Proceedings of the IEEE 1996 CUSTOM INTEGRATED CIRCUITS CONFERENCE

Town and Country Hotel
San Diego, California

May 5–8, 1996

The CICC '96 is sponsored by the IEEE Electron Devices Society with cooperation from the IEEE Solid State Circuits Council. Its goal is to provide a forum for manufacturers, circuit designers, CAD developers, and users of ASICs to present and discuss exciting new developments, future trends, and innovative ideas.

96CH35886

Papers have been printed without editing as received from the authors.

All opinions expressed in the Digest are those of the authors and are not binding on the Institute of Electrical & Electronics Engineers, Inc.

Additional copies may be ordered from:
IEEE Order Dept
445 Hoes Lane
Piscataway, NJ 08854
TEL: (Toll Free) (800) 678-4333

IEEE Catalog No. 95CH35886
ISBN 0-7803-3117-6 Softbound
ISBN 0-7803-3118-4 Casebound
ISBN 0-7803-3119-2 Microfiche
Library of Congress Catalog Card No. 85-653738

WELCOME TO CICC '96

Welcome to the 18th annual Custom Integrated Circuits Conference. The IEEE Electron Devices Society, in cooperation with the IEEE Solid State Circuits Council, cordially invites you to participate in CICC'96. The CICC provides the international forum for system designers and IC designers to meet together and share information about the most recent advances in technology and its application to real world problems. This year we convene in sunny San Diego at the Town and Country Hotel. The conference begins on Sunday, May 5 with a trio of Educational Session tracks, making up two full-day sessions and two half-day sessions. These sessions are led by design engineers and project managers currently working at the leading edge of their fields. The sessions provide practical, ready-to-use information that can be applied immediately or help prepare you for new opportunities. Areas covered include Topics in Advanced IC Design, Wireless, Advanced Technology Interconnect Issues and Topics in Communication Systems.

Allen Barlow
General Chairman

On Sunday evening we have our highly successful marketing sessions. Here a number of our exhibitors will present 14 papers in three tracks. These sessions give attendees an opportunity to hear important technical information on products they can buy today.

Our first plenary session on Monday morning kicks off the technical conference with a keynote address from Wally Rhines, President and CEO of Mentor Graphics. He'll share with us his perspectives and some challenges on "Laying the Groundwork for the 21st Century Design Environment".

Jake Buurma
Conference Chairman

Of course, the focus of the CICC has always been the presentation of exceptional technical papers that provide the latest, most significant developments in the IC industry. This year we have 113 papers in 20 sessions that cover the design and application of programmable logic devices, gate arrays, cell-based ICs and full-custom ICs in both Analog and Digital domains. Some of the important topics include wireless and digital communications, low-power design techniques, and tools for deep sub-micron design. Many sessions include invited and tutorial papers that provide a valuable perspective on the session topic.

Our exhibits will demonstrate the latest technology commercially available for IC design with Tuesday night featuring the CICC Hospitality Night hosted by our exhibitors. In the exhibits hall, you'll find hors d'oeuvres, refreshments and professional networking at its very best.

At our Tuesday luncheon Cdr. John Green, Science Advisor for Submarine Development at the Navy's Deep Submergence Unit, will discuss Deep Sea Exploration. Cdr. Green will talk about the vehicles and technologies involved in deep sea exploration and share results from some of the missions.

Robert Cordell
Technical Program Chairman

Tuesday evening features two spirited panel discussions with industry pundits who are sure to offer strong opinions and discernments on the following two topics: Deep Sub Micron Chip Design - What Business Issues will Dominate?; and Technical Challenges in One-Million-Gate Chip Designs. These panel discussions are always both entertaining and enlightening, and you'll come away with information and insights about the not-too-distant future.

We extend our thanks to the CICC Technical Program committee and our dedicated conference staff for all of their support. It's through their diligent efforts that we keep CICC the best place to discover the latest integrated circuit design techniques, learn of the newest product announcements and debate the most effective business strategies. Custom ICs are at the very center of a fast moving market with global competition. You can't afford to miss it!

STEERING COMMITTEE

Allen Barlow
AKM DesignTek

Jake Buurma
Toshiba America

Lauren Christopher
Thomson Consumer Electronics

Robert Cordell
Bellcore

Gregory Ledenbach
Sematech & Intel Corporation

Resve Saleh
Simplex Solutions

ORGANIZING COMMITTEE

GENERAL CHAIRMAN
Allen Barlow
AKM DesignTek
San Diego, CA

CONFERENCE CHAIRMAN
Jake Buurma
Toshiba America
San Jose, CA

TECHNICAL PROGRAM CHAIRMAN
Robert Cordell
Bellcore
Red Bank, NJ

EDUCATIONAL SESSIONS CHAIRMAN
Jerry Molnar
Mitel Semiconductor
Kanata, Ontario, Canada

EXHIBITS CHAIRMAN
Brian Fitzgerald
IBM Corporation
Waltham, MA

PANEL CHAIRMAN
Venu Gopinathan
AT&T Bell Laboratories
Holmdel, NJ

PUBLICITY CHAIRMAN
Jenny Ford
Motorola
Mesa, AZ

TREASURER
Gregory Ledenbach
Sematech & Intel Corporation
Austin, TX

CONFERENCE COORDINATORS
Melissa Widerkehr
Cathy Coyle
Widerkehr and Associates
Gaithersburg, MD

TECHNICAL PROGRAM COMMITTEE

Aitken, R., Hewlett Packard, Palo Alto, CA
Antao, B., Motorola Inc., Austin, TX
Au, K., Motorola Inc., Austin, TX
Avery, L., David Sarnoff Research Center, Princeton, NJ
Beunder, M., Compass Design Automation, San Jose, CA
Billig, P., Compass Design Automation, San Jose, CA
Brassington, M., Sun Microsystems, Mountain View, CA
Carley, L. R., Carnegie Mellon University, Pittsburgh, PA
Christopher, L., Thomson Consumer Electronics, Indianapolis, IN
Cravens, S., Symbios Logic Inc., Ft. Collins, CO
Dumlugol, D., Cadence Design Systems, San Jose, CA
Fang, P., Advanced Micro Devices, Sunnyvale, CA
Fiez, T., Washington State University, Pullman, WA
Fitzgerald, B., IBM Corporation, Waltham, MA
Ford, J., Motorola, Mesa, AZ
Frenkil, J., Sente Inc., Chelmsford, MA
Garrity, D., Motorola, Tempe, AZ
Gopinathan, V., AT&T Bell Laboratories, Holmdel, NJ
Green, D., Nippondenso America Inc., Carlsbad, CA
Gutierrez, G., Applied Micro Circuits Corp., San Diego, CA
Haque, Y., Maxim Integrated Products, Sunnyvale, CA
Hotta, M., Hitachi Ltd., Tokyo, Japan
Jou, S., National Central University, Taiwan, Republic of China
Kizilyalli, I., AT&T Bell Laboratories, Orlando, FL
Kumar, R., Cadence Spectrum Design, San Diego, CA
Ledenbach, G., Sematech, Austin, TX
Lee, B., Samsung Electronics Company, Suwon, Korea
Leung, R., LSI Logic Corporation, Milpitas, CA
Lipman, J., EDN Magazine, Livermore, CA
Lu, F., Broadcom Corporation, Irvine, CA
Maamari, F., AT&T Bell Laboratories, Princeton, NJ
Mackie, I., Advanced Hardware Architecture, Pullman, WA
Mayaram, K., AT&T Bell Laboratories, Allentown, PA
Milor, L., Advanced Micro Devices, Sunnyvale, CA
Molnar, J., Mitel Semiconductor, Ontario, Canada
Mori, S., Toshiba Corporation, Kawasaki, Japan
Mullinger, N., Mentor Graphics, Wilsonville, OR
Okada, K., Mitsubishi Electric Corporation, Hyogo, Japan
Onodera, H., Kyoto University, Kyoto, Japan
Oppold, J., IBM, Essex Junction, VT
Penning de Vries, R., Philips Semiconductors, Nijmegen, The Netherlands
Rich, D., AT&T Bell Laboratories, Allentown, PA
Sakurai, T., Toshiba Corporation, Kawasaki, Japan
Sapatnekar, S., Iowa State University, Ames, IA
Stamoulis, G., Intel Corporation, Santa Clara, CA
Starr, L., Ford Microelectronics, Inc., Colorado Springs, CO
Szorc, F., Delco Electronics, Kokomo, IN
Takada, M., NEC Corporation, Kanagawa, Japan
Tavana, D., Xilinx, San Jose, CA
Thorn, M., Texas Instruments Inc., Dallas, TX
Tu, P., Cadence Design Systems, San Jose, CA
Turner, J., Altera Corporation, San Jose, CA
van Bavel, N., , Austin, TX
Visweswariah, C., IBM TJ Watson Research Center, Yorktown Heights, NY
Weste, N., Macquarie University, Sydney, Australia
Whitaker, S., AKM DesignTek, San Diego, CA
Whitney, T., Actel Corporation, Sunnyvale, CA
Wright, J., American Microsystems Inc., Pocatello, ID

Technical Program Committee (Left to Right):
Front Row (Seated): K. Au, G. Gutierrez, R. Kumar, M. Widerkehr (conference manager), J. Molnar, R. Saleh, G. Ledenbach, R. Cordell, A. Barlow, J. Ford, D. Tavana, K. Okada, D. Dumlugol
Row 2: H. Onodera, R. Leung, J. Turner, B. Antao, L. Milor, T. Whitney, D. Green, N. van Bavel, R. Aitken, J. Oppold, T. Fiez
Row 3: S. Jou, M. Hotta, D. Rich, M. Brassington, D. Garrity, Y. Haque, F. Szorc, N. Weste, F. Lu, S. Mori, M. Takada, R. Penning de Vries
Row 4: T. Sakurai, L. Starr, F. Maamari, J. Wright, S. Whitaker, I. Mackie, J. Lipman, P. Billig, G. Stamoulis, I. Kizilyalli, S. Cravens, C. Visweswariah, R. Carley, M. Thorn
Not Shown: L. Avery, M. Beunder, L. Christopher, P. Fang, B. Fitzgerald, J. Frenkil, V. Gopinathan, B. Lee, K. Mayaram, N. Mullinger, S. Sapatnekar, P. Tu, C. Coyle (conference manager), P. Mahoney (conference manager)

CONFERENCE OVERVIEW

THE TOWN AND COUNTRY HOTEL						
	PRESIDIO	FRIARS/PADRE SIERRA	GOLDEN WEST	CALIFORNIA	TOWN & COUNTRY SAN DIEGO (EXHIBIT HALL)	DELMAR/HELIX/ SANTA FE (Registration)
SUNDAY, MAY 5						Sat. May 4: 6–9 pm Ed. Sess. Reg.
8:00 am–5:00 pm		Educational Session 1 Topics in Advanced IC Design	Educational Session 2 Wireless	Educational Session 3 Advanced Technology Interconnect Issues		7 am–2 pm Ed. Sess. Registration 3–9 pm Conference Registration
7:00 pm–9:30 pm		Exhibitor's Marketing Session #1	Exhibitor's Marketing Session #2	Exhibitor's Marketing Session #3		
MONDAY, MAY 6						
7:30 am–8:00 am						7:30 am
8:00 am–9:30 am	1. KEYNOTE					
9:30 am–10:00 am					9:30 am	
10:00 am–12:05 pm	2. High Speed Datacom	3. Process Technology and Manufacturing	4. Advances in Gate Arrays	5. DSP Computational Elements and Digital Filters	Exhibits	Registration
12 Noon–2:00 pm	"ALL CONFERENCE LUNCHEON" IN EXHIBIT HALL					
2:00 pm–4:35 pm	6. Analog Circuit Techniques	7. Low Power Circuit Design Techniques and Implementations	8. Advances in Programmable Logic	9. Innovations in Physical Design	3:00 pm	5:00 pm
TUESDAY, MAY 7						
7:30 am–8:30 am						7:30 am
8:30 am–11:30 am	10. RF Communications Circuits	11. Power Estimation and Analysis	12. New Architectures in Programmable Logic	13. Memories, I/Os, and Read Channels	10:30 am	Exhibits
12 Noon–1:30 pm	Conference Luncheon in Regency Hall					
2:00 pm–5:00 pm	14. Sensor Application and Interface Circuits	15. Device Modeling and Analog Simulation	16. Video, Image and Speech Digital Signal Processing	17. Critical Design Methods for the late 90's	Exhibits	4:00 pm
5:30 pm–7:30 pm	"Open Hospitality Night" in the Exhibit Hall				7:30 pm	
8:00 pm–10:00 pm (PANELS)			18. Deep Submicron Chip Design—What Business Issues will Dominate?	19. Technical Challenges in One-Million-Gate Chip Designs		
WEDNESDAY, MAY 8						
7:30 am–8:30 am						7:30 am
8:30 am–11:30 am	20. Data Converters	21. Simulations and Modeling of RF Circuits	22. Improvements in Test and Reliability	23. Cell and Macro Generation/Optimization	9:30 am Exhibits 12:30 pm	Registration 11:00 am

Author interviews are held in the Council/Chamber/Cabinet/Forum/Senate/Committee Rooms immediately following afternoon sessions.

EDUCATIONAL SESSIONS

Chair: **Jerry Molnar**
Mitel Semiconductor

Co-Chair: **Terri Fiez**
University of Washington

Michael Beunder
Compass Design

SESSION 1 **FRIARS/PADRE/SIERRA ROOM**

TOPICS IN ADVANCED IC DESIGN

E1.1 8:00 - 9:50 AM **High-speed Converter Design**
Rudy van de Plassche, Eindhoven University of Technology, Philips Research Laboratories

After the introduction of important converter specifications and specific problems encountered in high-speed designs, the full flash, two-step and folding and interpolation architectures will be described. The folding and interpolation architecture combines the advantages of a full flash converter with the small die size and low-power consumption of a two-step converter. Comparators are important parts of high-speed converters. Different architectures will be analyzed. Attention will be paid to the Bit Error Rate of a comparator and a design relation between technology and BER will be evaluated. Furthermore, in digitally sampled converters a relation between the minimum analog bandwidth and the third-order distortion of a converter can be determined. An example of an 8-bit A/D converter will be discussed. Finally, digital information can be converted into analog signals using a high-speed digital-to-analog converter. An example of a 1 Gsample low-distortion 10-bit D/A converter will be described.

E1. 2 10:10 - 12:00 NOON **Bipolar Analog for CMOS Designers**
William F. Davis, Private Consultant

With BiCMOS technology, CMOS designers are faced with bipolar device options to enhance their designs. This minority carrier device is uniquely different from the majority carrier CMOS counter part. The respective terminal characteristics are of sufficient difference to create bipolar design pitfalls for the unwary CMOS designer. However, these differences can also produce distinctive design advantages. This session will contrast the physical differences and highlight the resulting difference in terminal and circuit behavior. Examples include circuit functions where bipolar implementation (or BiCMOS combination) offer superior advantages over CMOS implementation. Bipolar circuit building blocks are also addressed, such as current sources (OTC, PAT, and otherwise), bandgap voltage references, current mirrors, multipliers, input and output stages, and split collector topologies. Critical bipolar layout issues are also addressed.

E1. 3 1:00 - 2:50 PM **Low Power Digital Design**
Anantha Chandrakasen, MIT

It is clear that batteries are a major factor in the weight and size of portable electronic systems and these, in turn, are directly impacted by the power dissipated by the electronic circuits. In this talk, a system level approach to power reduction will be presented which involves optimizing algorithms, architectures, logic design, circuit design, and technology. Potentially one of the most important techniques involves combining architecture optimization with voltage scaling, which allows a trade-off between silicon area and low-power operation. Several techniques will also be presented to minimize the switched capacitance including optimizing signal correlations, choice of data representation, minimizing spurious transitions, optimizing sequencing of operations, activity driven power down, transistor sizing, etc. Support circuitry for low-voltage operation including level-converters and DC/DC converters will also be presented.

E1. 4 3:10 - 5:00 PM **Structured Custom CMOS Design for Performance Oriented DSP Applications**
Neil Weste, Macquarie University

For reasons of design productivity, most CMOS digital chips being designed today are constructed using a combination of synthesis and automatic place and route software. In combination with sub-micron multi-level metal CMOS processes these systems produce large complex chips which are capable of operating at high speed. To all intents and purposes traditional "custom" design has been relegated to an art form. Having said all of this, there are still emerging applications (particularly in DSP) where for one reason or another a "structured custom" design might have significant benefit in a design. But in today's highly automated environments one is still faced with the problem of integrating these foreign modules into an otherwise smooth design flow. Of equal importance is the problem of maintaining the investment over subsequent process migrations. This session discusses a complete top down approach to custom DSP implementation. Topics include the decision to go custom, design partitioning, design flows, system issues (e. g. clocking, I/O interfaces, timing analysis), design metrics (e. g. power estimation, size, speed) process independence/migration and outsourcing. Detailed examples of real circuit implementations including an embedded processor core and FIR filters will be discussed.

SESSION 2 **GOLDEN WEST ROOM**

WIRELESS

E2. 1 8:00 - 9:50 AM **RF Systems Design**
Ken Hansen, Motorola

This talk provides the IC designer with a strong top down view of the issues related to implementing IC solutions in RF systems. The session begins with an overview of some of the most prominent types of wireless communications systems. Other topics to be covered include radio propagation, cellular system planning, transceiver block diagram, noise figure and sensitivity, intermodulation,
spurious responses and IF selection. The IC requirements in these areas will be emphasized. An example of a receiver will also be discussed in more detail.

E2. 2 10:10 - 12:00 NOON **RF Design Basics**
Ico Coulias, Wireless Microsystems

Selected topics on RF Integrated Circuit design will be reviewed. Particular emphasis will be placed on understanding basic RF concepts and nomenclature. Specific RF IC design examples include mixers, low-noise amplifiers, I/Q modulators, phase splitters, and AGC amplifiers. Block-level design and discrete-component impedance matching circuits will be discussed. The effect of device, layout, package, and test board parasitics on the electrical performance of the RF IC will be examined.

E2. 3 1:00 - 2:50 PM **Frequency Synthesis I (High Level)**
Tom Higgins, Hewlett Packard

There has been a dramatic reduction in the size and cost of frequency synthesizers over the last twenty years. What originally took a large array of mixers, filters, and other expensive devices can now be done with an IC, a VCO, and few other parts. Also, the recent availability of Fractional-N synthesis on the general market has allowed arbitrarily fine resolution to be achieved. This session will give an overview of frequency synthesis techniques, followed by a more detailed discussion of PLL-based (indirect) synthesizers. Both API- and Sigma Delta Fractional-N will be discussed. A performance comparison of the various methods will also be given. Topics to be covered include frequency synthesis techniques (direct, indirect, PLL based, brute force). A detailed look at indirect synthesis will include a review of PLLs, integer division, Fractional-N division as well Sigma Delta techniques. Performance comparisons between different frequency synthesis techniques will be discussed.

E2. 4 3:10 - 5:00 PM **Frequency Synthesis II (Circuit Level)**
Tad Kwasniewski, Carleton University

The evolution of the IC technology over the past decade resulted in the dominance of 'digital' CMOS process with niche applications still served by the bipolar or lately GaAs implementation technologies. Today's of-the-shelf synthesizers' chips integrate all functions but the VCO and external loop filter. The synthesizer, whether it is based on the DDS (Direct Digital frequency Synthesis) or IS (Indirect frequency Synthesis) concepts, remains a circuit whose performance is determined by the performance of high-precision, analog circuitry; e. g. noise, speed and power dissipation. Circuit performance requirements for the synthesizer building blocks such as phase detectors, loop filters, VCOs and prescalers will be discussed together with circuit techniques and implementation technologies used today in the synthesizer design. The digital radio and clock-recovery synthesizer applications will be discussed in detail.

SESSION 3 **CALIFORNIA ROOM**

ADVANCED TECHNOLOGY INTERCONNECT ISSUES

E3. 1 8:00 - 9:50 AM **Advanced Technology Interconnect Issues**
Gary McGuire, MCNC

The scaling of device dimensions with each new generation of CMOS and DRAM technology is creating increasing demands on the performance of existing materials. The recent Semiconductor Industry Association Roadmap highlights a number of materials and processing changes that are anticipated at 0. 25 micrometer design rules and beyond. One of the areas most impacted is interconnects associated with multilevel metallization schemes. Four and five levels of metallization will be utilized in the next few generations of technology to reduce the impact of scaling on the interconnect line width and spacing. The high current densities and resulting concern with electromigration will eventually result in a switch from the traditional Al/0. 5 % Cu metallization to Cu or other low resistivity metallization. Device performance will no longer be limited at the transistor level but by interconnect performance. As a result there will be a shift toward the use of low dielectric constant materials. The trends in device performance, materials selection and process integration will be discussed in a tutorial manner with the goal of providing the design engineer with an overview of these issues.

E3. 2 10:10 - 12:00 NOON **Design and Analysis of Submicron ULSI Interconnects**
Laszlo Gal, Motorola

The sustained growth of integrated circuit density and speed is a direct consequence of the transistor and interconnect scaling. While this growth was unconstrained by the metal interconnects for 0. 5 micron geometries and above, in deep submicron digital circuits on-chip delays, power dissipation, die area and die costs are dominated by the interconnects. Therefore optimizing the ULSI interconnects is becoming the number one priority of both the process developer and the CAD tool provider. In this course the trend of performance degradation, signal integrity and reliability issues of ULSI interconnects are analyzed and potential design and CAD solutions discussed. As device geometries approach 0. 15 micron the optimum solution will be the combination of increased number of interconnect layers and non-scaled metal lines. The session is concluded by discussing the need for replacing circuit simulation by electromagnetic simulation as on-chip clock frequencies reach 500 MHz.

TOPICS IN COMMUNICATION SYSTEMS

E4. 1 1:00 - 2:50 PM **High Speed (Broadband) ATM Networks**

Ness B. Shroff, Purdue University

ATM is fast becoming the likely transport standard for the Broadband Information Superhighway. Its ability to packetize ("containerize") many different types of information into standardized fix-sized chunks that can be handled and switched in the same way makes it attractive for use in networks that will support the many exciting, current and future, applications. ATM also makes "bandwidth-on-demand" sharing of network resources possible. This leads to efficient utilization of the network infrastructure at a much lower cost than techniques in which bandwidth must always be dedicated to each user whether it is needed all the time or not. The elements of the ATM cell will be described, along with how it is routed using Virtual Circuits and Virtual Paths. Basics of ATM switching techniques and some ATM switch examples will be covered. In addition, some of the open challenges that ATM faces, especially in the areas of control and traffic management, will be discussed.

E4. 2 3:10 - 5:00 PM **The Anatomy of a Full-Service Network Set-top Box**

Michael Adams, Time Warner Cable

This session will describe the development of a digital set-top for the Orlando Full Service Network (FSN). We will consider in detail the requirements for communications, MPEG audio and video decompression, graphics, computing and synchronization. The chosen solutions in these areas will be described, and the lessons we learned in the development process shared. Special attention will be given to the transport of digital data to the set-top over a hybrid fiber-coax network. The cost reduction of a digital set-top box to a $300 target will be discussed, and key functional blocks that can be integrated in hardware identified. The software architecture of the interactive set-top will also be covered in some detail.

SESSION QUICK INDEX

CONTENTS

TECHNICAL SESSIONS

SESSION 6— PRESIDIO ROOM
Monday Afternoon

SESSION 7— FRIARS/PADRE/SIERRA ROOMS
Monday Afternoon

SESSION 8— GOLDEN WEST ROOM
Monday Afternoon

SESSION 9— CALIFORNIA ROOM
Monday Afternoon

SESSION 17— CALIFORNIA ROOM
Tuesday Afternoon

SESSION 18— GOLDEN WEST ROOM
Tuesday Evening

SESSION 19— CALIFORNIA ROOM
Tuesday Evening

SESSION 20— PRESIDIO ROOM
Wednesday Morning

SESSION 21— FRIARS/PADRE/SIERRA ROOMS
Wednesday Morning

SESSION 22— GOLDEN WEST ROOM
Wednesday Morning

SESSION 23— CALIFORNIA ROOM
Wednesday Morning

LIST OF CONTRIBUTORS

Laying the Groundwork for the 21st Century Design Environment,
Wally Rhines, President and CEO, Mentor Graphics

Through most of the electronic design history, integrated circuit (IC) designers were specialists - memory designers, logic designers, analog designers, etc. Developers of embedded software were also frequently specialized by application, by processor, etc. Printed circuit board, multi-chip module and box designers tended to be more oriented toward "system design", involving integration of disparate technologies to meet system requirements. Challenges of "system on silicon" are changing roles. The next decade will be driven by "integrated system design" that requires new levels of specialization by those who create the hardware and software building blocks for reuse and by those system integrators who combine original design of hardware and software with reuse of major blocks like embedded core processors. The design team of the future will make product tradeoffs in hardware, software and system design in small teams. Product design success will require new approaches to the electronic design automation tools and the environment in which they are used.

Chair: Nicholas van Bavel
Co-Chair: German Gutierrez

Do not come to this session unless you want to see the latest developments in DSP and mixed-signal circuits to support high-speed data communications in CMOS!

Internet access, telecommuting, video on demand, and video teleconferencing are the killer applications that are driving an explosion in the demand for faster, cheaper, digital communications into the home and business. This session covers new innovations in DSP and mixed-signal circuits that are enabling system level solutions in a single chip that address this need. The papers in this session cover a wide spectrum of communications applications: wire-line transceivers, hard-disk read channels and ATM switches.

The first paper is an invited tutorial covering techniques used in full-duplex communications on unshielded twisted-pair copper cable - in particular the area of HDSL and fast-ethernet. The modeling of the cable is discussed as well as the effects of crosstalk. The basic building blocks of both applications are analyzed including ADCs, DACs, line drivers, echo cancellers, equalizers and timing recovery circuits.

The second paper focuses on the timing recovery problem for the Carrierless AM/PM passband modulation scheme. The nonlinear spectral-line method is analyzed and the implementation is described using high-speed analog circuits for squaring, bandpass filtering and PLL circuits. This chip works for 50 Mb/s ATM applications and is implemented in a 0.9 μm CMOS process.

The third paper describes the latest developments in mixed-mode DFE for read-channels and other communications problems that have severe intersymbol interference. A unique structure uses parallelism and look-ahead logic to improve the throughput to 55 Mb/s in a 2 μm process.

The final paper describes a Content Addressable Memory (CAM) chip for use in ATM cell processing. This chip uses analog comparison techniques to develop high-speed access times in the CAM with very small die area. At a clock rate of 40 MHz, 622 Mb/s operation is achieved with only 70 mW power dissipation at 3.3 V supply voltage.

Please join us for this exciting session to see where DSP and mixed-signal are having the most impact in data communications!

Integrated Circuits for Data Transmission over Twisted-Pair Channels

D. A. Johns[1] and D. Essig[2]

[1]Dept. of Elec. and Comp. Eng.
University of Toronto
Toronto, Ontario, M5S 1A4, CANADA
(johns@eecg.toronto.edu)

[2]Brooktree Corp.
9868 Scranton Rd.
San Diego, CA, 92121, USA
(essig@brooktree.com)

Abstract

This tutorial paper discusses typical architectures and future challenges in designing integrated circuits for data transmission over twisted-pair wire channels. To highlight the various architectural approaches, two main applications are addressed — HDSL and fast-ethernet. Following a discussion of twisted-pair cable modelling, common building blocks and possible trade-offs are described. Finally, future challenges facing integrated circuit designers are presented.

Introduction

Due to the abundance and low-cost of unsheilded twisted-pair (UTP) cables, there is great interest in transmitting high-speed data over UTP cable. However, there are certain challenges that face circuit and system designers to accomplish this task. This tutorial paper intends to highlight the various architectural approaches used for overcoming some of the challenging design obstacles.

While common circuit and architectural approaches can be used in a variety of applications, for illustrative purposes, it is useful to focus on specific applications. Specifically, two main applications are addressed here — High-bit-rate Digital Subscriber Loop (HDSL) and fast-ethernet. Other twisted-pair data-communication applications of interest but not specifically addressed in this paper include ISDN, ADSL (Asymmetric Digital Subscriber Loop), E1/T1, and emerging ATM standards.

HDSL and fast-ethernet make use of a variety of main blocks as shown in Fig. 1. The transmit D/A is used to convert digital levels to a suitable analog signal. A low-impedance line-driver is used to supply the necessary drive currents to the cable. The 2-4 wire hybrid allows both transmission and reception of data over a single cable. The receive A/D converts the received signal (plus some transmitted signal) to a digital signal. The echo canceler is used to eliminate residual transmit signal on the receive path due to a non-ideal 2-4 wire hybrid (as we shall see, practical hybrids only partially cancel the transmit echo). Finally, clock-recovery and equalization is performed on the received signal to determine the correct timing information and remove inter-symbol interference.

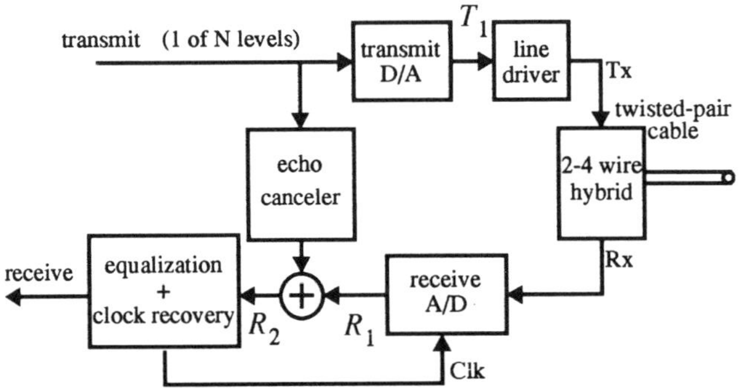

Fig. 1 The main blocks for twisted-pair data transmission

The organization of this paper is as follows. First, modelling of twisted-pair cables is discussed. Next, the two main applications of HDSL and fast-ethernet are described. Each of the blocks shown in Fig. 1 are then discussed in more detail. Finally, future challenges of some of these blocks are described.

Twisted-Pair Cable Modelling

A twisted-pair cable is modelled as a transmission-line and can be described using four primary constants — R, G, L, C [1,2]. These four constants are shown in Fig. 2 where (in terms of "per unit length"), R is the resistance, G is the conductance, L is the inductance and C is the capacitance. Typical values for a twisted pair cable are approximately

$$R(f) \approx \sqrt{f/4} \;\; \Omega/km$$
$$L \approx 0.6 \;\; mH/km$$
$$C \approx 0.05 \;\; \mu F/km \tag{1}$$

with $G \approx 0$ and therefore negligible. Note that R is shown as proportional to the square-root of frequency (where f is in units of Hz) due to the "skin-effect" while L and C are relatively constant at frequencies higher than $100kHz$.

A. Characteristic Impedance

The characteristic impedance of a transmission-line can be shown to equal

$$Z_o = \sqrt{\frac{R + j\omega L}{G + j\omega C}} \tag{2}$$

and thus, for high frequencies where $\omega L \gg R$ (i.e. $\omega \gg 2\pi \times 16kHz$ using the typical values above), we have

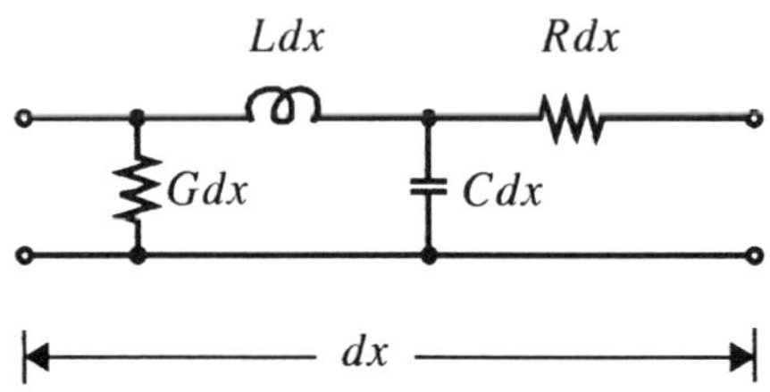

Fig. 2 A lumped-parameter model for a short section of cable.

2.1.1

0-7803-3177-6 $5.00 © 1996 IEEE

$$Z_{oh} \approx \sqrt{\frac{L}{C}} \quad (3)$$

In other words, since L and C are relatively constant, the characteristic impedance of the line is relatively constant at higher frequencies.

Using the typical values in (1), we have that $Z_{oh} \approx 110\ \Omega$ for frequencies greater than $100kHz$. Thus, when terminating the line, impedances around 110Ω should be used.

B. Transfer-Function

When properly terminated, the transfer-function of a twisted-pair cable is modelled by

$$H(d, f) = e^{-d\alpha(f)} e^{-jd\beta(f)} \quad (4)$$

where α and β are the attenuation and phase constants, respectively, while d is the cable length (in units of km). When $\omega L \gg R$, the attenuation and phase constants can be approximated by [1]

$$\alpha(f) \approx \frac{R(f)}{2}\sqrt{\frac{C}{L}} \quad (5)$$

$$\beta(f) = 2\pi f\sqrt{LC} \quad (6)$$

Combining (4), (5) and (6), we see that for a given length d, the cable's transfer-function phase is roughly linear with respect to frequency. The transfer-function gain (in dB) through a cable is given by

$$H_{dB}(d, f) = 20\log_{10}|H(d, f)| = \frac{-20}{\ln 10} d\alpha(f)$$

$$= -8.686d \times \frac{R(f)}{2}\sqrt{\frac{C}{L}} \quad (7)$$

Thus, the ***gain in dB is proportional to the square-root of frequency*** due to $R(f)$ which, in turn, is proportional to $\sqrt{f}$. This $\sqrt{f}$ response should not be confused with a pink-noise $\sqrt{f}$ response (in the pink-noise case, spectral density falls off at $-10\ \text{dB/decade}$). Specifically, consider the curves shown in Fig. 3 for an ideal $\sqrt{f}$ cable and a first-order lowpass filter where the 3dB frequency of each is at a normalized frequency of 1Hz. In the case of a cable, the gain falls off slowly when the attenuation is small and then rapidly as the attenuation increases.

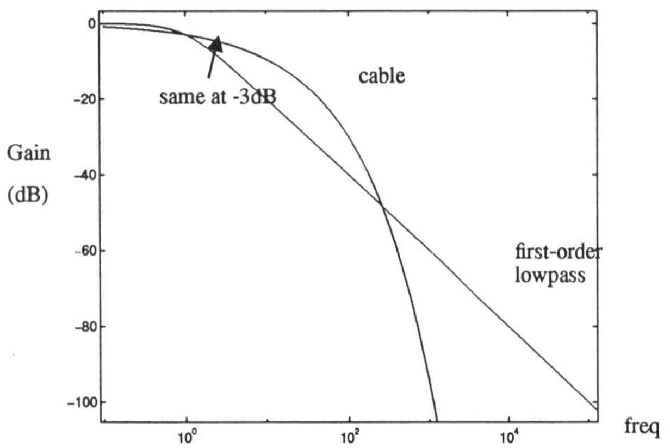

Fig. 3 Comparison of gain for an ideal cable and a first-order lowpass filter.

Using the values in (1), we can find the attenuation for a ***typical*** twisted-pair cable to be approximately

$$H_{dB}(d, f) \approx -0.02d\sqrt{f} \quad (8)$$

where d is in km and f is in hertz. Note that (8) is only useful in estimating the loss in a "typical" cable as it depends on the primary constants of the cable which in turn depend on the physical construction of the cable (such as wire gauge, twist lengths, insulation type and thickness, etc.).

C. Crosstalk

As shown in Fig. 4, there are two main classes of crosstalk — near-end crosstalk (NEXT) and far-end crosstalk (FEXT). Far-end crosstalk is due to the leakage of transmitted signals onto a nearby wire pair and interfering with a locally transmitted signal. Thus, in the case of FEXT, the interfering crosstalk and the desired transmitted signals are both attenuated by the cable length. Near-end crosstalk is a result of large transmitted signals immediately leaking onto a nearby wire pair where small received signals from a distant transmitter are present (the received signals are attenuated due to the length of the cable). Thus, when present, NEXT almost always dominates FEXT in terms of limiting performance. Unfortunately, NEXT cannot be cancelled similar to echo cancellation since the crosstalk interferer is typically not available.

A reasonable model for the squared-gain of NEXT is given by

$$|H_N(j\omega)|^2 = K_N|\omega|^{1.5} \quad (9)$$

where K_N is a constant determined by physical parameters of the cables. Note that the crosstalk coupling increases with frequency. Since the cable transfer-function decreases with frequency, we see that there will some frequency value where the signal-to-interference power becomes $0\ \text{dB}$. Thus, transmitting signal power above this frequency value is of little use and the lower frequency region of cables is where one should place most of the signal energy.

It should be mentioned here that a measured crosstalk gain will actually have many peaks and nulls over the frequency spectrum. However, the above relation is reasonable for modelling the envelope of the peaks in crosstalk gain.

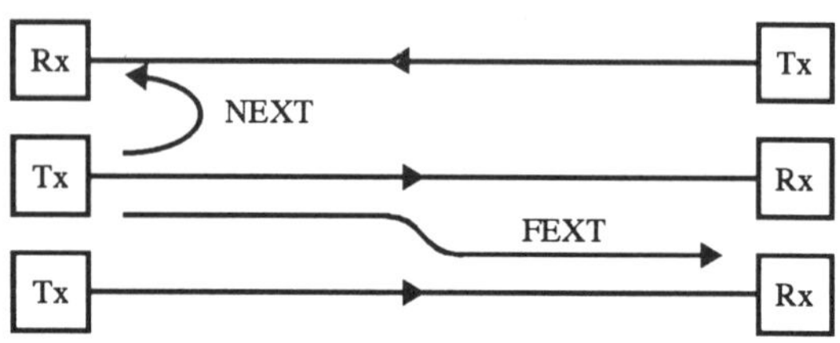

Fig. 4 Illustration of NEXT and FEXT.

HDSL Application

HDSL is a transmission standard for full-duplex delivery of 1.544Mb/s over long lengths (up to 4.5km) of existing telephone-loop UTP cables. To accomplish this goal, two pairs of cables are typically used along with a 4-level PAM code referred to as "2B1Q" (two bits - one quaternary). In other words, one of four levels are sent corresponding to 2 bits of information. As a result, each cable must operate at a rate of 386kS/s (kilo-Symbols per second) in dual-duplex mode (actual transmission rate is 392kS/s with overhead).

Detailed cable modelling in the HDSL environment was presented in [3] where it was shown that the $\sqrt{f}$ attenuation model is valid at both low and high frequencies and has a transition region where the attenuation falls off more slowly. Making use of (8) and letting $d = 4.5\text{km}$, we see that a 200kHz signal is attenuated by roughly 40dB. Thus, the high-frequency portion of a $5V_{pp}$ transmit signal would appear on the receive end as a $50mV_{pp}$ signal! Clearly, there is strong need for effective echo cancellation in HDSL applications and therefore, effective cancellation becomes one of the main design challenges.

When receive signals are small in relation to the transmit signals, it is not uncommon for the linearity of the echo path to limit performance of the overall system. Such a limitation occurs because the echo cancellation circuitry cannot typically deal with non-linearities and/or noise in the echo path. For example, when the receive signal is attenuated by 40dB, if the 2-4 wire hybrid performs only 6dB echo attenuation (a typical number due to the large load variations in HDSL applications), then to maintain a 40dB SNR receive signal, the linearity and noise in the transmit path should be better than 74dB. Referring to Fig. 1, the blocks that need to meet this 74dB requirement are the transmit D/A, the line-drive, the 2-4 wire hybrid and the receive A/D. This high linearity and noise requirement in the echo path is one of the most demanding aspects in realizing a fully-integrated HDSL transceiver.

Once the transmit echo signal is removed, the performance limit is mainly due to NEXT since HDSL signals are often sent in cable bundles where many nearby large transmit signals can couple into the received signal path.

Fast-Ethernet Application

Presently, ethernet accounts for approximately 85% of all local-area-network (LAN) implementations. At modest data rates, 10Mb/s ethernet is realized over UTP cabling using a 10Base-T standard. Within this LAN environment, there are two main types of UTP cables available — CAT3 and CAT5. The 10Base-T standard operates well over both types of cabling

CAT3 cables have poorer attenuation and crosstalk characteristics and are often referred to as "voice-grade" cables. The attenuation response of $100m$ CAT3 cables at $20°C$ is reasonably well modelled by

$$H_{\text{dB}}(f) = 2.32\sqrt{f} + 0.238f \qquad (10)$$

where f is in MHz [4]. Note that there is a term which is proportional to f in this model and should be taken into account during equalization. CAT5 cables are often referred to as "data-grade" as they have better attenuation and crosstalk characteristics. A reasonable model for $100m$ of CAT5 cable at $20°C$ is

$$H_{\text{dB}}(f) = 1.967\sqrt{f} + 0.023f + 0.05/\sqrt{f} \qquad (11)$$

with f in MHz. Note that this type of cabling follows the $\sqrt{f}$ attenuation curve much more closely.

Fast-ethernet refers to an emerging 100Base-T standard intended to upgrade ethernet networks to 100Mb/s over $100m$ of CAT3 and CAT5 cabling. Due to the two main types of UTP cabling available, three separate sub-standards have emerged within the 100Base-T standard.

100Base-T4 makes use of 4 pairs of either CAT3 or CAT5 cables where 2 pairs are used dual-duplex (with hybrids) while the remaining 2 pairs are broken down into one transmit and one receive pair. In other words, if pairs 1-4 are available, pairs 1,2 and 3 are used for transmission in one direction while pairs 1,2 and 4 are used for transmission in the other direction. To reduce radiated emissions, 8 binary bits are coded into 6 ternary (i.e. one of three levels) outputs. As a result, the data rate of 100Mb/s is reduced to 33Mb/s since it sent over 3 pairs and further reduced to 25MS/s on each pair due to the $6/8$ coding.

100Base-TX makes use of 2 pairs of CAT5 wiring each dual-duplex through the use 2-4 wire hybrids. Once again, a three level PAM code is used to reduce radiated emissions.

100Base-T2 is not yet finalized as a standard but is intended to make use of 2 pairs of either CAT3 or CAT5 cables each dual-duplex through the use of 2-4 wire hybrids. However, to accommodate the difficult radiated emission standards, it is likely a 5x5 code will be used where one of 5 PAM levels will be transmitted on each pair. The received constellation point is found by determining the received levels on both pairs. During regular transmission, only one of 16 possible symbols are sent (i.e. one of 4-levels on each pair). Therefore, for 100Mb/s, the transmission rate over each pair is 25MS/s since one symbol is sent for every 4 bits (i.e. effectively 2-bits on each pair).

To determine typical attenuation values for fast-ethernet applications, we make use of (8) and letting $d = 0.1\text{km}$, we see that 12.5MHz and 20MHz signals are attenuated by only about 7dB and 9dB, respectively. These values might be typical attenuation numbers for CAT5 wire as seen also through the use of (11). Similar values in CAT3 wiring can be found from (10) and correspond to 11dB and 15dB, respectively.

Transmit D/A

As mentioned above, the transmit D/A block typically includes filtering to shape the transmit spectrum. This filtering can be realized in both the digital and analog domains as

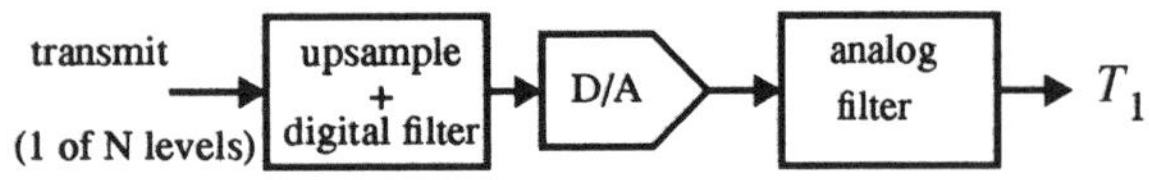

Fig. 5 Typical transmit D/A block.

shown in Fig. 5. Here, the digital transmit signal is upsampled and passed through a digital filter (these two operation are usually combined through the use of "poly-phase" filtering [5]). Such an approach requires a faster D/A converter but it reduces the requirements for the analog filter.

One of the challenges in HDSL realizations is ensure better than 72dB linearity in the overall echo path. Thus, it is important that the D/A converter have better than 12-bit linearity which is difficult to achieve with component matching. To overcome this challenge, one can use an oversampled 1-bit D/A converter as is commonly done in audio applications. For example, in [6,7], the transmit signal is upsampled by 16 times, applied to an 48 Tap FIR digital filter and upsampled once more by 4 (for an overall oversamplin-g-rate of 64 times). This digital signal is then applied to a delta-sigma DAC (with one-bit current output) to obtain an overall linearity and noise performance of at least 12-bits. The use of an oversampled one-bit DAC eliminates the need for accurate on-chip capacitor matching (to the 12-bit level). A similar 1-bit oversampled DAC approach was used in [8].

At higher speeds, such as in fast-ethernet applications, the linearity and noise requirements of the echo path are not nearly so severe. A typical requirement might be around the 40dB level. For example, the proposed transmit paths for 100Base-T2 suggests upsampling by 3, using a 75MHz 4-bit DAC and following it with a 3'rd-order lowpass filter. The design challenges here are relatively modest except for high-speed operation. One approach for adjusting a continuous-time pulse-shaping filter at high-speeds makes use of comparator outputs at specific sampling instances and is presented in [9].

Line Drivers

As seen in Fig. 1, a line-driver is used to supply the necessary drive current to the twisted-pair cable. These line-drivers are commonly realized as a voltage buffer having a low-impedance output.

As with other portions of the echo path, HDSL line-drivers need to have at least a 72dB performance. Such a requirement for a power amplifier makes this block one of the more difficult challenges in the analog interface (in fact, many systems use discrete amplifiers for this function). Various circuit architectures for linear line-drivers were developed for ISDN applications [10-12]. However, while the frequency requirement of ISDN drivers is only 40kHz, it is nearly 5 times higher for HDSL with a similar linearity requirement.

To achieve large output swings and low distortion, modern line-drivers make use of fully-differential structures.

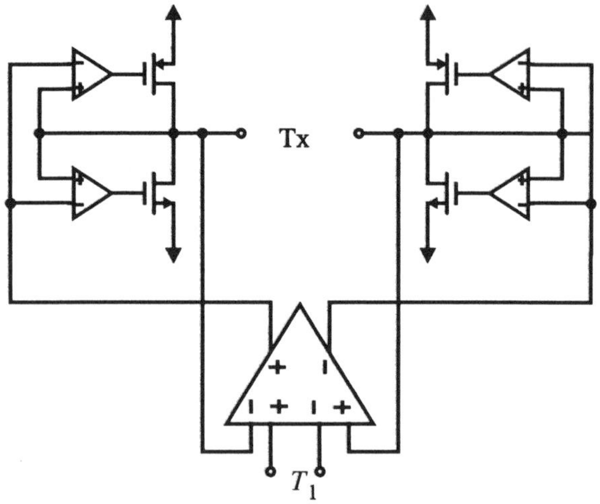

Fig. 6 A possible fully-differential line driver.

One possible architecture is shown in Fig. 6 [12]. Here, four large transistors are used with class-AB biasing and connected in a common-source configuration. As well, two feedback loops are used to obtain high linearity, speed and output swings. Another alternative is to not place immediate feedback around the four internal amplifiers and instead rely on the main feedback path from input to output. In this case, more elaborate compensation schemes are necessary such as the nested-Miller approach [13]. A challenge in these types of designs is ensuring that quiescent current is well controlled as well as the usual need for a common-mode feedback circuit. Note that it is imperative that the distortion of the final output transistor stage be modest since the gain of the open-loop amplification will be reduced at high frequencies and will therefore not help in improving linearity much. For this reason, the last stage is likely biased as class-AB with a relatively large quiescent current. An interesting approach for realizing a rail-to-rail output amplifier with class-AB biasing is described in [14].

In fast-ethernet applications, the linearity of the line-drivers is not as stringent as in the HDSL application since received signal levels are only around 10 dB below transmit levels. In addition, the echo cancellation of the 2-4 wire hybrids is improved since the impedance seen by the cable more controlled as compared to the HDSL case. As a result, line driver linearity need only be around the 35dB level. However, due to the high-speeds involved and relatively low impedances that must be driven, many fast-ethernet implementations make use of bipolar transistors through the use of a BiCMOS technology.

2-4 Wire Hybrids

The use of 2-4 wire hybrids allows one to transmit and receive simultaneously over a single cable (i.e. dual-duplex transmission). However, hybrids are also commonly used even when two separate cables are available to reduce the frequency of the transmitted signal on each pair. Such a reduction in signal frequency is often important to meet emission requirements and can be particularly beneficial for longer cable lengths. Specifically, when longer cables are used, the received signals radiate little energy and therefore one is most concerned with reducing the high-frequency content of the transmitted signal. The main disadvantage of a dual-duplex scheme is the need for excellent echo cancellation.

A possible single-ended hybrid configuration is shown in Fig. 7. Here, assuming the line impedance seen through the transformer equals R_T, then $V_2 = V_1$ and the transmit signal is cancelled. However, due to line impedance variations, the echo cancellation is only moderate. For example, a plot of various line impedances (including the transformer) for HDSL applications are shown in Fig. 8. Note that above $100kHz$, the majority of impedances are around 100Ω but there are lines which go as low as 40Ω and as well as ones that go over 200Ω. These large line impedance variations are due to un-terminated bridge-taps and one would not expect as large a variation in fast-ethernet applications.

Note the zero at dc and pole around $10kHz$ both due to the transformer. This low frequency pole results in a long impulse response in the echo path. To shorten the echo impulse response, the pole can be placed at higher frequencies (through the use of a different transformer) but then channel bandwidth is reduced.

Other variations on this hybrid circuit are to eliminate the R_1 network altogether and rely on the digital echo canceler to remove the transmit signal. Such an approach would require more dynamic range in the receive path (i.e. more bits on the A/D converter). Alternatively, the R_1 divide by 2 circuit might be made more complex so that better echo cancellation occurs thereby reducing the requirements on the receive path. Specifically, the two resistor R_1 circuit can be replaced with a circuit that better matches a nominal trans-

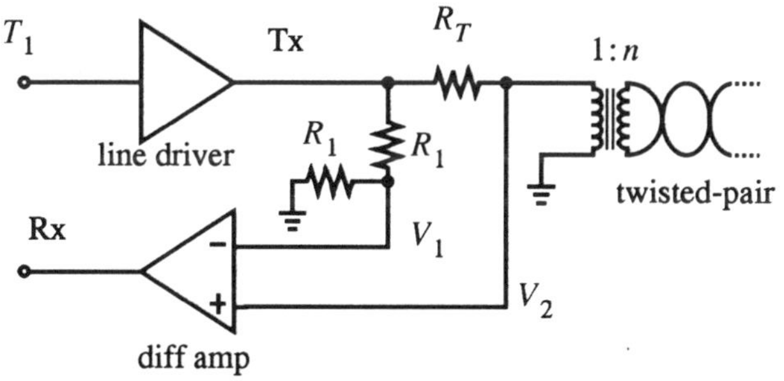

Fig. 7 A single-ended 2-4 wire hybrid interface circuit.

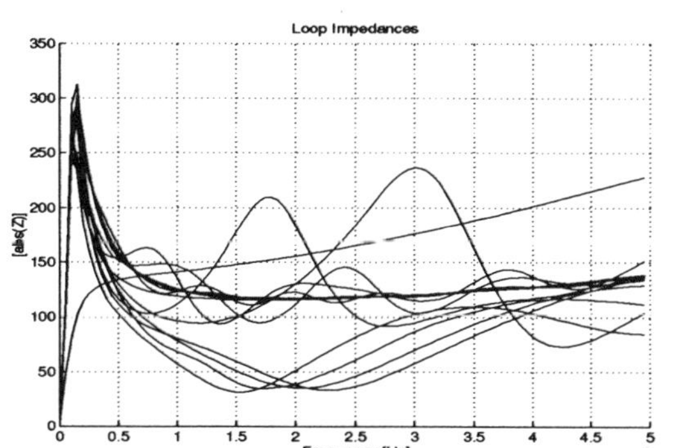

Fig. 8 Typical impedances seen on HDSL lines (transformer included).

fer-function from Tx to V_2. Finally, fully-differential realizations of this hybrid circuit are also commonly used.

Receive A/D

A typical receive A/D block is shown in Fig. 9. Note that to reduce analog complexity, the control signal for the voltage controlled amplifier is often derived from a digital signal after A/D conversion (as opposed to an analog AGC).

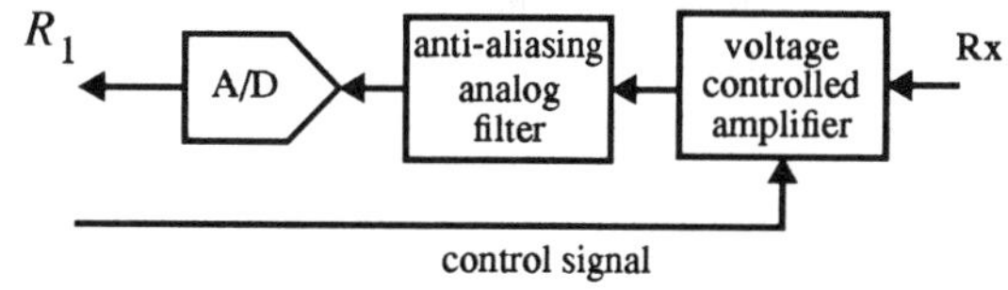

Fig. 9 Typical receive A/D block

The requirements of the anti-aliasing filter depends the sample-rate of the A/D converter. Thus, in lower speed applications, such as HDSL, oversampling can be used to ease these requirements at the expense of more digital filtering after A/D conversion. In fast-ethernet applications, this analog filter is most likely continuous-time due to its high-speed and moderate linearity requirements. For example, the proposed T2 standard suggests a 5'th-order continuous-time lowpass filter here while a 6-bit A/D is sampled at 3 times the symbol-rate (i.e. $75MHz$). One of the challenges here to keep the size and power consumption of this fast A/D converter low.

Echo Cancellation

Echo cancellation is typically realized as an adaptive transversal FIR filter making use of the LMS algorithm as shown in Fig. 10. Note that the input to the transversal filter is the transmit symbols and therefore full multiplications are not needed within the filter or in the LMS algorithm due to the limited number of transmit symbols.

Due to the high dynamic range requirement and low frequency pole (due to transformer coupling) in an HDSL echo path, the impulse response of the echo path in HDSL is quite long and requires around a 120 tap FIR adaptive filter [15]. Besides requiring many multiplications and additions, another disadvantage of a large FIR filter is that the coefficient bit size increases as the number of taps increases. Specifically, it can be shown that because of coefficient quantization, every four times increase in tap length requires another bit of coefficient accuracy. Thus, although an HDSL application needs only 13-bit dynamic range, coefficient accuracy in the echo canceler might be around 20 bits.

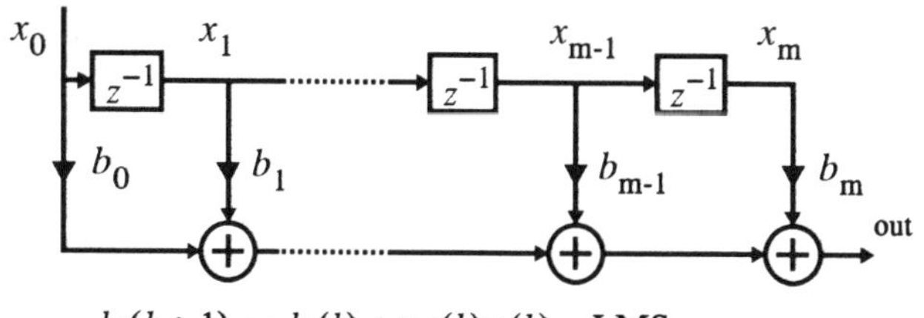

$$b_i(k+1) = b_i(k) + \mu e(k) x_i(k) \quad \text{LMS}$$

Fig. 10 An adaptive transversal FIR filter and LMS algorithm.

As a result, this echo cancellation circuitry can consume a large amount of silicon area.

In fast-ethernet applications, there are lower dynamic range requirements because of larger receive signals and thus a typical echo canceler might be around 30 taps. Also, it is sometimes possible to add some NEXT cancellation when one is transmitting on 2 (or more) cables in the same bundle. Specifically, if one is transmitting on cables 1 and 2, then the NEXT from cable 1 onto cable 2 can be cancelled by applying the transmit signal from cable 1 into the receiver circuitry of cable 2. Similarly, the NEXT from cable 2 to cable 1 can also be cancelled. Of course, the NEXT due to unrelated cables in the same bundle remains.

Equalization

Assuming digital equalization is performed, it is often accomplished through the use of a feed-forward equalizer (FFE) and decision-feedback equalizer (DFE) as shown in Fig. 11. While an FFE alone could be used, it is usually arranged that the FFE removes precursor intersymbol interference (ISI) while the DFE removes postcursor ISI.

One advantage of using a DFE is reduced noise enhancement since the input to the DFE is noise free (it is assumed to be the correct received signals). However, error propagation reduces the effectiveness of the DFE since if a wrong decision is made, this wrong decision remains in the tapped-delay line within the DFE until it is flushed out. Another advantage of using a DFE is that the multiplications within the DFE are trivial since the received signals take on a limited number of values (say one of four levels). In the case of the FFE, full multiplications are required. As a result, one would like to use a smaller number of taps in the FFE in comparison to the DFE. Fortunately, such an arrangement is often possible since the channel tends to have a longer tail following the main peak than preceding it. For example, one HDSL approach made use of a 19-tap FFE and a 128 DFE [8].

One issue that can make the DFE difficult to implement is that there should be a low latency through the DFE otherwise the first few postcursors cannot be removed. Such a low latency requirement can be difficult in a standard transversal filter where the output is formed as the sum of many signals. For example, in an m tap transversal filter as shown in Fig. 10, m signals need to be combined within one clock cycle. To overcome this large addition requirement one can use a transposed transversal filter as shown in Fig. 12 [16]. Here, the input signal is applied to many multipliers in parallel but there is a delay block between each adder. Thus, each adder has only two inputs and there is a delay between adders. It should be noted that the gradient signals are the same for the filters of Fig. 10 and Fig. 12 and therefore a separate delay line is needed in the case of the transposed filter to realize the gradient signals. Another disadvantage of the transposed structure shown in Fig. 12 is that the delay blocks need to be at least as large as the coefficient widths (say 20-bits wide) as opposed to only 2-bits wide in the transversal filter of Fig. 10.

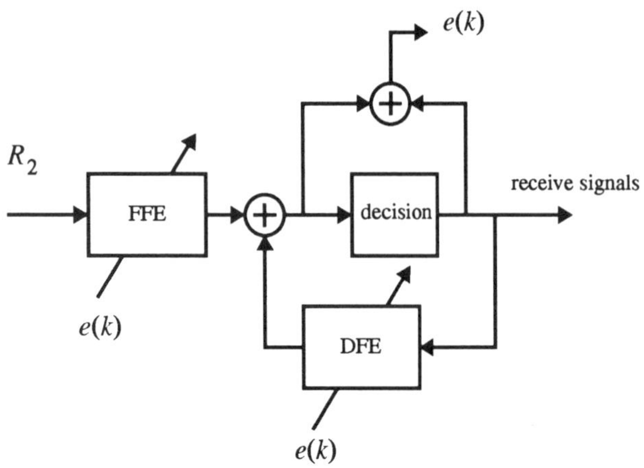

Fig. 11 A typical equalizer with feed-forward and feedback equalization.

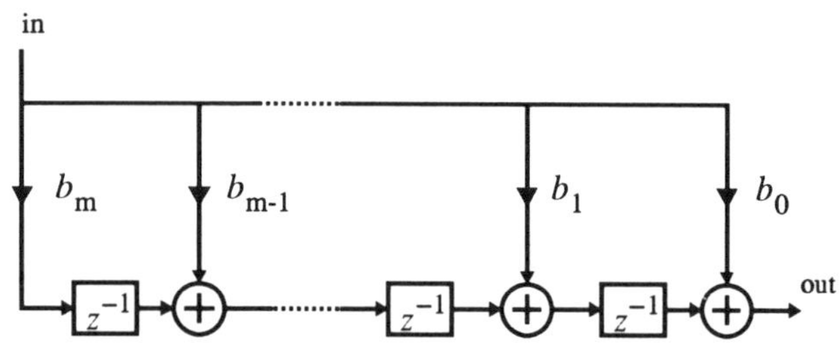

Fig. 12 Transposed transversal FIR filter.

One common alternative to the above architecture is to use a fractionally-spaced FFE such that both matched filtering and equalization is performed within the FFE [1]. A fractionally-spaced FFE operates at a higher rate than the data rate (say 3 times) and can be efficiently implemented as a polyphase filter [5] assuming downsampling is performed immediately following the FFE. Another variation is to include an error-prediction filter to aid in adaptation [6].

While digital equalization is common at HDSL rates, analog equalization is a serious alternative for applications such as fast-ethernet. The advantages of analog equalization are reduced size and power. As well, in applications such as 100Base-Tx, the cable transfer-function is reasonably well modelled with little variation once the cable length is known. As a result, a common analog equalization approach is to estimate the cable length based on the size of the received signal and thereby tune the analog filter appropriately. More complex analog equalizers have not found common usage mainly due to the difficulty in designing complex analog systems as well as their susceptibility to dc offsets [17].

Clock Recovery

A typical clock-recovery arrangement is shown in Fig. 13. If the complexities of the equalizer and echo canceler are relatively modest, the A/D can be operated at twice the symbol rate such that an early-late gate approach is used [1,18]. Specifically, the output of the equalizer is observed before and after the decision instant by $T/2$ (where T is the symbol rate) and the VCO is adjusted to make these two intermediate values equal. However, this approach requires that the echo canceler and the equalizer run at twice the

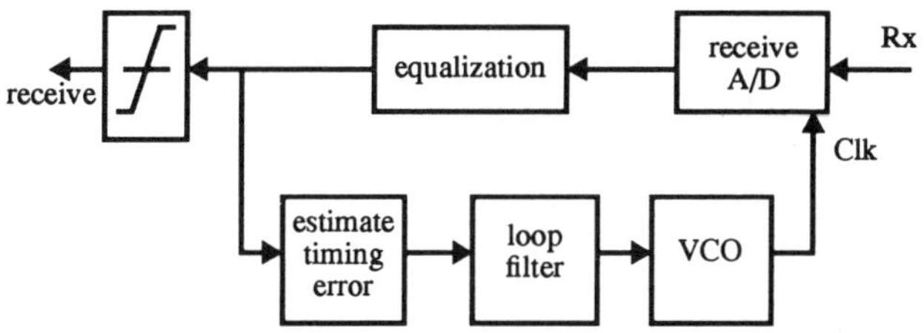

Fig. 13 A clock-recovery system (echo cancellation not shown).

symbol-rate.

In order that the rate of the equalizer or echo canceler not be increased, baud-rate timing recovery is often used [1,19]. This approach makes use of decision-directed timing adjustments that are performed in the digital domain and then indicate to the VCO whether the phase should be increased or decreased.

The VCO is often a frequency synthesizer that should display low jitter performance. While past approaches have relied on voltage-controlled crystal oscillators to maintain low jitter, a technique which is quickly gaining in popularity is to make use of oversampling techniques within the synthesizer as shown in Fig. 14 [20]. Here, a divide-by-N counter is set to one of three integer values by applying N to an oversampled delta-sigma modulator and quantizing its output. As a result, the frequency accuracy of the VCO can be set quite precisely and the jitter noise is mostly at high-frequencies allowing the loop filter to attenuate it significantly.

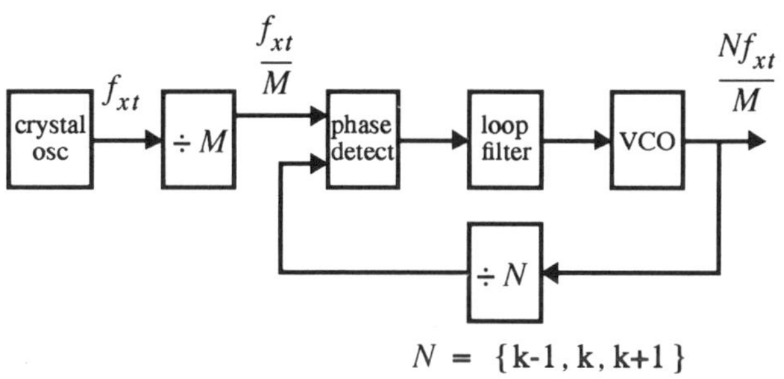

Fig. 14 A fractional-N frequency synthesizer.

Finally, it should be mentioned that low-frequency jitter is often tolerated by the addition of a FIFO which remains approximately half-full through the use of a feedback circuit (this techniques is also referred to as an elastic-buffer).

Future Challenges

While HDSL is presently delivered over two twisted-pair cables, there is much interest in single pair implementations. Realizing single-pair HDSL will effectively double the symbol-rate unless a new signalling scheme is adopted. In the fast-ethernet area, vendors are currently looking at designing 100Base-T2 chipsets that are cost competitive with TX and T4 solutions. As well, there is always interest in looking at other modulation schemes [21-23]. Finally, there is the usual push towards smaller geometry processes that allow more digital circuitry but will most likely challenge analog designers (such as only a 3.3V power-supply voltage being available).

A. Line-Drivers

In HDSL applications, maintaining the linearity of line-drivers will continue to challenge circuit designers. Furthermore, as power-supply voltage levels decrease with smaller device dimensions, producing the necessary output line voltage will be difficult. One method to allow lower voltage levels from the line-driver is to adjust the turns ratio of the line-transformer. Unfortunately, this also lowers the impedance that must be driven by the line-driver. Another approach is to use a current line-driver as shown in Fig. 15. Here, the line-driver approximates an ideal current source and needs only supply half the voltage as the approach shown in Fig. 7. However, this current line-driver must source twice as much current and it may be difficult to maintain the same linearity as the voltage approach.

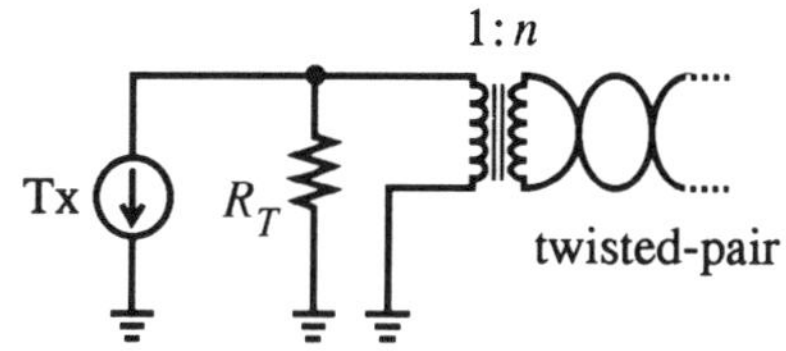

Fig. 15 A current driven line-driver.

B. 2-4 Wire Hybrids

If 2-4 wire hybrids can be realized with less echo return-loss, the linearity requirement of the echo path can be reduced (i.e. the D/A, line-driver, and A/D converter). Such extra echo cancellation may be necessary to realize single--pair HDSL. However, as the trend is towards more integration with less analog and discrete components, it is likely that this extra cancellation might cost extra A/D or D/A converters. One such hybrid echo cancellation circuitry is discussed in [24].

C. Echo Cancellation

As discussed above, the echo cancellation circuitry in HDSL applications can require over a 120 tap FIR filter. For higher data-rates, this already long filter length will further increase. As noted in [15], there is likely some benefit of creating hybrid FIR/IIR type echo cancellation circuits to reduce overall complexity. Thus, a challenge for circuit designers will be to realize effective adaptive IIR filtering.

Another interesting avenue of research is to realize efficient non-linear echo cancellation circuits such that the linearity requirements of the echo path can be relaxed. This challenge is complicated by the fact that the non-linearities being introduced into the echo path are not typically memoryless and are therefore better modelled using Volterra series expansions rather than Taylor series.

D. Equalization

It is likely that digital equalization will be used on channels having modest data-rates. To improve performance, it is sometimes advantageous to equalize to a lowpass response thereby treating the channel as a type of partial-response system (this approach is now adopted by many disk-drive manufactures and is known as PRML [25]). By reducing the channel boost at high-frequencies, less noise is amplified. However, now some type of maximum likelihood detector is needed and is most likely sub-optimal for complexity reasons [26]. While this maximum likelihood detection approach can ideally achieve an extra 3dB performance, other imperfections in the receive signal and modelling typically limit the improvement to much smaller values.

As mentioned above, for high-speed channels, such as fast-ethernet, analog equalization is a serious alternative. The two main advantages of analog equalization are lower power and less silicon area. The main challenge in analog equalization remains dealing with dc offsets [17] as well as realizing efficient programmable filter structures.

E. Clock-Recovery

Jitter is always an important performance criteria of clock-recovery. While most present clock recovery techniques adjust the sampling instance, it is also possible to perform fully digital clock recovery by re-sampling the digital signal (in other words, use sample-rate conversion techniques [5]). While the complexity of digitally re-sampling is usually considered too costly, there are advantages to be gained such as running the A/D converter off a crystal controlled clock having extremely low jitter. Another clock-recovery challenge is that of tracking high-frequency jitter since the latency of the clock-recovery loop often includes equalization and therefore has limited tracking speed. This latency can be shortened by not placing the equalizer and any extra filtering in the clock-recovery loop but implies that either an oversampled or analog equalizer is used. The clock-recovery can then be accomplished by digital re-sampling at the cost of extra circuit complexity.

Summary

This paper discussed modern architectures for realizing high-speed data transmission over twisted-pair cables and addressed two applications in particular — HDSL and fast-ethernet. With HDSL, the echo path linearity is seen to be one of the most challenging system requirements. In the fast-ethernet application, maintaining a low cost and power consumption is perhaps the biggest challenge. However, with the wide variety of applications requiring data transmission over twisted-pair channels, perhaps the biggest challenge is in deciding which applications to pursue.

References

[1] E.A. Lee and D.G. Messerschmitt, *Digital Communication*, Kluwer Academic Publishers, Norwell, Mass., 1994.

[2] W.D. Reeve, *Subscriber Loop Signalling and Transmission Handbook: Digital*, IEEE Press, Piscataway, NJ, 1995.

[3] J.J. Werner, "The HDSL environment," *IEEE Journal on Selected Areas in Communications*, vol. 9, pp. 785-800, Aug. 1991.

[4] G.H. Im, et al, "51.84 Mb/s 16-CAP ATM LAN standard," *IEEE Journal on Selected Areas in Communications*, vol. 13, pp. 620-632, May 1995.

[5] P.P. Vaidyanathan, *Multirate Systems and Filter Banks*, Prentice Hall Inc., Englewood Cliffs, New Jersey, 1993.

[6] D. Essig, et al, "An HDSL chipset for high-speed transmission over copper," *Proc. of IEEE Int'l Conf. on Comm.*, pp. 461-465, New Orleans, 1994.

[7] *HDSL Data Book*, Brooktree Corp., 1995.

[8] M.A. Kuczynski, et al, "A 1Mb/s digital subscriber line transceiver signal processor," *IEEE Int'l Solid-State Circuits Conf.*, pp. 26-27, San Francisco, Feb. 1993.

[9] A Shoval, W.M. Snelgrove and D.A. Johns, "A 100Mb/s BiCMOS adaptive pulse-shaping filter," in press, *IEEE Journal on Selected Areas in Communications: Special issue on Copper Wire Access Technologies for High Performance Networks.*

[10] F.N.L. Op'T Eynde, P.F.M. Ampe, L. Verdeyen, and W.M.C. Sansen, "A CMOS large-swing low-distortion three-stage class AB power amplifier," *IEEE Journal of Solid-State Circuits*, vol. 25, pp. 265-273, Feb. 1990.

[11] L. Tomasini, A. Gola, and R. Castello, "A fully differential CMOS line driver for ISDN," *IEEE Journal of Solid-State Circuits*, vol. 25, pp. 546-554, Apr. 1990.

[12] H. Khorramabadi, "A CMOS line driver with 80-dB linearity for ISDN applications," *IEEE Journal of Solid-State Circuits*, vol. 27, pp. 539-544, Apr. 1992.

[13] R.G.H. Eschauzier, R. Hogervorst and J.H. Huijsing, "A programmable 1.5V CMOS class-AB operational amplifier with hybrid nested Miller compensation for 120dB gain and 6MHz UGF," *IEEE Journal of Solid-State Circuits*, vol. 29, pp. 1497-1504, Dec. 1994.

[14] R. Hogervorst, J.P. Tero, R.G.H. Eschauzier, and J.H. Huijsing, "A compact power-efficient 3V CMOS rail-to-rail input/output operational amplifier for VLSI libraries," *IEEE Journal of Solid-State Circuits*, vol. 29, pp. 1505-1512, Dec. 1994.

[15] W.Y. Chen, J.L. Dixon, and D.L. Waring, "High bit rate digital subscriber line echo cancellation," *IEEE Journal on Selected Areas in Communications*, pp. 848-860, Aug. 1991.

[16] H. Samueli, B. Daneshrad, R.B. Joshi, and B.C. Wong, "A 64-tap CMOS echo canceller/decision-feedback equalizer for 2B1Q transceivers," *IEEE Journal on Selected Areas in Communications*, pp. 839-847, Aug. 1991.

[17] A. Shoval, D.A. Johns, and W.M. Snelgrove, "DC offset performance of four LMS adaptive algorithms," *IEEE Trans. on Circuits and Systems — II: Analog and Digital Signal Processing*, vol. 42, pp. 176-185, March 1995.

[18] A. Buchwald and K.W. Martin, *Integrated fiber-optic receivers*, Kluwer Academic Publishers, Norwell, Mass., 1995.

[19] K.H. Mueller and M. Muller, "Timing recovery in digital synchronous data receivers," *IEEE Trans. on Communications*, vol. 24, pp. 516-531, May 1976.

[20] T.A.D. Riley, M.A. Copeland and T.A. Kwasniewski, "Delta-sigma modulation in fractional-N frequency synthesis," *IEEE Journal of Solid-State Circuits*, vol. 28, pp. 553-559, May 1993.

[21] G.H. Im and J.J. Werner, "Bandwidth-efficient digital transmission up to 155 Mb/s over unsheilded twisted pair wiring," *Proc. of IEEE Int'l Conf. on Comm.*, pp. 1797-1803, Geneva, 1993.

[22] G. Cherubini, S. Ölçer, and G. Ungerboeck, "A quaternary partial-response class-IV system for 125Mb/s data transmission over unsheilded twisted-pair cables," *Proc. of IEEE Int'l Conf. on Comm.*, pp. 1814-1819, Geneva, 1993.

[23] W.E. Stephens and T.C. Banwell, "155.52 Mb/s data transmission on category 5 cable plant," *IEEE Communications Magazine*, pp. 62-69, April 1995.

[24] D. Mueller and A. Kaelin, "A hybrid HDSL echo canceler," *Int'l Symposium on Circuits and Systems*, Seattle, Washington, May, 1995.

[25] M.H. Shakiba, D.A. Johns and K.W. Martin, "A 200MHz 3.3V BiCMOS class-IV partial response analog Viterbi decoder," *Custom Integrated Circuits Conference*, pp. 567-570, Santa Clara, California, May, 1995.

[26] M.V. Eyaboglu and S. Qureshi, "Reduced-state sequence estimation with set partioning and decision feedback," *IEEE Trans. on Comm.*, vol. 36, Jan. 1988.

Timing Recovery in CMOS using Nonlinear Spectral-line Method

Un-Ku Moon, Angelo Mastrocola, Jeanann Alsayegh, and Scott Werner
AT&T Bell Laboratories, Allentown, PA 18103

ABSTRACT: **In a Carrierless AM/PM (CAP) passband modulation scheme where the waveform does not contain a baud rate spectral line due to zero mean-value of the data symbols, the nonlinear spectral-line method is applied to extract the symbol rate. It is achieved first by a squaring function which draws out the higher moments of the signal that are periodic at the symbol rate; then the signal is further processed through a bandpass filter and phase-locked loop (PLL) combination to recover signal timing. All necessary timing recovery functions are implemented in the analog (continuous-time) domain, and the recovered timing information is used by a receive equalizer. The timing recovery block implemented in 0.9μm CMOS includes a multiplier for the squaring function, a self-tuning bandpass filter, and a PLL using an external VCXO. The 51.84MHz recovered clock allows a BER of 10^{-10} [1].**

I. INTRODUCTION

This paper describes an IC implementation of timing recovery using the nonlinear spectral-line method [2]. This analog implementation of the timing recovery IC was fabricated in 0.9μm CMOS technology and fully verified to meet all the requirements for a system utilizing Carrierless AM/PM (CAP) modulation scheme [1]. The following sections will include a brief summary of the nonlinear spectral-line method (section II), the overall structure and some sub-block details of the IC implementation (section III) including the squarer, the self-tuned bandpass filter, and PLL, and some meaningful measurement results (sections IV).

II. NONLINEAR SPECTRAL-LINE METHOD

The key principle in nonlinear spectral-line timing recovery is that a memoryless nonlinearity, $f(\cdot)$, is applied to the incoming signal, $s(t)$, so that the expected value of the output, $E[r(t)]$, where $r(t)=f(s(t))$, is non-zero and periodic with the symbol period T [2-5]. This is due to the inherent fact that if we treat the phase of the signal (signals in modulation schemes such as QAM, PAM, and CAP) as a random parameter, the signal represents a cyclostationary process. In a simpler case where $E(s(t))$ is already non-zero and periodic with period T (i.e. cyclostationary), the necessary spectral-line at the symbol rate exists and a nonlinearity does not need to be applied. Such condition qualifies as *linear* spectral-line method [2].

Typical ways of applying a memoryless nonlinearity to the zero-mean input signal are using a class of absolute value circuits (i.e. rectifiers), a squarer, and even fourth-power circuits [6]. Regardless of the method used, it must elicit the cyclostationary behavior that is contained at the higher moments of the incoming signal. Perhaps the most common and clearly explained implementation is the squarer. The IC implementation uses this method, sometimes known as envelope-derived timing recovery [2-3].

In regard to understanding the functions of the memoryless nonlinearity in the context of timing recovery, it is generally "safe" to dismiss any distinction between different in-phase and quadrature data-transmission systems such as QAM and CAP which can fall under passband PAM [2], and further dismiss any differences between baseband and passband PAM [2-3]. This is well-justified given that some sort of band-limiting pre-filter normally exists (typically part of the transmission path), and the output of the nonlinear function, $f(\cdot)$, is filtered by a bandpass filter.

The timing recovery IC operates on a received passband signal

$$s(t)=\mathrm{Re}\left\{\sum_{n=-\infty}^{\infty} c_n g(t-nT)e^{(j\omega_c t)}\right\},$$

where $c_n=a_n+jb_n$ is the complex input data sequence and $g(t)$ is the filtered baseband pulse shape modulated by the carrier ω_c. The squared signal, $r(t)=s(t)^2$, has a desired timing tone (symbol rate ω_s) whose power is [3]

$$R(\omega)|_{(\omega=\omega_s)}=\sum_{k=0}^{N-1}|C(k\Omega)|^2G(k\Omega)G(\omega_s-k\Omega). \quad (1)$$

In (1), $C(k\Omega)=\sum_{n=0}^{N-1}c_n e^{(j\omega_c t)}$, $\Omega=\frac{\omega_s}{N}$, is the DFT of the complex input data sequence c_n with period N. $G(\omega)$ is the spectrum of $g(t)$.

The DFT is invoked because of the periodic nature of c_n. However, in the limiting case of perfectly random, uncorrelated c_n, $N\rightarrow\infty$ and $|C(\omega)|^2\rightarrow$ constant, and (1) approaches the form

$$R(\omega)|_{(\omega=\omega_s)}=\overline{c^2}\int_{-\infty}^{\infty}G(\omega)G(\omega_s-\omega)d\omega. \quad (2)$$

From Fig. 1 and 2, it is clear that the amount of overlap between $G(\omega)$ and $G(\omega_s-\omega)$ directly impacts on the strength of the desired timing tone. This spectral overlap is a function of the excess bandwidth parameter α, which can vary from 0 to 1 (0 to 100%). Note that in the extreme case of α=0, $G(\omega)$ and $G(\omega_s-\omega)$ are disjoint, and no timing tone is generated. In our applications, signals are transmitted with

0-7803-3177-6 $5.00 © 1996 IEEE

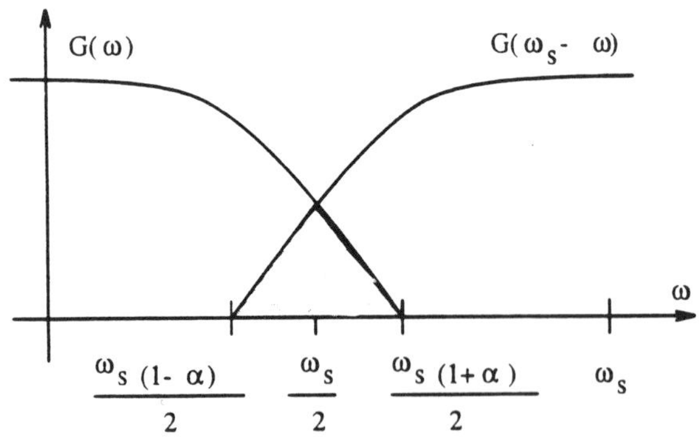

Fig. 1 Effect of excess bandwidth on symbol tone extraction

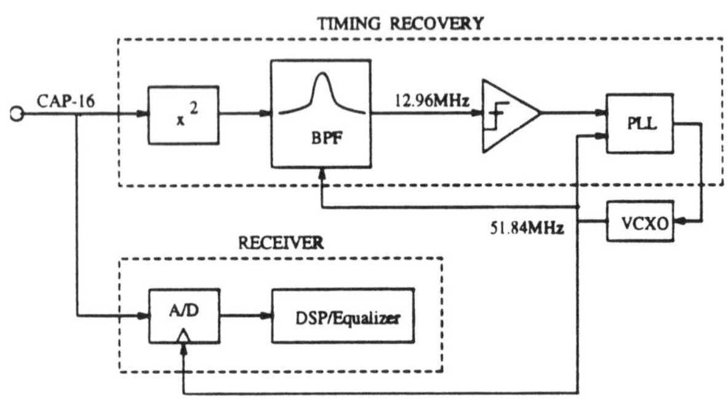

Fig. 2 Overall timing recovery structure

100% [1] and 50% excess bandwidth.

III. Timing Recovery Implementation

The overall structure of the IC implementation is shown in Fig. 2. This timing recovery is realized in the continuous-time (except for the inherent sampling nature of the phase detector) domain. To minimize sensitivity to the surrounding noisy environment, all circuits are fully differential wherever possible. As discussed in the preceding section, the squaring function is used to apply a memoryless nonlinearity to the incoming signal. And the subsequent bandpass filter eliminates a large amount of undesired tones (noise) surrounding the symbol frequency (12.96MHz). The input to the PLL is a slicer which turns this "noisy tone" into a "jittery clock" (which can have missing or extra edges) at this symbol rate; the PLL locks onto this jittery input clock and filters out its jitter using an external VCXO. After the symbol timing is recovered, the jitter-free clock is fed into an application-specific receiver which processes the incoming CAP signal.

A. Squarer (Multiplier)

The squaring function is implemented with a set of cross-coupled MOSFETs operating in triode [7], as shown in Fig. 3. The input and output buffers are source-follower stages to remove the resistive load requirement from the previous stage, and to have a large enough driving capability at the output. The op amp symbol represents a differential single-transistor gain stage with simple common-mode feedback.

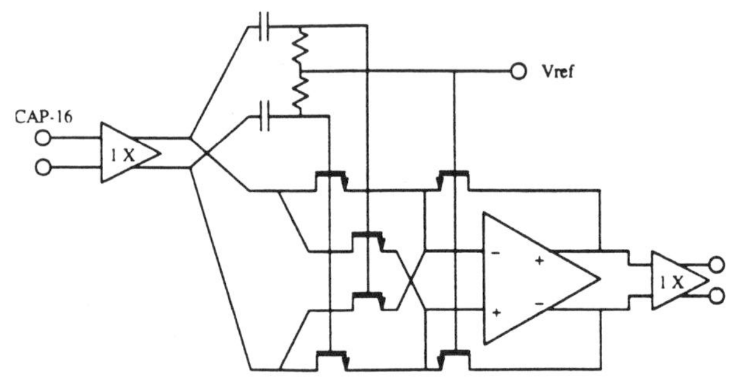

Fig. 3 Squaring function by using a multiplier

This portion of the design was kept simple to achieve a large bandwidth so that the squaring function would not be hindered by the limited bandwidth. The overall signal path has more than 100MHz bandwidth for either inputs (X or Y) according to simulation results. The bandwidth on the squaring/multiplying function is limited only by the input buffers and the cross-coupled MOSFETs (current output). The rest of the bandwidth limitation only applies as a spectrum shaping function, which only needs a 12.96MHz bandwidth in this application. For the set of cross-coupled MOSFETs alone, even a gigahertz range of input bandwidth can be achieved by placing an enhancement capacitor across the virtual grounds [8]. Lastly, the feedback resistors are implemented with triode MOSFETs to match the input triode devices for superior gain stability over process and temperature variation.

B. Self-tuned Bandpass Filter

The bandpass filter and the automatic tuning circuit structures are similar to Khoury's Gm-C filter design [9]. For this application, the two blocks used for voltage-controlled oscillator (VCO) and the bandpass filter are identical. For the VCO, the signal input is disabled, and for the filter, the positive feedback is disabled. This should allow a better matching between master (VCO) and slave (filter) in achieving an accurate automatic-tuning.

The VCO oscillates due to the positive feedback at the signal's zero-crossing [9]. This is done by the limited range in the transconductance of the positive feedback input. At larger signal swings of either polarity, the positive feedback is removed, until the signal swings back near zero. At zero crossing, a net positive feedback is realized. One can visualize the operation by picturing a pendulum swinging back and forth at its natural frequency ($\frac{G_m}{C}$) while a small horizontal force is applied in the direction of movement at the instance where the arm is lined up vertically.

Implanted in these VCO and filter biquad blocks is a small range of digital tunability should it be needed for improved tuning accuracy. Figure 4 displays the bandpass filter measured frequency response for a few digital settings. For the few dozen devices characterized, the center frequency of the

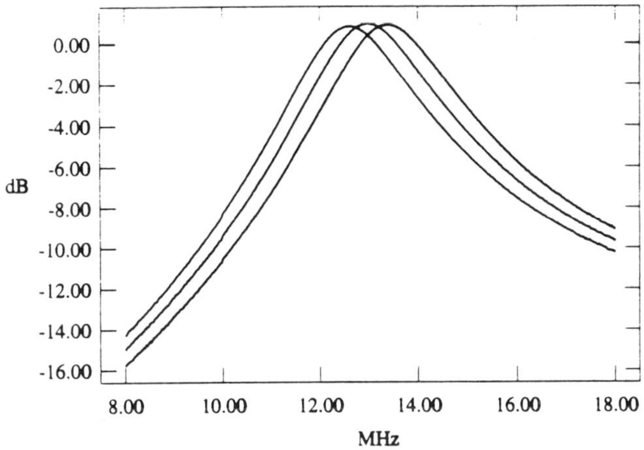

Fig. 4 Measured bandpass filter frequency response

filter at its default setting was within ±5% of 12.96MHz. The quality factor (Q) and the frequency accuracy (f_c) of the bandpass filter directly affects the final jitter of the recovered clock. Having a smaller Q requires a better PLL implementation (more noise gets through). However, a larger Q filters out more noise, but requires a more accurate f_c (or else we lose the timing tone). Such tradeoffs must be considered in optimizing performance over process, temperature, and matching tolerances. Reported discrete (manually-tuned) implementations have up to Q=50 [5], while it it impractical for IC design. Some pre-filtering techniques (before the squarer) can further aid in reducing the jitter [2-3],[5].

C. Phase-locked loop

The PLL requires a very narrow-band jitter transfer function; it must filter out the majority of the wide-band jitter still contained in the signal processed through the squarer and the bandpass filter. Use of a VCXO with a very narrow tuning range aids in this matter. An AT&T 154-type quartz crystal oscillator (or equivalent) with a tuning range of ±100ppm for 1-4volt input control range (5V supply) is used to make up the complete PLL.

As shown in Fig. 2, the input to the PLL is first squared-up by a high-speed slicer (zero-threshold comparator). The phase detector inside the PLL block is a simple XOR gate. It is important to use a non-edge-triggered phase detector because the input to the phase detector is very jittery with multiple edges (or no edge) per symbol period. Simulation results as well as lab measurements have verified this phenomenon. The buffered XOR output drives a charge-pump, which is filtered by external components. An all-digital implementation similar to [10] was also verified to function properly both in simulation and measurement, but the XOR and charge-pump combination was chosen for its simplicity and smaller chip area.

IV. Measurement Results

Presented in the following figures are the measurement results, verifying the intended and anticipated performance of the timing recovery IC. Extensive simulation results prior to fabrication have shown close agreement with the measurement. Simulations were performed using a combination of C code, MATLAB, and AT&T's version of SPICE (ADVICE). The channel to the timing recovery block was modeled with a pseudo-random number generator for the data sequence, in-phase and quadrature FIR filters for generating the CAP-16 signal, an FIR-modeled transmission line, an all-passive LC smoothing filter, and additive white Gaussian noise (AWGN) yielding a SNR of 20dB.

The measured spectrum of the incoming signal for a 50% excess bandwidth system is shown in Fig. 5. The lab measurements displaying the regeneration of the symbol tone by squaring and filtering is shown in Fig. 6.

With the help of a narrowly tunable VCXO, the measured jitter transfer function of the PLL demonstrates about 500Hz -3dB corner frequency. Fig. 7 displays the PLL transfer

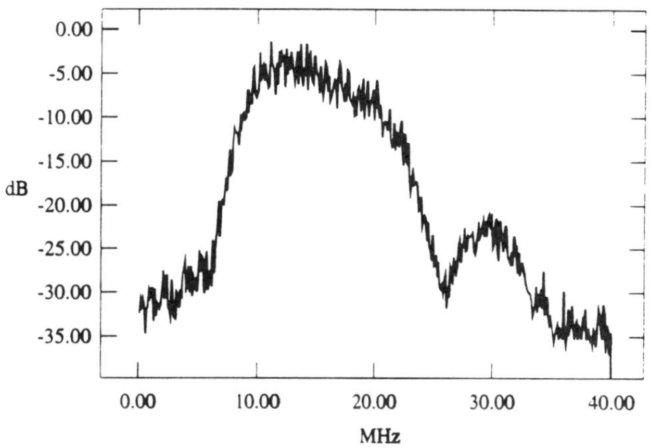

Fig. 5 Measured spectrum of incoming CAP-16 signal

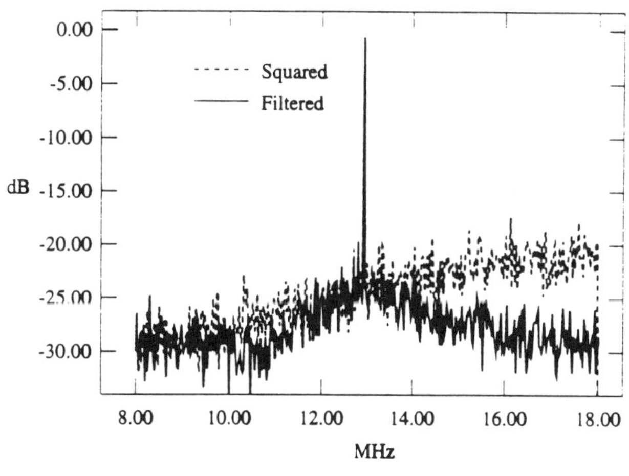

Fig. 6 Measured spectrum after squarer and filter

function and its measured output jitter spectrum, which follow a similar shape over frequency. Finally, shown in Fig. 8 and 9, respectively, are the recovered clock used by a CAP-16 receive equalizer and the constellation of the demodulated, received data that was sent over a maximum length (worstcase) test-link. The recovered clock, measuring about

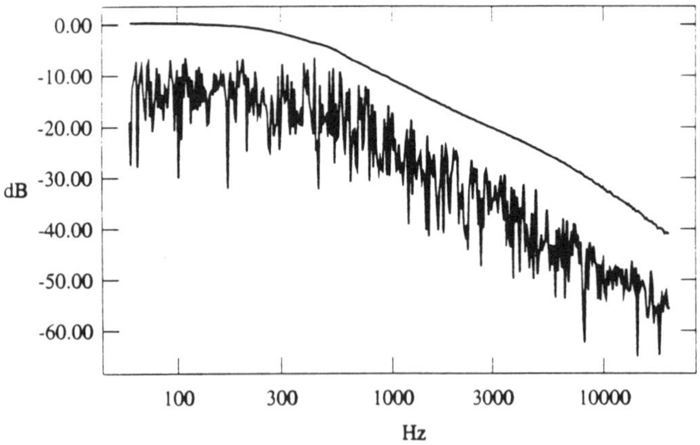

Fig. 7 Measured jitter transfer function and spectrum

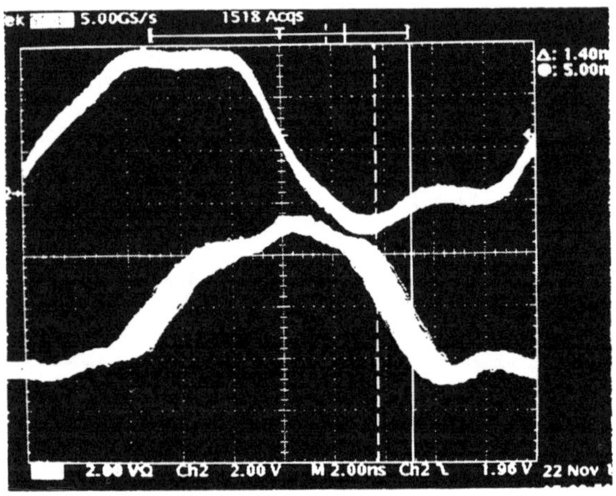

Fig. 8 Measured recovered clock at sample rate (51.84MHz)

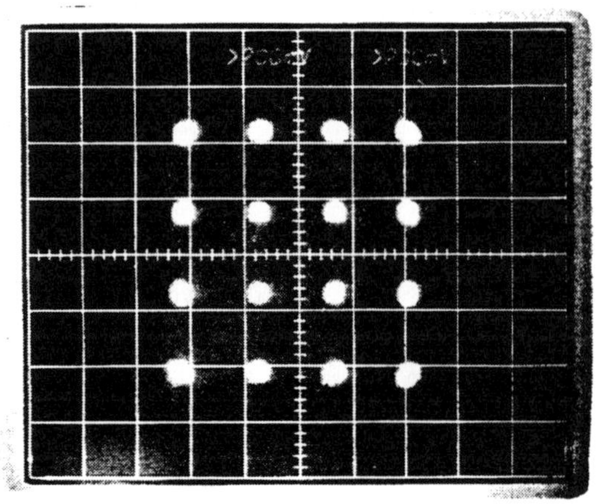

Fig. 9 Measured received CAP-16 constellation

1.4ns peak-to-peak jitter, meets the BER requirement of 10^{-10} for an application such as [1].

V. Conclusion

A timing recovery scheme using a nonlinear spectral-line method was presented. An all-analog implementation in 0.9μm CMOS technology successfully demonstrates the required performance for an application using Carrierless AM/PM [1]. Justification for the IC's applicability in all synchronous multi-level pulse amplitude modulation systems has been summarized. The three key functional blocks in this timing recovery structure, squarer, bandpass filter, and PLL, were described, and design tradeoffs were discussed. Some ways of improving the timing recovery process such as prefiltering before the squarer, and higher-Q with more accurate f_c of the post-filter have been mentioned.

VI. Acknowledgment

Authors are thankful for much help provided by Nat Dwarakanath and his team, Art Grandle and his team, Derrick Johnson, Jit Kumar and his team, Dale Nelson, Gerry Pepenella and his team, Jacque Ruch, Joe Trackim, J.J. Werner and his team, and Jay Zeman.

References

[1] G.H. Im, D.D. Harman, G. Huang, A.V. Mandzik, M.H. Nguyen, and J.J. Werner, "51.84 Mb/s 16-CAP ATM LAN standard," *IEEE J. Select. Areas Commun.*, vol. 13, no. 4, pp. 620-632, May 1995.

[2] E.A. Lee and D.G. Messerschmitt, *Digital Communication*, Second Edition, Boston: Kluwer Academic Publishers, 1994.

[3] R.D. Gitlin and J.F. Hayes, "Timing recovery and scramblers in data transmission," *Bell Syst. Tech. J.*, vol. 54, no. 3, pp. 569-593, March 1975.

[4] W.R. Bennett, "Statistics of regenerative digital transmission," *Bell Syst. Tech. J.*, vol. 37, pp. 1501-1542, Nov. 1958.

[5] L.E. Franks and J.P. Bubrouski, "Statistical properties of timing jitter in a PAM timing recovery scheme," *IEEE Tran. Commun.*, vol. COM-22, pp. 913-920, July 1974.

[6] J.E. Mazo, "Jitter comparison of tones generated by squaring and by fourth-power circuits," *Bell Syst. Tech. J.*, vol. 57, pp. 1489-1498, May-June 1978.

[7] B.-S. Song, "CMOS RF circuits for data communications applications," *IEEE J. Solid-State Circuits*, vol. SC-21, no. 2, pp. 310-317, April 1986.

[8] J. Crols and M.S.J. Steyaert, "A 1.5 GHz highly linear CMOS down-conversion mixer," *IEEE J. Solid-State Circuits*, vol. 30, no. 7, pp. 736-742, July 1995.

[9] J. Khoury, "Design of a 15-MHz CMOS continuous-time filter with on-chip tuning," *IEEE J. Solid-State Circuits*, vol. 26, no. 12, pp. 1988-1997, Dec. 1991.

[10] J.F. Ewen, A.X. Widmer, M. Soyuer, K.R. Wrenner, B. Parker, and H.A. Ainspan, "Single-chip 1062Mbaud CMOS transceiver for serial data communication," *ISSCC Dig. Tech. Papers*, pp. 32-33, Feb. 1995.

A Mixed-Signal Decision-Feedback Equalizer That Uses Parallelism

Ravinder S. Kajley, James E. C. Brown and Paul J. Hurst

Solid-State Circuits Research Laboratory
Department of Electrical and Computer Engineering
University of California, Davis, CA, 95616

ABSTRACT

A mixed-signal decision-feedback equalizer (DFE) that uses parallelism is described. The parallelism in the look-ahead DFE achieves an increase in the data rate over that of a conventional DFE. The DFE occupies 23 mm^2 in a 2-μm CMOS process, operates at 55 Mb/s and dissipates 450 mW.

INTRODUCTION

Decision-feedback equalization is a sampling detector that can be used when intersymbol interference (ISI) is present, such as in a disk-drive read channel [1,2,3] or other communication channels [4]. The decision-feedback equalizer (DFE) is used in combination with a linear forward equalizer to remove all ISI. Fig. 1 shows the block diagram of a complete hard-disk read channel that uses a DFE. The read channel contains many components in addition to the DFE: timing recovery, forward equalizer, automatic gain control (AGC), and preamp.

This paper describes a mixed-signal implementation of a look-ahead DFE for a disk-drive read channel. This DFE is an extension of a previous mixed-signal DFE design [5]. The key difference here is the use of parallelism in the look-ahead structure, which achieves an increase of the data rate of the DFE. The data rate of the new DFE is 55 Mb/s, while the DFE in [5] operates at 35 Mb/s. Both DFEs are fabricated in the same technology and use many of the same circuits.

DFE BACKGROUND

The forward equalizer in Fig. 1 is responsible for removal of precursor ISI. The DFE removes the remaining postcursor ISI by using a linear combination

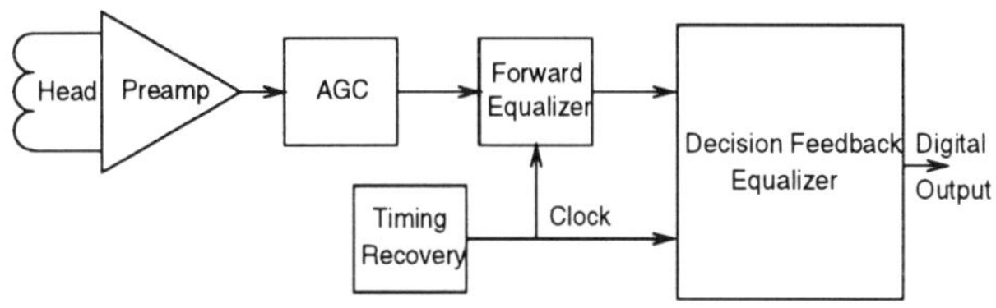

Fig. 1: Disk-drive read channel block diagram

of past decisions to cancel the interference [1,4]. The postcursor ISI, ISI^+, at time nT can be expressed as a weighted summation of past bits, plus an offset:

$$ISI^+(n) = h_{offset} + \sum_{k=1}^{M} h(k)\hat{y}(n-k) \qquad (1)$$

where $\{h(k)\}$ is the response of the disk channel plus forward equalizer sampled at bit times kT, $\{\hat{y}(n-k)\}$ is past binary data, and h_{offset} is the d.c. offset. M is the number of bits that contribute significant postcursor ISI.

Fig. 2 shows an adaptive 5-tap DFE, which can cancel the postcursor ISI. Four taps cancel postcursor ISI, and one tap cancels the d.c. offset. Based on simulations using a Lorentzian channel model [1] with $PW_{50} = 2T$ and appropriate forward equalization, a 5-tap DFE is sufficient to achieve MSE < –25 dB for a noise-free channel, where MSE = $E(e^2(n))$ is the mean-squared error and $e(n) = \hat{y}(n) - \tilde{y}(n)$ is the error across the slicer.

The coefficients d_k in Fig. 2 can be determined by adapting the DFE using an algorithm which minimizes the MSE. Once the DFE has adapted, the DFE coefficients are $d_k = -h(k)$ for $1 \le k \le M$ and $d_0 = -h_{offset}$. Here d_k is the coefficient at tap k, d_0 is the d.c. offset cancellation coefficient, and M is the number of taps. With these coefficients, the ISI^+ is canceled by the DFE.

The DFE coefficients can be adapted using the sign-LMS algorithm [4,6]:

$$d_k(n+1) = d_k(n) + \beta \text{sign}(e(n))\hat{y}(n-k) \quad \text{for } 1 \le k \le M$$

and $$d_0(n+1) = d_0(n) + \beta \text{sign}(e(n)) \, , \qquad (2)$$

This work was supported in part by a UC MICRO grant, co-sponsored by Allied Telesyn, Analog Devices, Hewlett-Packard, IBM, Level One Communications, National Semiconductor, Quantum, Seagate and Silicon Systems.

0-7803-3177-6 $5.00 © 1996 IEEE

where $\beta << 1$ is the gain of the update integrator and n is time. The sign-LMS update equation for each coefficient can be implemented by a L-bit up/down counter if $\beta = 2^{-L}$ because sign($e(n)$) $\hat{y}(n-k)$ is either +1 or −1. The sign-LMS update for coefficient d_1 is shown in Fig. 2.

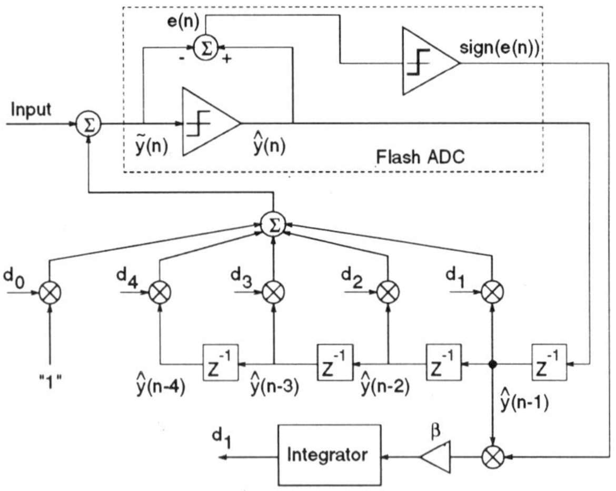

Fig. 2: DFE with sign-LMS update shown for coefficient d_1. Z^{-1} is a delay of a bit period T.

A MIXED-SIGNAL DFE ARCHITECTURE

Fig. 3 shows a block diagram of a mixed-signal DFE, which is practically identical to the design in [5]. (The only difference is that the ADC outputs a 2-level sign(e) here rather than the 3-level quantized error signal $q(e)$ in [5].) The look-ahead DFE that is described in the next two sections is a modification of this mixed-signal DFE and uses many of these circuit blocks. The input is an analog current coming from a preceding analog forward equalizer [7-11]. A current-to-voltage (I2V) converter feeds a small flash ADC which outputs the binary decision $\hat{y}(n)$ and sign($e(n)$) using only 3 comparators. Digital logic and 10-bit counters are used in the coefficient update. The digital output of the counters feed resistor-string 6-bit digital-to-analog converters (DACs) that drive current-output multipliers [12]. The multipliers perform the multiplication of the past binary decision $\hat{y}(n-k)$ by the analog coefficient value d_k. The current outputs are all summed at the I2V input simply by connecting them together.

LOOK-AHEAD DFE CONCEPT

The critical, speed-limiting feedback loop in Fig. 3 is through the I2V converter, flash ADC and d_1 multiplier. This feedback loop must have a total delay less than one bit period, T. Although shown as a separate delay element, most of the Z^{-1} delay in this critical loop is in the flash ADC. To reduce T and thereby increase the data rate of the DFE, parallelism can be used.

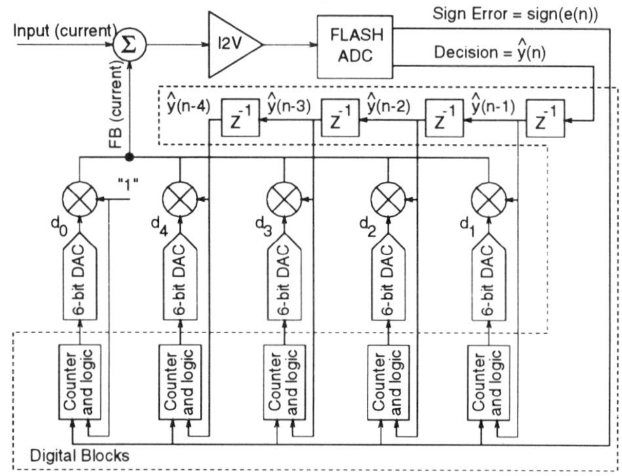

Fig. 3: Mixed-signal DFE showing analog and digital circuit blocks. The digital blocks are enclosed by the dashed line.

Parallelism is used in the look-ahead (LA) DFE to change the critical, speed-limiting feedback loop to being through the d_2 multiplier rather than the d_1 multiplier [13]. The LA DFE can be explained conceptually using Fig. 4. There are two parallel processing paths in the LA DFE. Each path is a copy of the DFE of Fig. 2 with the following changes. The upper path always assumes $\hat{y}(n-1) = +1$ and uses that value in its convolution calculation. The lower path always assumes and uses $\hat{y}(n-1) = -1$. Both paths use the same d_k coefficients and the same (correct) values of $\hat{y}(n-2)$, $\hat{y}(n-3)$ and $\hat{y}(n-4)$ in the convolution calculation. The two parallel paths produce inputs to the upper and lower ADCs that output two tentative decisions $\hat{y}_{+1}(n)$ and $\hat{y}_{-1}(n)$, where $\hat{y}_{+1}(n) = [\,\hat{y}(n)$, if $\hat{y}(n-1) = +1]$ and $\hat{y}_{-1}(n) = [\,\hat{y}(n)$, if $\hat{y}(n-1) = -1]$.

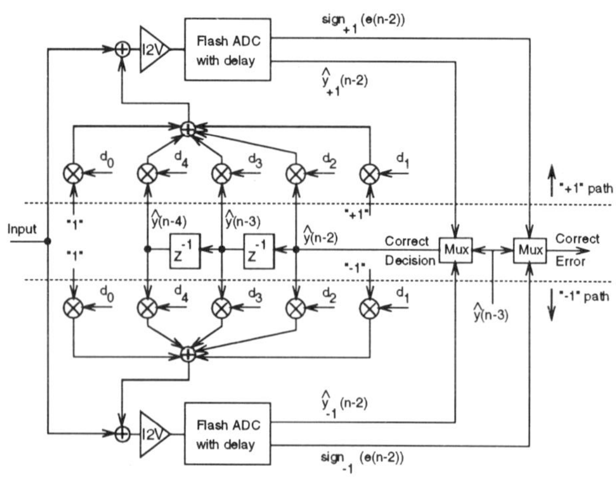

Fig. 4: LA DFE block diagram. LMS update hardware is not shown.

Selection between tentative decisions $\hat{y}_{+1}(n)$ and $\hat{y}_{-1}(n)$ could be made using $\hat{y}(n-1)$, if these values were available at time n. However, at time n, the tentative decisions and $\hat{y}(n-1)$ are essentially 'stuck' inside the ADC and are not available. Therefore, the selection is delayed and is made between the delayed tentative decisions $\hat{y}_{+1}(n-2)$ and $\hat{y}_{-1}(n-2)$, with the selection determined by the value of $\hat{y}(n-3)$. At start-up, a training sequence of known data is used for rapid and reliable convergence [4]. This sequence of known data initializes the selection process, giving known values for the select signal $\hat{y}(n-3)$ during training. At the end of the training data, the selection process can continue to operate using the correct delayed decision that is output by the LA DFE. The two paths also output tentative sign(e) signals, one of which is selected by the past decision $\hat{y}(n-3)$. The selected sign($e(n-2)$) signal is used to update the coefficients, using a delayed version of the sign-LMS coefficient update algorithm [14]:

$$d_k(n+1) = d_k(n) + \beta \mathrm{sign}(e(n-2))\hat{y}(n-k-2) \quad \text{for } 1 \le k \le M$$

and $$d_0(n+1) = d_0(n) + \beta \mathrm{sign}(e(n-2)) \,. \tag{3}$$

In the LA DFE, the critical, speed-limiting feedback loop is through the d_2 multiplier and now must have a total delay less than $2T$. More than half of this loop delay ($>T$) is in the flash ADC and the rest ($<T$) is in the mux, d_2 multiplier, and I2V.

LOOK-AHEAD DFE DESIGN

The two paths in the LA DFE in Fig. 4 use the architecture shown in Fig. 3. All analog circuits are fully differential to minimize the coupling of digital noise into the analog signal paths. An I2V converter followed by a flash ADC is used in each of the upper and lower paths. The ADC uses pipelined comparators to allow decisions to be made at a $1/T$ rate, with a delay less than $2T$ through the ADC. The comparators are as described in [15], except that a different input sampling network is used [16]. The critical feedback loop through the d_2 multiplier allows a total delay of $2T$ through the I2V, flash ADC, mux and multiplier. The I2V and multipliers are the same as those used in [5], and the DACs and ADCs use the same structures as in [5] but have undergone minor modifications.

The LMS update circuitry and DACs (not shown in Fig. 4) and decision delay line are shared by the two paths. The update circuitry can be shared because only the selected past decisions and error signal are used to update the coefficients. This sharing saves both power and IC area. To further reduce area, one up/down counter is shared for updating the five d_k coefficients. The logic design is such that two clock cycles are required to update a coefficient. Therefore, each coefficient is updated every 10th clock cycle, and its value is stored in a register after updating. This scheme slows convergence but does allow coefficient adaption. Since the main goal of this work is to demonstrate that parallelism in the analog signal paths can increase the data rate without a significant degradation of performance, speed of convergence was not a major concern. The LA DFE could easily be modified to have a dedicated counter for each coefficient.

The two parallel paths in the LA DFE must be identical in order to achieve the same performance as the conventional DFE in Fig. 2. In an IC implementation, the analog circuits in these paths will be mismatched to some extent. The effect of d.c. offset mismatch between the two paths and convolution-path gain mismatch were simulated [17]. The results of these simulations are that the LA DFE can tolerate convolution-path gain mismatch of 10% and large offsets with negligible degradation. In the LA DFE, coefficient d_0 cancels the average input-referred offset in the two paths, and d_1 cancels mismatch in the offset of the two paths in addition to ISI caused by $\hat{y}(n-1)$ [17]. To achieve good matching between the upper and lower paths, circuits that have similar functions (the I2Vs, flash ADCs and convolution multipliers) are placed close together with identical orientations.

MEASURED PERFORMANCE

The LA DFE was tested by generating a realistic DFE input signal that contains only postcursor ISI using software, sending that signal through an arbitrary waveform generator, and adding bandlimited noise to the signal before it enters the LA DFE. The test signal is generated using a Lorentzian channel model with $PW_{50}=2T$ and a 5-tap FIR forward equalizer.

A plot of measured bit-error rate (BER) versus signal-to-noise ratio (SNR) is shown in Fig. 5 for a data rate of 55 Mb/s. This data was gathered using coefficients that were found by adapting, but adaption was turned off when measuring BER. The sampling phase is adjusted by hand. For comparison, simulated BER performance of the DFE in Fig. 2 is also plotted. This early data shows that the LA DFE is functional and operates faster than the DFE in [5]. That DFE used similar circuitry, the same CMOS technology, and the architecture in Fig. 3. It had a maximum operating speed of 35 Mb/s. The parallelism is allowing faster operation.

The IC layout is shown in Fig. 6. The IC was fabricated in a 2-μm CMOS process. The total active area of the LA DFE is 23 mm^2. The measured performance is summarized in Table 1.

CONCLUSION

A mixed-signal look-ahead DFE was presented that increases the data rate by using parallelism. The area increase due to parallelism is much less than a factor of two because all blocks do not have to be duplicated.

ACKNOWLEDGMENT

The authors are grateful to the Mass Storage Group at National Semiconductor for performing microsurgery to correct wiring errors on the chip.

REFERENCES

[1] J. W. Bergmans, "Density Improvements in Digital Magnetic Recording by Decision Feedback Equalization," *IEEE Trans. on Magnetics*, pp. 157-162, May 1986.

[2] J. M. Cioffi, W. L. Abbott, H. K. Thapar, C. M. Melas, K. D. Fisher, "Adaptive Equalization in Magnetic-Disk Storage Channels," *IEEE Comm. Magazine*, pp. 14-29, Feb. 1990.

[3] W. L. Abbott, J. M. Cioffi and H. K. Thapar, "Channel Equalization Methods for Magnetic Storage," *ICC*, Boston, MA, June 1989.

[4] M. L. Honig and D. G. Messerschmitt, *Adaptive Filters*, Kluwer Academic Publishers, Boston, 1984.

[5] J.E. Brown, P. Hurst and L. Der, "A 35 Mb/s Mixed-Signal DFE for Disk-Drive Applications," *CICC*, Santa Clara, pp. 26.6.1-26.6.4, 1995.

[6] B. Widrow and S. D. Stearns, *Adaptive Signal Processing*, Prentice-Hall, New Jersey, 1985.

[7] J. Sonntag, et. al., "A High-Speed Low-Power PRML Read Channel Device", *IEEE Trans. on Magnetics*, pp. 1186-1195, March 1995.

[8] B. Rothenberg, S. Lewis and P. Hurst, "A 20 MS/s Switched-Capacitor FIR Filter in 2 μm CMOS," *ISSCC.*, San Francisco, pp. 210-211 and 369, Feb. 1995.

[9] R. Gomez, M. Rofougaran and A. A. Abidi, "A Discrete-Time Analog Signal Processor for Disk Read Channels," *ISSCC*, pp. 212-213, Feb. 1993.

[10] R. Yamasaki, T. Pan, M. Palmer, and D. Browning, "A 72 Mb/s PRML Disk-Drive Channel Using an Analog Sampled-Data Signal Processor," *ISSCC*, pp. 278-279, 1994.

[11] S. B. Lyle, G. Worstell, and R. R. Spencer, "An analog discrete-time transversal filter in 2μ CMOS," *Asilomar Conf. on Signals, Systems and Comp.*, pp. 970-974, 1992.

[12] K. M. Ware, H-S. Lee and C. G. Sodini, "A 200-MHz CMOS phase-locked loop with dual phase detectors," *IEEE J. Solid-State Circuits*, pp. 1560-1568, Dec. 1989.

[13] K.K Parhi, "Pipelining in algorithms with quantizer loops," *IEEE Trans. on CAS*, pp. 745-754, July 1991.

[14] G. Long, F. Ling and J. G. Proakis, "The LMS Algorithm with Delayed Coefficient Adaption," *IEEE Trans. on ASSP*, vol. 37, pp. 1397-1405, Sept. 1989.

[15] J. T. Wu and B. A. Wooley, "A 100-MHz Pipelined CMOS Comparator," *IEEE J. Solid-State Circuits*, pp. 1379-1385, Dec. 1988.

[16] L. Der, S. Lewis and P. Hurst, "A Switched-Capacitor Differencing Circuit with Common-Mode Rejection for Fully Differential Comparators" *Midwest Symp. on CAS.*, pp. 911-914, Aug. 1993

[17] R. Kajley, P. Hurst, and J.E. Brown, "A Mixed-Signal DFE Using a Look-Ahead Architecture," *Asilomar Conf. on Signals, Systems and Comp.*, pp. 1413-1417, 1994.

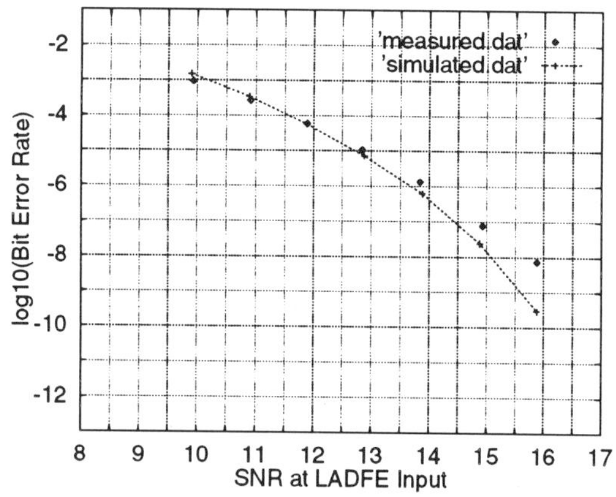

Fig. 5: Measured BER vs. SNR at 55 Mb/s, and simulated BER of a DFE.

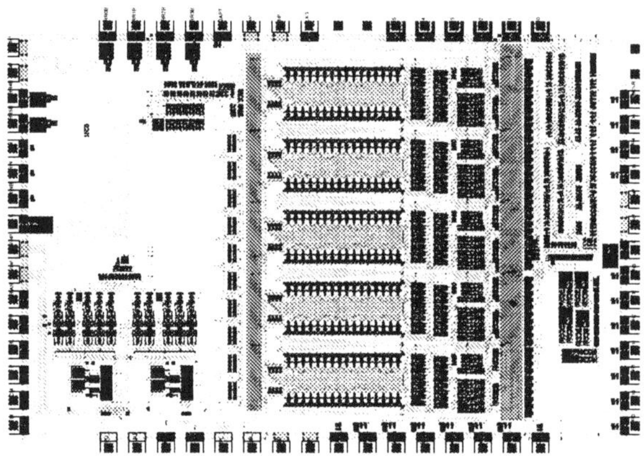

Fig. 6: Layout plot of the LA DFE IC.

Parameter	Value
Process	2 μm CMOS
Active Area	23 mm^2
Power Supply	5 V
Power Dissipation	450 mW @ 55 Mb/s
Data Rate	55 Mb/s
Number of taps	4 for ISI, 1 for d.c. offset

Table 1: LA DFE Performance Summary

A New CAM Macro for 622 Mbps ATM Cell Processing

Hideaki Odagiri, Noriaki Takahashi, Tsunaaki Shidei *,
Koji Takeshita * and Yutaka Kumagai *

OKI Electric Industry Co., Ltd.
Integrated Designing Division / VLSI R&D Center *
1-3-B23, Nakase, Mihama-ku, Chiba-shi, Chiba, 261-01, Japan
550-1, Higashiasakawa-cho, Hachioji-shi, Tokyo, 193, Japan *

Abstract

A CAM macro for 622 Mbps ATM cell processing has been developed with two built-in circuits, the Idle Word Detection Circuit and the Multiple Hit Detection Circuit. The macro is fabricated in a standard 0.5 um 3 layer CMOS technology, 40MHz operation and 70 mW power dissipation with 3.3 V power supply has been confirmed.

1. Introduction

In Asynchronous Transfer Mode (ATM) communication systems, an ATM cell is composed of a header field and an information field. Each connection to which an ATM cell belongs is identified by the Virtual Path Identifier/Virtual Channel Identifier (VPI/VCI) that is placed in the header of all ATM cells. At the User-Network Interface (UNI), the VPI/VCI is a 24 bit field in the ATM cell header, while at the Network-Node Interface (NNI), the VPI/VCI is a 28 bit field. The ATM communication system has to support all VPI/VCIs for maximum flexibility of ATM cell processing. Therefore, when VPI/VCI is used as the physical address itself, large amounts of memory, for example (2^{24} words or 2^{28} words) x (total amount of ATM cell processing bits required for all potential connections), will be needed in the system.

On the other hand, if the VPI/VCI is used as the logical address, the amount of memory can be reduced to correspond to the number of connections that can be handled simultaneously in the system. Using a Content Addressable Memory (CAM) is one solution for translating the logical address into the physical address (1). This will greatly reduce the amount of memory that the system uses (2). However, in a high speed ATM cell processing application (up to 622 Mbps), there are two unresolved problems. The first is how to select an inactive word of CAM for entering a new VPI/VCI, and the second is how to detect the duplicate VPI/VCIs that have been entered in CAM.

These problems could be solved by an external wired logic, but this will require a large circuit area and a processing delay. Therefore, these circuits would need to be embedded in a macro cell to increase the through-put of the system.

In this paper we will describe a new CAM macro for 622 Mbps ATM cell processing with 40 Mbps 16bit parallel operation in ATM communication systems. We developed two circuit architectures and hardware solutions that are embedded into the CAM macro, the Idle Word Detection Circuit and the Multiple Hit Detection Circuit.

2. Application and Configuration of the CAM Macro in ATM Cell Processing

In an ATM communication system, the CAM macro can be used in many functions. These functions include:

1) ATM Layer functions
(e.g., VPI/VCI translation, routing and switching)
2) Traffic control functions for each VPI/VCI
(e.g., Usage Parameter Control/Network Parameter Control and Congestion Control)
3) Operation and Maintenance functions for each VPI/VCI
(e.g., Detection and reporting of transmission quality degradation and transmission failure).

Fig. 1 shows the block diagram of an ATM cell processing circuit that executes these functions. The CAM macro shown in Fig. 1 includes the CAM to be stored with the VPI/VCI assigned to the connection and the RAM to be stored with the data for ATM cell processing of that connection. The ATM cell processing Arithmetic Logic Unit (ATM-ALU) shown in Fig. 1 uses the data that is read out from the RAM as the input, calculates the data, and outputs the results of the ATM cell processing.

0-7803-3177-6 $5.00 © 1996 IEEE

IEEE 1996 CUSTOM INTEGRATED CIRCUITS CONFERENCE

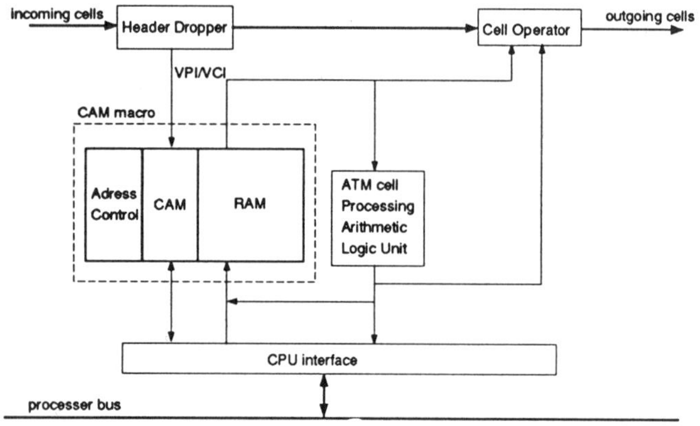

Fig. 1 Block diagram of ATM cell processing circuit

The operations of the ATM cell processing circuit shown in Fig. 1 are:

1) Connection Setup

When a connection request is accepted, the VPI/VCI assigned to the connection is stored in the CAM and the data assigned to the connection is stored in the RAM, through the CPU interface circuit.

2) Data Transfer

The VPI/VCI is extracted from every incoming ATM cell at the Header Dropper. In the CAM macro, the data corresponding to the VPI/VCI is read out from the RAM. Then the ATM-ALU calculates the data and outputs the results. Based on the results, the Cell Operator operates on the ATM cell (e.g., cell discard, modification of the header information). The output of the ATM-ALU is transferred to the system through the CPU interface circuit. The data in the RAM is updated with necessary.

3) Connection Release

When a connection release is ordered through the CPU interface circuit, the location of the VPI/VCI of the connection is identified with the search operation of the CAM, then the VPI/VCI in the CAM is deleted.

3. Circuit Design

When the system uses the CAM, there are two unresolved problems. The first is how to select an inactive word of CAM for entering a new VPI/VCI. Selecting an inactive word of the CAM is not easy, because inactive words of the CAM are scattered after many times of connection setup and release. The second is how to detect the duplicate VPI/VCIs that have been entered in the CAM. The duplicate VPI/VCIs may cause a system crash and must be avoided.

The circuits that solve these problems require the status of all words of the CAM. Therefore, the circuits need to be embedded in the CAM macro to reduce the total area of the circuits and to achieve the required high speed.

The Idle Word Detection Circuit and the Multiple Hit Detection Circuit are described as developed below:

3.1. Idle Word Detection Circuit

A Valid bit (Vbit) is placed in every CAM word and is used to indicate which CAM word is active or not. The Idle Word Detection Circuit (IWDC) selects one of the inactive words. When the status of the Vbits is updated, then another inactive word is selected.

Fig. 2 shows an example of the logic diagram of the IWDC, and Fig. 3 shows the transistor level circuit diagram corresponding to a part of the logic diagram shown in Fig. 2. This circuit uses a complementary Vbit output as the input and uses a dynamic CMOS logic circuit to obtain a high speed operation. This circuit consists of only three transistors for each CAM word (2 NMOS transistors for the complementary Vbit input and 1 PMOS transistor for pre-charge). Therefore, the minimum number of transistors are used and the circuit can be embedded in this macro.

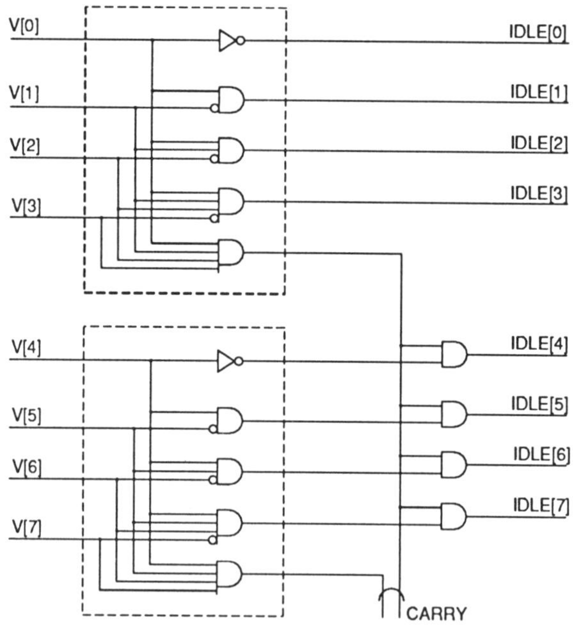

Fig. 2 Logic diagram of IWDC

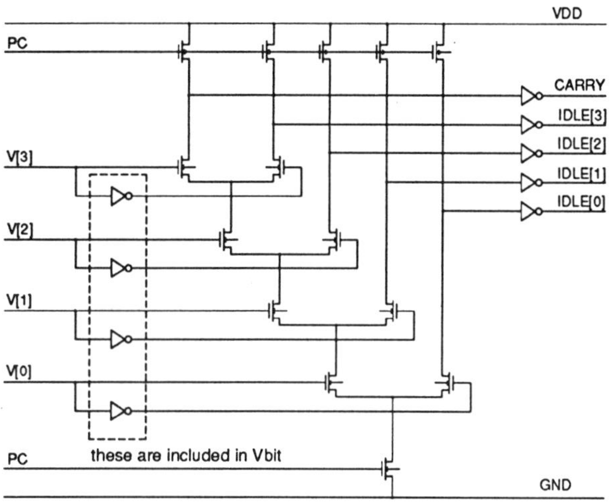

Fig. 3 Circuit diagram of IWDC

3.2. Multiple Hit Detection Circuit

The Multiple Hit Detection Circuit (MHDC) detects that the duplicate VPI/VCIs have been entered in the CAM in the search operation. Fig. 4 shows the circuit diagram of this circuit.

This circuit detects the on resistance difference of NMOS transistors. One is the reference (MN0) that is always on, the others are the NMOS transistors connected in parallel with the match lines (MN1-MN128) as the gate inputs. If zero or one of these transistors (MN1-MN128) is on, the combined on resistance of all transistors is greater than that of the reference transistor; and, if two or more are on, then the combined on resistance of all transistors is smaller than that of the reference transistor. Therefore, this difference of on resistance results in a voltage difference. This voltage difference (node BC-BCN) is amplified to a logic level at the next comparator stage.

Fig. 5 shows the simulation wave form of this circuit. This wave form shows that the difference voltage output of node BC-BCN obtains over 800mV amplitude at the worst condition of this circuit (only two transistors are on).

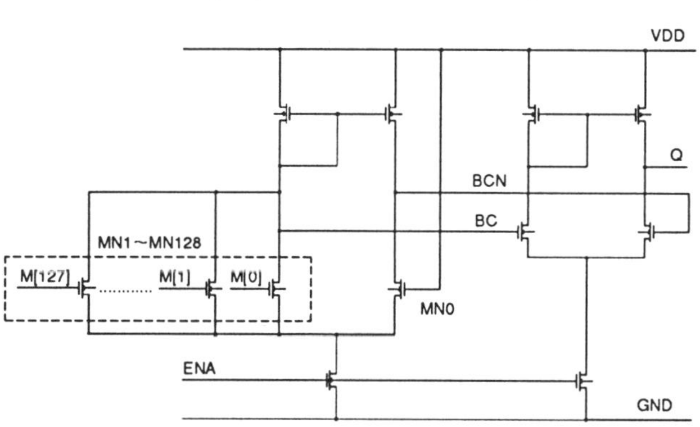

Fig. 4 Circuit diagram of MHDC

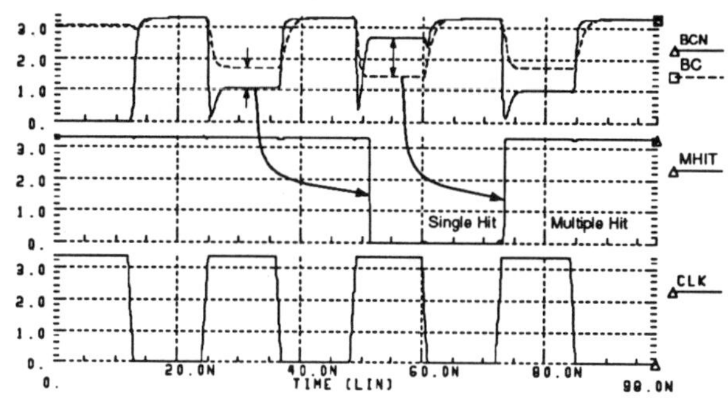

Fig. 5 Simulation result of MHDC

4. Implementation

This macro is composed of CAM, RAM, Bridge, CAM I/O, RAM I/O, and control circuits. Its block diagram and photomicrograph are shown in Fig. 6 and Fig. 7, respectively. The die size of this test chip (0.5um CMOS technology) is 4.3 x 8.0 mm^2. The area of this macro is 2,030 x 4,496 um^2, and 160,625 transistors are integrated in this macro.

The CAM is 33 bits x 128 words (one bit is used as the valid bit). The RAM is 64 bits x 256 words. To minimize the layout of interconnection between CAM and RAM, the height of each word of the CAM and the Bridge is the same as the height of 2 RAM cells.

The Bridge includes the Idle Word Detection Circuit (IWDC), the Multiple Hit Detection Circuit (MHDC), the No-Hit Detection Circuit (NHDC), the Full Detection Circuit (FDC) and the Empty Detection Circuit (EDC). The NHDC detects that the target VPI/VCI does not exist in the CAM in the search operation. The FDC detects that all VPI/VCIs stored in the CAM are active. The EDC detects that all VPI/VCIs stored in the CAM are inactive.

The control circuit includes a command decoder and a timing generator. A standard 0.5um SOG logic design method is used to allow flexible combinations of commands and operations based on the needs of each application.

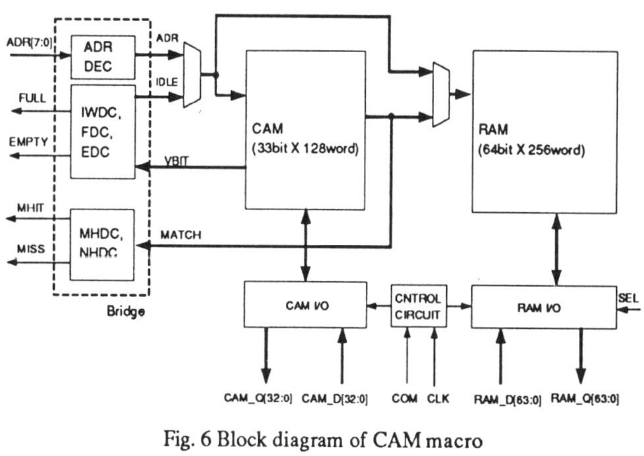

Fig. 6 Block diagram of CAM macro

Fig. 7 Photomicrograph of CAM macro

5. Experimental Results

Fig. 8 shows the shmoo plot of this macro operation at room temperature. It shows that this macro operates at 13.0 ns with a 3.3 V power supply, and demonstrates that this macro can operate at 40 MHz in the worst condition. The speed will satisfy 622 Mbps ATM cell processing with 16 bit parallel operation in the system.

Fig. 9 shows the wave form of the Multiple Hit Detection Circuit measured with an EB tester. In this case, the external clock frequency is 10 MHz and the power supply is 3.3 V.

Performance of this CAM macro is summarized in Table 1.

6. Conclusion

The CAM macro which integrates the CAM, RAM, and logical circuits for ATM Cell processing has been described. This new circuit adopted architecture and hardware solutions such as Idle Word Detection Circuit and Multiple Hit Detection Circuit.

The design is 33 bits x 128 words CAM and 64 bits x 256 words RAM in single-polysilicon, triple metal, 0.5 um CMOS technology. The measurement result confirms 40 MHz operation and 70 mW power dissipation at 3.3 V power supply.

7. References

(1) K.J.Schultz and P.G.Gulak, "CAM-based Single-Chip Shared Buffer ATM Switch", ICC '94 Conf. Rec., New Orleans, LO, May 1994, pp. 1190-1195

(2) H.Higuchi, S.Tachibana, M.Minami and T.Nagano, "A 2-ns, 5-mW, Synchronous-Powered Static-circuit Fully Associative TLB", 1995 Symposium on VLSI circuits, Kyoto, June 1995, pp. 21-22

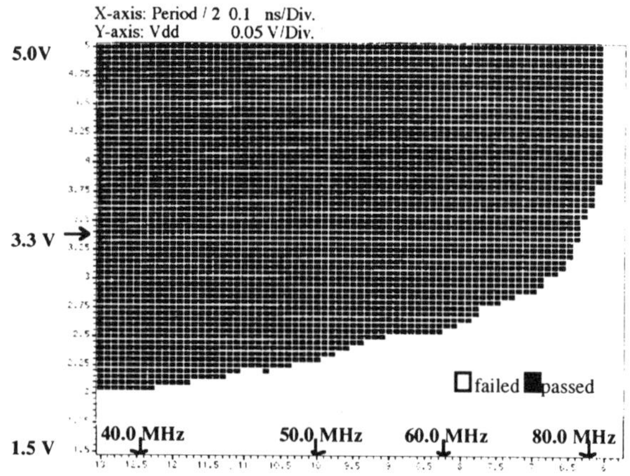

Fig. 8 Shmoo plot of CAM macro test chip

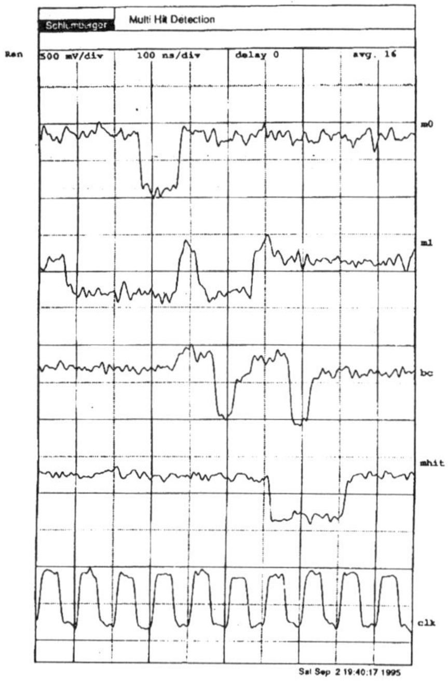

Fig. 9 EB Tester wave form of MHDC

TABLE 1. Characteristics of CAM macro

Technology	Single poly-Si Triple Metal 0.5 um CMOS
CAM Word Length	128 words
CAM Bit Width	33 bits
RAM Word Length	256 words
RAM Bit Width	64 bits
Macro Cell Area	2,030 x 4,496 um^2
Maximum Frequency	over 40 MHz
Power Consumption	70 mW (at 40 MHz)
Power Supply	3.3 V

Chair: Rene Penning de Vries
Co-Chair: Michiel Beunder

In this session, manufacturable process technologies illustrating trends in scaled CMOS, BiCMOS and CCD integrated circuits are featured along with methodologies for yield prediction in VLSI designs.

The first paper describes various predictive yield models that relate yield to product design characteristics. Using critical area, transistor density and minimum metal space with critical area as input, yield models are developed and compared with data.

In the second paper, a high performance 0.35 μm 33 volt CMOS technology with a gate oxide thickness of 65 Å and 50 ps ring oscillator delay for ASIC applications is described. This technology features n+-polysilicon gates, poly-buffered LOCOS isolation, and three levels of metal. Circuits that have been fabricated include a 256K SRAM development vehicle, internal cell designs and the ORCA-FPGA.

The following paper discusses a thin-film silicon on insulator (TFSOI) complementary BiCMOS technology based on a 0.5 μm CMOS flow. The process, optimized for low power, demonstrates ECL delays of 80 ps at 20 μA gate current, a 60% reduction over representative bulk BiCMOS results. Bipolar transistor noise figures, device matching, and flicker noise data confirm the suitability of this technology for RF and mixed-mode signal applications.

Finally, a new pixel technology for 1/4 inch PAL inter-line transfer charge coupled device (IT-CCD) is presented. The paper highlights a thin light shield and fabrication of an in-process micro-lens that allows smaller pixel sizes without sacrificing performance.

Yield Loss Forecasting in the Early Phases of the VLSI Design Process

Hans T. Heineken, Jitendra Khare and Wojciech Maly

Department of Electrical and Computer Engineering
Carnegie Mellon University
Pittsburgh PA 15213.

Abstract

This paper describes three new yield models. The first takes as input the critical area of a layout; the second approximates the critical area with the minimum spacing area between metal lines; and the third uses transistor density to model critical area. The models were developed and verified using manufacturing data.

1.0 Introduction

The success of an integrated circuit in the semiconductor market is determined not only by new technologies, innovative functionality, or enhanced performance, but also by manufacturing cost. If cost incurred by the design and fabrication of an IC exceeds market demand, failure is likely [1,2]. Traditionally designers have associated cost with die area; the underlying assumptions being that smaller dies are less prone to defect-related yield loss, and have a higher die count per wafer. Unfortunately, die area does not address the entire cost reduction problem. Fig. 1, [3], shows the yield vs. area plot for several designs, all fabricated on the same fab line during the same time period. Contrary to traditional yield models [4,5] the yield increases with die area.

The yield trend in Fig. 1 was previously described and empirically modeled in [3]. The objective of this paper is to expand and enhance the theory proposed in [3] and to propose new analytical yield models capable of forecasting the **yield of designs in the early phases of the design process**.

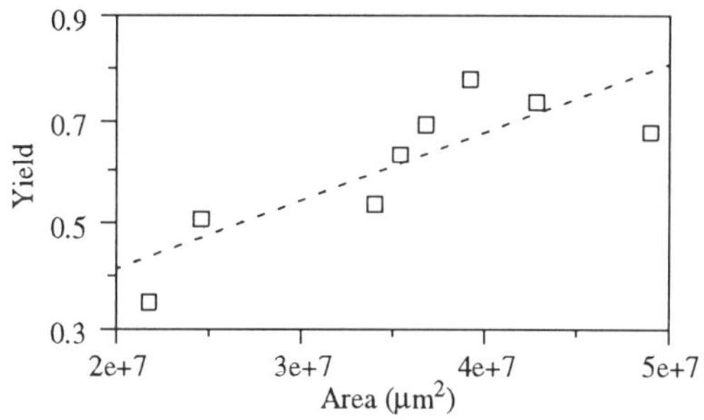

Fig 1. Measured yield vs. area for a number of ICs [3].

2.0 Yield Measurements

To develop the yield model described in this paper, an extensive experiment was performed to collect and analyze data from a variety of products. Data was extracted with the manufacturability analysis environment MAPEX [6] and is presented in Table 1.

The data was taken from ICs manufactured over a period of one year on the same multi-product fabrication line. The ICs averaged a volume of 33 lots per design and ranged in transistor count between 71k and 224k. Each of the designs used a number of design styles including RAMs, ROMs, standard cells, and custom cells. Additional data related to the ICs is given in Table 1. Note that three of the ICs in Table 1 are optical shrinks of the other ICs, e.g., design D2 is fabricated in 1.2, 1.0, and 0.8 μm minimum feature size.

Table 1. Design characteristics of products presented in Fig. 1.

Products		Chip area [cm2]	Min. feature size [μm]	Number of transistors (K)	Measured yield
No.	Design				
1.	D1	0.367	1.2	144.5	0.691
2.	D2	0.218	0.8	223.9	0.351
3.	D2	0.340	1.0	223.9	0.539
4.	D2	0.490	1.2	223.9	0.677
5.	D3	0.427	1.2	70.88	0.731
6.	D4	0.392	1.2	64.08	0.775
7.	D5	0.245	1.0	224.7	0.511
8.	D5	0.353	1.2	224.7	0.631

3.0 Yield Model

The "traditional" Poisson yield model uses the die area and the notion of point defects to calculate the yield of a die [4]. In its simplest form the yield model takes the following form: $Y = exp(-A_o D)$, where A_o is the die area and D is the defect density. Clearly this model is inadequate to explain the yield trend in Fig. 1. In this section a modification of this model is presented.

3.1 Defect Model

Yield loss is a by-product of a mismatch between an IC's specified and actual functionality as measured by a tester. In a mature process a dominant mechanism causing yield loss is defects deposited or formed on a particular layer of an IC [1]. The traditional yield model represents these defects as dimensionless points and assumes that any such defect in a design causes a functional failure in the design.

In this paper defects are assumed to be two-dimensional disks of extra or missing material embedded in a conducting, semiconducting, or insulating layer of an IC [7]. In addition, each defect is assumed to have the following parameters: (a) size distribution which specifies the variation in frequency between defects of different radii, (b) density, D_{oi}, which specifies the frequency of occurrence of defects of type i; and (c) location, (x,y), which specifies the spacial location.

A number of relationships exist that model the defect size distribution, i.e., the frequency of occurrence of defects, as a function of their radius [8,9,10]. The yield models in this paper assume a widely used defect size distribution $f_i(r)=k/r^p$, where r is the defect radius, and k and p are parameters of the model [8]. The defect density variation, $D_i(r)$, for defects of type i, can then be calculated by multiplying the defect size distribution, $f_i(r)$, by the defect density, D_{oi} [11]:

$$D_i(r) = D_{oi} k / r^p = K_i / r^p \quad (1)$$

where the constant K_i is substituted for the product $D_{oi}\, k$.

3.1.1

0-7803-3177-6 $5.00 © 1996 IEEE

3.2 Layout Sensitivity to Defects

Another shortcoming of the traditional yield model lies in the fact that defects forming on the layout of an IC do not always adversely affect the functionality of an IC. For example, their size and location determine whether they can span the spacing between non-equipotential conducting lines. Defects deposited in a densely populated region of a layout are naturally more likely to produce a short than defects deposited in sparsely populated regions.

A useful measure of an IC's sensitivity to spot defects is "critical area" [7,8,9]. The critical area, *Acr(r)*, for a defect of radius r is defined as the area of a die where if the center of a circular defect is deposited a fault occurs in the circuit. Fig. 2 shows the metal1 critical area vs. defect radius curves of the products listed in Table 1. The critical areas of the products were extracted using MAPEX [6].

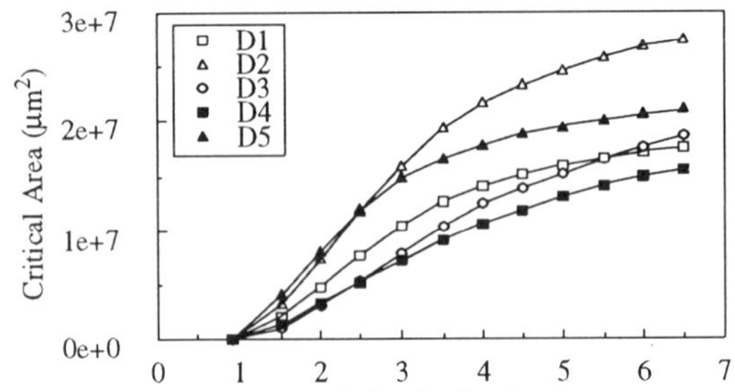

Fig 2. Metal1 critical area vs. defect radii for 1.2μm designs in Table 1.

3.3 Yield Model

Given the critical area of an IC and the defect density and size distribution for a given defect type *i*, the defect-related yield, Y_i, of an IC can be modeled using a Poisson based yield model [7]:

$$Y_i = exp\left(-\int_0^{\infty} Acr_i(r)\, D_i(r)\, dr\right), \quad Y_{chip} = \prod_{i=1}^{N} Y_i \qquad (2)$$

where N is the number of defect types.

4.0 Modified Yield Models

A number of problems arise in the application of yield model (2). One problems is the difficulty and cost of extracting accurate values of the defect densities and defect size distributions. Another problem is the cost of calculating the critical area of designs for multiple layers and for multiple defect sizes. In addition, the critical area can not be calculated until the latter stages of the design process, after layout is complete.

In this section three yield models are proposed that address these problems. They are analytically derived from the Poisson based yield model (2). The models are presented in the following way. First the model is described. Next parameter extraction is performed and the ability of the model to explain the yield trend in Fig. 1 is demonstrated. Finally the ability of the model to forecast yield is demonstrated on two additional products. These products were fabricated over six months towards the tail end of the period that the eight pervious products in Table 1were fabricated. The measured design and yield data of the products are given in Table 2.

Table 2. Design characteristics of two products.

Products No.	Products Design	Chip area [cm²]	Min. feature size [μm]	Number of transistors (K)	Measured yield
9.	D6	0.165	0.8	168.4	0.389
10.	D6	0.371	1.2	168.4	0.693

4.1 Single Layer Critical Area Based Model

Model: In an analysis of the designs listed in Table 1 it was observed that a correlation existed between the critical areas of the metal layers. This leads to the assumption that yield can be modeled as a function of a single critical area layer. The validity of this assumption is demonstrated in this section.

The critical area, *Acr(r)*, can be approximated by a piecewise linear function:

$$Acr(r) = m_j\left(r - r_j\right) + Acr_j \qquad (3)$$

where: Acr_j is the critical area at data point *j*,
r_j is the defect radius at data point *j*,
m_j is the critical area/defect radius slope between points *j* and *j +1*, (i.e., $m_j = (Acr_{j+1} - Acr_j)/(r_{j+1} - r_j)$).

Substituting (3) and (1) for *Acr(r)* and *D(r)* in model (2), and integrating with respect to defect density, the yield model takes the following form:

$$Y = exp\left[-\sum_{j=1}^{Ncr-1}\left[K\left(\frac{m_j r_j - Acr_j}{p-1}\right)\left(\frac{1}{r_{j+1}^{p-1}} - \frac{1}{r_j^{p-1}}\right) - \left(\frac{m_j}{p-2}\right)\left(\frac{1}{r_{j+1}^{p-2}} - \frac{1}{r_j^{p-2}}\right)\right]\right] \qquad (4)$$

where *Ncr* is the number of critical area data points. In this yield model the critical area points and their corresponding defect radii are the independent variables; *K* and *p* are the model parameters.

Parameter Extraction: To determine the validity of model (4) and its ability to explain the yield trend in Fig. 1, the yield data and the metal1 critical area data of the designs were applied. The model parameters, *K* and *p* were optimized to minimize the error between the measured and modeled yield. The optimal values are $K = 60.26$ and $p = 4.74$; error = 7.6% (error is calculated by summing |measured yield - modeled yield| and dividing by the sum of the measured yield). The measured yield and modeled yield are plotted in Fig. 3.

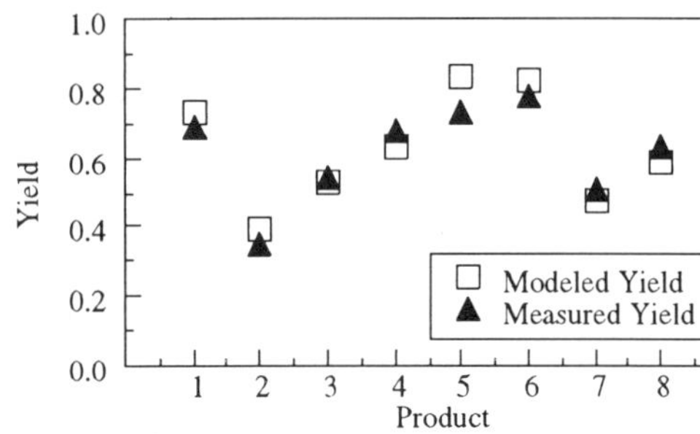

Fig 3. Measured and modeled yield for products using model (4).

3.1.2

Prediction: Using the values of p and K extracted above, and applying the critical areas of the products in Table 2, yield model (4) calculates the yield for product 9 to be 0.381 and for product 10 to be 0.621. Measured values are 0.389 and 0.693, respectively.

Conclusion: The ability of yield model (4) to explain the yield trend in Fig. 1 and to predict the yield of additional products validates the assumption that yield can be modeled by a single critical area layer. The model also demonstrates that the critical area of an IC is a good measure of the sensitivity of a design to defects, and can be used to model the yield of an IC.

4.2 Minimum Spacing Based Yield Model

Yield model (4) relates the yield of an IC to its critical area. Unfortunately, the extraction of multiple critical area data points might be computationally expensive. In this section, a modification of the yield model is provided that addresses this issue. In particular, model (4) is used to show that the sum of the spacing area between minimum spaced metal lines, a design attribute previously used to model yield, can be effectively used to approximate the critical area.

Model: In yield model (4) the critical area function was modeled as a piecewise linear function. This can be taken a step further: the critical area curve can be approximated by two linear functions, one modeling the initial rise in the critical area; the second modeling the critical area as it begins to saturate. Fig. 4 shows a two linear function fit of the critical area curve of design D1. The functions are fit using least squares fit.

Of the two linear functions only the first is of relevance from a yield perspective. This is a consequence of the high value of the parameter p extracted in section 4.1. This value of p, (p=4.74), is such that defects with radii larger than those covered by the first linear function, occur so infrequently that they have a negligible contribution to the yield. (Model (4) calculates the error in yield for D1, when ignoring defects with radii covered by the second linear function, to be less than 1%.)

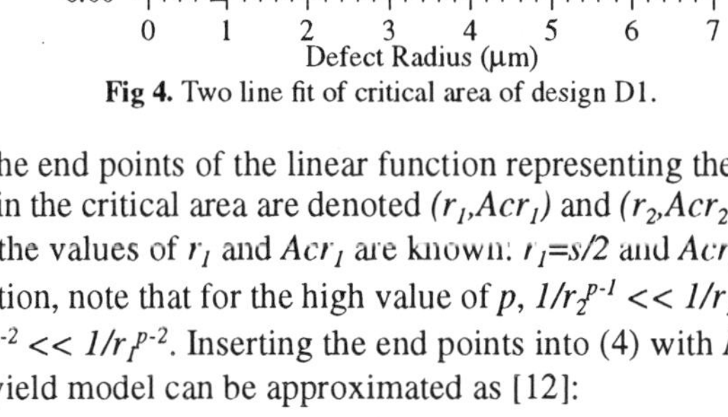

Fig 4. Two line fit of critical area of design D1.

The end points of the linear function representing the initial rise in the critical area are denoted (r_1,Acr_1) and (r_2,Acr_2). Note that the values of r_1 and Acr_1 are known: $r_1=s/2$ and $Acr_1=0$. In addition, note that for the high value of p, $1/r_2^{p-1} << 1/r_1^{p-1}$ and $1/r_2^{p-2} << 1/r_1^{p-2}$. Inserting the end points into (4) with $Ncr=2$, the yield model can be approximated as [12]:

$$Y = exp\left[-\frac{K \cdot m_1}{(p-1)\,(p-2)\,r_1^{p-2}}\right] \qquad (5)$$

where m_1 is the slope of the linear function.

To evaluate (5) the slope of the linear function, m_1, needs to be calculated. Note from Fig. 4 that the rise in critical area is well modeled by a linear function. Therefore in principle, any critical area point $(r,Acr(r))$ such that $r_1<r<r_2$ can be chosen to calculate the slope: $m_1=Acr(r)/(r-r_1)$.

Let r be chosen such that $r=s/2+dr$, and let dr be very small, $dr<<s/2$, (for example let dr be smaller than the grid resolution of the layout), then the critical area at r is the critical area between metal lines that are spaced a minimum spacing, s. Furthermore the critical area is proportional to the sum of all the area between minimum spaced lines, *Aminsp*:

$$Aminsp = Acr(r)\frac{s}{dr}, \qquad r = \frac{s}{2} + dr \qquad (6)$$

Inserting (6) into m_1, yield equation (5) can be expressed as:

$$Y = exp\left[-\frac{K \cdot Aminsp}{s\,(p-1)\,(p-2)\left(\frac{s}{2}\right)^{p-2}}\right] \qquad (7)$$

Parameter extraction: To test the validity of yield model (7) the design data in Table 1 was applied again. The model parameters K and p were optimized to minimize error between measured and modeled yield. The optimal values are K=54.77 and p=4.80; error = 5.0%. The measured yield and modeled yield are plotted in Fig. 5.

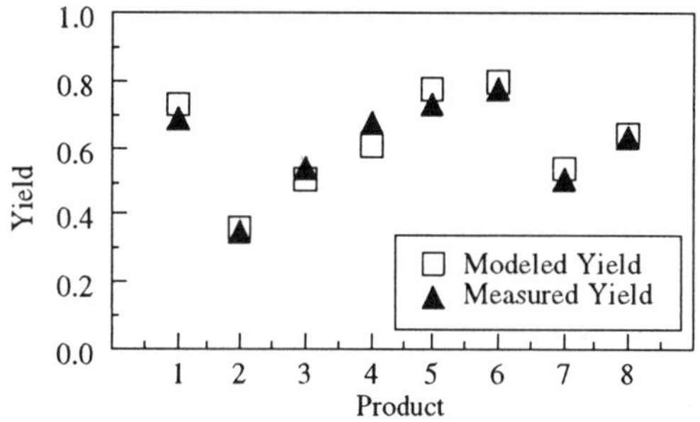

Fig 5. Measured and modeled yield for products using model (7).

Prediction: Using the values of p and K extracted above, and applying the design data of the products in Table 2, yield model (7) calculates the yield for products 9 and 10 to be 0.414 and 0.654, respectively. Measured yield values are 0.389 and 0.693.

Conclusion: The proximity of the modeled and measured yield data points in Fig. 5, shows that yield model (7) provides a good explanation of the yield trend of the products in Table 1. In addition model (7) is able to predict the yield of the products in Table 2.

The advantage of this model over others that use the sum of spacing area between minimum spaced metal lines, is that it uses both process and design attributes as independent variables. Consequently the yield of new designs can be accurately predicted.

4.3 Transistor Density Based Yield Model

In this section a yield model is developed that takes as independent variables design attributes that are more accessible in the earlier stages of the design process.

Model: The development of the yield model takes the following form: the critical area is used to define a new measure of the layout density. This measure is then correlated to the transistor density. The transistor density is then used to model yield.

In Fig. 6 the critical areas of two designs with equal area are drawn. Their critical area slopes are: m_{arr} and m_l. The first design with slope m_{arr} is an array of metal lines with minimum space and width. The layout density of this design is at a maximum, i.e., at any given defect size the critical area of this design is greater or equal to that of any other design of equal area. The second design is a "typical" design whose critical area is modeled as a linear function of defect radius as was done in Section 4.2 for yield model (5).

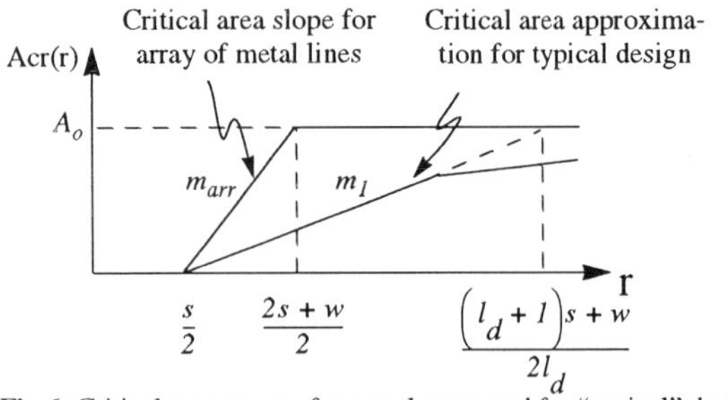

Fig 6. Critical area curves for metal array and for "typical" design.

The critical area slope, m_l, of a design can be expressed in terms of m_{arr}. In particular, a new layout density measure, l_d, is defined as $l_d = m_l/m_{arr}$. Inserting l_d into yield model (5), the yield model takes the following form:

$$Y = exp\left[-\frac{K \cdot l_d \cdot A_o'}{(p-1)\,(p-2)\,(s'+w')\left(\frac{s'}{2}\right)^{p-2}\lambda^{p-3}}\right] \quad (8)$$

In this equation, lengths and areas are expressed in terms of minimum feature size, λ. I.e., A_o', s' and w' are area, minimum spacing, and minimum width, respectively, in units of λ. This is useful since a number of the designs in Table 1 are direct shrinks of others.

The layout density l_d can be calculated from the critical area of a design. However, a more accessible means exists to calculate l_d. An analysis of the designs using MAPEX shows that layout densities of the products in Table 1 correlate to the square root of the transistor densities, T_d', of the products (T_d' is expressed in units of minimum feature size squares per transistor).

Inserting T_d' into (8), the yield model takes the following form:

$$Y = exp\left[-\frac{K_o N_{tr} / \sqrt{T_d'}}{(p-1)\,(p-2)\,(s'+w')\left(\frac{s'}{2}\right)^{p-2}\lambda^{p-3}}\right] \quad (9)$$

where N_{tr} is the number of transistors and K_o incorporates the constant K and the constant of proportionality between l_d and T_d'.

Parameter Extraction: The relevant data from the products in Table 1 was applied to yield model (9). The model parameters were optimized to minimize error between measured and modeled yield. The optimal values are $K_o = 352$ and $p = 4.75$; error =4.9% (Fig. 7).

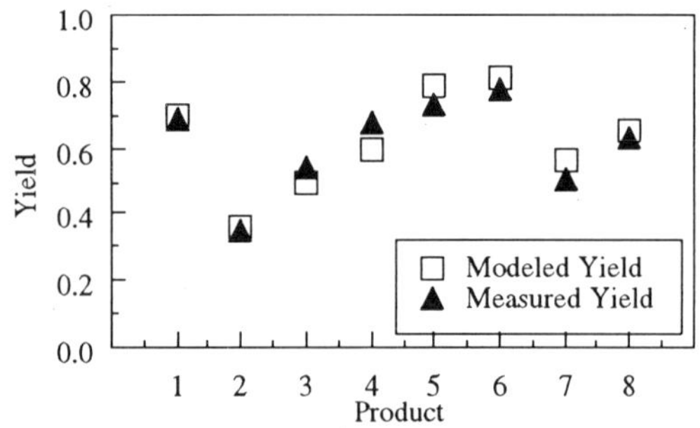

Fig 7. Measured and modeled yield for products using model (9).

Prediction: Again the products in Table 2 were used to determine the yield model's ability to estimate yield of new designs. Using the values of p and K_o extracted above, yield model (9) calculates the yield for product 9 to be 0.422 and product 10 to be 0.683. Measured yield values are 0.389 and 0.693.

Conclusion: Fig. 7 demonstrates that yield model (9) explains the yield trend in Fig. 1. This yield model is useful in that it relies on more accessible variables. The model takes as input the number of transistors in the design and the transistor density. The yield model also captures the effects of design shrinks on yield.

5.0 Conclusion

To perform cost analysis of an IC, an accurate yield model is needed to capture the effects of process and design features on the yield of the design. In addition, for cost analysis to be of maximum use, the yield model must be capable of forecasting the effects of design decision on the yield in the earlier stages of the design process. To this end three Poisson based yield models have been proposed and verified. The yield models were tested on several ICs. They were used successfully to explain the yield trend in several products as well to forecast the yield in two additional products.

Acknowledgments

The authors would like to thank Philips Research Labs for their support and in particular Frank Agricola and Eric van Utteren.

References

1. W. Maly, "Computer-aided design for VLSI circuit manufacturability," *Proc. of the IEEE*, vol. 78, no. 25, pp. 356-392, Feb. 1990.
2. W. Maly, "Cost of Silicon Viewed from Design Perspective," *Proc. 31st ACM/IEEE Design Automation Conf.*, pp. 135-142, June 1994.
3. W. Maly, H.T. Heineken, and F. Agricola, "A Simple New Yield Model," *Semiconductor International*, pp 148-154, July 1994.
4. R.M. Warner, Jr., "Applying a composite model to the IC yield problem," *IEEE J. Solid-State Circuits*, vol. SC-9, no. 3, pp. 86-95, June 1974.
5. T.L. Michalka, R.C. Varshney, J.D. Meindl, "A discussion of yield modeling with defect clustering, circuit repair, and circuit redundancy," *IEEE Trans. on Semi. Manufacturing*, vol. 3, no. 3, pp. 116-127, Aug. 1990.
6. H.T. Heineken and W. Maly, "Manufacturability Analysis Environment - MAPEX," *Custom Integrated Circuits Conference*, May 1994.
7. W. Maly and J. Deszczka, "Yield estimation model for VLSI artwork evaluations," *Electron. Lett.*, vol. 19, no. 6, pp. 226-227, March 1983.
8. C.H. Stapper, "Modeling of defects in integrated circuit photolithographic patterns," *IBM J. Res. Develop.*, vol. 28, no. 4, pp. 461-474, July 1984.
9. A.V. Ferris-Prabhu, "Modeling the critical area in yield forecasts," *IEEE J. Solid-State Circuits*, vol. 20, no. 4, pp 874-878, Aug. 1985.
10. C.H. Stapper, "Modeling of integrated circuit defect sensitivities," *IBM J. Res. Develop.*, vol. 27, no. 6, pp. 549-557, Nov. 1983.
11. J. Khare, D. Feltham, W. Maly, "Accurate estimation of defect-related yield loss in reconfigurable VLSI circuits," *IEEE Journal of Solid-State Circuits*, vol. 28, no. 2, Feb 1993.
12. H.T. Heineken, J. Khare, W. Maly, "Yield loss forecasting in the early phases of the VLSI Design Process," Carnegie Mellon Univ., CMU-SRC Tech. Rep. Dec.1995.

A Very High Performance and Manufacturable 3.3 V 0.35-μm CMOS Technology for ASICs

I.C. Kizilyalli, S.A. Lytle, B.R. Jones, E.P. Martin, S.F. Shive, A.L. Brooks, M.J. Thoma, R.W. Schanzer, J.W. Sniegowski, D.M. Wroge, R.W Key, J.W. Kearney, and K.R. Stiles

AT&T Bell Laboratories, 9333 S. John Young Pkwy., Orlando, FL 32819.

1. ABSTRACT

In this paper a manufacturable and high performance 0.35 μm CMOS ASIC technology optimized for 3.3 V operation will be presented. This CMOS technology features a 65 Å gate oxide, single n^+-polysilicon gate, and 3 levels of metal. An improvement of 1.6X in circuit performance and 1.56X in packing density is achieved over AT&T's previous generation 0.5 μm 3.3 volt CMOS technology [1] by device scaling, and aggressive isolation and interconnect design rules. The nominal ring oscillator delay time is 50 ps.

2. INTRODUCTION

This paper describes a manufacturable high performance 0.35 μm CMOS technology that is optimized for 3.3 V operation with a gate oxide thickness of 65 Å. Aggressive interconnect and isolation design rules are employed to achieve a factor of 1.56 and 5.3 improvement in packing density over AT&T's previous generation 0.5 μm [1] and 0.9 μm [2] CMOS technologies, respectively. Nominal gate delays of 50 *ps* for an unloaded ring oscillator and 94 *ps* for a 2-input NAND gate with a fan-out of 2, are measured. These improvements in performance and packing density have been obtained through the use of: (i) Poly-buffered LOCOS isolation scheme with a channel-stop implant which allows the active area to tub edge separation to be reduced below 0.55 μm (Fig. 1), (ii) a 65 Å gate oxide resulting in much improved transistor subthreshold characteristics and exceptional transistor current drive, (iii) hard-mask gate processing with L_{eff} control of $3\sigma \approx 0.06$ μm, (iv) three level metal routing with metal design rules optimized for ASIC applications (Fig. 2).

3. THE PROCESS

The starting wafers consist of 7.0 μm p- epi grown on a (100) p^+ Si substrate. Self-aligned N- and P-tubs are formed. The N-tub is defined by a deep phosphorus implant which determines the tub depth and an arsenic implant which serves as a channel-stop for PMOS devices. After N-tub drive-in, boron is implanted to form the P-tub. The P-tub is diffused, and boron is implanted a second time for punch-through suppression for NMOS transistor. A poly-buffered LOCOS (PBL) isolation scheme is used. The PBL stack consists of 200 Å of pad-oxide, 600 Å of amorphous silicon, and 2400 Å of Si_3N_4. Active areas are defined and etched into the PBL stack. Then boron is implanted as channel-stop into the P-tub regions. The energy and dose of this implant is critical in determining the active area to tub edge separation (Fig. 1) as well as the threshold voltage of the narrow-width NMOS transistor (narrow-width effect). The PBL stack, field-oxide (FOX) thickness, and FOX growth conditions that result in minimum birds beak length, minimum narrow-width effect, and negligible FOX thinning in between narrow PBL stack lines were determined by process simulation and experiments. A 3000 Å field oxide is grown at atmospheric pressure at 1050°C. Figure 3 shows a SEM micrograph of the as-grown FOX and the PBL stack. The width of the FOX and the PBL stack is 0.5 μm. The length of the birds beak is 0.07 μm and no PBL stack lifting is observed. This is true for PBL stack widths as narrow as 0.35 μm. Negligible FOX thinning is observed for as-grown FOX in between the two closely spaced PBL stack layers where the measured FOX thickness between two PBL stack lines spaced 0.5 μm apart is 2850 Å. As shown in Fig. 4, the simulated and measured field oxide thickness as a function of FOX width are in good agreement. Reducing the as-grown FOX thickness results in a much reduced surface topography, a shorter birds beak, negligible FOX thinning, reduced transistor δW, and no transistor narrow-width effect. After PBL stack removal, BF_2 is implanted through a sacrificial oxide to set the threshold voltage for the NMOS transistor. A patterned BF_2 implant adjusts the threshold voltage of the PMOS transistor along with added punch-through protection implants. A 65 Å gate oxide is grown. The gate oxide thickness is well-controlled with $\sigma = 1$ Å over 14 lots. Dielectric breakdown tests were performed to optimize the interaction of the isolation and gate oxidation processes. 2500 Å of heavily doped N^+ polysilicon serves as a gate electrode. The poly gate is patterned using a PECVD oxide hard-mask with an anti-reflective coating to obtain excellent gate linewidth control. Patterned arsenic and BF_2 LDD (lightly doped drain) rotated implants are used in NMOS and PMOS transistors, respectively. The NMOS LDDs are diffused before the PLDD implant. TEOS spacers are formed and source-drain (S/D) regions are implanted. Arsenic is used for NMOS S/D and BF_2 is used for PMOS S/D. The SEM photograph in Fig. 5 shows a cross section of a transistor. PETEOS/PEPTEOS/PEBPTEOS constitutes the dielectric 1 (D1). It is flowed in a conventional furnace at 875°C that also diffuses the S/D junctions. D1 is smoothed with a blanket wet etch-back step. Tapered contact windows are created using a combination of isotropic and anisotropic etches followed by a low temperature RTA. All three metal levels consist of Ti/TiN/AlCuSi/TiN. A RTA in N_2 of the Ti/TiN barrier layer of metal 1 (M1) affords low contact resistance by siliciding the junctions at the bottom of contact windows. TiN is used as an anti-reflection coating at all metal levels. PETEOS is used as interlevel dielectric for the dielectric 2 (D2) and dielectric 3 (D3) levels. Gap fill for both levels is accomplished by a deposition, argon/RIE etch and deposition process. D2 is globally planarized using a metal topography reduction (MTR) process whereby resist etch-back of a photoresist layer over another patterned photoresist layer is employed to eliminate the M1 topography. Tapered via profiles are obtained in D2 and D3 by a two step tapered via etch process. The first step is a conventional anisotropic RIE etch step that partially etches the patterned vias into the PETEOS layer. The second step is a resist etch-back step that further smooths the dielectric surface and etches through the remaining PETEOS to make contact to the

0-7803-3177-6 $5.00 © 1996 IEEE

IEEE 1996 CUSTOM INTEGRATED CIRCUITS CONFERENCE

underlying metal. Metal 2 (M2) is deposited and etched. The gap fill process for D3 is then carried out, and D3 is smoothed with a blanket RIE etch-back step. After tapered vias are etched, metal 3 (M3) is deposited and etched. Three level metal metrics are summarized in Table 1.

TABLE 1. Three Level Metal Metrics

METRIC	*Contact Size*
*W*1	0.56 μm
*W*2	0.64 μm
*W*3	0.64 μm
	Uncontacted Pitch
*M*1	1.02 μm
*M*2	1.20 μm
*M*3	1.36 μm
	Contacted Pitch
*M*1	1.60 μm
*M*2	1.76 μm
*M*3	1.76 μm

The passivation consists of a dual layer of plasma silicon nitride over PETEOS. This technology requires a total of 17 mask levels.

4. DEVICES, CIRCUITS and RELIABILITY

Using extensive process, device, and circuit simulations performance targets for the transistors and evaluation circuits were obtained. These targets were realized using lot experiments. In this technology the NMOS transistors are surface-channel devices and the PMOS transistors are buried-channel devices. Since it is economically desirable to extend the use of n^+-polysilicon gates, buried-channel PMOS transistors were chosen over surface-channel PMOS transistors. Careful analysis shows that buried-channel PMOS transistors can be utilized in a CMOS technology with no penalty in speed for supply voltages as low as 2.5 V [3] (see also data in [4]).

TABLE 2. AT&T CMOS Technology Comparison

PARAMETER	0.35 μm *CMOS*	0.5 μm *CMOS*
V_{DD}	3.3 V	3.3 V
t_{ox}	65 Å	90 Å
$V_{th}(NMOS)$	0.6 V	0.6 V
$V_{th}(PMOS)$	0.85 V	0.9 V
$I_{ON}(NMOS)$	600 $\mu A/\mu m$	425 $\mu A/\mu m$
$I_{ON}(PMOS)$	300 $\mu A/\mu m$	180 $\mu A/\mu m$
$L_{poly}(NMOS)$	0.36 μm	0.5 μm
$L_{poly}(PMOS)$	0.40 μm	0.6 μm
I_{off} *at* 125*C*	$< 1nA/\mu m$	$< 1nA/\mu m$
Ring Oscillator	50 *ps*	82 *ps*
2$-$*NAND FO*=2	94 *ps*	144 *ps*

Table 2 compares the nominal transistor characteristics and circuit performance of AT&Ts 0.35 and 0.5 μm CMOS processes. Typical triode saturation characteristics are shown in Fig. 6. In Fig. 7 we demonstrate that the subnominal gate length transistors have excellent subthreshold characteristics at 125°C. The target drive current (I_{ON}) for the NMOS transistor is 600 $\mu A/\mu m$, and is realized with margin without channel punch-through (I_{off}) as shown in Fig. 8. The dependence of NMOS narrow-width effect on the isolation channel-stop implant is demonstrated in Fig. 9. In this technology the threshold voltage of the NMOS is independent of the transistor width. Peak substrate current in the NMOS transistor is kept below 2000 $nA/\mu m$ by adjusting the LDD implant energy (Fig. 10) and the LDD implant dose. Design tolerance experiments, where transistor gate lengths are physically varied, were performed. Fig. 11 depicts ring oscillator delay time as a function of the NMOS I_{ON} from this experiment. Increasing the NMOS drive current clearly results in reduced ring oscillator delay time. A 50 ps delay time is achieved when the NMOS transistor provides $I_{ON} = 600\ \mu A/\mu m$. The measured 199 stage ring oscillator wave-forms are shown in Fig. 12. The delay per stage versus supply voltage is shown in Fig. 13. This 0.35 μm CMOS technology, at a supply voltage of 3.3 V, has a targeted ring oscillator delay time of about 50 ps. Equivalent speed to our 0.5 μm 3.3 V (82 ps) is achieved at $V_{supply} = 2.0\ V$. Furthermore, excellent speed characteristics are apparent for supply voltages as low as 1.5 V. Other circuits that have been fabricated include a 256K SRAM development vehicle, internal AT&T standard cell designs, and the ORCA-FPGA (Fig. 14).

5. CONCLUSIONS

In this paper a manufacturable and a very high performance 0.35 μm CMOS 3.3 V ASIC technology has been presented. This CMOS technology features a 65 Å gate oxide, single n^+-polysilicon gate, and 3 levels of metal. An improvement of 1.6X in circuit performance and 1.56X in packing density is achieved over AT&T's previous generation 0.5 μm 3.3 volt CMOS technology. Ring oscillator delay times of 50 ps are measured.

ACKNOWLEDGEMENTS

The use of the AT&T Manufacturing Line (OR-1) in Orlando (FL.) is gratefully acknowledged. The authors recognize significant contributions of R. Graver, M. Twifford, T.S. Yang, D. Ziger, F. Stevie, T. Giniecki, D. Vitkavage, K. Steiner, W. Jungling, A. Taylor, M. Cruz, E. Persson, J. Petrizzi, W. Cochran, A. Oates, P. Bechtold to this work, and thank K. Carlisle for carefully reviewing the manuscript. One of the authors (ICK) thanks T. Troiani and the Allentown Art Services for their timely support.

REFERENCES

[1] I.C. Kizilyalli and M.J. Thoma et al, IEEE Trans. Semiconductor Manufacturing, vol. 8, pp. 440-448, 1995.

[2] M.L. Chen et al, IEEE IEDM Tech. Digest pp. 256-259, 1986.

[3] I.C. Kizilyalli et al, IEEE Electron Device Lett., vol. 10. pp. 457-459, 1995.

[4] M.P. Brassington, IEEE CICC Tech. Digest, pp. 47-54, 1995.

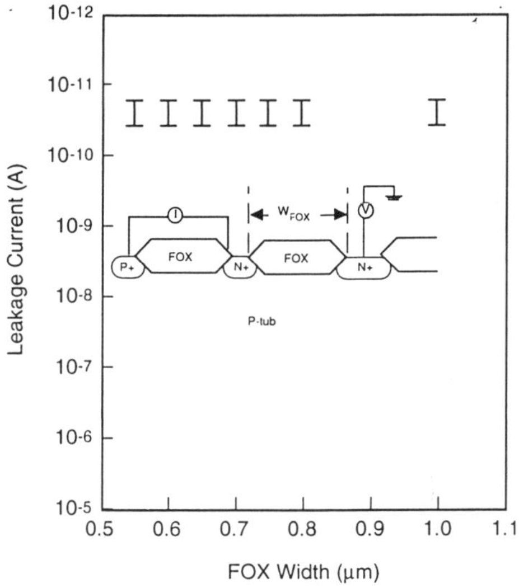

Fig. 1: Leakage current versus FOX width.

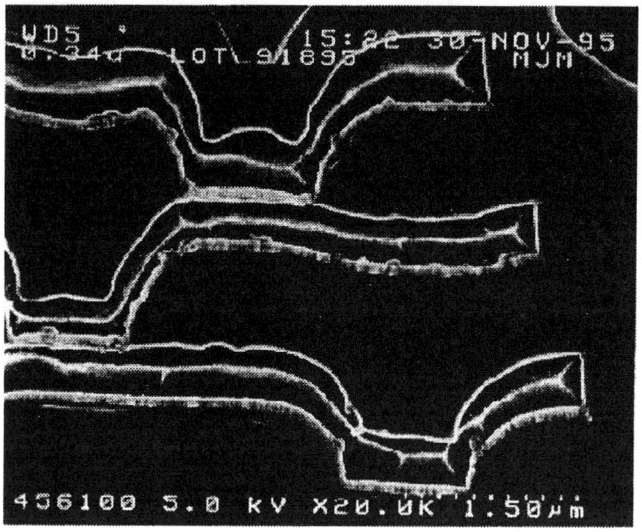

Fig. 2: SEM photograph of the 3 layer metal.

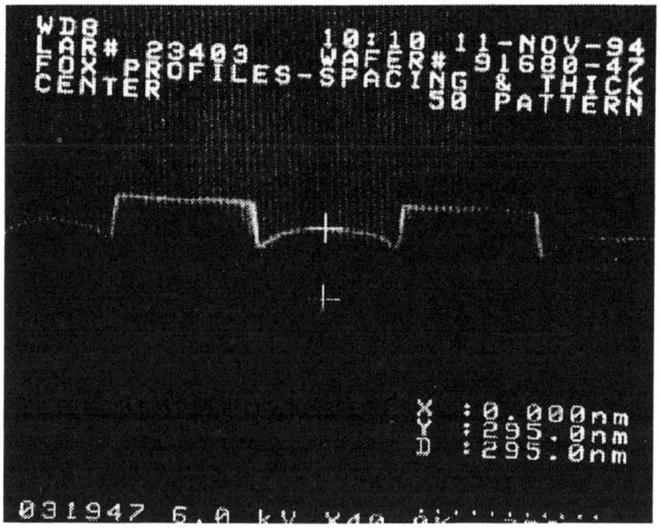

Fig. 3: SEM photograph of the FOX and the PBL stack.

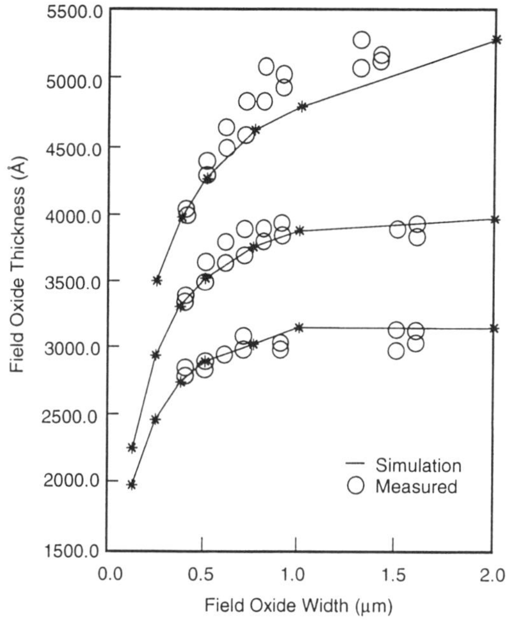

Fig. 4: FOX thickness versus FOX width.

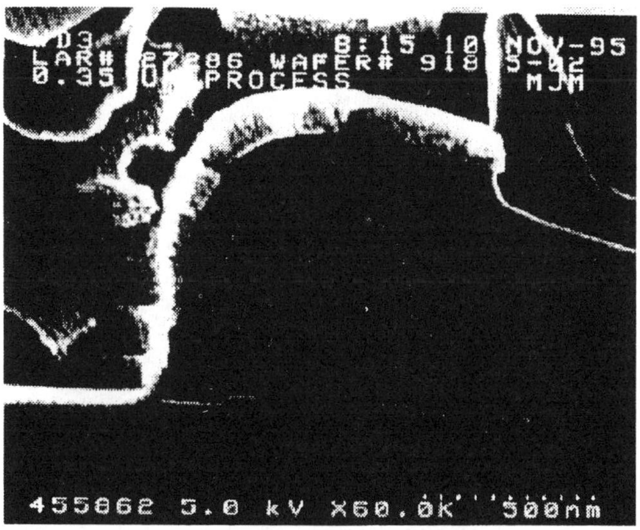

Fig. 5: SEM photograph of a transistor.

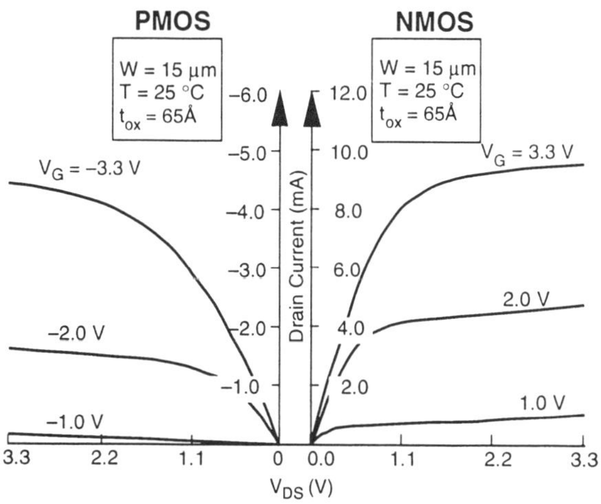

Fig. 6: NMOS and PMOS triode saturation.

AUA50731.combine-1

3.2.3

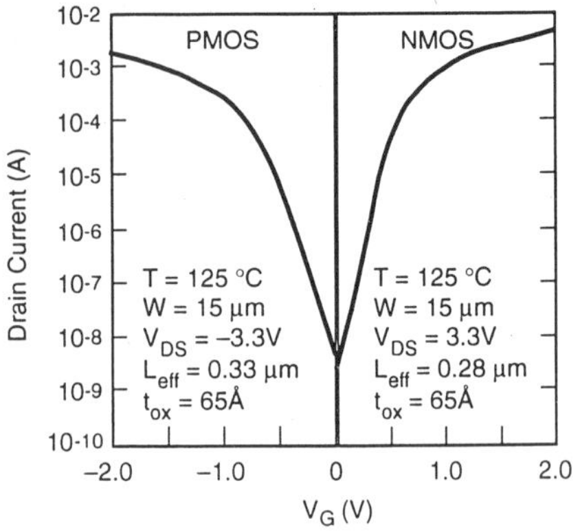

Fig. 7: NMOS and PMOS subthreshold characteristics.

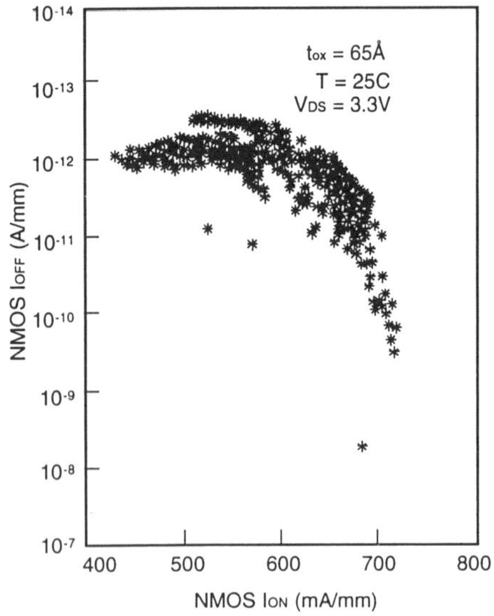

Fig. 8: NMOS I_{ON} versus I_{OFF}

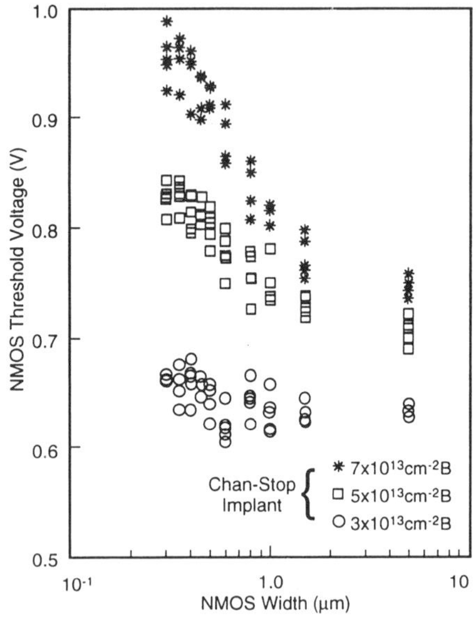

Fig. 9: NMOS narrow-width effect.

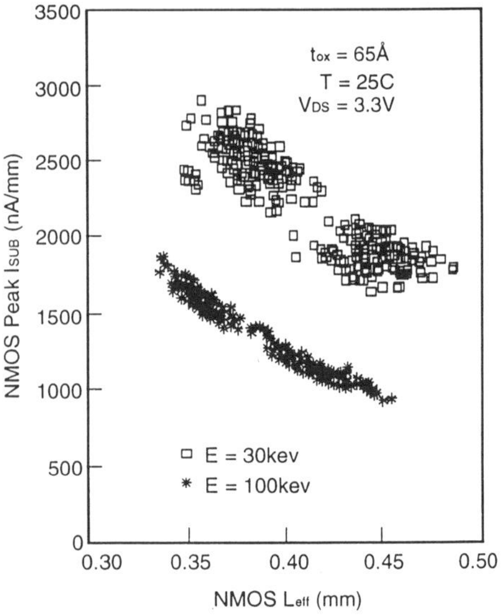

Fig. 10: NMOS I_{sub} versus L_{eff}.

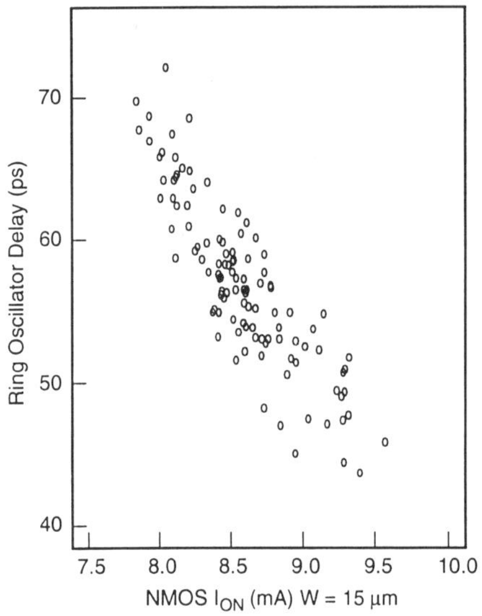

Fig. 11: Ring oscillator delay versus NMOS I_{ON}.

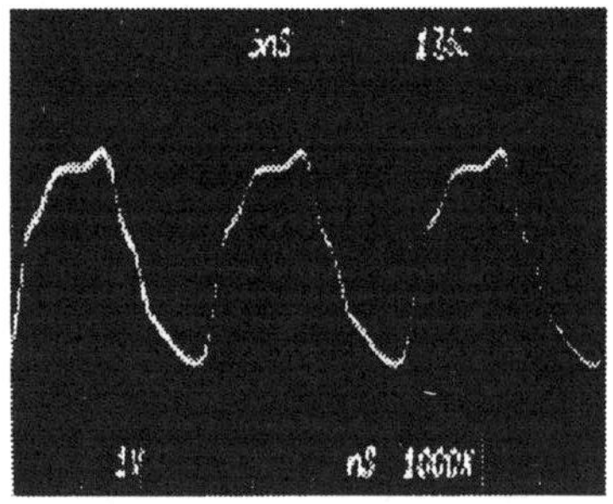

Fig. 12: Oscilloscope trace for a 199 stage ring oscillator circuit.

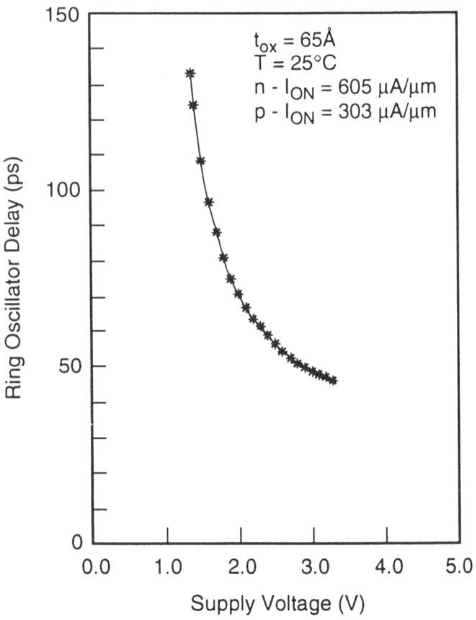

Fig. 13: Ring oscillator delay versus supply voltage.

Fig. 14: 0.35μm CMOS technology ORCA-FPGA chip.

TFSOI Complementary BiCMOS Technology for Low Power RF Mixed-Mode Applications

W. M. Huang, D. Ngo, J. Babcock, H.C. Shin, P. Welch, M. Racanelli, J. Foerstner. J. Ford, and S. Cheng

Advanced Custom Technologies, Motorola Inc.,
2200 W. Broadway Rd., Mesa, AZ 85202

Abstract

A Thin_Film_Silicon_On_Insulator Complementary BiCMOS technology has been developed for low power applications. The technology is based on a manufacturable, near-fully-depleted 0.5μm CMOS process with the lateral bipolar device integrated as a drop-in module for CBiCMOS circuits. Excellent low power performance is demonstrated through low current ECL and low voltage CMOS circuits. For the first time, good RF and analog performance of a TFSOI (C)BiCMOS technology is demonstrated. Device gain, noise figure, 1/F noise and matching characteristics comparable to bulk BiCMOS technologies is achieved.

Introduction

Interest in low power technologies for RF mixed-mode applications has been escalating in recent years, fueled by the growing portable wireless communication market. The use of (C)BiCMOS technology on SOI for low-power applications have been reported in the past [1-4]. The reduced collector-substrate capacitance of the bipolar devices and reduced source/drain capacitance of the CMOS devices in addition to the reduced sensitivity to alpha particles offered by the full dielectric isolation of the SOI substrate makes the technology attractive for low power applications. The use of the oxide isolated substrate also allows the (C)BiCMOS SOI process to be realized at a lower cost than bulk (C)BiCMOS process using trench isolation [5].

In this work, expansion of the TFSOI CBiCMOS process for RF mixed-mode applications will be described. Good RF and analog characteristics of the bipolar and CMOS devices will be discussed.

Process Integration

Separation_by_IMplantation_of-OXygen (SIMOX) substrate material with 0.1μm silicon thickness and a 0.4μm buried oxide is used for the process. The baseline 0.5μm CBiCMOS TFSOI process consists of Poly-Buffered-LOCOS (PBL) isolation, LDD CMOS devices with 105Å gate oxide, lateral bipolar devices with silicided polysilicon base contact and split-oxide spacers to allow independent control of the bipolar base and emitter widths [5].

The use of the oxide-isolated substrate in conjunction with the PBL isolation (Figs.1-3) allows the devices to be fully isolated without the need of well isolation for the MOS devices or trench isolation for the bipolar devices. The lateral bipolar device structure also offers reduction to the overall device area and thus parasitic capacitance by eliminating the need of the deep collector contact used by conventional vertical bipolar device on bulk silicon.

Device Characteristics

Table 1 listed the key device parameters for both the MOS and bipolar devices.

Typical Gummel and f_T characteristics for a NPN device with a peak current gain of 65 and f_T of 14 GHz is shown in Figs. 4 and 5. Figs. 6 and 7 shows the corresponding PNP device with peak current gain of 40 and f_T of 9 GHz. The use of the silicided top base contact is reflected in the high f_{MAX} the NPN device achieves. The low collector-substrate and collector base capacitance reflects the advantages of using the SOI substrate and the lateral device layout.

The bipolar device also shows good device characteristics for mixed-mode applications. Fig. 8 shows noise figure and corresponding gain for the lateral NPN device measured at V_{CE}=1V and a frequency of 900MHz. RF performance comparable to that of vertical NPN device on bulk silicon [6] is demonstrated. Device 1/*f* noise also shows comparable performance to bulk silicon despite the use of SIMOX SOI substrates (Fig. 9)

In addition to good noise and gain performance, the lateral bipolar device also shows less than 500μV V_{BE} mismatch and less than 5% Beta mismatch, suggesting that this technology is applicable for analog applications where device matching is important.

3.3.1

0-7803-3177-6 $5.00 © 1996 IEEE

The MOSFET output characteristics is shown in Fig. 10 for device with L_{eff} of 0.45μm. The 3V I_{DSAT} for the NMOS and PMOS devices are 310 μA/μm and 140 μA/μm respectively. The "kink" seen with the NMOS output characteristics is typical for a near-fully-depleted floating body SOI MOS device. The "kink" can be eliminated by using a body-tied device (Fig. 11). The body-tied device can be easily realized by a layout change which allows the body of the device to be connected to the source of the device or as a separate terminal, without extra processing. Such a device behaves similarly to a bulk silicon device except with a reduced junction capacitance and is recommended for analog application. NMOS device f_T of 16 GHz and f_{MAX} of 19 GHz is also demonstrated using a 15 finger 25μm wide device.

Circuit Performance

Speed-power advantages of this technology are demonstrated using both bipolar and CMOS circuits. Fig. 12 shows ECL gate delays for this technology and for a 0.4μm trench-isolated double-polysilicon self-aligned vertical NPN BiCMOS technology on bulk silicon [6]. At a gate current of 100μA, the 0.1x7.2μm^2 TFSOI circuit achieves a ECL gate delay of 45 psec, a 25% reduction in delay compared to the bulk circuit. As the gate current is reduced to 20μA, the 0.1x1.2μm^2 TFSOI circuit achieves a ECL gate delay of 80 psec, a 60% reduction over the bulk circuit.

Similar low-power advantages are demonstrated using a CMOS 8-bit CPU. Figs. 13 compares the power dissipation of the CPU built in bulk and TFSOI CMOS technologies with similar device threshold voltages and Leff [8]. A 30% reduction in power dissipation is realized using the TFSOI technology.

Conclusion

A modular 0.5μm CBiCMOS process on TFSOI featuring near-fully-depleted CMOS devices and lateral bipolar devices has been developed for low power applications. Expansion of the process for RF mixed-mode applications has been demonstrated with good noise figure, gain and device matching characteristics. Excellent speed-power improvement over 0.5μm bulk technology for both the bipolar and CMOS circuits are demonstrated.

Acknowledgments

The authors wish to acknowledge the ACT IDF for wafer processing support, MRST for SOI substrate support and special thanks to B. Fulton, and J. Jones wafer processing support.

References

[1] S. Parke, et al., "A versatile, SOI BiCMOS technology with complementary lateral BJT's," IEDM Technical Digest, p. 453, 1992.
[2] G.G. Shahidi, et al., "A novel high performance lateral bipolar on SOI," IEDM Technical Digest, p. 663, 1991.
[3] W.M. Huang, et al., "TFSOI BiCMOS technology for low power applications," IEDM Technical Digest, p. 449, 1993.
[4] J.-P. Colinge, "Recent advances in SOI technology," IEDM Technical Digest, p. 817, 1994.
[5] W.M. Huang, et al., "TFSOI complementary BiCMOS technology for low power applications," IEEE Transactions on Electron Devices, vol. 42, no. 3, p. 506, 1995.
[6] J. Kirchgessner, et al., "An advanced 0.4μm BiCMOS technology for high performance ASIC applications," IEDM Technical Digest, p. 97, 1991.
[7] J. Costa, et al., "Extracting 1/f noise coefficients for BJT's"," IEEE Transactions on Electron Devices, vol. 41, no. 11, p. 1992, 1994.
[8] W.M. Huang, et al., "TFSOI CMOS technology for sub-1V microcontroller circuits," to be published 1995 IEDM.

TABEL 1
DEVICE PARAMETERS

BIPOLAR		**NPN**	**PNP**
SOI Thickness	μm	0.1	0.1
Emitter Size	μm^2	0.1 x 7.2	0.1x7.2
Peak Current Gain		65	40
Peak f_T	GHz	14	9
f_{MAX}	GHz	22	
C_{EB}	fF	4.1	3.5
C_{CB}	fF	2.3	2.2
C_{CS} (SIMOX)	fF	2.2	2.2
BV_{EBO}	V	4.8	6.5
BV_{CBO}	V	>6	>10
BV_{CEO}	V	>3.5	>8

CMOS		**NMOS**	**PMOS**
SOI Thickness	μm	0.1	0.1
Gate Oxide	Å	105	105
Leff	μm	0.44	0.45
Threshold Voltage	mV	510	450
I_{DSAT} @ 3V	μA/μm	310	140
sub-Vt slope	mV/dec	67	68
BVDSS	V	3	5
Peak f_T	GHz	16 .	
f_{MAX}	GHz	19	

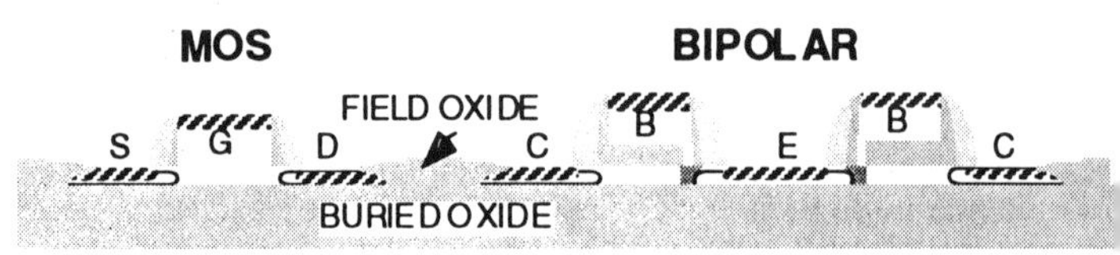

Fig. 1 TFSOI CBiCMOS process after spacer formation and self-aligned silicide formation showing the use of full oxide isolation between the MOS and the bipolar devices.

3.3.2

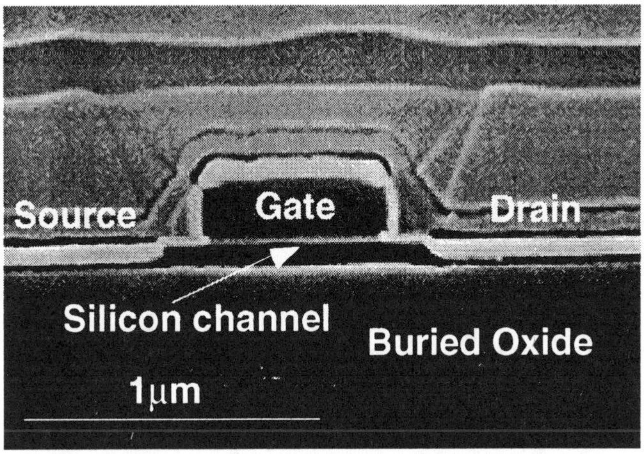

Fig. 2 SEM cross section of the completed MOS device.

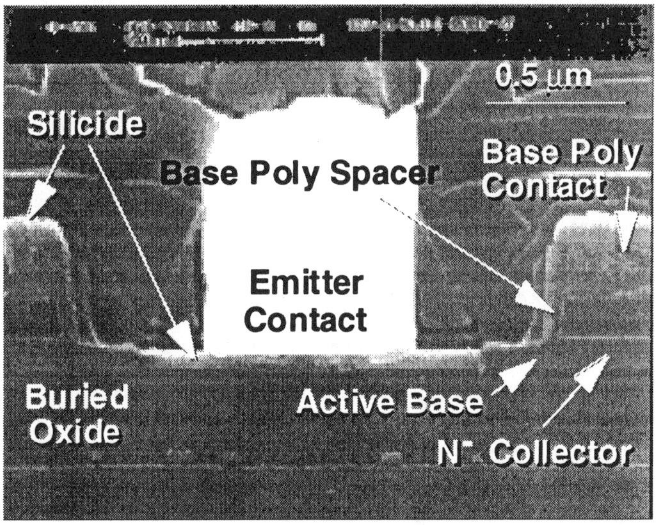

Fig. 3 SEM cross section of the completed bipolar device.

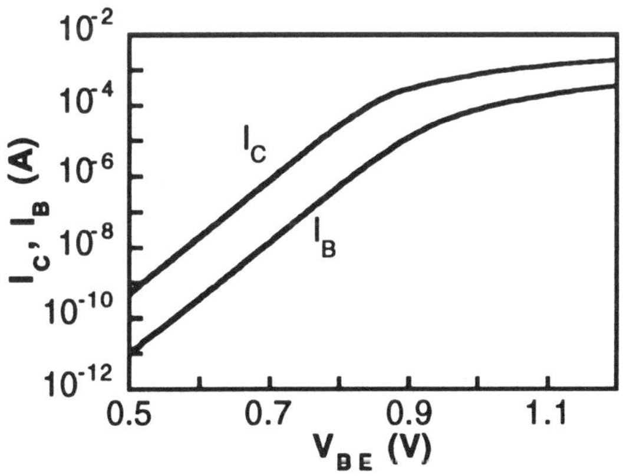

Fig. 4 Gummel plot for the 0.1x7.2μm^2 lateral NPN device.

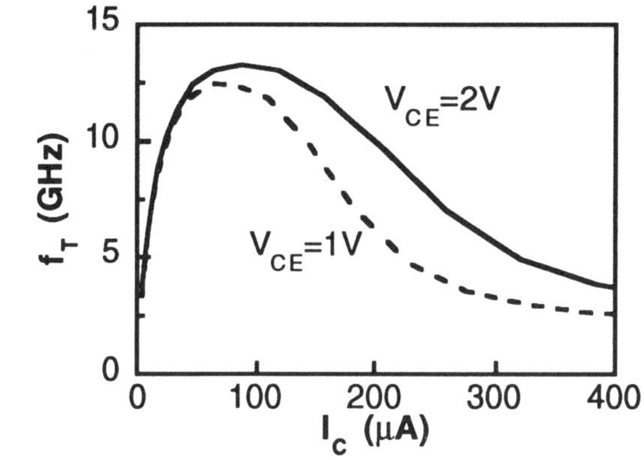

Fig. 5 f_T versus collector current @ V_{CE}=1 and 2V for the 0.1x7.2μm^2 lateral NPN device.

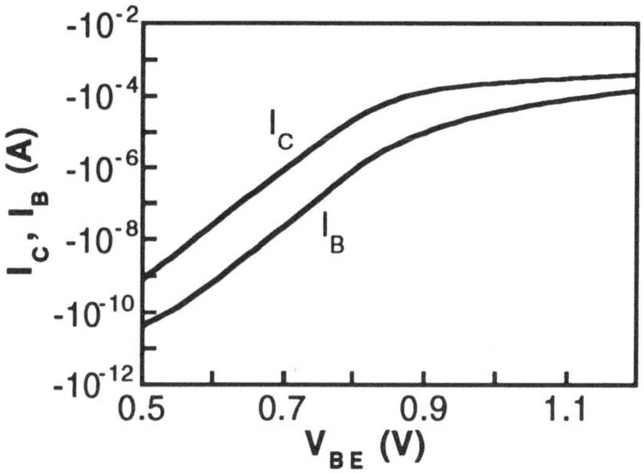

Fig. 6 Gummel plot for the 0.1x7.2μm^2 lateral PNP device.

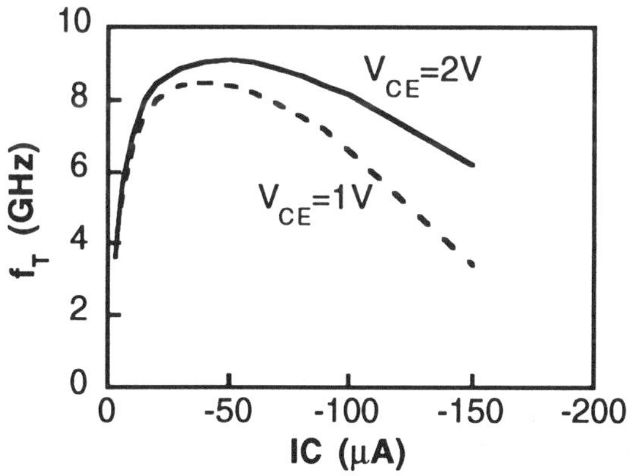

Fig. 7 f_T versus collector current @ V_{CE}=1 and 2V for the 0.1x7.2μm^2 lateral PNP device.

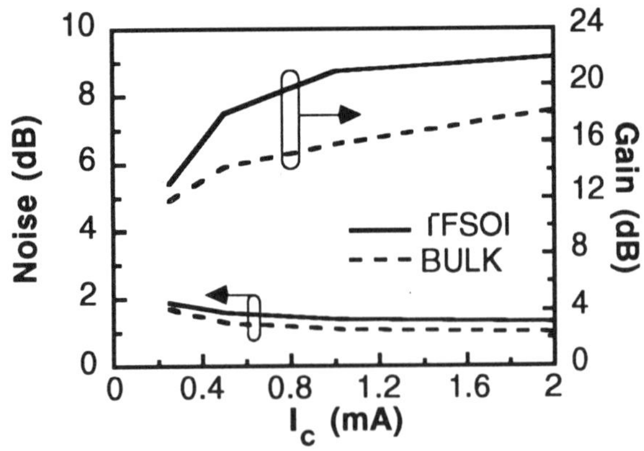

Fig. 8 Noise figure and gain measured at 900MHz with VCE=1V comparing the TFSOI lateral bipolar and bulk double-polysilicon self-aligned bipolar.

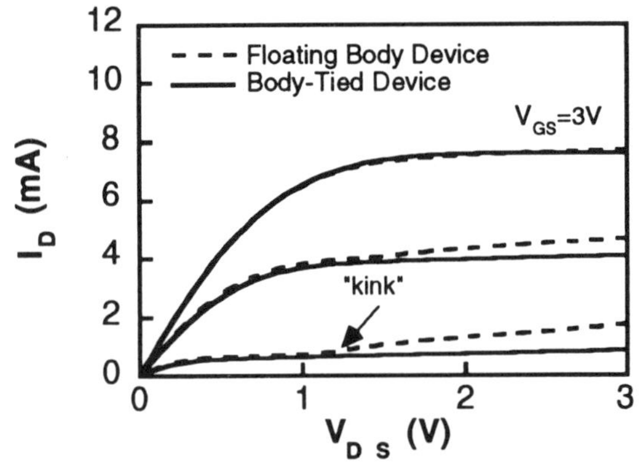

Fig. 11 Output characteristics for the standard floating body and the body-tied NMOS devices showing the elimination of the "kink" with the body-tied device.

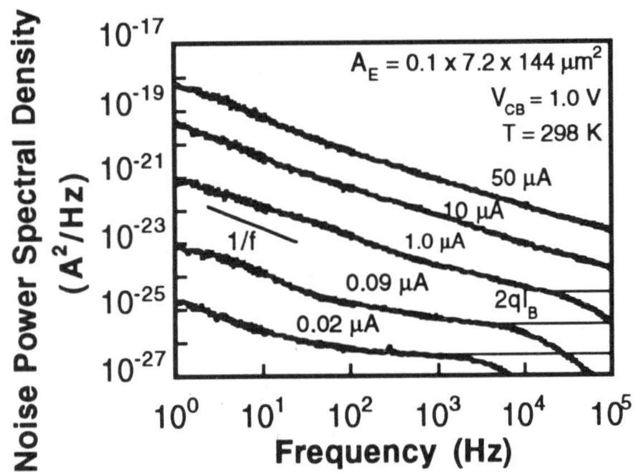

Fig. 9 Normalized noise spectrum versus frequency for different base currents for a TFSOI lateral bipolar device. 1/f noise performance comparable to bulk bipolar is demonstrated [7].

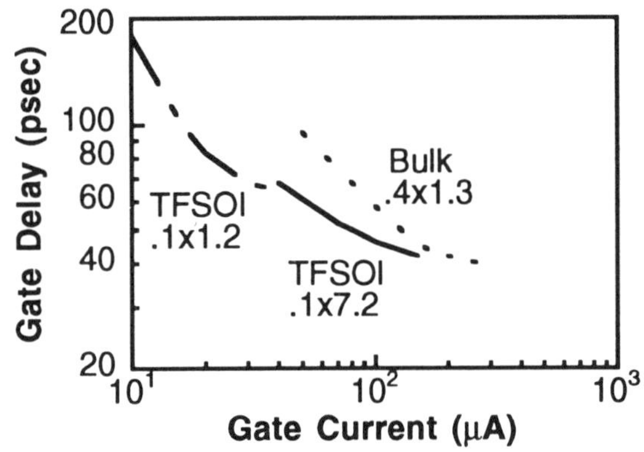

Fig. 12 ECL gate delay versus gate current for the TFSOI lateral bipolar and bulk double-polysilicon self-aligned bipolar.

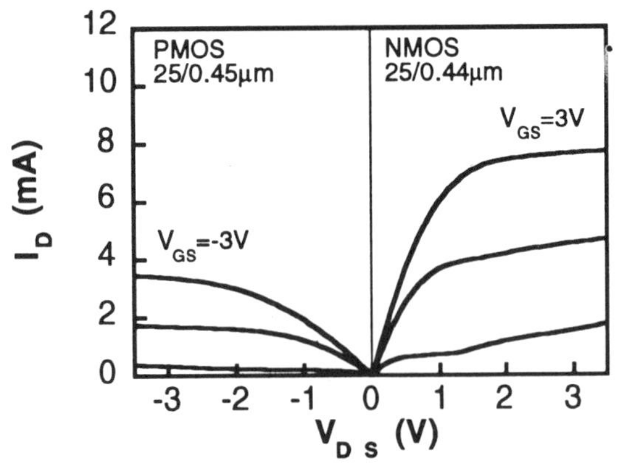

Fig. 10 Output charaacteristics for the PMOS and NMOS floating body devices.

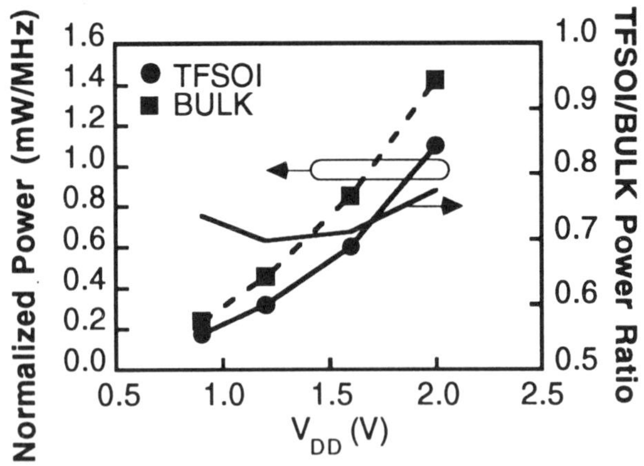

Fig. 13 8-bit CMOS CPU power dissipation comparing TFSOI and bulk CMOS technologies. The TFSOI process offer a 30% power reduction without changes to the circuit design.

High Sensitivity Pixel Technology for A 1/4-inch PAL 430k pixel IT-CCD

Akira Tsukamoto, Wataru Kamisaka, Hiroyuki Senda, Naoto Niisoe
Hiromitsu Aoki, Tomoko Otagaki, Yoko Shigeta
Masaji Asaumi, Yuji Miyata, Yoshikazu Sano
Toshihiro Kuriyama*, and Sumio Terakawa

Kyoto Research Laboratory
Matsushita Electronics Corporation
19 Nishikujo-Kasugacho, Minami-Ku, Kyoto 601, Japan

Abstract

A new pixel technology has been developed for a 1/4-inch 430k (PAL) pixel IT-CCD (Inter-line Transfer Charge Coupled Device). The new pixel has a thin light shield made of tungsten silicide film and an inner-layer optical micro-lens pre-defined by BPSG flow. This process technology can reduce the pixel size to 22.6μm^2 with improved sensitivity and reduced smear value. The sensitivity is increased by 30% and the smear value is reduced by 6dB compared to a conventional pixel technology. These characteristics are comparable to a conventional 1/3-inch 560k pixel IT-CCD with the pixel size of 33.6 μm^2(1).

Introduction

Nowadays, use of CCD image sensors adopted into electronic equipments is increasing. Fine pattern processing makes it possible to shrink their pixel sizes. As reducing the pixel size, the aperture size of the light shield and the channel stop region become so small that its low sensitivity and large smear value become major problems for CCD image sensors.

As shown in columns A and B of Table I, when reducing the pixel size to 67% by the conventional technology, the sensitivity decreases to 75% and the smear value increases to 200%.

To realize the high sensitivity while keeping the smear value low enough, we have developed an advanced pixel technology with a thin refractory metal light shield and an inner-layer micro-lens pre-defined by Boro-Phospho Silicate Glass (BPSG) flow.

* Picture Component Division, Semiconductor Group
Matsushita Electronics Corporation
1 Kotari-Yakimachi, Nagaokakyo-Shi, Kyoto 617, Japan

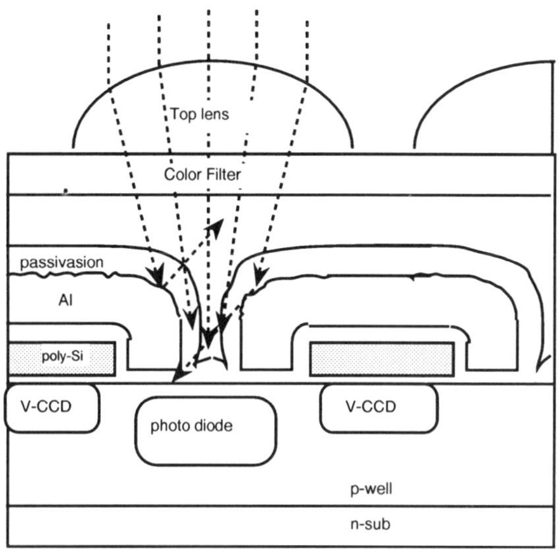

Fig.1 Cross-sectional view of a conventional pixel with thick aluminum light shield. Dotted lines indicate paths of incident light.

Pixel technology

Cross-sectional view of a conventional pixel with an aluminum light shield is shown in Fig.1. Because of high reflectivity of the aluminum surface, some of the incident light flux is scattered by the aluminum light shield resulting in decrease of the sensitivity and deterioration of the smear value. Since the light shield is too thick for the aperture size being used, the solid angle from the photo diode surface to the on-chip micro-lens becomes too small as reducing the aperture size of the light shield. This is the reason for decreasing the sensitivity.

For a photo diode to collect photon more efficiently, one may try to thin the light shield film and to place the on-chip micro-lens closer to the photo diode. However, using the conventional pixel technology, it is impossible to thin the aluminum light shield film further because wiring is made of the same layer of aluminum film as the light shield. Thinning the light shield would results in poor reliability of wiring. To be worse, transparent light flux through the

0-7803-3177-6 $5.00 © 1996 IEEE

IEEE 1996 CUSTOM INTEGRATED CIRCUITS CONFERENCE

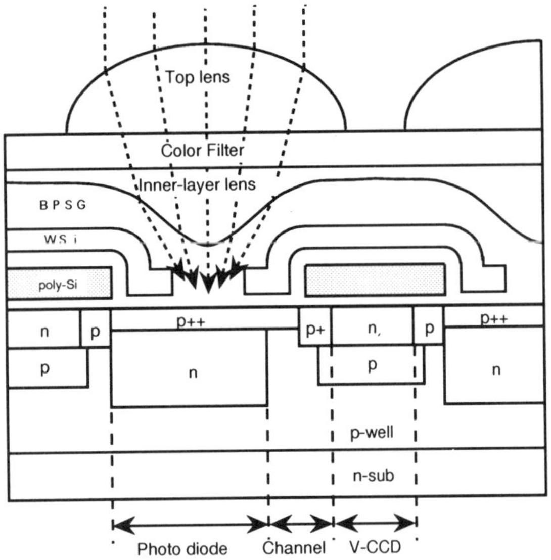

Fig.2 Cross-sectional view of a new pixel. Dotted lines indicate paths of incident light.

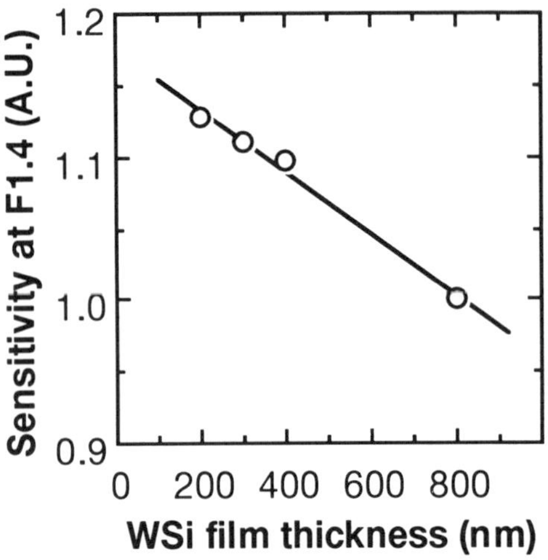

Fig.3 Dependence of the sensitivity at F-number of 1.4 on the film thickness of the light shield made of tungsten silicide. The variation of the aperture size of photo diode is within ±0.07μm The sensitivity is normalized by that at 800nm. This experiment was made on a 1/4-inch 430k pixel IT-CCD image sensor.

aluminum grain boundaries increases as thinning the aluminum light shield.

The new pixel has a light shield made of refractory metal allowing us to use BPSG flow which is necessary for forming bottom profile of the inner-layer micro-lens. The inner-layer micro-lens realizes a high efficiency for collecting photon, details of which are discussed below.

A. Photo diode and transfer gate

Fig.2 shows a cross-sectional view of the new pixel. The pixel consists of a photo diode, a transfer gate (Vertical CCD) region and a channel stop region. The shallow p++ layer of the photo diode suppresses the dark current generated in vicinity of the silicon substrate surface. The deep n-type region of the photo diode that is formed by a high energy ion implantation leads to a high charge saturation as well as high anti-blooming capability. The transfer gate region must be shielded against incident light to suppress the smear.

B. Tungsten silicide light shield

One element of the new pixel technology is to form a light shield with a thin refractory metal film. We chose tungsten silicide as a material of the light shield for its easiness in patterning and for its low transparency. The reflectivity of tungsten silicide is about a half of aluminum, which should be advantageous to reduce the smear value.

Fig.3 shows the experimental results of the sensitivity as a function of film thickness of the light shield made of tungsten silicide. The sensitivity decreases as increasing the film thickness. We chose the thickness of tungsten

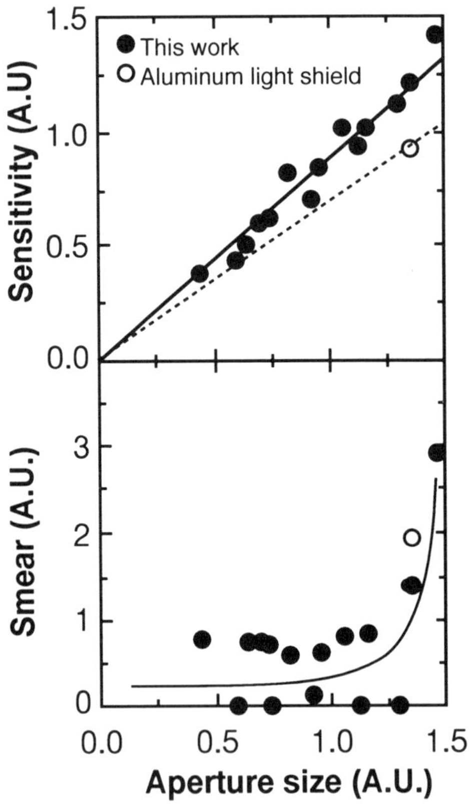

Fig.4 (a) Experimental results of the sensitivity on the photo diode aperture size of the light shield. (b) The dependence of the smear value on the aperture size. The aperture size is normalized by the optimum value. The sensitivity and the smear value are normalized at the optimum aperture size. These experiments were made on a 1/4-inch 430k pixel IT-CCD image sensor. Open dots indicate the sensitivity and the smear value with conventional pixel technology.

silicide film to be about 400nm considering the transparency. This is in contrast to our conventional pixel having an aluminum light shield with thickness of order of μm. The new pixel technology improves the sensitivity by 10% compared to our conventional one owing to the larger photon collection solid angle.

The experimental results of the aperture size dependence of the sensitivity and the smear value are shown in Fig.4. The sensitivity is in proportion to the aperture size. We optimize the aperture size of the light shield to maximize the sensitivity while keeping the smear value as small as possible. The aperture size is normalized by the optimum value. The smear value increases dramatically for the normalized aperture size greater than 1.3.

The results of the conventional pixel having an aluminum light shield are also shown. The new pixel indicates the sensitivity ~1.3 times larger than conventional one. This is because of the difference between their light shield thicknesses. These results are quantitatively consistent with the results of Fig.3.

C. Inner-layer Micro-lens

Another element of new pixel technology is an inner-layer optical micro-lens. To define the bottom profile of the inner-layer lens, a BPSG film is deposited over the light shield followed by high temperature reflow in a thermal furnace. Thus, the hollow of the BPSG film is self-aligned with respect to the photo diode aperture of the light shield. Next, filling the hollows with a high transparency organic resin that has a larger refractive index than the BPSG film. In this way, convex lens arrays are formed on photo diodes

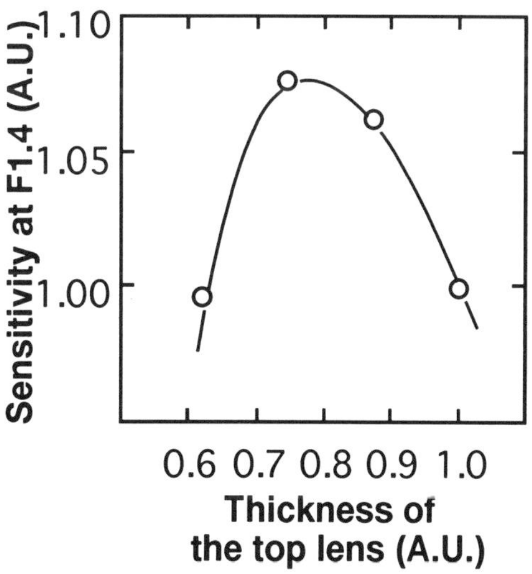

Fig.5 Simulation results of the sensitivity at F-number of 1.4 as a function of the top lens thickness with the inner-layer micro-lens structure. The thickness and the sensitivity are normalized by those of a new pixel without the inner-layer micro-lens.

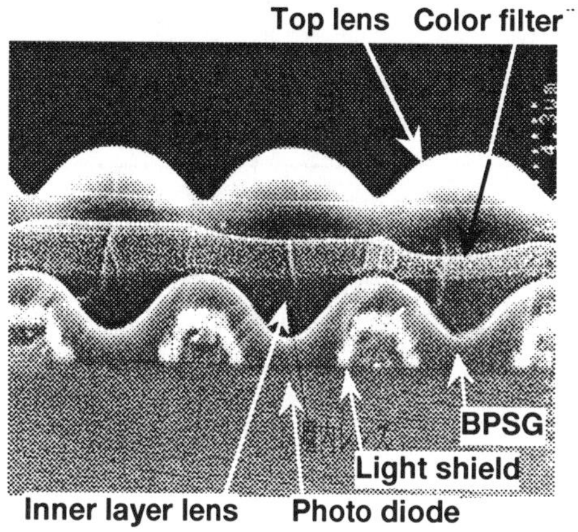

Fig.6 A cross-sectional SEM micrograph of a 1/4-inch 430k pixel IT-CCD image sensor.

in self-aligned fashion. Finally, an on-chip color mosaic filter with synthetic resin and a half-micron-spaced on-chip micro-lens (top lens) are formed to optimize the overall sensitivity with the inner-layer micro-lens(2)(3).

Fig.5 shows the simulation results of the sensitivity as a function of the top lens thickness for F-number of 1.4. The thickness of the top lens is optimized so that the sensitivity is maximized. With the inner-layer micro-lens structure, the optimum sensitivity is expected to be increased by 8% compared to one without the inner-layer micro-lens.

A cross-sectional SEM micrograph of a 1/4-inch 430k pixel IT-CCD image sensor is shown in Fig.6 where optical centers of the top lens and inner-layer lens are self-aligned exactly.

Device characteristics

Integration of the thin light shield and the inner-layer micro-lens results in a high sensitivity as well as a satisfactorily low smear value. Owing to the high photo sensitivity and monolithic low noise output amplifier, high signal to noise ratio (S/N) is realized.

Device characteristics of the 1/4-inch 430k(PAL) IT-CCD image sensors are summarized in Table I. The pixel size is 4.85(H) x 4.65(V) μm^2 for PAL format. Our new pixel technology increases the sensitivity by 30% and the smear value is reduced by 6dB compared to the conventional pixel technology. These results show that this pixel technology has achieved a remarkable improvement on the sensitivity and the smear value.

Conclusion

A new pixel technology has been developed for a 1/4-inch 430k (PAL) pixel IT-CCD (Inter-line Transfer Charge Coupled Device). The new pixel technology consists of two elements. (a) The thin light shield made of tungsten

Table I. Device characteristics

	A	B	C
Pixel size shrinkage	1.0	0.67	0.67
Pixel technology	Conventional	Conventional	This work
Light shield	Al	Al	WSi
Inner-layer lens	No	No	Yes
Device (format)	1/3-inch 560k EIS (NTSC)	1/4-inch 430k (PAL)	1/4-inch 430k (PAL)
Pixel size (μm^2)	5.7(H) x 5.9(V)	4.85(H) x 4.65(V)	4.85(H) x 4.65(V)
Color filter	Ye, Cy, Mg, G mosaic array	Ye, Cy, Mg, G mosaic array	Ye, Cy, Mg, G mosaic array
Sensitivity (F number:8) (mV)	220	140	210
Sensitivity(F: 1.4) (mV)	160	120	155 (50mV/lx:3200K)
Smear (V/10) dB	-83	-78	-85
S/N (dark) dB	-58	-58	-62
Carrier saturation voltage (mV)	600	400	600
Electronics shutter (sec)	1/60 - 1/10000	1/60 - 1/10000	1/60 - 1/10000
Lag (%)	Negligible small	Negligible small	Negligible small
Vsub,Reset voltage adjustment	Manual adjustment	Manual adjustment	Adjustment free
H-CCD driving voltage (V)	5.0	5.0	3.0

silicide with an optimum aperture size. (b) The inner-layer optical micro-lens pre-defined by BPSG flow. This process technology can reduce the pixel size to 22.6μm^2 while keeping the sensitivity and the smear value comparable to those of a conventional 1/3-inch 560k pixel IT-CCD with the pixel size of 33.6 μm^2.

Acknowledgment

The authors would like to express their gratitude to H.Esaki, Director of Kyoto Research Laboratory, and the technical staff of KRL and Picture Component Division for their supports and encouragements. S.Hashimoto is also acknowledged for his useful discussions and comments.

References

(1) S.Terakawa, Y.Sano, T.Imanishi, and Y.Hiroshima,"Advanced high performance CCD technology for a 1/4-inch 560k pixel IT-CCD image sensor," CICC, 1994, pp.479-483.

(2) K.Shimomura, T.Otagaki, Y.Tamagawa, M.Kobayashi, H.Aoki, Y.Sano, and S.Terakawa,"Novel synthetic aqueous photoresist for CCD micro color filter," Proceeding of SPIE, vol.2195, 1994, pp.823-833.

(3) Y.Sano, T.Nomura H.Aoki, S.Terakawa, H.Kodama, T.Aoki, and Y.Hiroshima,"Submicron spaced lens array process technology for a high photosensitivity CCD image sensor," IEDM meeting technical digest, 1990, pp.283-286.

Chair: John Wright
Co-Chair: Jeffery Oppold

Gate array usage has continued to climb and almost every engineer that deals with ASICs has contributed to the design of a gate array in one form or another. Even with all the familiarity of gate arrays new frontiers continue to be discovered. Each of these discoveries bring with it greater risk and greater reward. We hope with this session to bring you some aids in crossing these new frontiers and maybe some ideas on how you can do it with first pass success! We included some papers that have solved some of the traditionally difficult issues that we face in todays high speed low power environments. If these papers strike a cord for your application, let us know. If you have specifics you'd like to see addressed, let us know that also.

The first paper is an invited paper from IBM on "A Proven Methodology for Designing One-Million-Gate ASICs". How many of you have designed a million gate circuit? What are the odds that within three years you will have? This paper will give you insights in what to look out for and what tools you may need. Even if you are one of the few that have designed a million gate circuit, there are ideas in this paper that will help you the next time.

The next paper, titled "A High-Speed Low-Power 0.3μm CMOS Gate Array with Variable Threshold Voltage (VT) Scheme" presents a new twist on the approach to getting high speed and low power all at the same time.

Paper number three, titled "A Quarter-Micron SIMOX-CMOS LVTTL-Compatible Gate Array with an Over 2,000 V ESD-Protection Circuit" addresses several issues that have haunted SIMOX-CMOS products including 2,000 V ESD protection.

The last paper in the session, titled "A 1860kG CMOS Gate Array with GTL Input Flip-Flop Circuits" looks at some of the issues in transmitting data at 300MHz and what design techniques they used to make a reliable system.

We hope that you enjoy these papers and that you come away from the session with ideas you can implement in future designs.

A Proven Methodology for Designing One-Million-Gate ASICs

Ann Marie Rincon, Michael Trick, Thomas Guzowski

IBM Microelectronics
Essex Junction, Vermont 05452

Abstract

This paper describes the methodology used to design a family of ASIC chips in the 400K to one-million-gate range in a 0.5 micron technology [1]. Working first-pass hardware was produced at an average clock speed of 100MHz with over 99% testability. The methodology utilized a variety of third-party CAD tools from Synopsys and Cadence in combination with proprietary IBM tools.

Introduction

The last two years have seen a steady rise in the number of ASICs in the 400K gate range and above. In 1995, the one-million-gate ASIC became a reality.The high gate density of these ASICs and their implementation in a 0.5 micron technology, coupled with the need to support industry-standard design styles, presents several key challenges to the development of the design methodology.

Front-end methodology development focuses on the problem of prohibitively large development and verification times required to design and verify high gate-count ASICs. The industry-standard practice of verifying ASIC timing and testability through gate-level simulation is not viable for designs over 100K, given today's time-to-market requirements; therefore alternative verification strategies had to be developed.

The back-end, or layout, portion of the methodology was faced with the half-micron paradigm of wire capacitance and wire delay dominating the gate capacitance and gate delay in the customer's design. Front-end timing requirements had to be tightly coupled with the layout implementation to prevent costly iterations through the design and layout processes. Increased wire parasitics also increased the difficulty in balancing clock trees to minimize skew.

Rigorous design verification and checking in the release-to-layout and release-to-manufacturing procedures, along with timing-driven placement and placement-based optimization tools, helped solve these problems. This paper discusses these solutions and the entire methodology used to design high gate count ASIC chips (Figure 1). Specifically, we will discuss the factors that were key to the success of the methodology, cite case histories of designs that used the methodology, and what we learned from processing these designs.

I. Support for Edge-Triggered Designs

The first major step in the methodology provides an automated method to map edge-triggered, flip-flop-based designs into level-sensitive latch-based libraries.

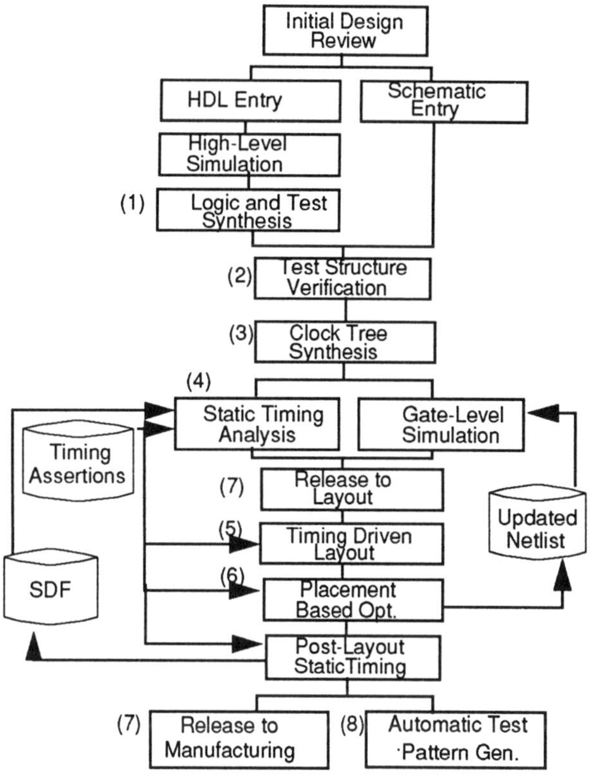

Figure 1 - Design methodology overview.

This was key in enabling customers to design in an edge-triggered fashion and leverage the high testability and automated pattern generation capability of full-scan design. Customer investment in existing software dictated that the mapping method be accomplished by using industry-standard tools. Success was achieved through a three-step process involving the Synopsys Design Compiler and Test Compiler products and the IBM synthesis product, Booledozer.

Synthesis tools cannot directly model the behavior of a level-sensitive latch, so a simplified set of latch models was created [2]. These "pseudo-latches" are modelled as traditional D-flip-flops (DFF), allowing the synthesis tool to infer them directly as Step 1 of the mapping pro-

0-7803-3177-6 $5.00 © 1996 IEEE

cess (Figure 2).

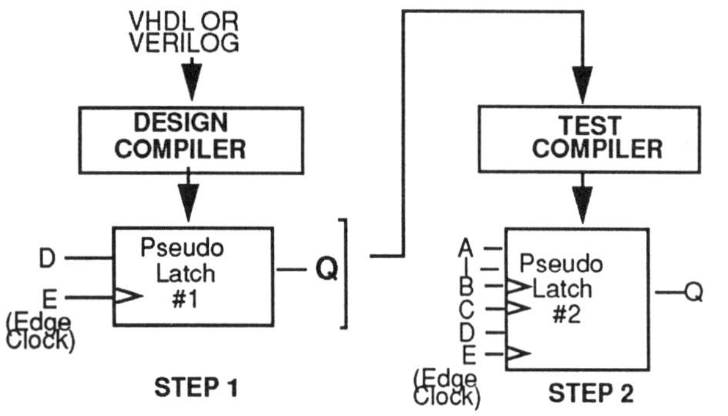

Figure 2 - Mapping to pseudo-latch.

Step 2 uses the Synopsys Test Compiler in "Auxiliary Clock LSSD" mode to map each DFF to another pseudo-model, representing its scan equivalent (Figure 2).

Test Compiler replaces the first pseudo-latch with a second pseudo-latch containing scan master and slave clock pins (C,B) as well as a scan test clock (A) and the scan-in pin (I). Auxiliary clock nets (A,B, and C) and scan input/output nets are created by the Test Compiler and connected to predesignated chip ports. If ports are not predesignated, they are added to the design by Test Compiler.

II. Signoff for Design-For-Test (DFT) Compliance

With scan clocks inserted and scan paths connected, the design proceeds into DFT verification, one of the key signoff phases of the methodology. Compliance with DFT requirements is checked using a test structure verification (TSV) tool (Figure 1 (2)). The test signoff requirement dictates that a design cannot contain any "fatal" DFT errors; i.e., those errors preventing automatic test pattern generation later in the process. Designs without TSV errors achieve, on average, greater than 99 percent stuck-fault test coverage.

The last phase of mapping is accomplished with a set of netlist "transforms" and the Booledozer synthesis tool (Figure 3). The design containing pseudo-latch #2 is first read into Booledozer; then the netlist transform replaces the specified pseudo-latches with a predetermined true full-scan latch from the ASIC library.

A series of clock splitters is also inserted into the design via the Booledozer tool. They split the single-edge clock (E) into the respective master (C) and slave (B) clocks used to drive the full-scan latch. After splitter insertion, the clock signal that previously drove the edge clock (E) in pseudo-latches #1 and #2 now drives the oscillator input (O) to the clock splitter. (The number of splitters inserted into the design and the number of latches driven by each splitter are determined as part of the clock-planning and clock-tree synthesis phase.) Reference to any E pin is dropped from the design representation at this point.

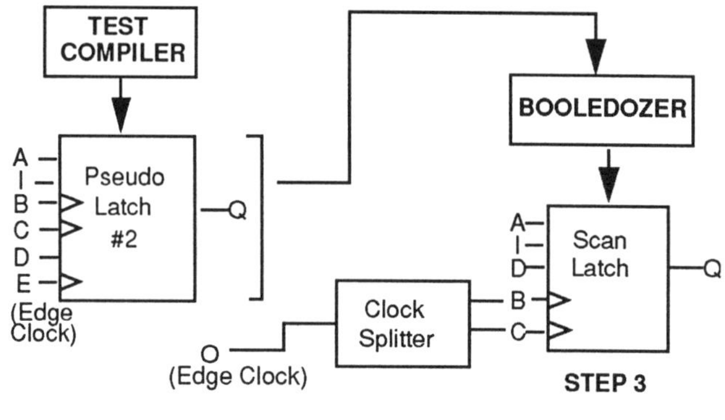

Figure 3 - Mapping to target latch.

III. Clock Planning and Clock-Tree Synthesis

Effective clock-tree synthesis (Figure 1 (3)) is the result of detailed clock planning performed early in the design cycle. A four-step process comprised of data gathering, tree calculation, customer review, and data revision, is used.

Data Gathering

The data-gathering phase involves gathering customer information that pertains to the design content, including die size/internal circuit area, circuit area devoted to RAMs, the number and types of technology cells driven by the clock, and the worst-case operating temperature and voltage. Performance objectives in the form of transition and latency targets for the repowering tree are also established by the customer.

Tree Calculation

Tree calculation is done with an IBM application called "Clockpro." Given a range of transition times, the design-content information from the data-gathering step and a set of driver cell types and strengths (Figure 4), Clockpro calculates all the possible repowering tree configurations that meet the specified transition-time target. Clockpro lists all the clock-tree combinations generated, sorted by latency, from the least to greatest amount of latency.

Terminator cells can also be factored into the clock repowering tree for nets with critical performance and skew criteria. A terminator is a single-pin cell that capacitively loads a net. Each terminator's capacitance can be set to one of four values. When these cells are seeded into a clock repowering tree, additional control over balancing can be accomplished by varying either the terminator load value or its location.

This tree calculation and synthesis method can be

4.1.2

used on any number of clock domains. Many of the ASICs produced using this method have multiple clock domains on a single chip.

```
# IBM clockpro, version 1.19.4 (Mar 9, 1995 09:13:56)
#
 Enter technology name (cmos4lp, cmos5l, cmos5s): cmos5l
 Enter timing rule path name: /rules/timing/cmos5l/v7.0/
 Enter design name: Virginia
 Enter Net Name to be repowered: edge_clock

 Enter Operating Temp (Celsius): <Best Case> <Worst Case>
0 125
 Enter Operating Voltage (Volts): <Best Voltage> <Worst Voltage>
3.8 3.0
 Enter number of loads: 1666
 Enter the area to be considered (sq mm): 23
 Enter the area to subtract (sq mm): 3.02

 Enter Slew Rate (nSec): <Slew_starting_rate> <Slew_increment>
                         <Number_of_increments>: 0.5 0.1 6
 Enter incoming signal transition to clock tree (LH/HL): LH
 Clock tree to invert resultant signal polarity? (Y/N) <default is N>: N
 Enter lead driver type of clock tree: (DRIVER/SPLITTER): splitter
 Enter latch (or original loads) data: <cell name> <input_pin>
                                       <output_pin> <drive strength>
weighted
# Enter the number and type of each latch (one per line), ending
# with a blank line.
# Weighted latch data: <number_of_this_type> <cell_name>
# <input_pin> <output_pin> <drive_strength>
850 LPH0001 B L2 a
418 LPH0001 B L2 b
30 LPH0001 B L2 c
151 LDM0001 B L2 a
214 LDM0001 B L2 b
3 LSC0001_G B L2 b
# Enter clock splitter data: <cell name> <input_pin> <ouput_pin>
 <drive_strength>: CLKSP_G OSC ZB d
# Enter inverting clock driver data: <cell_name> <input_pin>
 <output_pin> <drive_strength>: CLKI_G A Z d
# Enter non-inverting clock driver data: <cell_name> <input_pin>
 <output_pin> <drive_strength>: CLK_G A Z d
 Enter terminator cell type <cell_name_drive_strength> to add
# terminators to the tree: TERM_G_b
```

Figure 4 - Example clockpro input.

Several additional clock-tree features can be accommodated with this process, including early clock taps, heterogeneous sink distributions, and inter-tree balancing.

Customer Review

A family of clock trees with transition times bounding the supplied target is given to the customer for review and to allow the customer to make trade-offs between tree depth, latency, total cells added, and power consumption. Recommendations are also provided to the customer if certain tree configurations are considered to be more wirable than others. Customer review and critique drive the final selection of the clock tree(s) implemented in the design.

Data Revision

The clock-planning data is revised only when a satisfactory tree cannot be selected from those generated. In most cases, revisions are constrained to the choice of driver cell and drive strength.

When the final tree selection is complete, the output of the Clockpro program (Figure 5) is fed into the Booledozer tool, and used to drive a set of "build-tree" scenarios. The resulting output from this four-step process is the customer design with the repowering trees inserted.

```
# Resulting clock trees sorted by latency
#
(DataRev..:1.19.4 )
(Treenum..: 1 )
(Design...: "Virginia" )
(Netname..: "edge_clock" )
(Latency..: 4.085 Worst case, 1.129 Best case )
(Slew.....: 0.800 )
(Leaftype.: SPLITTER )
(Counts...: 4 levels, 151 Drivers and Splitters, 171 Terminators )
(Levels...:    4         3         2          1)
(Drivers..:    1         4         17         129)
(Cells....:    CLKI_a    CLKI_a    CLKI_a     CLKSPL_d)
(Inverting     I         I         I          I )
(Cmax.....:    1.546     1.546     1.546      0.705
(Cwire....:    1.206     1.150     0.758      0.220 )
(Cgate....:    0.290     0.348     0.693      0.480 )
(Area.....:    2.540     10.780    19.364     19.980 )
(Fanout...:    4         6         9          14 )
(Sinks....:    4         17        129        1666 )
(Terms....:    0         7         24         140 )
(Termcell.:    Term_b    Term_b    Term_b     Term_b )
```

Figure 5 - Clockpro output.

IV. Signoff with Static Timing Analysis

Once the clock trees are fully created, the design is ready to enter the second signoff step, static timing analysis (Figure 1 (4)).

For many years, timing signoff in the IBM ASIC design methodology was based on static timing analysis rather than gate-level delay simulation. As ASIC gate counts per die reached the 100-300K range in the late 1980s, the timing of all relevant paths through chips using delay simulation became impractical. As gate counts continued to expand into the 500K to one-million-gate range in the mid-90s, the move away from delay simulation became more critical for maintaining aggressive development schedules.

Static timing analysis allows all paths on the chip (under best-case and worst-case conditions) to be examined in a single timing run. To yield meaningful timing information about the design, careful attention must be given to the development of the "timing assertions" (such as those files that define expected arrival times, clock-phase relationships, false paths, etc.) that drive the analysis tool [4]. While the development of a meaningful assertion set requires up-front designer investment, once completed, these assertions are used for both pre- and post-layout timing signoff and for driving timing-driven layout (Figure 1 (5)), another key phase of the chip design cycle.

The elapsed clock time for a static timing run on a 850K-gate ASIC is approximately 90 minutes (using an IBM RISC System/6000, Model 590), compared to the days or weeks required to obtain timing information on an equal number of paths when using delay simulation.

In this methodology, the purpose of gate-level simulation is for functional verification only, and is owned entirely by the ASIC designer. Models are provided in the design kit for a variety of industry-standard simulators such as Verilog and VHDL-based simulators. Functional simulation was not repeated at the ASIC Design Center during the signoff process.

Methodology Checks and Release-to-Layout

When logic design and verification completes, the designer and the ASIC Design Center take the design through the release-to-layout (RTL) signoff procedure. Final netlist and assertion files are identified, time- and date-stamped, and rerun through the key signoff tools (TSV for testability and IBM's Einstimer static timing analyzer). An additional set of technology-specific methodology checks, called CMOSCHKs, is run to perform electrical rules checking (ERC) to ensure that library elements and test flags required for manufacturing test are included in the netlist and that I/O locations are properly assigned.

To exit the RTL signoff, the design is required to time without negative slacks, transition, or capacitance violations. This is key in enabling the timing-driven layout and placement-based optimization routines to achieve timing closure with minimum intervention by the logic designer.

The design must also be free of any "fatal" TSV errors (those errors preventing automatic generation of test patterns). Designs with "warning" errors are allowed to proceed to layout if the customer agrees to accept the lower test coverage resulting from the error condition.

V. Timing-Driven Layout

The timing-driven layout methodology (Figure 6) begins with an initialization step to place input receivers, output drivers, large macrocells, cores, memory arrays, and timing-critical logic. Next, timing-driven placement iterates through a constraint generation and placement loop until latch-to-latch timing meets the specification or ceases to improve.

When placement has been optimized, scan-chain connections are ordered to minimize wire length, the drive strengths of logic cells are optimized to improve performance, and clock trees are balanced to minimize skew. After optimization, the placement of objects does not change significantly, and the design can be wired. A final run through a suite of electrical and physical checks and test-pattern generation prepares the chip for manufacturing.

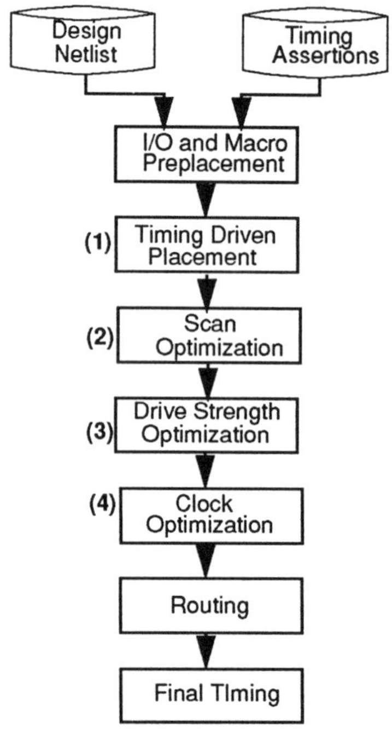

Figure 6 - Timing-driven layout process.

The timing-driven placement process (Figure 7) uses a static timing analyzer coupled with placement tools and a calculator for wiring resistance and capacitance. The initial placement step uses capacitance and region constraints generated from floorplanning tools.

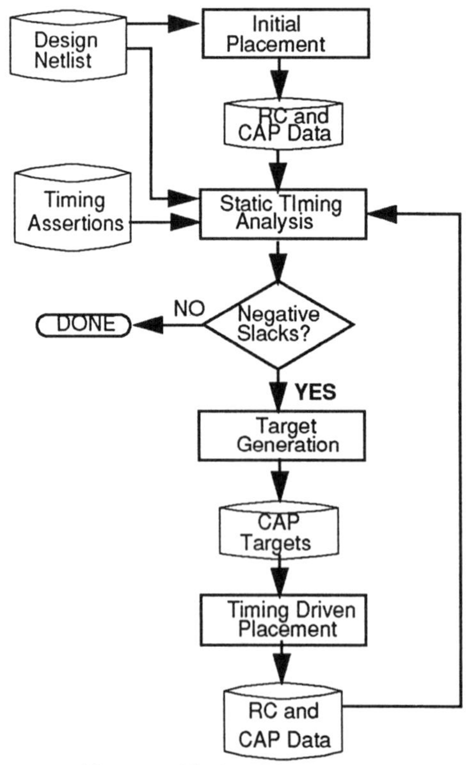

Figure 7 - Timing driven placement.

A rectilinear Steiner route for each net determines wiring parasitics after the initial placement step, and the timing analyzer calculates delays, based on the wiring parasitics, to report paths not meeting timing constraints. The target generator uses the timing report for each path in the design to optimize capacitance targets to improve timing. The timing-driven placement process is repeated until the placement tools cannot further improve the timing.

VI. Placement-Based Optimization

Scan-Chain Optimization

Scan-chain optimization (Figure 6 (2)) reduces wire length and capacitive loading on latches. Each latch on a design is connected serially in a scan chain of up to a few thousand latches. The scan-chain-optimization program uses a simulated annealing algorithm to swap connections and minimize total wire length.

Drive-Strength Optimization

Further refinement of timing is accomplished through drive-strength optimization (Figure 6 (3)) and buffer insertion. The static-timing tool contains two algorithms to optimize drive strength: one simply assigns each cell a strength that minimizes the delay on its output; the other algorithm assigns a drive strength to each cell to minimize the delay through the slowest, latch-to-latch path passing through the cell.

Clock Optimization

During placement, clock connections and high fanout repowering networks are ignored while the placement tools try to converge on latch-to-latch timing. This prevents clock nets from biasing latch placement. After placement, clock trees are optimized (Figure 6 (4)) to drive the latches with a minimum of wire, latency, and skew [8].

The clock-optimization algorithm traces each clock tree to identify equivalent nets and pins. It then creates clusters of equivalent objects at the leaves of the tree so that each cluster contains the same total (wire plus pin) capacitance. The driver of the net for each cluster is placed in the centroid of the cluster, and the algorithm continues recursively up the tree, clustering objects and placing the driver at the center of each cluster until all levels of the tree are balanced. The first two or three levels of the clock tree span large areas of the chip and use fat balanced wire to minimize skew, while leaf levels of the tree are balanced with dummy capacitive loads. The automatic clock-optimization tools balance delays to latch inputs to within a few hundred picoseconds. Manual tuning eliminates any remaining skew problems.

Using a clock tree instead of a fully connected grid of clock lines requires more sophisticated algorithms to reduce skew, but the tree structure is more flexible and efficient [9]. Using a tree structure instead of a clock grid allows for multiple clock domains covering overlapping regions of a chip. The clock-optimization tools adjust the tree to accommodate non-uniform distributions of latches, creating less blockage and using less power than a full grid of clock lines. Using a tree allows the addition of skew to steal time from slow paths and give the time to fast paths.

Routing

When placement, timing optimization, and skew minimization are complete, the physical designer uses a maze runner with rip-up and reroute capability to wire the design. Most large ASICs are wire-limited, not cell-limited; therefore, the placement tools are set to spread circuits evenly over the entire chip to minimize areas of high pin density. Bit-stacking, or arranging logic in regular structures, relieves congestion in very tightly connected logic structures like multiplexor trees and crossbar switches.

VII. Final Methodology Checks

When physical design completes (all wires are embedded, and all timing constraints are met), an exhaustive suite of checking programs ensures that I/O cells are in the correct positions, all required cells are in the netlist, pin usage is correct on every cell, NWell contacts occur at the proper frequency, gate-array filler cells are placed in every spare location, power busses are connected properly, test and boundary scan structures are correct, all electrical rules are followed, and the proper versions of all library cells are used. When these checks are complete, shapes are generated for each mask level, design-rule checking (DRC) is performed and the netlist extracted from the layout is compared to the schematic netlist.

VIII. Automatic Test Pattern Generation

Once layout is complete and all timing problems have been corrected, it is possible to generate test patterns for the design. Many ASIC customers are accustomed to generating functional test patterns for their designs; this is a complicated process that can take weeks or months to complete. By contrast, an automatic test-generation program can generate test patterns in a matter of hours for designs that pass test structure verification.

Boundary scan (either compliant with IEEE 1149.1 or a simplified IBM version) allows chips to be tested on low-cost, low-pin-count testers and is required on all IBM ASIC parts.

All RAMs embedded on an ASIC contain array built-in self-test (ABIST) that automatically generates and applies patterns to test the RAM as part of manufacturing scan test.

enhanced the clock-optimization methodology.

Case Histories

Case History #1

In 1994, Cray Research, Inc. requested that the IBM ASIC Design Center implement ten of Cray's chips in a half-micron process. The chips received contained up to 400K gates, were designed to run at 100MHz, and contained phase-locked loops to synchronize clock signals. Many of the chips also contained multiple memory arrays. The chip layout was flat (without hierarchy), and floorplanning tools were not used to help with the physical design. The logic designers at Cray provided lists of global nets that could be extended to allow other groups of tightly connected logic to congregate in small regions, minimizing delay between local circuits.

We then developed customer-specific algorithms for tracing logic that needed to be placed near the input/output drivers and receivers to minimize chip-to-chip timing problems. All the chips required exceptionally precise placement of the logic connected to off-chip drivers and receivers. We then wrote software to place that logic close to the edges of the chip and minimize off-chip delays.

Two smaller chips of less than 80,000 gates neither stressed the design tools from a data volume standpoint nor posed siginificant delay or wiring congestion problems. The small chips had very tight skew constraints so that hundreds of wires had to be manually balanced. The large chips posed different problems because it was often difficult to place logic on the chips so that both timing problems and wiring congestion could be simultaneously eliminated. One chip had a very tightly connected crossbar switch that made the chip extremely difficult to wire. Another had a configuration of register array macros that created a wiring bottleneck that was difficult to smooth out. A third part proved especially tricky because placements that met timing constraints could not be wired, while placements that minimized congestion and enabled complete wiring did not meet the 10-ns cycle time. Manual placement changes and buffer insertion were required to meet the final timing objectives. Many electromigration, transition-time, and capacitance violations were corrected manually if drive-strength optimization programs could not remove the violations. Many buffers were inserted for timing optimization.

As a result of the difficulties we experienced with the Cray chips and other early (1994) chip designs, the ASIC design group increased support for floorplanning tools, developed a timing-driven design methodology, wrote additional software to streamline the placement of logic near off-chip drivers and receivers, and

Case History #2

One of the largest ASIC chips we processed was a 66MHz video-processing chip containing 1.15 million gates of logic and a digital-to-analog converter. The netlist contained equal amounts of standard-cell components and register-array and RAM macros. The die was 17.3mm on a side with four levels of metallization. The chip was designed hierarchically to facilitate transmission and processing of 300 megabytes of netlist files.

Five months before the design was released to manufacturing, we began initial floorplanning based on partial sections of the netlist and diagrams of the functional blocks. For the next two months, additional pieces of the netlist were completed and the floorplan was refined. Some hierarchical blocks were merged to eliminate those with high aspect ratios that were unwirable. Some logic changed as latches were consolidated into register arrays. Two months before release, the logic was functionally complete, and timing assertions correctly specified arrival times and clock phases for each hierarchical design entity. One month before release, test structures were correct, negative slacks removed and clock trees balanced; and, during the final month, all electrical violations were removed, final wiring parsitics calculated, and drive strengths optimized. Two weeks after the final logic change, the chip was design-rule clean, the comparison between layout and logical netlist showed no errors, and the chip was released to manufacturing.

Lessons Learned

Data Management

Most designers think of problems with layout tools when contemplating the difficulties of designing large chips, but placement and wiring appears easy when compared to the difficulty of managing data and specifying a design for ASICs containing one million gates. There are tens of thousands of models and control files in a typical ASIC library. Library designers constantly enhance the library, creating a logistical nightmare for logic or physical designers who must use a consistent set of models on their designs.

It is difficult to specify a million-gate ASIC completely and accurately because there are many details in multiple clock domains and timing assertions that can affect a timing-driven physical-design methodology. It is also difficult for designers to correctly specify all test structures, test flags, and boundary-scan structures. When an error in a timing assertion, clock specification, test structure, or library model is found, costly iterations in processing the netlist can occur.

Rigorous checking minimizes the likelihood that a netlist will be rejected because it contains syntactic or structural errors. When the logic designer uses static timing analysis, test-structure verification, electrical-rule checks, and methodology checks in the design shop, errors can be eliminated before the netlist is sent to the physical design center. The checking programs indicate errors and allow the logic designer to fix problems in a matter of hours; if physical designers had to find the problems and send the netlist back to logic designers for repair, the same problem would take days to debug.

Library designers use a rigorous version-control mechanism to maintain coherent sets of models. New designs use the latest level of models, but many old designs are often enhanced with metal-only engineering changes. Inserting gate-array cells between standard-cell logic provides great flexibility for modifying a design without disturbing placement, timing, or wireability. Modifying only the metal levels decreases manufacturing time and mask-build costs, but this requires that old versions of models be used to ensure consistency of design.

Floorplanning

Logic synthesis uses an estimate of wire capacitance based on the circuit area and the fanout of a net. These estimates are pessimistic for most nets on a chip, but are optimistic for a small percentage of nets that must cross the chip. Without accounting for physical variation in the length of nets, the logic design does not achieve the highest attainable clock speeds for most paths. There are some paths that run slower than predicted because physical constraints elongate the nets.

Floorplanning allows the logic designer to take into account physical constraints on nets so the nets connecting tightly grouped clusters of logic are correctly anticipated as fast nets while cross-chip connections will be estimated as slow nets. This allows logic synthesis to pack more logic into local paths and decrease the number of latches required, but it does increase the number of re-drive buffers required on long paths.

It is very difficult to use net length estimates early in the logic synthesis process because synthesis often transforms a net and invalidates any parasitics estimated for the net. Area-based wire load models may be used for small blocks of logic; and, when the logic is fairly stable, incremental repowering of long global nets can reduce the risk of timing problems during layout.

Two industry-standard floorplanners and the IBM floorplanner are supported as part of the ASIC design flow. All three tools can generate capacitance targets, region constraints, and preplacement information in a format that guides the placement tools.

Deciding whether to perform the chip-layout process in a flat or hierarchical manner depends upon the capacity and run times of the tools used in the design methodology. A flat physical design approach becomes unwieldy for chips larger than 500K gates (approximately 100K-200K placeable objects). For large chips, placement, timing, and clock optimization each require hours of CPU time for a single run; wiring can sometimes run for many days. The hierarchical approach enables parallel design of each block on the chip and effectively isolates design changes in a single block to facilitate incremental processing of the design.

Designs smaller than 300K gates do not cause significant run time for any step in the design process; and, because a flat chip enables more streamlined power-bus routing and clock-tree optimization, it is quicker to process small chips as a single entity.

Data Volume

Early half-micron ASICs flushed out many problems with data capacity of the design tools. Hard-coded limitations in placement and clock-optimization tools forced recompilation of those programs. Files for the wiring tool exceeded one gigabyte until that program was changed to reduce disk usage by an order of magnitude. Run time for some translation steps grew exponentially, making those tools almost unusable. The entire physical design methodology was streamlined to minimize unnecessary file translation and to reduce the amount of data generated by the tools.

Despite efforts to increase efficiency of the tools, non-hierarchical 700K gate ASICs required over one gigabyte of memory for interactive wire embedding, while blockage maps exceeded one gigabyte of memory. Simulated annealing algorithms for placement, clock, and scan optimization required hundreds of megabytes of memory and ran for over 10 hours on an IBM RISC System 6000, Model 590 workstation with one gigabyte of real memory and two gigabytes of virtual storage. The design center maintained six machines of that capacity and a dozen smaller machines for smaller jobs and desktop display services.

The simple transmission, reading, and writing of design data for ASICs over 500K gates consumed valuable time in the product-development cycle. The typical size of a flattened EDIF file for a 500K design was over 100 megabytes. Compressing those files made transmission of the data to the ASIC Design Center more efficient, but required time and CPU cycles to "uncompress" before processing. Therefore, several tools in the Design Center were modified to read compressed data formats, eliminating many hours from the design cycle.

Customer Front-End Processing
A key lesson learned during the processing of the early ASICs was that the ASIC logic designer should synthesize the clock trees and repower the nets. Until mid-1995, all clock insertion was done at the ASIC Design Center, primarily because there was no organization trained to support the Clockpro and Booledozer tools in the customer's environment. As a result, a loop was created in the design process to receive the netlist at the ASIC Design Center for front-end processing and then return it to the customer for final prelayout timing and functional verification. The hand-off of design information from one organization to another for processing and reverification can add days or even weeks to a schedule, especially with design data in the 100- to 200-megabyte range.

In June 1995, the front-end processing tools were installed at a customer site for use on a family of ASICs in the 250K to 450K gate range. The resulting ease with which these designs passed through the release-to-layout procedure more than offset the resource expended on software support, documentation and customer training.

Summary

The ASIC design methodology described in this paper has been successfully used by both IBM and non-IBM designers to produce many large half-micron ASICs. Figure 8 illustrates that for 1994-1995, thirty-three percent of these ASIC designs were over one-half million gates, and seven percent were in the 750K-1000K gate range.

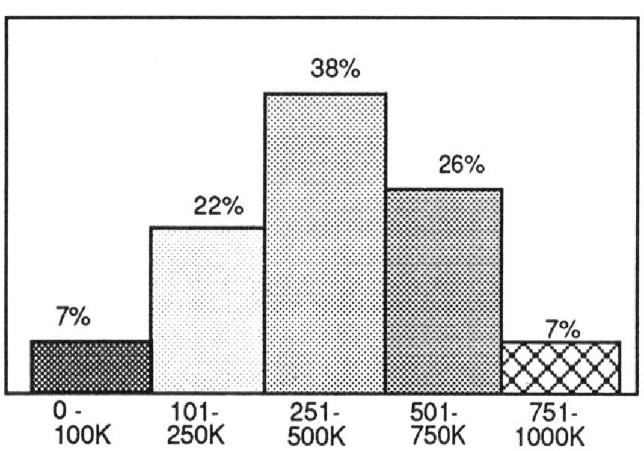

Figure 8 - Distribution of gate count for half-micron ASICs produced in 1994-1995.

The methodology combines industry-standard tools from companies such as Synopsys and Cadence for logic design, and proprietary IBM tools for design signoff. The latch-mapping techniques developed provide support for edge-triggered design and leverage the high testability of scan-based design.

Signoff is based on static timing analysis and design-for-testability verification. All ASICs designed using this methodology achieved over 99% test coverage (with most achieving over 99.5%) and performed at-speed or faster than the guaranteed performance.

References

(1) R. Gregor, C. Ng, J. Libous, E. Carter, R. Beaudoin, A. Chu, D. Grindel, J. Kinney, M. Lee, L. Mentes, J. Oppold, M. Russell, A. Secor, G. Yenik, "A One-Million-Circuit CMOS ASIC Logic Family," IEEE Custom Integrated Circuits Conference, 1993.

(2) S. Oakland, J. Monzel, R. Bassett, P. Gillis, "An ASIC Foundry View of Design and Test," International Test Conference, 1994.

(3) J.Panner, R. Abato, R. Bassett, K. Carrig, P. Gillis, D. Hathaway, T. Sehr, "A 300K-Circuit ASIC Logic Family CAD System," IEEE Custom Integrated Circuits Conference, 1990.

(4) J. Rowson, "EDA and Electronic System Design," IEEE 1994 Custom Integrated Circuits Conference, 1994.

(5) T. Guzowski, "Clockplanning Options and Requirements," IBM Application Note ASIC Products, 1995.

(6) M. Murakata, M. Murofushi, M. Igarashi, T. Aoki, T. Ishioka, T. Mitsuhashi, N. Goto, "Concurrent Logic and Layout Design System for High Performance LSIs," IEEE 1995 Custom Integrated Circuits Conference, 1995.

(7) M. Brassington, "Semi-Custom ASIC Technology Trends," IEEE 1995 Custom Integrated Circuits Conference, 1995.

(8) Keith M. Carrig,David J. Hathaway, Keith W. Lallier, Jeannie T. H. Panner, andTerrence W. Sehr, U.S. patent no. 5,339,253, "A Method and Apparatus for Making a Skew-Controlled Signal Distribution Network", filed June 14, 1991, issued August 16, 1994.

(9) Satyamurthy Pullela, Noel Menezes, and Lawrence T. Pillage, "Low power IC clock tree design," IEEE Custom Integrated Circuits Conference, 1995, pp. 263-266.

A High-Speed Low-Power 0.3μm CMOS Gate Array with Variable Threshold Voltage (VT) Scheme

Tadahiro Kuroda, Tetsuya Fujita, Tetsu Nagamatu, Shinichi Yoshioka, Toshikazu Sei*, Kenji Matsuo*, Yoichiro Hamura*, Toshiaki Mori*, Masayuki Murota, Masakazu Kakumu, and Takayasu Sakurai

Semiconductor Device Engineering Laboratory, Toshiba Corp., Kawasaki, Japan
* Semiconductor Group, Toshiba Corp., Kawasaki, Japan

Abstract

Circuit techniques for dynamically varying threshold voltage are introduced to reduce active power dissipation by 50% with negligible overhead in speed, standby power and chip area. No additional external power supply or additional step in process is required. A gate array with this scheme is fabricated in a 0.3μm CMOS technology whose performance is investigated. The gate array is best fit for multimedia portable applications that require low standby power dissipation and high performance.

1. Introduction

Lowering both of a supply voltage, V_{DD}, and a threshold voltage, V_{th}, enables high-speed and low-power operation. For example, in a 0.3μm CMOS technology, if V_{th} is reduced to 0.3V, V_{DD} can be decreased down to 2V while maintaining the speed at V_{th}=0.7V and V_{DD}=3V which is a typical operation condition for normal LSI's. In this case active power dissipation is reduced by more than 50% [1]. This approach, however, increases standby power dissipation due to the low V_{th} and therefore cannot be applied to battery-operated products.

Two circuit schemes to solve this problem were previously proposed. One is to employ two V_{th} ; low V_{th} for fast circuit operation and high V_{th} for small leakage current. This scheme, however, requires special circuits for latches and registers and therefore imposes area and speed penalties. The other scheme is to dynamically vary V_{th} through substrate-bias, V_{BB}, such that V_{th} is raised in a standby mode to cut off leakage. This scheme is called a Variable Threshold Voltage scheme (VT scheme). It doesn't require those special circuits, but imposes an area penalty for separating substrate from power supplies. A circuit proposed in [1] requires additional external power supplies and a circuit presented in [2] needs additional steps in process for triple-well structure.

In this paper, a new circuit for the VT scheme is proposed which employs MOS switches for an active mode and a Self Substrate Bias (SSB) circuit for a standby mode. It doesn't requir any additional external power supply nor additional step in process. It enables the VT scheme to be applied to logic gates, memory elements, and analog circuits with practically no penalty. Several new circuit techniques are introduced to internally generate voltages above V_{DD} or below GND, assure high reliability, and reduce area and power penalties.

2. Variable Threshold (VT) Scheme

The VT scheme is illustrated in Figure 1. When "SLEEP" is asserted ("1") for the standby mode, the SSB is activated and 100μA current is drawn from a substrate until V_{BB} reaches $V_{standby}$, typically -2V. V_{BB} is monitored by a Substrate Voltage Monitor (SVM) circuit which controls the SSB such that V_{BB} is set to $V_{standby}$. Consequently, V_{th} is raised to 0.7V to reduce the leakage current by four orders of magnitude smaller than that in the active mode. When "SLEEP" becomes "0" for the active mode, the SSB is disabled and MOS switches, MP1 and MN1, are turned on. Large current is injected into the substrate until V_{BB} reaches 0V. As a result, V_{th} is set to 0.3V for fast circuit operation under a low V_{DD}.

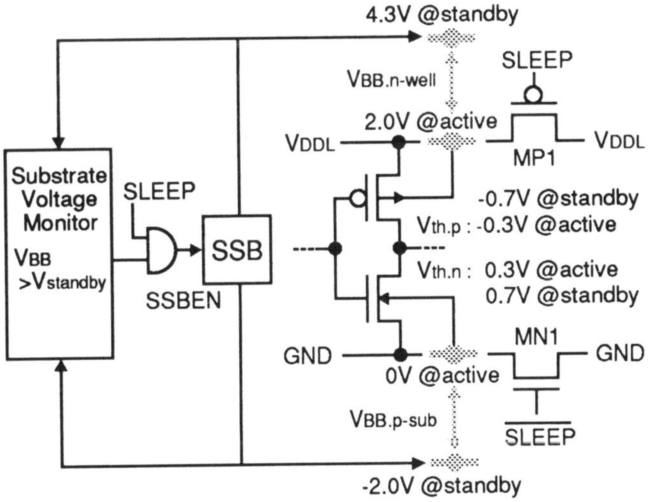

Fig. 1 Variable Threshold (VT) scheme.

0-7803-3177-6 $5.00 © 1996 IEEE

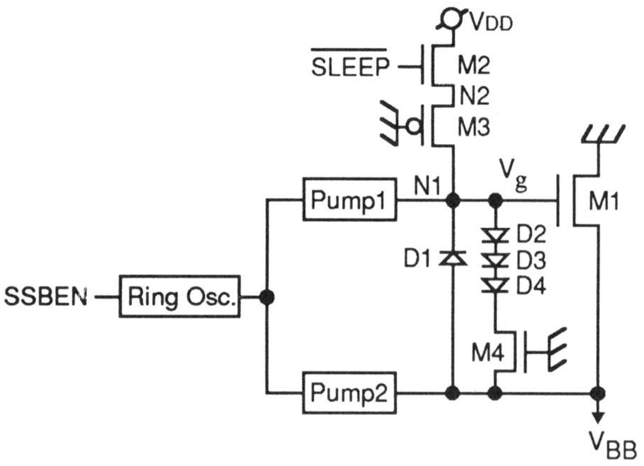

Fig. 2 Substrate Voltage Switch (SVS) circuit.

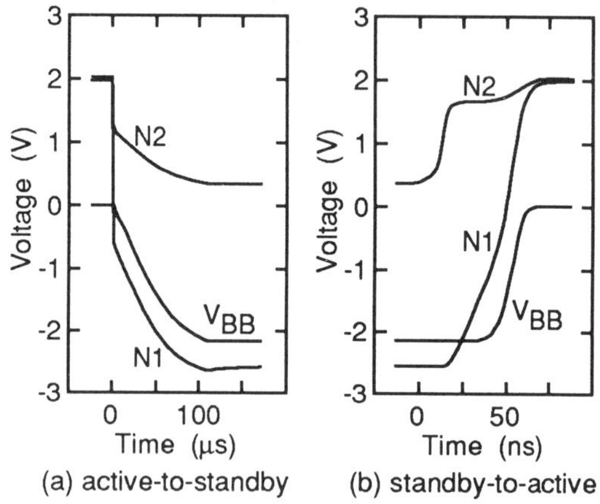

Fig. 3 Waveforms in SVS by SPICE.

An active to standby mode transition takes about 100μs, while a standby to active mode transition is completed in 0.1μs. This "slow falling asleep but quick awakening" feature is acceptable for most applications. The VT circuit works with 3.3V V_{DD} which is usually available on a chip for standard interfaces with other chips.

3. Circuit Techniques

Figure 2 depicts a Substrate Voltage Switch (SVS) circuit. Waveforms simulated by SPICE are presented in Fig. 3. In order to turn off the MOS switch M1 in the standby mode, its gate voltage V_g should be lower than V_{BB}. Two pump circuits are employed; one for V_g and the other for V_{BB}. Since parasitic capacitance at node N1 is much smaller than that in the substrate, V_g lowers much faster than V_{BB}. V_g is set to V_{BB}-0.8V by a clamp diode D1 and therefore M1 is turned off completely. In the active mode, on the other hand, the pump circuits stop, transistor M2 is turned on, V_g rises to turn on M1, and consequently, large current is injected to the substrate to raise V_{BB} very quickly. Clamp diodes D2, D3, and D4 limit V_g to V_{BB}+2.4V. Transistor M3 keeps the voltage at node N2 above GND. These clamping circuits prevent transistors from receiving high-voltage stress on their gate oxide. Clamping diodes are formed by using p+ and n+ diffusions both in an n-well.

To reduce area and power penalties, MOS transistors in subthreshold region are utilized. Figure 4 depicts the Substrate Voltage Monitor (SVM) circuit. The stacked pMOS transistors, M1 to Mn, operate in subthreshold region whose drain current is expressed as

$$I_{DS} = \frac{I_0}{W_0} \cdot W \cdot 10^{(V_{GS}-V_T)/S} \qquad (1)$$

where S is the subthreshold swing, V_T is the threshold voltage, I_0/W_0 is the current density to define V_T, and W is the channel width. By applying (1), the voltages at the intermediate nodes are given like

$$V_2 - V_1 = V_3 - V_2 = \cdots = V_n - V_{n-1}, \qquad (2)$$

and therefore divided-by-n voltages between V_{DD} and V_{BB} can be generated. By taking the k-th and the $(k+1)$-th intermediate nodes, hysteresis is introduced to the DC transfer characteristics of the SVM to improve noise immunity. The DC transfer characteristics of the SVM exhibit little dependence on process fluctuations and temperature. The SVM circuit dissipates 10pA DC current at room temperature and occupies 64μm by 215μm Si area. If resistors made by a well diffusion of about 1kΩ are employed instead, it would occupy 260mm by 260mm Si area for the same DC current, or dissipate 50μA DC current for the same Si area.

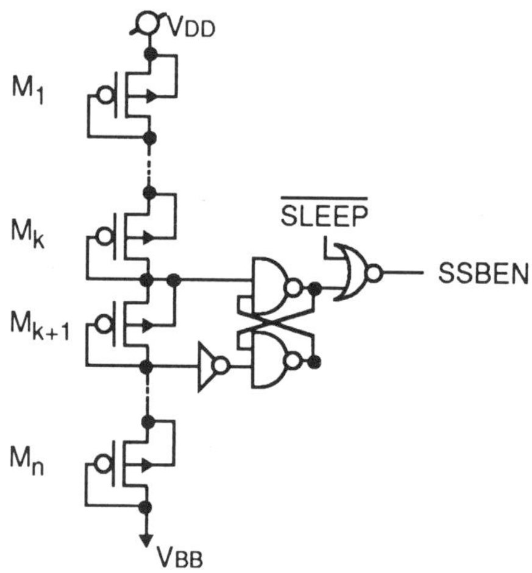

Fig. 4 Substrate Voltage Monitor (SVM) circuit.

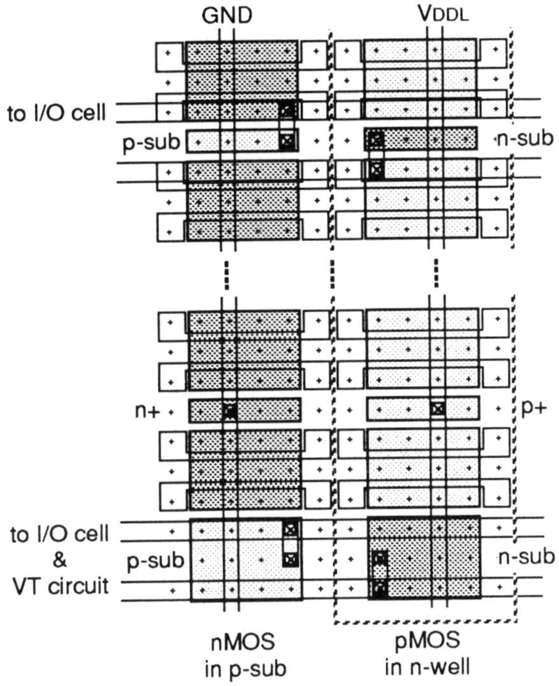

Fig. 5 Substrate-contact seperation in gate array.

4. Gate Array Design

The VT circuits are implemented in a custom macro in a peripheral I/O area. It occupies 0.86mm by 0.46mm Si area which is less than 0.4% in a 10mm-square chip. If experimental circuits are removed, the macro size is to be half. The M1 transistors in the SVS circuit are distributed by utilizing unused transistors in the I/O cells.

In a basic cell area substrate-contacts are separated from power/ground lines as illustrated in Figure 5. Substrate-contacts placed at the bottom edge of the basic cell area are connected to the pump circuit in the VT macro. In most of the basic cells substrate-contacts are replaced by source diffusions which stabilizes V_{BB}. In several rows substrate-contacts are placed and connected to the M1 transistors in the I/O cells by the second metal. The area penalty is therefore less than 1%.

Level shift circuit depicted in Fig. 6 is employed to transfer signals from internal to peripheral circuits, while conventional gates can be directly connected to transfer signals in the opposite way. Low-V_{th} transistors in the basic cells and high-V_{th} transistors in the I/O cells are used. They are placed close by a layout tool so that interconnections between them can be minimized. The level shift circuit can receive signals as low as 0.5V.

A gate array is designed which includes available gate counts to 188 thousand on a 8mm die. Test circuits are implemented such as gate chains, a 4K-bit SRAM, and a DLL.

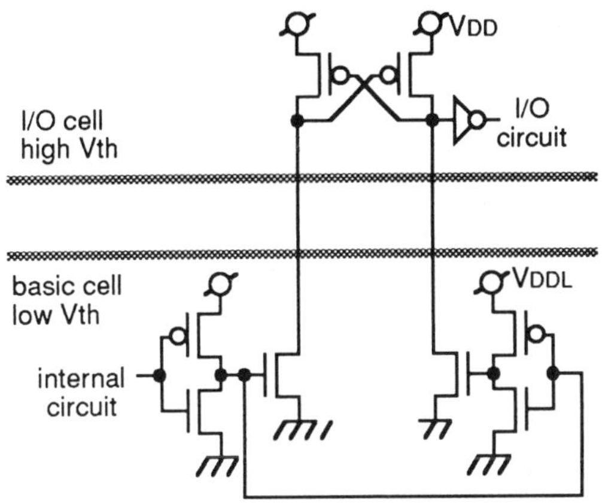

Fig. 6 Level shift circuit.

5. Experimental Results

The gate array is fabricated in a 0.3μm CMOS n-well double-metal technology. The low-V_{th} is 0.3V and the high-V_{th} is 0.7V. Chip photomicrograph appears in Fig. 7.

Figure 8 shows measured V_{BB} waveforms. Due to large parasitic capacitance in a probe card the transition takes longer time than anticipated; 500μs for the active to standby and 0.5μs for the standby to active. Measured chip leakage current at room temperature is about 100μA in the active mode and 0.1μA in the standby mode.

Figure 9 shows measured propagation delay time from an input to an output. The gate array functions under V_{DDL}'s as low as 0.5V. Figure 10 presents a measured shmoo plot of the SRAM. The same access time is obtained at V_{DD}=2V, V_{th}=0.3V as that at V_{DD}=3.3V, V_{th}=0.7V. Latchup effect due to V_{BB} noise was not observed.

6. Conclusion

Main features of the gate array is summarized in Table 1. The VT-scheme reduces active power dissipation by 50% with negligible overhead in speed, standby power and chip area. It requires no additional external power supply nor additional step in process.

Acknowledgments

The authors would like to acknowledge the encouragement of J. Iwamura, K. Maeguchi, T. Sakaue, K. Ichinose, O. Ozawa, and Y. Unno throughout the work.

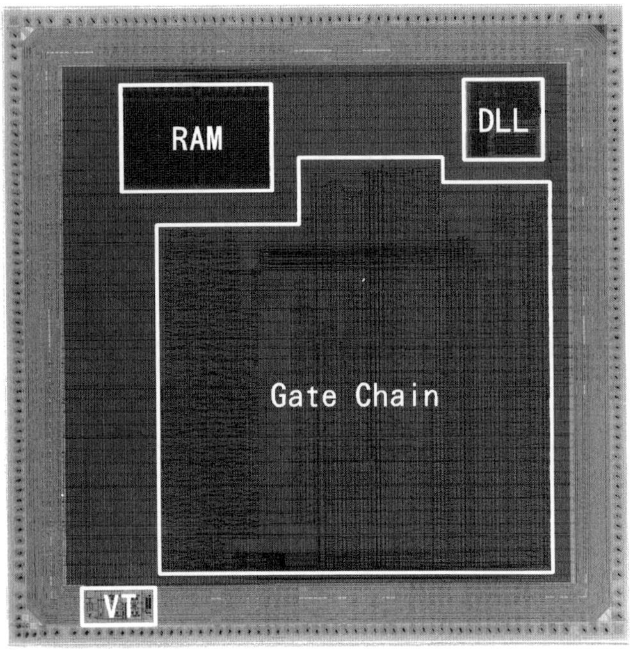

Fig. 7 Chip photomicrograph.

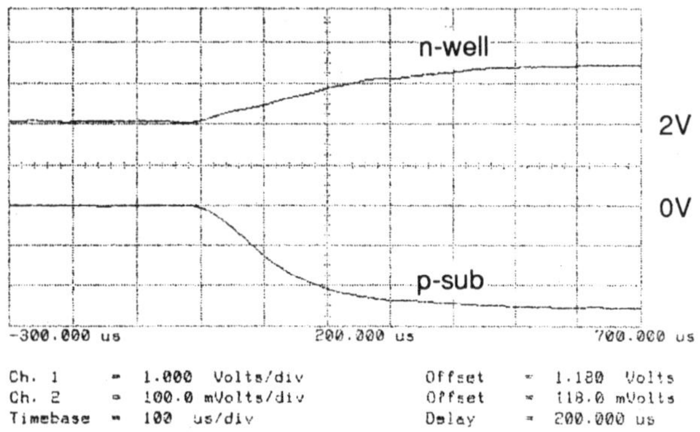

(a) active-to-standby

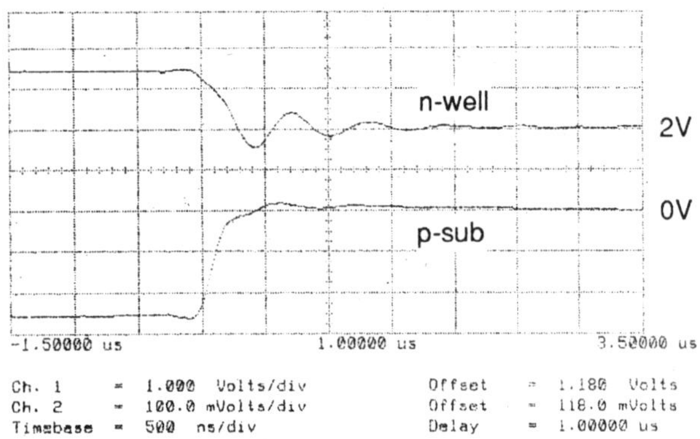

(b) standby-to-active

Fig. 8 Measured substrate voltage.

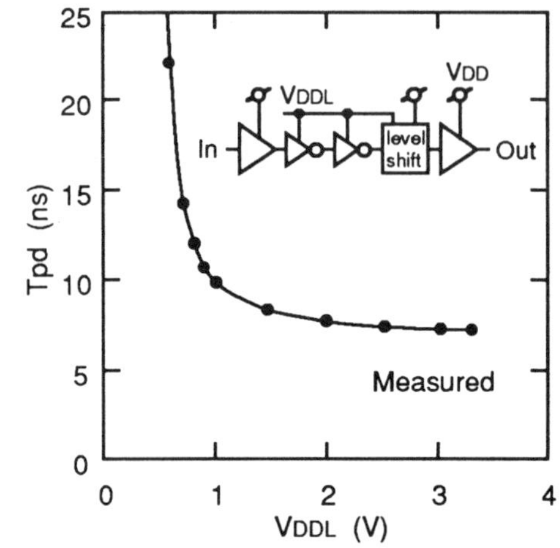

Fig. 9 Input-output propagation delay.

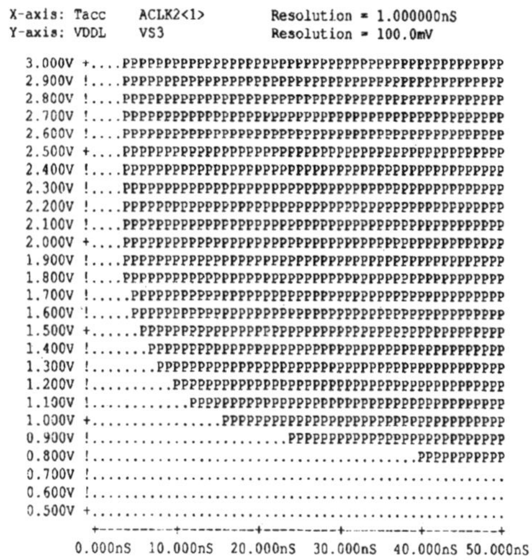

Fig. 10 RAM Schmoo plot.

Table 1 Main features.

Technology	0.3μm n-well CMOS 2-layer metallization
Gate array capacity	188k gates
Die size	8 x 8 mm^2
Supply voltage	3.3V, 2V
Leakage current	100μA at active mode 0.1μA at standby mode
Mode transition time	100μs for active-to-standby 0.1μs for standby-to-active

References

[1] Seta, K. et al.,"50% Active-Power Saving without Speed Degradation using Standby Power Reduction (SPR) Circuit," ISSCC Dig. of Tech. Papers, pp. 318-319, Feb. 1995.

[2] Kuroda, T. et al.,"A 0.9V 150MHz 10mW 4mm^2 2-D Discrete Cosine Transform Core Processor with Variable Threshold Voltage Scheme," to be published in ISSCC Dig. of Tech. Papers, Feb. 1996.

A Quarter-micron SIMOX-CMOS LVTTL-compatible Gate Array with an over 2,000 V ESD-protection circuit

Yusuke Ohtomo, Takeshi Mizusawa, Kazuyoshi Nishimura,
Hirotoshi Sawada and Masayuki Ino

NTT LSI Laboratories
3-1, Morinosato Wakamiya, Atsugishi, Kanagawa, 243-01, Japan

Abstract

A quarter-micron SIMOX-CMOS gate array with an LVTTL interface is described. The SIMOX-CMOS gates have the same delay at 1.2 V of supply voltage as that of the gates in 0.5-μm bulk CMOS at 3.3 V and reduce power consumption 87 %. The interface circuits in the array convert 3.3 V external signal from/to 2.0 V-1.2 V internal signal with little power penalty. An ESD-protection circuit for the chip shows over 2,000 V of ESD-hardness in the advanced SIMOX-CMOS having 50-nm-thick silicon film.

1. Introduction

Demands for LSIs that consume less power has made it necessary to lower LSI supply voltage. To enhance speed at low voltage, fully depleted SIMOX-CMOS technology is effective because it can reduce threshold voltage 0.1 V while keeping the same off current as that of bulk CMOS.

However, as the supply voltage is gradually lowered, advanced LSIs will continue to be used with previous-generation LSIs operated at high voltage on the same PC board. It is difficult to use quarter-micron SIMOX-CMOS LSIs at supply voltage of 3.3V. Even though parasitic bipolar effects have been repressed [1], the device has low drain-breakdown voltage. The key to open up the high-performance SIMOX-CMOS to wide use is to develop a circuit that converts high (3.3 V) voltage signal from/to low (< 2.5 V) voltage signal while maintaining device reliability. There are some reports [2], [3] of 5V/3.3V conversion; however DC current of the output buffer affects the power consumption. Another major issue related to using the SIMOX-CMOS is ESD protection [4], [5]. No protection circuit has been reported that satisfies the MIL standard.

This paper describes solutions to these problems for SIMOX-CMOS. The proposed circuits are evaluated with a quarter-micron SIMOX-CMOS gate array.

2. Architecture

An LVTTL compatible gate array using a low supply voltage of under 2.5 V requires another supply voltage of 3.3 V (Fig. 1). Below 2.5 V, an output buffer can not output the LVTTL high level of 2.4 V considering the 10 % drop of the supply voltage. The input buffer and internal gates are operated at the low supply voltage to decrease power consumption, and only the output buffer is operated using both voltages.

3. Internal Gate Performance

Figure 2 shows the advantages of using SIMOX-CMOS at low supply voltage. The experimental gate delay of a 2-input NAND with fan-out 3 and 1-mm metal wiring is compared between quarter-micron SIMOX-CMOS and bulk CMOS. The SIMOX-CMOS has 74 % of the gate delay at 2.0 V and 61 % of the gate delay at 1.2 V compared with that of the bulk CMOS. At 1.2 V the gate delay is 350 ps, which is the same delay in 0.5-μm bulk CMOS at 3.3 V. This indicates the internal gate performs the same function at the same speed using less than 13 % of the power consumption of 0.5-μm bulk CMOS gates. A test chip using 120K gates [6] correctly operated and showed a 97 % reduction in power consumption. The device features are summarized in Table 1.

4. LVTTL Compatible I/O

4-1. Output Buffer

A new LVTTL output buffer circuit was developed to make full use of the internal gate performance. The circuit shown in Fig. 3 (a) reduces power consumption at the I/O by cutting DC current. The PMOS Q1 and Q2 at the output stage require intermediate gate voltage (e.g. 3.3~1.0 V for Q1, 2.0~1.0 V for Q2) to avoid drain breakdown. Voltage generators 1 and 2 generate the gate voltages. The output low voltage of generator 1 is clamped at V_{g1L} (= 1.0 V) by Q7 and Q8. The latch made with crosscoupled PMOS Q3 and Q4 and NMOS Q5 and Q6 cuts the DC current of generator 1. Voltage generator 2 is used when VDD is over 1.5 V. The diodes in generator 2 clamp the output low voltage at 2^*V_j (= 1.0 V). The output of the generator 2 is pinned to VDD and the generator 2 is grounded by a metal layout when the chip is used at VDD below 1.5 V. Figure 3(b) shows the converted signals when the buffer drives

0-7803-3177-6 $5.00 © 1996 IEEE

15 pF of load capacitance. At VDD = 2.0 V, the DC current was 117 μA when outputting high but 0 μA when outputting low. At VDD =1.2 V, the DC current was eliminated because only the generator 1 worked.

4-2. Input Buffer

An LVTTL compatible input buffer circuit is shown in Fig. 4(a). The voltage conversion method is the same as a previously reported circuit [2]. However, the ESD-hardness is improved by adding NMOS Q7 and Q8, and a diode. The NMOSs raise the breakdown voltage from the PAD to VDD to a level higher than that of an input protection circuit shown in next chapter. The diode is used to protect feedback PMOS Q9 instead of the well-diode used in the bulk CMOS. The simulated waveform is shown in Fig. 4(b). The delay increased by the added NMOSs was only 40 psec.

5. ESD Protection Circuit

5-1. Architecture

The SIMOX-ESD protection circuit shown in Fig. 5 was designed based on the following concept: put lateral diodes where the well-diodes and the source/drain-substrate-diodes are in bulk CMOS. It makes paths for the ESD current to flow through like in the bulk CMOS. In addition, diodes are put between VDDO and VDD, and between GNDO and GND so each can use the others path. The voltage added to the protection circuit is reduced so that less heat is generated. To prevent current from flowing into the VDD when inputting 3.3 V, the gate of PMOS Q10 and Q11 in the input protection circuit are respectively connected to VDDO and VDD.

5-2. Physical Layout

The diodes in Fig. 5 require a large area and a broad I/O pitch in the physical layout. The area was reduced by using a pattern where the MOSFET and diode are merged as shown in Fig. 6. The n^+ (or p^+) area of the diode is eliminated and the total junction length of the 1024-μm diode could be squeezed into a 105-μm I/O pitch. To avoid local heating, a 74μm x 32μm silicon island is formed using the Diode/MOSFET merged pattern, and the island is connected to metal 2, 3, 4 of the power lines through the thermal via. The space Lj between the junction and diffusion contact is set at 0.9 μm to diffuse current from the 1.4μm pitch tungsten contact to the junction in the 50-nm-thick silicon film.

5-3. HBM ESD Results

The protection circuit was evaluated according to the Human Body Model (HBM) specified in Mil-Std 883D Method 3015.7. Both the positive- and negative-polarity HBM discharge are added between the input pin and VDD/GND, or the output pin and VDDO/GNDO. The experimental results are summarized in Table 2. Compared with the voltages (in parenthesis) before applying the diodes and PMOS input protection circuit, the failure voltage is improved over 1,500 V. The results show the protection circuit satisfies the MIL standard, which requires more than 2,000 V of ESD-hardness.

6. Conclusion

The two main issues associated with using the SIMOX-CMOS gate advantage have been discussed. To make the internal supply voltage low and to interconnect to the external high voltage (LVTTL) signal, an output buffer used crosscoupled PMOS was developed. Together with the input buffer it converted the 3.3 V signal from/to 2.0-1.2 V signal with little DC power consumption. The I/O buffer is protected by a SIMOX-CMOS ESD protection circuit using MOSFET merged with lateral diodes, which is the first protection circuit to satisfy the MIL standard. This technology opens up the possibility of the wide use of SIMOX-CMOS low-power, high-speed LSIs.

Acknowledgment

The authors would like to acknowledge K. Takeya, T. Sakai and T. Tsuchiya for their encouragement. They also thank T. Nikaido and T. Komatsu for helpful discussion and M. Yamada for help in making the circuit layout.

References

[1] T. Ohno, M. Takahashi, A. Ohtaka, Y. Sakakibara and T. Tsuchiya, "Suppression of the Parasitic Bipolar Effect in Ultra-Thin-Film MOSFETs/SIMOX by Ar Ion Implantation into Source/Drain Regions," Proc. IEEE IEDM, Dec. 1995.

[2] Y. Tanaka et al., "A 0.5 μm BiCMOS SOG with Selectable 5V/3.3V Operations," Proc. IEEE CICC, pp.14.4.1-14.4.4, May, 1991.

[3] H. Fukuda et al., "A BiCMOS Chanelless Masterslice with On-Chip Voltage Converter," IEEE ISSCC Digest of Technical Papers, pp. 176-177, Feb., 1989.

[4] M. Chan et al., "ESD Reliability and Protection Schemes in SOI CMOS Output Buffers," IEEE Trans. Electron Devices, vol. ED-42, No. 10, pp. 1816-1821, Oct., 1995.

[5] K. Verhaege, G. Groeseneken, J. P. Colinge and H. E. Maes, "Double Snapback in SOI nMOSFET's and its Application for SOI ESD Protection," IEEE Electron Device Lett., vol. 14, pp. 326-328, July, 1993

[6] M. Ino et al., "Quarter-micron CMOS/SIMOX Gate Array LSI," IEEE ISSCC Digest of Technical Papers, Feb., 1995.

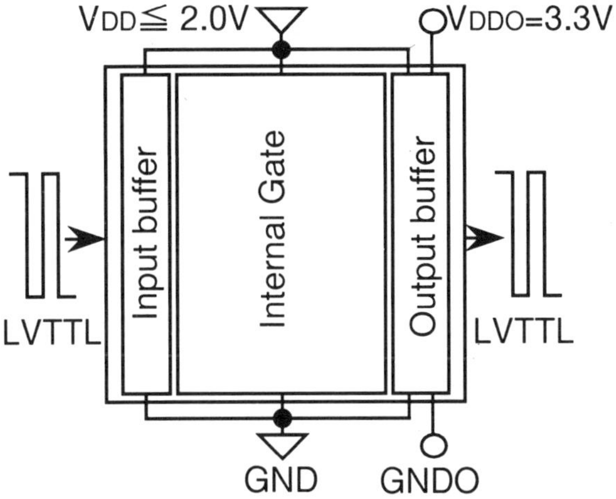

Fig. 1 Chip architecture.

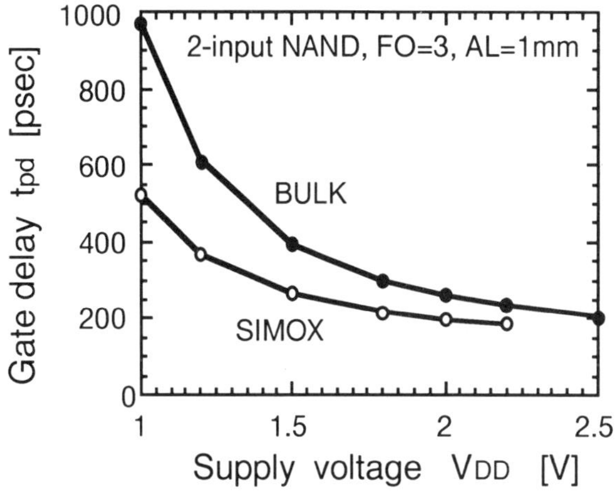

Fig. 2 Speed advantage at low supply voltage.

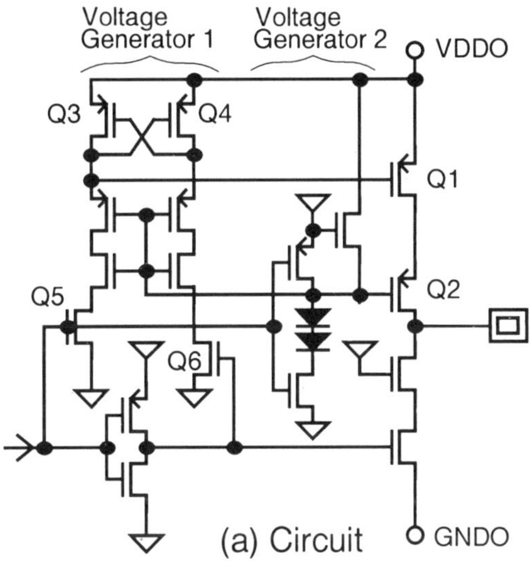

(a) Circuit

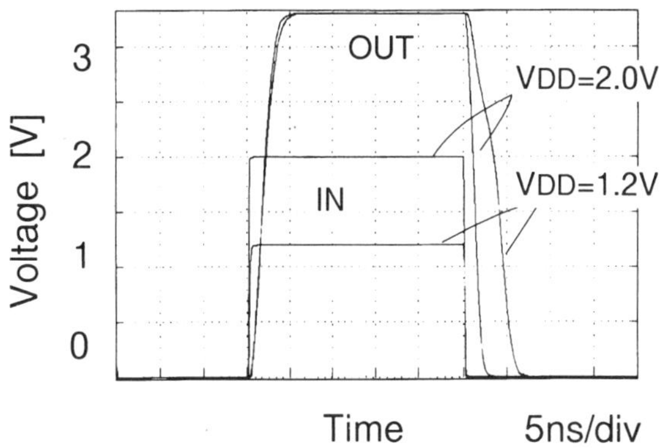

(b) Waveform

Fig. 3 Output buffer.

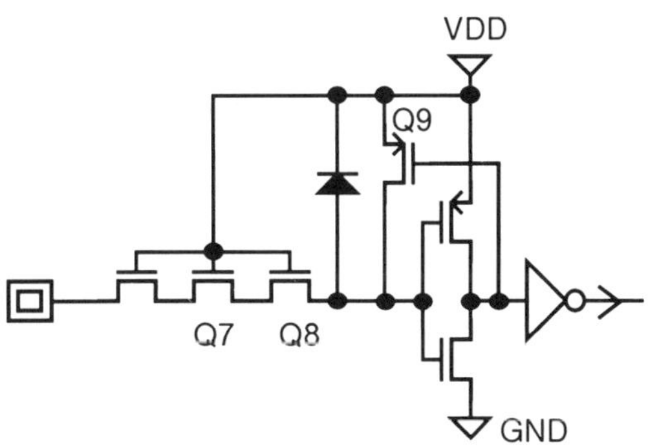

(a) Circuit

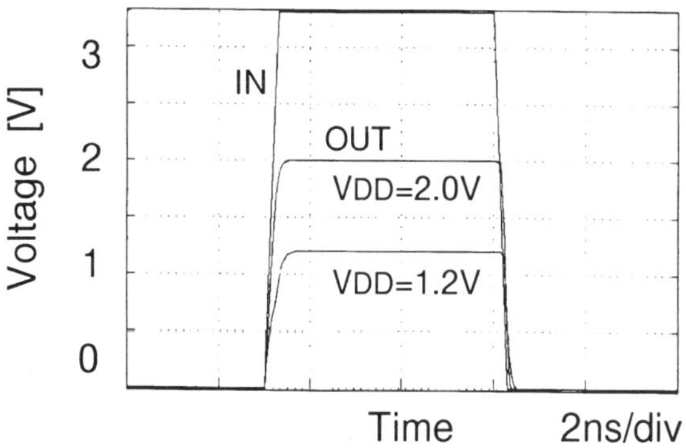

(b) Waveform

Fig. 4 Input buffer.

4.3.3

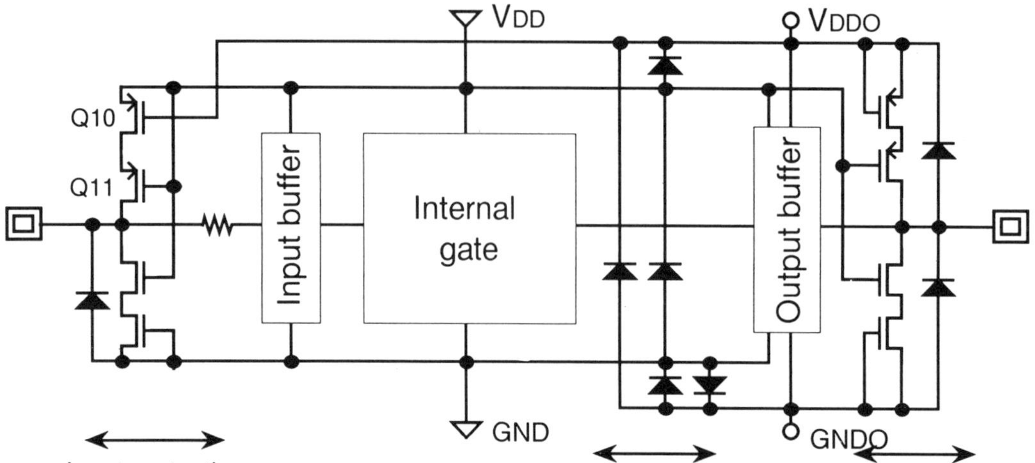

Fig. 5 SIMOX-ESD protection circuit.

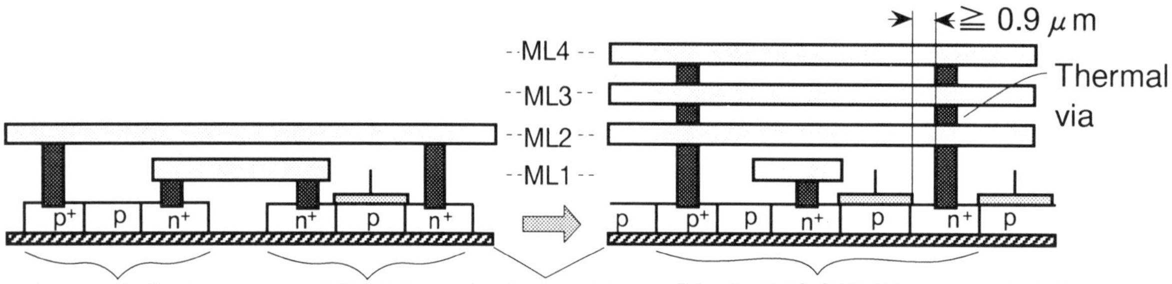

Fig. 6 SIMOX-ESD protection circuit pattern.

Table 1 LVTTL compatible device features

Transistor	
Gate Length	Ln = Lp = 0.25 μm
Threshold Voltage	Vtn = 0.2V, Vthp =0.3V
Tox	5nm
Saturation current	Idn=359, Idp=203 μA/μm
Off current	Ioffn=95, Ioffp=108 pA/μm
Breakdown voltage	Vsdn= 2.9V, Vsdp=3.8V
Gate array	
Size	10 mm□
Gate count	300Kgate
Metal pitches	1.4 μm for ML1 - ML4
I/O count	340 (105 μm I/O pitch)
Supply Voltage	3.3V, VDD(≦2.0V)

Table 2 HBM ESD results

		Positive	Negative
Input	VDD	2,000 V	> 3,000 V
	GND	2,600 V (300)	> 3,000 V
Output	VDDO	> 3,000 V (1,200)	4,300 V (1,800)
	GNDO	> 5,000 V (1,200)	> 3,000 V (1,500)

Values in parenthesis are for MOS protection circuit only.

A 1860kG CMOS Gate Array with GTL Input Flip-Flop Circuits

Katsuichi Tomobe, Toshiro Takahashi, Masatoshi Kawashima, Yasuo Sonobe, Tadashi Kiyuna*1, Sigeru Yamamoto*2

Device Development Center, Hitatchi Ltd.
2326 Imai Ome-shi Tokyo, 198 Japan
*1Hitachi Computer Engineering Co., Ltd.
*2 Hitachi VLSI Engineering Co., Ltd.

Abstract

A 1860kG CMOS gate array with a high speed I/O circuit using 0.35μm CMOS process technology, has been developed. 300MHz synchronous data transmission through a 30cm line has been achieved with a flip-flop circuit which can directly receive and store a low voltage swing signal. This circuit technique reduces the latency time of data transmission between 2 LSIs by as much as 1.7ns compared with conventional circuits.

Introduction

Machine cycles of RISC MPUs have improved rapidly in recent years, through the use of submicron CMOS technology [1][2][3]. However, it is difficult to design massive-parallel processors and high performance servers in such a way that the data transmission frequency between LSI chips increases proportionally to the MPU speed, because the physical size of the machines remains the same. Therefore, system performance is often limited by data bus throughput between MPUs, memories, cross-bar switching LSIs and other peripheral LSIs.

One approach to increase bus capacity is a wide data width bus like 16B, for example. But the package cost of LSIs becomes high because the technique requires many signal pins. Another approach is to use high speed asynchronous data transmission in which one LSI sends data with a clock to another LSI. However, this technique requires a long latency time for clock synchronization in the receiver LSIs. A small transmission latency between LSIs is important for system performance because high speed MPUs require large amounts of timely data. Reducing latency time is also effective to attain high data throughput because a synchronous high bus frequency can be achieved.

Low voltage swing I/O interfaces are widely considered for high speed data transmission because the reflection of signal waves is restrained by termination resistances. GTL is one type of standerd low voltage swing I/O interface [4][5]. But data transmission time through a signal line on a board and a cable still cannot be scaled down because physical size of the machine remains the same. We propose a new circuit technique for dramatically minimizing the delay time of an input buffer.

Input Flip-flop Circuit

A timing distribution schema of one cycle synchronous data transmission in a conventional I/O circuit is shown in Fig. 1. The delay time of the data tramsmission from a flip-flop of a driver LSI to a flip-flop of a receiver LSI is the sum of $t_{pd\text{-}FF}$(the delay time of the flip-flop of the driver LSI), $t_{pd\text{-}out}$(the delay time of the output buffer), $t_{pd\text{-}line}$(the data transmission time through a signal line), $t_{pd\text{-}in}$(the delay time of the input buffer), $ts_{etup\text{-}FF}$(the setup time of the flip-flop of the receiver LSI), and t_{skew}(clock skew). The delay time should be within one cycle time. Although, $t_{pd\text{-}line}$ is a major factor in latency, reducing $t_{pd\text{-}line}$ is difficult because the transmission line length is constant. The delay time of the input buffer, $t_{pd\text{-}in}$, is also an important factor because the buffer requires a relatively long delay time to amplify the low voltage swing input to a 3.3V CMOS level. Our approach is to eliminate $t_{pd\text{-}in}$, using a GTL input edge-trigger type flip-flop circuit which can directly receive and store a low voltage swing signal.

The circuit diagram of the GTL input flip-flop is shown in Fig. 2. Three p-channel MOS transistors, MP1, MP2, and MP3 make up the conventional

4.4.1

0-7803-3177-6 $5.00 © 1996 IEEE

GTL input buffer. MP4, MP5, and MP6 make up the memory. Two n-channel MOS transistors MN1 and MN2 act as common part for both the GTL input buffer and the memory. The input buffer and the memory are constituted within one circuit to store data with a low voltage swing signal. Therefore, delay time for amplifying low swing level signals can be eliminated.

Setup time of the flip-flop circuit is as small as -0.1ns from circuit simulation because the response from the clock is somewhat slow. On the other hand, the sum of $t_{pd\text{-}in}$ and $t_{setup\text{-}FF}$ is 1.6ns. As a result, latency time of the input flip-flop circuit has improved by 1.7ns compared with the conventional circuit. The delay time of the input flip-flop circuit is 1.3ns when it receives data and the clock signal simultaneously.

Experimental Results

The performance of the input flip-flop circuit has been examined by a 0.35μm CMOS gate array test chip. The evaluation circuit diagram is shown in Fig. 3. Both driver and receiver circuits are integrated in the same chip for measuring transmission frequency without clock skew. The conventional circuit consists of a driver flip-flop, a push-pull type GTL level output buffer, a transmission line, a GTL input buffer and a receiver flip-flop circuit. The GTL input buffer and the flip-flop circuit are replaced by a GTL (low voltage swing signal) input flip-flop circuit in the proposed circuit. All other conditions are kept the same. The outputs of receiver flip-flop circuits have been monitored through CMOS type output buffers. The transmission line consists of 20cm board lines and a coaxial cable. Line lengths has been changed by selecting the cable length.

Fig. 4 shows reciprocals of maximum synchronous transmission frequencies with various line lengths. A transmission delay time is reduced by 1.7ns by the GTL input flip-flop circuit. The total delay minus tpd-line has been improved by 50%. The maximum frequency of the proposed circuit for one cycle transmission is 340MHz through a 30cm line while that of the conventional circuit is 210MHz. Therefore, the bus latency time is improved by 37%. Measured waveforms of the GTL input flip-flop circuit at 300MHz is shown in Fig. 5. The input flip-flop correctly receives and stores data on the transmission line. 500MHz operation of the input flip-flop circuit has been observed.

Gate Array Features

The features of a CMOS channelless gate array and a photomicrograph of the test chip are shown in Table 1 and Fig. 6. 1860k logic gates, 752 I/O circuits and a PLL (phase-locked-loop) circuit are integrated in a 14.75mm by 14.75mm size chip using 0.35μm CMOS process technology. 5 metal layer wiring has been developed for high gate utilization. Wiring pitch of the first to third metal layers is as small as 1.4μm for the high gate count. Wiring pitch of the fourth and fifth metal layer is 2.8μm for small metal resistance and parasitic capacitance using thick metal and thick interlayer dielectric.

High speed switching time of gates is achieved by MOS transistors with 0.35μm gate length and 8nm gate oxide thickness. Delay times of various ring oscillators are shown in Fig. 7. Gate delay times of an inverter with a fanout 1 and a 2 input NAND gate with fanouts 2 plus 0.2mm metal lines are 45ps and 100ps, respectively. A PLL is embedded for minimizing the clock skew between LSIs [6]. The effective frequency range of the voltage-controlled oscillator is between 300MHz and 700MHz. The peak-to-peak jitter of the PLL is 0.10ns.

Conclusion

A 1860kG CMOS gate array with high speed I/O circuits has been developed using 0.35μm CMOS process technology. The GTL input flip-flop circuit can directly receive and store the low voltage swing signal. This circuit technique reduces the latency time of data transmission between 2 LSIs by as much as 1.7ns compared with conventional circuits. The maximum synchronous transmission frequency is 340MHz through a 30cm line while that of the conventional circuit is 210MHz. An embedded PLL creates a high frequency clock with 0.10ns peak-to-peak jitter. Short latency time between LSIs has been achieved for data transmission from 200MHz to 300MHz. The gate array is suitable for system controller LSIs, a cross-bar switching LSIs and other peripheral controller LSIs for massively parallel processor systems and high performance servers.

Acknowledgment

The authors would like to thank T. Hayashi and K. Koide for their helpful advice.

Reference

[1] David Bearden, et al., "A 133MHz 64b Four-Issue CMOS Microprocessor," ISSCC Dig. of Tech. Papers, pp.174-175, Feb. 1995.
[2] A. Charnas, et al., "A 64b Microprocessor with Multimedia Support," ISSCC Dig. of Tech. Papers, pp.178-179, Feb. 1995.
[3] William J. Bowhill, et al., "A 300MHz 64b Quad-Issue CMOS RISC Microprocessor," ISSCC Dig. of Tech. Papers, pp. 182-183, Feb. 1995.
[4] Bill Gunning, Leo Yuan, Trung Nguyen, Tony Wong., "A CMOS Low-Voltage-Swing Transmission-Line Transceiver," ISSCC Dig. of Tech. Papers, pp. 58-59, Feb. 1992.
[5] Junya Kudoh, et al., "A CMOS Gate Array with Dynamic-Termination GTL I/O Circuits," Proc. of ICCD, pp.25-29, Oct. 1995.
[6] Toshiro Takahashi, et al., "A CMOS Gate Array with 600Mb/s Simultaneous Bidirectional I/O Circuits," ISSCC Dig. of Tech. Papers, pp. 40-41, Feb. 1995.

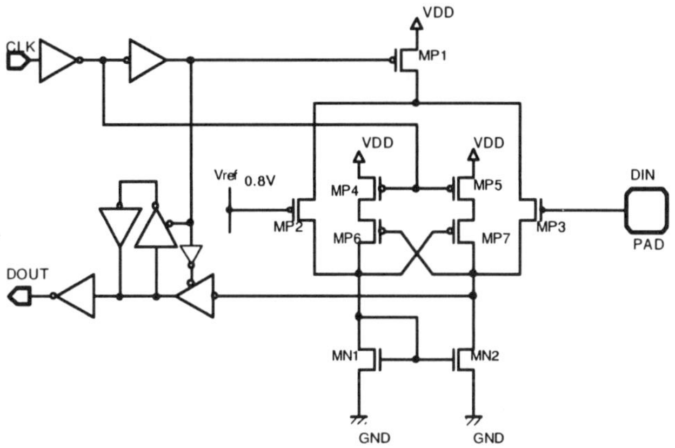

Fig. 2 Circuit Diagram of a GTL Input Flip-Flop (Edge Trigger Type)

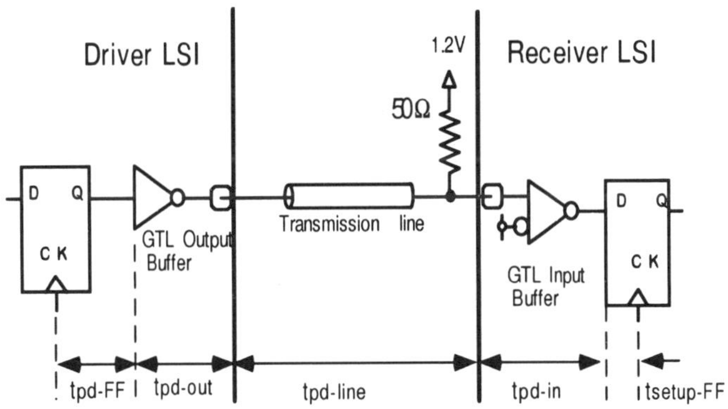

tcyc > tpd-FF + tpd-out + tpd-line + tpd-in + tsetup-FF + tskew

Fig. 1 Timing distribution

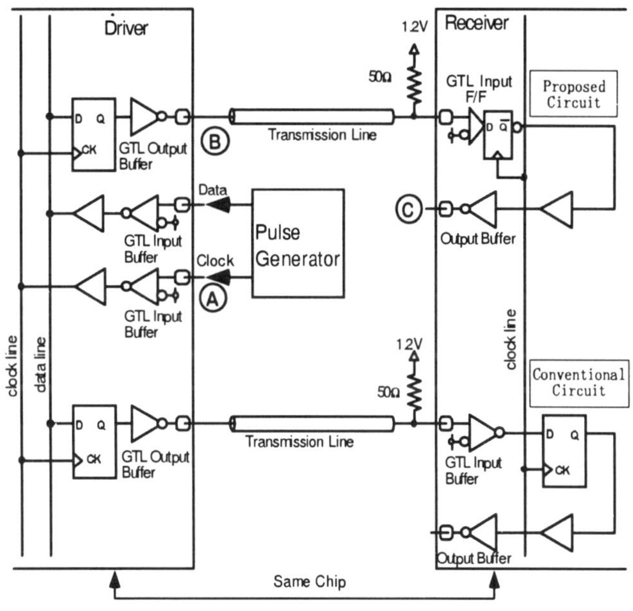

Fig. 3 Evaluation Circuit Diagram

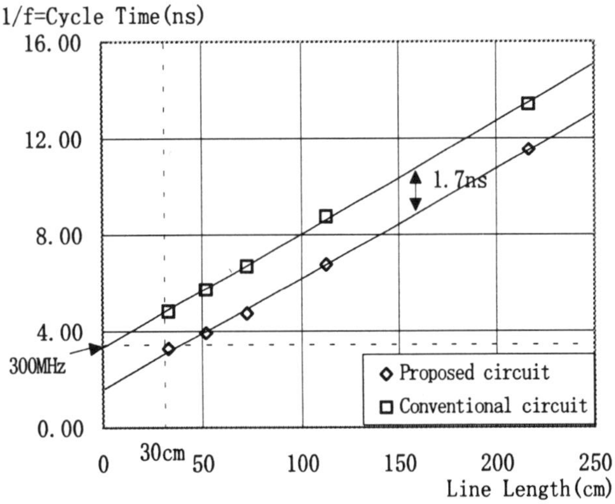

Fig.4 Reciprocals of maximum syncronous tramsmition frequencies with various line length

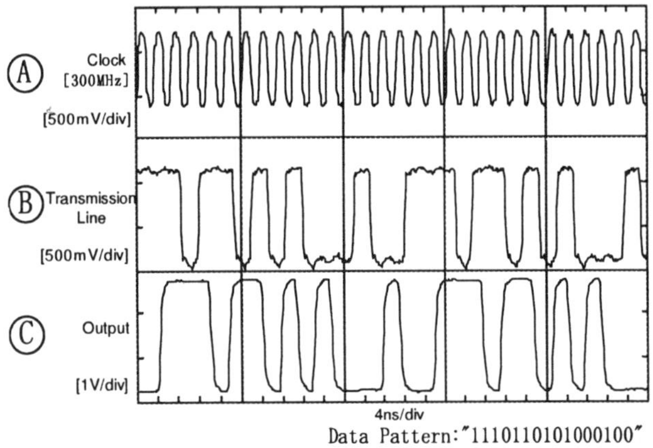

Fig. 5 Measured Waveforms(32.7cm line length)

Table 1 Gate array features

Raw Gate		1860kG
Number of I/O circuit		752
Chip size		14.75×14.75mm
Wiring grid pitch	M1~M3	1.4 μ m
	M4,M5	2.8 μ m
MOS transistor	Gate Length	0.35 μ m
	Gate Oxide	8nm
Power supply		3.3V
Package		400pin PGA

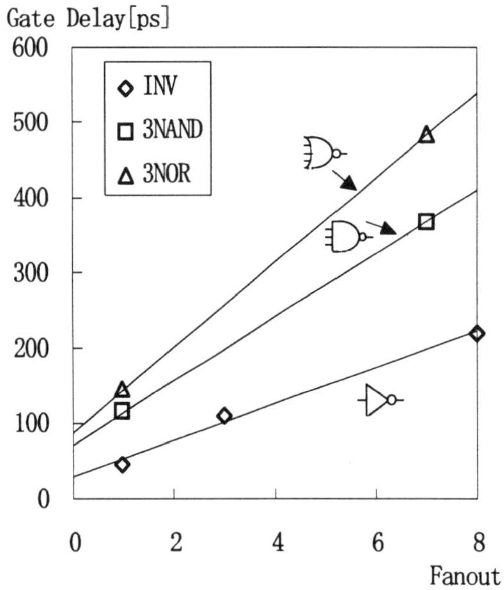

Fig. 7 Measured Gate Delay

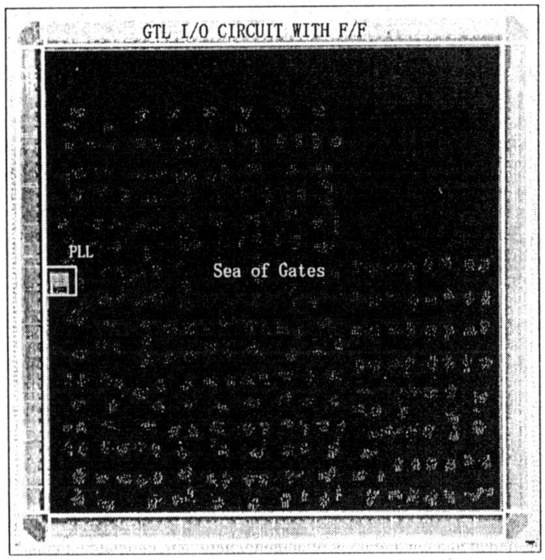

Fig. 6 Micrograph of Test Chip

Chair: Fang Lu
Co-Chair: Keisuke Okada

Digital signal processing ICs, no matter whether they are user programmable or application specific, have gained acceptance in a wider range of applications over the past few years.

Through the use of new circuit design techniques and architectural innovations, DSP ICs continue to improve their area and power efficiency while also increasing their performance. This progress allows more and more DSP functions to be integrated onto a single chip using mature and cost-efficient IC processes. This session presents the speed/area optimization techniques applied to several high-performance DSP computational building blocks and digital filters.

The first paper introduces several floating point datapaths developed for graphics and simulation applications. These CMOS datapaths can support multiple FP and integer formats without speed degradation, and they have been embedded in a 125 MHz, 290 MFLOPS Vector Pipelined Processor for use in supercomputers. Pipeline throughputs have been increased through architectural rearrangements. A clever on-line test technology has been implemented in these datapaths for improving circuit reliability under actual operating conditions. This technique can detect not only static faults but also delay faults, which have traditionally been difficult to identify.

By applying distributed arithmetic, the second paper proposes an efficient array architecture for the multiplication of complex numbers. Instead of using four real multipliers, this complex number multiplier only takes the area of two real multipliers and its speed is almost equivalent to a single real multiplier. The efficient design is derived by taking a close look at a real multiplier structure with data in the Offset Binary representation. Residue error compensation and the functions of various boundary cells are discussed in detail. Besides a test chip implementation, a VHDL module with generic parameters has been developed so that the complex multiplier module can be included in large designs.

The third paper describes a digital matched filter with a differential and pipelined structure for CDMA communication systems. A novel differential PN code scheme is adopted to reduce the number of multiplications and summations (M & S). This filter not only saves half of the hardware and power, but also improves the speed to a single operation time of M & S. These features make the filter more suitable for the high speed and low power requirements of personal communications. Besides an FPGA for verification on a filter building block, a 64-tap matched filter with 4-bit soft decisions has been designed using CMOS standard cells. Simulation results show that the PN code chip rate in this filter can achieve 45M chips/sec.

To perform ghost cancellation and channel equalization, the last paper presents a compact 10-bit 288-tap FIR filter using a structured architecture which employs an optimized partial product tree compression method. The new architecture is based on the addition of equally weighted partial products which result from 288 multiplications of the filter coefficients and the inputs. Optimized parallel compression schemes such as 4:2 and 5:5:2 compressors are used to perform decomposed partial product addition. The 288-tap FIR filter occupies only 7 mm x 9 mm of die area in a 0.6 μm CMOS technology.

Come join us and learn how these works achieve high computational density while maintaining excellent silicon and power efficiencies.

Floating Point Datapaths with On-Line Built-In Self Speed Test

Yasuhiko Hagihara*, Shigeto Inui*, Fuyuki Okamoto*
Masato Nishida**, Toshihiko Nakamura*** and Hachiro Yamada*
e-mail:[hag,inui,oka,yamada]@mel.cl.nec.co.jp
*Microelectronics Res. Labs, NEC Corporation
**Computers Division, NEC Corporation
***2nd Computer Engineering Department, NEC Kofu, Ltd.

Abstract

This paper describes floating point (FP) datapaths developed for graphics and simulation applications. The datapaths are fabricated using 0.35 μm CMOS technology and embedded in a 125 MHz, 290 MFLOPS Vector Pipelined Processor[1] for use in supercomputers. A new on-line test technique has been developed for the purpose of improving reliability under actual operating conditions. The technique makes it easy to detect not only static faults but also delay faults, which have traditionally been difficult to detect.

1. Introduction

High performance floating point (FP) datapaths are necessary both for graphics applications[2] (e.g., ray-tracing, 3D-rendering, etc.) and for simulations (e.g., calculating matrix eigen values, solving differential equations, etc.). We have developed and fabricated with 0.35μm CMOS technology an FP adder, an FP multiplier, and an FP divider and we have embedded them in a Vector Pipelined Processor. In order to improve performance, FP datapaths with the following characteristics are developed: 1) the FP adder adds three operands simultaneously, 2) the FP multiplier rounds the intermediate product of a Wallace tree, and 3) the FP divider determines four digits of a quotient per clock cycle and execute 4 divide operations every 12 clock cycles.

Graphics applications and simulations need to be executed in parallel, and reliability concerns make it necessary for LSIs contained in parallel processing systems to contain some kind of on-line test scheme. We have implemented into our FP datapaths a newly developed on-line test technique which makes it easy to detect not only static faults but also delay faults.

Our paper is organized as follows: Section 2 describes our newly developed high performance FP datapaths; Section 3 describes the on-line speed test technique; Section 4 describes processor characteristics; and Section 5 describes experimental results.

2. Floating Point Datapaths

A. Formats

We have introduced three 64-bit Floating-Point (FP) formats into the processor: 1) IEEE-754[3], 2) original format (A), and 3) original format (B), as seen in Fig. 1. IEEE-754 is the most popular format for PCs and workstations. The two original formats were necessary in order for the processor to be supercomputer compatible. As shown in Fig. 1, the IEEE-754 format has an 11-bit exponent, original format (A) has a 7-bit exponent, and original format (B) has a 15-bit exponent. Format (A) also has a characteristic radix-16 mantissa.

The processor also supports three integer formats and it contains an integer ALU capable of converting back and forth between integer and floating-point formats. A 32-bit floating point format number can be loaded by converting it to a 64-bit format and writing it into internal registers. Both the integer ALU and the FP divider are capable of converting back and forth between 32-bit and 64-bit FP formats. While the introduction of more than one floating point format generally tends to increase area and reduce operating speed, our newly developed design techniques, described below, have helped us overcome the problem of reduced speed.

B. Floating Point Adder

We employ an FP adder which executes 64-bit floating point addition/ subtraction operations, 32/64-bit integer addition/ subtraction operations, and various shift operations. The FP adder is fully pipelined, and the completion of any one of these operations is achieved in 3 cycles of 8nsec each. As seen in the FP adder block diagram (Fig. 2), the adder adds three operands simultaneously. The first pipeline stage consists of an unpacker, which separates operands into sign/ exponent/ mantissa parts, exponent comparators, and mantissa barrel shifters. For use in the simultaneous addition of three operands, it contains three exponent comparators and two mantissa barrel shifters. The second pipeline stage consists of a 3-input adder, priority encoders, and an exponent adder. The third pipeline stage consists of another exponent adder, a rounder, an exception detector, and a packer.

C. Floating-Point Multiplier

We employ an FP multiplier which executes 64-bit floating point multiplication operations and 32/64-bit integer multiplication operations. The FP multiplier is fully pipelined, and the completion of any one of these operations is achieved in 2 cycles of 8nsec each. As shown in the FP multiplier block diagram (Fig. 3), the first pipeline stage consists of an unpacker, an exponent adder, a radix-4 Booth recoder, and the first part of a mantissa Wallace tree. The second pipeline stage consists

0-7803-3177-6 $5.00 © 1996 IEEE IEEE 1996 CUSTOM INTEGRATED CIRCUITS CONFERENCE

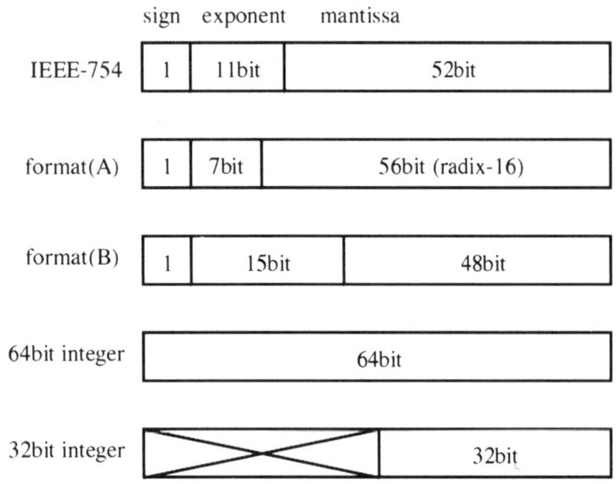

Fig. 1 Floating Point and Integer Formats

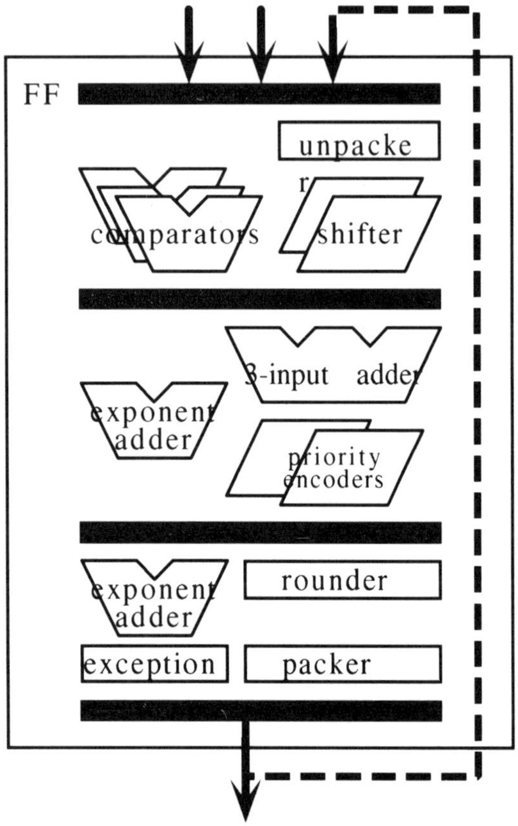

Fig. 2 Floating Point Adder Block Diagram

of the second part of the Wallace tree, mantissa adders, an exception detector, and a packer. In order to be able to handle multiplication of both signed and unsigned 64-bit operands, we designed the binary multiplier portion of the FP multiplier to be a signed 65-bit multiplier. The Wallace tree consists of five stages of 4-to-2 compressors. The first three of these stages are contained in the first

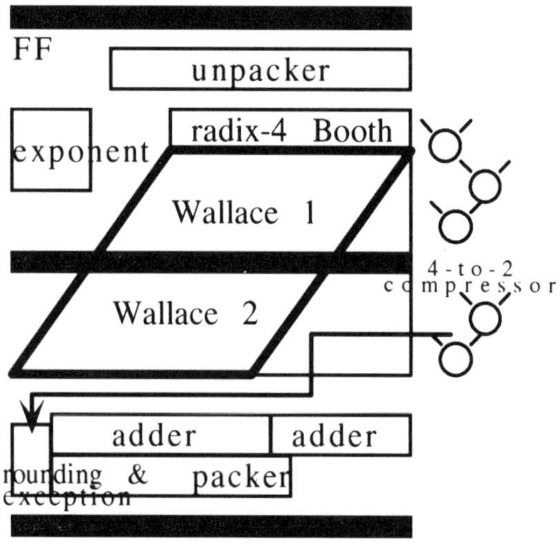

Fig. 3 Floating Point Multiplier Block Diagram

pipeline stage of the FP multiplier. Because IEEE-754 prescribes rounding in floating point calculations, its multiplication requires more logical gates than does that of the other two formats. To help accelerate IEEE-754 multiplication, a rounder in the exception detector rounds the output of the fourth 4-to-2 compressor stage of the Wallace tree. The rounder also bypasses adders, which calculate a mantissa product by adding Wallace outputs. These two look-ahead schemes decrease delay time in the second pipeline stage (which contains the maximum delay path of the FP multiplier) by, respectively 750psec and 1000psec each.

D. Floating-Point Divider

We employ an FP divider to execute floating point division in three formats and conversion between 32-bit and 64-bit FP formats. As shown in Fig. 4, the FP divider consists of the following 16 pipeline stages; (PRE)- (A0)- (A1)- (A2)- (A3)- (A4)- (A1)- (A2)- (A3)- (A4)- (A1)- (A2)- (A3)- (A4)- (A5)- (POST). The (PRE) stage consists of an unpacker, an exponent adder and a mantissa tripler, and the (POST) stage consists of an exponent adder, an exception block, and a packer. The (A0)-(A5) stages are each of the same structure, and (A1)-(A4) are used three times in each single divide operation. Each (A) consists of two redundant binary adders and two devices that we refer to here as detQs, each of which calculates two digits of a quotient. In short, four digits of a quotient can be calculated in a single clock cycle.

In order to improve FP divider throughput, we have developed the following techniques. The first is a fast detQ, in which partial remainders and partial quotients are expressed as redundant (signed-digit) binary numbers, which means that digits of the quotient can be calculated without the necessity of comparing all the bits of a divisor with those of the partial remainders. In the second

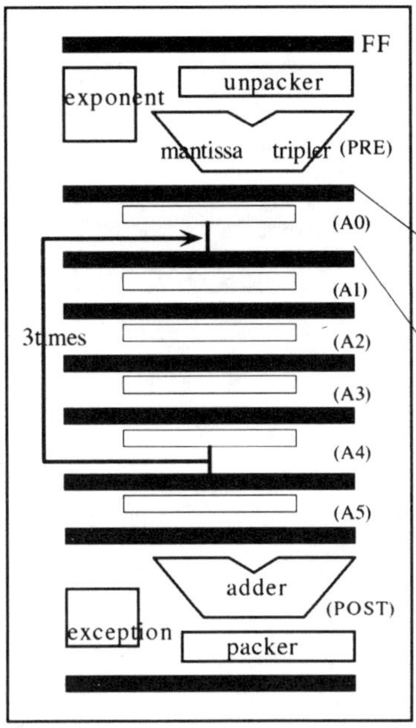

Fig. 4 Floating Point Divider Block Diagram

technique, the introduction of a four stage block, (A1)-(A4), allows us to perform four divide operations every 12 clock cycles. As a result, we have been able to reduce throughput significantly, to three, which is only one-fifth that of a conventional 200MHz divider, which determines two digits of a quotient per clock cycle.

E. Floating-Point Accumulation

Fast accumulation is necessary in such graphics applications as DCT and FFT, as well as in simulations. We achieve fast accumulation with a newly developed FP adder which adds three operands simultaneously. With this adder, our FP accumulator is able to calculate a vector product, Σ (Ai * Bi) [i=1..N], in N-cycles of multiply operations and (N/2+2) cycles of addition operations. The number of operations is thus reduced to 75% of that of conventional FP accumulators with a 2-input multiplier and a 2-input adder.

3. BISST

The increased clock frequency of recent high performance LSIs has increased the difficulty of testing. Specifically,

1) VDD/VSS dynamic bounce noise has increased, because complicated functionality has increased power dissipation;
2) coupling noise between signal nets has increased, because fine, multi-layer wiring technology has increased the coupling capacitance between signal nets, and coupling capacitance increases wiring delay, and
3) the adoption of advanced circuit technologies (e.g. pass-transistor logic, dynamic circuits, etc.) into an LSI makes it difficult to determine its longest delay path

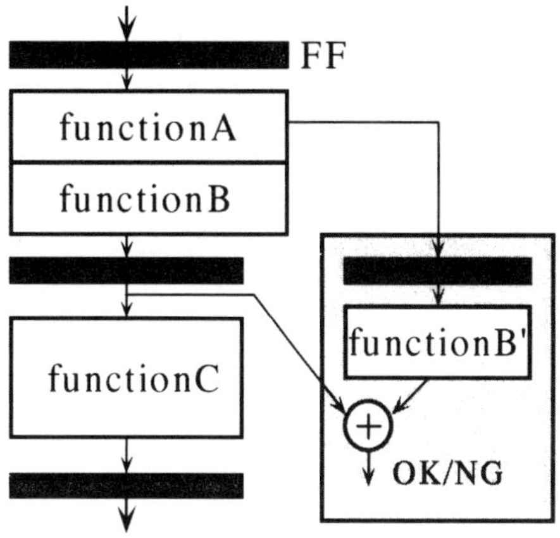

Fig. 5 BISST Scheme

because the various circuit technologies depend on such operating parameters as temperature, supply voltage, transistor deviation, etc.

In response to these difficulties, we have developed a BISST (Built-In Self Speed Test: see Fig. 5). FunctionA and functionB are serially connected in a pipeline stage. To help determine the speed of critically slow path in that pipeline stage, we add 1)flip-flops, which hold functionA outputs, and 2) a functionB', which generates expected outputs of functionB. These are then compared with the actual outputs of functionB. If functionB outputs are different from those expected, either functionB or functionB' must contain a fault. At low frequencies this is an indication of a static fault; at higher frequencies, it indicates a delay fault. As may be seen in the design illustrated in Figure 5, the flip-flops in front of functionB' ensure that it will not contain the critical path of this circuit, i.e. if a delay fault is detected, it will almost certainly have occurred in functionA or functionB.

Because functionB' creates overhead in both area and delay time, we want to make it as small as possible. A modulo-3/modulo-7 ALU can be used for it if functionB is an adder, subtracter, multiplier, or divider; a parity generator can be used if functionB is a selector. If functionB is a random logic block, we have to develop adequate techniques, or functionB' cannot be made smaller; it must be the same as functionB. While aliasing will not occur, in this case, the aliasing probability is 33% for modulo-3 ALUs and 50% for parity generators. Aliasing can be reduced, however, by increasing the number of points of comparison per function, i.e. by increasing the minuteness of the comparison.

4. Implementation

Figure 6 is a photo of the Vector Pipelined Processor, which may be seen to consist of three FP datapaths, an integer unit, vector registers, a crossbar to connect datapaths and registers, and controllers to decode instructions and schedule them to be executed. Table 1 summarizes FP datapath characteristics. Table 2 lists LSI features.

	Latency	Throughput	Format	etc.
FP Add	3	1	3 Integer/ 3 FP	3 operand
FP Mpy	2	1	3 Integer/ 3 FP	
FP Div	16	3	3 FP	12 cycle/ 4 operation

Table 1: Performance of Floating Point Datapaths

Technology	0.35μm, 4 metal CMOS
Number of Devices	4 million transistors
Chip Size	305 mm^2
Clock Frequency	over 125 MHz@3.3V
Processing Performance	~290 MFLOPS Add 125MFLOPS Mpy 125MFLOPS Div 41MFLOPS
I/O Band Width	32 Gbit/s 512signal I/O Load x2, Store x2

Table 2: Characteristics of the Chip

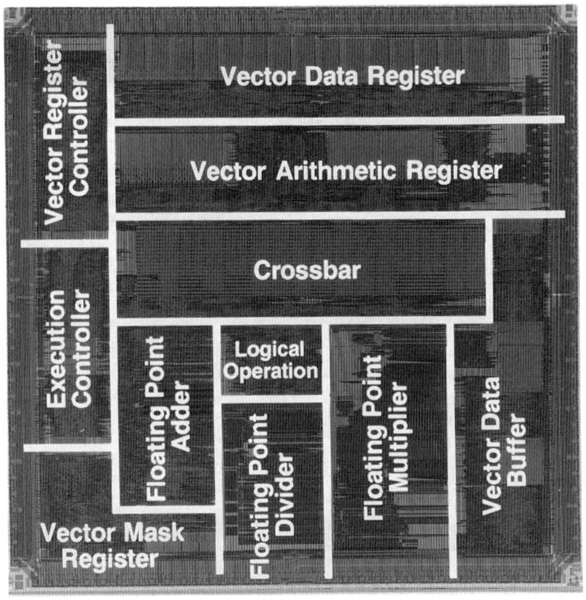

Fig. 6 VPP Chip Photo

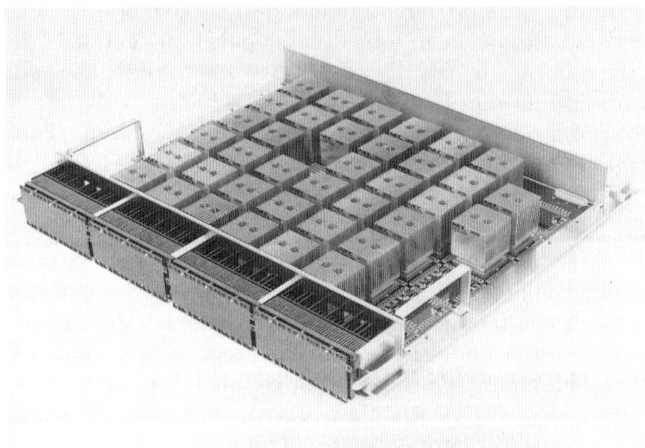

Fig. 7 Evaluation Board Photo

5. Evaluation

Figure 7 shows a photo of a evaluation board of the processor. Measurements were made on a combined 256Mbyte SRAM board and 8 VPP board (Fig. 7), which also included the BISST. Test programs included: specFP, Linpack, etc. The board and the processor were confirmed to operate at over 125MHz with a 3.3V power supply. As we considerably increase the frequency, decrease power supply voltage or increase noise magnitude, the BISST indicates the occurrence of a delay fault. The LSI is regarded as operating at 133MHz with a 2.9V power supply, and at 160MHz with a 3.3V power supply.

6. Conclusion

High performance floating point datapaths which supports three FP formats have been developed and been implemented in a Vector Pipelined Processor that has been verified to operate at over 125MHz under actual operating conditions. A BISST scheme has been confirmed as an effective speed test technique.

Acknowledgments

The authors would like to thank Dr. Abe, Dr. Fukuma, Mr. Takada, Mr. Ohwada, Mr. Furui, Mr. Nishi, Mr. Nakazaki and Mr. Inoue for their encouragement, and would also like to express their appreciation to Ms. Shijo, Mr. Kawaguchi, Ms. Endo, Mr. Osada, Mr. Nakazato, Mr. Wadasaki, Mr. Tagaya and many many persons who have supported the VPP project.

References

(1) F. Okamoto, Y. Hagihara, C. Ohkubo, Y. Sekine, N. Nishi, H. Yamada and T. Enomoto, "A 200MFLOPS 100MHz 64-bit BiCMOS Vector-Pipelined Processor (VPP) ULSI," ISSCC Digest of Tech. Papers, Feb., 1991.

(2) L. Kohn, S. Fu, "A 1,000,000 Transistor Microprocessor," ISSCC Digest of Tech. Papers, Feb., 1989.

(3) "IEEE Standard for Binary Floating-Point Arithmetic," IEEE Std. 754-1985.

A Complex Array Multiplier Using Distributed Arithmetic

Shousheng He and Mats Torkelson
Department of Applied Electronics, Lund University
S-22100 Lund, Sweden
email: he@tde.lth.se; torkel@tde.lth.se

Abstract– **The design of an efficient array architecture for the multiplication of complex numbers applying the distributed arithmetic is presented. The complex multiplier takes area just over that of two real multipliers and its speed is almost the same as a single real multiplier. The texture of the design is obtained by an in-depth examination of a real multiplier structure with data in the Off-set Binary representation. Residue error compensation and the functional requirement of various boundary cells, such as negative weight addition, are discussed in detail. VHDL module with generic parameters has been written and successfully simulated, which enable the complex multiplier module to be included in large designs with required word-lengths for both operands. A testing chip has been implemented with a standard library in 0.8μ CMOS process and fabricated.**

I INTRODUCTION

In a vast majority of computational tasks, the high complexity of VLSI implementation is largely due to the requirement of multiplicative operations. Although in some applications, such as digital filters, fixed coefficients multipliers can be optimized quite efficiently with Canonical Signed Digit (CSD) representation, in most cases full multipliers are indispensable. In many applications, such as the Discrete Fourier Transform (DFT), computation involves complex numbers is required. A straightforward implementation of a complex multiplication will need 4 real multiplications and 2 real additions:

$$\begin{aligned} Z &= AB = (A_R + jA_I)(B_R + jB_I) \\ &= (A_R B_R - A_I B_I) + j(A_R B_I + A_I B_R) \end{aligned} \tag{1}$$

Employing polynomial reduction technique[1, 2], the multiplication requirement can be reduced to 3, with 5 additions:

$$\begin{aligned} Z = \ & B_R(A_R - A_I) + A_I(B_R - B_I) \\ & + j[A_R(B_R + B_I) - B_R(A_R - A_I) \end{aligned} \tag{2}$$

In the DFT computation, complex multiplication is done with one of the operands, e.g. B, as the constant "twiddle factor". So two additions involving the twiddle factor in eqn. (2) can be precalculated, leads to a 3-multiplication, 3-addition scheme. This has been frequently quoted as the yard stick of fundamental requirement for the complex multiplication[3]. In hardware implementation, however, it has the disadvantage of requiring 3 real coefficients instead of usual 2, which adds a great deal of communication load. The delay due to the cascading real multiplications and additions are also excessively large.

In this paper we are to consider yet another efficient approach applying the Distributed Arithmetic (DA) on bit-level operations, so that the complex multiplier takes the area roughly equivalent to 2 real multipliers with the speed as that of a single multiplier in parallel implementation. Bit-parallel complex multiplier with DA approach has been proposed in the context of a radix-2 butterfly[4], where only the implementation of basic cell in the array was discussed. The details of residue error compensation and the boundary cell functionality of the implementation had been overlooked, which have been proven to be of significant importance if a more efficient and precise realization of the algorithm is to be sought.

In the following discussion, the two operands of multiplications, whether real or complex, are distinguished as coefficients B, which has a predefined value, and data A, whose value is unknown. All the arguments are expressed as fixed point fractional, in the range of $(-1, 1)$.

II A MULTIPLIER WITH OFF-SET BINARY REPRESENTATION

We begin with a multiplier for real numbers with the data A in the Off-set Binary (OB) representation, where logical 0 is interpreted as -1[5]. An N-bit data in the 2's complement (2C) coding can be easily converted into the OB coding by the Most-Significant Bit (MSB) inversion[6]:

$$A = -a_0 + \sum_{n=1}^{N-1} a_n 2^{-n}, \qquad a_n \in \{0, 1\} \tag{3}$$

$$A' = \sum_{n=0}^{N-1} \alpha_i 2^{-n-1}, \qquad \alpha_n = \begin{cases} -1, & a_n = 0 \\ 1, & a_n = 1 \end{cases} \tag{4}$$

A small residue error of weight 2^{-N} is introduced by the conversion:

$$A = A' - 2^{-N} \tag{5}$$

It should be also noticed that each α_n in the OB coding has a weight one bit lower than a_n in the 2C coding.

Consider the product of the N-bit A and an M-bit coefficient B:

$$\begin{aligned} AB &= (A' - 2^{-N})B \\ &= \sum_{n=0}^{N-1} \alpha_i 2^{-n-1} B - 2^{-N} B \end{aligned} \tag{6}$$

5.2.1

0-7803-3177-6 $5.00 © 1996 IEEE

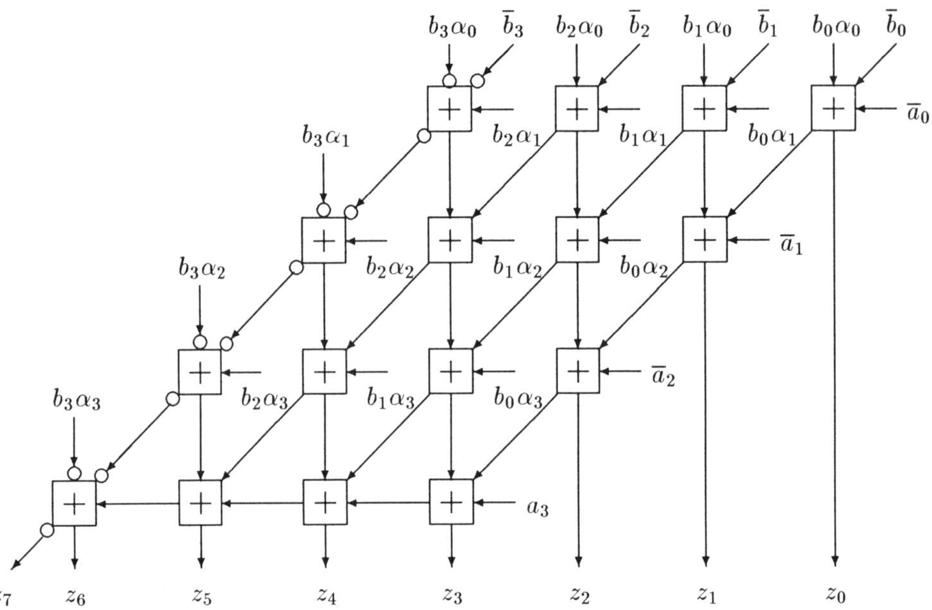

Figure 1: An $N = M = 4$ array multiplier with data A in OB coding

For an OB multiplier, explicit conversion of the data representation is not necessary. The summation term in eqn. (6) can be implemented in bit modules similar to that in a 2C multiplier, only with a XNOR gate in place of the AND gate. The residue error introduced, expressed in eqn. (5), can be compensated with an subtraction of 2^{-N} weighted coefficient, as demonstrated in eqn. (6).

Fig. 1 shows a parallel $N \times M = 4 \times 4$ multiplier with data A in the OB coding, using an array multiplier structure with carry-save adders[7], where the "bit product" can be logically implemented as $b_i\alpha_j = \overline{b_i \oplus a_j}$. Since each Partial Product (PP) of B in the OB array multiplier will be either added or subtracted according to α_n in eqn. (4), an extra Least Significant Column (LSC) of full adders, to the right of the main array, is necessary for the addition of the complement bit in case of subtraction. At the other end of the PP summation, care has to be taken to maintain the correct sign representation in 2C coding where the two PP's are aligned one bit apart, due to their weighting difference. Although this can be accomplished by either sign extension, or an extra summation of a correction factor[8], the most efficient approach is using negative weighted addition[7, 9]. A compact structure can be obtained with merged Most Significant Column (MSC) cells, where adders with 2 negative weighted inputs are used, and merged Least Significant Row (LSR) cells, shown in Fig. 1. The carry output with negative weight propagates downward to the partial sum of higher significance. It is not necessary to use negative input adders in the Most Significant Row (MSR) cells as in a ordinary array multiplier, since the MSB of the OB coded data is not solely negative, but is used to decide whether to add or subtract the PP of the corresponding weight.

The cost of this efficient implementation, where all inputs of all elementary bit adders are utilized, is a small error of weight $2^{-(N+M-2)}$ introduced in some input combination due to the over-range at the LSB cell in the LSR[1]. This error, however, is well bellow the rounding/truncation error for reasonable N and M in ordinary applications.

III A COMPLEX ARRAY MULTIPLIER WITH DISTRIBUTED ARITHMETIC

The Distributed Arithmetic (DA) technique of [10] can be applied to calculate a inner product by summation of a series of Partial Inner Product (PIP) from a look-up table. In the case of a complex multiplication, which can be considered as a 2-point inner product, the look-up table has only two PIP words, in the forms of sum and difference respectively,

$$\begin{aligned} B_S &= B_R + B_I \\ B_D &= B_R - B_I \end{aligned} \tag{7}$$

Now consider the complex multiplication of eqn. (1). Replacing B in the PIP form and A in the OB representation, after simplification, the real and imaginary parts of eqn. (1) are given as

$$\begin{aligned} Z_R &= \sum_{n=0}^{N-1} (PB_S + QB_D)\alpha_{Rn} 2^{-n-1} - B_D 2^{-N} \\ Z_I &= \sum_{n=0}^{N-1} (QB_S + PB_D)\alpha_{In} 2^{-n-1} - B_S 2^{-N} \end{aligned} \tag{8}$$

[1] Since error compensation subtracts $2^{-N}B$, with a fixed complement "1" in the LSB position, over-range happens when both $a_0 = b_0 = 0$. This error can be eliminated when still an extra column LSC adders are used for the rounding of the output result.

5.2.2

where P, Q are selecting functions determined by the n-th bit-pair of A in the OB form,

$$P = (\alpha_{Rn} - \alpha_{In})(2\alpha_{Rn})^{-1}$$
$$Q = (\alpha_{In} + \alpha_{Rn})(2\alpha_{In})^{-1} \quad (9)$$

Comparing the real and imaginary parts of eqn. (8) with eqn. (6) respectively, it is clear that 2 basic functors of the OB multiplier can be combined to form a complex multiplier with just an additional decoding circuit for $\{\alpha_{Rn}, \alpha_{In}\}$ bits to address the PIP of B. The arithmetic operation at the bit level can be substituted with logic operations, which can be easily implemented with an XOR gate and a commutator.

$$P = a_{Rn} \oplus a_{In}$$
$$Q = a_{Rn} \overline{\oplus} a_{In} \quad (10)$$

Applying the DA technique, a bit-parallel complex multiplier can be built upon the structure of a dual OB array multiplier, with some additional decoding and multiplexing circuitry. The global organization of a $N \times M$-bit complex multiplier in Fig. 2 demonstrates the geometrical relation of the main array cells and the boundary cells required for the implementation.

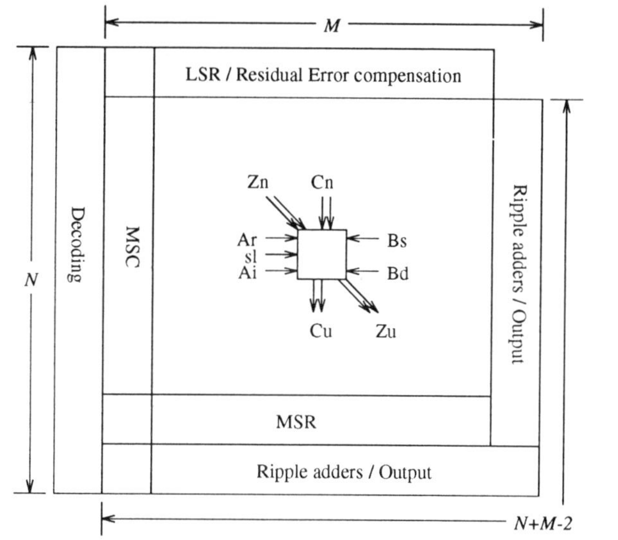

Figure 2: Global structure of bit-parallel complex multiplier using DA

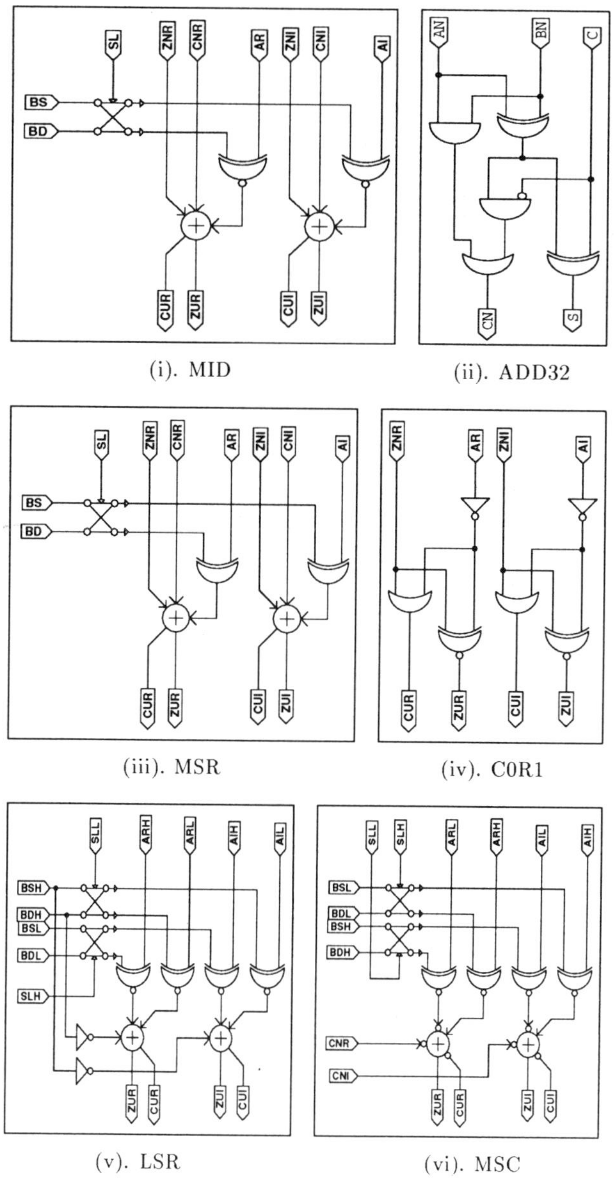

Figure 3: Basic cells in the array complex multiplier

The main array contains $(N-3)(M-2)$ basic cells of type MID, as shown in Fig. 3(i), with a dual carry-save adder to implement the bit operation specified in eqn. (8), where the decoding XOR is extracted out to the left most position, to be shared among a complete row of cells. Just like in typical layout of array multipliers, the summations propagate diagonally, the carries vertically, as shown in Fig. 2. The MSR in the array contains cells of MSR type, Fig. 3(iii), which is similar to MID, only with the data bits inverted.

The LSR cells each contains two commutators with a dual adder, as in Fig. 3(v), where the residue error compensation, expressed in eqn. (8) is accomplished with an implicit subtraction by adding inverted coefficient inputs in the PIP form. The MSC cells contain also two commutators and use adders with two negative inputs, as in Fig. 3(vi). Note that adders with negative inputs are not symmetrical and the outputs/inputs have different logic relations, as depicted with primitive gate implementation of the ADD32(3 input/2 negative), Fig. 3(ii). The LSB outputs are thrown away, since they are always zero due to the fact that the LSB operation in eqn. (8) has always two terms in PIP form of eqn. (7). Taking this factor into account, the carries of the LSB cell in the Least Significant Column (LSC) have values of tautological "1", which helps in logic reduction of the second dual ripple adder in the LSC, as shown in Fig. 3(iv). The carries of the MSB in the MSC are also dropped off since they are not required in an in-range operation, as will be discussed in section IV. Those boundary cells at the corners will need combined modification to comply with both row and column cell properties, as already discussed above.

IV Constraints for in-range fractional computation

In fixed point arithmetic operations, one of the rudimental requirement is that the output data should be *in-range* within a given word-length, so that it can be represented without any ambiguity and is valid for further processing. A complex multiplication can be considered as a product of two vectors in the complex plane, which is quite different from the multiplication of real numbers, where no over-range consideration is necessary provided the input operands are of in-range fractional. For general complex multiplications, even if all the real/imaginary parts of the two operands are in-range, i.e. $-1 < A_R, A_I, B_R, B_I < 1$, output over-range can still occur. This can be clearly seen if the complex numbers are expressed in a polar form, as shown in Fig. 4:

$$Z = Z_v e^{j\phi} = AB = A_v e^{j\alpha} B_v e^{j\beta} = A_v B_v e^{j(\alpha+\beta)} \tag{11}$$

where the Z_v, A_v, B_v represent the norms, or the vector lengths in the complex plane, of the corresponding complex numbers.

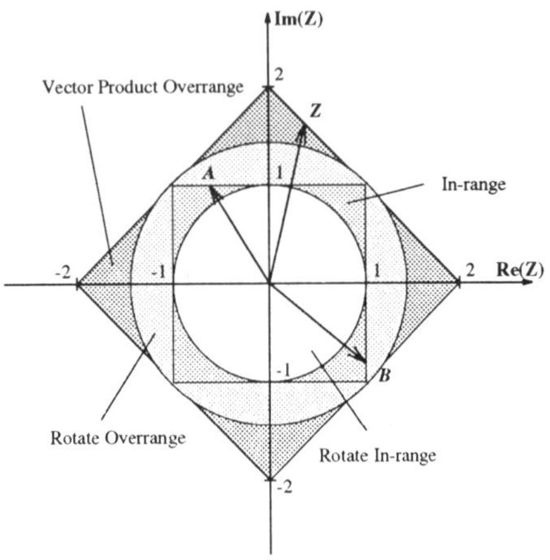

Figure 4: Range constraint for complex multiplication

In such application as the DFT transform, the DFT coefficients are *rotation* factors, which have the property that $B_v = 1$. Thus if a constraint is put onto the input data so that $A_v < 1$, i.e. the vector is inside the region of unit circle in Fig. 4, an in-range output can be guaranteed. Otherwise, a factor of $1/\sqrt{2}$ should be applied to one of the operands, which represents approximately 3dB loss in precision.

V Conclusion

Bit-parallel complex multiplication with the DA technique has been discussed. A compact architecture has been presented for the array implementation based on the structure of an OB multiplier. The functionality of boundary cells has been discussed in detail. The constraint for in-range operation has been specified for DFT computation so that the precision can be fully utilized within a given word-length. VHDL modules with generic parameters for word-length requirement have been written and undergone extensive simulation. The correctiveness of the architecture has been verified. A primitive 16×12 bit-parallel complex multiplier has been implemented using a standard cell library in 0.8μ CMOS process[11] and fabricated. The core takes area approximately $1.8 \times 1.9\text{mm}^2$. Its layout is shown in Fig. 5. Full testing of the chip should have been finished by the time of the publication of this paper.

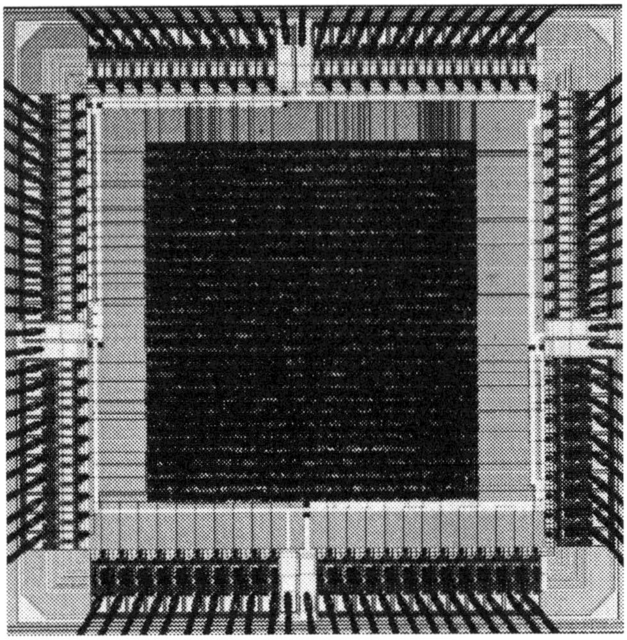

Figure 5: Chip die photo for the complex multiplier

References

[1] D.E. Knuth. *The Art of Computer Programming*, volume 2. Addison-Wesley, 2nd edition, 1981.

[2] H. J. Nussbaumer. *Fast Fourier Transform and Convolution Algorithms*. Springer-Verlag, 1981.

[3] P. Duhamel and M. Vetterli. Fast Fourier transforms: a tutorial review and a state of the art. *Signal Processing*, 19(4):259–299, 1990.

[4] I. R. Mactaggart and M. A. Jack. A single chip radix-2 FFT butterfly architecture using parallel data distributed arithmetic. *IEEE J. Solid-State Circuits*, sc-24(3):368–373, June 1984.

[5] C. F. N. Cowan, S. G. Smith, and J. H. Elliott. A digital adaptive filter using a memory-accumulator architecture: theory and realization. *IEEE Trans. Acoust., Speech, Signal Processing*, ASSP-31(3):541–549, June 1983.

[6] S. G. Smith. Efficient serial/parallel inner-product computation. *Electronics Letters*, 22(14):750–752, July 1986.

[7] I. Koren. *Computer Arithmetic Algorithms*. Prentice-Hall, 1993.

[8] A. Clements. *The Principles of Computer Hardware*. Oxford University Press, 2nd edition, 1991.

[9] L.R. Rabiner and B. Gold. *Theory and Application of Digital Signal Processing*. Prentice-Hall, Inc., 1975.

[10] A. Peled. On the hardware implementation of digital signal processors. *IEEE Trans. Acoust., Speech, Signal Processing*, ASSP-24(1):76–86, Feb. 1976.

[11] Austria Mikro System International. *0.8-Micron Standard Cell Databook*, 1993.

A Pipelined Digital Differential Matched Filter FPGA Implementation & VLSI Design

Kuang-Chan Liu, Wun-Chang Lin and Chorng-Kuang Wang

Department of Electrical Engineering
National Central University
Chung-Li, Taiwan, ROC
E-mail:ckwang@mbox.ee.ncu.edu.tw

Abstrct— A digital matched filter having a differential and pipelined structure is proposed for CDMA (Code Division Multiple Acess) communication systems. A novel differential PN (Pseudo-Noise) code scheme is adopted to reduce the number of multiplication and summation (M & S). The PDDMF (Pipelined Digital Differential Matched Filter) not only saves half of the hardware and power, but also improves the speed to a single operation time of M & S. These features make the PDDMF more suitable for personal communication high speed and low power requirements than the conventional methods. The FPGA of the DDMF (Digital Differential Matched Filter) verifies the matched filter using the proposed differential PN code scheme. Moreover, a 64 taps length PDDMF with 4 bits soft decision using 5V $0.8um$ CMOS standard cells is designed. The active chip area is 2.6*2.6 mm^2. Simulation results demonstrate that the chip rate of the PN code in the PDDMF can achieve 45M chips/sec.

I. Introduction

Matched filters are widely used for code acquisition in modern direct sequence spread spectrum communication systems because they offer significant advantages in the speed of synchronization.

However, the primary limitation of the conventional digital matched filters (in Fig.1) is that as the number of stages increases, the number of multiplication and summation remarkably increases. The complexity of the multiple inputs summation circuits does as well. These restrictions constrain the matched filters applications in VLSI implementation owing to their low speed, huge hardware and power consumption. The number of stages of commercial digital matched filters is mostly in the range of 8 to 64; the chip rate is limited to 30M chips/sec.

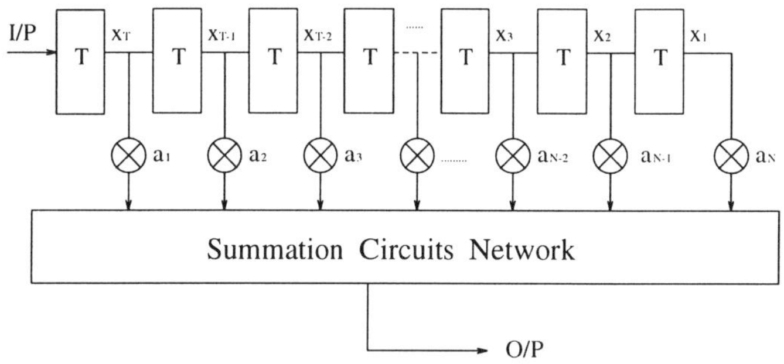

Figure 1: Architecture of conventional matched filter.

This paper is organized as follows. Section II presents a new PDDMF architecture in addition to a novel differential PN (Pseudo-Noise) code scheme. A comparison of the number of M & S of the conventional matched filter with that of the PDDMF is then made. Section III provides the FPGA experimental results, and a 64 taps length PDDMF with 4 bits soft decision using 5V $0.8um$ CMOS standard cells is also designed. Finally, Table I summarizes the characteristics and performances of the new PDDMF.

II. Pipelined Digital Differential Matched Filter

PDDMF - Pipelined Structure Using Differential PN Code Scheme

The output of the conventional digital matched filter (Fig.1) can be expressed as follows:

$$f_c(T-1) = a_N x_0 + a_{N-1}x_1 + .. + a_2 x_{T-2} + a_1 x_{T-1}$$
$$f_c(T) = a_N x_1 + a_{N-1}x_2 + ... + a_2 x_{T-1} + a_1 x_T \quad (1)$$

5.3.1

0-7803-3177-6 $5.00 © 1996 IEEE

where N is the length of PN code, a_i, i=0 to N, is the PN code coefficient and x_i, i=0 to T, is the received signal. The number of M & S of a conventional matched filter M_c is

$$M_c = N = 2^r - 1 \quad (2)$$

where N is the length of PN code and r is the order of the PN code generator polynomial [1]. An increase in the number of stages obviously causes the number of M & S to significantly increase; the related hardware and power do as well.

To reduce the number of M & S, the differential PN code scheme is proposed. An accumulator is placed at the output of the matched filter (Fig.2) and accumulates the previous output f(T-1). The output function of the digital matched filter can then be modified as follows:

$$f_D(T) = [b_{N+1}x_0 + b_N x_1 + b_{N-1}x_2 .... + b_2 x_{T-1} + b_1 x_T] + f_D(T-1)$$

$$= [-a_N x_0 + (a_N - a_{N-1})x_1 + (a_{N-1} - a_{N-2})x_2 + ... + (a_2 - a_1)x_{T-1} + a_1 x_T] + f_c(T-1) \quad (3)$$

where b_i, i=0 to N+1, is the multiplication coefficient of the matched filter with differential PN code scheme, and a_i, i=0 to N, is the PN code in the conventional matched filter. The $f_D(T)$ in (3) is identical with the $f_c(T)$ in (1) if

$$b_1 = a_1$$
$$b_i = a_i - a_{i-1}, for \quad i = 2 \quad to \quad N$$
$$b_{N+1} = -a_N$$

The multiplications coefficients of the PDDMF architecture can be modified to $b_i = a_i - a_{i-1}$. Since the PN code a_i can be +1 or -1, b_i must therefore be 0, 2 or -2. Multiplication and summation (M & S) for those $b_i = a_i - a_{i-1} = 0$ is unnecessary. The number of M & S is therefore significantly reduced for PDDMF with long stages.

Fig.2 shows the PDDMF with a differential PN code scheme and a pipelined structure. Fig.3 illustrates the associated data flow. In PDDMF, the input data are first broadcasted simutaneously to each multiplier and then summed by the following pipelined structure. The output function of the proposed PDDMF is the same as (3).

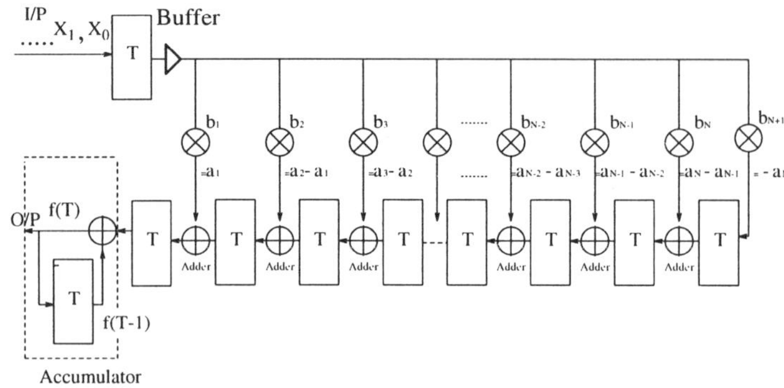

Figure 2: Architecture of the proposed PDDMF (Pipeline Digital Differential Matched Filter).

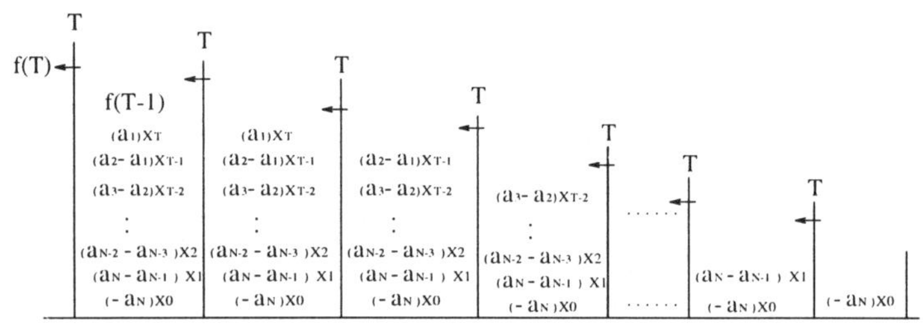

Figure 3: Data flow of the proposed PDDMF.

The PDDMF takes advantages of the proposed differential PN code scheme and the pipelined structure. The differential PN code scheme eliminates the number of M & S, and thus saving the hardware and power. On the other hand, the pipelined structure simplifies the multiple inputs summation circuits and makes the throughput of the PDDMF independent of the PN code length. Moreover, it improves the speed of the PDDMF to the operation time of a single M & S.

M & S Comparison between the PDDMF and the Conventional MF

According to the run property [1] of a PN sequence with length $N = 2^r - 1$, where r is the order of polinomial generator. The total number of run in a PN sequenc is

$$N_{run} = 2 + 2 \cdot (1 + 2 + 2^2 + 2^3 + ... + 2^{r-3}) = 2^{r-1}$$

Here, a run is a subsequence of identical symbols within the PN seqence, and the length of the subsequence is the length of the run.

The condition where $b_i = a_i - a_{i-1}$ is not equal to zero (i.e., the nonzero multiplication coefficient) in the PDDMF only takes place between adjacent two runs.

Therefore, the number of nonzero coefficients is 2^{r-1}-1. If b_1 and b_{N+1} are considered, the number of M & S in the PDDMF is

$$M_D = 2^{(r-1)} \tag{4}$$

rather than $M_c = 2^r$-1 of the conventional matched filter. The ratio of M & S of the conventional MF to the PDDMF can be expressed as follows:

$$R_{M\&\ S} = \frac{M_D}{M_c} = \frac{2^{(r-1)}}{2^r - 1} \tag{5}$$

From (5), for PN codes with a sufficiently long length, the ratio $R_{M\&\ S}$ approaches to 1/2. That is, the PDDMF saves half of the hardware and power in comparison with the conventional MF.

III. FPGA Implementation & Chip Design

A. FPGA Implementation

Fig.7 shows an FPGA implementation of a matched filter with differential PN code scheme- DDMF (Digital Differential Matched Filter). The prototype of length 7 DDMF with soft decision is realized by an FPGA, and the soft decision is made by an 8 bits ADC. An 8 bits DAC is connected to the DDMF to show the correlation output.

The PN code in the DDMF is (1, 1, 1, 0, 1, 0, 0). If 0 and 1 are converted to 1 and -1, the coefficients of the DDMF will be

$(b_1, b_2, b_3, b_4, b_5, b_6, b_7, b_8)$
$=(a_1, a_2 - a_1, a_3 - a_2, a_4 - a_2, a_5 - a_4$
$, a_6 - a_5, a_7 - a_6, -a_7)$
$=(1, 0, -2, 2, -2, 0, 0, 1)$

FPGA Experimental Results

Fig.8 shows the FPGA experimental results of the DDMF with a length of 7. In Fig.8, the peaks of matched filter correlation outputs can still be detected when the input signal is embedded with AWGN (Addative White Guassian Noise) of SNR (Signal to Noise Ratio) - 3 dB.

B. 64 Taps 4 Bits Soft Decision PDDMF Chip Design

In IS-95 based CDMA systems, the 9.6KHz input data are spread to a signal band 1.2288MHz. Because the SNR of the spreading signal is very low, the 4 bits soft decision matched filters are adopted to increase the reliability of data detection. In the

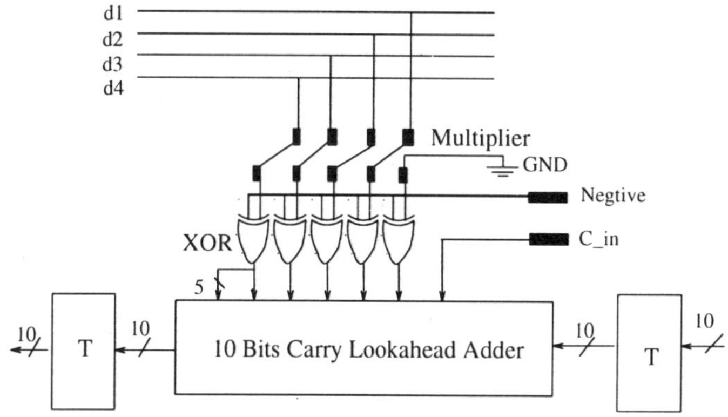

Figure 4: Circuit design for M & S of 2, -2.

64 taps PDDMF, the number of multiplication coefficients where $b_i = a_i - a_{i-1} = 0$ is 31. Only 32 M & S need to be multiplied, rather than 64 in the conventional matched filter.

Multiplier and Adder

Fig.4 displays the circuit of two's complementary M & S of 2 or -2. If the multiplication coefficient is bi=-2, the pin *Negative* and *Cin* are set to logic 1; otherwise they will be set to logic 0 for bi=2. When the pin *Negative* and *Cin* are set to logic 1, the input data are shifted to the left and converted to their two's complement. Such an operation causes the input data multiplied by -2. Finally, the summation is performed by the 10 bits carry lookahead adder.

Verilog Simulation Results

The 64 taps PDDMF with 4 bits soft decision using 5V 0.8um CMOS standard cells is designed as follows. The active chip area shown in Fig.6 is 2.6*2.6 mm^2. The Verilog simulation results (in Fig.5) demonstrate that the chip rate of the PDDMF can achieve 45MHz.

	Conventioal matched filter	PDDMF	Ratio
number of M & S (one sample)	$M_c = 2^r - 1$	$M_D = 2^{(r-1)}$	$\frac{M_D}{M_c} = \frac{1}{2}$
number of M & S (M samples)	$M_{cM} = (2^r - 1)*M$	$M_{DM} = 2^{(r-1)}$	$\frac{M_{cM}}{M_{DM}} = \frac{1}{2*M}$
Summation Circuit Signal Delay	$T_{d\_c} = N * t_{adder}$	$T_{d\_D} = t_{adder}$	$\frac{T_{d\_D}}{T_{d\_c}} = \frac{1}{N}$

Table 1: The charcteristics of PDDMF in comparison with the conventional MF.

IV. Conclusions

A new architecture - PDDMF (Pipeline Digital Differential Matched Filter) with a novel differential PN code scheme is proposed. The theoretical analysis indicates that the PDDMF can save the hardware and the power for the matched with M samples to 1/(2*M). In particular, the PDDMF shortens the signal delay in summation circuit to 1/N of the conventional MF, where N is the level of the summation circuits. Consequently, the PDDMF speed can be improved to one operation time of the single M & S. The FPGA hardware implementation and the 5V 0.8*um* CMOS standard cells chip design of the DDMF and the PDDMF architectures respectively verify the predicted performaces.

References

[1] R. E. Ziemer and R. L. Peterson " Digital Communication and Spread Spectrum Systems", chap. 8, pp. 386.

[2] Fang Lu, and Henry Samueli, "A 200M Hz CMOS Pipelined Multiplier-Accumulator Using a Quasi-Domino Dynamic Full-Adder Cell Design ", *IEEE Journal of Solid-State Circuits*, vol. 28, no. 2, pp. 123-132, Feb, 1993.

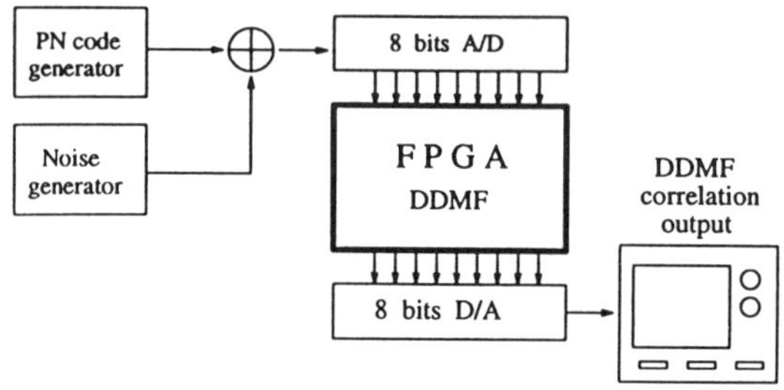

Figure 7: The block diagram of the DDMF FPGA prototype .

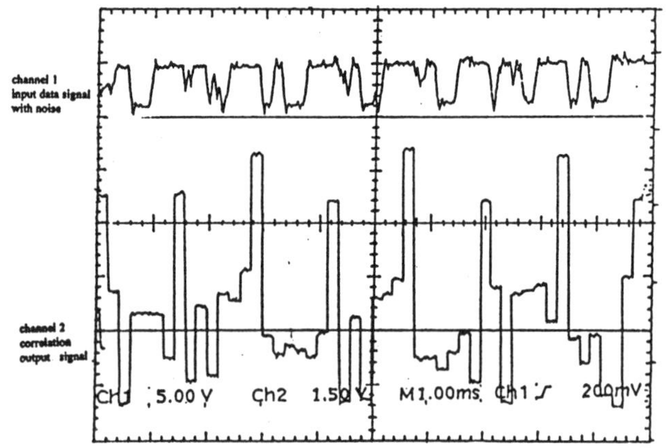

Figure 8: The correlation output of the FPGA DDMF of 7 taps with input SNR of -3dB.

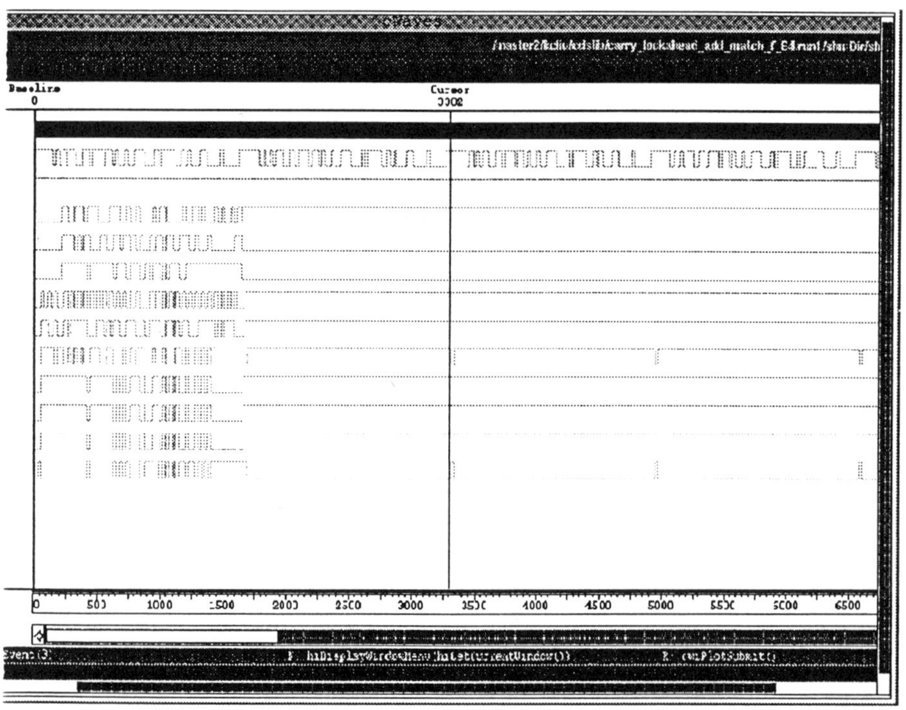

Figure 5: The Verilog simulation results of the 64 taps PDDMF using 5V 0.8*um* CMOS standard cells .

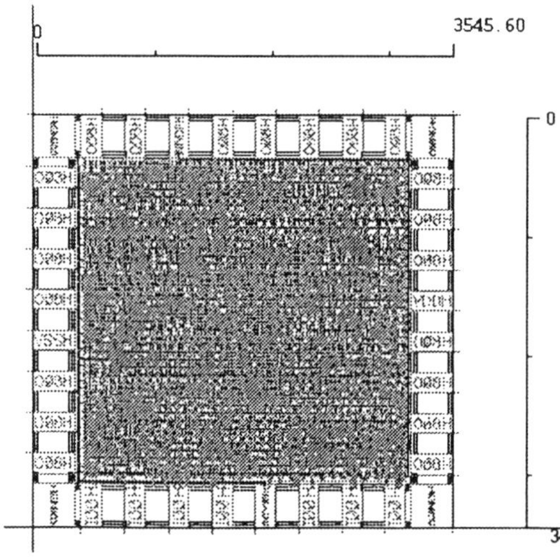

Figure 6: The 5V 0.8*um* CMOS standard cells layout of the 64 taps PDDMF. The active chip area is 2.6*2.6 mm^2.

5.3.4

Structured Design of a 288-tap FIR Filter by Optimized Partial Product Tree Compression

Jun Rim Choi, Seong Wook Jeong, Lak Hyun Jang and Jin Ho Choi

ASIC Center, LG Electronics Research Center
16, Woomeon-Dong, Seocho-Gu
Seoul, Korea 137-140
Tel : +82-2-526-4522
Fax : +82-2-575-5104
e-mail : jrc@asic.goldstar.co.kr

Abstract

A compact 10-bit, 288-tap FIR filter is designed by adopting structured architecture which employs optimized partial product tree compression method. The new architecture is based on the addition of equally weighted partial products which result from 288 multiplications of the filter coefficients and the inputs. The 288 multiplication and 287 addition operations are decomposed to add 1440 partial products to meet the tight timing requirement. Optimized parallel compression schemes such as 4:2 and 5:5:2 compressors are used to perform decomposed partial product addition. The completed 288-tap FIR filter occupies 7x9mm^2 of Silicon area which consists of 385754 transistors in 0.6μm triple-metal CMOS technology.

Introduction

Finite impulse response (FIR) filter has been used for a video rated digital filtering such as ghost cancellation and channel equalization. However, a major drawback is the large number of arithmetic operations needed for its implementation, which prevents from providing a one-chip solution for most filtering applications. Thus, various methods have been proposed to reduce the computational complexity of multiplications and additions.[1]-[4] An efficient architecture and optimized circuits are the keys to design a compact video-rate FIR filter that meets the timing constraint. Since most conventional video-rate FIR filters consist of parallel multipliers and adders, the critical signal path lies in the parallel multiplier. The parallel multiplier can be categorized in three domains; Booth encoder, partial product summation tree, and the final adder. The size introduced by the partial product summation tree is relatively large compared to that of the other components and the delay of the final adder is unavoidable. Therefore, we need to optimize the partial product summation tree for compact design and restrict using the final adder for high speed processing.

In this work, we propose an optimized parallel architecture for partial product tree compression. The new architecture features a partial product decomposition method to achieve high density and speed. A 10-bit, 288-tap FIR filter is designed by adopting the new structured architecture and optimized circuits.

Decomposition of Partial Products

A 288-tap FIR filter is described as follows;

$$y[n] = w[n] + \sum_{k=0}^{287} h[k] \cdot x[n-k] \tag{1}$$

The FIR function requires 288 multiply-add-accumulate (MAC) operation as shown in Fig. 1.

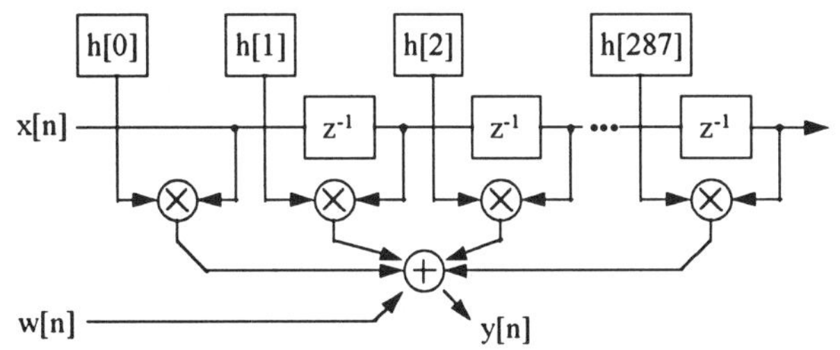

Fig.1. Functional Diagram of a 288-tap FIR filter

To perform 288 operations, we divided the 288-tap operation module into nine 32-tap sub modules for distributed computing as shown in Fig. 2. The nine 32-tap sub modules generate sums and carries respectively which are added altogether in the second adder to yield the final

5.4.1

0-7803-3177-6 $5.00 © 1996 IEEE

sum and carry. Thus the slow final adder is used only once at the end of the 288-tap filter to save the processing time. The post-processor rounds the output and checks the overflow. The clock of the 32-tap sub module is multiplied by four using a phase locked loop (PLL) to save the Silicon area and to enhance the operating speed. Since the 32-tap sub module performs 8-tap MACs in four clock cycles, we need a hardware to add 40 partial products generated from eight 10-bit multipliers within a multiplied clock cycle.

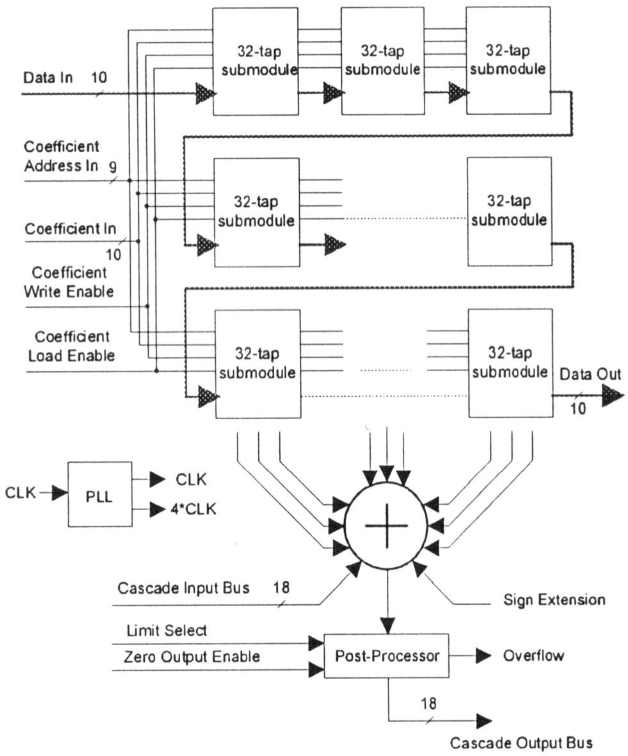

Fig. 2. Block diagram of a 288-tap FIR Filter

According to the modified Booth algorithm (MBA)[5], multiplier components can be divided into three components : partial product, sign conversion bit, sign extension constant as shown in Fig. 3. Instead of adding all the components together, we decomposed the multiplier components of the same weight to save the unnecessary adder bits generated when adding the partial products of different weight. Each partial product is decomposed and added in a 32-tap sub module as shown in Fig. 4. Sums, carries and the sign bits are grouped respectively to keep the proper weights. The sums, carries and the sigh bits are added in each sub module using the carry save adder[5]. Since the final adder used in most multipliers is the critical path and the processing time increases as the number of bits increases, it is used at the end of processing when 288 MAC operations are completed. The sign extension constant of the 288 multipliers can be treated as a sign extension constant of the 288 multipliers, thus the constant becomes -349184 X 288 = -100564992. It is to be added at the end of the adder tree.

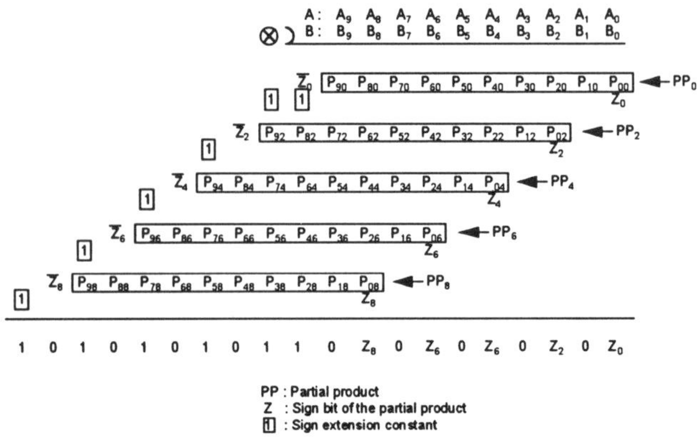

Fig. 3. 10X10-bit multiplier components from modified Booth sign extension algorithm

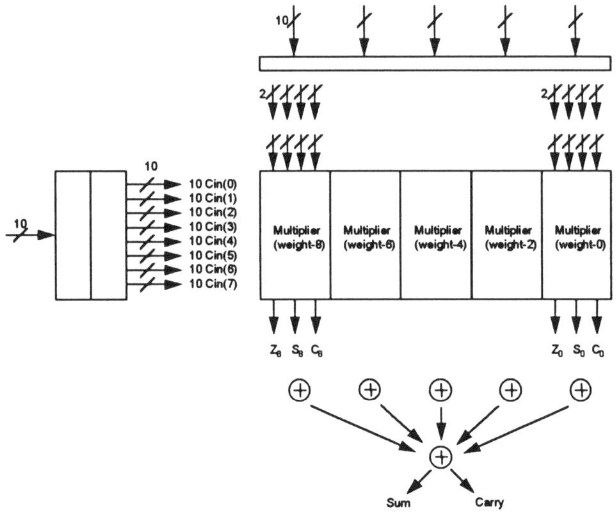

Fig. 4. Decomposition of the partial products of the same weight

The decomposed partial products and the sign bits in a sub module are added in parallel to generate sums and carries respectively as shown in Fig. 5 and Fig. 6. The eight partial products of the same weight are again decomposed into 3 and 5 partial products to save the number of parallel adders. The 3:2 compressor is well known and is used widely as proposed by Wallace[6]. The 5:5:4 compressor is proposed by Song[7] and it is optimized for high density parallel adder. The 5:5:4 compressor uses 6 full adders and it requires 4 full adder processing time as shown in Fig. 7. The sums and carries are generated 2-bits each. Since the 5:5:2 compressor adds each 2-bits of

5.4.2

the 5 partial products in parallel, its hardware is optimized at the expense of longer timing than 3:2 compressor. The generated 4 sums and carries are compressed again using the 4:2 compressor to yield the final sum and carry.

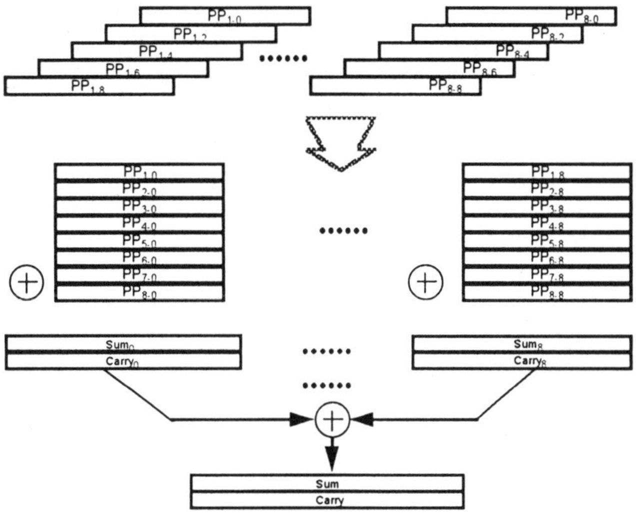

Fig. 5. Decomposition of the partial products and their sums and carries in 8-tap sub module

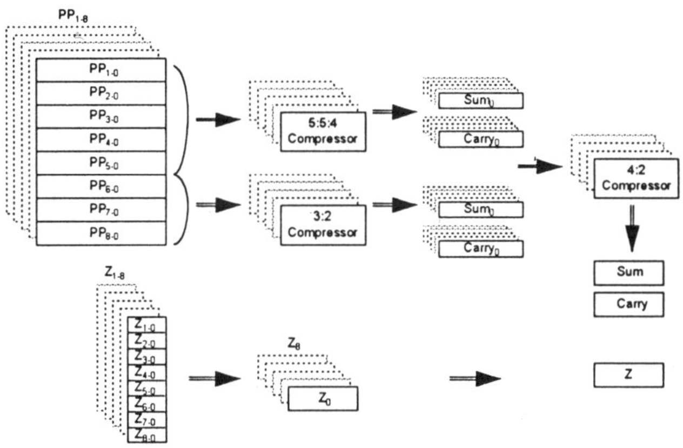

Fig. 6. Addition of multiplier components to yield sums and carries

Design of an Optimized Fill Adder

A full adder is an important cell in the parallel compression architecture proposed in this paper. Several different circuit configurations have been proposed to exploit exclusive operations of a full adder. Among different configurations complementary pass transistor logic family provides compact silicon area and high speed compared to conventional CMOS logic. However, its complementary characteristic requires increased wiring, thus it is not the most suitable for the design of a full adder. Pass transistors in combination with transmission gates provide the minimum number of gates and they consume less power at the expense of weak load driving capability.

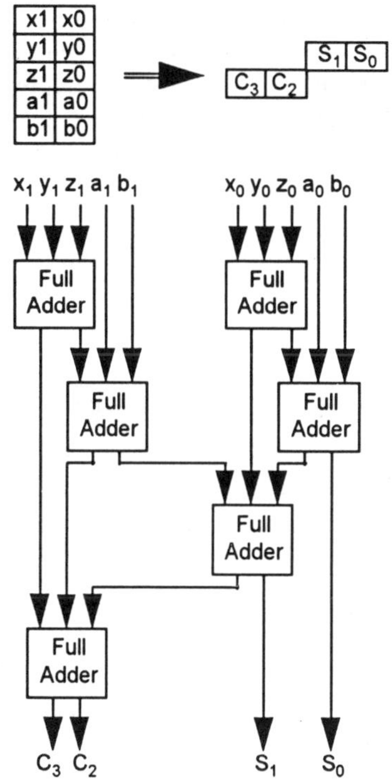

Fig. 7. Architecture of a 5:5:4 Compressor

Fig. 8. shows a carefully designed full adder schematic which exploits exclusive operation of pass transistor logic. Its driving capability is enhanced by adding buffers at the expense of area and timing delay. The only drawback is the glitch which is generated because of delayed timing at the input. It consists of 24 transistors and has 3 gate delays.

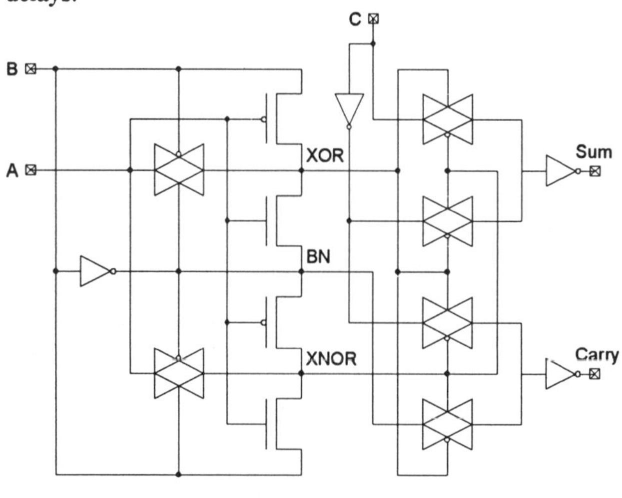

Fig. 8. Full adder circuit using pass transistor logic

Design of a 288-tap FIR filter

The 288-tap FIR filter is designed in 0.6μm triple metal CMOS technology. It is compatible with LG Semiconductor's 0.6 μm high density 3.3V library. It consists of nine 32-tap sub modules and each sub module consists of the PLL, memory bank, Booth encoder,partial product generator, partial product tree compressor and the final adder. Its characteristics are summarized in Table 1 and the chip layout is shown in Fig. 9.

Table 1. Characteristics of a 288-tap FIR filter

Number of Taps	288
Supplied Voltage	3.3V
Technology	0.6μm standard cell, triple metal
Chip Size	7X9 mm^2
Number of Transistors	385754
Operating Frequency	15MHZ external, 60MHz internal
Power Consumption	3.5W at 60MHz

Fig. 9. The 288-tap FIR Filter Chip Layout

Conclusion

A compact 10-bit, 288-tap FIR filter is designed by adopting structured architecture which employs optimized partial product tree compression method. The new architecture is based on the addition of equally weighted partial products which result from 288 multiplications of the filter coefficients and the inputs. Optimized parallel compression schemes such as 4:2 and 5:5:2 compressors are used to perform decomposed partial product addition. The completed 288-tap FIR filter occupies 7x9mm^2 of Silicon area which consists of 385754 transistors in 0.6μm triple-metal CMOS technology.

Acknowledgments

The authors would like to thank the process engineers of LG Semiconductors who fabricated the chip and Mr. Dong Myung Park and Dr. Sung Ho Kang who encouraged us to develop the chip.

References

[1]. J.W. Adams and A.N.Wilson, Jr., "A new approach to FIR digital filter with fewer multipliers and reduced sensitivity," IEEE Trans. Circuits and Systems, vol. CAS-30, pp.277-283, May, 1983.

[2] G. Privat and L. Paris, "Design of digital filters for video circuits", IEEE Journal of Solid State Circuits, vol. 21, pp441-445, June, 1986.

[3]. T. Miyazaki et. al., "Video rate FIR filter structures for a silicon compiler," *Systolic Array Processors* (Mc. Canny, Mc. Whirter, and Swartzlander), Prentice Hall, 1990.

[4]. E. Bidet, at.al., "FDF, a 512-tap FIR filter using a mixed temporal-frequential approach," 1995 IEEE Custom Integrated Circuits Conference, Santa Clarara, 1995.

[5]. E. E. Swarzlander, *Computer Arithmetic*, vols. 1-2, IEEE Computer Society Press, 1990.

[6]. L.S. Wallace, "A suggestion for fast multipliers," IEEE Trans. on Computers, vol. EC-13, pp. 14-17, Feb, 1964.

[7]. P. Song and G. De Michelli, "Circuitand architecture trade-offs for high speed multiplication," IEEE Journal of Soilid State Circuits, vol. 26, no. 9, Sept. 1991.

Chair: David Rich
Co-Chair: Terri Fiez

Mixed-mode integrated circuits are becoming increasingly common. Analog circuit design often dominates the performance limits that can be achieved. In this session, advances in the design of continuous time filters, current sources, operational amplifiers, and analog to digital converters will be reported. Both CMOS and BiCMOS technologies are represented in these papers.

The first three papers of this session deal with the design of high performance continuous time filters. Siniscalchi et al. present a programmable Gm/C based 4th order filter with a gain and bandwidth that can be programmed over a 10X range. The wide tuning range is achieved using bipolar current splitting pairs in the transconductance blocks. Chang et al. describe a 14th order Gm/C based bandpass filter that achieves a 75dB dynamic range. The V-I converter of the transconductor uses conventional source degeneration with a new MOSFET based Gm tuning technique to achieve the high linearity. Ruymbeke et al. present a BiCMOS realization of a log domain continuous time filter. The log domain technique has the potential to yield integrated filters with lower power consumption or higher speeds than linear methods allow. This paper reports one of the first monolithic realizations of a log domain filter. Despite the strong nonlinear behavior of the integrators, 1% distortion can be achieved. For a 30kHz cutoff frequency the 3rd order filter consumes 45μA.

Innovations in linear analog circuit design are presented in the next two papers. Although the current source described by You et al. has a compliance voltage slightly smaller than that of a single transistor current source, it has an output resistance 25 times larger. An active error amplifier is used to achieve this result. Burger et al. present a CMOS folded cascode operational amplifier with a very high gain bandwidth product and high DC gain in a 0.7μm CMOS process. A systematic design approach for transistor sizing which allows for good gain bandwidth product and phase margin is presented.

Finally, Goes et al. describe a 4 bit multiplying DAC that is a building block for a pipelined A/D converter. Gain and nonlinearity errors are reduced using an analog self-calibration technique. The analog part of the calibration circuit requires a pulse counting reference DAC, a high accuracy comparitor, and a calibrating DAC. 14 bit performance is reported for an MDAC with an 8 bit linearity before calibration.

High-Precision, Programmable 1-10MHz Bandwidth, 0-20dB Gain Communication Channel for Digital Video Applications

Patrick Siniscalchi, Adam Wyszynski*, and Davy Choi

Texas Instruments, Inc.
P.O. Box 655303, MS 8213, Dallas, TX 75265
e-mail: pps@msg.ti.com

* currently with Cirrus Logic, Inc.
5068 W. Plano Pkwy., Plano, TX 75093

Abstract

The analog signal processing portion of a digital video demodulation scheme is built in a 5V BiCMOS process. The channel is both cutoff frequency and gain programmable from 1→10 MHz and 0→20dB, respectively. The filter is realized as a 4th order Gm/C Butterworth lowpass and includes in-package trim for accurate control. A programmable gain amplifier is placed in front of the filter for better S/N performance (>40dB) and lower intermodulation distortion (>-52dB). The PSRR of this single-ended channel is better than 40dB. The channel dissipates 250mW.

1. Introduction

Digitally encoded video signals can be transmitted using either MCPC (multiple channel per carrier) mode, where all channel data are multiplexed and transmitted on a single carrier, or SCPC (single channel per carrier) mode, which allows each channel to be transmitted on its own subcarrier. Since both SCPC and MCPC modes utilize different carrier frequencies and different modulation techniques, the cutoff and gain requirements for the channel must vary, with the variation occurring as lower gain for higher cutoff and higher gain for lower cutoff. A practical digital video receiver is shown in Fig. 1. The system consists of an RF preamp, two mixers, a quadrature local oscillator (LO), two programmable, continuous-time filters, two ADCs, and a demodulator. After down-conversion, the quadrature data signals are filtered, amplified, and converted to digital form. The wide programmability of the data channel filter is mandatory due to the variety of data rates and modulation schemes. The accuracy (less than 5% for the cutoff and 1dB for the gain) is required for acceptable bit-error-rate (BER). A monolithic realization of a data filter channel, where both the cutoff and gain can be programmed over a 10X range, is presented in the following. It should be noted that wide programmability combined with the need for accurate control, plus the requirements for low channel noise and intermodulation distortion, produce a challenging set of design constraints for the data channel.

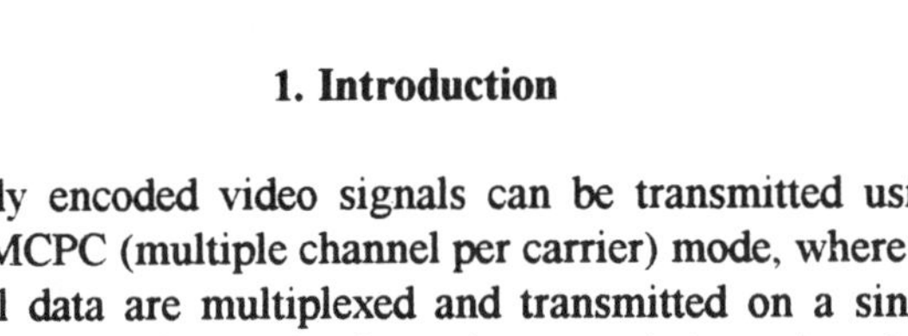

Fig. 1. Digital video demodulation system

2. Channel Overview

The block diagram of the data filter channel appears in Fig. 2. It consists of a single-to-differential converter (S/D), a programmable gain amplifier (PGA), widely-programmable continuous-time lowpass filter, a fixed gain amplifier (FGA) and a differential-to-single-ended converter (D/S). This configuration results in superior signal-to-noise (S/N) performance (over 12dB improvement) as compared to an arrangement where the placement of the PGA and FGA are reversed. The external ac coupling capacitors eliminate any dc voltage offsets between the two phases of the differential signal. This is essential for good distortion performance through the D/S circuit. The single-ended I/O complicated the design and results in some degradation in distortion and power-supply rejection ratio (PSRR) performance.

The chip is built in a 7 GHz , 0.8 μm BiCMOS process. The use of a high-performance process was necessitated by the requirement that the frequency response, out to 2X the cutoff frequency, should be nearly ideal. This dictated that all circuits in the signal path must be very wideband.

3. Programmable Filter

The Butterworth transfer function is chosen for its

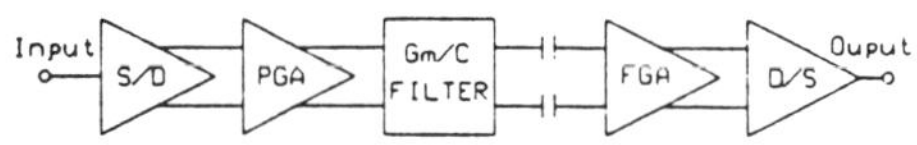

Fig. 2. Analog processing data filter channel.

0-7803-3177-6 $5.00 © 1996 IEEE

IEEE 1996 CUSTOM INTEGRATED CIRCUITS CONFERENCE

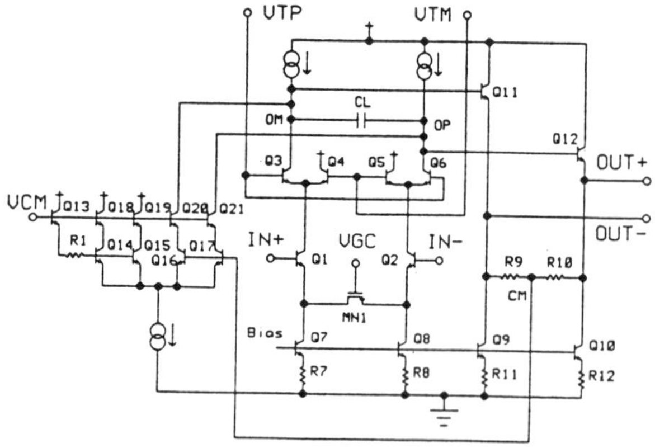

Fig. 3. Widely tunable transconductance stage.

maximally-flat passband magnitude response. Sufficient attenuation is achieved from a 4th order filter using g_m-C techniques and composed of two cascaded biquad sections. A main design issue in the filter is the wide programming range of 1 to 10 MHz. This requires a transconductor (G_M) whose gm can be varied > 18X to compensate for capacitor variations and to allow further minimization of BER by dynamically modifying the cutoff frequency as needed.

A realization of a widely-tunable transconductor is shown in Fig. 3. It consists of a bipolar differential pair, Q_1 and Q_2, degenerated by a triode region MOS device, MN_1. For $g_{mQ1} >> g_{ds1}$, the transconductance of the input stage is $g_m \approx g_{ds1}$. The wide tunability is achieved by the current splitting pairs Q_3-Q_4 and Q_5-Q_6, which were chosen over a Gilbert cell for their lower noise and higher output impedance. Dual-input G_M stages were obtained by duplicating Q_1,Q_2, and MN_1 and their associated current sources. The different g_m/C of each integrator was achieved with equal transconductance among all G_M stages and ratioed capacitors. The worst case phase error of any integrator is < 0.75° at its unity-gain frequency.

Since the g_m of the input stage is the g_{ds} of MN_1, it is heavily dependent upon the common-mode (c-m) level of the input. This fact drives the need for a very good common-mode feedback (CMF) loop that can keep variations in the c-m voltage to less than 10mV if the cutoff and gain accuracies are to be preserved. The CMF loop consists of transistors Q_{13}-Q_{21} in Fig. 3 in a cascoded, dual-differential pair arrangement. If the detected output c-m level, CM, rises above(or falls below) the reference c-m voltage, VCM, minus 1 V_{BE}, then Q_{16} and Q_{17} will draw more (or less) current from nodes OP and OM, thus stabilizing the output c-m voltage at OP and OM to be equal to VCM. The cascode devices, Q_{20} and Q_{21}, increase the output impedance of the CMF loop so that it will not degrade the transconductor's output impedance. The CMF loop contains a single gain stage and employs current-mode operation, so it exhibits excellent high frequency response (GBW > 50 MHz). The load capacitor, C_L, can be split into two capacitors to ground, which will provide increased phase margin in the feedback loop.

4. Filter Tuning/Programming

A. MOS g_{ds} Tuning

The triode region MOS device in the transconductor replaces a laser-trimmed resistor used in (1). Previously, in (2) , to compensate for process and temperature variations in g_{ds}, a PLL was used to control the MOS gate voltage, VGC. A significantly more compact circuit that performs the MOS g_{ds} tuning appears in Fig. 4. This circuit combines a replica of a G_M input stage with an 'error amplifier' to control VGC. A bandgap-derived voltage reference that is differential about the input c-m voltage is applied to the replica input stage so that MN_1 is biased as it is in the G_M stage. A temperature-independent reference current is sourced into the drain of the MN_1. This forces the current in Q_2, and consequently MP_2, MP_1 and Q_1, to $I_B - I_{REF2}$. However, to satisfy KCL at the MOSFET drain, VGC is adjusted until $I_{REF2}/2$ flows through MN_1. At that point the r_{ds} of the MOSFET is given by

$$r_{ds} = \frac{V_{REF}}{\left(I_{REF2}/2 \right)} . \qquad (1)$$

The PMOS current mirror acts as an error amplifier by forcing the same current through Q_1 and Q_2. VGC is then applied to all MOSFET 'resistors' in all G_M stages.

B. G_M Tuning

With the MOSFET tuned to a specific resistance, the G_M stage must be adjusted for capacitance variations. A quasi-open loop approach was chosen, where fuse trimming is performed during in-package testing to adjust the cutoff frequency to a precise value. The capacitor process variation is trimmed by adjusting a temperature-

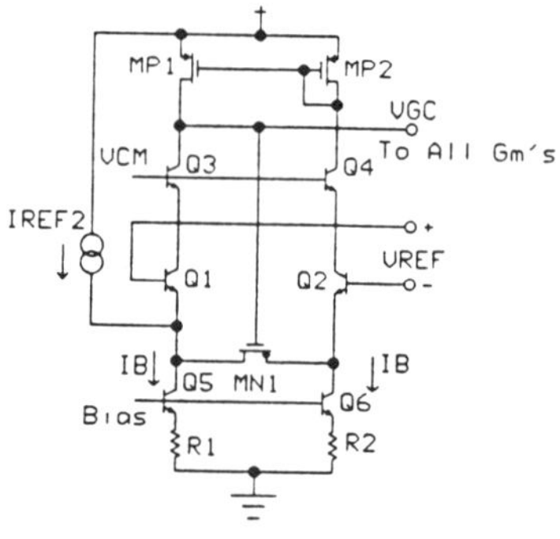

Fig. 4. MOSFET g_{ds} tuning circuit.

6.1.2

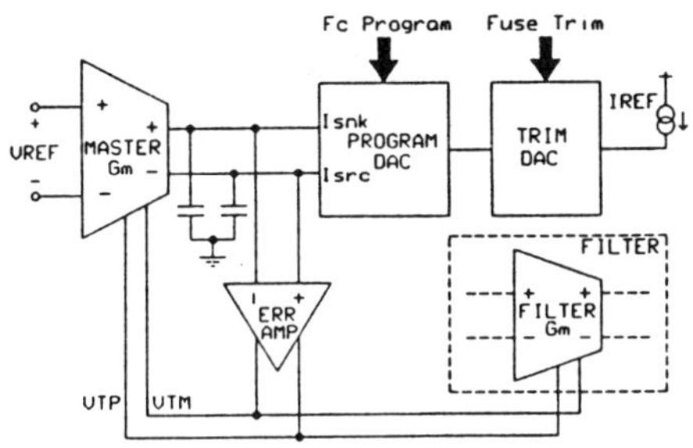

Fig. 5. GM stage tuning and programming circuit.

independent reference current that is applied to a tuning loop for the G_M stage. The control loop, shown in Fig. 5, corrects for any temperature variations. For cutoff frequency programming, a DAC is used to modify the trimmed reference current prior to the tuning loop.

The tuning loop consists of a differential reference current (modified by the Trim and Program DACs), a master G_M stage with a differential temperature-independent voltage reference as its input, an error amplifier, and compensation capacitors. The error amp adjusts the current splitting pairs in the master G_M stage until its output current is equal to the modified reference current. The gm of the master G_M cell is

$$gm = I^*_{REF} / V_{REF} \qquad (2)$$

where I^*_{REF} is the trimmed and programmed reference current. The differential control voltages, VTP and VTM, are also applied to all the G_M cells in the filter and since the master G_M cell is a replica of the filter G_M stages, their transconductances are all equal.

5. S/D, Programmable Gain Amp, & D/S

Consider the amplifier shown in Fig. 6, which is a BiCMOS, folded-cascode version of (3). It combines the S/D with the PGA, resulting in lower power, smaller area, reduced distortion and noise levels, and a wider frequency response. Its linearity is enhanced by the two operational transconductance amplifiers (OTAs) that increase the effective g_m of Q_1 and Q_2 by the voltage gain of the OTA, so that R_1 appears to be driven from a nearly ideal voltage source. Current-splitting pairs were again chosen for gain control. The PGA gain can be approximated as

$$Av \approx \frac{R_2}{R_1}\frac{1}{1+\exp\left(\frac{V_2-V_1}{V_T}\right)} \approx \frac{R_2}{R_1}\exp\left(\frac{V_1-V_2}{V_T}\right)$$

$$= \frac{R_2}{R_1}\exp\left(\frac{I_{prg}R_4}{V_T}\right) = \frac{R_2}{R_1}\exp\left(\frac{mR_4}{R_3}\right) \qquad (3)$$

where $\exp((V_2-V_1)/V_T) >> 1$ is assumed, R_3 and R_4 are on-chip resistors, $R_2 = R_9+R_{10}$, and $I_{prg} = mV_T/R_3$. I_{prg} is a PTAT current controlled by a linear DAC that provides for temperature independent gain control in linear dBs.

In order to minimize response inaccuracies in the filter transfer function, the PGA is designed for a bandwidth of over 200 MHz. The folded cascode design provides for high frequency performance, as well as high output impedance ($r_o >> R_2$) which reduces gain errors. Finally, the dc level of the output is controlled by a CMF loop which is necessary to minimize any dc input variation to the filter that would result in a channel gain error.

As with the S/D and PGA, the FGA and D/S functions were combined into one block. The FGA is similar in design to the PGA. However, both the current splitting pairs and the CMF loop are absent. The output buffer is a simple emitter follower stage.

6. Measured Results

A partial die photo showing both I and Q data filter channels and their associated tuning circuitry appears in Fig. 7. It can be seen that the tuning and programming circuitry occupies approximately the same area as the filter. This overhead in area was justified by the tight accuracy and wide programmability requirements.

The performance results are summarized in Table 1. Fig. 8 shows the amplitude response of the channel for the extremes, 1MHz cutoff with 20dB of gain and 10MHz cutoff with 0dB of gain. Fig. 9 shows the difference between the actual amplitude response of the channel and an ideal Butterworth filter at 10MHz. Deviations from ideality are believed to be due to inexact cancellation of parasitic zeros and low dc gain in the G_M stage at the higher and lower cutoff frequencies, respectively. The data channel THD for the 10MHz-0dB case is shown in Fig. 10 for an input at 2MHz and $300mV_{p-p}$. The even order harmonics are created during the differential to single-ended conversion process.

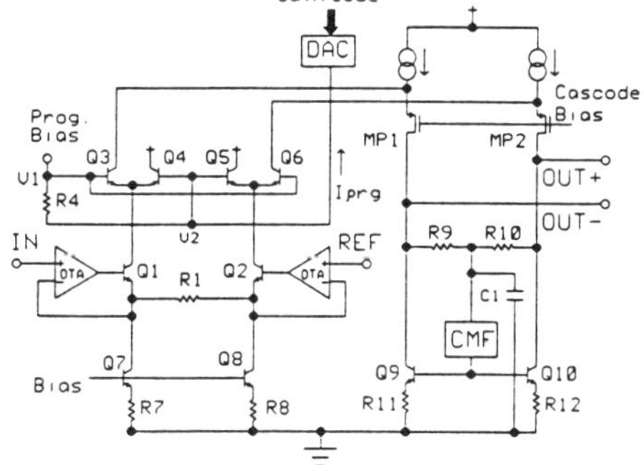

Fig. 6. Highly-linear programmable gain amplifier.

Fig. 7. Die photo showing I/Q channels and tuning circuitry.

Table 1
Performance Results

Cutoff Frequency	1-10 MHz
Channel Gain	0-20 dB
Initial Cutoff Accuracy	< 2 %
Cutoff Variation (0-70 ° C)	< 3%
Amplitude Response Accuracy (dc to f_c)	< 0.25 dB
I-Q Cutoff Matching	< 2 %
I-Q Gain Matching	< 1 dB
THD (Vout = 300mV)	< 0.4 %
IMD (fc=1MHz, A_v=20dB, fin = 9 & 10MHz @ 25mV ea.)	< 0.25 %
Output Noise (integrated to 100MHz)	< 300 μVrms
PSRR (to 1 MHz)	> 40 dB

7. Conclusions

A dual, programmable cutoff frequency and gain data channel has been presented that offers precise control over a 10X range for both cutoff frequency and gain. Circuits were shown that allow the channel to maintain an approximately ideal 4th order Butterworth response while accurately controlling the bandwidth from 1 to 10 MHz. Though the I/O of the channel is single-ended, the levels of intermodulation distortion, THD and PSRR were shown to be moderately low. An overview of the digital video application, as well as the factors that drive the need for such an accurate, programmable channel, were discussed.

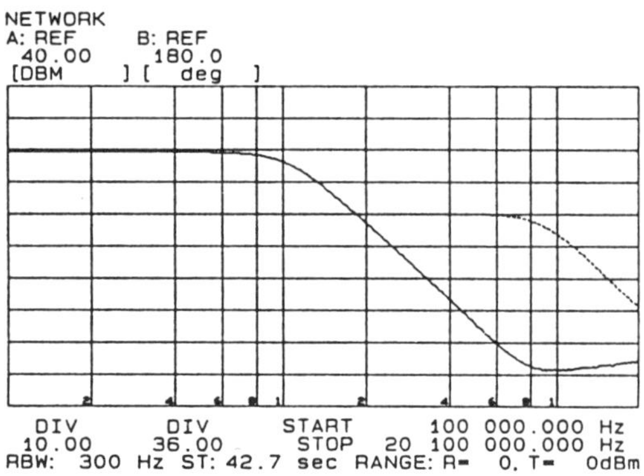

Fig. 8. Amplitude response of the data filter channel.

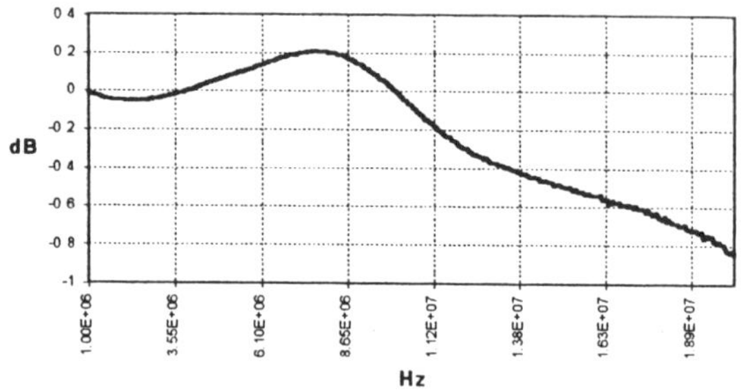

Fig. 9. Channel response accuracy to ideal 4th order Butterworth.

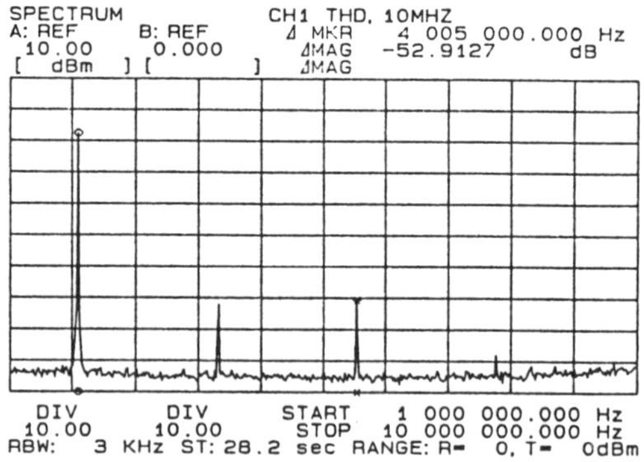

Fig. 10. Data filter channel THD for fc = 10MHz.

Acknowledgments

The authors wish to gratefully acknowledge M. Aragon for the serial interface design, V. Gopinathan and G. Mayfield for design support, D. Hart for chip testing, O. Valdez and G. Wakeman for layout design, and B. Krenik and M. Chiang for their encouragement and management support.

References

(1) R. Richetta, et al., "A 16 Mbytes/s PRML Read/Write Data Channel," IEEE ISSCC Digest of Technical Papers, pp.78-79, Feb. 15, 1995.

(2) V. Gopinathan, Y. P. Tsividis, K.-S. Tan, and R. K. Hester, "Design considerations for high-frequency continuous-time filters and implementation of an antialiasing filter for digital video," IEEE Journal of Solid State Circuits, pp. 1368-1378, Dec. 1990.

(3) A. Wyszynski, "High-Frequency Linear Tunable Single-Ended Voltage to Differential Current Converter," in Proc. IEEE 38th Midwest Symp. Circ. Syst., Aug. 1995.

A Highly Linear CMOS Gm–C Bandpass Filter for Video Applications

Z.Y. Chang, D. Haspeslagh, J. Boxho and D. Macq

Alcatel Bell

Francis Wellesplein 1. B–2018 Antwerp, Belgium

FAX: 32–3–2409947, TEL: 32–3–2407686

Abstract

A 14th order CMOS transconductance–C (Gm–C) bandpass filter for video applications is described. By using highly linear Gm–C integrators, the filter achieves 75 dB dynamic range over 700kHz noise bandwidth. The measured IM3 @ 600kHz is –61dB for a 4Vpp input signal. On–chip automatic frequency tuning provides more than 300% center frequency range of the filter with 1% frequency accuracy. The 0.7 um CMOS filter measures 4.8 mm^2 and consumes 70 mW from a single 5 V power supply.

I. Introduction

Integrated continuous–time transconductance–C (Gm–C) filters have been widely adopted for high frequency applications such as digital video, IF filters, etc [1–3]. Gm–C filters offers many advantages over other continuous–time filters such active–RC and MOSFET–C in terms of low–power and high frequency capability. However, due to the openloop operation nature Gm–C filters generally perform poorly as far as linearity is concerned. Many techniques have been proposed in the literature to improve the linearity performance [4–7]. Most of them require accurate matching of MOSFET devices for non–linearity cancellation. Due to process variations and component mismatching, reported practical results are limited to the 40–50 dB linearity range. For FDM–based ADSL (Asymmetrical digital subscriber loops) applications [8], bandpass filters with a linearity in excess of 60dB are required. To fulfill such high linearity requirements, a new approach to linear Gm–C integrators is developed. By using the new integrators, a 14th order Gm–C bandpass filter is designed in a 0.7um CMOS technology, achieving 75 dB dynamic range over 700kHz bandwidth. Powered with a single 5V supply, the measured IM3 @300kHz and @600kHz is only 0.025% and 0.1%, respectively for a 4Vpp input signal.

II. Linear Gm–C integrator

Since all continuous–time filters can be synthesized using integrators, the filter linearity performance is determined by the integrator linearity. Fig.1 shows the new linear integrator based on source degeneration with linear resistors. The input voltage–to–current V–I conversion is realized by means of a super Gm input stage in combination with passive resistors. Due to the input local negative feedback and the use of linear resistors a very linear V–I conversion is obtained without any matching requirements. However, as linear resistors can not be tuned in practice, a way has to find to vary the output integrating current so that the integrator time constant can be tuned. The technique presented uses three MOS transistors M1–M2–M3 operating in their linear regions. The large signal transconductance Gm of the integrator can be tuned by the control voltages V_{C1} and V_{C2} at the gates of the three transistors. By choosing W/L(M2,M3) = 2W/L(M1), it can be calculated that the Gm is linearly proportional to the differential control voltage as given by:

$$G_m = \frac{1}{R}\left(\frac{1}{2} + \frac{V_{cd}}{V_{cm}-V_T-V_A}\right) \qquad (1)$$

where $V_{cm} = (V_{c1}+V_{c2})/2$, $V_{cd} = V_{c1}-V_{c2}$ represents the common–mode and differential control voltage, respectively, V_T is the threshold voltage and V_A is the dc bias voltage at the node A. For a practical case where V_{cm}=3V, V_T=1V and V_A=1V, the Gm value can be tuned from 28% to 72% of its maximal value (i.e. 1/R) for a control voltage range from –0.5V to +0.5V. It should be noted that in contrary to existing

6.2.1

0-7803-3177-6 $5.00 © 1996 IEEE

tuning techniques, the above approach does not change the integrator biasing condition and therefore a stable performance is obtained over the whole tuning range.

Since the input V–I conversion is linear the integrator linearity performance is determined by the three tuning transistors. It can be shown that as long as the voltage across M1 is differential, no distortion will be generated by the tuning transistors [6]. The requirement of differential voltage swing across M1 is realized by designing the tuning transistors with a large W/L ratio and the use of regulated cascode output stages to limit the voltage swing at the node A and B. Simulations show that by choosing W/L=10u/5u, the third harmonic distortion HD3 for a 4 V_{pp} input differential signal at 500kHz is –65dB. It is interesting to note that owing to the fact that the tuning transistors experience only a small voltage swing, even better linearity performance can be obtained as compared to MOSFET–C filters where the tuning MOSFET experiences the full signal swing.

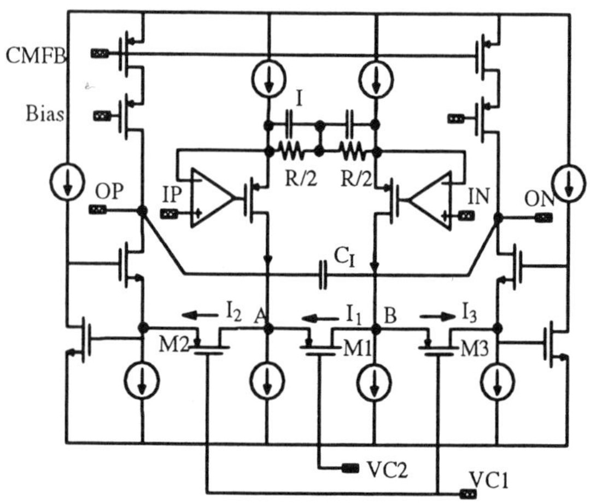

Fig.1. Linear Gm–C integrator

III. Bandpass filter design

Using the above Gm–C integrator design technique, a 14th order Chebyshev bandpass filter with a transition band of 138kHz to 838kHz and an in–band ripple of 1.5 dB is designed. The target filter dynamic range is 72dB, with < 0.1% total intermodulation distortion (IM3). The frequency accuracy must be within ±1%. On–chip frequency tuning is thus a must.

To obtain a passband frequency response with lower sensitivities to individual component values, the Gm–C bandpass filter topology is derived by applying the well–known lowpass to bandpass transformation (2) to a doubly–terminated passive RLC lowpass prototype.

$$s_L = \frac{(s^2 + \omega_o^2)}{s\ \omega_c} \qquad (2)$$

In expression (2), s_L is the normalized low pass Laplace variable, ω_o is the center frequency and ω_c is the band–width of the band pass filter to be designed. Since expression (2) transfers an integrator transfer function $1/s_LC$ into the transfer function of a resonator $s\omega_cC/(s^2+\omega_o^2)$, the bandpass filter topology consists of coupled resonators. The filter center frequency is given by the resonators frequency and the filter bandwidth is determined by the amount of coupling between resonators.

A Gm–C realisation of coupled resonators is shown in Fig.2. Multiple input stages Gm–C integrators are used which allow easy dynamic range scaling and output stages design. The resonator is composed of two Gm–C integrators (e.g. Ing1, Ing2) in a negative feedback loop. The filter quality factor is realised by the coupling between the resonators. The center frequency ω_o and bandwidth ω_c of the filter are given by:

$$\omega_o^2 = \frac{G_m^2}{C_1 C_2} \qquad (3.a)$$

$$\omega_c/C = \frac{G_{m1}}{C_2} \qquad (3.b)$$

where G_m and G_{m1} represent the transconductance of the resonator and the couplings branch. As expected the filter characteristic is completely determined by the G_m/C ratio's. The choice of Gm values and integrating capacitors are mainly determined by the requirements concerning the input/output dynamic range scaling, noise, Gm/C matching and power consumption. Since in practice capacitor matching is better than Gm matching, only four different Gm values (i.e. 0.015mS, 0.02mS, 0.0225mS and 0.3mS) are used in the complete 14th order bandpass filter which contains 14 multiple inputs Gm–C integrators.

Fully differential topology is used for the bandpass filter to obtain the well–known advantages. Thanks to the use of input degeneration resistors in the integrator (see Fig.1), a very simple common–mode feedback is possible as illustrated in Fig.2. The output common–mode voltage of the Gm–C integrator Ing3 is sensed at the middel of the input degenration resistor (see Fig.1) of the integrator Ing4 and is set to the common–mode voltage Ref through the common–mode feedback control loop.

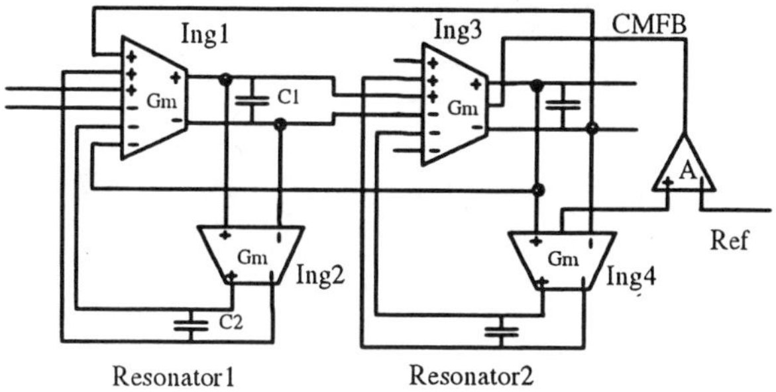

Fig.2. Gm–C realisation of coupled resonators

IV. Frequency tuning

Since the bandpass filter frequency characteristic is determined by the G_m/C ratio's, any process variation and temperature dependencies can be compensated for by tuning the G_m value of the integrators. As explained above, the G_m of the new integrator can be tuned over a large range by controlling the gate voltages of the three tuning transistors M1, M2, M3. The required control voltages V_{c1} and V_{c2} are generated in the tuning circuit shown in Fig.4 which is based on the charge balancing principle as proposed in [4]. A constant reference voltage V_{REF} is applied at the input of a reference Gm integrator and a switched–capacitor negative resistor network. In the steady state condition, the integrator output current must be equal to the negative resistor current, giving the following tuning condition:

$$\frac{G_m}{C_R} = f_{Clk} \tag{4}$$

The reference clock frequency f_{Clk} for the SC circuit is derived from the master system clock and is chosen to be 1.536Mhz to eliminate possible in–band (i.e. 138kHz to 838kHz) clock feedthrough.

The frequency accuracy of the bandpass filter is determined by the matching of the G_m/C ratio's of the tuning circuit and the main filters and the accuracy of the tuning circuit itself. To obtain good G_m/C matching the reference Gm value is chosen to be the middel Gm value of the bandpass filter. In addition, relative large device dimensions are used and well–known layout techniques such as common centroid input are employed. The accuracy of the tuning circuit itself is mainly limited by the offset within the tuning loop. Taking into account the offset effect the tuning condition (4) becomes:

$$\frac{G_m}{C_R} = f_{Clk}\left[\frac{1+\frac{V_{osA}}{V_{REF}}}{1-\frac{V_{osG}}{V_{REF}}}\right] = f_{Clk}\left(\frac{1+\varepsilon_A}{1-\varepsilon_G}\right) \tag{5}$$

where V_{osA} and V_{osG} are the offset voltages of the amplifier A and integrator G_m, respectively. To minimize this offset error, a large reference voltage V_{REF} is desirable. For a practical case where $V_{osA} = V_{osG} = 10mV$, an error of 1% is possible by choosing $V_{REF} = 2V$.

The stability of the tuning loop is concerned with the proper choice of the two time constants: C_I/G_m and R_2C_2. It can be shown that by designing $C_I/G_m >> R_2C_2$ the loop stability can be guaranteed. A large integrating capacitor C_I is also desirable to obtain a relaxed output dynamic range requirement for the amplifier A. It is interesting to note that no external capacitors are required for the tuning circuit.

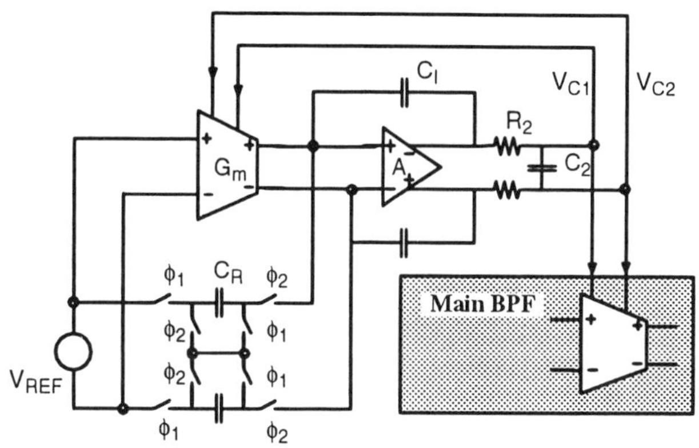

Fig.4. Principle of automatic frequency control

Since the quality factor Q of the bandpass filters is relatively low (i.e. $Q=\omega_o/BW\approx0.5$), no Q–tuning is applied. To obtain a well controlled filter quality factor, the integrator phase lead error is minimized by using the regulated cascode output stage (see Fig.1) while the phase lag error is compensated for by adding small capacitors in parallel with the input degeneration resistors (i.e. input compensation) and small resistors in series with integrating capacitors (i.e. output compensation).

V. Experimental results

The complete bandpass filter is fabricated together with an analog front–end chip for ADSL applications [8] in a 0.7 μm CMOS technology. Fig.5 shows the microphotograph of the filter. The filter active area measures $4.8mm^2$ and the associated frequency tuning circuit (not shown) is 1mm^2. The filter is powered by a single 5V supply and consumes 70mW. The measured frequency characteristics is shown in Fig.6. The accuracy of the center frequency is within ± 1% over the ambient temperature range of 0°C to 70°C and power supply range 4.7V to 5.3V. The high frequency band edge peaking is less than 2dB in the worst case. The frequency tuning range is measured to be slightly over 300%. The linearity performance is given in Fig.7 where the intermodulation distortion IM3 v.s input signal is shown. As can be seen for input signal up to 4Vpp, the IM3 is lower than –72dB @300kHz and –61dB @ 600kHz. The total equivalent input noise over 700kHz bandwidth is 0.25mV$_{rms}$. The filter achieves thus a dynamic range of 75dB with 0.1% IM3 linearity.

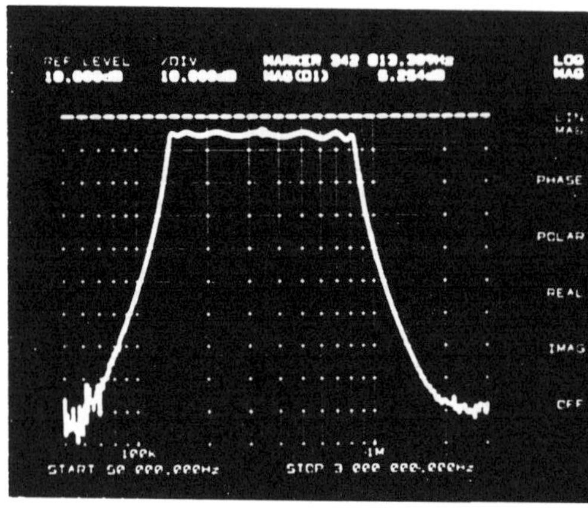

Fig.6. Measured frequency response

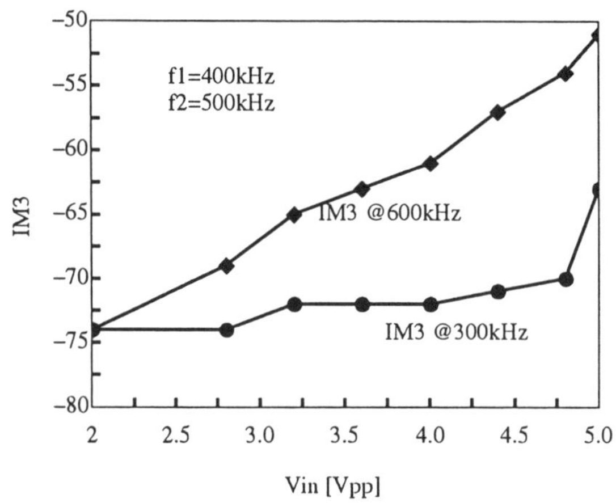

Fig.7. Intermodulation distortion IM3 v.s input amplitude

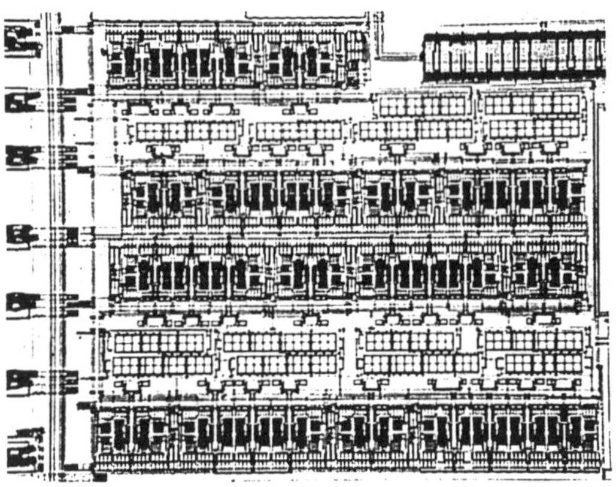

Fig.5. Microphotograph of the bandpass filter

VI. Conclusions

This paper describes a highly linear CMOS Gm–C bandpass filter used for video–frequency applications. By combining the conventional source degeneration with the new Gm tuning technique, a very highly linear Gm–C integrator is developed. Based on the new linear Gm–C integrators, a 14the order high–frequency bandpass filter is designed achieving 75 dB dynamic range over 700kHz noise bandwidth. The measured IM3 @ 600kHz is only –61dB for a 4Vpp input signal. On–chip automatic frequency tuning provides more than 300% center frequency range of the filter with 1% frequency accuracy under all operating conditions. The 0.7 um CMOS filter measures 4.8 mm^2 and consumes 70 mW from a single 5 V power supply.

References

[1]. V. Gopinathan, Y. Tsividis, K.S.Tan and R. Hester, "Design considerations for high–frequency continuous–time filters and implementation of an anti–aliasing filter for digital video" *IEEE J. Solid–State Circuits*. Vol.25 pp. 1368–1378, Dec 1990

[2]. J. M. Khoury, "Design of a 15–Mhz CMOS continuous–time filter with on–chip tuning" *IEEE J. Solid–State Circuits*. Vol.26, pp. 1988–1997, Dec. 1991

[3]. J. Silva–Martinez, M. Steyaert and W. Sansen, "A 10.7Mhz 68–dB SNR CMOS continuous–time filter with on–chip automatic tuning" *IEEE J. Solid–State Circuits*. Vol. 27, pp. 1843–1853, Dec. 1992

[4]. Y. P. Tsividis and J.O. Voorman, *Integrated continuous–time filters*. New York: *IEEE Press, 1993*

[5]. Y.P. Tsividis, M. Banu and J. Khoury, "Continuous–time MOSFET–C filters in VLSI" *IEEE JSSC*. Vol.21,pp. 15–30, Feb. 1986

[6]. Y. P. Tsividis, Z. Czarnul and S.C. Fang, "MOS transconductors and integrators with high linearity" *Electron. Lett., Vol.22, pp. 245–246, Feb.27, 1986*

[7]. F. Krummenacher and N. Johl, "A 4–MHz CMOS continuous–time filter with on–chip automatic tuning," *IEEE J. Solid–State Circuits*. Vol.23, pp.750–758, June 1988

[8]. Z. Y. Chang, D.Macq, D. Haspeslagh, P. Spruyt and B. Goffart, "A CMOS analog front–end circuit for an FDM–based ADSL system," Proceedings ISSCC'95, San Francisco. pp.330–331

A BiCMOS Programmable Continuous-Time Filter using Voltage-Companding

Gilles van Ruymbeke, Christian Enz, François Krummenacher, Michel Declercq
Swiss Federal Institute of Technology (EPFL), Electronics Laboratory, ELB-Ecublens, CH-1015 Lausanne, Switzerland
Phone: +41 21 693 46 09, Fax: +41 21 693 36 40, E-mail: ruymbeke@leghp1.epfl.ch

Abstract- **This paper presents a BiCMOS realization of a programmable continuous-time filter demonstrating the feasibility of the voltage companding technique. The circuit implements four low-pass filters (order 2,3,4,5) and one band-pass filter (order 2) by interconnecting a set of identical non-linear cells. The filter is obtained by a proper component substitution transforming a g_m-C structure into its log-domain equivalent. The g_m-C structures are derived from LC ladder prototypes which are synthesized using the image-parameter method. The programmable filters can additionally be frequency tuned from 10 kHz to 100 kHz with a very low THD despite their class AB operation (1% for a modulation index of 3 corresponding to an input current of 30 μA).**

I. Introduction

Programmable and tunable analog integrated filters are useful components of many systems. This paper describes an original synthesis method based on the image-parameter and implemented using voltage-companding integrators.

II. Principle of the voltage-companding filter

The general structure of a linear filter based on a voltage-companding or log-filter is presented in Fig. 1 [1][5]. The latter is embedded between the input and output of a simple current mirror. The input consists of a linear V-to-I converter having a transconductance g_{m0} and a diode connected transistor which generates the input compressed voltage V^*_{in}. The log-filter delivers a compressed voltage V^*_{out} which is then expanded into a current by the output BJT. The current is then transformed back to a voltage through the linear conductance of value g_{m0}. The compressed voltages at the input and output of the log-filter are related to the input and output voltages of the linear system by the following relation:

$$V^*_k = U_T \ln(g_0 V_k / I_S) \tag{1}$$

which also holds for the internal voltages V^*_k of the log-filter and the internal voltages V_k of the equivalent linear filter.

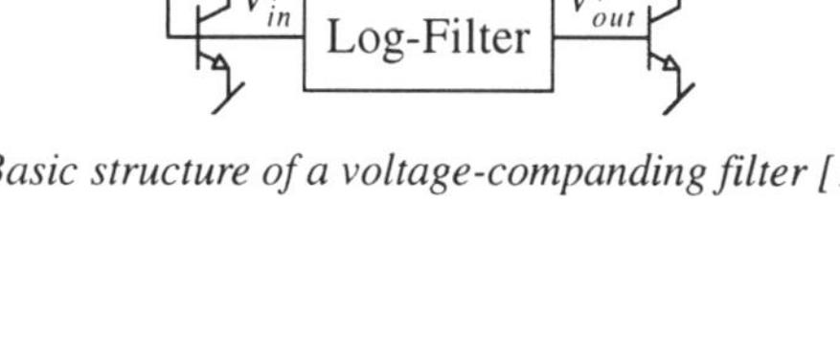

Fig. 1: *Basic structure of a voltage-companding filter [1][5].*

III. Linear- to log-domain mapping of g_m-C filters

This section illustrates how the voltage-companding filter shown in Fig. 1 can be derived from an existing g_m-C filter using a simple component substitution. Fig. 2 presents a 2nd-order low-pass LC ladder prototype filter and the corresponding g_m-C structure. Note that the LC ladder of Fig. 2 a) corresponds to the lowest order low-pass filter that has been implemented.

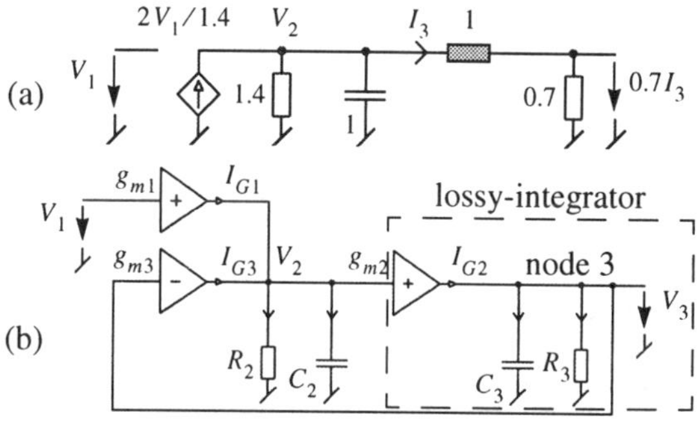

Fig. 2: *LC ladder prototype (a) and linear g_m-C filter (b).*

Since the log-filter should be derived from the existing g_m-C structure by simple component substitution, the circuit topology of both realizations should be identical. The transformation should additionally preserve all the capacitors (assumed to be linear) as well as the linear relation between the same input and output voltages. These three constraints are used to define the linear- to log-domain mapping by writing the sum of the currents at node 3 of the lossy integrator of Fig. 2 b):

$$C_3 (dV_3/dt) = g_{m2} V_2 - V_3/R_3 \tag{2}$$

As shown hereafter, the transformation consists simply of replacing the linear transconductors and resistors in the g_m-C filter by respectively an exponential transconductor and a current source, while the capacitors and the input and output voltages remain unchanged. Applying this transformation to the g_m-C filter of Fig. 2 b) results in the voltage-companding filter presented in Fig. 5, where all the voltages and currents are identified by a star. The current in capacitor C_3 of the log-filter shown in Fig. 3 is then given by (3) where I^*_{G2} is the current sourced by the nonlinear transcondutor and I^*_{R3} is the current sinked by the current source.

$$C_3 (dV^*_3/dt) = I^*_{G2} - I^*_{R3} \tag{3}$$

Differentiating the definition of the compressed voltage given by (1), leads to (4) where V_k and V^*_k are the voltages at node k of the linear g_m-C and the log-filter respectively.

0-7803-3177-6 $5.00 © 1996 IEEE

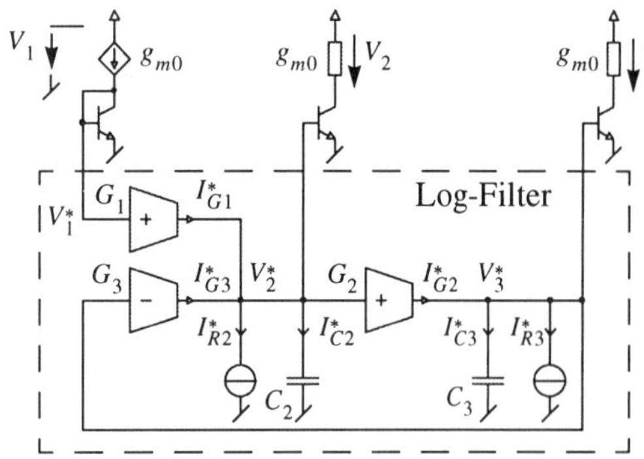

Fig. 3: *Log-domain filter corresponding to the filter of Fig. 2.*

$$\frac{dV_k^*}{dt} = \frac{U_T}{V_k} \cdot \frac{dV_k}{dt} \tag{4}$$

Substituting (4) into (3) leads to:

$$C_3\frac{dV_3}{dt} = \frac{V_3 I^*_{G2}}{U_T} - \frac{V_3 I^*_{R3}}{U_T} \tag{5}$$

The components related to currents I^*_{G2} and I^*_{R3} can be characterized by identifying each term of (5) and (2):

$$I^*_{G2} = \frac{g_{m2} U_T}{V_3} \cdot V_2 \qquad \text{and:} \qquad I^*_{R3} = \frac{U_T}{R_3} \tag{6}$$

Eqn. 6 states that I^*_{R3} indeed corresponds to a constant current source while I^*_{G2} corresponds to the current delivered by a nonlinear transconductor having a transconductance controlled by its output voltage V_3. This transconductor can easily be realized by using a simple bipolar device as shown in Fig. 4 (the bias is not shown for simplicity). Neglecting the base current, the output current is given by:

$$I^*_{G2} = I_S \exp\left[\frac{V_2^* + V_S^* - V_3^*}{U_T}\right] = g_m U_T \frac{\exp\,[V_2^*/U_T]}{\exp\,[V_3^*/U_T]} \tag{7}$$

which obviously results in the definition of I^*_{G2} given by (6). The transconductance g_m appearing in (7) is defined by (8) and sets the integrator time constant C_3/g_m which should be made equal to the corresponding integrator in the g_m-C structure.

$$g_m = \frac{I_S}{U_T} \exp\,[V_S^*/U_T] \tag{8}$$

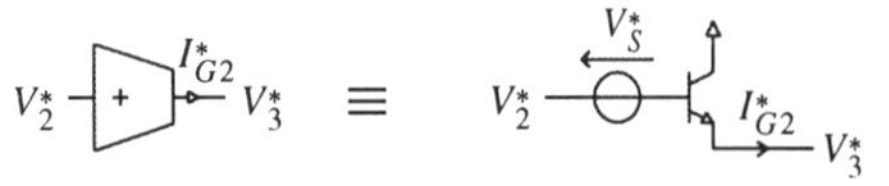

Fig. 4: *Exponential transconductor.*

It should be note that the current source I^*_{R3}, simulating the loss, can be used to tune the quality factor Q of the filter. It can also be used in an oscillator to compensate for the loss by simply inverting the current which would correspond to a negative resistance.

IV. Programmable filters using image-parameter method

The image-parameter synthesis method was used many years ago to design filters graphically. Nevertheless, this method is exact and allows to build a filter by cascading elementary sections such that the image impedance at each port are matched two by two following the domino rule. The resulting image attenuation is obtained by summing the image attenuation of each section as illustrated in Fig. 15. Although *m-derived* type sections can be implemented thanks to the voltage-companding technique [1], the circuit presented in this paper only uses *k-constant* type sections. The actual attenuation is the result of the image attenuation and an additive bandpass ripple introduced by the mismatch between the image impedance at the ends of the image chain and the resistive termination [1]. A canonical g_m-C implementation based only on *k-constant* type sections and resistive terminations is presented in Fig. 5[7]. All the low-pass filters realized within this work result from an iteration of the central canonical section. A bandpass filter is obtained by coupling the central sections used for the 4th- and 5th-order low-pass filters to realize two antiresonant circuits.

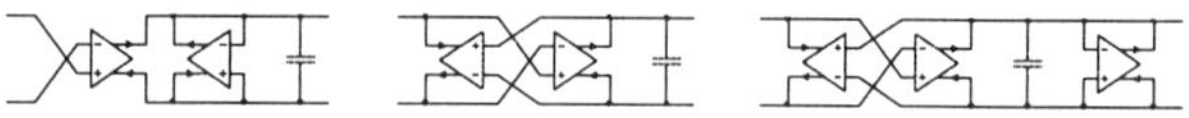

Fig. 5: *A canonic programmable filter structure (k-constant)*

V. Realization and measurement of the programmable filter

The Operational Transconductance Exponentiator (OTE) is shown in Fig. 7. The lower part of the schematic corresponds to the circuit proposed by Seevinck [4]. The main disadvantage of the original circuit lies in its sensitivity to base currents. The upper part of Fig. 7 has been added to compensate for the base currents of the middle range BJTs, while the base currents of the grounded NPNs are compensated by the circuit of Fig. 8. In order to maintain all the currents positive, an input signal conditioner that operates basically as a current rectifier is added in front of the filter. It is a variant of the circuit proposed in [6].

The five filter configurations shown in Fig. 6 have been integrated in a 2 µm BiCMOS process (c.f. Fig. 9). The measured transfer functions are shown in Fig. 11. The middle traces of Fig. 10 correspond to the measurement of the compressed voltages obtained for a sinusoidal input (upper trace). Despite the strong distorted aspect of these compressed voltages, the output signal (lower trace) remains sinusoidal, which demonstrates the global linear behavior. Fig. 12 shows that the cut-off frequency can be tuned linearly over at least one decade of frequency. The measured THD is plotted in Fig. 13 versus the modulation index m defined as the ratio of the differential input current to the bias current. The THD remains smaller than 1% for m lower than 3 (corresponding to a 30 µA differential current) even though the filter operates in large-signal and in class AB. Fig. 14 represents the measured output power of the fundamental and the in-band 3rd-order intermodulation product referred to the input power of a duotone input signal of same amplitude. The 3rd-order intercept point referred to the input of the filter corresponds to a modulation index of about 1.8. This demonstrates that the log-domain technique can achieve intermodulation performance similar to classical linear filters. The quiescent current consumption of the 3rd-order low-pass filter with a cut-off frequency fixed at 30 kHz is about 45 μA at 4 V supply.

VI. Conclusion

The realization and measurement of a programmable BiCMOS voltage-companding filter have been presented. Thanks to the image-parameter method synthesis, the integrated filter can be programmed by combining 5 elementary sections in order to achieve four different low-pass characteristics (of order 2 to 5) and one 2nd-order bandpass function. All the measurement have been performed on the 3rd-order low-pass filter. The THD is smaller than 1% for a modulation index smaller than 3 and the 3rd-order intercept point referred to the input corresponds to a modulation index of 1.8. These results tends to demonstrate that despite the strong nonlinear behavior of the integrators, satisfactory distortion and small intermodulation products can be achieved.

Acknowledgments

F. Yang is greatly acknowledged.

VII. References

[1] G. van Ruymbeke, “Filtres continus intégrés programmables,” PhD Dissertation, EPFL, N° 1397, 1995.

[2] F.Krummenacher and G. van Ruymbeke, “Integrated Selectivity for Narrow-Band FM IF System”, *IEEE JSSC*, vol. SC-25, no. 3, pp. 757-760, June 1990.

[3] C. C. Enz,” Low-Power Log-Domain Continuous-Time Filters: an Introduction”, ESSCIRC’95 WorkShop on LP and LV.

[4] E. Seevinck, “Companding current-mode integrator: A new circuit principle for continuous-time monolithic filters,” Electron. Lett., vol. 26, pp. 2046-2047, Nov. 1990.

[5] D. R. Frey, “Log domain filtering: An approach to current mode filtering,” in IEE Proc., pt. G, Dec. 1993, pp. 406-416.

[6] D. R. Frey, “Current mode class AB second order filter,” Electron. Lett.,1994, Vol. 30, No. 3.

[7] G. van Ruymbeke and F. Krummenacher, “A Programmable Continuous-Time Filter”, ISCAS’93, pp. 1263-1266.

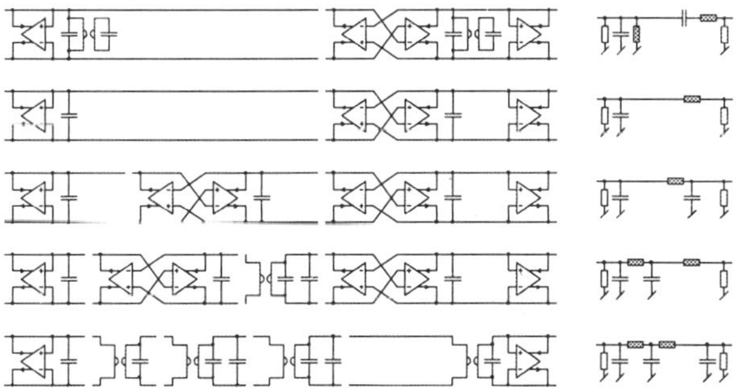

Fig. 6: *The band-pass and the four low-pass filters.*

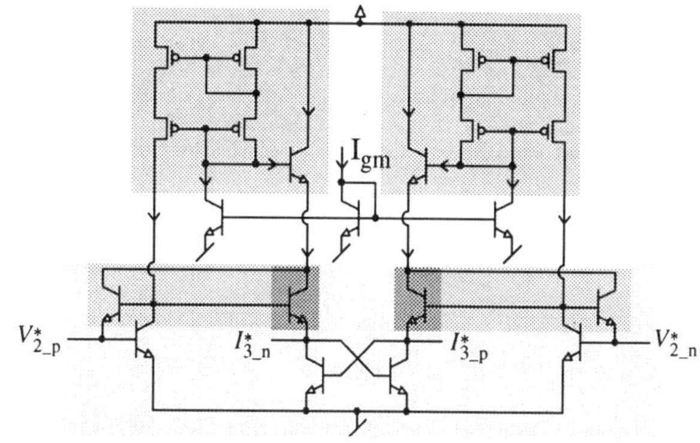

Fig. 7: *OTE schematic with base current compensation.*

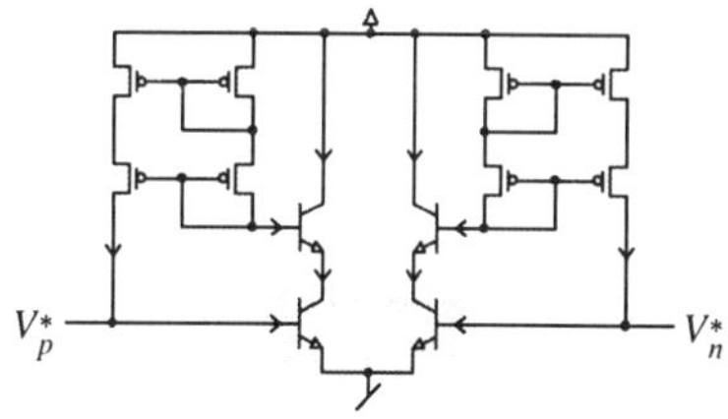

Fig. 8: *Base-current compensation of grounded NPN transistors.*

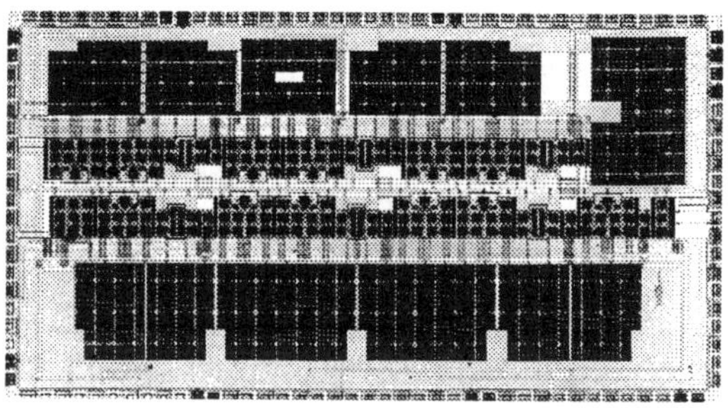

Fig. 9: *Microphotograph of the log domain filter circuit*

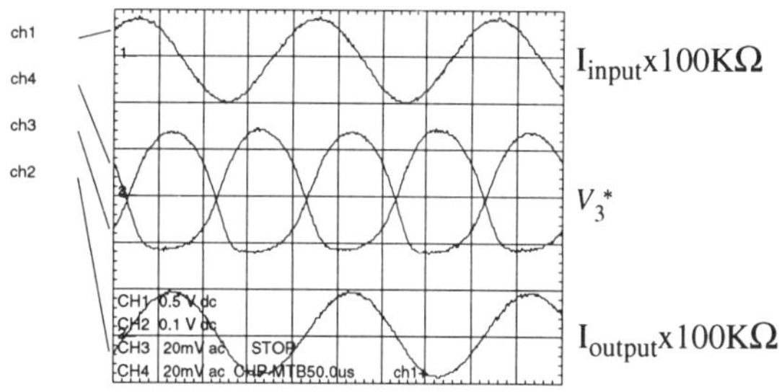

Fig. 10: *Measured voltages: input voltage, internal compressed voltages and output voltage.*

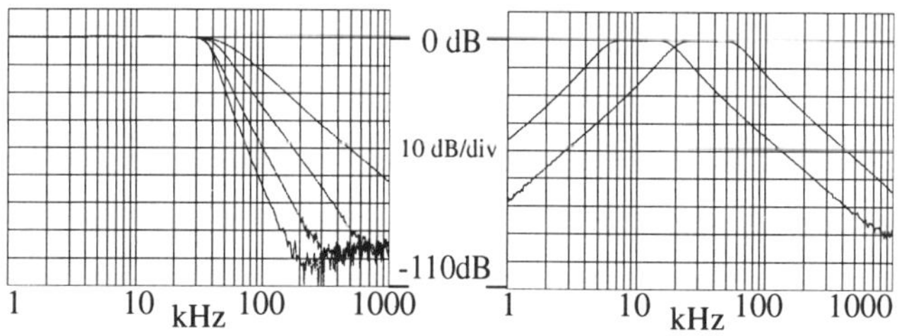

Fig. 11: *Measured characteristics of the 2nd, 3rd, 4th, 5th-order low-pass and the 2nd-order band-pass configurations.*

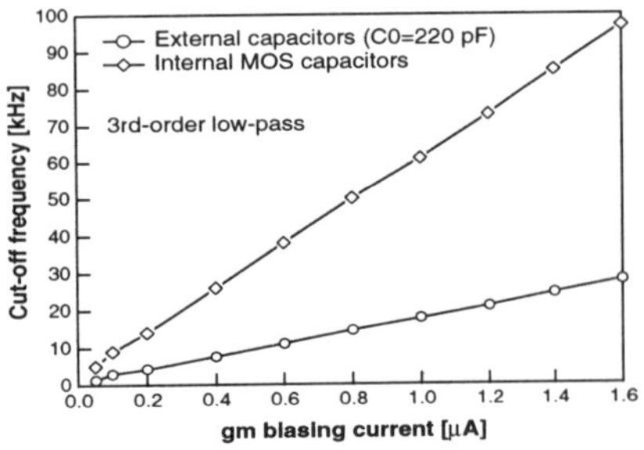

Fig. 12: *Measured frequency tuning versus* Ig_m

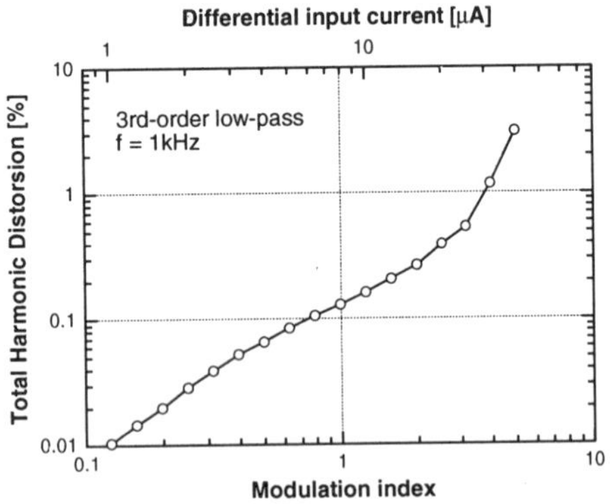

Fig. 13: *Measured THD versus modulation index.*

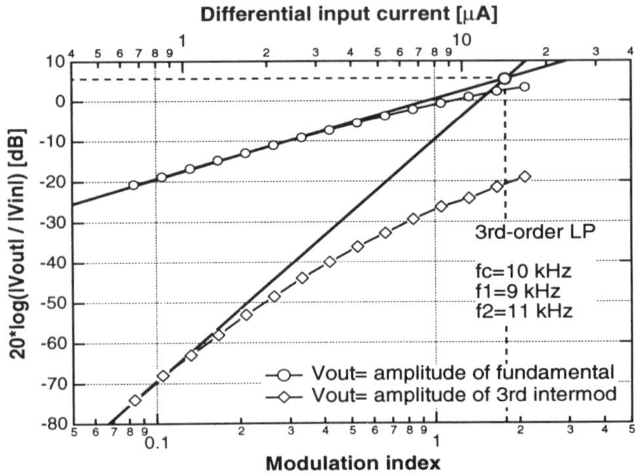

Fig. 14: *Measured power of the fundamental and the in-band 3rd-order intermodulation product referred to the input power versus the modulation index.*

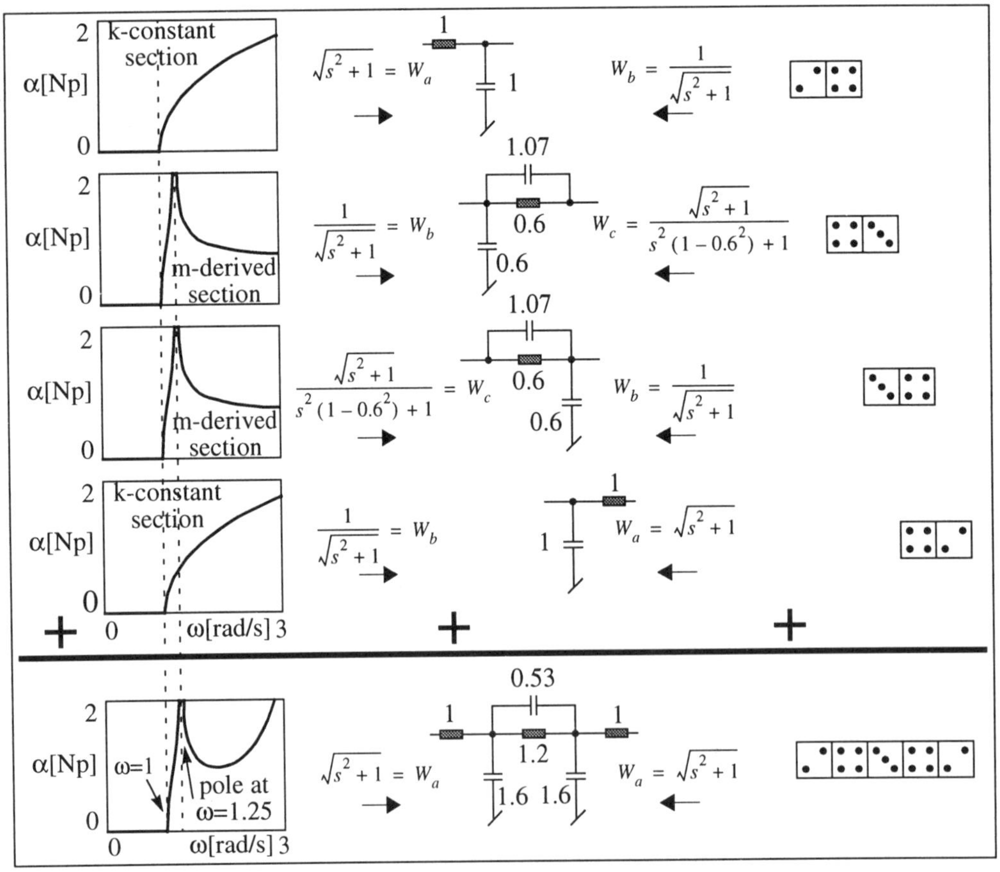

Fig. 15: *Images-parameters method synthesis*

An Improved Current Source for Low Voltage Applications

Fan You, S.H.K. Embabi, J. F. Duque-Carrillo(*) and Edgar Sánchez-Sinencio

Department of Electrical Engineering, Texas A&M University
College Station, Texas 77843-3128, USA

(*)Dept. of Electronics, University of Extremadura, Badajoz, Spain

Abstract

A new current source which is well suited for low voltage applications is proposed. Measured compliance voltage is slightly smaller than that of a single transistor. Its output resistance is a factor of 25 larger than that of a single transistor current source and was measured at 8MΩ. The use of the new current source improves the common-mode input range and the common-mode rejection ratio of fully-balanced and single-ended differential amplifiers.

I. Introduction

The demand for analog circuits which can operate at low voltage is an established fact and does not need any further justification. Traditional current sources (see the first three types in Table I) suffer from a trade off between the output resistance and the compliance voltage (the minimum voltage required for the current source). Other topologies such as the generalized cascode [1] and the active regulated cascode [2] have improved R_{out} with a relatively low compliance voltage of $2V_{dssat}$. This, however, may still be too large to be tolerated in low voltage circuits. For example, in differential amplifiers, the compliance voltage of the tail current source must be minimal to maximize the input common-mode range (CMR). Krenik et al. proposed a current source which has a high output resistance and its compliance voltage is one V_{dssat} [3]. Yet, their circuit can only be used for biasing differential amplifiers.

In this paper, we are proposing a general purpose current source structure which offers both the low compliance voltage and the high output resistance.

II. Current Source Structure

A. Circuit Description

Fig. 1 shows the conceptual schematic of the proposed low voltage (LV) current source. A constant current, I_B, is injected into M1 which is mirrored to the output transistor M2. A change in V_x (the output voltage of the current source) will force a similar change in V_y because of the virtual short circuit provided by the error amplifier A. This implies that the drain-to-source voltage of M1 tracks that of M2. Since the drain current of M1 is fixed and is equal to I_B, the amplifier adjusts the gate voltage to compensate for the change in the drain-to-source voltage of M1 and M2. This results in a constant output current. The amplifier provides a negative feedback to boost the effective resistance of M2.

B. Circuit Analysis

Fig. 2 shows the small signal equivalent circuit for the current source in Fig. 1. The analysis shows that the output resistance (seen at the drain of M2) is given by

$$R_{os} = \frac{A_L^-}{A_L^- - A_L^+} r_{o2} \approx -R_{oB} \quad (1)$$

where A_L^- is the gain of the negative feedback loop, which consists of the error amplifier and M1. A_L^+ is the gain of the positive feedback loop which includes the amplifier and M2. From (1) it can be shown that the output resistance is negative and is equal to R_{oB} (the resistance of the reference source I_B). The gain of the positive loop is reduced when the output of the current source is loaded by a low resistance (e.g. in Fig. 6). This implies that the negative feedback will dominate to guarantee the stability of the circuit.

The proposed realization of the current source is shown in Fig. 3. The error amplifier is implemented as a single differential amplifier. The reference current is provided by a cascode source. Thus, the output resistance is:

$$R_{os} \approx -\frac{g_{my}}{g_{ox} g_{oy}} \quad (2)$$

where g_{my} and g_{oy} are the transconductance and output conductance of My and g_{ox} is the output conductance of Mx. Note that if a regulated cascode or an active regulated cascode current source is used to implement the I_B source, the output resistance of the LV current source will be in the order of g_m^2/g_o^3, while, its compliance voltage is slightly less than one V_{dssat}.

C. Experimental Results

The LV current source in Fig. 3 has been fabricated in $2\mu m$ CMOS Orbit process (via MOSIS). The same circuit has been used as a simple current source by connecting nodes Z and Y and connecting node V to V_{ss} (to disable the error amplifier A). The aspect ratios of M1 and M2

0-7803-3177-6 $5.00 © 1996 IEEE

are $200\mu m/3\mu m$. Fig. 4 illustrates the measured current sourced by both circuits. Note that the slope of curve A (of the simple mirror) is larger than that of curve B (of the LV source). The output resistance of the simple current source is 320kΩ, while, that of the new LV source is 8MΩ (a factor of 25 improvement in the resistance). The positive slope of curve B implies that the resistance of the LV current source is negative, which was predicted by the analysis (see (1)). The compliance voltage varies from 120mV at 20μA to 300mV at 90μA (see Fig. 5). In comparison, the compliance voltage of the simple current source is 200mV at 20μA and 400mV at 90μA.

III. Using The LV Current Source for Low Voltage Differential Amplifiers

The proposed current source is well suited for low voltage differential amplifiers. The increased output resistance of the source with its small compliance voltage improve the CMRR and the input CMR.

A. Single-Ended(SE) Differential Amplifiers

(i) Circuit Analysis

A detailed small signal analysis for the SE amplifier, shown in Fig. 6(a), has been carried out. The mismatching between Mi1 and Mi2 is accounted for by assuming that $g_{mi1} = g_{mi}(1+\epsilon_i/2)$ and $g_{mi2} = g_{mi}(1-\epsilon_i/2)$. Similarly, for the load transistors M3 and M4, $g_{m3} = g_{ml}(1+\epsilon_l/2)$ and $g_{m4} = g_{ml}(1-\epsilon_l/2)$. The common-mode gain for the SE differential amplifier has been derived in terms of its systematic and random components A_{cs} and A_{cr}, respectively.

$$A_{cs} = \frac{-g_{os}}{2g_{ml}} \tag{3}$$

$$A_{cr} = \frac{\epsilon(g_{oi}+g_{os})}{g_{oi}+g_{ol}} \tag{4}$$

where g_{oi} and g_{ol} are the output conductances of Mi1(Mi2) and M3(M4), respectively. To obtain the simple expression for the random common-mode gain, we have assumed that $\epsilon_i = \epsilon_l = \epsilon$. The common-mode gain components of the SE differential amplifier using the simple current mirror have been compared with those of a SE differential amplifier with a LV current mirror (see Table II). For the case of the simple current mirror, g_{os} is comparable to g_{oi} and g_{ol}. Hence, the random common-mode gain is in the order of ϵ. If ϵ is 1%, the systematic common-mode gain dominates over the random gain. For the LV current source, g_{os} is in the order of g_o^2/g_m. Thus, the random common-mode gain is the dominant component (see Table II). The use of the LV current source reduces the common-mode gain by about 4 (if $g_m/g_o = 25$ and $\epsilon = 1\%$). This translates into a 12dB increase in the CMRR.

(ii) Experimental Results

The measured output voltage of the SE differential amplifier versus the common-mode input is shown in Fig. 7. Curve A corresponds to the SE differential pair with the simple mirror, while, B corresponds to that with the LV source. The slopes of the linear portions of the curves can be used as a measure for the common-mode gain. From the figure it is obvious that the LV source reduces the common-mode gain by a factor of 5, which means that the CMRR improves by 14dB (as predicted by the analysis). It is also important to mention that ϵ was measured to be slightly less than 1%.

B. Fully-Balanced (FB) Differential Amplifiers

(i) Circuit Analysis

First, recall the following equations for the differential and common-mode output voltages of a FB pair.

$$v_{od} = A_{dd}v_{id} + A_{dc}v_{ic} \tag{5}$$

$$v_{oc} = A_{cd}v_{id} + A_{cc}v_{ic} \tag{6}$$

where v_{id} and v_{ic} are defined as the differential and common-mode input voltages, respectively. Both types of common-mode gains, A_{dc} and A_{cc} have been derived. The first is only due to random mismatch and is given by:

$$A_{dc} = \epsilon\,\frac{g_{oi}g_{ol} + g_{os}(g_{oi}+g_{ol})}{g_{ol}(g_{oi}+g_{ol})} \tag{7}$$

Table II shows that the common-mode gain A_{dc} of the FB differential amplifier with the LV current source is a factor of 3 smaller than that of the same amplifier with a simple current mirror. This implies that the CMRR of the FB differential amplifier, which is defined as A_{dd}/A_{dc}, gains an additional 10dB if the proposed LV current mirror is used instead of the simple one.

The common-mode gain A_{cc} consists of the following systematic and random components,

$$A_{cs} = \frac{g_{os}}{2g_{ol}} \tag{8}$$

$$A_{cr} = \frac{\epsilon g_{ol}}{2(g_{oi}+g_{ol})} \tag{9}$$

Assuming that $\epsilon < 1\%$, the systematic component dominates when both types of current sources are used (see Table II). Yet, in the case of the simple mirror, the value of A_{cc} is higher than that of the LV current mirror by a factor of 10 which corresponds to 20dB reduction in A_{cc}.

(ii) Experimental Results

Fig. 8 shows the measured differential output voltages (V_{od})of the FB differential amplifier with the simple mirror (curve A) and with the LV mirror (curve B). Two observations can be made. First, the input CMR of the amplifier with the simple mirror is between 250mV and 650mV, while that of the LV mirror is between 250mV and 870mV. A 200mV increase in the CMR is significant for low power supply voltages. For the 2V supply, used in our measurements, the CMR improves by 10%. To compare the values of A_{dc} one can measure the slopes of both curves in the valid input range. While it is difficult to measure exact values for the slopes because of the measurement's resolution, one can show that the LV current source results in a reduction of A_{dc} by a factor of 6 (15dB). Curves C and D in Fig. 8 show the measured common-mode outputs (V_{oc}) of the FB amplifiers with the simple and the LV sources, respectively. The dramatic improvement of A_{cc}, which is predicted by the analysis, has been confirmed by comparing the slopes of curves C and D. The improvement of A_{cc} is measured to be 31dB. The fact that the common-mode output voltages are well controlled through the use of the LV current source, relaxes the requirement of the common-mode feedback loop needed to stabilize the common-mode component in the feedback amplifier.

IV. Conclusion

A new low voltage current source has been proposed. This source offers both high output resistance (8M Ω) and low compliance voltage (200mV). We have demonstrated that the proposed current source improves the input common-mode range and the common-mode rejection ratio of low voltage differential amplifiers. The analysis has shown, however, that the CMRR of single-ended and fully-balanced differential amplifiers will be eventually limited by the process mismatching when the output resistance of the tail current source is increased. The new LV current source provides very good control over the common-mode output of the fully-balanced amplifiers, which allows for simple common-mode feedback circuits.

References

(1) P.J. Crawley and G.W. Roberts, "High-Swing MOS Current Mirror with Arbitrarily High Output Resistance," Electronics Letters, Vol. 28 No. 4 pp. 361-363, 1992.

(2) Teresa Serrano and Bernabe Linares-Barranco, "The Active-Input Regulated-Cascode Current Mirror," IEEE Trans. on CAS, vol. 41. No. 6, pp. 464-467, 1994.

(3) W. Krenik et. al., "High Dynamic Range CMOS Amplifier Design in Reduced Supply Voltage Environment," Tech. Dig. of Midwest Symposium on Circuits and Systems, 1988.

TABLE I

COMPARISON BETWEEN CURRENT SOURCE TOPOLOGIES

Current Source Type	R_{out}	$V_{compliance}$
Simple	r_o	V_{dssat}
Cascode	$r_o^2 g_m$	$2V_{dssat} + V_T$
Regulated Cascode	$r_o^3 g_m^2$	$2V_{dssat} + V_T$
Generalized Cascode [1]	$r_o^2 g_m$	$2V_{dssat}$
Active Reg. Cascode [2]	$Ar_o^2 g_m$	$2V_{dssat}$

r_o is the output resistance of a MOS transistor and V_{dssat} is the drain-source saturation voltage.

TABLE II

IMPACT OF THE SIMPLE AND LV SOURCE ON A_c, A_{dc}, A_{cc}.

		Simple CS $g_{os} = g_o$	LV CS $g_{os} = g_o^2/g_m$	$\frac{A_{simple}}{A_{LV}}$
SE	A_{cs}	$-g_o/2g_m$	g_o^2/g_m^2	$g_o/\epsilon g_m$
	A_{cr}	ϵ	$\epsilon/2$	($\approx 12dB$)
	A_c	$-g_o/2g_m$	$\epsilon/2$	
FB	A_{dc}	1.5ϵ	0.5ϵ	3 ($\approx$ 10dB)
	A_{ccr}	0.5	g_o/g_m	
	A_{ccr}	$\epsilon/4$	$\epsilon/4$	$g_m/g_o \approx 10$
	A_{cc}	0.5	g_o/g_m	(20dB)

Each entry gives the order of magnitudes of A_c, A_{dc} and A_{cc} in terms of ϵ and g_m/g_0 ($\epsilon < 1\%$, $g_m/g_o = 25$)

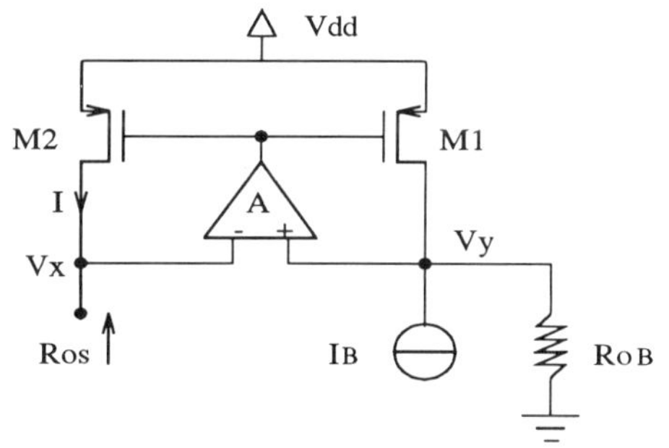

Fig. 1. A conceptual schematic of the low voltage current source.

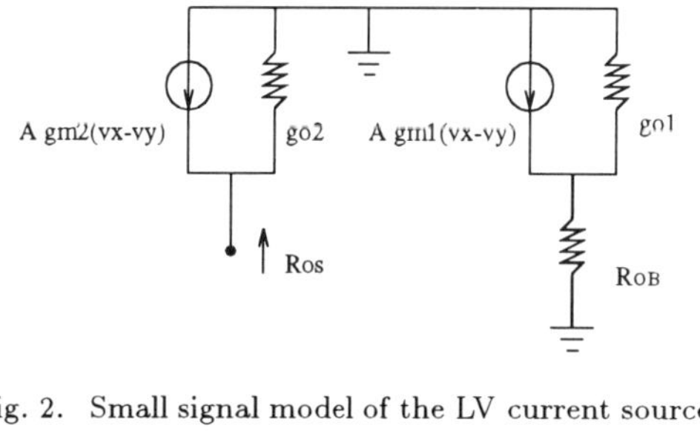

Fig. 2. Small signal model of the LV current source.

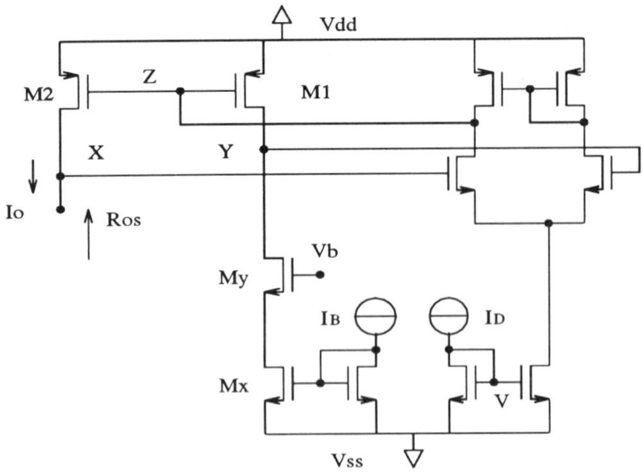

Fig. 3. Full implementation of the LV current source.

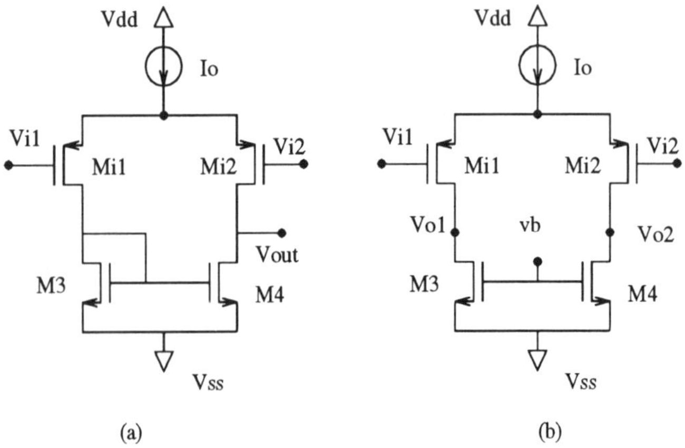

Fig. 6. Differential Amplifiers, (a) single-ended and (b) fully-balanced.

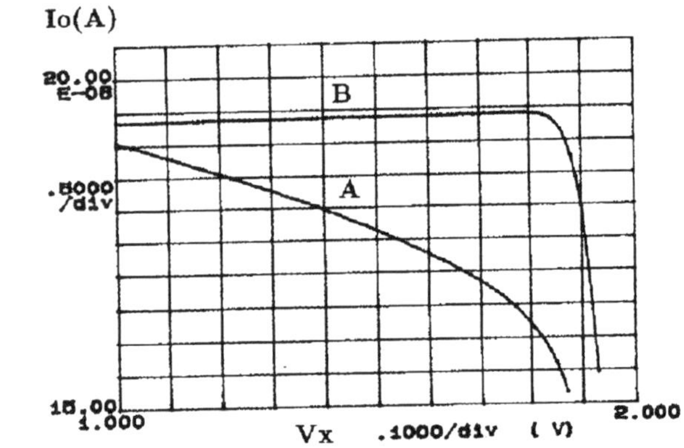

Fig. 4. Measured output current of the simple (curve A) and LV (curve B) current source.

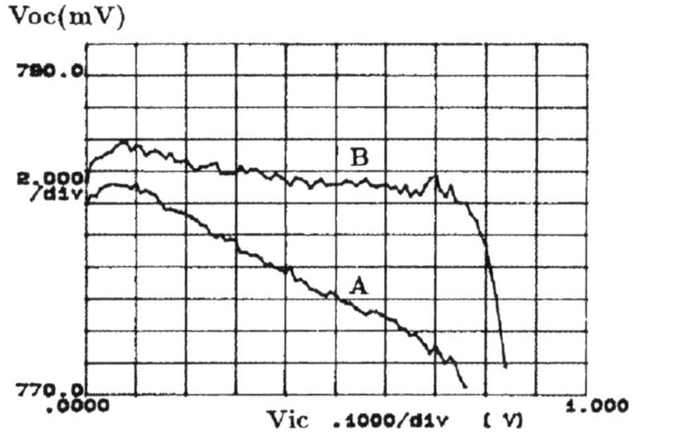

Fig. 7. Measured output voltage vs common-mode input voltage for a SE amplifier with simple (curve A) and LV (curve B) current source.

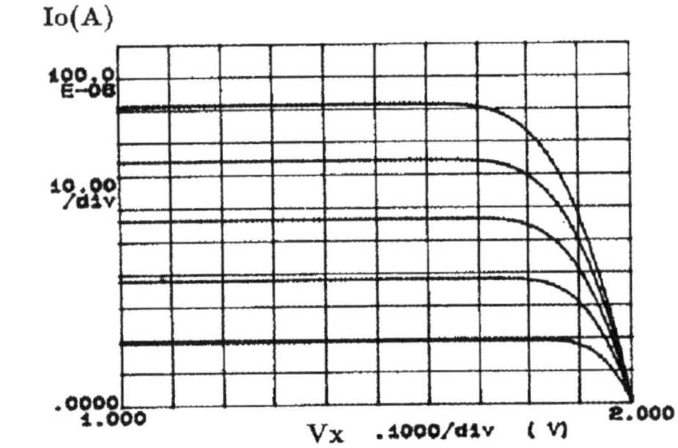

Fig. 5. Measured I-V characteristics of the LV current source.

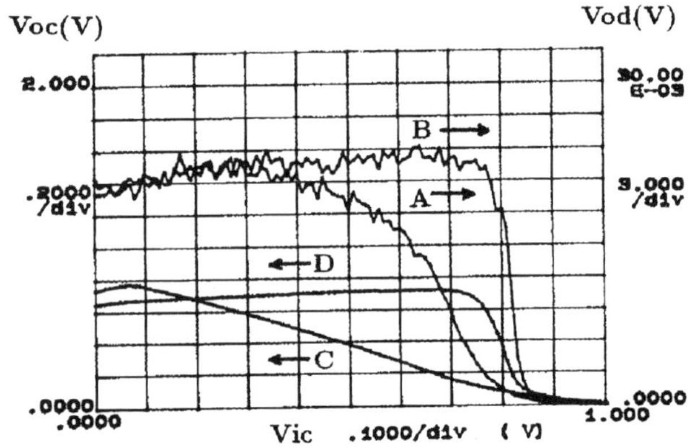

Fig. 8. Measured Voc and Vod vs common-mode input for the FB amplifier with simple (curves A and C) and LV (curves B and D) current sources.

6.4.4

A 100 dB, 480 MHz OTA in 0.7 μm CMOS for Sampled-Data Applications*

Thomas Burger and Qiuting Huang

Swiss Federal Institute of Technology, Integrated Systems Laboratory
CH-8092 Zürich, Switzerland

Abstract

A fully differential operational transconductance amplifier with a gain-bandwidth of 480 MHz at 52 degrees of phase margin and a DC-gain of 100 dB has been implemented in a 0.7 μm CMOS process. The circuit utilizes the regulated cascode gain enhancement technique and has a switched-capacitor dynamic common mode feedback structure. It has been designed for a +/- 3 V output swing at a single supply of 5 V. The power consumption is 145 mW.

1 Introduction

Operational amplifiers are a critical element in analog sampled-data circuits, such as switched-capacitor filters, delta-sigma modulators and pipelined A/D converters. The latter circuits, especially bandpass delta-sigma A/D converters, require clock frequencies in excess of 40 MHz to provide cost-effective solutions for wireless communications and video applications. In a wireless application, such as the GSM transceiver, A/D conversion at the IF frequency, using a bandpass delta-sigma converter, offers a way to improve the level of integration [1]. A higher sampling frequency enables A/D conversion at a higher IF, 21.4 MHz for example, which improves image rejection and relaxes the RF filter requirement. A higher sampling frequency also improves resolution for a given channel bandwidth. As the frequency requirement of these sampled-data circuits becomes higher and higher, it almost directly translates to higher and higher frequency requirement for the op-amp. A high unity-gain bandwidth (GBW) is essential for accurate dynamic charge transfer in an SC circuit in a short sampling period, to realize high Q and precise poles, whereas the precise phase requirement of ideal integrators normally requires that the DC-gain of the amplifier be higher than 80 dB [2],[3].

In addition to GBW and DC-gain, which often pose conflicting requirements on the amplifier design, adequate slew rate, sufficient phase margin and sufficient output signal swing are all important parameters that affect the overall trade-off named performance.

[1]*This work has been funded by FSRM

In this work, we are seeking a systematic design approach to reach high speed while maintaining high DC-gain, good phase margin and output swing. In the following, we shall first describe how the use of amplifier-regulated cascode technique frees the DC-gain consideration from the GBW optimization, before describing how the latter can be optimized using the available current and silicon area, subject to system constraints such as load capacitance, slew rate, output swing and phase margin.

2 Gain enhancement

The folded cascode OTA is one of the best known structures for high frequency application [4]. Its implementation in modern CMOS processes with submicron channel length devices results in a DC-gain that is not sufficient for many applications. The regulated cascode structure [5] is an excellent way to overcome this problem. It utilizes an additional gain stage, another OTA for example, to increase the cascoding effect of transistor

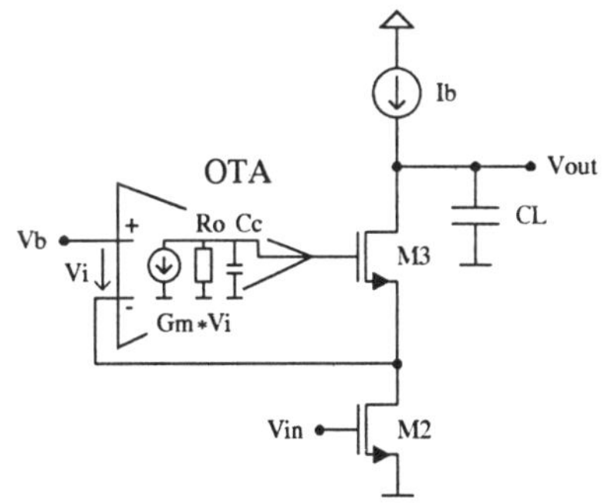

Figure 1: Regulated cascode gain enhancement

M3 as shown in figure 1. The total output resistance as well as the DC-gain are increased by the gain $G_m * R_o + 1$ of the regulating OTA. If the GBW of the additional gain stage is designed such that its unity-gain frequency lies between the -3 dB and the unity-gain frequency of the original cascode, a first order frequency roll-off results as shown in figure 2. In implementations using this gain enhancement technique DC-gains of 90 dB have been reported [6], [7]. Our experience shows that more

6.5.1

0-7803-3177-6 $5.00 © 1996 IEEE

IEEE 1996 CUSTOM INTEGRATED CIRCUITS CONFERENCE

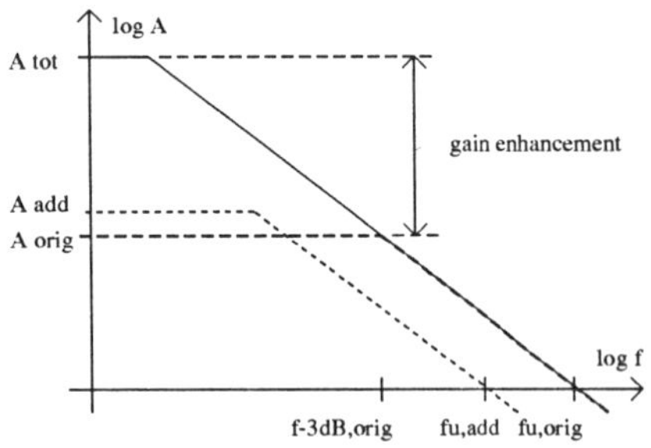

Figure 2: Gain Bode plots illustrating gain enhancement

than 80 dB gain can be easily realized using minimum length ($\approx 1\ \mu m$) transistors, even at relatively high current levels. Assuming minimum length transistors and allowing much more current while being able to maintain relatively high gain, makes the maximization of speed a more straightforward task.

3 Gain-bandwidth optimization

When an OTA is optimized for a specific application the requirements at the system level usually dictate the load capacitance and the minimum acceptable phase margin and output voltage swing. Optimizing for speed to enable the system to run at a high clock rate, is then a question of sizing the transistors

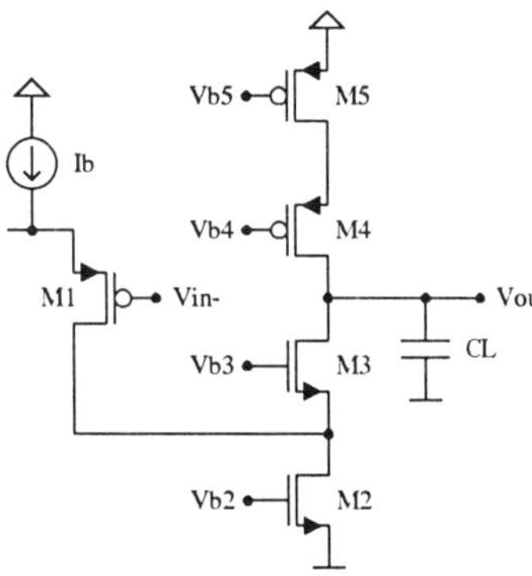

Figure 3: Half circuit of fully differential folded cascode OTA

correctly and providing sufficient current so that the slew rate and GBW are maximized. The slew rate is a linear function of current whereas the unity gain-bandwidth GBW is related to bias current and the width W_1 of the input transistor M1 (see fig. 3) by

$$GBW = \sqrt{2K'I_{d1}W_1/L_1}/C_L \qquad (1)$$

For a given load capacitance both increasing the current and W_1 increases GBW. The limit to the achievable GBW, however, is set by the acceptable phase margin which is determined by the nondominant pole P_3 associated with the source node of M3,

$$P_3 = \frac{g_{m3} + g_{mb3}}{C_{db1} + C_{gd1} + C_{db2} + C_{gd2} + C_{gs3} + C_{sb3}} \qquad (2)$$

If the saturation voltages of M2 and M3 are fixed by a given output swing requirement, then the two transistor widths must scale with the bias current linearly, and so do the parasitic capacitances associated with these two transistors. On the whole, a smaller output swing allows a larger overall saturation voltage of M2 and M3 ($V_{dsat2} + V_{dsat3}$), so that both transistors can be narrower to reduce the parasitic capacitance, which increases P_3, allowing a higher GBW to be realized. For a given output swing, different ratios between the two saturation voltages may give different values of P_3. Part of the GBW optimization is to find the best saturation voltage ratio between M2 and M3. To maintain a constant phase margin PM, the width W_1 for the input transistor is determined for a given bias current by the nonlinear equation:

$$\frac{g_{m1}(W_1)}{P_3(W_1)C_{out}} - \frac{cot(PM)}{sin(PM)} = 0 \qquad (3)$$

where C_{out} is the sum of C_L, C_{db3} and C_{db4}. The solution to (3) enables us to plot the achievable unity-gain bandwidth as a function of bias current in figure 4 for different ratios between the saturation voltages of M2 and M3. In the calculation, we have assumed that the bias currents of M1 and M3 are equal, $C_{out} = 6pF$, $PM = 60°$ and $V_{dsat2} + V_{dsat3} = 1\ V$. It can be shown through more detailed examination of (3) that at lower current levels, GBW increases as the cubic root of the bias current. At higher current levels the GBW becomes constant. This constant is only determined by V_{dsat2} and V_{dsat3}. This is shown in figure 4. Interesting to note is that the saturation voltage ra-

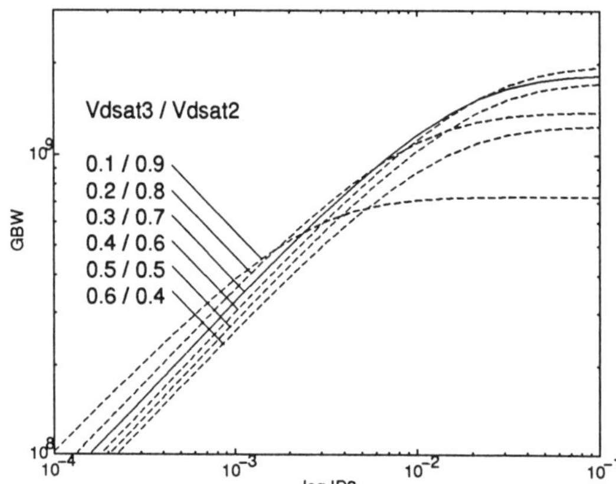

Figure 4: GBW vs. I_{D3} and different ratios V_{dsat3}/V_{dsat2}

tio for which the highest GBW can be achieved varies with the bias current. This means that at low current levels it is better to have transistor M3 much wider than M2. Over a wider current range the best V_{dsat} ratios lie between 0.4/0.6 and 0.3/0.7. The latter (solid line) was used in the design we implemented. The above analysis assumes that the input capacitance of the OTA, which is proportional to W_1, is unrelated to the load capacitance. In practical switched capacitor applications, however, the input capacitance often forms part of the load capacitance, which means that the optimum W_1 may be lower than indicated in fig. 4.

The most important conclusions to draw from the above analysis, is that the power consumption increases no slower than the cubic function of the desired unity gain bandwidth. Beyond a certain current level, GBW can only be increased by either reducing the output swing or reducing the phase margin. Figure 5 depicts the GBW curves for two V_{dsat} ratios and

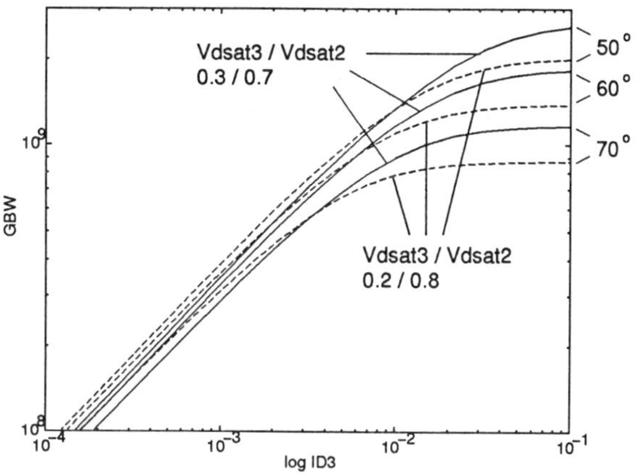

Figure 5: GBW vs. I_{D3} and different phase margins

phase margins of 70, 60 and 50 degrees. The same general behaviour of GBW versus I_{D3} as found in figure 4 appears here. As expected the curves for the lower phase margins reach higher GBW values at high current levels. Again, there is a dependence of the optimum GBW on the ratio V_{dsat3}/V_{dsat2}.

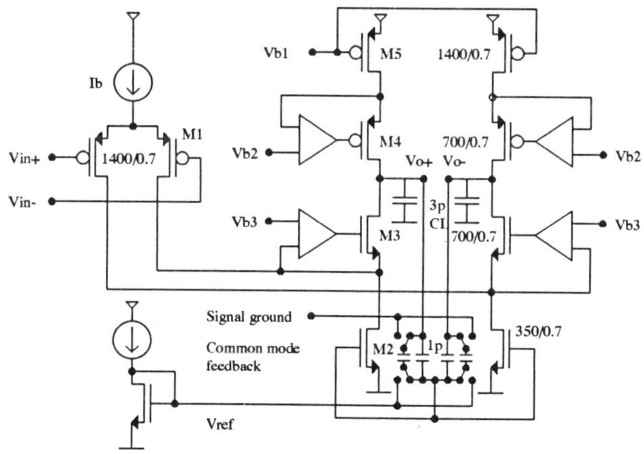

Figure 6: Main OTA schematic

4 Implementation

To validate the optimization procedure described above and to demonstrate that very high bandwidth and very high DC-gain can be simultaneously achieved without sacrificing output swing and phase margin, we implemented a fully differential folded cascode amplifier with amplifier-regulated gain-enhancement and a dynamic common mode feedback. Figure 6 shows the main OTA. For implementation the design parameters of section 3 have been set to $C_{out} = 6pF, I_{D3} = I_{D1} = 5mA$, $(V_{dsat2} + V_{dsat3}) = 0.9V$, $V_{dsat3}/V_{dsat2} = 0.3/0.7$, and $PM = 60°$ resulting in $W_1 = 2900\,\mu m, W_2 = 343\,\mu m, W_3 = 935\,\mu m$. The value of W_1 corresponds to a C_{gs1} that was considered too large for application of the OTA in SC circuits, since in such an application the OTA input capacitance is present as a part of the output load. Therefore the value of W_1 was set to $1400\,\mu m$ resulting in a simulated GBW of 580 MHz at a phase margin of 63° for the actual implementation.

The common mode feedback circuit utilizes a summing network directly placed at the circuit output leading back to the gate of transistor M2 and its counterpart. Two switched capacitors provide the necessary offset voltage to establish the common mode output voltage. The common mode feedback circuit uses the same compensation capacitance as the differential circuit and was designed to have equal low frequency gain and gain-bandwidth as the differential circuit.

The regulating OTA's of figure 7 have been designed with a PMOS input differential pair for the low input voltage level

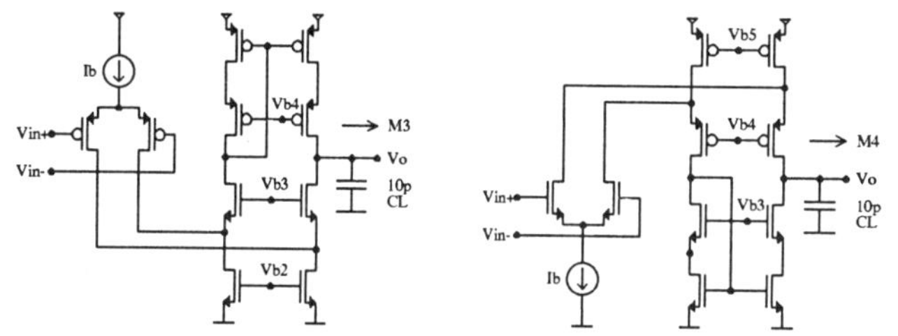

Figure 7: Regulating OTA schematics

at the source of transistor M3 and a NMOS input for the high input voltage level at the source of M4. To guarantee a fast settling behaviour of the whole OTA as required by sampled-data applications the regulating OTA's have been built with a GBW of 60 MHz (PMOS input) and 80 MHz (NMOS input) consuming 10 % of the main amplifiers current each. The OTA's are compensated with a large output capacitance of 10 pF such that the unwanted pole-zero pair (doublet) in the frequency response [6] is kept closed.

Figure 8 shows the circuit implementation in a $0.7\,\mu m$ CMOS process with linear capacitance option. While the regulation OTA's are relatively small compared to the main amplifier, the compensation capacitances of these OTA's consume considerable chip area. For measuring purposes large output buffers have been added to the circuit implementation.

5 Measurements

Frequency response measurements have been performed using a differential sine source and operating the OTA in a feed-

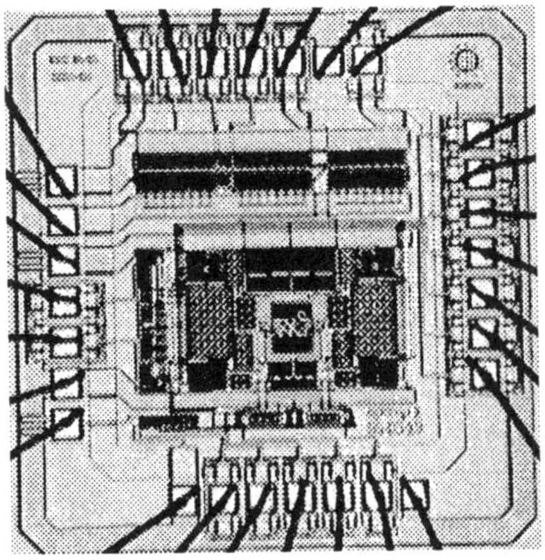

Figure 8: Photomicrograph of the OTA with output buffers

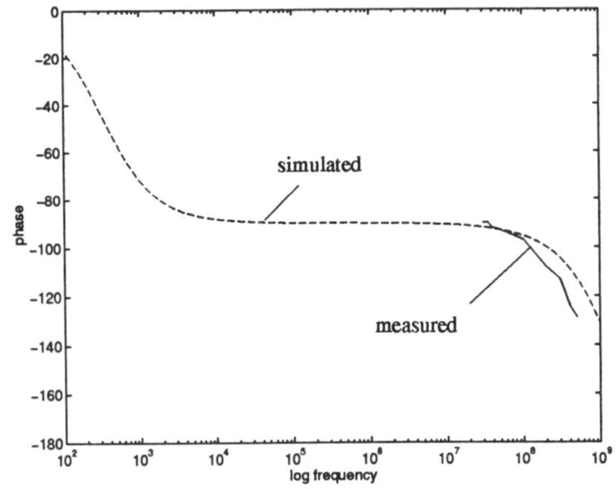

Figure 10: Simulated and measured phase response

back network to obtain an amplifier with a gain of 10. For low to medium frequency range the differential input of the OTA was fed to an instrumentation amplifier providing an amplified OTA input voltage. At high frequencies the differential input to the feedback network of the test circuit has been taken for measurement directly. Figures 9 and 10 show the simulated and the measured amplitude and phase response of the OTA. The DC-gain was measured to be 102 dB. The measurement of unity gain frequency gave a value of 480 MHz at a phase margin of 52°. Table I summarizes the OTA characteristics.

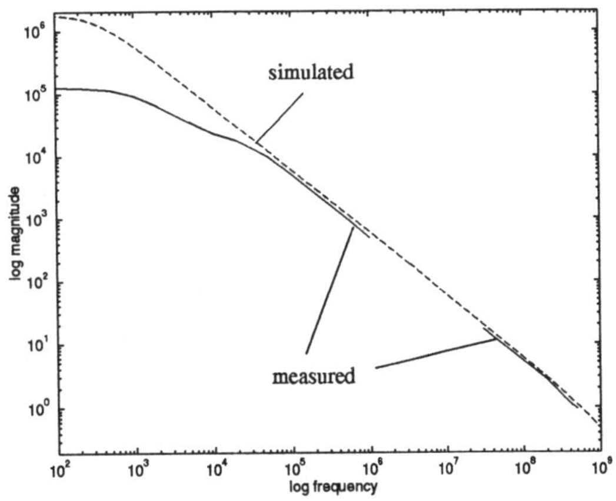

Figure 9: Simulated and measured frequency response

TABLE I : OTA characteristics	
DC-gain	102 dB
Unity-gain frequency	480 MHz
Phase margin	52 deg.
Slew rate	1600 V/μs
Settling to 0,1 % for ΔVin=1V	6.7 ns [2]
Differential output swing	+/-3 V
Power supply	5 V
Power dissipation	145 mW

6 Conclusions

It has been shown that using a modern submicron CMOS process it is possible to build an OTA having both a very high unity gain bandwidth and DC-gain. For the design of such a circuit individual sizing of every transistor is important to improve GBW and phase margin. A systematic design approach for GBW optimization of the folded cascode structure using convenient design parameters is described.

References

[1] F.W. Singor and W.M. Snelgrove. Switched-Capacitor Bandpass Delta-Sigma A/D Modulation at 10.7 MHz. *IEEE Journal of Solid-State Circuits*, 30(3), March 1995.

[2] K. Martin and A.S. Sedra. Effects of the Op Amp Finite Gain and Bandwidth on the Performanc of Switched-Capacitor Filters. *IEEE Trans. on Circuits and Systems*, CAS-28(-):822–829, August 1981.

[3] W. Sansen, Q. Huang, and K. Halonen. Transient Analysis of Charge Transfer in SC Filters - Gain Error and Distortion. *IEEE Journal of Solid-State Circuits*, SC-22(2):268–276, April 1987.

[4] F. Op't Eynde and W.Sansen. Design and Optimization of CMOS Wideband Amplifiers. In *Proceedings of the IEEE Custom Integrated Circuits Conference*, pages 2571–2574, June 1989.

[5] B.J. Hosticka. Improvement of the gain of mos amplifiers . *IEEE Journal of Solid-State Circuits*, SC-14:1111–1114, December 1979.

[6] K. Bult and G. Geelen. A Fast-Settling CMOS Op Amp for SC Circuits with 90-dB DC Gain. *IEEE Journal of Solid-State Circuits*, 25(6):1379–1384, December 1990.

[7] M. Helfenstein, Q. Huang, and G.S. Moschytz. 90dB, 90MHz, 30mW OTA with the Gain-Enhancement Implemented by One- and Two Stage Amplifiers. In *Proceedings of the IEEE 1995 International Symposium of Circuits and Systems*, pages 1732–1735, May 1995.

[2] Simulation result

A CMOS 4-bit MDAC with Self-Calibrated 14-bit Linearity for High-Resolution Pipelined A/D Converters

J. Goes, J. C. Vital and J. E. Franca

Instituto Superior Técnico, IST Centre for Microsystems
Integrated Circuits and Systems Group
Av. Rovisco Pais,1-1096 Lisboa Codex—PORTUGAL

Abstract— This paper presents an integrated 4-bit MDAC for high-speed high-resolution pipelined ADCs which employs a code-by-code analogue self-calibration technique. Measured results from the prototypes fabricated in a 1.0 μm CMOS technology show that the proposed self-calibration technique corrects the MDAC linearity and the interstage gain to the 14-bit level, while allowing conversion rates in the MHz range.

I. INTRODUCTION

High-resolution analogue-to-digital conversion with sampling rates in the MHz range is required in a broad area of applications with enhanced performance, ranging from high quality imaging systems to wireless communications. The most attractive solutions currently available for implementing this type of analogue-to-digital converters (ADCs) employ pipelining as a means of relaxing the speed requirements of the analogue components. Such architectures use a cascade of stages comprising a low-resolution flash quantizer and a low-resolution multiplying digital-to-analogue converter (MDAC), which computes and amplifies the residue to be quantised by the following stages [1]. As a consequence of the residue amplification, the precision requirements are more critical in the front-end stage of the pipeline, which must exhibit the overall accuracy of the converter, and are progressively relaxed towards the last stage. Without using neither trimming nor self-calibration techniques, the overall resolution of these ADCs is limited by the linearity and gain errors of the first MDACs, which are naturally bounded at the 10-bit level by the component matching accuracy provided by most CMOS processes available today. As trimming is expensive, self-calibration techniques must be considered for extending the resolution of such high-speed pipelined ADCs above 10-bits.

Self-calibration techniques based on the measurement and digital storage of the errors for later correction have been reported and which can be classified into two main groups, depending on whether the calibration is performed in the analogue or in the digital domain. The proposed analogue techniques require separate calibration DACs and precision analogue components [2-4] and, additionally, may need a reference feedforward technique to compensate for gain errors in the MDACs. On the other hand, digital calibration techniques do not require fancy analogue circuitry but put an extra burden in the digital part and require additional stages or additional resolution in each calibrated stage to overcome the loss in effective resolution produced by digital truncation errors during calibration [5, 6]. Furthermore, interstage gain proration techniques may be necessary to overcome the effects of interstage gain errors [7].

This paper describes an integrated CMOS 4-bit MDAC for high-resolution pipelined ADCs which uses an analogue code-by-code self-calibration technique based on a high-precision reference DAC [8, 9]. This solution is capable of simultaneously calibrating both the D/A conversion linearity and the gain of the low-resolution MDAC by employing just a single, compact sub-binary capacitor-array. Measured results from the 1.0 μm CMOS 4-bit MDAC prototypes show levels of accuracy compatible with 14 bits of resolution, while allowing sampling rates in the MHz range.

II - SYSTEM ARCHITECTURE

Block Diagram

The integrated 4-bit MDAC was designed to be used in the front-end stage of a 15-bit high-speed pipelined ADC, as illustrated in the block diagram of Fig. 1. The 4-bit resolution was carefully selected by following a systematic design for optimisation strategy [10]. Besides the MDAC, the analogue part of the self-calibrated circuit includes three extra blocks to implement the proposed calibration technique: a high-precision pulse-counting reference DAC, a high-accuracy comparator (HAC) and a calibrating DAC which shares the same virtual ground of the MDAC. The digital part is composed by a random access memory (RAM) and by a successive approximation register (SAR). In addition to the self-calibrated 4-bit MDAC, the front-end stage of the pipelined ADC includes a sample-and hold circuit and a 4-bit flash quantizer.

The system operates in two main modes: the calibrating mode, where the errors in the MDAC are digitised and stored for subsequent correction, and the conversion mode where the S/H, the flash quantizer and the calibrated MDAC provide the functions required in the front-end stage.

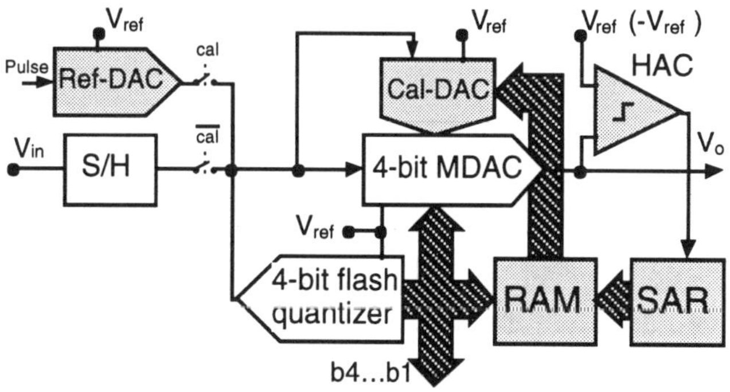

Fig. 1: Block diagram illustrating the use of the self-calibrated MDAC in the front-end stage of a pipelined ADC

6.6.1

0-7803-3177-6 $5.00 © 1996 IEEE

IEEE 1996 CUSTOM INTEGRATED CIRCUITS CONFERENCE

MDAC and Errors to be Calibrated

Since 1-bit redundancy and digital error correction are employed between stages, only the errors that are introduced in the residue amplification performed by the MDAC must be calibrated. The MDAC architecture originally proposed in [11] is schematically illustrated in Fig. 2. It is composed of a binary-weighted capacitor-array associated with the input bits b_1 through b_N, two additional unit capacitors C' and C'' and an operational amplifier (opamp). Assuming a sufficiently high DC gain of the opamp, and $V_{ref}^- = -V_{ref}^+ = -V_{ref}$, the output voltage of the MDAC at the end of the residue amplification phase is given by

$$v_{RA} = \frac{1}{C'+C''}\left[v_{in}(C'+\sum_{i=1}^{N} C_i) - V_{ref}\sum_{i=1}^{N} C_i \cdot b_i\right], \tag{1}$$

where v_{in} is the voltage stored in the capacitor-array in the sampling phase and bits b_i, $i = 1,...N$, assume the values +1 or -1, respectively, according to logic "1" or logic "0". If the capacitance values satisfy the condition

$$C' = C'' = C,\ C_i = 2^{i-1}C,\ i = 1,\ldots,N, \tag{2}$$

expression (1) can be rewritten as

$$v_{RA} = 2^{N-1}(v_{in} - V_{ref}\sum_{i=1}^{N} 2^{i-N-1} b_i), \tag{3}$$

which corresponds to the correct output in the residue amplification phase. Due to the inherent component tolerances in the fabrication process, condition (2) will not be satisfied in practical situations. In order to reflect this, let the unit capacitance be defined as

$$C = (C'+C'')/2, \tag{4}$$

and the remaining capacitance values as

$$C' = (1+\varepsilon')C, \quad C_i = (2^{i-1}+\varepsilon_i)C, \quad i = 1,\ldots,N. \tag{5}$$

With the above definitions, expression (3) can be rewritten as

$$v_{RA} = 2^{N-1}\left\{v_{in}\left[1+\frac{1}{2^N}\left(\varepsilon' + \sum_{i=1}^{N}\varepsilon_i\right)\right] - \frac{V_{ref}}{2^N}\sum_{i=1}^{N}\left(2^{i-1}+\varepsilon_i\right)b_i\right\} \tag{6}$$

By comparing (6) with (3) one can conclude that the mismatches between the capacitors produce not only a gain error that is evident in the multiplying factor affecting v_{in} but also a code dependent nonlinearity error affecting the second term in (6). If these errors are present in the front-end stages of high-resolution pipelined ADCs, they can produce large discontinuities in the conversion characteristic and, therefore, must be compensated by using self-calibration [4, 7].

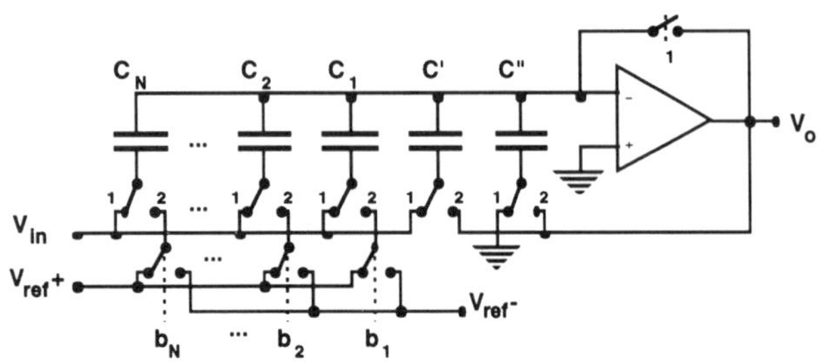

Fig. 2: Low-resolution MDAC architecture [11].

III - The Self-Calibration Technique

Calibration of the Gain Error of the MDAC

The basic MDAC configurations employed in the calibration of the gain error are shown in Fig. 3. Here, the calibrating capacitor-array is represented in a simplified form using two variable capacitors $C_{g.cal1}$ and $C_{g.cal0}$, which represent the groupings of all the capacitors associated with calibrating bits at logic "1" and logic "0", respectively, during gain calibration. The calibration process is started by producing at the output of the pulse-counting reference DAC the voltage $V_{ref}^- + 2\,\mathrm{LSB}$, as shown in Fig. 3(a), where the LSB weight is given by $(V_{ref}^+ - V_{ref}^-)/2^N$. This voltage is sampled by the capacitors of the MDAC as well as by those in the calibrating DAC, as indicated in Fig. 3(b). Note that the grouping $C_{g.cal0}$ samples the symmetric of the input voltage, which is available with no extra hardware in fully-differential structures commonly used in high-resolution ADCs. Assuming again, for the sake of simplicity, that the condition $V_{ref}^- = -V_{ref}^+ = -V_{ref}$ holds, and considering a residue amplification of code "00...0" with $C_{g.cal0}$ and $C_{g.cal1}$ respectively connected to $+V_{ref}$ and $-V_{ref}$, as illustrated in Fig 3(c), the output voltage is given by

$$v_0 = -V_{ref} + V_{ref}\frac{2^{-N+2}}{C+C'}\left(\sum_{i=1}^{N} C_i + C + C_{g.cal1} - C_{g.cal0}\right). \tag{7}$$

Ideally, the output voltage would be exactly $+V_{ref}$. However, in practical situations the value $C_{g.cal1} - C_{g.cal0}$ must be adjusted to minimise the difference between V_o and $+V_{ref}$. This is performed by repeating this sequence for every bit of the calibrating code, starting from the most significant, and comparing the resulting output voltage to $+V_{ref}$. The SAR is used for this purpose. At the end of this operation, the calibrating code ensures that the gain of the residue amplification is equal to 2^{N-1}. The gain-calibrating code is stored in the RAM and is used in the next calibration steps

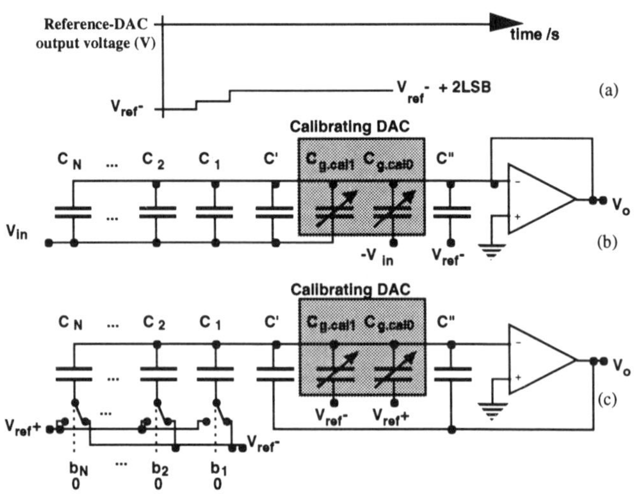

Fig. 3: Configurations of the MDAC for gain error calibration: a) Input voltage generation, b) sampling , c) residue amplification.

6.6.2

for non-linearity correction as well as during the normal conversion mode.

Calibration of the Nonlinearity of the MDAC

The nonlinearity calibration process for a generic MDAC input code k is illustrated in Fig. 4. First, the pulse-counting reference DAC generates the voltage $V_{ref}^{-} + k \times \mathrm{LSB}$. This voltage is sampled by the MDAC and by the calibrating DAC with $C_{g.cal0}$ and $C_{g.cal1}$ configured according to the gain-calibrating code, as shown in Fig. 4(b). Then, a residue amplification is performed with the configuration presented in Fig. 4(c). In this case, $C_{d.cal0}$ and $C_{d.cal1}$ represent the groupings of all the capacitors associated with calibrating bits at logic "0" and logic "1", respectively, during the calibration of the DAC nonlinearities. The output voltage, given by

$$v_o = \frac{-1}{C'+C''}\left[V_{ref}\left(1-k\times 2^{-N+1}\right)\left(\sum_{i=1}^{N} C_i + C' + C_{g.cal1} - C_{g.cal0}\right) + V_{ref}\left(\sum_{i=1}^{N} C_i b_i + C'' + C_{d.cal1} - C_{d.cal0}\right)\right], \quad (8)$$

should be exactly $-V_{ref}$ with ideal capacitance values and no calibrating arrays. In practical situations, the gain factor is already calibrated by the term $C_{g.cal1} - C_{g.cal0}$ found in the previous step, and the term $C_{d.cal1} - C_{d.cal0}$ can be adjusted to compensate the code-dependent nonlinearity. This is performed in the configuration of Fig. 4 (c) by testing each bit of the calibrating DAC, starting from the most significant, and comparing the resulting output voltage to $-V_{ref}$. The SAR is again used to find the calibrating code which minimises the final error. This calibrating code is stored in the RAM cell addressed by code k, and will be fetched whenever this code is used in a residue amplification during conversion mode. This procedure must be repeated for every possible MDAC input code.

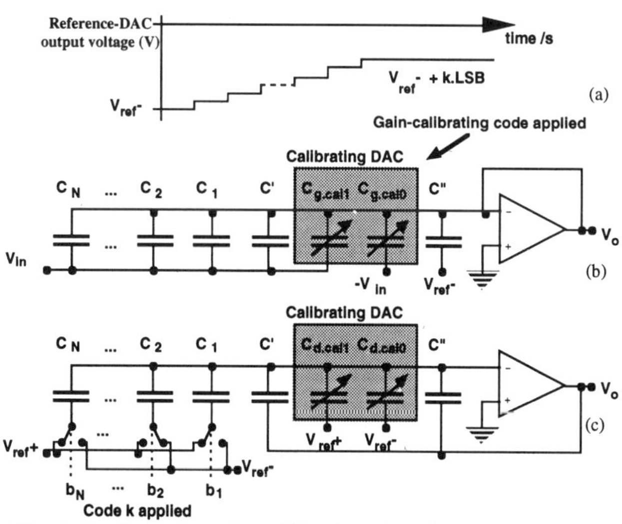

Fig. 4: Configurations for calibration of nonlinearity for code k: a) Input voltage generation, b) sampling , c) residue amplification.

Conversion Mode.

In this mode, the RAM already holds the calibrating-codes for the gain error and code-dependent MDAC nonlinearities. The gain error calibrating code is applied to the calibrating DAC whenever the sampling phase is enabled and, during residue amplification, the calibrating code is fetched from the RAM addressed by the input-code of the MDAC. The configurations represented in Fig. 5 illustrate this operating mode. The output voltage resulting from the residue amplification is given by (9), where the errors already discussed are appropriately compensated.

$$v_o = \frac{v_{in}\left(\sum_{i=1}^{N} C_i + C' + C_{g.cal1} - C_{g.cal0}\right) - V_{ref}\left(\sum_{i=1}^{N} C_i b_i + C_{d.cal1} - C_{d.cal0}\right)}{C'+C''} \quad (9)$$

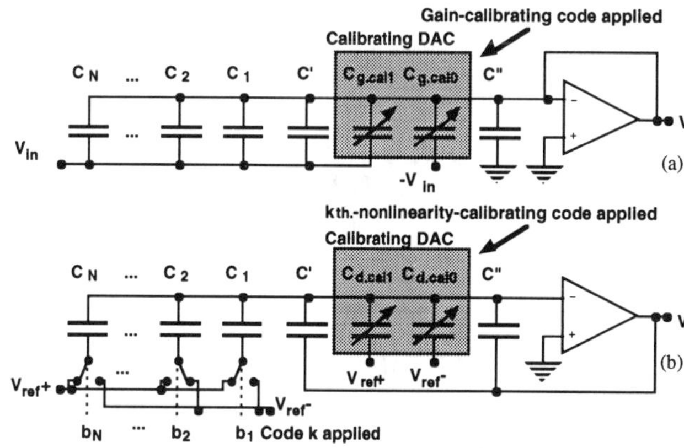

Fig. 5: Conversion mode: a) sampling, b) residue amplification.

IV. Integrated MDAC Prototype

In order to prove the feasibility of the proposed self-calibration technique, the 4-bit MDAC prototype was designed in a 1.0 µm CMOS technology for 15-bit linearity and 5 MS/s and includes only the analogue part of the block diagram of Fig.1. The S/H, the 4-bit flash quantizer and an additional 5-bit flash quantizer following the output of the MDAC were included for testing purposes. For improved performance fully differential structures were adopted. Fig. 6 shows the scheme of the pulse-counting reference DAC, which is based on a low gain-sensitivity integrator [12]. The capacitance values of 4 pF and 64 pF for the input and feedback capacitors, respectively, insure that sampled thermal noise is bellow the 16-bit level. Although this structure would reach the target 15-bit of linearity by using an opamp with only 65 dB of DC-gain, the same high-speed high-gain opamp was used for this reference DAC and for the MDAC. This fully differential folded cascode OTA with gain boosting obtained from regulated cascodes exhibits simulated 106 dB DC-gain and 100 MHz of gain-bandwidth product when loaded with 9.6 pF. The MDAC employs a unit capacitor of C=0.5 pF due to thermal noise constraints. A 12-bit sub-binary calibrating-DAC implemented with a capacitor divider array was used, as shown in Fig. 7 [4] . The adopted sizing of C1'=20fF, C2'=35.2fF, C3'=66.2fF, C4'=124.4fF, Cd1=30fF, Cd2=20fF does not significantly increase the load of the opamp. The flash quantizers and the S/H were designed with classic structures [11]. Fig. 8 shows a microphotograph of the complete die, where the the blocks Ref-DAC, MDAC, Cal-DAC and HAC occupy 1.4 mm^2 and dissipate 60 mW. The measured residue amplification characteristics are represented in Fig. 9, before and after calibration. Fig. 10 shows the magnified middle-points of the various segments of the residue amplification characteristic. The initial nonlinearity of the MDAC was at the 8-bit level and the gain was 7.984. After calibration, the observed nonlinearity is aproximately ±2 LSB at the 15-bit level and the gain was corrected to 7.997. These results are compatible with an effective 14-bit overall converter resolution.

V. CONCLUSIONS

A 1.0 μm CMOS integrated MDAC for a 15-bit high-speed ADC was described which uses an analogue self-calibration technique based on a high-linearity reference-DAC. The experimental results show that the adopted self-calibration technique is capable of calibrating both MDAC nonlinearities and interstage-gain errors to the 14-bit level.

ACKNOWLEDGEMENTS: This work was partially supported by the CEC under ESPRIT Project 8795 (AMFIS) and by JNICT under the Project PCERN/P/FAE/122/94 (ACTIX).

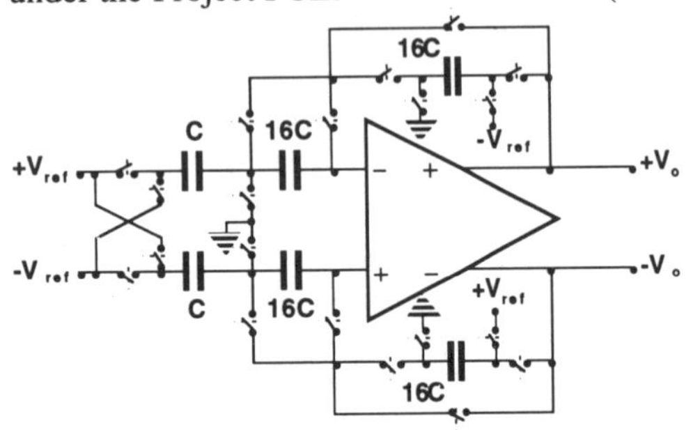

Fig.6: Pulse-counting reference DAC.

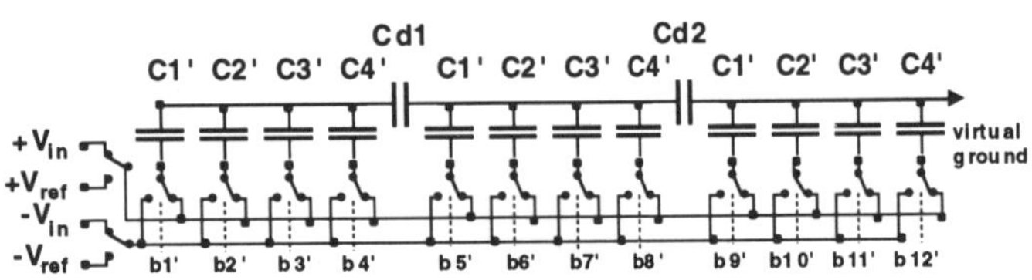

Fig. 7: The structure of the 12-bit calibrating-DAC.

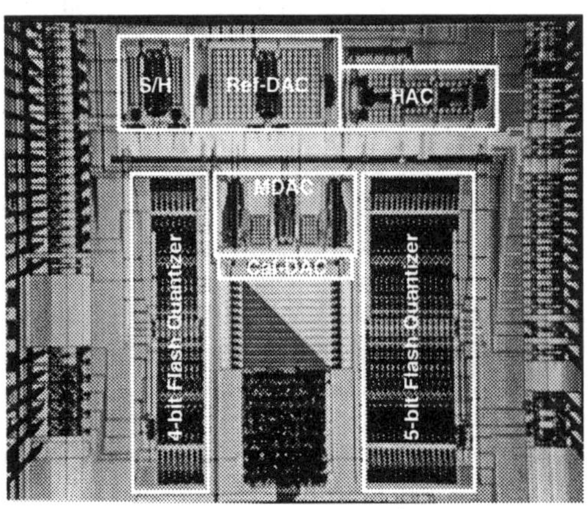

Fig. 8: Chip Microphotograph

REFERENCES

[1] S. H. Lewis, P. R. Gray, "A Pipelined 5-Msample/s 9-bit Analog-to-Digital Converter", *IEEE Journal of Solid-State Circuits*, Vol. SC-22, No. 6, pp. 954-961, Dec. 1987.

[2] H.-S. Lee, D. A. Hodges, P. R. Gray, "A Self-Calibrating 15 Bit CMOS A/D Converter", *IEEE Journal of Solid-State Circuits*, Vol. SC-19, No. 6, pp. 813-819, Dec. 1994.

[3] S.-H. Lee, B.-S. Song, "A Direct Code Error Calibration Technique for Two-Step Flash A/D Converters", *IEEE Transactions on Circuits and Systems*, Vol. 36, No. 6, pp. 919-922, Jun. 1989.

[4] Y.-M. Lin, B. Kim, P. R. Gray, "A 13-b 2.5-MHz Self-Calibrated Pipelined A/D Converter in 3-μm CMOS", *IEEE Journal of Solid-State Circuits*, Vol. 26, No. 4, pp. 628-636, Apr. 1991.

[5] S.-H. Lee, B.-S. Song, "Digital-Domain Calibration of Multistep Analog-to-Digital Converters", *IEEE Journal of Solid-State Circuits*, Vol. 27, No. 12, pp. 1679-1688, Dec. 1992.

[6] S.-H. Lee, B.-S. Song, "Simplified Digital Calibration for Multi-Stage Analog-to-Digital Converters", *Proc. IEEE ISCAS*, pp. 1216-1219, 1993.

[7] S.-H. Lee, B.-S. Song, "Interstage Gain Proration Technique for Digital-Domain Multi-Step ADC Calibration", *IEEE Trans. CAS-II: Analog and Signal Processing*, Vol. 41, No. 1, pp. 12-18, Jan. 1995.

[8] J. Goes, J. Franca, N. Paulino, J. Grilo, G. Temes, "High-Linearity Calibration of Low-Resolution Digital-to-Analog Converters", *Proc. IEEE ISCAS*, pp. 345-348, 1994.

[9] J. Goes, J. Vital, J. Franca, "An Analogue Self-Calibration Technique for High-Resolution Video-Rate Pipelined A/D Converters", *Proc. IEEE 38th. Midwest Symp. on CAS*, 1995.

[10] J. Goes, J. Vital, J. Franca, "Optimum Resolution-per-Stage in High-Speed Pipelined A/D Converters using Self-Calibration", *Proc. IEEE ISCAS*, pp. 525-528, 1995.

[11] B.-S. Song, S.-H. Lee, M. F. Tompsett, "A 10-b 15-MHz CMOS Recycling Two-Step A/D Converter", *IEEE Journal of Solid-State Circuits*, Vol. 25, No. 6, pp. 1328-1338, Dec. 1990.

[12] K. Nagaraj, J. Vlach, T. R. Viswanathan, K. Singhal, "Switched-Capacitor Integrator with Reduced Sensitivity to Amplifier Gain", *IEE Elec. Letters*, Vol. 22, No. 21, pp. 1103-1105, Oct. 1986.

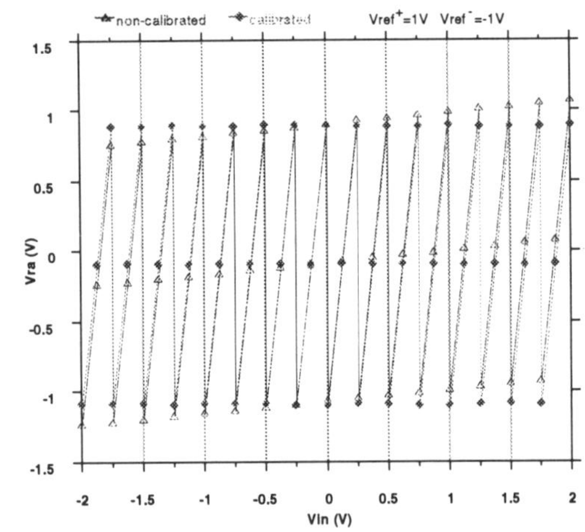

Fig. 9: Measured residue amplification characteristics

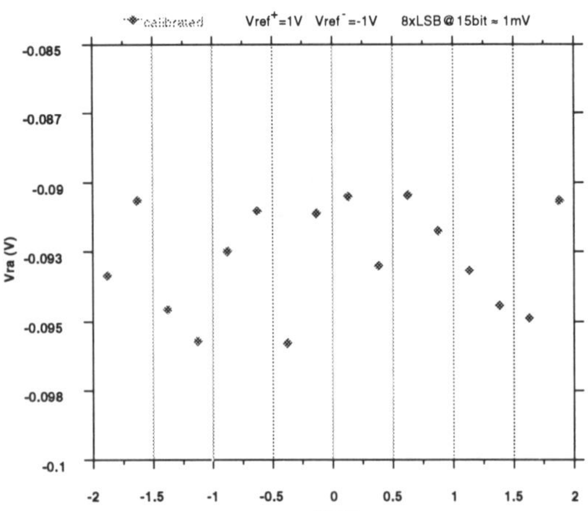

Fig. 10: Details of the middle-points of the calibrated residue amplification characteristic

6.6.4

LOW POWER CIRCUIT DESIGN TECHIQUES AND IMPLEMENTATIONS

SESSION 7

Chair: Michael Brassington
Co-Chair: Doug Garrity

The progressive scaling of supply voltages for deep submicron technologies, is being outpaced only by the exploding global demand for low power, portable electronic systems such as cellular telephones, laptop computers, PDAs, etc. This trend is forcing the development of electronic circuits which can operate at very low voltage and consume negligible power. In some cases the constraints imposed by these requirements are forcing the development of radically new circuit design concepts. Several such approaches are presented in this session.

A range of design techniques and implementations are discussed for low power digital, analog and mixed signal applications. Examples include power management by an embedded micro-controller, adiabatic circuits, 0.9V SRAM and low power telecommunications circuitry. In addition, two micro-power circuits are described for rechargeable lithium-ion battery protection.

The first paper, highly efficient power management for a DSP application is achieved by the incorporation of an embedded processor which is used to control sleep modes throughout the integrated circuit. A 90% reduction in power consumption is achieved.

Adiabatic circuits have been proposed as an inherently low power design methodology. The second paper in this session provides a valuable evaluation of the characteristics arithmetic units which incorporate adiabatic circuitry.

This paper is followed by the description of a 4Kb SRAM design which combines the use of sub-blocked array architectures, selective use of NOR/NAND based decode logic and glitch-free sense amps to achieve functionality of 12MHz at 0.9V and power consumption of only 22uW.

The fourth paper describes a singly chip circuit design for a high performance, low voltage baseband processor intended for use in cordless telephones. This design operates down to 2.7V.

In line with the growing popularity of rechargeable lithium ion batteries for portable applications, the final two papers address the need for in-pack, electrical over-stress protection circuits which consume the absolute minimum power in stand-by mode.

Power Management Technique for 1-V LSIs using Embedded Processor

Satoshi SHIGEMATSU, Shin'ichiro MUTOH and Yasuyuki MATSUYA
NTT LSI Laboratories
3-1, Morinosato Wakamiya, Atsugi-shi, 243-01 Japan

Abstract

A new power management technique is proposed for low-power, high-speed LSIs. This technique reduces the power consumption and enhances the performance of an LSI by using an embedded small processor to control the sleep modes and the processing of the LSI. Using this technique for a low-power DSP, the total power is reduced to about 10 % of one without this technique, while maintaining the speed performance in 1-V LSIs.

Introduction

There has recently been an increasing demand for small long-life high-performance portable devices such as cellular phones and notebook computers. Reducing the supply voltage to 1 V (one battery cell) is the most effective way to extend battery life, downsize the equipment, and assure the reliability of circuits in sub-quarter micron devices. In low voltage circuits, the threshold voltage V_t should also be reduced for high-speed operation. However, lowering V_t exponentially increases the standby leakage current. To solve this problem, circuit schemes having two modes (active and sleep) have been proposed to control the leakage current in the circuit; the standby leakage current is reduced in the sleep mode [4-6]. These modes are controlled by an external control signal (sleep signal) in all schemes, however there has not been any discussion about this signal yet. Any scheme designed to lower the standby power cannot work satisfactorily without effective control of the sleep signal. Therefore, management of the sleep and active modes is the most important issue in low-voltage and low-V_t LSIs. This paper proposes a new power management technique in which these modes are effectively managed using an embedded small processor. This produces low-power high-performance LSI at low voltages.

Low-voltage and low-V_t schemes

Conventional circuits such as those used to implement MPUs and DSPs use the power management technique in which the clock is stopped to reduce the standby power. However, in the low-V_t circuit, this technique cannot reduce the standby power satisfactorily because a large amount of the leakage power is wasted even though the circuits stop. Therefore, many circuit schemes have been studied to try and reduce this leakage power.

We have already proposed a multi-threshold-voltage CMOS (MTCMOS) circuit scheme to lower standby leakage power of the low-V_t circuit [1]. This MTCMOS circuit is shown in Fig. 1. The logic gates are composed of low-V_t MOSFETs to enhance speed at low voltages. A

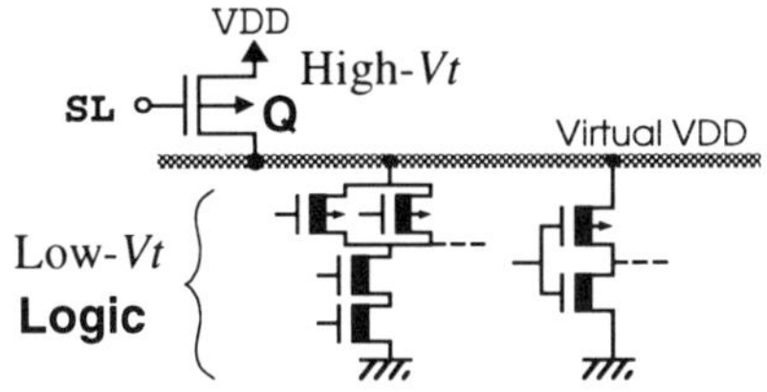

Fig. 1 MTCMOS circuit scheme

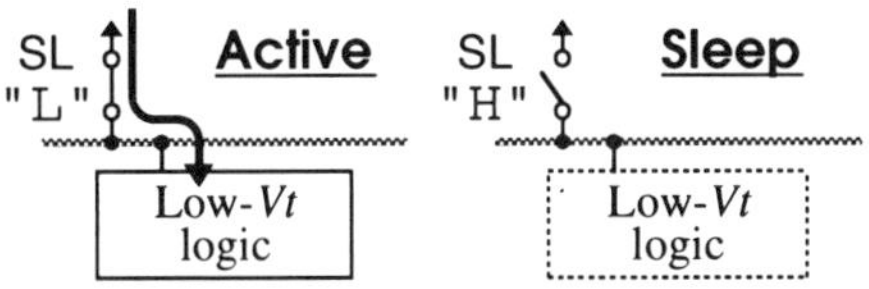

Fig. 2 Modes in MTCMOS

high-V_t MOSFET, Q, connects the power supply to the logic core, and this Q is controlled by a sleep signal (SL), which switches between the active and sleep modes. Fig.2 shows these modes. In the active mode, Q supplies power to the logic core. In the sleep mode, Q suspends the power supply to cut the standby leakage current. The MTCMOS circuit scheme produces low-power, high-speed operation at voltages as low as 1 V. Other circuit schemes also use the active and sleep modes.

The management of these modes is very important for applying low-V_t schemes to real devices, but no effective management technique has been devised so far.

Power management technique

A. Low-V_t LSI architecture

The most effective way to manage low-V_t schemes is to extend the sleep period as much as possible, because leakage power is reduced during this time in all schemes. In our new power management technique, a small embedded power management processor (PMP) controls the low-V_t circuit to extend the sleep period. The PMP is composed of only "high-V_t " MOSFETs for low leakage power. Fig. 3 shows a block diagram of a low-voltage, high-speed LSI with the processor. The application circuit is composed of a low-V_t MOSFET with the low-V_t scheme to lower the standby power. The PMP is embedded to control the low-V_t circuit using the sleep signal. It also has a slow interface to communicate with external devices, and has a sleep control circuit that produce several sleep signals for blocks in the low-V_t circuit independently. The PMP can control some blocks carefully according to their status and can distribute some operations between itself and the low-V_t circuit, so it is embedded on the low-V_t circuit chip. A common memory is used to transfer

7.1.1

0-7803-3177-6 $5.00 © 1996 IEEE
IEEE 1996 CUSTOM INTEGRATED CIRCUITS CONFERENCE

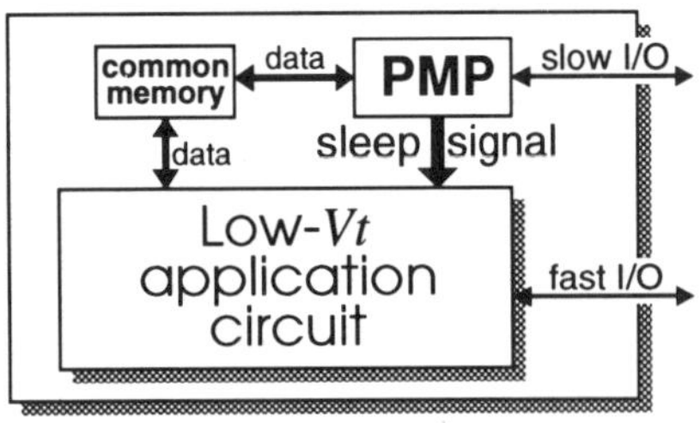

Fig. 3 Block diagram of low-voltage, high-speed LSI

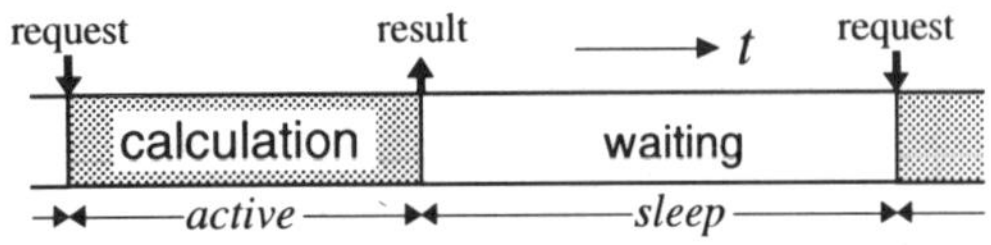

Fig. 4 Operations of low-Vt LSI

data between the PMP and the low-V_t circuit. In this architecture, the PMP masters the low-V_t circuit to "find" and "create" the sleep period in order to extend it. This capability is the main feature of our technique.

B. Finding the sleep period

It is necessary to find a period when the low-V_t circuit can sleep, and then to change the mode rapidly.

Fig. 4 shows a very common operation in portable devices. In this operation, calculations in the low-V_t circuit are started by external request signals from a keyboard, a timer, or an I/O device. After the calculation, the low-V_t circuit outputs the result and waits for the next request signal. During this waiting period, the low-V_t circuit can sleep. In our technique, the PMP always watches the low-V_t circuit and the external devices to find out when this period is. If the PMP detects a transition of the state, then it must switch the mode of the low-V_t circuit rapidly to reduce the power and suppress any delay in calculation.

When the PMP detects a request from the external devices, it must wake the low-V_t circuit up rapidly. Fig. 5 shows the wake-up protocol. The PMP wakes the low-V_t circuit according to following sequence:

1. The PMP watches the request signals and receives the data instead of the low-V_t circuit, thus allowing the low-V_t circuit to sleep until data collection is complete.
2. When the PMP has collected enough data for calculation, it sends the data and some information, which the low-V_t circuit uses, to the common memory.
3. The PMP sends signals to switch the mode of the blocks to be used for calculation in the low-V_t circuit from sleep to active.
4. The blocks in the low-V_t circuit are powered, and prepare to begin the calculation.
5. The PMP sends a reset or interrupt signal to the low-V_t circuit to start the calculation.
6. The low-V_t circuit gets the data from the common memory, and starts the calculation.

When the PMP detects completion of the calculation, it must put the low-V_t circuit to sleep rapidly. Fig. 6 shows protocol to do this. The PMP puts the low-V_t circuit to sleep according to the following sequence:

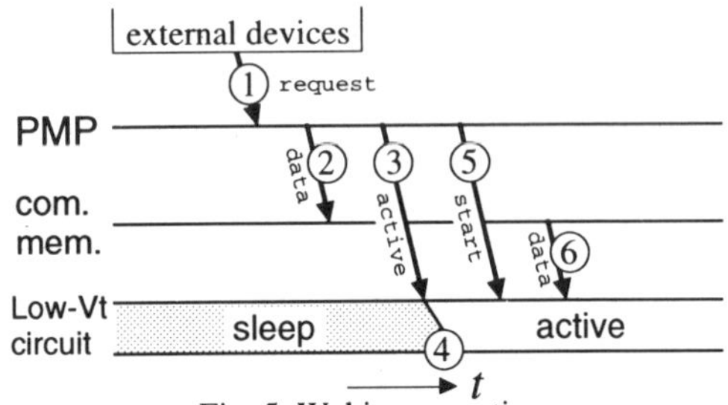

Fig. 5 Waking operation

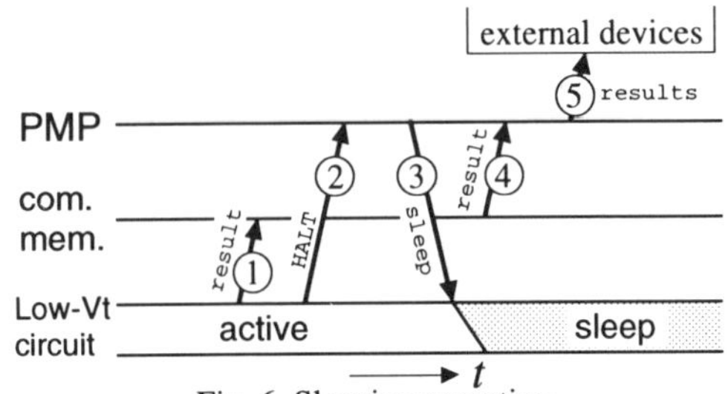

Fig. 6 Sleeping operation

1. When the calculation is finished, the low-V_t circuit sends the result and some information to the common memory.
2. The low-V_t circuit halts, and a "halt" signal is sent to the PMP.
3. The PMP sends sleep signals to switch the mode of the blocks used from active to sleep, and the low-V_t circuit goes to sleep.
4. The PMP gets the result from the common memory.
5. The PMP sends the result through the slow interface instead of the low-V_t circuit.

Thus it is rather complicated to control the sleep signal because a sleep signal controller must carefully watch the state of both the external devices and the low-V_t circuit (e.g. running, halting, or sleeping). Therefore, the best solution is to use the small embedded processor PMP as the sleep controller on the same chip. Moreover, this PMP can effectively determine the sleep period according to the state of the system because it can perform complex operations and can make sleep signals for several blocks independently.

C. Creating the sleep period

The PMP can reduce the leakage power by determining the sleep period and rapidly putting the low-V_t circuit to sleep. In order to reduce the power consumption further, it is necessary to create a new sleep period.

From the analysis of the calculation period (see Fig. 7(a)), we found that there are two kinds of operations: one where the calculation quantity is light and another where it is heavy. The power leaks during the active period regardless of the calculation quantity. In our technique, the new sleep periods are created by distributing these light and heavy operations to the PMP and the low-V_t circuit.

While the PMP is controlling the sleep signal, it can also execute light operations such as controlling the external devices and communicating with them through the slow interface because it basically operates only during

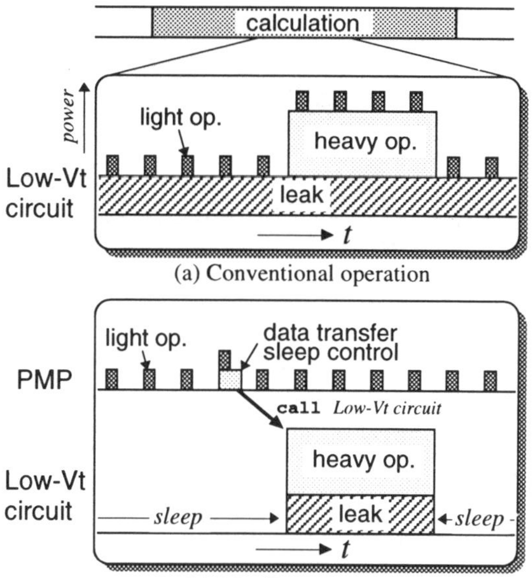

(a) Conventional operation

(b) Distributed operation

Fig. 7 Operation during calculation

the mode transition. Therefore, heavy operations, such as voice CODEC, and light operations, such as data collection, can be distributed to both the low-V_t circuit and the PMP, and the low-V_t circuit can sleep when the PMP is executing light operations.

In the conventional operation shown in Fig. 7(a), all of the low-V_t circuit, which comprises low-V_t MOSFETs, is used even for light operations. Hence LSI power is wasted due to leakage current because the mode of the low-V_t circuit is always active. In the distributed operation shown in Fig. 7(b), the PMP is used for light operations with low power consumption, and the low-V_t circuit is used only for heavy operations. Therefore, the active period of the low-V_t circuit can be minimized. In this technique, the power consumption of the PMP must be smaller than the leakage power of the low-V_t circuit. This is not a problem because the PMP operates very slowly and the leakage power of the low-V_t circuit increases because V_t is reduced to increase the speed for high-speed application circuits such as MPUs and DSPs.

In this distributed operation, the common memory is used to communicate between the low-V_t circuit and the PMP, and can absorb a difference in speed between the two. This memory has two ports whose sleep modes are controlled independently in order to put the port which is not in use to sleep.

The power consumption is further reduced because this technique shortens the active period in which the leakage current is being wasted. Moreover, the low-V_t circuit can be designed for high-speed by, for instance, reducing the V_t and narrowing the gate of the MOSFET. Therefore, using this technique, which finds and creates sleep periods, lower-power and higher-speed low-V_t circuits can be built.

Experimental results

In order to confirm the effectiveness of this power management technique, it was applied to a 1-V DSP [3].

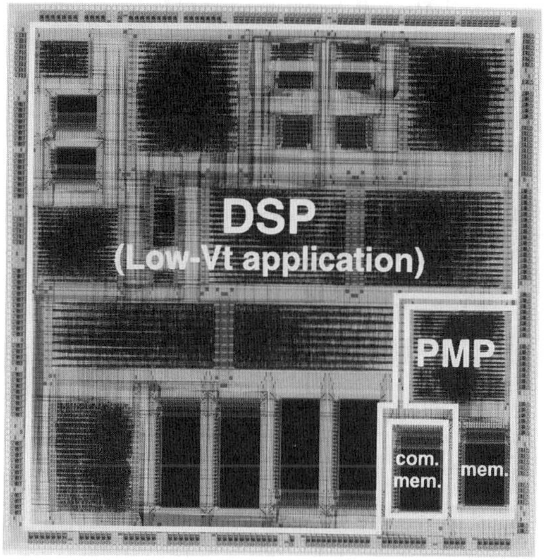

Fig. 8 1-V DSP chip microphotograph

TABLE I Characteristics of 1-V DSP

Process technology	0.5 μm MTCMOS
Die size	15 mm x 15 mm
Transistor count	Logic 267 K, RAM 123 K
Operating frequency	13.2 MHz at 1.0 V
Power consumption	2.2 mW/MHz at 1.0 V
Standby power	350 μW (active) at 1.0 V
	600 nW (sleep) at 1.0 V

A. 1-V DSP

The DSP chip was fabricated using the 0.5-μm MTCMOS process. The chip microphotograph is shown in Fig. 8. The operating frequency of 13 MHz and MAC performance of 26 MOPS were confirmed. The energy is 2.2 mW/MHz at 1.0 V. The details of the DSP are summarized in Table I.

B. PMP

The PMP was embedded on the DSP chip, and was composed of only high-V_t MOSFETs. The PMP's core area is less than 2.8 % of the chip area, as shown in Fig. 8. It was designed as a 16-bit micro-processor for power management, and has simple instructions like an 8-bit micro-processor. The details of the PMP are summarized in Table II. An operating cycle of 3.9 MHz was confirmed at 1.0 V. The operating current dependence on the supply voltage is shown in Fig. 9. The energy is only 57 μW/MHz at 1.0 V, and the power consumption of the PMP is less than 10 % of the leakage power of the DSP core. These results confirmed that the PMP is a low-power processor and it can be used for the new power management technique.

C. Effectiveness of power management

Fig. 10 shows the power consumption as a function of the intermittent ratio, which is the waiting/calculation ratio. It shows the effectiveness of finding the sleep period. At the intermittent ratio of 100, the power is reduced to

TABLE II Characteristics of PMP

Process technology	0.5 µm CMOS
Die size (core)	2.7 mm x 2.4 mm
Transistor count	31 K Tr.
Operating cycle	3.91 MHz at 1.0 V
Power consumption	56.8 µW/MHz at 1.0 V
System clock	4 clock / cycle
Instruction issue	1 instruction / cycle
Sleep control	16 blocks

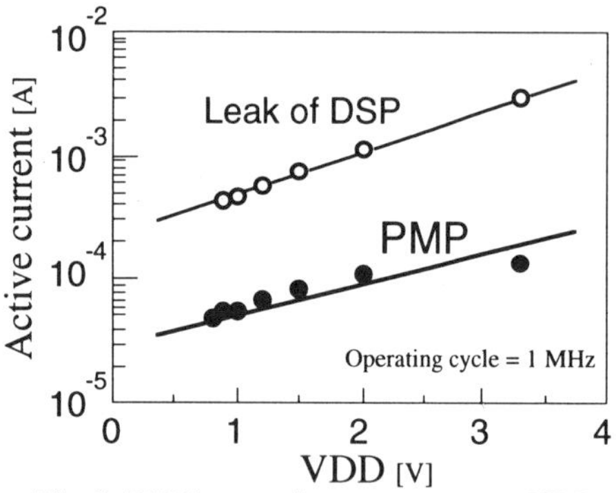

Fig. 9 PMP's operating current versus VDD

about 1/20. This means that finding the sleep period is very effective in lowering the power, especially for portable devices with large intermittent ratios.

To confirm the effectiveness of creating the sleep period, the distribution technique was applied to a calculation equivalent to the one for half-rate voice decoding. In this application circuit, the data transmission frequency is 5.6 KHz, the frame period, which represents a unit of calculation, is 40 ms, and 140000 MAC operations are executed in each frame by the DSP. The data transmission and data decoding are distributed to the PMP and the DSP. Fig. 11 shows the effectiveness of the distribution dependence on the V_t of low-V_t MOSFETs. The distributed operation is much lower than the two operations — i.e., in which the calculations are not distributed and in which the calculations are not distributed but the DSP sleeps without I/O circuit — for any V_t This means this technique can lower the V_t of LSIs while maintaining low power, and so low-power, high-performance circuits can be realized.

The total effectiveness of the new power management technique is shown in Fig. 12. This application circuit

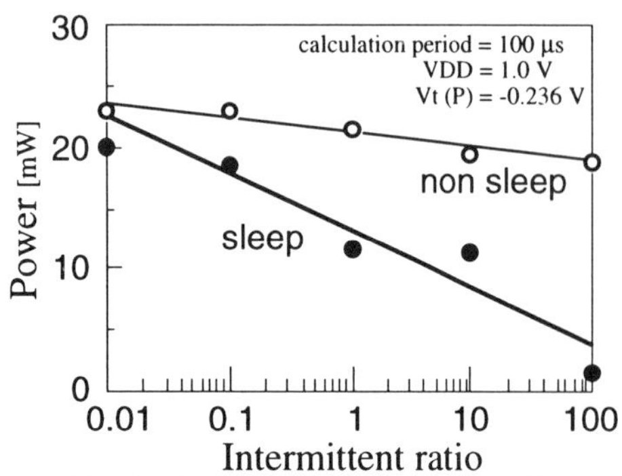

Fig. 10 Power versus intermittent ratio

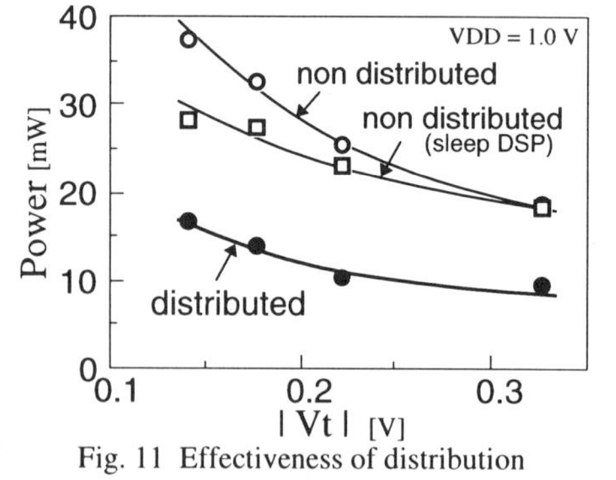

Fig. 11 Effectiveness of distribution

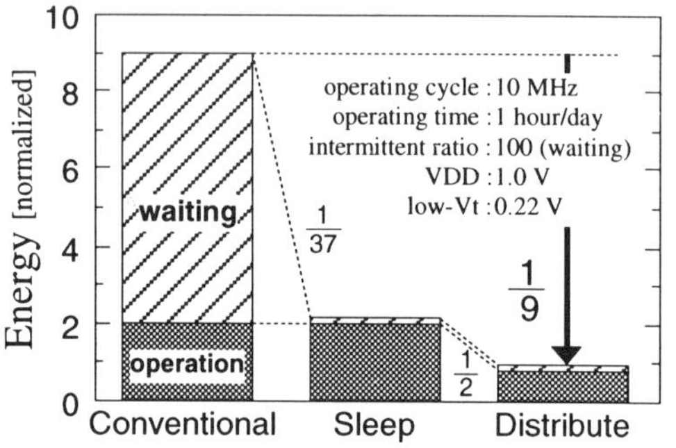

Fig. 12 Total power management effect

is a common portable device that is generally used for one hour a day and operates at 10 MHz at 1.0 V. Using the finding technique, the power required for waiting is reduced to about 1/40. The operation power is decreased to one half by the distribution. Thus this new power management technique reduces the total power of the low-V_t circuit to about 10 % of one which is not managed.

Conclusion

An effective power management technique is proposed for low-power, high-speed LSIs. The embedded power management processor PMP was used to enhance the sleep period in which leakage power is suppressed by low-V_t circuit schemes. The PMP finds and creates the sleep period by controlling the sleep signal and the distribution of the calculation. Using this technique for a low-power DSP, the total power is reduced to about 10 % of one not using this technique, while maintaining the speed performance in 1-V LSIs.

Acknowledgment

The authors would like to thank S. Horiguchi, H. Fukuda and J. Yamada for their support and comments.

References

[1] Mutoh, S. *et al.,IEEE J.Solid-State Circuits,* vol.30, no.8, pp. 847-854, Aug. 1995.

[2] Mutoh, S. *et al.,ISSCC Dig.Tech.Papers,* FA 10.4, Feb. 1996.

[3] Burr, J. *et al.,ISSCC Dig.Tech.Papers,* pp.84-85, Feb. 1994.

[4] Horiguchi, M. *et al.,IEEE J.Solid-State Circuits,* vol.28, no.11, pp. 1131-1135, Nov. 1993.

[5] Takashima, D. *et al.,IEEE J.Solid-State Circuits,* vol.29, no.4, pp. 441-447, Apr. 1994.

[6] Seta, K. *et al.,ISSCC Dig.Tech.Papers,* pp. 318-319, Feb. 1995.

Implementing and Evaluating Adiabatic Arithmetic Units*

Micah C. Knapp Peter J. Kindlmann Marios C. Papaefthymiou

Department of Electrical Engineering
Yale University
New Haven, CT 06520

Abstract

In recent years, several adiabatic logic architectures have been proposed for low-power VLSI design. However, no work has been presented describing the implementation and evaluation of nontrivial adiabatic circuits. We have evaluated a specific adiabatic architecture and used it in the design of low-power arithmetic units. We investigated implementation issues specific to adiabatic system development and performed a systematic comparison of our designs with corresponding CMOS circuits. In this paper we describe our adiabatic designs, discuss implementation issues at the logic and architectural level, and report our empirical findings.

1. Introduction

Low-power circuit design methodologies based on the thermodynamic principle of adiabatic changes have received considerable attention in recent years. Adiabatic circuits achieve low energy consumption by maintaining small potential differences across their devices while they are conducting and by allowing the energy stored in their capacitors to be recycled. Energy is supplied periodically to the gates of an adiabatic circuit by means of slowly changing clocking waveforms. These waveforms are driven by oscillators which recycle the energy stored in the circuit's capacitors. Despite the development of several adiabatic logic architectures, there exist no reports on the difficulties associated with the development of nontrivial adiabatic circuits or on the actual power savings that have been achieved by nontrivial adiabatic circuits.

We recently performed an empirical study of several adiabatic logic structures that have been proposed in the literature. We focused our investigation on the logic architecture described in [4] and we used it to design adiabatic adders and multipliers with low energy dissipation. In this paper we describe our designs and discuss related logic and architectural level implementation issues such as transistor sizing, data synchronization, latency, and throughput. We also present an empirical comparison of our designs with corresponding CMOS designs in terms of their energy consumption. When simulated at operating frequencies of 10MHz and 33MHz, our arithmetic units achieved energy savings factors of 4 and 3, respectively.

The remainder of this paper has five sections. In Section 2 we briefly present a few adiabatic logic structures, and in Section 3 we discuss the logic-level implementation issues associated with the architecture we used in our designs. In Section 4 we describe our designs and discuss related system-level implementation issues. In Section 5 we present the results of our empirical comparison. We conclude with a discussion of possible target systems for adiabatic implementation.

2. Adiabatic logic families

Researchers have proposed several adiabatic logic architectures. A family of structures based on reversible logic has been described in [6, 7]. An architecture inspired by bipolar transistor structures has been proposed in [3]. Two fairly simple adiabatic architectures based on diodes have been presented in [1, 2], and an extension of these structures that uses cross-coupled p-type transistors instead of diodes has been described in [4]. Inverters from the last three logic families are shown in Figure 1 and are named by the number and types of devices needed to generate them.

Our empirical study covered the structures in Figure 1 and focused on the 2N2P architecture with two NMOS and two PMOS transistors in its inverter. All of these adiabatic structures are dynamic, that is, the input clock provides the energy that drives the output loads of each gate. The adiabatic logic families we studied require either two or four clock phases for their operation, with relative shifts of 180° or 90°, respectively. These clocks can be sinusoidal and can be driven from free-running oscillators.

Our simulations of simple adiabatic structures and [1, 2, 4] indicated that the most important operational characteristics for the successful utilization of adiabatic logic architectures are a constant load presented by the gates to the clock, the reduction or elimination of floating output nodes, the ability to generate a signal and its complement on the same clock phase, the reduction of the number of clock phases needed for correct operation, the ability to drive a high output to V_{dd} and a low output to V_{ss}, predictable operation under normal operating con-

*This work was supported in part by a Young Investigator Award from the US Army Research Office under Grant No. DAAH04-95-1-0281.

0-7803-3177-6 $5.00 © 1996 IEEE IEEE 1996 CUSTOM INTEGRATED CIRCUITS CONFERENCE

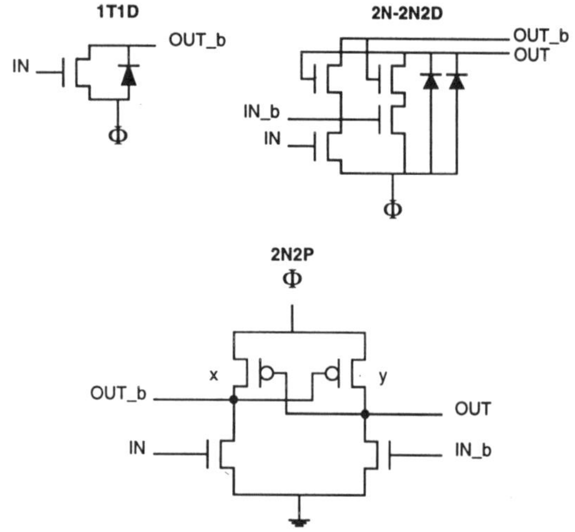

Figure 1: Adiabatic inverters with and without diodes.

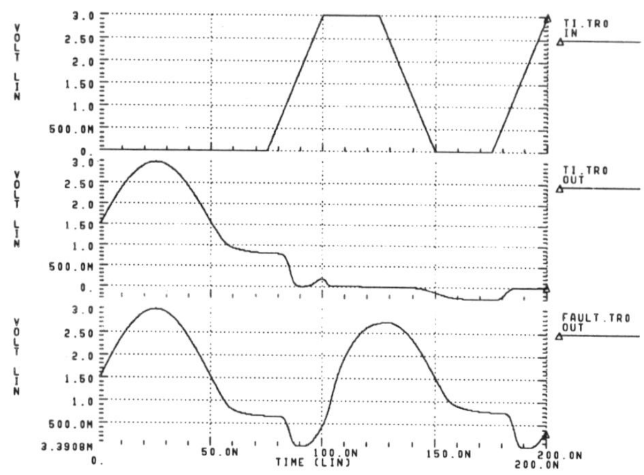

Figure 2: Fault in an incorrectly ratioed 2N2P inverter.

ditions, and the reduction or elimination of diodes. No single logic family covered all of these requirements, but we found that the 2N2P devices covered more than the other structures.

3. Logic-level issues concerning 2N2P

The 2N2P logic family is very similar to CMOS CVSL logic [5] and thus possesses several of the positive characteristics of conventional CMOS. Since 2N2P gates use ground nodes and avoid diodes, they can drive their output loads to V_{dd} and V_{ss}. Also, for at least some portion of the clock cycle, the outputs of the 2N2P gates are not floating and, therefore, are much less susceptible to noise. The 2N2P logic gates generate both the output signal and its complement on the same clock phase and can use the differential signal pairs to generate fairly complex logic functions such as XOR in one gate. Another benefit of generating both a signal and its complement from the same gate is that the load "seen" by the input clock is relatively the same regardless of what output level is being driven. This balanced load prevents the clock oscillator from changing frequency due to shifts in the capacitance of the circuit. Even though they are connected to the ground, the 2N2P logic gates can recycle a large percentage of the energy used to charge their output nodes.

Unfortunately, the 2N2P logic family has some potentially serious problems. The logic gates cannot hold an output value for more than a quarter of the clock cycle and, therefore, require four clocks to build logic circuits. Moreover, the 2N2P logic gates use a ratioed gate structure such that the PMOS and NMOS transistors must be sized according to the output load and to each other. If an incorrect ratio is used, a logic fault may occur which, in addition to driving an incorrect output value, causes the gate to consume significant amounts of energy.

We looked closely at the logic fault that can occur with 2N2P, and discovered that it is caused by the manner in which the logic gates evaluate their output levels. We describe the events which force a logic fault to occur in terms of the 2N2P inverter (as shown in Figure 1). When the input clock Φ reaches the PMOS threshold voltage, both PMOS transistors x and y turn on and start conducting. The drain of one of the two PMOS transistors, for example x, is connected to ground through a tree of NMOS gates, and its output OUT_b is thus forced to "low". Due to the cross-coupling of OUT and OUT_b, transistor y remains on and turns x off when the voltage of signal OUT exceeds the PMOS threshold voltage. The problem arises when the PMOS transistors are too strong with respect to the NMOS tree, in which case the output OUT_b that should be forced to ground through the NMOS tree is driven to the PMOS threshold value via x. When this occurs, the output OUT_b is forced to the wrong state.

Figure 2 illustrates the logic fault that can occur with a 2N2P logic inverter. The top graph shows the input data. The middle graph shows the output waveform of a correctly functioning 2N2P inverter, where the PMOS width is 20λ, the NMOS width 3λ, and the output load is 100fF. The bottom graph shows the output of a 2N2P inverter exhibiting a functional fault (the second output should be at "0" instead of "1"). In this case, the PMOS width is 30λ, the NMOS width is 3λ, and the output load is 100fF.

To prevent this logic fault, we can recommend, that the NMOS logic trees be kept fairly small, that is, not many NMOS transistors in series between the output and ground, and that the PMOS transistors are kept fairly "weak" with respect to the NMOS transistors. Unfortunately, "weak" PMOS transistors connect the output loads to the input clock through a large resistance, and therefore, increase the energy consumed by the gate. These factors must be considered and balanced correctly to build a functional, low-power 2N2P logic gate. The 2N2P logic gates must be used with caution, and large circuits built with this logic family must be thoroughly simulated with accurate load estimates to ensure a functional circuit.

4. Systems implementation issues

We have used 2N2P logic gates to design two adiabatic circuits of medium complexity, a 4x4 bit carry lookahead

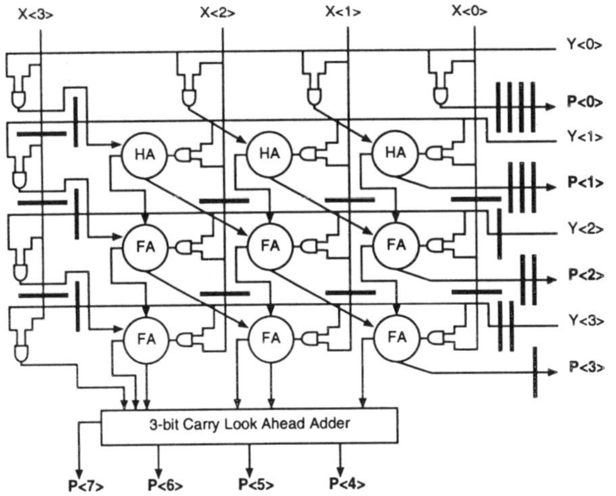

Figure 3: 4x4 bit adiabatic multiplier.

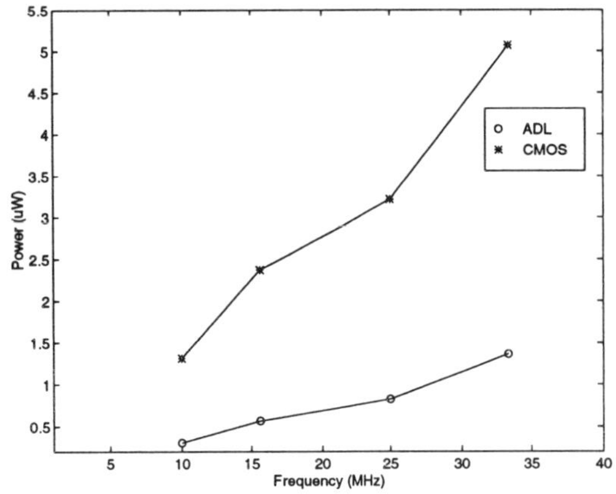

Figure 4: Power vs frequency for 4x4 bit adders.

adder with about 200 transistors and a 4x4 bit multiplier (shown in Figure 3) with about 500 transistors. Both circuits assume parallel data in and out, have been extensively simulated using Hspice and function correctly. We have also designed larger circuits, an 8x8 bit carry lookahead adder with about 700 transistors and an 8x8 bit multiplier with approximately 2500 transistors, which also appear to be functionally correct based on preliminary simulations. For all designs we used MOSIS $1\mu m$ process technology with $V_{dd} = 3$ Volts and minimum-size transistors ($l/w = 3\lambda/2\lambda$, where $\lambda = .5\mu m$).

We have identified three important system design issues with 2N2P that are mostly applicable to system design with any adiabatic logic family. The first issue is that adiabatic circuits have limited potential to achieve both significant energy savings and high performance. One of the largest performance limiting factors is that the dynamic gates in an adiabatic circuit can only perform one logic evaluation per clock cycle. Since the gate clock frequencies must be slow to achieve savings in energy consumption, the overall circuit speed can be *very* slow. For example, consider an adiabatic circuit consisting of four cascaded logic gates. The latency of this circuit is 100ns, assuming a 10MHz gate clock frequency. The upper bound on the throughput is 10MHz (that is, new input data every 100ns), since each gate in the circuit can analyze new data only once per cycle. Since 2N2P circuits have large latencies and relatively low throughputs, each gate must perform as complicated a function as possible in order to increase system performance. In Section 3, however, we saw that a long and weak NMOS tree can lead to logic faults with 2N2P. Thus, the complexity of the functions that can be performed reliably by any gate without excessively increasing its size and energy consumption is limited.

Another critical system issue that stems from the dynamic nature of the adiabatic gates is data synchronization. Since every gate introduces a phase delay from input to output, two signals that originate from different clock phases must be synchronized in order to be used as inputs to the same adiabatic gate. This requirement leads to the insertion of data buffers, each of which shifts a signal by one phase (a quarter clock period). A large number of additional gates are thus added into the circuit just to maintain data synchronization. For example, our 4x4 bit adiabatic multiplier requires the addition of 25 data buffers to maintain data synchronization. These buffers are denoted as thick black lines in the block diagram of our 4x4 adiabatic multiplier shown in Figure 3. An adiabatic logic family that only requires two clocks (where each gate introduces a half clock period delay) would require fewer data buffers but would get less work done per period, since only two logic functions could be performed per clock period.

Every adiabatic logic family we investigated uses mostly analog signals to perform digital functions, thus allowing the logic to recycle energy and limit the power dissipated by the circuit. Unfortunately, it is impossible to simulate such systems with simple digital simulators. More complex analog simulators such as Hspice must be used to not only test critical circuits and special "corner cases", but also to test the functionality of the circuit. For more complex circuits, the computing requirements can be very high. For example, simulations of our 4x4 multiplier with 256 input test vectors take on the order of 4 hours to complete using Hspice on a state of the art PowerPC-based workstation with 64MB main memory. Since adiabatic systems must be thoroughly tested to ensure correct functionality, large scale development of adiabatic systems can be very expensive in terms of time and resources.

5. Systems evaluation

We compared the energy consumption of our adiabatic arithmetic units with that of equivalent CMOS circuits that we designed for minimum energy consumption. The input data rates for the CMOS circuits were identical to those for the adiabatic circuits, and therefore, our comparison was based on energy consumed per computation (the circuits did equal amount of computing during equal amounts of time). We reduced the amount of glitching in the CMOS circuits using standard CMOS design techniques. Since the energy consumed per operation in a CMOS circuit can be approximated by $C \cdot V_{dd}^2$, we used minimum sized transistors in all CMOS gates in order to minimize the overall circuit capacitance.

Figures 4 and 5 show the energy consumption per clock

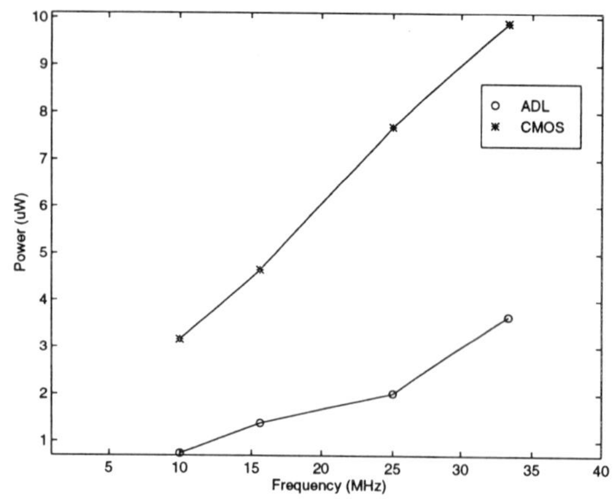

Figure 5: Power vs frequency for 4x4 bit multipliers.

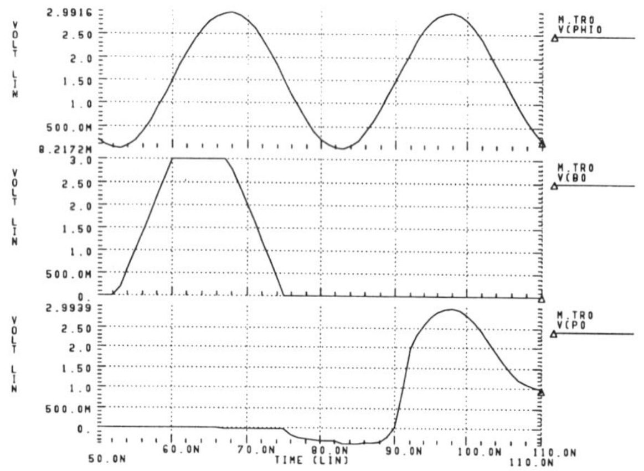

Figure 6: Latency in 4x4 bit adiabatic multiplier.

cycle for our 4x4 carry lookahead adders and 4x4 multipliers, respectively. The graphs were obtained from Hspice simulations of 256 test vectors at data rates of 10, 16, 25, and 33MHz. Our adiabatic circuits exhibit significant savings in energy consumption, especially at low clock frequencies. At 10MHz the relative savings factor is 4, and at 33MHz the relative savings factor is 3. Preliminary simulations of our 8x8 bit designs indicate power savings similar to those achieved with the 4x4 bit circuits.

Figures 6 and 7 illustrate the performance issues in adiabatic systems design. Figure 6 illustrates the latency of our 4x4 bit adiabatic multiplier running at 33Mhz. The top graph shows one of the four gate clocks, the middle graph shows the input, and the bottom graph shows the output which is lagging five phase delays after the input signal. When compared with CMOS, adiabatic circuits suffer from very long latencies. In Figure 7, the top graph shows the input and the bottom graph shows the output for the CMOS multiplier. In this circuit, the input/output latency is 8ns and the effective throughput can reach 125MHz. Our 4x4 adder has a latency of 3 clock phases, which is still much more significant than that of the CMOS adder. For the 8x8 bit designs, the differences are even more significant. The latency of our

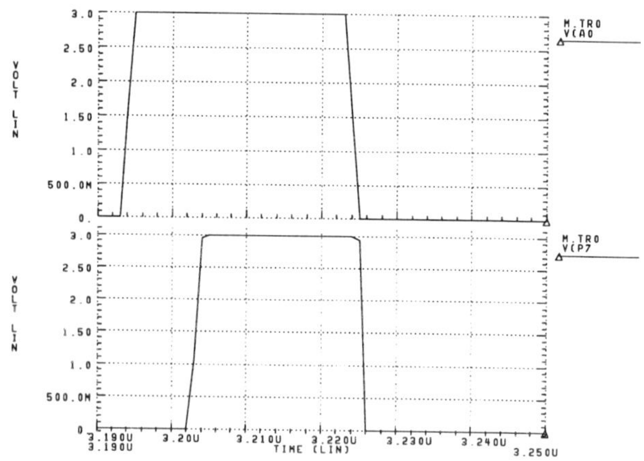

Figure 7: Latency in 4x4 bit CMOS multiplier.

adiabatic adder is 4 clock phases, and the latency of our adiabatic multiplier is 12 clock phases.

6. Conclusion

Large scale system development using adiabatic technologies is more complex than conventional CMOS circuit development because of data synchronization and simulation issues. Moreover, adiabatic circuits have large latencies due to the dynamic nature of their gates.

Another subtle yet important performance issue with adiabatic systems is that although it is possible to build adiabatic memory elements, these elements have a latency of one clock cycle. Therefore, they can only be clocked at a maximum frequency of half the gate clock frequency. The implication of this fact is that the clock frequencies specified by default for adiabatic and CMOS circuits are not directly comparable, since the former usually refers to *gate* frequencies, and the latter refers to *system* frequencies.

Therefore, it seems that adiabatic technology is more suitable for low-speed, combinational circuits.

References

[1] J. Denker, S. Avery, A. Dickinson, A. Kramer, and T. Wik. Adiabatic computing with the 2N-2N2D logic family. In *1994 International Workshop on Low Power Design*, April 1994.

[2] A. Dickinson and J. Denker. Adiabatic dynamic logic. In *CICC*, 1994.

[3] R. Hinman and M. Schlecht. Recovered energy logic: A highly efficient alternative to today's logic circuits. In *IEEE Power Electronics Specicalists Conference Record*, pages 17–26, 1993.

[4] A. Kramer, J.Š. Denker, B. Flower, and J. Mulrony. 2nd order adiabatic computation with 2N-2P and 2N-2N2P logic circuits. In *1995 International Workshop on Low Power Design*, April 1995.

[5] N. Weste and K. Eshraghian. *CMOS VLSI Design*. Addison-Wesley, Reading, Massachusetts, 1985.

[6] S. Younis and T. Knight. Practical implementation of charge recovering asymptotically zero-power CMOS. In *Research in Integrated Systems: Proceedings of the 1993 Symposium*, March 1993.

[7] S. Younis and T. Knight. Asymptotically zero energy split-level charge recovery logic. In *Proceedings of the 1994 International Workshop on Low Power Design*, March 1994.

A 0.9V, 4K SRAM For Embedded Applications

James S. Caravella
Motorola, Inc., Semiconductor Products Sector
Advanced Design Technology, Low Power Circuits Research Lab
2100 E. Elliot Rd. MD EL 712
Tempe, AZ 85284
caravell@adtaz.sps.mot.com

Abstract:

A 4Kb SRAM design is presented with functionality of 12MHz at a supply voltage of 0.9 volts with a RMS run power (1MHz) of 22μW. The circuit operates at maximum frequency of 38MHz at a supply voltage of 1.6 volts with a RMS run power (1MHz) of 32μW. The design utilizes a sub-blocked array architecture as well as selective use of NOR/NAND based decode logic. The sense amplifier design is a low voltage, glitch-free design to conserve power.

I. Introduction

Battery operated systems are an ever increasing portion of the worldwide electronics market. The key to any battery operated system is low power consumption. Battery life has and will continue to be an increasingly important specification and could mean marketing life or death to a battery operated product. Advanced technologies as well as low power circuit techniques are together the key to reduce system power consumption.

Power consumption in conventional, static CMOS logic can be expressed as:

$$P = C \times (Vdd)^2 \times f \qquad (1)$$

where C is the average switched load and parasitic capacitances, Vdd is the supply voltage, and f is the operating frequency of the circuit (neglecting short circuit currents, device leakages, etc.). From equation (1), power consumption is a function of the square of the supply voltage, therefore, the most dramatic reduction in power can be achieved by reducing the power supply. If a digital system's supply voltage was dropped from a typical 5 volts to a typical 1.2 volts, while maintaining system speed, the system would achieve a factor of 17 reduction in power consumption. Since, the most readily available batteries are alkaline type, which have an initial potential of 1.6 volts and has an end of life potential of 0.9 volt, battery operated systems must be able to accommodate these aggressive supply voltage targets. For systems to accommodate these supply voltage targets, new circuit techniques and process technologies must be developed.

This paper will discuss a 4K bit SRAM that has been fabricated and tested below supply voltages of 0.9 volt. Specific circuit techniques as well as architectural techniques will be presented.

II. Architecture

The bitcell for a standard CMOS SRAM consists of cross coupled inverters and is accessed via two n-channel transistors, whose gates form a common word line. Data is read from the cell by discharging one of the bitlines through the n-channel transistor (wordline select transistor) and into the bit cell. The bitcell must be designed to be able to sink this discharge current without causing a disruption of data in the cell and must do so over process, temperature and supply voltage. Typically, the current sink limit is determined by the wordline select transistor, since it must have a smaller W/L ratio than that of the bitcell n-channel device to insure data integrity in the bitcell. Therefore, the discharge rate of the bitlines depends on the wordline select transistors and is defined by equation (2) below, assuming a discharge of the bitline from 90% of Vdd to 10% of Vdd:

$$t_{HL} \cong \tau_d\left(\frac{2(Vtn - 0.1Vdd)}{(Vdd - Vtn)} + ln\left[19 - \frac{20Vtn}{Vdd}\right]\right) \qquad (2)$$

where τ_d = Cbitline/[K' W/L (Vdd-Vt)]. From equation (2), if Vdd is reduced, the technology could help to offset the reduction in current sink capability by decreasing the threshold voltage of the device (Vt). CMOS technology has limits to which Vt can be scaled [1] and the reduction in Vt could not offset the reduction in Vdd. For example, if W/L is constant, and Vdd is reduced from 5 volts, with a Vt ~1 volt to Vdd of 0.9 volt with a Vt ~ 0.4 volt, the time required to discharge the bitlines would be increased by a factor of 3.6. Another option to increase the current sinking capability of the bitcell would be to increase the bitcell size and also the wordline select transistors. Increasing the cell size is not a viable option since this would cause the overall memory size to grow and thus would increase cost. Another option is to reduce the capacitance that the bit cell needs to discharge. This can be achieved by reducing the number of bitcells sharing a given bitline.

For a 4K bit memory, with an 8 bit data word, the standard array organization is 64x64 bitcells. Assume Cj is the parasitic capacitance (junction and metal) associated with a single bitcell load on a bitline and Cch is the parasitic capacitance (channel, fringe, metal) associated with a single bitcell on the word line. For the above mentioned standard array organization, the total bitline capacitance is 64*Cj and the total word line capacitance is 64*Cch. If the array of bitcells

0-7803-3177-6 $5.00 © 1996 IEEE

IEEE 1996 CUSTOM INTEGRATED CIRCUITS CONFERENCE

were to be divided into four sub-arrays of 32x32 bitcells, the total bitline and wordline capacitance would be halved from that of the 64x64 array. More importantly, the sub-array would achieve lower power consumption per read/write than that of the 64x64 array, assuming only one of the four sub-arrays are active. The total capacitance per read/write that would need to be discharged/charged is 1024*Cj + 32*Cch for the sub-array architecture and 4096*Cj + 64*Cch for the 64x64 array. Therefore, for identical bitcells, not only is the sub-array architecture faster, but also uses considerable less power. The only concern would be the area penalty by going to a sub-array architecture due to the increased overhead of decode logic, control logic and routing. The memory presented here uses the sub-array architecture and a block diagram can be found in Fig 1.

The idea of sub-arraying has been well documented for fast SRAM designs and recent research indicates that sub-arraying the memory has benefit for low power low voltage designs as well [2], [3].

III. Circuit Designs & Design Considerations

A. Address Buffers: The address buffers used for this design are minimum sized, static NAND gates. The NAND function is based on a memory select signal (RAMS) and the address bit. This gating of the address into the memory is done to conserve power when the memory is not being used.

B. Control logic: The control logic resides in the center of the 2x2 array of memory and provides clock buffering and distribution. Like the address buffers, the system clock is gated by RAMS to conserve power.

C. Column Decode: 1:4 muxing is done at the bitlines via a single n-channel transistor, therefore, the column decode must be a 4:1 decode based on two address bits. The decode circuit is dynamic NAND based and can be found in Fig 2. The NAND-type structure was chosen due to its low power consumption, that is, only the decoded row transitions. Sizing of the devices in this circuit was done in the following manner: First size the inverter to drive the known load capacitance, then size the stacked n-channel devices to provide the necessary discharge time of node A. Likewise, the precharge device, M3 should be sized to provide for node A to Likewise, the precharge device, M3 should be sized to provide for node A to be charged to Vdd during the low clock cycle. The feedback/latch transistor, M4, was sized to a W/L of 0.3. This device must be large enough to properly "hold" a Vdd at node A. From a low voltage operation standpoint, a dynamic NOR based decode would provide lower delay times through the decoder due to the limited amount of stacking of devices. This type of decode was not used here since the

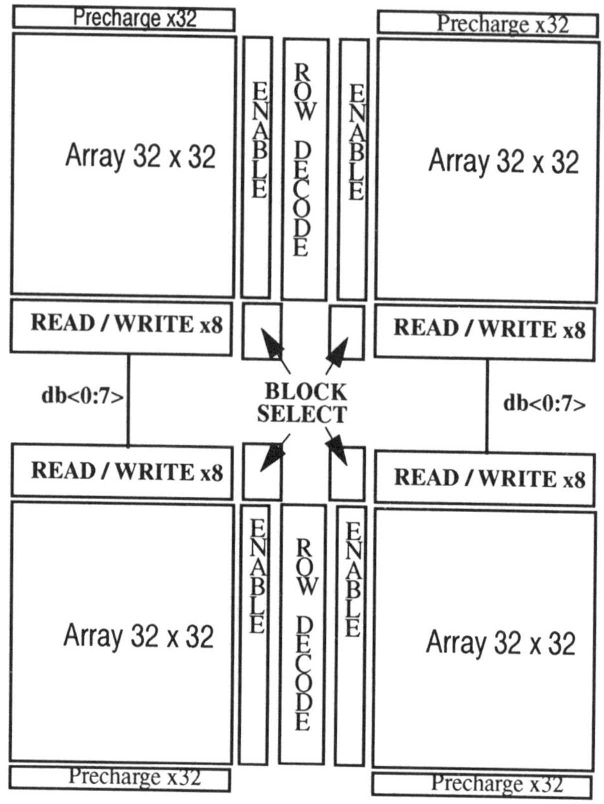

Figure 1: "Memory Architecture"

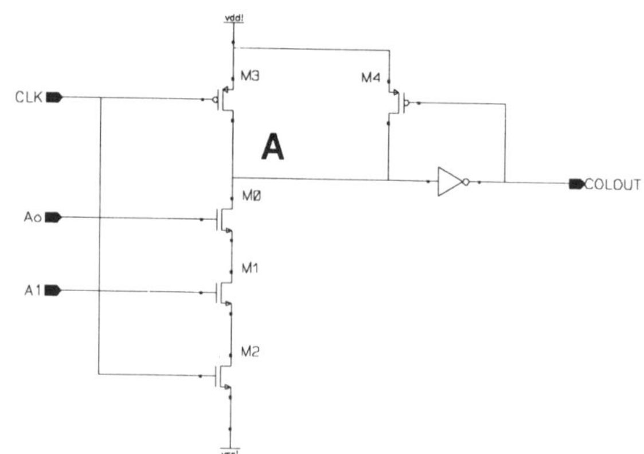

Figure 2: "Column Decoder"

NAND structure met the delay requirements (column decode is not in the worst case access delay path) and did so at a much lower power consumption.

D. Row Decode: As noted in Fig. 1, the row decoders are common among two array quadrants. The row decode delay is in the critical access delay path and therefore, dynamic NOR based decode logic was used for these decoders. Fig. 3 contains a schematic of the row decode. Since there is a single row decode per wordline, there are 32 row decodes that are shared among two array quadrants. If no additional decoding was done, for each read/write, 31 of these decoders would transition (precharged/dynamic) and only one (the decoded address) would not. An additional level of decoding is done to reduce the overall power consumption of these row decoders. The addition decode is done by a decode circuit as found in Fig 2. The row decoders are grouped in groups of eight (3 addresses) and are activated (CLK signal) by the secondary decode (2 addresses). This method provides a balance between minimal delay and high power consumption. Designing of the row decoder follows from that of the column decoder.

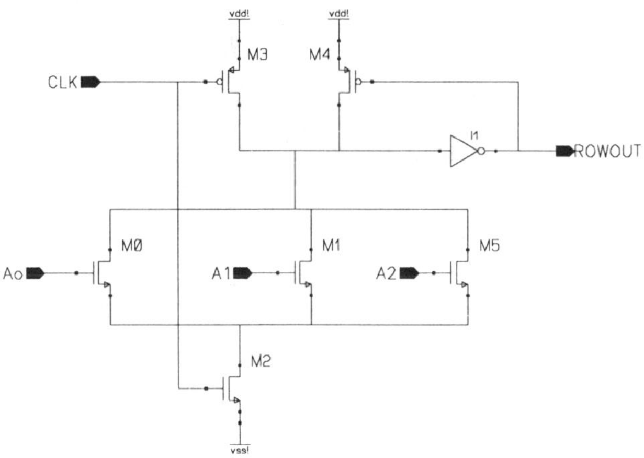

Figure 3: "Row Decoder"

E. Enable: The enable circuit isolates the row decoders from the array's wordline. The enable circuit can be found below in Fig. 4. ROW_OUT signal is the output of the row decoders, the BL_SEL signal is the output of the block select circuit, and the RBL_SEL is the output of the coarse row decode (1 block of 8 row decoders). In order to isolate the row decoder (ROW_OUT) from the wordline (WL), the delay through the NAND gate must be less than the delay through the decoder. If the NAND is slower than the row decoder, glitches could happen on the wordline. Glitches on the wordline could cause data corruption in the bitcells. Transistors M1, M2 and M3 are sized to provide necessary drive to charge the wordline.

F. Sense amplifier: Standard differential amplifiers no longer function at supply voltages near 2*Vt. The tail current source has problems staying saturated, likewise, the current mirror load cannot stay saturated to provide high output resistance. In addition, the common mode input range suffers as a result of a low Vdd to Vt ratio. As a result of many of the afore mentioned problems, this type of amplifier suffers from low gain at low supply voltages. Fig. 5 contains a simplified schematic of the sense amplifier used in this SRAM design. Additional devices are present in the design to ensure glitch-free operation, but are not shown in Fig. 5.

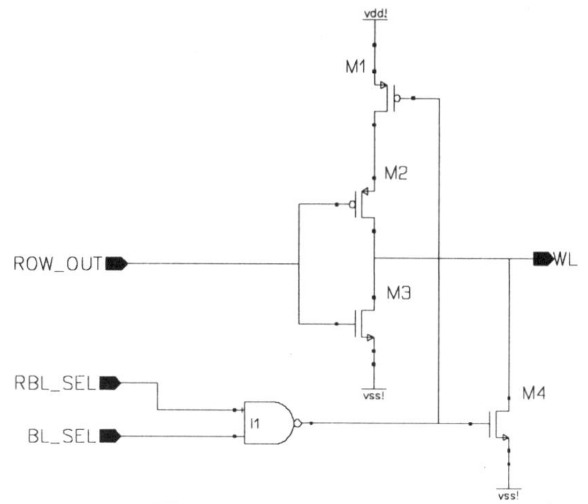

Figure 4: "Enable Circuit".

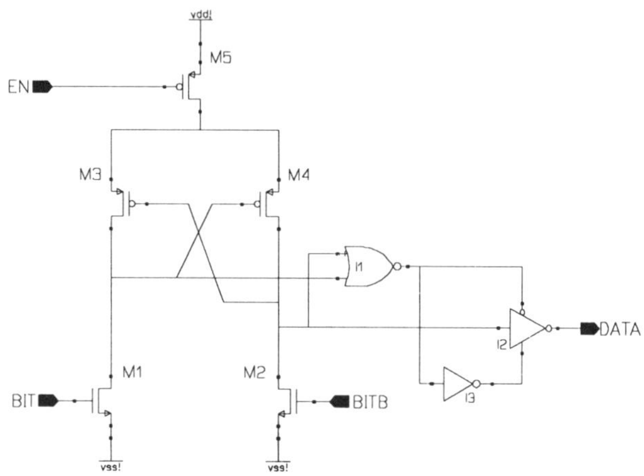

Figure 5: "Sense Amplifier"

The EN signal found in Fig. 5 is the complement of the precharge control signal. Therefore, during the precharge state of the clock, M5 is off and both M1 and M2 are conducting. The outputs of the amplifier are at Vss. When a read occurs, M5 feeds current into the sense amplifier. A ΔV on the bitlines is amplified due to the g_m of M1/M2 and the positive feedback load M3/M4. Both M3/M4 are saturated at turn-on, thus, providing a high output resistance at the output nodes of the sense amplifier. The positive feedback of M3 and M4 help to maintain the gain of the amplifier as the supply voltage is lower. Second level sensing is done via a NOR gate. If either of the outputs of the sense amplifier goes to Vdd, the output of the NOR gate will go to Vss, thus enabling the tristate buffer which buffers the sense amplifier. This secondary sensing eliminates false bus transitions and thus saves system power.

IV. Experimental Results

The RAM described within this paper has been fabricated using a low threshold voltage CMOS technology, using 105A gate oxides, L of 0.65μm, and salicided poly. The design was characterized over supply voltage and a plot of access time (clock period for this design) vs. the RMS run current can be found in Figure 6. The RMS current was measured given a read and write pattern of a checkerboard across the entire memory at 1MHz. The tester load on the DUT was a 50Ω transmission line environment, terminated to Vdd/2. .

Vtn ~ 530mV (W=1.6μm) Vtp ~ -480mV (W=1.6μm)

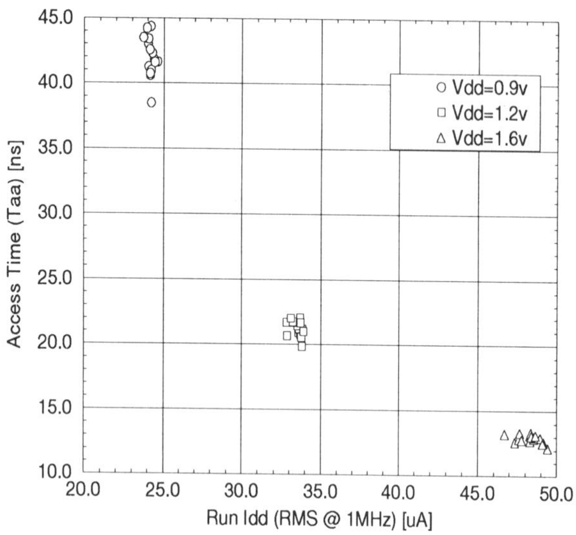

Figure 6: "Taa vs. Run Idd Current (RMS @ 1MHz)"

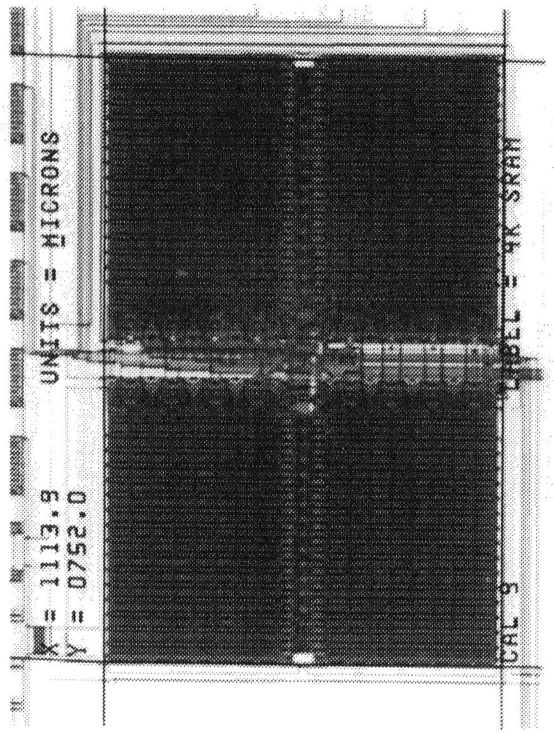

Figure 7: "Mircophotograph Of The 4Kb SRAM"

The access time in Fig. 6 includes delay due to the bidirectional pad ring I/O. This I/O delay is typically 10ns at Vdd of 0.9 volt and typically 5ns at Vdd=1.6v.

A table of the minimum voltage for given frequency of operation can be found in Table 1. Clock frequencies below 500KHz were not attempted.

Table 1: Minimum Supply Voltage For Given Frequency Vtn ~ 550mV, Vtp ~ -490mV

Vddmin @ F=500KHz	VDDmin @ F=1MHz	VDDmin @ F=2MHz	VDDmin @ F=4MHz
580mV	590mV	620mV	690mV

The size of the memory is 1113μm x 752 μm as noted in Fig. 7.

Conclusions

The memory presented in this paper has shown functionality at supply voltages below 0.9 volt. The design uses a sub-array architecture, selective use of NAND/NOR decode logic and a glitch-free low voltage sense amplifier. The area penalty for the sub-array architecture is on the order of 16% over that of a full custom, single array memory and is equivalent in area to a well established compiled single array memory. Performance varies from 500KHz at Vdd=580mV to 50MHz at Vdd=1.8 volts. Extremely low supply voltage operation (>1 volt) is demonstrated with a high level of performance and low power dissipation. This design is a low voltage/low power SRAM module for the Motorola, 68HC08 family of microcontrollers.

Acknowledgments

The author wishes to acknowledge Ken Papworth of Motorola Advanced Custom Technologies (ACT), for all the testing and characterization done on this memory. An acknowledgement to the low power process group of ACT for all device and process development work. Also an acknowledgment to the members of the Motorola, Low Power Circuits Design Lab as well as Jim Feddeler, Jeff Jorvig and Dan Ong, of Motorola CSIC Design for the many technical discussions, and logistical help pertaining to this design.

References

[1] E. Takeda, et al.; IEEE Electron Devices, vol. 32, 1985, p322.

[2] D. Wong, et al.; "An 11-ns 8Kx18 CMOS Static RAM with 0.5-μm Devices," IEEE Journal of Solid-State Circuits, vol 23, Oct. 1988.

[3] A. Chandrakasan, et al; "A Low Power Chipset for Portable Multimedia Applications", ISSCC 1994, p64

A 2.7V CMOS SINGLE CHIP BASEBAND PROCESSOR FOR CT2/CT2+ CORDLESS TELEPHONES

C. Crippa, G. Nicollini, P. Confalonieri, S. Pernici, A. Mecchia, P. Rizzo, F. Adduci, E. Viani, I. Bietti, A. Nagari, and C. Dallavalle
SGS-THOMSON Microelectronics, Agrate Brianza, Milano, Italy.

A. Leblond - Dassault Automatismes et Telecommunications, Plaisir, France.

P. Busserolle - Dassault Electronique, Saint-Cloud, France.

ABSTRACT

A low voltage, low power CMOS single chip baseband processor for CT2 and CT2+ cordless telephones is presented. The chip integrates a complete voiceband codec, a tone generator, a G721 ADPCM coder/decoder, a Burst Mode Logic controller for CT2/CT2+ framings, and an I/Q baseband signal generator. It can be easily interfaced with standard microcontrollers through a parallel interface. It can operate from a 2.7V minimum supply with operative and stand-by power consumptions of 35mW and $25\mu W$, respectively. Maximum operative supply is 5.5V. Chip area is $55.5mm^2$ in a 0.8μ N-well CMOS process.

I. INTRODUCTION

During the last ten years, cordless telephones have gained widespread user acceptance as a wireless counterpart to the standard wired ones. The handset typically operates within 50 to 100 meters of the user "residential" base station, which connects to the public switched telephone network (PSTN). With the advent of digital cordless telephony, cordless systems with enhanced functionality (CT2, CT2+, DECT, PHS) have been developed that can support higher data rates and more sophisticated applications such as wireless PBXs and public-access Telepoint systems. This paper presents a CMOS baseband processor which performs all the baseband functions needed in CT2 and CT2+ applications in a single chip.

II. GENERAL ARCHITECTURE

The chip functional block diagram is shown in Fig. 1. In transmission, the Acoustic Front-End (AFE) converts the human voice coming from the microphone (in case of handset) or from PSTN (in case of residential, PBX, and public base stations) to a digital signal which is compressed to 32Kbps by the G721 ADPCM encoder. Then the Burst Mode Logic (BML) controller multiplexes this signal in the B-channel with signaling information in the D-channel. The resulting signal is digitally filtered by a Gaussian filter contained in the I/Q generator and converted in two analog quadrature modulating signals (GMSK modulation). In reception, the BML demultiplexes the voice B-channel and the signaling D-channel. The B-channel is expanded by the ADPCM decoder, converted from digital to analog in the AFE and then sent to the earpiece or loudspeaker receivers. As other features, the B-channel signal can also be sent to the PCM or ADPCM interfaces to be used by external devices such as static answering machines, or for ISDN applications. Master clocks for each functional block are generated by integer divisions of a 18.432MHz one pin quartz oscillator output.

A) ACOUSTIC FRONT-END

The AFE block diagram is shown is Fig. 2. TX section consists of a programmable microphone preamplifier, with two switchable high impedance balanced inputs, followed by an antialiasing 8th order SC band-pass filter, and a successive approximation 14 bit linear A/D converter. RX section is made of a linear 14 bit D/A converter followed by an 8th order SC bandpass filter with $\frac{\sin x}{x}$ compensation, and two programmable attenuators with output stages capable to directly drive loads as heavy as 200Ω or 400nF. TX and RX sections can also be connected by sidetone or by amplified listening paths. Moreover, the AFE includes a tone generator which can be directed to TX and RX, enabling a DTMF signal on the sending B-channel and a tone signal for warning or calling informations on the acoustic receivers. The gains of each section are software programmable on wide ranges to allow the use of low cost microphones, earpiece capsules, loudspeakers, and buzzers. Bias circuit, voltage references, and common mode generator are made internally, avoiding any external component. The AFE design structure rely on a fully differential approach, that allows to get much better performance in presence of "digital noise" coming from the other functional blocks or external disturbances arising from the radio frequency circuits. Three couple of power supply pins are provided, so that internal analog, digital, and driver amplifier supply lines can be completely separated.

B) ADPCM PROCESSOR

The ADPCM block implements the compression and decompression functions of a digitized voice signal in full compliance with CCITT G721 recommendation [1]. The ADPCM coder compresses a PCM (linear 14 bit

0-7803-3177-6 $5.00 © 1996 IEEE

or A/mu law companded 8 bit) stream into a 4 bit ADPCM pattern at 32Kbps rate to be properly formatted by the BML and I/Q generator for transmission. The ADPCM decoder realizes the reverse operation (decompression) for reception. The coding and decoding algorithms are executed within one 8KHz frame duration. The ADPCM block contains about 7000 gates, 1Kbit RAM, and 23Kbit ROM. Nearly half of the random logic was auto-routed, while the remaining part was realized with a datapath generator to achieve a higher density and less power consumption.

C) BML CONTROLLER

The BML block diagram is shown is Fig. 3. The main task of the BML controller is the formatting of voice and signaling data into three different multiplexes called MUX1, MUX2, and MUX3 [2]. MUX1 is used to exchange voice and signaling informations in the B and D channels, respectively. MUX2 and MUX3 are used during the set-up of the handset/base station radio link, and contain preambles (series of 1 and 0) to gain bit synchronization and special patterns to establish frame synchronization in the SYNC channel. Bit and frame synchronizations are extracted by means of a digital PLL. Moreover the BML performs the CRC and bit parity check on the transmit and receive data streams. The BML contains about 10000 gates and was entirely auto-routed.

D) DIGITAL INTERFACE

The Digital InterFace (DIF) is devoted mainly to PCM and ADPCM interfaces. It contains a timeslot assigner logic to enable the connection to external PCM and/or ADPCM busses, and can be programmed to work in delayed or non-delayed data mode with different data format. Moreover it synchronizes the data transfer among the other blocks. As a last feature it includes a comfort noise generator for the receive AFE channel that can be activated when the radio link between base station and handset is lost.

E) I/Q GENERATOR

The I/Q generator is the baseband section of an I/Q modulator. It implements a Gaussian Minimum Shift Keying (GMSK) modulation with normalized bandwidth of the Gaussian filter set to 0.5. The I/Q generator has a digital part that starting from the BML output data stream generates the samples of the I and Q waveforms, and an analog one that performs a digital to analog conversion of these samples. The I/Q generator block diagram is shown is Fig. 4. The digital part is made of a finite state machine, that select the appropriate waveform starting from the input data, followed by a ROM, that contains the samples for all the possible I and Q waveforms, and a counter that scans the samples of the selected waveform. The other block is the timing generator for offset and gain correction of the I and Q outputs. The analog part includes two differential 8 bit, 1.152MHz D/A converters, followed by two differential 2nd order continuous time reconstruction filters with a nominal cutoff frequency of 250KHz, and a 5 bit A/D converter used to cancel the offset and perform gain matching correction on the I and Q differential outputs. The D/A converters are realized with a common resistive array for the three most significant bits, and two separate differential capacitive arrays for the remaining five bits. The A/D converter is realized with four differential capacitive arrays (I offset, Q offset, I and Q gain matching) which share a common comparator and logic during a successive approximation algorithm. A dedicated couple of power supply pins was provided for I/Q to reject as much as possible the high frequency noise coming from the other blocks, especially the 18.432MHz oscillator and the 9.216MHz ADPCM.

F) CRYSTAL OSCILLATOR

All the clock and synchronization signals needed by each functional block are generated by division from a 18.432MHz one pin quartz crystal oscillator output. This oscillator (Fig. 5) is made of a simple non inverting differential input stage with a capacitive feedback C_1 to realize, taking into account the capacitance at the amplifier output C_2 and the total capacitance at the input pin C_t (included that one across the quartz electrodes C_{op}), an input impedance $Z_{in} = R_{in} + jX_{in}$ that satisfies the following relations at the resonant angular frequency ω_o :

$$R_S + R_{in} < 0 \qquad X_S + X_{in} = 0$$

where $R_S + jX_S = R_S + j(\omega L_S - \frac{1}{\omega C_S})$ is the linear equivalent circuit of the mechanical quartz oscillation. The gain stage transconductance g_m must be chosen close to $g_{m_{opt}}$, i.e. that one that leads to the absolute maximum value for R_{in} to assure a minimum start-up time constant [3] :

$$g_{m_{opt}} = \omega_o \frac{C_t(C_1 + C_2) + C_1 C_2}{C_1}$$

The critical transconductance for oscillation $g_{m_{crit}}$, which is much less than $g_{m_{opt}}$, results:

$$g_{m_{crit}} = \omega_o \frac{[C_t(C_1 + C_2) + C_1 C_2]^2}{Q C_1 C_S (C_1 + C_2)}$$

where Q is the quartz quality factor.

A single couple of power supply pins was used for the

oscillator, the ADPCM, and the BML controller.

III. EXPERIMENTAL RESULTS

The chip was integrated in a 0.8μ N-well single-poly double-metal CMOS process with implanted capacitors. Its microphotograph is shown in Fig. 6. Table I lists the main characteristics of each functional block. The total stand-by power consumption is around $25\mu W$ at 2.7V supply. Tables II and III summarize measured AFE and I/Q generator performance, respectively.

REFERENCES

[1] CCITT G721 Recommendation, Blue Book, Melbourne '88.

[2] ETSI I-ETS 300 131 R2, June '94.

[3] E. A. Vittoz, M. G. R. Degrauwe, and S. Bitz: "High-Performance Crystal Oscillator Circuits: Theory and Application", IEEE JSSC, June '88.

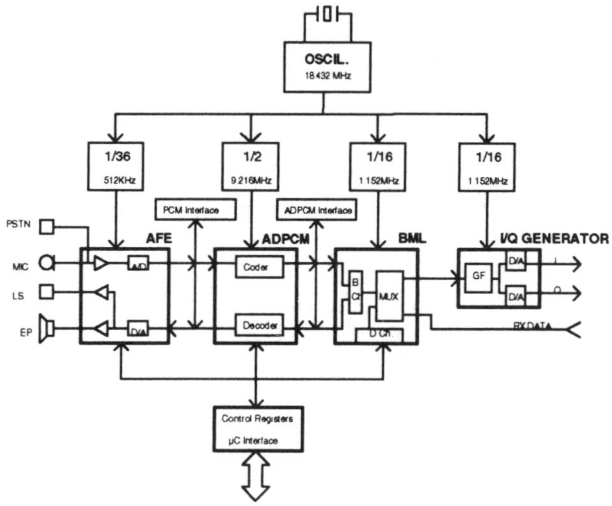

Fig.1: Chip architecture

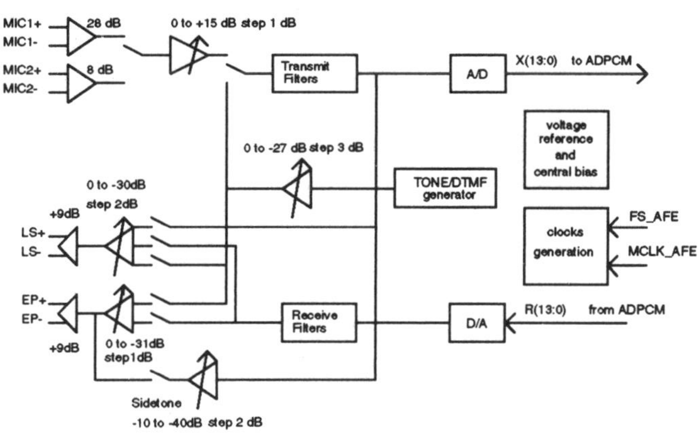

Fig.2: AFE block diagram

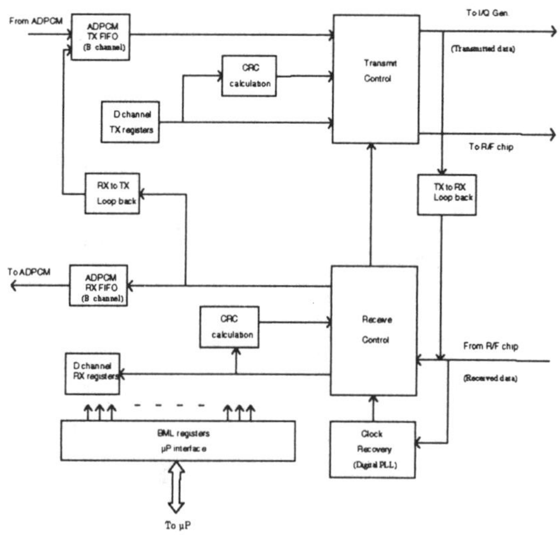

Fig.3: BML block diagram

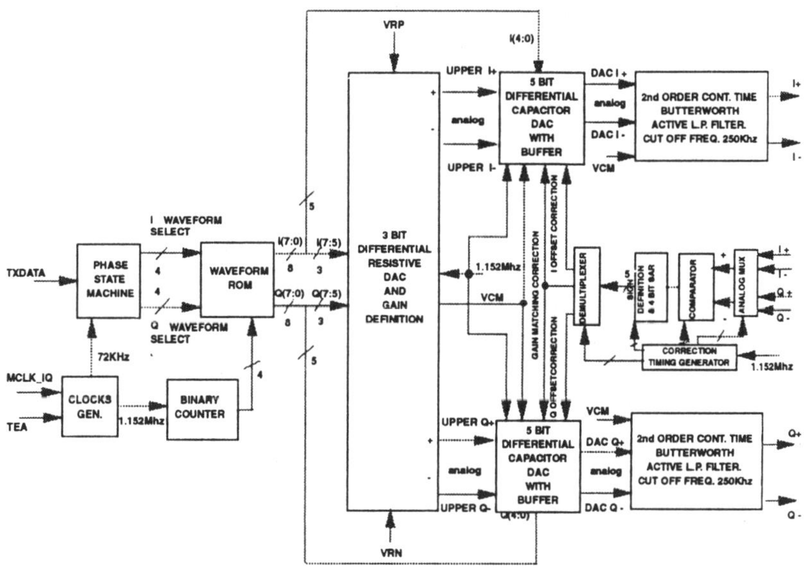

Fig.4: I/Q generator block diagram

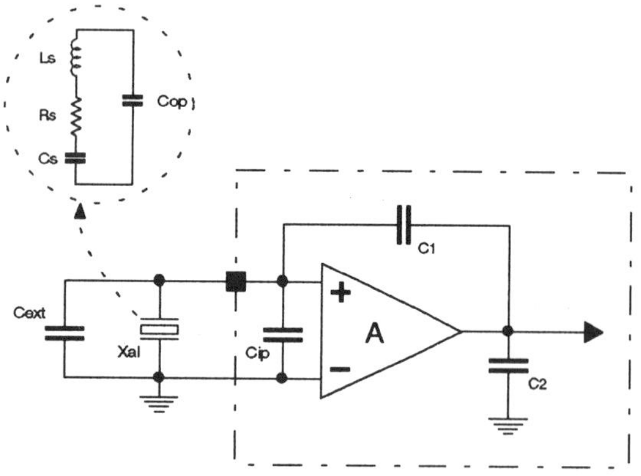

Fig.5: One pin crystal oscillator

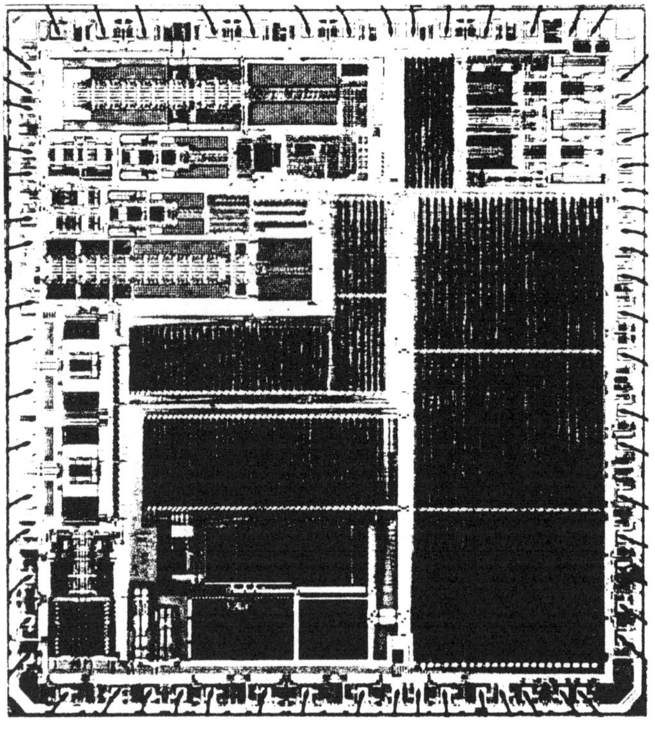

Fig.6: Chip microphotograph

	AREA (mm²)	POWER (mW)
AFE	11.9	13.3
ADPCM	7.8	9.0
BML	11.0	6.3
I/Q	3.3	3.5
Oscillator	0.4	2.7
DIF	3.2	0.1
Interconnections	4.3	
Wrap-around/Scribe-line	13.6	
Total	55.5	34.9

Table I: Chip characteristics

	Transmit	Receive
Resolution	14b	14b
Sample rate	8KHz	
S/(N+THD) (dB)	72	73
ICN (dBmOp)	-78	-79
THD (dB)	-72	-73
Abs. gain error (dB)	<0.25	<0.20
Gain tracking error(dB)	<0.05	<0.05
PSR (dB)	-65	-65
Crosstalk (dB)	-80	-85
DC offset (mV)	< 0.02	<8.0

Table II: AFE performance

Resolution		8b
Sample rate		1.152MHz
S/(N+THD) (dB)		48
THD (dB)		-63
Peak spurs (dB)	f<2MHz	-65
	f>2MHz	-73
DC offset (LSB)		<0.5
Gain error (dB)		<0.5
Gain matching (dB)		<0.05
Phase matching (deg.) (at 18KHz)		<1.0

Table III: I/Q generator performance

A Micropower Safety IC for Rechargeable Lithium Batteries

Troy Stockstad, Tom Petty
Amplifier and Power Design Group, Analog Division, Motorola, Inc.
Tempe, AZ

Renwin Yee
Colorado Springs, CO

Abstract

A rechargeable lithium battery safety IC designed for monitoring cell voltage and current in the battery is presented. The circuitry is capable of protecting batteries of one to four series connected cells via a programmable circuit architecture with a cell voltage limit accuracy of +/- 1%. The circuitry has an average current drain of 25μA. The Li safety IC with integrated series FETs has an area of 18.4 mm^2.

Introduction

Lithium rechargeable cells are very attractive for portable applications because of their lighter weight and higher energy density than NiCd cells. In the past, lithium batteries have not been widely used in the commercial market due to potential safety hazards. However, improved cell design, coupled with overcharge protection circuitry in the battery, can protect the cells within the battery pack from abuse.

The normal operating voltage range of lithium ion rechargeable cells currently available is approximately 4V to 2.5V. The 4V limit is referred to as the "over voltage" limit, while the 2.5V limit is the "under voltage" limit. There is variability in the exact limits specified for each manufacturer's cells depending upon the chemistry being used. If the voltage of a lithium ion cell exceeds the over voltage limit, the cell may become unstable. A cell which is discharged below the under voltage limit will degrade in terms of cycle life performance. In order to maintain the integrity of the cells, electronics can be used to interrupt currents through the battery when either an over voltage or under voltage condition is present.

In this paper, a lithium battery safety IC which is placed in the battery pack with the cells to monitor cell voltage limits and current is described. The IC uses a battery measurement scheme which "polls" cell voltages, utilizing a duty cycle approach to minimize current drain. A pair of multiplexers selects cell voltages to transfer to a single trimmed comparator circuit for checking the voltage limits. This method differs from other design approaches that continuously monitor each cell in the battery, requiring a precision reference for each cell, and operating at higher drain current. Features of the safety IC include programmable +/- 1% voltage limit accuracy, programmable over charge and over discharge current limits, and circuitry regulated to 10V for driving series FETs. The circuit architecture allows the battery pack designer to choose the cells voltage limits and current limits with external resistors. Finally, a cell balancing circuit will be described which corrects for individual cell voltage differences within each battery, thus maintaining the efficiency of the battery and extending its cycle life. The circuit has an awake mode current drain of 25μA while monitoring four cells, with a sleepmode state that draws less than 10nA.

Lithium Safety IC Circuit Requirements

To protect battery cells from over charge and over discharge conditions, two FETs are placed in the series path of the cells to interrupt current, in the event of a cell reaching an under or over voltage limit or if a current limit condition is detected. Any one cell in the battery reaching a over voltage or under voltage limit causes the corresponding charge or discharge path FET to be turned off. Fig. 1 illustrates a battery pack with a Li safety IC.

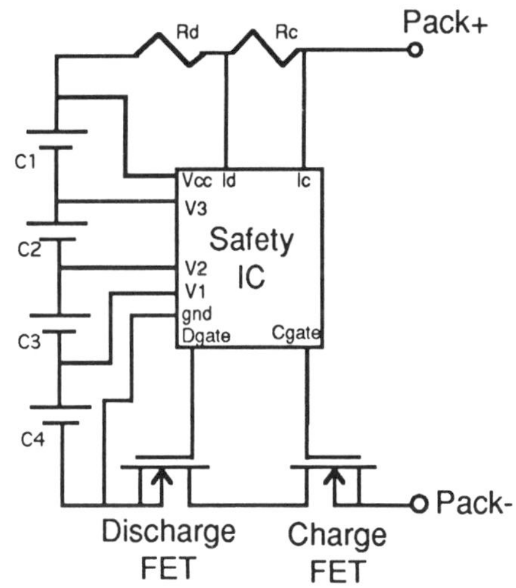

Fig. 1. Lithium Battery Pack.

One of the key parameters of the safety IC is to obtain as low as possible drain current on the battery while monitoring the cells. Ideally the safety circuitry would draw little (< 1μA) or no current when an under voltage cell is measured, since continued current drain degrades the cell cycle life. Another important specification is the voltage limit accuracy of the safety monitoring circuitry. The over voltage limit accuracy must be 1% or better to

0-7803-3177-6 $5.00 © 1996 IEEE

get the maximum capacity out of the cells while still maintaining safety. Other desirable features would be current limit circuitry which provides a time delay before deactivating the series FETs (for capacitive load purposes, for example), and sufficient drive voltages for the series FETs to keep the Rds(on) of the devices low.

Circuit Implementations

The design method chosen in this design for protecting the cells relies upon taking high precision measurements of cells periodically, and remaining in a low current drain state the remainder of the time (1). The cell voltages are measured approximately 1.5 milliseconds out of every 1.4 seconds. By "polling" the cells and using multiplexers to level shift cell voltages to a comparator, only one reference is required to be trimmed, and the drain current of that reference is effectively only 1/1000th of its steady state value. Fig. 2 illustrates the block diagram of the safety IC. The analog multiplexers connected to the individual cells level shift the precision comparator to each cell. The cell positions which are measured are determined by the user by selecting the voltage levels of the program pins, PROG1 and PROG2.

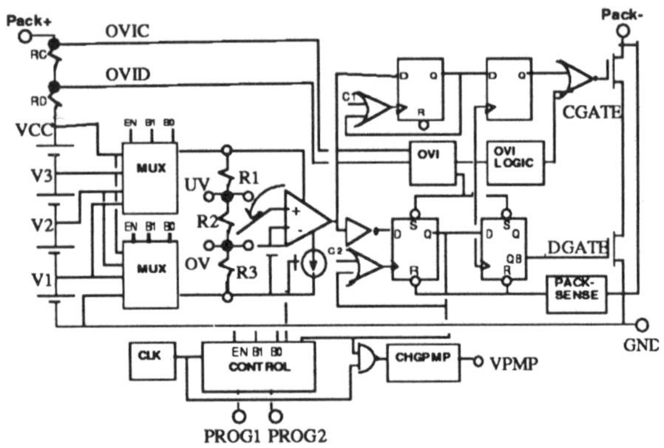

Fig. 2. Li Safety IC Block Diagram.

All cells are first measured to check for the over voltage limit. Referencing Fig. 2, the input to the trimmed "bandgap" comparator is switched to the OV node on the resistor string of R1, R2, and R3, which are external to the IC. Placing R1, R2, and R3 external to the chip allows the voltage limits to be easily adjustable. If any one cell is over voltage (output of the comparator is high), the first D flip-flop is latched to a one by clock C1. The over voltage information is propagated to the driver of the charge FET, CGATE, after all cells have been measured. The over voltage limit where the comparator changes state is given by

$$Vov = \frac{R1+R2+R3}{R3} * 1.23V \qquad (1).$$

The input of the comparator is then switched to the UV point on the external program resistors, R1, R2, and R3, and all cells are measured for under voltage. Each measurement in clocked into the first under voltage D flip-flop, and similarly, propagated to the output driver of the discharge FET, DGATE. The under voltage cell limit is given by

$$Vov = \frac{R1+R2+R3}{R2+R3} * 1.23V \qquad (2).$$

When an under voltage cell is measured, the biasing of the IC is disabled. This stops the oscillator, and puts the IC into "sleep mode". While in sleep mode, the only drain current on the IC is due to leakage currents. The IC "wakes up" when a voltage drop is measured across the parasitic back gate diode of the discharge protection FET. The back gate diode is the conduction path of charge current if the battery is placed in a charger.

"Bandgap" Comparator

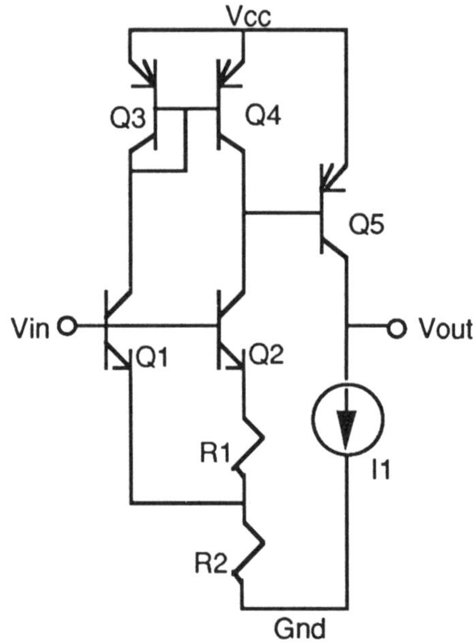

Fig. 3. "Bandgap" Comparator.

A simplified view of the bandgap comparator circuitry is shown in Fig. 3. The comparator is similar to a Brokaw reference (2), except the input terminal, made up of the bases of transistors Q1 and Q2, is not fed back to the output. Transistor Q5 provides current gain to the difference of collector currents from Q2 and Q4, which switches Vout from a low to high logic level as the input voltage becomes greater than the bandgap voltage of silicon. The switch point of the comparator is stable over temperature. The comparator is trimmed to have its switch point at input Vin within +/- 1% of 1.23V by adjusting the value of R2. R2 is trimmed by shorting out zener diodes across segments of the resistor, a method referred to as zener zapping (3). Using the bandgap comparator reduces circuitry by combining the comparator and reference function into one compact circuit.

Multiplexers

The control logic drives all the multiplexing switches and provides the system clocks at the appropriate times during polling. An on chip oscillator designed to run at 11kHz provides the system clock. Fig. 4 shows the storage oscilloscope photograph of the top pin of the analog multiplexer, MUX+, as the IC samples the voltages of four series cells.

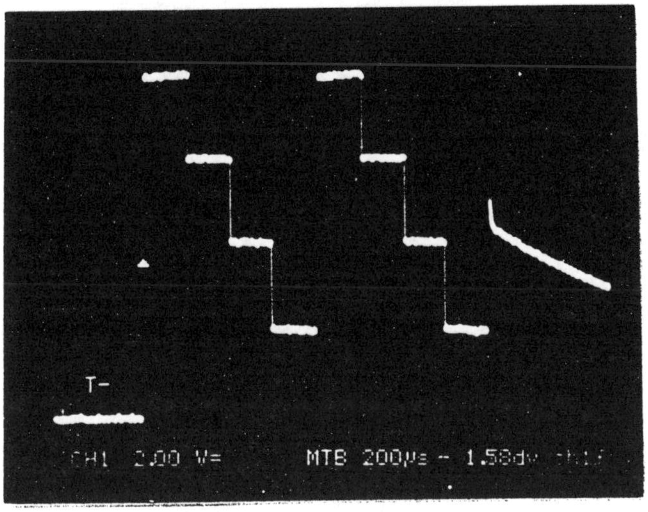

Fig. 4. Photograph of Cell Voltage "Polling".

Oscillator

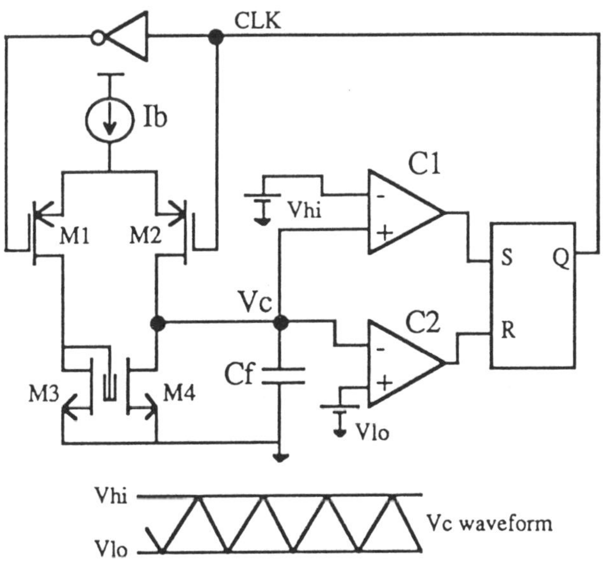

Fig. 5. Oscillator Circuitry.

The oscillator used for the control logic, shown in Fig. 5, was chosen for its ease of integration and predictable startup conditions. The design is guaranteed to start when the bias Ib sources current. The differential input stage, made up of transistors M1 - M4, controls the path Ib flows through, either sinking out of Cf through M4, or sourcing into Cf through M2, depending on the value of logic level CLK. The comparators C1 and C2 are used for switching the state of the SR latch when the excursions of the capacitor voltage cross the voltages Vhi and Vlow. The oscillation frequency can be shown to be

$$f_{osc} = \frac{I_b}{2(V_{hi} - V_{low}) * C_f} \quad (3).$$

Integrated LDMOS devices with a total series resistance of 350mΩ are on the initial version of the IC. A second generation safety IC has also been developed without on chip power FETs, since many users require a series resistance lower than 100mΩ. The FETs have their gates driven by circuitry which is powered by a charge pump regulated to 10V. The charge pump is required for applications where batteries contain only one or two series cells, since the battery voltage itself may be too low to sufficiently drive the series FETs.

The current limits for the IC are asynchronous, so that over current conditions can be immediately interrupted as they are sensed. The current limit is accomplished by detecting the voltage drops across Rc and Rd, shown in Fig. 2. A ΔVbe comparator is used to signal when the IR drop of the current sense resistor becomes greater than the desired limit. When over current conditions are detected, the IC's logic shuts off the appropriate FET to interrupt current.

Cell Balancing

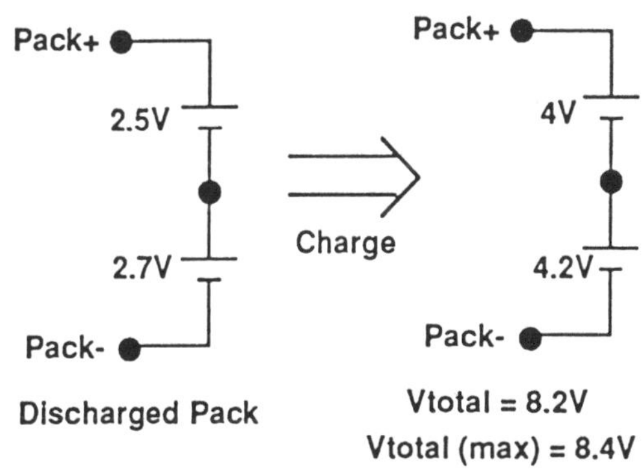

Fig. 6. Unbalanced Battery Pack.

The lithium safety IC also provides cell balancing to the battery pack. Fig. 6 illustrates an unbalanced battery pack. If cells are unbalanced (i.e. different voltages) when the battery pack is inserted into a charger, the full capacity of all the cells is not realized. This is due to the requirement that charging must terminate when any one cell reaches an upper voltage limit. If a method is used to keep the cell voltages in a battery at the same level, the battery will charge to a higher capacity and realize greater cycle life.

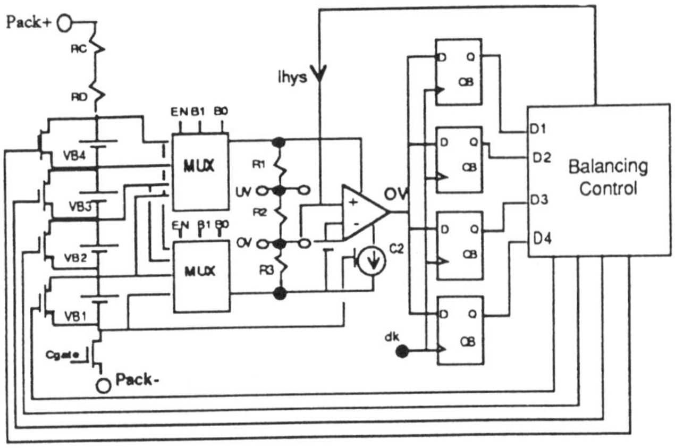

Fig. 7. Cell Balancing Circuitry.

Fig. 7 shows a block diagram of the balancing circuitry of the lithium safety IC. As over voltage measurements are taken by the bandgap comparator, the output result from each cell's over voltage comparison is clocked into a cell-specific flip flop. If one cell reaches the over voltage limit before the other cells, the over voltage cell is discharged by an internal FET. This cell is discharged until it falls a "hysteresis" voltage below the over voltage limit. The hysteresis voltage is set by the program resistors, R1 and R2, and the hysteresis current, which is 2µA. The equation for the hysteresis voltage is

$$V_{HYS} = I_{HYS} * [R1 + R2] \quad (4).$$

Measured Results

Parameter	Minimum	Typical	Maximum
Voltage limit accuracy	+/- 1%	+/- 0.5%	
Awake Drain Current, Vcc		4.5µA	6µA
Awake Drain Current, V3		2nA	10nA
Awake Drain Current, V2		16µA	20µA
Awake Drain Current, V1		2.2uA	5uA
Charge Pump Voltage	9V	10V	11V
Sleep Mode Current, all pins		< 1nA	10nA
Series FET On Resistance		350mΩ	400mΩ

Table 1. Measured Data for Lithium Safety IC. Data taken with four 4V cells in battery.

The IC's drain current is a function of the number of cells and cell voltages. The measurements given here are for a battery with four 4V cells. A photograph of the safety IC with integrated LDMOS power devices is shown in Fig. 8. The die is packaged in a 20 pin SOIC.

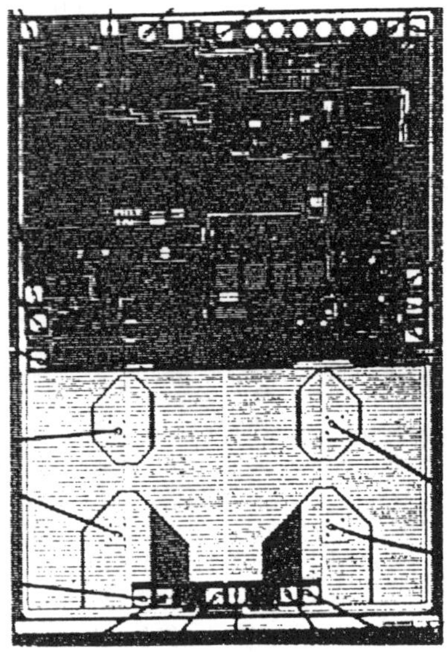

Fig. 8. Die Photograph.

Summary

A lithium rechargeable battery safety IC has been described which monitors lithium cells in a battery pack. The cells are protected with two series FETs which are controlled by the safety circuitry. The Li safety IC "polls" the cell voltages, checking each cell for an over voltage limit and an under voltage limit, and monitors battery current. The IC precisely controls the cell voltages and protects them from possible abuse while only requiring 25µA of drain current.

Acknowledgments

Special thanks to Don Zobel, Monty Palmer, and Marv Bellefeuille for their valuable help.

References

[1] R.L. Vyne, T.D. Petty, R. Koda, "Sleep-mode Amplifier", IEEE Bipolar Circuits and Technology Meeting, pp.285 - 288, 1991.
[2] A. Paul Brokaw, "A Simple Three-Terminal IC Bandgap Reference", IEEE J. Solid-State Circuits, vol. SC-9, pp. 388-93, Dec. 1974.
[3] George Erdi, "A Precision Trim Technique for Monolithic Analog Circuits," IEEE J. Solid-State Circuits, vol. SC-10, pp.412 -6, Dec. 1975.

Micro Power Protection Chip for Rechargeable Lithium-ion Batteries

Greg Smith
National Semiconductor Corporation
Tucson Design Center - Lithium-ion Products Group
Tucson, AZ 85715

Abstract

Lithium-ion (Li+) rechargeable batteries have gained the attention of portable electronic equipment manufactures because of a 70% increase in specific energy (150 W-hr/kg) over nickel based chemistries (1), (2). Unlike NiCd or NiMH cells, Li+ cells require in-pack electronics for protection against inadvertent electrical over-stress. The topic of this paper is the micro-power design techniques utilized for an in-pack protection IC that is powered continuously from a single Li+ cell without significant discharge of the cell.

Introduction

Figure 1 illustrates the basic configuration of the protection IC which resides in the battery pack. Li+ cells supply about three times the voltage/cell than NiCd cells. Applications for single cell voltage Li+ packs are cellular phones and audio/video equipment ranging from personal cassette players to video recorders. Single cell voltage packs may have 2 or more cells connected in parallel. The back to back power MOSFETs block current into or out of the pack once the protection IC has detected that the cell has been overcharged (overvoltage), overdischarged (undervoltage), or has had excessive load or discharge current.

The choice of power MOSFETs must satisfy the need for small size, low on-resistance, and the BV_{DS} necessary to stand off all improper voltage chargers when the MOSFETs are off. Dual low threshold (<2V) N-MOSFETs with low on-resistance (<0.05Ω) are available in 8 pin SOIC packages. The circuit described was also manufactured in an 8-pin SOIC so that the total active pack electronics is limited to two small surface mount packages. The IC's supply is the cell and therefore must operate over the full range of cell voltages. The design is capable of operation between 1V and 5.5V, which well exceeds the cell's range of operation, even during extreme operating conditions.

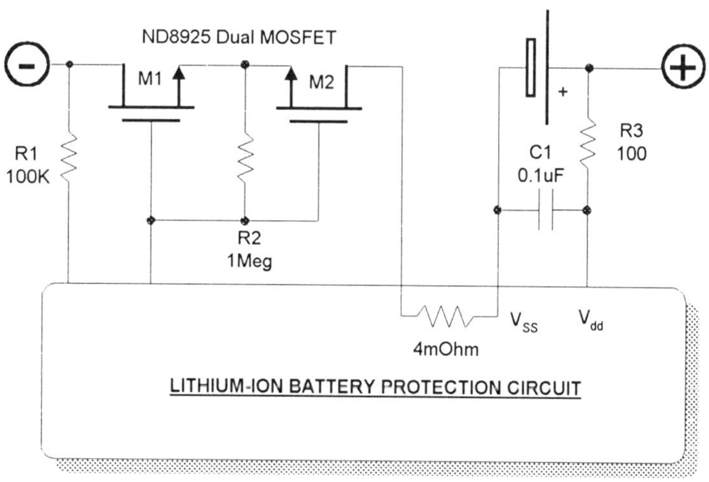

FIGURE 1. 3.6V BATTERY PACK CONFIGURATION

7.6.1

0-7803-3177-6 $5.00 © 1996 IEEE

Circuit Description

A. Cell Voltage Detection.

The maximum cell voltage is the most critical measurement made by the IC. A sampling technique is used where the cell voltage is attenuated once a second by a pair of precision series capacitors. The attenuated voltage is compared with a reference signal to determine if the cell exceeds an overvoltage limit. This technique is also ideally suited to multiple series cell packs because the same attenuator network could be used to sample the other cells in a time division multiplexing scheme that requires very little extra circuitry. The attenuator ratio is modified to also test for minimum cell voltage and other intermediate voltage points if desired. Figure 2 illustrates the basic switch capacitor attenuator in the pre-measurement (nulled) switch position. The reference voltage is also attenuated for the following reasons; 1) the capacitive loading at the comparator's inputs are now balanced and result in differential cancellation of charge injection, 2) the common mode signal is platformed to an optimum level for worst case low cell voltage operation of the comparator, and 3) the multi-level attenuation is obtained by simply switching in extra capacitance in either of the bottom capacitors, C_B and C_D. The primary reason that this technique is used is that it consumes no significant power from the battery and is inherently accurate (≈0.2%) using poly-ONO-poly capacitors. A sensitivity analysis shows that the change of attenuation ratio is linear with changes in the value of the bottom capacitor of each attenuator leg. This attenuator ratio is fine tuned at the package level final test operation by adding or subtracting bottom capacitance, C_B and C_D. The switches that control the binarily weighted trim components of capacitors, C_B and C_D are driven from an SRAM configured EEPROM register that is serially loaded from a single pin on the package (3). This trimming operation allows the part to assume a custom termination voltage for each customer while maintaining less than ±25mV total error over temperature. Modification of the overcharge termination voltage allows the IC to adapt to developing changes in battery chemistries. Programmable termination voltages allows the user to determine the proper tradeoff of total cell energy charge cycle versus the total number of useable charge cycles.

The bandgap reference for the voltage measurement system is shown in Figure 3. Rather than employ >10MegΩ of resistance to force the quiescent current of the reference down to the desired level (<1μA), the voltage measurement system was designed to be powered down except for the actual measurement period. The power-on/power-off cycle of 5mSec/1Sec reduces the power-on quiescent current of 50μA to an average current of 250nA. Poly resistors were employed for their excellent linearity. The T-network connected resistors reduces the area required by the resistors. Both the voltage reference amplifier's input offset voltage and the comparator's input offset (Figure 2) are auto-zeroed as part of the 1Hz powerup

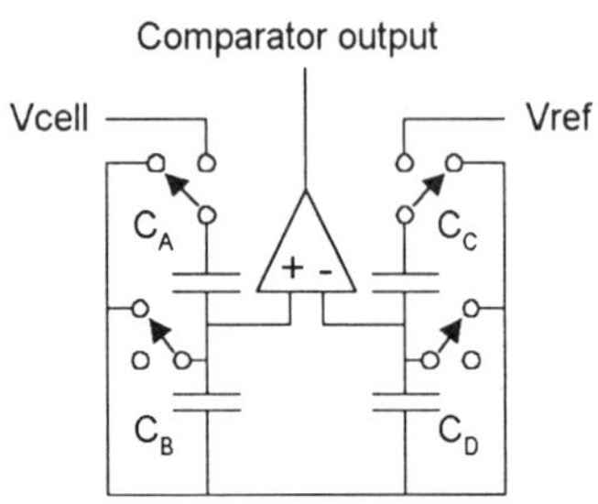

FIGURE 2. ATTENUATOR

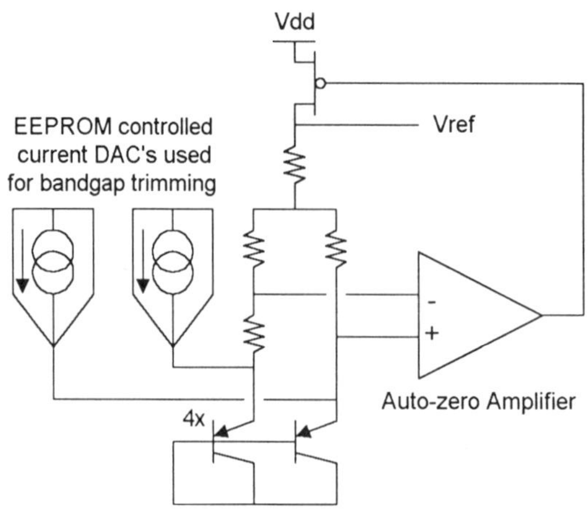

FIGURE 3. BANDGAP REFERENCE

7.6.2

sequence. The hold capacitors for the offset cancellation are located at the current mirror of the input stage (4). Recent results of the cell voltage samples are stored by the logic. Multiple consecutive fault readings are required before the IC turns off the power MOSFETs to avoid false termination due to transient events.

B. Pack Overcurrent Detection

The battery pack current measurement system utilizes a 4 milliohm resistor that is integrated in the molded plastic package's copper alloy leadframe. The 8 mil by 8 mil cross-section of the resistor provides an extremely robust resistor capable of handling 60 Amp, 6mSec current pulses that could occur during a short circuit of the battery pack. An accurate current measurement system is continuously active if the pack's current is above a threshold level as determined by the voltage drop across the power MOSFETs. The method for measuring the pack's current by the MOSFETs drop alone has been used by earlier protection IC's, yet this method is inherently inaccurate due the variation of the MOSFETs on-resistance with Vgs, temperature, and the characteristics of different MOSFET models. For this design, the voltage across the MOSFETs is crudely measured, but only for information about the pack's external bias (load or charger) and to provide the accurate current measurement with a power-up signal when the magnitude of the current is past roughly the mid-band of allowable pack current. The crude MOSFET voltage measurement circuit is similar to the accurate current measurement circuit with the difference that it not trimmed and operates at lower current drain. The inputs of the accurate current measurement system consist of a pair of level shifting substrate PNP's operating with EEPROM trimmed emitter currents. The predictable offset due to the mismatch of emitter current density provides a reference voltage against which the voltage across the 4 milliohm resistor is compared. Therefore the IC's trip point for the battery pack current is equal to $\Delta V_{BE}/R_{SENSE} = [(kT/q)*\ln(I_{E1}/I_{E2})]/R_{SENSE}$. For example, a programmed over-current trip point of 4 Amp requires a ΔV_{BE} of 16mV and an emitter current ratio of 1.87. The temperature coefficient of resistance of the copper alloy resistor is ≈3000ppm/°C and therefore is first order canceled by the temperature coefficient of ΔV_{BE}. Data on the parasitic substrate PNP's performance indicates that the ΔV_{BE} remained constant for a given ratio of emitter currents over the 6 decade current range of 100pA to 100μA. There was no significant matching penalty for biasing the PNP's down to 50nA. The program range of trip currents for the battery pack is 1A to 9A with 0.25A resolution. The direction of current in the sense resistor can be of either polarity; charging or discharging. The accurate current measurement system switches the polarity of the resistor's inputs based on the current direction information from the crude MOSFET voltage measurement system. Thus the need for two accurate comparator stages are avoided. This coarse/accurate pair of comparators provides an order of magnitude accuracy improvement over simply using the MOSFET voltage for detecting unsafe battery current levels. The bias for the accurate comparator is only 500nA, however even this current is turned off except when the pack current exceeds mid-band. An over-current event is allowed to occur for a delay period (≈4mSec), thereby allowing transient currents to be tolerated. If a current exists in the pack that exceeds both the time and programmed current limits, the power MOSFETs are turned off interrupting the current flow in the pack. The MOSFET voltage detection circuit is used to determine when the pack's operating conditions are safe for the MOSFETs to resume conduction again.

C. System Clock

Approximately 50% of the chip area is logic for the timers, FSM's and test blocks. Because even the most important events in the operation of the battery pack are measured in milliseconds, the system clock is only 10khz. A clock/counter delay scheme was selected over the 'analog' techniques seen on existing protection chips that used small currents charging large external capacitors for a single ramp period. It was believed that this analog method is less accurate, more costly, consumed more power and is less reliable than digital means. The clock circuit is RC based with enhancements on existing micropower designs (5). At this low clock frequency, the ≈1K gates of synchronous logic dissipates only a few μA's. The clock circuit itself dissipates less than 0.3μA. The basic clock operation is illustrated in figure 4. The transistors in this circuit operate in weak inversion. The Id/Vgs of MOSFETs in weak inversion operate with I-V relationships similar to that of BJT's. The voltage across the resistor is proportional to absolute temperature (PTAT). The duty cycle is ideally 50% as the capacitor is alternately charged with I and -I currents. The clock period is effectively set by the RC product. In order to control the period for 5% accurate timing purposes, four bits of EEPROM trimming is employed. The EEPROM control switches that select various segments of the R and C timing elements. The clock circuit also provides the PTAT bias currents for the various PNP input comparators on the chip. In addition, the clock circuit generates the power on reset (POR) signal to the logic. The POR circuit is a replica of the clock loop without a capacitor and is biased to switch H-L and L-H at a Vdd potential greater than that required for the clock to operate. The clock therefore is assured to be operating whenever POR goes high so that false state events are avoided in the logic. POR switches at Vdd ≈ 1V and has 100mV of hysteresis.

Conclusion

The 3mm^2 Lithium-ion protection IC was fabricated in 1μ CMOS with an EEPROM module. Poly-poly capacitors provided the low power and analog accuracy required without the need for special resistors. The key performance element for this particular application after accuracy and low cost is power dissipation. The quiescent supply current of 10μA is low enough to be well below the typical self-discharge of the Li+ cell. Power reduction methods that where employed included reduced clock speed, weak inversion operation of analog components, extensive power-down sequencing and shared functionality. Next generation chip designs will reduce the power even lower through reduction of the clock speed and duty cycle of analog circuit operation. Table 1 illustrates the measured accuracy capabilities of the circuits as evaluated in Nov. 1995. The author wishes to acknowledge the invaluable design contributions of Dan Bellemare, Yu Cheung, Mike Henry, David Kunst, Suzy Lillis, and Stu Shacter.

Table 1.

Parameter	Trim range	Trim resolution
max Vcell	4.0 - 4.5V	± 20mV
max I charge	1 - 9Amp	± 0.25Amp
max I discharge	1 - 9Amp	± 0.25Amp

References

(1) David Linden, editor, *Handbook of Batteries,* 2nd edition, McGraw Hill, 1994, Chapter 36

(2) J, Broadhead and M. Farley, editors, *Journal of Power Sources, Proceedings of the Seventh International Meeting on Li Batteries,* Boston MA, May 1994, page 151

(3) D. Oto, et al, "High Voltage Regulation and Process Conditions for High Density 5V Only EEPROM", IEEE Journal of Solid Sate Circuits, vol. SC-18 #5, Oct 83, pages 532-53899

(4) G. Temes, P. Deval, and V. Valencia, "SC Circuits: State of the Art Compared to SI Techniques", International Symposium on Circuits and Systems, Chicago 1993

(5) P. Kakela, T. Rahkonen, J. Kostamovaara, "A Micropower RC Oscillator Chip for Consumer ASIC Applications, journal unknown, ©1991 IEEE, page 278-281

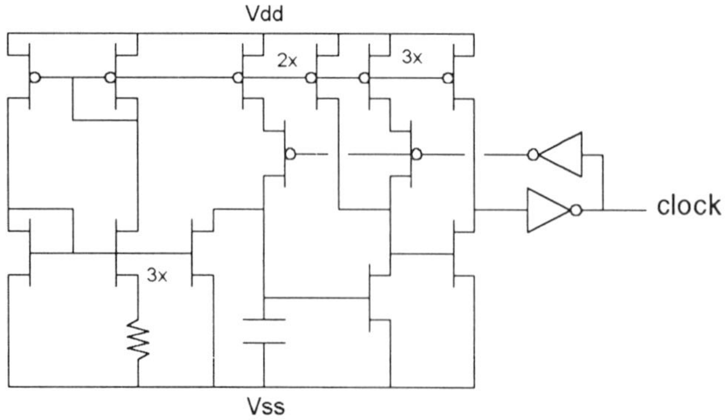

FIGURE 4. SYSTEM CLOCK

Chair: John Turner
Co-Chair: Danesh Tavana

The use of programmable logic is becoming more widespread with consumption doubling every 2.5 years. Programmable logic devices can be found in many applications such as telecommunications, computers, industrial control and logic emulation hardware. The number of programmable logic product offerings continues to expand with no apparent convergence in device architecture. The success of a device architecture is very dependent upon the routing resources available to interconnect logic blocks. Teams of researchers from industry and academia are pursuing new methods aimed at producing flexible and efficient routing structures for interconnecting logic blocks. Papers in this session look at advances in interconnect schemes that address problems encountered with the routing of signals in programmable logic devices.

The session opens with an invited paper from Hewlett Packard describing the development of an FPGA architecture targeted for use as a reconfigurable processor that is capable of accelerating specialized computational tasks. Routing resources to support multi-chip designs as well as rapid compile time make it possible for very large systems to be built from hundreds of FPGA chips.

The second paper offers insight into the interconnect flexibility necessary to route signals between logic and memory arrays. The effects of the connection block and the number of embedded memory blocks as well as interconnect delay is considered.

The third paper represents experimental work on self clocked FPGA circuits aimed at avoiding traditional clock distribution problems. Routing and logic area comparisons are made with respect to a synchronous device of equivalent capability. Prototype versions of the self clocked logic block have been fabricated and performance assessment suggests that greater system level performance is achievable.

The fourth paper describes time-multiplexed routing resources as a means to reduce the routing channel size and the number of logic blocks necessary to implement a design. Two SRAM switches per connection are time-multiplexed to dynamically change the configuration pattern during run time.

The final paper of the session presents a new multiplexer decoding switch that enhances signal routability. This scheme makes possible a reduction in SRAM cell count per switch while providing increased routing resource. No additional circuitry is required creating minimum impact to die area when compared to traditional decoding methods.

An FPGA for Multi-Chip Reconfigurable Logic

by R. Amerson, R. Carter, W. Culbertson, P. Kuekes, G. Snider

Hewlett-Packard Laboratories

L. Albertson

Hewlett-Packard California Design Center
Palo Alto, California 94304

Abstract

The Plasma chip, designed specifically to address issues important to custom computing machines (CCM), completes a 100% fully automatic place and route in approximately three seconds. Plasma FPGA's using 0.8 micron CMOS are packaged in large multichip modules (MCMs). Plasma introduces some innovative architecture concepts including hardware support for large multiported register files.

Introduction

Plasma, an acronym for "Programmable Logic And Switch Matrix," is the key component of a large custom computing machine, Teramac[1]. Its unique marriage of architecture and compiler give it this remarkable performance. Unlike commercial field programmable gate array (FPGA) chips, Plasma incorporates features making it ideal for building large systems with hundreds of chips that can readily be configured to solve new problems.

A. Design Needs

CCM requirements differ from the traditional FPGA market needs. They require rapid compile time since they may implement hundreds or thousands of designs over their lifetimes, most of which are unproved and need to be compiled for bug fixes, performance enhancement, and algorithm experiments. Hundreds or thousands of hours for a compile is prohibitive.

The Teramac system uses several commercial FPGAs to implement random logic for control interface and state machines. These chips should be cost effective, allow implementation of a variety of general purpose circuits, handle interface to other system components such as the bus, and operate at system speed. Compile time is not an issue since once a satisfactory design is compiled, it need not be compiled again. We are happy to trade compile time for the other features needed. One hour for a single chip is quite acceptable since the same design is loaded into multiple chips.

The target market for commercial vendors values lower cost above high pinout. Customers are willing to trade long compile times during design for lower cost, higher density chips. Modifying a design by hand is an acceptable process.

B. Compiler

The desire for fast compilation of large designs inspired separating the problem of mapping a user circuit onto programmable hardware into smaller independent subproblems that could be attacked with O(n) algorithms. This separation guided the hardware in directions that allowed for fast compilation. As implementation problems developed or the need for choosing parameters arose, the compiler then became a tool for exploring alternatives and refinement.

Before any hardware was designed, we created a compiler to place and route a million gates using a variety of hypothetical interconnection schemes. Not surprisingly, the interconnect topologies that the compiler found easy to use had a very large number of wires available. This need drove the architecture of the chip in which three-fourths of the active area and nine-tenths of the programming bits are devoted to interconnect.

C. Rent's Rule

Richard Rent observed that the number of signals crossing the boundaries of a partition of a logic design has a power law dependence on the number of gates in the partition[2]. The exponent was found to range from 0.5 to 0.7 for a variety of designs. Our experiments with a number of custom designs confirmed this. Teramac follows Rent's rule with an exponent of approximately two-thirds.

Commercially available chips have fewer pins and less interconnect than Rent's rule predicts are needed. Since these parts did not meet our requirements we created a custom FPGA to ensure that we had enough wires on the chip and enough I/Os off the chip.

User Visible Chip Features

Plasma is a routing-rich FPGA consisting of lookup tables (LUT) with configurable latches and registers on

0-7803-3177-6 $5.00 © 1996 IEEE

IEEE 1996 CUSTOM INTEGRATED CIRCUITS CONFERENCE

their outputs, interconnected by crossbar switches. It has many features making it ideal for reconfigurable computing applications: each chip has 336 programmable signal pins; placement and routing are insensitive to signal pin assignments; compilation time is measured in seconds; multiported register files are efficiently supported; raw die are available.

A. Programmable Atomic Logic Elements (PALE)

The basic primitive for implementing logic is the PALE. It consists of a six-input, two-output universal gate, each output of which is followed by a pair of latches that can be configured as an edge triggered flip-flop. The LUT can be configured to implement any six-input boolean function. The PALE structure is shown in Figure 1. Each LUT output drives a PALE output block, which drives the PALE output.

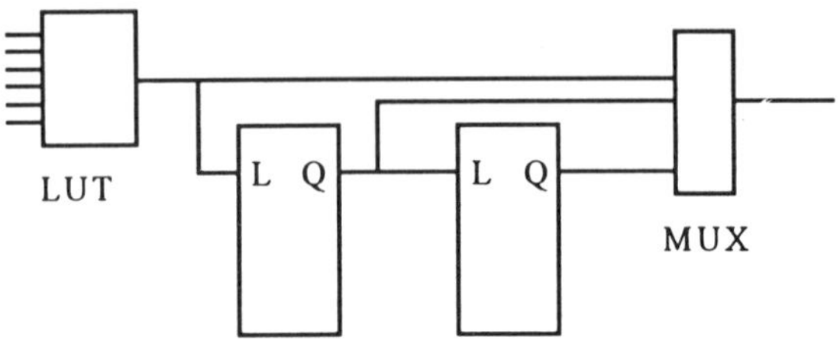

Figure 1: 1/2 PALE Logic

The output block (Figure 2) can be configured to pass its input directly to its output or pass the input to the output via a level- or edge-triggered register. Each latch has two user clocks for control and a configuration clock used for scan. The user can select either clock. The separate configuration clock allows scan without destroying the contents of either latch during both scan-in and scan-out.

Plasma is hierarchically structured. At the bottom level are PALEs. Sixteen PALEs connect into a crossbar. Each PALE has eight connections into the crossbar– six input

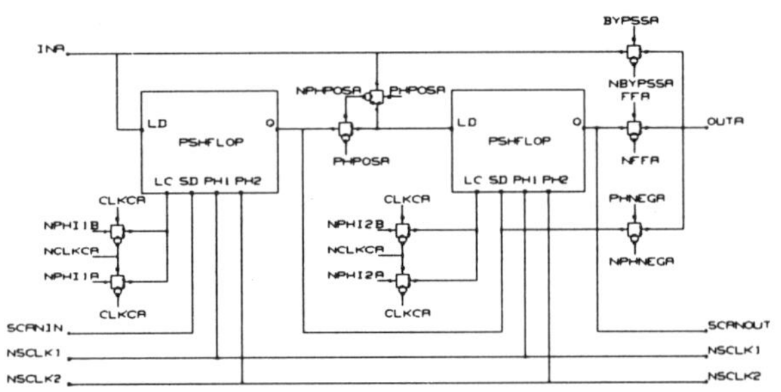

Figure 2: Output Block

and two output. The sixteen PALEs make a total of 128 lines. For space reasons, the PALE outputs are partially populated, driving half of the crossbar lines.

Tri-state logic is not directly supported since errors in user designs could unintentionally damage the chip. Instead, a straightforward mapping to an AND-OR structure gives the equivalent for most cases. It also allows detecting the case of multiple outputs enabled simultaneously, alerting users to design bugs.

Rent's rule projects between 32 and 56 wires needed to connect to the central crossbar from each hextant for exponents of 0.5 to 0.7. Additional wires are needed to connect between PALEs within the hextant that do not connect outside. Empirical studies found few designs requiring over 100 lines, so 100 was felt to be adequate for most designs.

B. Interconnect

Plasma contains 256 PALEs, organized into sixteen groups, called hextants, of sixteen PALEs each. Eight PALEs bound each side of the hextant, and eight hextants bound each side of a central crossbar (Figure 3). These two levels of crossbars allow (1) interconnections of PALEs within the same set; (2) interconnections of PALEs in different sets; and (3) interconnections of PALEs and I/O signal pins. I/O signal pins are connected to the top and bottom of the central crossbar.

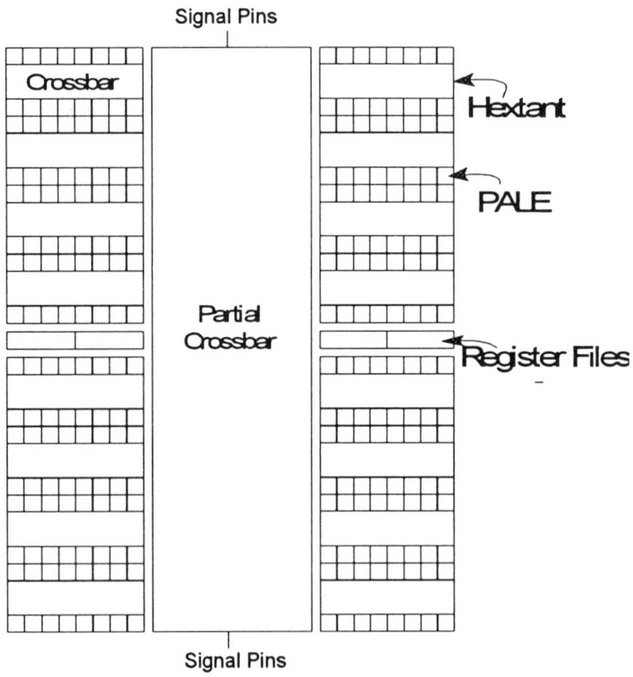

Figure 3: Plasma Block Diagram

The sixteen hextant crossbars connect into a central crossbar that is 400 lines wide and 1600 lines deep. This crossbar is one-quarter populated because of space consid-

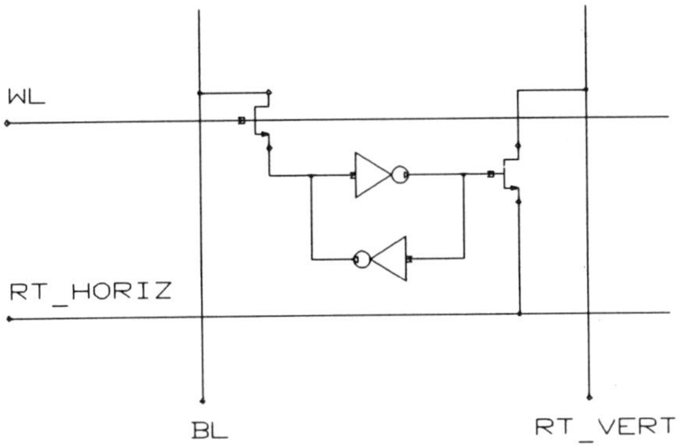

Figure 4: Five Transistor Memory Cell

erations. Each of the 100 wires from a hextant connects to 100 of the 400 central crossbar lines. The central crossbar is divided into four sections with buffers between to speed signals. Each line segment connects 100 hextant crossbar lines. Alternatively, one could view the central crossbar switch array as sixteen, interleaved, fully populated, 100X100 crossbar arrays.

The dense population of switches in the crossbars slows the chip. It speeds the compiler significantly, since a routing path is nearly always available. Knowing the difficulty of routing commercial chips having significantly less connectivity, we considered a trade of compile speed for execution speed a good choice.

C. *Configuration Memory*

Plasma contains nearly 400,000 bits of configuration memory integrated throughout the silicon layout. The configuration memory is a considerable percentage of the total chip area. It is used to store the configuration of both the PALE LUTs and the crossbar routing resources. The memory cell design is a 5-T memory cell as shown in Figure 4, and was used to save a routing channel, normally taken up by the extra bit line of a standard 6-T memory cell. An n-channel transistor is used for crossbar switches instead of a full transmission gate. Both of these design trade-offs were made early in the design to conserve area and sacrifice speed and noise margin. In order to gain back some of the lost performance, when using only an n-channel transistor, no signal ever goes through more than two switches without being buffered. The 5-T memory cells are designed so that it is difficult to read 0 and write 1. The noise margin of the cell is improved by having a slow word line turn-on to avoid flipping the cell during a read cycle. Extensive corner case SPICE simulations and functional first silicon proved the robustness of the memory cell design.

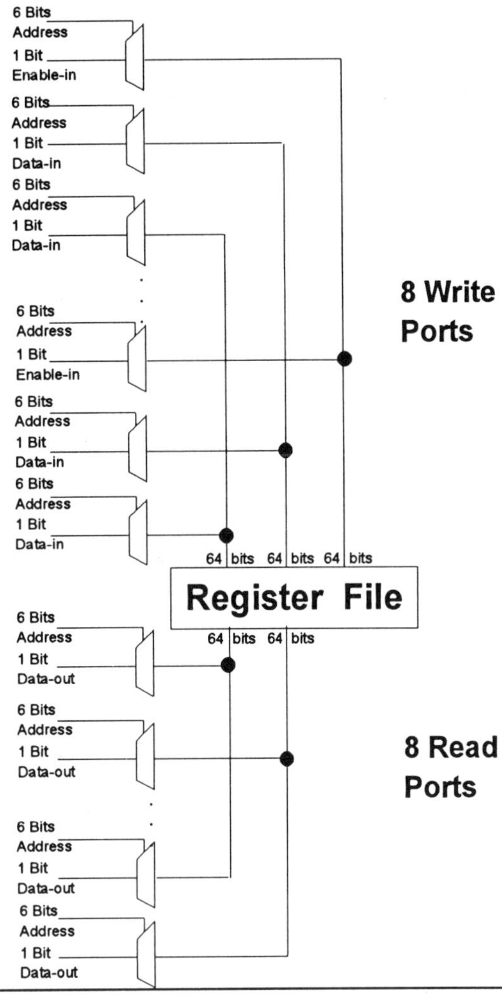

Figure 5: Plasma Register File

The number of bits of configuration memory could be greatly reduced from the 1000 per PALE Plasma has. Each of the PALE inputs connects to only a single line of the 100 in the hextant crossbar at any one time, but 100 bits of configuration memory are used to express this connection. Obviously, only seven bits are required, a 93% reduction. But, using only seven configuration bits would require a 100-1 multiplexor to select the correct line. Such large multiplexes are quite slow and require significant chip area. The number of transistors used would not decrease significantly.

D. *Register Files*

Custom computing designs frequently have need for large numbers of registers. Building these from scratch using gates (viz. LUTs) is expensive. The large decoders used to select the register accessed by each port require a minimum of one output per register per port. Building a 64-deep register file with two read and one write port would require 256 LUTs to drive the row lines. A 64 wide OR gate is required for each bit (column) of each read port.

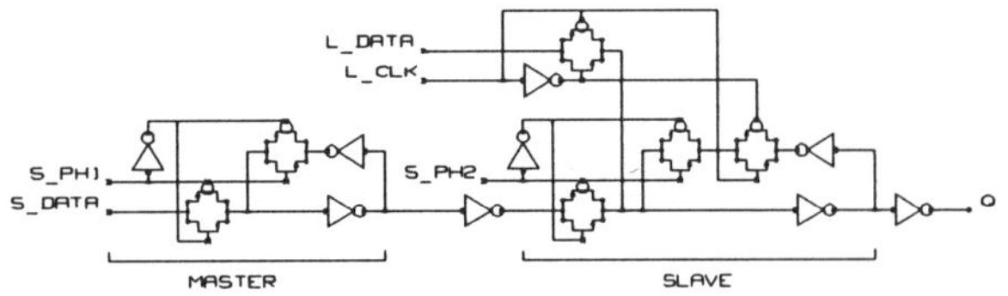

Figure 6: Shift Latch Configuration

The chip does not support wire-OR or tri-state logic. Using the six input PALEs of Plasma, a 64 deep, 32-bit wide, register file requires approximately 1200 PALEs.

Plasma takes a novel approach to reducing this count. Because the PALE LUTs are large trees of pass transistors, they can be connected "backwards" to build large decoders. The six input PALE conveniently converts into a 6-64 decoder. By taking a slice of PALEs from each of eight hextants, a register file primitive is constructed (Figure 5). Each Plasma contains four, 2-bit wide, 64-deep, 16 port register files[3]. Eight ports are configured as read ports and eight as write ports. A write port has both an address and write enable; a read port has only an address.

D.1 Average gate count

For designs using this feature, extremely high gate efficiency can be achieved. One test design, a sort engine for 64 elements, achieved an equivalent gate count of 60,000 for a single Plasma, nearly 240 gates per PALE. A more typical number for random logic is eight to ten gates per PALE, approximately 2000 gates per chip. These numbers are achieved by 100% automatic place and route..

System Visible Chip Features

A. Scan

The Plasma chip is designed to allow debugging of a user's custom design to be easy and natural. The complete internal state of the chip is available on a user scan chain. This includes all flip-flops and register files as well as all the I/O pins of the chip. Both D and Q of every flip-flop are simultaneously observable. The inputs and outputs of all LUTs are observable and the user interface recreates signals that exist on the original schematic but were subsumed into lookup tables.

The shift-latch configuration (Figure 6) used in the PALE storage module has two functions: a full master/slave flip-flop used for scan, and a data latch for user circuits. The slave portion of the flip-flop is shared to implement both functions. The system logic clocks are stopped before examining the scan chain and the user may change the state of the machine before resuming clocking by simply changing the state of the latch via the scan chain. This architecture allows the user full peek and poke capability of his user circuit while debugging. This scan capability uses different commands to the chip than the normal read and write configuration commands allowing very fast debugging.

B. Plasma Controller

B.1 Programming

Programming of the Plasma chip is accomplished using a configuration scan chain. This scan chain is separate from the user scan chain so that fast reading and writing of the user state can take place. This allows simple algorithms for examining and setting the state of all user visible signals in the chip to provide excellent debug capability.

The configuration scan chain, controlled by the chip controller, writes and reads the underlying configuration bits. The read allows verification that a configuration was received correctly. Configuration requires first halting the clocks and placing the chip in configure mode. A separate high frequency clock is used for configuration. A complete configuration consists of nearly 400,000 bits. The configuration bits are stored in a large SRAM structure underlying the crossbars and PALEs. Each row may be read and written independently, allowing minor modifications to a design without completely downloading a new configuration.

Plasma uses a parent-child scheme on the scan chain signals. A header record tells the chip whether this configuration data was intended for it. If not, the remaining data is passed to the child pin. This allows the chips to be connected serially for system configuration, reducing the signals that must be connected to the master system. A broadcast command can address all chips on a given chain.

B.2 Implementation

The Plasma controller consists of several small state machines, each controlling various aspects of the chip programming and scan. The "message" state machine inspects the incoming serial bit stream looking for configuration or scan commands. The "parent/child" state machines control how the data is to be routed within the system, whether it is to be used locally or passed through. Additionally, the controller contains the configuration "Address Generator" and controls various other signals required for reading and writing configuration data such as bitline precharge and the serial to parallel read/write control signals.

The chip controller was implemented in standard cell technology. Standard cells were chosen so that the design of the controller could easily be modified with minimal impact to the overall chip schedule. It was physically implemented in the center of the Plasma chip, to minimize the lengths of the connections to the rest of the chip. The Plasma controller consists of less than 1000 gates and consumes an area of about 0.7mm^2.

C. Clock Generators

C.1 Logic Clock

The internal logic clock for the Plasma chip is a non-overlapping two phase clock. Its frequency is set by the single phase clock signal at the pin interface to the chip.

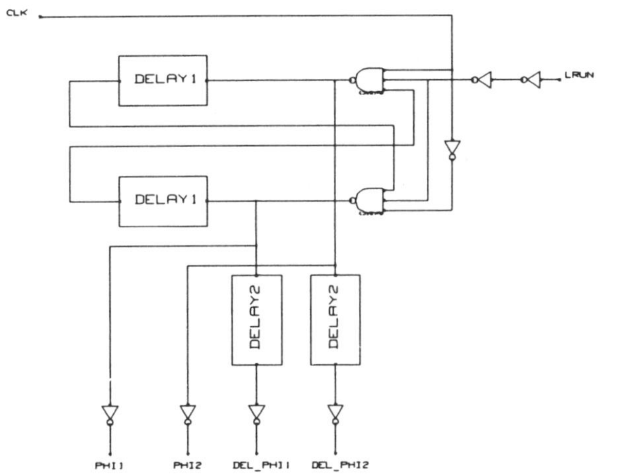

Figure 7: Logic Clock Generator

Because of simultaneous switching buffer concerns with the chip, the logic clock generator (Figure 7) additionally produces a delayed version of the two phase clock. The delayed phases are routed to the eight hextants closest to the center of the chip.

C.2 Configuration Clock

The configuration clock for the Plasma chip produces both single phase and a two phase clocks, which are used to control the configuration and scan of the chip. The controller is clocked with a single phase of this clock, and the two phase version is used to clock the scan registers. The frequency of this clock is roughly ten times higher than that of the logic clocks.

C.3 Clock Tree Routing

Both the logic clocks and configuration clock were physically implemented in the center of the chip, to minimize the lengths of the clock tree to the rest of the chip. The clock trees are buffered going into the right and left sections of the chip, then at the interface to each hextant and PALE and again within each PALE.

D. Chip Interface

D.1 Pinouts

A CCM requires greater connectivity than is provided by commercially available FPGA's. Plasma's 336 signal I/O pins, 84 per side, are roughly double that which can be obtained from an off-the-shelf part. In addition, placement and routing within Plasma is insensitive to signal pin assignments because all I/O pins have equal connectivity into the chip's central crossbar.

D.2 Padring

The Plasma padring is a custom designed 408 pin staggered padring, with each side consisting of 102 pads. Two large VDD/GND connections per side, allow excellent current carrying capability for the core logic. Two large, and five additional pairs of VDD/GND power pins separately supply all of the off-chip pad drivers.

Circuit Design Considerations

A. Simultaneous Switching Outputs (SSO)

FPGAs used for custom computing have a high probability of simultaneous switching problems. A user may design a massive data path with most of the signals switching from 1 to 0 (or 0 to 1) at the same clock edge. The resulting collapse of the internal power supply can cause configuration bits to change, leaving a random design in the chip. Not only would the resulting design be incorrect, but the chip outputs could be shorted together, resulting in damage to the chip.

We took significant precautions to reduce the danger of simultaneous switching problems. Great care was taken in the design of the buffer arrays: central crossbar buffers, peripheral crossbar buffers, and PALE buffers.

A.1 Buffer Design

The buffers for Plasma were designed with small devices to keep switching currents small, at the expense of sacrificing speed. Break-before-make design of the buffers minimizes crossover currents when switching between logic states (Figure 8). Peripheral crossbar buffers and I/O pads are inverting so alternate buffers in a path switch to the opposite state. Logic states are positive true in the peripheral crossbars and negative true in the central crossbar.

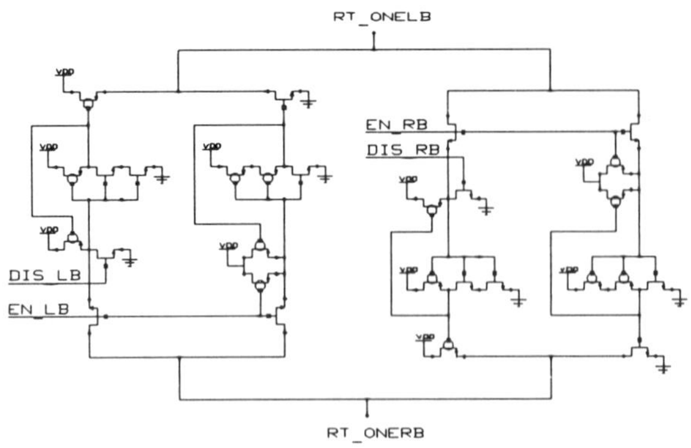

Figure 8: Break-before-make Buffer Design

A.2 Power Busing

The power busses within Plasma are designed using the lowest resistivity metal layer, to meet the instantaneous current requirements when switching all flip-flops simultaneously. In all cases, the power buses are wide and are routed directly over high current buffers to minimize voltage drops when switching. Wide internal VDD/GND power rings between the core circuitry and the pads provide good power connections from the core to the power pads. Eight large VDD/GND connections, each with three bonding wires, allow better current carrying capability and lower inductance between the pads and the internal power ring.

Dirty VDD/GND power pins, and internal power rings within the I/O pads separately supply all the off-chip pad drivers. Plasma is unusual because the on-chip core SSO is much more severe than I/O pad SSO

A.3 Memory Cell Precharge

To assist in keeping the configuration bits from changing logic states during normal operation of the chip, the configuration cell bit lines are precharged high. They are continually precharged high at all times except during reading or writing of the configuration memory. This is done to take advantage of the difficulty of writing a “1” into the configuration memory cells. Many conventional designs precharge the bit lines only immediately before accessing the cells.

A.4 Logic Clock Skewing

To prevent all 512 flip-flops from switching at the same clock edge, the logic clocks are deliberately skewed up to seven nanoseconds by the clock distribution tree. The inner eight sets of hextants are skewed to clock later than the outermost eight hextants to deal with the worst case hold time as seen from outside the Plasma chip. All hold times are kept negative because the best case buffer switching times are much less than the clock skew. To account for skewing, additional time is added to the non-overlap time of the clock. SPICE shows that all hold times can be met. Because the Plasma compiler computes the clock period required by each individual design, the skewing of clocks in the chip causes no setup time problems.

A.5 Global Drive Enable

A potential problem arises with “Global Drive Enable,” the signal that enables all buffers out of tristate after a configuration has been loaded. When all buffers on the chip are enabled at once, a significant current surge can occur. Our solution created five individual “Global Drive Enable” signals each separated by one hundred nanoseconds, routed so that four tied to one hextant in each quadrant and the fifth enabled the pad drivers. The timing is controlled by the chip controller.

B. Electromigration

Because of simultaneous switching concerns, electromigration had to be looked at very carefully. In the case of Plasma, we not only had to size the buses to carry the current, but also had to determine the correct amount of contacts to use when changing metal layers and what the best contact pattern would be for the flow of current.

Results

FIRST SILICON WORKED! The 16.2 mm by 16.2 mm three million transistor Plasma chip (Figure 9) required no modification. Test designs performed as predicted due to extensive peer review and conservative design.

The chip performs at its projected speed. Simultaneous switching has not been an issue. Most designs have compiled with no user intervention, though quite a bit more compiler work was required to bring the compiler to accomplish this.

A. Performance

Designs contained in a single Plasma have run at speeds in excess of 10MHz. The Teramac compiler calculates the correct maximum clock speed based on an individual design.

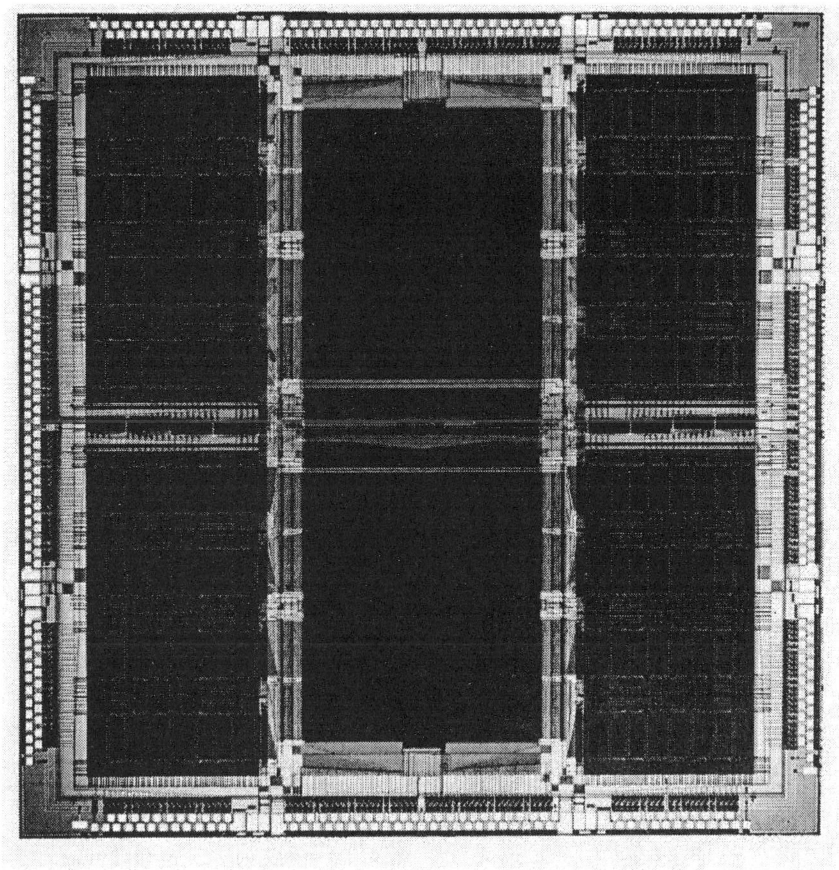

Figure 9: Plasma Chip Photograph

Conclusions

The Plasma architecture simplifies the compiler such that the resulting speed of compilation, tens of minutes rather than tens of hours, allows another layer of abstraction in defining highly parallel custom computers. The speed of compilation and execution along with easy access to many large memory banks and multiported register files provided by Teramac will allow fruitful methods for the simple specification and synthesis of custom computers.

Acknowledgments

Lyle Albertson, Sue Blockstein, Brian Jung, and Tom Meyers did the detailed design and verification of Plasma. Betty Kitajima and Cindy Laird did the custom artwork layout. Peter Maxwell developed the chip yield model for Plasma. Greg Snider was the compiler writer and chief architect. Arnie Berger designed the Logic Board and managed the system hardware design. Andy Blasciak designed and implemented the control board. Martin Guth perfected the MCM wire-bonding process. Dan Kary developed the control and user interface software. Wulf Rehder developed test algorithms. Bruce Culbertson created the system test code. Dick Carter defined the user interface, created numerous custom computing designs, and improved the compiler. Rick Amerson provided management and contributed to the Plasma architecture. Phil Kuekes was project manager and has contributed to both the MCM design and the architecture. Barry Shackleford and Bob Rau originally conceived Teramac.

References

[1] R. Amerson, R. Carter, W. Culbertson, P. Kuekes, G. Snider. "Teramac -- Configurable Custom Computing", Proceedings of the 1995 IEEE Symposium on FPGA's for Custom Computing Machines.

[2] B. Landman and R. Russo, "On a Pin vs. Block Relation- ship for Partitions of Logic Graphs," IEEE Transactions on Computers, December 1971.

[3] Greg Snider, Philip Kuekes, W. Bruce Culbertson, Richard J. Carter, Arnold S. Berger, Rick Amerson, "The Teramac Configurable Compute Engine," Field Programmable Logic and Applications, Will Moore and Wayne Luk (Eds). Springer, 1995

Memory/Logic Interconnect Flexibility in FPGAs with Large Embedded Memory Arrays

Steven J.E. Wilton, Jonathan Rose, and Zvonko G. Vranesic
Department of Electrical and Computer Engineering
University of Toronto
Toronto, Ontario, Canada, M5S 1A4
{wilton,jayar,zvonko}@eecg.toronto.edu *

Abstract

As the capacities of field-programmable gate arrays (FPGAs) grow, it becomes desirable to create FPGAs with embedded memory arrays. This paper examines the flexibility of the interconnect structure that joins memory and logic. For architectures with only a few memory arrays, we find that both the routability and the delay of circuits are insensitive to the memory/logic interconnect flexibility, which implies that this interconnection can be made very inflexible. This is in contrast to the logic connection block flexibility, which has been shown to require high flexibility [1]. For architectures with more arrays, the memory/logic interconnect flexibility requirements increase and approach those of logic interconnect.

1 Introduction

As FPGAs grow to encompass entire systems, it is becoming clear that an efficient way to implement large blocks of memory is needed. FPGAs with embedded memory blocks are now available from several vendors [2, 3], in contrast to the fine-grain LUT RAM available earlier [4, 5]. Both types of memory provide higher integration, faster memory access and reduce the demand for FPGA I/O pins compared to off-chip memory implementations.

In our previous work [6, 7], we examined FPGA memory architectures in isolation and showed how to provide the ability to implement many different numbers and shapes of user memories. This paper explores the structure of the interconnection between memory and logic. We are interested in finding the minimum memory/logic interconnect flexibility (the cost of the interconnect) that provides both routability and high speed. To do this we employ an experimental approach in which benchmark circuits are implemented on many different interconnect architectures. The following section describes our base architectural assumptions. Section 3 describes how we created enough circuits to test the architecture, and the implementation procedure. The subsequent sections give results and conclusions.

*This work was supported by the Natural Sciences and Engineering Research Council of Canada, MICRONET, and the Walter C. Sumner Foundation.

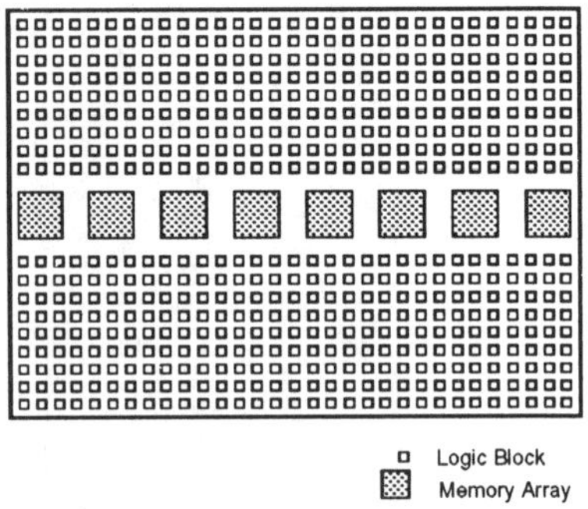

Figure 1: FPGA with Embedded Memory

2 Architectural Framework

We assume that an FPGA with embedded memory consists of distinct logic and memory resources, as illustrated in Figure 1. The memory resources consist of a set of identical arrays that can be combined to implement the required user memory configuration, similar to [6]. The number of bits in each array is fixed, but the aspect ratio can be configured by the user. In the results presented in Section 4, we have assumed that each array has two kilobits, and has a configurable data width of 1, 2, 4, or 8 (similar to [2] and [7]). Unlike [6], here we assume that each memory array has separate input and output data ports (as well as an address port).

The logic resources of the FPGA are assumed to consist of five-input lookup tables, interconnected using symmetric horizontal and vertical channels like those of [1] and [4]. At the intersection of each horizontal and vertical channel is a switch block. The flexibility of each switch block, F_s, is defined as the total number of possible connections offered to each incoming wire [1]. In this paper, we assume $F_s = 3$, which is consistent with previous work. All segments are of length 1; that is, segments only connect neighbouring switch blocks. Each pin of each lookup table can be connected to two channels; within each channel, each pin can be connected to

0-7803-3177-6 $5.00 © 1996 IEEE

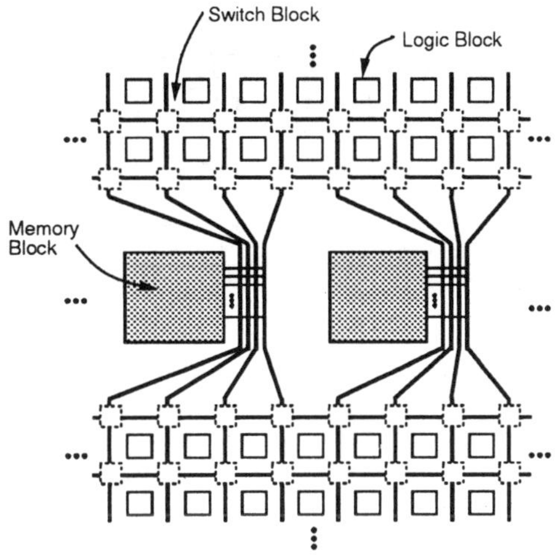

Figure 2: Memory/Logic Routing Structures

F_c tracks. In Section 4, we vary F_c between W and $\frac{W}{2}$, where W is the number of routing tracks per channel.

The memory arrays are assumed to be positioned in a single row in the middle of the chip, as shown in Figure 1, with logic blocks above and below the memory arrays. This is similar to the Altera FLEX10K architecture [2] and the Actel 3200DX architecture [3].

Figure 2 shows the assumed interconnect structure between the logic and memory portions of the FPGA. In the figure, each solid line is a channel of W parallel routing tracks. The vertical tracks in the top half of the logic are connected to those in the bottom half, and the pins on each memory array are connected to one or more of these vertical tracks. Note that in Figure 2, the vertical tracks are shown to "bend" around the memory blocks; in multi-metal layer implementations, the vertical tracks might simply cross on top of the arrays.

We define the flexibility of the memory/logic interconnect, F_{lm}, as the number of vertical tracks to which each memory pin can connect. If there are r logic blocks per memory block in the horizontal dimension, the maximum possible F_{lm} is rW (in Figure 2, $r = 4$). In Section 4, we will vary F_{lm} from 1 to its maximum value and examine the effects on routability and delay.

3 Architecture Evaluation Method

This section describes how we investigate the effect of varying the amount of flexibility in the memory/logic interconnect structure, F_{lm}. The general approach is to map, place and route benchmark circuits on FPGAs with different architectural parameters. We measure the effect on routing demand (the number of tracks needed to route, W) and the speed of the resulting circuits. The following subsections describe the development of appropriate benchmark circuits, and the implementation procedure.

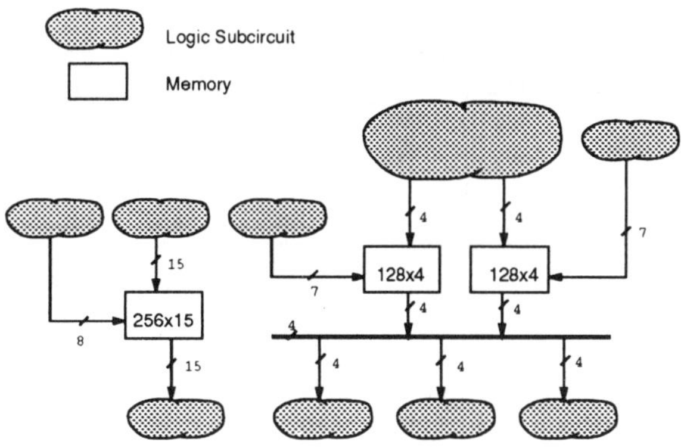

Figure 3: Example Generated Circuit

3.1 Benchmark Circuit Generation

It is difficult to apply the traditional method of placing and routing 10 to 20 benchmark circuits [1, 8] in this analysis because circuits typically have only a few memories each, and so hundreds of such examples may be required to properly exercise the memory-logic connection architecture. Because it isn't feasible to gather that many benchmark circuits, our approach is to study the types of memory configurations found in systems, and then to develop a stochastic memory configuration generator based on that study.

A memory configuration is the set of all memories required by a circuit, and consists of the number of memory *clusters* (clusters are groups of memories which share data input or data output subcircuits), the number of memories within each cluster, and the width and depth of each memory.

The generator chooses the memory configuration parameters independently from probability distributions based on statistics gathered from a set of 171 custom ASIC designs. (Although we were able to obtain data on the memories in these circuits, the detailed circuits were not available).

Next, the logic sub-circuits needed to supply address and data as well as consume the data are generated and attached to the memories. The memory/logic interconnect pattern is chosen from a set of commonly occurring patterns in our example ASIC designs using measured distributions. The sub-circuits are randomly chosen from a collection of 38 MCNC circuits, which have been optimized using SIS [9] and technology-mapped to 5-input lookup tables using FlowMap [10]. The actual construction of the circuits is non-trivial and beyond the scope of this paper. Full details of the memory circuit study and construction of the generator are described in [11].

Figure 3 illustrates an example generated circuit. It has three memories, two of which receive data from a single logic source and drive a common data bus. Interconnections between the various logic sub-circuits and external input/output pins are not shown in this figure.

The circuits are generated without accounting for the

Arch	5-LUTs Used		Arrays Used		Total Terminals		Memory Terminals		r		Aspect Ratio	
	Avg	St.Dev	Avg	St.Dev	Avg	St.Dev	Avg	St.Dev	Avg	St.Dev	Avg	St.Dev
2	382	192	1.63	0.48	2094	1033	28.7	11.7	10.4	3.43	1.00	0
4	502	307	2.61	1.13	2761	1649	47.4	23.3	5.98	2.28	1.01	0.05
8	574	340	3.93	2.27	3164	1838	69.3	41.6	3.50	0.85	1.53	0.96
16	638	360	5.46	3.97	3543	1968	97.8	73.8	3.05	0.23	4.64	3.44

Table 1: Circuit and Implementation Statistics

number of memory bits or number of independent arrays in the FPGA architecture. Once generated, the circuits that do not fit in the target architecture (in terms of total number of bits or number of independent arrays) are discarded. In the next section, we present results for architectures with 2, 4, 8, and 16 memory arrays; for each architecture we generate circuits until we have 200 circuits that fit. Table 1 shows statistics from the 200 circuits generated for each architecture (r and Aspect Ratio will be discussed in Section 3.2).

This stochastic generation approach was also used in [6] which studied only the memory portion of an embedded-memory FPGA.

3.2 Circuit Implementation Procedure

Each benchmark circuit is "implemented" in each FPGA using the following procedure: First, each memory in the memory configuration must be implemented using one or more of the physical memory arrays. For example, a 4Kx2 user memory can be implemented using four 2-kilobit arrays each configured as a 2Kx1 memory with appropriate decoding. We call this process the logical-to-physical mapping. Because there can be many ways to perform this mapping, it is an optimization problem. We use an algorithm described in [11].

Next, the mapped circuits are then placed on an appropriately-sized FPGA using a simulated annealing-based placement program, and the detailed routing is performed using a multi-pass maze router (both tools are described in [11]). The routing is repeated for different values of W (the number of tracks per channel) to determine the minimum W that gives a 100% routable solution. The router considers all input pins of a lookup-table equivalent, as well as all address pins of a memory array. Data pins are also considered equivalent with the constraint that a pin assignment for a specific bit in the data-out port fixes the corresponding assignment in the data-in port (and vice versa).

The size of the FPGA used in the place and route step depends on the number of lookup tables in the circuit to be implemented, as well as the number of memory arrays to be included. For a given number of arrays, we place the required number of logic blocks above and below the memory row, creating a roughly square chip. This will result in different values of r (the number of logic blocks per memory block in the horizontal dimension) for different circuits. If, however, the resulting r is less than 3, we force r to be 3, resulting in a non-square aspect ratio. For some circuits, the number of input/outputs will determine the chip area; for these circuits, a square array with the required number of input/output blocks will be used. Table 1 shows the average values of r and the average aspect ratio (ratio of horizontal logic blocks to vertical logic blocks) for circuits implemented on the four architectures.

Finally, a timing analyzer is used to measure the critical path through the circuit. The timing analyzer finds each net delay by constructing an RC-tree. The memory access time is measured from a modified version of CACTI, a detailed cache access time model [12]. The CLB delay is assumed to be constant. A 0.5um CMOS process was assumed throughout.

4 Experimental Results

4.1 Effect of F_{lm} on Routability

Figure 4-a shows the effect of changing F_{lm} on the minimum track count required for 100% routability (averaged over 200 stochastically generated circuits). The results are shown for architectures with 2, 4, 8, and 16 two-kilobit blocks. The right-most point on each line corresponds to architectures with $F_{lm} = rW$; as discussed in Section 2, this is the maximum possible F_{lm}. As the graph shows, for architectures with fewer than 8 arrays, the measured minimum W is not sensitive to F_{lm}. This is contrast to [1] which shows that circuit routability for logic is severely hampered by low values of the connection block flexibility, F_c.

As the number of memory arrays increases above 4, the number of tracks needed for complete routing *does* increase significantly for low F_{lm} ($F_{lm} = 4$ is sufficient). There are several reasons for this. First, the circuits generated for these architectures tend to have larger memory requirements, and thus, use more memory pins (this is shown in Table 1). For small values of F_{lm}, the nets connected to each array have fewer routing options than nets connected to only logic elements. The more of these nets there are, the more sensitive the overall circuit routability will be to F_{lm}.

A second effect is from nets that connect to more than one memory array. Memory arrays are connected together when they are combined to form larger user memories, or when independent user memories share a common data bus. The more arrays in a circuit, the more of these memory-to-memory nets there are. The number of routing paths for these nets is very dependent on F_{lm}.

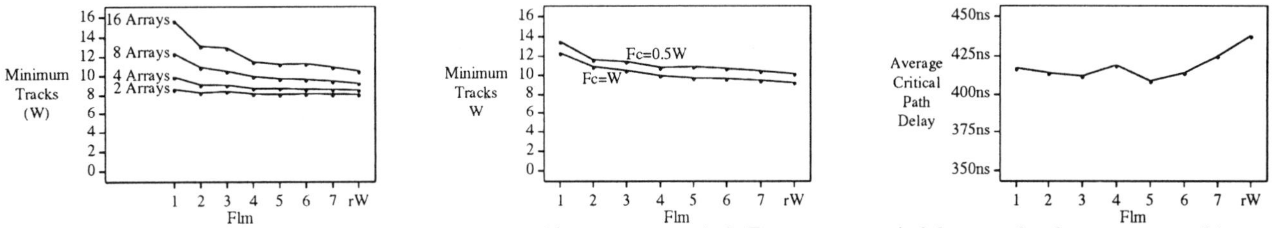

a) $F_c = W$ b) 8 arrays, varied F_c c) delay results for 8-array architecture

Figure 4: Effects of F_{lm} on routability and delay

A final reason may be due to the way in which the memory connections are "spread" among the logic blocks, as determined by the parameter r. For architectures with only a few arrays, the value of r is significantly larger than for architectures with many arrays, as shown in Table 1, column 10. The higher r is, the more "spread out" the memory pins are. It is possible that a concentration of the low-flexibility memory pins in a single channel decreases routability.

4.2 Effect of F_c on choice of F_{lm}

Figure 4-a assumed that the connection block flexibility, $F_c = W$; Figure 4-b shows how the results (for 8 arrays) affected by F_c. Although the track requirements go up as F_c is decreased, the dependence on F_{lm} remains unchanged.

4.3 Circuit Delay

Although the main motivation behind using a small value of F_{lm} was to reduce the area required for the memory/logic interconnect block, removing switches generally increases the speed of a circuit implemented on an FPGA. In this case, however, the memory/logic interconnect block does not represent a major portion of the critical path delay, so we would expect the speed to be roughly independent of F_{lm}. Figure 4-c shows the average critical path delay of the 8 array architecture. As the graph shows, for low flexibilities, the delay actually drops slightly as F_{lm} increases; this is because a low F_{lm} often results in circuitous routes for some nets.

5 Conclusions

In this paper, we have investigated the effects of the memory/logic connection block flexibility on the overall routability and speed of a circuit implemented on an FPGA with embedded memory arrays. We have found that the routability of circuits with memory is not strongly affected by a low memory/logic interconnect flexibility, especially for architectures with fewer than 8 embedded arrays. For such architectures, the logic routing resources are flexible enough to compensate for the low flexibility in the memory/logic interconnect. This becomes less true as the number of arrays (and hence the number of connections to memory) is increased.

The delay of such circuits was also shown to be roughly independent of the flexibility of the memory/logic interconnect. This is primarily because this interconnect block represents only a small portion of the critical path of a circuit.

References

[1] J. Rose and S. Brown, "Flexibility of interconnection structures for field-programmable gate arrays," *IEEE Journal of Solid-State Circuits*, vol. 26, pp. 277–282, March 1991.

[2] Altera Corporation, *Datasheet: Flex 10K Embedded Programmable Logic Family*, 1995.

[3] Actel Corporation, *Datasheet: 3200DX Field-Programmable Gate Arrays*, 1995.

[4] Xilinx, Inc., *The Programmable Logic Data Book*, 1994.

[5] AT&T Microelectronics, *Product Brief: AT&T Optimized Reconfigurable Cell Array (ORCA) Series Field-Programmable Gate Arrays (FPGAs)*, April 1993.

[6] S. J. E. Wilton, J. Rose, and Z. G. Vranesic, "Architecture of centralized field-configurable memory," in *ACM/SIGDA International Symposium on Field-Programmable Gate Arrays*, pp. 97–103, 1995.

[7] T. Ngai, J. Rose, and S. J. E. Wilton, "An SRAM-Programmable field-configurable memory," in *IEEE Custom Integrated Circuits Conference*, pp. 499–502, May 1995.

[8] J. L. Kouloheris and A. E. Gamal, "PLA-based FPGA area versus cell granularity," in *1992 Custom Integrated Circuits Conference*, pp. 4.3.1–4.3.4, 1992.

[9] E. Sentovich et al, "SIS: A system for sequential circuit analysis," tech. rep., Electronics Research Laboratory, University of California, Berkeley, May 1992.

[10] J. Cong and Y. Ding, "FlowMap: An optimal technology mapping algorithm for delay optimization in lookup-table based FPGA designs," *IEEE Transactions on Computer-Aided Design*, vol. 13, pp. 1–12, June 1994.

[11] S. J. E. Wilton, *Architecture of Field-Configurable Memory*. Ph.D. thesis, University of Toronto, in preparation.

[12] S. J. E. Wilton and N. P. Jouppi, "Cacti: An enhanced cache access and cycle time model," in press.

A Self Clocked FPGA for General Purpose Logic Emulation

Dana L. How

Information Systems Laboratory, Stanford University, Stanford, CA 94305-4055

Abstract

This paper describes the architecture of an entirely self clocked FPGA. Such an FPGA would avoid clock distribution problems, facilitate circuit composition and increase the amount of hardware concurrency, as well as reliably operate at its maximum speed without timing analysis. These unique advantages would be of great use in many functional simulation and logic emulation applications targeting larger designs requiring many FPGAs.

Introduction

Major applications for current FPGAs include rapid prototyping and logic emulation. Unfortunately, current synchronous FPGA CAD tools require extra effort to be spent, by either the user or the CAD tools, on performance-limiting timing and clock distribution problems, resulting in slower design iterations. This is particularly noticeable in multi-FPGA simulation engines [7] where clocks and data must be distributed across several levels in an interconnect hierarchy.

This paper presents a self clocked logic block for use in an entirely self clocked FPGA. The built-in local handshaking control in such an FPGA entirely eliminates clock distribution problems, and it facilitates circuit composition and vastly increases the amount of hardware concurrency [2]. Additionally, a circuit implemented in a self clocked FPGA would reliably operate at its maximum speed without the intervention of timing analysis programs to determine safe clock rate limits. Finally, not only does self clocking reliably adapt better to the variable routing delays in FPGAs, but in an optimized implementation the accepted incremental cost of configurability may even overshadow the incremental cost of self clocking.

Many issues in circuit design and design automation must be investigated before a self clocked FPGA emulation system may be implemented. This paper focuses on the design of the self clocked logic block and accompanying routing network, with secondary attention on chip integration issues. Considerations in developing a CAD system for a self clocked FPGA are overviewed. Finally, results from the four CMOS test chips fabricated through MOSIS are presented and analyzed.

Approach

The self clocked logic block is based upon a *micropipeline* [2], illustrated in Figure 1. Each of the C-elements in the figure changes its output state only when its inputs differ, and then to the value of the input from the left. The micropipeline is considered empty when all the C-elements are in the same state. When the leftmost request R_i transitions to the other state, the C-element it drives will react to the difference at its inputs and change its output. This change will propagate through the delay element and serve as the acknowledgment A_o to the previous stage, as well as the request to the next, where the process will repeat. Note that to prevent overrun, a particular C-element will not accept a new transition from its left unless its neighbor to the right has accepted its last transition. Therefore the circuit acts as an elastic buffer or FIFO for transitions arriving from the left.

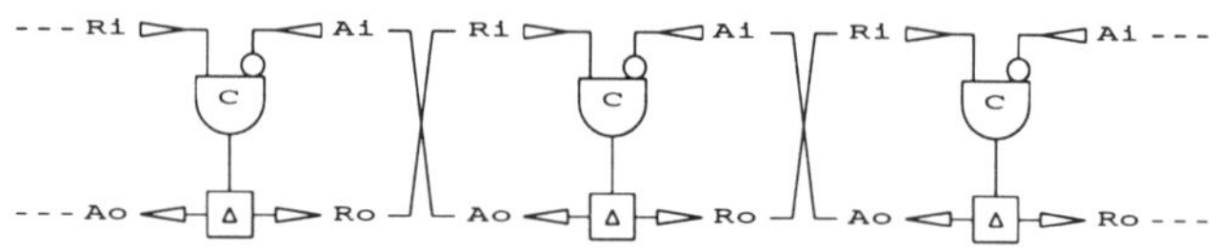

Figure 1: Micropipeline control structure.

With this micropipeline control structure, the outputs from the C-elements can be used to clock double-edge-triggered latches in a pipelined datapath flowing to the right *if* the delay elements are properly configured. Unfortunately, this approach not only bases the circuit timing on delay elements which must have worst-case values, but it also imposes a *bundling constraint*: the delay on each request signal must be no less than the delays on all of its associated inter-stage data signals. This is difficult to guarantee when routing in FPGAs [12].

To avoid these complications of separate control and data an integrated signaling scheme must be used. For

0-7803-3177-6 $5.00 © 1996 IEEE

IEEE 1996 CUSTOM INTEGRATED CIRCUITS CONFERENCE

example, the forward request lines could be replaced by pairs, one of which could indicate a new **0** and the other a new **1,** as shown in the top half of Figure 2, but this protocol is combinationally ambiguous and requires a decoder with extra state. Fortunately, it is simple to devise a protocol for the signal pair such that for each request exactly one of the pair has a transition *and* the data value is directly available, and this protocol is shown in the bottom half of Figure 2. Here the signals are called D and F: to send a different value, flip D to the new value; otherwise flip F to send the same value "fresh." This technique, here called LETS — "Level Encoded Transition Signaling," was first presented in [6].

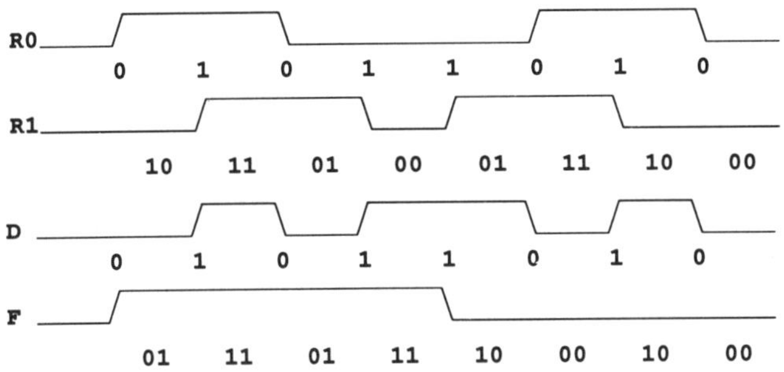

Figure 2: Identical data encoded with Standard (top) and Level Encoded (bottom) Transition Signaling.

LETS turns out to be the best signaling protocol to use in a self clocked FPGA. This is not only because signals entering logic blocks do not need to be decoded, but also because each new data value requires exactly one transition, rather than the two transitions that would be required by conventional 4-phase dual rail. This reduces the number of trips through the slow, SRAM-programmable routing network without imposing any need for transition signaling converters around an inherently level-sensitive look-up table in the logic block core.

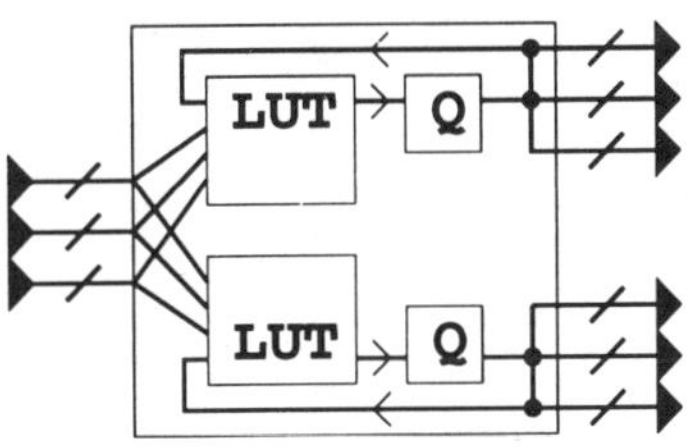

Figure 3: Logic block architecture.

Finally, the logic block architecture, shown in Figure 3, was chosen for simplicity and easy comparison to other architectures. It consists of two 4:1 SRAM look up tables, with three shared inputs driving two outputs registered in self clocked latches; each signal comprises three wires. A better solution would use [10] to tune this architecture to the surrounding routing structure, presented next.

Routing

In this work segmented routing channels [8] interconnect the array of logic blocks. Normally such channels require far too many switches to be implemented using SRAM, but the number of switches can be reduced. Because the three wires comprising a self clocked signal follow the same path, the extra wires do not require additional bulky SRAM bits to determine their routing. Of course, with individual acknowledgments all nets must be decomposed into two-point connections, and this increases the SRAM requirement for some nets.

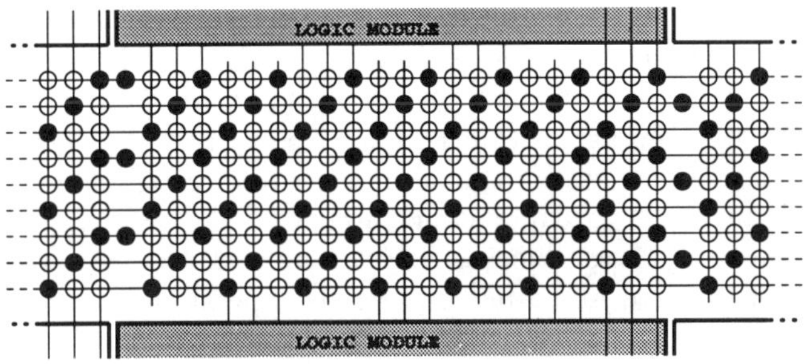

Figure 4: Depopulated segmented routing channel.

Furthermore, the proposed logic block exhibits a very useful symmetry: with no special clocking pins, all signals occur in interchangeable triplets. For example, there are three interchangeable inputs to the look-up tables, and each output has three acknowledged ports. Any routing channel passing by the logic block need only be connectable to *one* signal in each of these triplets. Consequently each 3×3 submatrix in the switch matrix connecting the logic block to the routing channel will only need programmable switches along the diagonal, depopulating 2/3 of the switch matrix at no cost in routing flexibility, as shown in Figure 4.

Synthesis

Clearly an FPGA composed of these logic blocks could implement various self clocked structures directly [1]; this will not be addressed further in this paper. However, synchronous logic easily maps onto such an FPGA. Figure 5 shows a trivial synchronous system on the left, and the same system mapped onto three self clocked logic blocks on the right. The flip-flops have disappeared as separate components since each self clocked logic block already has RS flip-flops at the outputs, and an extra buffer has been inserted to satisfy the second bullet below.

CAD tools for mapping synchronous designs onto these logic blocks must include these considerations:

- For optimum performance, each cycle must contain some logic blocks with *bubbles,* useless data already consumed by the logic blocks' fanout. (Other logic

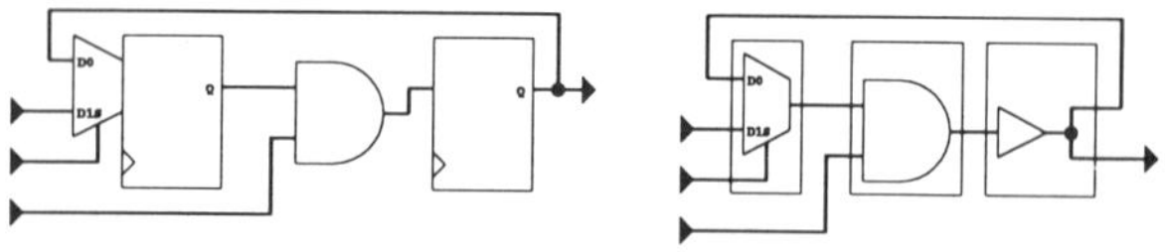

Figure 5: Mapping synchronous logic.

blocks contain useful *tokens*). The proportion of bubbles which optimizes performance and utilization depends on the timing properties of the logic block [4].

- Any cycle of logic blocks must include at least three logic blocks [1] and an even number of tokens.
- Intersecting cycles must have the same number of tokens. Thus shorter cycles may severely limit the utilization of logic blocks in longer cycles, making it advantageous to insert extra logic blocks in the former.

It is unclear at this time how much perturbation of a synchronous design is required in order to *optimize* the performance — throughput and latency [3] — of its mapping onto these logic blocks. Such optimization is, however, not required for correct and reliable operation.

Implementation

Figure 6 shows the implementation of part of the logic block, with three input signals on the left and an output signal with a fanout of three, hence the three ackowledgments on the right. When all new inputs have arrived and all the driven logic blocks have acknowledged the current logic block output, the C-element changes and the XOR gate it drives goes to **1**. The **LUT** Look Up Table contains eight SRAM bits; the three data inputs at the top select one of these bits. When the *GO* pulse arrives, the selected SRAM bit determines whether the pulse is passed through to the *L* or *H* signal. The **SWAP** block and the two RS latches convert the *L* or *H* pulse into the proper *D* and *F* protocol; the completion of this conversion ends the *GO* pulse and then *L* or *H*. The NOR gate, D latches and XOR gate at the bottom detect the fall of *L* or *H* and produce an acknowledgment to the left. Note that in this circuit *GO* is *level*-sensitive and indicates when the logic block is computing; thus it changes twice as each data value passes through the logic block. Although this logic block is self clocked in its external behavior, clearly its internal design is neither delay- nor speed-insensitive; rather, it makes conservative assumptions about relative gate delays.

The logic block includes other functional, configuration and testability features:

- State feedback supports flip-flops with enable, JK flip-flops, and other memory elements.
- Multiple outputs ease the implementation of functions which share inputs, such as full adders.
- Input deletion allows the logic block to be configured for functions of fewer than three inputs. Complete input deletion indicates an unused logic block.
- Output deletion allows the logic block to be configured for smaller fanouts. SRAM bits control deletion.
- Initialization loads the flip-flops and C-elements from programmable bits, providing some of the functionality of set and preset inputs and specifying the initial locations of data tokens.
- Single-stepping allows the logic block to be slowed down reliably for testing and debugging.
- Microprobing allows internal state to be read out. This provides full observability of the internal states in an array of logic blocks.

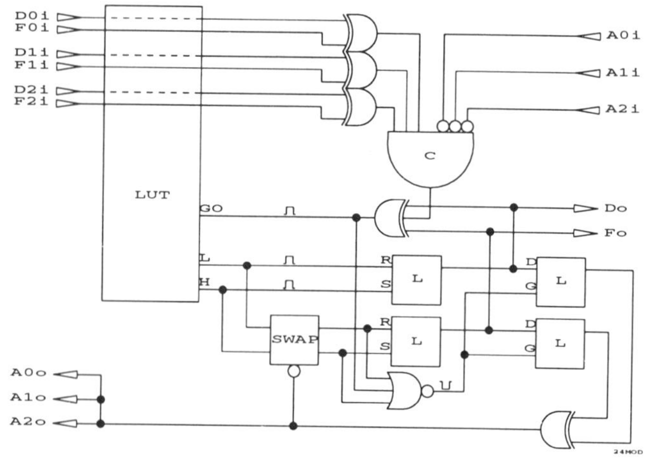

Figure 6: Overview of self clocked logic block.

Figure 7: 4 × 6 array micrograph.

Results

Two preliminary versions of the logic block and a preliminary 2×2 array were fabricated through MOSIS in 1μm 3M CMOS, and these three test chips verified the correctness of the approach. A larger 4×6 array, shown in Figure 7, with a programmable I/O architecture and numerous testing and debugging features was then fabricated through MOSIS in 0.8μm 3M CMOS.

Table 1: Measured performance.

	4×6 array	XC4003 -5
Logic block delay	10–17ns	≈8ns
Chip performance	33MHz	35MHz

Table 1 shows the performance of a 4 bit counter mapped into this final chip and the PREP Average Internal Performance of the Xilinx 4003 [9]. Although the two architectures differ too much for such a simple comparison, the self clocked FPGA does achieve comparable internal performance in an equivalent process. In general this is because inferior routing, clock skew and *worst* case margins prevent Xilinx's faster logic block from being fully used. Finally, for the self clocked FPGA, achieving 33MHz is not limited to small designs; performance depends more on token distribution than on circuit size.

Not only does the self clocked array offer internal performance comparable to a commercial FPGA, but it can be used to construct a multi-FPGA system featuring improved system-level performance. This results from eliminating clock skew, which can consume 4–8% of the clock period, as well as the advantages of *average* case performance and automatic pipelining. For reference, [5] shows a more than 2× performance gain from average case performance in a system with more general asynchrony, while the detailed analysis in [13] shows that together average case performance and automatic pipelining yielded a 15–20% performance gain in a system utilizing the same type of self clocking used here. Moreover, self clocking is less susceptible to the irregular delays of programmable routing, but without building a complete emulation system this advantage cannot be quantified.

Table 2: Area comparison.

Clocking	Logic area	Routing area	4×6 core area
Synchronous	0.28mm^2	0.27mm^2	11.5mm^2
Self clocked	0.47mm^2	0.40mm^2	18.4mm^2
	+67%	+50%	+60%

Self clocking typically incurs an area overhead of +100% [11, 15]. To determine this overhead in the self clocked FPGA, the sizes of various components were compared to their counterparts in a synchronous equivalent, constructed by removing the self clocked features and adding clock configuration circuitry. As Table 2 shows, the area overhead of self clocking amounted to +60%. The more area-efficient implementation is primarily because three-wire signaling can be incorporated into a segmented routing network, where the SRAM cell size limits the wire pitch, with an area increase of +50% instead of +200%.

Conclusion

When used in a multiple-FPGA system such as a logic emulator, the self clocked FPGA here presented offers many ease-of-use and performance advantages, in particular, more than 20% extra system-level performance in at most 60% additional area. Of course, more work on CAD tools and inter-FPGA routing would be needed to build a complete multi-FPGA system.

Acknowledgments

Many thanks to my advisor, Prof. Abbas El Gamal, and my officemates for their support and insight.

References

[1] M. R. Greenstreet, T. E. Williams and J. Staunstrup, "Self-Timed Iteration," *VLSI '87,* C. H. Séquin (Ed.), Elsevier Science Publishers, 1988, pp. 309–322.

[2] I. Sutherland, "Micropipelines," *Communications of the ACM,* V. 32, N. 6, June 1989, pp. 720–738.

[3] T. E. Williams, "Latency and Throughput Tradeoffs in Self-Timed Speed-Independent Pipelines and Rings," CSL-TR-90-431, Computer Science Laboratory, Stanford University, August 1990.

[4] M. R. Greenstreet and K. Steiglitz, "Bubbles Can Make Self-Timed Pipelines Fast," *Journal of VLSI Signal Processing,* V. 2, N. 3, November 1990, pp. 139–148.

[5] T. H.-Y. Meng, R. W. Brodersen and D. G. Messerschmitt, "Asynchronous Design for Programmable Digital Signal Processors," *IEEE Transactions on Signal Processing,* V. 39, N. 4, April 1991, pp. 939–952.

[6] M. E. Dean, T. E. Williams and D. L. Dill, "Efficient Self-Timing with Level-Encoded 2-phase Dual-Rail (LEDR)," *Advanced Research in VLSI 1991,* pp. 55–70.

[7] S. Walters, "Computer-Aided Prototyping for ASIC-Based Systems," *IEEE Design and Test of Computers,* June 1991, pp. 4–10.

[8] V. P. Roychowdhury, J. W. Greene and A. El Gamal, "Segmented Channel Routing," *IEEE Transactions on CAD of Integrated Circuits and Systems,* V. 12, N. 1, January 1993, pp. 79–95.

[9] "Benchmark Suite #1," Version 1.2, PREP, March 28, 1993.

[10] J. L. Kouloheris, *Empirical Study of the Effect of Cell Granularity on FPGA Density and Performance,* Ph.D. Thesis, Stanford University, July 1993.

[11] S. B. Furber, P. Day, J. D. Garside, N. C. Paver and J. V. Woods, "A Micropipelined ARM," *VLSI '93,* T. Yanagawa and P. A. Ivey (Eds.), Elsevier Science Publishers, 1993, pp. 211–220.

[12] E. Brunvand, "Using FPGAs to Implement Self-Timed Systems," *Journal of VLSI Signal Processing,* V. 6, N. 2, August 1993, pp. 173–190.

[13] M. A. Franklin and T. Pan, "Clocked and Asynchronous Instruction Pipelines," *Proceedings of the 26th ACM/IEEE Symposium on Microarchitecture,* Austin, TX, December 1993.

[14] *The Programmable Logic Data Book*, Xilinx, 1994.

[15] K. van Berkel, R. Burgess, J. Kessels, M. Roncken, F. Schalij and A. Peeters, "Asynchronous Circuits for Low Power: A DCC Error Corrector," *IEEE Design & Test of Computers,* V. 11, N. 2, Summer 1994, pp. 22–32.

Time-Multiplexed Routing Resources for FPGA design

Chih-chang Lin, Douglas Chang†, Yu-Liang Wu‡, and Malgorzata Marek-Sadowska

ECE Dept.
Univ. of California
Santa Barbara, CA 93106

CS Dept.†
Univ. of California
Santa Barbara, CA 93106

CS/CE Dept.‡
Chinese Univ. of Hong-Kong
Shatin, Hong Kong

Abstract

We propose a time-multiplexed routing architecture for SRAM based FPGAs. This can be implemented by having two programmable SRAMs for each routing connection. The goal of this approach is to alleviate the on-chip routing bottleneck, and to increase the range of circuit sizes which can be accommodated on a single chip. We consider a Xilinx 4000 style architecture with and without time-multiplexed routing. Our experimental results show that time-multiplexed routing can reduce the channel density by 30%. Also, sharing permutation equivalent LUTs between the time phases can result in a 14% reduction of the number of LUTs required to implement a design.

1 Introduction

Field Programmable Gate Arrays (FPGAs) are arrays of pre-fabricated logic blocks, e.g., Look-Up-Tables (LUTs), and wire segments with user reprogrammability. In addition to their economical cost for low volume production, FPGA technology also offers rapid prototyping, hardware reuse, and faster solutions for many originally software-oriented applications (e.g., logic emulators) [1].

In the past, SRAM FPGA reprogrammability was used statically. That is, the FPGA was configured only once for a specific circuit design. Currently, there is growing interest in the dynamic reconfiguration of FPGAs. In this type of FPGA reuse, the operations implemented by the FPGA are tightly coupled, and the FPGA is reconfigured many times during the system operation [2]. One problem with dynamic reconfiguration is that the reconfiguration of FPGAs is typically slow, in the range of tens of milliseconds [3]. However new FPGA architectures have been proposed that reduce the reconfiguration time to tens of nanoseconds [4, 5, 6]. This is done by having multiple FPGA programmings on-chip, thus eliminating the need for down-loading a programming from off-chip.

There are three factors limiting the circuits that can be implemented on an FPGA. They are the fixed number of pins on an FPGA package, the fixed amount of routing resources, and the fixed number of LUTs on an FPGA. Recently, several works [7, 8] applied the concept of time-multiplexing on FPGA I/O pins to reduce the first limiting factor. In these techniques, multiple logical connections between FPGAs are transmitted over a single physical connections in a time-multiplexed, pipelined fashion.

In this paper, we address the other two limiting factors by having two SRAMs associated with each programmable routing switch. Thus the FPGA has two routing structures on-chip instead of the customary one. The two routing structures are time-multiplexed, and we dynamically switch between them during run-time. Thus the number of nets that need to be routed at a single time is reduced by half. Our experiments show that this leads to a 30% reduction in the size of the routing channel.

We also found that a high percentage of the LUTs in a mapped circuit are permutation-equivalent functions. P-equivalent functions are functions that are equivalent up to a permutation of the inputs. LUTs that are P-equivalent are said to be in the same P-class. We use this fact to implement LUT sharing between the two routing phases. We show experimentally that LUT sharing leads to a 14% reduction in the number of LUTs required to implement a design.

In section 2, the architecture used to perform the experiments is described. Section 3 describes the software algorithms for phase assignment, placement, pin assignment, and routing. Section 4 discusses our results and conclusions follow in section 5.

2 Architecture

The investigated architecture is a 2-D array of logic blocks (L), Connection boxes (C), and Switch boxes (S), arranged like the Xilinx 4000 series (see Figure 1). Each L block contains a 4-input LUT and a flip-flop. The output of the LUT is available directly at pins 6 and 7, and via the flip-flop at pin 5. The pins 0, 1, 2, and 3 are the inputs to the LUT. Pin 4 is used for the clock. This LUT is similar to the one used in [9].

The main difference between this time-multiplexed

0-7803-3177-6 $5.00 © 1996 IEEE

IEEE 1996 CUSTOM INTEGRATED CIRCUITS CONFERENCE

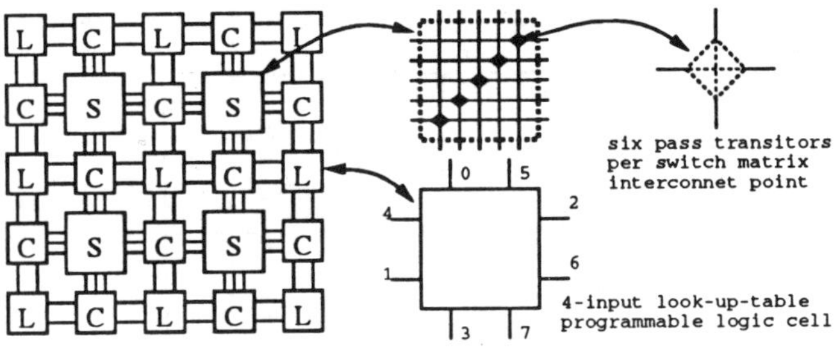

Figure 1: FPGA routing architecture.

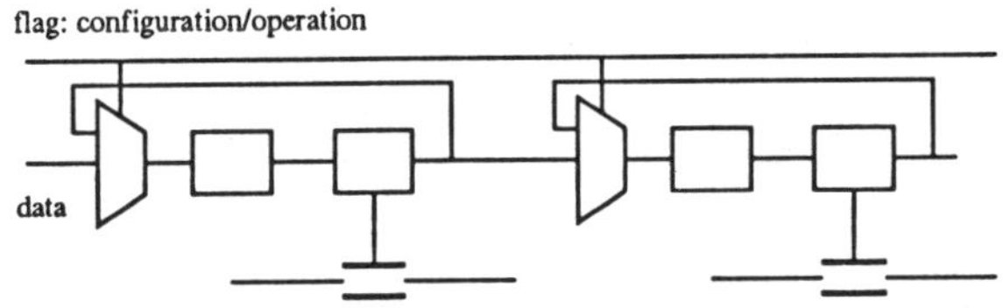

Figure 2: 2 SRAMs per switch.

architecture and conventional ones is that each programmable switch in the C and S blocks has two SRAMs to control it as shown in Figure 2. During configuration mode, data is shifted into the SRAMs. During operation mode, the two SRAM bits are shifted in a loop by the phase clock, so that each bit alternately controls the switch. Thus this architecture has two, time-multiplexed routing structures.

During one system clock, the phase clock will clock twice thus creating two phases, phase 1 and phase 2. During phase 1, the first routing structure will be implemented, and during phase 2, the second.

3 Overview of our system

To show how time-multiplexed routing can improve FPGA routability and reduce the number of LUTs needed to implement a circuit, we created a software system to implement combinatorial logic on this architecture. The input is a technology mapped netlist. This is a directed acyclic graph (DAG) where each node has at most four incoming edges. Each of these nodes will become a LUT in the FPGA. From this netlist, the nodes are assigned to an initial phase and placement. An initial pin assignment is also made. Then simulated annealing is used to improve the phase assignment, placement, and pin assignment. Finally detailed routing for each phase is performed. These steps are discussed in more detail below. For simplicity, the discussion of LUT sharing is postponed until section 3.3.

3.1 Initial phase assignment, placement, and pin assignment

The two routing structures enable the nets to be routed in phase 1 or phase 2. Thus only about half the nets need to be routed at one time. Since the output signal of a node is not computed until all of its input signals are ready, the netlist topology shows the natural dependency between the nets. Thus, to decide which phase each net should belong to, we need to cut the given DAG such that all the edges cross the cut line in one direction. This will give us a phase assignment for the nodes which also gives us the phase assignment of the nets.

The initial phase assignment algorithm starts with the primary inputs in phase 1, and all other nodes in phase 2. Then it considers all nodes in phase 2 that have all their inputs in phase 1. These are the phase 2 boundary nodes and they are the nodes that could potentially be assigned to phase 1 from phase 2. (There is also a corresponding set of phase 1 boundary nodes.) When a node is assigned to phase 1 that will force certain nets to be routed in phase 1. We pick the node that minimizes the difference in the number of nets to be routed in each phase. We iterate this process until all nodes are in phase 1. Then we pick the best cut (closest to equal numbers of nets in both phases) to be the initial phase assignment. Note that each net is viewed as a 2 pin net; multi-pin nets are decomposed into 2-pin nets.

The initial placement of the nodes is then done by a breadth-first traversal of the netlist and filling the chip from left to right. This is done because the output pins (pins 6 and 7) of our LUT are on the right and bottom of the LUT, so the natural computation flow is from left to right (or top to bottom). The initial pin assignment is done by considering each placed LUT individually and finding the best assignment assuming all the pins on other LUTs are fixed. There are four nets to be assigned to the four input pins. The cost for assigning a net to each pin is calculated based on the best routing length. Then using minimum weighted matching [10], an assignment with minimum cost can be found. For the output pin, the net is assigned to the pin that results in the minimum cost.

3.2 Simulated annealing

We use the cost function

$$\alpha \sum_i d_i + \beta * cf,$$

where α and β are constants, d_i is half the perimeter of the bounding box containing the pins of net n_i, and it is used to approximate the length of the routed net. cf is the routing congestion factor. To compute cf, let $area_i$ be the area of the bounding box of net n_i.

Table 1: Number of P-classes in MCNC benchmarks.

name	*#LUT*	*#P-classes*	*#P-classes / #LUTs*
C432	63	25	39.7%
C499	106	7	6.5%
C880	107	38	35.5%
C1908	106	61	57.5%
alu2	119	63	52.9%

Then n_i adds 1 to the congestion within its bounding box. Overlapping bounding boxes will create areas of high congestion. cf is the size of the area of congestion which is greater than a given congestion threshold. cf is used to approximate the number of nets that need to be routed within the same area.

Two costs are maintained, one for each phase. Simulated annealing is used to minimize the maximum of the two costs. In each iteration, the annealing will either move a phase 1 or phase 2 boundary LUT to the other phase, or swap the position of two LUTs or two I/O pads. After moving LUTs or I/O pads the pins assignments are updated to minimize d_i using minimum weighted matching. Thus this combines phase assignment, placement, and pin assignment into the annealing.

The detailed routing remains the same as in the single-phase conventional architecture, except that here we only need to route approximately half of the nets at a time. To complete the experiment, we adopted the one-step router described in [11].

3.3 Sharing of logic blocks

If a LUT in phase 1 is in the same P-class as a LUT in phase 2, then they can share the same physical LUT. The potential for sharing is quite high. As seen in Table 1, the number of P-classes is low relative to the number of LUTs. Thus the number of different LUTs used is low relative to the total number of LUTs used, and the potential for sharing is high.

The sharing can be used for two purposes. One is to reduce the total number of physical LUTs needed to implement a circuit. Sharing can also be used to provide more freedom for placement, thus improving the final routing. To maximize the sharing possibilities, the initial phase assignment can be tuned to try to split each P-class equally into both phases. The simulated annealing placement is also modified to allow P-equivalent nodes in different phases to be shared or unshared. These changes integrate the extra freedom given by LUT sharing into the system.

4 Results and Discussion

We performed two experiments. First we compared the channel density (after detailed routing) of the FPGA with and without time-multiplexed routing. Second we attempted to maximize the sharing

Table 2: Using time-multiplexing to reduce channel density.

name	*# LUT*	*channel density*	*phase 1, 2 density*
C432	63	8	5, 6
C499	106	9	6, 6
C880	107	8	6, 6
C1908	106	9	6, 6
alu2	119	10	7, 7
total	501	44	31

Table 3: Using time-multiplexing to reduce #LUTs.

	original		*time-multiplexed*	
name	*#LUT (chip size)*	*channel density*	*#LUT (chip size)*	*phase 1,2 density*
C432	63 (8×8)	8	53 (8×8)	6, 6
C499	106 (11×11)	9	92 (10×10)	7, 7
C880	107 (11×11)	8	92 (10×10)	6, 6
C1908	106 (11×11)	9	94 (10×10)	7, 8
alu2	119 (11×11)	10	99 (10×10)	7, 8
total	501	44	430	35

of LUTs on the time-multiplexed FPGA. We mapped ISCAS89 and MCNC91 benchmark circuits to 4-input LUT blocks by using SIS [12] LUT optimization script to obtain a set of test suites.

Table 2 shows the channel density of the FPGA with (column 4) and without (column 3) time-multiplexed routing. For example, circuit C499 has a channel density of 9 when implemented on an FPGA without time-multiplexing. With time-multiplexing, C499 has a channel density of 6 for both phase 1 and phase 2. In general we show a 30% reduction in channel density. Table 3 shows the amount of LUT sharing possible. In some cases the LUT sharing can result in a smaller chip needed to implement the circuit. For example, C499 has 106 LUTs. This is implemented on a chip size of 11×11 with channel density 9. However with time-multiplexing, the number of LUTs needed is reduced to 92. C499 then is implemented on a chip size of 10×10 with channel density of 7 for both phases. In general we show a 14% reduction in number of LUT needed to implement a circuit.

Although the time multiplexed routing allows a larger range of circuits to be implemented, it adds certain overhead to an FPGA in terms of area, power, and performance. If we represent the extra area required by the extra number of SRAM bits needed, we can see that the time-multiplexed routing FPGA requires $2(6W|S| + 4W|C|) + 16|L|$ bits as opposed to $6W|S| + 4W|C| + 16|L|$ bits for a normal FPGA, where W is the track width, $|S|$, $|C|$, $|L|$ are the number of switch boxes, connection boxes and logic blocks in the FPGA. The extra bits used for time-multiplexed routing could instead be used to create a normal FPGA with channel width $W', W < W' < 2W$ or increasing $|L|$, $|S|$, and $|C|$. However the normal FPGA created will be be less flexible than the time-multiplexed one because it will not have the ability to trade-off LUTs and routing resources as the time-multiplexed chip.

The critical delay of the chip will increase due to the time used for switching phases. However since the routing is split into two phases, the critical path in each phase could be shorter than the critical path for the circuit. The power dissipated by this chip will be higher due to the switching needed between phases. However if a given switch is open or closed in both phases, then the SRAMs controlling the switch will not change. Thus to minimize power dissipation, the routing in each phase should be as similar as possible. We are currently working on this topic.

5 Conclusion and future work

We proposed an FPGA with time-multiplexed routing resources. Our experimental results show that time-multiplexed routing can reduce the channel density by a factor of 30%. Also, sharing P-equivalent LUTs between the time phases can result in a 14% reduction of the number of LUTs required to implement a design. This routing resource allows us to trade-off the number of LUTs and channel density to implement a circuit. Several new optimization problems arise due to the new routing structure. For example, in the technology mapping stage, instead of finding a solution that minimizes the number of LUTs, as it is done in the normal approach, we may want a solution, with perhaps more LUTs, but having more P-equivalent LUT pairs in different phases. Also we believe that the channel density can be further reduced by allowing the detail routings in both phases to work together instead of independently. For example, by adding a rip-up and re-route stage to the router that also can change the phase in which a net is routed.

Although we just considered a two-phase FPGA time-multiplexing routing on a Xilinx 4000 style architecture, it can be extended to an arbitrary multi-phased time multiplexed routing on an arbitrary FPGA architecture. In particular, the area overhead for time-multiplexed routing will be reduced as the number of bits for the LUTs increases relative to the routing. Thus using LUTs with more than 4 inputs may be of interest.

ACKNOWLEDGMENTS

This work was supported under NSF grant MIP9419119 and MICRO/XILINX/ACTEL.

References

[1] S. D. Brown, R. J. Francis, J. Rose, and Z. G. Vranesic, *Field-Programmable Gate Arrays.* Boston, Mass.: Kluwer Academic Publishers, 1992.

[2] M.J. Wirthlin and B.L. Hutchings, "A dynamic instruction set computer," *IEEE Symposium on FPGAs for Custom Computing Machines*, 1995.

[3] B. K. Fawcett, "Taking advantage of reconfigurable logic," *ACM/SIGDA Int. Workshop on FPGAs*, 1994.

[4] N.B. Bhat, K. Chaudhary and Ernest S. Kuh, "Performance-oriented fully routable dynamic architecture for a field programmable logic device," Memorandum No. UCB/ERL M93/42, University of California, Berkeley, 1993.

[5] A. DeHon, "DPGA-coupled microprocessors: Commodity ICs for the early 21st century," *IEEE Workshop on FPGAs for Custom Computing Machines*, pp. 31–39, 1994.

[6] D. Jones and D. M. Lewis, "A time-multiplexed FPGA architecture for logic emulation," *IEEE Custom Integrated Circuits Conference*, pp. 24.2.1–24.2.4, 1995.

[7] J. Babb, R. Tessier and A. Agarwal, "Virtual wires: Overcoming pin limitations in FPGA-based logic emulators," *IEEE Workshop on FPGAs for Custom Computing Machines*, pp. 142–151, 1993.

[8] J. Li and C.K. Cheng, "Routability improvement using dynamic interconnect architecture," *IEEE Symposium on FPGAs for Custom Computing Machines*, 1995.

[9] G. Lemieux and S. Brown, "A detailed router for allocating wire segments in FPGAs," *Proc. ACM/SIGDA Physical Design Workshop*, 1993.

[10] T. Lengauer, *Combinatorial Algorithms for Integrated Circuit Layout.* Applicable theory in computer science, New York, NY: John Wiley & Sons, 1990.

[11] Y.L. Wu and M. Marek-Sadowska, "Orthogonal greedy coupling – a new optimization approach to 2-D FPGA routing," *Proc. ACM/IEEE Design Automation Conf.*, pp. 568–573, 1995.

[12] "SIS: A system for sequential circuit synthesis," Report M92/41, University of California, Berkeley, 1992.

NON-ORTHOGONAL DECODING: AN ARCHITECTURAL ELEMENT FOR REPROGRAMMABLE INTERCONNECT OR LOGIC

Adi Srinivasan

Aptix Corporation
2880 N. First St., San Jose, CA 95134.

Abstract

A reprogrammable logic or interconnect architecture termed non-orthogonal decoding[1] is described, which increases routability with no additional circuitry and negligible added layout area, compared to an existing reprogrammable routing architecture, the orthogonal decoding architecture introduced by Aptix.

An orthogonal decoding architecture's routing channel box is modified to illustrate non-orthogonal decoding. The former is shown to be replaceable, without restrictions, by non-orthogonal decoding, which is also shown to increase routability. A generalization of the 2-d decoding space to higher dimensions and the application of non-orthogonal decoding to FPGAs are discussed.

1. Introduction

Routed reprogrammable interconnect or logic architectures using SRAM to control routing switches have relatively large chip areas compared to alternatives [1]. In such cases die size reduction can be achieved with a smaller SRAM cell, or a reduction in the SRAM to switch count ratio.

To reduce die size in a reprogrammable interconnect architecture using one SRAM cell per pass transistor switch [2], two methods were described in [3]; a 9T high voltage SRAM was replaced by a smaller 5T cell with no process changes and large read and write margins, and an architecture which decoded SRAM data to control switches reduced the SRAM to switch ratio. The 1024 pin switch-box in [3] termed FPIC2, uses a 32 x 32 array of L-shaped pin-stubs, each spanning a single horizontal and vertical interconnect channel (a channel box), to implement passive bi-directional paths.

In non-orthogonal decoding, a set of switches (no restriction on geometric distribution) in a physical *track space* is mapped to a virtual ***D*** dimensional *decoding space* addressed by ***D*** sets of SRAMs; each switch's state is decoded by a ***D***-wide *and.* ***D*** need not be the track-space dimension (generally 2) and can be > 3. For simplicity, in this paper ***D*** is 2 except in Sec. 3.1. While implementations of non-orthogonal decoding may appear physically similar to orthogonal decoding, the former is an extensive generalization of the latter, yielding significant routability increases.

1. Covered by U.S. patent 5,406,138 issued Apr 11, 1995.

2. Non-orthogonal decoding

2.1. Orthogonal decoding review

Given N independently controllable switches in a channel box 2^N patterns (termed maps) are specifiable, but a significant number of these are not useful in practice [3]. To eliminate infrequently used maps, [3] motivates the introduction of a prototypical routing constraint (C) of the form: in a set S of N_S switches, turn on at most n_S switches (where typically $n_S << N_S$). If the constrained switches comprise K sets, S1 through SK, then the K sets must be disjoint, and a constraint C_{Si} exists for each Si, $1 \leq i \leq K$.

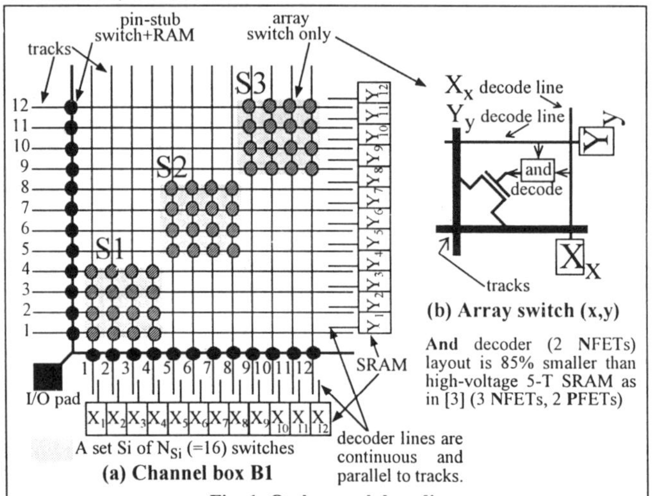

Fig. 1 Orthogonal decoding

Fig. 1a shows the example channel box of [3] with K=3, N_{Si}=16, n_{Si}=1, $\forall$ i $\in$ [1,K] using *orthogonal decoding* — an architecture in which array switches are controlled by orthogonal decoder lines (a switch is on only if both its decoder lines are high) driven by SRAM. In a set Si, while > n_{Si} (=1) switches may be on, at most 1 net can exist [3]. If v_x and h_y refer to vertical & horizontal tracks respectively, with x & y as numbered in Fig. 1a, switches are identified by their 2-d track (v_x,h_y) or and-decoder (X_x,Y_y) addresses. Set notation is used to indicate the switches on one or more tracks or decoder lines, e.g. $v_3=\{(X_3,Y_y) \mid 1 \leq y \leq 4\}$ or $Y_y=\{(v_x,h_y) \mid x,y \in [1,4]\}$. Given the switch distribution of Fig. 1a, the decoding is completely described by

$$\forall i \in [1,K] \text{ and } \forall x,y \mid (v_x,h_y) \in Si,\ (v_x,h_y) = (X_x,Y_y) \qquad (1)$$

0-7803-3177-6 $5.00 © 1996 IEEE

IEEE 1996 CUSTOM INTEGRATED CIRCUITS CONFERENCE

2.2. Non-orthogonal decoding, 1 direction constrained.

From Fig. 1a or (1) it is clear that in a set Si, all switches on at most one horizontal or vertical track can be independently controlled, but can create only 1 net. The number of nets can be increased by the introduction of a decoding (i.e. physical) constraint in either direction, e.g. "vertical" X, as below

$\forall\ s_a \equiv (v_{xa}, h_{ya}),\ s_b \equiv (v_{xb}, h_{yb}) \in X_x,\ s_a \neq s_b \Leftrightarrow v_{xa} \neq v_{xb}\ \&\ h_{ya} \neq h_{yb}$ ($D1_X$)

i.e. 2 switches controlled by a vertical decode line X_x must be on different horizontal *and* vertical tracks.

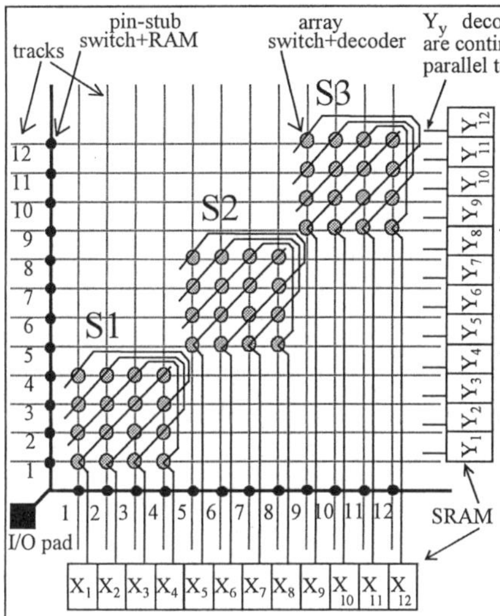

(a) Channel box B2

	X_1	X_2	X_3	X_4
Y_1	v_1,h_1	v_2,h_1	v_3,h_1	v_4,h_1
Y_2	v_2,h_2	v_3,h_2	v_4,h_2	v_1,h_2
Y_3	v_3,h_3	v_4,h_3	v_1,h_3	v_2,h_3
Y_4	v_4,h_4	v_1,h_4	v_2,h_4	v_3,h_4

(b) Decoding space of S1

Fig. 2 Non-orthogonal decoding, X constrained.

Fig. 2a shows a *non-orthogonal* (X lines aren't orthogonal to Y lines) distribution satisfying $D1_X$. If at most one X is on at a time, Y decoder lines may be used to create up to 4 independent nets per set Si, e.g. if $(X_2,Y_3) \equiv (v_4,h_3)$ is on, the other 3 switches on X_2 (see Fig. 2b) may be independently turned on by Y_1, Y_2 or Y_4. Thus all $n_{Si}=4$. If at most one Y is on at a time, multiple X lines may be on; the controlled switches would all be on one track h_3, resulting in at most one net, i.e. $n_{Si}=1$. The usage mode with $n_{Si}=4$ is preferred.

If a constraint $D1_Y$ like $D1_X$ is applied to horizontal decode lines Y only, the decoding space for S1 could be derived from Fig. 2a by swapping X and Y, and replacing v_x,h_y with v_y,h_x.

2.3. Non-orthogonal decoding, both directions constrained.

An obvious extension to the above scheme is to apply $D1_X$ and $D1_Y$ simultaneously to increase the number of possible nets. However, an additional decoding constraint is needed to eliminate degenerate decoder addresses.

The relationship between (v,h) and (X,Y), denoted $f(v,h) \rightarrow (X,Y)$, must be an injection, i.e. a one-to-one mapping. (D2)

The mapping requirement ensures no two (X,Y) address the same switch. The one-to-one requirement ensures that while some (X,Y) may not correspond to any switch, no two (v,h) map to the same (X,Y) address.

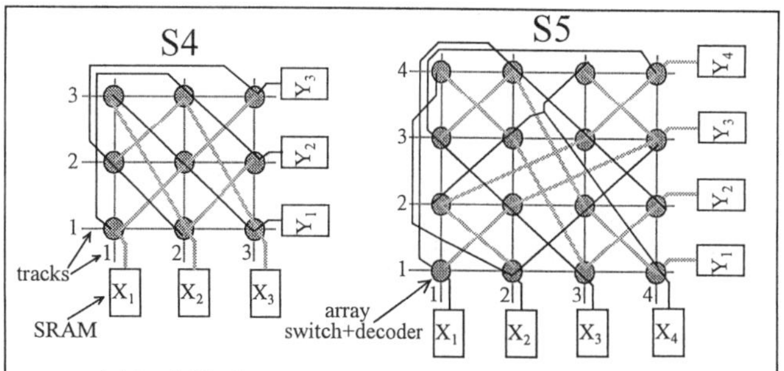

(a) 3 x 3 block (c) 4 x 4 block

	X_1	X_2	X_3
Y_1	v_2,h_2	v_1,h_3	v_3,h_1
Y_2	v_1,h_1	v_3,h_2	v_2,h_3
Y_3	v_3,h_3	v_2,h_1	v_1,h_2

(b) Decoding space of S4

	X_1	X_2	X_3	X_4
Y_1	v_2,h_4	v_3,h_2	v_1,h_3	v_4,h_1
Y_2	v_4,h_2	v_1,h_4	v_3,h_1	v_2,h_3
Y_3	v_1,h_1	v_4,h_3	v_2,h_2	v_3,h_4
Y_4	v_3,h_3	v_2,h_1	v_4,h_4	v_1,h_2

(d) Decoding space of S5

Fig. 3 Non-orthogonal decoding, X and Y constrained.

Examples satisfying $D1_X$, $D1_Y$ and D2 are shown in Fig. 3, with one set of switches in each case for clarity. With only a D1 constraint in effect, "diagonal" (X_x in Fig. 2a) decoder lines always offer a solution. If $(a_1,a_2,\ldots,a_n)$ denotes a *cyclic permutation*, i.e map a_i to a_{i+1} $(0 < i < n)$ and a_n to a_1, then the permutations of $\{v_1,v_2,v_3,v_4\}$ obtained by scanning "down" any X column of Fig. 2b are generated by $\sigma_V=(v_1,v_2,v_3,v_4)$, i.e. σ_V^4 (identity) for X_1, σ_V for X_2, σ_V^2 for X_3 and σ_V^3 for X_4, form the cyclic group of order 4.

With D1 and D2 in effect the cyclic permutation described above (applied in both directions) always works for square blocks of odd "width" e.g S4 in Fig. 3a. In S4, $\sigma_V=(v_1,v_2,v_3)$ applied to $\{v_3,v_2,v_1\}$ and $\sigma_H=(h_1,h_2,h_3)$ applied to $\{h_1,h_2,h_3\}$ generate the mapping in Fig. 3b. Lack of space prohibits giving the lengthier solution for any square block of even width; however, cyclic permutations of the type above (diagonal decoder lines) never work in this case. One solution is shown for a 4 x 4 block in Figs. 3c & 3d.

With D1 and D2 applied, while the maximum number of nets per set is the same as with D1 alone, the number of choices increases. In S5 (Fig. 3c) $n_{S5}=4$, as with any n_{Si} in B2 (Fig. 2a). If the number of choices available for *event k* (i.e. creation of the kth independent net) in set S is denoted by S_k, then in S1 (Fig. 2a) $S1_1=16$, $S1_2=3$, $S1_3=2$ and $S1_4=1$, while $S5_1=16$, $S5_2=6$, $S5_3=2$ and $S5_4=1$.

The increase in choices for event 2 may be seen from Fig. 3d; if event 1 in S5 uses $(X_2,Y_3) \equiv (v_4,h_3)$, the choices for event 2 are any of the 6 remaining switches in column X_2 or row Y_3. The choices for event 3 are the 2 remaining switches in the X_2 column or Y_3 row in which event 2 occurred. Thus after the first event in S5 the router has 2 groups of three switches available for event 2; for subsequent events it is restricted to the group of event 2.

For a square set of n^2 switches Snxn, with D1 and D2 in effect,

$Snxn_1 = n^2$, $Snxn_2 = 2(n - 1)$, $Snxn_j = 2(n - j + 1), j \in [3,n]$ (2)

An additional usage mode exists for blocks of width > 3, involving the simultaneous use of 4 switches. If the 4 switches are chosen so that exactly 2 or 0 of them are on any (X or Y) decoder line, they can be turned on using 2 X and 2 Y lines, e.g. in Fig. 3d (v_2,h_4), (v_4,h_1), (v_4,h_2), (v_2,h_3) are turned on by X_1, X_4, Y_1, Y_2, creating 2 nets. Some combinations exist which each generate 4 nets; e.g. 4 such combinations exist in Fig. 3d — $\{X_1, X_2, Y_1, Y_3\}$, $\{X_1, X_2, Y_2, Y_4\}$, $\{X_3, X_4, Y_1, Y_3\}$ and $\{X_3, X_4, Y_2, Y_4\}$. Though useful, nets of this type, termed *quad nets*, are ignored in the following analysis, which considers only independently creatable nets.

2.4. Non-orthogonal decoding in rectangular blocks

Though previous examples showed non-orthogonal decoding in square blocks of switches it is generally applicable to a rectangular block of x=n by y=m switches Snxm, where arbitrarily $m \leq n$. Fig. 4a shows an example with n=4, m=3 and $D1_X$ in effect.

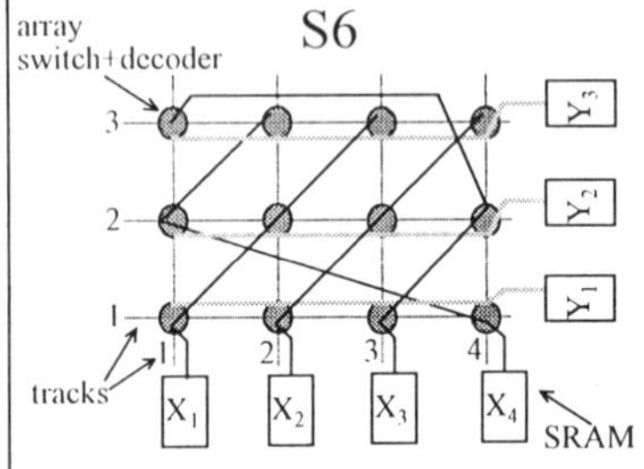

(a) 4 x 3 block

	X_1	X_2	X_3	X_4
Y_1	v_1,h_1	v_2,h_1	v_3,h_1	v_4,h_1
Y_2	v_2,h_2	v_3,h_2	v_4,h_2	v_1,h_2
Y_3	v_3,h_3	v_4,h_3	v_1,h_3	v_2,h_3

(b) Decoding space of S6

Fig. 4 Rectangular block, X constrained.

In general n X and m Y SRAM suffice and D1 is satisfiable only when applied to the shorter decoding lines (X lines in Fig. 4a); the cyclic permutation described in 2.3 works, resulting in diagonal decode lines similar to Fig. 2a. The *maximum number of independent nets* $m_{Snxm} = m$ and the total number of combinations of independent nets or *maps* is

$$M_{Snxm} = 1 + n \sum_{j=1}^{m} \binom{m}{j} = 1 + n(2^m - 1) \quad (3)$$

For a constant number of switches nm, M_{Snxm} in (3) is maximized by maximizing m, i.e. m=n; the number of SRAM needed = (n+m) is also minimized by m=n. Thus blocks of aspect ratio m/n equal or tending to 1 are preferred. By comparison if orthogonal decoding is applied,

$m_{Snxm} = 1$ and $M_{Snxm} = 1 + nm$ (4)

With $D1_X$, $D1_Y$ and D2 in effect in Snxm, additional circuitry must finally be introduced. The D1 and D2 constraints cannot be satisfied unless (n — m) SRAM are added, "squaring" the decoder space to n x n, which generates n(n—m) unused (X,Y) addresses. Fig. 5a shows a solution for S4x3 with D1 and D2 in effect.

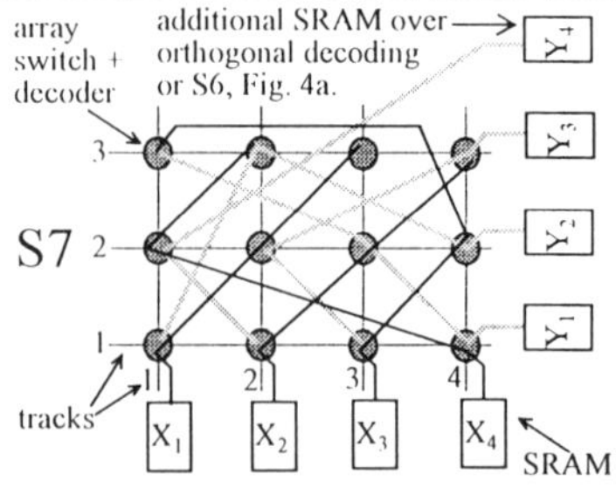

(a) 4 x 3 block

	X_1	X_2	X_3	X_4
Y_1	—	v_3,h_2	v_1,h_3	v_4,h_1
Y_2	v_1,h_1	—	v_4,h_2	v_2,h_3
Y_3	v_2,h_2	v_4,h_3	v_3,h_1	—
Y_4	v_3,h_3	v_2,h_1	—	v_1,h_2

(b) Decoding space of S7

Fig. 5 Rectangular block, X and Y constrained.

The choice of nets in S7, Fig. 5a is similar to that for S5, Fig. 3c. Now $S7_1=12$, $S7_2=5$ and $S7_3=1$ or 0. The number of choices for events 2 and 3 merit explanation: if event 1 uses $(X_1,Y_3)\equiv(v_2,h_2)$, the remaining 4 switches on X_1 or Y_3 are independently addressable, but there is always a 5th choice at (X_A,Y_B) where X_A is the X of the unused address in Y_3, and Y_B is the Y of the unused address in X_1. (X_A,Y_B) — in this case $(X_4,Y_1)\equiv(v_4,h_1)$ — can be turned on without creating the usual dependent nets at (X_1,Y_1) and (X_4,Y_3) as they are unused. If (X_A,Y_B) is on, a 3rd switch cannot be independently used, so $S7_3=0$; any other event 2 yields $S7_3=1$. Note that an alternate solution to Fig. 5b is obtained by replacing all (v_x,h_4) in Fig. 3d by — (unused). This method is generally applicable for deriving the decoding space of Snxm from Snxn.

In general, for Snxm with D1 & D2 in effect, $m_{Snxm} = m$ and

$$M_{Snxm} = 1 + n\left(2 \sum_{j=1}^{m} \binom{m}{j} - 1 + n - m\right) = (1 + n(2^{m+1} + n - m - 3)) \quad (5)$$

For a constant number of switches nm, M_{Snxm} in (5), as in (3), is maximized at m=n, which also minimizes the SRAM count.

2.5. Area increase due to non-orthogonal decoding.

Non-orthogonal decoding subject to D1 adds no circuitry for square or rectangular switch blocks; however, decoder lines must be re-routed, causing an area increase ≈ 0-5%. The increase is small because the SRAM are not laid out peripherally as in Figs. 2-5, but are spatially distributed among the switches. The orthogonal decoding of [3] requires convoluted decoder line crossovers in layout; with non-orthogonal decoding, careful placement of SRAM in the switch area can exploit this condition resulting in a minimal area increase. Non-orthogonal decoding in both directions adds no circuitry for the preferred square blocks, though re-routing decoder lines causes an area increase ≈ 10%.

2.6. Routability increase due to non-orthogonal decoding

Let N(B) denote the set of 2-wire nets routable in a channel box B, and B3 be the channel box derived by replacing each of S1—S3 in B2 (Fig. 2) by S5. Then $N(B1) \subset N(B2) \subset N(B3)$, i.e. non-orthogonal decoding increases routability over orthogonal decoding. The switch box flexibility measure F_S [1,6] (used to determine routability and #paths-using-N-switch-boxes=$F_cF_S^N$ in [6]) is assumed invariant [1] over all switch-box events, which is not true for the switch-blocks of Figs. 1-5. If $F_{Si}(j)$ is the switch-box flexibility of set Si at event j, enumerated $F_{Si}=\{F_{Si}(1),F_{Si}(2),...,F_{Si}(j)\}$, then in Fig 1a $F_{S1}=\{4,0,0,0\}$, in Fig. 2a $F_{S1}=\{4,1,1,1\}$, and $F_{S5}=\{4,2,1,1\}$ which provide some measure of the improvement. The number of sets b in a channel box B is chosen so $n_B \ge \sum_{i=1}^{b} n_{S_i}$; event 1 typically occurs in all sets before event 2 in any, making it difficult to compute an effective F_{Si}.

However the measure M_{Snxm} in (3), (4), & (5) makes the effect of dependencies explicit by summing over all combinations, and provides an effective comparison of the potential of each decoding scheme. In orthogonal decoding of an nxm block, M_{Snxm} is linear in m, while it is exponential for both types of non-orthogonal decoding. Also, the maximum number of independently creatable nets m_{Snxm} increases from a constant, 1, to linear in m, typically allowing m times more nets in the channel-box as a whole. Thus a layout change with no added circuitry results in significantly increased numbers of possible nets.

3. Further applications of non-orthogonal decoding

3.1. Decoding spaces with dimension > 2.

While the preceding discussion was restricted to 2-d (X & Y) decoding spaces, higher dimensional decoding spaces are possible; e.g. if the and-decode of Fig. 1b is replaced by a 3-input-and with Z_z SRAM & decoder lines added, then with $D1_X$, $D1_Y$, $D1_Z$, and D2 in effect, a mapping $f(v,h) \rightarrow (X,Y,Z)$ can be found to implement a 3-d decoding space.

For example, in a 27 switch block Q, laid out as 3 square blocks of width 3 in a 9x9 track grid, 3-d decoding is implemented with 9 SRAM as 3 S4-type blocks Q1, Q2, Q3 sharing $X_{1\text{-}3}$ & $Y_{1\text{-}3}$, with Z_i addressing all switches in Qi, i $\in$[1,3]. Ref. [5] discloses more details.

3.2. Non-orthogonal decoding in FPGAs.

In an FPGA connection box with flexibility F_C and W tracks, 100% routability is feasible with low F_S values, if $F_C > W/2$ [6]; attaining these F_C values requires highly populated pin-stubs. When a logic block pin can connect in a connection-box, it is generally connectable to all W tracks [7]. The number of on switches along any 1 pin-stub is typically small.

A routing constraint C_{Spi} on Spi, the set of switches on pin-stub i, is satisfiable for some n_{Spi} ; if n_{Spi} is small non-orthogonal decoding allows a high switch population to be addressed by a small number of SRAM. If, as in Fig. 6, $n_{Spi} \le N_{Spi}^{1/2}$, the preferred square array is used; otherwise an nxm (Section 2.4) array with n (i.e. the "longer" side) $=n_{Spi}$ uses n+m SRAM in a manner similar to Fig. 6a.

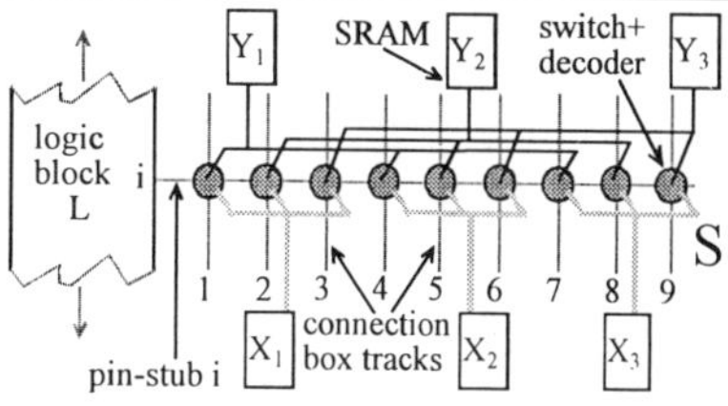

	X_1	X_2	X_3
Y_1	v_1,h_i	v_4,h_i	v_7,h_i
Y_2	v_2,h_i	v_5,h_i	v_8,h_i
Y_3	v_3,h_i	v_6,h_i	v_9,h_i

(a) FPGA connection-box CB1 **(b) Decoding space of S8**

Fig.6 Non-orthogonal decoding in FPGA connection box.

In either case n_{Spi} switches are independently usable,though $> n_{Spi}$ switches can be on. For a square array, this results in $2N_{Spi}^{1/2}$ $(=2n_{Spi})$ SRAM controlling N_{Spi} switches; a considerable area reduction for large N_{Spi} ($\approx$ W).

4. Summary

A new reprogrammable logic or interconnect architecture, non-orthogonal decoding, has been presented. An existing orthogonal decoding architecture is shown to be replaceable by non-orthogonal decoding with no restrictions or added circuitry, resulting in significantly increased routing resources, while adding minimal layout area.

The architecture is shown to be generally applicable to switch blocks of varied aspect ratio. A general analysis of 2-d decoding shows the preferred switch-block aspect ratio to be unity. Non-orthogonal decoding is shown to significantly increase the number of independent nets creatable over orthogonal decoding. An extension of the non-orthogonal decoding space to > 2 dimensions reduces the SRAM/switch ratio further. Finally, an application to FPGA connection boxes results in an SRAM-count reduction.

Non-orthogonal decoding can thus be applied to existing orthogonal decoding or other reprogrammable logic or inter-connect architectures to increase routability or reduce SRAM count, reducing area and improving performance.

Licenses for use of this architecture may be obtained from Aptix Corporation.

5. References

[1] S. D. Brown, R. J. Francis, J. Rose and Z. G. Vranesic, "Field Programmable Gate Arrays," *K.A.P.*, 1992.

[2] R. Guo, H. Nguyen, A. Srinivasan, H. Verheyen, H. Cai, S. Law and A. Mohsen, "A 1024 pin universal interconnect array with routing architecture," *Proc. of the IEEE Custom Integrated Circuits Conference*, pp 4.5.1-4.5.3, May 1992.

[3] R. Guo, H. Nguyen, A. Srinivasan, Q. Nasir, H. Cai, S. Law and A. Mohsen, "A novel reprogrammable interconnect architecture with decoded RAM storage," *Proc. of the IEEE Custom Integrated Circuits Conference*, pp 9.5.1-9.5.3, May 1994.

[4] A. Srinivasan, R. Guo and H. Cai, "A reprogrammable interconnect architecture using fewer storage cells than switches," *U.S. Patent 5,319,261*, Dec 1993.

[5] A. Srinivasan and R. Guo, "Programmable interconnect architecture using fewer storage cells than switches," *U.S. Patent 5,406,138* Apr 1995.

[6] J. Rose and S. Brown, "The Effect of Switch Box Flexibility on Routability of Field Programmable Gate Arrays," *Proc. of the IEEE Custom Integrated Circuits Conference*, pp 27.5.1-27.5.4,May 1990.

[7] M. Khellah, S. Brown and Z. Vranesic, "Minimizing Interconnection Delays in Array-based FPGAs," *Proc. of the IEEE Custom Integrated Circuits Conference*, pp 9.2.1-9.2.4, May 1994.

Chair: Frank Szorc
Co-Chair: Sachin Sapatnekar

Place and route algorithms have been a topic of research through the years. This research has fragmented into several niche areas over time, due to the various implementation technologies used: standard cell, gate array, structured datapath and analog. This session will highlight recent results in several niche areas of component place and route for high-performance ICs.

Placement algorithms rely on a partitioning of the components into clusters in an effort to minimize both overall area and wire length. The first paper discusses a ratio-cut top-down clustering technique which is used to rearrange clusters in module migration algorithms. This technique outperforms the classic Fiduccia-Mattheyes and Cheng-Wei algorithms by 27% to 47% in overall net length. The algorithm avoids local minima and improves the overall size of the circuit.

Accurate estimates of interconnect length are a critical need for timing-driven layout and synthesis tools used for sub-micron design. The second paper describes a new model of interconnect which improves the mean and variance of net lengths for large designs given a circuit netlist as input. This improved information can then be put to use by the appropriate place and route tools.

Gate array routing algorithms have developed into their own niche market, and are especially critical for very large gate arrays, ie, 1-million gate designs. These large arrays are especially prone to local routing congestion, causing massive amounts of rip-up and reroutes to occur to complete the array routing. The rip-up and reroute phase can take many hours of cpu time. A new rip-up and reroute algorithm is described which runs up to 3 times faster than the fastest commercial routers.

Automatic routing of analog signals has always faced different and complex challenges as compared to digital signals. In particular, the manufacturability of analog circuits is highly dependent on the component placement and bounds and matching on routing parasitics. A new performance-based routing algorithm for analog circuits is described which improves both manufacturability and reliability.

Routing of multiple metal layers in switchboxes and channels has steadily improved over the years through the use of heuristic methods. Although efficient heuristics exist, they do not always produce an optimum solution. A new algorithm is described in the final paper for exact, multi-layer planar routing. This algorithm relies on the use of zero-suppressed BDDs to achieve this result.

This is the session to attend if you are interested in learning more about how design tools will allow you to place and route your next design more efficiently.

A Stable Partitioning Algorithm for VLSI Circuits*

Jong-Sheng Cherng and *Sao-Jie Chen*

Department of Electrical Engineering
National Taiwan University
Taipei, Taiwan, R.O.C.
E-mail: csj@cc.ee.ntu.edu.tw

Abstract

In this paper, a novel module migration based two-way partitioning algorithm is proposed to enhance the stability and quality of partitioning result. The proposed algorithm intensifies the capability of escaping from local optimal by releasing the size constraint temporarily and controlling the migration direction. And a circuit clustering procedure is incorporated into the algorithm to further improve the partitioning quality. Compared with the Fiduccia and Mattheyses [3] and Cheng and Wei [1] algorithms, the experimental results of our proposed algorithm show a significant improvement in most cases and outstanding performance in particular with large size circuits.

1. Introduction

With its increasing complexity, a VLSI circuit may contain millions of transistors. As a result, the size of problem has become so large that hierarchical approaches become essential to reduce the design complexity. Partitioning plays an important role in the hierarchical design of many physical design problems, such as circuit packaging, circuit layout, etc.. In partitioning electrical circuits, since the interconnection of nets among modules is the most essential, the main objective of many circuit partitioning algorithms is to keep the number of nets connecting different partitions minimal.

Since the problem of partitioning with size constraints is *NP-complete*, optimal solution is hard to obtain when the problem size is large and various heuristics have been developed. Kernighan and Lin [5] proposed a well-known module interchanging strategy for two-way partitioning which has become the basis of most iterative improvement partitioning algorithms. Fiduccia and Mattheyses [3] improved this algorithm by reducing time complexity to be in linear proportion to the pin number by employing a bucket list data structure. However, the Kernighan and Lin based algorithms have been observed to be highly sensitive to the choices of initial partition [6] [9]. As a consequence, many trials are needed to guarantee the quality of solutions. Also, the size of each partition must be predefined in the Kernighan and Lin based algorithms. A module migration process will be interrupted when the size constraints are broken, so the freedom of exploring the solution space is limited by this rule. The concept of ratio cut proposed by Wei and Cheng [8] [10] has received much attention recently. Since this ratio cut approach removes the constraint on subset size and tends to identify natural clusters in the circuit, it generates better result.

On the other hand, clustering techniques have been found to be useful in reducing the problem complexity and producing stable partitioning result [2] [4] [7] [9]. A cluster is a group of highly connected components in a circuit. The goal of clustering algorithms is to identify the natural clusters in a circuit. Generally speaking, the clustering algorithms are used as a preprocessing step for partitioning. That is, clustering is first applied on the original network, and then partitioning is applied on the clustered network. There are approaches to achieve partitioning with this two-level scheme. For example, Cheng and Wei [1] [9] combined the ratio cut scheme [8] with the Fiduccia and Mattheyses [3] algorithm to obtain a stable performance two-way partitioning algorithm. At the first level of their approach, some highly connected components from the original circuit are clustered into groups by a ratio cut based top-down clustering technique to reduce the problem complexity. Then the Fiduccia and Mattheyses algorithm is applied to rearrange the groups into two subsets with prespecified sizes at the second level. Similarly, Shin and Kim adopted a bottom-up clustering technique to cluster highly connected components at the first level, then a gradual constraint-enforcing partitioning algorithm was used to complete the second level partition. The algorithm starts with a relaxed size constraint on subsets and then gradually tightens the size constraint until the required constraint is met. Cong and Smith [2] also presented a parallel bottom-up clustering algorithm to generate stable partitioning results. Additionally, Hagen and Kahng [4] used a random-walk method on the network to extract the global relations among modules and to discover natural clusters, then the Fiduccia and Mattheyses algorithm was incorporated to complete the second level partitioning.

2. Ratio Cut for Clustering

Since clustering-based two-level partitioning algorithms can produce partitioning results stabler than direct partitioning, we used the ratio cut technique as a preprocessing step for clustering the original network. The ratio cut method removes the size constraints on partitions and identifies natural clusters in a circuit, so it is suitable for ratio cut to achieve a good clustering result. According to Wei and Cheng [8] [10], the ratio of a cut is defined as $RC = \frac{C_{AB}}{|A|*|B|}$. The set of nodes V of a network $H = (V, E)$ is divided into two disjoint subsets $A \subset V$ and $B = V - A$ with $A \neq \varnothing$. The cut-size C_{AB} is the number of nets connecting the partitions A and B. The ratio cut is the cut that generates the minimum ratio among all cuts in the network. However, finding the ratio cut belongs to the class of *NP-complete*

*This work was supported by the National Science Council, R.O.C., under Grants NSC85-2221-E002-050 and NSC85-2215-E002-019

0-7803-3177-6 $5.00 © 1996 IEEE

IEEE 1996 CUSTOM INTEGRATED CIRCUITS CONFERENCE

problems. The algorithm in [8] is adopted as our basic top-down clustering tool to recursively divide the circuit into small, highly connected groups. Next, a module migration based algorithm is applied to the contracted network, which nodes are groups of modules generated during the clustering procedure.

3. Module Migration Partitioning Algorithm

After clusters are formed by using a ratio cut based top-down clustering technique, a partitioning algorithm must be applied to rearrange the clusters into two subsets with prespecified sizes. For that purpose, we will develop a module migration partitioning algorithm. In a given circuit, if the connections among the modules are localized, some highly connected groups exist such that the connections among these modules are heavier than others, and dividing these groups into different partitions will increase the number of nets cut. Therefore, if we can move some modules to make these groups not to be cut, the net cut number will be fewer. This concept will be used as the main guideline in our proposed approach. An overview of our algorithm is shown as follows:

Algorithm : *Module Migration Partitioning*

1) Obtain an initial partitioning.
2) Set migration direction as from the left partition to the right partition.
3) Randomly select a module from the left partition.
4) Move the selected module and update associated information.
5) If the total accumulated size *TAS* of moved modules exceeds *GSLB* and the reversing point is found, then go to Step 7; otherwise, go to Step 6.
6) Select from the left partition a module having a maximal gain and the strongest connection to previously moved modules, then go to Step 4.
7) Set migration direction as from the right partition to the left partition.
8) Randomly select a module from the right partition.
9) Move the selected module and update associated information.
10) If current partitioning result satisfies the size ratio and generates a minimal net cut, then output this best result and go to Step 12; otherwise, go to Step 11.
11) Select from the right partition a module having a maximal gain and the strongest connection to previously moved modules, then go to Step 9.
12) Update the *GSLB* value.
13) Assign the current best result as a new initial partitioning and repeat Steps 2 to 12 until the iteration count is reached.
14) Output the current best result and use a self-adjusted probabilistic function set to update the values of starting *GSLB* and net weighting. Repeat Steps 13 and 14 until the number of trials is reached.
15) Report the best partitioning result.

First, an initial partitioning solution is generated as a starting point, then we attempt to move the modules in highly connected groups of one partition into the other one to improve the existing solution. This is achieved by fixing the migration direction and choosing modules to be moved with great care. This "forward" migration process is continued until the assumed group is swept into the other side. Next, the migration direction is reversed and the modules are moved to keep the size ratio fall into an acceptable range. We call a *pass* is done when the size ratio falls below the acceptable range during the "backward" migration. The best feasible solution will be recorded any time it occurs and used to form a new starting point for the next *pass*. The above *pass* operation will be repeated

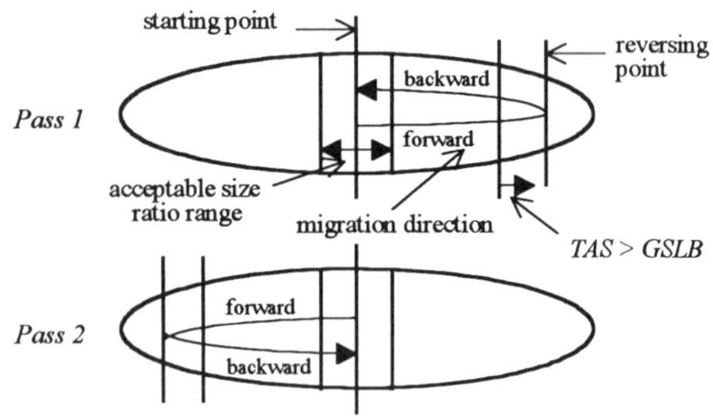

Fig. 1: *Pass* operation of the module migration algorithm.

iteratively to improve the result and the iteration count can be determined in advance. Note that we have relaxed the size constraints temporarily during a *pass* process. The basic operation of a *pass* (Steps 2 to 12 in the above Algorithm) is shown in Fig. 1, where the migration direction of each *pass* is changed alternatively.

We hope only modules in one highly connected group are moving during forward or backward migration. Thus it is undesirable to choose a module which has no connection to the previously moved modules as the next target to move. Therefore, a module m_i chosen for migration must have a maximal gain (here gain means the decrement of the net cut number if m_i is moved into the opposite partition) and have the strongest connection to previously moved modules. In Step 5 of the above Algorithm, *GSLB*, which is the *guessed size lower bound* of the maximal module set in the highly connected group that we want to move, is devised to help moving modules efficiently. Heuristically, we hope modules in larger groups be first moved into one side, and then modules in smaller groups. Therefore, the value of *GSLB* is set to be high at the first *pass*, and will be gradually decreased at the following *passes*.

Backward migration is essential to search acceptable size ratio results and to find better solution for each *pass*. Naturally, the change of migration direction occurs when modules in a highly connected group have just been moved into one side. By direct observation, when such a group is moved into the opposite side, the net cut number will fall into a minimal point. If we still continue the moving operation unaware, the net cut number will certainly increase. Under this consideration, when the accumulated size of moved modules is greater than the given *GSLB*, we begin to keep record of the decreasing net cut number and it is time to reverse the migration direction when this value begins to increase. Since the moving gain of candidate modules in a module migration based algorithm can reflect this value changing, we need only to check the "moving gain" at the implementation, and this is one of the factors that makes our module migration based algorithm efficient.

Since weighting nets (especially those multi-pin nets) in a partitioning graph and setting the *GSLB*'s starting value are two important parameters in our algorithm, all these parameter settings must be managed efficiently. To reduce the sensitivity to the variety of parameters, we hope the algorithm has the ability to search for an optimal parameter range automatically. Therefore, we need a set of probabilistic functions to aid the algorithm in finding a better value range for each of these parameters iteration by iteration. The function of such a parameter management scheme is briefly described as follows: if the current value of a certain parameter used in the algorithm produces most of times good (bad) results, the probability that setting the next iteration value of this parameter close to its current value must be kept high (low).

Since the result from just a single run of the proposed module migration partitioning algorithm is very stable, we have only to apply the algorithm just five times to obtain a good result on the contracted network. Finally, the last run of the proposed algorithm will be applied to the original flattened network to fine tune the partitioning result. In next section, experiments on partitioning benchmarks are performed to verify the superiority of our proposed algorithm.

4. Experimental Results

Our clustering-based module migration (CMM) algorithm was coded in C language and implemented on a Sun SPARC workstation. For comparison, we implemented the Fiduccia and Mattheyses (FM) algorithm [3] and the Cheng and Wei (CW) algorithm was from [1]. We tested our algorithm using 18 circuits as listed in Table 1. The runs of the CW and our CMM algorithms are set to be 20. The *g* value of the CW algorithm is set to be 50 based on [1]. We set the run of the FM algorithm to be 100 for meaningful comparison. By experimental results, we demonstrate that our partitioning algorithm outperforms the FM and CW algorithms in terms of the minimal net cut number and the average net cut value.

In Table 1, we compare our result to the FM and CW results with the size of each partition being allowed to have 1% deviation from exact bipartition. For most circuits, the CW algorithm performs better than the FM algorithm in terms of the average net cut value, but the situation is opposite in terms of the best minimal net cut number. However, the proposed CMM algorithm generates better results than those of the FM and CW algorithms for both terms. For instance, in the example of *industry2*, although much improvement to the FM algorithm is achieved by the CW algorithm (1098 $\rightarrow$ 938), the minimal net cut number is still high compared with our result (197). The average net cut value also drops from 2028.53 to 1139.90 to 241.65. These are 82% and 79% improvements for the minimal net cut number, and 88% and 79% improvements for the average net cut value, over the FM algorithm and the CW algorithm, respectively. On average, our module migration partitioning algorithm shows a 27% and 26% improvements over the best results and a 51% and 45% improvements over the average results achieved by the FM algorithm and the CW algorithm, respectively, for all the 18 test circuits. For huge circuits like *avg.small* and *avg.large*, although our algorithm spends approximately three times of run time than the CW algorithm, it is worthy for us to achieve 58% and 64% improvements for the minimal net cut number, and 50% and 53% improvements for the average net cut value over the CW algorithm, for the circuits *avg.small* and *avg.large*, respectively. From Table 1, we also find that the CMM and CW algorithms have the same time magnitude, and the average time used in the FM algorithm is much less than others. Although the FM algorithm works very fast for each run, more than 100 runs are required to derive the same solution quality as our CMM algorithm. Therefore, our approach is superior to the FM method with the whole performance consideration.

In next experiment, we relax the deviation such that each partition is allowed to have a up to 10% deviation from bipartition. Table 2 presents the minimal net cut number and the average net cut value in this interval. On average, the CMM approach yields 23% and 47% improvements in terms of the minimal net cut number and the average value compared to the FM approach. For the average net cut value, the improvement of the CMM approach over the CW approach ranges from 1% (*primGA1*) to 43% (*industry2*), with an average of 9%. For the minimal net cut number, the CMM method outperforms the CW method on most circuits, and about an average of 8% improvement is obtained.

Since we have put a rather large tolerance on the size constraints (1 : 1.222) in Table 2, all of the three algorithms have more chances to search for better partitionings, therefore, they generate similar solutions for some circuits. But for large circuits like *industry2*, *industry3*, *avg.small*, and *avg.large*, the CMM algorithm obtains outstanding performance over other algorithms not only in Table 2 but also in Table 1. Additionally, the stability of the FM method is worse than the others when comparing the differences between average values and best values for circuits. According to the experimental results of both tables, we show that in the case where a strictly balanced partition is required, both the FM and CW algorithms have a poor performance. On the contrary, the behavior of the proposed CMM algorithm is very stable all the way. This shows that for two-way partitioning, our module migration algorithm is a better choice than the FM and CW algorithms.

5. Conclusion

A new two-way partitioning algorithm has been developed, where we adopted an efficient module migration scheme to release the size constraints temporarily and to control the migration direction. Experimental results obtained indicate that our proposed module migration partitioning algorithm improves the unstable property of conventional module migration based algorithms and outperforms the Fiduccia and Mattheyses algorithm [3] and the Cheng and Wei algorithm [1] in terms of the minimal net cut number and the average net cut value with the prespecified size constraints. Furthermore, our heuristic algorithm generates even better performance on large circuits than the others with reasonable runtime. Especially, our algorithm is very stable even when the deviation is quite small, that is, the solution quality of our proposed algorithm is less sensitive to the choice of the deviation.

Acknowledgements

The authors are thankful to Professors C. K. Cheng and C. W. Yeh for their supplying us both the testing benchmarks and the ratio cut program.

References

[1] C. K. Cheng and Y. C. Wei, "An improved two-way partitioning algorithm with stable performance," *IEEE Trans. Computer-Aided Design*, vol. 10, no. 12, pp. 1502-1511, Dec. 1991.

[2] J. Cong and M. Smith, "A parallel bottom-up clustering algorithm with applications to circuit partitioning in VLSI design," in *Proc. ACM/IEEE Design Automation Conf.*, 1993, pp. 755-760.

[3] C. M. Fiduccia and R. M. Mattheyses, "A linear-time heuristic for improving network partition," in *Proc. ACM/IEEE Design Automation Conf.*, 1982, pp. 175-181.

[4] L. Hagen and A. Kahng, "A new approach to effective circuit clustering," in *Proc. IEEE Int. Conf. Computer-Aided Design*, 1992, pp. 422-427.

[5] B. W. Kernighan and S. Lin, "An efficient heuristic procedure for partitioning graphs," *Bell System Technical Journal*, vol. 49, no. 2, pp. 291-307, Feb. 1970.

[6] T. K. Ng, J. Oldfield, and V. Pitchumani, "Improvements of a mincut partition algorithm," in *Proc. IEEE Int. Conf. Computer-Aided Design*, 1987, pp. 470-473.

[7] H. Shin and C. Kim, "A simple yet effective technique for partitioning," *IEEE Trans. VLSI System*, vol. 1, no. 3, pp. 380-386, Sept. 1993.

[8] Y. C. Wei and C. K. Cheng, "Towards efficient hierarchical designs by ratio cut partitioning," in *Proc. IEEE Int. Conf. Computer Aided Design*, 1989, pp. 298-301.

[9] Y. C. Wei and C. K. Cheng, "A two-level two-way partitioning algorithm," in *Proc. IEEE Int. Conf. Computer Aided Design*, 1990, pp. 516-519.

[10] Y. C. Wei and C. K. Cheng, "Ratio cut partitioning for hierarchical designs," *IEEE Trans. Computer-Aided Design*, vol. 10, no. 7, pp. 911-921, Jul. 1991.

Table 1

Partitioning results allowing 1% deviation from bipartition.

Example				FM			CW			CMM			Improvement (%)			
	# nets	# modules	# pins	min	avg	sec/run	min	avg	sec/run	min	avg	sec/run	I_0	I_1	I_2	I_3
19ks	3343	2684	10387	137	181.32	11.14	154	184.45	65.35	117	125.85	74.58	15	31	24	32
19kstw	3731	3079	11238	98	161.89	13.53	103	151.60	86.23	90	101.35	119.11	8	37	13	33
primGA1	904	752	2707	48	77.62	1.60	51	70.90	13.59	47	48.75	15.99	2	37	8	31
primGA2	3029	2907	10959	161	281.88	16.30	169	266.85	73.72	134	139.80	72.58	17	50	21	48
primSC1	904	752	2707	50	80.45	1.59	50	71.75	13.23	47	49.15	16.67	6	39	6	31
primSC2	3029	2907	10959	172	286.55	16.00	172	249.50	72.79	133	135.10	69.88	23	53	23	46
test02	1866	1663	6220	139	238.50	2.65	108	163.25	28.24	90	99.80	82.12	35	58	17	39
test03	1699	1607	5866	90	132.21	3.70	71	112.20	27.66	56	58.50	39.99	38	56	21	48
test04	1738	1515	5998	78	92.82	1.35	45	73.15	25.56	44	44.00	88.22	44	53	2	40
test05	2910	2595	10121	71	113.53	2.46	64	92.25	52.36	51	51.60	106.09	28	55	20	44
test06	1745	1752	6728	68	88.80	4.88	70	85.20	34.15	60	67.55	39.20	12	24	14	21
8870	349	286	1138	16	41.12	0.35	18	34.95	4.22	14	14.90	9.01	13	64	22	57
5655	843	801	2755	54	84.03	1.20	58	76.65	13.45	49	51.45	19.61	9	39	16	33
industry1	2594	2271	7721	34	78.33	7.73	34	77.95	54.51	24	27.10	52.69	29	65	29	65
industry2	13915	12142	47232	1098	2028.53	10.58	938	1139.90	191.24	197	241.65	377.37	82	88	79	79
industry3	21966	15059	65356	360	688.98	56.97	409	804.50	726.43	272	290.85	530.90	24	58	33	64
avg.small	22124	21918	76231	336	638.33	509.13	501	615.65	2246.57	211	307.90	6456.26	37	52	58	50
avg.large	25384	25178	82751	432	825.57	634.45	536	711.05	2544.80	193	336.90	7775.00	55	59	64	53
Average													27	51	26	45

Note: $I_0 = (\min_{fm} - \min_{cmm}) / \min_{fm}$ $I_1 = (\text{avg}_{fm} - \text{avg}_{cmm}) / \text{avg}_{fm}$

$I_2 = (\min_{cw} - \min_{cmm}) / \min_{cw}$ $I_3 = (\text{avg}_{cw} - \text{avg}_{cmm}) / \text{avg}_{cw}$

Table 2

Partitioning results allowing up to 10% deviation from bipartition.

Example	FM		CW		CMM		Improvement (%)			
	min	avg	min	avg	min	avg	I_0	I_1	I_2	I_3
19ks	133	178.41	118	123.00	112	121.80	16	32	5	1
19kstw	97	167.32	90	96.50	79	88.00	19	47	12	9
primGA1	45	67.80	42	42.65	42	42.30	7	38	0	1
primGA2	169	280.22	120	122.60	120	121.95	29	56	0	1
primSC1	44	71.28	42	42.95	42	42.50	5	40	0	1
primSC2	158	282.02	119	124.65	119	123.05	25	56	0	1
test02	128	192.59	81	98.50	75	85.35	41	56	7	13
test03	69	131.98	56	61.10	55	56.15	20	57	2	8
test04	44	48.34	44	44.95	42	42.80	5	11	5	5
test05	42	60.70	42	43.20	42	42.00	0	31	0	3
test06	60	85.02	60	69.60	60	62.50	0	26	0	10
8870	14	22.47	14	15.10	14	14.35	0	36	0	5
5655	52	69.23	49	53.10	47	48.25	10	30	4	9
industry1	25	70.52	20	24.80	20	22.35	20	68	0	10
industry2	1429	1991.48	346	422.90	178	239.85	88	88	49	43
industry3	261	653.89	211	252.45	190	220.20	27	66	10	13
avg.small	341	638.75	270	373.20	191	286.85	44	55	29	23
avg.large	449	815.09	212	353.40	160	319.20	64	61	25	10
Average							23	47	8	9

Standard Cell Interconnect Length Prediction from Structural Circuit Attributes

Hans T. Heineken and Wojciech Maly
Department of Electrical and Computer Engineering
Carnegie Mellon University
Pittsburgh PA 15213.

Abstract

A new interconnect model is proposed that predicts the distribution parameters of net lengths. The model takes as input a standard cell netlist and provides as output estimates of the mean and variance of net length on a net by net basis. The model was developed and verified on designs produced with two different place and route algorithms.

1.0 Introduction

The semiconductor industry continues to scale feature sizes. Scaling improves the performance of transistors and decreases die size allowing for increased levels of integration. However, scaling increases parasitic resistance and degrades the RC delay in interconnects [1]. In addition, high levels of integration lead to longer interconnect lines. Consequently scaling increases the contribution of interconnects to delay and **increases the uncertainty in predicting worst case delays in critical paths**. Therefore a need exists for an accurate interconnect length model.

Numerous interconnect length distribution models exist, many based on Rent's rule [1,2,3,4,5,6,7]. This rule relates the number of pins on a logic block to the number of gates in this block raised to the Rent's constant. However, the inability to derive Rent's constant up front severely limits the practicality of the model. In addition the model provides little feedback as to how a particular interconnect is affected by structural changes to a circuit.

Pedram et. al. [8,9] and Hamada [10] provide models that better account for the effects of structural changes on the interconnect length. They examine the neighborhood population of an interconnect, where the neighborhood population is the number of cells encountered within a distance of one from the cells on the interconnect. Given the neighborhood population, Pedram averages the interconnect length over all possible placements of the cells within the neighborhood. Hamada considers the interconnect's location within a neighborhood, assumes a Weibull interconnect length distribution and solves a neighborhood density function to find the parameters of the distribution.

Jyu [11] collects a vast array of structural characteristics from a standard cell netlist and uses linear regression to build a model. The accuracy of the model, however, depends on classifying nets into short, medium and long nets; categories that are difficult to determine before layout.

In this paper the interconnect length distribution is postulated to be a gamma distribution. The objective is to create a model capable of predicting distribution parameters, mean and variance, on a net by net basis. The model emphasizes the prediction of interconnect lengths in the critical paths of circuits. Consequently, only local interconnects are modeled.

The model is envisioned to work in the design environment pictured in Fig. 1. In this environment, a database of completed designs will be analyzed by a tool set called MAPEX [12]. This tool set allows users to extract design attributes from netlists and layouts and to analyze and model relationships between attributes. Models will be incorporated in the design environment and employed on new designs. Upon completion of a design, model parameters will be re-tuned.

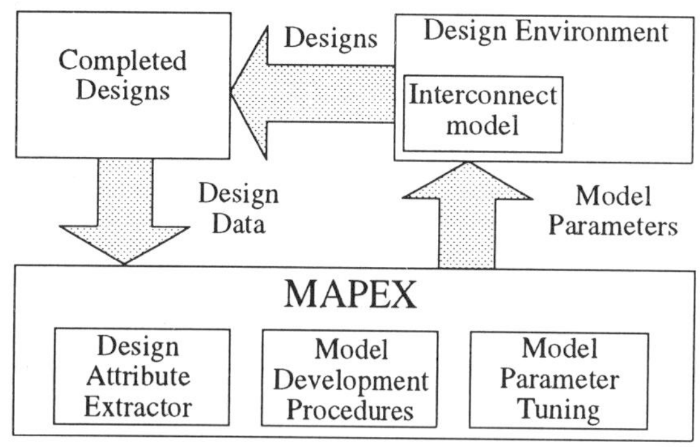

Fig 1. Interconnect model environment.

Within the design environment the interconnect model is envisioned to work on circuit paths that have been identified as possible critical path candidates. The model will estimate length distribution parameters for each net in a path, providing direct feedback to designers as to how their design decisions affect performance.

2.0 Methodology

An extensive experiment was conducted to develop the interconnect model described in this paper. The methodology by which the experiment was conducted was as follows:

1. First a set of representative standard cell netlists was selected for analysis. The netlists were placed and routed with commercial tools; two layouts were generated for each netlist.
2. Data was extracted from the layouts and netlists using the analysis environment MAPEX. Data included a listing of all nets and their physical and structural attributes, e.g., length, fanout, neighborhood. Data also included attributes of the circuit as a whole, e.g., number of cells, number of interconnects.

3 a. A set of structural attributes deemed to affect net lengths was selected. Nets with similar structural attributes were grouped together.

b. Models were postulated for the distribution parameters of net length for each group of nets.

c. Models were analyzed for correctness. The model development process iterated to 3a. until the required level of accuracy was reached.

3.0 Designs

The interconnect model was developed and verified using a set of representative standard cell circuits. Table 1 lists the circuits.

0-7803-3177-6 $5.00 © 1996 IEEE

IEEE 1996 CUSTOM INTEGRATED CIRCUITS CONFERENCE

Table 1. Design Data

Circuits	Number of Cells	Number of Interconnects
1	2751	3403
2	4720	6448
3	3210	3703
4	382	535
5	3464	3722
6	1855	2144
7	1572	1859
8	5598	5836
9	3440	3467
10	2021	2826

4.0 Model Development

4.1 Attributes of Interest: Definitions

In developing the interconnect model a select set of structural attributes was postulated to have an affect on net length. These attributes were broken into two categories: circuit attributes - those related to the circuit as a whole; and net attributes - those specific to nets. Examples refer to Fig. 2.

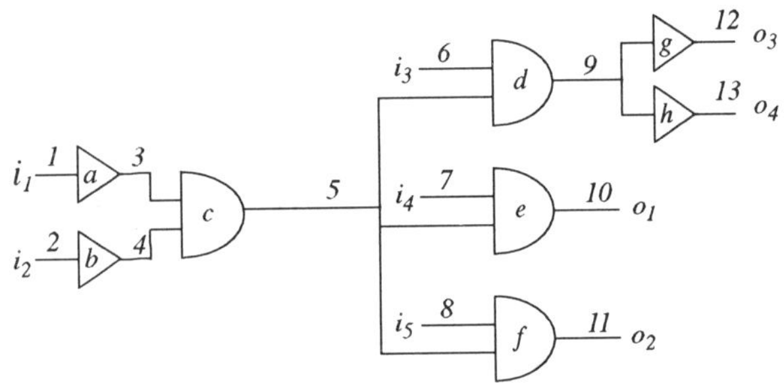

Fig 2. Example circuit used to define circuit and net attributes.

Circuit Attributes:

- *Interconnect* - any multi-terminal equipotential region connecting inputs and outputs of cells in the layout. E.g., *Interconnects* = 1, 2, 3, ...
- *Net* - a two terminal portion of an interconnect connecting the output of a cell with the input of a cell on the same interconnect. (Nets are labeled by the cells they connect - net *cd* connects cells *c* and *d*.) E.g., *nets* of interconnect 5 = {*cd, ce, cf*}.
- *Ncell* - number of standard cells in circuit. E.g., *Ncell* = |{*a, b, c, d, e, f, g, h*}| = 8.
- *Nint* - number of interconnects. E.g., *Nint* = |set of all interconnects| = 13.
- *Nnet* - number of nets. E.g. *Nnet* = |set of all nets| = 16.
- *Distance(cell1,cell2)* - the number of cells traversed in shortest path between cells *1* and 2. E.g., *Distance(a,g)* = 2.
- $Nngh(cell)_i$ - neighborhood population at level *i*: the number of cells residing within a distance *i* from cell in the circuit.

  E.g., $Nngh(d)_1$ of cell *d* = |{*c, e, f, g, h*}| = 5.
  $Nngh(d)_2$ of cell *d* = |{*a, b*}| = 2.

The definition of neighborhood population is similar to that given in [9]. However, the purpose of collecting neighborhood data is slightly different. In this paper neighborhoods are employed as used in [13], where neighborhoods are analyzed to identify clusters in a circuit. However, instead of finding clusters, the neighborhood populations are summed over all cells to find a figure of merit of the "congestion" of a circuit.

- $Angh_i$ - neighborhood population at level *i* summed across the circuit. E.g., $Angh_1 = 22$.

Net Attributes:

- *length* - length of net (μm).
- *size* - number of cells attached to the interconnect from which net is taken. E.g., *size* of net *cd* = |{*c, d, e, f*}| = 4.
- *attint* - number of interconnects attached to the cells of a net excluding the interconnect from which the net is taken. E.g., *attint* of net *cd* = |{3, 4, 6, 9}| = 4.
- *attint_nc* - similar to *attint* but excludes interconnects common to both the input and output cells of the net. E.g., *attint_nc* of net *cd* = |{3, 4, 6, 9}| = 4.

4.2 Model Selection

4.2.1 Net Groupings

Each net in a netlist has a set of attributes associated with a net. The values of these attributes differ from one net to the next in the netlist. Depending on the attribute the difference may or may not be reflected in net length.

Those attributes that affect a net's length are defined as a net's *context*. The sum of the associated values of a net's context, i.e., the sum of the values of attributes that affect a net's length, is defined as a net's *weight*. Nets with similar weights are grouped together and models are developed to estimate net length distribution parameters for each group.

An analysis of the designs showed that an appropriate net weight is the following:

$$Net\ weight = size + attint\_nc$$

4.2.2 Model Tuning

To determine if the net length can be modeled as a function of net weight, the net lengths were plotted against net weight, (Fig. 3).

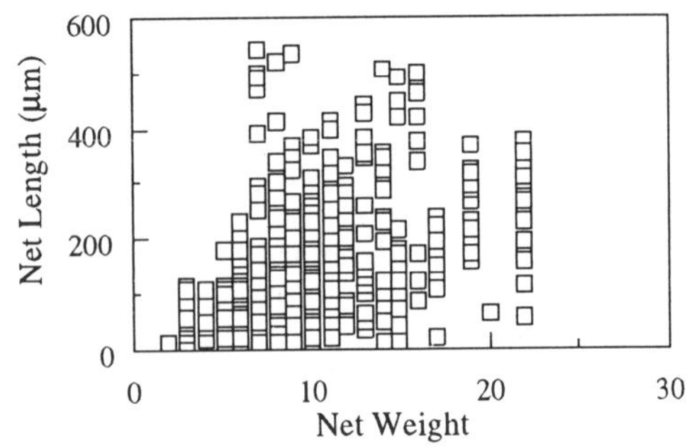

Fig 3. Net length vs. net weight.

Each set of nets with the same net weight formed a unique group of nets. An analysis showed that the distribution of net lengths within each group could be modeled by a gamma distribution.

Only local nets were considered. (Local nets are defined

in this paper as those with a net weight less than or equal to 12. In the designs used in this experiment, nets with weights greater than 12 constituted less than 5% of the interconnects.)

Fig 4 shows a plot of mean net length in each group vs. net weight.

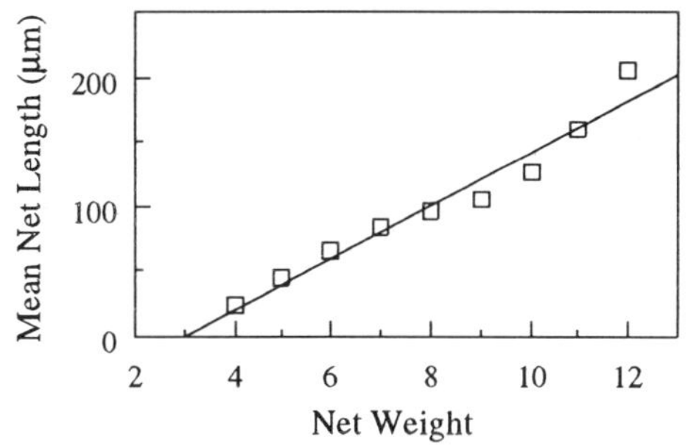

Fig 4. Mean net length vs. net weight.

Fig. 4 shows clearly that the mean net length can be well modeled as a linear function of net weight:

$$Mean\ Length \propto m \cdot Net\ \ Weight$$

where m is the slope of the linear function. The linear function is shown in Fig 4 as a solid line and is fit using least square fit. The correlation for this circuit is high (0.983). The error is less than 6%. (The error is calculated as: $error = |l_m - l|/l_m$, where l_m is the mean measured net length and l is the mean modeled net length.)

Table 2 gives the slopes of the linear functions, correlations, and errors for the circuits in Table 1. The high correlations and small errors show that the postulated net context is applicable and that mean net lengths for a given circuit can be well modeled by a net's weight.

Table 2. Slopes, correlations, and errors of mean and variance net length vs. net weight.

Circuits	Slope: Mean Net Length vs. Net Weight	Correlation: Mean Net Length vs. Net Weight	Error: Mean Net Length vs. Net Weight (%)	Slope: Variance Net Length vs. Net Weight	Correlation: Variance Net Length vs. Net Weight	Error: Variance Net Length vs. Net Weight (%)
1	22.7	0.973	5.94	4070	0.971	1.47
2	15.9	0.981	0.61	2860	0.941	3.06
3	17.0	0.924	1.28	3340	0.907	3.46
4	11.5	0.977	1.38	753	0.979	1.24
5	16.9	0.937	5.16	3610	0.906	9.76
6	13.9	0.976	5.34	1690	0.953	1.90
7	13.7	0.947	2.05	1230	0.493	7.80
8	17.4	0.932	2.21	4050	0.571	7.80
9	14.6	0.984	5.97	3490	0.962	8.78
10	12.4	0.970	2.03	1350	0.595	2.89

Modeling the mean value is not sufficient for net length prediction. Another required parameter is variance. An analysis of the nets showed that the variance can also be modeled as a linear function of net weight (Fig. 5). The slopes, correlations, and errors are listed in Table 2.

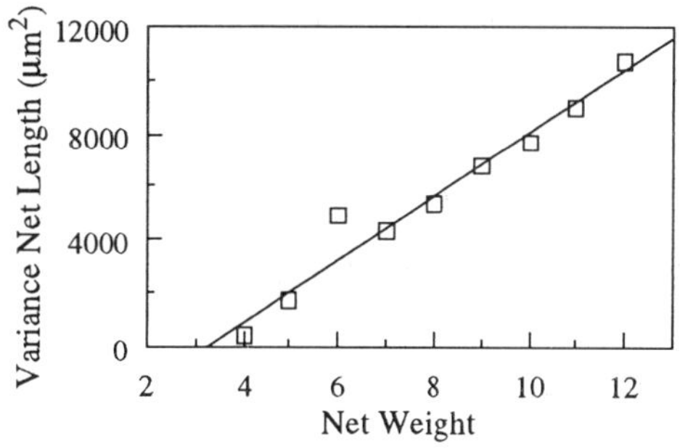

Fig 5. Variance net length vs. net weight.

It was observed that incremental changes to the circuit topology did not affect the slopes. The slopes were therefore assumed to be a function of circuit attributes (as opposed to net attributes). An analysis with MAPEX showed that the slopes for the circuits in Table 1 were affected by the overall interconnect density or congestion of a circuit. A good measure of the congestion of the circuit was the neighborhood population of the circuit, $Angh_i$. In particular, a correlation existed between the slopes and the neighborhood populations at level i=3: $Angh_3$. Fig. 6 shows the relationship.

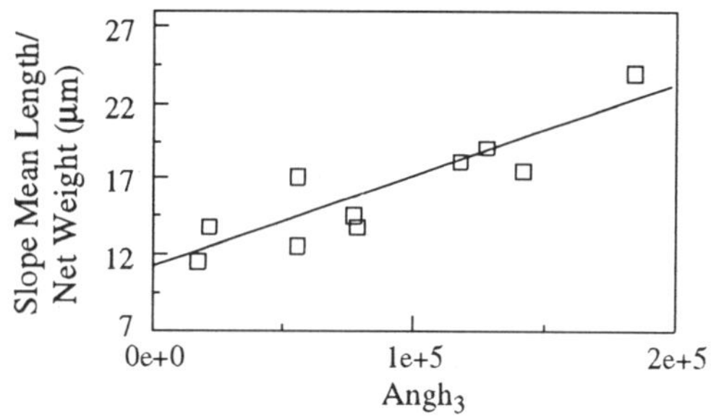

Fig 6. Slope of mean net length/net weight vs. $Angh_3$.

The slope of variance of net length vs. net weight is similarly affected by the congestion of the circuit. However, in addition, the size of the circuit and the number of interconnects in the circuits contributed to the slope. An analysis of the circuits showed that a correlation existed between the slope of the variance and $(Angh_3\ Ncell\,/\,Nint)$.

Applying the above developed relationships for mean and variance of net lengths, the net length model can be described as a function of net and circuit attributes in the following way:

$$Mean(length) = a_m + b_m\,(Angh_3)\ Net\ weight$$

$$Variance(length) = a_v + b_v\,(Angh_3\ Ncell\,/\,Nint)\ Net\ weight$$

where a_m, b_m, a_v, b_v are model parameters.

The model parameters were tuned to the designs and the following values were derived: 11.3, 4.95×10^{-5}, 1028, and 2.57×10^{-2}, for a_m, b_m, a_v, b_v, respectively. Fig. 7 shows the modeled and measured mean net length for one of the circuits. Table 3 lists the errors for the modeled mean and variance of net length for the circuits in Table 1. The average error in mean net length is 8.64%. The average error in variance is 24.5%.

To check the postulated theory of the model, a second design environment with a different placement algorithm was used to generate layouts of the circuits in Table 1. Attributes were extracted from the netlists and layouts and the parameters of the model were re-tuned. The values of the parameters were: 12.3, 4.65x10^{-5}, 647, and 2.38x10^{-2}, for a_m, b_m, a_v, b_v, respectively. The results showed an average error for mean net lengths of 10.9% and for variance of net lengths of 31.2%.

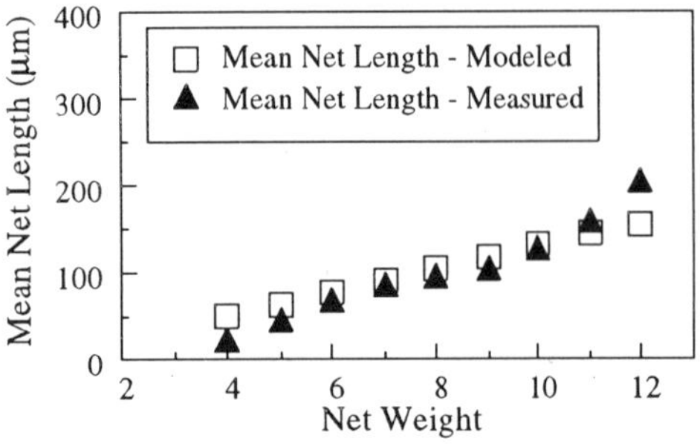

Fig 7. Measured and Modeled net length vs. net.

Table 3. Errors in modeled mean and variance net length.

Circuits	Error: Mean Net Length (%)	Error: Variance Net Length (%)
1	6.40	19.7
2	20.3	10.2
3	19.7	31.4
4	0.44	37.5
5	0.64	35.7
6	5.13	46.0
7	8.02	12.2
8	0.96	22.3
9	15.7	7.65
10	9.11	22.3

4.2.3 Net Length Distribution

Fig. 8 shows the histograms for measured net lengths for two different net weights (5 and 9). On top of these histograms the gamma distribution is drawn. The gamma distribution was derived using the mean and variance of the net lengths predicted by the model.

5.0 Conclusions

To decrease the uncertainty in predicting worse case delays in critical paths, an accurate empirical interconnect length model has been proposed. The model takes as input a standard cell netlist and provides as output estimates of the distribution parameters of net length on a net by net basis. The model was developed and verified on two place and route algorithms.

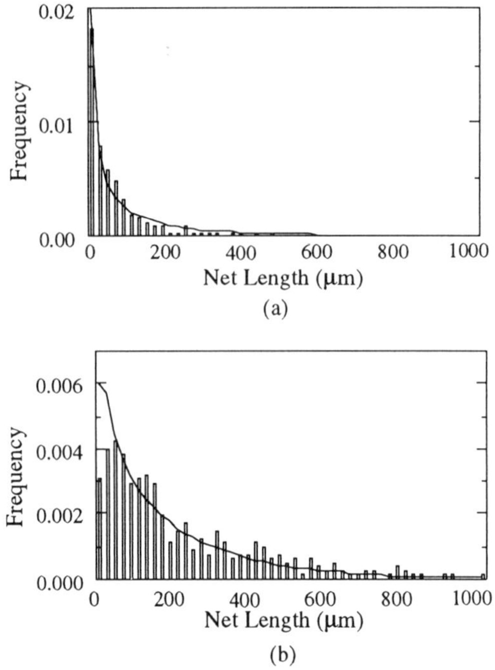

Fig 8. Net Length histograms and gamma distributions for net weights (a) 5 and (b) 9.

References

1. H.B. Bakoglu, "Circuits, interconnections, and packaging for VLSI," *Addison-Wesley Publishing Company,* 1990.

2. B.S. Landman and R.L. Russo, "On a pin versus block relationship for partitions of logic graphs," *IEEE Trans. on Circuits and Systems*, vol. cas-20, pp. 1469-1479, Dec. 1971.

3. W.E. Donath, "Placement and average interconnection lengths of computer logic," *IEEE Trans. on Circuits and Systems*, vol. cas-26, no. 4, pp. 272-277, April 1972.

4. W.E. Donath, "Wire length distribution for placements of computer logic," *IBM Journal of Research and Development*, vol. 25, no. 3, pp. 152-155, May 1981.

5. M. Feuer, "Connectivity of random logic," *IEEE Trans. on Computers*, vol. C-31, no. 1, pp. 29-33, Jan. 1982.

6. C.V. Gura and J.A. Abraham, "Average interconnection length and interconnection distribution based on Rent's rule," *Proc. 26th Design Automation Conference*, pp. 574- 577, June 1989.

7. J.E. Cotter and P. Christie, "The analytical form of the length distribution function for computer interconnections," *IEEE Trans. on Circuits and Systems*, vol. 38, no. 3, pp. 317-320, March 1991.

8. M. Pedram and B. Preas, "Accurate prediction of physical design characteristics for random logic," *Proc. Int. Conference on Computer Design*, pp. 100-108, Oct. 1989.

9. M. Pedram and B. Preas, "Interconnection length estimation for optimized standard cell layouts," *Proc. Int. Conference on Computer-Aided Design*, pp. 390-393, Nov. 1989.

10. T. Hamada, C.-K. Cheng, and P.M. Chau, "A wire length estimation technique utilizing neighborhood density equations," *Proc. 29th Design Automation Conference*, pp. 57-61, June 1992.

11. H.H. Jyu, "Statistical delay model based performance analysis and optimization in logic circuit design," *Ph.D. Thesis*, Princeton University, Nov. 1994.

12. H.T. Heineken and W. Maly, "Manufacturability analysis environment - MAPEX," *Proc. Custom Integrated Circuits Conference*, pp. 309-312, May 1994.

13. J. Garbers, H.J. Promel, and A. Steger, "Finding clusters in VLSI circuits," *Proc IEEE Int. Conference on Computer-Aided Design*, pp. 520-523, Nov. 1990.

A New Rip-up and Reroute Algorithm for Very Large Scale Gate Arrays

Hiroshi Shirota, Satoshi Shibatani, and Masayuki Terai
System LSI Laboratory
Mitsubishi Electric Corporation
4-1 Mizuhara, Itami, Hyogo 664, Japan

Abstract

A fast rip-up and reroute algorithm for large scale gate arrays is reported. The algorithm combines 'global' and 'local' rip-up and reroute processes to efficiently eliminate the unconnects introduced by an initial routing process. The global process reduces the local wire congestion by ripping up and rerouting global paths. The local process eliminates the unconnects, mainly caused by routing order dependency, by ripping up and rerouting local paths. The effectiveness of our method is demonstrated by our experimental results on industrial sea-of-gates (SOG) circuits and a well-known benchmark circuit.

1. Introduction

In recent years, the scale of application specific integrated circuits (ASICs), especially gate arrays, has grown significantly. As the process dimensions scale into deep submicron, 0.35µm-0.5µm, over 1 million-gate (MG) SOGs have been developed (e.g., [1]). The gate arrays are characterized by the quick turn around time (QTAT), so that a fast automatic routing program for the gate arrays has been required.

The most important goal of the routing program for the gate arrays, where the size of the routing region is fixed, is 100% completion of the connections (i.e., the inter-cell required routing). In general, the routing program can be divided into two processes: an initial routing process and a rip-up and reroute process. A usual method is to perform the initial routing process in two steps: a global routing and a detailed routing[2, 3]. The detailed routing can be classified as: maze routers[4, 11], channel routers[2, 3], and so on. The initial routing process often leaves some unconnects for the following two reasons. (1) Routing order dependency; Most initial routing processes use a sequential approach, i.e., all the connections are routed one by one. Therefore, the results of the initial routing process are dependent on the order in which the connections are routed. Some unconnects are often left over because the previously routed connections block the routing of the connections yet to be tried. (2) Local wire congestion; For dense large scale gate arrays, the degree of congestion is not even over the whole chip after the initial routing process. The unconnects may occur in local congested regions.

The function of the rip-up and reroute process is to route the unconnects that remains after the initial routing process. It rips up the previously completed connections to eliminate a blocking situation, and then routes the blocked connections and all the ripped up connections. For dense gate arrays, the time required to complete the rip-up and reroute process is long because of the large number of unconnects. A major part of the time required for routing is consumed by the rip-up and reroute process. Our experience shows that the rip-up and reroute process of a dense large scale gate array requires more than ten hours. In the practical design of a SOG circuit, it is necessary to rerun routing several times for various placements to find a successful result from the standpoint of resultant circuit performance and routing completion. It is essential to develop a rip-up and reroute method that attains a high completion ratio and a short run-time.

We consider a rather general form of the rip-up and reroute problem containing a set of connections, most of which are already routed by an initial router and the remainings are left unrouted. Up to now, many rip-up and reroute algorithms have been proposed[5-10]. They can be classified into two main approaches: a weak modification and a strong modification.

The weak modification[5, 6] relocates part of the existing wiring by a shove-aside operation, a layer switching operation, and so on, while maintaining all existing connections. This strategy tends to be quicker than the strong modification but not as successful due to its strong local bias. This strategy is rarely used alone. It is usually used in conjunction with a strong modification(e.g., [5, 6]).

The strong modification[5-10] rips up the blocking wires and routes the blocked connections and then routes all the ripped-up connections. It usually uses maze router[4] for routing. This technique can be classified into two approaches. One approach maintains design rules while routing[5, 7]. The other approach allows illegal conditions[6, 8, 10]. After it routes all unconnects by allowing nets to intersect (crossover), or violate design rules with previously routed wires, it rips up and reroutes illegal nets in a certain order until all intersections and violations are eliminated. However, because both of them can not determine which wires are the best ones to be ripped up, we must rip up a great number of wires by trial and error to obtain a higher completion ratio and then reroute them with the time-consuming maze router. Therefore, as it is, this strategy is not applicable to the large scale routing problems.

Among the experimental examples in literature[5-10], the biggest rip-up and reroute problem [8] contains only 400 x 304 = 121,600 grid points per layer, as far as the authors know. However, for a 1 MG gate array, which contains about 9,000 x 6,000 = 54,000,000 grid points per layer, the routing region is 2 orders of magnitude larger than the largest problem in literature[8], so it is difficult for existing algorithms to be applied to the 1 MG gate array in terms of the processing time. The traditional remedy for such a large scale gate array is to partition the routing region into many small subregions and perform a rip-up and reroute process for each subregion (e.g., [10]). This method tends to be quicker and can efficiently eliminate the violations mainly caused by the routing order dependency. However, it spends much time eliminating violations in the local wire congested subregions. The reason is that we must repeat the rip-up and reroute process many times while shifting a subregion of interest over the entire region to propagate the violations toward the less congested subregions and finally eliminate all the violations. Therefore, this method is slow especially for large dense chips where the results of the initial routing process usually have local wire congestion.

In this paper we propose a new rip-up and reroute algorithm for large scale gate arrays which is effective even in cases where the routing region has local wire congestion. This algorithm is based on a strong modification method and allows design rule violations in its intermediate solution. Our algorithm consists of two rip-up and reroute processes: a global rip-up and reroute process and a local rip-up and reroute process. The global process rips up the prerouted wires that go through the congested subregions and reroutes the ripped-up connections around them to reduce the local wire congestion. The two main features of the global process are a "coarse" maze router and a congestion-directed cost function for picking up the nets to be ripped up. The local process builds a local routing window centered on each violation, and then rips up and reroutes the existing connections within each window to eliminate the violations mainly caused by the routing order dependency. Unlike the existing method[9] which rips up and reroutes the prerouted connections only within the small subregion, the local wire congestion is reduced by the global process and the local process only has to eliminate the violations mainly caused by the routing order dependency. Therefore, because it is not necessary to repeat the local process

0-7803-3177-6 $5.00 © 1996 IEEE

many times to reduce the local wire congestion, our algorithm is faster than the existing method especially for dense gate arrays, whose initial routing results often have the local wire congestion. Our experimental results on 1 MG SOG and MCNC's PRIMARY1-GA benchmark circuit support the effectiveness of our rip-up and reroute algorithm.

The remainder of this paper is organized as follows: In section 2, we define the routing problem. In section 3, the 'global' and 'local' rip-up and reroute algorithm is outlined. In section 4, the results of our implementation are presented.

2. Problem description

A routing region is an area inside rectangular boundaries that have n layers available for routing. For each layer, there are non-uniform pitched grid lines in horizontal and vertical direction superimposed over the entire region (See Fig. 1). A cell terminal and an obstacle (i.e., intra-cell routing pattern) are located inside the region on arbitrary layers and have rectilinear shapes. A net is a set of terminals to be connected. A route of a net connects all the terminals of the net, and is a set of wire segments on the grid lines and/or vias on the grid points. Each via connects segments on the neighboring layers. On each routing layer, a segment is allowed to run either horizontally or vertically. Design rules are following different types of constraints: segment width, via size, segment-to-segment spacing which is the minimum allowable edge-to-edge separation, segment-to-via spacing and via-to-via spacing. In automated routing, we must check to see whether objects on adjacent grids satisfy the design rules or not. For the maze router, such a check can be done on each grid point in the wave expansion phase by the method of literature [11]. Most nets are already routed by an initial routing process and the remainings are left unrouted. The prerouted nets can be ripped up and rerouted. The problem is to find routes for all the nets within the given routing region without causing any design rule violations.

The whole routing region is divided into a two-dimensional array of subregions each of which is h grid high and w grid wide. For every subregion, the congestion value is calculated and fragments of grid line for the "coarse" maze router are selected, as will be described later. We also define a local routing window centered on each violation of interest, which is n grid high and m grid wide. The local process performs the rip-up and reroute of existing wiring only within the window. Figure 2 shows an example of a subregion and a local routing window. In the sections that follow, examples are restricted to a single layer model.

3. Algorithm

The entire algorithm is described in Subsection 3.1 and the "coarse" maze routing algorithm is described in Subsection 3.2.

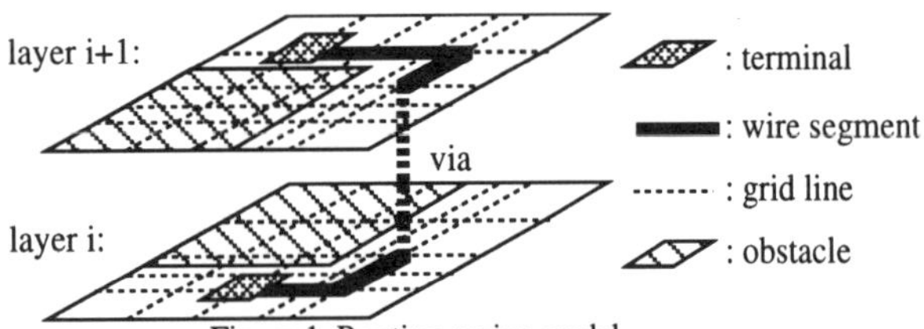

Figure 1. Routing region model.

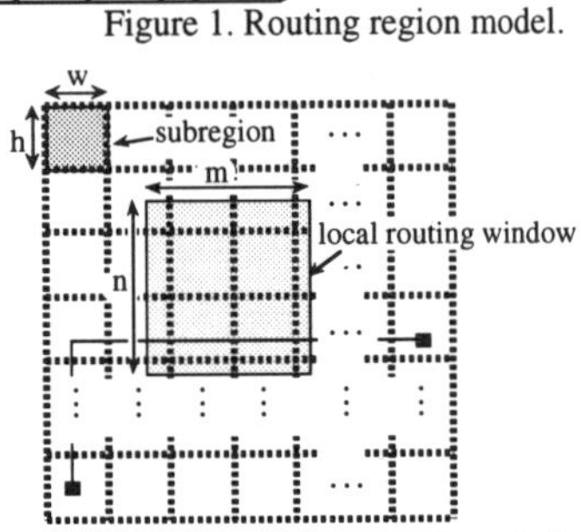

Figure 2. Subregion and local routing window.

3.1 The rip-up and reroute algorithm

The algorithm consists of two main processes: a local rip-up and reroute process that eliminates the violations mainly caused by routing order dependency and a global rip-up and reroute process that reduces local wire congestion. Figure 3 shows the overall flowchart of the algorithm.

As a preprocessing step, "routing of unconnects with violations" generates an initial wiring for all the unconnects by the fast "coarse" maze router described in the following Subsection 3.1. It completes all the routing allowing for some resulting violations and thereby detects where are the blocking spots.

Then a local phase eliminates the violations. The local process builds a local routing window centered on each violation and rips up and reroutes the prerouted wires within the window using a statistical optimization technique[10]. The local process is iterated until either all the violations are eliminated or all the subregions, includes violation, are processed for preset times. If the local phase fails to eliminate all the violations, a "global and local" phase takes over.

In the "global and local" phase, we define the congestion of subregion as the number of violations in it. The reason for this definition is that most violations caused by the routing order dependency have been eliminated by the local phase, and the primary contributor to the remaining violations is considered local wire congestion. To eliminate all the violations while reducing the local wire congestion, the "global and local" phase combines (1) the global rip-up and reroute process and (2) the local rip-up and reroute process. (1) The global process rips up a net that goes through the subregion of the highest congestion and reroutes it around subregions of high congestion. The two major features of the global process are a rerouting method using the fast "coarse" maze router and a "rip-up measure" that is used to select a net to be ripped up. We should give higher "rip-up measure" to the net with higher probability of being able to successfully reduce the congestion of the subregion of interest. There are two useful guidelines for computing the "rip-up measure". The first is: Rip-up the wiring that causes the congestion of the subregion. The second is: Rip-up the wiring that has high probability to be able to reroute around the subregion. As a representation of these guidelines, we introduce the following factors related to the rip-up measure of a net: the number of violations it involves in the subregion, the number of its terminals in the subregion, the half perimeter length of the minimum bounding box of the net's routing, and the number of rip-up times. The reason of introducing the number of rip-up times is to prevent a net from being ripped up many times. Then, the local process is invoked again to eliminate the violations within the subregions

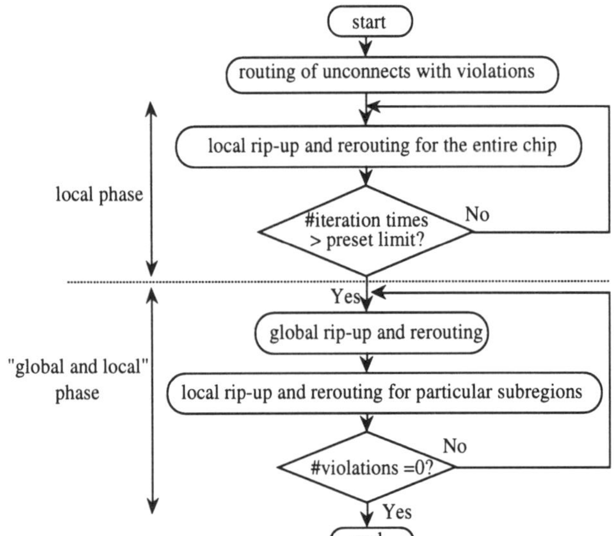

Figure 3. Overall flowchart of rip-up and reroute.

where the path, of the net of interest, which existed before and exists after the execution of the last global process, passes through. The chance that the local process will be successful is rather high, because the number of existing wires within the congested subregion has been reduced by the global process. The "global and local" phase is iterated until either all the violations are eliminated or the number of iteration times exceeds a preset limit. Consequently, because the algorithm eliminates violations by the local process while reducing the local wire congestion by the global process, it can efficiently eliminate violations even in the routing region with the local wire congestion.

3.2 The "coarse" maze routing algorithm

In this subsection, we explain the "coarse" maze routing. The two major features of the router are a fast wave expansion method based on some selected fragments of the grid line and a cost definition.

Note that the routing region is divided into small subregions. For each subregion, we single out 1 to 2 promising horizontal and vertical fragments of grid line. These fragments are a subset of the fragments each of which runs across a subregion (See Fig. 4(a)). We compute the total number of vacant grid points on each fragment. Each vacant grid point is not occupied by route, terminal or obstacle. Then, we select the fragments with the greatest number of vacant grid points on them for each subregion. However, we cannot find a path on these selected fragments, because the fragments in adjacent subregions do not cross each other. Therefore, we extend all the selected fragments to the opposite end of the adjacent subregions. These extended fragments form a fragmentary-coarse(FC) grid map for the "coarse" maze router (See Fig. 4(b)). We generate a minimum-cost path on the FC grid map with a maze routing algorithm. We expand the wavefront only on the FC grid map from the source terminals until the target terminal is reached. However, the wavefront can not reach to the target terminal in cases where no extended fragments of FC grid map cross the terminals, so that within all the subregions that include the terminals, we expand the wavefront on all the grid lines. The cost of a path is computed based on four measures: a wiring cost, a via cost, a violation cost and a congestion cost. In the cost of path calculation the wiring cost is proportional to the wire length (in grids), the via cost is proportional to the number of the via along the path and the violation cost is proportional to the number of the violations along the path. These are similar to the measures used by most maze routers. The congestion cost is proportional to the number of the existing violations within the subregions along the path. Note that the estimation of the congestion for a subregion is based on the number of existing violations in it. Therefore, with these measures, we can give a priority to the path that go around congested subregions. This method is quite fast because the sum of grid points over all the FC grid map is much fewer than that of grid points over all the grid lines. Figure 5 shows an example of wires on the FC grid map.

A similar approach has been presented in Codar[6]. It selects some coarse grid lines for an entire routing region and finds a path on them. However, due to the following two features, it is fair to consider our "coarse" maze router to be a new method. (1) Codar is suitable for the small routing problems and is difficult to apply to the large routing problems which are the focus of our algorithm. Larger routing region makes the length of coarse grid lines longer, therefore, more blockages are on the coarse grid lines and more violations are introduced on the Codar's resulting path. However, it is no problem for our algorithm, because it selects fragments of grid line for each subregion and the length of them is independent of the size of the routing region. (2) Codar introduces some violations in cases where there are no legal solutions, then following process eliminates all the violations. However, because it does not consider the location of the resulting violations, it may introduce many violations within the congested areas and it may spend much time to eliminate the violations within the congested areas. Because our algorithm gives priority to the path that go around the congested areas in wave expansion phase, it introduces few violations within the congested areas. The violations within less congested areas can be eliminated faster.

4. Evaluation

We have implemented the rip-up and reroute algorithm in C programming language as part of the HGALOP (high-speed gate array layout program) system[3] for CMOS SOGs. The system is installed on 133 MIPS SPARC station 20 which has 600 MB of virtual memory. We have run the rip-up and reroute program with the following three environmental parameters from our experience: (1)The size of a subregion is set to as same as that of global cell[2] which is the basis of the congestion estimation for the global routing. (2) The size of a local routing window is set to 3 by 3 subregions. (3) The iterative times of executing the local rip-up and reroute process at the local phase is set to 3.

The circuits used in our evaluation are shown in Table 1. The PRIMARY1-GA in Table 1 is a benchmark circuit provided by MCNC which uses double-metal layers for routing. The other circuits in Table 1 are CMOS triple-metal-layer SOGs. The results for the circuits produced by our rip-up and reroute algorithm are shown in Table 1. Figure 6 shows the routing result of circuit A in Table 1. Placement and initial routing have been performed by the HGALOP system. However, because HGALOP system can easily complete all the routing requirements of PRIMARY1-GA within the given base array, we have placed some cells of PRIMARY1-GA randomly to introduce some unconnects. The required routing is completed in all the circuits by our algorithm.

Then, in order to compare our rip-up and reroute algorithm with other approaches, we have applied the following three algorithms to all the circuits in Table 1 after connecting all the unconnects, introduced by the initial routing, with the "coarse" maze router: (1) a flat rip-up and reroute algorithm[10] which is applied to the entire chip, (2) the local rip-up and reroute algorithm which is our local phase described in Subsection 3.1 and (3) our 'global' and 'local' rip-up and reroute algorithm described in Subsection 3.1. However, we have not been able to apply method (1) to the circuits in Table 1 except PRIMARY1-GA, because it requires over 600 MB of virtual memory for the circuits. Method (1) required 70 minutes to eliminate all the violations in PRIMARY1-GA. Figures 7 and 8 depict the reduction of the number of violations while applying the method (2) and (3) versus the CPU time in cases of the circuit A and B in Table 1, respectively. These experimental results show the following: (i) The

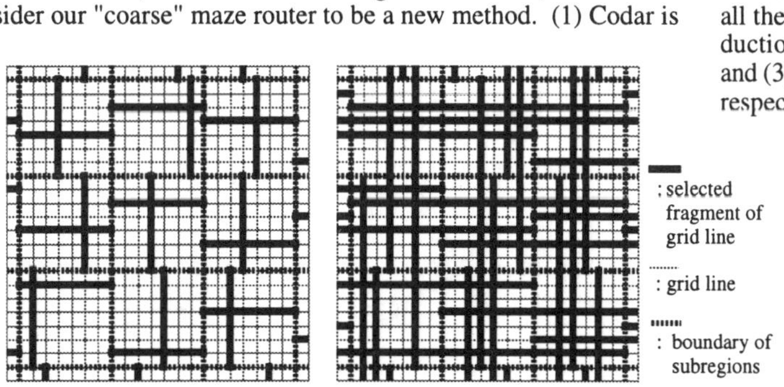

(a)Independent fragments of grid line. (b)Extended fragments of grid line.

Figure 4. An example of FC grid map.

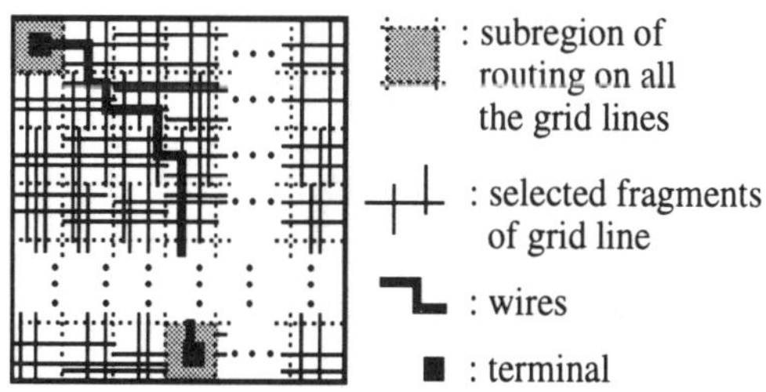

Figure 5. An example of wires on the FC grid map.

method(1) is not applicable to large size gate arrays, because it requires too much memory and time. With our implementation, we have been able to run only the smallest routing problem (PRIMARY1-GA) with 600 MB of virtual memory. The processing time is about 4 times longer than that of method(3). (ii) It is difficult to eliminate all the violations of dense gate arrays by the local rip-up and reroute alone (method(2)). (iii) Our algorithm (method(3)) which integrates the 'global' and 'local' rip-up and reroute, efficiently, eliminates all the violations introduced by either local wire congestion or routing order dependency.

Finally, we compare the rip-up and reroute algorithm to two commercially available routing tools A and B developed by a CAD vender. Tool A is considered to be one of the top of the field. The circuits we have used for this comparison are the same as in Table 1. After routing all the unconnects of the circuits using the "coarse" maze router with violations, we have run tools A and B to eliminate the resulting violations. The results are provided in Table 2. Our rip-up and reroute algorithm is faster than that of vendor's tools A and B. Our algorithm is about 3 times faster than that of tool A. Therefore, we can say that our algorithm is more efficient.

5. Conclusion

In this paper, we have reported an efficient rip-up and reroute algorithm for large scale gate arrays, which combines global rip-up and reroute and local rip-up and reroute. It achieved, on average, more than 3 times faster execution than two commercially available routers.

Table 1. Results of our rip-up and reroute algorithm.

Circuit		A	B	C	D	PRIMARY1-GA
Base array size (raw-gates)		400K	400K	400K	1000K	0.18K
#of nets		28K	38K	48K	142K	0.9K
#of grid points per layer		21M	21M	21M	57M	7.8K
#of uncon. of the initial routing		66	97	106	173	59
#of ripped up nets	the local process	46,371	60,949	99,697	211,320	28,267
	the global process	22	17	60	74	11
CPU time [min]	the routing of uncon.	2.39	4.58	7.88	6.77	0.26
	the rip-up & reroute	38.37	34.95	67.70	240.59	23.24
#of resulting violations		0	0	0	0	0

Table 2. Results of two CAD vendor's tools for the same circuits in Table 1.

Circuit		A	B	C	D	PRIMARY1-GA
Tool A	CPU time[min]	92.23	100.80	217.16	703.77*	fail**
	#of resulting violations	0	0	0	59	-
Tool B	CPU time[min]	185.25	169.71	284.71	514.28*	fail**
	#of resulting violations	0	0	0	138	-

*) The program has stopped because of running out of memory.
**) We don't know the reason. It may be bug.

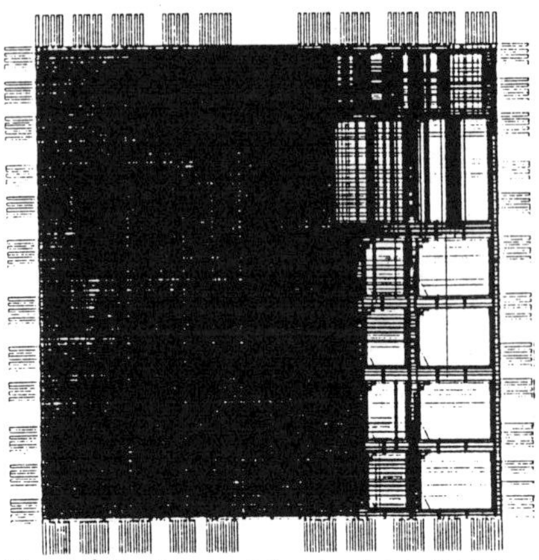

Figure 6. Routing result for circuit A in Table1.

Acknowledgement

The authors wish to thank Dr. K. Sato of Mitsubishi Electric Corporation for his encouragement and valuable comments.

References

[1] N. Ikeda, A. Ishibashi, H. Maeno, S. Matsue, K. Asahina, T. Arakawa, and S. Kato, "0.5μm 1M gate CMOS SOG," CICC, 1993, no.23.2.

[2] B. S. Ting, and B. M. Tien, "Routing techniques for gate array," IEEE Trans. CAD, vol. CAD-2, no.4, pp.301-312, 1983.

[3] M. Terai, K. Takahasi, H. Shirota, and K. Sato, "A new efficient routing method for channel-less sea-of-gates arrays," CICC, 1994, no.28.3.

[4] C. Y. Lee, "An algorithm for path connections and its applications," IRE Trans. Electronic Comput., vol. EC-10, pp.346-365, 1961.

[5] H. Shin, and A. Sangiovanni-Vincentelli, "Mighty: a rip-up and reroute detailed router," ICCAD, 1986, pp. 2-5.

[6] P. S. Tzeng, and C. H. Sequin, "Codar: a congestion-directed general area router," ICCAD, 1988, pp.30-33.

[7] M. Raith, and M. Bartholomeus, "A new hypergraph based rip-up and reroute strategy," 28th DAC, 1991, pp.54-59.

[8] E. Rosenberg, "A new iterative supply/demand router with rip-up capability for printed circuits boards," 24th DAC, 1987, pp.721-726.

[9] A. Fujimura, "Automating the layout of very large gate arrays," VLSI SYS. Apr 1988, pp.22-27.

[10] Y. L. Lin, Y. C. Hsu, and F. S. Tsai, "A detailed router based on simulated evolution," ICCAD, 1988, pp.38-41.

[11] M. H. Arnold, and W. S. Scott,"An interactive maze router with hints," 25th DAC, 1988, pp.672-676.

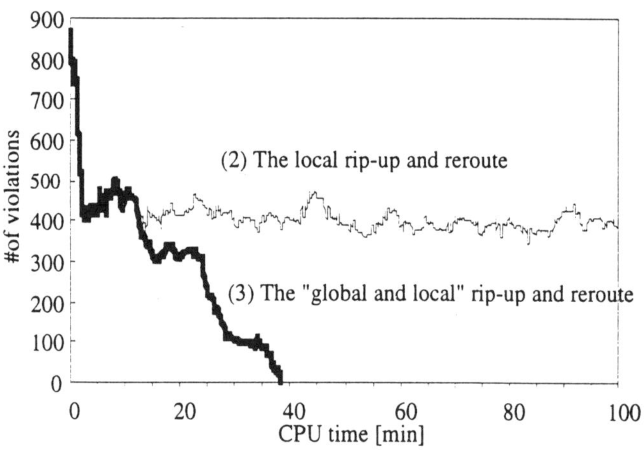

Figure 7. Results for circuit A(#of violations vs. CPU time).

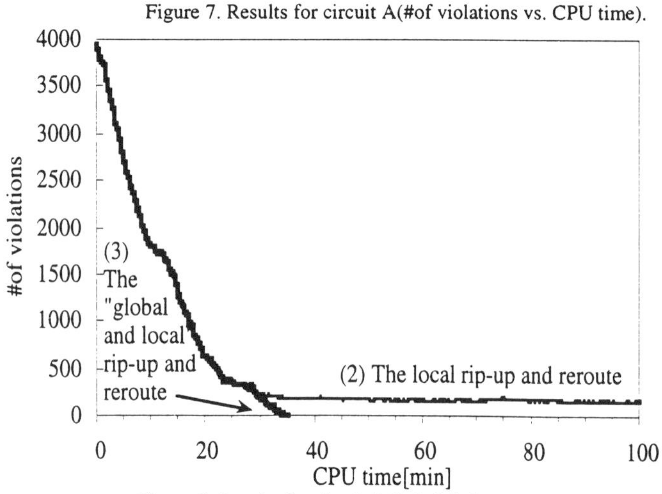

Figure 8. Results for circuit B(#of violations vs. CPU time).

Analog Routing for Manufacturability

K. Lampaert, G. Gielen*, W. Sansen

Katholieke Universiteit Leuven, Dep. Elektrotechniek, ESAT-MICAS
Kardinaal Mercierlaan 94, B-3001 Heverlee, Belgium
Tel.: +32/16/321077 - Telefax: - +32/16/321986
*research associate of the National Fund of Scientific Research

Abstract

The goal of a performance-driven routing tool is to route an analog circuit such that the performance degradation caused by layout parasitics remains within the specification margins imposed by the designer. For a given set of circuit specifications, several valid routing solutions can be found. In this paper, we propose an algorithm that selects the solution that additionally maximizes the yield and the testability of the resulting layout. Initially, the circuit is routed with a cost function designed to enforce all performance constraints. After all nets have been routed, the layout parasitics are extracted and the performance of the circuit is verified. In a second phase, nets are ripped up and rerouted to optimize the yield and the testability of the layout. During this process, care is taken not to introduce performance constraint violations. An industrial example is presented to demonstrate the effectiveness of the approach.

1. Introduction

Routing is an important problem in analog design automation. The performance of an analog circuit is critically dependent on the parasitics introduced during the routing phase and it is therefore necessary to control their influence and to keep the performance degradation within the limits imposed by the designer and his application. Routing tools which try to accomplish this are called performance-driven routers.

Several approaches to performance-driven routing have been proposed. In [1] sensitivities of performances to wiring parasitics are computed and a set of bounding constraints for these parasitics is determined. By enforcing this fixed set of parasitic constraints during routing, a layout respecting the high level performance constraints can be obtained. This algorithm was extended with a technique for dynamically updating the bounds in [2] and support for RF constraints was added in [3]. In [4] a technique for parasitic-bounded routing was presented. In this approach, parasitically nonviable paths are pruned as the search progresses.

In our router, we have implemented a performance-driven algorithm similar to the ones mentioned. A line-expansion algorithm with a cost function designed to minimize performance degradation is used as basic routing mechanism. To evaluate the performance degradation induced by a partial path, we use a linear approximation based on precalculated sensitivities. These sensitivities are also used to determine the net routing order and to control a rip-up and reroute mechanism.

Experiments with this algorithm showed that it guarantees a solution respecting the performance constraints if one exists. Moreover, in all but the tightestly constrained cases, several valid solutions could be found by varying the control parameters of the algorithm. Although these solutions are equivalent in terms of performance degradation, their quality differs substantially if yield and testability are taken into account. These observations suggest an extension of the performance-driven routing algorithm to also include yield and testability effects.

In the past, a number of routing algorithms have been proposed to increase circuit yield [5, 6]. However, these algorithms were designed for digital channel routers only and did not consider the testability of the circuit. In this paper we present a performance-driven routing tool which, while bounding the performance degradation within the allowed margins, also minimizes the defect sensitivity and increases the testability of an analog circuit. In section 2 we give an overview of our routing tool that includes these manufacturability considerations. In section 3 we derive a model to extract the number of bridging faults between two circuit nodes for a given layout. In section 4 we discuss the power-supply current monitoring technique used for testing analog circuits, its relation to the layout of a circuit and the modifications to the routing algorithm necessary to improve its effectiveness. Experimental results are presented in section 5 and conclusions are given in section 6.

2. Overview

The flow diagram of the routing tool is shown in Fig. 1. The input of the tool consists of a netlist file, placement information, a set of circuit specifications and a technology file. Before the actual routing tool is started, a circuit analysis tool is called. This tool

0-7803-3177-6 $5.00 © 1996 IEEE

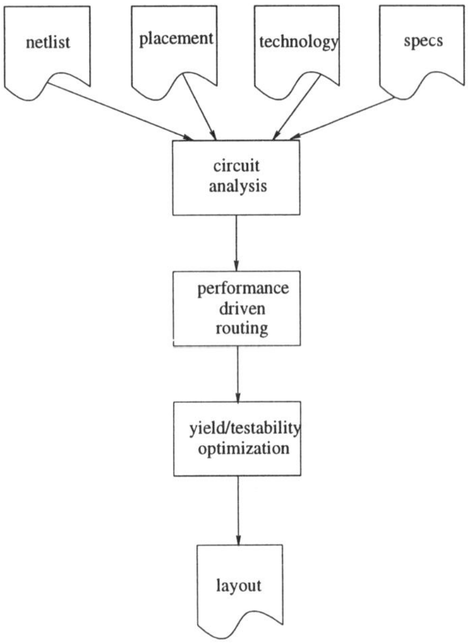

Fig. 1: Program flow of routing tool

interfaces to a numerical simulator and determines the performance sensitivities and operating point information about the circuit.

After circuit analysis, a performance-driven routing algorithm is used to wire the layout such that the performance degradation introduced by layout parasitics remains within the specifications imposed by the designer. The line expansion algorithm [7] is used as the basic path finding mechanism. The router starts with a collection of source cells, out of which the best cell is chosen and expanded in all possible directions and on all possible layers. These new cells are checked for design rule violations and the design rule correct cells are added to the routing cell collection. Out of the new collection, the best cell is chosen again and the process of cell selection, expansion, design rule check is repeated until one of the partial paths hits a target terminal. The cost function used to select the best cell for expansion is comprised of two terms :

$$C(cell) = F(source, cell) + P(cell, target) \quad (1)$$

where $F(source, cell)$ is the cost of the path from the source to the cell and $P(cell, target)$ is an estimation of the cost from the cell to one of the target terminals. $F(source, cell)$ is a measure of the performance degradation induced by the partial path:

$$F(source, cell) = \sum_{i=1}^{N_s} \left(\sum_{j=1}^{N_p^k} S_j^i p_j \right) \quad (2)$$

where N_s is the number of performance specifications, N_p^k the number of parasitics associated with net k and S_j^i the sensitivity of the ith performance to parasitic p_j. This cost function minimizes the performance degradation introduced by the net. The order in which the nets are routed is based on a heuristic estimation of their impact on the performance. This basic routing mechanism has been augmented with a rip-up and rerouting mechanism to ensure a fully functional layout as a starting point for the yield/testability optimization.

If the performance-driven routing phase was succesful and there is enough performance margin left, a yield and testability optimization loop is entered. During this loop, nets are removed and rerouted until the available performance margin is consumed or no further yield/testability improvement is found. The algorithm used for the rerouting of nets differs in two ways from the one used during initial performance-driven routing.

- Instead of minimizing performance degradation, partial paths that introduce specification violations are detected during the search and pruned from the search heap. During rerouting of a net, the geometry of all other nets is known. Therefore, the additional performance degradation introduced by a partial path can be computed exactly and if it exceeds the available performance margin, the partial path can be removed from the search heap. In this way, it is guaranteed by construction that no specifications are violated during yield/testability optimization.

- The cost function is changed to favor paths that optimize the manufacturability of the circuit. In equation (1), $F(source, cell)$ is replaced by:

$$F(source, cell) = \sum_{i=1, i \neq k}^{n} n_{f,ki} P_{ki} \quad (3)$$

with n the number of nets and k the net which is rerouted. In this equation, $n_{f,ki}$ denotes the expected number of bridging faults between circuit node k and i and P_{ki} is a factor indicating the probability that a bridging fault between node k and i will remain undetected by the power-supply current monitoring test technique for analog circuits [9]. Equation (3) is crucial for our routing approach and will be explained in the next sections.

3. Yield estimation

In this section, we derive a model to extract the expected number of bridging faults between two nodes of a circuit. This number is calculated by combining the defect density and the critical area, i.e. that portion of the circuit active area in which the occurrence of a defect results in a failure.

Defects have been classified in two categories in the literature [8]. First are *dielectric pinholes*, very small

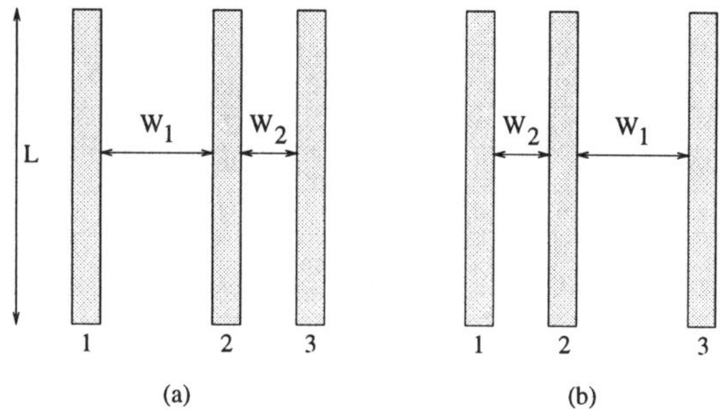

Fig. 2: Layout fragments

defects which often occur in chip insulators. Their occurrence can result in a short between wires at different routing levels. The critical area associated with these defects is the overlap region between two wires. The expected number of faults caused by dielectric pinholes n_f^{ph} can thus be determined as the total overlap area A_{ov} between wires at different photolithographic levels multiplied by the average pinhole density $\overline{D_{ph}}$:

$$n_f^{ph} = A_{ov}\overline{D_{ph}} \tag{4}$$

The second class of defects are photolithographic defects, for which the defect size becomes of importance. These defects can cause shorts between wires on the same routing level. The expected number of faults for two parallel conductors of length L, separated by a narrow slit of width w can be calculated by combining the critical area in function of defect size and a defect size distribution with the following integral:

$$n_f^{pd} = \int_0^{\infty} A(\chi)D(\chi)d\chi \tag{5}$$

The result was computed in [8] :

$$n_f^{pd} = \frac{L\chi_0^2\overline{D_{pd}}}{2w} \tag{6}$$

where $\overline{D_{pd}}$ is the average photolithographic defect density and χ_0 a process-dependent parameter. By combining equations (4) and (6), the expected number of faults between any two circuit nodes can be extracted from the layout.

4. Testability

Based on the expected number of faults, it is impossible to distinguish between the two layout fragments depicted in Fig. 2. In both cases, the expected number of faults is equal to $L\chi_0^2\overline{D_{pd}}(\frac{1}{2W_1} + \frac{1}{2W_2})$. However, from a testability point of view, it is possible that faults between nodes 1 and 2 are easier to detect than faults between nodes 2 and 3 and in that case (b) is to be preferred over (a). To enable the router to make intelligent decisions in cases like this, we need a measure of detectability to weigh the faults. Such a measure is developed in this section for the power-supply current monitoring technique [9].

We denote the power-supply current (I_{PS}) spectrum by the m-dimensional vector $\phi = (\phi_0, \phi_1, ..., \phi_{m-1})^T$, where ϕ_0 is the RMS value and $\phi_1, ..., \phi_{m-1}$ are the first $m-1$ harmonics. During the design stage of the circuit, the mean vector μ_0 and the covariance matrix Σ_0 of ϕ for the fault-free circuit are estimated by performing a Monte-Carlo analysis. During the testing stage of the circuit, a time-domain measurement of the power-supply current is performed, followed by a discrete frequency transformation to obtain a sample of ϕ. It can be shown that, in the assumption of Gaussian distributions of ϕ, the vector $(\phi-\mu_{(0)})^T\Sigma_{(0)}^{-1}(\phi-\mu(0))$ is distributed according to the χ^2-distribution with m degrees of freedom and that the circuit under test is fault-free with probability $(1-\alpha)$ under the following condition :

$$(\phi - \mu_{(0)})^T\Sigma_{(0)}^{-1}(\phi - \mu(0)) \geq \chi^2(\alpha) \tag{7}$$

where α is a user-supplied constant. The effectiveness of the above technique strongly depends on how separable the I_{PS} spectrum distribution of the fault-free circuit is from the I_{PS} spectrum distributions of all possible types of faulty circuits. In a circuit with n nodes, there are in general $\frac{1}{2}n(n-1)$ types of shorts possible between the nodes. Denote by $\mu_{(k)}$ the mean vector of the RMS value and the harmonics of the I_{PS} spectrum for the kth type of faulty circuit. The statistical distance between the spectrum of the good circuit and that of the circuit with fault k can be defined as [10]:

$$d_{0,k} = (\mu_{(0)} - \mu_{(k)})^T\Sigma_{(0)}^{-1}(\mu_{(0)} - \mu_{(k)}) \tag{8}$$

The smaller the value of $d_{0,k}$, the higher the chance that fault k will remain undetected during testing. We can thus put P_{ki} in equation (3) equal to the inverse of the statistical distance between the I_{PS} spectrum of the good circuit and that of the circuit with nodes k and i shorted.

The cost function term (3) can thus be interpreted as the sum of the expected number of faults between net k and all other nets i, multiplied by the probability that they will remain undetected during testing. Minimizing this term for each net means optimizing the yield and the testability of the overall circuit.

5. Experimental results

The CMOS operational amplifier shown in Fig. 3 was used to test this algorithm. During circuit analysis, Monte-Carlo analysis was used to determine the statistical distance between the I_{PS} spectrum of the good circuit and that of all possible faulty circuits.

The circuit was then routed using the described algorithm. The resulting layout is shown in Fig. 4. The performance and yield/testability characteristics for the circuit after layout are given in Tab. 1. The first column

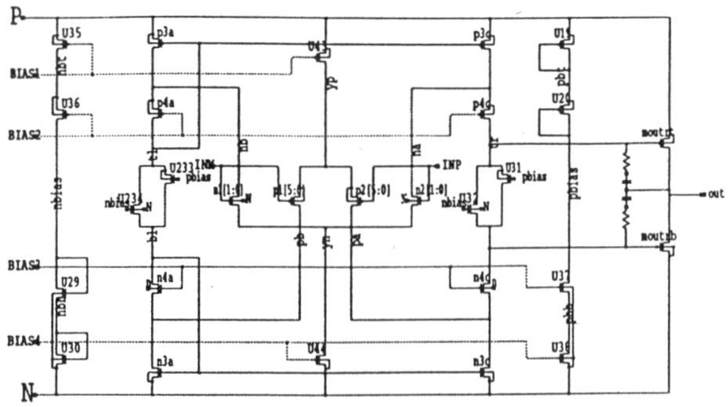

Fig. 3: Test circuit : CMOS operational amplifier

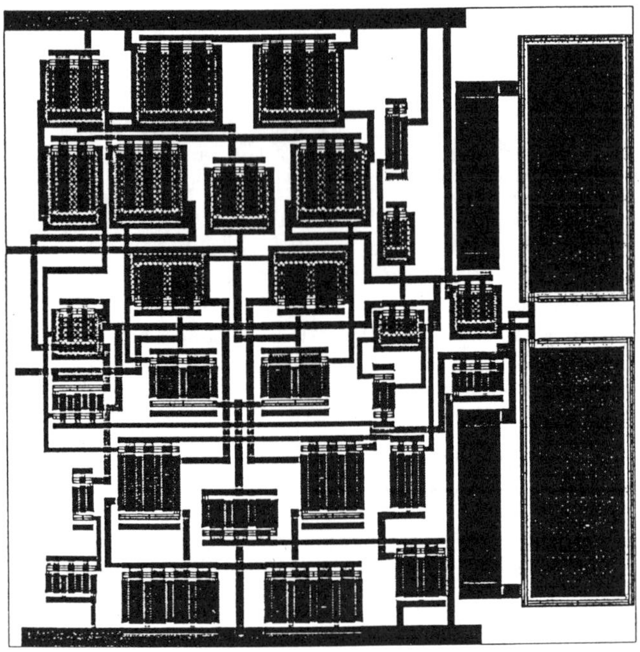

Fig. 4: Layout of the amplifier optimized for yield and testability

gives the results after the performance driven stage of the algorithm and the second column gives the results after yield/testability optimization. The performance characteristics given are the unity-gain bandwidth (UGBW) and the phase margin (PM). To evaluate the testability of the layout, a large number of circuits was tested using the I_{PS} monitoring technique. Every type of fault was represented in the test set with a probability calculated from the layout using the technique described in section 3. The test error rate (TER) is defined as the percentage of faulty circuits that are classified as good circuits by the test algorithm. It can be seen from Tab. 1 that the test error rate is significantly improved by the algorithm, while the performance remains within the specifications.

Performance	Spec	Stage 1	Stage 2
$UGBW(MHz)$	>7	7.13	7.06
$PM(^\circ)$	>70	70.3	70.2
Test Error Rate(%)		0.21	0.03

Table 1: Performance and Test Error Rate for the opamp after performance-driven routing (Stage 1) and after yield and testability optimization (Stage 2).

6. Conclusions and Future Work

A yield/testability optimizing router was presented. A circuit is routed in two steps. During the first step, the circuit is routed using a performance-driven routing algorithm. After the performance of the circuits has been verified, the remaining performance margin is used to optimize the yield and the testability of the layout. An industrial circuit was used to demonstrate that the yield and the testability of a layout is signicantly improved while the performance remains within the specifications.

Acknowledgments

Parts of this research are supported by ESA-ESTEC, the CEC ESPRIT project ADMIRE and Rockwell Semiconductor Systems.

References

[1] E. Malavasi, A. Sangiovanni-Vincentelli, "Area Routing for Analog Layout," IEEE Trans. Computer-Aided Design, vol. 12, pp. 1186-1194, Aug. 1993.

[2] E. Malavasi, A. Sangiovanni-Vincentelli, "Dynamic Bound Generation for Constraint-Driven Routing," Proc. IEEE CICC, pp. 477-480, May 1995.

[3] E. Charbon, G. Holmlund, B. Donecker, A. Sangiovanni-Vincentelli, "A Performance-Driven Router for RF and Microwave Analog Circuit Design," Proc. IEEE CICC, pp. 383-386, May 1995.

[4] B. Basaran, R. Rutenbar, L. R. Carley, "Latchup-Aware Placement and Parasitic-Bounded Routing of Custom Analog Cells," Proc. ACM/IEEE ICCAD, pp. 415-421, Nov. 1993.

[5] A. Pitaksannonkul, S. Thanawastien, C. Lusinsap, J. A. Gandhi, "DTR: A Defect-Tolerant Routing Algorithm," Proc. 26th ACM/IEEE DAC, pp. 795-798, June 1989.

[6] S.-Y. Kuo, "YOR: A Yield-Optimizing Routing Algorithm by Minimizing Critical Areas and Vias," IEEE Trans. Computer-Aided Design, vol. 12, pp. 1303-1311, Sep. 1993.

[7] W. Heynes, W. Sansen, H. Beke, "A Line-Expansion Algorithm for the General Routing Problem with a Guaranteed Solution," Proc. ACM/IEEE DAC, June 1980.

[8] C. Stapper, "Modeling of Defects in Integrated Circuit Photolithographic patterns," IBM J. Res. Develop., vol. 28, pp. 461-475, July 1984.

[9] G. Gielen, Z. Wang, W. Sansen, "Fault Detection and Input Stimulus Determination for the Testing of Analog Integrated Circuits Based on Power-Supply Current Monitoring," Proc. ACM/IEEE ICCAD, pp. 495-498, Nov. 1994.

[10] T. Anderson, "An Introduction to Multivariate Statistical Analysis," John Wisley and Sons, 1958.

Exact Multi-Layer Topological Planar Routing

Olivier Coudert
Synopsys, Inc., 700 East Middlefield Rd.
Mountain View, CA 94043

C.-J. Richard Shi
University of Iowa, ECE Department
Iowa City, Iowa 52242

Abstract—This paper describes an exact algorithm for multi-layer topological planar routing in switchboxes and channels. Using recent developments in set covering resolution, this method produces in a few minutes the optimum solutions for routing problems that were previously solved by heuristics.

I Introduction

A *routing region* (Fig. 1) is an area enclosed by an external boundary, with some blocks inside the boundary (called holes), and pins on the external and internal boundaries. A routing area is a *switchbox* when there is no hole, and a *channel* when pins are located on the top and bottom of the boundaries. A *net* is a set of pins to be connected without going through any boundary. A *planar subset* is a set of nets that can be routed in one layer without crossing each other. The maximum k-layer problem consists of routing as many nets as possible with k planar subsets. The minimum α-cover problem consists of routing at least a fraction α of the nets with as few planar subsets as possible.

These problems are motivated by MCM, PCB and performance-driven IC routing. For submicron VLSI designs, wire delays dominate gate delays, and the routing area exceeds the area used by transistors. Thus multi-layer routing technology is of increasing importance in such a high performance design. Polynomial algorithms have been developed for special cases, but the general problem is NP-complete [2].

The approach we propose to multi-layer topological routing is made of two steps:

(a) compute all the maximal planar subsets.
(b) solve a set covering problem.

Not only is the set covering problem NP-complete, but also the number of maximal planar subsets can be exponential w.r.t the number of nets. However we show that these two steps can be solved *exactly* with a low computational cost in practice.

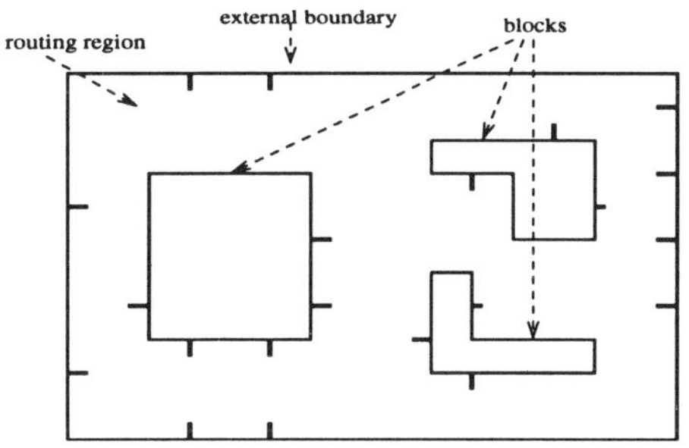

Figure 1: A routing region.

In this paper, step (a) is addressed in Section II for switchboxes or channels only. Section III shows how step (b) can be solved using a new lattice based paradigm. Section IV explains how this paradigm can be efficiently implemented with ZBDDs (Zero suppressed Binary Decision Diagrams). Experimental results are discussed in Section V.

II Maximal Planar Subsets for Switchboxes and Channels

Two nets are *compatible* iff they can be routed in a single layer without crossing. In the case of switchboxes or channels, sets of mutually compatible nets (called clusters) are exactly the planar subsets. Consider Fig. 2, where pins with the same number form a net. Nets 1, 2 and 3 form the cluster $\{1,2,3\}$, and these three nets can be routed in a plane without crossing each other. The maximal clusters are $\{1,2,3,4\}$, $\{1,2,5\}$ and $\{3,4,6\}$. The minimum layer routing is $\{\{1,2,5\},\{3,4,6\}\}$.

If we see the set of nets V as a set of vertices, and the set of compatible pairs E as a set of edges, then a cluster is a clique of the undirected graph (V, E). Thus step (a) amounts to computing all the maximal cliques of this graph.

0-7803-3177-6 $5.00 © 1996 IEEE

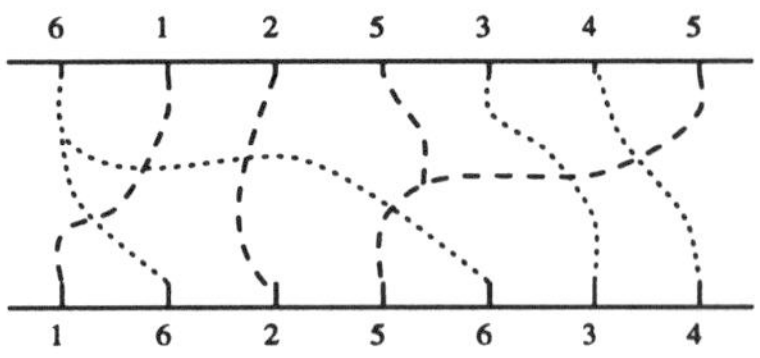

Figure 2: A topological routing channel.

Given the set U of *uncompatible* pairs, a cluster is a set of nets that does not contain any elements of U. Thus we can express the set of all clusters *Cluster* by:

$$\begin{aligned} \mathit{Cluster} &= \mathit{NotSupSet}(2^V, U), \quad \text{where} \\ \mathit{NotSupSet}(X, Y) &= \{x \in X \mid \forall y \in Y,\ y \not\subseteq x\}. \end{aligned}$$

The set of all maximal planar subsets *MaxPlanSet* is:

$$\begin{aligned} \mathit{MaxPlanSet} &= \max_{\subseteq} \mathit{Cluster}, \quad \text{where} \\ \max_{\subseteq} X &= \{x \in X \mid \forall x' \in X, x \subseteq x' \Rightarrow x = x'\}. \end{aligned}$$

III Solving Set Covering

Let X and Y be two sets, and R a binary relation on $X \times Y$. An element y of Y *covers* an element x of X iff $x\ R\ y$. The *set covering problem* $\langle X, Y, R\rangle$ consists of finding a minimum subset S of Y such that any x of X is covered by some y of S. Step (b) amounts to solving:

$$\langle V, \mathit{MaxPlanSet}, \in\rangle. \tag{1}$$

The classical approach consists of building the corresponding covering matrix, and of solving it using the Quine–McCluskey procedure [1]. The size of the covering matrix is in $O(|V| \times |\mathit{MaxPlanSet}|)$, which limits this approach since $|\mathit{MaxPlanSet}|$ can be exponential (> 100000 in some of our examples) w.r.t. $|V|$.

To overcome this limitation, we use the results developed in [3]. The method consists of transforming the original set covering problem (1) into an isomorphic set covering problem (2) expressed on a complete lattice. Here, an obvious translation is:

$$\begin{aligned} &\langle X, \mathit{MaxPlanSet}, \subseteq\rangle, \\ \text{where} \quad X &= \{\{v\} \mid v \in V\}. \end{aligned} \tag{2}$$

X and *MaxPlanSet* are subsets of the lattice $(2^V, \subseteq)$. Applying the fundamental result of [3, Theorem 8], the following reduction rules:

$$\begin{aligned} \langle X, Y, \subseteq\rangle &\rightarrow \langle X, \max_{\subseteq} \tau_X(Y), \subseteq\rangle, \\ \langle X, Y, \subseteq\rangle &\rightarrow \langle \max_{\subseteq} \tau_Y(X), Y, \subseteq\rangle, \quad \text{and} \\ \langle X, Y, \subseteq\rangle &\rightarrow \langle X - E, Y - E, \subseteq\rangle \quad \text{with } E = X \cap Y, \end{aligned}$$

where τ_X and τ_Y are defined from 2^V into 2^V by:

$$\begin{aligned} \tau_X(y) &= \bigcup_{\substack{x \in X \\ x \subseteq y}} x \\ \tau_Y(x) &= \bigcap_{\substack{y \in Y \\ y \supseteq x}} y \end{aligned}$$

are applied on (2) to yield a new set covering problem, say (3), which is isomorphic to the cyclic core of (1). From the solutions of (3), one can recover the solutions of (1) using basic set operations [3]. The interest of such an approach is that the reduction to the isomorphic cyclic core (3) can be done with ZBDDs in an *implicit* way, i.e., with a complexity that does not depend on the number of elements of X and Y.

IV Algorithms

ZBDD (Zero suppressed Binary Decision Diagram) is a graph data structure to represent subsets of 2^V in a compact and canonical way. We have shown that steps (a) and (b) can be solved by using some complex set operations defined on 2^V. Indeed, these set operations can be efficiently implemented with ZBDDs, so that their costs are not related to the size of the sets they operate on, but to the size of the ZBDDs representing these sets. In practice, very large sets can be represented with small ZBDDs (e.g., the ZBDD of 2^V has size $O(|V|)$), which enables us to solve effectively the multi-layer problems.

Given $V = \{v_1, \ldots, v_n\}$, let a *combination* be a subset of V, and X be a set of combinations, i.e., $X \subseteq 2^V$. The set X can be decomposed w.r.t to an element v_k in two unique sets X_{ε_k} and X_{v_k}, where: X_{ε_k} is the set of combinations belonging to X that do not contain v_k; X_{v_k} is made of the combinations belonging to X that contain v_k, from which v_k has been removed. For instance, with $X = \{\{\}, \{v_1, v_2\}, \{v_1, v_3, v_5\}, \{v_4\}\}$ and $k = 1$, we have $X_{\varepsilon_1} = \{\{\}, \{v_4\}\}$ and $X_{v_1} = \{\{v_2\}, \{v_3, v_5\}\}$.

If X is recursively decomposed w.r.t. the elements of V by applying the decomposition

$$X = \mathit{vertex}(v_k, X_{\varepsilon_k}, X_{v_k}),$$

function *NotSupSet*(X, Y);
if $X = \{\}$ **or** $\{\} \in Y$ **or** $X = Y$ **return** $\{\}$;
if $X = \{\{\}\}$ **or** $Y = \{\}$ **return** X;
let $T1 = NotSupSet(X_{v_k}, Y_{\varepsilon_k}) \cap NotSupSet(X_{v_k}, Y_{v_k})$
    $T0 = NotSupSet(X_{\varepsilon_k}, Y_{\varepsilon_k})$ **in**
    return $vertex(v_k, T0, T1)$;

function *MaxTauRow*(X, Y);
if $Y = \{\}$ **or** $X = \{\}$ **return** $\{\}$;
if $Y = \{\{\}\}$
  if $\{\} \in X$ **return** $\{\{\}\}$ **else return** $\{\}$;
if $X = \{\{\}\}$ **and** $\{\} \in Y$ **return** $\{\{\}\}$;
let $X1 = NotSubSet(X_{\varepsilon_k}, Y_{\varepsilon_k})$
    $T1 = MaxTauRow(X1 \cup X_{v_k}, Y_{v_k})$
    $T0 = MaxTauRow(X_{\varepsilon_k} - X1, Y_{\varepsilon_k} \cup Y_{v_k})$ **in**
    return $vertex(v_k, NotSubSet(T0, T1), T1)$;

function *MaxSet*(X);
if $X = \{\}$ **or** $X = \{\{\}\}$ **return** X;
let $T1 = MaxSet(X_{v_k})$
    $T0 = MaxSet(X_{\varepsilon_k})$ **in**
    return $vertex(v_k, NotSubSet(T0, T1), T1)$;

function *MaxTauCol*(Y, X);
if $Y = \{\}$ **or** $X = \{\}$ **return** $\{\}$;
if $Y = \{\{\}\}$
  if $\{\} \in X$ **return** $\{\{\}\}$ **else return** $\{\}$;
if $X = \{\{\}\}$ **return** $\{\{\}\}$;
let $Y0 = NotSupSet(Y_{v_k}, X_{v_k})$
    $T0 = MaxTauCol(Y0 \cup Y_{\varepsilon_k}, X_{\varepsilon_k})$
    $T1 = MaxTauCol(Y_{v_k} - Y0, X_{\varepsilon_k} \cup X_{v_k})$ **in**
    return $vertex(v_k, NotSubSet(T0, T1), T1)$;

Figure 3: Set operations needed to solve steps (a) and (b) implicitly.

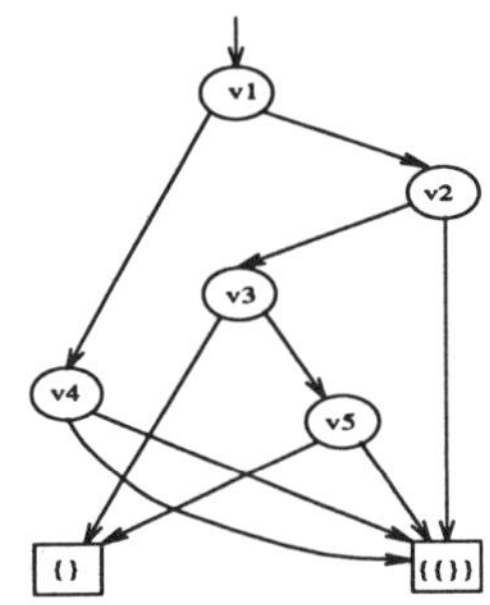

Figure 4: A ZBDD.

one obtains a binary tree whose leaves are $\{\}$ and $\{\{\}\}$. When representing sets of *sparse* combinations, most of the right branches (which denote the subsets of combination where v_k's do not occur) points to $\{\}$. This suggests the following compression scheme [4]: (1) isomorphic subgraphs are shared in memory; (2) a vertex whose right branch points to $\{\}$ is removed and replaced with the vertex pointed by its left branch. The resulting graph is the ZBDD of X, and is canonical w.r.t. to the ordering of the elements of V. Fig. 4 shows the ZBDD of the set X given above.

Fig. 3 describes the complex set operations needed to solve steps (a) and (b) in terms of basic set operations (union, intersection, *vertex*, etc). *MaxSet*(X) evaluates $\max_\subseteq X$, *MaxTauRow*(X, Y) evaluates $\max_\subseteq \tau_Y(X)$, and *MaxTauCol*(Y, X) evaluates $\max_\subseteq \tau_X(Y)$, with both X and Y subsets of the lattice $(2^V, \subseteq)$. Function *NotSubSet*, defined as $NotSubSet(X, Y) = \{x \in X \mid \forall y \in Y,\ x \not\subseteq y\}$, can be implemented in a way similar to *NotSupSet*. All these operations use a divide-and-conquer strategy on the decompositions of the arguments' ZBDDs w.r.t. their top labels v_k.

V Experimental Results and Conclusion

Table 1 presents some experimental results on multilayer topological routing problems. The CPU time includes reading the uncompatible pairs, building the ZBDD of the maximal planar subsets, and solving the covering problem.

Computing the ZBDD of the maximal planar subsets does not take more than 15 seconds. Note that the size of the ZBDD (i.e., its number of vertices) remains small, even when the number of maximal planar subsets is large.

The cyclic core is implicitly computed using the ZBDD based method presented in Section III and IV. The set covering problem is reduced by a factor > 20. The cyclic core is then explicitly solved using the minimizer SCHERZO [3].

The difference between the greedy solution [2] and the optimum solution is not significant. However, the interest of this exact approach is as follows:

- Our formulation only uses two aspects of the

example			max planar subset		CC		Sol		
Name	**#net**	**#uncomp**	**|ZBDD|**	**#plan**	**row**	**col**	**greedy**	**exact**	**CPU**
bus	24	133	62	45	6	8	9	9	0.13
ex1	21	77	71	50	9	14	9	7	0.11
ex3a	44	176	149	495	16	41	11	10	0.44
ex3b	47	283	139	7685	31	804	10	9	1.85
ex3c	54	336	224	3329	32	621	13	12	1.46
ex4b	54	298	524	6885	31	535	13	11	2.05
ex5	64	405	282	61269	47	6749	9	9	135.36
ex5b	64	427	5988	44938	46	4773	11	10	38.05
deut	72	763	196	101427	50	5777	18	16	18.49
exam1	200	17124	734	6711	107	1512	135	126	51.02
exam2	250	26081	1061	19409	146	1897	150	141	159.53
exam3	300	36801	2294	100520	228	13316	176	162	597.53

For each **example**, **#net** is the number of nets, and **#uncomp** the number of uncompatible pairs.
|ZBDD| is the size (number of vertices) of the ZBDD representing the **#plan** maximal planar subsets.
The cyclic core **CC** has size **row** × **col** (note that the size of the explicit initial covering problem is **#net** × **#plan**).
Sol gives the **greedy** and **exact** solutions for the minimum 1-cover problem.
The **CPU** time to solve exactly the problem is given in seconds on a 60 MHz SuperSparc (85.4 SpecInt).

example	k = 1		k = 2		k = 3		k = 4		k = 5	
bus	10/10	41%/41%	14/14	58%/58%	17/17	70%/70%	19/19	79%/79%	20/20	83%/83%
ex1	8/8	38%/38%	11/12	52%/57%	14/15	66%/71%	16/17	76%/80%	17/19	80%/90%
ex3a	20/21	45%/47%	26/27	59%/61%	30/31	68%/70%	33/34	75%/77%	35/36	79%/81%
ex3b	15/15	31%/31%	25/26	53%/55%	32/33	68%/70%	37/37	78%/78%	39/40	82%/85%
ex3c	20/20	37%/37%	30/31	55%/57%	34/36	62%/66%	38/40	70%/74%	41/43	75%/79%
ex4b	22/22	40%/40%	31/31	57%/57%	37/38	68%/70%	41/43	75%/79%	44/46	81%/85%
ex5	20/20	31%/31%	31/32	48%/50%	38/42	59%/65%	45/49	70%/76%	50/55	78%/85%
ex5b	20/20	31%/31%	31/32	48%/50%	39/41	60%/64%	46/48	71%/75%	51/54	79%/84%
deut	17/17	23%/23%	28/29	38%/40%	36/39	50%/54%	42/46	58%/63%	46/50	63%/69%

The first column under each k describes the number of nets routed by greedy vs optimum algorithm.
The second column describes the percentage of nets routed by greedy vs optimum algorithm.

Table 1: Experimental results: minimum 1-cover problem and maximum k-cluster problem.

problem: (1) whether two nets can be routed in the same layer or not, which is a local information; (2) for switchboxes and channels, a planar subset is a set of mutually compatible nets. Since our method does not exploit the special structure of the problem, it can be extended to handle the general case.

- The method can handle practical design constraints, like the physical capacity constraint, i.e., the maximal number of nets that are allowed to be routed in one layer.

References

[1] R.K. Brayton, G.D. Hachtel, C.T. McMullen, A.L. Sangiovanni-Vincentelli, *Logic Minimization Algorithms for VLSI Synthesis*, Kluwer Academic Pub., Dordrecht, 1984.

[2] J. Cong, M. Hossain, N. A. Sherwani, "A Provably Good Multi-Layer Topological Planar Routing Algorithm in IC Layout Designs", *IEEE Trans. CAD*, **12**-1, pp. 70–78, Jan. 1993.

[3] O. Coudert, "2-Level Logic Minimization: An Overview", *Integr.*, **17**-2, pp. 97–140, Oct. 1994.

[4] S. Minato, "Zero-Supressed BDDs for Set Manipulation in Combinatorial Problems", *Proc. 30th DAC*, Dallas, TX, pp. 272–277, June 1993.

[5] C.-J. Shi, J. A. Brzozowski, "A Framework for the Analysis and Design of Algorithms for a Class of VLSI-CAD Optimization Problems", *Proc. ASP-DAC*, pp. 67–74, Chiba, Japan, Aug. 1995.

Chair: Don Green
Co-Chair: Rick Carley

The burgeoning growth of commercial, RF wireless communications is fueling the demand for low cost, compact, and power efficient circuitry. The development of Si ICs for analog and digital functions has proceeded rapidly. Although small scale integration has been achieved for some RF functions, implementations of RF circuits in UHF and microwave frequency bands still rely heavily on discrete active and passive components with microstrip interconnect on a circuit board. RF transceiver size reductions, for products such as cellular telephones, have largely resulted from shrinking component package sizes for RF components, not from increases in RF integration level. Similarly, RF transceiver cost reductions are largely the result of higher production volumes and economies of scale, not reduced component count.

This session presents seven papers, reporting work on overcoming some design barriers for monolithic integration of RF functions.

- The first paper discusses optimization of monolithic (passive) inductors on Si ICs with acceptable Q-factors in the 1-3 GHz range.
- The second and third papers report results for 1.8 and 2.5 GHz monolithic filters using passive inductors with active elements for tuning and Q-enhancement in Si BJT and BiCMOS technologies.
- The fourth paper demonstrates 1 GHz CMOS mixer performance using a 0.7μm CMOS process.
- The fifth paper shows the integration of RF signal processing circuitry to linearize a transmitter.
- The sixth paper demonstrates the feasibility of a bipolar only sampled data delta-sigma modulator.
- The final paper is a harbinger of future integration trends, presenting an entire 900 MHz, spread-spectrum, RF transmitter IC, fabricated in a 1μm CMOS process.

Continuing improvements in IC transistor performance at high frequencies, along with the development of new circuit and system level design approaches, can be expected to substantially alter the realization of RF functions for wireless communication products in the near future.

Modeling, Characterization and Design of Monolithic Inductors for Silicon RFICs.

John R. Long and Miles A. Copeland*

University of Toronto
Toronto, Ontario, Canada, M5S 1A4.
*Carleton University, Ottawa, Canada.

Abstract

The results of a comprehensive investigation into the characteristics and optimization of inductors fabricated in the top-level metal of a sub-micron silicon VLSI process are presented. A computer program which extracts a physically-based model of microstrip components which is suitable for circuit (SPICE) simulation has been used to evaluate variations in metallization, layout geometry and substrate parameters upon inductor performance. 3-D numerical simulations and experimental measurements of inductors were also used to benchmark the model accuracy. It is shown in this work that low inductor Q is primarily due to the restrictions imposed by the thin interconnect metallization available in most VLSI technologies, and that computer optimization of the inductor layout can be used to achieve a 50% improvement in component Q-factor over unoptimized designs.

Introduction

Optimization and refinement of silicon circuit designs incorporating on-chip inductors and transformers has been identified by others [1] as lacking in the present state of the design art. In order to achieve this goal, a circuit model for the inductor is required for the efficient simulation and optimization of RFICs. A computationally efficient, scalable, lumped-element model can be applied to an arbitrary configuration of microstrip lines, such as a spiral inductor[2,3]. The accuracy of such a lumped-element model is evaluated in this paper for variations in metallization thickness, layout geometry and substrate parameters, using a combination of 3-D numerical simulations and experimental measurements. The results of this investigation highlight areas where process improvements and parameter optimization can be applied to maximize the RF performance of silicon-based inductor designs.

The Monolithic Inductor Model

A spiral microstrip transmission line is the most popular method of realizing an on-chip inductor using a single layer of metallization, because the spiral configuration allows a long transmission line to be used without consuming a large chip area. The spiral inductor structures, while compact and more space efficient than a straight wire inductor, are more difficult to analyze. The accurate characterization of these structures at microwave frequencies requires an analysis of the fringing fields, parasitics, groundplane effects and most importantly for silicon IC design, an analysis of the effect of the conductive substrate on the performance of the component. These effects cannot be fully analyzed to yield closed form expressions which adequately predict the inductor behaviour.

The physical length of a microstrip line on an RFIC is typically less than 1mm, and so a lumped element model can be used to approximate the electrical behaviour of each conductor segment of a spiral inductor. However, the analysis of the electric and magnetic coupling between a group of conducting lines still requires some numerical computation. A computer program, GEMCAP2 [2,4], was used to extract the electrical parameters for a lumped-element circuit model from the inductor specifications. The parameters of the lumped-element model for each segment are computed from the layout geometry, and the substrate and metallization properties, as described in a previous publication[4]. The lumped-element circuit model can then be used in a time domain (e.g., SPICE) or frequency domain (e.g., Touchstone) simulation along with other active and passive circuit components. The substrate and metallization parameters that are required for the extraction of the lumped-element circuit parameters are listed in Table 1, for the BiCMOS process [5] used to fabricate the inductors described in this paper. It should be noted that the lumped-element model is not limited to modeling spiral inductors, but can also be applied to the analysis of other microstrip-based components such as monolithic transformers.

Table 1: Microstrip Substrate Parameters for a VLSI-BiCMOS Process[5].

Parameter	Value
Substrate resistivity, ρ	10 Ohm-cm
Substrate thickness, b_2	380 μm
Silicon dielectric constant	11.7
Oxide thickness, b_1	5 μm
Oxide dielectric constant	3.9
Metal resistivity	0.03 Ohm-μm

Model Validation

Ohmic losses of an inductor fabricated on a silicon IC are caused by dissipation of energy in the semi-conducting substrate and by the ac resistance of the conductor

0-7803-3177-6 $5.00 © 1996 IEEE

IEEE 1996 CUSTOM INTEGRATED CIRCUITS CONFERENCE

metallization (which includes the skin effect). The quality factor or 'Q' of the inductor is given by the ratio of the inductive reactance to the total dissipation. The Q-factor obtained from a three-dimensional electromagnetic field simulation of a spiral inductor is a means of verifying the predictions of the lumped-element model. A commercial simulator for planar structures, EM-Sonnet, was used for the three-dimensional numerical simulations. A 4.5 turn square spiral 5nH inductor with a 10μm wide line and a 5μm line spacing was selected for this comparison. A small ratio of linewidth to line spacing was chosen to ensure that the memory requirements and simulation time for the numerical computations were not excessive. The substrate and metallization parameters from the BiCMOS process (Table 1) were assumed for the simulations. Excellent agreement is seen (refer to Figure 1) between the inductor Q-factor simulated using Sonnet (3D-EM) and the lumped-element model (GEMCAP2) over the frequency range of interest. The Q-factor predicted by GEMCAP2 for the same planar spiral on a lossless substrate (Ideal) is also plotted on the Figure for comparison. The slight downward curvature is the result of capacitive parasitics which cause self-resonance. Figure 1 demonstrates that the effect of the substrate conduction on the inductor Q-factor is significant at higher frequencies, but that operation close to the ideal curve is obtained for frequencies below approximately 1.5GHz. The ability to predict the substrate effects on the inductor performance is key to the design and realization of inductors with acceptable performance in silicon technology. When substrate effects are properly accounted for in a circuit simulation, the inductor design can be optimized to minimize these losses and maximize the circuit performance.

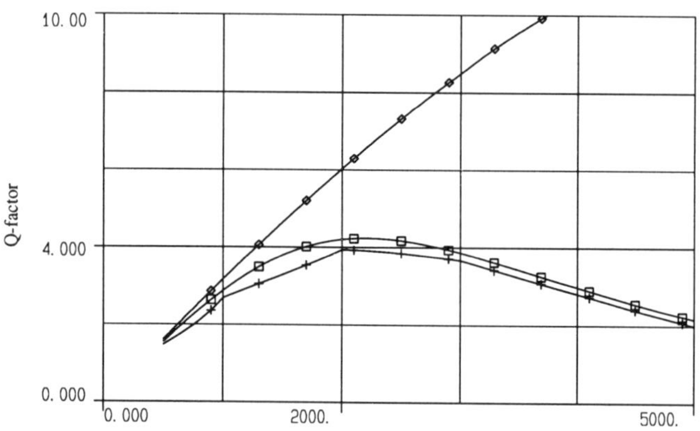

Figure 1: Simulated Q-factor for 5nH inductor (N_t=4.5, W=10μm, S=5μm)

Experimental Measurements

The effect of the layout geometry and the substrate parameters upon the inductor performance have been investigated experimentally. The results of the experimental measurements are useful as a benchmark to confirm the predictions of the lumped-element (i.e., GEMCAP2) computer model and also to provide insight into the possible optimization of different aspects of the inductor performance, such as the Q-factor.

Thin metallization in most VLSI processes limits the quality factor of microstrip inductors, because energy is dissipated by the finite resistivity of the metallization as well as in the conductive substrate. The quality of the spiral inductor, or Q-factor, could therefore be improved by increasing the conductor thickness. The measured quality factor for an 1.8nH spiral inductor fabricated with 1μm and 3μm thick metal layers is plotted in Figure 2. This inductor consists of 3.5 turns of 15μm wide top-level metal in the BiCMOS process with a line spacing of 1.5μm. The measured behaviour of the inductor fabricated with thicker metal shows close agreement with the predictions of the lumped-element circuit model. The Q-factor initially rises with frequency as the reactive component of the impedance increases, peaks, and then decreases due to the increasing energy dissipation at higher frequencies. The inductor Q-factor at 3GHz improves from 5 to 10 when the thickness of the aluminium metal layer is increased by from 1 to 3μm. The improvement in inductor Q is less than a factor of 3 because of the inductance is inversely proportional to the thickness of the conductor. Hence, the inductance decreases with increasing metal thickness, lowering the Q. The frequency dependence of the conductor resistance, or the skin effect, is more pronounced as the thickness of the metallization is increased, which also contributes to higher dissipation and a lower Q-factor at RF.

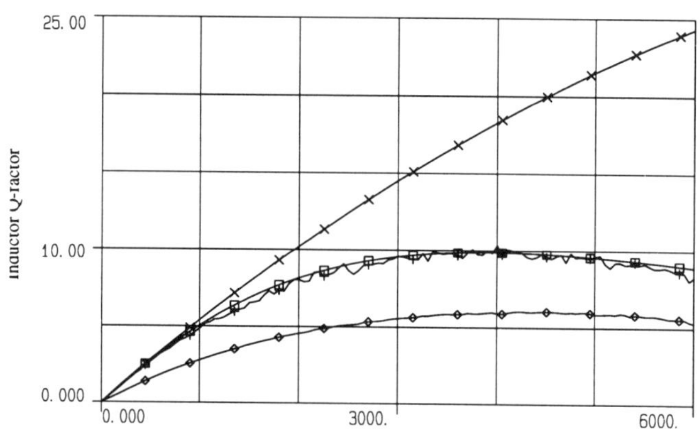

Figure 2: Q-factor for different metal thickness (1.8nH inductor).

The substrate conductivity also introduces a significant degradation in the component Q, as indicated by the large difference in Q-factor between the measured performance and that predicted for an ideal substrate from simulation. However, these results indicate that inductors with a Q-factor of 10, which is suitable for many RFIC applications, could be fabricated in silicon production technologies if a low resistivity metal of adequate thickness were available.

The effect of the spacing between conductor lines on inductor Q-factor has also been investigated. Narrow line spacings were found to increase the magnetic coupling between windings, which causes an increase in the inductance and the Q-factor for a given layout area. However, the interwinding capacitance also increases as the conductors are brought closer together. For the thin metallization used to fabricate on-chip inductors, the effect of the interwinding capacitance was found to have a negligible effect upon the performance.

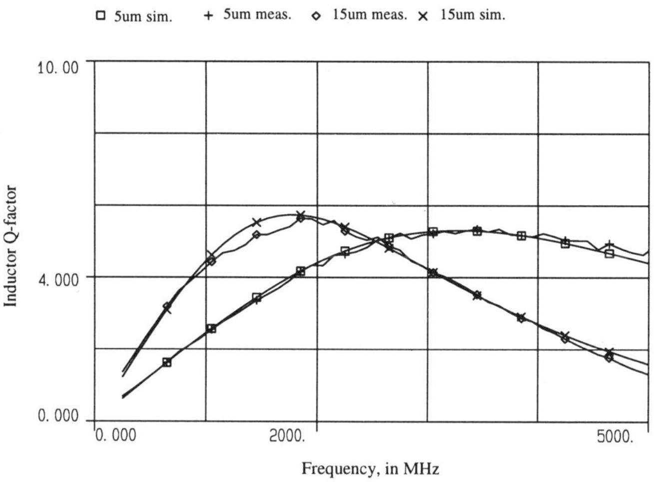

Figure 3: Q-factor as a function of metal width (L= 5nH).

The measured behaviour of the spiral inductor and the performance predicted by the lumped-element computer model for conductor widths of 5 and 15μm are shown in Figure 3. A layer of top-level aluminium 1μm thick was used to fabricate a 4.5-turn spiral with an inductance of 5nH in each case. The outside dimensions of the inductor were increased in order to maintain the inductance value constant in both cases. The larger surface area of the inductor with wide conductor metallization results in higher parasitic capacitances, which lowers the inductor self-resonant frequency. Thus, the peak of the inductor Q-factor shifts to a lower frequency as the conductor width increases. However, this can be used to advantage in the optimization of a given inductor for maximum Q within a desired range of frequencies.

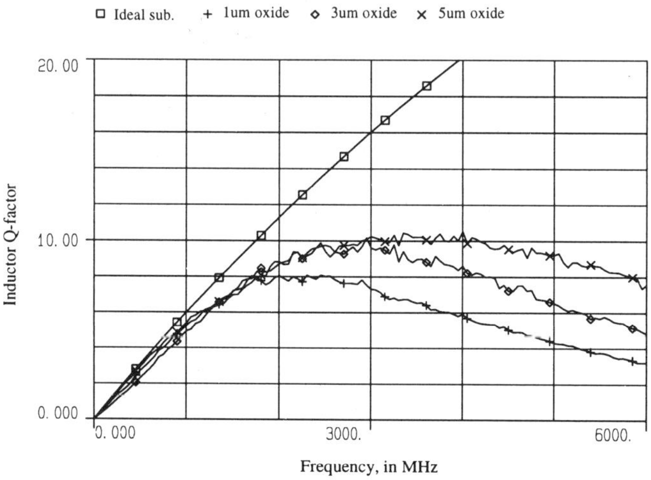

Figure 4: Q-factor as a function of oxide thickness (L= 1.8nH).

The dielectric oxide which separates the top level metal and the substrate is a key parameter influencing the inductor performance. The effect of changes in the oxide thickness upon the component Q-factor is illustrated in Figure 4, for the 1.8nH inductor discussed previously (refer to Figure 2, metal thickness of 3μm). A thicker oxide layer reduces the parasitic capacitance of the structure, which improves the inductor self-resonant frequency. This is seen as a broadening of the inductor Q-factor with frequency as the oxide thickness increases in Figure 4. Performance closer to that predicted for an ideal (i.e., non-conductive) substrate can be achieved when the oxide thickness is maximized.

The measured behaviour of the 5nH spiral inductor is shown in Figure 5 for two different substrate conductivities (10 and 0.01 Ohm-cm). A layer of aluminium 1μm thick and 10μm wide was used to fabricate the 4.5-turn spiral in both cases. At low frequencies, the resistive component of the impedance (Re[Z1]) is close to the dc resistance of the inductor, which is about 8.5 Ohms. As the frequency increases, the resistive component increases, primarily due to dissipation in the conductive substrate. As seen in the Figure, the dissipation of the inductor fabricated on a low resistivity substrate (0.01 Ohm-cm) is substantially greater than for the substrate resistivity normally used in the BiCMOS process (10 Ohm-cm). The peak Q-factor is reduced by a factor of 2 because of the increased dissipation, which makes the inductor unsuitable for most RF circuit applications.

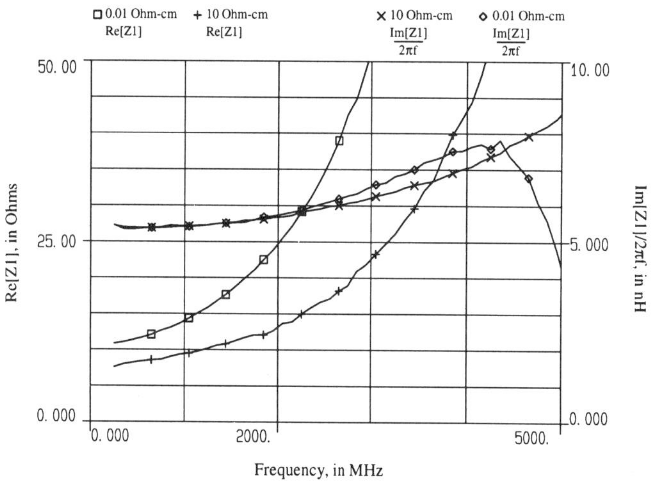

Figure 5: Measured one-port impedance versus substrate resistivity (L= 5nH).

In Figure 5, the inductance, which is given by the reactive portion of the measured impedance (Im[Z1]) divided by the frequency, is 5nH in each case at low frequencies. The parasitic capacitance between the metal and substrate is larger when a highly doped substrate is used to fabricate the inductor, which leads to a lower self-resonant frequency. The reactance of the inductor fabricated on the 0.01 Ohm-cm substrate rises to a peak at 4GHz and then falls rapidly as the inductor approaches self-resonance. However, the low frequency inductance is unaffected. Current flow in the substrate beneath the spiral would cause negative mutual coupling and thus a reduction in the low frequency inductance of the spiral. These results indicate that the substrate current that is induced by the magnetic field is small.

Design Rules

Design of an inductor to fit a particular application requires extensive simulation work in order to determine the number of turns, linewidth and line spacing necessary to achieve the best performance within a given technology. Extraction of a lumped-element component model is faster than other simulation methods, making this type of optimization more practical.

The results of this study have shown that a silicon technology with a substrate resistivity in the 10 Ohm-cm range and a 5μm thick oxide between metal and substrate can be used to fabricate inductors with acceptable Q-factors in the 1-3 GHz frequency range. Use of the minimum space that the process allows between conductors reduces the chip area required to fabricate the inductor and results in the highest component Q. The metallization thickness has a large effect upon the inductor quality factor, and a Q-factor of 5 is achievable with 1μm thick aluminum topmetal. However, a Q-factor of 10 can be achieved if the metal thickness can be increased to 3μm.

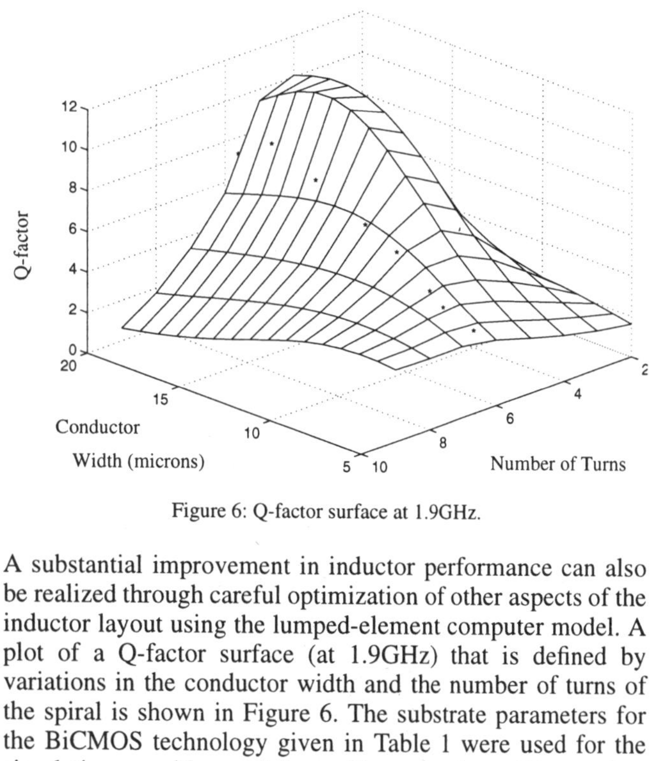

Figure 6: Q-factor surface at 1.9GHz.

A substantial improvement in inductor performance can also be realized through careful optimization of other aspects of the inductor layout using the lumped-element computer model. A plot of a Q-factor surface (at 1.9GHz) that is defined by variations in the conductor width and the number of turns of the spiral is shown in Figure 6. The substrate parameters for the BiCMOS technology given in Table 1 were used for the simulations, with a 1.5μm line spacing. The points corresponding to an inductance of 5nH are plotted onto the surface with an asterisk. These points indicate that an optimum geometry can be identified which maximizes the inductor Q-factor for a given inductance. The spiral geometry which yields the highest Q-factor at 1.9GHz can also be identified from this Figure. The three-dimensional plot is a powerful design aid which can be used to visualize the effects of the many parameters which influence the inductor performance. A fast and efficient computer-based model is required in order to generate these types of design aids.

Conclusions

A lumped-element computer model adequately models the performance of inductors fabricated in a silicon VLSI process for a wide range of layout geometries. Both the measured and simulated results indicate that the quality factor of the inductor is dominated by losses in the metallization at lower frequency, and can be improved through the use of thicker metal in fabrication. Losses introduced by the conductive substrate tend to dominate performance at higher frequencies. However, these losses do not significantly degrade the performance in the low GHz frequency range, making these devices potentially useful in many RFIC applications. Dielectric losses can be minimized by proper selection of the conductor width, spacing and metal thickness for a given set of substrate parameters. The lumped-element modeling approach is computationally efficient, and can be used to develop optimized inductor and transformer designs for various RF circuit applications.

Acknowledgements

This work was supported by Micronet, the Natural Sciences and Engineering Research Council of Canada (NSERC) and the Telecommunications Research Institute of Ontario (TRIO). The authors appreciate fabrication support and services provided by Northern Telecom Ltd. and Bell-Northern Research Ltd. Special thanks to Anand Veluswami for his assistance in the preparation of graphics for this manuscript.

References

[1] P.R. Gray and R.G. Meyer, "Future directions in silicon ICs for RF personal communications", Proc. of the Custom Integrated Circuits Conference, Santa Clara, 1995, pp. 83-89.

[2] G.G. Rabjohn, Monolithic Microwave Transformers, M.Eng. thesis, Carleton University, April 1991.

[3] E. Frlan, S. Meszaros, M. Cuhaci and J. Wight, "Computer-Aided Design of Square Spiral Transformers and Inductors", in *Proceedings of the IEEE MTT-S*, Long Beach CA., June 1989, pp. 661-664.

[4] J.R. Long and M.A. Copeland, "Modeling of monolithic inductors and transformers for silicon RFIC design", Proc. IEEE MTT-S International Symposium on Technologies for Wireless Applications, Vancouver, Canada, February, 1995.

[5] R. Hadaway et al, "A Sub-Micron BiCMOS Technology for Telecommunications", *Journal of Microelectronic Engineering*, vol. 15, pp. 513-516, 1991.

1.8 GHz Tunable Filter in Si Technology

Spyros Pipilos[1], Yannis Tsividis [1,2] and Josef Fenk [3]

[1] Division of Computer Science, National Technical University of Athens, 15773 Zographou, Greece
[2] Department of Electrical Engineering, Columbia University, New York, NY 10027, USA
[3] Semiconductor Division, Siemens AG, 81541 Munich, Germany

Abstract

An active filter using integrated inductors has been implemented in Si bipolar technology. The filter has a second-order bandpass response, with quality factor tunable through Q enhancement and center frequency tunable through reactance multiplication. For a nominal center frequency of 1.8 GHz and a quality factor of 35, the filter has a 1 dB compression dynamic range of 40 dB, and draws 8 mA from a 3 V supply.

1. Introduction

There are points in a wireless communications system where RF frequency filtering with only moderate dynamic range is needed. Such is the case, for example, after the transmitter up-conversion mixer, where the transmitted signal has to be filtered in order to suppress the image and carrier signals. Conventional transconductance-capacitance filters do not perform adequately in the GHz range, because parasitic capacitances cannot be tuned out; in addition, their power dissipation may be prohibitive (1). One is then naturally lead to the classical use of inductors in lieu of active devices, to resonate with capacitances and provide zero phase shift at the resonant frequency. In standard Si technology, the quality factor (Q) values achievable in practical inductors in the 1–2 GHz range are small (2). To increase the value of Q, and to make it tunable, active Q enhancement has been proposed (3-8). It has also been proposed (9, 5, 6) to employ reactance multiplication (10) for frequency tuning. Both types of tuning are necessary for eliminating the effects of tolerances and environmental variations on the filter response, a task which should eventually be done by an on-chip automatic tuning system (not considered in this paper). With these enhancements, of course, the active circuits are back in the filter. However, the main role in resonance is still played by the inductor (in addition to the capacitor); the active circuits are only used to improve and adjust the performance. Thus the advantages of LC circuits can, at least partly, be maintained. In this paper, we report on results from an experimental chip employing the above techniques.

2. Basic Operation of the Circuit

The circuit used is shown in Fig. 1, and is a variation of circuits proposed (but not built) in (5, 6). Ignoring R_s for the moment, the elements L and C form parallel resonant tanks, which are the loads to the moderately emitter-degenerated differential pair Q1A, Q1B. This stage has a second-order bandpass response. By adjusting the tail current I_G of this stage, its effective trasnconductance G_{m1}, and thus the filter gain, can be varied.

To vary the resonant frequency, one could use a varactor diode if the process used makes such a diode possible with good performance (low losses and good linearity). In our case, we opted instead for reactance multiplication: a very small resistance R_s samples the current in each capacitor, and feeds a portion $G_{m2}R_s$ of that current back to the tank, where G_{m2} is the transconductance of the pair Q2A-Q2B. The effective capacitance thus becomes:

$$C_{eff} = (1 - G_{m2}R_s)C \tag{1}$$

and can be tuned by varying the tail current I_F of this stage. Thus, the center frequency ω_o of the filter becomes tunable.

The resonant circuit discussed so far is lossy, mainly due to the inductor and capacitor series resistances, including the current-sampling resistance R_s. The following analysis is very approximate due to the low Q values involved. Near the resonant frequency, the effect of these series resistances can be represented by an equivalent conductance G parallel with a lossless LC tank (11, 6). The Q of the circuit is then $Q_o = \omega_o C_{eff}/G$ (11), and is lower than the inductor quality factor. To enhance the value of Q, the losses are replenished by energy supplied from the active circuit consisting of Q3A, Q3B, Q4A, Q4B, and the two resistors R_Q. This circuit

0-7803-3177-6 $5.00 © 1996 IEEE

IEEE 1996 CUSTOM INTEGRATED CIRCUITS CONFERENCE

can be viewed as a resistive voltage-to-current converter, followed by a current mirror; it is a linearized transconductor (13), with an effective transconductance which will be denoted by G_{m3}. In the positive feedback connection shown, the stage forms a negative conductance of value $-G_{m3}$, which appears in parallel with G; thus the effective Q is now $\omega_o C_{eff}/(G - G_{m3})$, which can be written as follows:

$$Q_{eff} = Q_o/(1 - G_{m3}/G) \tag{2}$$

This can be varied by varying G_{m3} through the bias current I_Q of the stage. Thus, the filter Q becomes tunable and can be increased above the value of the inductor quality factor. Because tuning ω_o through C_{eff} also affects Q, the proper way to tune the circuit is to adjust ω_o through I_F first, and then adjust Q through I_Q. Depending on the accuracy and Q value required, it may be necessary to go through one more iteration. An automatic tuning circuit to achieve this is being planned.

In order to drive the package parasitics and the measurement circuitry, on-chip buffers were included; as seen in Fig. 1, each of these consists of two emitter followers in cascade.

3. Dynamic Range Issues and Design Considerations

A calculation gives the following mean square noise voltage across the tanks:

$$\overline{v_o^2} = \eta \frac{2kT}{C} \frac{Q}{Q_o} \tag{3}$$

where k is Boltzmann's constant, T is the absolute temperature, and η is a factor somewhat larger than 1 which depends on the amount of frequency tuning. For fixed Q and $\overline{v_o^2}$, we can use a capacitor inversely proportional to Q_o. Thus all impedance levels will be proportional to Q_o, which implies a proportionate advantage in power dissipation. This effect does not have its analog in $Gm - C$ filters, although the proportionality to the final quality factor, Q, is observed in those, too (1). A more detailed comparison of $Gm - C$ filters to the above approach can be found in (7). To keep Q_o as large as possible, R_s should be kept as small as possible. However, this means that for sufficient tunability G_{m2} in (1) will have to be large, which will increase the power dissipation.

If a gain larger than 1 is present between input and output, the subcircuit potentially most prone to distortion in Fig. 1 is the Q-enhancing circuit, since it handles the full output signal. Special attention must be paid to linearizing this circuit. Various possibilities exist (1); the topology adopted was chosen as a compromise between linearization, noise, power consumption, and tunability.

4. Experimental Chip

The circuit of Fig. 1 has been fabricated using the Siemens 0.8 μm bipolar production technology B6HF, with f_t of 25 GHz (12). The chip microphotograph is shown in Fig. 2. As seen, slightly more than half of the chip is occupied by the inductors. The inner part of the inductor was not filled, in order to achieve a satisfactory compromise between inductance value, quality factor, and self-resonance frequency. The inductors were laid out in 3d-level metal, and their inside terminal was contacted through multiple vias to a 2nd-level metal bridge. These inductors have a nominal value of 4 nH, and a quality factor of 6.5 at 1.8 GHz. The test chip was designed for large tuning range. In order to achieve this without too much power dissipation (see previous section), we had to use a rather large value (20 ohm) for R_s; these resistors were implemented using $p+$ diffusion. The active chip area was 0.4 mm^2.

5. Measurements

The filter, with its on-chip buffers, was ac-coupled on a printed circuit board using capacitors chosen for low series inductance between two baluns with proper termination, as shown in Fig. 3. *With the exception of the power dissipation, the results given below are for the entire combination of filter, internal buffers, and external circuitry.* They can thus be used as very conservative estimates of the performance of the filter by itself. We estimate the losses of the test circuitry to be 15 dB. Unless stated otherwise, all measurements are at room temperature.

By adjusting the bias currents as explained in Sec. 2, the results of Figs. 4 and 5 are obtained. It is seen that both center frequency and Q can be adjusted over a wide range; it fact, Q was found to be easily adjustable from 3 to 350. In the following, unless mentioned otherwise all measurements reported are for a center frequency of 1.8 GHz and a Q of 35. We have performed two-tone third-order intermodulation (IM3) distortion measurements, with the two tones and IM3 products in-band. *For the entire measurement setup of Fig. 3*, the input and output IM3 intercept points (11) were -16 dBm and -9 dBm, respectively. The output noise power spectral density at the center frequency was -131 dBm/Hz, and the integrated noise power was -52d Bm (550 μV rms). The "spurious-free dynamic range" (which takes as maximum input signal level the one producing IM3 product power equal to the noise power) and the "1-dB compression dynamic range", (which takes as maximum input signal level the one producing 1 dB compression at the output) are plotted in Fig. 6 and 7 as a function of Q. If a smaller tuning range is needed it should be possible to decrease the value of R_s and thus increase the dynamic range further.

The power dissipation is a function of center frequency and Q. For a center frequency of 1.8 GHz and a Q of 35, the filter (excluding buffers) draws a current of 8 mA. Keeping Q and w_o constant at 1.8 GHz and 25, respectively, the 1 dB compression dynamic range varied by about 0.5 dB for power supply voltages varying between 2.8 and 4.5 V, and by 6 dB for temperature variations between -10 and $+50^0C$.

6. Conclusions

A fully integrated, Si-technology second-order bandpass active RLC filter with a nominal center frequency of 1.8 GHz has been presented. The filter quality factor, which by itself would have been unacceptably low due to the high losses of the on-chip inductors, has been increased using parallel-mode Q enhancement circuitry; the same circuitry is also used for making the Q tunable from 3 to 350. Reactance multiplication is used to provide center frequency tunability from 1.6 to 2 GHz. Good performance is maintained for supply voltages between 2.8 and 4.5 V, and temperatures between -10 and $+50^0$C. At its nominal center frequency of 1.8 GHz and for a filter Q of 35, the filter has a 1 dB compression dynamic range of 40 dB, and draws a supply current of 8 mA. The performance of this filter makes it suitable for applications where the signal to be filtered is of large, stable amplitude, as is the case in the synthesizer of certain cellular radios.

Acknowledgements

The authors thank Messrs. F. Dielacher, S. Heinen, H. Herrmann, R. Marcus, and W. Veit for helpful discussions and support during the realization and evaluation of the chip.

References

(1) J. O. Voorman, "Continuous-time analog filters", in Y. P. Tsividis and J. O. Voorman, eds, *Integrated Continuous-Time Filters*, IEEE Press, New York, 1993.

(2) N. M. Nguyen and R. G. Meyer, "Si IC-compatible inductors and LC passive filters", *IEEE J. Solid-State Circuits*, vol. 25, pp. 1028-1051, August 1990.

(3) R. A. Duncan, K. W. Martin, and A. S. Sedra, "A Q-enhanced active-RLC bandpass filter", *Proc. 1993 Int. Symp. on Circuits and Systems*, Chicago, pp. 1416-1419, May 1993.

(4) Y. P. Tsividis, "Integrated continuous-time filter design", *Proc. 1993 Custom Int. Circuits Conference*, San Diego, pp. 6.4.1-6.4.7, May 1993.

(5) S. Pipilos and Y. Tsividis, "RLC active filters with electronically tunable center frequency and quality factor", *Electronics Letters*, vol. 30, pp. 472-474, March 1994.

(6) S. Pipilos and Y. Tsividis, "Design of active RLC integrated filters with application in the GHz range", *Proc. 1994 Int. Symp. on Circuits and Systems*, London, pp. 645-648, May-June 1994.

(7) W. B. Kuhn, F W. Stephenson, and A. Elshabini-Riad, "Dynamic range of high-Q OTA-and enhanced-Q LC RF bandpass filters", *Proc. 1994 Midwet Symp. on Circuits and Systems*, pp. 767-771, 1994.

(8) R. A. I. Duncan, *Active-RLC Filters in Silicon*, Ph.D. Dissertation, University of Toronto, 1995.

(9) R. A. I. Duncan, private communication.n.

(10) B. S. Camies, *Principles of Frequency Modulation*, Iliffe, London, 1959.

(11) H. L. Kraus, C. W. Bostian, and F. H. Raab, *Solid-State Radio Engineering*, Wiley, New York, 1980.

(12) H. Klose, R. Lachner, K. R. Schoen, R. Mohnkopf, K. H. Malek, M. Kerber, H. Braun, A. V. Felde, J. Popp, O. Cohrs, E. Benagnolli, and P. Sehrig, "B6HF: A 0.8 micron 25 GHz/25 ps bipolar technology for mobile radio and ultra fast datalink IC products", *Proc. 1993 Bipolar Circuits and Technology Meeting*, Minneapolis, pp. 125-127, 1993.

(13) W. J. A. de Heij, E. Seevinck, and K. Hoen, "Transconductor and integrator circuits for integrated bipolar video frequency filters", *Proc. ISCAS'89*, pp. 114-117.

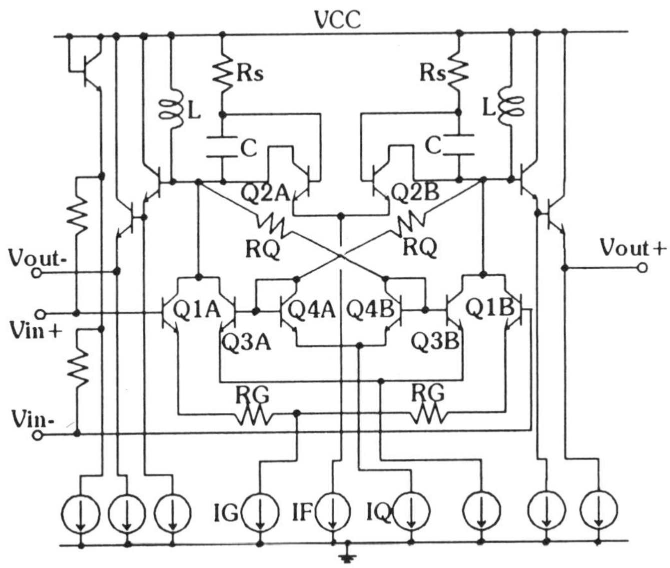

Fig. 1. Filter schematic diagram, including output buffers.

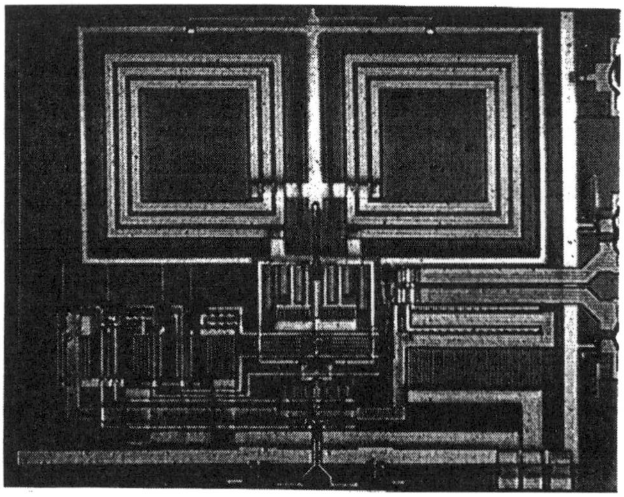

Fig. 2. Chip microphotograph.

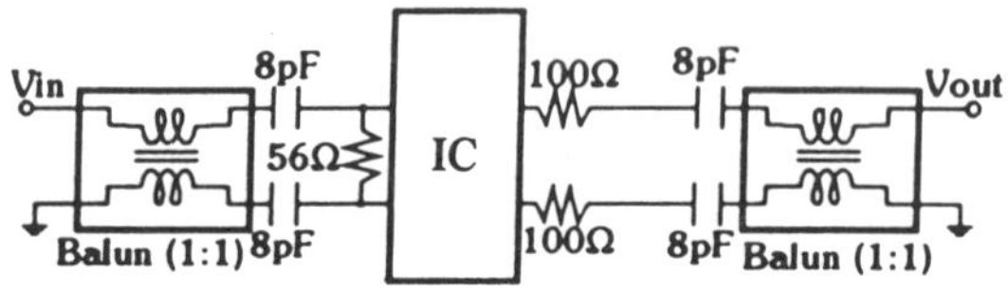

Fig. 3. Test setup.

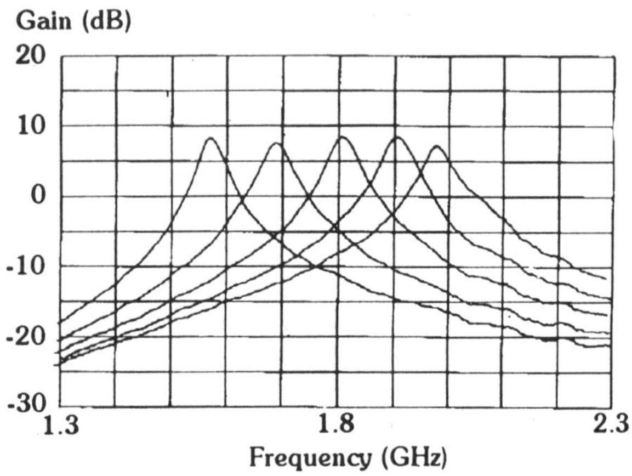

Fig. 4. Effect of varying the center frequency, while keeping Q constant, for the system of Fig. 3.

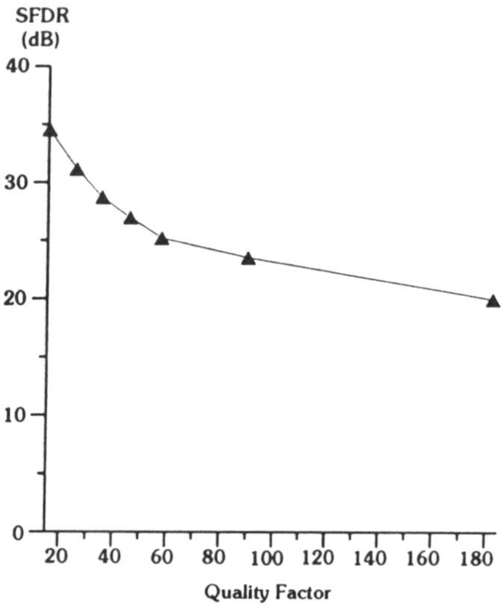

Fig. 6. Spurious-free dynamic range vs. Q for the system of Fig. 3, for a center frequency of 1.8 GHz.

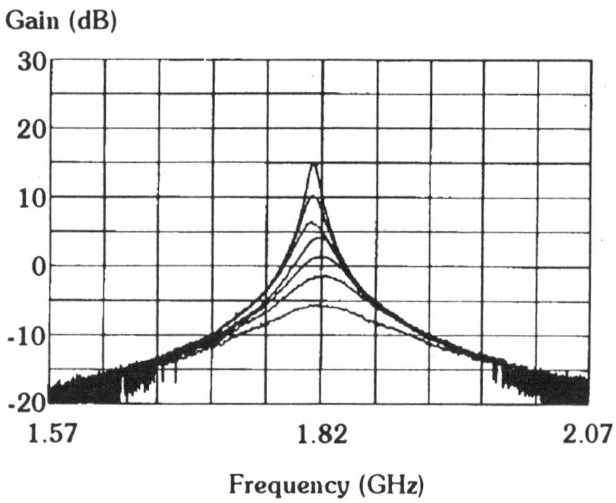

Fig. 5. Effect of varying Q, while keeping the center frequency constant, for the system of Fig. 3.

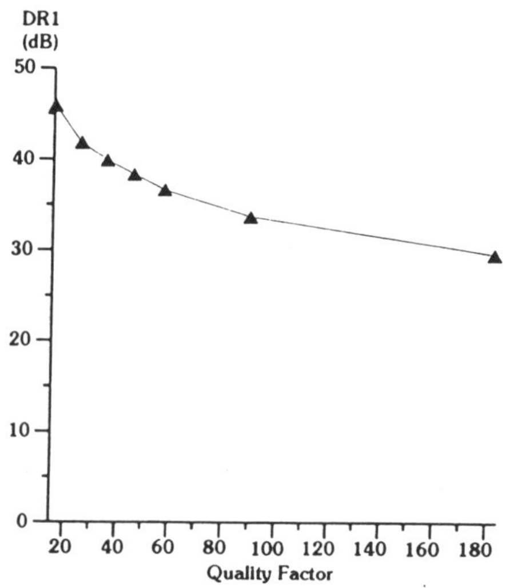

Fig. 7. 1-dB compression dynamic range vs. Q for the system of Fig. 3, for a center frequency of 1.8 GHz.

A 2.5GHz Monolithic Silicon Image Reject Filter

Jose Macedo (1), Miles Copeland (1), Peter Schvan (2)

(1) Department of Electronics, Carleton University, 1125 Colonel By Drive, Ottawa, Ontario, Canada, K1S 5B6

E-mail: jmacedo@doe.carleton.ca

(2) Bell-Northern Research Ltd., M.S.014, P.O.Box 3511, Station C, Ottawa, Ontario, Canada, K1Y 4H7

Abstract

A low power 2.5GHz monolithic silicon bipolar image rejection filter for superheterodyne receiver application is presented; which uses an on-chip inductor and a negative resistance circuit to realize a notch filter with better than 50dB rejection (measured) at the image frequency. Using 0.8micron BiCMOS technology [1] a tuned amplifier was fabricated with 1.9GHz passband, a deep notch at 2.5GHz to attenuate the image frequency (assuming 300MHz intermediate frequency (IF)), and 3.2mA current consumption at +3v. A second chip incorporated a varactor to add notch frequency tuning capability over a 239MHz range.

Introduction

With the growing market for wireless personal communication systems (PCS) there is a need for low power, low cost radio receivers. Low cost can be achieved by integrating the required functions as much as possible, thus minimizing the number of off-chip components. Currently superheterodyne receivers are the state-of-the-art in mobile communications. However, most implementations still require several chips and off-chip components, such as passive filters. A superheterodyne receiver requires image rejection to suppress the image frequency which is located two IFs away from the desired radio frequency [2]. The required image rejection varies from 60 to 90dB depending on the application (ideally, infinite rejection would be desirable!). Currently, off-chip filters are used to provide image rejection, thus increasing receiver cost. Schultes [3] states that the technological and economical challenge in heterodyning is in the development of low-cost, high-performance image rejection and IF filters. Hence, it is highly desirable to develop monolithic image rejection filters, the motivation of the present work.

In Fig. 1 the desired radio frequency (RF) to which the receiver is tuned is 1.9GHz and the IF is 300MHz. A high side local oscillator (LO) of 2.2GHz is used and therefore the unwanted image is located at 2.5GHz as shown. Hence, an image reject filter is included after the low noise amplifier (LNA) and before the mixer. This filter must suppress the undesired 2.5GHz image.

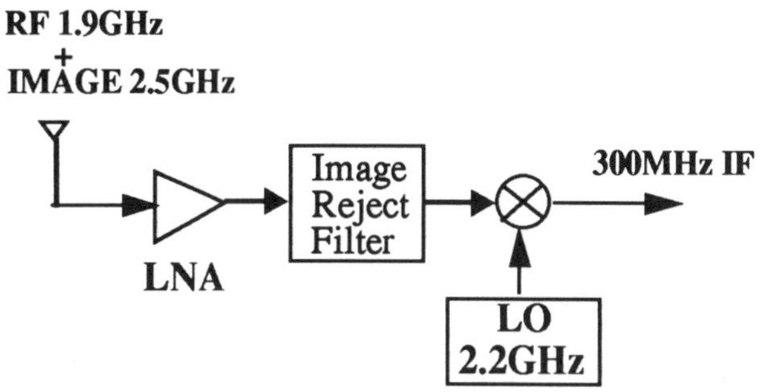

Fig.1: Superheterodyne receiver block diagram

In principle, image rejection with monolithic circuits can be realized with an image reject mixer, a highly selective bandpass filter or a notch filter.

The classical image reject mixer relies on phase cancellation and ideally would eliminate the need for RF filtering. However, in this method the practical difficulty lies in realizing networks with accurate phase shifts. D.Pache et al. [4] report a BiCMOS image reject mixer with 25-35dB image rejection, W. Baumberger [5] achieves 33dB rejection using GaAs technology. Clearly, in practice, an image reject mixer can achieve a maximum of about 30 dB rejection, and thus additional filtering is still necessary in the front end receiver to obtain acceptable image rejection.

The second possible method to reject the image is to pass the desired signal and not the image. However, to realize a low power tuneable monolithic bandpass filter with high selectivity, i.e. high Q, in the GHz range is extremely challenging, due to stability and linearity considerations.

Using a notch filter centered at the image frequency is the method proposed here. As will be shown in the next section, with the availability of monolithic inductors it is possible to implement such a filter with a simple circuit which is attractive for low power operation. A rejection better than 50dB

0-7803-3177-6 $5.00 © 1996 IEEE

IEEE 1996 CUSTOM INTEGRATED CIRCUITS CONFERENCE

has been obtained. The circuit is suitable for integration with an LNA and a mixer to realize the architecture shown in Fig. 1 with low power consumption. This notch filter could also be used in conjunction with an image reject mixer to obtain a monolithic receiver with very high image rejection (70-80dB). It is also possible to have an additional off-chip input filter in front of the LNA to increase image rejection.

Circuit Description

Consider first the series resonator shown in Figure 2. It uses an inductor (L2) in series with the base of an emitter follower (Q2) whose emitter is loaded with a capacitor (C2). The input impedance Zb looking directly into the base of Q2 is given by [6]:

$$Zb = \frac{1}{j\omega} \times \left(\frac{1}{C_\pi} + \frac{1}{C2}\right) - \frac{g_m}{\omega^2 \times C_\pi \times C2} \qquad (1)$$

where $C\pi$ is the emitter-base capacitance of transistor Q2 and g_m is the transconductance of Q2 which depends on the DC bias current of the transistor.

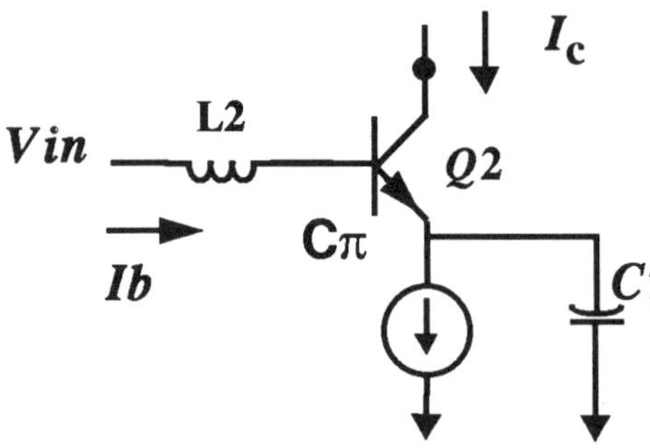

Fig.2: A Series Resonator

The input impedance Zin of Fig.2 is thus seen to be:

$$Zin = Zb + j\omega L2 \qquad (2)$$

The negative resistance seen at the base of Q2 (see Equation 1 above) compensates the losses of the monolithic inductor L2 and the intrinsic base resistance of Q2, dramatically increasing the quality factor Q (the monolithic inductor by itself has a Q of approximately 5 at 2GHz). Clearly, there is an LC series resonator formed by L2, $C\pi$ and C2. Hence, the resonance frequency is determined by the following formula:

$$fnotch = \frac{1}{2\pi \times \sqrt{L2 \times \frac{C_\pi \times C2}{C_\pi + C2}}} \qquad (3)$$

This series resonator is now used in combination with a tuned amplifier to obtain the completed circuit schematic shown in Fig.3. The completed circuit consists of a cascoded amplifier (Q1, Q3) to which a series (notching) resonator has been added (L2, Q2, C2). The cascoded configuration provides improved frequency response [7]. The resistor at the emitter of Q1 provides degeneration for improved linearity. The cascoded amplifier is tuned to the passband at 1.9GHz using an LC tank (L3, C3) at the collector of transistor Q3. The cascoded amplifier output is buffered with a pair of emitter followers to enable it to drive 50 ohm loads for testing purposes.

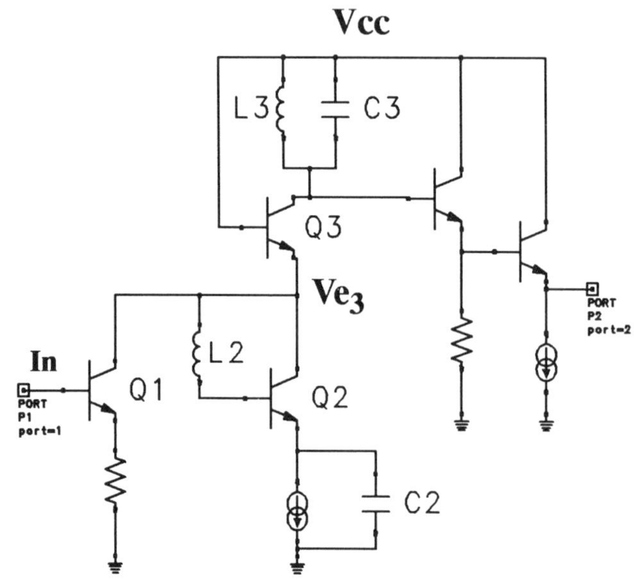

Fig.3: Image Reject Filter schematic

The series resonator creates a zero impedance at the emitter of Q3 when its Q is tuned to infinity. This zero impedance path shunts the input signal current from Q1, and thus there is no output signal on the LC tank (L3, C3) at the notch frequency.

The series resonator effect, via the base current through L2 in Figures 2 and 3, is augmented by the connection of the collector of Q2 as shown in Fig. 3. It can be shown that the effect of this connection is to reduce the impedance Zin of equation (2) by dividing it by the term $(1 - j\omega_T/\omega)$ where ω_T is the transit frequency in rad/sec. This is because the collector current is related to the base current by the term $-j\omega_T/\omega$.

Stability of the circuit in Figure 3 can be an issue. If we treat the node at the emitter of Q3 as a current summing node, we can construct the block diagram shown in Figure 4. Here, Ze3 is the emitter impedance of the cascode Q3 in Figure 3, Zs is the series resonator impedance as discussed above, and Ve_3 is the emitter voltage of Q3. The voltage caused in Ze3 by the input current is opposed by the voltage caused by the current in the series resonance.

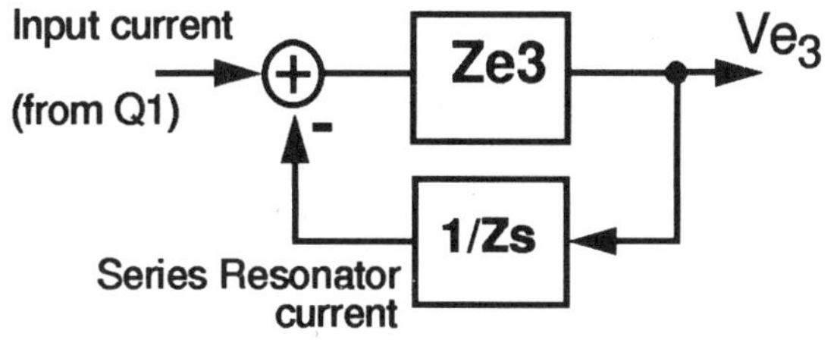

Fig.4: Block Diagram view of circuit

In this diagram the loop gain is given by Ze3/Zs. The relative phases and magnitudes of Ze3 and Zs then become important for stability. The circuit can be shown to be stable as long as Ze3 is not too large and does not have too large a capacitive component. Thus the size and transit frequency (f_t) of cascode Q3 must be carefully chosen.

In summary, the completed circuit consists of a tuned amplifier, implemented with the bipolar subset of NORTEL's 0.8 micron BiCMOS technology, with 1.9GHz passband and a notch at 2.5GHz. This assumes a superhet receiver as shown in Fig.1. The notch Q can be tuned by adjusting the current trough transistor Q2 in Fig.3.

Circuit Layout

Figure 5 shows a microphotograph of the layout. Clearly visible are the two inductors fabricated in low resistance aluminum metal, one for the notch resonator (at the left) and one for the output LC tank. The pads have been arranged to allow for wafer probing using coaxial microwave probes as well as for wirebonding into a package. The active area is approximately 0.365 x 0.767 mm^2.

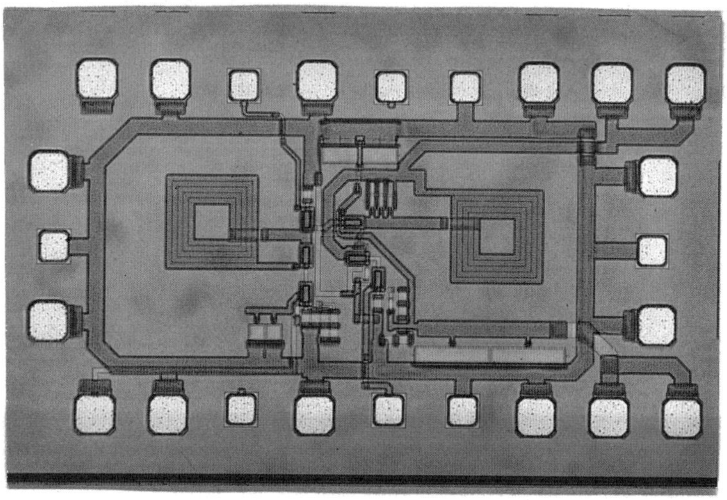

Fig. 5: Microphotograph of Notch Filter Chip

Simulation and Experimental Results of a 2.5GHz image reject filter

The circuit of Fig.3 was simulated with typical models of NORTEL's 0.8 micron BiCMOS process [1]. Using HP/EESOF's Libra simulator the frequency response shown in Figure 6 was obtained. It exhibits an 11.5dB gain at the passband and a deep notch at 2.67GHz.

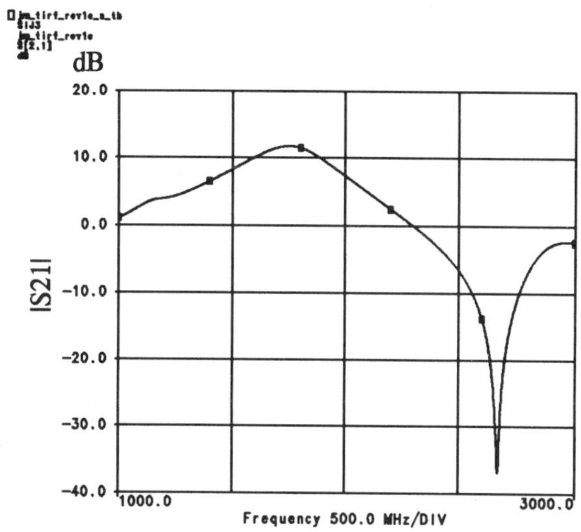

Fig. 6: Notch Filter Simulated Frequency Response

Figure 7 shows the frequency response (S21) of a packaged device measured with an HP8753B network analyzer. A "Bias-T" network connected to the base of Q1 was used to DC bias Q1 to a 2mA DC collector current. The notch Q was adjusted by means of the current source at the emitter of Q2. The plot clearly shows the notch at 2.52GHz with -47.7dB loss (marker 2). The passband gain is 12dB at 1.8GHz. Thus, the measured rejection is 59.7dB. The notch width was 33MHz at a depth of 40dB from the passband.

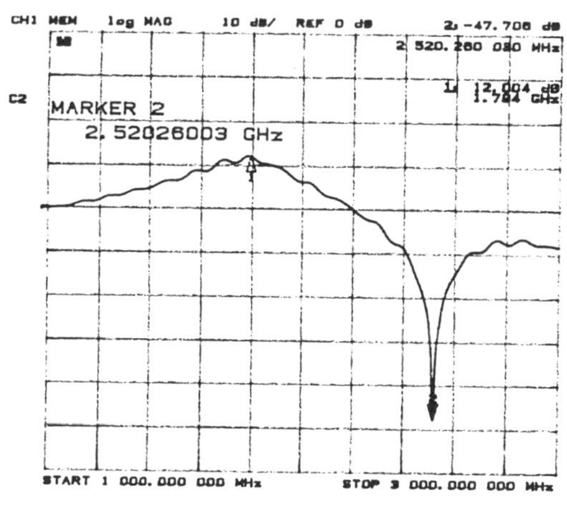

Fig.7: Notch Filter Measured Frequency Response (S21)

The following table summarizes the obtained results.

Table 1: Notch filter results (0.8 micron BiCMOS technology):

Parameter	Sim'n, Libra	Sim'n, Hspice	Meas't (pkg)
Gain @ 1.8GHz	11.5dB	12.3dB	12dB
NF @ 1.8GHz	11.3dB	11.1dB	11.4dB
Input IP3	-3.2dBm	---	-7.8dBm
Notch Frequency	2.67 GHz	2.65 GHz	2.52 GHz
DC Current	2.8mA	2.72mA	3.2mA
Power Supply	3v	3v	3v

Simulations with HP/EESOF Libra were done including bondwires but without layout parasitic capacitances. Hspice simulations included bondwires and layout parasitic capacitances. For this reason, Hspice yields a slightly lower notch frequency than Libra.

As shown in Table 1 good agreement is observed between simulation and measurement results. The differences are due to modelling inaccuracies as well as process variations. There are inaccuracies in the modelling of the wirebonds which connect Vcc to the L3, C3 tank of Figure 3 (wirebonds were approximated to 1nH). Another source of inaccuracy occurred because the package was not modelled.

The simulated and measured noise figures are very close. The noise figure results in this implementation show that a preliminary LNA stage is required, as shown in figure 1. A monolithic front end incorporating LNA, image reject notch filter and mixer is currently in fabrication. The BiCMOS process used [1] has a bipolar f_T of 11GHz and performance improvements in a faster process are expected

The measured third order intercept point (IP3) is 4dB worse than simulated. This indicates that there are non-linearities present in the circuit have not been accounted for in the simulation. The simulation shows the notch at 2.67GHz while the actual measurement was 2.52GHz. This error appears to be systematic. System design centering optimization has not been completed yet.

To compensate for the frequency error and allow notch frequency tuneability a second chip was fabricated in which capacitor C2 of Figure 3 is now replaced with an on-chip varactor. In this case the measured notch frequency tuning range was 239MHz, from 2.490GHz to 2.729GHz, corresponding to a varactor control voltage of 0v and 3v respectively.

Conclusions

With a novel combination of on-chip inductors and negative resistance techniques a monolithic image rejection filter for superheterodyne receiver has been realized. A first 2.5GHz notch filter has been fabricated with NORTEL's 0.8micron BiCMOS process. This circuit requires Q tuning to provide sufficient negative resistance to obtain a deep notch. In order to compensate for process variations and to allow for possible frequency tuning a second fabricated circuit included an on-chip varactor, giving a measured notch frequency tuning range of 239MHz. Additional work is required to finalize the design centering optimization and to develop the electronic tuning circuitry.

Acknowledgements

This work was supported by the Natural Sciences and Engineering Research Council of Canada (NSERC), NORTEL's Telecom MicroElectronics Center, Micronet and the Telecommunications Research Institute of Ontario (TRIO). Thanks also to Mr. John R. Long for providing the inductor layout and model.

References

(1) R. Hadaway et al, "A sub-micron BiCMOS technology for telecommunications", Journal of Microelectronic Engineering, vol.15, pp. 513-516, 1991

(2) R.S.Carson, "Radio Communications Concepts: Analog", Wiley, 1990, Chapter 8.

(3) G. Schultes, P. Kreuzgruber, A. L. Scholtz, "DECT Transceiver Architectures: Superheterodyne or Direct Conversion?", Proceedings of the 43rd Vehicular Technology Conference, Secaucus, NJ(USA), May 18-20, 1993, pp.953-956.

(4) D. Pache, J.M.Fournier, G.Billiot and P.Senn, "An Improved 3V 2GHz BICMOS Image Reject Mixer IC", IEEE 1995 Custom Integrated Circuit Conference

(5) W. Baumberger, "A single chip image rejecting downconverter for the 2.44GHz band", 1993GaAs IC Symposium

(6) G. D. Vendelin, A. Pavio, U.L. Rohde, "Microwave Circuit Design Using Linear and Nonlinear Techniques", New York, N.Y.: Wiley, 1990, Chapter 6.

(7) A.S.Sedra, K.C.Smith, "Microelectronic Circuits", Saunders College Publishing, 1991, Chapter 7.

A 1 GHz CMOS Upconversion Mixer

Peter Kinget and Michiel Steyaert

Katholieke Universiteit Leuven, ESAT-MICAS
Kard. Mercierlaan 94, B-3001 Heverlee, Belgium

Abstract

A high frequency mixer topology is presented for the realization of a 1 GHz upconversion mixer in a standard 0.7 μm CMOS technology. The high output bandwidth is achieved by the development of an nMOS only current amplifier to convert the modulated current of an nMOS transistor in the linear region to an output voltage.

Introduction

Modern wireless telecommunication systems require fully integrated RF receivers and transmitters. For economic reasons the realization of these building blocks in a CMOS technology is preferred since the analog building blocks can then be integrated with the digital signal processing circuits on a single chip. The RF upconversion mixer is a key building block for the realization of a fully integrated transmitter.

A first possibility for the realization of the upconversion mixer is to use a switching mixer [7]. Due to the use of a square wave local oscillator (LO) signal, a large signal at third harmonic of the LO frequency is generated. In the full architecture of the transmitter, this requires an extra filter block after the switching mixer to suppress this harmonic signal in order to meet the out-of-band signal specifications. However the specifications of this filter block will be very hard to meet with fully integrated filters.

A linear mixer topology is thus preferred for the upconversion mixer since almost no out-of-band harmonics are generated. Traditionally a MOS version of the Gilbert bipolar multiplier topology is used [1]. Due to the square-law current relationship of the MOS transistor they have only a small input range and require large supply voltages. These problems can be overcome by using complex predistortion circuits or by using topologies based on the exact square law relationship between gate voltage overdrive and drain current [2,3]. However for high speed sub-micron transistors the quadratic current region is almost inexistent. Moreover, those structures are very sensitive to transistor mismatches and the RF performances are limited.

As a result, for CMOS analog multipliers operating at RF frequencies, the linear resistance type architecture is preferred [4,5]. In a downconversion mixer the mixed signal is a low frequency signal and only a low output bandwidth is required which is easier to achieve. Consequently, high performance downconversion mixers for RF frequencies up to 1.5 GHz are possible [4]. However, for the realization of the upconversion path, a very high output bandwidth at LO frequencies must be achieved. In [5] CMOS RF circuits have been demonstrated for upconversion to RF frequencies below 100 MHz.

In this paper an upconversion mixer circuit is presented operating up to 1 GHz. The circuit is fully compatible with a standard 0.7 μm CMOS process. The current of a MOS transistor in the linear region is modulated by the LO signal applied at the gate and by the differential baseband signal applied at the drain and source. The high frequency output path is realized with a current amplifier which has an nMOS only transistor structure so that very high bandwidths are achieved. The circuit is designed and characterized for the operation in an embedded RF application.

Circuit topology and operation

Mixing principle

In figure 1 the topology of the mixer is presented. The operation of the mixer is based on the modulation of the current of the nMOS transistor M1 in the linear region by the gate and drain-source voltages. For a gate voltage of $V_1 + v_{in1}$ a drain voltage $V_2 + {}^{v_{in2}}\!/_2$ and a source voltage $V_2 - {}^{v_{in2}}\!/_2$ the current is given by:

$$i_{DS} = \beta_1 \cdot (v_{in1} \cdot v_{in2}) + \beta_1 \cdot (V_1 - V_T - V_2) \cdot v_{in2}. \quad (1)$$

The LO signal is connected to the gate of M1 as v_{in1} and a baseband signal is used for v_{in2}. The current then contains frequency components around the LO due to the first term in eq. (1) and components in the baseband due to the second term. The differential input voltage v_{in2} is connected to the gates of M2 and M3 which operate as source followers and copy the voltage variations at their gates to the drain and source of M1.

HF current buffer with low-impedance input

The key issue in this design is the nMOS only transistor structure M2, M4 and M6 (M3, M5 and M6) in fig. 1. This structure operates as a current amplifier in unity feedback that copies the signal current to the output and has a low-impedance input. A small current flowing into M2 (M3) results in a large gate voltage swing for M4 (M5). Due to the feedback all signal current of M1 will flow into M4 (M5) and is copied with M6 (M7) to the output. The feedback loop through transistors M4 and M5 boosts the effective g_m

0-7803-3177-6 $5.00 © 1996 IEEE IEEE 1996 CUSTOM INTEGRATED CIRCUITS CONFERENCE

of M2 and M3 so that they act as very good LF source followers with low distortion and so that the input impedance of the current amplifier is lowered. The allowed differential swing for input 2 is extended by adding a level-shifter, implemented with C_{LS} and R_{LS} and a current source, in the feedback-loop.

By developing an nMOS only current buffer structure, extremely high speed performance has been achieved. The GBW of the current amplifier is determined by the g_m of M4 and the capacitive load at the gate of M4. The second pole of the feedback loop is determined by the g_m of M2 and the capacitive load at the source of M2. When M2, M4 and M6 have equal sizes, the capacitive load at the gate of M4 is more than twice the load at the source of M2, and M2 and M4 have equal g_m. The second pole is as a result located at least at twice the GBW and a good loop stability for unity feedback is guaranteed.

The sizing of C_{LS}, M2, M4, M6 and the pMOS current source M8 is optimized for high GBW for the current amplifier and as a result a high BW for the mixer. In table 1 the sizes and biasing conditions of the circuit are summarized. In the present realization the drain junction of the pMOS current source M8 represents a large capacitive loading which can only be reduced by using a large gate-overdrive for the current sources. This sets the required supply voltage to 5 V. The total current consumption of the circuit is 17 mA.

LF output suppression and Conversion Gain

At the output a bias circuit with a low frequency feedback is implemented that acts as a coil. In the full circuit schematic of figure 2 the biasing circuits are located on the left and right hand side. They short-circuit the baseband frequency components in the output current, that are generated by the second term in (1) and keep the DC level of the output to V_{REF}. The high frequency signal are short-circuited in C_{LF} and no feedback occurs.

The high frequency differential output current can be converted to a voltage by the differential load R_L of the next stage. The conversion gain of the mixer is determined by:

$$G = \beta_l \cdot V_{LO\,rms} \cdot R_L \qquad (2)$$

with V_{LOrms} is the rms value of the local oscillator signal. The circuit is designed as a building block for a single chip transmitter. The output will be connected directly to the next stage without leaving the chip and the capacitve load at the ouput is low. Therefor a on-chip impedance R_L of 500 Ω can be used so that a conversion gain of -10 dB is achieved for a 0 dBm LO signal.

Experimental results

A test-chip has been fabricated in a 0.7 μm CMOS technology. Figure 3 is a microphotograph of the chip. The area occupied by the circuit is 0.25 mm^2. The free area between the bonding pads and the circuit has been filled with decoupling capacitors for the power supply. The total chip area is 0.5 mm^2.

The speed performance of the mixer is measured with a Microtech Cascade probe station for the RF input and outputs. For test purposes both outputs are loaded with a load impedance of 50 Ω and a single output is measured. The mixer is designed for a 500 Ω on-chip load so that the measured output strength is reduced by 26 dB. A fixed 20 kHz sine is applied as baseband signal and the LO frequency is swept. In figure 4 the measured frequency response is displayed. For a 0 dBm LO signal the measured conversion gain, recalculated for a differential output load of 500 Ω, is -11 dB. This agrees very well with the predicted value of eq. 2.

The output spectrum for a double sideband modulation of a 0 dBm 1 GHz LO signal is displayed in figure 5. The output spectrum is measured with an external balun for the differential to single ended conversion. The modulating signal is a 20 kHz 2Vpp differential sine. All the unwanted signals, such as the LO feedthrough and the distortion components are below -30 dBc, which is as good as the performance of bipolar RF mixers [6]. The distortion components in the output spectrum are due to the distortion introduced by the output resistance of M4 and M5 on the baseband signal. The distortion performance can be further improved without degrading the speed performance by adding cascode transistors to M4 and M5.

Conclusion

In this paper a CMOS realization for a 1 GHz RF upconversion mixer is demonstrated. The high bandwidth is achieved by an nMOS transistor in the linear region for the modulation and an nMOS only transistor structure to output the modulated current to a load resistor. The unwanted signals in the measured output spectrum are below -30 dBc.

References

[1] D. Soo and R. G. Meyer, " A Four-Quadrant NMOS Analog Multiplier", *IEEE Journal of Solid-State Circuits*, pp. 1174-1178, Dec 1982.

[2] K. Bult and H. Wallinga, "A CMOS Four-Quadrant Analog Multiplier", *IEEE Journal of Solid-State Circuits*, pp. 841-848, June 1986.

[3] H-J Song and C-K Kim, "An MOS Four-Quadrant Analog Multiplier Using Simple Two-Input Squaring Circuits with Source Followers", *IEEE Journal of Solid-State Circuits*, pp. 841-848, June 1990.

[4] J. Crols and M. Steyaert, "A Full CMOS 1.5 GHz Highly Linear Broadband Downconversion Mixer", *Proc. ESSCIRC'94*, pp. 248-251, Sept. 1994.

[5] Bang-Sup Song, "CMOS RF Circuits for Data Communications Applications," *IEEE J. of Solid State Circuits*, vol. SC-21, no.2, pp.310-317, Apr. 86.

[6] T. Stetzler, I. Post, J. Havens and M. Koyama, "A 2.7 V to 4.5 V Single-Chip GSM Transceiver RF Integrated Circuit", *ISSCC Digest of Technical Papers*, pp. 150-151, Feb. 1995.

[7] I.D. Robertson, "Mixer designs for GaAs technology" in *GaAs technology and its impact on circuits and systems*, D. Haigh and J. Everard, Eds., London: Peter Pelgrinus Ltd., 1989, pp. 281-312.

Figures and Tables

	Size	Bias Current
M1	16/0.7	
M2, M4, M6	80/0.7	4 mA
M8	50/0.7	4.1 mA
R_{LS}	10 kΩ	100 μA
C_{LS}	280 fF	

Table 1: Component sizes and bias currents.

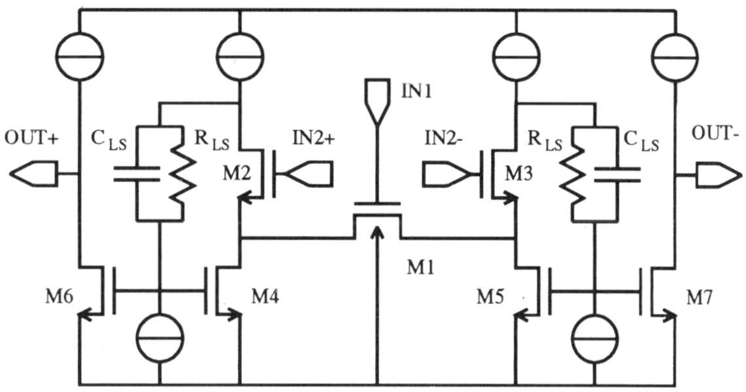

Figure 1: The nMOS only mixer topology.

Figure 3: Chip microphotograph.

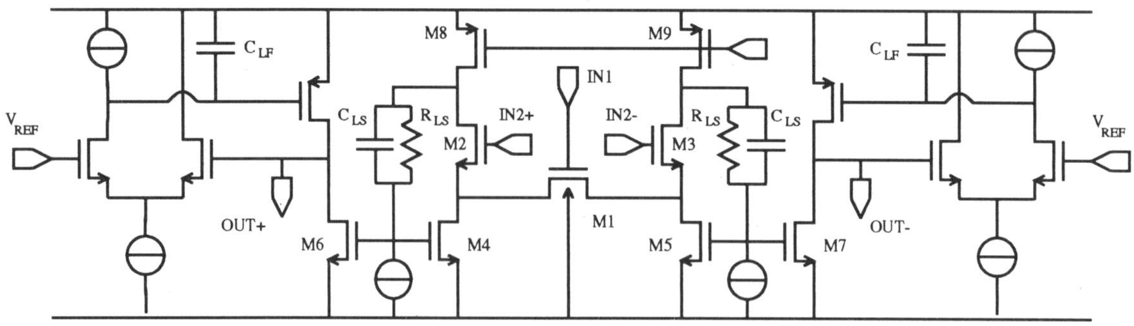

Figure 2: Total circuit diagram, including the bias circuitry for the output on the right and left side. The baseband frequency components in the output signal are short-circuited through the feedback.

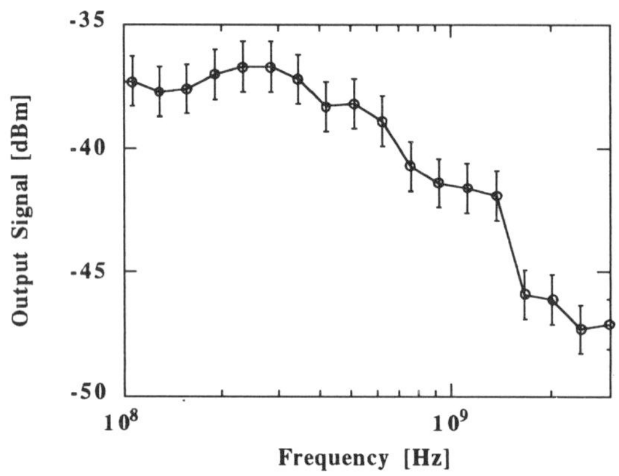

Figure 4: Measured frequency response using a Cascade probestation with bondpad calibrations. The error in the measured single ended gain is +/- 1 dB. Only a single ouput is measured which results in an output signal reduction of 6 dB for a 50 Ω load.

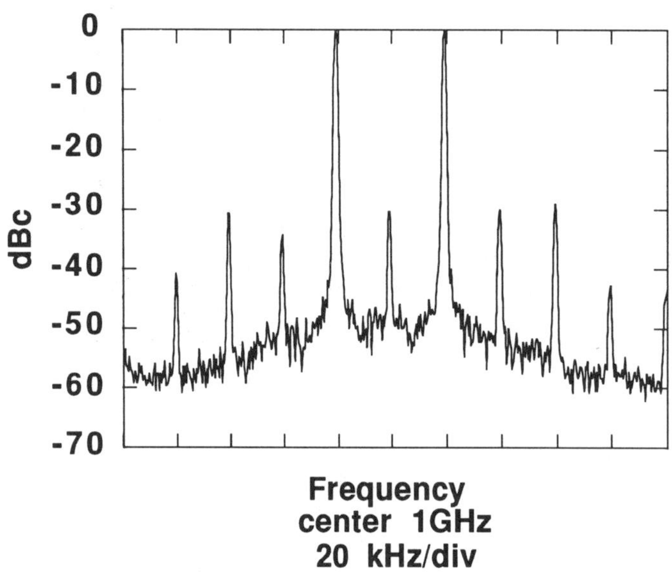

Figure 5: Measured double sideband output spectrum for a sine modulation signal of 20 kHz. and a 1 GHz LO.

Linearized High Efficiency Class E Power Amplifier for Wireless Communications

Tirdad Sowlati[1], Yuriy Greshishchev[1], C. Andre T. Salama[1], Gord Rabjohn[2] and John Sitch[2]

[1] Department of Electrical and Computer Engineering
University of Toronto
Toronto, Ontario, M5S 1A4, Canada

[2] Bell Northern Research
P.O. Box 3511, Station C
Ottawa, Ontario, K1Y 4H7, Canada

Abstract

A linearized high efficiency Class E power amplifier for wireless communications is presented in this paper. It consists of a Class E power stage and a phase correcting feedback loop. The Class E power amplifier operating at 835 MHz delivers 443 mW to the load at a supply voltage of 2.4 V. The phase correcting feedback loop comprises novel limiting amplifier, phase detector and phase shifter. The phase correcting feedback reduces the 30° phase distortion of the Class E amplifier down to 4°, resulting in a 20 dB reduction in adjacent channel interference. The total efficiency of the linearized Class E power amplifier is 65%.

1. Introduction

One of the challenging blocks in designing the handset for wireless communication transceivers is the power amplifier. Among the specifications that it must meet are high efficiency and capability of operating at a low supply voltage. In the North American Digital Cellular standard, where a linear modulation scheme is used, the power amplifier must also be linear. Otherwise, the signal spectrum is degraded and severe adjacent channel interference occurs. There is a trade-off between efficiency and linearity in power amplifiers: high efficiency can be achieved at the expense of the amplifier working in the nonlinear region, and linear amplification results in a low or moderate efficiency.

In order to exploit the limited power resource (battery) of the handset and achieve linearity at the same time, a high efficiency linearized power amplifier must be used. Class B, Class C and Class F amplifiers have been widely suggested in this application [1-3]. Recently, it has been shown that Class E power amplifiers have considerable advantages for low voltage wireless transceiver applications [4] and a linear transmitter architecture using a Class E power amplifier has been proposed [5]. The large envelope variations in the Class E results in AM-to-PM distortion which can be cancelled out by using a phase correcting feedback loop as shown in Fig. 1. In this architecture, the limiting amplifier limits the output voltage of the Class E amplifier to a fixed value. The phase detector extracts the phase error signal which is then low-pass filtered, amplified and fed to the phase shifter. The phase shifter changes the phase of the signal applied to the power amplifier to compensate the phase distortion generated by the power amplifier.

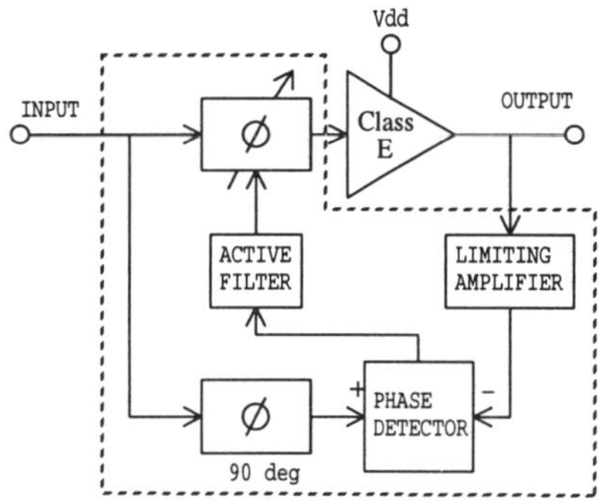

Fig. 1: Block diagram of the linearized Class E power Amplifier

This paper discusses the design and implementation of a linearized high efficiency Class E power amplifier using the aforementioned architecture. The circuits were fabricated in a 0.8 μm GaAs MESFET technology. A number of novel circuit design solutions are introduced to minimize power, allow the use of low supply voltages and improve circuit performance.

2. Circuit Design

In this section, the design of the building blocks in Fig. 1 will be described. The circuits were designed and simulated using harmonic balance analysis in the "HP Microwave and RF Design System". A custom large signal MESFET model was used in the simulations. In order to accurately model the high frequency parasitics, the spiral inductors were represented by measured S-parameters data, and the critical interconnections were modeled by distributed transmission lines.

A. Class E Power Module

The Class E power module, shown in Fig. 2, consists of a Class F driver stage and a Class E power amplifier. It was designed to operate at 835 MHz and deliver 500 mW to the 50 Ω load at a supply voltage of 2.4 V. The design methodology is similar to the one used in an earlier publication [4]. However, the circuit was designed for a higher output power and efficiency. Transistors M1 and M2 in Fig. 2 are interdigitated depletion mode MESFETs. The former has a

0-7803-3177-6 $5.00 © 1996 IEEE

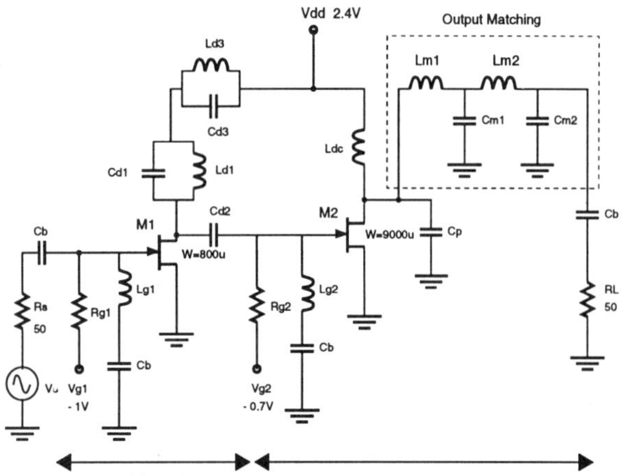

Fig. 2: Circuit diagram of the Class E power module

total width of 800 μm and 8 fingers, and the latter has a total width of 9 mm and 80 fingers. The output matching circuit transfers the 50 Ω load to the optimum load (8.9 Ω) at the drain of the power transistor M2. The power dissipation in M1 and M2 are 35 mW and 130 mW respectively, and the power added efficiency is 73%.

B. Limiting Amplifier

The limiting amplifier, shown in Fig. 3, consists of two basic parts: a symmetrical serial-parallel MESFET limiting cell (M1-M4 in Fig. 3) and a saturating gain stage (M5-M12 in Fig. 3). In the limiting amplifier, the phase variation with amplitude must be kept as small as possible since it cannot be cancelled out in the phase correcting feedback loop and produces a residual phase error. Therefore, the limiting amplifier was developed to have very low phase error. Unlike conventional limiting schemes using a directional coupler, this novel limiting cell prevents the undesirable attenuation of low signal levels due to low resistance of M1 and M2 in triode region and high resistance of M3 and M4. The input current is limited by the current source M1 (or M2

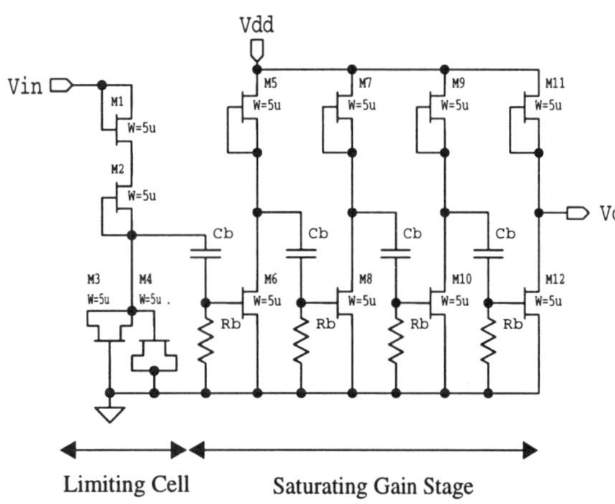

Fig. 3: Limiting amplifier circuit diagram

depending on the polarity of the input voltage) resulting in a low power dissipation.The maximum allowable input voltage which can be applied to the limiting cell is constrained by the breakdown voltage of the MESFET.

The limiting gain stage is based on inverters with equal MESFET widths, allowing the use of a single power supply. The gain stage is designed with AC coupling (C_b-R_b) which also serves as the inverter load. This approach results in better performance than a source follower interstage connection in terms of power dissipation, phase error and signal swing.

C. Phase Shifter

In the present design, the phase shifter can have a square output waveform, but must have a low duty cycle variation versus phase changes. The phase shifter, shown in Fig. 4, consists of two varactor-tuned-LC circuits at the drain and the source of the transistor M1. It offers a larger range of phase deviation, a more constant sensitivity and a better tolerance to process variations compared to single varactor-tuned circuits. To cancel the amplitude variations due to phase changes, a 3-stage saturating gain amplifier (M6-M11 in Fig. 4) is used. The gain stage is similar to the one described in Subsection B.

The same phase shifter design was also used to achieve the 90° constant phase shift between the two inputs of the phase detector (see Fig. 1) under zero phase error condition in the loop.

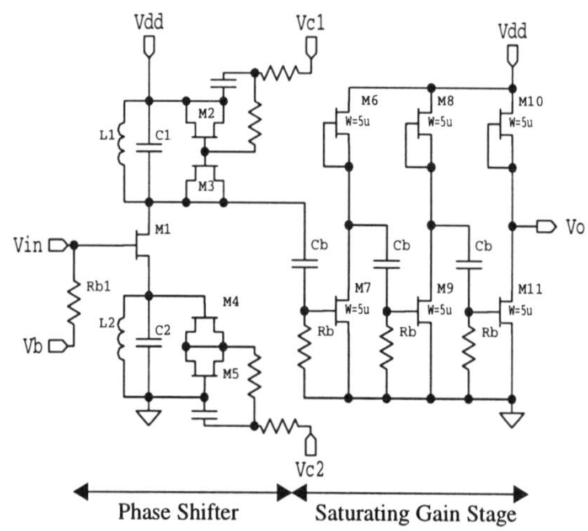

Fig. 4: Phase shifter circuit diagram

D. Phase detector

The phase detector is a modified Gilbert Cell based on the mixer configuration reported in [6]. As shown in Fig. 5, two resistors are introduced at the sources of transistors M3 and M6 to reduce the quiescent current to 100uA and to provide symmetrical drain current variation versus Vrf. Low-pass filter capacitors Co are used to improve the dynamic performance of the switches M1-M2 and M4-M5. The differential amplifiers DA1 and DA2 serve as single-ended-

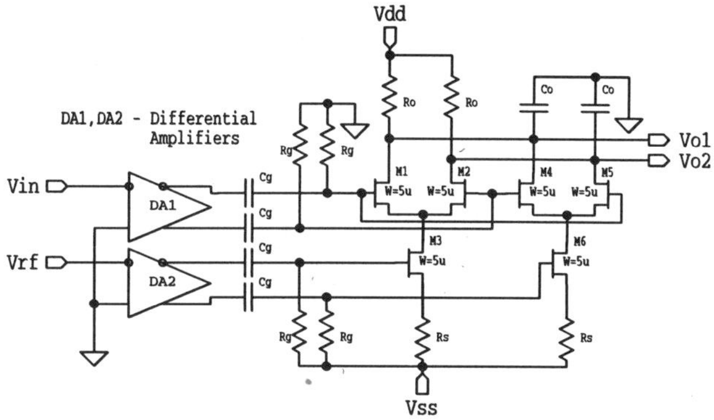

Fig. 5: Phase detector circuit diagram

to-differential converters. The quiescent current, the resistive loads of the differential amplifiers and the Cg-Rg values were optimized considering the effect of their parasitic capacitances to the substrate. These parasitics are the major power demanding factor in the circuit.

3. Implementation and Experimental Results

The circuits mentioned in Section 2 were fabricated in a 0.8 μm depletion mode GaAs MESFET technology. The depletion mode FET transistors in this technology have a threshold voltage of –1V and a cut-off frequency of 20 GHz. The substrate thickness is 625 μm and the wafer is not thinned after processing. There are also no via holes available.

Fig. 6 shows the micrograph of the GaAs Class E power module. The output matching network was realized on an alumina substrate with a thickness of 1.02 mm, a dielectric permittivity of 9.8 and a loss tangent of 0.0002 at 1 GHz. The output matching network was then wire-bonded to the GaAs power module.

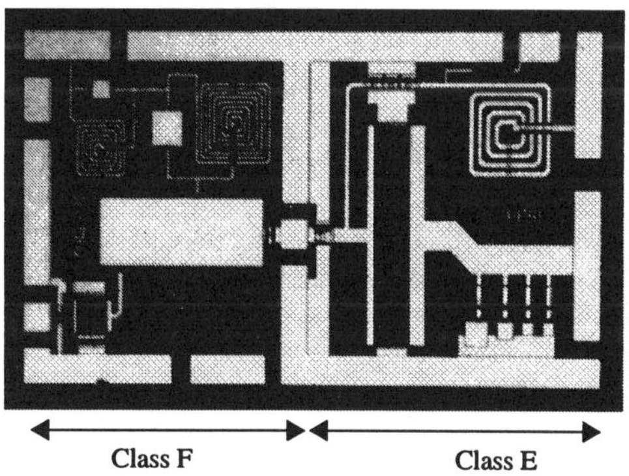

Fig. 6: Micrograph of the Class E power module

The micrograph of the phase correcting feedback loop is shown in Fig 7. The dimensions of the fabricated chip are 1.7×2.4 mm^2. Interconnections were mostly done in the second metal layer with airbridges to reduce parasitics.

The limiting cell DC characteristics illustrated in Fig. 8, are symmetrical and in good agreement with simulations. From 0 V up to 1.5 V, the limiting cell has the same output

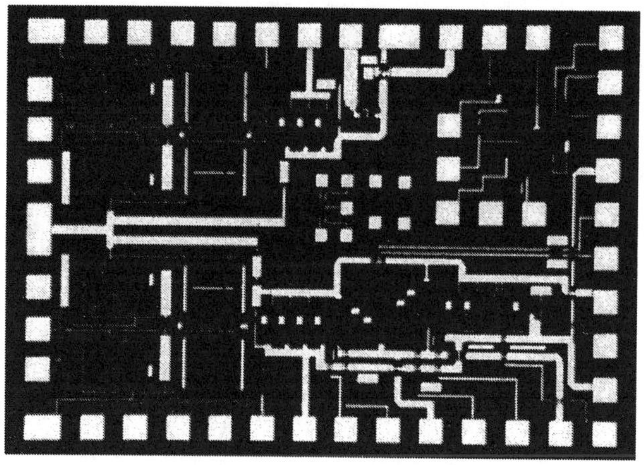
Fig. 7: Micrograph of the phase correcting feedback

voltage as a conventional diode limiter with a 2 KΩ equivalent series resistance. Beyond 1.5 V, the limiting capability increases drastically due to the current saturation in M2 (refer to Fig. 3), and the output voltage is almost constant.

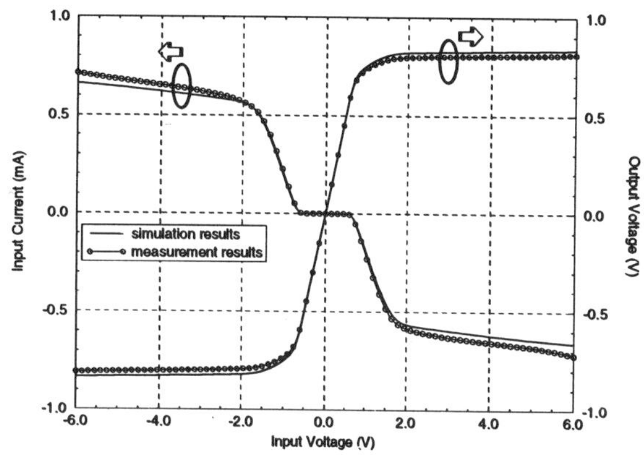

Fig. 8: DC characteristics of the limiting cell

Fig. 9 shows the measured phase shifter performance with the control voltage applied to both Vc1 and Vc2 inputs (refer to Fig. 4). The phase range is ± 55°, the phase variation in the

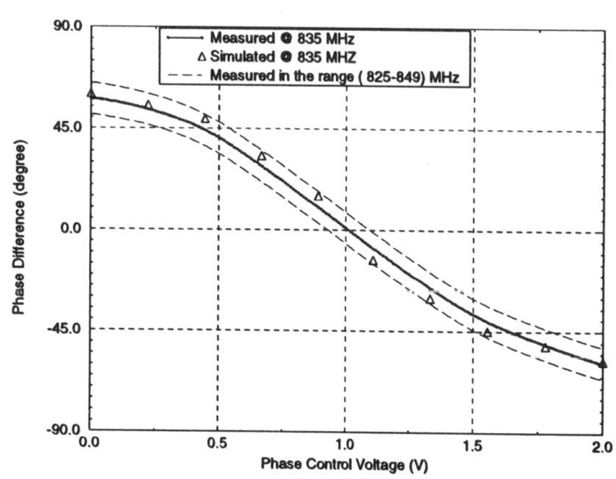

Fig. 9: Measured characteristics of the phase shifter

frequency range of 825 MHz to 849 MHz is less than 7.5° and the sensitivity is 80°/V. The phase detector overall sensitivity was measured to be 10 mV per degree.

Fig. 10 shows the phase offset of the limiting amplifier measured at the output of the phase detector. Measured results are very close to simulation down to 0.6 V. Below this voltage, the inaccuracy of the MESFET model results in a predicted lower phase offset. The measured offset phase error is less than ± 2.5° for Vin > 0.3 V, and less than ± 0.25° for Vin > 0.8 V confirming the excellent performance of the limiting amplifier.

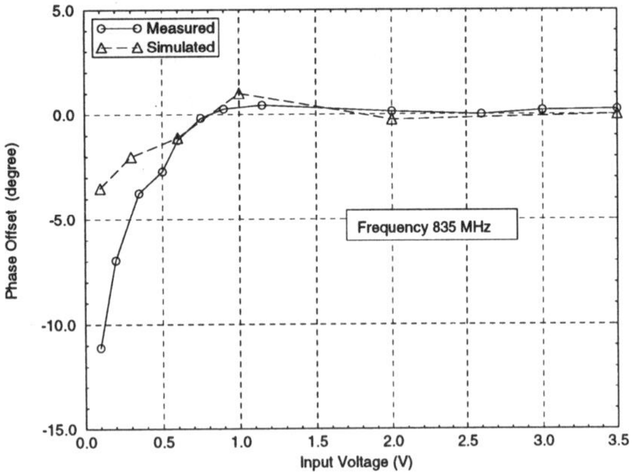

Fig. 10: Phase offset of the limiting amplifier

The power dissipation of the phase correcting feedback loop (two phase shifters, phase detector and limiting amplifier) is 20 mW at supply voltages of Vdd=3 V and Vss=-1.5 V. The active filter in Fig. 1, implemented using a silicon micropower OpAmp, has a power dissipation of 1.5 mW.

The Class E amplifier was characterized with and without linearization feedback. As shown in Fig. 11(a), the amplifier delivers 443 mW to the load at a supply voltage of 2.4V and the output voltage varies linearly versus the supply voltage. However, the power amplifier without linearization has a phase error of about 30°, as illustrated in Fig. 11(b). The addition of the phase correcting feedback, with a loop gain of 20, reduces this phase error down to 4°, as illustrated on the same figure. Consequently, the linearization of Class E power amplifier reduces the phase nonlinearity and results in a 20 dB reduction in adjacent channel interference. The power added efficiency of the power amplifier both before and after linearization, as shown in Fig. 11(b), is higher than 65%.

4. Conclusion

A linearized high efficiency Class E power amplifier for wireless communications was presented. The design and implementation of the Class E power module, limiting amplifier, phase shifter and phase detector of the phase correcting feedback loop were discussed. The performance of the designed circuits show that the linearized Class E

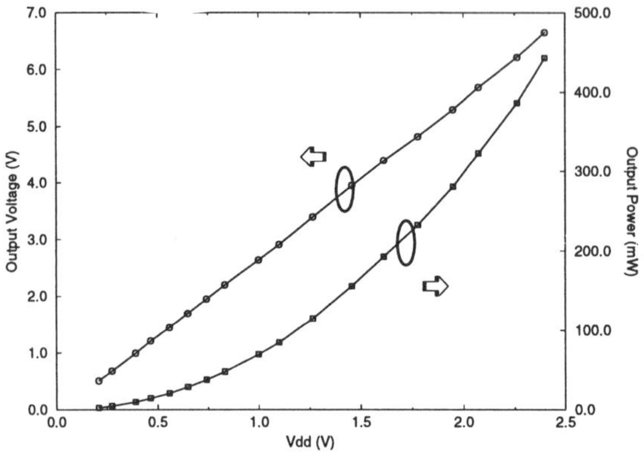

a)

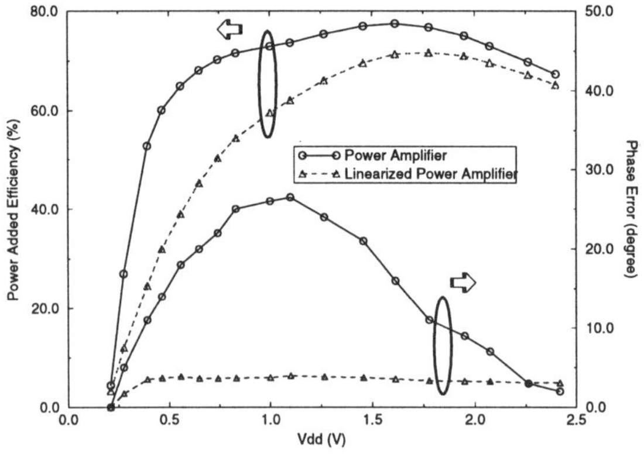

b)

Fig. 11: Characteristics of the Class E power amplifier at 835 MHz
a) Output voltage and output power
b) Power added efficiency and phase error

power amplifier provides a suitable solution to achieve the high efficiency (65%) and linearity required in wireless transmitters using the North American Digital Cellular standard.

Acknowledgments

This work was supported by the Natural Sciences and Engineering Research Council of Canada and Micronet.

References

[1] A. Lohtia, P. A. Goud and C. G. Englefield, "Power Amplifier Linearization using Cubic Spline Interpolation", IEEE Vehicular Technology Conf., Proceedings, pp676-679, 1993.

[2] S. A. Hetzel, A. Bateman and J. P. McGeehan, "A LINC Transmitter", IEEE Vehicular Technology Conf., Proceedings, pp133-137, 1991.

[3] K. Chiba, T. Nojima and S. Tomisato, "Linearized Saturation Amplifier with Bidirectional Control for Digital mobile Radio", IEEE Global Telecommunication Conf., Proceedings, pp1958-1962, 1990.

[4] T. Sowlati, C. A. T. Salama, J. Sitch, G. Rabjohn and D. Smith, "Low Voltage, High Efficiency Class E GaAs Power Amplifiers for Wireless Transmitters", IEEE J. Solid-State Circuits, vol. 30, pp. 1074-1080, 1995.

[5] T. Sowlati, Y. Greshishchev, C. A. T. Salama, J. Sitch and G. Rabjohn, "Linear Transmitter Design using High Efficiency Class E Power Amplifier", IEEE Personal Indoor and Mobile Radio Communications Symposium, Tech. Digest, pp. 1233-1237, 1995.

[6] A. Boveda, F. Orgitoso, and J. I. Alonso, "A 0.7-3 GhZ GaAs QPSK/QAM direct modulator", IEEE J. Solid-State Circuits, vol. 28, pp. 1340-1349, 1993.

A Bipolar Sampled-Data Bandpass Delta-Sigma A/D Modulator

Theodore Varelas, Seyfi S. Bazarjani, and W. Martin Snelgrove

Department of Electronics, Carleton University, Ottawa Canada K1S 5B6

E-mail: vth@doe.carleton.ca

Abstract

A second-order bandpass delta-sigma A/D modulator has been implemented in the bipolar subset of a 0.8μm BiCMOS process [1]. This prototype demonstrates the feasibility of a bipolar-only delta-sigma modulator based on sampled data techniques. The circuit was clocked at 250 MHz and demonstrated a 40dB SNR in a 1 MHz bandwidth for an input signal at 62.5 MHz. The circuit dissipates 350 mW when it operates from a 5V power supply.

Introduction

Digitizing a radio signal at an intermediate frequency (IF) stage of a radio receiver allows quadrature demodulation to be performed digitally, eliminating any phase and amplitude imbalance between the in-phase (I) and quadrature (Q) channels. Direct conversion at IF is done either by using a wide band video or ΔΣ analog to digital converter (A/D) or bandpass delta-sigma modulator (BP-ΔΣ).

Monolithic BPΔΣ modulators have been reported that digitize signals with a bandwidth up to 200kHz, for use in GSM, IS-54, or similar radio applications. They use switched capacitor (SC) techniques [2],[3],[4] or continuous-time circuits with off-chip inductors as resonators [5]. Increasing demand for wider bandwidth applications requires higher oversampling or higher order design to achieve the required performance. Digitizing at a higher IF is also desirable in order to simplify the radio.

The SC approach has proven to be a very robust and linear technique that has a manufacturing advantage because it does not require any tuning but it has demonstrated limited speed, clocking at 10MHz to 100MHz. Continuous-time circuits that use G_m-C or LC based resonators are on their way [6] [7] [8] to addressing the need for wider bandwidth A/Ds.

In this paper, we present a second order BPΔΣ modulator. The modulator is designed using only npn bipolar devices. It demonstrates the feasibility and high speed of a sampled-data design style based entirely on a track and hold cell.

Architectures for Bandpass ΣΔ modulator

A second-order modulator is the simplest circuit that demonstrates bandpass operation, and so is useful to test new circuits and structures. It has limited linearity, though enough for radio receivers that include good filtering and low-speed AGC.

The second order ΔΣ bandpass modulator noise transfer function can be derived from the first order lowpass modulator noise transfer function by mapping the zeros of the transfer function form DC to $f_s/4$. This is achieved by a $z^{-1} \rightarrow -z^{-2}$ transformation. The stability and SNR characteristics of the resulting bandpass modulator will be identical to that of low pass prototype [9]. Based on this transformation the noise transfer function of the second-order bandpass ΔΣ modulator is $H_q(z) = (1+z^{-2})/z^{-2}$ and the signal transfer function is $H_s(z) = z^{-2}$.

The z-domain block diagram of the above second order modulator is shown in Figure 1. Note that there are two full delays from input to output and two full delays in the loop, and also note that we never have to sum more than two voltages at any node. This implementation makes possible the use of the 2-operand summer/subtracter cell described in [10] to sum the D/A feedback signal with the loop voltage.

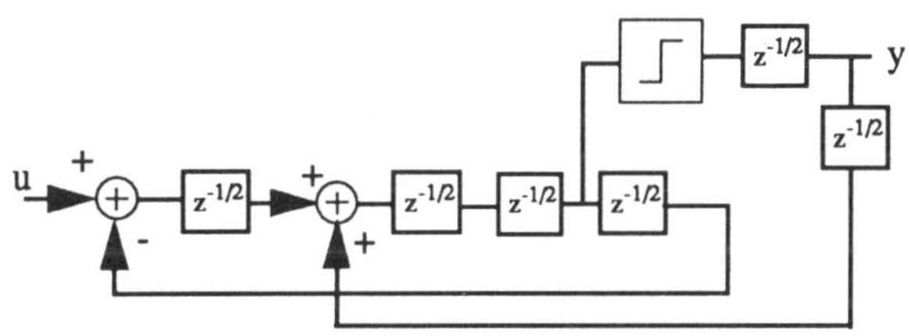

Fig 1. A second order bandpass ΣΔ modulator

Modulator implementation

The basic building block of the modulator is a bipolar track and hold circuit. Any track and hold circuit with a 50% duty cycle holds the sampled data for half of a clock period and therefore introduces half a clock delay. We can represent this action in the z-domain by saying that the track/hold circuit has a transfer function of $T(z) = Az^{-0.5}$ where A is the gain of the track/hold. Because we always use the half-delays in pairs, variation in clock timing from a 50% duty cycle is unimportant.

The modulator in Figure 1 can be implemented by cascading four track/hold circuits in a master-slave configuration.The complete circuit implementation is shown in

0-7803-3177-6 $5.00 © 1996 IEEE

IEEE 1996 CUSTOM INTEGRATED CIRCUITS CONFERENCE

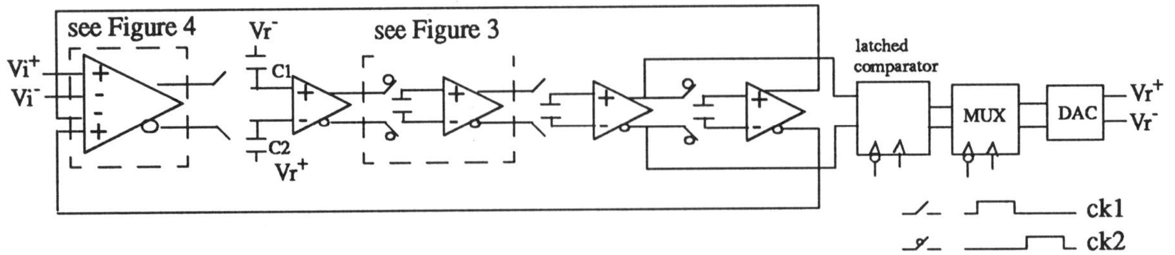

Fig 2. Simplified block diagram of our second-order modulator

Figure 2. For simplicity the track/hold circuits are presented as a pair of switches charging a capacitor followed by an amplifier. The quantizer is a master-slave latch D-type flip-flop providing a full clock delay for the quantizer feedback signal. The quantizer is followed by a multiplexer that controls the one bit DAC. During ck2 it passes the quantizer levels and during ck1 it inverts the quantizer levels. In this way it provides signal levels at the bottom plate of C1 and C2 that add to the loop signal by KVL, thus avoiding the need to sum signals at a virtual ground as is done in switched-C circuits.

The track and hold circuit design

A diode-bridge based track/hold circuit was used for this design. The diodes were npn devices that had their bases tied to their collectors. Since the diode bridge is a single-ended circuit and the modulator requires a fully balanced signal a quasi-differential circuit was used, (Figure 3) consisting of two bridges followed by a differential output buffer in a unity gain configuration [11].This way we reduce even-order distortion. The track/hold shown corresponds to the switches, sampling capacitor and buffer of Figure 2.

The input impedance of the buffer was designed to be fairly high to reduce droop caused by the input current to the output buffer that discharges the hold capacitor. Simulations predicted that the track/hold would have a small-signal bandwidth of 550 MHz and a small- signal gain of -0.5 dB in the hold mode. In simulation it also exhibits 8 bits linearity for signals up to 50MHz when clocked at 500 MHz. Its power consumption is 55mW when operated from a 5V power supply.

Summation/subtraction implementation

Our $\Delta\Sigma$ implementation emulates the classic SC-based designs. In a differential SC approach the number of switches required to perform a summation is eight. Following the same approach would increase the power consumption of our design drastically; it would also require an op-amp with a unity-gain frequency of nearly 2GHz. Instead we prefer to use the SC approach shown in Figure 2 to sum the signal coming from the DAC. Because the DAC signal is balanced and has a set value we do not use any additional switches but we take advan-

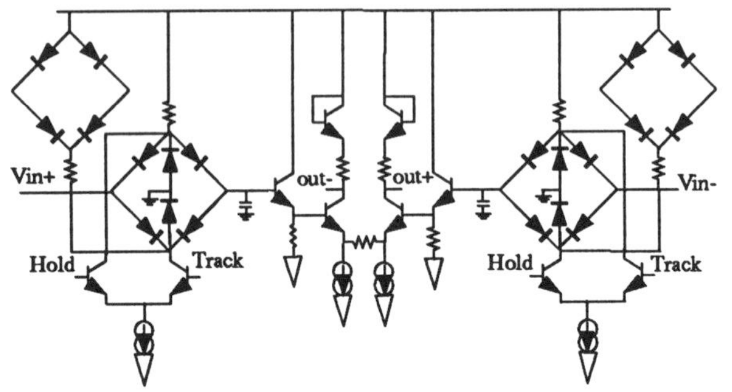

Fig 3. Schematic of the track/hold circuit.

tage of the clock phasing by multiplexing back at the bottom plate of capacitor C1 and C2 the right DAC level to perform the summation. This method is not applicable for the loop filter feedback.

To introduce the input signal to the loop a continuous-time summation was used as shown in Figure 4. A common-emitter differential pair with separate parallel inputs provides current summation at the load resistors. The differential pair was emitter-degenerated to achieve unity gain. This stage allows intermodulation between the input and loop signals and is a candidate for redesign.

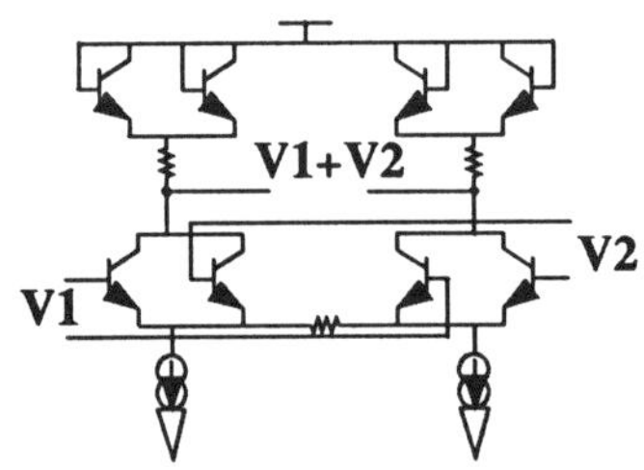

Fig 4. A current summing summation

The comparator and the multiplexer are ECL based cells and their detailed design will not be discussed in this paper. They were optimized to meet the speed requirements of the modulator while trying to minimize power. Their total power consumption is 64 mW.

Experimental Results

A photomicrograph of the modulator is shown in Figure 5. The active area of the test chip is 975υmX525υm. All the bitstreams for the results presented here were captured by a logic analyzer and a Hanning-windowed 32768 point FFT was performed. A typical output spectrum of the modulator when clocked at 250 MHz is shown in Figure 6. Figure 7 is an in-band plot of the same measurement. Based on this measurement the SNR was 41dB for a 1MHz bandwidth.

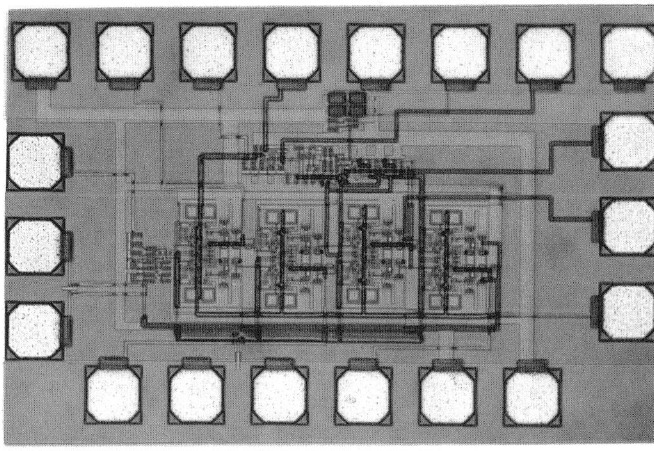

Fig 5. Photomicrograph of the modulator.

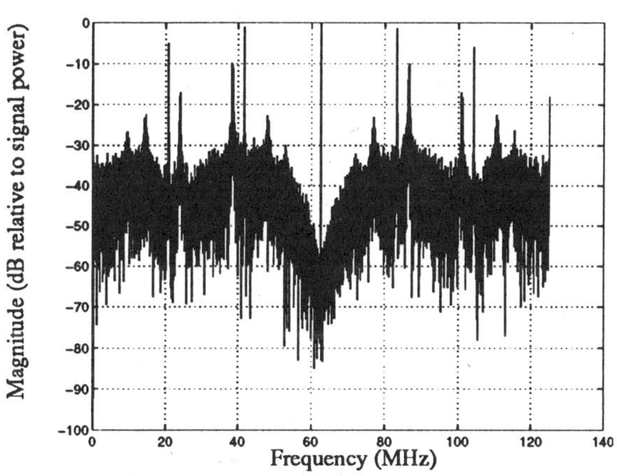

Fig 6. Output spectrum of the second order bipolar BPΔΣ for a full-scale tone at fs/4. Clock frequency is 250 MHz

Measurements were taken for different input signal levels. The SNR at each signal level is plotted in Figure 8 (solid line) and is compared with the SNR predicted for an ideal second order bandpass sigma-delta modulator (dashed line). The experimental SNR is 20dB lower than the idealized model predicts. As was noted in above, the track and hold gain is less than unity and that reduces the in-band noise attenuation; the flat passband noise in Figure 7 is also characteristic of this. Figure 9 plots simulated SNR versus the percentage of gain reduction below unity of the track and hold cell, for a half-scale tone input at 62.5 MHz and for a clock frequency of 250 MHz and a signal bandwidth of 1 MHz. In a wider band the actual performance is closer to the ideal.

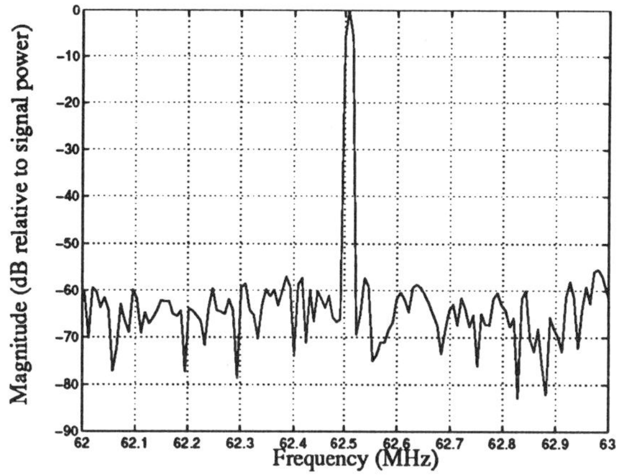

Fig 7. In-band-plot of Figure 6. The abscissa contains131bins

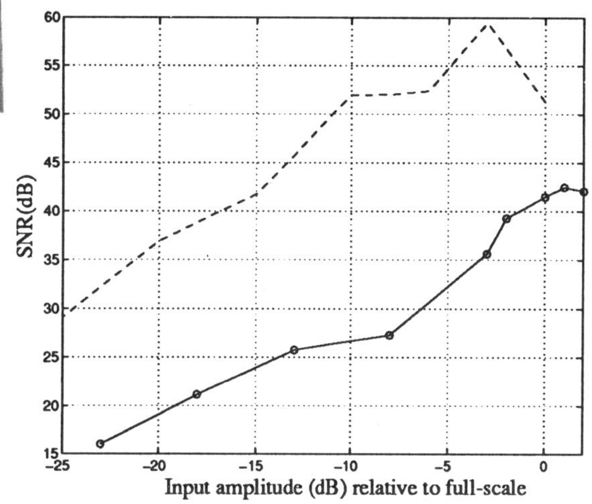

Fig 8. SNR for 1MHz bandwidth and a clock frequency of 250MHz.

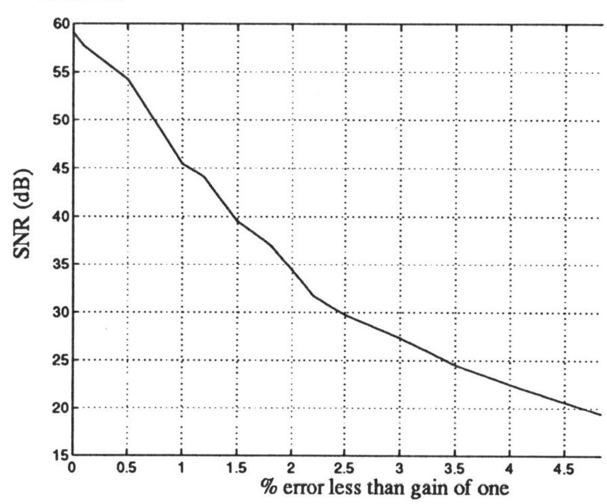

Fig 9. SNR (dB) versus percentage of gain reduction of the track/hold

Conclusions

The design and experimental evaluation of a fully bipolar second-order bandpass delta-sigma modulator were presented. The prototype design demonstrated the feasibility of a bipolar-only sampled data ΔΣ modulator. It can be an alternative to continuous time solutions to provide the high oversampling necessary for wide band applications. Because this is a sampled data technique it does not require any tuning to achieve the desired frequency of operation. It is important though to note the sensitivity of the structure to the gain of the track and hold cell. Other bipolar track and hold circuits are under evaluation in an effort to provide a cell with more nearly unity gain. Also, we would like to note that fourth order design would show less tonal behavior and higher SNR and make this approach useful for wide band applications.

Acknowledgments

Fabrication of the chip was provided by Northern Telecom. Financial support for this work was provided by the OCRI/NSREC high speed industrial chair and TRIO. The evaluation boards were made by Philip Lauzon. I would also like to thank Dr. Omid Shoaei and Ash Swaminathan for useful discussion on the ΔΣ theory.

References

[1] R. Hadaway et al, "A sub-micron BiCMOS technology for Telecommunications" J. of Microelectronics Engineering, vol.15, pp. 513-516, 1991.

[2] Lorengo Longo and Bor-Rong Hong, "A 15b 30KHz bandpass sigma-delta modulator", ISSCC-93 Digest, pp.226-227.

[3] Frank W. Singor and W. Martin Snelgrove, "Switched- capacitor bandpass delta-sigma A/D modulation at 10.7MHZ", IEEE J. Solid-State Circuits, vol 30, no 3, pp. 184-192, March 1993.

[4] Philsar Electronics, "10-bit high speed bandpass sigma-delta A/D converter with digital I/Q", advanced data sheet, December 1995.

[5] G. Troster, et al, "An interpolative bandpass converter on a 1.2 sum BiCMOS analog/digital array", IEEE J. Solid-State Circuits, vol 28, no. 4, pp. 471-477, April 1993.

[6] Omid Shoaei, "Continuous-time delta-sigma A/D converters for high speed applications" Ph.D. Thesis, Carleton.

[7] W. Gao,W. Martin Snelgrove, TheodoreVarelas, S. J. Kovacic, and D. L. Harame,"A 5-GHz SiGe HBT return-to-zero comparator" Proc. of IEEE 1995 Bipolar/BiCMOS Ciruits and Technology Meeting,pp. 166-169, Oct. 1995.

[8] J. F. Jensen, G. Raghavan, A.E. Cosand and R.H. Walden "A 3.2GHz second-order delta-sigma modulator Implemented in InP HBT technology" IEEE J. of Solid State Circuits, vol.30, no. 10, October 1995.

[9] S. Jantzi, R. Schreier and M. Snelgrove, "The Design of Bandpass Delta-Sigma ADCs," chapter in Delta-Sigma Data Converters, IEEE Press, to appear.

[10] Seyfi S. Bazarjani, Theodore Varelas and W. Martin Snelgrove, "Sample-and-Hold Based bandpass ΔΣ modulators.

[11] W. T. Colleran and A.A. Abidi, "A 10-b, 75-MHz two dtage Pipelined Bipolar A/D Converter", IEEE J. of State Circuits, vol. 28, no. 12, pp. 1187-1199, December 1993.

A 900 MHz CMOS Frequency-Hopped Spread-Spectrum RF Transmitter IC

A. Rofougaran, G. Chang, J. J. Rael, M. Rofougaran,
S. Khorram, M-K. Ku, E. Roth, A. A. Abidi, and H. Samueli

Electrical Engineering Department
University of California
Los Angeles, CA 90095-1594

Introduction

This paper reports on the first monolithic CMOS RF transmitter, which is part of a complete low-power transceiver operating in the 902-928 MHz ISM frequency band [1]. The system can transmit data at rates up to 160 kb/s by frequency-shift keying a carrier which is periodically hopped according to a pseudorandom code across 26 MHz. The key elements to accomplish this in the transmitter are: an *agile* baseband *frequency synthesizer*, a low-noise 915 MHz *local oscillator*, *upconversion mixers*, some needed *RF filters*, and a *power amplifier*. All these elements are integrated on a 1-μm CMOS IC (Fig.1(a)).

Transmitter Architecture

At the core of the transmitter architecture is the agile frequency synthesizer. Hopping frequency synthesizers are conventionally based on a phase-locked loop (PLL) with a digitally-controlled variable divider. Finely separated frequencies may be synthesized from a fixed crystal-based reference by operating the divider in fractional-N mode. This approach suffers from two limitations: the finite PLL unity-gain bandwidth, often about one-tenth of the reference frequency, results in a long loop settling-time when the divider modulus is switched; and it is difficult to make a VCO with a wide tuning range, whose output spectrum is also acceptably free of harmonics and spurious frequencies over that range. We have instead adopted a solution based on direct-digital frequency synthesis (DDFS) to overcome both these problems. A 12-bit input frequency control word sets the accumulation rate in a register, which addresses ROMs containing coefficients of the trigonometric sine and cosine waveforms (Fig.2) to produce digital-domain quadrature outputs at the programmed frequency. The accuracy of ROM coefficients determines the harmonic and spurious frequency content, guaranteed to be -72 dBc or less in this implementation. The DDFS output is followed by a 10-bit D/A converter (DAC) to obtain an acceptably low harmonic distortion and noise at the synthesizer output. We have previously reported measurements on a prototype of this agile frequency synthesizer [2]. Frequency-shift keyed modulation is superimposed on the hopping carrier by simply adding or subtracting a small, fixed offset to the frequency-control word.

The quadrature outputs of the DDFS/DAC are upconverted by quadrature phases of a fixed frequency 915 MHz local oscillator (LO), and added or subtracted to select either the upper or the lower side-frequency. In this way, any frequency may be synthesized in the 902-928 MHz band, to within the resolution of the 12-bit frequency control word. DAC and subsequent nonlinearities will set the important spurious frequencies in the output spectrum. This arrangement requires a *fixed-frequency* 915 MHz LO, which may be entirely integrated and then embedded in a wideband PLL to reduce phase noise.

Each upconversion mixer in the quadrature channels comprises a four-FET commutating switch. With a sufficiently large LO voltage applied at the FET gates, this mixer upconverts a complex baseband spectrum with little intermodulation distortion. An on-chip bandpass filter suppresses signal energy upconverted by the 3rd LO harmonic.

The local oscillator consists of a pair of cross-coupled *LC* oscillators, which synchronize into quadrature phases [3] (Fig.3). A source-follower FET with a voltage-controlled PMOS resistor in series with its gate synthesizes a load which appears inductive at high frequencies. This resistor tunes the LO frequency. The LO output phase noise spectral density is -85 dBc/Hz at a 100 kHz offset from the carrier.

The UCLA Low-Power Wireless Transceiver Project is supported by the US Advanced Research Projects Agency, and by a consortium of semiconductor companies under the State of California MICRO Program. For more information, e-mail abidi@icsl.ucla.edu

0-7803-3177-6 $5.00 © 1996 IEEE

IEEE 1996 CUSTOM INTEGRATED CIRCUITS CONFERENCE

A prototype of the on-chip power amplifier has been previously published [4]. Under control of a digital word, it delivers power in the range of –15 dBm (30 μW) to +15 dBm (30 mW) to an antenna load in increments of 1 dB. The power amplifier comprises an inductively loaded common-source FET preamplifier, which drives a binary-weighted array of FETs whose drains tie together into the off-chip load (Fig.4). The array of FETs is biased close to threshold, and attains a peak power-conversion efficiency of about 40% when amplifying a single-tone input. Two halves of the amplifier drive the antenna in balanced mode.

Circuit Design

The frequency spectrum of the discrete-time DDFS analog output contains images of the main tone around the clock frequency. As shown in Fig.1(b), the single-sideband upconversion mixer selects the images ❷ and ❸ separated from the main tone ❶ by exactly the clock frequency, while suppressing the other image in much the same way that it cancels the unwanted sideband of the main tone. These images must not radiate from the antenna. They may be removed after the DAC with a lowpass filter of sufficiently high order, but with a 20 MHz cutoff frequency, it is very difficult to build an active filter with a 60 dB linearity commensurate with the DAC output.

Instead, the DDFS/DAC clock frequency is raised to 80 MHz, and the upconverted tone and the image frequencies are applied to the power amplifier with no prior filtering. All harmonics and intermodulation tones that nonlinearity in the power amplifier creates are also separated at least by 80 MHz, and fall well into the stopband of the three-resonator, passive, off-chip dielectric resonator filter connected to the amplifier output (Fig.1(b)). With a stopband attenuation of 40 dB or greater, this passive filter rejects the images without distorting the signal. Owing to the *sin x/x* rolloff in the spectrum of the sampled waveform, the two main image tones, ❷ and ❸, are at least 15 dB lower than the main tone, and do not overload the power amplifier to the point that it suffers a loss in efficiency.

The quasi-passive DAC core consists only of capacitors and switches (Fig.5(a)), and readily operates with a three-phase clock at 80 MHz. In the earlier prototype the op amp charge-to-voltage buffer required two additional clock phases, and its settling time became the main bottleneck to higher clock rates. We have reconfigured it to re-use two of the DAC clock phases, and now this too clocks at 80 MHz.

The DAC, through the four-FET upconversion mixer, must drive the input capacitance of the on-chip power amplifier with a 0.5V full-scale signal. This capacitance is too large a load for the op amp, so a second open-loop buffer consisting of a differential pair degenerated by a polysilicon resistor is used (Fig.5(b)). The spurious-free dynamic range at the DDFS/DAC output, prior to the second buffer, is about 60 dB [2]. Although the buffer drains a relatively small current, it adds –45 dB 3rd harmonic distortion to the full-scale DDFS output, and is the weak point in overall system linearity.

The two four-FET switch mixers are shorted together at the outputs (Fig.6) to select one sideband. The sidebands upconverted by the 3rd harmonic of the LO must be adequately suppressed, otherwise they will alias as spurious tones within the ISM band after suffering power amplifier nonlinearity. A low-Q bandpass filter, consisting of an on-chip 50 nH spiral inductor in series the input capacitance of the power amplifier, adequately suppresses signals at 2.7 GHz. The spiral inductor is suspended above a cavity in the substrate (Fig.7) [5], which greatly reduces the parasitic capacitance and increases its self-resonance to beyond 2.7 GHz.

All the above circuits operate on a 3V power supply.

Experimental Results

The entire transmitter is fabricated in a 1-μm CMOS, double-metal technology with a linear capacitor. It occupies an active area of 6×3.8 mm (Fig.8) mainly consumed by the large on-chip inductors. The circuit, except the LO, drains 55 mA when operating at speed. The LO with spiral inductor loads drains 10 mA, while in this version with simulated inductors the current drain is 30 mA.

The transmitter output spectra are evaluated at a mid-scale power amplifier setting of about +5 dBm (3 mW). A 10 MHz DDFS/DAC tone shows several harmonics in the upconverted spectrum at about –50 dBc, which is also the LO leakage level (Fig.9). The unwanted sideband is roughly at –43 dBc. This performance is either at par or superior to sideband selection in previously reported integrated high-f_T bipolar transmitter ICs for GSM [6-8], and implies an on-chip quadrature inaccuracy of less than 1°.

The transmitter is also evaluated by hopping the DDFS in the lower-half ISM band, and measuring the average levels of the unwanted sidebands lying in the upper-half band, which will interfere with other users in that part of the band (Fig.10). All the unwanted sidetones are 46 dB lower than the wanted signal, except one which is only 40 dB lower. This is an acceptable sideband suppression in a microcell-based system where transmitted power is controlled over a 30 dB range. No off-chip filters are inserted at the power amplifier output for these measurements.

These results show that with the appropriate mix of architecture and circuits, it is possible to attain a a level of integration and performance in a modest 1-µm CMOS technology comparable to more advanced RF silicon technologies, with no apparent untoward effects of parasitic coupling through the substrate or other observable detriments.

References

1 J. Min, A. Rofougaran, H. Samueli, and A. A. Abidi, "An All-CMOS Architecture for a Low-Power Frequency-Hopped 900 MHz Spread-Spectrum Transceiver," presented at *Custom IC Conf.*, San Diego, CA, pp. 379-382, 1994.

2 G. Chang, A. Rofougaran, M. K. Ku, A. A. Abidi, and H. Samueli, "A Low-Power CMOS Digitally Synthesized 0-13 MHz Agile Sinewave Generator," presented at *Int'l Solid State Circuits Conf.*, San Francisco, pp. 32-33, 1994.

3 A. Rofougaran, J. Rael, M. Rofougaran, and A. A. Abidi, "A 900 MHz CMOS LC Oscillator with Quadrature Outputs," to be presented at *Int'l Solid State Circuits Conf.*, San Francisco, CA, 1996.

4 M. Rofougaran, A. Rofougaran, C. Olgaard, and A. A. Abidi, "A 900 MHz CMOS RF Power Amplifier with Programmable Output," presented at *Symp. on VLSI Circuits*, Honolulu, pp. 133-134, 1994.

5 J. Y.-C. Chang, A. A. Abidi, and M. Gaitan, "Large Suspended Inductors on Silicon and their use in a 2-µm CMOS RF Amplifier," *IEEE Electron Device Letters*, vol. 14, no. 5, pp. 246-248, 1993.

6 W. Veit, J. Fenk, S. Ganser, K. Hadjizada, S. Heinen, H. Herrmann, and P. Sehrig, "A 2.7V 800MHz-2.1GHz Transceiver Chipset for Mobile Radio Applications in 25 GHz f_T Si-Bipolar," presented at *Bipolar Circuits & Technology Mtg.*, Minneapolis, pp. 175-178, 1994.

7 T. Stetzler, I. Post, J. Havens, and M. Koyama, "A 2.7V to 4.5V Single-Chip GSM Transceiver RF Integrated Circuit," presented at *Int'l Solid State Circuits Conf.*, San Francisco, pp. 150-151, 1995.

8 C. Marshall, F. Behbahani, W. Birth, A. Fotowat, T. Fuchs, R. Gaethke, E. Heimerl, S. Lee, P. Moore, S. Navid, and E. Saur, "A 2.7V GSM Transceiver IC with On-Chip Filtering," presented at *Int'l Solid State Circuits Conf.*, San Francisco, pp. 148-149, 1995.

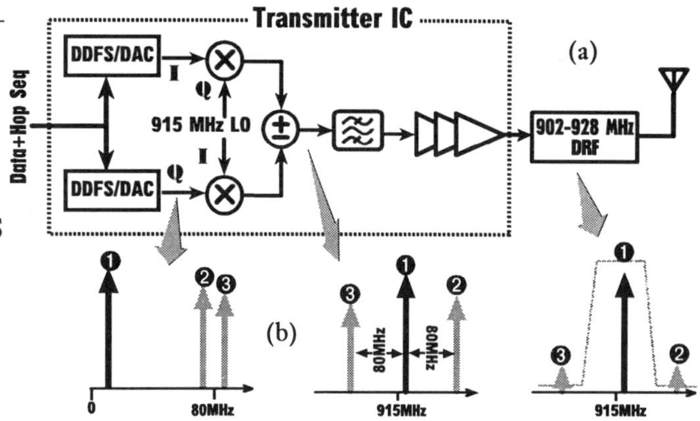

Fig.1 (a) Block diagram of transmitter IC. (b) Frequency spectrum of a single tone generated by DDFS in various parts of the transmitter. The output dielectric resonator filter acts as an image-suppression mixer.

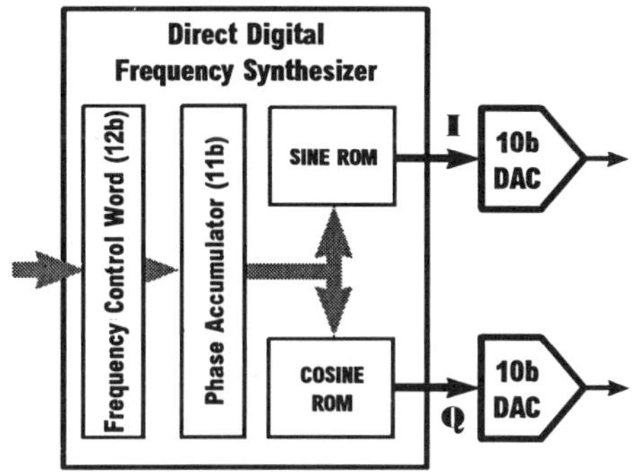

Fig.2 Block diagram of DDFS/DAC unit

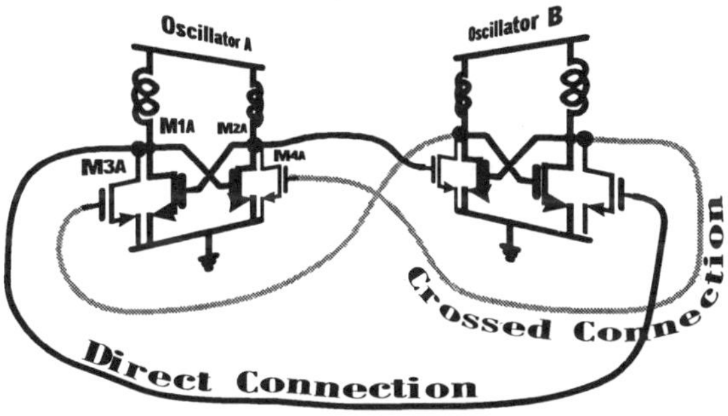

Fig.3 LC oscillator which produces balanced outputs in quadrature.

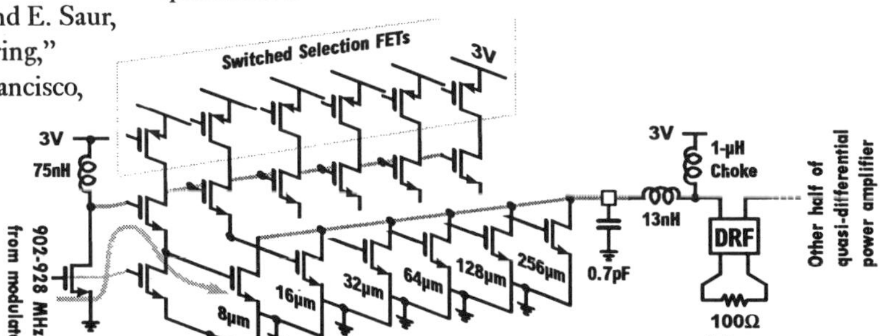

Fig.4 Power amplifier consists of binary-weighted array of FETs.

10.7.3

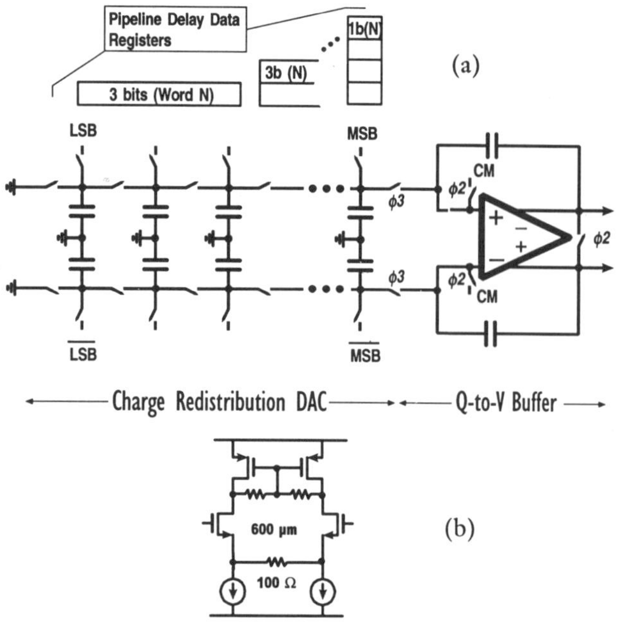

Fig.5 (a) Quasi-passive 10-bit pipelined charge redistribution DAC; (b) Additional buffer to drive power amplifier through upconversion mixers.

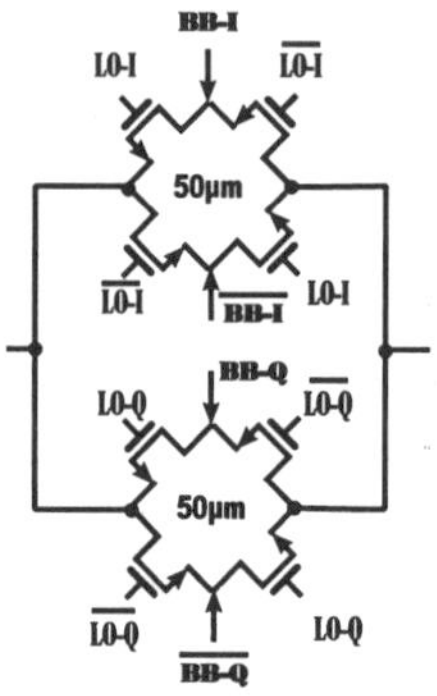

Fig.6 Four-FET switch upconversion mixers in I and Q channels sum outputs as currents.

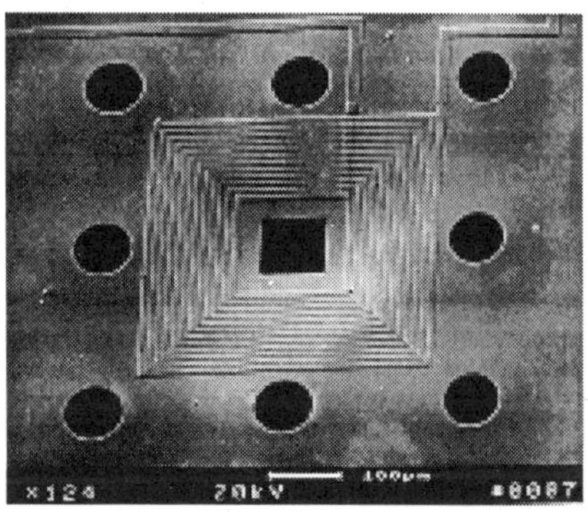

Fig.7 A large-inductance spiral suspended above a cavity on a CMOS substrate. The etchant isotropically excavates a cavity under the exposed holes surrounding the spiral.

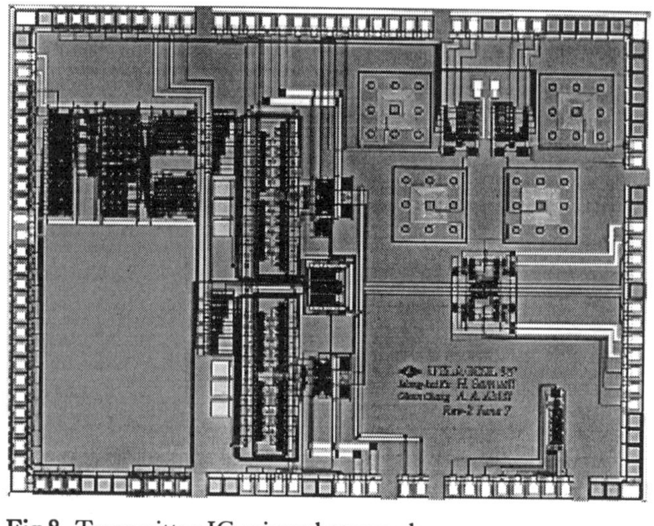

Fig.8. Transmitter IC microphotograph

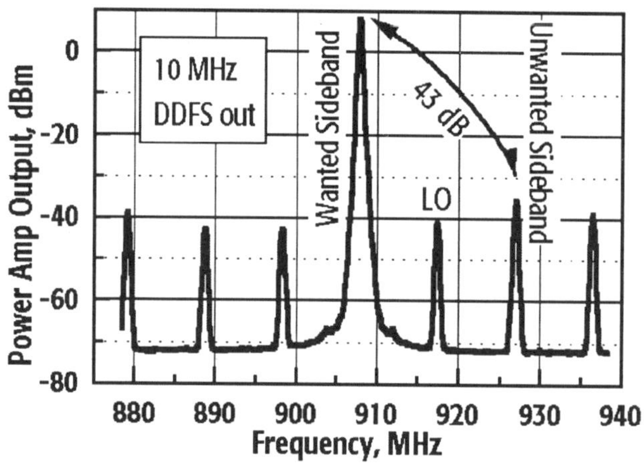

Fig.9 Measured output spectrum from power amplifier, with single 10 MHz tone generated by DDFS.

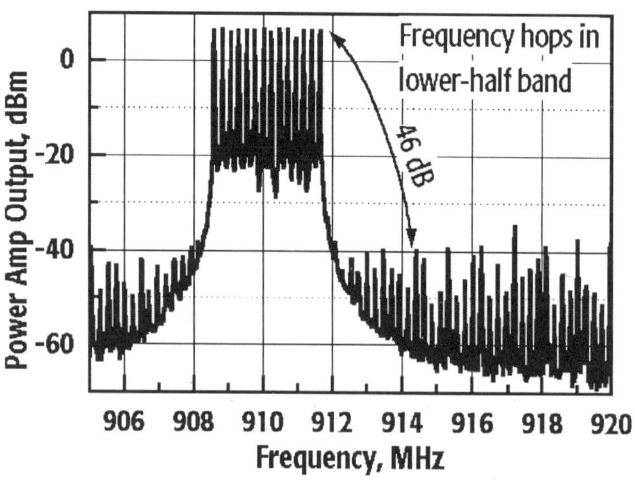

Fig. 10 Measured output spectrum from power amplifier, with DDFS hopping over 14 frequency slots in lower half of 902-928 MHz band.

Chair: Dundar Dumlugol
Co-Chair: Kartikeya Mayaram

Keeping the power dissipation within bounds is rapidly becoming one of the main challenges in contemporary digital design. This requires a "power conscious" design methodology that addresses dissipation at every level of the design hierarchy together with accompanying design automation tools. An intensive effort in the recent years has resulted in both academic and industrial tools and methodologies.

A tutorial paper in this session focuses on the analysis facet of the low power design process. The paper starts with a discussion of power analysis at the circuit and logic design abstraction levels for which mature commercial solutions are available. However, in the author's opinion, it is the high-level power prediction tools, including "power exploration" at the RTL level and above, that will have the greatest impact on power reduction.

In the following paper, an efficient and accurate Monte-Carlo approach to the problem of estimating average node switching probabilities in sequential circuits is presented as the basis for power estimation. Error bounds together with experimental results are discussed.

The next paper presents a non-parametric technique for statistical power estimation which has advantages over traditional techniques because it does not require parametric distribution functions. Lower and upper bounds of the average power are derived together with a stopping criterion for a desired percentage error for the estimated power with a specified confidence level.

An Automatic Test Generation (ATG) method to generate the input patterns for maximum power is the subject of the following paper. Benchmark results of the ATG method are compared to the traditional simulation method both in terms of speed and performance.

A combined AC and transient power distribution analysis is the topic of the next paper which proposes a method that efficiently derives both steady state and transient responses from a series of fixed-time point transient impulse responses.

Exploring the Power Dimension

Jan M. Rabaey

Department of EECS, University of California at Berkeley

1. Introduction

Keeping the power dissipation within bounds is rapidly becoming one of the main challenges in contemporary digital design. Design experience and research in the early 1990s has amply demonstrated that doing so requires a "power conscious" design methodology that addresses dissipation at every level of the design hierarchy [Chan95]. Evidently, this cannot be achieved if no accompanying design automation environment is available. Spurred by this observation, we have seen an intensive effort in the domain of design automation for low power in recent years that has resulted in both academic and industrial tool environments [Sin95].

In this tutorial, we only focus on the analysis facet of the low power design process. Not only is the availability of analysis, simulation, and prediction tools the "conditio sine qua non" for low power design, it also represents the most understood and established component of the low-power design methodology. The paper starts with a discussion of power analysis at the circuit and logic design abstraction levels. This domain is becoming relatively mature with multiple entries commercially available. In our opinion, it is the high-level power prediction however that will have the most impact on the reduction of power dissipation. We have therefore included an in-depth discussion on the efforts and results in the so-called "power exploration" domain, what includes power prediction and analysis at the RTL level and above.

2. Power Analysis at the Circuit and Logic Levels

With the reemergence of power as an important parameter in the design process, it became rapidly clear that no adequate and efficient tools were available for the analysis or prediction of the dissipation of a complex design. The only option available was to run SPICE simulations on the transistor-level circuit diagram. This solution is rendered inefficient by the extra requirement that power simulations are performed over a large sequence of input patterns. In fact, dissipation is not only a function of the signal probabilities (i.e. the probability of being 0 or 1) of the input patterns, but also of the spatial and temporal correlations of excitations. For instance, the signals "0101" and "0011" have the same signal probabilities but display a considerably different transition pattern. As a result, simulations have to be performed over a meaningful set of realistic input patterns. Consequently, techniques have been devised in years to expedite the analysis process, some of which are discussed in the paragraphs below.

2.1 Sources of Dissipation

To understand the effectiveness and the accuracy of a power estimator, we should first have a clear insight in the sources of dissipation in a digital circuit and how they are addressed in the estimator. Power dissipation sources in CMOS circuits can be classified in three different groupings [Rab95]:

- The charging and discharging of circuits capacitors during transitions. Some of these (dis)charges, called glitches or dynamic hazards, are non-intentional and are caused by timing effects.
- Short-circuit currents, which are due to the dc path between the supply rails during transitions.
- Static power consumption. This can be consequence of the chosen circuit topology (such as the pseudo-NMOS gate) or by leakage effects.

The dynamic dissipation, represented by the first bullet, represents the dominant source of dissipation in standard CMOS circuits, ignoring the latter factors can drastically impact the accuracy of the power estimation.

2.2 Power Analysis at the Circuit Level

Obvious ways to expedite the power analysis at the circuit level are to simplify the transistor model, or/and the data representation.

- *Timing simulators* use an event-driven simulation mechanism and simplified, linearized transistor models to provide a relatively accurate transient analysis at a fraction of the computational cost of a SPICE-like circuit simulation. It furthermore models all dissipation sources in an adequate way, although the simplified transistor model is rather inaccurate in modeling leakage effects. A prime example of a timing simulation-based power estimator is the PowerMill tool of Epic Design Technology [Den94].
- When even more efficient simulation is desired, we can simplify the transistor model further to a simple resistive switch and used a discrete data representation {0,1,X}. This class of simulators, called *switch simulators*, has the disadvantage that

0-7803-3177-6 $5.00 © 1996 IEEE

IEEE 1996 CUSTOM INTEGRATED CIRCUITS CONFERENCE

the short-circuit and leakage current fractions are ignored. Glitch effects are accounted for in a pessimistic fashion as it is assumed that every glitch is rail-to-rail which is rarely the case in reality. The major plus-point is the reduction in simulation speed with orders of magnitude compared to SPICE. An example of switch-level power estimator is the IRSIM-CAP simulator [Lan94], which is a modification of the popular IRSIM tool (developed at Stanford).

To put things in perspective, we have compared SPICE, timing simulation, and switch simulation for two simple benchmark circuits — an 8-bit adder and an 8-bit TSPC shift register. The errors with respect to SPICE and the speed-up are shown in Table 1. It can be observed that the timing results are in close correspondence with SPICE for both circuits (error < 7%). Switch-level analysis displays a large error of 27% for the adder, where the effects of glitches and short-circuit currents are considerable.

Table 1. Comparison between circuit-level simulation, and timing and switch-level power analysis.

	Adder		Shift Register	
	% Error	Speedup	%Error	Speedup
Timing	6	15	7	3.7
Switch	27	60	4	22

2.3 Power Analysis at the Logic Level

Power prediction at the logic level has been the topic of extensive research in recent years, spurred by the popularity of gate-level design methodologies. From an accuracy perspective, power analysis at the logic level only addresses the dynamic component of the power budget, similar to the switch-level. Moreover, all capacitance is lumped into the output nodes of the gate and is relatively unknown, as it depends upon precise information from the cell-library, fan-out structure, and interconnect.

Once again, extensive simulation of the logic structure can be used to generate the necessary signal statistics. This is the approach followed in the Entice-Aspen tool of Motorola [Geo94]. The Aspen environment combines the statistics, obtained from an ordinary logic simulator, with cell characterization data generated by Entice. Using this approach, it is even possible to estimate (roughly) some of the direct path and static power components. For a 16-bit adder, a relative error of 10.1% was reported with respect to SPICE for a speedup of 4 orders of magnitude [Geo94].

On the other hand, the logic level has the advantage that the output data statistics of a gate are a well-defined function of the input statistics. It should therefore be possible to compute the switching statistics of every node in a network as a function of the input statistics. For instance, given a 50% chance of "1" inputs at the input of a 2-input NAND static CMOS gate, the chances of a 0->1 transition to occur at the output equals 3/16. This observation gives rise to an alternative power-analysis technique, called *probabilistic estimation*. In this approach, switching statistics for the various nodes in the network are obtained using a static evaluation and no time consuming simulation is needed. While simple at first glance, the evaluation of these statistics is complicated by correlations between signals, caused by reconvergence. Feedback in sequential circuits and real delay models make the evaluation even more complex. A number of efficient probabilistic evaluation techniques have been proposed. For a detailed description, please refer to [Ped96]. The jury is still out regarding the applicability and the accuracy of the probabilistic techniques.

3. Early Exploration

The central foundation of a sound system-level design methodology for low power is the capability to identify potential design bottlenecks in the early phases of the design process. At this phase, absolute accuracy in the power prediction is not the focal issue. The main goals are to break down the power dissipation of a design under consideration into its main components, to identify the dominant consumption sources and to evaluate the impact of design decisions on the power budget. In other words, a relative accuracy should be maintained between the various design components and between various versions of the same design. For design exploration to be useful, it is furthermore required that the process should be simple and require a minimum of user input, in terms of component models and system description — both are in fact relatively unknown, or undefined at this early phase of the design process.

Similar to the power analysis at the circuit and logical levels, power estimation requires a prediction of two components: the physical capacitance that is being switched C, and the average rate of switching (switching activity α) as described in (EQ 1).

$$P = \left[\sum_{\text{all hardware resources}} \alpha_{res} C_{res} \right] V_{DD}^2 \qquad \textbf{(EQ 1)}$$

While capacitance estimation has been explored before in the realm of area prediction, it is the statistical analysis that makes power prediction a new challenge. In the subsequent sections, we will describe techniques to evaluate both factors at various stages of the design process.

3.1 Switching Activity

Switching activity can translate into many views, depending upon the abstraction level or the application domain envisioned. When the only known fact about an application is its behavioral description, activity is almost synonymous to complexity — the more complex the description, the higher the probable power dissipation. When more details regarding the implementation structure are becoming available, the activity model can be refined.

Activity Estimation at the Behavioral Level

In the conception phases of a design, the only source for the activity analysis is the behavioral description of the design, i.e. the functional code that specifies the required operation. This could come in the form of a textual program (e.g. using C, C++, or VHDL) or in a schematics flow-chart (e.g. a dataflow graph or a state-chart). Such a description does not imply an implementation architecture, and hence does not allow for an estimation of the architecture-dependent switching effects. It however allows for a fairly detailed analysis of the so-called **algorithm-inherent activity**.

> **The algorithm-inherent activity of a design is the switching activity that has to occur in any implementation of the design, independent of the implementation style chosen.**

For instance, if an application requires the execution of N multiplications per execution, these multiplications have to be performed independent upon the fact that the chosen architecture is a single-treaded Von-Neuman structure, a super-scaler machine or an array-processor. Arithmetic operations, input-output operations, and basic memory accesses are the most important members of the inherent-activity class.

The extent of the inherent activity can be determined using a number of techniques:

- *Dynamic profiling* tallies the occurrence-rates of basic operations during a typical execution of the algorithm for a set of input-excitation vectors, defined by the designer. This approach has the advantage of being readily available. Profilers such as *gprof* and *quantify* [Qua93] provide ample information on the breakdown of the computational complexity of a code-fragment into its components. The main problem with the current incarnation of these tools is that the granularity of the profiling is confined to the subroutine level. For power-analysis purposes, a breakdown up to the operator level is necessary. Profilers that do so (e.g. the Pixie profiler from MIPS Computer Systems) are operating at the assembler level, and are hence implementation sensitive.

Example: Profile of VSELP speech coder

Fig. 1. shows the dynamic profile constructed for a VSELP coder using the Quantify profiler, augmented with an operator tallying capability. It reveals that only a small percentage of the code is responsible for most of the computational complexity, and presumably also for power dissipated in the datapath components. Any low-power strategy should obviously focus on bringing these components down.

- The *static profiling* approach determines the inherent activity of the code by analyzing the program structure. The occurrence of each operation can be tallied by code traversal. Combining this information with the (known) statistics of the input signals (e.g. repetition rate) allows for a detailed prediction of the activity. This has a number of obvious pitfalls. First, it only works with deterministic programs. The presence of data-dependent cycles and branches makes it impossible to predict exactly the number of operations. This problem can be resolved when branch and loop statistics are available such as obtained from dynamic profiling. In that case, an average operation count can be determined. Secondly, it requires the availability of special code-analysis tools, which in general are not available. The static profiling approach has a number of strong arguments in its favor as well. It has the advantage of being fast — it requires only a single perusal of the syntax graph of the code —, it is relatively independent upon the excitation vectors, and it presents substantially more information, that could be used to improve the power profile. Examples are the amount of concurrency available in the various parts of the algorithm, the partitionability, the regularity, etc. [Rab95].
- While the previous bullets deal with the occurrence rate of the operators, algorithmic profiling can also yield information regarding the data statistics of the algorithm, i.e. the mean, variance and correlation coefficients of the various data signals. For instance, Fig. 2. shows how the transition probability for the various bits of a two's complement data word is impacted by the mean value and the autocorrelation coefficient ρ. Positive correlation between consecutive samples can lead to a substantial reduction in signal activity. The data statistics can help to predict the power-savings of architectures that preserve signal correlations. This is not the case in general-purpose programmable architectures, that tend to multiplex all data-streams over time-shared hardware

once per frame	LPC Analysis		(2.1%)
	Subframe Processing		(95.9%)
	once per subframe	Compute Weighted Input Speech	(3.9%)
		Compute Lag	**(21.9%)**
		Filter Basis Vectors for Codebook #1	**(23.2%**
		Orthogonalize Filtered Basis Vectors	(0.8%)
		Search Codebook #1	**(4.1%)**
		Filter Basis Vectors for Codebook #2	**(23.3%)**
		Orthogonalize Filtered Basis Vectors	(2.0%)
		Search Codebook #2	**(4.1%)**
		Quantize Gains	(3.6%)
		Update Filter State	(1.7%)

Fig. 1. Dynamic profiling of VCELP. The top three functions take 77% of the total compute time. Dot-vector products account for 70% of the total time.

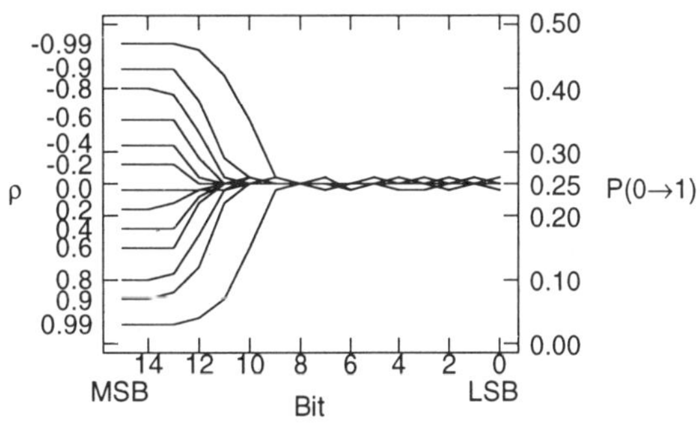

Fig. 2. Impact of temporal correlation between consecutive samples in two's complement notation.

resources. Under those circumstances, all data can be modeled as uniform, white noise ($\rho = 0$).

The presented estimations have to be treated with care as they do not take into account the **architecture-induced switching overhead**. This includes the power consumed in interconnect, controllers, registers, and clock, as well as the extra switching introduced by hardware sharing. For most applications, the overhead takes more than 80% of the total power budget. We introduce a *power-efficiency measure* η:

$$\eta = \frac{\text{algorithm-inherent dissipation}}{\text{total dissipation of implementation}} \quad \text{(EQ 2)}$$

η is a quality-measure for an implementation from a power perspective. Obviously, in an power-conscious application, η should be as close as possible to 1.

The overhead power is hard to determine at the behavioral level as it requires a knowledge of implementation details. However, it is possible to get a reasonable idea of the overhead that will be incurred for a known architectural target, based on properties of the application-at-hand and empirical information about previous implementations.

EXAMPLE: Prediction of Controller Overhead

A controller-overhead prediction model was build for the HYPER behavioral synthesis environment [Meh96]. The model uses information about the application, the architecture, and the layout tools to predict the power consumed in the controller FSM. The resulting model and its accuracy is shown in Fig. 3. The average and maximum errors of this model are 11.5% and 41.1%, respectively.

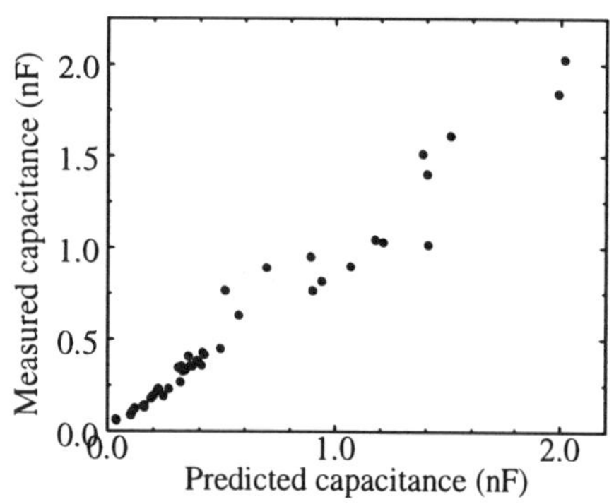

Fig. 3. Predicted versus measured switching capacitance in controllers of synthesized DSP processors.

Activity Estimation at the Structural Level

Once the precise structure of an architecture is known, more precise activity measures for both control and data signals can be obtained. Again, a static and a dynamic approach is conceivable. The latter is easily accomplished. Augmenting a VHDL or VERILOG simulation with a set of activity monitors at the inputs and outputs of the hardware modules produces switching activity (for control busses) and mean, variance, and correlation measures (for data busses).

While a simulation-based approach has the advantage of being simple and easily implemented, it has the pitfall of being pattern-dependent — and hence might miss some potentially harmful sequences. A probabilistic approach can be envisioned that computes the output statistics of a module given the input statistics and the functional description of a module [Lan94]. It can be seen that this evaluation can be of an exponential complexity, especially if one takes spatial correlations between input signals into account. Still, it is possible to construct a simple and efficient power estimation tool by using simplified propagation rules — for instance, by ignoring the spatial correlations. *Watt Watcher*, an architectural power estimation tool offered by Sente Inc., is based on this concept . It performs a power analysis of a circuit of 120,000 gates in 34 minutes [Sen95].

3.2 Physical Capacitance

The physical capacitance presents the weighting coefficient for the switching capacity in (EQ 1). The exact determination of its value during design exploration is complicated by a number of issues:

- Most often, the hardware resource has not been designed yet at this phase in the design process.
- Even if it has been, no simple power macro-model is available. For example, the availability of a transistor or gate netlist of a module such as a multiplier or a memory does not automatically present a model that is useful at the higher design abstraction level. In fact, very often the trees tend to obstruct the forest.

The promotion of power considerations to the prime scene has prompted the emergence of macro power models in the recent years, some of which are detailed below.

- The PDA model presents the power consumption of a module as a function of a number of complexity parameters such as the wordlength or the size (of a memory). For example, the energy consumption of an adder can be presented as

$$E = C_u N V^2_{DD} \quad \text{(EQ 3)}$$

  where N is the bit-width of the inputs and C_u the capacitive weighting factor. The latter can be obtained from analysis of a multiplier instance at the logic or circuit levels. The disadvantage of this approach is that C_u is a function of the applied input patterns. White noise (or random) data patterns are typically used during the characterization process. This tends to lead to pessimistic results.
- This deficiency has been addressed by the DBT and ABC models of Landman [Lan94], who expresses the capacitance as a function of an activity parameter that depends upon the chosen data representation. For instance, the bits in a two's complement data stream typically fall into two classes: non-correlated or white noise (lsb's) and correlated (msb's) (see also Fig. 2.). A meaningful activity factor is here the number of bits in the lsb-region N_u. The energy of the adder of can now be rewritten:

$$E = (C_u N_u / N + C_c (1 - N_u)/N) N V_{DD}^2 \quad \textbf{(EQ 4)}$$

This can be extended to more complex modules. Instead of representing a module by a single switching capacitance, an array of capacitances is needed. The value of these weights are defined during a characterization phase, where instances of the modules are simulated using well-defined data patterns, followed by a simple regression-based parameter extraction. Models can also be developed for other number representations such as sign-magnitude, or pure binary control data, represented by enumerated data-types [Lan94].

- An alternative approach towards building power-models for building blocks was offered by Svensson [Sve96], who models switching capacitance analytically by calculating the input and output capacitance of each stage in a functional block, and determining the switching activity at the input and output of each stage as a function of the input. A stage is a single PMOS pull-up, NMOS pull-down configuration. A bit slice may have one or more of these stages. This approach has the advantage of generating capacitance models without requiring extensive simulations. Svensson's model for a single pull-up/pull-down stage of logic can be generalized to

$$C_S = \alpha_{in} C_{in} + \alpha_{out} C_{out} \quad \textbf{(EQ 5)}$$

where α_{in} is the probability of input transitions, α_{out} is the probability of output transitions and C_{in} and C_{out} is the input and output capacitance, respectively. To get the total capacitance of a module requires summing the capacitance switched in each stage. Therefore,

$$C_{ST} = \sum_{j}^{stage} \alpha_{in,j} C_{in,j} + \alpha_{out,j} C_{out,j} \quad \textbf{(EQ 6)}$$

Analytical models have been provided for a wide range of modules, going from arithmetic, random logic, interconnect, and memory.

- On a completely different note, researchers at Princeton [Tiw94] have built instruction-level power models for microprocessors. Instead of looking into the architecture of a processor to understand the source of dissipation, the processor is modeled as a black-box, and the energy consumed for a given assembly-level instruction is measured. This is accomplished by putting the instruction in an infinite loop, and by observing the average current consumed during its execution. It is observed that this information on itself is not sufficient to accurately predict the overall energy consumed by an assembly-level program. The change of state between two instructions incurs some extra overhead. This can be adequately modeled by measuring the power consumed by a pair of instructions put in a loop, in excess of what is predicted by the per-instruction consumption. A complete model of a processor then includes power models per instruction and overhead-factors for pairs of instructions. This approach was successfully used to predict the power consumption of programs running on a number of processors such as the Intel 486DX2, a Fujitsu SPARCLite and a Fujitsu DSP processor.

Example: Average Cost of Instructions on DSP

The average current consumed by some basic instructions on the Fujitsu DSP processor is given in Table 2. Observe that this does not include the overhead-factor, which is defined on a per-pair base and is of the same order of magnitude.

Table 2. Average base cost per instruction (in mA)

	LDI	LAB	MOV1	MOV2	ASL	MAC
range	15.8-22.9	34.6-38.5	18.8-20.7	17.6-19.2	15.8-17.2	17.0-17.4
average	19.4	26.5	19.8	18.4	16.5	17.2

3.3 Power Exploration - Examples

To bring the techniques, presented in the previous sections into perspective, we present two examples of early power exploration environments. The presentation will include examples of other frameworks as well.

Example 1: Very-Early Prediction using PowerPlay

PowerPlay [Lid96] offers power estimation based solely upon the manipulation of power, timing, and area models of functional blocks. Each model is parameterized (for example, by bit-width) and is scalable with supply voltage and technology. Existing hardware models are shared among all users, and new models are easily created and integrated.

The main attraction of PowerPlay is that it offers a spreadsheet like environment for back-of-the-envelope computations in a format that is universally available. As the world-wide web (WWW) has become the de-facto standard for information gathering and displaying, it is the most natural choice for a design exploration environment. Besides the relatively seamless data gathering and push-button tool launching capabilities (through hyperlinks -textual pointers to files or programs) that a WWW browser provides, the WWW is publicly accessible and in widespread use, encouraging shared library models and design reuse. Another advantage is a seamless integration of modeling and documentation.

The example of shows the power model of a multimedia terminal, the infopad, as shown using the Netscape WWW browser. Notice that the (hierarchical) model includes power models for ASICs, processors, FPGAs, radio modem, and display.

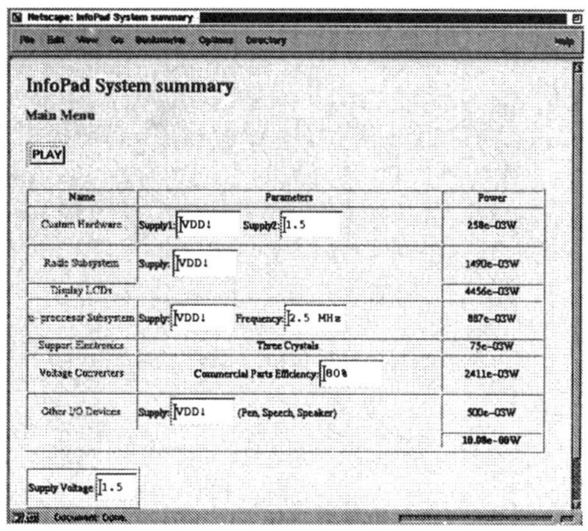

Fig. 4. Power model of Infopad Multimedia terminal, as offered by PowerPlay.

Example 2: Architectural Power Estimation using SPA

The SPA [Lan94] tool combines RTL-level simulation with black-box models of the composing modules (using the ABC and DBT models, described earlier) to provide rather accurate predictions at the architectural levels. More precisely, signal statistics are gathered through a VHDL simulation of the architecture over a meaningful set of excitation vectors. Monitors are automatically attached to the signals and the inputs and outputs of the modules. The obtained statistics are combined with the pre-characterized module models to yield the final energy (or power) predictions. Since the interconnect capacitance is generally not known at this point in the design process, a simple model is used that estimates the average wire length from the active area of the design.

Shown in Fig. 5. is the SPA-estimated breakdown of the power dissipation of an 48-processor multiprocessor chip, while running different applications. The accuracy of this simulations has proven to be within 10% of the results obtained from an extracted layout using switch-level simulation, yet the results are obtained in minutes of simulation time. Notice furthermore that this multiprocessor architecture achieves a high level of power-efficiency as almost 50% of the power is dissipated in the data paths (or $\eta \sim 0.5$).

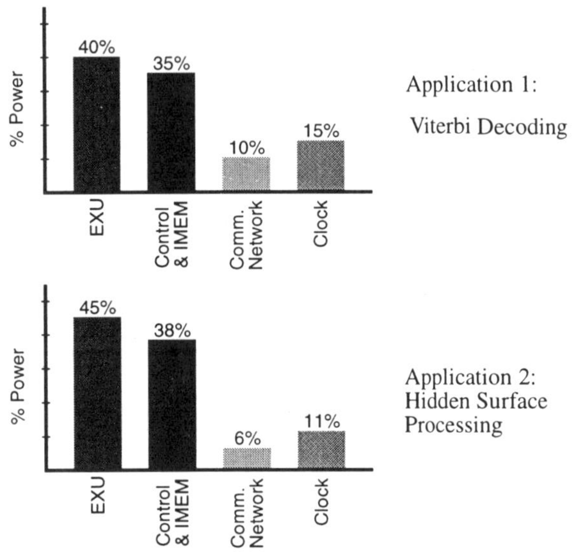

Fig. 5. Breakdown of power-consumption in 48-processor multi-processor chip for rapid-prototyping of DSP, as obtained by SPA. The breakdown is shown for two different applications running on the processor array.

4. Summary and Perspective

Design automation for the analysis and minimization of power has received ample attention in recent years, yet is still in an infantile state. Tools for analysis at the circuit and logic level are now commercially available. Yet the true applicability and value of these techniques has still to be proven, especially at the logic level where the bounds on the obtained accuracy are not completely clear.

Power analysis at the higher level of abstraction is of crucial importance if a "power-conscious" design methodology is ever to take hold. Innovation at the architectural and behavioral levels requires first of all that the sources of dissipation are well understood and that the opportunities for minimization become clear. It is the authors belief that a number of the techniques presented in this tutorial (and others that are currently emerging) will make it into the commercial market place in the near future. Power considerations are becoming an integral part of the overall design conception process and are bound to do so in the foreseeable future.

5. References

[Cha95] A. Chandrakasan, *et al*, *Low Power Digital CMOS Design*, Kluwer Academic Publishers, Boston, MA, 1995.

[Den94] C. Deng, "Power Analysis for CMOS/BiCMOS Circuits", in *Proc. 1994 Int. Workshop for Low Power Design*, pp. 3-8, April 1994.

[Geo94] B. George et al., "Power Analysis and Characterization for Semi-Custom Design", in *Proc. 1994 Int. Workshop for Low Power Design*, pp. 215-218, April 1994.

[Lan94] P. Landman, *Low-Power Architectural Design Methodologies*, Ph. D. Thesis, Electronic Research Laboratory, University of California, Berkeley, August 1994.

[Lid96] D. Lidksy et al., "Early Power Exploration: A Worldwide Web Application", submitted for 33rd Design Automation Conference, Las Vegas.

[Meh96] R. Mehra et all, "Algorithm and Architectural Level Methodologies for Low Power", in *Low Power Design Methodologies*, Ed. J. Rabaey and M. Pedram, Ch. 11, Kluwer Academic Publishers, 1996.

[Ped96] M. Pedram, "Logic Synthesis for Low Power", in *Low Power Design Methodologies*, Ed. J. Rabaey and M. Pedram, Chapter 6, Kluwer Academic Publishers, Boston, MA, 1996.

[Pow91] S. Powell and P. Chau, "A Model for Estimating Power Dissipation in a Class of DSP VLSI Chips", IEEE Transactions on Circuits and Systems, pp. 646-650, June 1991.

[Qua93] Quantify Documentation, Pure Software Inc, CA.

[Rab95] J. Rabaey et al., "Design Guidance in the Power Dimension", Proc. ICASSP 1995, Detroit, April 1995.

[Rab96] J. Rabaey, *Digital Integrated Circuits: A Design Perspective*, Chapters 4 and 7, Prentice-Hall, 1996.

[Sen95] Watt Watcher Product Sheet, Sente Corp., Chelmsford, MA.

[Sin95] D. Singh et al., "Power Conscious CAD Tools and Methodologies: A Perspective", Proceedings of the IEEE, pp. 570-594, April 1995.

[Sve96] C. Svensson *et al*, "Low Power Circuit Techniques," in *Low Power Design Methodologies*, Ed. J. Rabaey and M. Pedram, pp.37-64, Kluwer Academic Publishers, Boston, MA, 1996.

[Tiw94] V. Tiwari et al., "Power Analysis of Embedded Software: A First Step towards Software Power Minimization", IEEE Trans. on VLSI Systems, pp. 437-445, Dec. 1994.

6. Acknowledgments

The author would like to acknowledge the contributions of David Lidsky, Renu Mehra, George Varghese, and Sharad Malik.

A Monte-Carlo Approach for the Accurate and Efficient Estimation of Average Transition Probabilities in Sequential Logic Circuits

Georgios I. Stamoulis

Intel Corp. 2200 Mission College Blvd. Santa Clara, CA 95008

Abstract: This paper presents an efficient and accurate Monte-Carlo approach to the problem of estimating average node switching probabilities in sequential circuits which are used in average power estimation and reliability analysis of these circuits. Specific error bound for the proposed estimation method are given at a certain level of confidence. This method is based on the analysis of *paths* in the State Transition Graph (STG) of the circuit and is validated by both theoretical analysis as well as experimental results.

1 Introduction.

There has recently been much interest in the simulation of VLSI circuits for the estimation of their average power dissipation and their susceptibility to cumulative degradation phenomena such as electromigration and hot-carrier degradation. For the simulation of combinational circuits a number of methods have been introduced in the past. Some of them approached the problem at the transistor level and estimated the average current drawn by each individual gate [1], [2]. Others proposed a gate level simulation, using the average switching activity in the circuit nodes as a measure of the average power dissipation of the circuit [3], [4], [5], [6], [7]. Sequential circuits were considered as a separate problem in [8], [9], [10], as the combinational approaches were not readily extendable. However, the complexity of the derived algorithms was high and no simulation results for large sequential circuits were given, primarily due to the exponential complexity of the problem.

In this paper we describe a computationally efficient method for obtaining the average switching activity for synchronous sequential circuits with high accuracy by using Monte-Carlo simulation at the logic gate level. The accuracy obtained is to within 10% of the actual switching probability value for a particular node at a 95% level of confidence or better. This is achieved by introducing the notion of "paths" in the State Transition Graph (STG) of the circuit which efficiently scans the sample space that has both the primary inputs and the initial state of the latches as variables. This results in a two step approach, in which the average switching probability being estimated along a single "path" in the first step, and over several "paths" in the second. The validity of this method has been explained theoretically and validated by experimental results. The acceptable error can be set by the user at different values and at different confidence levels, resulting in a different number of required samples in each case. Another significant attribute of this approach is that it can be efficiently applied to even the largest benchmark circuits (ISCAS 89 [11]) and that it can be coupled with any of the existing methods for estimating the average power dissipation of combinational circuits either at the gate or at the transistor level. As an example, the simulator presented in [12] has been used to estimate the average power dissipation of the ISCAS 89 benchmark circuits. The results presented have been geared towards high accuracy rather than speed. Even so, the simulation times are in the 30 minute range for even the largest circuits.

This research was supported in part by NSF Grant No. MIP-9357581 and by the Semiconductor Research Corporation under Contract SRC 93-DP-109.

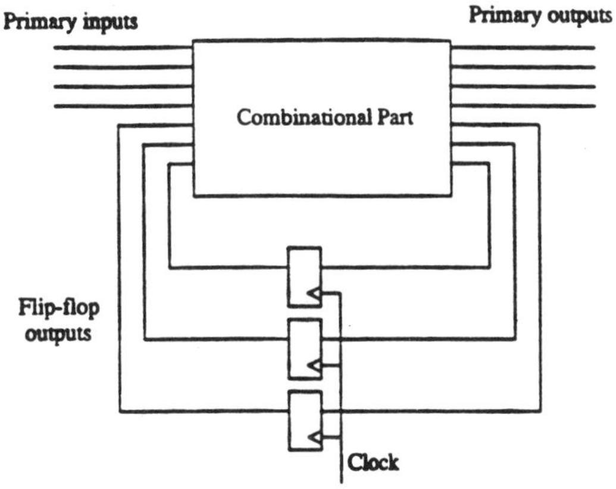

Figure 1: The sequential circuit model used.

2 The new approach.

The circuit model that will be used in this paper is the one in Fig. 1. The nodes of interest are the ones termed "flip-flop output nodes" and the analysis will focus on determining the accurate switching probabilities for these nodes. We assume: a) the circuit can be at any state after power up with equal probability, b) there are no glitches at the outputs of the latches, and c) all latches reach a steady state before their next state is allowed to enter into the combinational logic part. All of the above are reasonable expectations from a well-designed sequential circuit. Furthermore, the above assumptions permit the estimation of the power dissipated by the latches separately from the rest of the circuit (the combinational part) and limit the number of times the output of a latch can switch in one clock cycle to at most once. The latter is a critical part of the ensuing study.

0-7803-3177-6 $5.00 © 1996 IEEE

IEEE 1996 CUSTOM INTEGRATED CIRCUITS CONFERENCE

2.1 Initial State Probability Analysis.

The dependence on the initial state probabilities could be alleviated if it can be proven that the stochastic processes which describe the state of the flip-flops in the time domain are stationary. Fortunately, by applying Borovkov's theory of renovating events [13] we can prove that this is true if the latches can be driven to a known state by a sequence of input vectors, which is the definition of initializable circuits [14]. Let a system be described by the stochastic recurrence:

$$W_{n+1}^{Y} = h(W_n^Y, \xi_n) \quad n = 0, 1, \ldots \tag{1}$$

where ξ_n is the driving sequence. The theorem stated in [13] is the following:

Theorem: If there exists a sequence of events $\{A_n\}$ compatible with the shift and satisfying the condition $P^O(A_0) > 0$, a function Φ and an integer m, such that, on A_n, $W_n^Y = \Phi(\xi_n, \ldots, \xi_{n+m-1})$, for all $Y \in \mathcal{Y}$, then there exists a stationary sequence $\{Z \circ \theta^n\}$, solution of expression (1), and such that, for all $Y \in \mathcal{Y}$, the sequence converges with strong coupling to $\{Z \circ \theta^n\}$. (In [13])

Strong coupling between $\{X_n\}$ and $\{Z \circ \theta^n\}$ implies that $\{X_{n+k}\}_{n \geq 0}$ converges in variation to $\{Z \circ \theta^n\}$. Convergence in variation implies convergence in distribution. For the analysis in this paper, expression (1) is the definition of the next state of a sequential circuit, $\{\xi_n\}$ is a stationary and ergodic process as we have assumed that the circuit inputs are driven by random bit patterns, the sequence $\{A_n\}$ is an initialization sequence, which due the random bit pattern applied has a non-zero probability of occurring and drive the circuit to a certain state independent from previous states (i.e. the $W_n^Y = \Phi(\xi_n, \ldots, \xi_{n+m-1})$ condition is satisfied). All the above directly imply that, if the sequential circuit is initializable (i.e. there exists a vector sequence that can drive it to a known state irrespective of the present state), then the random processes describing the state of every node in the circuit, including the states of the latches, converges in distribution to a stationary process. Furthermore, states that are N clock ticks apart become independent as $N \to \infty$, which means that the node processes are also mean ergodic. Stationarity and ergodicity in combinational circuits is a trivial subproblem of the initializable sequential circuits. In the case of a non-initializable sequential circuit, no such convergence can be proven in the general case.

Consequently, for the initializable circuits, the average switching probability converges to an average value which is independent of the initial state of the circuit and the pathwise average is equal to the time average due to ergodicity. However, the number of samples required is not readily available. By defining the notion of "paths" in the State Transition Graph (STG) of the circuit, a new perspective appears. Instead of running one long simulation with a given initial condition (time average) which may take a long time to converge, we sample through simulations with different initial conditions (pathwise average). As the process is stationary and ergodic, the initial state probability will not bias the estimated average value and, thus, equiprobable initial states can be safely used.

A path is defined as a sequence of adjoining transitions in the STG. The state from which the first transition in a path originates from is termed the head of the path, while the state at which the last transition arrives to is termed the tail of the path. In the random search that is proposed by this paper the sampling is performed one path at a time. The transition probability samples are the average transition probabilities calculated by the analysis of a "random" path, a path in which the head is randomly selected and so are the vectors applied to the primary inputs. A study on the effect of the path head assignment probabilities to the calculated transition probabilities is presented in Section 3, in which it is shown that considering all states originally equiprobable is a valid assumption. The transition probability estimation is performed in two steps: a) estimation of the pathwise transition probabilities, and b) estimation of the overall transition probabilities.

2.2 Step 1: Average along one path.

A random path is created by first randomly choosing a state as the head of the path and subsequently applying random vectors to the circuit's primary inputs. The transition probabilities observed in this path, call it i, for the output of flip-flop j, are given by the following expressions:

$$plh_{ij} = \frac{Number\ of\ L-H\ transitions\ in\ path\ i}{Total\ number\ of\ clock\ cycles} \tag{2}$$

$$phl_{ij} = \frac{Number\ of\ H-L\ transitions\ in\ path\ i}{Total\ number\ of\ clock\ cycles} \tag{3}$$

Since the following analysis is valid for both plh_{ij} and phl_{ij}, the symbol p_{ij} will refer to either of the two quantities. The number of vectors needed for achieving a maximum error ϵ with level of confidence $(1-a)$, as given by basic statistics, is:

$$N \geq \left(\frac{t_{a/2}\sigma}{\eta\epsilon}\right)^2 \tag{4}$$

where $t_{a/2}$ is the Student-t coefficient for the level of confidence $(1-a)$, σ is the variance of p_{ij} and η its mean. It is useful to make here certain remarks regarding the worst case conditions for N. The random variable p_{ij} follows the Bernoulli distribution. This means that $\eta = p_{ij}$ and $\sigma^2 = p_{ij}(1-p_{ij})$. Based on this observation it can be deduced that as $\eta \to 0$, N becomes increasingly larger for constant ϵ. This becomes a problem when η goes below 0.05 since the number of vectors required for adequate approximation becomes prohibitively large. However, a larger amount of error can be tolerated for low transition valued nodes since they contribute little to the overall power dissipation. Furthermore, the error accumulated in this step is compensated for in the second step of the estimation. Another way of expressing the observed transition probabilities is by the following expression:

$$\eta_{i,obs} = \eta_i + E_i(N) \tag{5}$$

where $\eta_{i,obs}$ is the observed average from the sample, η_i is the true path mean and $E_i(N)$ is the error, which is a random variable with zero mean and variance given as a function of N:

$$\sigma_E = \frac{t_{0.05}\sigma}{4\sqrt{N}} \tag{6}$$

The maximum error observed is for $\sigma = 0.5$ which is the maximum variance that the random variable representing low to high or high to low switching might attain.

2.3 Step 2: Average over all paths.

The second step in the average switching probability estimation algorithm would be to average over all paths, with the switching probability estimate of each path being one measurement. In this case the random variable can take any real value in the interval [0,0.5].

In addition to the error due to the sampling of the switching probabilities of the paths, there is an additional term due to the error in the estimation of the average switching probability for each path (recall Eq. 4). However, the average switching probability for a node j is given as:

$$p_{j,obs} = \frac{\sum_i p_{ij,obs}}{N} = \frac{\sum_i p_{ij}}{N} + \frac{\sum_i E_{ij}}{N} \tag{7}$$

In the analysis of the pathwise sample we have concentrated just on the first term of the right hand expression in Eq. 6. The second term is the error term that introduces additional error in our calculations. Since this term is a sum of zero-mean normal variables, its contribution to the value of the estimated overall mean switching probability is zero. However, it contributes an additional absolute error ϵ_1, which is bound by:

$$\epsilon_1 < \frac{t_{a/2}\sigma_E}{\sqrt{N}} \tag{8}$$

where σ_E is given by Eq. 5.

At this point there is a tradeoff to be made between the speed of the simulation and the accuracy that can be achieved. The number of vectors required depends on the probability density function of the transition probabilities along a path of fixed length. However, for realistic density functions the number of paths required is manageable. It is also important to stress that an accurate estimate for the high switching probability nodes is highly desirable, while for nodes with low transition probability a rough estimate should be sufficient since they contribute very little to the overall power dissipation, a notion shared with other work in the area [6].

3 Initial state assignment.

A critical part of the average switching probability analysis of sequential circuits is the assignment of the initial state probabilities, the probabilities by which the heads of the paths are selected.

In the path approach described in Section 2, the head of a path is the only state directly influenced by the state probability assignment, while the rest of the states in the path are influenced by the random vectors applied to the primary inputs of the circuit. Thus, the state of the circuit, after the application of the first random vector, is determined exclusively by the circuit structure. Furthermore, the number of random input vectors in a path is such that any bias induced by the uniform probabilities assumed for path head states is minimized. This is true for all sequential circuits in which the primary inputs can drive the flip-flops to a known state (initializable circuits). For non-initializable circuits, the uniform probability assumption for the initial states is warranted as the flip-flops can be in any state after power-up. In order to test this approach we simulated all of the ISCAS89 benchmark circuits by the method proposed in this paper, with the probability that a given flip-flop is at logic high value in the head state of a path ranging from 0 to 1 in increments of 0.1. A notion of convergence was also introduced: a switching probability value has "converged" if the difference between the minimum and the maximum values over all the runs was less than a preset threshold t_{conv}. The length of the paths was chosen as 1000, the number of the sampled paths was 1000 as well, and t_{conv} was set to 0.001. Table 2 shows the number of nodes that converged and did not converge for a sample of the ISCAS89 benchmark circuits, along with and indication of whether the circuit is initializable or not according to [14]. It is noteworthy that out of the 28 circuits tested the 25 that converged were also deemed initializable and the three that did not converge were considered non-initializable by [14], exactly as predicted by the theoretical analysis of Section 2. A χ^2 test shows that there is significant evidence to support that initializable circuits converge (i.e. there is no bias from the uniform selection probability of the initial states) at a confidence level greater than 99.99%. This result conforms with the hypothesis stated in the beginning of this section.

4 Experimental results.

The approach described in this paper has been applied to the ISCAS89 benchmark circuits with the path length set to 400 and the path number set to 2000 in order to achieve high accuracy even for very low switching probabilities. With this setup, the switching probability of a node, which has an actual switching probability of 0.01, can be estimated to within 5% at 95% confidence level. The results of this estimation along with the combinational part of the circuits were then analyzed by the simulator described in [12] to estimate the average power dissipation. The runtimes for both the switching probability estimation program and the average power simulator are shown in Table 3. The power estimates have been omitted since they are implementation dependent.

5 Conclusions. In this paper a method for the accurate estimation of the average switching probability of flip-flop outputs in a general sequential circuit along with its error bounds has been presented. This approach introduces the notion of paths in the state transition graph of the sequential circuit and employs them in a Monte-Carlo logic simulation approach to estimate the average switching probability. Furthermore, the possible bias on the results by the initial state selection has been alleviated, as shown by theoretical and experimental results, in initializable circuits and special guidelines set forth for non-initializable ones. It should be noted that this approach achieved high accuracy in relatively little time and with a very small memory overhead.

Acknowledgements: The author would like to acknowledge the contribution of Dr. Mark S. Andersland to the theoretical development of this paper and the contribution of Mr. Shashank Goel to the implementation of the algorithms. Furthermore, the author would like to thank Dr. Resve Saleh for his encouragement and support.

References

[1] Stamoulis et al., "Improved Techniques for Probabilistic Simulation Including Signal Correlation Effects," *30th DAC Proceedings*, pp. 379–383, 1993.

[2] Li et al., "A Probabilistic Timing Approach to Hot-Carrier Effect Estimation," *ICCAD 1992 Proceedings*, pp. 210–213, 1992.

[3] Tsui et al., "Efficient Estimation of Dynamic Power Consumption under a Real Delay Model," *ICCAD 1993 Proceedings*, pp. 224–228, 1993.

[4] F. Najm, "Transition Density: A New Measure of Activity in Digital Circuits," *IEEE Transactions on Computer-Aided Design*, pp. 310–324, 1993.

[5] Devadas et al., "Estimation of Power Dissipation in CMOS Combinational Circuits Using Boolean Function Manipulation," *IEEE Transaction on CAD*, pp. 373–383, 1992.

[6] M. Xakellis, *Estimating Node Transition Densities with Statistical Simulation Techniques.* Master's thesis, UIUC, 1993.

[7] Burch et al., "A Monte-Carlo Approach for Power Estimation," *IEEE Transactions on VLSI*, vol. 1, pp. 63–71, 1993.

[8] Ghosh et al., "Estimation of Average Switching Activity in Combination and Sequential Circuits," *29th DAC Proceedings*, pp. 253–259, 1992.

[9] Monteiro et al., "A Methodology for Efficient Estimation of Switching Activity in Sequential Logic Circuits," *31st DAC Proceedings*, pp. 12–17, 1994.

[10] Tsui et al., "Exact and Approximate Methods for Calculating Signal and Transition Probabilities in FSMs," *31st DAC Proceedings*, pp. 18–23, 1994.

[11] Brglez et al., "Combinational Profiles of Sequential Benchmark Circuits," *ISCAS 1989 Proceedings*, pp. 1929–1934, 1989.

[12] G. Stamoulis, *Probabilistic Simulation for Reliability and Average Power Estimation.* PhD thesis, UIUC, 1994.

[13] F. Baccelli and P. Brémaud, *Elements of Queueing Theory.* Springer-Verlag, 1994.

[14] Wehbeh et al., "On the Initialization of Sequential Circuits," *ITC 1994 Proceedings*, pp. 233–239, 1994.

Table 1: Additional error due to the error in the estimate of the pathwise average (length of paths=400, number of paths=2000)

Transition probability	Additional error (%)
0.01	5.37
0.10	0.54
0.20	0.27
0.30	0.18
0.50	0.11

Table 2: Analysis of the effect of the initial state probability.

ckt name	# of FFs	conv'd	not conv'd	init
s208	8	8	0	Yes
s208.1	8	5	3	No
s298	14	14	0	Yes
s344	15	15	0	Yes
s349	15	15	0	Yes
s400	21	21	0	Yes
s420	16	16	0	Yes
s420.1	16	5	11	No
s444	21	21	0	Yes
s832	5	5	0	Yes
s838	32	32	0	Yes
s838.1	32	5	27	No
s953	29	29	0	Yes
s1196	18	18	0	Yes
s1488	6	6	0	Yes
s1494	6	6	0	Yes
s35932	1728	1728	0	Yes

Table 3: Simulation times for the sequential and the combinational parts.

ckt name	# of FFs	seq	comb
s208	8	100.8	0.3
s420	16	185.5	0.9
s510	6	189.8	0.9
s641	19	351.6	3.1
s953	29	173.5	1.9
s1423	74	199.3	10.4
s5378	179	481.3	11.8
s9234	228	456.0	69.0
s13207	669	775.1	102.5
s15850	597	808.5	185.0
s35932	1728	1468.5	778.5
s38584	1452	1383.4	847.4

Nonparametric Estimation of Average Power Dissipation in CMOS VLSI Circuits *

Li-Pen Yuan, Chin-Chi Teng, and Sung-Mo Kang
Dept. of Electrical and Computer Engineering and Coordinated Science Lab.
University of Illinois at Urbana-Champaign
1308 W. Main St., Urbana, IL 61801
{lpyuan, teng, kang}@uivlsi.csl.uiuc.edu

Abstract

In this paper, we present a new statistical technique for estimation of average power dissipation in digital circuits. Present statistical techniques estimate the average power based on the assumption that the power distribution can be characterized by a preassumed function. Large error can incur when the assumption is not met. To overcome this problem, we propose a nonparametric technique in which no distribution function needs to be assumed. A distribution-independent upper and lower bound of the average power are derived from the Kolmogorov-Smirnov theorem. A stopping criterion is designed based on the bounds for a desired percentage error with a specified confidence level. Since it does not resort to the assumption of any particular distribution function, the technique can be applied to all the circuits irrespective of their power distributions.

I. Introduction

Current approaches for estimation of average power dissipation in CMOS circuits can be roughly divided into two categories: probabilistic and statistical techniques. In probabilistic approaches, signal probabilities and transition densities are propagated from the primary inputs throughout the circuit[1]. Average power measure can be obtained by summing the product of load capacitance and transition density over all the internal nodes of the circuit. Although such an approach can estimate the average power in short CPU time, accuracy is lost because spatial signal correlations are neglected. Signal correlations can be considered in probabilistic approaches only at the cost of high computational complexity[2]. In statistical approaches, randomly generated input patterns are applied to the circuit and the energy dissipated per clock cycle is sampled using a circuit simulator. Based on the central limit theorem, the average power can be calculated from the sample mean within a desired percentage error ϵ for a given confidence level $1-\alpha$. Although for most circuits it can provide fast and accurate estimation of average power, it incurs large error in the cases with non-normal power distribution[3].

Since there is no way of knowing *a priori* the power distribution of a circuit, the required assumption of a particular power distribution discourages the use of the current statistical technique. To overcome this problem, we propose a new statistical technique for average power estimation. This technique is *nonparametric*, meaning that no particular distribution function or parameter needs to be assumed. With use of this technique, statistical conclusions can be drawn only by analyzing the power sample data with distribution-independent statistics. Owing to this feature, the proposed technique can be applied to circuits with arbitrary power distributions.

The rest of the paper is organized as follows. In Section II we show how to express the power consumption of a circuit as a random variable so that statistical techniques can be applied. In Section III we use a distribution-independent theorem, namely the Kolmogorov-Smirnov theorem, to find the confidence band of the cumulative distribution function(cdf). Next we show how to design the stopping criterion to terminate the random simulation based on the confidence band. We implemented the proposed technique and tested it on a set of benchmark circuits. The results are discussed in Section IV, followed by concluding remarks in Section V.

II. Problem Formulation

Consider a digital circuit with n primary inputs, each of which may possess a different signal probability. The logic state of primary input i can be treated as a random variable $\boldsymbol{R}_i$ with its probability density function(pdf) $f_{r_i}(x)$ defined as

$$f_{r_i}(x) = \begin{cases} P_i, & x = 1 \\ 1 - P_i, & x = 0 \end{cases} \tag{1}$$

where P_i is the signal probability of primary input i. Assume spatial independence among all the input signals, i.e., the signal value of each primary input can be determined independently, the input pattern of the circuit can be represented as a random variable(RV) $\boldsymbol{V}$ whose pdf is defined as

$$f_v(v) = f_v(v(x_1, x_2, \ldots, x_n)) = \prod_{i=1}^{n} f_{r_i}(x_i), \tag{2}$$

where x_i is the logic state of primary input i. In static CMOS circuits, a gate dissipates power only when the gate output makes a switching. Here the power due to leakage currents is ignored. To study

*This research was supported by Joint Services Electronics Program (N00014-95-J-1270) and Semiconductor Research Corp. (SRC95DP109).

0-7803-3177-6 $5.00 © 1996 IEEE

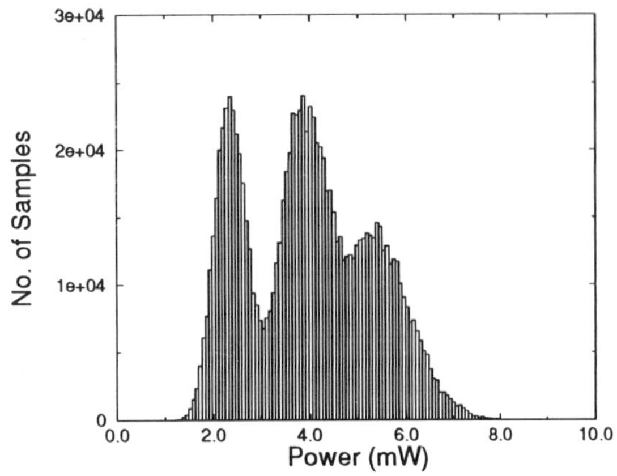

Figure 1: Power distribution of circuit apex6.

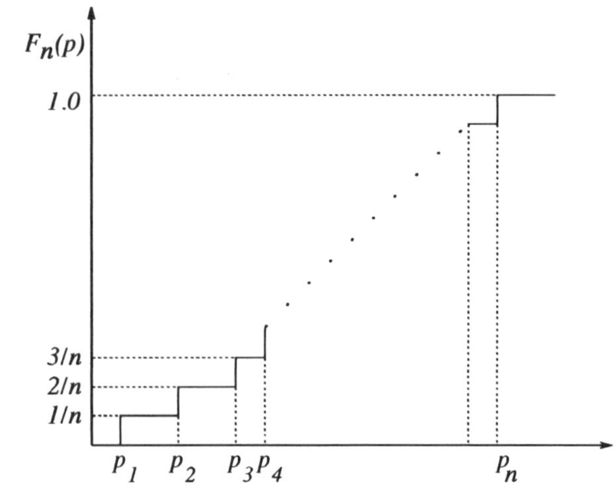

Figure 2: Construction of sample cdf $F_n(p)$.

the power dissipation behavior of the circuit, we use $\boldsymbol{V}_1$ and $\boldsymbol{V}_2$, two independently and identically distributed RV's, to represent two consecutive random input patterns. The power dissipation of the circuit with m gate output nodes can be determined by $\boldsymbol{V}_1$ and $\boldsymbol{V}_2$ as

$$\boldsymbol{P} = \frac{V_{DD}^2}{2T} \sum_{i=1}^{m} C_i \boldsymbol{n}_i(\boldsymbol{V}_1, \boldsymbol{V}_2), \quad (3)$$

where RV $\boldsymbol{n}_i$ is the number of transitions occurring at the ith gate, C_i is the load capacitance at node i, and T is the clock cycle time. $\boldsymbol{n}_i$ is a function of $\boldsymbol{V}_1$ and $\boldsymbol{V}_2$. Since $\boldsymbol{P}$ is a function of $\boldsymbol{n}_i$, $i = 1, \cdots, m$, it is also a random variable and possesses a distribution function. It should be noted that since the number of transitions at each gate can only take positive integers, $\boldsymbol{n}_i$ has a discrete distribution. So does $\boldsymbol{P}$ according to (3). For practical circuits, the sample space of $\boldsymbol{V}_1$ and $\boldsymbol{V}_2$ is usually large enough so that the difference between two adjacent observable values of $\boldsymbol{P}$ is very small. Therefore, we can assume that $\boldsymbol{P}$ has a continuous cdf without causing much error.

Depending on the function and the implementation, circuits can have various power distribution. While most of the circuits are observed to have a single-modal power distribution, there are many circuits showing bimodal or even multimodal power dissipation behavior. Fig. 1 shows the power distribution of a benchmark circuit apex6 after 1 million random simulations. It is clearly observed that the power distribution of apex6 is far from the normal or even single-mode. For cases like this, the existing statistical technique[3] tends to produce large error in estimating the average power dissipation because the sample mean of $\boldsymbol{P}$ is explicitly assumed to be normally distributed. Furthermore, it is also hard to detect the existence of such a non-normal distribution unless numerous simulations are conducted to extract the distribution. In order to overcome this difficulty, in next section we introduce a nonparametric technique.

III. Estimation of Average Power

A. Kolmogorov-Smirnov Statistic

For a given circuit, let $F(p)$ be the cdf of $\boldsymbol{P}$. Suppose that the random variables $\boldsymbol{P}_1, \ldots, \boldsymbol{P}_n$ form a set of random samples from $F(p)$, and let $p_1, \ldots, p_n$ denote the observed values of $\boldsymbol{P}_1 \ldots, \boldsymbol{P}_n$. A sample cdf $F_n(p)$ is constructed from the values $p_1, \ldots, p_n$ such that for any p $(0 < p < \infty)$ the value of $F_n(p)$ represents the proportion of observed values in the sample which are less or equal to p, as depicted in Fig. 2. $F_n(p)$ can be regarded as the cdf of a discrete distribution which assigns probability $1/n$ to each of the n values $p_1, \ldots, p_n$. Let $\boldsymbol{D}_n$ be the maximum difference between the sample cdf $F_n(p)$ and the real cdf $F(p)$, i.e.,

$$\boldsymbol{D}_n = \sup_{0<p<\infty} |F_n(p) - F(p)|. \quad (4)$$

If the random samples $p_1, \ldots, p_n$ are drawn from $F(p)$, the Kolmogorov-Smirnov theorem[4] states that $\sqrt{n}\boldsymbol{D}_n$ has the following limiting distribution:

$$\begin{aligned} H(t) &= \lim_{n\to\infty} \Pr(\sqrt{n}\boldsymbol{D}_n \le t) \\ &= 1 - 2\sum_{i=1}^{\infty}(-1)^{i-1}e^{-2i^2t^2}, \quad \forall t > 0. \quad (5) \end{aligned}$$

Note that although the equality in (5) only holds when $n \to \infty$, the approximation that $\Pr(\sqrt{n}\boldsymbol{D}_n \le t) \approx H(t)$ is valid for n is as small as 80. Since the number of samples used in our simulation is much larger than 80, this approximation will not pose a problem for our purpose. Equation (5) can be used to find the confidence band of the unknown cdf $F(p)$ and can be used as a guide for estimating the average power dissipation. For a given set of sample data $p_1, \ldots, p_n$, suppose c is the observed value of

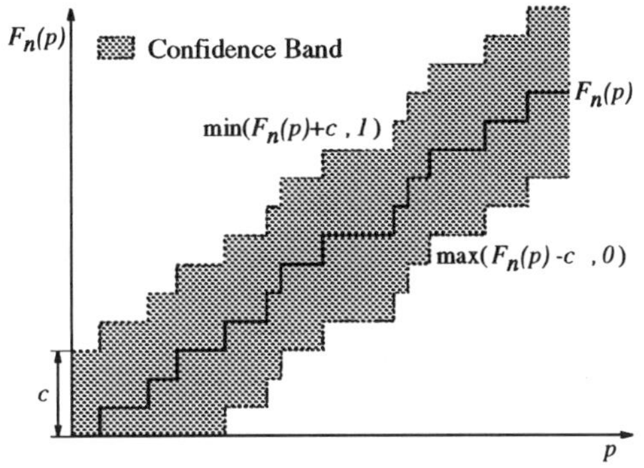

Figure 3: $(1-\alpha)$ confidence band of $F(p)$.

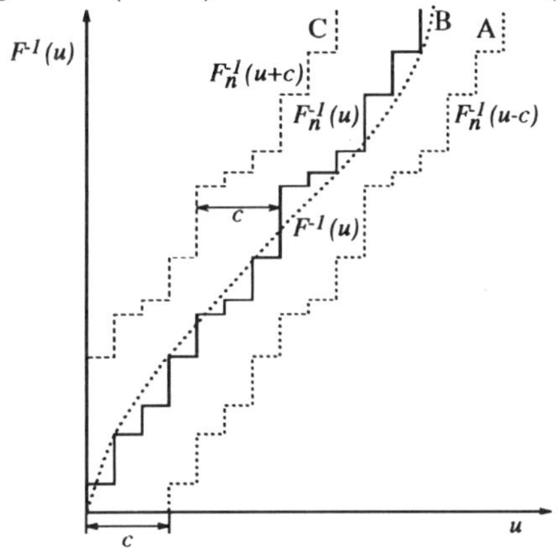

Figure 4: Lower and upper bound of average power.

$\boldsymbol{D}_n$, we can find a corresponding value α such that $H(\sqrt{n}c) = 1-\alpha$. From (4) and (5) we know

$$\Pr(F_n(p) - c \leq F(p) \leq F_n(p) + c, \forall p) = 1-\alpha. \quad (6)$$

Therefore, the curves $[\max(F_n(p)-c, 0), \min(F_n(p)+c, 1)]$ encompass a *confidence band* [5] for $F(p)$ with $1-\alpha$ confidence, as shown in Fig. 3.

B. Bounds of Average Power and Stopping Criterion

The confidence band generated by (6) is distribution-independent and is the key result for finding the bounds of the average power and designing the stopping criterion to terminate the random simulation. To begin with, recall that the average power μ_p of a circuit is the mean of $\boldsymbol{P}$:

$$\mu_p = E[\boldsymbol{P}] = \int_0^{\infty} pf(p)dp, \quad (7)$$

where $f(p) = dF(p)/dp$. Since the unknown cdf $F(p)$ is a non-decreasing function, its inverse function F^{-1} is well defined. Let $u = F(p)$, by substituting $f(p)dp$ by du and p by $F^{-1}(u)$, the integration in (7) can be performed on the domain of variable u as

$$\mu_p = \int_0^1 F^{-1}(u)du. \quad (8)$$

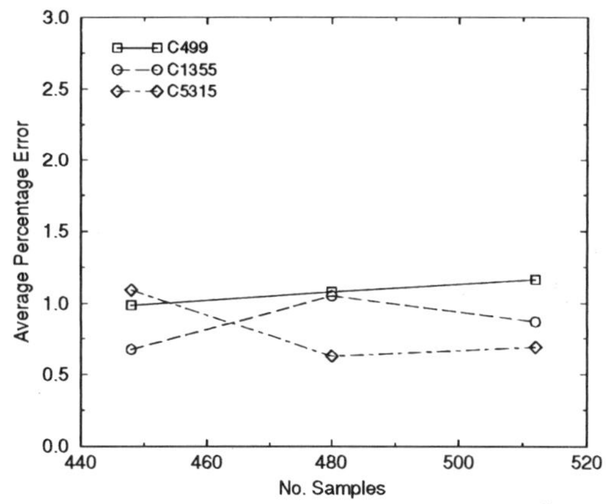

Figure 5: Comparison results between $F^{-1}(u)$ and $F_n^{-1}(u)$ for different power sample sizes.

If curve B in Fig. 4 represents $F^{-1}(u)$, μ_p is actually the area between the u axis and curve B. By making the approximation that $F^{-1}(u_i) \approx F_n^{-1}(u_i)$, where $u_i = F_n(p_i), i = 1, \ldots, n$, (6) can be transformed to the following inequality:

$$\int_0^1 F_n^{-1}(u-c)du \leq \mu_p \leq \int_0^1 F_n^{-1}(u+c)du, \quad (9)$$

with $1-\alpha$ confidence. (9) provides an upper and a lower bound of the average power. Graphically, it can be easily shown in Fig. 4 that $\int_0^1 F_n^{-1}(u-c)du$ is just the area between curve A and the u axis, while $\int_0^1 F_n^{-1}(u+c)du$ is just the area between curve C and the u axis. Let $\mu_{pL} = \int_0^1 F_n^{-1}(u-c)du$, $\mu_{pU} = \int_0^1 F_n^{-1}(u+c)du$, and $\bar{\mu}_p$ be the sample mean of n independent simulations, (9) can be recast as

$$\frac{\mu_{pL} - \bar{\mu}_p}{\bar{\mu}_p} \leq \frac{\mu_p - \bar{\mu}_p}{\bar{\mu}_p} \leq \frac{\mu_{pU} - \bar{\mu}_p}{\bar{\mu}_p} \quad (10)$$

For a user-specified percentage error ϵ and confidence level $1-\alpha$, the circuit simulation is stopped when the following criterion is satisfied:

$$\max\left[\frac{\bar{\mu}_p - \mu_{pL}}{\bar{\mu}_p}, \frac{\mu_{pU} - \bar{\mu}_p}{\bar{\mu}_p}\right] \leq \epsilon. \quad (11)$$

By (11), we can bound $|\mu_p - \bar{\mu}_p|/\bar{\mu}_p$ to the specified percentage error and guarantee the obtained sample mean $\bar{\mu}_p$ is close enough to μ_p. It is noteworthy that no assumption on the distribution of $\boldsymbol{P}$ is made in deriving (11). The stopping criterion is thus valid for all the circuits.

IV. Experimental Results and Discussion

The nonparametric technique proposed in Section III has been implemented using a variable delay circuit simulator and used to estimate the average power for a set of benchmark circuits. Since in deriving (9) we approximated $F^{-1}(u_i)$ by $F_n^{-1}(u_i)$ for

Circuit Name	SIM (mW)	LB (mW)	UB (mW)	$\bar{\mu}_p$ (mW)	No. Samples
C499	5.845	5.574	6.106	5.867	2016
C880	2.907	2.756	3.042	2.901	3616
C1355	5.843	5.550	6.036	5.841	2016
C1908	5.357	5.107	5.621	5.375	3008
C2670	7.331	6.995	7.624	7.362	2272
C3540	15.275	14.497	15.999	15.260	3200
C5315	21.357	20.299	22.076	21.366	2048
C7552	33.309	31.666	34.640	33.181	2656
apex6	3.906	3.728	4.099	3.923	3072
dalu	4.736	4.538	5.012	4.774	4224
x3	4.251	4.064	4.437	4.277	2208

Table 1: Power simulation results for benchmark circuits.

$i = 1, \ldots, n$, we need to know first if the approximation is valid for the sample size of interest. In Fig. 5 we show the average percentage error ϵ_{av} calculated by

$$\epsilon_{av} = \frac{1}{n} \sum_{i=1}^{n} \frac{|F_n^{-1}(u_i) - F^{-1}(u_i)|}{F^{-1}(u_i)} \times 100\% \qquad (12)$$

as a function of sample size n for several benchmark circuits. In (12) we approximate $F^{-1}(u)$ by the $F_n^{-1}(u)$ constructed from 1 million random samples. It is observed that for the sample sizes of interest, the average percentage error is smaller than 1.2%. Thus, the approximation is good enough for our purpose. By conducting the same analysis, similar conclusion can be drawn for all other test circuits.

Table 1 shows the simulation results for the set of benchmark circuits. For each circuit, every primary input had a signal probability of 0.5. The maximum error allowed was specified as 5% with 0.99 confidence. In this table, SIM is the average power obtained by calculating the sample mean of 1 million random samples, and represents the best estimate we can get; LB(UB) is the lower(upper) bound of the average power calculated by (9); $\bar{\mu}_p$ is the sample mean obtained from taking the average of the sample set whose size is listed in the last field. For all the cases, the technique produced very accurate estimate of the average power. It should be noted that for circuits with multimodal power distribution such as apex6, our technique can still be applied. Another distinguished property of the technique is that it is *dimensionally-independent*[3], i.e., the number of samples needed to achieve the specified accuracy is independent of the circuit size. Thus, this technique is very useful for average power estimation of very large circuits.

In another experiment, we studied the convergence speed of the technique under different simulator initializations by conducting 1000 simulation runs for each circuit. The minimum, maximum, and average number of samples required to achieve 10% maximum error with 0.99 confidence are shown in Table 2. As seen from the table, among all the simulations, no run estimated the average power with error larger than 10%. This result agrees with the expected 0.99 confidence and shows the superiority of the technique.

Name	Min	Max	Avg	>10% Err
C499	448	512	480	0.0
C880	672	2240	889	0.0
C1355	448	512	482	0.0
C1908	640	1056	694	0.0
C2670	480	608	547	0.0
C3540	672	896	747	0.0
C5315	448	512	485	0.0
C7552	608	704	649	0.0
apex6	672	960	732	0.0
dalu	832	1376	1054	0.0
x3	512	576	539	0.0

Table 2: Statistics from 1000 runs.

V. Conclusion

We have proposed a novel nonparametric technique for estimation of average power dissipation of digital circuits. The technique has the following features:

1. Average power can be estimated without any assumption on power distribution. Therefore it can be applied to circuits with arbitrary power distribution.
2. It is dimensionally independent. This feature makes it suitable to handle very large circuits.
3. Along with average power estimate, we also acquire certain knowledge about the cdf of power dissipation. This information may be very important in the design of highly reliable circuits.

REFERENCES

[1] F. N. Najm, "Transition density: a new measure of activity in digital circuits," *IEEE Trans. Computer-Aided Design,* vol. 12, no. 2, pp. 310-323, 1993.

[2] C.-Y. Tsui, M. Pedram, and A. M. Despain, "Efficient estimation of dynamic power consumption under a real delay model," *IEEE Int. Conf. Computer-Aided Design,* Santa Clara, CA, 1993, pp. 224-228.

[3] R. Burch, F. N. Najm, P. Yang, and T. Trick, "A Monte Carlo approach for power estimation," *IEEE Trans. VLSI Systems,* vol. 1, no. 1, pp. 63-71, 1993.

[4] M. H. DeGroot, *Probability and Statistics,* Menlo Park, CA: Addison-Wesley, 1986.

[5] J. D. Gibbons, *Nonparametric methods for quantitative analysis,* Columbus, OH: American Sciences Press, 1985.

Maximum Power Estimation for Sequential Circuits Using a Test Generation Based Technique *

Chuan-Yu Wang, *Kaushik Roy*, and *Tan-Li Chou*
Electrical and Computer Engineering, Purdue University, West Lafayette, IN 47907, USA.

Abstract

With the high demand for reliability and performance, accurate estimation of maximum instantaneous power dissipation in CMOS circuits is essential to determine the IR drop on supply lines and optimizing the power and ground routing. However, the problem of determining the input patterns to induce maximum current, and hence, the maximum power, is NP-complete. *In this paper, we present an* Automatic Test Generation (ATG) *based technique to efficiently estimate maximum power dissipation in sequential circuits. The technique can generate tight lower bounds of maximum instantaneous power within very short CPU time compared to random simulation based techniques. In addition, we also generate the measure of the quality or effectiveness of our approach from a statistical point of view. Experiments were performed on* ISCAS-89 *sequential circuit benchmarks. Results show that the ATG-based estimation is superior to traditional simulation-based technique in both speed and performance. For large circuits, the ATG approach is on an average* 29% *better and* 37099% *faster than simulation based technique.*

1 Introduction

Maximum instantaneous power dissipation in VLSI circuits has a great impact on circuit's reliability and the design of power and ground lines. Accurately estimating the peak power is essential in synthesizing highly reliable systems. However, accurate estimation maximum power in CMOS circuits involves searching for input patterns to maximize the switching activities in the circuits, and hence, is CPU time intensive. For large-scaled circuits, the only feasible approach is to generate tight upper and lower bounds of the maximum instantaneous power within reasonable amount of CPU time such that the gap between the bounds is as narrow as possible.

In the past, several pattern-dependent approaches have been proposed for estimating maximum power in combinational circuits [3, 4, 5]. However, their applications are limited to very small circuits because of the CPU-time consumption. In [3], Devadas et. al. maximized the power dissipation in CMOS circuits by solving a weighted *max-satisfiabiliy* problem, which has the complexity exponentially proportional to the number of primary inputs (PI's). In [4, 5], a loose upper bound is initially generated by propagating signal uncertainty through the circuit. The upper bound is then successively made tighter by a branch-and-bound strategy. However, without an efficient process to generate reliable lower bounds, estimate of the maximum power can not be obtained with the upper bounds alone.

For sequential circuits, the estimation problem is more complicated due to the fact that the sequential relationship between the PI's (Primary Inputs) and states must be considered. Given a sequential circuit with n PI's and m state bits, the complexity of searching the state and inputs which induce maximum switching activity is $O(2^{(2n+m)})$. Consequently, for large-scaled sequential circuits, the estimation of peak power by directly searching the input space can quickly become intractable. As demonstrated in our experiments, the pattern-dependent estimation (based on logic-simulation) is not only intolerably time-consuming, but the quality of the estimates generated are also not acceptable. In [9], Manne et. al. addressed the the problem of estimating maximum *sustained* power dissipation of sequential circuits provided that the *State Transition Graph* (*STG*) has been given. Exhaustive search is conducted in the transformed STG to identify the loop along which the average power dissipation is maximized. However, the CPU time for identifying the loop is proportional to the size of the given STG. The technique is, hence, only applicable to sequential circuits of very small size.

Motivated by these, we tackle the problem of efficiently generating good estimates (tight lower bound) of the maximum *instantaneous* power dissipation for large-scaled sequential circuits. We solve the problem based on *ATG*-based technique, which was originally developed by us for estimating combinational circuits in [1]. For a given sequential circuit, the sequential relationship between states and inputs can be derived from structure of the circuit. By unrolling the sequential circuit in time, we translate the relationship into boundary conditions. The constraint is then attached to the original problem, which formulates the maximizing of power consumption in combinational circuits. In the resultant problem, the objective function denoting instantaneous power in sequential circuit can be maximized efficiently based on *Automatic Test Generation* technique.

In addition, the quality (tightness) of the estimates generated are measured by a *Monte Carlo* based technique. However, unlike combinational circuits [6], the *distribution* of the instantaneous power of sequential circuits is not known. In this paper, the *Tchebycheff* inequality is proposed to generate the estimate of the quality measure.

The paper is organized as follows. Section 2 introduces main ideas of the ATG-based maximum power estimation. Section 3 demonstrates how to apply the ATG-based estimation to sequential circuits. Section 4 illustrates pattern-dependent estimation for sequential circuits, and how to measure the quality of estimates statistically. Section 5 presents the implementation and experimental results. Conclusions are given in Section 6.

2 ATG-Based Estimation

In CMOS circuits, power dissipation is dominated by the dynamic power consumption. Hence, the instantaneous power dissipation of a *combinational* circuit due to two consecutive input vectors is proportional to: $P_c = \sum_{for\ all\ gates} T(g) * C(g)$, where $C(g)$ denotes the output capacitance of gate g, and $T(g)$ is a binary variable which indicates whether gate g switches or not corresponding to the two input vectors. $T(g)$ equals 1 if gate g switches, and is 0 if gate g does not switch. Since the capacitive load at a gate output can be approximated by the fanout factor of the gate, we can re-express the instantaneous power dissipation as :

$$P_c = \sum_{for\ all\ gates} (g(I_1) \oplus g(I_2)) * F(g) \tag{1}$$

where I_1, I_2 denote two input vectors for two consecutive time frames, $g(*)$ denotes the *Boolean* function of gate g in term of PI's, and $F(g)$ denotes the fanout of gate g. Subjected to the structure of the given circuit, P_c is the objective function to be maximized.

To maximize P_c, instead of searching for optimal I_1 and I_2 directly, we greedily assign transitions to the gates with large

*This research was supported in part by ARPA (F33615-95-C-1625) and NSF CAREER award (9501869-MIP)

0-7803-3177-6 $5.00 © 1996 IEEE

IEEE 1996 CUSTOM INTEGRATED CIRCUITS CONFERENCE

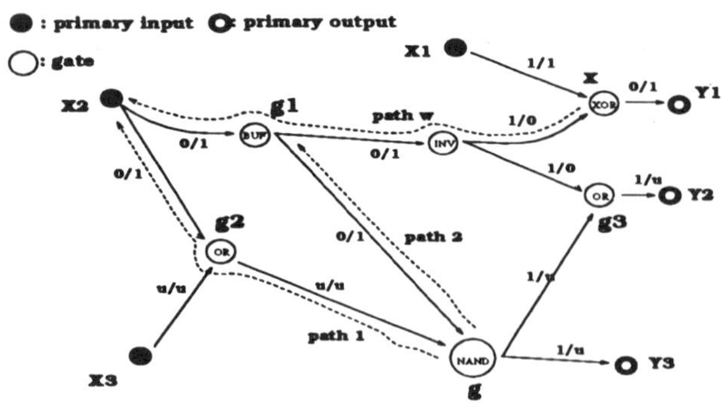

Figure 1: Justifying a switching

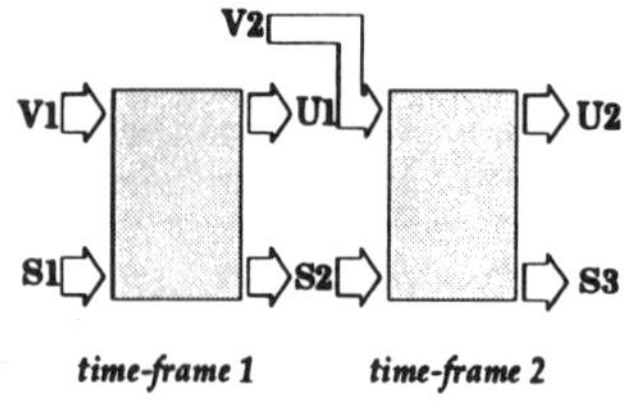

Figure 2: Unrolling the sequential circuit

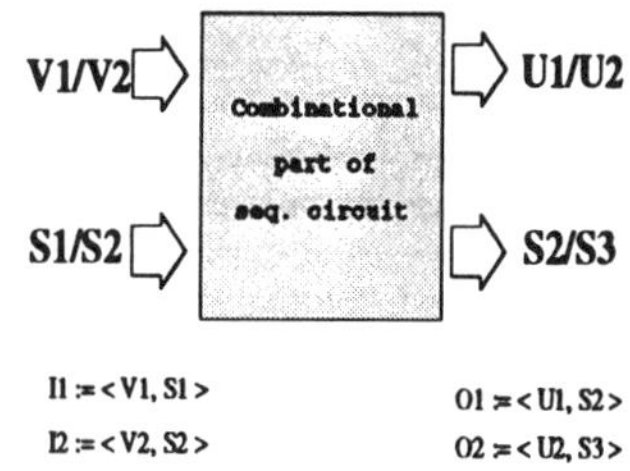

Figure 3: Overlapping two time frames

fanout ($F(*)$). The gates are first sorted in non-increasing order by their fanout. Then, in each iteration, we select a gate g which is untried and currently has the the largest fanout $F(g)$, to make it switch (i.e. $g(I_1) \oplus g(I_2) = 1$). Either rising transition ($g(I_1)/g(I_2) = 0/1$) or falling transition ($g(I_1)/g(I_2) = 1/0$) is assigned to g. The assignment is then verified by the *justification* mechanism. The *justification* mechanism in our algorithm includes two processes - *backtracing* and *implication*, which work similarly to the *9-Valued D algorithm* (an *Automatic Test Generation* algorithm) [7]. In 9-Valued D algorithm, to detect a stuck-at fault at a gate's output, a *sensitized path* is generated to propagate the fault *forward* from the gate to a *primary output*. The path generated is then verified by the *Implication* process. Similarly, in ATG-based estimation, to justify a rising (0/1) or falling (1/0) transition assigned to a gate, a path is generated to propagate the transition *backward* from the gate to a *primary input*. In our algorithm, *Backtracing* process is used to build such a path by recursively assigning switching to one of the fanins gate by gate, until a PI is met. The path is then verified in the same way as in 9-Valued D algorithm. A gate may take the logic value 0, 1 or u (unknown) in a time frame. For each gate, the logic values during time-frame 1 and 2 are packaged into a composite logic value. We define the *state* of the circuit as a combination of the most recently updated composite values of all gates. If the justification of assigning transitions to a gate fails, the state of the circuit will be recovered from the incorrect decision (*Backtracking*), and next gate will be tried. The gates in the circuit are assigned and justified one by one until all gates have been processed. By the above procedure, ATG-based estimation implicitly searches the solution space, and the vector pair I_1/I_2 for stressing the circuit is generated as a side effect.

A transition assigned to the output of a gate is said to be *justified* if either one of the following paths can be generated:

1. *The Primary Path* : a path through which the transition can be propagated backward from the assigned gate to a PI, which has the composite logic value not *in conflict* with the transition. Two composite logic values are in conflict if they have 0 and 1 at the same position. For example, $u/1$ and $0/0$ are in conflict.

2. *The Secondary Path* : a path along which the transition can be propagated backward from the assigned gate to a *primary path*, which was built to justify a transition earlier.

For example, as shown in Figure 1, since we can assign transitions to g_2 and x_2 under current state of the circuit, path 1 is a primary path to justify the falling transition at gate g. On the other hand, since g_1 is switching, we can also backtrace the transition at g to the primary path w, which was generated for justifying the transition at the output of gate x. Hence, a secondary path can be also generated (path 2). Either path 1 or path 2 justifies the transition at gate g.

3 Estimation for Sequential Circuit

In a sequential circuit, outputs of the latches are defined as *pseudo-primary inputs*, and inputs to the latches are defined as *pseudo-primary outputs*. For a given sequential circuit, after combining all the *primary inputs* and *pseudo-primary inputs* into a single input vector, the process of maximum power estimation turns out to be searching for two consecutive input vectors, which stimulate maximum switching activity in the circuit. Compared to the problem described by equation 1, the only difference is that the input vectors I_1 and I_2 are now correlated. We will derive and formulate the correlation next. Details about maximizing power consumption will be given in Section 3.2

3.1 Derivation of the Correlation

To incorporate the correlation into the ATG-based estimation, we unroll the sequential circuit into two consecutive time-frames. As shown in Figure 2, primary input vector V_1 makes the circuit to transit from State S_1 to S_2 in time-frame 1. V_2, along with the state vector S_2, makes the circuit to transit to State S_3 in time-frame 2. Here U_1 and U_2 denote the primary output vectors of time-frame 1 and 2, respectively. In addition, in each time-frame, each gate is associated with a 3-valued-logic value - 0, 1, or u (unknown). In Figure 3, by considering time-frame 1 and 2 together, we package the 3-valued-logic values of two time-frames into 9-valued-logic (composite logic value). If a gate g takes the value 1 in time-frame 1 and value u in time-frame 2, the composite logic value of g will be $1/u$. Each gate is then associated a stack to store the composite logic values which have been assigned to the gate. It can be seen from Figure 3 that, V_1 and S_1 can be packaged into a single vector I_1, since they are in the same time frame. We denote the packaging as $I_1 := <V_1, S_1>$. Similarly, other pairs of vectors in the same time frame can be packaged as $I_2 := <V_2, S_2>$, $O_1 := <U_1, S_2>$, and $O_2 := <U_2, S_3>$. If there are n PI's and m state bits in the sequential circuit, it can be seen from Figure 3 that the elements in vector I_2 and O_1 must be made equal from the $(n+1)$th to $(n+m)$th position.

Based on Equation 1 and the above discussion, the problem of maximum power estimation can be described as :

$$Maximize \;\; P_s = \sum\nolimits_{for\ all\ gates} (g(I_1) \oplus g(I_2)) * F(g)$$

$$Subject\ to\ \Phi \equiv \{O_1^j = I_2^j \;/\; j = n+1, n+2 \ldots, n+m\}$$

where I_2^j and O_1^j denote the jth element of vector I_2 and O_1 respectively, $F(g)$ is the fanout factor of gate g, and Φ is the constraint set to the objective P_s. Since O_1 is the corresponding output vector for I_1, Φ represents the *temporal correlation* between the consecutive input vectors I_1 and I_2.

3.2 Maximization

To maximize P_s for a given sequential circuit ζ, transitions (i.e. $g(I_1) \oplus g(I_2) = 1$) are greedily assigned to gates having large $F(*)$. The gates in ζ are first sorted by the number of fanout ($F(*)$) in non-increasing order. Rising or falling transitions ($g(I_1)/g(I_2)$ $= 0/1$ or $1/0$) are then assigned to the gates one by one in the sorted order. However, as derived in Section 3.1, the *justification* mechanism have to cope with the boundary condition provided by constraint set Φ. The second element of the composite logic value of any pseudo-primary input must be always equal to the first element in the composite logic value of the corresponding pseudo-primary output. To achieve this, during the justifications of switchings, O_1^j and I_2^j ($j \in \{n+1, n+2..., n+m\}$) must be updated simultaneously. Suppose g is the switching gate which currently needs to be justified. Before assigning the transition to g, the kth element ($k \in \{n+1, n+2..., n+m\}$) of I_1/I_2 has the composite value a^k/b^k, and the kth element of O_1/O_2 has the composite value b^k/c^k. We describe the modifications of the original *justification* mechanism as follows :

(i) During *backtracing* or *backward-implication* process, if I_2^k is updated as the value λ, the composite value λ/c^k is assigned to the kth element of O_1/O_2. The assignment will be verified immediately by *backward-implication*. If the verification fails, the *primary* or *secondary path* built is abandoned and other paths are tried for gate g.

(ii) In *forward-implication* process, if O_1^k is updated as the value β, the composite value a^k/β is assigned to the kth element of I_1/I_2. The assignment is then verified immediately by *forward-implication*. Similarly, as above, if the verification fails, other paths are tried to make gate g switch.

By Rule **(i)** and **(ii)**, satisfication of the constraint set Φ can be guaranteed when P_s is being maximized.

For example, let sequential circuit ζ has only one state bit (i.e. m=1), and the circuit shown in Figure 1 is the combinational part of ζ. Suppose PI X_3 is the pseudo-primary input and Y_3 is the corresponding pseudo-primary output. To justify the falling transition at gate g, Path 1 (a *primary path*) implies PI X_3 must take the value $0/u$. However, driven by gate g, Y_3 is forward-implied to take the value $1/0$. There will be conflict when we *backward-imply* the value $0/u$ from the pseudo-primary output Y_3. According to Rule **(i)**, Path 1 must be abandoned since it violates the constraint set Φ. It can be seen from Figure 1 that, *backtracing* the falling transition at g through Path 2 can be verified. No conflict will be generated if both X_3 and Y_3 take the value $1/0$. Hence, Path 2 (a *secondary path*) successfully justifies the falling transition at g in ζ.

4 Pattern-Dependent Estimation

In estimating maximum power for combinational circuits, pattern-dependent estimation has been used to serve as a comparison version for the ATG-based estimation. In [1], we proposed to simulate the circuit and monitor the maximum value of the instantaneous power generated from simulation, which is defined as the *simulated maximum power*. The simulation stops if (1) the most updated simulated maximum power is satisfactory, or (2) the pre-set upper bound of the number of simulations has been reached. To determine if a simulated maximum power (X) is satisfactory, we associate each X with a measure of quality, and pre-specify a minimum requirement of the quality before simulating the circuit. We view the instantaneous power as a random variable. By the *normality* assumption of the distribution, the quality measure of X is computed by integrating the *density function* of power from 0 to X. The density function of power can be derived from the *mean* and *deviation* of the instantaneous power. Therefore, we apply *Monte Carlo* technique to estimate the *mean* and *deviation* of the instantaneous power while stressing the circuit.

In estimating the maximum power of sequential circuits by *Monte Carlo* based simulation, there are two major differences from maximum power estimation for combinational circuits. *First*, the *warm-up* period (the period after which sampling should start) must be considered in estimating the *mean* of instantaneous power. Otherwise, the estimate of the mean will be biased by the selection of initial states [2]. *Second*, the distribution of the instantaneous power of sequential circuits is now unknown.

Given a sequential circuit the cycle length of the warm-up period for average power estimation can be computed from the number of PI's plus the number of state bits (i.e. (n+m)) [2]. To solve the second problem, let P_s denote the instantaneous power of the circuit, and X is the *simulated maximum power*. By *Monte Carlo* based simulation, the mean μ and deviation ϵ of P_s can be estimated with pre-specified precision. Let p_s denote a random sample of P_s. The probability of the event $\{X \geq p_s\}$ (i.e $Pr\{X \geq p_s\}$), which represents the quality of X, needs to be computed without knowing the distribution of P_s. To achieve this, we introduce *Tchebycheff Inequality* [8] as follows: $Pr\{|p_s - \mu| \geq \alpha\} \leq (\epsilon^2/\alpha^2)$, where $Pr\{*\}$ is the traditional probability operator, p_s denotes a random sample of P_s, μ is the mean of P_s, ϵ is the deviation of P_s, and α is an arbitrary positive real number. No matter what the distribution of P_s is, the above inequality is always valid. If we let $\alpha = X - \mu$, the above inequality can be transformed into: $Pr\{p_s \leq (X - 2 * \alpha)\}$ + $Pr\{p_s \geq X\} \leq (\epsilon^2/(X-\mu)^2)$. Since $Pr\{p_s \leq (X - 2 * \alpha)\} \geq 0$ and we are only interested in the probability of the event $\{X \geq p_s\}$, the following inequality can be derived :

$$Pr\{X \geq p_s\} \geq 1 - (\epsilon^2/(X-\mu)^2) \qquad (2)$$

By Inequality 2, as shown in Figure 4, given a simulated maximum power X, $\gamma(X) = 1 - (\epsilon^2/(X-\mu)^2)$ is a conservative estimate of the quality of X ($Pr\{X \geq p_s\}$). Let Q_m denote the user-specified minimum requirement of the quality of X. The simulator will stop and report the most updated X as a satisfactory estimate once $\gamma(X) \geq Q_m$ is achieved.

The flowchart of pattern-dependent estimation is given in Figure 5. The simulator monitors the simulated maximum power X while estimating the mean (μ) and deviation (ϵ) of instantaneous power. We denote the ith sample of μ and ϵ as μ_i and ϵ_i, respectively. The procedure to generate μ_i and ϵ_i of power is implemented as follows :

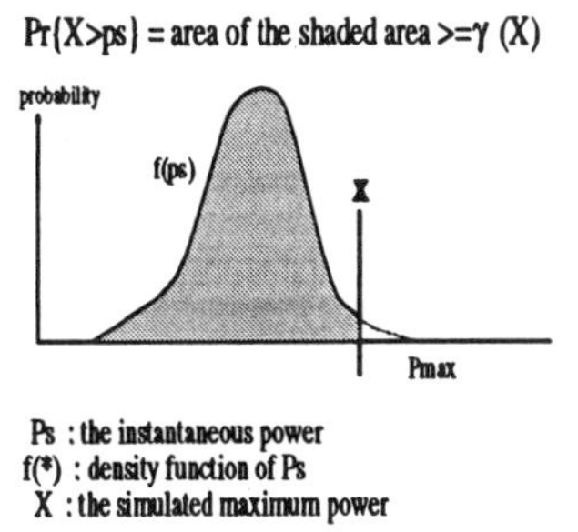

Figure 4: Measuring the quality of X

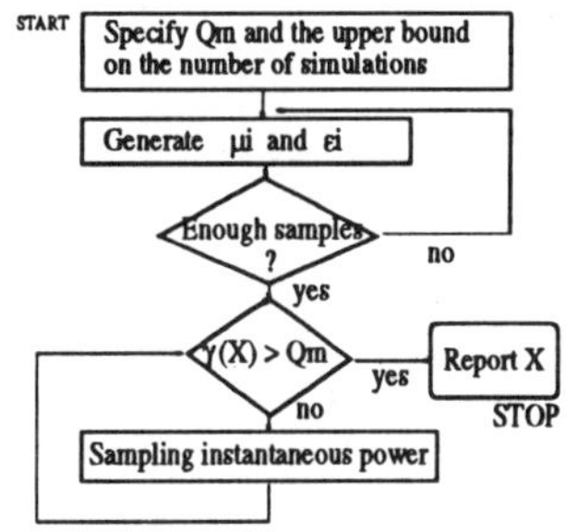

Figure 5: Flowchart of pattern-dependent estimation

(i) Assign *state bits* randomly to generate an *initial state*.
(ii) Warm-up the circuit according to the cycle length calculated as in [2].
(iii) Make t samples of the instantaneous power. In our experiment, t = 300.
(iv) Compute the sample mean of the t samples, which is the value of sample μ_i. Compute the deviation of the t samples, which is the value of sample ϵ_i.

By repeating the above procedure, to estimate μ with the pre-specified precision, sample μ_i (i=1,2,..,30) are collected and analyzed to compute the number of samples of μ required [6]. The estimate of μ can then be obtained by further simulations. Experimentally we found that the deviation of sample ϵ_i (i=1,2,..,30) is small enough to be ignored. The arithmetic mean of ϵ_i (i=1,2,..,30) should be a good estimate of ϵ. After the estimates of μ and ϵ of instantaneous power are generated, the two criterions for stopping the simulation are checked immediately. If $\gamma(X) \geq Q_m$ or the number of simulation runs exceeds the pre-specified bound, the simulater stops and reports X. Otherwise, the simulation continues until one of the criterions is met.

5 Experimental Results

The proposed ATG-based estimator (ATG_P), along with the pattern-dependent estimator (SIM_P), are implemented in C on a HP 715/50 workstation.

The primary inputs to a circuit are modeled as stochastic processes each associated with a *signal probability* (the probability of taking the logic value of ONE) and *activity* (average number of transitions per unit time). The input signals for simulation are generated based on the given signal probabilities and activities. For maximum power estimation using SIM_P, the PI's were specified to have signal probability and activity of 0.5 and 0.9, respectively, in order to ensure high switching activity at the inputs. We assume that the high switching activity at the inputs also produce high switching activity at the internal nodes of the circuit [1].

In SIM_P, with 95% *confidence*, we estimate the mean of power (μ) within the specified 5% error. To compare the performance of ATG_P and SIM_P, the minimum requirement Q_m is set as 0 in SIM_P. On the other hand, to make the estimate X as good as possible to compete with ATG_P, we didn't set the bound on the number of simulation runs. The simulator stops and reports the most updated simulated maximum power X after the estimate of μ with pre-specified precision is generated.

We show the experiment results on $ISCAS-89$ benchmark circuits in Table 1. The resultant maximum instantaneous power X's are measured according to Equation 1: $P_s = \sum_{for\ all\ gates} (g(V_1) \oplus g(V_2))F(g)$, where $F(g)$ denotes the fanout of gate g. It can be seen from Table 1 that, due to the variation of warm-up periods, the required number of simulation runs are different for each circuit. In Table 2, the quality measure $\gamma(X)$ of each X is calculated according to Inequality 2. For each circuit, we also show the number of gates, and the mean and deviation of instantaneous power.

Judging from X, $\gamma(X)$, and the CPU time consumed for the experiments shown in Table 1 and 2, the ATG-based estimation is far superior to the traditional simulation-based technique for estimation of maximum power. In Table 1, for example, for benchmark $s38417$, ATG_P improves the maximum power generated from SIM_P by 61% with 0.3% cpu time. For $s38584$, ATG_P improves the maximum power by 15% with 0.5% cpu time. Within a very small amount of CPU time, the ATG-based estimator generates estimates with the quality which cannot be achieved by simulating the circuit for a very long period of time.

6 Conclusions

In this paper, to estimate the maximum power consumption in sequential circuits, we propose an Automatic Test Generation (ATG) based technique. Experiments show that this approach is far superior to the traditional pattern-dependent simulation-based techniques. Within very small amount of CPU time, our approach can generate estimates of very high quality. In estimating the maximum power of very-large-scale circuits, like microprocessor, our approach may be the only existing way to generate reliable estimates.

References

[1] C.Wang and K.Roy, "Maximum Power Estimation for CMOS Circuits using Deterministic and Statistic Approaches", Ninth International Conference on VLSI Design (1996), to appear.

[2] T.Chou and K.Roy, "Statistical Estimation of Sequential Circuit Activity", IEEE/ACM International Conf. on CAD, pp.34-37, 1995.

[3] S.Devadas, K.Keutzer and J.White, "Estimation of Power Dissipation in CMOS Combinational Circuits", IEEE Custom Integrated Circuits Conf., pp. 19.7.1-19.7.6, 1990.

[4] H.Kriplani, F.Najm, and I.Hajj, "Maximum Current Estimation in CMOS Circuits", 29th ACM IEEE Design Automation Conf., pp.2-7, 1992.

[5] H.Kriplani, F.Najm, P.Yang and I.Hajj, "Resolving Signal Correlations for Estimating Maximum Currents in CMOS Combinational Circuits", 29th ACM IEEE Design Automation Conf., pp.2-7, 1992.

[6] R.Burch, F.Najm, P.Yang and T.Trick, "A Monte Carlo Approach for Power Estimation", IEEE Trans. VLSI Systems, Vol.1, No. 1, pp.63-71, March, 1993.

[7] M.Abramovici, M.Breuer and A.Friedman, Digital Systems Testing and Testable Design, Computer Science Press, 1990.

[8] A.Papoulis, Probability, Random Variables, and Stochastic Process, McGraw-Hill, Inc., 1991.

[9] S.Manne, A.Pardo, R.Bahar, G.Hachtel, F.Somenzi, E.Macii and M.Poncino "Computing the Maximum Power Cycles of a Sequential Circuit", 32nd Design Automation Conf., pp.23-28, 1995.

Table 1: Results of ATG_P and SIM_P for $ISCAS$-89 benchmarks

Ckt.	Max. Power X		CPU Time (sec)		Num.
	$ATG.$	$SIM.$	$ATG.$	$SIM.$	sim.
s208	135	138	0.24	51.8	13590
s298	201	195	0.61	30.7	13410
s344	185	187	0.39	70.7	14790
s349	180	188	1.42	68.6	14790
s382	266	226	1.96	42.8	14790
s386	268	269	0.33	77.6	12630
s510	244	241	3.58	89.7	14790
s526	298	297	0.74	72.4	14790
s820	612	624	4.66	155.7	14580
s832	619	633	6.83	169.6	14580
s1196	594	634	10.6	321.6	16380
s1238	610	649	1.63	311.0	16380
s1423	784	710	4.33	673.9	28020
s1488	938	923	15.4	187.0	12810
s1494	938	926	3.46	188.9	12810
dsip	4097	3994	40.4	16042.0	99270
s5378	2998	2114	23.0	3446.1	49320
s9234	4709	3006	54.5	9689.5	58800
s13207	6525	3508	188.6	13000.9	148200
s15850	6734	4059	706.2	20402.9	130650
s35932	19872	18066	718.3	466808.9	357990
s38417	18946	11802	592.1	160249.9	338460
s38584	17472	15136	1294.2	249051.4	298980

Table 2: Quality of estimates of ATG_P and SIM_P for $ISCAS$-89 benchmarks

Ckt.	Num.	$\gamma(X)$ (%)		mean (μ)	dev. (ϵ)
	gates	$ATG.$	$SIM.$	of power	of power
s208	104	95.28	95.60	51.2	18.2
s298	119	99.15	99.07	53.4	13.6
s344	160	92.63	93.00	108.0	20.9
s349	161	91.17	92.87	109.3	21.0
s382	158	99.49	99.20	67.7	14.1
s386	159	97.45	97.48	132.7	21.6
s510	211	78.03	76.86	127.7	54.5
s526	193	99.00	98.99	78.8	21.9
s820	289	94.34	94.75	299.1	74.4
s832	287	94.36	94.82	300.2	75.7
s1196	526	79.52	89.74	497.2	43.8
s1238	508	78.15	89.00	514.8	44.5
s1423	657	90.45	85.23	405.9	116.8
s1488	653	87.04	86.40	312.6	225.1
s1494	647	87.17	86.66	310.3	224.8
dsip	2097	97.97	97.53	2981.0	159.8
s5378	2779	99.56	98.04	1330.1	109.7
s9234	5597	99.62	98.10	1647.4	187.1
s13207	7951	99.96	99.80	1098.5	106.7
s15850	9772	99.65	98.29	1887.4	283.3
s35932	16050	87.56	75.66	13539.4	2233.0
s38417	22397	99.63	97.76	7011.1	716.9
s38584	19407	92.76	84.36	10161.4	1967.0

Combined ac and Transient Power Distribution Analysis*

Xun Yang
Electrical and Computer Engineering
The University of Texas at Austin
Austin, TX 78712
yang@cad.ece.utexas.edu

Byron Krauter
IBM Corp.
11400 Burnet Road
Austin, TX 78758
krauter@austin.ibm.com

Lawrence T. Pileggi
Electrical and Computer Engineering
Carnegie Mellon University
Pittsburgh, PA 15213
pileggi@ece.cmu.edu

Abstract

The increased use of noise sensitive circuits and power up or sleep modes, and the high cost of on-chip decoupling have combined to make the VLSI CMOS power distribution problem both a global chip and a local circuit concern. Extensive simulation is required to evaluate the worst case power rail sag due to the enormous possibilities in power up sequences. Moreover, VLSI CMOS power distribution networks are non-ideal in that they fail to provide a low impedance path from the power supply all the way to the circuit terminals over a large range of frequencies [2,3]. *Resonant peaks typically litter the impedance spectrum, even at the chip level* [10], *making both transient and steady state analysis necessary.*

This paper proposes an analysis formulation that efficiently derives both steady state and transient responses from a series of fixed-time point transient impulse response solutions (frequency-shifted moments). This formulation is used to efficiently evaluate worst case power supply sags. Practical examples involving VLSI CMOS power distribution networks are shown which demonstrate that phase-shifts across the chip further complicate the calculation of the worst case voltage sag.

1: Introduction

Because it is otherwise intractable, the VLSI power distribution problem is invariably modeled as an equivalent (albeit sometimes large) linear circuit. Switching activity is modeled with independent sources, and details concerning power rail structure and quiet circuit characteristics are coalesced into larger macro models. One favorable consequence of this approach is that total circuit response can be readily obtained from the circuit impulse responses.

For example, assume the switching activity on the VLSI CMOS chip can be modeled with M current sources. The total response $V(t)$ at time t is given by the sum of the individual responses to each of the M sources:

$$V(t) = \sum_{k=1}^{M} V_k(t) \tag{1}$$

where $V_k(t)$ is the individual response due to the k-th current source. These individual responses include both the steady state and transient responses

$$V_k(t) = V_k^{ss}(t) + V_k^{tr}(t) \tag{2}$$

Because of the synchronous nature of computers, the steady state response $V_k^{ss}(t)$ in (2) can be expressed as a complex Fourier series:

$$V_k^{ss}(t) = \sum_{n=-\infty}^{\infty} I_k(jn\omega_0) \cdot Z_k(jn\omega_0) \cdot e^{jn\omega_0 t} \tag{3}$$

In (3), $\omega_0 = 2\pi(T_{cycle})^{-1}$, $I_k(jn\omega_0)$ is the n-th series coefficient of the periodic steady state signal $i_k^{ss}(t)$, and $Z_k(jn\omega_0)$ is transfer impedance of the network. The latter can be found using the transfer function $H_k(s)$ of the impulse response $h_k(t)$:

$$Z_k(jn\omega_0) = H_k(s = jn\omega_0) \tag{4}$$

The transient response, $V_k^{tr}(t)$, can be specified in terms of a convolution of $i_k^{tr}(t)$ with the impulse response $h_k(t)$:

$$V_k^{tr}(t) = \int_0^t i_k^{tr}(\tau) \cdot h_k(t-\tau)\, d\tau \tag{5}$$

Because of unique relationship between time domain, frequency domain, and the highly synchronous operation of VLSI chips, we can accurately formulate the VLSI CMOS power distribution noise using a series of frequency-shifted moments taken about a single expansion point (hence from fixed-time step, transient simulations of the circuit impulse responses). Furthermore, since transient power distribution response is required for only a few cycles, and steady state impedances can be computed via Padé approximations using a limited number of circuit moments, only a small amount of information is required. Moreover, this information can be reused to efficiently evaluate worst case power supply sag.

2: Power Rail Analysis Using Shifted Moments

We describe our formulation approach following a brief review of frequency shifted moments.

2.1: Shifted Moments and Related Procedures

In [1], the generating matrices of Asymptotic Waveform Evaluation (AWE) [9] are shown to resemble those used in fixed time step numerical integration. In particular, shifted moments [8] computed about $s = \alpha$, correspond to backward Euler integration with time step $\Delta t = \alpha^{-1}$. Indeed, the passive elements in the companion models are identical, and the sources differ only by a factor α. These small differences are depicted in Fig.1 for both L and C.

The recursive procedure followed in moment computations is also identical to the recursive iterations followed in

* This work was supported in part by IBM, the Semiconductor Research Corporation under contract 95-DJ-343, and The Advanced Research Projects Agency.

0-7803-3177-6 $5.00 © 1996 IEEE

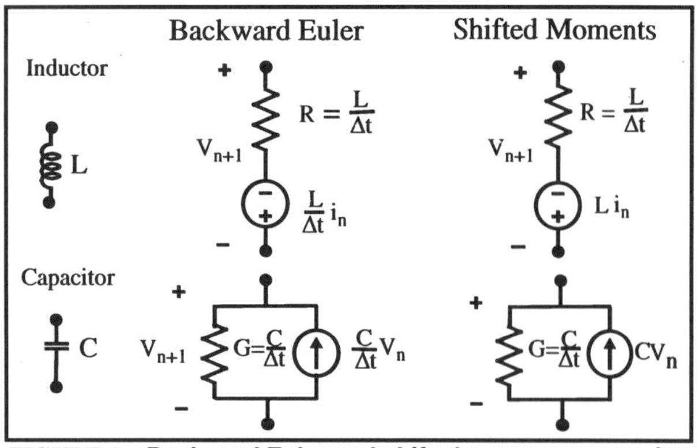

FIGURE 1: **Backward Euler and shifted moment companion models for expansion about $\alpha = \Delta t^{-1}$.**

the power method for computing the dominant eigenvalue(s) of a matrix. Given a matrix A, the power method is defined by the iteration [7]

$$\mathbf{x}^{new} = \mathbf{A}\mathbf{x}^{old} \quad (6)$$

Except for special starting vectors, the iterations converge to an eigenvector corresponding to the largest or dominant eigenvalue of the matrix **A**. As this convergence occurs, all non-dominant eigenvalue information is lost due to the finite precision of floating point numbers [6].

Just like moment computations, the power method is also altered to extract additional eigenvalues. The straight power method uses the matrix **A** in (6) and is equivalent to moment computation about s=0. The inverse power method uses $\mathbf{A}^{-1}$ and corresponds to derivative computations or expansions about $s = \infty$. The shifted power method replaces the matrix **A** with $(\mathbf{A} - \alpha\mathbf{I})^{-1}$ and is equivalent to shifted moment computation about $s=\alpha$. The shifted power method converges to the eigenvalue closest in magnitude to α.

2.2: VLSI CMOS Power Distribution Noise

On-chip steady state noise analysis in VLSI CMOS design is concerned with the forced circuit response at integer multiples (or harmonics) of the operating frequency (i.e. dc, f_c, $2f_c$, $3f_c$, .. where f_c is 1/Tcycle). If we can ignore the dc case for now -- which we recommend be covered in a separate and less complicated analysis -- the frequency range of interest covers the low order harmonics of the operating frequency only. Therefore, an appropriate expansion point, when considering the eventual convergence of shifted moment computation on the poles closest in magnitude to the expansion point, is $\alpha = |jn\omega_c|$ where $\omega_c = 2\pi f_c$ and $1 \le n \le 4$. For n = 2, this corresponds to a time step $\Delta t = 0.08 \cdot T_{cycle}$ (see Fig.2).

On-chip transient noise analysis in VLSI CMOS design is concerned with noise pulses that might disrupt circuit operation during the course of a single clock cycle. Clearly the minimum step size must be less than a single cycle, and for chips with half cycle paths, clearly less than a half cycle. But should the minimum step size be tenths, hundredths, or even thousandths of a cycle? Model uncertainties in global timings and switching factors, and the smoothing effects of model coalescence suggest that anything less than tenths of a cycle would be meaningless. Therefore, $\Delta t \approx 0.1 \cdot T_{cycle}$ would be an appropriate time step for a global (and even a local) transient noise analysis.

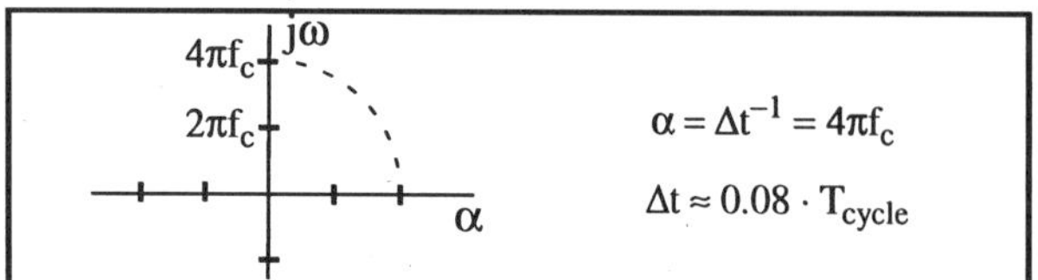

FIGURE 2: **Expansion point closest to 2nd harmonic.**

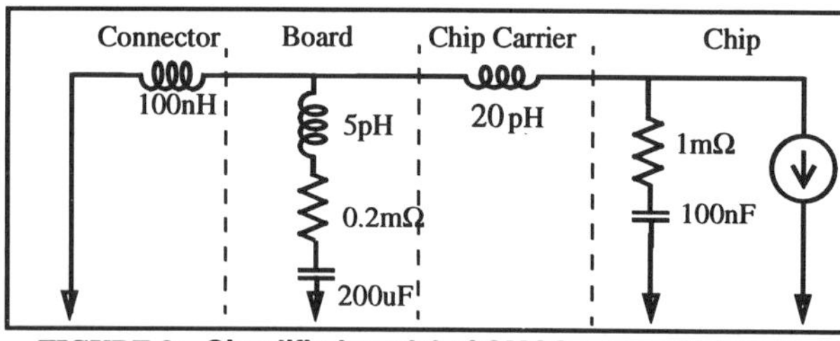

FIGURE 3: **Simplified model of CMOS power distribution.**

Because the time step required for transient noise analysis is comparable to that required for ac analysis (again see Fig.2), a single expansion point is sufficient for most VLSI CMOS power distribution analyses. The following examples will demonstrate this point.

3: Examples

Because of their high clock rates and abundant decoupling (both on and off chip [4]), today's microprocessors generally operate above the natural (or resonant) frequencies of their power distribution network. These networks look capacitive at the frequencies of interest, and therefore, steady state driving point and transfer impedances are monotonically decreasing. Moreover, these impedances can be determined from a single expansion point taken above the natural frequency of the packaged chip.

Consider the chip, board, and connector model depicted in Fig.3. While this first example was intentionally made simple so that the circuit poles could be evaluated exactly, it is, nevertheless, representative (with respect to the chip and chip carrier characteristics [5]) of today's large microprocessor designs that operate at clock rates exceeding 100 MHz.

The exact poles and the dominant poles computed using 3rd order AWE approximations for expansions about s=0 and s=1e9 sec^{-1} are shown in Table 1. Note, a 3rd order AWE approximation using circuit moments taken about s=0 accurately computes the low frequency poles of this circuit, but

TABLE 1. Exact and dominant poles for circuit in Fig.3.

Exact Poles	3rd Order AWE $s = 0\ sec^{-1}$	3rd Order AWE $s = 1e9\ sec^{-1}$
-9.990e2 - j 2.235e5	-9.990e2 - j 2.235e5	-4.353e2
-9.990e2 + j 2.235e5	-9.990e2 + j 2.235e	-
-2.400e7 - j 6.322e8	-7.148e9	-2.400e7 - j 6.322e8
-2.400e7 + j 6.322e8	-	-2.400e7 + j 6.322e8

misses the high frequency poles. When the same approximation is applied to frequency-shifted moments taken about s=1e9 sec^{-1}, just the opposite occurs. The high frequency poles are accurately determined, while the low frequency poles are completely missed.

When these dominant poles and their residues are used to compute transfer and driving point impedances (see (4)), these same errors persist. The driving point impedances of this circuit, computed using the exact and approximate transfer functions, are depicted in Fig.4 and Fig.5. When s=0 is used as the expansion point, this impedance is accurately computed at low frequencies only. When s=1e9 is used as the expansion point, this impedance is accurately computed for the higher frequencies only.

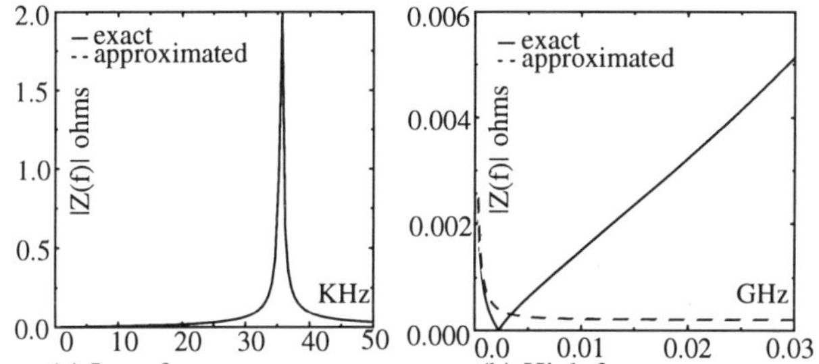

(a) Low frequency range (b) High frequency range

FIGURE 4: **Frequency response using 3rd order Padé approximation with s = 0 sec^{-1}.**

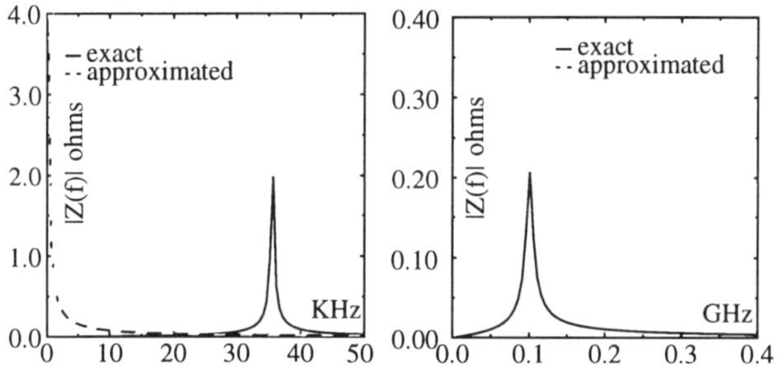

(a) Low frequency range (b) High frequency range

FIGURE 5: **Frequency response using 3rd order Padé approximation with s = 1e9 sec^{-1}.**

3.1: Complete Analysis with Shifted Moments

For our second example, we created a model that was again representative of today's large VLSI CMOS microprocessor designs. First and foremost, our assumptions included a 200 MHz internal clock rate and peak switching currents exceeding 40 amps. Because on-chip decoupling is frequently used in these designs, our example also assumed an on-chip decoupling capacitance on the order of the total quiet circuit capacitance. Finally, we assumed the chip was flip chip mounted in a BGA package using ground planes and decoupling capacitors.

Our example is depicted in Fig.6, and a simple macro model that includes both chip and package features is given in Fig.7. Note that the chip area has been divided into a 3x3 region (with nodes 1 through 9 representing on-chip locations), and source substitution has been applied to create a simple circuit model with a common ground node. That is, all impedances model a portion of the combined power-ground loop.

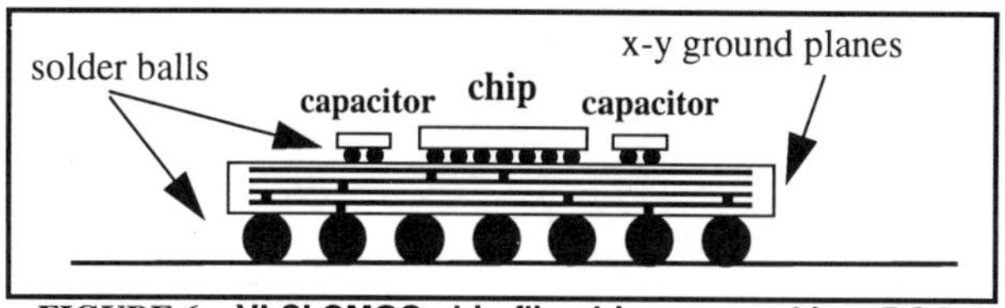

FIGURE 6: **VLSI CMOS chip flip-chip mounted in a BGA package with internal ground planes and capacitors**

In subcircuit *a*, R_1 and L_1 model the lossy path available through the on chip power distribution network, and L_2 models the low loss parallel path available through the package ground planes. In subcircuit *b*, the inductance L_3 models the impedance getting off the chip to the card, (or to adjacent decoupling capacitors on the package), while the two RC networks, R_3C_3 and R_4C_4, model the total on-chip capacitance, (inherent plus added on-chip decoupling) with two time constants. Finally, the independent current source I_1 models the switching activity in that region of the chip. Note, because of the full chip area connection of flip chip assembly, the center of the chip is modeled with the same subcircuits as the perimeter sections.

The passive circuit parameters for our example are as follows: R1 = 750mΩ, L1 = 20 pH, L2 = 200 pH, R2 = 9.0 mΩ, C2 = 8.9 nF, R3 = 18.0 mΩ, C3 = 13.3 nF, and L3 = 500 pH. The nine independent current sources were all assigned the same current waveforms. This signature is depicted in Fig.8. With a supply voltage of 3.3 Volts and the current signatures depicted in Fig.8, the minimum and maximum chip powers are 26.4 and 66.0 watts respectively.

To evaluate worst case power supply sag (which coincides here with a 29.6 watt increase in chip power), using transient simulation only, one must first model the minimum switching activity until a steady state condition is achieved. Because settling time can be long (in some cases even hundreds of cycles), this procedure is inefficient when multiple power up sequences must be evaluated. These inefficiencies can be avoided by using shifted moments to evaluate worst case power supply sag.

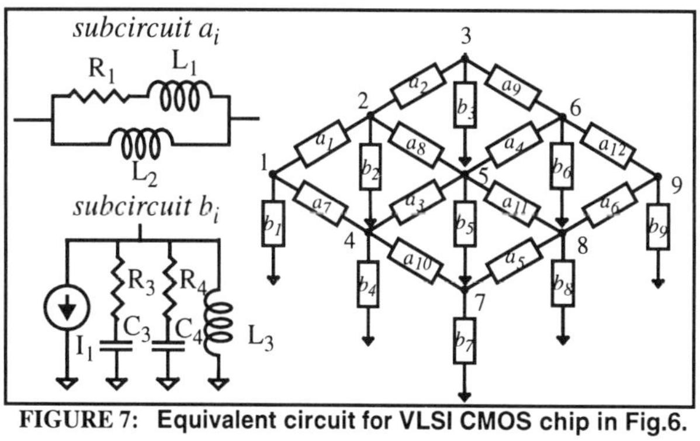

FIGURE 7: **Equivalent circuit for VLSI CMOS chip in Fig.6.**

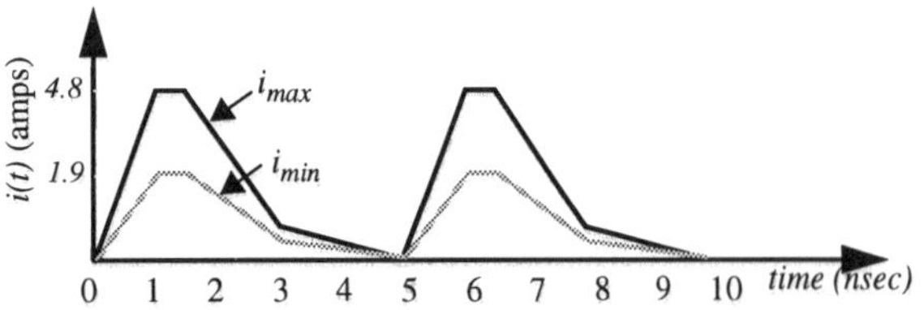

FIGURE 8: Current signatures for model in Fig.7.

Depicted in Fig.9 is the worst case sag at node 1 (computed using both shifted moments and SPICE) when all nine independent sources switched from the i_{min} to i_{max} waveforms in Fig.8. Although the results match well, the SPICE simulation took longer to compute because a 300 nsec settling time was required for the circuit to reach steady state conditions. While separate ac and transient analyses could be performed and these results combined, this approach would require several additional, very costly factorizations of the circuit equations (one each for every harmonic considered). To make matters worse, because it's an ac analysis, these additional factorizations would involve complex matrices.

The shifted moment solution, however, was obtained using only 60 circuit moments (60 time domain points from a circuit simulation) taken about a s=2e9 sec^{-1} expansion point (or 0.5 nsec time step). Although these moments had to be computed separately for each of the nine source locations, only one LU factorization of the circuit equations was required, and these same moments were subsequently reused to evaluate other switching currents.

3.2: Phase Shifts and Multicycle Effects

It is generally taken for granted that simultaneous switching (i.e. minimum to maximum switching activity in a single cycle) causes the worst case power supply sag. However, because of increasing chip sizes, decreasing cycle times, and the introduction of on chip decoupling capacitance, power supply disturbances can take more than a cycle time to propagate across chip. This effect is demonstrated in Fig.10 where the impulse response due to a source at node 1 is observed at nodes 1, 5, and 9. Clearly, for a 5 nsec cycle time, a power supply distrubance originating at node 1 would arrive at node 9 sometime in the subsequent cycle, and simultaneous switching would not produce the worst sag at this node.

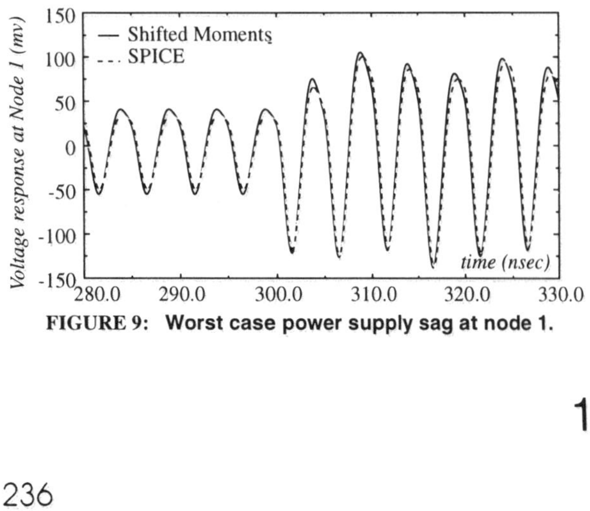

FIGURE 9: Worst case power supply sag at node 1.

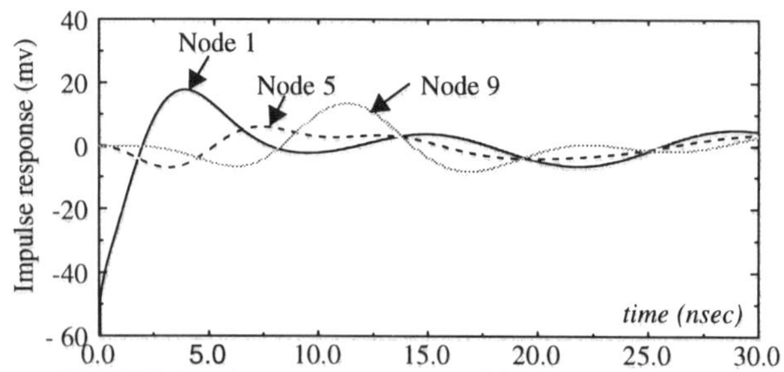

FIGURE 10: Impuse response with source at node1.

4: Conclusions

Because of their unique relationship to both time and frequency domain simulation, frequency-shifted moments can be used to compute both steady state and transient responses in VLSI CMOS power distribution networks. Furthermore, since a limited number of circuit moments are required, and these same moments can be reused to evaluate all possible switching patterns, shifted moment power rail analysis offers an efficient means to evaluate power up sequences for worst case power supply sag.

References

[1] M.M.Alaybeyi, J.Y.Lee, and R.A.Rohrer. Numerical Integration Algorithms and Asymptotic Waveform Evaluation (AWE). *Proceedings of the IEEE International Conference on Computer-Aided Design (ICCAD)*, pp. 76-79, Nov. 1992.

[2] H.B.Bakoglu. *Circuits, Interconnections, and Packaging for VLSI.* Addison-Wesley Publishing Company, 1990

[3] E.E.Davidson. Electrical Design of a High Speed Computer Package. *IBM Journal of Research and Development,* Vol. 26, 349-361, May 1982.

[4] D.Dopperpuhl et al, A 200 MHz 64b Dual Issue CMOS Microprocessor, *IEEE Journal of Solid-State Circuits*, Vol. 27, Nov. 1992.

[5] R.Evans and M.Tsuk. Modeling and Measurement of a High-Performance Computer Power Distribution System, *IEEE Transactions on Components, Packaging, and Manufacturing Technology-Part B*, Vol. 17, no. 4, Nov. 1994.

[6] P.Feldman and R.W.Freund, Efficient Linear Circuit Analysis by Pade Approximation via the Lanczos Process, *IEEE Trans. on Computer-Aided Design,* Vol. 14, May, 1995

[7] W.W.Hager. *Applied Numerical Linear Algebra,* Prentice Hall, 1988.

[8] X. Huang, Pade Approximation of Linear(ized) Circuit Responses, Ph.D. Disseration, Carnegie Mellon University, Nov. 1990.

[9] L.T. Pillage and R.A. Rohrer, Asymptotic Waveform Evaluation for Timing Analysis, *IEEE Trans. on Computer-Aided Design,* April 1990.

[10] R.F.Sechler and B.Krauter. Resonant Noise in VLSI CMOS. *IBM Technical Report*, TR 51.0481, Oct. 1988.

Chair: Telle Whitney
Co-Chair: Danesh Tavana

As the use of programmable logic arrays continues to increase, the demand for higher complexity parts, flexible routing architectures and additional features continues to drive the development of architecture innovations. A new and exciting generation of programmable logic architectures is emerging based on advancing technologies. This session includes products based on flash, antifuse, and SRAM. These architectures cover the spectrum of volatile, non-volatile, one time programmable and in system reconfigurable. These devices include novel features such as embedded systems and distributed blocks. Each of these architectures provide possibilities for exciting and novel applications.

The session opens with a paper describing a fine-grained architecture based on flash technology. The fine-grained nature of the chip allows it to maintain the flexibility associated with gate array designs.

The next two papers describe new architectures using a reconfigurable logic array. The first architecture is a medium-grained SRAM based product developed by IBM based on a grid based routing system. The part allows dynamically re-programmability with clock rates up to 33 MHz. The subsequent paper describes AT&T's new generation programmable device based on the same overall architecture with optimized multiplication and distributed memory features.

Altera describes their newest architectural entry with densities up to 100,000 gates based on embedded memory. We finish up with Xilinx's newest programmable device based on a metal-to-metal antifuse and a sea-of-gates architecture.

A HIGH DENSITY FLASH MEMORY FPGA FAMILY

Robert Lipp, Rick Freeman, Tim Saxe

Zycad Corporation

GateField Division

Fremont, CA 94538

ABSTRACT

An NMOS transistor switch, direct coupled to a Flash memory cell without the use of a sense amp, is the basis for a high density FPGA. The switch is capable of transmitting the full digital switching voltage; novel circuit design techniques prevent the generation of hot electrons which would otherwise deprogram the memory during normal operation. The small programmable Flash switching element facilitates use of a fine grain architecture optimized for design techniques based upon high level design and synthesis. At 0.8μ technology, the largest member of the family is rated at 100,000 gates. Future scaling to 0.35μ technology will make practical a 400,000 gate device.

INTRODUCTION

Field Programmable Gate Arrays (FPGAs) have become increasingly indispensable in modern systems. The two primary FPGA technologies are SRAM and antifuse based. SRAM based products have the advantage of using a standard CMOS process, but have the drawbacks of limited density due to the large SRAM cell, limited performance due large interconnect loading from the large NMOS switch and the long lines required for these large chips, poor results with synthesis tools, and volatility - the requirement to reload the pattern every time the device is powered up. Antifuse devices are non-volatile, but one time programmable; they require non-standard processing and, despite the small antifuse size, have limited density due to extensive programming and testing overhead.

Flash memory is defined by two aspects, bulk erase and selective programming. Typically, these are combined into a single floating-gate transistor structure. The most common structures employ Fowler-Nordheim (FN) tunneling global erase with channel-hot-electron (EPROM) programming. Non-volatile Flash memory has previously not been thought suitable for FPGAs. Flash memory elements operate only over limited voltages, typically less than 1.5v, across the source-drain, to prevent the generation of hot electrons that might program an erased device during readback. This low voltage requires all Flash based memories to use relatively large sense amps for reading.

FPGAs can be defined as fine grained and coarse grained. A core logic cell which directly implements only one simple logic or register function is typically considered fine grained. Logic cells that implement two or more simple logic or register functions are considered coarse grained.

Fine grained architectures require more programmable interconnections and therefore require more interconnect switches, placing greater demands upon the base technology. However, they are considered superior when they can be implemented with small, low impedance switches for several reasons: they are more compatible with high level design (HDL) tools and synthesis, and they provide more predictable results in density and performance. High level design tools and synthesis have been optimized for fine grained architectures, such as conventional gate arrays, and are very difficult to optimize for the typical complex logic/register coarse grained cell. Additionally, it is very difficult for coarse grained architectures to have the right mix of logic and register cells, and to have them in the right physical locations.

DESCRIPTION OF FLASH SWITCH CELL

A novel device structure has been developed in which the floating gate of each Flash memory element is shared with the floating gate of an NMOS switching transistor (Fig. 1).

Instead of a sense amp detecting the state of the memory element, the floating gate directly modulates the switch transistor. The EPROM is used to add electrons to the floating gate, discharging it and turning the switch transistor off. The floating gate is charged by removing electrons via the FN tunnel, turning the switch transistor on. The identical combination Flash-memory/NMOS-switch structure is used for all programming functions: interconnect routing and programming the logic cell.

0-7803-3177-6 $5.00 © 1996 IEEE

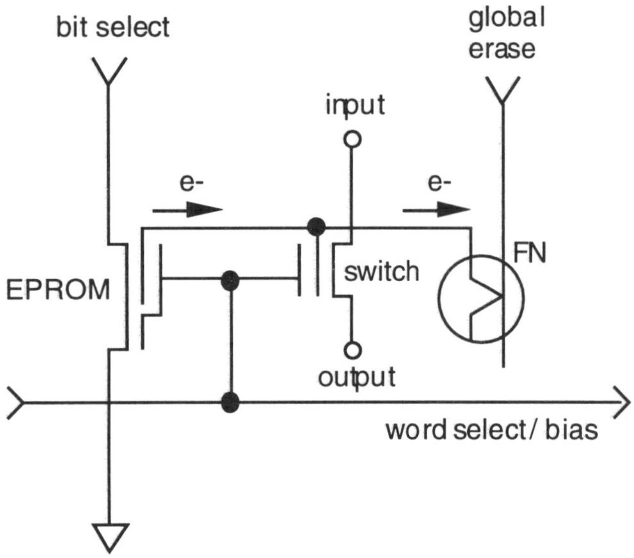

Schematic

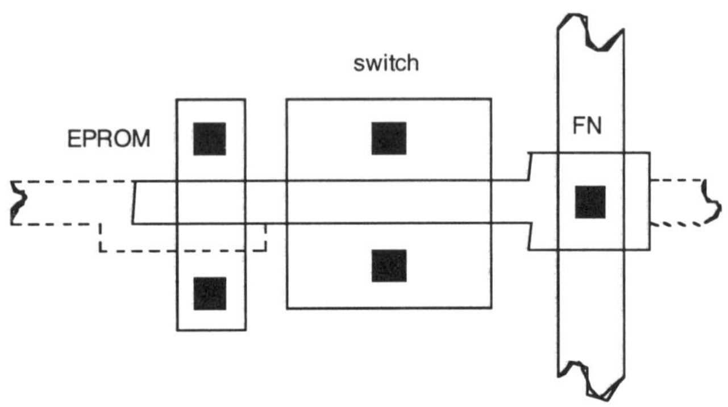

Layout

Figure 1: *Flash FPGA Memory/Switch Cell*

The Flash memory cell employed uses conventional FN tunneling global erase with EPROM programming. Note the use of a split gate EPROM transistor and the physical separation of the FN tunnel away from the EPROM. This eliminates two of the major problems of Flash memories, high voltage source engineering of the EPROM and control of over-erase. In fact, the floating gates of all memory elements are deliberately over-erased by charging them to 4v by FN tunneling. They are then selectively programmed off by individually discharging them through the EPROM to -4v. During operation, the switch transistor is biased such that the voltage is +6.5/-1.5v for on/off switch transistors respectively. This permits balancing and minimizing oxide field stresses, during storage as well as during operation when the device is biased properly. Performance is enhanced with no compromise in either retention or reliability. When switching a high voltage, the floating gate voltage is further bootstrapped to 7.5v. Bootstrapping is a singularly effective technique for enhancing performance while maintaining reliability. As the gate voltage tracks the increase in source, drain and channel voltage, the gate oxide is never exposed to the higher 7.5v fields. The overall 9v operating range (-1.5v to 7.5v) provides excellent on/off drive for the native transistor NMOS switches, including the ability to switch a full rail 5v logic signal without signal degradation. Self loading (RC delay) is reduced by 50% as compared to using a conventional NMOS transistor to switch the signal.

PHYSICS OF SWITCHING RAIL-TO-RAIL SIGNAL WITH A FLASH CELL

The work function of silicon is 3.1ev. Any electron with an energy in excess of 3.1ev is sufficiently hot to penetrate the oxide and be captured by the floating gate. To assure long operating life, effectively no electrons must be allowed to acquire this energy in the vicinity of the floating gate. Accelerating an electron across a 5v source-drain voltage is obviously out of the question. The source-drain voltage must remain below 3.1v. In actual fact, the voltage must remain considerably below 3.1v so the tail of the energy distribution, including electrons further energized by the higher gate voltage, remains below the work function. In practice, most Flash memory designs use a conservative source-drain voltage in the 1.5v range.

The maximum source-drain voltage on the FPGA switches is constrained to conservative 1.2v by a combination of resistive and capacitive division. A simple resistive divider consisting of the driver and switch impedances would be sufficient to manage the relative voltage drops. However, the driver performance would be unacceptably degraded. Instead, the driver impedance, capacitive load and input risetime are dynamically ratioed to the switch impedance and capacitive load so as to prevent more than 1.2v from ever appearing across the switch (Fig. 2).

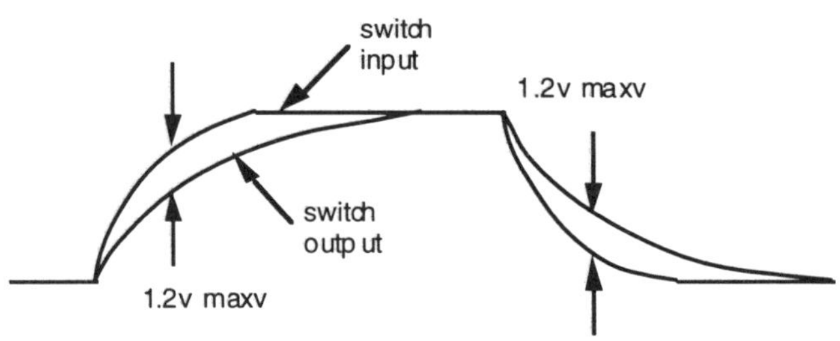

Figure 2: *Switch Input-Output Voltage Relationship*

These constraints required careful architecture design and layout, including doubling switches on critical timing paths. With this technique, performance degradation is negligible. In the final design, switches are physically connected in a such a fashion that programming cannot violate these restrictions.

Logic Cell

The small, low impedance Flash switch facilitates implementation of a fine grained cell and architecture. The logic comprises three inputs and two identical outputs, with programmable inversion on all inputs (Fig. 3).

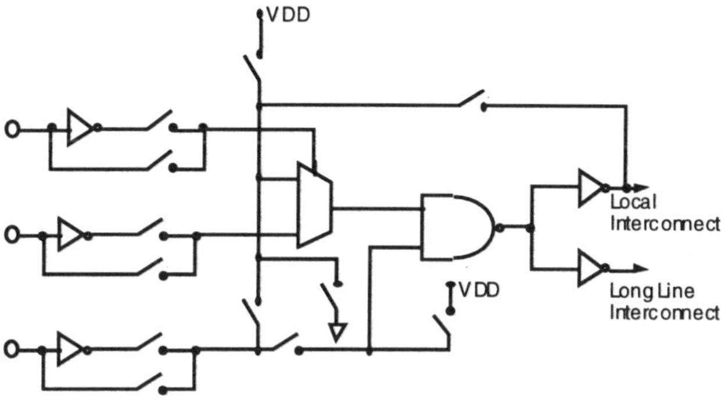

Figure 3: *Logic Cell*

The cell can be configured as any two input cell, all common three input cells, or a latch with clear. More complex cells, such as flip-flops, are created by combining two or more cells. Programmable inversions on the inputs eliminate the need to waste logic cells by using them as inverters. Software automatically handles cell merging and inverter reduction, including timing constraints, in netlist optimization. This simple logic cell is fleshed out to an highly optimized synthesis library consisting of 322 elementary library elements (core cells only), none more complicated than a flip-flop with clear and set. More complex logic macros (optimized counters, adders, etc.) can be readily created and added to the synthesis library as desired. The logic cell is nominally rated at a capacity of 5 gates. Synthesis, on the average, achieves a packing of 2.8 gates/cell.

IO Cell

The input-output (IO) cell is a flexible tri-state bi-directional cell. There are a total of 684 IO library cells with the following programmably selectable features:

6/12ma output drive

25/50/100ma/ns skew control

open drain/source

CMOS/TTL input levels

pull-up

Architecture

The logic cell is combined with 150 Flash switches to implement the basic logic tile. Fourteen of these switches program the logic function, while the balance provide for randomly interconnecting the various tiles though the wiring grid.

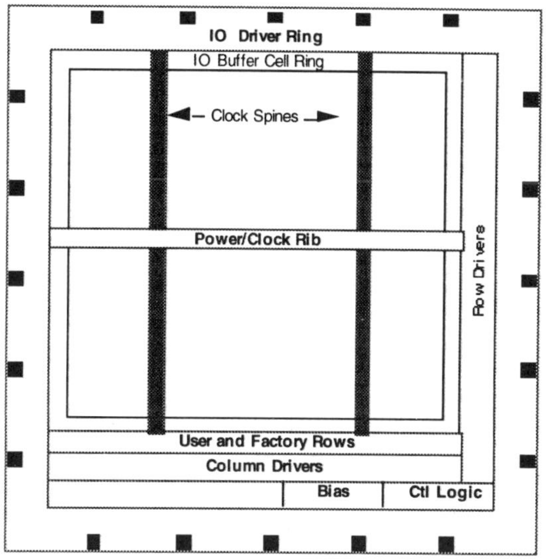

Figure 4: *Chip Architecture*

The FPGA is tiled uniformly with logic tiles. The interconnect structure is arranged into 8x8 groups of supertiles which uniformly overlay the logic tiles. A low skew (0.5ns) global interconnect structure comprised of four independent signals further uniformly overlays the entire chip. The uniformity of the array is a major advantage in optimizing place and route, assuring high quality, predictable results in tile usage and timing.

The cells are interconnected with a multi-level interconnect structure. Each cell is connected to its eight nearest neighbors by a high performance local interconnect structure. The local interconnect has fixed fanout loading and is driven by one of the two identical cell outputs. This local interconnect is used for interconnecting multi-tile logic cells as well as closely spaced logic blocks, providing functions such as fast carry which might otherwise need special purpose routing. The local interconnect employs an expensive, full crossbar interconnect scheme. But as up to 60% of the interconnections are made on this network, it is very effective.

For random intermediate and long routing, staggered length lines are used. Buffering is handled automatically by the router for long and heavily loaded nets.

Software

The FPGA is designed and defined to look like a conventional gate array instead of a conventional logic block array style FPGA. DeskTop Foundry ™ (DTF) software library kits are available for the major design systems. The logic library is essentially identical to those provided by major conventional gate array suppliers. Conventional gate array netlist and timing interfaces are used. With a typical tile utilization of 80% and gate utilization of 2.8 gates/tile, the FPGA achieves an average of 40-45% gate utilization, in line with the high end of conventional gate arrays. As most users will use this software infrequently, the interface is highly intuitive and the operations highly automated. The system will determine what needs to be done and then does it, but under full user control. Place and route (P&R) is designed to be fully automatic, with 100% completion. The user controls the process by 'effort level' and constraint files: pinout, path timing, and even preplacements, if desired.

The DTF minimizes the netlist size by automatically pruning the netlist of constant propagation, and unnecessary buffering and inversions, relying upon the cell input inversions and automatic routing buffering to accomplish these latter functions. Timing is back annotated to the original netlist, using the original user defined names. Performance can be analyzed using any conventional timing analyzer. Should performance be inadequate, the timing constraint file can be updated to include more paths as desired and P&R rerun to try to meet these updated requirements.

Programming

Programming is accomplished on the ASICmaker™, a special programmer which provides the required high voltages and programming waveforms. The part must be powered down during programming as the bulk erase initially turns on all switches, shorting all internal nodes together. The programming bit pattern is downloaded to this programmer through a standard RS-232 port. Programming time ranges from 30-60 sec., depending upon the size of the array. In-system programming is available through the PC card edge connector or through a special 14 pin connector.

Two rows of memory elements are dedicated to programming factory and user features.

Special Programming Features:

Programmable security bit to prevent readback

Internal part identification and serialization

User defined internal part ID

Reliability and Endurance

Conventional Flash memories have proven high reliability. This Flash FPGA is based upon conventional Flash processes and benefits from the learning curve applied to these processes. Endurance is not a major issue for FPGAs as compared to memories which may require one million cycles, or more. The biggest obstacle to providing high endurance FPGAs is the cost involved in qualifying the product due to the relatively long test times involved. Endurance specifications therefore have been set by market requirements rather inherent material limits:

10 cycles - manufacturing applications

100 cycles - prototype applications

5000 cycles - emulation applications

Summary

A family of high density Flash memory based FPGAs ranging from 9,000 to 100,000 gates has been created and manufactured. The problems of switching full digital signal levels without soft programming or the use of sense amps has been solved. Demonstrated reliability is in line with mainstream Flash memories. The FPGA is scaleable with future Flash process improvements to a capacity of 400,00 gates on 0.35μ technology.

References

(1) [GateField 1995] Zycad Corporation, *"GF100K Series DATAbook"*, 1995

(2) B. Britton, D. Hill, W. Oswald, N. Woo, S. Singh, Optimized Reconfigurable Cell Array Architecture for High-performance Field Programmable gate Arrays", *1993 IEEE Custom Integrated Circuits Conference*, pp.7.2.1-7.2.5.

(3) A. El Gamal, J. Greene, J. Reyneri, E. Rogoyski, K. A. El-Ayat, A. Moshen, "An Architecture for Electrically Configurable Gate Arrays", *IEEE Journal of Solid State Circuits, vol.24, no.2, pp.394-398, April 1989*

(4) J. Rose, R. Francis, D. Lewis, P. Chow, "Architecture of Field Programmable Gate Arrays: The Effect of Logic Block Functionality on Area Efficiency", *IEEE Journal of Solid State Circuits, vol. 25, no.5, Oct. 1990, pp.1217-1225*

(5) Y. Ikawa, et. al, "A One Day Chip: An Innovative IC Construction Approach Using Electrically Reconfigurable Logic VLSI with On-Chip Programmable Connections", *IEEE Journal of Solid State Circuits, vol. SC-21, no.2, pp.223-227, April 1986*

(6) G. Verma, N. Mielke, "Reliability Performance of ETOX Based Memories", *Proc. IRPS, pp. 158-166, 1986*

An SRAM-Based FPGA Architecture

Scott Gould, Brian Worth, Kim Clinton, Eric Millham,

Fran Keyser, Ron Palmer, Steve Hartman, Terry Zittritsch

IBM Microelectronics
Essex Junction, Vermont 05452

Abstract

An SRAM-based FPGA architecture has been developed using a licensed AT6000 architecture base [1]. The logic-cell architecture exploits an efficient, medium-grained, fixed library cell that implements most frequently used synthesis functions. An internal routing structure enables dense designs using a highly connected grid-based routing system and a dedicated I/O routing structure that supports the highest I/O counts available. Dynamic reconfiguration is retained with an underlying SRAM structure like the AT6000.

Introduction

Our challenge was to build upon the AT6000 architecture to make it competitive with the densest and fastest mature FPGA technologies available and retain the features that give the AT6000 architecture advantages for certain types of applications. Attractive in the AT6000 are its flip-flop density, homogenous array, and dynamic "on-the-fly" reconfigurability. The areas we focused on for improvement were the logic-cell functionality, routing structure, I/O resources, and the global-clock and reset structure.

Designing an FPGA is an exercise of complex trade-offs between the implementation cost of a feature and its perceived benefit in terms of density, function, or performance. For developing the IBM FPGA, we used three design-analysis tools to help speed the benefits analysis: Cadence Corporation's Figaro Architects' Workbench, an auto place-and-route tool for FPGAs that runs against a parameterized model of the architecture; Emerald [2], a similar tool produced by the University of Washington; and IBM's logic-synthesis tool called Booledozer.

I. Array Structure

This FPGA family of devices utilizes a medium-grained fixed library cell that is matrixed in an unbroken symmetrical array and looks like the familiar sea-of-gates approach. The array is divided hierarchically into sectors. Each sector contains an array of eight-by-eight cells and a hierarchy of buses. Repeaters at the sector boundaries repower signals traversing sectors and switch signals from one level of busing hierarchy to another. The array is surrounded by a flexible I/O busing network and multifunction I/Os. Figure 1 is a photograph of our initial chip (IBM10016), which is a 56x56 logic-cell array.

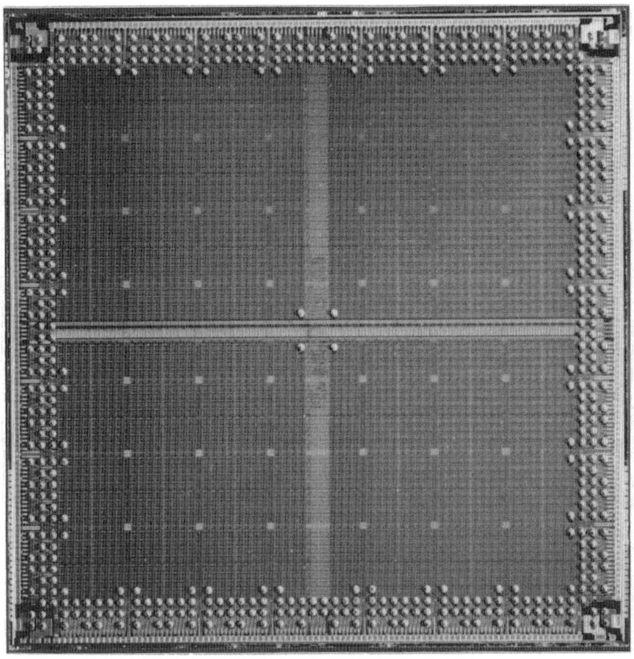

Figure 1. IBM10016 device.

II. Logic Cell

The goal of the logic-cell architecture was to have a high-performance cell with the most useful logic capability using a minimum number of SRAM configuration bits and silicon area. The optimal area usage and logic-cell functionality point for lookup tables (LUTs) is between three and four inputs [3]. However, the usage distribution of possible LUT functions is nonuniform[4]. Table 1 summarizes data taken from the Ref. 4 material and shows experimental results on designs using a fixed-library technology mapper Booledozer. Over 90 percent of the logic is implemented, using only a small set of the 65,536 possible functions for a four-input LUT. Therefore, to minimize the required number of configuration bits, the logic cell is designed to implement the most useful three- and four-input logic functions. The trade-off between the number of configuration bits and logic functionality is investigated

0-7803-3177-6 $5.00 © 1996 IEEE

IEEE 1996 CUSTOM INTEGRATED CIRCUITS CONFERENCE

using Emerald. Table 2 lists the initial-device array characteristics showing the efficient number of configuration data bits for the logic-cell density.

Table 1. Function usages for 3 and 4 input LUTs.

3-input	% usage	4-input	% usage
bc +d	35	ad +bc	26
bc	33	cd	25
bcd	22	bc +d	12
~bd + bc	6	bcd	11
b ⊕ c	3	abcd	7
		ab + c + d	5
		abc + d	4
		c ⊕ d	3

Table 2. IBM10016 device characteristics.

	Usable Gates	Logic Cells	Flip-Flops	Cell Delay (avg wc)	Configuration Data Bits
IBM 10016	16K	3136	3136	2 ns	192509

Medium-Grained Cell

The medium-grained logic cell(Fig. 2) has eight configuration bits allocated for logic: M inverter select(1), N and O inverter select(1), P inverter select(1), I and J programmable Nand/Nors(1), Fout mux(2), and Gout mux(2). The tri-state control uses a decoded state on the Lout mux. The M-O-P mux output is selected when both the C port on the Fout mux and the P port on the Gout mux are decoded. The logic cell efficiently uses the eight configuration bits and takes advantage of nonuniform usage distribution of functions used by synthesis to implement 60 functions and 181 permutations. In addition to increased logic capabilities, the AT6000 architecture enhancements include feedback within cell for counter design, F mux clock input for clock gating, and significant increases in the bus access to the cell for routing.

III. Routing

To investigate optimal routing resource allocation, we used the place-and-route algorithms in the CAD software. The routing resources in the Figaro Architects' Workbench were modelled with emphasis on flexibility. The following structures were parameterized and explored: number and type of buses, bus segmentation, local-bus-to-logic-cell connectivity, bus-to-bus connectivity, bus-to-bus switch connectivity, and logic-

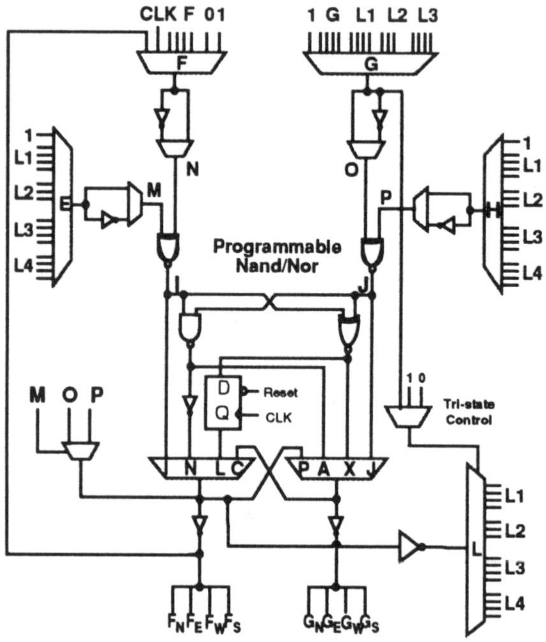

Figure 2. Logic cell.

cell-to-logic-cell connectivity. Changing any combination of these parameters can be accomplished in minutes. Test suites of designs were run and analyzed by looking at many statistics such as net delay and resources used. This parameterization enabled us to investigate an enormous number of routing structures in a very short time. Table 3 shows the results achieved when the routing resources are combined with the efficient logic cell.

Table 3. PREP benchmark design results [5].

PREP	1	2	3	4	5	6	7	8	9
IBM 10016	133	67	68	34	40	63	76	76	105

Interconnect Structure

Within the array, four types of routing buses exist (differing in quantity, length, connectivity, and speed): direct, local, express, and super. The performance data for some of the routing buses is shown in Table 4.

Table 4. Interconnect performance.

Direct	Local (4)	Local (8)	Express (8)	Express (16)
1 ns	1.7ns	2.2ns	1.1ns	1.4ns

Direct buses are used for the fastest adjacent cell-to-cell communication.

Local buses are used for localized logic cluster connections that are not possible with direct bus connections.

They span each sector in both the horizontal and vertical directions and connect to each cell along a cell row or column. Half of the local buses contain a programmable jumper halfway across the sector. The local buses are divided in half, effectively doubling their usefulness for short connections. A programmable bus turn is provided, enabling a signal to make a 90-degree turn at the intersection of horizontal and vertical local buses.

Express buses are used for longer connections; they span either one or two sectors and do not connect directly to the cells. Programmable bus turns are also incorporated into the express bus design, allowing 90-degree turns at regular intervals across the array.

The super bus spans the entire array and connects to the express and local buses at selected repeaters. Unlike the express and local buses that are broken at the repeaters, the super bus switches into the sectors without a break in the line.

Table 5 shows the increased bussing quantity over the AT6000 base. The super bus is an additional resource not found in the AT6000.

Table 5. Routing resource comparison of IBM10016 vs AT6000 base.

	Direct	Local	Express	Super
IBM10016	1X	4X	2X	"+"

IV. Input/Output

Two I/Os are coupled to each logic cell at the array boundary (Fig. 3). A variety of connections to and from the I/Os and the logic cells provide a fast and flexible architecture. Dedicated buses enable I/O signals to route without using the internal array routing, unlike the AT6000.

The I/Os connect to the cell array through extension buses that utilize the ports of the sector repeaters at the array boundary. Signal propagation in to or out of the array can be along the local, express, or super buses. In addition to the extension buses, each I/O port can connect to the logic cells along the array edge through the direct buses, providing the fastest path to the adjacent logic cells on the I/O periphery, as in the AT6000.

To ease I/O placement and routing problems, a new bus has been introduced. It runs around the chip and has programmable connections to the extension buses. The I/O bus is broken and rebuffered with I/O repeaters every 16 I/Os. This scheme is very flexible and allows all I/Os to be completely connected to the array.

The IBM10016 supports up to 448 user I/Os; the largest device in the family supports up to 704 user I/Os, exceeding all other industry FPGA devices.

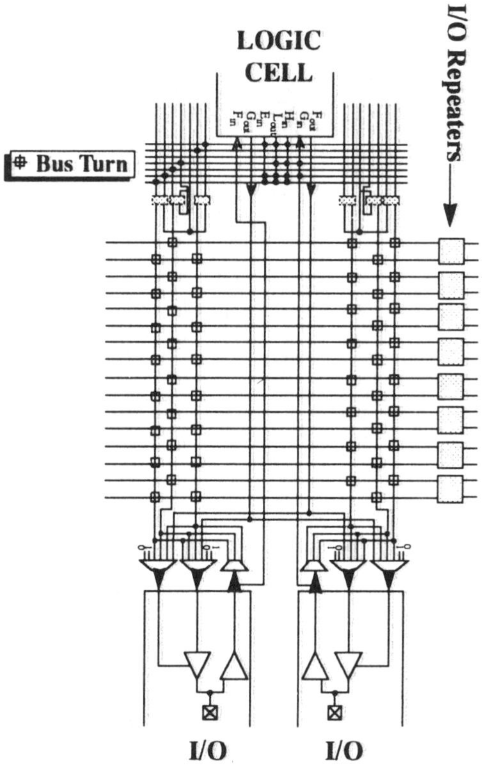

Figure 3. I/O connectivity to array through I/O busing.

V. Clock and Reset Structure

Figure 4 illustrates the flexibility of the clock and reset distribution across the array. Six primary inputs are available to bring clocks onto the clock network (three for reset). These six I/Os are not dedicated for clocking but may be used as standard I/O when not used as a clock. The six global clocks are routed horizontally across the array to the inputs of a mux located at each column; the mux can choose which of the global clocks to apply. As the global-column clocks pass each sector, a mux within each sector can choose from up to four of the global column clocks located two to the left or two to the right of the column. The resulting sector-column clock is then presented to each logic cell flip-flop.

A local bus is connected to an extra input of the global mux along each column of the array. When this input is selected, the signal on that local bus is fed into the clock network, enabling a clock to be gated, divided, or generated in any way, and allowing it to be placed back onto the clock network and routed to each flip-flop. The reset structure is identical.

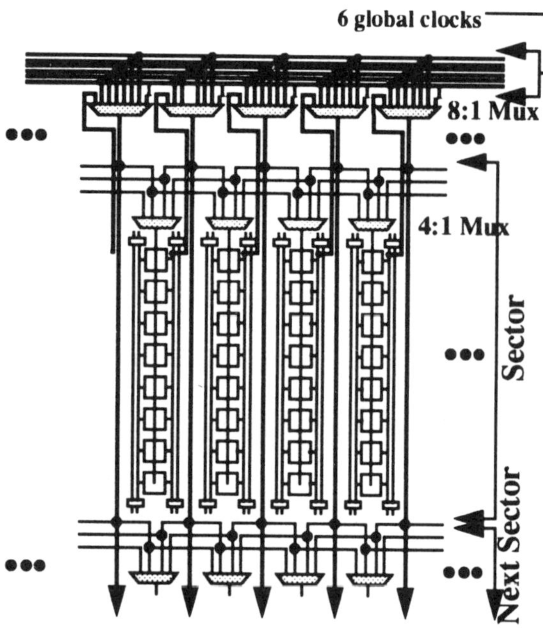

Figure 4. Clock and reset distribution; separate structures are used for clock and reset.

VI. Dynamically Reprogrammable SRAM Structure

Most SRAM-based FPGA's serialize incoming programming data and load the memory as a scan string, requiring a reset to disable the device while configuring. The serial-scan method disallows any real-time circuit updates.

Partial configuration, termed "windowing", is enabled in both the IBM10000 and the AT6000 architectures without interrupting unaffected areas. Special addressing circuitry accelerates writing the SRAM, allowing configuration clock rates of up to 33 MHz in parallel configuration modes.

The combination of loading speed, windowing capability, and a non-disruptive configuration cycle make up the features of dynamic reconfiguration. This opens up many possibilities for the logic designer, enabling fast reprogrammability and removing the limitations of serial scan-chain-based SRAM configuration methods.

Summary

An SRAM-based FPGA architecture has been developed using a licensed AT6000 architecture base. The logic-cell architecture exploits an efficient, medium-grained, fixed library cell that implements most frequently used synthesis functions. An internal routing structure enables dense designs using a highly connected grid-based routing system and a dedicated I/O routing structure that supports the highest I/O counts available. Dynamic reconfiguration is retained with an underlying SRAM structure like the AT6000.

Most FPGA studies use LUTs to investigate area and function trade-offs, assuming all possible functions are equally useful. The IBM10000 architecture leverages the nonuniform usage distribution of functions used by synthesis to reduce configuration bits and silicon area. This architecture trades off configuration bits for highly used synthesis functions to provide an efficient FPGA technology solution.

Future FPGA architectures and architecture enhancements are planned, using the same Figaro Architects' Workbench and Emerald development tools described in this paper. A field-programmable SRAM enhancement is planned for the IBM10000 family of FPGA devices.

Acknowledgments

The authors acknowledge Joy Barber, Wayne Beebe, Allan Bertolet, Sally Botala, Susan Bueti, Sam Bueti, Christine Fuller, Eberhard Gramatzki, Wendell Larsen, Tim Reny, Chris Ro, and Gulsun Yasar for their dedication and support to develop and implement this architecture.

References

(1) Atmel, Configurable Logic Design & Application Book, 1994 1995.

(2) D. Cronquist, L. McMurchie, "Emerald - An Architecture-Driven Tool Compiler for FPGAs," FPGA '96 1996 ACM Fourth International Symposium on Field-Programmable Gate Arrays, 1996.

(3) J. Rose, R. Francis, P. Chow and D. Lewis, "Architecture of Programmable Gate Arrays: The Effect of Logic Block Functionality on Area Efficiency," IEEE Journal of Solid State Circuits, Vol. 25, No 5, October 1990, pp. 1217-1225.

(4) L. Trevillyan, "An Experiment in Technology Mapping for FPGAs using a Fixed Library," IWLS Workshop Notes, pp. P9c1-P9c6, 1993.

(5) Programmable Electronics Performance Corporation, Benchmark Suite #1 Version 1.3, April 11, 1994.

A New Generation of ORCA FPGA with Enhanced Features and Performance

T. Ngai, S. Singh, B. K. Britton, W-B. Leung, H. Nguyen,
G. P. Powell, R. Albu, W. B. Andrews, J. He, C. W. Spivak

AT&T Microelectronics
1247 S. Cedar Crest Blvd.
Allentown, PA 18103-6209

Abstract

This paper describes the new AT&T Optimized Reconfigurable Cell Array (ORCA) 2CA and 2TA series of Field-Programmable Gate Arrays (FPGAs). Both series are based on the ORCA 2C/T architecture, but migrated to the advanced AT&T 0.35µ CMOS processes. These two processes are individually optimized for 5V and 3.3V operations. In addition, architectural innovations are incorporated into the new series to enhance both performance and functionality. These include:

- Efficient support of parallel multiplier
- Efficient support of synchronous single-port and dual-port memories
- Streamlined creation of large memory using port-enable control and internal tristate buffers
- Easy system integration with the 5V-tolerant input/output buffers, found on the 3.3V 2TA series.

As the result, the ORCA 2CA/2TA arrays are significantly smaller and 15%/30% faster (PFU speed) than the corresponding 0.5µ 2C/T arrays.

1. Introduction

The speed and gate density of FPGAs have grown rapidly for the past few years, but the demands of today's FPGA applications have evolved as well. On top of demanding faster, cheaper and higher density FPGAs, new requirements are posted by today's applications, such as better memory support (synchronous memories and multi-port memories such as register files), more arithmetical functions such as multipliers, and a fast 3.3V FPGA series for low-power applications. In order to fulfill these requirements, AT&T has re-engineered the second generation ORCA 2C/T series [ATT95a] [brit94] with new functionality, then individually optimized this architecture for two advanced processes: 0.35µ/3.3V and 0.35µ/5V CMOS. The resulting new series, called ORCA 2CA and 2TA, are 100% bitstream upward-compatible from the ORCA 2C/T series, even with the added functionality.

In this paper we present the details of implementing the ORCA 2CA/2TA series. In the next section we describe the new features for the new series. Section 3 describes the 0.35µ/3.3V and 0.35µ/5V processes used to implement the ORCA 2CA/2TA series and the advantages of these process versus the 0.5 µ process used for the original ORCA 2C/T series.

2. New Features

In this section we describe the new features of the ORCA 2CA/2TA series. They include a 4x1 multiplier, synchronous single- and dual-port memories, memory write-enable/port-enable control, and the 5V-tolerant input/output buffers for the 3.3V devices. These new 2CA/2TA series are referred as the 'A' version for the rest of the paper.

2.1 Multiplier

Multipliers are very common components in DSP applications. Because of their complexity, any reasonable sized multiplier has tended in the past to consume a large number of FPGA logic cells and the resulting performance was usually poor.

The 'A' version incorporates an innovative solution to implement fast parallel multipliers very efficiently. Figure 1 shows how the ORCA logic block, called a Programmable Functional Unit (PFU), can be used as a 4x1 multiplier. X_{0-3} and Y_n are the unsigned 4-bit and 1-bit data. P_{in0-3} and P_{out0-3} are the input and output partial products. C_{in} and C_{out} are the carry-in and carry-out signals. As shown in the diagram, the four AND gates and the 4-bit adder that are required for a 4x1 multiplier are found in one FPU, which translates to compactness and speed. Since the fast carry routing is used

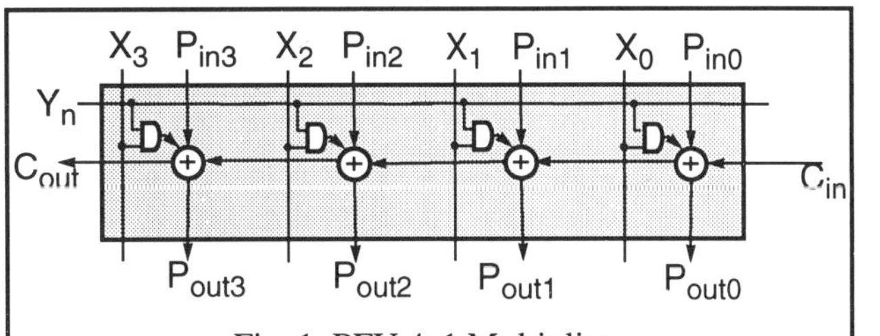

Fig. 1. PFU 4x1 Multiplier

12.3.1

0-7803-3177-6 $5.00 © 1996 IEEE

IEEE 1996 CUSTOM INTEGRATED CIRCUITS CONFERENCE

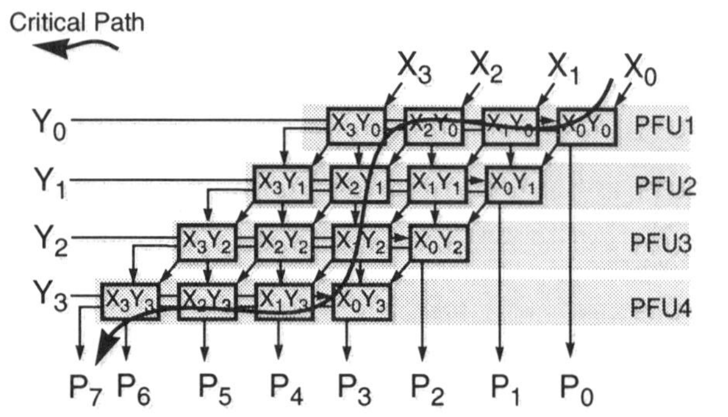

Fig. 2. Cascading 4 PFUs to Form a 4x4 Multiplier

for much of the critical path, this also translates to increased speed.

Each PFU multiplier can be easily stacked together to form deeper and/or wider multipliers. Figure 2 demonstrates how a 4x4 multiplier is implemented using four PFU multipliers. With the abundant 2CA/2TA routing resources, this multiplier can easily be packed into four neighboring PFUs. As the result, the critical path goes through only three segments of PFU to PFU direct routing resources, and the resulting 4x4 multiplier is fast. Speed results are summarized in Section 3. If an even higher throughput rate is required, this multiplier can be pipelined using the four PFU FFs that are available.

2.2 Distributed Synchronous Memory

Most digital systems require some form of memory. They range from small registers for state machine designs, to fast FIFOs for communication circuits, to large RAMs for general data storage. In order to accommodate these different requirements, the logic units of some SRAM-based FPGAs can be converted into distributed asynchronous memories and combined appropriately to satisfy various memory requirements. With increasing clock speed requirement of applications, such as ATM, the asynchronous-memory approach becomes more difficult to use because of all the address, data and control setup/hold time requirements.

PFUs are enhanced in the 'A' version to support 16x4 synchronous memory operations, as shown in Figure 3. All the incoming address, data and write enable signals are now latched by the clock while memory read operations remain asynchronous. Then, if the write enable is active, an internal write strobe is generated from the same clock edge to write the latched data into the latched address location. All latches become transparent for the next half clock cycle to sample the incoming signals. This design simplifies memory timing requirements: all setup/hold times are now referred to one clock edge. The output data can be latched by the configurable PFU FFs to meet the timing requirement of the next stage easier.

In order to boost performance for applications that require simultaneous reading and writing, the read and write addresses are individually decoded so that a different read address can be supplied to read memories during the half clock cycle when the write address is latched and memories are written. This approach requires an external address multiplexer to multiplex the read/write addresses between the first and second half clock cycles, but it allows the memory to be written and read from different addresses during every clock cycle.

The next section describes a dual-port synchronous memory mode which can be simultaneously read and written using two clocks that are asynchronous to each other.

2.3 Dual-Port Synchronous Memory

Figure 4 shows how a 16x2 dual-port memory can be constructed from the available 16x4 single-port memory. In

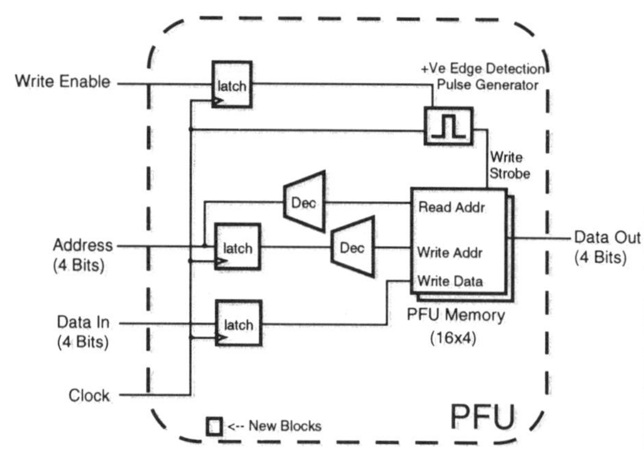

Fig. 3. PFU Synchronous 16x4 Memory

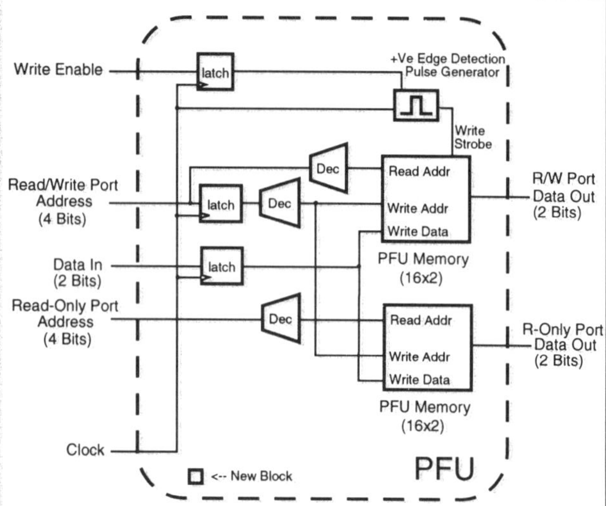

Fig. 4. PFU Synchronous 16x2 Dual-Port Memory

this mode, the 16x4 single-port memory is divided into two banks. The write address and data buses of the two banks are connected together so that any arriving data is written into both banks. The same address bus is also connected to the read address of the first bank, as in the single-port mode. However, another address bus is connected to the read address of the second bank so that it can be accessed independently. As a result, each PFU can be constructed as a 16x2 dual-port memory, with one port for both reading and writing and the other port for reading only.

The synchronous memory and the dual-port memory features make each PFU memory more flexible and easier to implement. The port enable function to be described next is used to make it easier to create larger memories from the individual PFU memories.

2.4 Port Enable

When a memory deeper than 16-bit words is required, multiple PFU memories can be grouped together to form it [Xili95]. For example, when a 64x4 memory is needed, four 16x4 PFU memories are needed, as shown in Figure 5a (to simplify the figure, only the read enable and write strobe signals of the PFU memories are illustrated). The upper two bits of address are decoded externally to control the read enable inputs, which in turn control the read-data tristate buffers of each PFU. The same decoded signals are logic-ANDed with the write enable line to turn on one of the PFU write strobes for writing. Since each of these write strobes requires one AND gate (one look-up table), PFUs are consumed for this function.

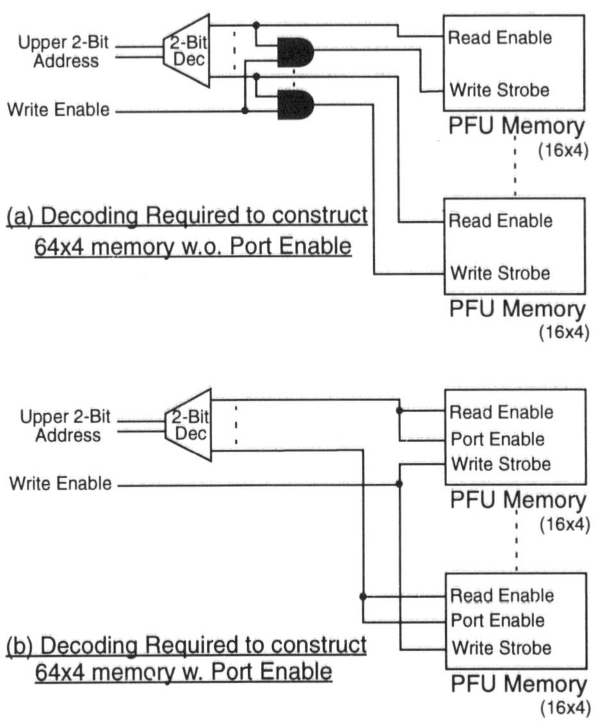

Fig. 5. Forming Deeper Memory W./W.O. Port Enable

In the 'A' version, a port enable pin is added to absorb this AND gate into the PFU memory to simplify the external decoding requirement. Figure 5b shows the simplified decoding with the port enable pin. The port enable and write enable inputs are internally ANDed together to replace the original write strobe function. This enhancement also speeds up the write enable path and reduces the number of routing nodes required.

2.5 5V Tolerant I/O buffers

Since not all the components in a typical digital design are readily available in 3.3V versions yet, one of the design objectives for the 3.3V 'A' version was to make sure that the I/O buffers can communicate with 5V devices. To achieve this, additional 5V power supply pins are required to ensure proper biasing of the transistors in the proximity of the I/O pads. All other programmable features such as selection for input levels, input speed, input float value, output drive, output speed, output sense and 3-state sense are maintained. Care was taken to ensure that the I/O characteristics for both the 5V and 3.3V devices remain compatible with the PCI specification [ATT95b].

3. Process Technology and Result

Until now, not many 3.3V SRAM-based FPGAs have been available. For the ones that are available, they are usually designed with the same process technology as the 5V devices. However, this compromises the performance of the 3.3V devices as shown in Table 1.

5V 0.35 μm	5V 0.35 μm operated at 3.3 V	3.3V 0.35 μm
75 ps	106 ps	59 ps

Table 1. Technology comparison (delay) for a typical gate delay

As shown in the first two columns, when using 5V technology operated at 3.3V, the typical gate delay degrades as much as 40%. The 'A' version 3.3V and 5V devices are designed with individually optimized 0.35μ CMOS processes with 3 levels of metal. Table 2 summarizes the key parameters of these technologies.

	3.3V/0.35μm	5V/0.35μm
N-Channel L' (μm)	0.32	0.46
Tox (A)	65	115
Metal 1 Pitch (μm)	1.04	1.04
Metal 2 Pitch (μm)	1.20	1.20
Metal 3 Pitch (μm)	1.36	1.36

Table 2. Key Technology Parameters for 0.35μ 3.3V/5V CMOS Processes

With the optimization of these two processes, as shown in the L' and Tox numbers in Table 2, the typical gate delay of this 3.3V technology is actually 20% faster than its 5V counterpart, as shown in Table 1. However, this speed advantage does not necessary apply in FPGA designs. In fact, it was found that the performance of the 3.3V 'A' version is only slightly faster than the 5V counterpart on the average. The main reason is that the driving capability of the N-channel devices (used as the MUXes and programmable switches) is greatly reduced by the body effect under 3.3V operating conditions. Table 3 shows some PFU and routing timing numbers for the four different processes. The new 2CA series (0.35μ/5V) is 18-31% faster than the 2C series (0.5μ/5V) and the 2TA series (0.35μ/3.3V) is 29-47% faster than the 2T series (0.5μ/3.3V). More importantly, the new 2TA 3.3V series is as fast as the 2CA 5V series, so that the same application can achieve equal or better speed in 3.3V FPGAs than in 5V FPGAs, with greatly reduced power consumption.

	5V 0.5 μm (2C)	3.3V 0.5 μm (2T)	5V 0.35 μm (2CA)	3.3V 0.35 μm (2TA)
LUT4 Delay (ns)	2.8	3.1	2.3	2.2
Carry-in to Carry-out(ns)	1.6	1.9	1.1	1.0
PFU to PFU Routing (ns)	1.1	1.4	0.9	0.9

Table 3. Worst Case Timing Numbers of the different ORCA Series

The above speed improvement is the result of circuit optimization in the advanced 0.35μ processes. Table 4 illustrates how the architecture innovation in the 'A' version improves performance when implementing an asynchronous parallel 8x8 multiplier and a 128x8 synchronous RAM. The first column shows the delay numbers of this two circuits in the current 2C/0.5μ technology. If the 'A' version architecture was implemented in the same 2C/0.5μ technology, these two circuits are 32-62% faster, as shown in the second column. With the advanced 0.35μ technology, the 'A' architecture becomes even faster.

	2C in 0.5 μm	If 2CA in 0.5 μm	2CA in 0.35 μm
Async. 8x8 Parallel Multiplier (ns)	75	51	40
128x8 Synchronous RAM[a](ns)	50	19	16

Table 4. Worst Case Timing Numbers of the different ORCA Series

a. Clock Period; Reading and writing are possible in each clock cycle.

All timing numbers shown in this paper are based on worst-case operating conditions. Other benchmark and performance results will be available during the conference. Shown in Figure 6 is a chip microphotograph of the 15K-usable gates device in 0.35μ process technology.

4. Conclusions

In this paper, we have described the new features and advantages of the AT&T ORCA 2CA/2TA 0.35μ 5V/3.3V series of FPGAs. By optimizing the 2C architecture in the advanced 0.35μ processes with enhanced functionality, the new high-capacity ORCA series is now capable of implementing a yet wider range of very high performance applications in both 5V and 3.3V.

Acknowledgments

The authors would like to thank C.T. Chen for his leadership and technical advice throughout the design, J. Hoff for his contribution in writing this paper and all the test, layout, and product engineers for their excellent support.

References

[ATT95a] AT&T Field-Programmable Gate Arrays Data Book, April 1995

[ATT95b] AT&T Preliminary Data Sheet, ORCA 2T15, Sept. 1995

[BRIT94] Barry K. Britton, Yaw T. Oh, William Oswald, Ho T. Nguyen, Satwant Singh, Chong Lee, Wai-Bor Leung, Carolyn Spivak, Jim Steward and C. T. Chen, "Second Generation ORCA Architecture Utilizing 0.5u Process Enhances the Speed and Usable Gate Capacity of FPGAs", IEEE International ASIC Conference and Exhibit, Sept. 1994, pp. 474- 478

[Xili95] The Programmable Logic Data Book, Xilinx Co., San Jose, CA., 1995

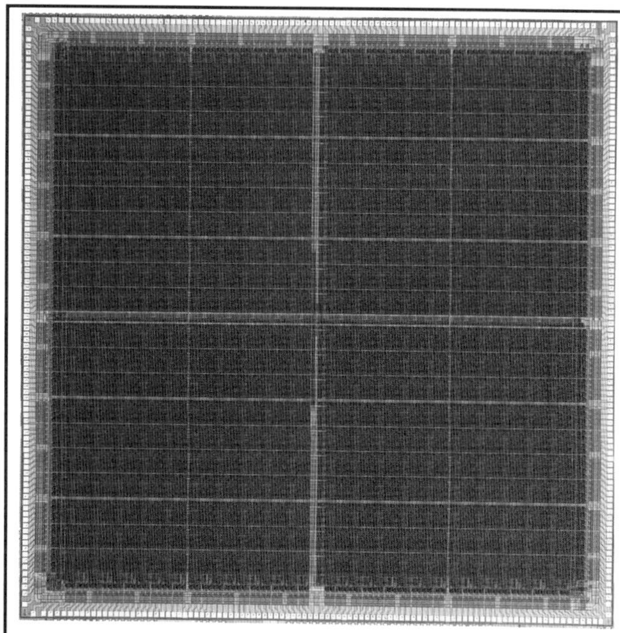

Fig. 6. Chip Microphotograph of the 15K-usable Gates Array in 0.35-Micron Process Technology

A High Density Embedded Array Programmable Logic Architecture

Srinivas Reddy, Richard Cliff, David Jefferson, Christopher Lane, C.K. Sung, Bonnie Wang, Joseph Huang, Wanli Chang, Todd Cope, Cameron McClintock, William Leong, Bahram Ahanin, John Turner.

ALTERA Corporation, 2610 Orchard Parkway, San Jose CA 95134-2020

Abstract

An SRAM based embedded array programmable logic architecture with densities ranging from 10,000 to 100,000 gates is discussed in this paper. An embedded array is incorporated into this architecture to implement megafunctions like microprocessors, FIFOs and multipliers efficiently. A multidimensional interconnect scheme is featured to achieve flexible routing between logic blocks, the embedded array and I/O pins. The first member of the family is currently available with a gate density of 50,000 gates.

Introduction

As programmable logic densities increase, more complex functions are being implemented on these devices. Most programmable logic devices use fine grained or coarse grained architectures for implementing the logic. Dual granularity has been shown by Cliff et al. (1) to be an efficient way to build a programmable logic device. Fine granularity logic blocks provide an efficient way to implement logic, and interconnect overhead is reduced by coarse granularity blocks which can be used for large fanout nets.

The FLEX 10K architecture is the industry's first embedded array programmable logic architecture using an embedded array to implement megafunctions and a dual granularity approach to implement the logic array. An improved flexible interconnect scheme with FastTrack continuous routing provides predictable interconnect delays, and easy placement of logic for the software. A powerful I/O structure enables the user to have individual tristate control and open collector option on each user I/O pin. In-Circuit Reconfigurability (ICR) via an external EPROM, and programming through the Joint Test Action Group (JTAG) port are provided. The architecture was developed to integrate seamlessly into ALTERA's MAX+PLUS II software design environment.

Architecture

The FLEX 10K architecture uses two components that are used to build logic: 1) Logic Array Block (LAB) and 2) Embedded Array Block (EAB). The LABs are used to implement logic functions like 8 bit counters, address decoders and small state machines. The EABs are used to build memory blocks and megafunctions like microprocessors, multipliers and DSP functions. Figure 1 shows an arrangement of LAB and EAB blocks in FLEX 10K devices.

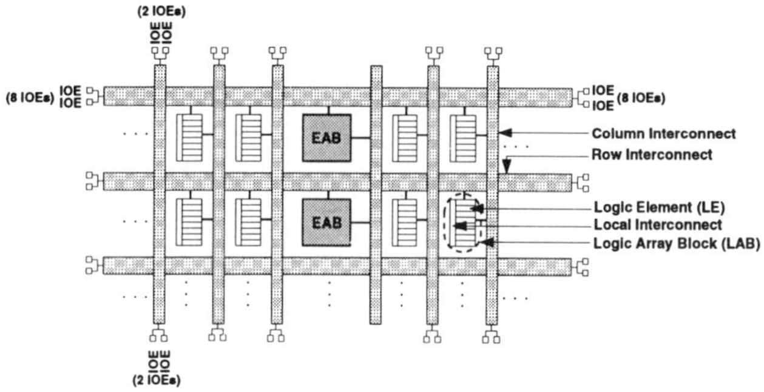

Figure 1: FLEX 10K Architecture showing the arrangement of LABs and EABs

Logic Array Block

The logic array block consists of 8 Logic Elements (LE) which are interconnected through a fully populated interconnect matrix, allowing any LE in the LAB to connect to any other LE in the same LAB. Rose et al. (2) has shown that a 4 input Lookup Table (LUT) is optimum for obtaining maximum logic density, and this has been proven by the existing FLEX 8K architecture. The LE in this architecture consists of a 4 input LUT followed by a register as shown in figure 2.

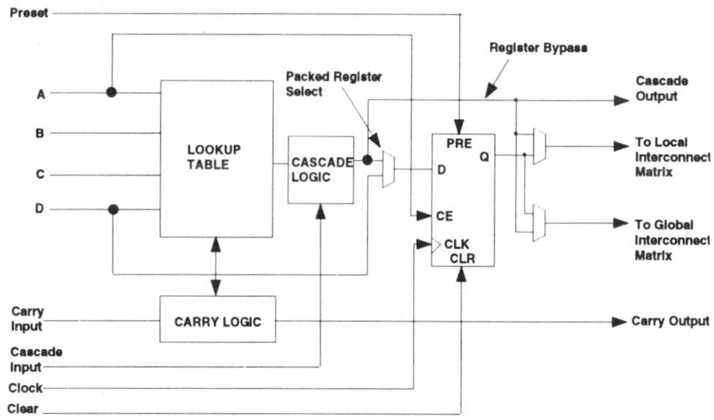

Figure 2: Logic Element block diagram

Each LE has four outputs: 1) cascade output providing a fast means of building wide fan-in functions 2) Carry output to provide a fast carry function between LEs 3) Global output to connect to global interconnect resources and 4) Local output to connect to other LEs in the same LAB. Clocks and clears for the register in the LE can be derived from any of a total of 6 dedicated input pins, or they can be asynchronously generated using other LEs. A clock enable signal can be derived by using one of the look up table inputs. A powerful new feature called register packing has been implemented in this architecture to increase

0-7803-3177-6 $5.00 © 1996 IEEE

IEEE 1996 CUSTOM INTEGRATED CIRCUITS CONFERENCE

capacity. LEs which use only 3 of the 4 available LUT inputs for logic, and not the register, can use the remaining LUT input as data to the register. Previous architectures wasted the register portion of the LE when only combinatorial functions were performed in the LE. The placement software by automatically recognizing unused registers in LEs is able to effectively increase the capacity of the device.

Embedded Array Block

The Embedded Array Block (EAB) is a new approach to providing RAM and logic in programmable logic devices. Similar to embedded gate array structures used in gate array devices, the EAB facilitates the building of megafunctions like microprocessors, FIFOs, RAM, ROM etc. (3). It consists of a RAM array with input and output registers and programmable options for varying width and depth. Logic can be implemented in the EAB by programming it with a read only pattern during configuration and using the EAB as a large LUT for performing combinatorial functions. The EAB can be compared to a heterogenous array that can implement logic (4). Complex functions can be built inside such arrays, with easy placement by the routing software. Such complex blocks can be defined as software macros that can be explicitly directed to be placed in the EAB.

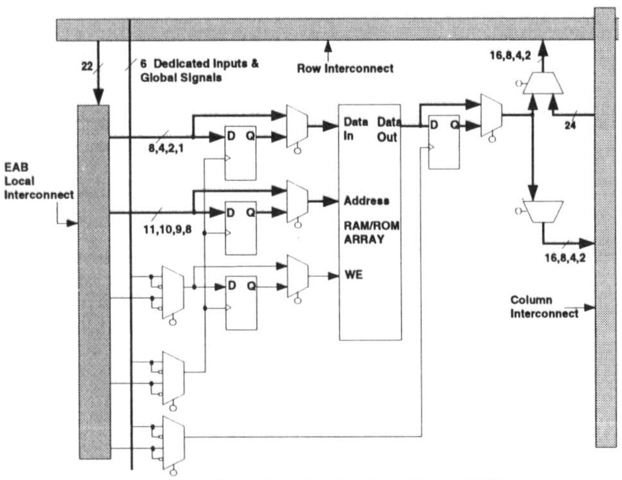

Figure 3. Embedded Array Block (EAB)

The EAB architecture has advantages over other architectures that implement RAM as blocks of small distributed RAM using LUTs. These distributed RAM arrays using LUTs, reduce the logic capacity of the device and have less predictable timing as the size of the RAM increases. The EABs on the other hand function as large blocks of self timed RAM with predictable timing and built-in address decoding without using any LUTs. SRAM arrays in other architectures consume LUTs to perform the address decode, which is a further reduction in density when memory blocks are built. The EAB enhances density by not only providing dedicated resources for megafunctions, but it does so without any or negligible impact on the logic capacity of the device using LABs.

A large fraction of user designs are synchronous in nature and the EAB can implement these synchronous functions efficiently by using programmable registers on the address, data and control inputs and data outputs of the EAB. These registers support extensive pipelining by using multiple clock sources as the register clocks to improve performance. Clocks to the EAB register can be from dedicated inputs or locally generated clocks from a LE, and registers on the inputs and outputs of the EAB can be clocked independently. A Write Enable (WE) signal is provided to synchronize the writing of data into the EAB when it is used as a synchronous RAM block, a FIFO, or a register file. Asynchronous operation of the EAB as a RAM array or as a logic array is also supported, by bypassing the input and output registers. Any combination of modes (registered or not registered) is possible on the inputs and outputs providing the user with complete flexibility.

Multi-Configurable Partioning of EABs

Each EAB has a capacity of 2K bits and when used as a RAM/ROM block can be programmed in 256x8, 512x4, 1024x2 or 2048x1 configurations. Multiple EABs from different rows can be cascaded to build wider and/or deeper RAM. To support this multi-configurable capability, a programmable decoding scheme is used for the addressing and steering of byte wide output data to the correct format. The addressing is divided into two components, a fixed portion (the X decode which is the word line) that is used for all configurations, and a variable Y decode which is the write column address, controlled by configuration RAM bits which vary the depth and width of the RAM. Mapping is also required for the data input to the array and the output to support the different widths.

Interconnect

This architecture uses the FastTrack Interconnect structure to connect LEs and I/Os. This is a global routing structure, present in both the horizontal and vertical dimensions of the chip, providing predictable timing on both axes. It is an enhancement over the FastTrack scheme employed in the previous FLEX 8K architecture. Increased classes of connections have been provided in the FastTrack Interconnect to account for the increased routability required when progressing to a higher density family of parts. A three layer metal process is used, which makes it possible to optimize the layout for the FastTrack Interconnect. As shown in figure 1, the FastTrack Interconnect consists of 1) a global horizontal routing channel servicing a row of LABs and the 16 I/O pins on the edges of each channel 2) a vertical routing channel allowing a column of LABs to communicate among themselves and the 2 I/O pins on the top and bottom of each column of LABs. All horizontal and vertical tracks are run in metal and driven by large dedicated drivers. No intermediate switches, which tend to degrade performance, are required to reach a destination in the same row. The vertical tracks are muxed to a subset of horizontal tracks in every LAB and can access the LAB inputs through the horizontal tracks. The vertical tracks are lightly loaded and this vertical to horizontal track connection is buffered with no loss in performance.

The LABs in a row get their inputs from the row interconnect channel. Signals from the row interconnect channel switch into the LAB or EAB inputs through a partially populated mux, which allows any horizontal row interconnect track to access any of the inputs in a LAB or EAB. The vertical interconnect channel can access the LAB or EAB by connecting to the row interconnect channel at every LAB and EAB in a row, and through the row channel to the LAB and EAB inputs. This scheme provides a predictable way by which LABs can communicate across row and column boundaries. For a LAB to connect to another LAB diagonally across the whole chip would entail one horizontal and one vertical track delay.

Routability among the LEs in a column or a row is increased by sharing horizontal and vertical tracks between adjacent pairs of LEs, using a scheme called "driver swapping". LEs in pairs of LABs can drive each other's vertical and horizontal tracks, in addition to their own. This doubles the number of vertical and horizontal tracks that a given LE can drive, resulting in increased routing flexibility. Carry and cascade connections between LEs and LABs tend to lock down cells and generally make routing more difficult. The sharing of resources between columns increases the flexibility of routing, especially in cases where cells are locked down due to carry and cascade chains. "Driver swapping" also increases the number of vertical I/O's that a LE in a given column can connect from 4 to 8, resulting in more flexible pin assignments.

The global routing resources described, simplify the software task of fitting a design. Signals which have a fanout of one, which are typically a large portion of a design, can be placed and routed within a single LAB. Due to the interconnect scheme with global row and column interconnects, the placement of high fanout nets is still easy, since any signal can access any LAB with a minimum of delay. Further if all destinations for a high fanout net are in the same row, then only one horizontal wire can be used. Dedicated low skew wires are available in every horizontal channel that can be used for signals driving high fanout nets, which are typical in bus intensive designs.

Global Clock Distribution

There are two dedicated high speed clocks which are provided for generating low skew signals. As programmable logic chips get larger, clock slew for global signals becomes an issue. Especially prone to hold time violations are I/O input registers, which tend to have very fast data path delays and very slow clock delays. One way to achieve a negative hold time at these registers is by slowing the data path using a delay chain. This fixes the problem, but at the expense of increased setup time for the register, which slows down performance.

The clock signals generated from the two dedicated clock pins use a direct drive technique to minimize clock delay and achieve negative hold times and fast setup times. The pad is routed directly to the four corners of the chip and individually buffered to drive the I/O's. The clock signal is treed into the core of the chip where it drives dedicated signal lines from both sides of the core into each LAB and the EAB. The low skew clock ensures a negative hold time for both I/O and LE registers with data to these registers being input from a pin. This balanced clock network also guarantees negative hold times between I/O registers and LE registers.

I/O Element

The I/O element (IOE) in this architecture features a bidirectional I/O buffer with a register that can be used as an input register or an output register. An input register at the I/O is useful when a fast setup time is required for data entering the device and an output register is useful when a fast clock to output time is needed to output data from the device. The pin associated with each IOE can be configured as an input pin, output pin, bidirectional I/O pin or as an open drain output pin. A programmable slew rate control for reducing output switching noise is provided for every output buffer. A network of peripheral signals control the functionality of the IOE. The IOE register's controls, like the clear and clock enable (CE), are generated from this peripheral bus. The clock for the register is generated from this peripheral bus and the two dedicated input clocks. By using these two dedicated clocks a very fast setup time and clock to output time for the I/O register can be achieved. The output enable(OE) for the output buffer is also generated from this peripheral signal bus. In order to increase the number of unique output enables and clock enables available at each pin, two separate control signals are provided for each IOE. These two signals, one of which controls the output enable and the other the clock enable, are derived from a separate LE for each IOE. This makes it possible to generate an OE or a CE signal with unique logic on every IOE. Alternatively, global clear and tristate functions are available to the user to clear all registers and tristate all pins respectively from two user I/Os, reducing the load on the peripheral bus which would otherwise have to be used to implement these functions.

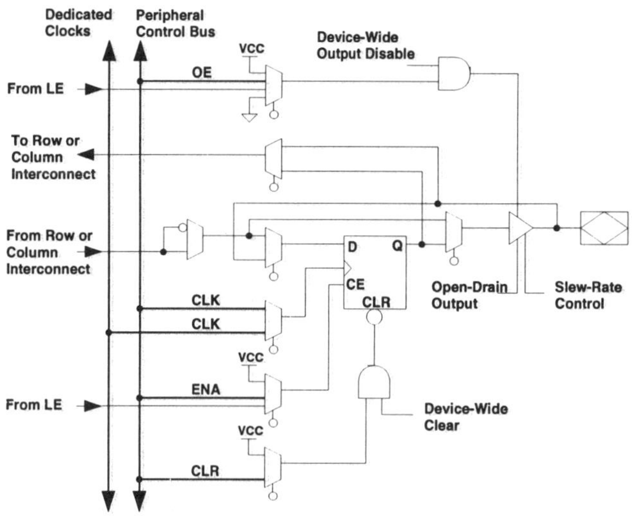

Figure 4. IO Element

Flexible Pin Assignments

IOEs can drive the row and column interconnects through programmable connections. Input routing flexibility is increased by having the ability to connect the IOE inputs on the row interconnect to the column interconnect. This vastly increases routability by providing the ability for inputs on any row to connect to LABs in any other row. Most users require the flexibility to change pin assignments as the design progresses and to be able to preassign pins. Preassigned pin assignments are handled in this architecture by the wealth of interconnect described and the flexibility of these interconnect resources. "Driver Swapping" allows more routing flexibility for locked down cells, the vertical to horizontal and horizontal to vertical connections in a LAB make it possible to route vertical and horizontal pins to anywhere on the chip thus helping preassigned and incremental pin assignment changes.

Pin Migration

The FLEX 10K architecture is designed such that migratability among different family members is possible. A user who requires a higher density part due to design changes can switch in the higher density member of the family without a complete redesign of the board. All power, ground, I/Os, and programming pins are positionally fixed for a given package making the migration possible. Migration is possible within a given package type, for all family members available in that package.

Performance

The first member of the 10K family is the 10K50 and some key performance statistics are listed.

16 bit counter fmax	- 98 MHz
Synchronous Read cycle time for EAB	- 11.2ns
Synchronous Write cycle time for EAB	- 8.3ns
Clock to data output valid for EAB	- 5.0ns

Conclusion

The FLEX 10K architecture is a high density programmable logic architecture featuring an embedded array for implementing megafunctions. Increased routing resources provide more flexibility to make it a highly routable architecture with pin migratability among different members of the family possible.

References

(1) R. Cliff et al., "A Dual Granularity and Globally Interconnected Architecture for a Programmable Logic Device", in Proc. CICC, May 1993, pp. 7.3.1 - 7.3.5.

(2) J.Rose, R.J. Francis, P. Chow & D. Lewis, "The effect of logic block complexity on area of programmable gate arrays", in Proc. CICC, May 1989, pp. 5.3.1-5.3.5.

(3) T. Ngai, J. Rose & S. J. E. Wilton, "An SRAM-Programmable Field-Configurable Memory", in Proc. CICC, May 1995, pp. 24.3.1-24.3.4

(4) J. He & J. Rose, "Advantages of Heterogeneous Logic Block Architecture for FPGAs", in Proc. CICC, May 1993, pp. 7.4.1-7.4.5

(5) R. Patel et al., "A 90.7MHz - 2.5 million transistors CMOS CPLD with JTAG boundary scan and in-system programmability"", in Proc CICC, May 1995, pp. 24.5.1 - 24.5.4

(6) K. Sawada et al., "A 5ns Port-Configurable Embedded SRAM with 0.5um CMOS Gate Array", in ISSCC Digest of Technical Papers, Feb. 1990, pp. 226-227.

(7) FLEX 10K Embedded programmable Logic Family Datasheet, July 1995.

The Architecture and Test Structures of the XC8100 FPGA Family

B. Fawcett, E. Goetting, D. Schultz, D. Parlour, S. Frake, P. Costello, R. Eccles, B. Leone, D. Marquez, S. Stokes, M. Palczewski, W. Peterson, T. Gannon, W. Hart, K. Look, M. Voogel, G. West, V. Tong, A. Chang, D. Chung, W. Hsieh, and W. Carter

Xilinx Inc.
San Jose, CA 95124 USA

Abstract

The XC8100 FPGA family is based on a new metal-to-metal antifuse technology and a "sea-of-gates" type architecture. Programmable interconnect elements are stacked vertically between metal layers and above the fine-grained logic cells, resulting in small die sizes and low cost. The basic cell is designed specifically for technology-independent design. Each cell can be configured to implement combinatorial, sequential, or three-state buffering functions. Other architectural features include a flexible set of high-drive buffers, output slew rate controls, and PCI-compatible I/O structures. A patented programming and test structure addresses the problem of providing for 100% post-program verification without test vectors.

Introduction

The evolution of Field Programmable Gate Array (FPGA) technology has been characterized by continuing increases in capacity and decreases in cost per gate. In an FPGA, both capacity and cost are direct functions of the silicon area required to implement the logic cells and their programmable interconnections. Traditionally, the logic elements and programmable interconnect have competed for the same silicon area. In the XC8100 FPGA architecture, the programmable interconnect is vertically stacked above the logic cells using the MicroVia™ technology; the new MicroVia technology combines a three-layer-metal CMOS process, a metal-to-metal antifuse programming element, and a sea-of-gates architecture. The combination of these three elements allows the development of FPGAs with extensive routing resources and high gate utilization in a very small die size.

Unlike SRAM-based FPGAs, where all the logic and routing resources of the device can be thoroughly tested by the manufacturer through configuration and reconfiguration of the programming cells, a one-time-programmable (OTP) antifuse-based FPGA must include special test structures to facilitate testing of the device both before and after programming of the antifuses. Older OTP FPGA technologies have to be verified with test vectors designed for the particular application after personalization. The XC8100 FPGAs address this problem through the use of extensive on-chip test circuitry and a novel method of post-programming verification that does not require test vectors.

To cope with increasing FPGA capacity, and the resulting increase in circuit complexity, as well as increasing time-to-market pressures, system designers are adopting design methodologies based on hardware description languages (HDLs), logic synthesis, and highly-automated design implementation tools. The XC8100 FPGA architecture and development tools are designed to be compatible with standard ASIC design methodologies, including the efficient implementation of HDL code.

MicroVia Technology

Antifuses are programming elements that are non-conductive until programmed. Older antifuse-based FPGAs use a 'dielectric-type' antifuse, wherein an oxide forms an insulator between a polysilicon layer and the silicon die. A portion of the antifuse is made from the same silicon layer as the transistors on the die, so the resulting FPGA architectures must be 'channeled', with the interconnect and its programmable switches running in channels between the logic cells. The new MicroVia antifuse is much different, using amorphous silicon as the insulator between two metal layers. When combined with three-layer-metal processes, the MicroVia antifuses can reside between the second and third metal layers. Thus, both the programming elements and the interconnect routing can reside 'above' the logic cells, further decreasing die size. In other words, the FPGA's routing resources - both wires and programmable interconnections - are stacked above the "gates", as in a 'sea-of-gates' gate array. In fact, the XC8100 architecture improves on the traditional sea-of-gates concept, in that it allows full usage of the logic cells regardless of interconnect requirements (i.e., cells are not used for routing). Because the routing no longer takes up die area, the XC8100 achieves the seemingly-contradictory combination of extensive routing on a very small die.

The MicroVia antifuse is formed by etching a small cylindrical hole through the isolation oxide between the second and third layers of metal. Chemical vapor deposition

0-7803-3177-6 $5.00 © 1996 IEEE

is used to place a thin layer of amorphous silicon at the bottom of the hole on top of the titanium-tungsten barrier in the metal 2 layer. The hole is then filled with the normal metal 3 material. Normally, the amorphous silicon forms an insulating layer with multi-megaohm resistance. However, when a high enough voltage is placed across the amorphous silicon barrier, a highly-conductive filament forms wherever the amorphous silicon is thinnest, and becomes a permanently conductive 'via' between the metal 2 and metal 3 layers. The resistance of amorphous silicon antifuses is about 50 Ω, (as compared to 100 - 500 Ω for polysilicon antifuses). The antifuse between metal 2 and metal 3 looks very similar to the via between metal 1 and metal 2, and can sit above the transistors that form the logic cells. Two additional mask layers are required for the MicroVia technology as compared to standard CMOS processes. One mask is used to deposit the amorphous silicon. The other additional mask involves implanting extra dopant in the 'high-voltage' circuits used to program the antifuses.

The XC8100 Architecture

The XC8100 FPGA architecture has three main configurable resources: configurable logic cells (CLCs), input/output (I/O) cells, and interconnections.

The CLCs provide the functional elements for constructing the design's combinatorial and sequential logic. While other antifuse-based FPGAs have fixed cell structures, the XC8100 architecture features a programmable cell, allowing a wide range of logic functions to map efficiently into the cells. Programmable choices in the CLC include input inversions, type of combinatorial function, synchronous logic internal feedback path selection, three-state enable, and cascade enable. Each CLC has four inputs and one output, plus a special cascade input and output. Fig. 1 shows the different possible implementations of a single CLC; each cell can be used to implement combinatorial logic, a latch, or a three-state buffer. The cascade inputs and outputs use dedicated routing to the nearest cells, extending along an entire row, and are used to construct wide combinatorial functions and flip-flops. The XC8100 software automatically configures the cell based on the user's design netlist.

For combinatorial logic, the cell has two programmable functions - AND or Sum-of-Products (SOP). Using inversion control, the SOP format can be used to implement multiplexer, XOR, and XNOR functions.

For synchronous logic, configuring the cell-internal feedback path implements a latch with data and control inputs. The inputs can be configured as active High or active Low. Two cells can be used to implement a latch with an asynchronous clear input. Two adjacent latches are combined through the cascade connection to create a D-type flip-flop, with one latch acting as the "master" and the second as the "slave". Flip-flops can be clocked on the positive or negative clock edge. Neither the feedback connections in the latches nor the master-slave connections use general routing segments nor have programming elements in their paths. This results in flip-flops with excellent performance and metastability behavior.

The MicroVia technology allows an abundant amount of routing, both in terms of wire segments and programmable

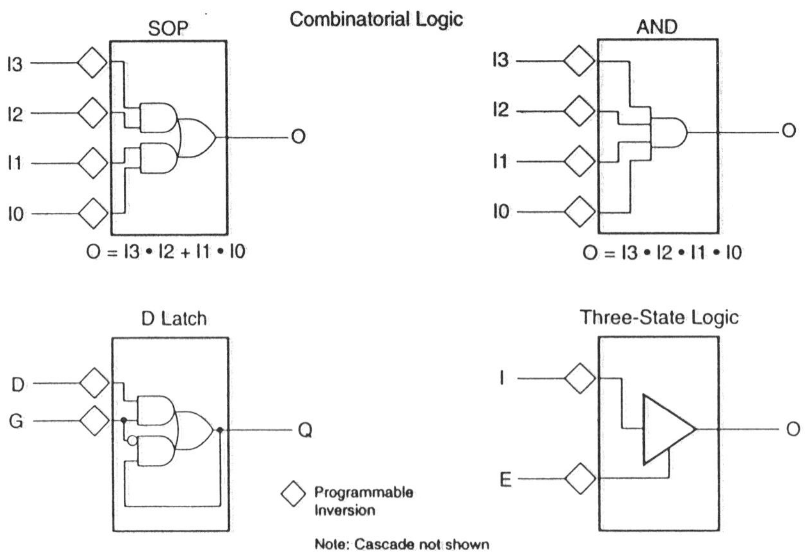

Fig. 1: Core functions that can be implemented in an XC8100 Configurable Logic Cell

12.5.2

interconnections, providing the routing software with many alternatives to insure fully-routed designs. For purposes of establishing an interconnect hierarchy, cells are grouped into 2-by-4 blocks. There are four types of interconnect in the horizontal direction. Single-block length wires extend for one block only, and are primarily intended to connect cells within a block. Double-block length wires connect left and right to adjacent blocks. Quad-block length wires connect left and right for a length of four blocks. Horizontal long lines extend the entire width of the device, and are optionally driven by "row buffers". There are two wire types in the vertical direction: double-block length and quad-block length. In addition, twelve vertical long lines at the right and left edges can connect with any horizontal wire, providing added flexibility for I/O connections.

Special high-drive buffers and nets are provided for the distribution of clocks and other time-critical or high fanout control signals (Fig. 2). Four high-drive buffers (BUFEDGE) each drive one of four vertical long lines running down the two sides of the device. Two horizontal long lines along each row of CLCs can be driven by any of those four buffers. Futhermore, each row of CLCs also has four dedicated buffers (BUFROW), each driving a horizontal longline, and each of which can be driven by the global buffers or other general-purpose interconnect signals. These buffers can be used independently or together to distribute clocks and/or control signals throughout the array.

Each I/O cell includes a programmable output section that drives a pad, and a programmable input section that receives data from the same pad. Aside from being connected to the same pad, the input and output sections of an I/O cell have no common logic. Inputs can be globally programmed for TTL or CMOS input thresholds. The output buffer is three-statable. As with the CLCs, all signals entering the cell from within the array are invertible; for I/O cells, this includes the output data and three-state enable signals. The "SoftEdge™" slew rate controls pioneered in the XC4000H FPGA family allow the choice of resistive or capacitive modes for the pull-down operation. In resistive mode, each output buffer can sink up to 24 mA. Each I/O pad can be programmed with or without an optional pull-up resistor. IEEE 1149.1 JTAG boundary scan test circuitry supports both the programming and testing of the FPGA device.

Unlike most other FPGA architectures, XC8100 FPGA utilization is not dependent on the mix of combinatorial, sequential, and buffering functions in the end application. Depending on the input netlist, any CLC can implement a combinatorial, synchronous, or three-state buffer function. The XC8100 software has the flexibility to choose the best

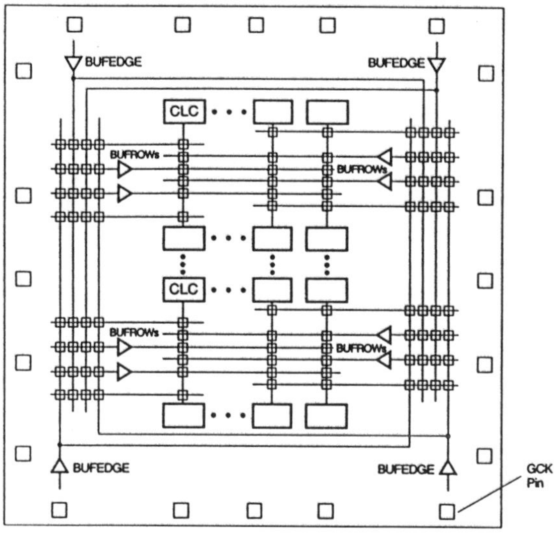

Fig. 2: XC8100 FPGA global buffer configuration

cell structure depending on the function to be implemented. Since any structure can be placed anywhere in the device, this flexibility eases placement and facilitates high-level floorplanning on a hierarchical basis. The objective is to allow the user to design entirely at a high level, without regard to the resources available on the device, and still achieve near-100% gate utilization. Thus, designs can be entered in HDLs with little concern for the FPGA architecture. Due to the programmable CLC and the abundance of routing resources, XC8100 devices often will replace competitive FPGAs claimed to be 30% to 50% larger.

Programming and Test Structures

Five pins are required for the programming interface; the TDI, TDO, TCK, and TMS JTAG pins are used to address and control the programming elements during programming, verification, and testing, and the V_{PP} pin provides the high voltage during programming operations and is used to measure antifuse resistance during verification.

There are several unique capabilities provided in the XC8100 architecture aimed at addressing the issues of testability, accurate programming, and programming yield. Special test structures included in the architecture include small metal-connect transistors that can be used to link metal segments without programming the antifuses, and Configuration Control Units (CCUs), used for both programming and test purposes (Fig. 3). The CCUs are connected in a scan daisy-chain, and allow for the random addressing of any CLC, wire segment or antifuse.

The metal-connect transistors are used to test the interconnect lines for shorts or opens prior to programming the

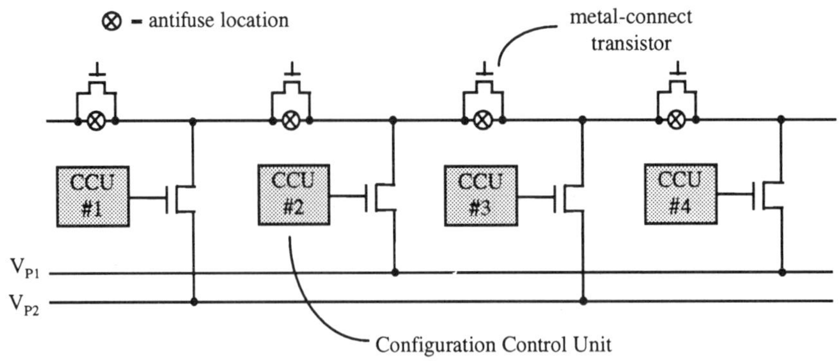

Fig. 3: Programming and test circuitry includes metal-connect transistors and Configuration Control Units (CCUs)

device. Similarly, the CCUs have connections to the Configurable Logic Cells (not shown in Fig. 3) as well as the interconnect segments, allowing for the full testing of all the logic in the CLCs prior to device programming.

The CCUs are used for programming as well as testing purposes. For example, to program the antifuse above CCU #2 in Fig. 3, the pass transistors controlled by CCU #1 and CCU #2 are turned on, the pass transistors associated with all other CCUs are turned off, and the programming voltage is applied between the V_{P1} and V_{P2} signal lines. Each antifuse can be individually selected for programming, and no more than two pass transistors reside along a signal path used to program an antifuse, regardless of the size of the array.

Once an antifuse is programmed, the same CCU structures can be used to test the resistance of the antifuse link. The resistance on the path from V_{P1} to V_{P2} through a pass transistor, the antifuse, and a second pass transistor can be measured, and the known resistances of the pass transistors and metal segments subtracted from the total to yield the programmed antifuse's resistance.

Once programming is completed, the CCUs are used to verify the integrity of signal nets in the design, in an operation unique to the XC8100 architecture. The CCUs can be used to selectively pull-up or pull-down metal segments. Thus, a net can be tested by pulling down just one segment of that net, and then sampling the state of all metal segments, testing for opens and shorts along that net's signal path.

Using these test structures, pre-programming tests include testing all the metal interconnect, verifying the functionality and speed of all the CMOS logic on the device (including programming circuitry, the configurable logic and I/O cells, and the internal buffers), and insuring that all the antifuses are in the correct (off) state. Two post-programming tests are performed by the programming hardware: the resistance of each programmed antifuse is measured, and all nets are tested to ensure that each is programmed exactly as specified in the design's netlist. Thus, complete post-program testing is performed without the need to generate test vectors.

XACT*step* Series 8000 Software

Because the XC8100 architecture is gate array-like, the design flow is ASIC-like. As with all Xilinx architectures, the XC8100 uses the standard Xilinx library conventions. However, the new XC8100 architecture necessitated a new placer, router, and other development software.

There are several new features in the Series 8000 software. All logic and nets, including hierarchy, are maintained by the back-end XC8100 tools, so this information is consistent for simulation and debugging. In other words, every logic gate in the design corresponds to one or more CLCs in the FPGA, and every net in the design corresponds to a net in the FPGA. The new PowerMaze™ router routes 1000's of gates per second. Series 8000 software can support incremental design within the synthesis environment. Most incremental design tools rely on matching symbolic names, and do not work in a synthesis environment where net names and instance names change from run to run. The Series 8000 software features the first incremental tools that match the topology of the two designs rather than the symbolic names, thereby providing incremental design capability in a synthesis environment.

There are five initial members in the XC8100 FPGA family ranging from about 600 to 8700 usable gates; larger devices are in development. Hundreds of test cases and real-world designs, entered with VHDL or Verilog and usually designed for another FPGA or ASIC architecture, have been retargetted to the XC8100 devices, and all have been implemented successfully.

Chair: Iain Mackie
Co-Chair: Brian Fitzgerald

This session focuses on a number of issues of interest to systems designers as well as those focused on the minutiae of the solutions. Discussions of different facets of RAM design and applications are followed by an exploration of some issues in IO and the session is rounded off by a review of a disk channel PRML read circuit.

First the architecture and implementation of a very high throughput SRAM buffer is discussed in some detail. The next subject is the management of the some trade offs that are presently inherent in embedding DRAMs in logic designs with a continued focus on performance and throughput. The third paper moves the review to the systems design domain with a performance evaluation of an architecture with DRAM embedded in the same chip as a processor. The final RAM paper in the session focuses on a CAM structure with embedded logic elements intended to solve a problem in Image Processing. At this point we are starting to complete the cycle with logic embedded in memory embedded in logic.

The session next moves to consider IO issues with two papers based on work done by one core team at AT&T. In the first of these papers a mechanism to manage noise caused by simultaneously switching outputs is reported upon. The second paper discusses high speed communication between chips using capacitive coupling without the need for coding for low bit error rates.

The final paper in the discussion moves to an implementation of a PRML (Partial Response Maximum Likelihood) circuit focused on disk read channels.

A 35 Gbit/s Throughput 64 kbit CMOS Buffer SRAM

Jonas Alowersson and Per Andersson

Department of Computer Engineering, Lund University, P.O. Box 118, S-221 00 Lund, Sweden
Phone: +46-46-222 4940, Fax: +46-46-222 4714, email: jonasa@dit.lth.se, pera@dit.lth.se

Abstract

A 64 kbit 0.8-µm pure CMOS buffer memory with 256 bit word-length and 3.6 ns cycle time, allowing 35 Gbit/s throughput, is presented. The memory consumes 1.5 W at the maximum frequency. The short cycle time is achieved through the use of a synchronously pipelined address decoder with one internal level of latches. The address decoder, based on TSPC latches, is described in detail.

Introduction

The remarkable speed improvement of CMOS circuits that has taken place over the last few years, with chips designed in 0.8-µm pure CMOS running at clock cycles of 2 ns or less, calls for the development of embedded memory structures that can match these clock speeds. A special demand exists for the class of memories commonly referred to as buffer memories, i.e., memories that serve as temporary storage between a data producer and a consumer. The premises for these memories differ from those for ordinary memories in that the throughput, as opposed to the access time, is the primary performance criterion. To increase the throughput, we may increase the word-length or shorten the memory cycle time or do both.

In this paper, we describe an experimental 64 kbit buffer memory, implemented in a 0.8-µm pure CMOS process [11], demonstrated at 275 MHz. The memory has an aggressive word-length of 256 bits, allowing a throughput of 35 Gbit/s when operating as a buffer memory with alternating write and read cycles. The power consumption of the memory core is estimated to 1.5 W at the maximum operating frequency.

The basic technique that allows us to design such a memory is pipelining. Pipelining of the read access is an attractive design approach for high-throughput memories with reasonable power consumption [3]-[5]. In [3], a 512 kbit pipelined memory in pure CMOS runs at 2 ns cycle and a power consumption of 3.5 W at 200 MHz, and in [4] a 72 kbit 0.8-µm BiCMOS memory reaches 3 ns cycle with 1.39 W power consumption at 333 MHz. Non-pipelined, memories which can match these times do exist, e.g., 1.5 ns access time for 64 kbit [1] and 256 kbit [2] BiCMOS memories, but they suffer from very high power consumption, e.g., 35 W for [2].

The bit-line and sensing circuits in our memory are built using techniques already described in [8] - [10]. This paper therefore primarily deals with the pipelined address decoder, which is based on TSPC latches [6]. In an earlier, three-ported 2 kbit memory design, employing a similar decoder [7], the decoders alone consumed 3 W out of a total of 4 W. Furthermore, the

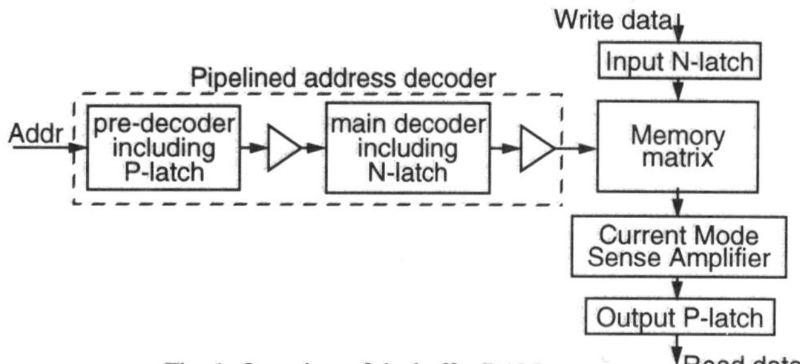

Fig. 1: Overview of the buffer RAM structure.

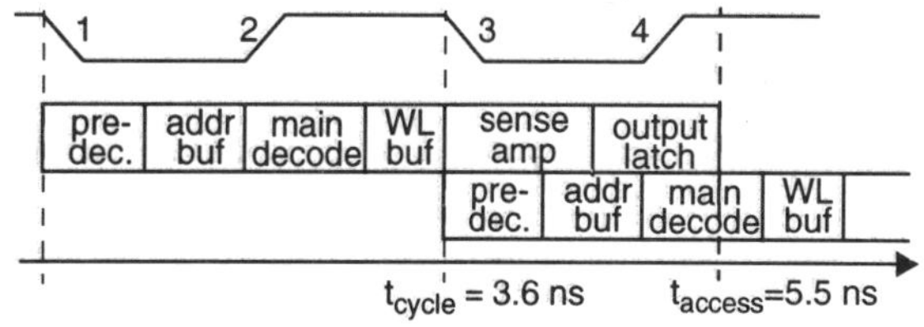

Fig. 2: The pipelined memory access.

critical delay path is in the decoder, not in the RAM cell matrix or sense circuitry. For this memory, we therefore restructured the address decoder in such a way that its power consumption could be reduced by an order of magnitude without speed degradation.

Memory Architecture

The memory block structure is similar to the one earlier reported in [7], see Fig. 1. The pipelined read access is illustrated in Fig. 2. Clocked synchronization points are:

1. Pre-decoder evaluation starts at falling clock edge 1.
2. Main decoder evaluation starts at rising clock edge 2.
3. Sense amplifier evaluation starts at falling clock edge 3. Simultaneously, output latch opens.
4. Output latch closes at rising clock edge 4.

Common power saving techniques, such as the divided word-line scheme, are not applicable to a memory of this type since the word length equals the row length. Thus, efforts had to be made to minimize the power consumed in each bit-line circuit. In modern memories, it is customary to reduce bit-line voltage swing at read cycles. In [9], we have shown how reduced bit-line voltage swing can also be applied at write cycles, leading to a decrease in write power by more than 80%. The low-power write scheme described in that paper is based on a combination of low common-mode bit-line voltage and relatively strong access transistors. The cell nodes will then in principle copy the bit-line voltages when the word-line is raised to V_{DD}. To prevent the cell from being erased at read, the word-lines are only raised to about 2/3 V_{DD} at read accesses.

0-7803-3177-6 $5.00 © 1996 IEEE

As sense amplifier we apply a clamped bit-line sense amplifier (CBLSA) [8], modified as shown in [10]. The CBLSA operates with bit-lines clamped to a reference voltage close to 0 V. As the difference in bit-line voltage between write and read cycles is minimal, it allows to fully take advantage of the low power write scheme in [9].

Pipelined Address Decoder

In order to make the most of synchronously pipelined operation (i.e., not wave pipelining), we want to insert one level of latches within the address decoder structure. That is a more aggressive use of pipelining than e.g., what is found in [4]. In an earlier design [7], we exploited that precharged multiple input nor-gates can be used as input stages of TSPC N-blocks at low delay penalty. Thus, an entire 5 to 32 decoding took place in one level of nor-gates, with little more delay than the inherent latch delay of a TSPC-block. However, the prize for this was an excessive power consumption in the address decoder. The main reason for this is, that in a dynamic nor-decoder, all but one of the dynamic nodes are discharged each clock cycle. Furthermore, these decoder gates had to be made with large devices, creating high capacitances in these dynamic nodes. The simple decoding structure in [7] could not be scaled to a 256 row memory with reasonable power consumption.

An attractive alternative in terms of power consumption would be to use serial instead of parallel input transistors in the decoder gates, i.e., a nand-decoder. That would lead to considerably increased delay, so we decided to keep the nor-structure, but made the following modifications aiming to decrease power consumption without sacrificing speed performance:

- Pre-decoders in address input buffers
- Altered decoder output timing to allow asymmetrical word-line buffering
- Two-level main decoder with logic in both stages of the TSPC-latch.

In Fig. 3, the logical structure of the resulting decoder is shown, and Fig. 4 shows a transistor schematic of the critical path through the decoder.

Pre-decoders consisting of two-input nand gates were added. The address input latches are made as TSPC p-blocks, and the nand-function is added to them simply by replacing the input transistor by two parallel transistors, T1 and T2 in Fig. 4. Pre-decoding of pairs of address signals into four-bit fields reduces the number of input transistors in each main decoder gate by one half, reducing the dynamic node capacitance and thereby in effect the power consumption by 50%. The number of signals distributed through the main decoder is not increased by this modification, as without the pre-decoders both the inverse and non-inverse of each address line had to be distributed. The fan-out of each pre-decoder gate equals one fourth of the number of decoder gates, which is a 50% reduction

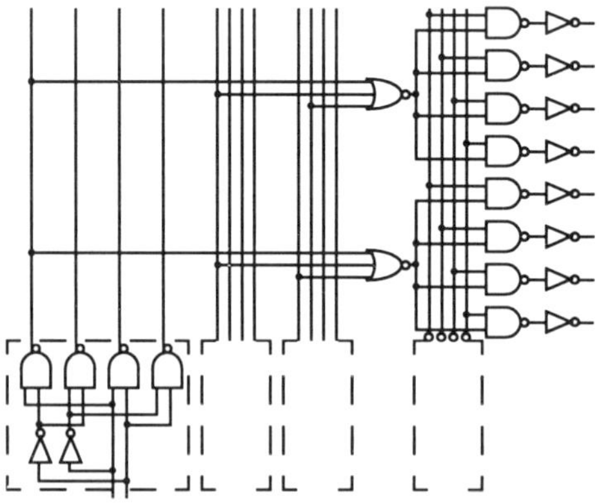

Fig. 3: Logical structure of 8 to 256 address decoder

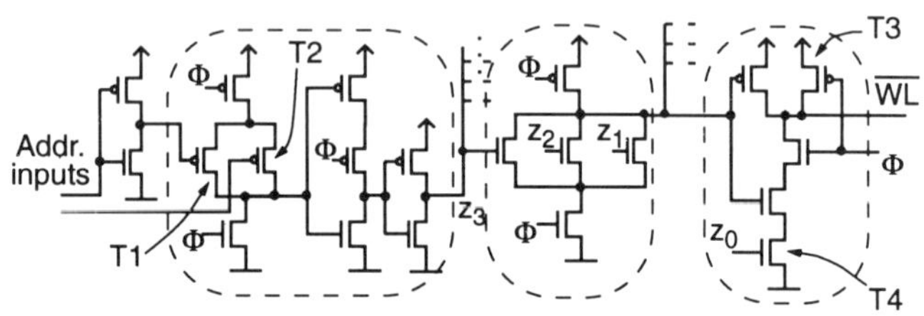

Fig. 4: Transistor schematic of critical path through decoder. T1-T4 mark the low-power features explained in the text.

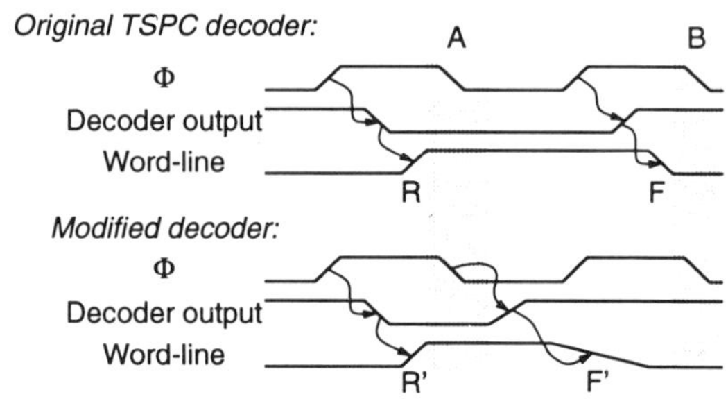

Fig. 5: Curve forms illustrating the timing difference between the original TSPC decoder and the modified decoder

compared to the fan-out of the address buffers without pre-decoders. Thus the power consumption in the address input buffers is also reduced. The small delay penalty of nand-function latches compared to plain latches is compensated for by the resulting reduced load at the pre-decoded address lines.

The second modification arises from the inherent latch property of the CBLSA, see Fig. 5. The word-line voltage is of interest only around and shortly after the falling clock edge, the regions A and B. After that, the sense amplifier has latched and cannot alter state until after it has been precharged again. To secure that only *one* word-line is high in these regions, the word-line buffer must be dimensioned so that rising edge R lies before region A *and* falling edge F lies before region B. We redesigned the main decoder, by adding the transistor T3 in Fig. 4. The latch stage is then turned into a nand-function between the clock and the nor-decoder output, and de-selection of word-lines starts during clock low instead of clock high. The word-line buffer can then be optimized for fast rise time and much slower fall time. The tapering factor of the buffer stages

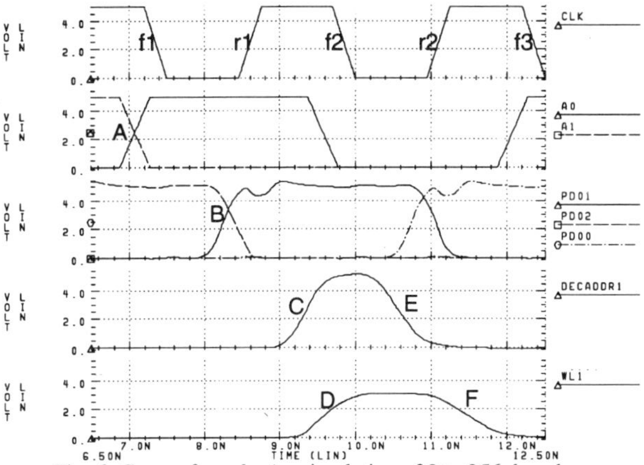

Fig. 6: Curves from *hspice* simulation of 8 to 256 decoder.

can be increased without rise time penalty (but with high fall time penalty), which leads to lower load on the decoders. Thus, the decoder can be built using smaller devices, resulting in a dynamic power reduction by as much as a factor of five.

The second stage of the main decoder offers one more opportunity. By inserting a third device in its pulldown-chain, T4 in Fig. 4, controlled by the pre-decoded two least significant address bits, this stage can perform a second level of decoding. This pulldown chain is in TSPC-latches normally not dimensioned for very high speed, as that would only lead to unnecessarily large output glitches while waiting for the input stage to evaluate. Thus, the device T4 can be added without increase of delay. The result is a two-level main decoder, where each nor-gate drives four nand-gates. The total load of all nor-gates is constant, but the number of input transistors in each nor-gate is decreased from four to three, so the net result is a further reduction in power consumption.

To evaluate the effects of the modifications described above, they were first applied to redesign the 5 to 32 decoder in our first design example [7]. This resulted in a decrease in simulated power consumption at 300 MHz from 1.05 W to 85 mW. The complete 8 to 256 decoder for the 64 kbit experimental memory has, according to *hspice* simulations, including word-line buffers driving 256 memory cells each, a power consumption of 460 mW when running at a 400 MHz.

Figure 6 shows a *hspice* simulation of the 8 to 256 decoder. The first panel shows the single clock, running at 400 MHz. The second shows the address inputs, where a new address is presented at *A*. Due to the inverters before the pre-decoder, this must take place at least one inverter delay before falling clock edge *f1*. The third panel shows the pre-decoded address lines after buffering. These signals are ready at *B*, which must occur before rising clock edge *r1* in order to secure correct evaluation of the main decoder. The curves shown in the figure are those for the least significant two bits, as these are the most heavily loaded (going to 256 nand-decoders instead of 64 nor-decoders, and thus are the last to be finished. The fourth panel shows main decoder output, at *C* the decoder has evaluated,

Fig. 7: Micrograph of 64 kbit experimental memory chip.

Table 1:
Characteristics of experimental 64 kbit SRAM

Organization	256 w × 256 b
Measured maximum clock frequency	275 MHz
Measured chip power at 275 MHz (including test circuitry)	3 W
Simulated SRAM power at 275 MHz	1.5 W
Cell size	15.6 × 17.0 µm
Chip size	5.9 × 6.2 mm
Technology	0.8 µm CMOS

and the fifth panel shows the corresponding word-line rising after buffering at *D*. The edge at *D* must come before falling clock edge *f2*, as the sense amplifier is activated at that time. De-selection of the word-line starts by the decoder output going low at *E*, triggered by falling clock edge *f2*, and reaches completion after buffering at *F*, well before the next sense amplifier activation at falling clock edge *f3*. The asymmetric word-line buffering mentioned above is evident in the figure from the difference in delay *r1* to *D* and *f2* to *F*.

Chip Design

To verify these circuit techniques an experimental chip has been designed and manufactured. A micrograph of the chip is shown in Fig. 7, and some characteristics can be found in table 1. To facilitate verification, the test circuitry shown in Fig. 8 is included on the chip. The data generator is a simple LFSR that produces a pseudo random sequence of input patterns. The data verifier is an identical LFSR that produces a replica of that pattern sequence, with a delay specified from the control unit. It compares these patterns with read data from the memory, and flags if there is a mismatch. Similarly, the address generator contains two LFSRs, which produces sequences of write and read accesses with relative delay matched to the data producer and verifier. For each *two* memory cycles, *one* input pattern is generated and written and *one* output pattern is read and checked. Thus, the test circuitry only needs to operate at half the memory clock frequency.

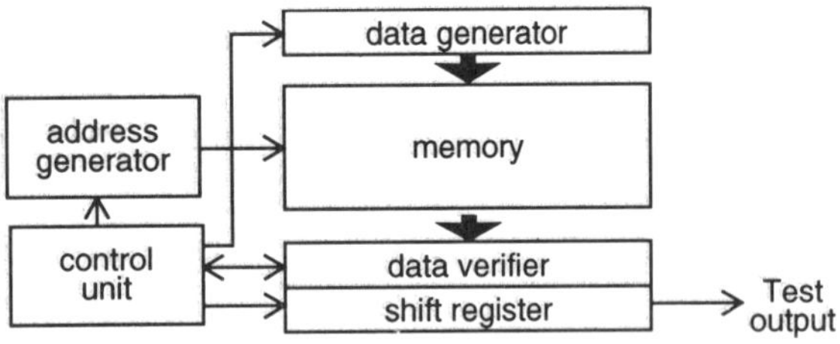

Fig. 8: Block diagram of on-chip test circuitry.

As typical in TSPC-designs, the clock load is high. Two clocks must be supplied, one at up to 400 MHz for the memory, and one at half the frequency for the test circuitry. The main clock is loaded by 60 pF, and the test circuit clock by 45 pF. Both clocks come from a central buffer that can be seen in the lower mid part of the chip micrograph. The main clock buffer is a six stage tapered CMOS buffer. After four stages, the signal is tapped to a frequency divider, producing a clock at half the main clock frequency. This clock is then buffered through two more stages to create the test circuit clock. The total widths of PMOS/NMOS devices in the output stages are 9.2 mm/3.2 mm for the main clock and 4.6 mm/1.6 mm for the test circuit clock.

Power is supplied through a large number of pads to minimize bond wire inductance. Most of these pads are at the vertical edges of the chip in Fig. 7, directly connected to large decoupling capacitors seen just inside the pads. These areas are covered with a sandwich structure of all available layers, with metal 2 and poly connected to V_{DD} and metal 1 and substrate to ground. Frequent contact cuts from the metal layers to the lower levels minimize the series resistance that would otherwise reduce the efficiency of the decoupling. Since both V_{DD} and ground are present in metal layers, this structure is also used for power distribution along the vertical edges of the chip. Apart from these two large structures, decoupling capacitors are spread around the chip wherever space allows. Totally, almost 2.5 nF of decoupling capacitance is supplied on the chip.

Measured Performance

Preliminary measurements show that the fabricated memory is functional at least up to 275 MHz clock frequency. The exact power consumption of the memory cannot be measured, as the on-chip test circuitry consumes almost as much power as the memory core, but the measured total chip power consumption of 3 W at 275 MHz indicates that the memory power is about 1.5 W, which is consistent with simulation results.

The measurements also show that there is a lower bound on clock frequency, not encountered in the earlier simulations. After the chip was fabricated, further simulations have been done, applying the actual process parameters measured from the wafers. These simulations indicate that this is caused by difficulties in controlling the word-line voltage at read. Due to the low power write scheme, this must be approximately 2/3 V_{DD}, but the exact value is, for the first generation word-line drivers that are used in this chip, highly process dependant. With the actual process parameters of the test chip, the word-line voltage at read will at lower frequencies become too high, causing the selected cells to be erased. We are planning a new version of the chip, which will include a circuit that automatically adjusts the word-line voltage to the actual process and supply voltage.

Conclusion

We have shown how a medium sized buffer memory with very high throughput rate and yet reasonable power consumption can be designed in pure CMOS. As a design example, a 275 MHz 64 kbit experimental SRAM with 256 bit words has been designed and manufactured in a 0.8-µm CMOS process. In terms of cycle time it is comparable to memories designed in high-end CMOS or BiCMOS technologies at much higher processing cost and to bipolar SRAMs with much higher power consumption. Furthermore, its very long word length does not lead to excessive bit-line power consumption. The circuit techniques applied does not need any non-standard processing techniques. The clocking scheme is compatible to TSPC, making it straightforward to use the memory as a building block in high performance chips. The area overhead for the proposed circuit techniques is not significant.

References

[1] H. Nambu *et al.*, "A 1.5ns, 64Kb ECL-CMOS SRAM," in *Dig. Tech. Papers, Symp. VLSI Circuits*, pp. 11-12, May 1991.

[2] N. Tamba *et al.*, "A 1.5-ns 256-kb BiCMOS SRAM with 60-ps 11-K logic gates," *IEEE J. Solid-State Circuits*, vol. 29, pp. 1344-1352, November 1994.

[3] T. Chappell *et al.*, "A 2-ns cycle, 3.8-ns access 512-kb CMOS ECL SRAM with a fully pipelined architecture", *IEEE J. Solid-State Circuits*, vol. 26, pp. 1577-1585, November 1991.

[4] K. Yokomizo and K. Naito, "Design techniques for high-throughput BiCMOS self-timed SRAM's," *IEEE J. Solid-State Circuits*, vol. 28, pp. 484-489, April 1993.

[5] K. Suzuki *et al.*, "A 500 MHz, 32 bit 0.4 µm CMOS RISC processor," *IEEE J. Solid-State Circuits*, vol. 29, pp. 1464-1473, December 1994.

[6] J. Yuan and C. Svensson, "High-speed CMOS circuit technique," *IEEE J. Solid-State Circuits,* vol 24, pp 62-70, February 1989.

[7] J. Alowersson, "A CMOS circuit technique for high-speed RAMs," in *Proc. 6th IEEE Intl ASIC Conf. and Exhibit*, pp. 243-246, September 1993.

[8] T. Blalock and R. Jaeger, "A high-speed clamped bit-line current-mode sense amplifier," *IEEE J. Solid-State Circuits,* vol. 26, pp 542-548, April 1991.

[9] J. Alowersson and Per Andersson, "SRAM cells for low-power write in buffer memories," in *Dig. Tech. Papers, IEEE Symp. Low Power Electronics*, pp. 60-61, October 1995.

[10] J. Alowersson and Per Andersson, "A 622 MHz current mode sense amplifier," unpublished

[11] *CYB process design rules,* Austria Mikro Systeme International GmbH, Unterpremstätten, Austria

This work was supported by the Swedish National Board for Technical Development (NUTEK) under contract no. 93-02633.

High-speed / High-band Width Design Methodologies for on Chip DRAM Core Multimedia System LSIs

Takahiro Tsuruda, *Mako Kobayashi, Masaki Tsukude,
Tadato Yamagata, and Kazutami Arimoto

ULSI Laboratory , Mitsubishi Electric Corporation *Daioh Electric Corporation

Abstract

Recently, as multimedia LSIs have developed, the demand for high-speed /high-band width LSIs which integrate the DRAM core and logic elements (CPU etc.) have been strongly required. However, the high-speed / high-band width operation induces the large switching noise. This noise degrades a DRAMs operating margin, and especially the data retention characteristics. In this paper, we analyze the noise transmission model and propose DRAM and logic compatible design methodology maintained the reliability of high-speed /high-band width system LSIs. And we obtained good experimental results on the test device.

1.Introduction

In the DRAM core implemented on system LSIs , there are three main problems as follows:

(1) High speed switching in logic sections heats up the die and increases the junction leakage current, degrading the static refresh characteristics of DRAM. This heat problem has been solved by cooling systems and package technology.

(2) The noise induced by large size buffer switching, in particular output buffers connected to large load capacitances degrades the dynamic refresh characteristics of the DRAM drastically. The dynamic refresh characteristics depend on the subthreshold leakage current of the access transistor and correspond to word line and substrate noise.

(3) As a DRAM requires a boosted potential, the logic section cannot use the optimum transistor for the high speed switching in order to maintain the reliability of the gate oxide film.

Here, we focus on the latter two problems and (2) in particular. We examined noise reduction techniques from the view point of circuits and device structures, and propose the best design approach for high-speed /high-band width system LSIs with a DRAM core.

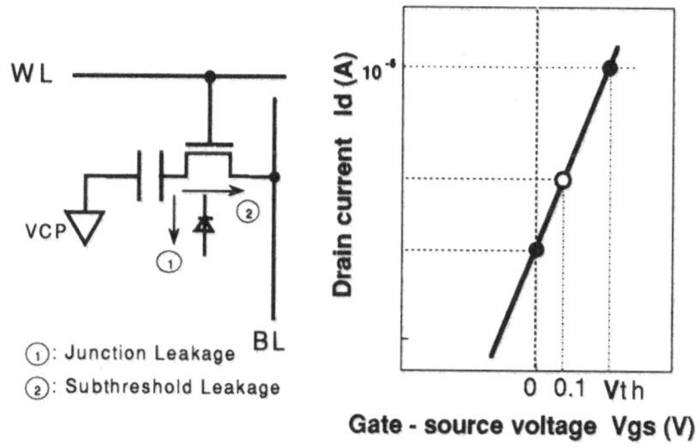

Fig. 1 DRAM cell leakage mechanism

2. The noise analysis for DRAM core implemented on system LSI

2.1 Degradation mechanism of dynamic refresh

In the logic section, a noise which level is lower than the CMOS threshold level is not a severe problem. However, in a DRAM core, noise on the word line increases the subthreshold leakage current of access transistors, exponentially. Fig. 1 shows the leakage mechanism of DRAM cell. If 0.1 V of noise appears on the word line, the leakage current increases ten times and the dynamic refresh time degrades, drastically. Fig. 2 shows the noise level on DRAM array GND level depends on the distance (Dis) between a DRAM core and a switching

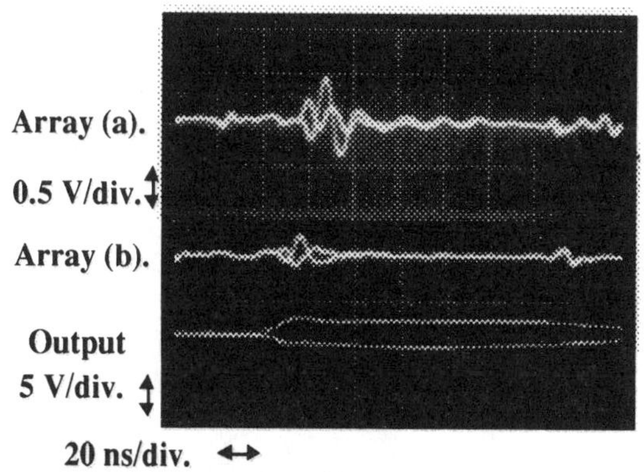

Fig.2 The waveforms of the noise level on DRAM array GND when the buffer is switching. Arrays (a) [Dis=100um] and (b) [Dis=7100um] are shown in Fig.11.

0-7803-3177-6 $5.00 © 1996 IEEE

IEEE 1996 CUSTOM INTEGRATED CIRCUITS CONFERENCE

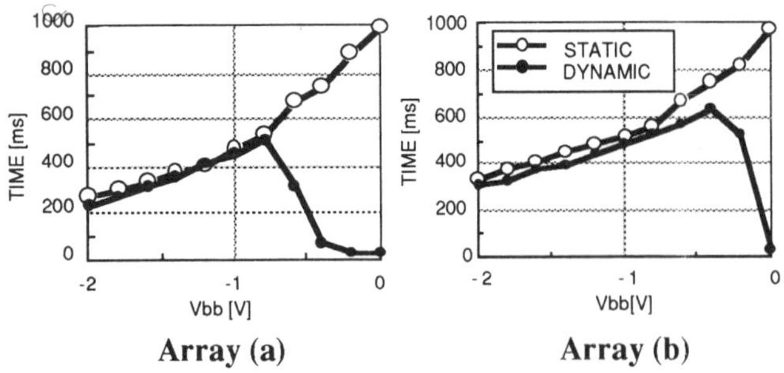

Fig. 3 Refresh time charactaristics depend on Dis.

buffer. The overshoot of DRAM array GND synchronizing with buffer switching increases the subthreshold leakage current. Fig. 3 shows DRAM core refresh time characteristics dependence on the Dis. The static refresh time does not depends on Dis, on the other hands dynamic refresh time strongly depend on Dis. Also, when the buffer does not switch, both the static and the dynamic refresh time become almost the same. So, it's clear that the buffer switching noise level corresponding to Dis is in proportion to the dynamic refresh time characteristics. It cannot be helped that the noise will occur at the switching buffer. Therefore, we discussed how to provide the noise transmission. To suppress the switching noise, the separated VDD/GND line scheme is well used in the on chip analog/digital LSI. In fact, the separated VDD/GND scheme for DRAM core improves the noise level as shown in Fig.4. However, the separated VDD/GND line scheme does not provide the sufficient dynamic refresh time. Because the additional noise paths exist. There are large junction capacitance (Cj) between the buffer transistor and substrate, therefore buffer - Cj - substrate - DRAM noise path (Cj path) is existing. Fig. 5 shows the waveforms of substrate noise through Cj path by buffer switching. Fig. 6 shows the

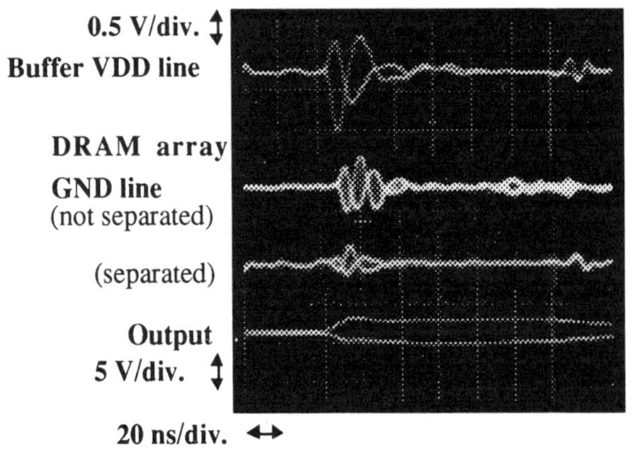

Fig. 4 Not separated and separated DRAM array GND line noise waveforms from VDD line through the Cdec path.

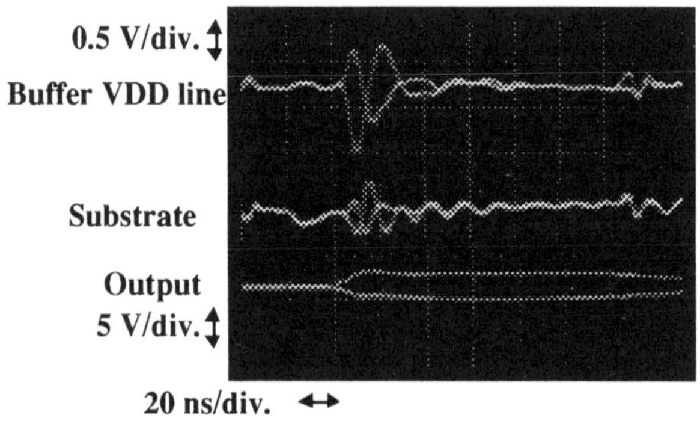

Fig. 5 Waveforms of substrate noise through the Cj path.

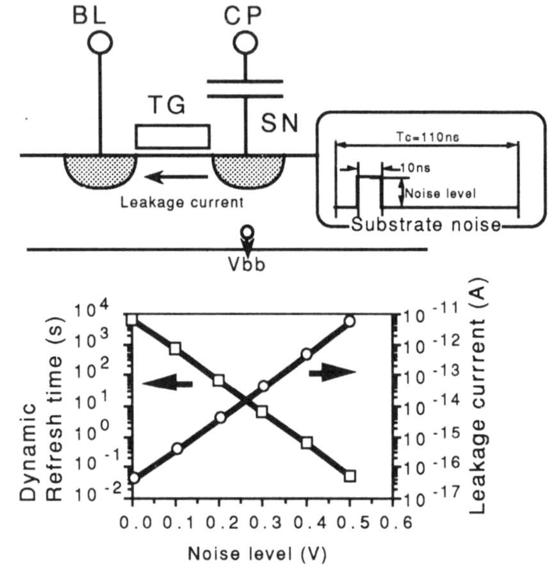

Fig. 6 Simulated dynamic refresh time and leakage current versus the substrate noise level.

simulated dynamic refresh time characteristics and leakage current versus the substrate noise level.

2.2 Noise immunity design proposal

Fig. 7 shows cross sectional view of the output buffer with adopting the triple well structure. To prevent the switching buffer noise being transmitted towards the substrate (Cj path), the bottom n-well must be biased by a stable VDD line. The potential of the p-well (a) and (b) is biased at Vbb. The noise is well suppressed by this triple well structure named noise barrier well. To prevent the switching buffer noise being transmitted to the p-well of the DRAM core through the Vbb line, the filter element can be inserted at the Vbb line and the clean Vbb level is suppled to the p-well (b). Fig. 8 shows dynamic refresh time characteristics for DRAM core in the three types of substrate structure: (A) conventional twin well struc-

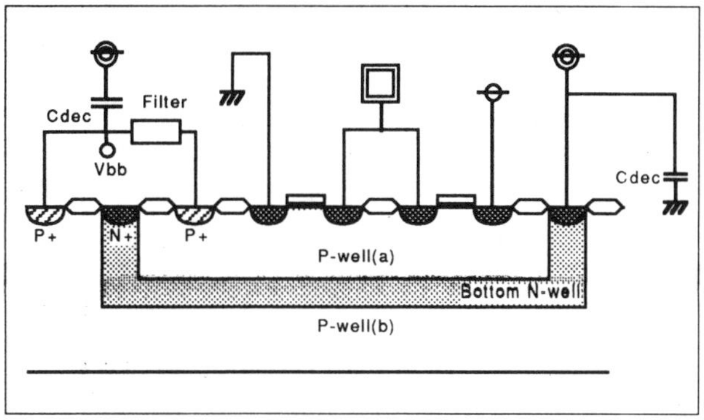

Fig. 7 Cross sectional view of output buffer with noise barrior well.

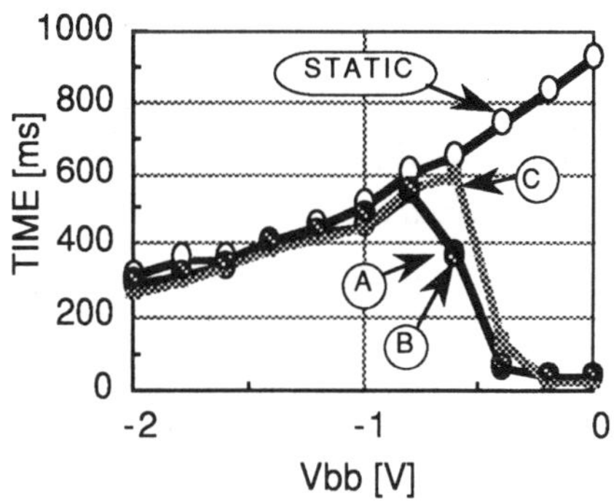

Fig. 8 Dynamic refresh time characteristics (A) conventional twin well structure, (B)normal triple well with bottom n-well biased by buffer VDD line, (C) noise barrier well with another biased clean VDD line.

ture, (B)normal triple well with bottom n-well biased by buffer VDD line, (C) noise barrier well with another biased clean VDD line. The case (C) has the best result of the three types.

It is difficult to prevent the noise entirely with the above mentioned solutions. To get longer refresh time characteristics, a boosted sense GND technique (BSG) [1] and a negative word line technique [2] are reported already. Specially a negative word line technique is one of good methods to improve the refresh time characteristics for low voltage operation. Fig. 9 shows the principle of this one. This technique provides a

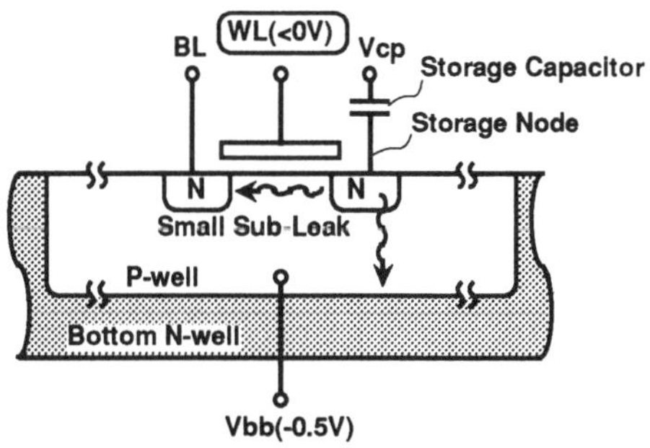

Fig.9 Negative word line technique.

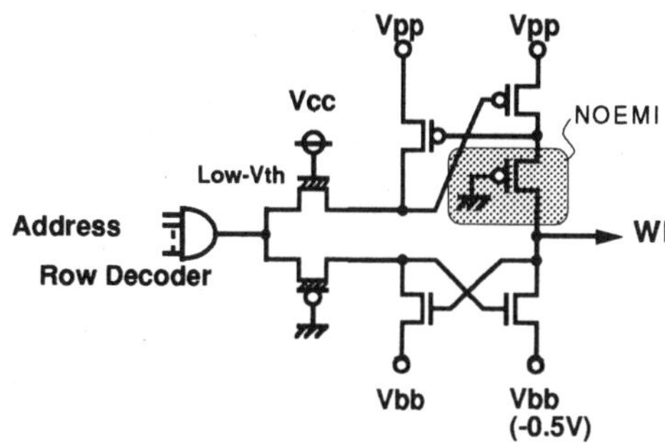

Fig.10 The negative word driver circuit.

negative Vgs which can bring noise free dynamic refresh characteristics, and enables the adoption of a lower Vth (0.6V) for the access transistor with a negligible subthreshold leakage current. Fig. 10 shows the negative word driver circuit. The hatched p-MOS transistor operates as the normally on enhancement MOSFET insertion (NOEMI) [3] to relax the TDDB and maintain reliability. Most gate oxide electric field are limited below Vcc.

3. Design methodology for on chip DRAM core system LSI

We fabricated a test device with a DRAM core and a large switching buffer using a 16 Mb DRAM process with the all of above described noise immunity design technique. Fig. 11 shows chip photography of the test device. Fig. 12 shows the results of static and dynamic refresh time characteristics of test device in page mode. The dynamic refresh time has the same value as the static refresh time characteristics at the minimum Dis condition.

Fig. 13 shows a block diagram of the VDD/GND line architecture of the on chip DRAM core system LSI with the pro-

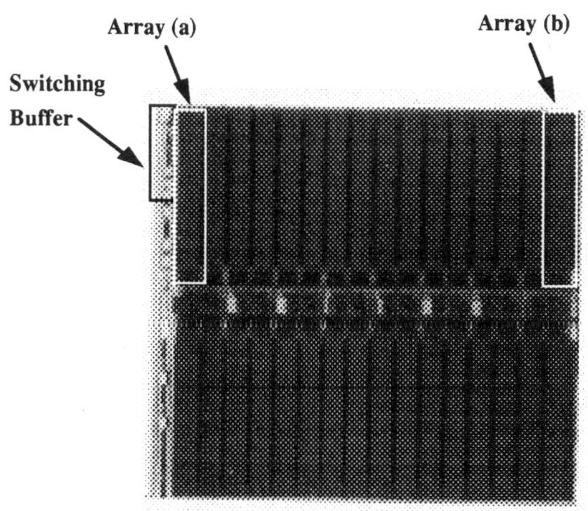

Fig. 11 Chip photography of test device.

posed circuit design methodology. A I/O element, logic element and DRAM control element, DRAM array element, have the individual VDD (2.5 V) / GND line. The 'H' level of word line is VDD not using the boosted potential. The power supply level of the DRAM core is scaled down to internal VDD (1.8V) by the voltage down convertor. The boostless word line scheme bring the following merits. This scheme enables the application of the optimum transistor to obtain high speed switching at the logic elements and the DRAM control element while maintaining the reliability of the gate oxide film. As the DRAM array element operates at the lower voltage, the DRAM core characteristics don't degrade at all. In the sense amplifier operation, there are no speed delays using a overdrive sensing technique. The static/dynamic refresh time does not become worse because the reduction of signal charge and the decrease of leakage current due to the scaled down of the p-n junction bias voltage are canceled. A soft error rate is no problem because a triple well structure is used.

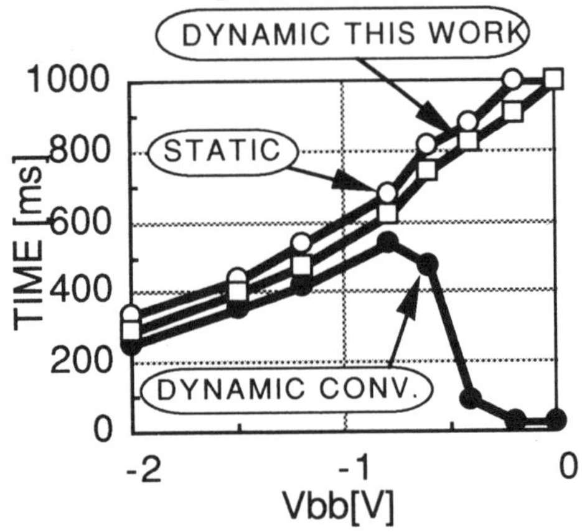

Fig.12 Experimental results of static and dynamic refresh time characteristics of test device.

Conclusion

We have analyzed the noise transmission model and discussed about the design methodologies for improving the noise margin of DRAM core for the system LSIs. The switching buffer noise protection technique for the DRAM core is indispensable. To say nothing of separated VDD/GND line scheme and longer distance between the DRAM core and the logic element, the noise barrier well structure to stop noise transmission through the substrate are very useful in obtaining a long dynamic refresh time. The wider noise margin DRAM core is provided with a negative word line scheme to maintain reliability. The boostless DRAM core and logic elements architecture can use the optimum transistor and realizes the high-speed / high-band width performance.

We believe the proposed design methodology will be useful for the on chip DRAM system LSI applied to the future multimedia system.

References

[1] M. Asakura et al. "A 34ns 256 Mb DRAM with Boosted Sense-Ground Scheme," in ISSCC Dig. Tech. Papers, Feb., 1994, pp. 140-141.

[2] T. Yamagata et al. "Circuit Design Techniques for Low Voltage Operating And/Or Giga-scale DRAMs." in ISSCC Dig. Tech. Papers, Feb., 1995, pp. 248-249.

[3] T. Sakurai et al. "Hot-Carrier Generation in Submicrometer VLSI Emvironment," IEEE J. Solid-State Circuits, vol. SC-21, No.1, Feb. 1986.

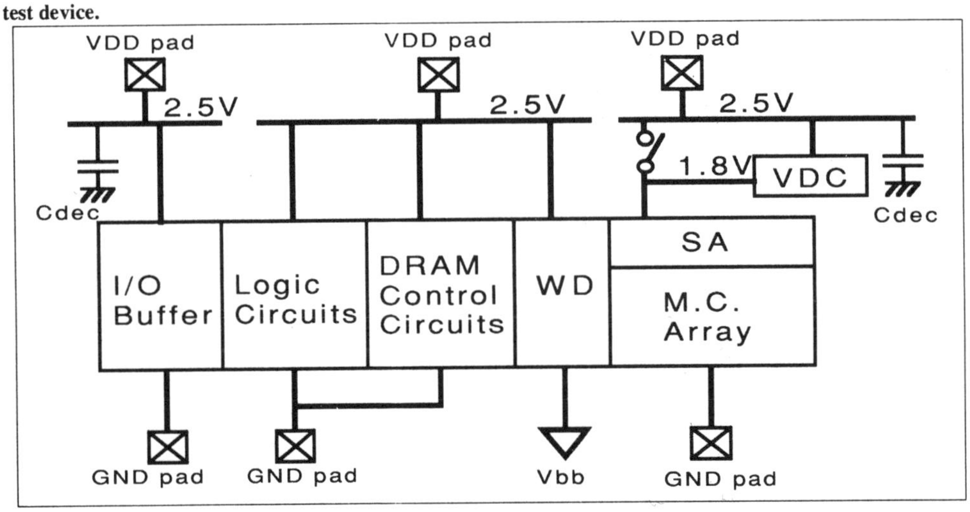

Fig. 13 Block diagram featuring the VDD/GND line architecture of on chip DRAM core system LSI.

Performance Evaluation of a Microprocessor with On-chip DRAM and High Bandwidth Internal Bus

Shunichi Iwata, Toru Shimizu, Jiro Korematu, Katsumi Dosaka, Hideo Tsubota, Kazunori Saitoh

Mitsubishi Electric Corporation
4-1 Mizuhara, Itami, Hyogo 664, Japan

Abstract

The M32R/D is a 32-bit microprocessor with on-chip 2M-byte (16M-bit) DRAM. It has a 32-bit RISC CPU, 32-bit × 16-bit multiply and accumulator, 2M-byte DRAM, 2K-byte cache memory, and a memory controller. The CPU, DRAM, and cache memory are connected with a 128-bit 66.6MHz internal bus. This wide internal bus allows high speed transfer of instructions. The use of a 128-bit bus resulted in a 13% improvement in performance as compared with 32-bit bus.

Introduction

A. Conventional integration approach

This chip is designed for multimedia applications, such as PDA, set-top box, network controller, printer, HDD controller, car navigation system, etc. A common factor of these applications are is that all require high performance, low power consumption, low cost, and large-capacity memory. One of the conventional solution for meeting these requirements is to integrate many peripheral circuits on a chip. However, in this approach, CPU and large memory are kept separate and need to be connected with an external bus. High performance is necessary for high bandwidth of the CPU-memory bus. High bandwidth can be obtained by wide bus width and high frequency. However, these factors increase the number and signal transition frequency of I/O pins and bus lines. Thus, high bandwidth results in an increase of power consumption and system cost.

B. M32R/D approach

Our approach integrates CPU and large-capacity memory (=DRAM) on a chip instead of peripheral circuits which are application specific. The peripherals are integrated on another chip as an ASIC (=I/O ASIC). The combination of the M32R/D, I/O ASIC and program ROM is capable of coping with a wide range of applications. Figure 1 illustrate a typical system configuration that effecting uses the M32R/D chip.

C. Advantages of DRAM integration

The main advantage of integrating the CPU and DRAM is the feasibility of using a wide and fast internal bus; the M32R/D has a 128-bit 66.6MHz internal bus. We can implement a narrow and slow external bus because the external bus transactions are reduced; the M32R/D has a 16-bit 16.6MHz external bus. The wide and fast internal bus leads to high performance while narrow and slow external bus reduces power consumption and cost.

This paper describes the implementation of the 128-bit wide internal bus and the result of its performance evaluation.

M32R/D Overview

Table 1 shows an overview of the chip characteristics. This chip is a 32-bit microprocessor with on-chip 2M-byte(16M-bit) DRAM. It includes a 32-bit RISC CPU, a 32-bit × 16-bit multiply and accumulator (MAC), a 2M-byte DRAM, a 2K-byte cache memory, an external bus interface unit (BIU), a memory controller, and some peripherals (see Figure 2). The 32-bit RISC CPU has a 5-stage pipeline structure and supports 83 instructions of 16 and 32 bits wide instruction formats. This chip integrates 17 million transistors onto a 19.9mm x 7.7mm die by using a typical 0.45 μm DRAM process, double metal CMOS technology (Figure 3). The CPU, DRAM, cache and BIU are connected with a single 128-bit 66.6MHz internal bus. The external bus is 16 bits wide and operates at 16.6MHz. The internal 66.6MHz clock is generated from the 16.6MHz bus clock by a 4x PLL. The typical power dissipation is 700 mW from 3.3V power supply at 66.6MHz.

128-bit internal bus implementation

A. Connection between CPU and DRAM

The M32R/D is a successor of the 16M-bit cache DRAM[1],[2]. We connected the instruction queue (I-queue), 128<-->32 selector and BIU buffer with the 128-bit bus between the DRAM array and the SRAM array (see Figure 4). The I-queue consists of two 128-bit entries, and holds the 32-bit instruction data for the CPU. When the CPU reads and writes an operand data, the 128<-->32 selector adjusts the data transfer between the 128-bit bus and the CPU. The BIU buffer is a 128-bit data buffer and supports burst transfers on 128-bit boundary data. This is effective for high speed data transfers from the external ROM to the cache and from the DRAM to the outside.

The DRAM and SRAM arrays are organized in four block (see Figure 5). The 128-bit bus is divided into four 32-bit buses. Each 32-bit bus is placed between the DRAM array and the SRAM array in each of the segments. The I-queue, 128<-->32 selector, and BIU buffer are also divided into 32-bit parts and placed between each of the DRAM array and the SRAM array segments. The CPU, MAC, memory controller, and other logic circuits are placed in the center of the chip. Divided buses have reduced the silicon area needed to connect the logic circuits in the middle.

B. Cache mode

The cache is direct mapped with 128-bit block size and follows a write-back policy. The cache has two caching modes: 1) "On-chip DRAM caching mode", see Figure 6, where the cache is used as an unified cache for caching on-chip DRAM. Thus, that the on-chip DRAM is used as the main memory. 2) "External ROM caching mode", see Figure 7, where the cache is used as an instruction cache for caching external memory. In this caching mode, an external ROM is used as program memory and the internal DRAM as data memory.

C. Bus operation

The 128-bit internal bus operates at 66.6MHz and transfers 128-bit data between the CPU and memory in one cycle. On a cache hit, the number of access cycles is 1 (=no wait cycles). In operand access operations, a 32-bit data is transferred between the CPU and the SRAM in one cycle through the 128<-->32 selector. In instruction fetch operations, 128-bit instruction codes are fetched into the I-queue in one cycle. The CPU is able to get 4 ~ 8 instructions (16/32-bit instruction format) in one cycle. On a cache read miss, the data transfer from the DRAM to the CPU and the update of the cache data takes 5 cycles (=4 wait cycles). This update includes a dirty data write-back and a new data load from the DRAM; the write-back and load are done in parallel through a write buffer. In this case, instruction transfer rate is 0.8 ~ 1.6 instructions/cycle. On a cache write miss, the data transfer from the write buffer to the CPU takes 1 cycles (=no wait cycles). Then, the data transfer from the write buffer to the DRAM is performed.

Evaluation

The 128-bit internal bus is wide enough to reduce the instruction and data collision. We have evaluated the performance of the M32R/D with a Verilog-HDL model. We used the Dhrystone benchmark program. The distributions of the executed operation for the Dhrystone program is as follows:

Branches	9%
Loads	25%
Stores	14%
Others	52%

Figure 8 shows the relative performance of the execution time, and bus collision time versus bus width for the on-chip DRAM caching mode. In this case, cache miss rate is 2%. The line graph shows the relative performance. The bar graph shows the relative execution time, and the parts of dark color show the collision time percentage of the total execution time. The bus collision reduces from 14.1% to 2.5% as the bus width is increased from 32 bits to 128 bits. The performance improvement is 13%. Figure 9 shows the relative performance and execution time versus bus width for the external ROM caching mode. In this case, the cache hit rate is 100% since all instructions are in the SRAM and all data are in the internal DRAM. The bus collision reduces from 42.5% to 20.9%. The performance improvement is 9%.

Conclusion

The M32R/D implements a 128-bit 66.6MHz internal bus connecting the CPU, on-chip DRAM, and on-chip cache. The use of a 128-bit internal bus reduced instruction and data collision, and the result of simulation using the Dhrystone benchmark program shows a 9 ~ 13% improvement in performance as compared to a 32-bit internal bus design.

Acknowledgement

The authors thank Y.Horiba, K.Nagasawa, H.Nishiumi, O.Tomisawa, and M.Yamada for support and encouragement.

References

[1] A.Yamazaki *et al.*, "A Concurrent operation CDRAM for Low Cost Multimedia", Symp. on VLSI Cir. of Tech. Papers, pp. 61-62, June 1993.

[2] K.Dosaka *et al.*, "A 90-MHz 16-Mb System Integrated Memory with Direct Interface to CPU", Symp. on VLSI Cir. of Tech. Papers, pp. 19-20, June 1995.

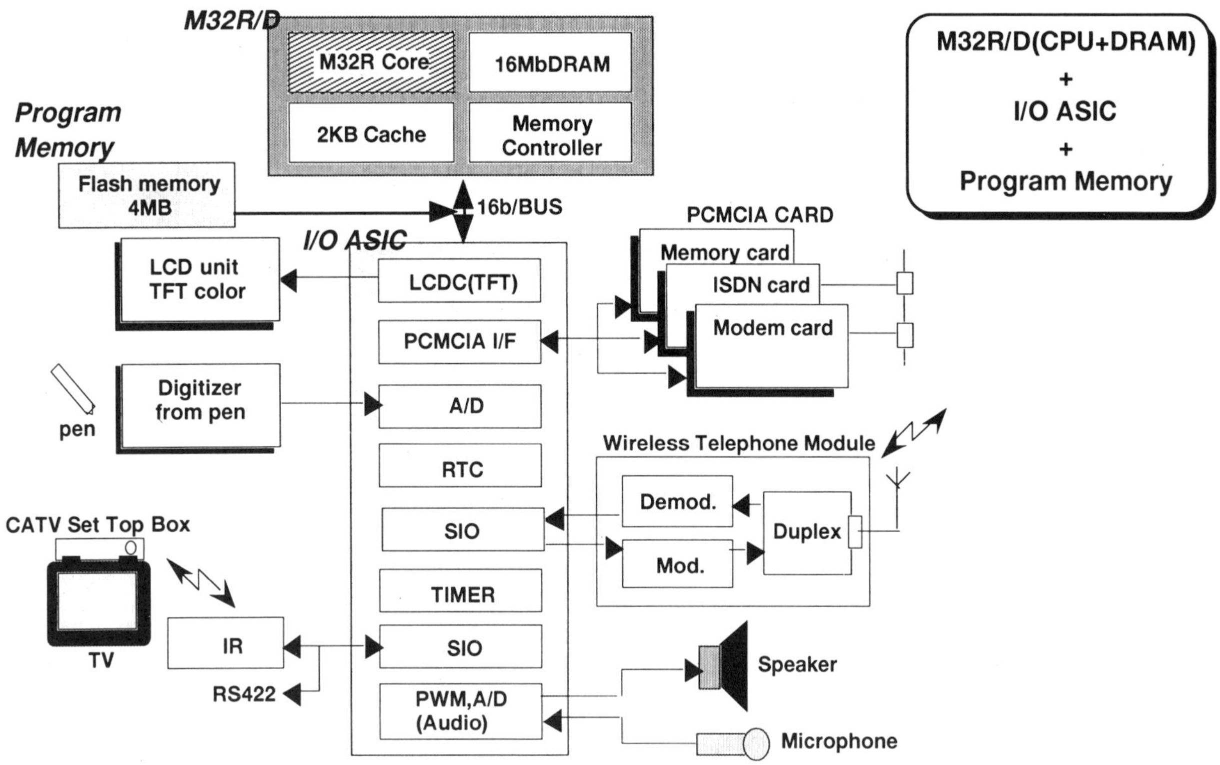

Figure 1. System configuration based on the M32R/D concept

Table 1. M32R/D Features

CPU core:	Small 32-bit RISC Architecture
VAX MIPS:	52.4 MIPS @ 66.6MHz (Dhrystone V2.1)
Memory:	2M byte DRAM with 2K byte Cache
Peripheral:	32b x 16b Multiply & Accumulator
	Memory Controller, etc.
External Bus:	24-bit address, 16-bit data
Clock:	66.6MHz (internal) / 16.6MHz (external)
Voltage:	3.3V
Power:	700mW(Typ.) / ≤ 2mW (Stand-by)
Die Size:	19.9mm x 7.7mm
CPU Size:	5.7 sq. mm
Process:	0.45μm CMOS, 2 metal layers
Package:	80-pin plastic QFP

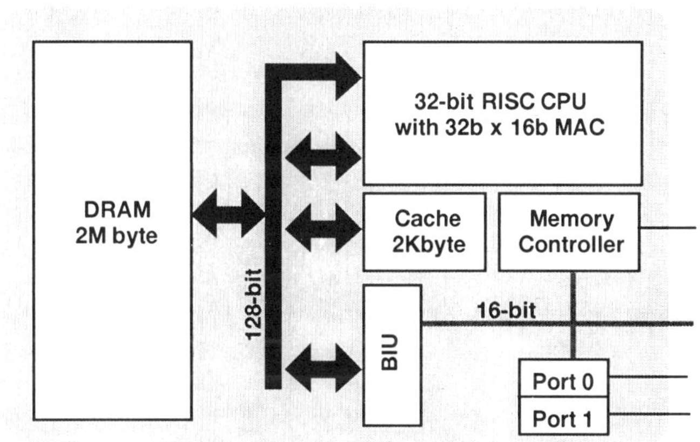

Figure 2. Block Diagram

Figure 3. Die Microphotograph

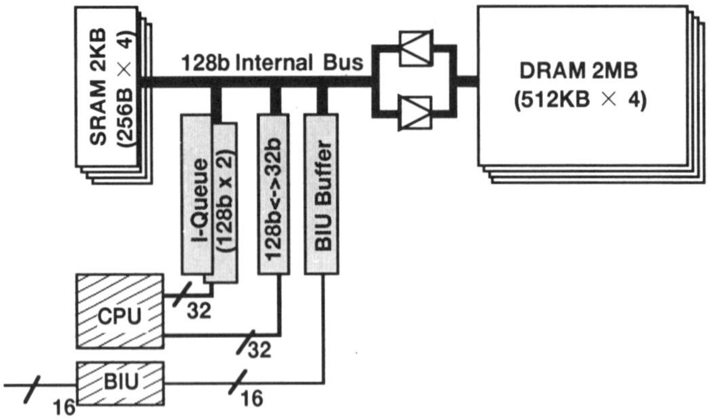

Figure 4. 128-bit internal bus

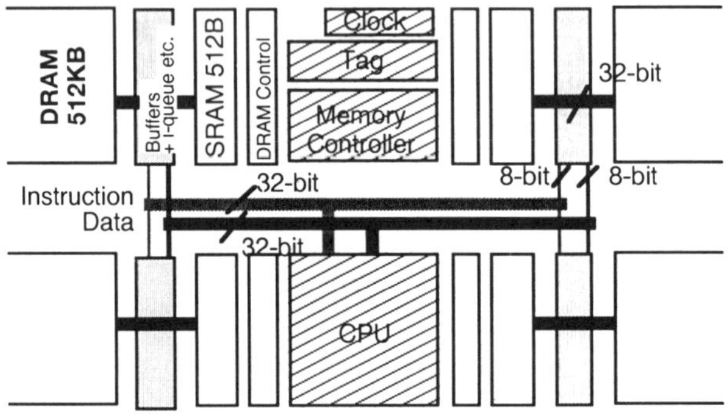

Figure 5. Floor Plan

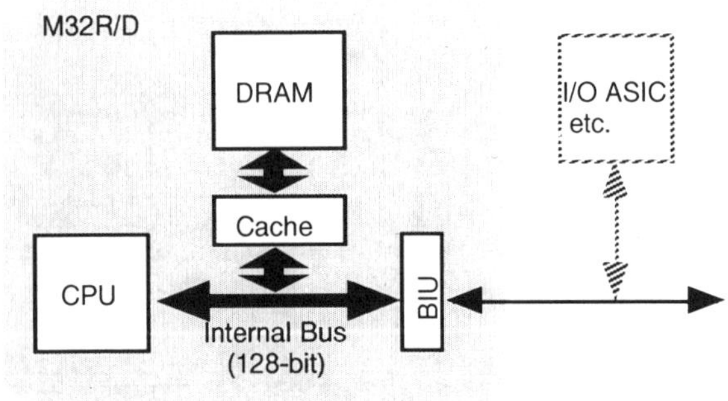

Figure 6. On-chip DRAM caching mode

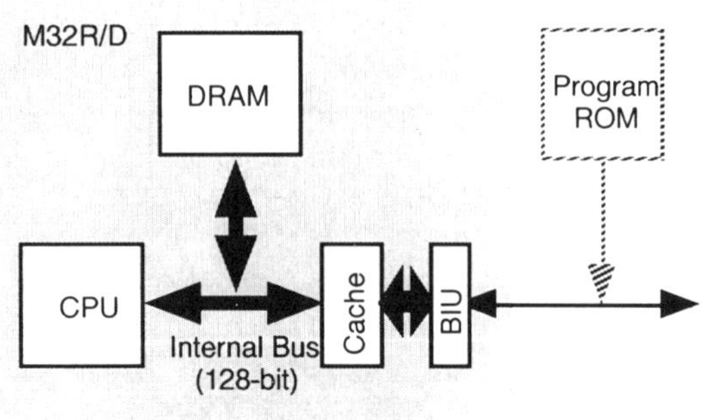

Figure 7. External ROM caching mode

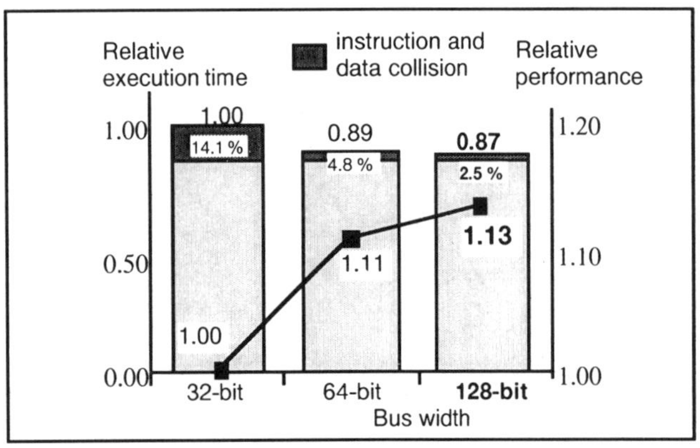

Figure 8. Performance evaluation: On-chip DRAM caching mode

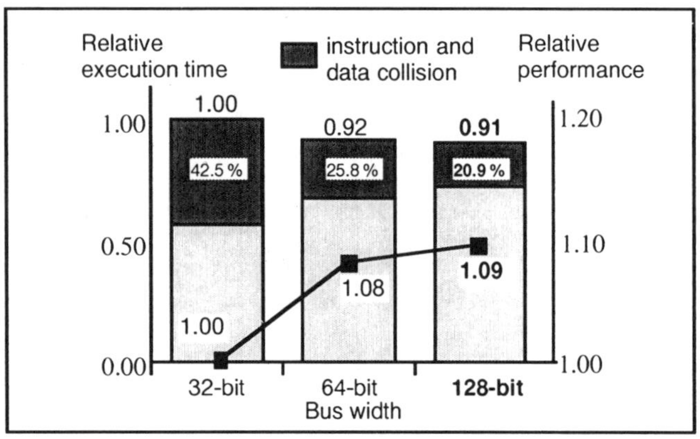

Figure 9. Performance evaluation: External ROM caching mode

A 336-kbit Content Addressable Memory for Highly Parallel Image Processing

Takeshi Ogura* Mamoru Nakanishi** Tatsuo Baba** Yasuji Nakabayashi** Ryota Kasai*

* NTT LSI Laboratories ** NTT Electronics Technology
Atsugi, Kanagawa, 243-01 Japan

Abstract

This paper describes a 336-kbit (4 kwords x 84 bits) Content Addressable Memory (CAM) LSI with dedicated functions for highly parallel image processing. This LSI can operate as a SIMD PE array, and is a key component of a CAM-based system, HiPIC (Highly-parallel Integrated Circuits and System). New hit-flag generation and multiple-response-resolution schemes are also incorporated in this LSI. A total of 4.2 million transistors are integrated using 0.5 μm CMOS process technology. The CAM LSI and HiPIC concept described here will make a significant contribution to the development of real-time image processing systems.

1. Introduction

A Content Addressable Memory (CAM) can not achieve only parallel search but also various parallel data processing. Therefore, many studies (1)-(3) have been performed based on CAM applications, and many CAM LSIs have been developed (4)-(8). There are, however, very few practical parallel data processing systems based on CAM. The main reason is that there are not many practical CAM-based hardware algorithms and system models, and large capacity CAM LSIs tuned for parallel data processing are difficult to develop.

In this paper, the concept of Highly-parallel Integrated Circuits and System (HiPIC), as a CAM-based system model, for image processing is described first. Then, CAM LSI architecture dedicated to the HiPIC is proposed. The HiPIC is an application-specific system, which achieves high performance and flexibility. The CAM LSI architecture is tuned for parallel data processing. Based on the proposed architecture, a fully parallel 336-kbit CAM LSI has been successfully developed using 0.5 μm CMOS process technology.

2. Highly-parallel Integrated Circuits and System (HiPIC) for Image Processing

A HiPIC consists of a highly-parallel PE (Processing Element) array, a reconfigurable logic element, RISC/DSP, and memory, as shown in Fig. 1. It may be implemented in a chip or a MCM (Multiple Chip Module) or a board. The highly-parallel PE array, shaded block in the figure, executes SIMD (Single instruction, Multiple Data stream) processing for high volume image data. The reconfigurable logic element controls the PE array and the interfaces to image data and the outer processor. The RISC/DSP controls the whole, and executes serial data processing.

The performance of a HiPIC strongly depends to the data processing ability of the highly-parallel PE array. To achieve high performance, it is most important to realize sufficient parallelism in the PE array. Therefore, the PE must be simple. Moreover, to avoid the so-called von Neuman bottleneck, the logic-in-memory type PE is the most suitable. The CAM LSI described here is representative PE array of a HiPIC.

Some highly-parallel image processing hardware algorithms based on the HiPIC concept have been studied (9), (10). An object tracking LSI (9), which contains 9-kbit CAM, strongly demonstrated that the HiPIC concept is effective for real-time image processing.

3. CAM LSI Architecture for HiPICs

Dedicated CAM functions for parallel processing

A CAM word can operate as a SIMD type PE by using dedicated functions for parallel processing. The dedicated

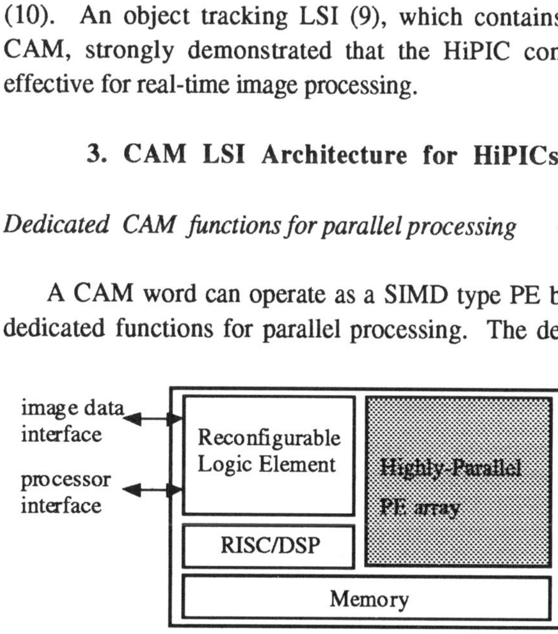

Fig. 1 HiPIC : Highly-parallel Integrated Circuits and System.

0-7803-3177-6 $5.00 © 1996 IEEE

IEEE 1996 CUSTOM INTEGRATED CIRCUITS CONFERENCE

functions are 1) maskable OR-search and 2) partial & parallel-write. The PE operations are carried out by an iterative bit-serial operation. In the maskable OR-search, search results for specified bit-positions are accumulated through OR logic. In the partial & parallel-write, data are written into specified bit-positions of multiple word locations in parallel. The SIMD-type PE operation is carried out in a bit-serial and word-parallel manner. The number of execution cycles is not affected by the data volume.

The CAM LSI, which supports dedicated functions for parallel processing, consists of four function blocks: the cell array, blocks for word and bit operations, and a control block. A block diagram of this CAM LSI is shown in Fig.2.

The cell array is composed of a 4 kwords x 72-bits writable CAM block and a 4 kwords x 12-bits read-only CAM block. Therefore, totally 336-kbit CAM is achieved. Each writable CAM cell circuit has capability of partial & parallel-write operation. The read-only CAM block also can achieve maskable search for fixed stored data. The read-only CAM block is composed of ROM (Read Only Memory) cells and a maskable address decoder. In the read-only CAM block, word-address data are stored. Complement data of the word-address data are also stored in an additional ROM. These data are used for generating single- and multiple-hit flags in word operation block.

The word operation block generates single- and multiple-hit flags using read-out data from the cell array. The single-hit flag indicates that only one word is selected or not selected in a search operation. On the other hand, the multiple-hit flag indicates that plural words are selected or not selected. In the word operation block, a response flag register is also

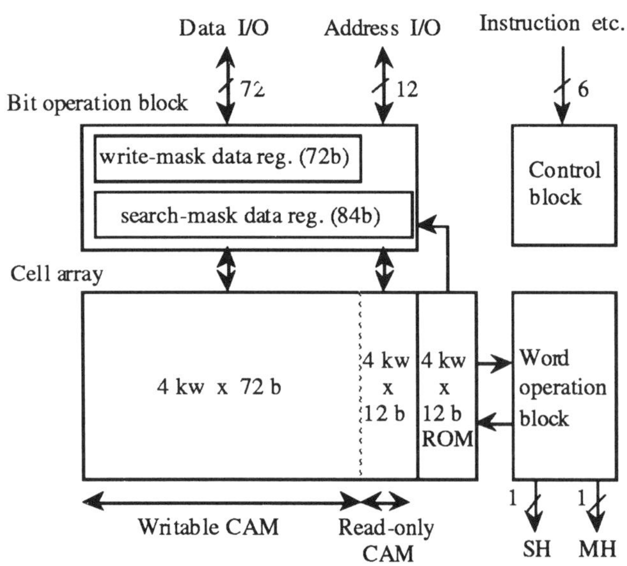

Fig.2 CAM Block Diagram. SH: single-hit flag, MH: multi-hit flag.

provided for each word. The accumulation of the results of the bit-serial search for parallel processing is carried out by this register.

The bit operation block specifies search bit positions in 84-bit-length. In a write operation, partial bit positions in 72-bit-length are also specified by this block.

This CAM LSI supports a powerful but simple instruction set for parallel processing. The instruction set contains 16 instructions classified into three modes: search, read, and write, as listed in Table I. Each operation is finished in one cycle. The parallel processing can be carried out by a combination of these instructions.

Single- and multiple-hit-flag generation

The usefulness of singe- and multiple-hit flags has been well known for about 30 years (11). Formerly developed CAM LSIs, however, can generate only hit flag, which indicates that selected words exist or not. This is because it is difficult to quickly generate single- and multiple-hit flags by small hardware.

This CAM LSI can quickly generate single- and multiple-hit flags using small and regular-structured hardware, ROM, where address-data and their complement data are stored. This ROM outputs "0", when no word-line is driven. When a single word-line is driven, the stored data are output. If plural word-lines are driven, logical OR of stored data of same bit-line is output. Therefore, by examining the output address-data and their complement data, single- and multiple-hit flags can be generated. The flag generation logics are expressed as

SH = &i (Ai@Bi) (1)

MH =(Ui (Ai) + Ui (Bi))&(!SH) , (2)

where SH is a single-hit flag, MH is a multiple-hit flag, & is AND, U is OR, @ is exclusive-OR, ! is negation, and Ai and Bi are output datum from read-only CAM and additional ROM, respectively.

TABLE I
Instruction Set of the 336-kbit CAM LSI

Mode	Operations
Search (3 instructions)	@fully parallel maskable search @maskable OR-search for parallel processing
Read (4 instructions)	@hit-flag, true and complement address read @address & data read at single hit @data read using address the same as RAM
Write (6 instructions)	@data write into matched words in parallel @data write using address the same as RAM @search-/write-mask-data write
NOP & Test (3 instructions)	

13.4.2

Multiple-response-resolution scheme

The multiple-response-resolution (MRR) becomes necessary only at the last stage of parallel processing in a CAM LSI for data read. The MRR is essential for a CAM LSI, but doesn't affect the performance of the parallel processing. Therefore, the best way is not to incorporate costly dedicated MRR circuits in a CAM LSI, unlike formerly developed CAM LSIs.

In this CAM LSI, the MRR can be realized without costly MRR circuits. The new MRR scheme features the ordered retrieval of fixed address-data stored in the read-only CAM block. The ordered retrieval is carried out an iterative maskable search for address-data. Key data for the iterative maskable search are generated using SH, MH and additional information, X. An X sensed in a given bit position indicates a mixture of 1's and 0's in that position for all selected words. The X can be sensed at off-chip by read-out address-data and their complement data.

An example of the MRR routine is shown in Fig. 3. The key data generation algorithm is the same as the proposed one (12). The number of search operation cycles necessary for the MRR is 2m–1, where m is the number of selected words. For this MRR scheme, only regular-structured hardware, ROM, is necessary. Therefore, this scheme is suitable for a large-bit-capacity and high-speed CAM LSI.

4. Implementation

Fabricated Devices

The 336-kbit CAM LSI was fabricated using 0.5 μm CMOS process technology with double-aluminum layers for interconnection. The CAM cell circuit is composed of twelve transistors, as shown in Fig. 4. The designed cell occupies 9.3 x 15.2 μm. A search operation is achieved by detecting whether a precharged match line has discharged through the cell or not. In a write and read operation, the electrical path to Vdd is cut by transistor Tr. 1. In this way, the bit line can be driven even if many cells are activated simultaneously. As a result, parallel-write operation up to 2 kwords can be achieved in one cycle. In a partial-write operation, writing into all cells connected to the same bit line is prohibited if the signal PWj is set at a low level.

A photomicrograph of the LSI is shown in Fig. 5. A total of 4.2 million transistors have been integrated on 16.5 x 16.5-mm chip. The effective area is 12.2 x 13.8 mm. The operation frequency of the CAM LSI is designed at the typical 66 MHz. This CAM LSI can also be used as an embedded macro block with conventional logic gates.

This LSI was tested using the evaluation board shown in Fig. 6. The evaluation board is composed of several FPGAs and a CPU-daughter board. The test sequence for this CAM LSI is generated by the FPGAs. Test results are reported through the CPU. Power consumption at 40-MHz operation is 3.6 W. The features of the 336-kbit CAM LSI are summarized in Table II.

Libraries for topdown system design

For rapid system design and development, libraries for the CAM LSI are strongly required. For the study of CAM-based hardware algorithms, a finite word-length library, written in C language, is supported. Performance evaluation of the proposed hardware algorithms, such as the number of processing cycles, and the accuracy of image processing, can be achieved by instruction-level simulation. Moreover, a

(a)

ID	address of selected words
a	0 0 1 0 1 0 0 1 0 1
b	0 0 1 0 1 0 1 0 1 1
c	0 0 1 1 0 0 1 0 0 1
d	0 0 1 1 0 0 1 0 1 1
e	1 0 0 1 1 0 0 0 0 1

search #	key & mask data	search result	output address true / complement	selected words
1	x x x x x x x x x x	P	1 0 1 1 1 0 1 1 1 1 1 1 1 1 1 1 1 1 1 0	a,b,c d,e
2	0 x x x x x x x x x	P	0 0 1 1 1 0 1 1 1 1 1 1 0 1 1 1 1 1 1 0	a,b,c d
3	0 x x 0 x x x x x x	P	0 0 1 0 1 0 1 1 1 1 1 1 0 1 0 1 1 1 1 0	a,b
4	0 x x 0 x x 0 x x x	1	0 0 1 0 1 0 0 1 0 1	a
5	0 x x 0 x x 1 x x x	1	0 0 1 0 1 0 1 0 1 1	b
6	0 x x 1 x x x x x x	P	0 0 1 1 0 0 1 0 1 1 1 1 0 0 1 1 0 1 1 0	c,d
7	0 x x 1 x x x x 0 x	1	0 0 1 1 0 0 1 0 0 1	c
8	0 x x 1 x x x x 1 x	1	0 0 1 1 0 0 1 0 1 1	d
9	1 x x x x x x x x x	1	1 0 0 1 1 0 0 0 0 1	e

(b)

Fig.3 An example of the MRR routine: (a) address of selected words and (b) MRR routine. x in key & mask data: don't care; P in search result: multi hit; 1 in search result: single hit.

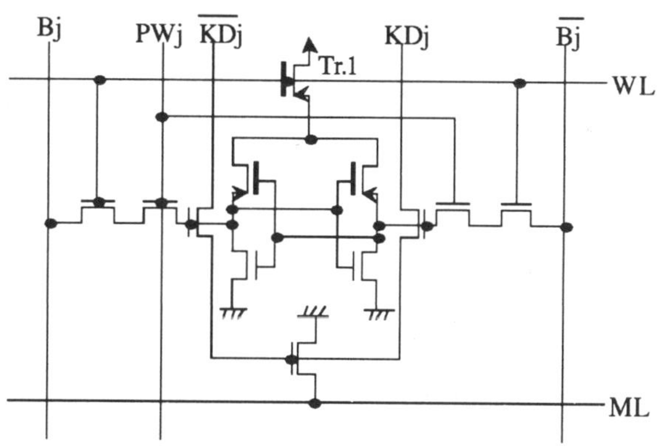

Fig.4 Cell circuit. B: bit line; WL: word line; KD: key-data line; ML: match line; PW: partial-write signal line.

Verilog library is also supported for hardware level design. With these libraries, it becomes easier to develop the HiPIC using this CAM LSI.

5. Conclusion

A 336-kbit (4 kwords x 84 bits) CAM LSI with dedicated functions for highly-parallel image processing has been developed. This LSI can operate as a SIMD PE array, and is a key component of a HiPIC (Highly-parallel Integrated Circuits and System). New hit-flag generation and multiple-response-resolution schemes are incorporated in this LSI. A total of 4.2 million transistors are integrated using 0.5 μm CMOS process technology. The immediate goal is to develop a HiPIC for real-time image processing using this CAM LSI. The CAM LSI and HiPIC concept described here are expected to contribute significantly to the development of real-time image processing systems.

Acknowledgment

The authors wish to thank Drs. Osamu Karatsu and Yasuo Ohmori for supporting this work.

References

(1) J. Naganuma and T. Ogura, "An associative processor for logic programming languages," Proc. 14th Annu. Hawaii Int. Conf. Syst. Sci., Jan. 1991, pp. 229-236.
(2) A. Krikelis, "Computer Vision Applications with the Associative String Processors," IEEE J. Parallel and Distributed Computing, Vol. 13, No. 2, Oct. 1991, pp. 170-184.
(3) T. Higuchi et al., "The IXM2 Parallel Associative Processor for AI," IEEE Computer, Vol. 27, No. 11, Nov. 1994, pp. 53-64.
(4) T. Ogura, S. Yamada, and T. Nikaido, "A 4-kbit Associative Memory LSI," IEEE J. Solid-State Circuits, Vol.SC-20, No. 6, Dec. 1985, pp. 1277-1282.
(5) H. Kadota, J. Miyake, Y. Nishimichi, H. Kudoh, and K. Kagawa, "An 8-kbit content-addressable and reentrant memory," IEEE J. Solid-State Circuits, Vol. SC-20, No. 6, Dec. 1985, pp. 951-963.
(6) H. Yamada et al., "Real-time String Search Engine LSI for 800 -Mbit/sec LANs," Proc. IEEE Custom Integrated Circuits Conference, May 1988, pp. 21.6.1-21.6.4.
(7) T. Ogura, J. Yamada, S. Yamada, and M. Tanno, "A 20-kbit Associative Memory LSI for Artificial Intelligence Machines," IEEE J. Solid-State Circuits, Vol. 24, No. 4, Aug. 1989, pp. 1014-1020.
(8) T. Yamagata et al., "A 288-kb Fully Parallel Content Addressable Memory Using a Stacked-Capacitor Cell Structure," IEEE J. Solid-State Circuits, Vol. 27, No. 12, Dec. 1992, pp. 1927-1933.
(9) Y. Fujino, T. Ogura, and T. Tsuchiya, "Facial image tracking system architecture utilizing real-time labeling," Proc. SPIE Visual Comm. and Image Processing '93, Nov. 1993.
(10)M. Nakanishi and T. Ogura, "Real-time CAM-based Hough transform Algorithm and Its Performance Evaluation," unpublished. [Proc. 49th Annual Convention IPS Japan, Sept. 1994.]
(11)R. R. Seeber and A. B. Lindquist, "Associative Memory with Ordered Retrieval," IBM Journal, Jan. 1962, pp. 126-136.
(12)M. H. Lewin, "Retrieval of Ordered Lists from a Content-Addressed Memory," RCA Review, June 1962, pp. 215-229.

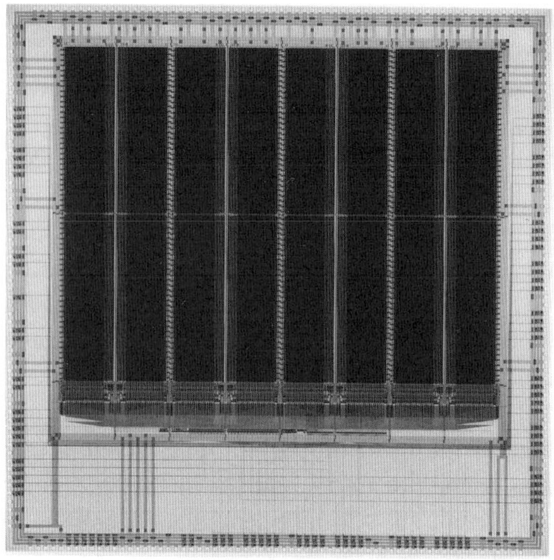

Fig. 5 Photomicrograph of a 336-kbit CAM LSI.

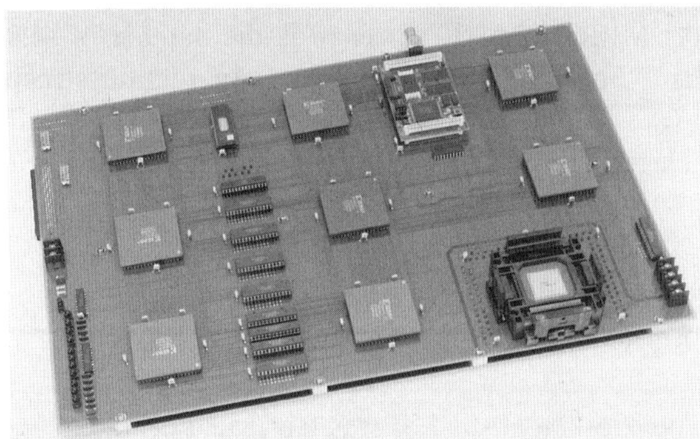

Fig. 6 Photograph of an evaluation-board.

TABLE II
Features of the 336-kbit CAM LSI

Configuration	4 kwords x 84 bits
Instruction set	16 instructions
Operation frequency	66 MHz (typical)
Supply voltage	3.3 V
Power dissipation	3.6 W at 40-MHz operation
I/O interface	I/O common, TTL compatible
Number of pins	92
Package	304-pin QFP
LSI process technology	0.5 μm CMOS with double-aluminum layers
Number of transistors	4.2 million
Chip size	16.5 x 16.5 mm (effective area 12.2 x 13.8 mm)

Forming Damped LRC Parasitic Circuits in Simultaneously Switched CMOS Output Buffers

Thaddeus Gabara†*, Wilhlem Fischer†, John Harrington§, William Troutman†

† AT&T Bell Laboratories, 600 Mountain Ave. 2A-232, Murray Hill, N.J. 07974
* Tel. (908) 582-2554 FAX (908) 582-5192 gabara@research.att.com
§ AT&T Bell Laboratories, 1247 S. Cedar Crest Blvd., Allentown, Pa 18103

ABSTRACT

Measurements of a 0.5μm CMOS testchip using several techniques have demonstrated a reduction in the generation of ground bounce. These techniques are: an automatic transistor sizing method that compensates for process, temperature, and supply voltage variations; a self-adjusting internal capacitive load that counteracts the increased switching rate of faster parts; and an integrated resistive element inserted directly into the power and ground leads that dampens the RLC oscillations. Comparison measurements between a conventional buffer and the new buffer have demonstrated that the amplitude and duration of the generated ground bounce has been reduced 2.5X and 2X, respectively. A single external resistor is required to set a reference current.

INTRODUCTION

Ground bounce is the result of exciting the parasitic RLC network existing in the power/ground network of a packaged device. Wide bus widths in IC packages increase the seriousness of this problem. Single ended outputs are the major contributors to the ground bounce effect: they generate large current fluctuations which energize the RLC network in the VDD or VSS power leads during simultaneous switching events.

The situation is further compounded by the process, voltage and temperature (PVT) variations that prevent the current fluctuations from remaining constant. This causes the ground bounce to vary over a wide range. It would be desirable to maintain the ground bounce constant, preferably, at its minimum value. However, in order to do so, one must be able to accurately control the PVT parameters: the solution is to use automatic PVT compensation techniques and RLC circuit damping techniques in the CMOS circuit.

This paper will describe an equation modeling the ground bounce problem, circuit techniques to minimize its effect, and measured results of a 0.5μm CMOS testchip.

SOURCE OF PROBLEM

Fig. 1 depicts a wide array of N-output buffers, each driving a capacitive load C_{pad} through the inductance L_{pad}. The box connected to the gates contains the pre-drive circuitry to disable or enable the output devices. L_{pwr} and L_{grd} are the parasitic inductances of each lead in the power and ground paths, respectively. C_{coup} is the on-chip coupling capacitance.

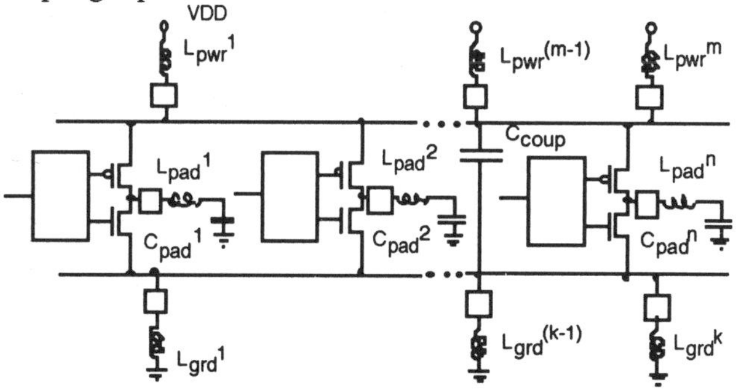

Figure 1. Approximate model of "N" output buffers with parasitics.

Two equivalent circuits of Fig. 1 are illustrated in Fig. 2 assuming all outputs are driven low.

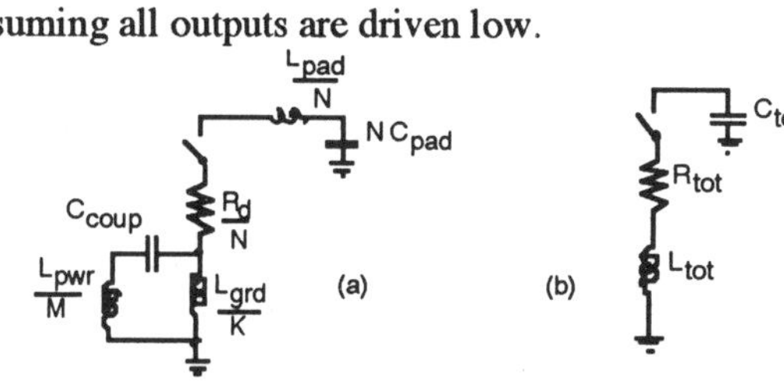

Figure 2. (a) Equivalent LRC circuit; (b) Simplified LRC circuit.

The resistance R_d is the on resistance of a single device;

$$R_d = 1/(2B(V_{gs}-V_t)) \quad (1)$$

since there are N devices which are activated, the total resistance is

$$R_{tot} = R_d/N \quad (2)$$

the total capacitance is equivalent to N capacitors in parallel;

$$C_{tot} = C_{pad} * N \quad (3)$$

The capacitor, C_{coup}, AC couples both power and ground equivalent inductances together while the equivalent interconnect inductance is in series;

$$L_{tot} = L_{pad}/N + \{ L_{pwr}/M \parallel L_{grd}/K \} \quad (4)$$

Since M and K are typically less than N, the inductance in the supply leads can dominate the total inductance.

The ground bounce effect occurs when the energy stored in capacitance, C_{tot}, is transferred into magnetic energy in the parasitic inductance of the loop ground path, L_{tot}, after passing through the series resistance, R_{tot}. This effect can be modeled by [1];

0-7803-3177-6 $5.00 © 1996 IEEE

IEEE 1996 CUSTOM INTEGRATED CIRCUITS CONFERENCE

$$I = -V/\beta L_{tot} \; e^{-\alpha t} \sin \omega t \qquad (5)$$

where V is the initial capacitance voltage and,

$$\beta = \sqrt{(\omega_0^2 - \alpha^2)} \qquad (5a)$$

$$\alpha = R_{tot}/2L_{tot} \qquad (5b)$$

$$\omega_0 = \sqrt{(1/L_{tot} C_{tot})} \qquad (5c)$$

The value of the resistance, R_{tot}, will determine whether an oscillatory waveform will be generated at the power/ground nodes of the chip.

$$R_{tot} < 2 \sqrt{(L_{tot}/C_{tot})} \text{ ; under-damped} \qquad (6a)$$

$$R_{tot} >= 2 \sqrt{(L_{tot}/C_{tot})} \text{ ; critically or over-damped} \qquad (6b)$$

Without compensation, this R_{tot} can vary by a factor of three over the PVT spread. In fact, this spread is a primary component affecting the difficulty of this problem [2]. At one end called WCS (Worst Case Slow; long channel, low voltage, high temperature), the resistance is larger such that (6b) may be satisfied, so the circuit is either critically damped or over damped. These devices, however, perform slower and cause large propagation delays. At the other end called BCF (Best Case Fast; short channel, high voltage, low temperature), the resistance is smaller causing an under-damped waveform as in (6a). Although these devices can in principle transfer data faster, the simultaneously switched noise or ground bounce prevents reliable data transfer. Due to the PVT spread, the amount of ground bounce is difficult to predict.

ATOMIC CELL SIZING TECHNIQUE

A digital transistor sizing technique [3] can compensate for the PVT spread and maintain the effective resistance of the output buffer in (1) constant. A constant resistance maintains α constant as in (5b); thus, the oscillatory waveform remains constant in (5).

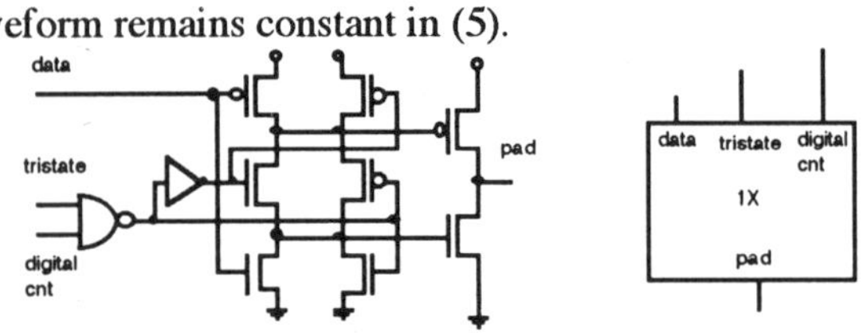

Figure 3. Atomic cell; circuit (left), block diagram (right).

This transistor sizing technique has been applied to a CMOS output buffer. The final and pre-drive stage of the CMOS buffer is called an atomic cell (AC). Fig. 3 illustrates the AC of an output buffer along with a block diagram representation. The digital control signal that is applied to the NAND gate enables/disables the output of that particular AC. This cell also contains two additional input signals; tristate - used to tri-state the buffer, and data - the signal which will be transferred to the pad. A buffer will consist of several of these ACs placed in parallel where the individual leads of tristate, data, and output are connected together as shown in Fig. 4; however, the digital control signals can be individually controlled to size the buffer. Thus, the weight of the parallel digital input controls can be adjusted to perform Atomic Cell Sizing (ACS).

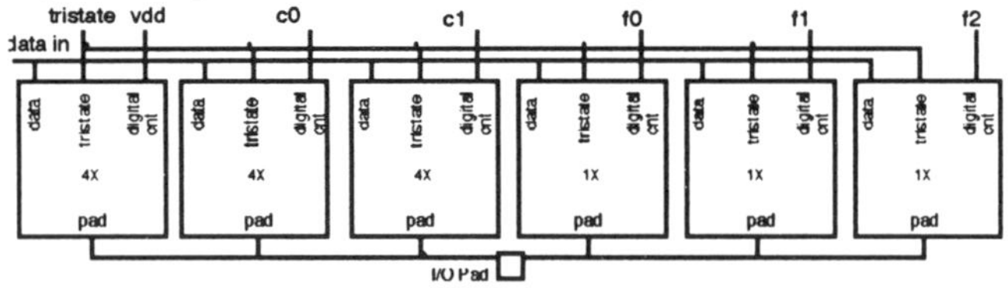

Figure 4. Combining coarse and fine atomic cells together to form an atomic cell sized output buffer.

The buffer shown in Fig. 4 consists of three fine and three coarse AC's placed in parallel. One coarse cell is equivalent to four fine cells. Note that the left most coarse cell has a solid input of VDD applied to the digital control. This insures that the composite size of this buffer can never decrease below that of a coarse cell or 4 AC's. For example, when all digital inputs are at a "1", the buffer is fully enabled such that the total transistor width of the final output stage is at a maximum of 15 AC's. When all digital inputs are at a "0", the total transistor width is at a minimum or 4 AC's, since one cell is always enabled.

REFERENCE CIRCUIT

The reference circuit that generates the digital control weight is illustrated in Fig. 5 and requires a single *external* resistor to generate a reference current. This current is mirrored by MPCUR into a digitally controlled N-channel transistor array MN1-6. These transistors mimic the effective resistance of a scaled version of the output buffer. The effective width of these devices is adjusted by the coarse and fine digital control bits (c0, c1, f0, f1, and f2) until the voltage drop across the transistor array matches that of the resistive divider. This insures that the array maintains a constant impedance over all PVT conditions.

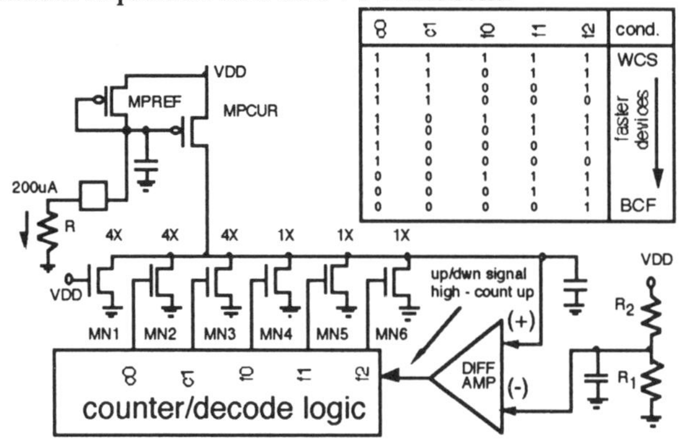

c0	c1	f0	f1	f2	cond.
1	1	1	1	1	WCS
1	1	0	1	1	
1	1	0	0	1	
1	1	0	0	0	
1	0	1	1	1	
1	0	0	1	1	
1	0	0	0	1	
1	0	0	0	0	
0	0	1	1	1	
0	0	0	1	1	
0	0	0	0	1	BCF

Figure 5. Reference circuit for the atomic cell sizing technique; inset, digital sequence.

The diff amp generates an UP/DOWN signal which increments/decrements a counter whose decoded output generates a digital weight that is provided back to the transistor array. The possible digital weights are indicated in the inset of Fig. 5. Note that a thermometer scale is used. At WCS, all control bits are at a logic "1", while at BCF, all control bits except one fine bit are at a logic "0".

Finally, note that in the array, device MN1 is always enabled. This device corresponds to the one coarse AC which is always enabled as illustrated in Fig. 4. The final generated digital weight is doubled latched (not shown) and sent to all the output buffers in the entire chip. It is this digital control signal which controls the effective impedance of all the output buffers that insures that the PVT variation is compensated over all conditions.

CAPACITIVE INSERTION

In addition to transistor sizing, the digital weight is used to adjust the capacitance of a node [4]. Capacitance is added to the pre-drive node of the output buffer as the chip progresses from the WCS to BCF conditions. The addition of this capacitance to the pre-drive node slows the charging of this node and provides an additional control over the slew rate of the output waveform generated at the pad.

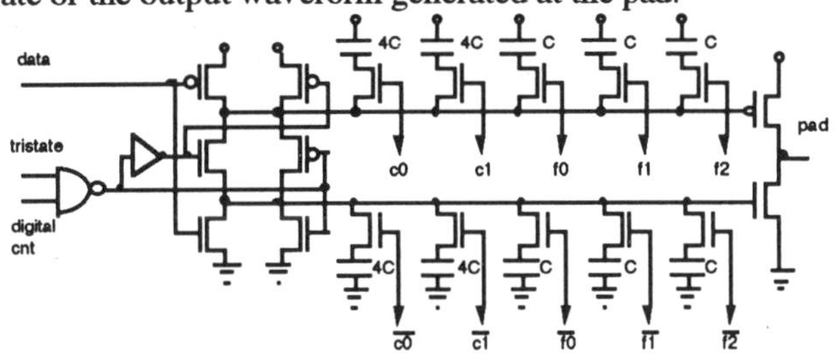

Figure 6. Adding digitally adjustable capacitive load to the pre-drive nodes of an atomic cell.

Fig. 6 illustrates the method of the capacitive insertion into the pre-drive node of a AC. Although the ACS technique connects several ACs in parallel, with a common output, data input and enable signal, the internal pre-drive nodes are NOT connected together. Therefore, each pre-drive node must contain this adjustable capacitive load circuit. The coarse and fine control bits insert a capacitance which is weighted 4X and 1X, respectively.

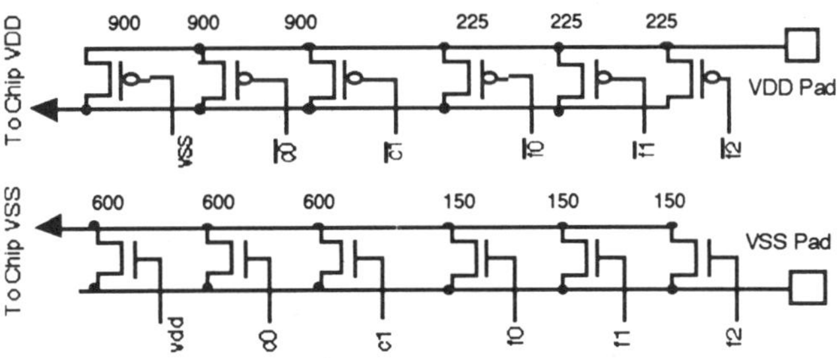

Figure 7. Power (top) and ground (bot) integrated resistive supply impedances. Bonding pad on right; chip core to left.

DAMPED LRC CIRCUIT

An integrated resistance is placed in series with the inductance which is in the VDD and VSS leads of Fig. 1; this resistance can be used to further dampen the oscillations. This achieves a faster buffer which satisfies the condition in (6b). Active circuit elements are used to form these resistances as shown in Fig. 7. These resistances are formed by transistors and will also be controlled by the same digital weight that has already been generated by the reference circuit. At WCS, all the devices will be enabled to minimize the resistance so a maximum specified voltage drop is satisfied. At BCF, the impedance is similar to the impedance at WCS. When several resistive power/ground pads are used, the overall impedance is their parallel combination. Resistive damping allows an independent ground bounce control over and above that of the control incorporated in the output buffer.

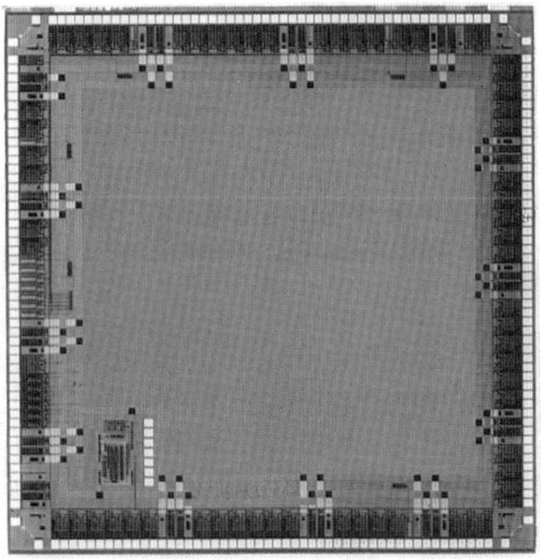

Figure 8. Micrograph of the ground bounce testchip.

MEASURED RESULTS

A 0.5μm CMOS testchip (6000μm X 6000μm), packaged in a 208 pin MQUAD package, has been measured comparing the ground bounce generation of ACS buffers versus Conventional buffers. The testchip is shown in Fig. 8. Either 2, 4, 8, 16, or 32 output buffers can be switched simultaneously; furthermore, a SELECT signal is used to enable either the ACS or Conventional buffer set allowing a comparison evaluation to be made. The ACS buffer was designed to match the delay of the Conventional buffer at WCS conditions. The micrograph of these two buffers is given in Fig. 9.

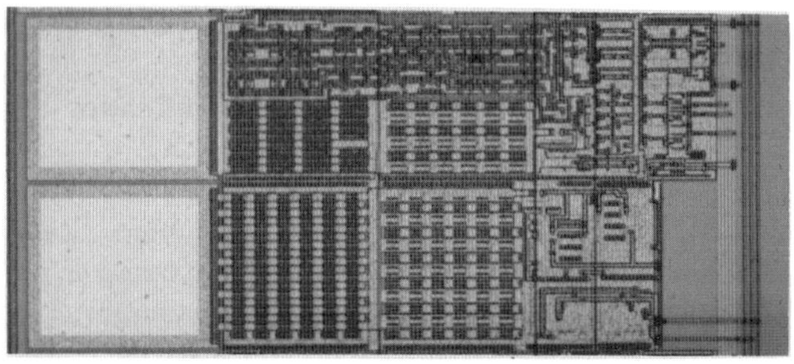

Figure 9. Micrograph of the ACS (top) and the Conventional Buffer (bottom).

Power can be supplied to the chip using either one of two sets of power/ground pads: a direct connection (metal resistance); or through a resistive component (active device resistance, R_{pwr} and R_{grd}). Fig. 10 shows a micrograph of both sets of these cells. The measured results of both sets of power/ground pads allows the direct observation of how this damping resistance in the parasitic RLC circuit reduces the simultaneous switching noise. Furthermore, the same package was used in all measurements to guarantee that the characteristics of the chip is not a factor in the measurements.

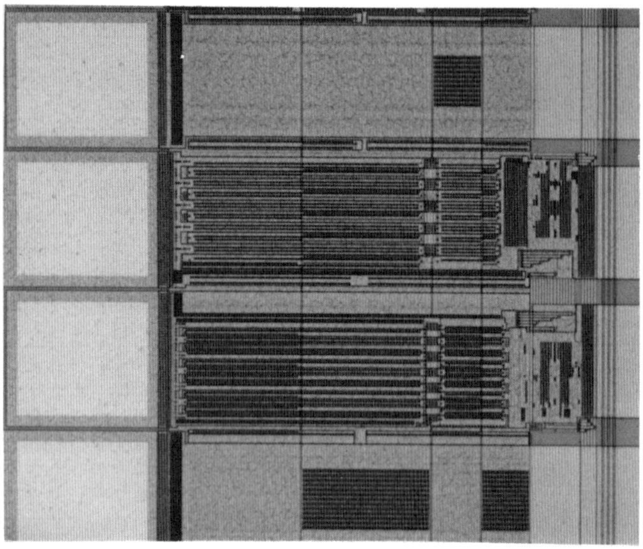

Figure 10. Micrograph of the Power Supply Cells: (top to bottom); Conventional VSS, Resistive VSS, Resistive VDD, Conventional VDD.

The measured ground bounce results for the Conventional Buffer are shown in Fig. 11. Fig. 12 depicts the same measurements using the ACS buffer with the active resistance in the VDD and VSS pads. The top waveform in both figures is an output waveform measured with no chip activity. The measured delay difference of the rising and falling edges of these two signals is within 200psec. The second, third, fourth, fifth, and sixth waveforms are the ground bounce measurements when 32, 16, 8, 4 and 2 buffers switch a capacitive load of 50pF simultaneously (VDD = 3.6V). Note that for the 32 bit case, the ACS buffer reduces the magnitude of the ground bounce by a factor of 2.5X. Moreover, the ringing of the oscillations for the ACS buffers dampens out 2X faster.

CONCLUSION

Several circuit techniques have been described which decrease the generated ground bounce of packaged CMOS devices. These basic techniques are PVT compensation, pre-drive capacitive loading, and resistive damping of the parasitic RLC tank circuit. A 0.5μm CMOS testchip containing both a conventional buffer and one incorporating the above techniques has been tested. The WCS delay of both buffers were identical. Measurements have demonstrated that the new buffer has: a 2.5X reduction in the amplitude of the generated ground bounce; and a 2X reduction of the duration of the ground bounce oscillations when compared to a comparable Conventional buffer.

ACKNOWLEDGEMENTS

The authors appreciate the comments of Mike Brown, Bernie Morris, and Mark Strauss and the support of Al Dunlop and Paul Henry.

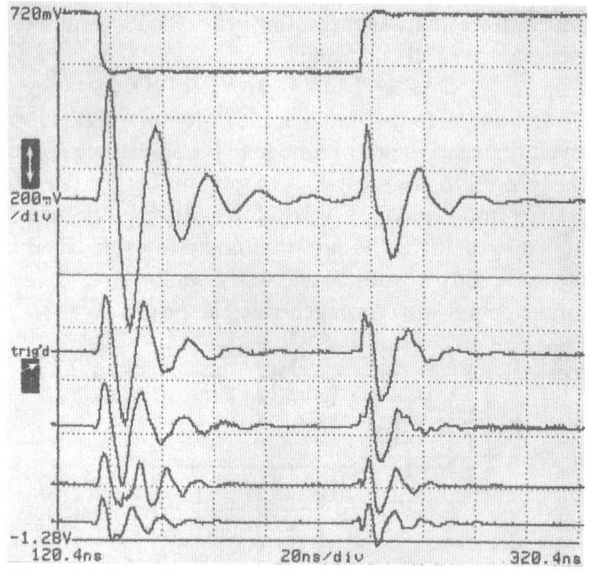

Figure 11. Ground bounce measurements of the Conventional Buffers.

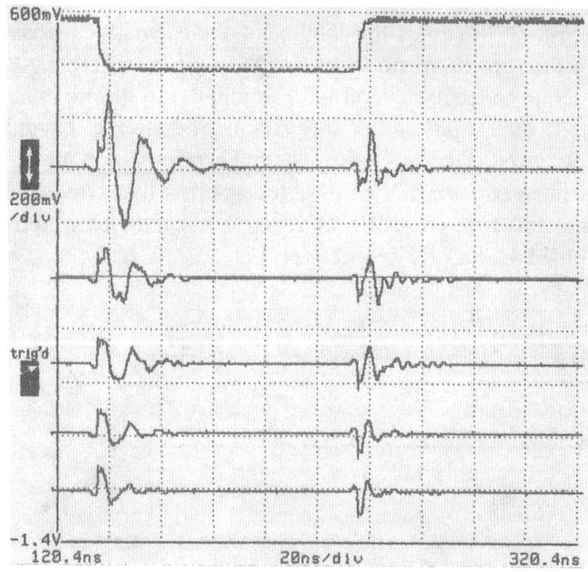

Figure 12. Ground bounce measurements of the ACS Buffers.

REFERENCES

[1] J. Groszkowski, "Frequency of Self-Oscillations", MacMillan Company, 1964.

[2] T. Gabara and D. Thompson, "Ground Bounce Control in CMOS Integrated Circuits", ISSCC, pp.88-89, Feb. 1988.

[3] T. Gabara and S. Knauer, "Digitally Adjustable Resistors in CMOS for High-Performance Applications", IEEE J. Solid-State Circuits, vol.27, no. 8, pp.1176-1185, August 1992.

[4] J.L. McCreary and P.R. Grey, "All-MOS charge redistribution analog-to-digital conversion techniques," IEEE J. Solid-State Circuits, vol. SC-10, no. 6, pp. 371-379, Dec. 1975.

Capacitive Coupling and Quantized Feedback Applied to Conventional CMOS Technology

Thaddeus Gabara and Wilhlem Fischer
AT&T Bell Laboratories, 600 Mountain Ave. 2A232, Murray Hill, N.J. 07974
Tel. (908) 582-2554; FAX (908) 582-5192; gabara@research.att.com

ABSTRACT

An on-chip capacitor which is formed under a bonding pad is used to block the DC level of an external input signal. Quantized Feedback using a self-triggered decision circuit is used to establish local DC voltage levels in the receiver and eliminate the "zero wander" effect. Measurement of a 0.5μm CMOS chip has demonstrated that: a BER (Bit Error Rate) Test for a 2^{31}-1 sequence showed no errors at 600Mb/s; the circuit detected a low frequency signal of 1Kb/s indicating that coding is not required; and the input signal can be DC biased anywhere, limited only by the ESD diodes, between VDD and VSS without affecting the final recovered CMOS waveform.

INTRODUCTION

Capacitive coupling has played a role in high speed transfer of signals between chips. This technique offers DC isolation between circuits; however, the data pattern is usually coded to prevent the loss of low frequency data. The receiver typically consists of a trans-impedance amplifier whose output would exponentially decay to a constant voltage if given a sufficient amount of time after detecting an edge transition. In order to minimize this effect, the RC time constant is made large and the input pattern is coded. However, the value of the capacitor can be very large (0.1μF) and would have to be a surface mount capacitor. Since discrete components use board area, the number of I/O (Input/Output) that can be capacitively coupled would be severely limited. Thus, when this technique is used, it is typically applied to one or a few of the high speed ports of a chip.

Salzman and Knight [1-2] have applied capacitive coupling to MCMs (Multi Chip Modules) to transfer signals between chips in their system. A metal plate is formed on the surface of two chips which are then placed facing one another with an air gap forming the dielectric. This capacitor transfers signals between the two chips. The receiver is a trans-impedance amplifier [2] followed by a register triggered by one of two inverter paths each with a different gate threshold.

Our technique forms the entire capacitor on one chip where the capacitor (1.6pF) can be formed under the bonding pad. This capacitor DC couples the external signal into the chip. A DC level is immediately established on the second plate of the capacitor independent of the data pattern. The receiver operates similar to the concept of Quantized Feedback [3-9] which was applied to the transfer of single ended data. This technique can detect signals for frequencies approaching DC since only edge transitions are necessary to operate the QFR (Quantized Feedback Receiver). Thus, coding of the input data is not required.

A description of the formation of the capacitance will be described in the 0.5μm CMOS technology. The basic CMOS version of the QFR will then be described. The on-chip coupling capacitor and the QFR are combined together to form the CCQFR (Capacitively Coupled Quantized Feedback Receiver). Measured results of a 0.5μm CMOS CCQFR will be described.

COUPLING CAPACITANCE

Several methods which use capacitive coupling are indicated in Fig. 1. The conventional method is illustrated in Fig. 1a. The dotted lines indicate the boundary of the IC chip. An external, discrete capacitor with approximate dimensions of 1mmX2mm is soldered directly onto the PCB (Printed Circuit Board). An ohmic path connects the second plate of the capacitor directly to the trans-impedance amplifier located on Chip A.

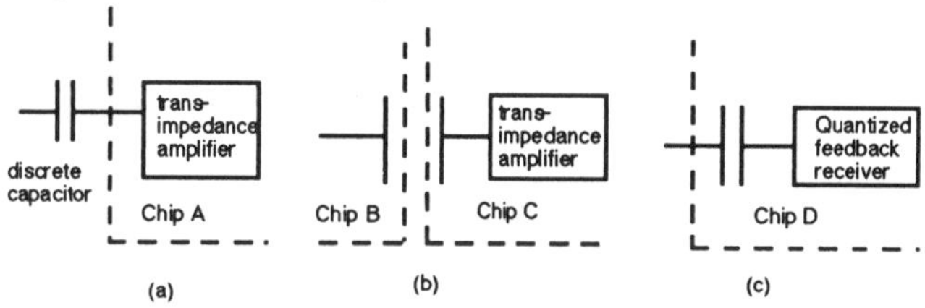

Fig. 1. Several ways of interfacing capacitive coupled signals to a chip.

Fig. 1b illustrates a recent technique of capacitively coupling signals between Chips B and C for MCMs. The first plate of the capacitor is on chip B, the second plate of the same capacitor is on chip C. These two chips face one another to form the capacitor. A trans-impedance amplifier detects the signal which is then captured by a register.

Fig. 1c illustrates the method that will be described in this paper. A DC path exists from the PCB to the first plate of the capacitor while the second plate is DC connected to the QFR. Note that the entire capacitor is formed entirely on Chip D.

In each of these methods, the capacitance is partitioned in different ways. In Fig. 1a, the entire capacitor is off-chip. In Fig. 1b, the capacitor is split between the two chips. In Fig. 1c, the entire capacitor is formed on one chip. The formation of an on-chip capacitance is prior art and has been done in many analog designs where a capacitor blocks the

0-7803-3177-6 $5.00 © 1996 IEEE

IEEE 1996 CUSTOM INTEGRATED CIRCUITS CONFERENCE

DC components of two circuit structures formed on one chip. However, to the best of our knowledge, no one has used a capacitor as illustrated in Fig. 1c, which blocks the DC component from circuits which are off-chip.

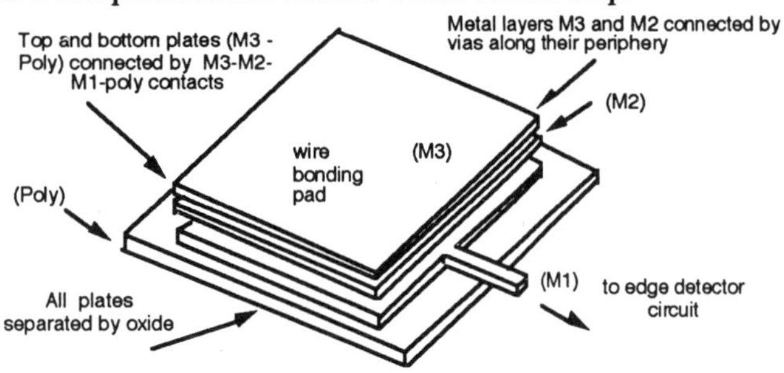

Figure 2. On-chip coupling capacitor under bonding pad.

ON-CHIP CAPACITANCE

Several methods of forming on-chip capacitors exist in the MOS technology. The oxide layer between conducting layers can be used as an insulator. Fig. 2 depicts a way of forming a capacitor in a processing technology using three metal layers. The entire capacitance can be formed under the bonding pad; this is a very efficient use of silicon area. The first plate is formed using the M3, M2 and poly silicon layers. The two top layers of M3 and M2 are connected along their periphery via contacts, and these contacts extend to the poly silicon layer. The M1 layer is sandwiched between the poly and M2 layers and forms the second plate of the capacitor.

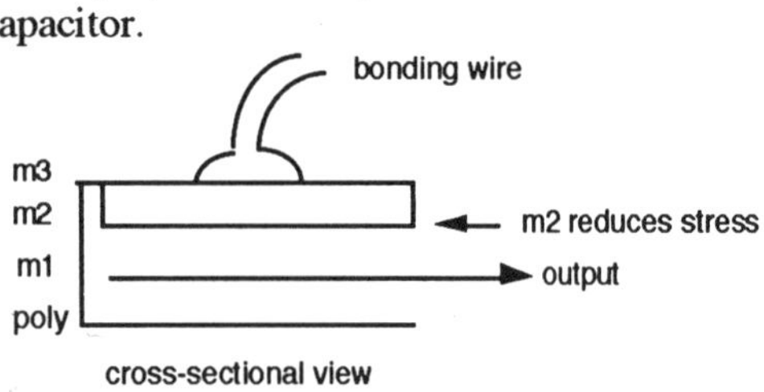

Figure 3. Cross sectional view of the capacitor in Fig.2.

A cross-sectional view of the on-chip coupling capacitor is given in Fig. 3. A M2 layer has been interposed between the M3 and M1 layer to prevent fractures from penetrating down to the M1 layer during the wire bonding process. Those fractures that do occur between the M3 and M2 layers are stopped at the M2 layer. And since the M2 layer has an ohmic connection to M3, these fractures do not deteriorate the integrity of the structure allowing a compact capacitor to be formed directly under the bonding pad. The actual coupling capacitance is the parallel combination of two capacitances. The first is between the M1 plate and the M2 plate above, while the second is between the M1 plate and the poly silicon layer below. The capacitor under the bonding pad optimizes silicon area usage and achieves a capacitance value of about 1.6pF.

QUANTIZED FEEDBACK

When a signal is passed through a high-pass filter, the signal experiences the effect known as "zero wander". This makes the detection of the signal more difficult since the effective "zero" of the signal drifts causing a loss of noise margin. Furthermore, this effect deteriorates as the bit activity rate decreases: this necessitates the need for coding the input signal.

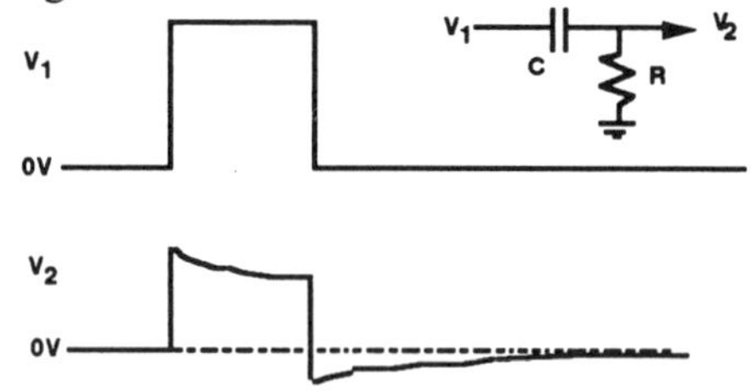

Figure 4. Response of a high-pass circuit to a single pulse.

Fig. 4 illustrates an RC high-pass network and two corresponding timing diagrams. The input waveform (V_1) is a single pulse, while the output waveform (V_2) is shown below. The leading edge of the pulse of V_1 causes the leading edge at V_2, however after the occurrence of this edge, due to the RC relaxation of the circuit (inset), the top of the V_2 waveform decays. When the falling edge of V_1 occurs, the voltage of the V_2 signal drops below 0V (dotted line). Since the output signal decreased below 0V, any decision circuit attempting to capture this pulse will lose noise margin due to this "zero wander" effect. Long-line and transatlantic cable transmission systems have experienced problems detecting signals due to this effect during the first half of this century. Various techniques have been proposed to remedy this situation and most fall into the category of quantized feedback [3-6].

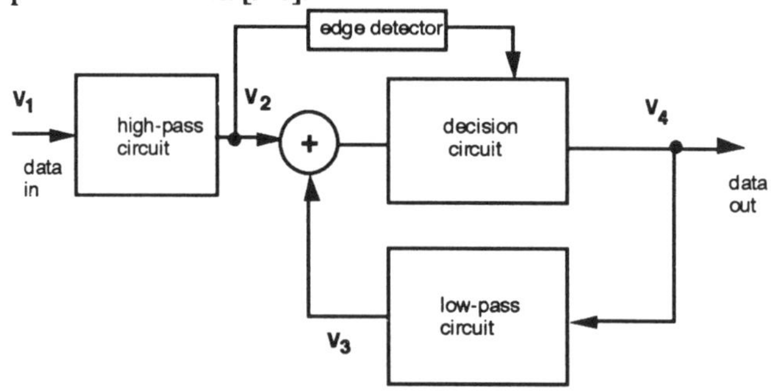

Figure 5. Block diagram of the quantized feedback technique.

The basic idea of quantized feedback is to introduce a complementary exponential signal back into the decaying exponential signal. This would compensate for the exponential decay of the waveform and generate a clean digital signal at its output [5-9]. Fig. 5 illustrates the block diagram of the basic concept of quantized feedback while the corresponding timing diagrams are given in Fig. 6. The signal V_1 (containing both DC and AC) passes through the high-pass circuit generating V_2. The edges of V_2 are sensed by the edge detector which then generates an equivalent of a clock pulse that is applied to the decision circuit. In previous designs, this clock signal was an independent signal which was not derived from the data signal. This is a new feature of our quantized feedback technique. The output of the decision circuit generates a clean digital signal which is then fed back through a low-pass filter to generate V_3 which is then added to V_2. The low-pass feedback circuit has an exponential constant which matches the time constant of the high-pass filter to insure that the summation of the two signals will generate a digital waveform at the input of the decision circuit. The signal at V_4, except for the possibility of amplitude scaling, is the digital representation of the signal at V_1. Thus, quantized feedback can be used in conjunction with the on-chip capacitance (described earlier) to eliminate the "zero wander" effect of the signal on the second plate of this capacitor.

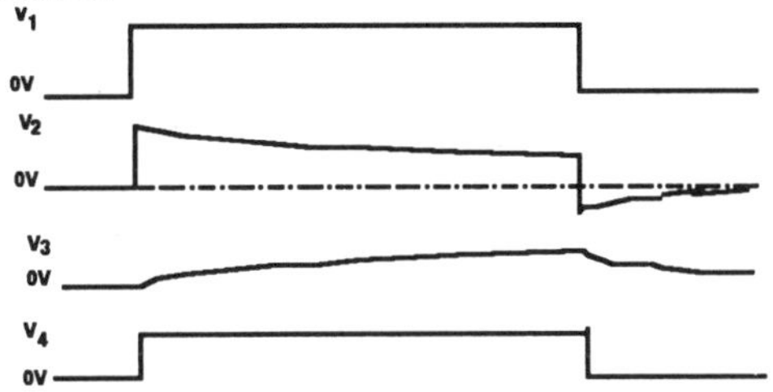

Figure 6. Waveforms for the block diagram of Fig.5.

CMOS CIRCUIT IMPLEMENTATION

Fig. 7 illustrates the CMOS circuit where the function of the edge detector, decision circuit, low pass circuit and the resistive portion of the high pass circuit are combined into one circuit structure. A balanced or differential signal (IN and $\overline{\text{IN}}$, 180 degrees out of phase with one another) is used to activate this cell. The balanced input signal offers a high common mode noise rejection.

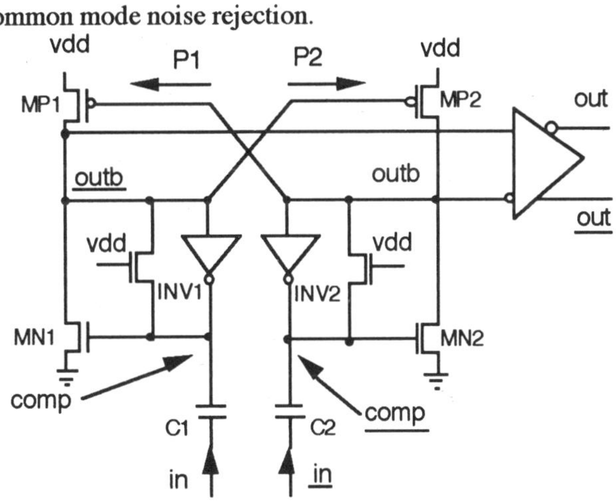

Figure 7. Balanced self-clocked CMOS quantized feedback circuit.

The input signal sees a high pass circuit consisting of the on-chip capacitance (C1 or C2) and the output impedance of the inverter (INV1 or INV2). Both inverters, INV1 and INV2, have a long N channel device connected across them which insures that the voltage swing of the signals at COMP and $\overline{\text{COMP}}$ remains bounded. The edge detector and decision circuit consists of the cross-coupled RAM-like cell consisting of devices MN1, MN2, MP1, and MP2. The positive feedback from each half of the RAM cell (arrowed paths, P1 and P2) cross couples, amplifies and latches the detected edge of the signal at OUTP and $\overline{\text{OUTP}}$. The low pass circuit (feedback path) is formed primarily of the output impedance of the inverter (INV1 or INV2) and the capacitor (C1 or C2). The equivalent location of the summing block of V_1 and V_2 in Fig. 5 is the node (COMP or $\overline{\text{COMP}}$) in Fig. 7. Thus, a local DC signal is immediately established (barring the delay around the feedback path) on the second plate of the capacitors C1 and C2. Note that the exponential coefficients of both the high and low pass filters are matched since they are the same network. A differential amplifier is used to generate the signals OUT and $\overline{\text{OUT}}$.

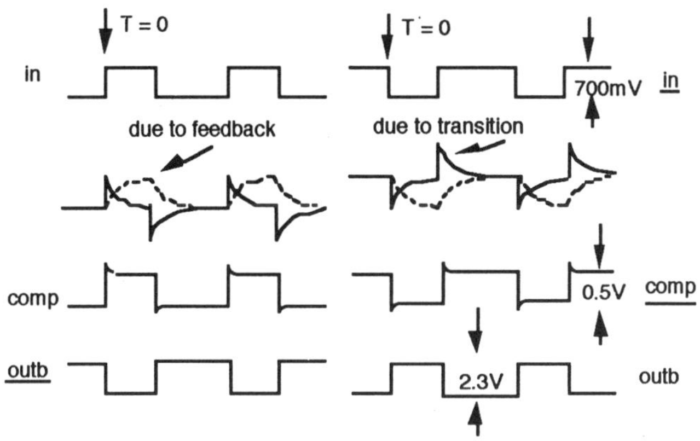

Figure 8. Timing diagrams for the circuit in Fig.7.

The timing waveforms for Fig. 7 are given in Fig. 8. The waveforms are split into two columns where T = 0 indicates the same time point. The top row gives the signals IN and $\overline{\text{IN}}$ each with a voltage swing of 700mV. The second row indicates the signal after passing through the high pass circuit (solid line) while the dotted line corresponds to the signal which is fed back via the low pass circuit. The third row gives the composite of the two signals of the row above or COMP and $\overline{\text{COMP}}$. The last row indicates the signal which is generated at OUTB and $\overline{\text{OUTB}}$.

EASEMENT OF STANDARDS INTERFACE

Note that since the off-chip inputs are AC coupled to the quantized feedback circuit via the on-chip capacitors C1 and C2, the absolute DC values of the input signals are immaterial since they are filtered out. This has advantages since several standards (GTL, PECL, HSTL, LVDS) can be detected by this CCQFB circuit. Several chips each generating different voltage swings can interface directly to a single chip which uses these CCQFR input buffers. Because the DC portion of the input signal is filtered out, these input buffers can operate over a large common mode voltage range.

MEASURED RESULTS

A micrograph of the 0.5μm CMOS testchip is

illustrated in Fig. 9. This chip has CCQF Receivers and open drain N-channel output buffers. A micrograph of the CCQFR buffer is illustrated in Fig. 10. Since this is a balanced input pad, two bonding sites are shown. The capacitors are formed entirely under each bonding pad. The quantized feedback circuit is the circuitry on the right.

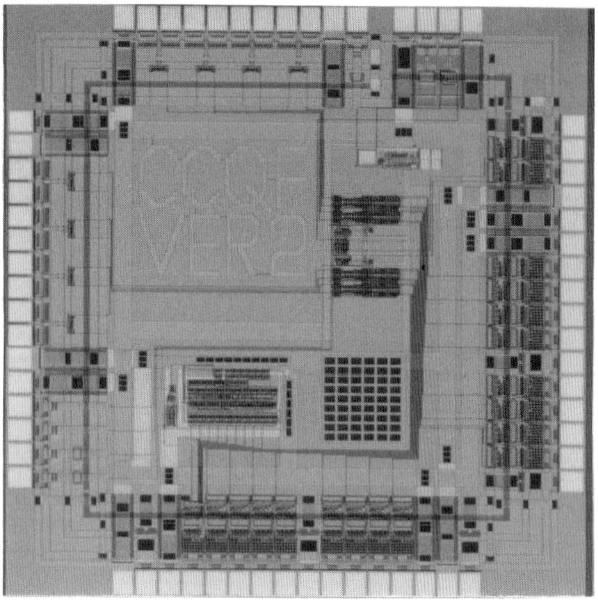

Figure 9. Micrograph of the 0.5μm CMOS testchip.

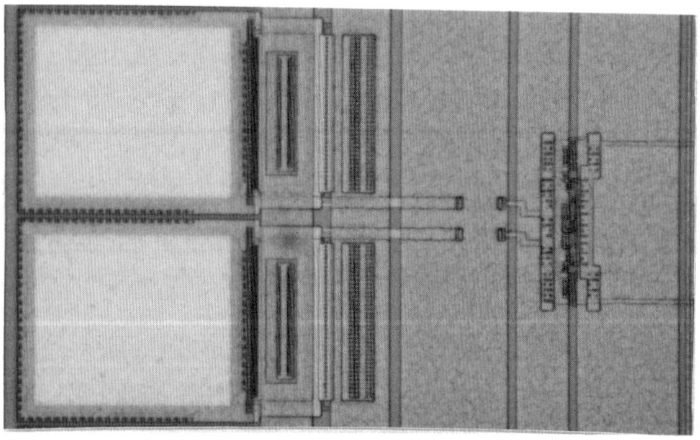

Figure 10. Micrograph of a Capacitive Coupled Quantized Feedback Buffers (capacitance under bonding pad).

The eye diagram for a 2^{31}-1 pseudo-random input sequence is illustrated in Fig. 11. The input signal swing was 700mV. At 600Mb/s and a supply voltage of 3V, no errors were detected. The DC level of the input signal was varied from VSS to (VDD - 0.7V) without detecting errors during the BER measurement.

Another test was made to evaluate the ability of the CCQF receiver to detect low frequency signals; this would demonstrate that coding of the data is not required. An input signal of 700mV was applied at 1Kb/s (1 millisecond period), and the signal was successfully detected indicating that coding of capacitively coupled signals is NOT required as long as the Quantized Feedback technique is used.

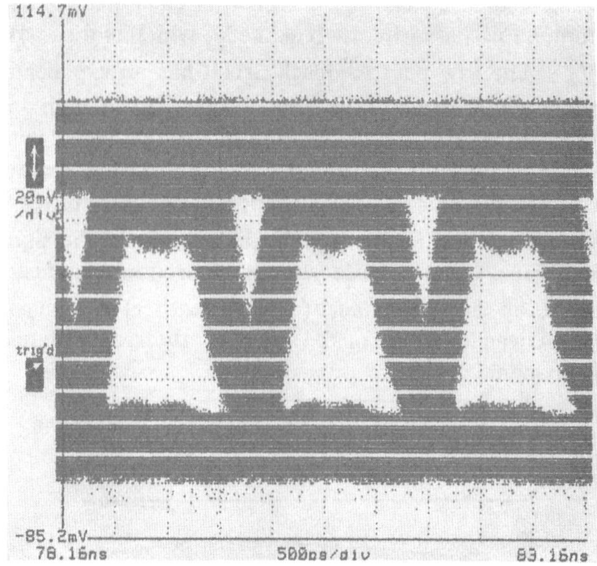

Figure 11. Eye diagram at 600Mb/s for a 2^{31}-1 pseudo-random sequence.

CONCLUSION

A new technique of forming an on-chip capacitance under the bonding pad has been described. Also the technique of Quantized Feedback was introduced which establishes a local DC value back into an AC coupled signal. A balanced receiver CMOS circuit consisting of the on-chip coupling capacitances and the quantized feedback technique has been described. This composite circuit can extract off-chip signals over a wide range of DC voltages and frequencies. The BER measurements (2^{31}-1) operating at 600Mb/s were error-free.

ACKNOWLEDGEMENTS

The authors also appreciate the discussions with Bernie Morris, Al Dunlop, and Paul Henry.

REFERENCES

[1] D. Salzman and T. Knight, "Capacitive Coupling Solves the Known Good Die Problem", IEEE Multi-Chip Module Conference MCMC-94, pp. 95-99, March 15-17, 1994, Santa Cruz, CA

[2] D. Salzman, T. Knight and P. Franzon, "Application of Capacitive Coupling to Switch Fabrics", IEEE Multi-Chip Module Conference MCMC-95, pp. 195-199, Jan 31, 1995, Santa Cruz, CA

[3] R. Mathes, "Signal Correcting System", 1919, US patent 1,295,553.

[4] W. Martin, "Correcting Circuits for High Distortion Lines", 1920, US patent 1,347,049.

[5] Le Roy MacColl," Signaling Method and Apparatus", 1936, US patent 2,056,284.

[6] M. Flavin, A. Gerrish, R. Howson and J. Salz, "Synchronous DC Restoration in VSB Transmission" MM 64-5311-10.

[7] W. R. Bennett, "Synthesis of Active Networks", Symposium on Modern Network Synthesis, Polytechnic Inst. of Brooklyn, pp.45-61, April 13-15, 1955.

[8] AT&T Blue Book, "Transmission Systems for Communications", 4th ed, 1971, Bell Telephone Laboratories, pp. 656.

[9] M. Steinberger and Y. Chang, "A Quantized Feedback Circuit for High Speed Digital Interfaces" unpublished.

A single chip HDD PRML channel

T. W. Ahn, G. Chen, B. Hua, C. J. Kim, D. Knee, A. Kondo, B. Kuo,
I. J. Lan, M. Ng, F. Shen, N. H. Yeh, and J. K. Young

Samsung Semiconductor, Inc., San Jose, CA

Abstract

A PRML (Partial Response Maximum Likelihood) channel chip employing a mixed-signal 0.8 μm BiCMOS process for hard disk drive (HDD) applications is described. This chip performs the complete write, read, and servo functions using only five external components with a data rate of up to 120 Mb/sec. Under normal operating conditions the power dissipation is less than 800 mW. The on-chip circuitry includes ADC, VGA, FIR filter, Viterbi decoder, two PLL's, etc.

Introduction

As areal densities of hard disk drives increase at a 60% annual rate, channels using the PRML concept have become a requirement in all the new drive developments to ensure system integrity (1), (2). A single chip PRML channel is especially attractive due to the reduced overall chip count and the simpler PC board inter-connections in the highly cost-sensitive HDD environment (3), (4). Class 4 partial response (PR4) is adopted in this chip for its close spectral matching with the magnetic recording channel characteristics. The Viterbi algorithm is incorporated in the maximum likelihood sequence detector. A mixed-signal 0.8 μm BiCMOS process with f_t equal to 15 GHz is used to make the chip achieve a data rate of 120 Mb/sec.

Description

The block diagram of the PRML chip is illustrated in fig. 1. The interface with the HDD controller is through a bi-directional eight bit parallel port labeled NRZ. The channel chip is differentially coupled to the write/preamplifier via WDP/WDN for write data and VINP/VINN for read data.

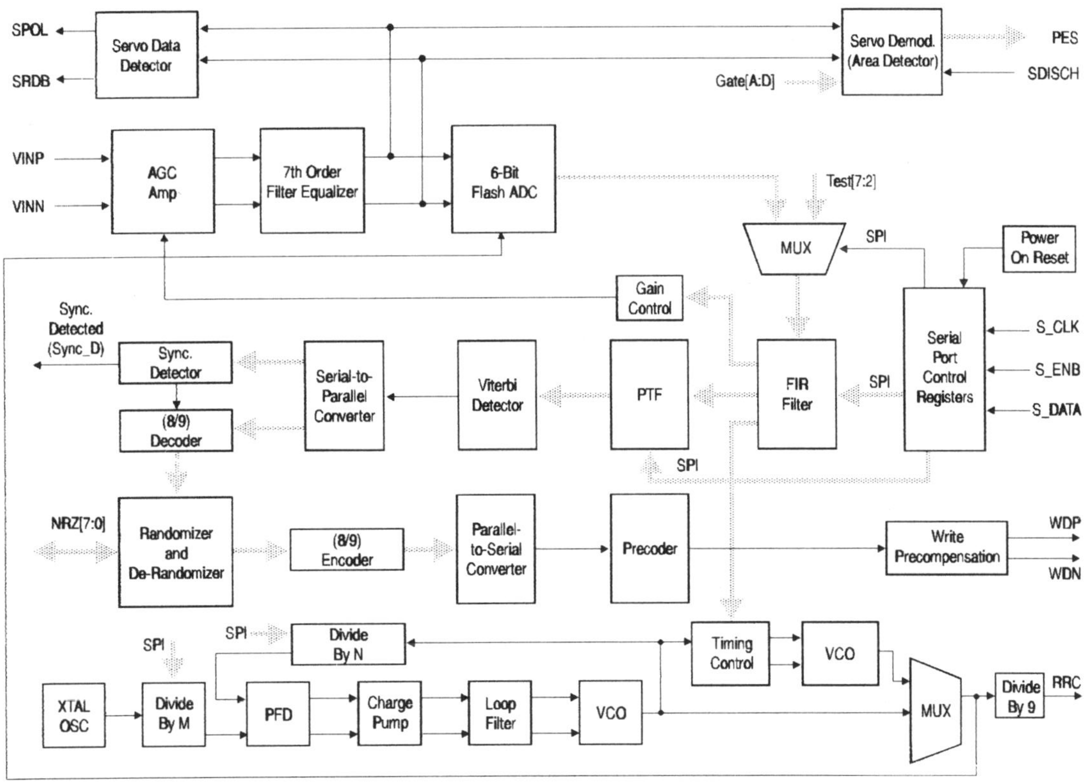

Fig. 1 Block diagram.

0-7803-3177-6 $5.00 © 1996 IEEE

A. Write channel

During the write mode, the data from the controller is first randomized with an internally generated pseudo-random sequence which has a generator polynomial of $x^7 + x + 1$. The pseudo-random sequence is preset to 7f and aligned with the first bit of the sector data. The randomized data passes through a (8/9,0,4/4) block code encoder to introduce a maximum run length constraint. This ensures reliable gain control and timing recovery in the subsequent read mode since no more than four consecutive zeros can occur in the bit stream. It also limits the Viterbi decoder memory depth and the resulting data propagation delay. The write data is precoded by a $1/(1 \oplus D^2)$ circuit to avoid error propagation. Transition shifts in the magnetic recording channel are pre-compensated by a 5 bit programmable delay line. This delay control may shift the transitions in multiples of 200 ps to eliminate the nonlinear interaction of adjacent bits.

B. Read channel

In the read mode, the differential signal from the preamplifier first enters an AGC amplifier. The AGC amplifier has a minimum bandwidth of 105 MHz and 28 dB gain control range. The continuous time filter is a seventh-order equi-ripple filter with two independent zeroes. The filter cut-off frequency is programmable from 10 MHz to 40 MHz. The two zeros may boost the amplitude up to 15 dB at the cut-off frequency. This continuous time filter serves both as an equalizer to the incoming signal and as an anti-aliasing filter for the ADC. All the analog circuitry including the ADC employs differential signal processing to maximize the common mode rejection.

The 6 bit flash ADC uses metal resistor ladders and has a conversion rate of at least 140 MHz. The ladder impedance is approximately 100 ohms per leg. Overflow and underflow are controlled by hard limiting the output. The FIR filter has three taps with the center tap fixed at 128, and both side taps are 7 bits programmable. The combined continuous time filter and the FIR filter outperforms a 5-tap FIR filter in equalizing a magnetic recording channel. An optional DFE based pole tip filter is utilized to compensate the trailing pole tip undershoot from a thin film head. Two Viterbi decoders designed for 1-D channels perform the maximum likelihood sequence detection of the interleaved even and odd bit streams separately (5). The (8/9) block decoder and de-randomizer restore the NRZ data.

C. Servo

A servo data detector and a servo demodulator are also integrated into this chip. The servo data detector detects the peaks and uses a hysteretic threshold level to qualify the peaks. The hysteresis range is adjustable from 10% to 90% of the peak amplitude. This ensures the integrity of alternate polarities for the servo transitions. In the servo demodulator, either an areal detection of the embedded servo burst or the head position error signal (sum and difference of the burst area) is directed to the output. The number of cycles in the burst used for areal integration is also programmable.

Features

Many features are designed into the PRML chip: an automatic insertion and stripping of the sync pattern (see fig. 2); a digitally controlled zero-phase re-start/PLL/VGA; an adaptive FIR filter; adaptive zero controls for the continuous time filter; channel quality monitor; multi-level power management, etc.

The intrinsic sync pattern generated in this chip is 24 bits long and is extremely robust against any bit slip as plotted in fig. 3. It provides a byte synchronization for the (8/9) block decoder. The qualification level for the sync pattern is programmable from a complete match of

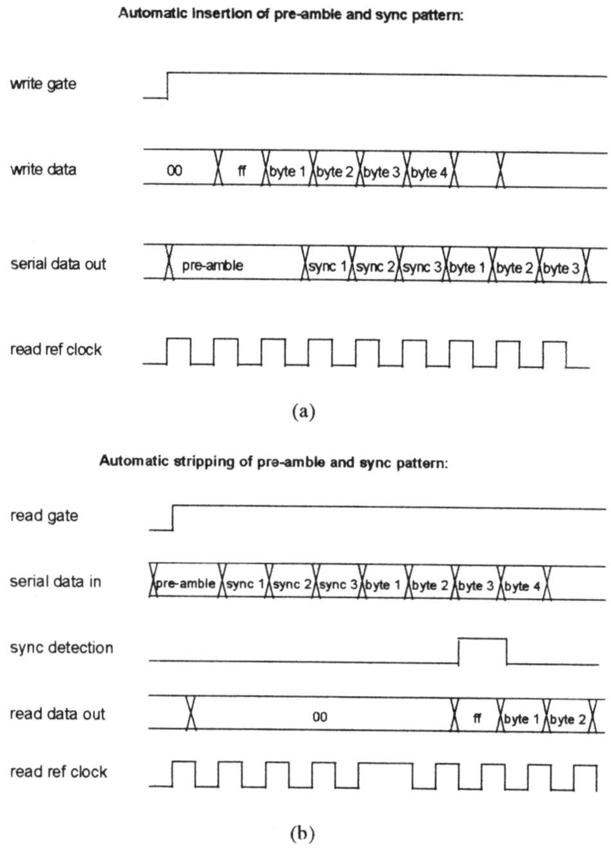

Fig. 2 Automatic insertion (a) and stripping (b) of sync pattern.

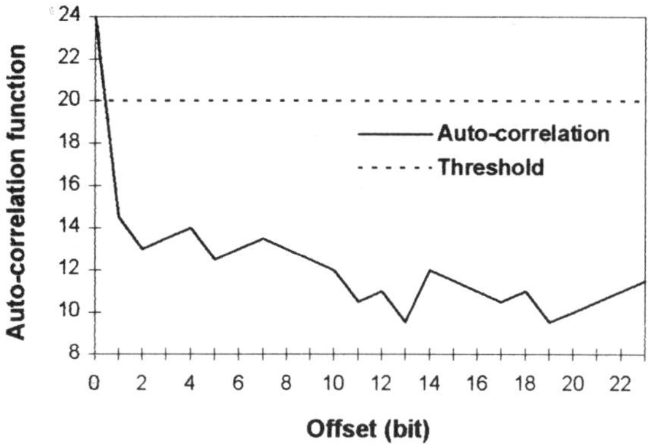

Fig. 3 Auto-correlation function of the 24 bit sync pattern.

24 bits down to only 17 bits. The default is set at 20 bits.

The digitally controlled PLL and AGC employ separate one shot current pulses to charge and discharge the VCO and AGC capacitors at every other bit clock cycle. These current pulses have a fixed duration of 5.8 ns and amplitudes proportional to the digital error signals. Programmable scaling factors for amplitudes of the one shot current pulses furnish frequency independent loop gain controls for both PLL and AGC. Thus there is no need to re-program the loop gain values as the data rate changes from i.d. to o.d. in zone bit recording.

The FIR filter coefficients and the continuous time filter zeros can be adaptively optimized using a built-in zero forcing sign algorithm circuit. The optimized coefficients may be read out through the serial interface port for future re-loading. The channel quality is monitored by accumulating the square error of the equalized signal samples. Results are re-scaled and stored in a 14 bit register. A threshold detector which decodes the tri-level data using two fixed thresholds is also built into the chip. Discrepancies between the Viterbi detector and the threshold detector can be counted and monitored. When the Viterbi decoder is working hard to correct the threshold detector errors, a higher count in this counter is usually observed, indicating a decreased channel margin.

There are seven power down modes in this chip. An additional level of power down mode may automatically disable the digital clocks when any block in the data path is not in use. All the testing and quality monitoring circuits have separate powerdown controls so that they only consume power in the diagnostic mode. The maximum power dissipation of this chip is 1.7 W. By applying appropriate power down modes, the power dissipation during normal operation may be reduced below 800 mW in the read mode.

Testability

Each analog block in this chip can be characterized separately using the analog test port. By-pass mode and probing point selection are controlled by internal programmable registers allowing access to the I/O's of the individual analog block.

In the digital section, partial scan is employed to ensure chip testability with minimal die size overhead. A 99.23% fault coverage is achieved with 50% scan insertion and 241,000 test vectors generated by ATPG. A high speed DAC allows monitoring of the ADC, FIR, pole tip filter or Viterbi decoder output at the operating clock frequency. Various key signals are also available for probing through an eight bit test port. A built-in serial port interface is used to load and download the 32 control registers which provides programmability for most functions of the chip.

Characterization

This chip resides in a 100 pin TQFP. Only five external components are needed to function in a HDD environment: Two capacitors for the AC-coupling from the preamp; two pull up resistors for the PECL write data output; and one 12.1 kΩ bias resistor. The total die size is approximately 55 mm^2 as shown in fig. 4. Substrate guard bands are placed between the digital and the analog blocks to reduce the crosstalk and isolate the noise. Multiple VDD and ground pickups ensure the stability of the supply voltage and the signal ground on all parts of the circuitry.

Fig. 4 A fully integrated PRML chip.

A built-in state machine controls the write and the read operation sequences. All the PLL and AGC loop gains are switched automatically from the acquisition mode to the tracking mode in read. During the acquisition mode, the digital timing and gain error signals are derived from the ADC output to minimize the loop propagation delay. Larger loop gains are employed to improve the acquisition speed without sacrificing the loop stability. In the tracking mode the error signals are derived from the pole tip filter output with reduced loop gains to ensure the accurate tracking of the sampling time and amplitude. The ADC sampling point is deliberately skewed to compensate the group delay introduced by the asymmetric coefficients of the FIR filter so that the transition of the PLL from the acquisition mode to the tracking mode is seamless. Fig. 5 plots the sampled data at the output of pole tip filter in the read process from the acquisition to the tracking mode.

The FIR filter coefficients are updated every eight clock cycles by averaging the signed errors. Although the coefficient convergence may take more than 1000 clock cycles as illustrated in fig. 6., the residual coefficient fluctuations are small at the end of the adaptation period, which is usually one data sector or approximately 5000 bits. The FIR filter coefficient adaptation is recommended for the initial factory calibration and optimization. During normal operation only re-loading of the already optimized coefficients is needed.

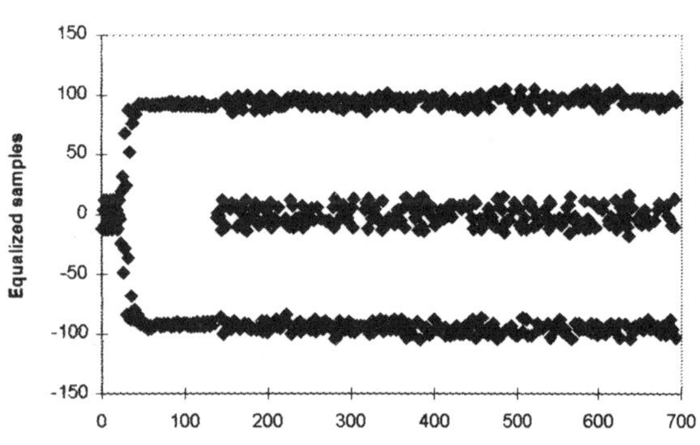

Fig. 5 Samples of equalized signal with seamless transition from acquisition mode to tracking mode.

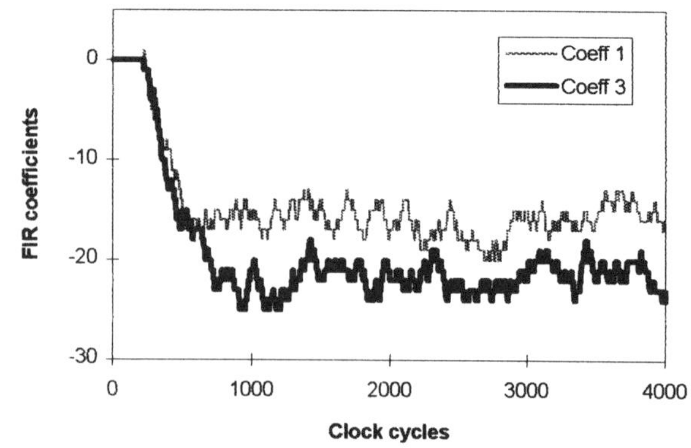

Fig. 6 Adaptive control of FIR filter coefficients.

Summary

A PRML channel chip employing a 0.8 μm BiCMOS process has been developed. Built-in adaptive circuitry allows the self-adjustment of filter parameters to match precisely with head and media characteristics. High fault coverage testing ensures excellent quality control in the mass production environment. Simple user interface enables easy firmware control in the read, write, and servo modes for HDD applications.

Acknowledgment

The authors would like to thank Dr. Yong Eue Park for the support and encouragement of this work.

References

(1) J. D. Coker, et al., "Implementation of PRML in a rigid disk drive," IEEE Trans. Mag., vol. 27, pp. 4538, 1991.

(2) R. D. Cideciyan, et al., "A PRML system for digital magnetic recording," IEEE J. Sel. Areas in Comm., vol. 10, pp. 38, 1992

(3) R. A. Philpott, et al., "A 7 Mbyte/sec, mixed-signal, magnetic recording channel DSP using partial response signaling with maximum likelihood detection," IEEE J. Solid-State Circuits, vol. 29, pp. 177, 1994

(4) R. A. Richetta, et al., "A 16 MB/s PRML read/write data channel," ISSCC Digest of Technical Papers, pp. 78, Feb., 1995

(5) R. W. Wood and D. A. Petersen., "Viterbi detection of class IV partial response on a magnetic recording channel," IEEE Trans. Comm., vol. COM-34, pp. 454, 1986

Chair: Larry Starr
Co-Chair: Sterling Whitaker

Processing information contained in sound, image, motion, physical and mechanical forces provides challenges for integrated circuit designers. This session examines applications and techniques for obtaining high-sensitivity measurements using sigma delta A/Ds, precision amplifiers and other specialized circuits.

The session begins with a review of a current detector circuit that allows accurate signal extraction in a high noise environment. Nano-amper signal sensitivity has been demonstrated over 100Hz to 30KHz bandwidth. The second paper examines a telemetry system circuit for orthopedic implant strain monitoring that exhibits automatic offset cancellation, a ratiometric transfer function, while minimizing component count and power dissipation. The third paper moves into the area of sound sensing by detailing an automatic gain control amplifier designed using a Bipolar/JFET process. The hearing aid application requires the circuit to run on a 1V battery with minimal power dissipation (0.3mW). Paper four explores the use of a Sigma Delta A/D circuit for image sensing, that overcomes previously reported high fixed pattern noise. At 3.3V, the dynamic range is >83db, the dissipation is <1mW with a fixed pattern noise of approximately 1%. The fifth paper describes a resistive bridge sensing application. Use of the sensing circuit in high volume automotive applications requires the circuit to have efficient calibration and test features to minimize costs. The sixth paper looks further into automotive applications by describing a motion detector that can be used for angular rate detection. The interface circuit, processed using standard CMOS with a high voltage module, estimates a linearity error of less that 0.2%. The final paper looks at a monolithic surface micromachined lateral accelerometer with on chip A/D. The common centroid layout of the co-fabricated element and interface circuit results in notable improvements to cross axis motion sensitivity.

The seven papers in this session explore many analog circuit techniques in quest of accurate sensing in diverse applications. Please join us in the review of these leading edge circuits.

A Low Noise Readout Detector Circuit for Nano-Ampere Sensor Applications

Donald C. Thelen Jr. and Dahlon D. Chu*

Microelectronics Research Center
2650 Yale SE, University of New Mexico

*Custom Microcircuits, Sandia National Laboratories
Albuquerque, NM 87185

ABSTRACT

A readout detector integrated circuit (IC) has been developed which is capable of detecting nano-ampere photo-current signals of interest in a high (micro-ampere) background illumination or DC noise level (SNR=92dB). The readout detector sensor IC processes transient signals of interest from a separate photodiode pixel array chip. Low noise signal conditioning, filtering, and signal thresholding implement smart sensor detection of only "active pixels." This detector circuit can also be used to perform signal conditioning for other sensor applications that require detection of very small signals in a high background noise environment.

INTRODUCTION

Biological image systems demonstrate that image processing can be done very efficiently if raw pixel information is processed in parallel to extract only relevant information. This parallel preprocessing dramatically reduces the amount of information that the CPU has to process. Even with the relatively low bandwidths of biological hardware, they are much faster than the inherently serial computation of most digital computers. Reducing the amount of information sent to the CPU also reduces the power consumed by the pad drivers, and makes it easier to keep digital switching noise off the sensitive analog pre-amplifiers. The application of the image processor dictates the relevant information to be collected and processed. Image processing pertaining to edge extraction, motion sensing, texture, or intensity changes is usually concerned with only one or two criteria. For example, finger print analysis could ignore motion, temporal intensity changes and average intensity. Missile tracking can be done without trying to figure out what shape the target is. Paint job inspection may only be interested in texture, and locating lightning strikes might only pay attention to temporal changes in intensity. In our satellite communications application, we need to constantly monitor pixels for flashes of light which have significant energy in the 100Hz to 30KHz bandwidth. When such a signal is detected, the data from that pixel is amplified, low pass filtered, and routed to an AD converter for processing by the CPU. The signals of interest may be as small as 1nA, and have a background illumination as large as 20uA. Sixty-four signal processors must fit on one chip and consume less than two watts for the 64 by 64 array. The chip must also operate in a satellite radiation environment.

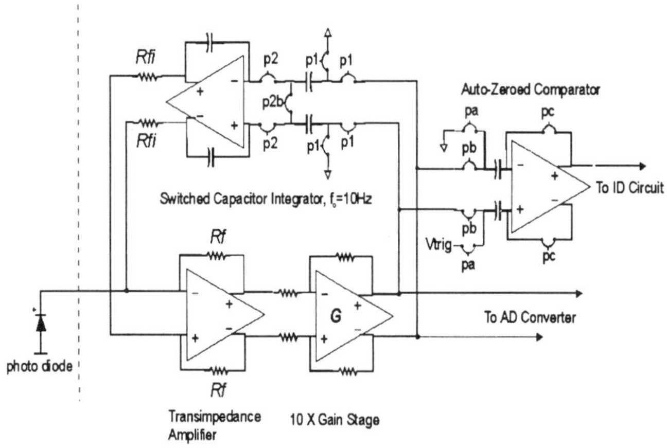

Figure 1. A single pixel readout detector signal processing module.

CIRCUIT DESCRIPTION & SYSTEM REQUIREMENTS

The system requirements of the satellite based sensor resulted in several technical challenges for the implementation of the readout detector electronics.

1) Input referred noise less than or equal to the photodiode shot noise in the hundreds of pico-ampere range.
2) On-chip 100Hz to 30KHz band pass filtering of photodiode signal.
3) Ability to adjust and disable the low frequency (100Hz, 200Hz, 300Hz, etc.) high pass pole.
4) Hold the DC 'background' signal (static background image) subtraction for several seconds.

Figure 1 shows a block diagram of a single pixel readout detector circuit. A transimpedance amplifier converts the photodiode current to voltage which then goes through a gain stage also used as a low-pass filter. Its bias current is set to produce a -3dB 30KHz roll-off. Next, DC and low frequency (< 100Hz) noise components of the signal are removed by a switched-capacitor integrator ('background' subtractor

0-7803-3177-6 $5.00 © 1996 IEEE

IEEE 1996 CUSTOM INTEGRATED CIRCUITS CONFERENCE

circuit) feeding back to the input of the transimpedance amplifier through resistors R_{fi}. Finally, an autozeroed comparator with an adjustable trigger level detects signals of interest. Tripping the comparator causes the background subtract circuit to hold its present value, and permits an ADC to digitize data from this activated pixel. Note that the integrator will not alias signals being fed to the ADC because the S-C integrator clock has been disabled by the comparator. The transfer function of the signal loop is:

$$\frac{V_{out}}{I_{in}} = \frac{sGR_f}{s + GAR_f / R_{fi}} \tag{1}$$

The poles of the circuit loop are located at:

$$\omega_{p1} = \frac{GAR_f}{R_{fi}}, \text{ and } \omega_{p2} = 2\pi(30KHz)rads$$

It must be pointed out that the transfer functions given herein are approximate because the integrator really should be modeled by a z transfer function. The gain stage is set at $G=10$, $R_f = R_{fi} = 50Kohm$, and A is the gain of the switched capacitor integrator. The unity gain frequency of the integrator is set to 10Hz by using a 16pF to 1pF capacitor ratio with a 1KHz sampling clock. This sets the high pass pole of the overall loop to 100Hz. Since the bandwidth of the transimpedance amplifier is also tuned to 30KHz, much of the noise generated by the transimpedance amplifier is aliased to lower frequencies. The 1KHz clock was chosen to keep the switched capacitor ratios small, and is skewed to a 99 percent duty cycle to maximize the time that the circuit looks for signals of interest.

The initial version of the readout detector circuit uses a low noise opamp transimpedance amplifier with p-well feedback resistors. Let's consider what happens to an input signal as it goes through the voltage opamp transimpedance amplifier. First, the difference between input current and feedback current is converted to a voltage on the differential pair. The differential pair is a transconductor which produces a current proportional to its differential input voltage. This current is converted to a voltage by cascode mirroring to a high impedance output. Finally, the output voltage is buffered to lower the impedance, and converted to a current by the feedback resistor in the transimpedance circuit. It would be much simpler if we could just connect the photodiode to a resistor directly, and then buffer the output. This however, can cause low pass filtering of the input current by a parasitic capacitance from the photodiode and the input lead capacitance. A current buffer is needed between the photodiode and the resistor to keep the input voltage more constant, thereby pushing the parasitic pole up in frequency, and letting the output voltage have large voltage swing [1].

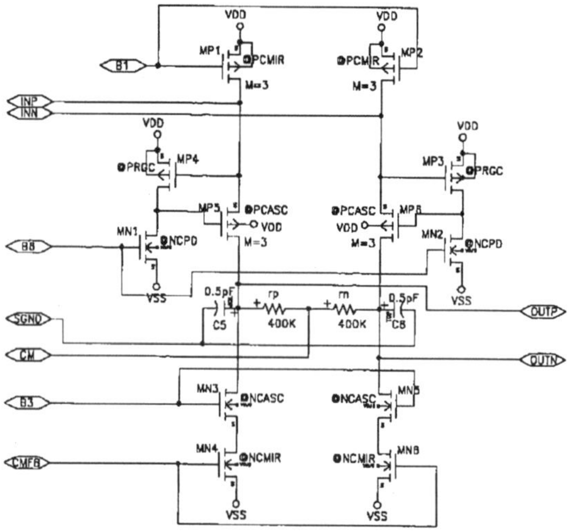

Figure 2. Schematic of an improved current buffer transimpedance amplifier.

An improved current buffer transimpedance amplifier has been designed and fabricated as shown in Figure 2. The noise analysis in the next section shows the superiority of this circuit given our constraints on power consumption and area.

RANDOM NOISE

There are two main sources of noise in the main feedback loop. First, the integrator's noise which is mostly thermal noise (kT/C) from the switches is added to the input through the feedback resistors R_{fi}. Second, the input referred noise of the transimpedance amplifier is added directly to the input signal. The transfer function for the input referred noise of the integrator to the output of the gain stage is:

$$\frac{V_{out}}{V_{noise}} = \frac{GAR_f / R_{fi}}{s + GAR_f / R_{fi}} \tag{2}$$

To refer the integrator input referred noise to the input of the overall loop, we divide (2) by (1):

$$Input\ Referred\ kT/C\ Noise = \frac{A / R_{fi}}{s} \tag{3}$$

Therefore, to minimize the effect of kT/C noise from the integrator, we would like to make the integrator gain A as small as possible, and the feedback resistor R_{fi} as large as possible. We must remember however that the pole at GAR_f/R_{fi} must remain at 100Hz. Dynamic range of the background and input signals also affect the choice of values here. The background current which can be subtracted is limited by V_{intmax}/R_{fi} and the signal is limited by V_{outmax}/R_f, where V_{intmax} is the maximum integrator output voltage, and V_{outmax} is the maximum transimpedance amplifier output.

As pointed out in the last section, the transimpedance amplifier was built two different ways. The first version used a voltage opamp and feedback resistors, R_f. We have found that the input referred current noise associated with the feedback resistors, is 3 orders of magnitude smaller than the noise from opamp. Therefore, the dominant noise source of the transimpedance amplifier is the input referred voltage noise of the opamp divided by the feedback resistor R_f. Making the input referred thermal noise of the opamp low requires the differential pair to have a large W/L and large bias current, and the current mirrors to have a small W/L. Input referred flicker noise requires a large W•L area for the differential pair, and a large L for the current mirrors.

On the other hand, the current buffer shown in Figure 2 has no differential pair, and no feedback resistors. The input referred current noise of this circuit is the mean square sum of the current noise in transistors MP1, MP2, MN4, and MN6. Because thermal noise current is proportional to $1/g_m$, we must make W/L small, and the drain current small to minimize thermal noise. To minimize the flicker noise, the channel lengths of MP1, MP2, MN4, and MN6 must be made long, and their drain currents small. Therefore, optimizing noise performance also complements our efforts to minimize area and power consumption to fit 64 channels on a chip. However, there is a tradeoff with having too small of a bias current in the buffer. If the input current coming out of INP exceeds the bias current sourced from MP1, the input current signal will be clipped. We also want to minimize the input impedance of this current buffer circuit. In our application, the input current is unidirectional (we expect pulses of light, not darkness) so the input stage can actually be used in a class AB fashion. Current coming into the INP pin increases the V_{gs} of MP5, and can be much larger than the bias current. The input impedance of this circuit is lowered by using a regulated gate cascode, which can allow for even more current to flow into the input before clipping. Another benefit of the current buffer is the lack of a complex pole caused by a parasitic input capacitance.

CORRELATED NOISE

Unwanted signals can be added to the photodiode signal of interest. Signals from neighboring circuits can couple through the power supplies, the substrate, and from parallel or overlapping interconnect. Capacitive coupling from parallel and overlapping interconnect were avoided by careful layout, and shielding of sensitive nodes. We also used differential circuits to reject cross talk through the power supplies. A separate power supply connected only to the substrate keeps voltage drops on the main supply lines from modulating the substrate and causing cross talk.

Figure 3. Die photo of the 3 pixel readout detector integrated circuit.

Because the input pads are low impedance, the voltage swings will be small, and minimal coupling between neighboring pixels will occur.

OFFSETS, LEAKAGE, CHARGE INJECTION, AND NONLINEARITY

Close attention was also paid to the cell and chip layout using matched and common centroid layout techniques to reduce input offsets. The ability to hold the background level for several seconds was achieved by using a differential integrator. While the integrator is holding the background current, any leakage current on the switches between the sampling capacitor and the opamp will be seen as a common mode signal, and rejected by the common mode feedback circuit. Charge injection was also reduced to a common mode signal by using delayed switches where the voltage on the switch is input signal voltage dependent [2].

We used p-well resistors in the transimpedance amplifier. These resistors are known to have several problems including a non-linear voltage to current relationship, distributed parasitic capacitance, large process variation, and large temperature coefficient. Using differential circuits cancels the odd order non-linearities of the p-well resistors. The distributed capacitance does not cause sufficient phase shift to cause instability in the opamp type transimpedance amplifier. The gain variation caused by the large resistor process variation and temperature coefficient was compensated for by setting the comparator reference with an external current source and a matched p-well resistor.

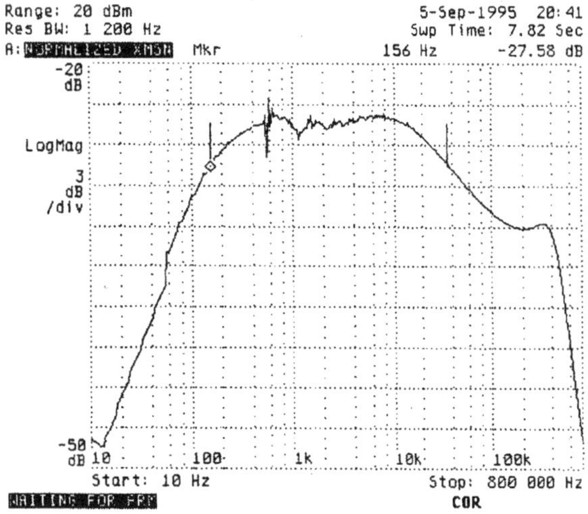

Figure 4. Frequency response of bandpass filter function in the readout detector channel.

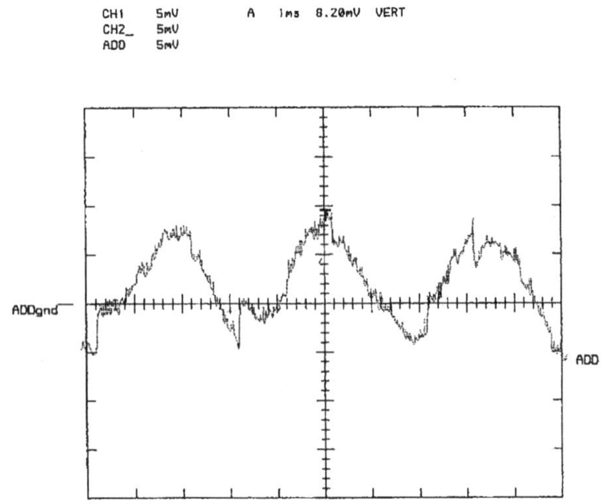

Figure 5. Output signal voltage response from a 4nA input sinusoid with a 1μA DC offset at 300Hz.

MEASURED RESULTS

The chip shown in Figure 3 was fabricated using Orbit Semiconductor's 2μm p-well process. Figure 4 shows the frequency response of the readout detector measured using a HP 3589A spectrum/network analyzer. A bandpass filter response is shown with -3dB high and low pass corner frequencies at 150Hz and 35KHz respectively. The flatband response shown is at -24dB because the input voltage signal is divided by a 1.2 megaohm resistor to generate a current input signal.

Figure 5 shows the detector output for a 4nA input signal with a 1μA DC offset. Note that the 1μA DC 'background noise' has been filtered out. The graininess of the scope plot is due to the operation of the background subtractor and actual system noise.

Input referred noise of less than 250pA rms over a 10Hz to 30KHz bandwidth has been measured in the current buffer transimpedance amplifier. This current buffer-resistor circuit consumes much less power than the voltage mode transimpedance amplifier, i.e. less than 2μA compared with 750μA. This is partially due to the elimination of having an output buffer stage on the new current amplifier.

SUMMARY

We have developed a low noise readout detector circuit suitable for sensing applications that require detection of nano-ampere signals. Moreover, in our satellite communications application, the circuit is able to discriminate transient light pulse signals having a characteristic amplitude and bandwidth out of a large (static) background image. Special circuit techniques were developed to maximize signal sensitivity and minimize noise source contributions. Measurements taken from the chip demonstrate nano-ampere signal sensitivity over a 100Hz to 30KHz bandwidth.

ACKNOWLEDGMENT

The authors would like to thank R. Kay, A. Medina, J. Falls, and M. Knoll from Sandia for system concept development and measurements.

REFERENCES

[1] H. W. Klein and M. E. Robinson, "A 0.8nV/√Hz CMOS preamplifier for magneto-resistive read elements," *IEEE J. Solid-State* Circuits, vol. 29, pp. 1589-1595, Dec. 1994.

[2] K. L. Lee and R. G. Meyer, "Low-distortion switched-capacitor filter design techniques," *IEEE J. Solid-State Circuits*, vol. SC-20, pp.1103-1113, Dec. 1985.

AN INTRACORPORAL TELEMETRY FOR STRAIN GAGE TRANSDUCER

J.B. Begueret, R. Benbrahim, Zhiqun Li, F. Rodes, J.P. Dom

Laboratoire de Microélectronique - IXL - ENSERB
351 cours de la Libération - 33405 TALENCE - FRANCE

Abstract

A new strain gage bridge to PPM (Pulse Position Modulation) converter has been developed. It consists of a chopper amplifier and a voltage to time converter associated in a feed-back loop. This topology exhibits mainly an automatic offset cancellation capability, a ratiometric transfer function and an accuracy of 10 bits. Therefore, this converter is insensitive to offset drifts, and is well suited for long term monitoring of orthopedic implants.

Introduction

Long term monitoring of strain in orthopedic implants, calls for an intracorporal circuitry with low component count, low power drain, high accuracy and low drift characteristics. We have designed and tested a new topology which fulfills these above mentioned requirements.

It consists of a strain gage bridge to PPM converter, which exhibits the transfer function of a special instrumentation amplifier, but with fewer components, and an automatic offset cancellation capability. This circuitry is powered by an 1 MHz inductive supply. Consequently, a voltage regulator, especially designed to eliminate the effects of the 1 MHz radiations, provides a 4 V supply. Hence, this arrangement have a low drift characteristic, and is well suited for long term monitoring.

Principle of operation

The diagram of the intracorporal circuitry is shown in Fig. 1 and Fig. 2. It includes a strain gage bridge to PPM converter which modulates the amplitude of a 180 MHz Radio Frequency (RF) transmitter.

A. Converter

The principle of operation of the strain gage bridge to PPM converter can be explained qualitatively with the diagram of Fig. 1 and the theoretical chronograms of Fig. 3. Assuming that the strain gage is stretched (causing Δl deformation), it produces on the wheatstone bridge output a positive differential voltage :

$$V_{AB} = \frac{V_{DD}}{1+\frac{X}{2}}\left(\frac{X}{4}\right) \quad \text{with} \quad X = \frac{\Delta R}{R} = F\frac{\Delta l}{l} \qquad (1)$$

Furthermore, if we also assume that the switches S_1 and S_2 are on position 1, then the amplifier's output (V_{OA}) settles to a high level voltage (V_{OAH}) and the comparator's output (V_{OC}) is high. Meanwhile, the output of the inverter forces the integrator's input to zero, and, as the integrator's positive reference (non inverted input of A_3) is biased at $V_{DD}/2$, a negative input current is pulled from the summing node of A_3, causing the integrator's output (V_{OI}) to ramp up toward the comparator threshold, equal to the high level amplifier's output voltage (V_{OAH}). As soon as the integrator's output (V_{OI}) reaches this high level threshold voltage, the comparator's output (V_{OC}) goes down to zero, while the output of the inverter rises to V_{DD}. The result is a sign inversion of both :

- the slope of the ramp generator
- the input voltage applied to the amplifier, due to the commutation of switches S_1 and S_2

Therefore, as shown in Fig. 3, the amplifier's output (V_{OA}) settles to its low level value (V_{OAL}), while the integrator's output (V_{OI}) ramps down toward the new comparator low level threshold voltage, equal to V_{OAL}. And, as soon as the integrator's output (V_{OI}) reaches this low level threshold voltage, the comparator's output (V_{OC}) goes high : the initial state is found again. Therefore, the above sequence repeats itself indefinitely. Consequently, the circuit of Fig. 1 is a free running oscillator, where the integrator's output (V_{OI}) ramps up and down symmetrically inside the high and low level thresholds of the comparator, producing a "double slope ramp". Meanwhile, the outputs of the comparator (V_{OC}) yield two square waves with a 50% duty cycle.

In order to produce the narrow pulses of the PPM signal, the output of the inverter drives a standard monostable multivibrator.

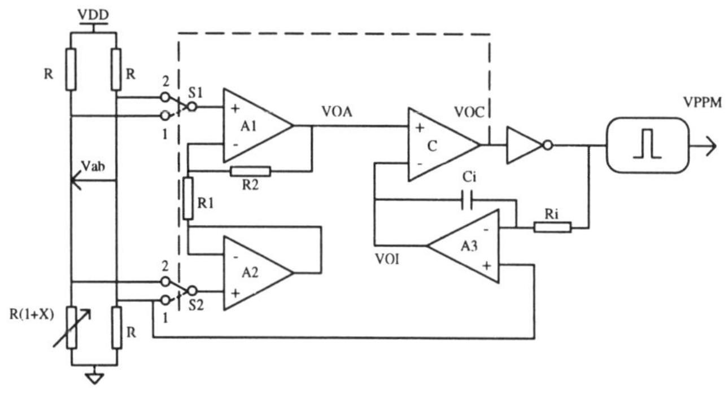

Fig. 1 : Diagram of the intracorporal strain gage bridge to PPM converter

0-7803-3177-6 $5.00 © 1996 IEEE

B. Radio Frequency transmitter

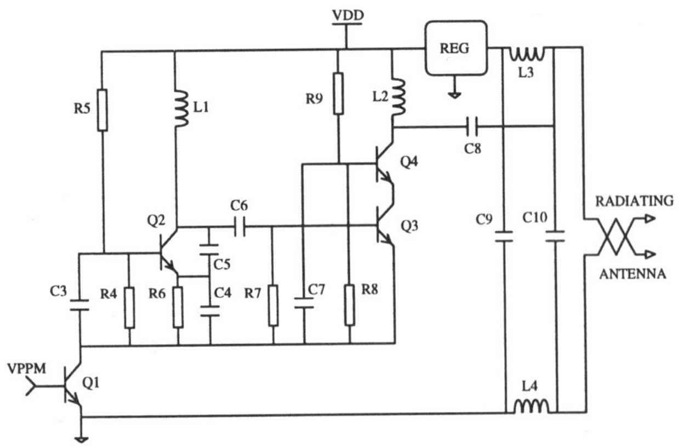

Fig. 2 : Diagram of the intracorporal RF transmitter

This PPM signal drives the switching transistor Q_1, which turns on or off the supply voltage of the RF transmitter, producing a 100% Amplitude Modulation (AM) of the 180 MHz carrier. The oscillator made with Q_2, L_1, C_4, C_5, R_4, R_5 and R_6 is a common base Colpitts [1].

This RF transmitter has the particularity of using its own power supply wires (coming from the inductive power supply) as a radiating antenna. Therefore, a coupling network (C_8, C_9, C_{10}, L_3, L_4) has been designed in order to allow the RF signal and the DC power supply current to share the same wires. To avoid frequency shifts caused by antenna deformations, produced by the patient's movements, we used an isolation circuit between the RF oscillator and the antenna, made with a class C cascode amplifier (Q_3, Q_4, L_2, R_7, R_8 and R_9).

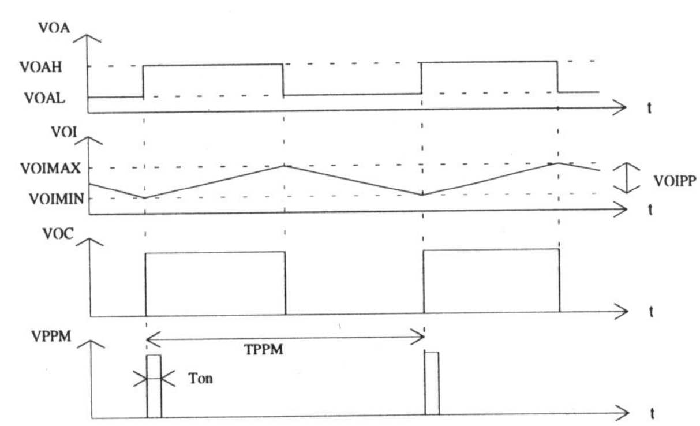

Fig. 3 : Chronograms of the strain gage bridge to PPM converter

Quantitative approach

The transfer function and the most important properties of this strain gage bridge to PPM converter can be derived by a quantitative analysis, where we have considered the influences of the offset voltages V_{OS1}, V_{OS2} and V_{OSC} of respectively the amplifiers A_1, A_2 and the comparator C.

The PPM output period (T_{PPM}) can be computed knowing that it is the time taken by the integrator's output (V_{OI}) to ramp up and down. Therefore, the positive peak and the negative peak of the ramp must be determined in order to find the peak to peak ramp amplitude.

The positive peak ramp voltage can be calculated when the switches S_1 and S_2 are on position 1, yielding a high level voltage on the amplifier's output (V_{OAH}). The positive peak ramp voltage (V_{OIMAX}) is equal to the high level threshold of the comparator, and is given by :

$$V_{OIMAX}=\frac{V_{DD}}{2}+V_{OS2}+V_{OSC}+\left(1+\frac{R_2}{R_1}\right)\left[\frac{V_{DD}}{1+\frac{X}{2}}\cdot\frac{X}{4}+V_{OS1}-V_{OS2}\right] \quad (2)$$

Similarly, the negative peak ramp voltage (V_{OIMIN}) can be calculated when the switches S_1 and S_2 are on position 2, yielding a low level voltage on the amplifier's output (V_{OAL}). The negative peak voltage of the ramp (V_{OIMIN}) is given by :

$$V_{OIMIN}=\frac{V_{DD}}{2}+V_{OS1}+V_{OSC}-\frac{R_2}{R_1}\left[\frac{V_{DD}}{1+\frac{X}{2}}\cdot\frac{X}{4}+V_{OS2}-V_{OS1}\right] \quad (3)$$

At this stage we can note that the expressions of V_{OIMAX} (2) and V_{OIMIN} (3) are heavily plagued by the offsets V_{OS1}, V_{OS2} and V_{OSC}.

The peak to peak ramp amplitude (V_{OIPP}) is the difference between the positive and the negative peak voltage of the ramp :

$$V_{OIPP}=\left(1+2\frac{R_2}{R_1}\right)\frac{V_{DD}}{\left(1+\frac{X}{2}\right)}\left(\frac{X}{4}\right) \quad (4)$$

It is important to note that the peak to peak ramp amplitude given by (4) is no longer affected by the offset voltages of the amplifiers (A_1, A_2) and of the comparator (C). These offset voltages have been canceled by the cyclic inversion of the input signal performed by the switches S_1 and S_2.

Therefore the circuit section including the switches S_1 and S_2, the amplifiers (A_1, A_2) and the comparator (C), have properties similar to the ones of a chopper amplifier or an AC amplifier.

Finally, we can compute the PPM period (T_{PPM}) knowing that the output voltage slope of the Miller integrator is constant and given by :

$$\left|\frac{\Delta V_{OI}}{\Delta t}\right|=\frac{V_{DD}}{2R_iC_i} \quad (5)$$

Substituting in (5) Δt by $T_{PPM}/2$ and ΔV_{OI} by V_{OIPP} given by (4), yields the complete transfer function of this strain gage bridge to PPM converter :

$$T_{PPM}=4R_iC_i\left(1+2\frac{R_2}{R_1}\right)\frac{\frac{X}{4}}{\left(1+\frac{X}{2}\right)} \quad (6)$$

Circuit design

The 5 kΩ strain gage has been chosen to minimize the consumption of the bridge. A standard variation of the strain gage causes a deviation of ±20 Ω. This deformation implies a voltage $V_{AB}=\pm 4$ mV at the input of the instrument amplifier. Consequently, given the low level input voltage, special three stages operational amplifiers, including cascode and level shifting stages [2], have been designed, as shown in Fig. 4, with high open loop gain in order to assure a sensibility of few μvolts at the input. To minimize the input equivalent noise, very large PMOS transistors have been choose in the input differential pair. Table I presents the specifications required for the amplifiers.

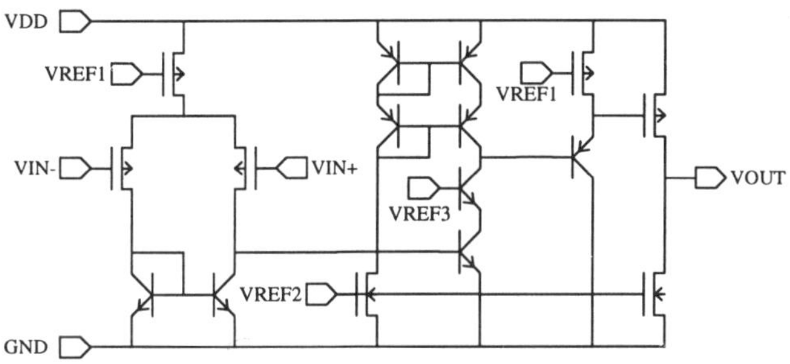

Fig. 4 : Schematic of the operational amplifier

Table I : Main amplifier specifications

Parameter	Value
gain	145 dB
GBW	6.3 MHz
PM	50°
Power dissipation	280 μW
Supply voltage	4 V
PSRR	-80 dB
CMRR	-80 dB

In the same way, the gain and the slew-rate of the comparator have been theoretically performed to have the required accuracy [3].

In order to minimize the fluctuations caused by the inductive power supply, the 4 V voltage regulator [4] has been designed with the characteristics shown in Table II :

Table II : Measured characteristics of the voltage regulator

Characteristics	Conditions	TYP
Line regulation	6 V<V_{IN}<12 V	0.08%
Load Regulation	100 μA<I_L<14 mA	14 ppm/mA
T.C.		170 ppm/°C
I_Q	V_{OUT}=4 V	0.53 mA
1 MHz Rejection		-43 dB

Furthermore, when considering the size minimization, the term $(1+2R_2/R_1)$ of (6), identical to the one of a classical instrument amplifier (IA), is obtained by a special circuitry with much lower component count : 2 operational amplifiers and 2 precision resistors are only necessary for the amplifier section of the PPM converter. On the other hand, the classical IA requires 3 operational amplifiers and 7 precision resistors [5].

The complete circuit has been designed and successfully tested in a single chip ASIC (2 μm BiCMOS technology) which integrates all the active circuitry of the strain gage bridge to PPM converter of Fig. 1 and Fig. 2. The microphotography of this ASIC is shown in Fig. 5.

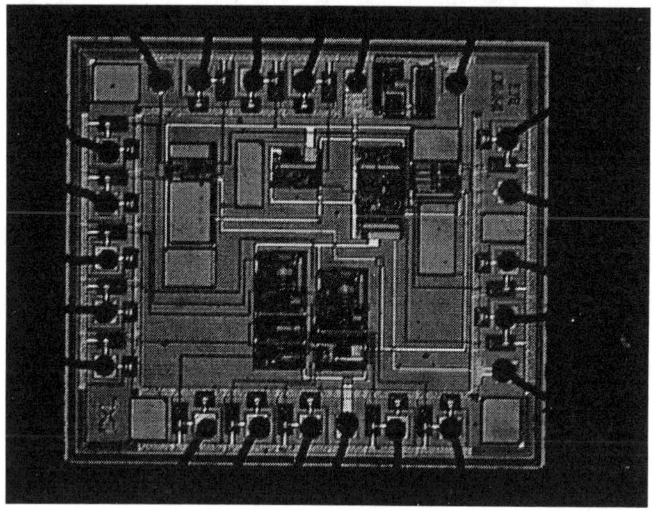

Fig. 5 : Micro-photography of the ASIC chip integrating all the active circuitry of the strain gage bridge to PPM converter (size 1.9 mm x 2.3 mm)

We designed an hybrid circuit allowing all the intracorporal circuitry (the ASIC and the associated external components) to fit into an implant cavity measuring 20 mm x 14 mm x 5 mm.

Results and discussion

The transfer function (6) clearly shows some advantages of this converter :

1- The offset voltages of the amplifier and of the comparator are canceled, and therefore, this PPM converter is insensitive to offset drifts, and is well suited for long term monitoring.

2- The PPM period is independent of the power supply V_{DD}. Therefore, this converter is ratiometric : it can reject the power supply variations. This feature is very attractive for an intracorporal application, where very often, the supply voltage can fluctuate.

3- The PPM period is independent of the pulse duration T_{on}, hence the "double slope" topology eliminates this source of error, encountered with single slope versions.

4- The PPM period is defined between two consecutive transitions of the PPM output signal (V_{PPM}) having the same sign. Consequently, the choice of the RF receiver is not critical : a receiver having an AM demodulator with different rise and fall time can be used.

5- The automatic offset cancellation of this PPM converter is performed by a "chopping frequency" which is also the PPM output frequency (see Fig. 3). Therefore, no intermodulation can be generated between the chopping frequency and the PPM frequency.

Fig. 6 is an oscilloscope screen plot of the 4 most significative waveforms of the converter : Channel 1 represents the amplifier's output voltage V_{OA}, Channel 2 the integrator's output voltage V_{OI}, Channel 3 the comparator output voltage V_{OC} and Channel 4 the PPM output signal V_{PPM}.

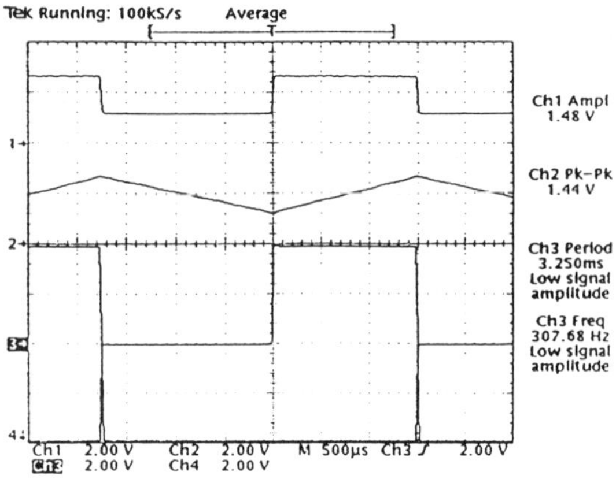

Fig. 6 : Oscilloscope screen plot of the strain gage bridge to PPM converter

One can notice that these experimental waveforms are very close to the theoretical ones of Fig. 3.
When producing theses waveforms, the PPM converter draws only 3 mA, when powered by the 4 V voltage regulator.

Fig. 7 is an experimental plot of the transfer function of the strain gage bridge to PPM converter which yields the PPM period (T_{PPM}) versus the strain gage resistance variation. Within any stretch, the strain gage resistance is R_0=5015 Ω and so, the reference PPM period is T_0=3.33 ms. For the complete deviation of the strain gage, the sampling frequency is between 200 Hz and 600 Hz. This experimental plot is in good accordance with the theoretical transfer function of the PPM converter given by (6). The obtained accuracy is 10 bits and the precision is 0.1%/month.

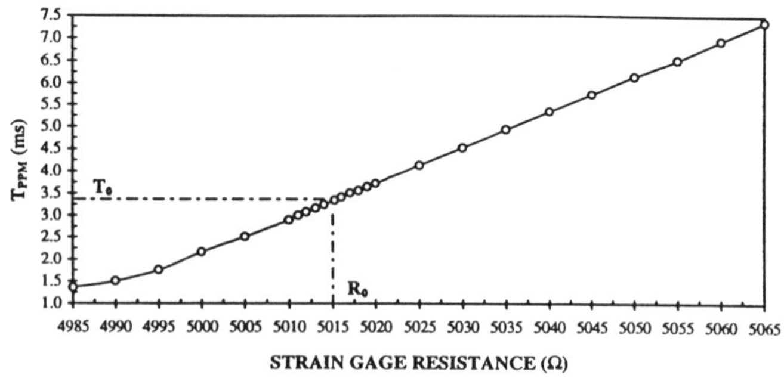

Fig. 7 : Experimental plot of the transfer function of the strain gage bridge to PPM converter

Conclusion

We designed and tested an intracorporal telemetry system using a strain gage bridge to PPM converter, associated with a RF transmitter, optimized for long term monitoring of stainless steel orthopedic implants.

This solution proved to be a good compromise in terms of accuracy, drift, power consumption and complexity.

Moreover, the strain gage bridge to PPM converter exhibits some specific advantages such as :

- automatic offset cancellation.
- ratiometric conversion law.
- low power consumption.
- low component count.
- small size circuitry using low cost technology.

Therefore, this PPM converter can find applications in other low power telemetry systems.

Bibliography

[1] W. H. Hayward, *Introduction to radio frequency design*, Prentice-Hall, New Jersey, 1982, pp. 265-279.
[2] R. Gregorian, G. C. Temes, *Analog MOS integrated circuits for signal processing*, Wiley & Sons, New York, 1986.
[3] P. E. Allen, D. R. Holberg, *CMOS analog circuit design*, Saunders, HBJ, New York, 1987, pp. 322-349.
[4] A. C. van der Woerd, W. A. Serdijn, R. H. van Beynhem, and R. J. H. Janse, "Low-power current mode 0.9 V voltage regulator," *IEEE J. Solid-State Circuits*, vol. 29, no. 9, pp. 1138-1141, September 1994.
[5] A. Basak, *Analogue electronic circuits and systems*, Cambridge, New York, 1985, pp. 80-85.

A BiFET Automatic Gain Control Amplifier for Hearing Aids

Frederick E. Sykes,
Stephen Armstrong.
Research and Development
GENNUM Corporation
P O. Box 489, STN A, Burlington,
Ontario, CANADA

ABSTRACT

This paper describes an audio band automatic gain control (AGC) amplifier implemented in Bipolar/JFET(Vp=0.3V) technology and is targeted for high performance hearing aid applications. A unique AGC topology was used where theAGC function is split into feedback and feedforward sections. The amplifier provides variable compression ratios from 1:1 to 4:1 compression, 4 pole bandsplit filter, and dualtime constant average signal detection. The single IC consumes only 0.3mW from a 1V supply.

INTRODUCTION

A significant number of hearing impaired people suffer from what is known as loudness growth [1]. That is the inability to hear quiet sounds and yet have no increase in the tolerance level of loud sounds, resulting in a reduction in the dynamic range of hearing. An AGC circuit can be used to compress the input signal into the remaining dynamic range of the person. But each person is unique in their particular hearing loss with the reduction in dynamic range often frequency specific, with different and varying degrees of loudness growth existing for low frequency and high frequency sounds. The transition from the different sensitivities can be rapid or gradual over the frequency range. The proposed AGC provides better matching to the specific persons loss by providing two controls over compression ratio, one for each frequency band, in order to compensate for the loudness growth. Since a complete hearing aid system is required to fit completely in the ear canal, small size and minimum use of externals is also a requirement.

A unique approach to splitting the AGC function into feedforward and feedback sections is presented which minimizes the circuit complexity, number of external components required, and the power dissipation. Audible temporal artifacts are minimized via a special twin average detector working with the two AGC sections. All circuitry has been optimized for 1V operation with very low operating current.

ARCHITECTURE

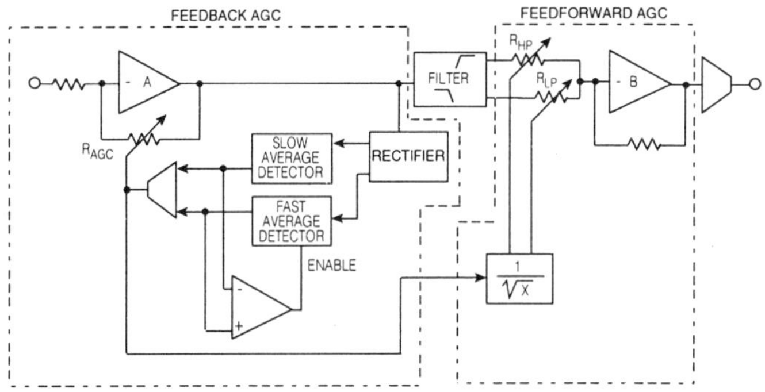

Figure 1. AGC architecture with feedback and feedforward components

The architecture of the automatic gain control amplifier* for customized loudness growth compensation is shown in Fig 1. The system is composed of an audio signal path and AGC control path. The audio signal progresses through a single ended amplifier with a current controlled variable resistor feedback element followed by a filter that splits the audio signal into low frequency and high frequency bands. In each band there is another AGC amplifier whose compression ratio is externally adjustable. The frequency bands are summed and sent to the output speaker driver.

The front end of the AGC control loop consists of a halfwave rectifier, two average detectors, and a current controlled variable resistor. The control from the front end AGC is used to control the band specific AGCs composed of a $1/\sqrt{x}$ block and two more current controlled variable resistors [2].

The front end AGC amplifier is low noise to allow AC coupling directly to the microphone. This amplifier has a fixed 2:1 compression ratio to improve the maximum input signal handling and minimize noise components from the following filter, resulting in a wide dynamic range system. Feedback control is used to improve AGC precision. The variable resistor is implemented with schottkys that provide over 2 decades of controllable impedance range.

* patent pending

0-7803-3177-6 $5.00 © 1996 IEEE

IEEE 1996 CUSTOM INTEGRATED CIRCUITS CONFERENCE

The filter is a unity gain 4 pole state variable filter with highpass and low pass outputs.

The rectifier is a half wave rectifier with current output,which is integrated to form the average of the input signal. A fast average detector is used to rapidly adjust the gain in the presence of intense but short duration transients in the input, the slow average detector is used to slowly adjust gain to track the environmental level of the input. This topology, which is called a twin average detector* provides smooth gain adjustment to adapt to the environment while still providing the protection to the user from intense short duration sounds[3].

The average detectors are each composed of an integrator and voltage to current (V/I) converter. The V/I converter takes the voltage from the integrator capacitors, converts them to current and provides amplification to set the proper loop gain. A V/I exists for each average detector and control is determined by a decision making block. The output currents are summed and fed into the variable resistor block where this control current defines the impedance.

This control current is also fed into the $^{1}/_{\sqrt{x}}$ block where two new control currents are generated for the high pass and low pass AGC variable resistors. The compression ratio is set via an external potentiometer for each frequency band. The adjustment range of 1:2 expansion to 2:1 compression, added to the fixed 2:1 compression of the first amplifier results in an overall system compression ratio of 1:1 to 4:1 compression independently controllable in each frequency band as illustrated in Fig 2.

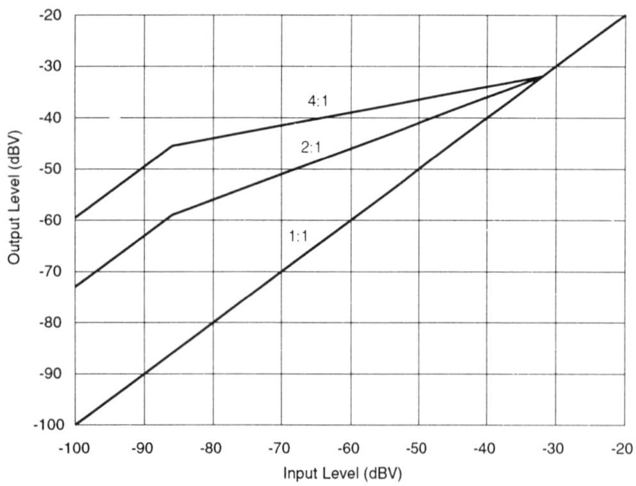

Figure 2. Input/Output response for various choices of compression ratio

The second amplifier sums the frequency bands and provides amplification in front of the output driver amplifier.

* patent pending

CIRCUITS

Front End Amplifier

The front end amplifier is an inverting amplifier with AGC obtained via the adjustment of the feedback resistor. The feedback resistor is made from a bias network of back-to-back schottky diodes, where adjustment of the bias current controls the terminal resistance. The maximum amplifier gain is 23dB and the minimum gain is -4dB, providing a compression range of 27dB. As seen in Fig 2 , the amplifier is linear until the lower threshold of -86dBV is reached at which point 2:1 compression starts and continues until the -32dBV input level. Compression is therefore active over a very wide input range which encompasses most of the sound levels of speech encountered by a hearing aid user.

This amplifier is also effective at reducing the dynamic range requirements of the following blocks, thc most critical one being the filter which is inherently noisy.

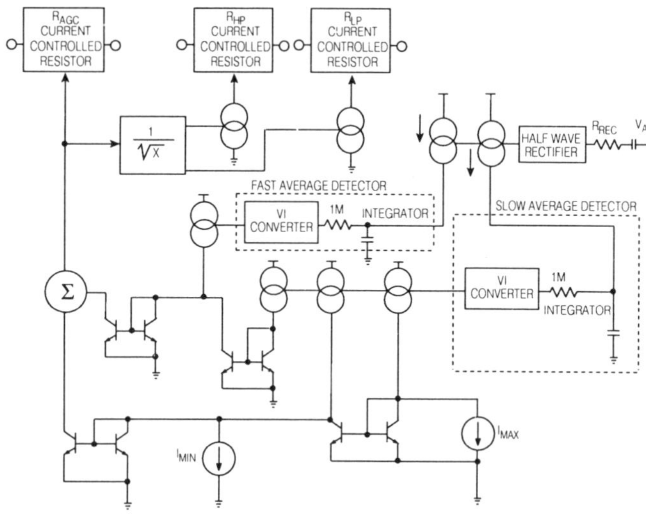

Figure 3. AGC control loop including feedback & feedforward components

Feedback AGC control

The complete AGC control circuit composed of the feedback and feedforward sections is shown in Fig 3. The feedback AGC control block is composed of the rectifier, integrators, V/I converters, threshold references, and the R_{AGC} variable resistor.

The rectifier senses the output from Stage A and converts the negative cycles into positive current pulses, which are injected onto the grounded time constant capacitor. The voltages on these capacitors are converted to a current through a V/I converter, which has an input impedance of 1MΩ. This converts the capacitor voltage to a nearly DC current that is proportional to the average energy of the input signal.

For a sine wave input the average current is given by

$$I_{AVG} = K \frac{V_{AOUTpk}}{\pi R_{REC}} \qquad (1)$$

The loop gain K is chosen to generate the desired I/O curve. Having two current outputs integrated on capacitors of different sizes results in a fast average and slow average detector. The 1MΩ input impedance with the capacitor values defines the time constants of the fast and slow detectors to be 10ms and 220ms respectively. Average detection is chosen over peak detection because of the theory that perceived loudness of a sound is based on its RMS energy [4]. Since speech and music have highly variable ratios of RMS to peak energy, a peak detector becomes a very poor detector of loudness. An average detector is a good practical approximation to an RMS detector. A current comparator is used to sense the output of the fast V/I and compare it to the slow V/I output. If at any instant in time the fast detector current exceeds the level of the slow detector current by 6dB, the AGC control will be transferred to the fast detector. This will only happen if a large rapidly rising transient sound occurs. Once the transient disappears, control is rapidly returned, as defined by the fast release time of 10ms, to the slow detector. This approach designated "Twin Average Detection", provides undistorted signal amplification based on the long term average of the sound environment, while still providing protection from loud sounds that significantly exceed the ambient level for short periods of time.

The current references of I_{MIN} and I_{MAX} define the lower and upper compression thresholds. I_{MIN} also sets the gain at inputs below the lower threshold.

Feedforward AGC Control

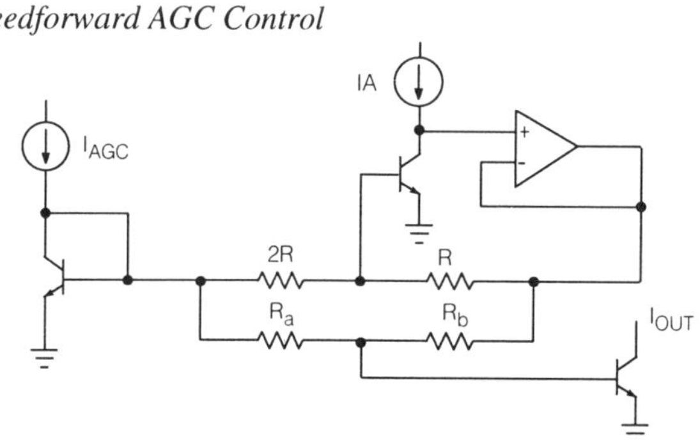

Figure 4. 1/X circuit to create variable compression or expansion functions

The $^1/_{\sqrt{X}}$ block, also called the compander circuit, in Fig 4, takes the control current of the front end AGC as an input and has the following current output:

$$I_{OUT} = \frac{Ia^m}{I_{agc}^{m-1}} \quad \text{where } m = \frac{3}{2}\,\frac{Ra}{(R_a + R_b)} \qquad (2)$$

This controls another pair of variable impedance blocks (R_{HP} and R_{LP}), one for each frequency band. Since these resistors are the input resistors to the virtual ground of an inverting amp, the bands are summed in inverse proportion to R_{HP} and R_{LP}. This gives the gain change for either band to be:

$$\Delta Gain = (\Delta V_{AOUT})^{(\frac{2}{N} - 1)} \qquad (3)$$

where N is the compression ratio and is related to the resistor ratio in the $^1/_{\sqrt{X}}$ block by m=2/N. The $^1/_{\sqrt{X}}$ block uses the exponential relationship of the transistors to implement the variable power function. ∞:1 compression is easily obtained by changing the feedback resistor to 2R, but to reduce current consumption a practical limit of 4:1 was placed on this implementation.The use of a single AGC control loop for all three variable resistors has dramatically simplified the circuitry as well as prevented temporal artifacts in the audio output.

Filter

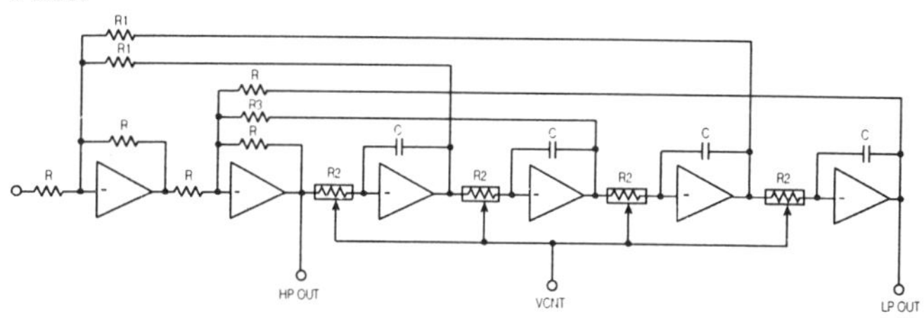

Figure 5. 4th order Linkwitz-Riley Filter

The schematic for the 4th order state variable filter is shown in Fig 5. A Linkwitz-Riley transfer function is used to provide the lowpass and high pass outputs with an adjustable crossover frequency [5]. Crossover adjustment is done by using JFETs for the four integrator resistors with a common gate control voltage [6]. The integrator caps are included in the IC layout. The low Vp JFETs facilitate operation at 1V with a control range of 2 octaves. The highpass output noise is 24μV, the low pass is 12μV but by having the filter follow Stage A which has 23dB of gain, the input referred noise is not dominated by the filter.

EXPERIMENTAL RESULTS

The die layout incorporating the AGC system as well as an output stage and regulator is shown in Fig 6. Transient performance is illustrated by Fig 7 where the input started at a -65dBV 1kHz level and is followed by a 1s burst at -40dBV. Note the characteristic envelope of the Twin Average detectors. This bump is where control is returned back to the slow detector from the fast detector.

Fig.8 and 9 show some of the possible frequency characteristics that can be obtained by adjusting the compression ratio in each frequency band. Fig. 8 is the frequency responses for the input levels of -85dBV, -65dBV, -45dBV, and -30dBV, with the setup conditions of 4:1 compression in the high band and1:1 compression in the low band. Fig. 9 has the same input levels with the setup conditions changed to 1:1 in the highband and 4:1 in the low band. Fig. 8 represents a response known as TILL, Treble Increases at Low Levels. Fig. 9 represents a BILL response, or Bass Increases at Low Levels.

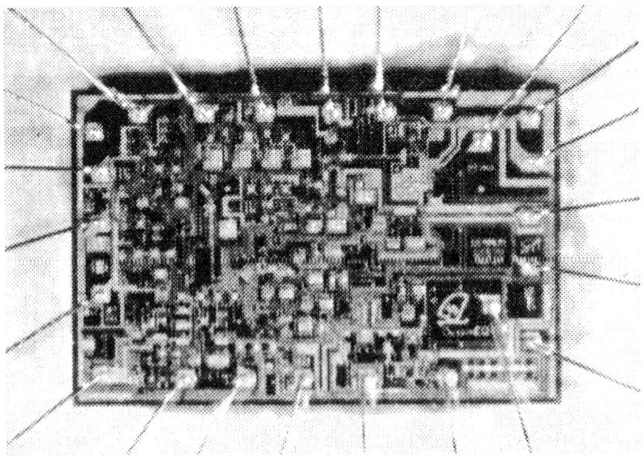

Figure 6. Die layout of AGC amplifier system

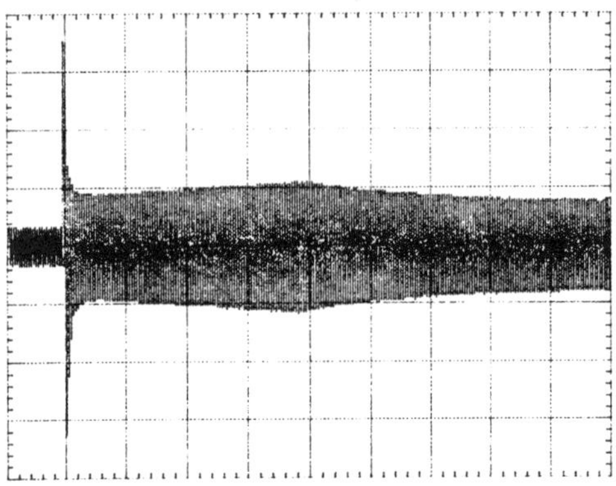

Figure 7. Transient envelope of AGC in response to a 1kHz burst of 1s duration

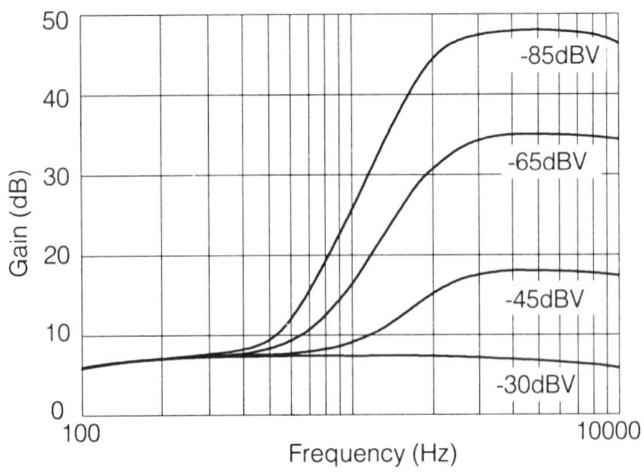

Figure 8. Measured frequency response for various input levels. Low band 1:1, High band 4:1 compression

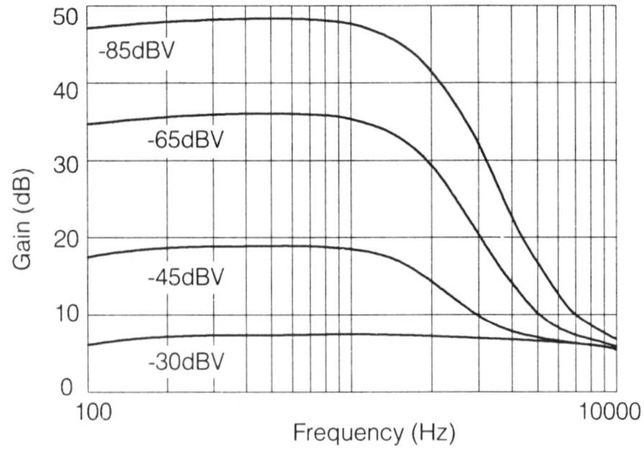

Figure 9. Measured frequency response for various input levels. Low band 4:1, High band 1:1 compression

CONCLUSIONS

A hearing aid AGC amplifier has been developed in a Bipolar/JFET process. The input referred noise (at maximum gain) is 3μVRMS, maximum system gain is 49dB with a 40.5dB compression gain range (at 4:1 compression), the compression ratio is independently adjustable in the low and high frequencies from 1:1 to 4:1. The amplifier can run off a 1V battery and consumes only 0.3mW. The approach of splitting AGC into front and backend compression has improved system flexibility and at the same time reduced circuit complexity.

ACKNOWLEDGEMENTS

The author would like to thank Ron Csermak for his technical guidance.

REFERENCES

[1] Mead Killion, Selda Fikret-Pasa, "The 3 types of sensorineural hearing loss: loudness and intelligibility considerations",The Hearing J.,vol 46, no11, pp.31-36, November 1993.

[2] Stephen Armstrong, Frederick Sykes, Ron Csermak, "Multi-channel synchronous companding system", Patent Pending.

[3] Stephen Armstrong, "Twin Average Detector", Patent Pending.

[4] Steve Armstrong, "The dynamics of compression: some key elements explored", The Hearing J.,vol. 46, no 11, pp. 43-47, November 1993.

[5] S.H/ Linkwitz, "Active Crossover networks for non coincident drivers", JAES, vol 24, no. 1, pp. 2-8, Jan 1976.

[6] Hans Wallinga, Klaas Bult, "Design and analysis of CMOS analog signal processing circuits by means of a graphical MOST model", IEEE J. Solid State Ccts.,vol 24, no 3, pp.672-679, June 1989.

A 128×128 Pixel CMOS Area Image Sensor with Multiplexed Pixel Level A/D Conversion

David X.D. Yang Boyd Fowler Abbas El Gamal

Information Systems Laboratory, Stanford University, Stanford, CA 94305-4055

Abstract

A 128×128 pixel CMOS area image sensor with a sigma-delta A/D Converter shared within each group of 2×2 pixels is described. Each pixel comprises a photodiode and 4 MOSFETs and occupies 20.8μm×19.8μm with a fill factor of 30% in a 0.8μm three layer metal one layer poly CMOS process. At 3.3V, the dynamic range is > 83dB, the dissipation is < 1mW and the fixed pattern noise is ≈1%.

Introduction

In digital imaging applications it is desirable to integrate A/D conversion with an area image sensor on the same chip. Such integration can significantly reduce power consumption and system cost. A number of approaches for integrating A/D conversion with image sensing have been developed [2, 3, 5, 4]. These approaches use CMOS technology because integration of sensing, A/D conversion and digital circuits on the same chip is difficult to accomplish with other image sensing technologies such as CCD. In [4] we proposed a pixel level A/D conversion approach using sigma-delta ($\Sigma\Delta$) modulation. The main advantage of such an approach is the use of very simple A/D conversion circuitry operating at low frequencies—in the kilohertz range—independent of the size of the sensor array. The low frequency also results in low power dissipation. An array of 64×64 pixels was fabricated in a 1.2μm two layer metal CMOS process. Each pixel consisted of a phototransistor and a first order $\Sigma\Delta$ A/D converter implemented using 22 MOS transistors. This approach had a large pixel size of 60μm×60μm and a low fill factor of 3%, relative to other image sensor approaches. In addition, the sensor suffered from unacceptably high fixed pattern noise of 10% due to beta variation of the phototransistors and MOS transistor mismatch in the $\Sigma\Delta$ A/D converters.

In this paper we describe an area image sensor that overcomes these limitations. To reduce pixel size and improve fill factor we share the $\Sigma\Delta$ A/D converter among four neighboring pixels through multiplexing. In addition more transistors are saved by using an analog shift register for implementing the D/A converter. In the original design 8 MOS transistors were used for the D/A converter compared to only 3 for the new design. Finally, by using a photodiode instead of a phototransistor, fixed pattern noise is reduced. These improvements result in a reduction of the pixel size by a factor of 4, an increase in the fill factor to 30%, and a reduction of fixed pattern noise to 1%. Sharing circuit components in a pixel block not only reduces pixel size but also slightly increases the A/D converter uniformity. These improvements make our sensor approach comparable in density and image quality with other CMOS image sensors with integrated A/D conversion while achieving significantly lower power.

The remainder of this paper is organized as follows: first we describe the chip architecture of the 128×128 area image sensor. Then we discuss the pixel block design and implementation. Finally we present measurement results from our area image sensor.

Chip Architecture

Figure 1 shows a functional block diagram of the area image sensor. The chip consists of an array of 64×64 multiplexed pixel blocks, each consisting of a group of 4 nearest-neighbor pixels sharing a single A/D converter. Each pixel block is connected to a bit line using a pass transistor. The bit lines are "read" using a 6:64 row address decoder and an array of 64 digital sense amplifiers. The topology is identical to read only memory (ROM) circuits. Each pixel block converts the analog light intensity into a digital code. The entire system is synchronous and after each clock pulse every pixel block produces one bit of data. This generates a two dimensional array of bits or a "bit plane". One frame of video data consists of L bit planes, where L is the oversampling ratio.

Since we have selected to use pixel-level A/D conversion, each A/D converter must be constructed using a very small amount of silicon area. The A/D converter's size is inversely proportional to the number of pixels in

0-7803-3177-6 $5.00 © 1996 IEEE

IEEE 1996 CUSTOM INTEGRATED CIRCUITS CONFERENCE

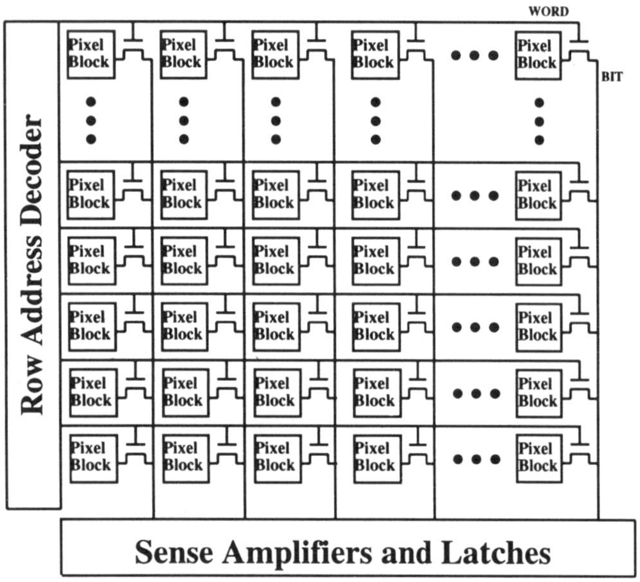

Figure 1: Image Sensor Chip Functional Block Diagram

the sensor, and to the fill factor of each pixel. Conventional Nyquist A/D conversion techniques require too much silicon area. We found that oversampled A/D conversion methods provide a reasonable compromise between size and functionality. Specifically, we chose first order one-bit $\Sigma\Delta$ A/D conversion because it is simple, and robust to process variation and system noise.

Figure 2 shows a first order one-bit $\Sigma\Delta$ A/D converter [1]. It consists of an integrator and a one bit quantizer. The $\Sigma\Delta$ A/D converter generates data at L times the Nyquist rate. The oversampled data is decimated to the Nyquist rate using a digital low pass filter. The resulting SNR of the Nyquist rate data can be computed as

$$\mathrm{SNR} = 9\log_2 L - 5.2\mathrm{dB}.$$

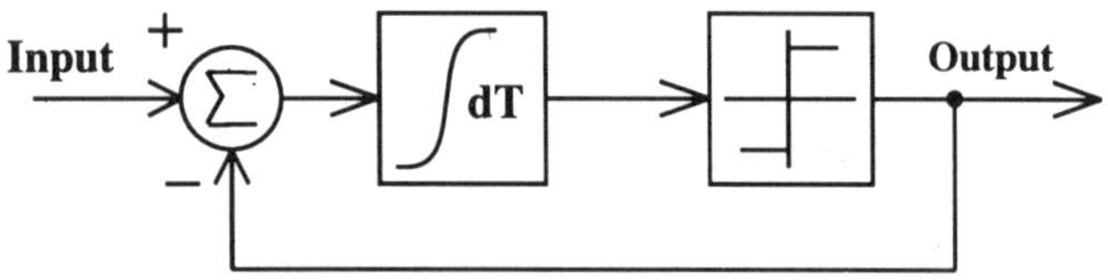

Figure 2: $\Sigma\Delta$ A/D converter.

Pixel Design

A circuit schematic of the multiplexed pixel block is given in Figure 3. Each photodiode in the multiplexed pixel block is formed by using an n+ diffusion in a p substrate. The photodiodes are reverse biased and exposed to light, while the rest of the circuitry is covered with the third layer of metal to reduce the chance of photon induced latch-up.

The multiplexed pixel block operates as follows. The pixels are selected one at a time using the select signals **S0, S1, S2,** and **S3**. (Figure 4 shows the control signal waveforms when pixel D0 is selected.) $\Sigma\Delta$ A/D conversion is performed on the photocurrent generated by each pixel. This current is integrated by the photodiode junction capacitance. The junction voltage of the selected photodiode is quantized using a comparator with regenerative feedback clocked via **CK**. The comparator is biased subthreshold in order to reduce power and noise, increase gain, and reduce any leakage current in the D/A converter. The quantized value is converted to a current using a 1 bit D/A converter and fedback to the photodiode. The 1-bit D/A is implemented by an analog shift register similar to a 3-phase CCD transfer structure. After a sequence of 3 control pulses **DUMP, STORE, CK** (as shown in Figure 4), a fixed amount of charge is dumped on the junction capacitance if the output of comparator **FEEDBACK** is low. The voltage of **STORE** can be used to control the amount of charge transferred to the junction capacitance. Therefore, electronic shuttering can be performed by adjusting **STORE**'s voltage. At the completion of each clock cycle a single bit is generated and read out using the bit line **BIT**.

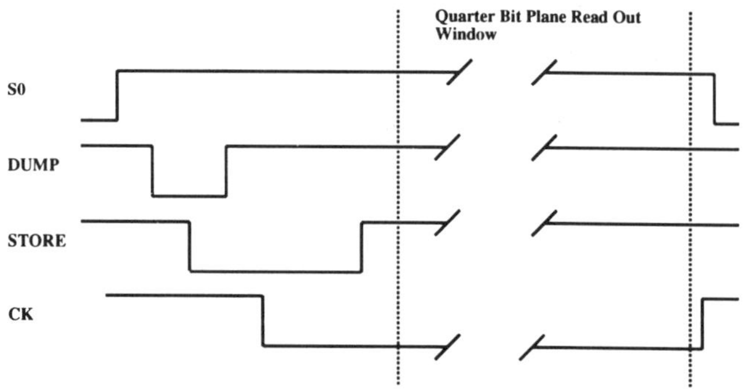

Figure 4: Control Signal Waveforms.

The area image sensor chip operates as follows: after an image is focused on the chip, the $\Sigma\Delta$ A/D converters are globally clocked at a rate F_s well above the Nyquist image frame rate, $2F_d$. First the top left pixel (D0) in each pixel block is selected by asserting **S0**. At the end of the clock cycle the outputs of the $\Sigma\Delta$ converters form a 64×64 array of bits, referred to as a "quarter bit plane", which is read out one row at a time. The quarter bit planes corresponding to D1, D2, and D3 are then read out in the same manner. The four quarter bit planes form a single "bit plane" of the image. The image is fully captured using a number of bit planes determined by the target SNR [4]. SNR degradation due to charge injection

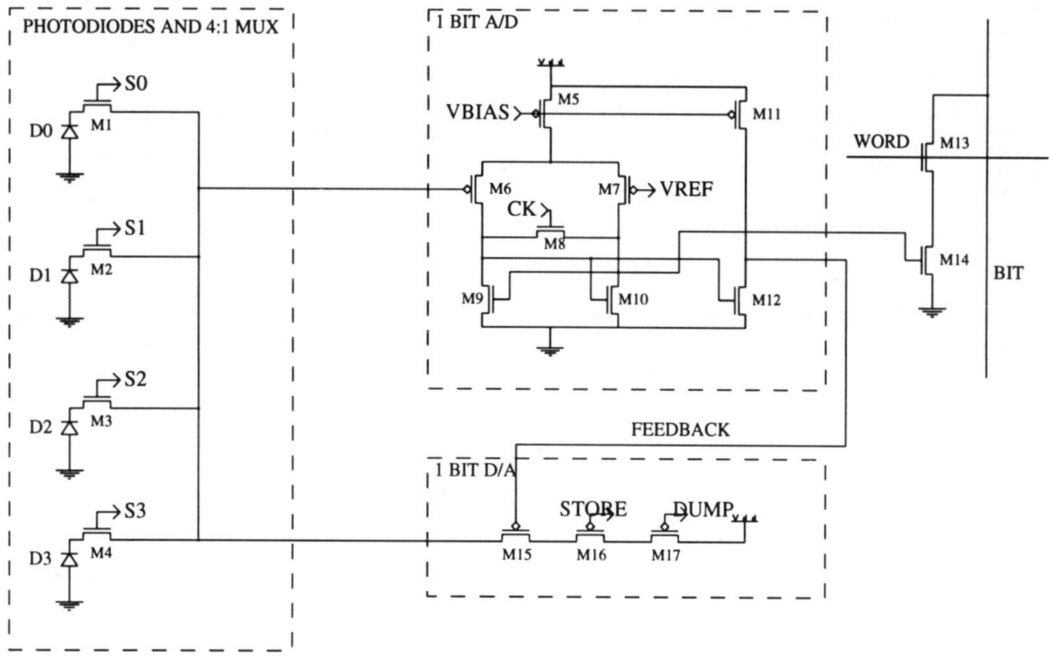

Figure 3: Circuit Schematic of Multiplexed Pixel Block.

of the digital circuitry in close proximity to the analog sensors is negligible since the frequency of operation—kilohertz range—is very low.

Measured Results

To characterize the new area image sensor we designed and fabricated a 128×128 pixel sensor using a 0.8μm three layer metal single layer poly CMOS technology. Figure 7 shows the die photo. Figure 5 shows the measured output from a single pixel's $\Sigma\Delta$ converter as a function of time. To reconstruct the digitized pixel value a decimation filter [1] is used. This is performed externally in software. For example, the lower graph in Figure 5 shows 3 'ones' in 8 clock cycles; if we use a counter as a simple decimation filter, the pixel output is 3/8 of the maximum output value.

Figure 6 was obtained by the sensor in room light. The images are focused onto the chip using an 8-mm lens.

In the process of switching between the multiplexed pixels, charge sharing occurs. This effect has been simulated and results show that charge sharing does not smear the image, on the contrary it performs slight edge enhancement.

The sensor achieves a dynamic range[1] greater than 83dB. This is the case since the magnitude of the feedback D/A converter can be varied by a factor of 40dB, and the maximum measured SNR at a frame rate of 30Hz and an oversampling ratio of 64 is approximately 43dB. The fixed pattern noise was calculated to be approximately 1% using measured results from the D/A converter and photodiode.

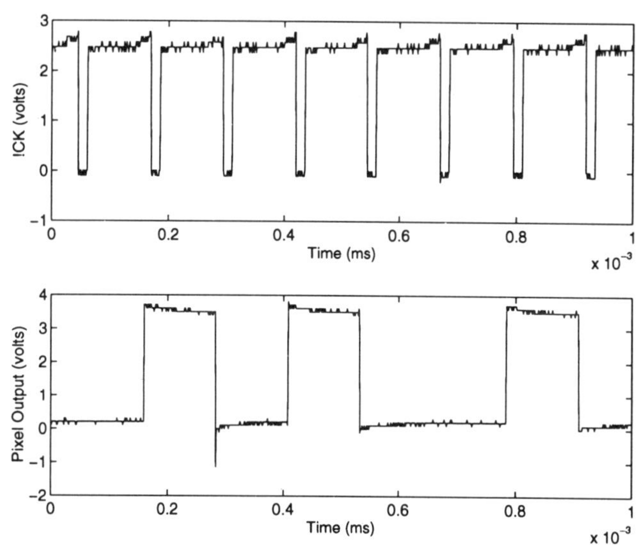

Figure 5: Single pixel $\Sigma\Delta$ A/D converter output from HP54601A. The top graph is **CK** bar and the bottom graph is the pixel output verses time.

[1] The ratio of the maximum non-saturating photocurrent to the dark current.

Figure 6: Image Obtained by the Sensor.

The sensor, without pads, consumes less than 1 mW. The static current of each pixel is approximately 1nA, and the dynamic current for the entire array is less than 200μA. We summarize the main characteristics of the chip in Table 1.

Table 1: Area Image Sensor Characteristics

128x128 Area Image Sensor Characteristics	
Technology	0.8 μm, 3-layer metal, 1-layer poly, nwell CMOS
Die Area	3337 μm × 3200 μm
Pixel Area	20.8 μm × 19.8 μm
Transistors per pixel	4.25 (17 per four pixels)
Fill Factor	30%
Package	65 pin PGA
Supply Voltage	3.3 v
Dynamic Range	83 dB
Dissipation w/o Pads	$<$ 1 mW
Fixed Pattern Noise	$\approx$ 1%
Measurement Temperature	21°C

Conclusion

We have demonstrated a CMOS area image sensor with multiplexed pixel level A/D conversion. With a fill factor of 30% and a pixel size of 20.8μm×19.8μm, our sensor is comparable in density and image quality with other CMOS image sensors with integrated A/D conversion while achieving significantly lower power. Additionally, our sensor topology scales well with CMOS technology advances. For example, the speed of each A/D converter is independent of the sensors size.

Acknowledgments

We thank M. Godfrey for his contribution to the test bed, and H. Elabd for his contribution to our new D/A converter circuit. We also thank B. Wooley and M. Horowitz for their support and encouragement, DARPA for funding fabrication through MOSIS, and Hewlett Packard, Intel, AMD, Interval Research and MSIS for their support of the Information Systems VLSI lab. In addition, we thank Hewlett Packard for its financial support to David Yang.

Figure 7: 128 × 128 Image Sensor Chip Micrograph.

References

[1] J.C. Candy and G.C. Temes, editors. *Oversampled Delta Sigma Data Converters*. IEEE Press, 1992.

[2] P. Denyer, D. Renshaw, G. Wang, M. Lu, and S. Anderson. On-Chip CMOS Sensors for VLSI Imaging Systems. In *VLSI-91*, 1991.

[3] R. Forchheimer, P. Ingelhag, and C. Jansson. MAPP2200, a second generation smart optical sensor. *SPIE*, 1992.

[4] B. Fowler, A. El Gamal, and D. X. D. Yang. A CMOS Area Image Sensor with Pixel-Level A/D Conversion. In *ISSCC Digest of Technical Papers*, San Fransico, CA, February 1994.

[5] S. Mendis, S.E. Kemeny, and E. Fossum. A 128×128 CMOS Active Pixel for Highly Integrated Imaging Systems. In *IEEE IEDM Technical Digest*, San Jose, CA, December 1993.

Readout Electronics with Calibration and On-Line Test for Resistive Sensor Bridges

O. Machul, D. Hammerschmidt, W. Brockherde, and B.J. Hosticka

Fraunhofer Institute of Microelectronic Circuits and Systems
D-47057 Duisburg, Finkenstr.61, Germany

ABSTRACT

In this paper a digitally programmable readout electronics for resistive sensor bridges is presented together with an efficient method for calibration based on the principle of piecewise linear interpolation. The calibration enables a temperature compensation in the range between -40°C and +120°C. The architecture also allows an automatic calibration process with a minimum expenditure of external hardware. In addition the system contains an implemented function for monitoring the resistive sensor bridge condition during operation. For automotive applications an integrated voltage regulator is available so that the chip can optionally operate either at a stabilized 5V power supply voltage or use an unregulated voltage in the range between 7.5V and 16V.

INTRODUCTION

Nowadays the total costs of sensor systems result from the NRE costs and recurring costs for manufacturing, testing, and calibration. Testing and calibration costs can make up to 50% of the total cost. To make such systems more attractive for industry these costs must be minimized. To keep down the costs for the testing and calibration of sensor systems it is necessary to provide efficient automated adjustment for compensation of nonidealities and cross-sensitivities towards the measured quantity. The proposed architecture of the sensor system with such features is briefly described in the next section.

SENSOR SYSTEM ARCHITECTURE

The system shown in Fig.1 contains a temperature sensor based on the bipolar bandgap principle, an oversampled third-order ΣΔ-modulator for digital conversion of temperature, and a programmable three-stage switched-capacitor (SC) amplifier readout electronics followed by a smoothing 3rd order Sallen-Key lowpass filter in order to obtain a continuous-time output voltage. The first SC-amplifier features a programmable input that cancels the bridge offset which is typically in the order of maximal signal bridge magnitude. In this version the temperature dependence of the bridge offset has not been compensated.

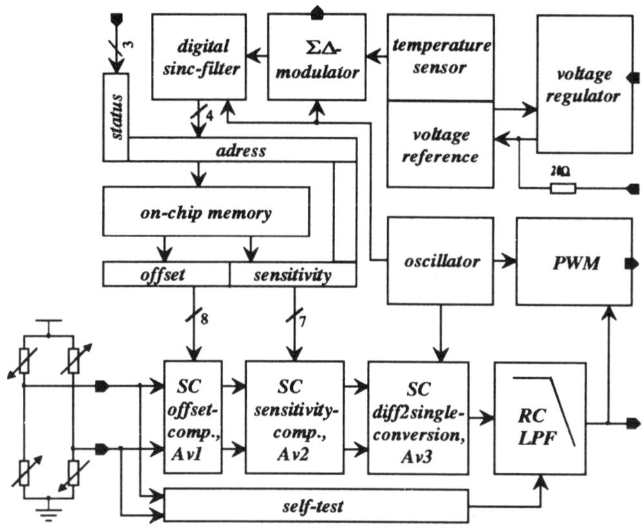

Fig.1: Block diagramm of the sensor system

The temperature-dependent sensitivity which is subject due to processing variations can be adjusted using the second SC-amplifier. All amplifiers feature capacitive resetting in order to obtain open-loop gain enhancement and sample-and-hold behaviour [1]. Furthermore, they employ the principle of correlated double sampling to achieve low frequency noise and offset reduction. Fig.2 shows the signal and noise transfer functions obtained by applying the above mentioned principles.

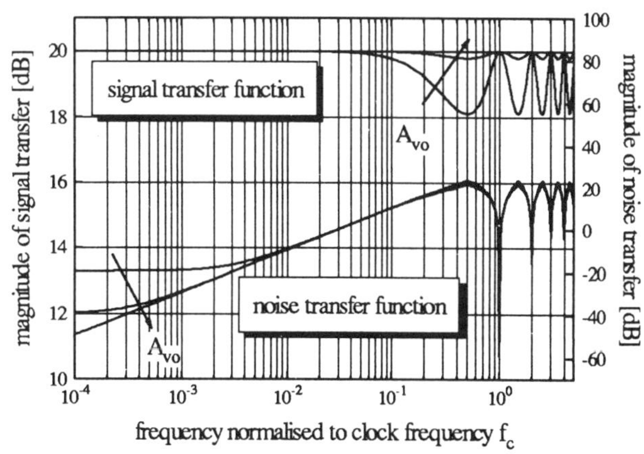

Fig.2: Signal- and noise transfer functions of SC-amplifiers using correlated double sampling and capacitive resetting

0-7803-3177-6 $5.00 © 1996 IEEE

IEEE 1996 CUSTOM INTEGRATED CIRCUITS CONFERENCE

These transfer functions are calculated for three different op amp open-loop gains featuring of 40, 60, and 80dB. It should be noticed that the noise transfer function rises up to the half clock frequency with 20dB per decade. Furthermore, Fig. 2 shows that the maximum suppression of low frequency noise and offset is limited by the open-loop-gain of the op amp used. The effect of capacitive resetting is not visible in Fig. 2 due to scaling effects, but it can be shown that the amplification error of the closed-loop gain is proportional to $1/A_{vo}^2$, where A_{vo} is the op amp open-loop gain.

A fully differential signal path technique has been used for the first two SC-amplifiers while the third one provides a single-ended output. A 3rd order Sallen-Key lowpass filter based on distributed RC-elements [2] follows immediately after the just mentioned differential-to-single conversion. This kind of improved smoothing filter offers a steeper stopband roll-off, higher crosstalk insensitivity, and a smaller chip area.

A serial scan-path-interface improves the testability of the entire sensor system and it is also used for the calibration process. During the calibration process the on-chip memory is operated in the write mode. The use of EEPROM can cause problems at high temperatures due to an unrealiable storage. For this reason we have choosen single-shot programmable zener-zapping diodes which provide a high temperature stability. A well-known problem of this type of storage is the high programming current requirement. This programming current has been reduced by applying layout modifications.

Additionally the system contains a circuit for monitoring the sensor bridge condition during system operation. The differential-mode signal of the sensor bridge output gives information about the sensed quantity whereas the common-mode signal contains information about the bridge bias. This can be used for self-test of the sensor bridge to detect faults. If the sensor is in a perfectly sound condition, the bridge bias and bridge resistance define a useful signal region. If in the case of a faulty bridge the common-mode signal leaves this region the analog output voltage is clamped at ground due to an internal modification of the low impedance output stage and this signals fault-detection. The evaluation of this information is obtained using an SC-network which generates the common-mode signal based on the differential sensor bridge output.

Furthermore, the sensor system contains analog and digital interfaces. The analog interface provides a continuous-time rail-to-rail output at low impedance while the digital interface yields a serial pulse-width-modulated, TTL-compatible signal. A voltage regulator allows the system user to employ either a stabilisized 5V power supply voltage or an unregulated voltage in the range between 7.5V and 16V.

Principle of Temperature Compensation

The temperature compensation of the sensor bridge requires a temperature sensor with thermal coupling to the bridge sensor, an on-chip memory used as an look-up table, and a piecewise-linear interpolator in order to save memory space [3]. We have adopted a novel approach for this purpose which is based on the interpolating behaviour of an oversampled $\Sigma\Delta$-modulator.

The oversampled modulator converts the baseband signal of the temperature sensor into a digital pulse-density signal with good linearity and an extremely high signal-to-noise ratio. A digital sinc-filter acting as lowpass filter transformes this bit stream into an oversampled 4 bit representation for addressing the following look-up table. This table contains sixteen values for sensitivity spaced over a wide temperature range between -40°C and +120°C. These values which can be adjusted with a resolution of 7 bit correspond to the break-points of the piecewise linear function. The look-up table converts the lowpass-filtered signal into digital control signals that modify the programmable amplification factors of the three-stage SC-amplifier circuit at oversampling rate. Finally, an analog decimation lowpass-filter with distributed RC-elements is used to provide a continuous-time output voltage that corressponds to amplified and temperature-compensated differential-mode sensor bridge signal in the baseband.

Calibration Process

One of the salient points of our approach is efficient calibration with only a few external hardware components. Fig. 3 shows a simplified block diagram of the required test board for a parallel calibration process.

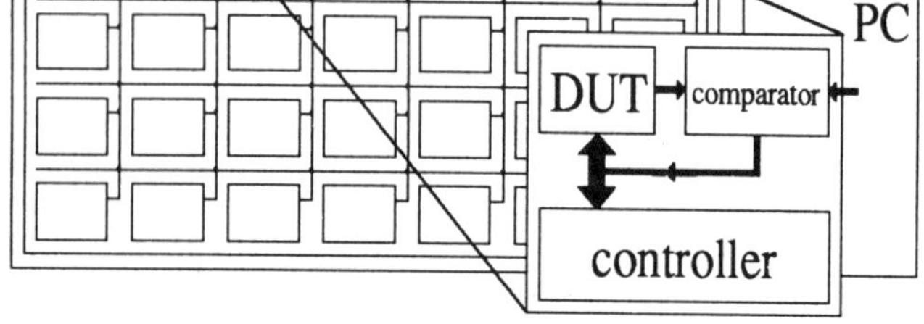

Fig.3: Parallel calibration process of many sensors

With the help of a personal computer (PC) and the HPIB-interface it is possible to control the sensed quantity, value of disturbance, and the desired output voltage applied to several devices under test (DUT's). The determination of the required calibration values is the task of the controller. First the controller sets the aquidistantly spaced points of the look-up table defined function using an IIR lowpass-filter with a precision of 10 bits which is implemented in controller software. The external output of the pulse-density-modulated temperature signal serves as the filter input. The four most significant bits represent the memory address at which the neccessary value for compensation must be stored. The remaining six least significant bits determine the distance from the actual address to the next one. The amplification needed for compensation must be determined at a point where ideally the six least significant bits are zero. If the controller detects such a point a high-precision comparator compares the analog output signal with the desired output voltage at this temperature. In the case of negative comparison result the controller modifies the amplification until the analog output matches the given output voltage. An efficient algorithm for modification of these values is the well-known successive approximation. Afterwards the established value is transfered into the memory using the serial scan-path interface. Finally the controller waits for the next point to establish another value of the desired look-up-table. This just described operation is reapeated until all breakpoints of the desired lock-up table defined function are established.

MEASURED RESULTS

The measured results of some important functions are shown in the following figures. Fig. 4 shows the baseband frequency spectrum of the oversampled ΣΔ-modulator for

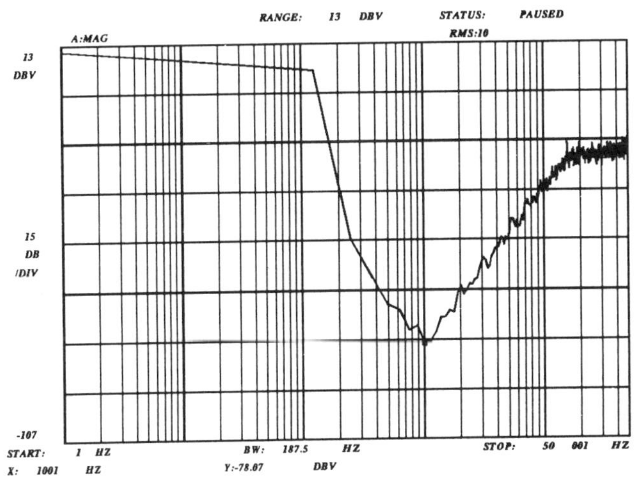

Fig. 4: Frequency spectrum of the ΣΔ-modulator

the digital conversion of the temperature signal at room temperature (T=27°C). The single bit stream output was measured using a dynamic signal analyser. The modulator yields a signal-to-noise ratio of 96dB which corresponds to a resolution of more than 15 bits in a baseband of 100Hz. Better results can be achieved, when the output is sampled and transfered to a PC where the data can be converted into the frequency domain using the discrete Fourier transformation. Fig. 5 shows the transfer function of the 3rd order Sallen and Key lowpass filter featuring distributed RC-elements which acts as analog decimation filter. This filter exhibits a 3dB cut-off frequency of 6.3kHz and a stopband rejection of 88dB at the system clock frequency. In comparison to the same filter type based on lumped RC elements an increase of 18dB has been achieved for rejection the stopband frequencies.

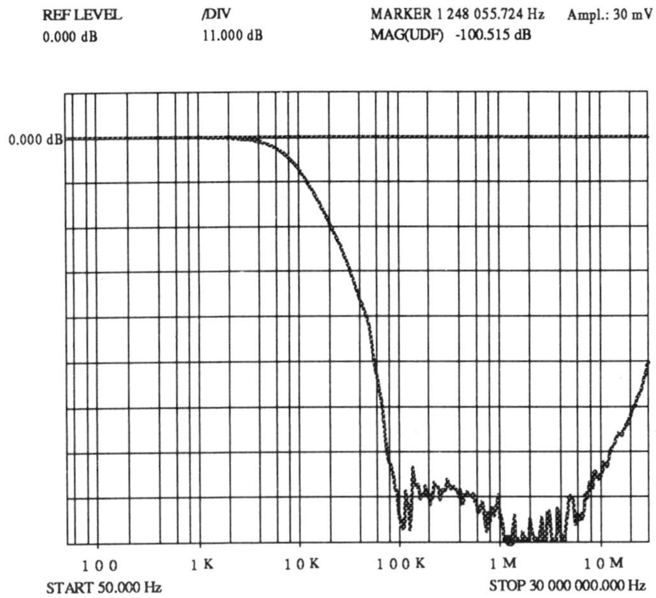

Fig. 5: Transfer function of distributed lowpass filter

The system draws a current of 12mA when supplied with a stabilized 5V power supply voltage. When using a unregulated voltage in the range between 7.5V and 16V as power supply voltage the system consumes 3mA more. The implemented voltage regulator yields a maximum efficiency of 66%. Noise measurements of the the three-stage SC-amplifier including analog lowpass filtering results in an equivalent minimum resolution of 10 bit. Finally for verification of the interpolation method an exponentially damped sinusoidal function has been programmed into the on-chip memory which can be normally approximated only using high order polynomial functions. This function has been chosen to demonstrate the high programming flexibility for various kinds of nonlinear functions that can be compensated. Fig. 6 shows the result of the verification.

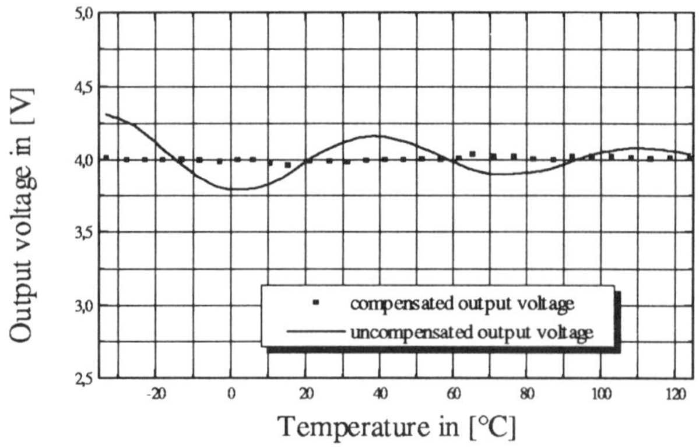

Fig. 6: Verification of the interpolation mechanism

Solid trace depicts the above mentioned programmed function contained in the on-chip memory which now simulates an uncompensated sensor signal as it appears at the output amplified by the electronics readout. The temperature varied between -40°C and +120°C. The dotted trace shows the corrected output voltage after the temperature compensation has been carried by the on-chip interpolation mechanism. The measurement proves that the temperature compensation can be better than 100ppm/°C.

The chip photomicrograph of the readout electronics system for resistive sensor bridges is shown in Fig. 7. The readout electronics has been fabricated in a standard 2 μm n-well silicon-gate CMOS technology. The die area of the fabricated chip is 26.79mm².

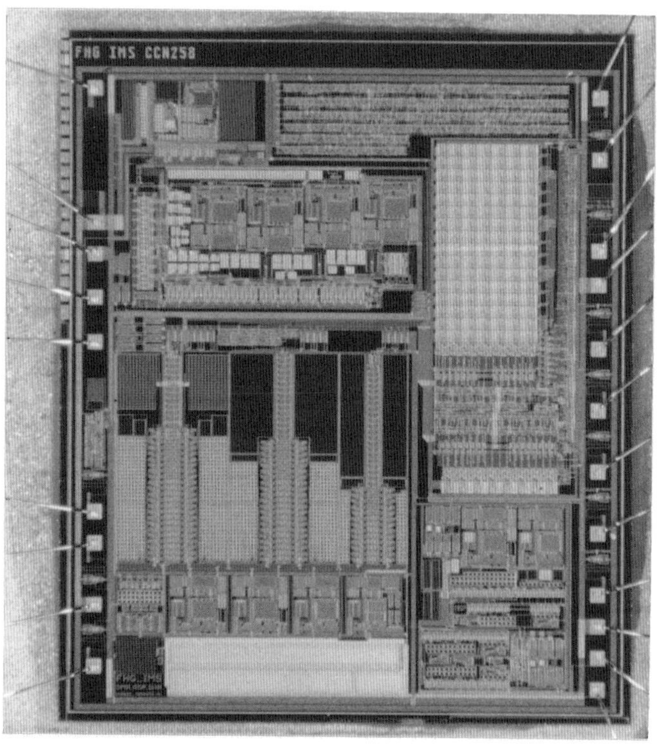

Fig. 7: Chip photomicrograpn

The following TABLE I summarizes the measured data.

TABLE I
MEASUREMENT RESULTS

Parameter	Value
Power supply voltages:	
- regulated	5V ± 5%
- unregulated	7.5 ... 16V
Oscillator frequency	1 MHz
System clock frequency	250kHz
System bandwidth	6.3kHz
Measured temperature range	-40°C ... +120°C
Min. equivalent resolution	10bit
Max. supply current	15mA
Chip area (2 μm CMOS)	26.79mm²

SUMMARY

This work has presented a readout electronics for resistive sensors featuring an efficient calibration method. In addition a high programming flexibility has been achieved due to the used calibration principle. This architecture allows an automatic parallel calibration process to minimize the total cost of the sensor system.

REFERENCES

[1] K. Martin, L. Ozcolak, Y.S. Lee and G.C. Temes, "A differential switched capacitor amplifier", IEEE J. Solid-State Circuits, vol. 22, no.1, pp. 104 -106, February 1987.

[2] D. Hammerschmidt, O. Machul, W. Brockherde, B.J. Hosticka, "Antialiasing and Smoothing Filters for Sampled Data Signals using Distributed RC Elements", ESSIRC'95, pp.198 -201

[3] P. Malcovati, C. Azeredo Leme, P.O'Leary, F. Maloberti and H. Baltes, "Data Conversion and Programmable Calibration for Smart Sensors", ESSIRC'93, p.25-28

An Angular Rate Sensor Interface IC

Seyed R. Zarabadi, Pedro E. Castillo-Borelly, and Jack D. Johnson

Delco Electronics Corporation
Kokomo, Indiana 46904

Abstract

This paper describes an interface IC that maintains the vibration of a surface micro-machined yaw rate sensor and processes its angular rate signal. In addition, it provides the necessary electronics to compensate for the nonuniformities, parameter shifts with temperature, and offset and sensitivity variations of the sensor. The IC has been fabricated in a 1.2μm CMOS process that includes a high voltage MOS transistor. Experimental results of several key circuit blocks and the angular rate sensor system indicate proper operation over temperature and supply voltage.

Introduction

Consumer demand for improved vehicle safety has driven several vehicle manufacturers to develop yaw control systems. Yaw rate is the angular rate of rotation about a vehicle's vertical axis, in other words, it is a measure of the turning of the vehicle to the left or the right. Automotive yaw control systems compare the driver's desired turning rate to the actual turning rate of the vehicle providing continuous feedback to maintain the vehicle on the driver's desired heading. The system includes a steering wheel angle sensor that provides a signal indicating the driver's desired turning rate. A yaw rate sensor is used to measure the actual turning rate of the vehicle. These two inputs as well as lateral acceleration are used by a microprocessor-based yaw control system to determine whether the vehicle is heading in the direction that the driver desires [1].

Delco's yaw rate sensor module is currently optimized for use in automotive applications. Changes to the system specifications in areas such as bandwidth, range, and scale factor can allow the sensor to be optimized for other applications such as video camera stabilization and virtual reality systems.

This paper describes an interface IC for an automotive yaw rate sensor. Delco's surface micro-machined angular rate sensor used with this interface IC is then briefly described. Next, the system block diagram (Fig. 1) and several key circuit blocks of the interface IC will be presented. We conclude the paper by providing some experimental data indicating the performance of the interface IC and the yaw rate system.

System Description

The complete sensor module includes a sensor IC, an interface IC and external components on a printed circuit board. Table 1 shows the expected performance of the module for a 50 Hz bandwidth. The bandwidth of the system can be tailored for a specific application by simply exchanging one of the external components. It should be noted, however, that the total noise is directly proportional to the system's bandwidth. In addition, the sensitivity and scale factor of the system can be adjusted by programming the interface logic.

Table 1 -- Yaw rate module expected performance

Parameter	Performance
Temperature	-40°C to +85°C
Operating Voltage	5V
Sensing Range	± 75 deg/sec
Bandwidth	50Hz
Operating Current	25mA
Zero Rate Output	2.5V
Scale Factor	30mV/(deg/sec)
Zero Rate Output Drift (over time and temperature, incl. noise)	± 3 deg/sec
Output Rate Nonlinearity Error	< 0.2%

The sensor IC used to detect rate consists of a vibrating ring gyroscope [2] plus sense electronics [1]. For normal operation, the ring requires 5 input and output signals (See Fig. 1): 0° and 45° drive, 0° and 45° sense, and the ring bias. The 0° drive and sense are used to maintain ring oscillation at the desired flexural mode. The 45° drive is used in the force-to-rebalance scheme to null out the signal detected at the 45° sense electrode as well as to cancel the initial zero bias. The rate information is determined from the amount of 45° drive needed to null out the signal detected at the 45° sense pick up. The ring bias is a high voltage (Vring) signal needed to improve the signal pick off by the capacitive electrodes (0° and 45° sense.) In addition to these signals, 4 balance voltages (V1 - V4) are needed to counteract the spring and mass nonidealities introduced during the manufacture of the ring.

The ring itself is surface micro-machined by electroplating a ring onto a fully processed CMOS wafer. To function properly, the ring requires a high vacuum environment which is achieved by bonding a silicon wafer on top of the micro-machined ring thus hermetically sealing the sensor.

0-7803-3177-6 $5.00 © 1996 IEEE

IEEE 1996 CUSTOM INTEGRATED CIRCUITS CONFERENCE

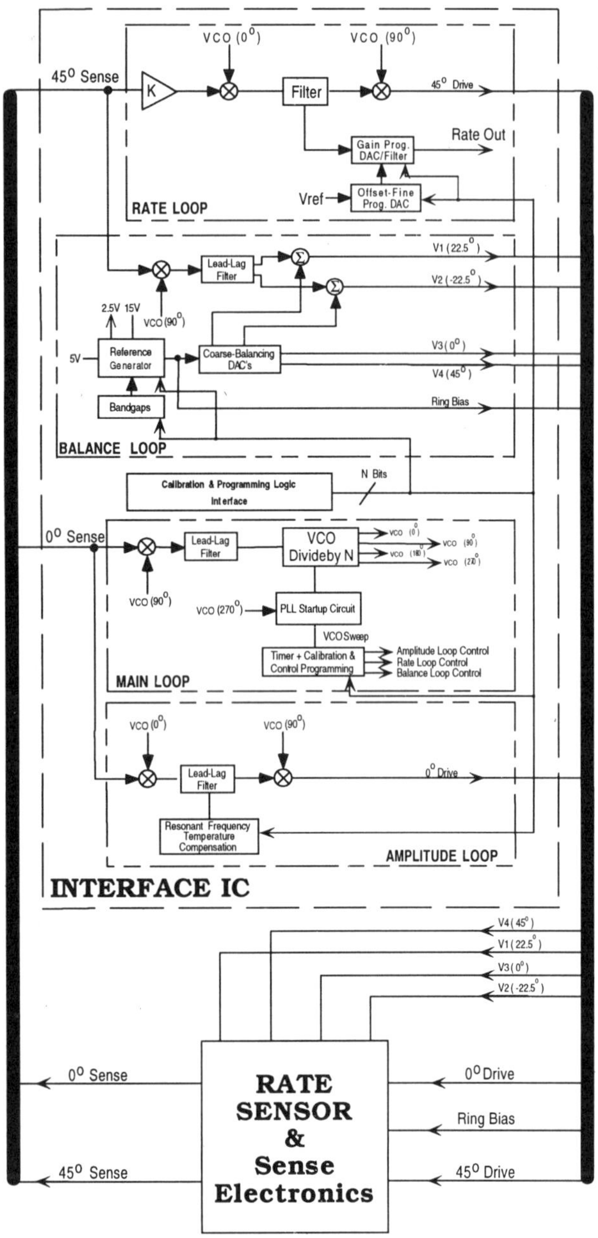

Figure 1 -- Yaw rate system block diagram

Interface IC Description

The interface IC was designed to be used with an angular rate sensor for automotive applications. Most yaw rate systems use rate sensors that are either bulky [3] or difficult to electronically compensate for the manufacturing-induced nonuniformities, hence requiring special trimming equipment that results in high system cost. The presented IC can be used with almost any vibratory rate sensor to maintain vibration and to process the angular rate signal. It is also capable of compensating nonidealities introduced during manufacturing and the parameter shifts with temperature of a surface micro-machined angular rate sensor.

The block diagram of the yaw rate system (shown in Fig. 1) includes the rate sensor and interface IC's. The sensor IC contains electronics to nondestructively sense and amplify the signals put out by the sensor. The interface IC comprises four feedback loops, a reference generator block, calibration and programming logic circuits, and miscellaneous circuitry. These four loops and several key circuits used in this IC are described below.

Main Loop

The main loop is a type II phased-lock loop designed to lock to the sensor's desired resonant frequency. It comprises a startup circuit that sweeps the VCO input until the sensor's desired resonant frequency is captured. This loop generates four logic signals that have 90° phase increment with respect to each other. They are used by the other loops for modulation and demodulation purposes. In addition, this loop also generates control logic signals used throughout the IC.

Amplitude Loop

The purpose of the amplitude loop is to maintain a constant drive voltage level to the sensor, provide ratiometric rate output voltage, and compensate for the sensor's resonant frequency shift with temperature. The reference voltage for setting sensor's drive amplitude is a product of the power supply voltage and a voltage whose temperature dependency matches the sensor's temperature coefficient. The linear voltage multiplier, shown in Fig. 2, is used to generate the reference voltage to the amplitude loop. The currents V_{ddk}/R, $VB(T)/R$, V_{ref}/R are generated from the power supply, temperature circuit, and a bandgap reference, respectively. Note that the voltage multiplier in Fig. 2 is linear with respect to V_{dd} and $VB(T)$ and, to first order, is insensitive to process and temperature variations [4].

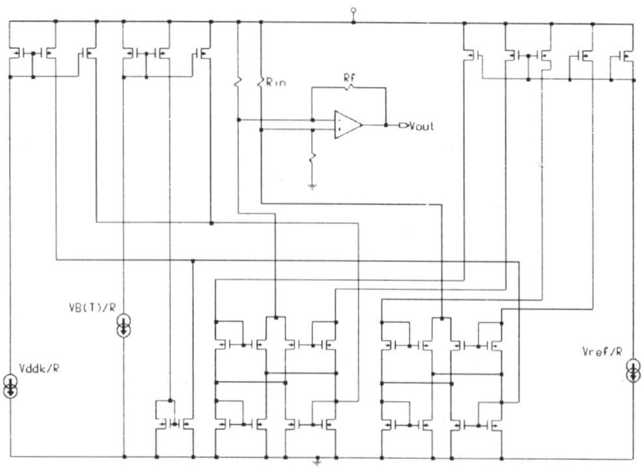

Figure 2 -- Simplified schematic of the linear voltage multiplier

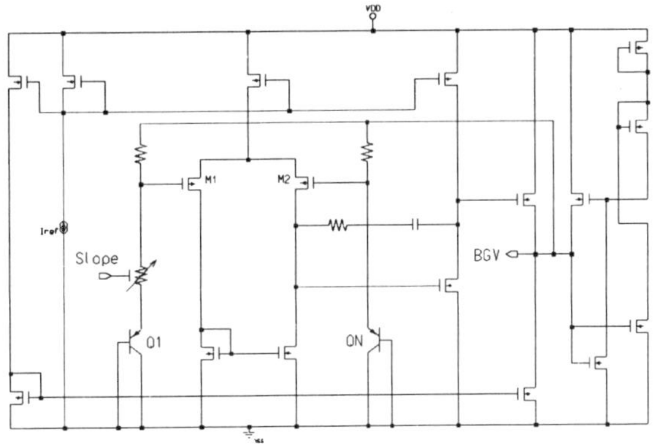

Figure 3 -- Simplified schematic of the bandgap reference generator

Balance Loop

Due to imperfections introduced during fabrication, it becomes necessary to compensate for the mass and stiffness nonuniformity of the rate sensor. Each sensor has two resonant frequencies at the desired vibration modes. The equations for the two resonant frequencies are:

$$\omega_1 = \sqrt{\frac{K_1}{m_1}}, \quad \omega_2 = \sqrt{\frac{K_2}{m_2}},$$

where K_i and m_i are the spring constant and lumped mass, respectively, corresponding to each normal mode. Rate sensitivity is reduced if the two frequencies are unequal. Imperfections that cause the resonant frequencies to be different are compensated for by the balance loop. This is accomplished by adjusting the effective ring stiffness using an electrostatic attractive force between the fixed balancing electrodes and the moving ring [2].

The spring constants of the ring are proportional to the balance electrode voltages, which usually have values ranging from the lowest IC potential to the ring bias voltage. Initially, balance voltages are programmed to their room temperature value and the balance loop will adjust these voltages to compensate for temperature and aging effects on the ring spring constants. In order to minimize silicon area and maximize reliability, the high voltage circuits are minimized.

Figure 4 shows the simplified schematic of a low current, high voltage calibration DAC. Voltages Vm and Vp are applied to the sensor's fixed electrodes to equalize the desired normal mode frequencies. Initially b1-bN are permanently programmed at the factory to compensate for nonuniformities of the sensor with Vbalf at ac ground. In the field, Vbalf will adjust Vm and Vp to compensate for temperature or aging that might be affecting the equal normal modes. Four high voltage devices (MU1, MU2, MB1, MB2) are used in this calibration DAC. The current drain from Vring is less than 50μA.

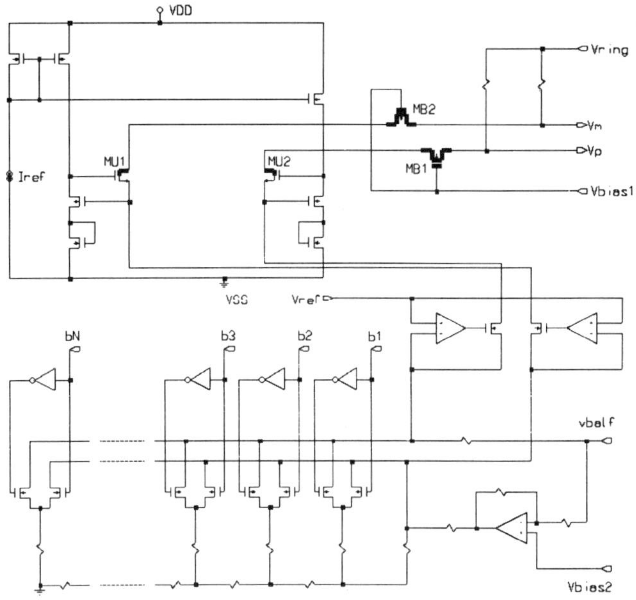

Figure 4 -- Simplified schematic of the high voltage balance DAC

Rate Loop

The rate loop is used to control the bandwidth of the rate sensor. Without this feedback loop the output rate bandwidth is given by: $BW = \omega/_{2Q}$, where ω and Q are the sensor's resonant frequency and quality factor, respectively. By feeding back a certain percentage of the rate signal, the nodal motion of the ring, representing the angular rate, is forced to zero. The closed loop bandwidth and damping is controlled by appropriately choosing the open loop gain and compensation circuits. The output rate is passed through a programmable gain block to adjust for the desired scale factor. The sensor's zero bias is compensated for by adding or subtracting a percentage of a stable reference signal to the output rate signal. This changes the sensor's vibration pattern such that no motion is detected at the 45° sense pickoffs [2].

The stable reference is generated by the bandgap circuit of Fig. 3. The zero TC of the bandgap voltage is obtained by programming.

Interface Logic

The interface IC contains logic circuitry designed to facilitate testing and to provide the flexibility needed to operate with different sensor IC's. This feature allows the interface IC to have the capability to set the proper balance voltages, to compensate for the difference in gain, offset, and scale factor in each chip, and to adjust the temperature characteristics of appropriate reference voltages to match the sensor's own temperature dependence. The inclusion of an on-chip non-volatile memory block allows these configuration settings to be made permanent.

Summary of Experimental Results

The angular rate sensor IC has been integrated in a 1.2μm CMOS process with an MOS high voltage module. Figure 5 shows a photomicrograph of the IC. The sensor's polarization and balance voltages are generated with high voltage MOS devices. Bandgap reference circuits have been used in the generation of stable high voltages. Figure 6 is a typical DC response of the multiplier. Each curve represents the product of one varying input (x axis) with other input fixed for each curve. The y-axis is the multiplier output. Linear regression by the least-squares method has been used to estimate a linearity error of less than 0.2%. Figure 7 shows a typical bandgap dc response with temperature for different programmed settings. The bandgap voltage error is less than 3mV over a 140°C temperature range (~18mV/°C). Figure 8 shows the response of the high voltage DAC vs. Vbalf fine input for three different codes. Typical measured INL and DNL are less than 0.15% of the full scale.

Figure 9 represents the typical frequency response of the rate output showing a 50Hz bandwidth. The system exhibits a phase lag of 30 degrees at this frequency. Figure 10 shows the measured noise performance of the system, indicating less than 1 deg/sec peak noise level. The measured non-linearity of the angular rate output is less than 0.2%.

Figure 5 -- Photomicrograph of IC

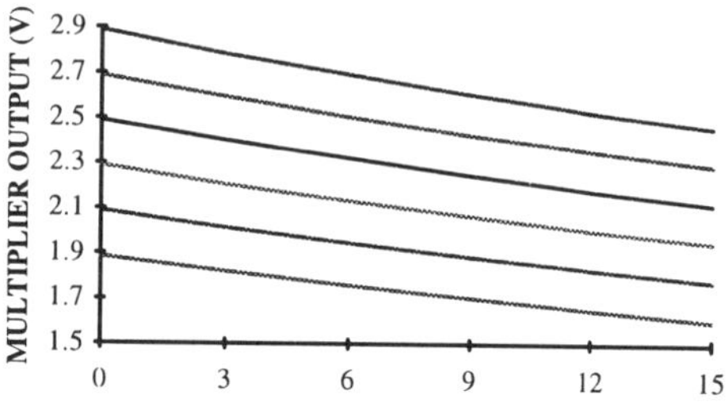

Figure 6 -- Multiplier output vs. Bandgap settings

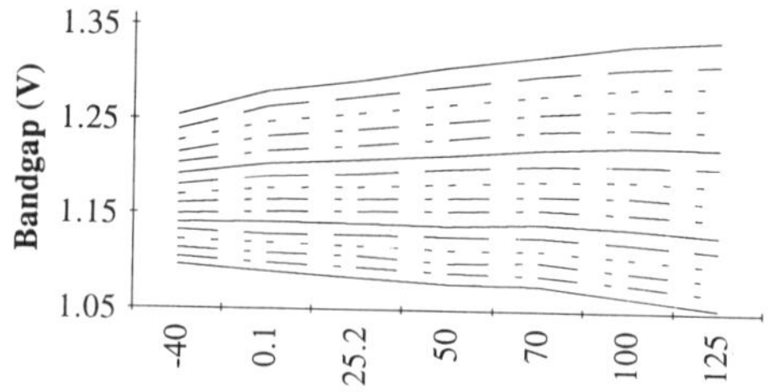

Figure 7 -- Bandgap output (V) vs. Temperature (°C) for different programming settings.

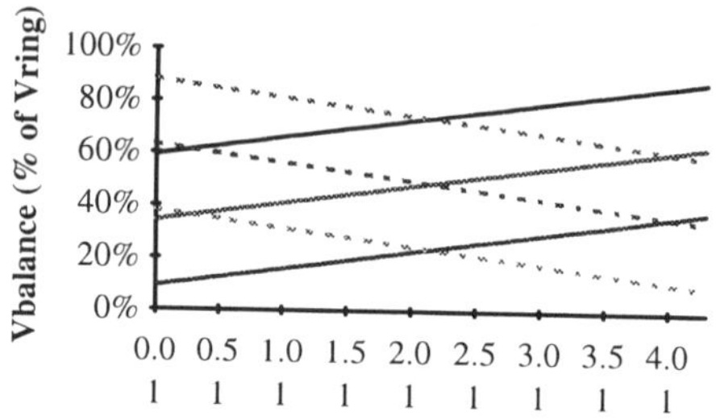

Figure 8 -- Fine DAC output vs. Vbalf input

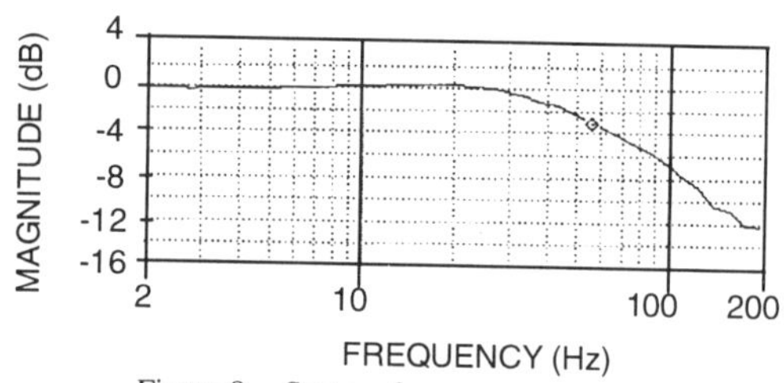

Figure 9 -- System frequency response

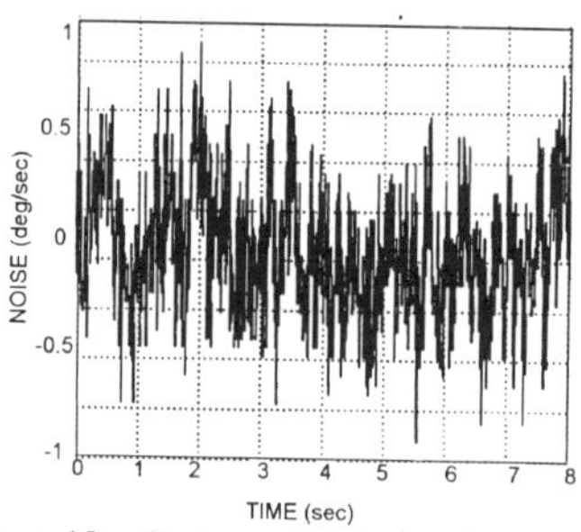

Figure 10 -- System noise in the absence of rate

Acknowledgments

We would like to thank M. Chia, F. A. Liem, G. Cook, G. Woodward, and K. Hong for their invaluable contributions.

References

[1] J. D. Johnson, S. R. Zarabadi, and D. R. Sparks, "Surface mircomachined angular rate sensor," *SAE Technical Paper Series, Sensor and Actuators* (SP-1066), 1995, pp. 77 - 83.

[2] M. W. Putty and K. Najafi, "A micromachined vibrating ring gyroscope," *Solid States Sensor and Actuators Workshop*, June 1994, pp. 213 - 220

[3] B. Johnson, "Vibrating rotation sensors," *Solid States Sensor and Actuators Workshop*, June 1994, pp. 41 - 47

[4] E. Seevinck and R. J. Wiegerink, "Generalized translinear circuit principle," *IEEE J. Solid State Circuits*, Vol. 26, No. 8, August 1991, pp. 1098 - 1002.

A Micromachined Fully Differential Lateral Accelerometer

Mark Lemkin
Berkeley Sensor & Actuator Center

Bernhard E. Boser
Electronics Research Laboratory

Department of Electrical Engineering and Computer Sciences
University of California, Berkeley, California 94720-1770

Abstract

A monolithic surface micromachined lateral accelerometer with on-chip A/D converter is described. The device features a fully differential capacitive interface between the sensing element and on-chip electronics to minimize the systematic offset and errors due to switch charge injection. The common centroid layout of the sensor interface reduces cross-axis sensitivity and susceptibility to angular accelerations. The measured noise-floor of the device is 500 µg/√Hz and the full-scale range is +/- 3.5 g (1g = 9.8m/sec^2).

I. Introduction

Surface micromachining has become an established technology for the co-fabrication of mechanical sensing elements with electronic interface and signal processing circuits on a single die [1]. In addition to monolithic integration, key advantages of this technology include small size, batch fabrication, and low cost. Applications of this technology include inertial sensors for the automotive market or computer I/O devices such as head-mounted displays.

In this paper, a monolithic accelerometer with on-chip ΣΔ ADC is described. The electronic circuitry is implemented in a standard 3µm BiCMOS process. The sensing element consists of a proof-mass anchored to the substrate with a compliant suspension. Acceleration of the die results in a proportional displacement of the proof-mass with respect to the substrate that is sensed capacitively.

The proof-mass and suspension are fabricated from a 2µm thick polysilicon film anchored 1.6µm above the substrate. This material has excellent mechanical properties, experiences no wear or fatigue, and can be processed with standard IC fabrication equipment and techniques. However, the limitation to planar structures combined with the small size of micromechanical devices imposes several design challenges. For example, the sensitivity of the accelerometer described here is only 0.3 fF/g; the nominal value of the sense capacitor is 84 fF. By comparison, a bulk micromachined sensor with substantially larger dimensions and fabricated on two chips achieves a sensitivity of 80 pF/g with a 50 pF sense capacitor [2].

The smaller sensitivity of the micromachined device is in part due to limitations on the achievable spring constant and hence displacement of the proof-mass due to acceleration. Size, robustness, and cross-axis sensitivity impose a lower practical limit on the value of spring constants achievable in micromechanical systems. In this design, the sensitivity of the sensor to undesired off-axis inputs is eliminated to first order with a common centroid layout of the capacitive pickup. The performance of the resulting device is compatible with challenging applications such as active automobile suspension or head-tracking for virtual reality displays which are not constrained to movement in only one direction.

Other work has utilized a differential sensor with single ended output for the sensing element [2, 3, 4]. The single ended interface leads to systematic offsets and distortion due to switch charge injection. The latter is of particular concern in this design because of the very small sense capacitance of the micromachined element. Both problems are eliminated with a differential sensor interface and fully differential electronics throughout. This approach also improves power supply rejection and reduces drift of the sensor. Because of the monolithic integration, more complicated wiring of the differential common centroid arrangement does not add substantially to the fabrication complexity or parasitic capacitance of the sensor.

In Section II design considerations of the micromachined sensing element are discussed. It is demonstrated how poor intrinsic rejection of off-axis inputs by the mechanical element is compensated with an appropriate layout of the capacitive interface. Section III describes the electronic position measurement interface and force-feedback loop of the sensor. Experimental results are reported in Section IV.

II. Sensing Element

In micromechanical accelerometers a proof-mass with mass m is attached to the substrate via a compliant suspension with spring constant k. When the substrate undergoes acceleration a, the proof-mass exerts a force $F = ma$ on the suspension. For frequencies below the mechanical resonance of the spring-mass system, this force causes the suspension to deflect a distance x given by

$$x = \frac{F}{k} = \frac{m}{k}a = \frac{1}{\omega_n^2}a \qquad (1)$$

In this equation, ω_n is the natural frequency of the sensor in the direction of applied acceleration. It should be minimized in order to maximize the displacement of the sensor for a given input, and hence its sensitivity. This can be achieved either by increasing the mass m of the mechanical sensing element, or by

0-7803-3177-6 $5.00 © 1996 IEEE

IEEE 1996 CUSTOM INTEGRATED CIRCUITS CONFERENCE

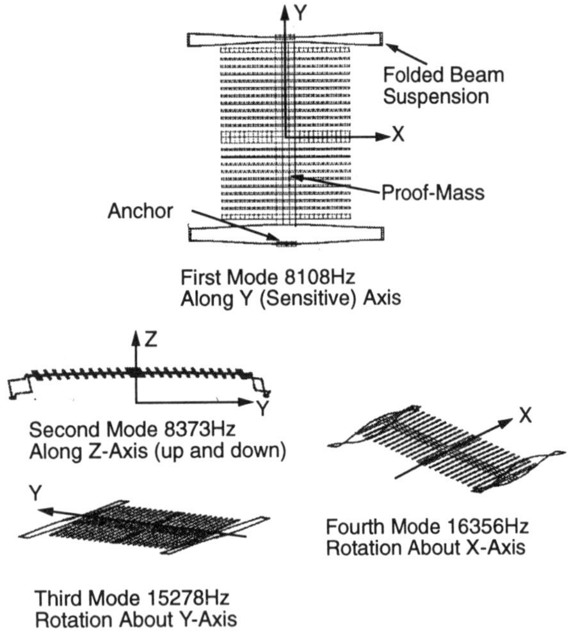

Figure 1: Motion of the sensing element in the four most sensitive directions. All but the first mode are undesired.

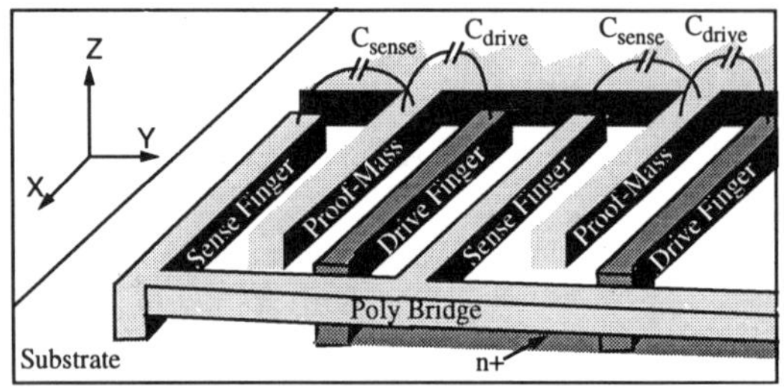

Figure 2: Segment of surface micromachined comb finger capacitor array

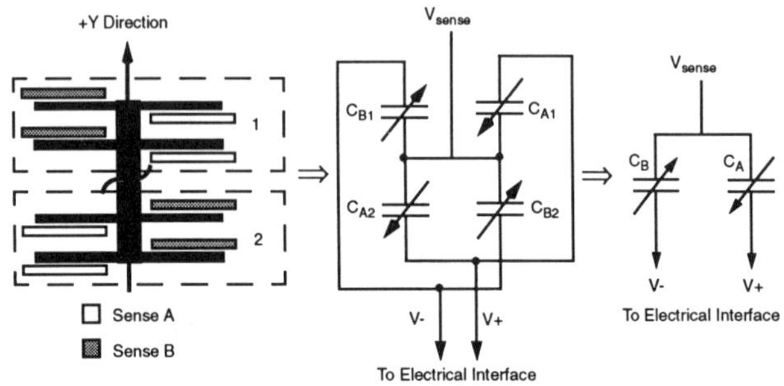

Figure 3: Common centroid schematic of proof-mass. Drive capacitors not shown for clarity

lowering the spring constant k. Fabrication constraints, such as warpage of the polysilicon film and patterning accuracy, limit the size of the proof-mass and constrain both the length and width of the suspension. Practical considerations such as shock survival set further limits on natural frequency. In principle, the resonant frequency also sets the useful bandwidth of the sensor. However, this limitation is overcome in this design with a force-feedback loop described in Section III.

A top view of the sensing element is shown at the top of Figure 1. A proof-mass with attached fingers for capacitive sensing of displacements in Y direction is suspended by four single-fold beams anchored to the substrate. The folds release stress in the polysilicon film and reduce the longest extension of the suspension. The mass of the sensing element including fingers is 0.2 μgram and the spring constant of all four beams combined is 0.5 N/m. The size of the complete structure is 400μm x 400μm and is limited in this design primarily by warping of the polysilicon film.

In a lateral Y-axis accelerometer, a low spring constant calls for long and narrow support beams. Since in this technology, the minimum width of beams (2μm) is close to the polysilicon film thickness (1.6μm), the resonant frequencies in the desired Y- and undesired Z-directions are very close. Finite element simulations with ABAQUS [5] shown in Figure 1 predict a Z-axis resonance at a frequency that is only 3% above the desired first mode. Consequently, inputs along the Y- and Z-axes result in approximately identical displacements of the proof-mass. Angular inputs about either the X- or the Y-axis result in slightly smaller motion of the proof-mass since the resonant frequencies of these modes are approximately twice that of the desired mode.

The poor cross-axis rejection of the mechanical element is compensated by the rejection in the capacitive and electronic position sensing interface. Figure 2 shows a section of the proof-mass with attached fingers along with the stationary position sense fingers. Several sense fingers are connected with a polysilicon bridge, rather than anchored individually. This layout minimizes the number of anchors and hence the bottom plate parasitic to the substrate. Also shown in the figure are drive fingers that are used for electrostatic force-feedback, described in Section III. Motion of the proof-mass in the Y-direction results in an increase or decrease of the sense capacitance, C_{sense}. Boundary element simulations with FASTCAP [6] predict a per tine sensitivity of 0.02 fF/μm, 1.39 fF/μm, 0.11 fF/μm, respectively for displacements in X, Y, and Z direction, suggesting an order of magnitude rejection of Z-axis inputs.

Complete first-order rejection of Z-axis and angular inputs is achieved by splitting the sense fingers into four groups, as is indicated in Figure 3, and measuring the difference between $C_A=C_{A1}+C_{A2}$ and $C_B=C_{B1}+C_{B2}$. Displacements along the Y-axis increase C_B but decrease C_A, hence resulting in a differential output. Z-axis inputs, on the other hand, cause all capacitors to change by the same amount, thus producing a common-mode signal that is rejected by the electronic circuitry. Rotations about either the X- or Y-axis, as well as translations in X-direction cause differences in individual capacitor values, but because of the common-centroid layout the sums $C_{A1}+C_{A2}$ and $C_{B1}+C_{B2}$ are unaffected to first order.

III. Electrical Interface

Mechanical acceleration sensing elements can either be operated open-loop, or be placed into a force-feedback loop that

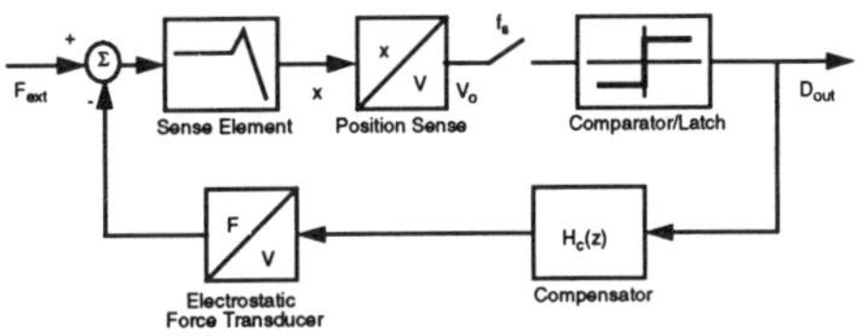

Figure 4: Sigma-Delta Force-Feedback Loop

senses displacement of the proof-mass from its nominal position and generates a restoring force that opposes the motion. Closed-loop operation has several advantages that are particularly important for miniature micromachined sensors. Moreover, the circuitry needed for the feedback loop comes virtually for free because of monolithic integration.

Closed-loop operation improves the dynamic range of the sensor by nulling proof-mass displacement even for large inputs. Important performance parameters including sensitivity or linearity are determined by the force transducer in the feedback loop, rather than the sensing element itself. Consequently, sensitivity to processing parameters and spring non-linearity are substantially reduced in a high-gain feedback loop [1]. Moreover, the bandwidth of a closed-loop system is determined by the loop bandwidth and hence decoupled from the natural frequency ω_n, which now can be optimized for maximum sensitivity alone. Feedback also eliminates gain-peaking in under-damped sensors.

A block diagram of the sensor embedded in a feedback loop is shown in Figure 4. Electrostatic actuation is used to counteract external acceleration signals acting on the sensor. Problems arising from the nonlinear relationship between voltage and force in electrostatic actuators are eliminated by quantizing the feedback signal to one Bit, encoding only the sign of the displacement of the proof-mass from its nominal position. Since in this case only two different force levels must be generated, linearity is assured. The entire system operates like a ΣΔ modulator, with the proof-mass acting as a second-order noise shaping filter [2, 7, 8].

The force-feedback accelerometer operates from the four-phase clock with period 2μs illustrated in Figure 5. In the first half of the cycle, the position of the proof-mass is determined by measuring the difference between C_A and C_B. The sign is obtained with the comparator during the "latch" phase, and finally a force-feedback pulse is generated. Because of capacitive feed-through it is important to separate the sense and feedback phases. A return-to-zero code eliminates signal-dependent distortion. The time between position measurement and force-feedback adds a delay to the loop transfer function which is compensated by lead filter $H_c(z)$ to maintain stability [7].

The position measurement interface is shown in Figure 5. The stationary sense fingers of the accelerometer are connected to the input of a fully differential amplifier. During the auto-zero phase, all capacitors are initialized to a constant voltage. Then a voltage V_{sense} is applied across C_A and C_B causing a

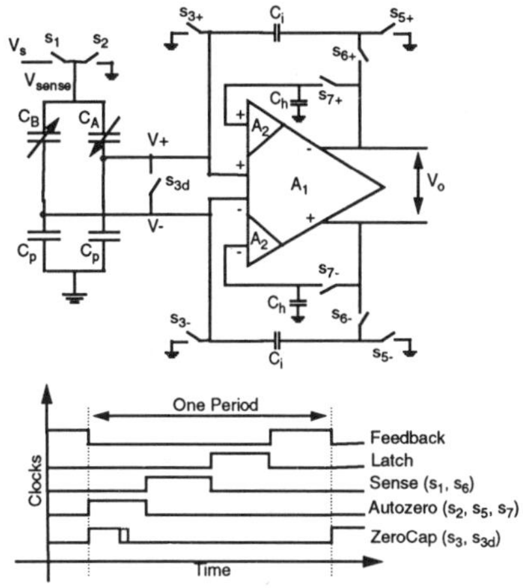

Figure 5: Timing diagram and integrator schematic

displacement charge to flow onto the integrating capacitors C_i. Only differential signals due to mismatch of C_A and C_B are amplified, while common-mode signals are rejected by the common-mode feedback loop at the output of the differential amplifier. Since the position measurement interface is followed by a comparator, the circuit is insensitive to variations of V_{sense}, representing another important advantage of force-feedback [2].

Owing to the small sense capacitors of only 42 fF each, combined with the relatively large switches in a 3μm technology, switch charge injection represents a potentially large source of error in this accelerometer. The fully differential design combined with bottom-plate sampling rejects these errors to first order. Still, opening switches s3+ and s3- during the auto-zero phase results in a 150mV step at the amplifier input due to charge injection and kT/C noise. The auxiliary gain stage A_2 stores this voltage and the amplifier offset and 1/f noise on the hold capacitors C_h. After the signal on C_h has settled, switches s7 are opened. Since these error sources are virtually constant over the period of one cycle, charge injection errors, kT/C noise and 1/f noise are attenuated by the open loop gain of the auxiliary amplifier, A_2. Offsets caused by mismatch between s3+ and s3- are eliminated by switch s3d, which is opened slightly later. After cancellation, sampling noise on the hold capacitors C_h and broad-band noise from the amplifier are the main noise contributors in the position measurement interface [9].

In practice, the preamplifier between the position measurement amplifier and the comparator contributes additional noise. Maximizing the amplifier output V_o minimizes this contribution. This, however, calls for small integrating capacitors and is at odds with the bandwidth requirement of the circuit which demands a value of C_i that is comparable to the total capacitance at the input of A_1. In this design, two stages of preamplification provide a gain of approximately 50 before the comparator.

The comparator output is fed to an off-chip digital decimation filter. Since the sampling rate must be chosen to be much higher than the resonant frequency and the bandwidth of interest in most sensing applications is much less than 1kHz, the

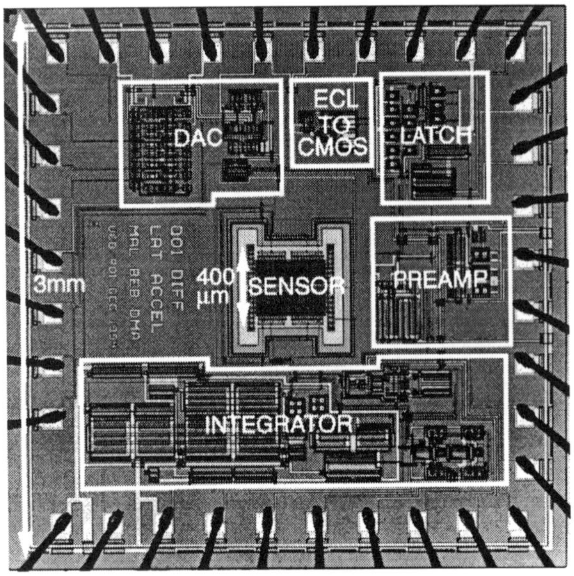

Figure 6: Die photo of fabricated device

oversampling rate is very large and quantization noise is no issue.

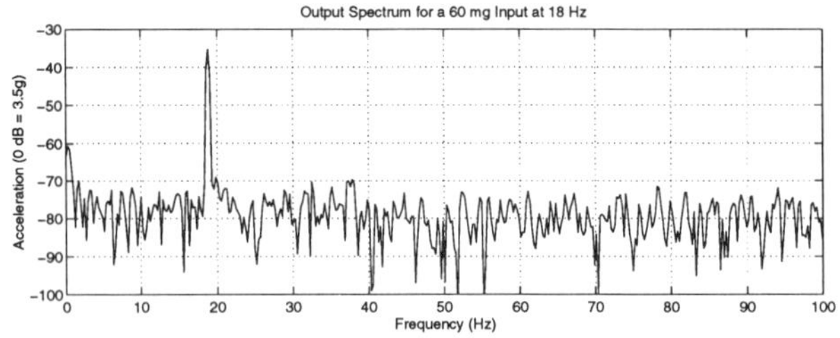

Figure 7: Power spectral density for a 60 mg input at 18Hz

Table 1: Accelerometer Specifications

Parameter	Value
Process	3µm BiCMOS
Power Supply	3.7 mA @ 5V
Die Size	3mm x 3mm
Sample Rate	500 kHz
Full Scale	+/- 3.5 g
Noise Floor	500 µg/sqrt(Hz)
Resonant Frequency	8.1 kHz
Proof-Mass	0.2 µg
Sense Capacitor (C_A=C_B)	42 fF each
Sensitivity (C_A=C_B)	32 fF/µm each

IV. Results

Figure 6 shows a die photograph of the lateral accelerometer. For evaluation, the device is mounted on a shaker table and excited with a sinusoidal acceleration signal. A reference accelerometer mounted on the shaker table is used for calibration. The one-bit output from the comparator is filtered and decimated with a programmable digital signal processor and the output sent to a computer for performance evaluation.

Output noise from the sensor was measured at 500 µg/√Hz. The full scale input range of the device is +/- 3.5g. Figure 7 shows the output power spectral density for a 60 mg input signal at 18 Hz. The increase in noise at low frequencies is due to 1/f noise in the preamp. A second version of the device, currently in fabrication, corrects this problem by using bipolar rather than MOS transistors as input devices in the preamp. Power consumption is 3.7mA at 5V with a sample rate of 500 kHz. Transverse attenuation has been measured at 45 dB for Z-axis and 43 dB for X-axis accelerations. Actual attenuation is better than these values, however cross-axis accelerations of the shaker table used for testing preclude more accurate measurements.

V. Conclusions

A differential capacitive sense interface eliminates errors due to charge injection in a monolithic surface micromachined accelerometer. Despite poor rejection of cross-axis acceleration signals by the mechanical sense element that are a consequence of the small size of the device, good overall performance is achieved because of the rejection of undesired signals by a common-centroid sensor interface.

Acknowledgments

The authors would like to thank Analog Devices for device fabrication and the California PATH program and ARPA for supporting this project. Discussions and contributions by Professors Auslander, Howe, and Pisano are gratefully acknowledged.

References

[1] Boser, B. E., Howe, R. T, "Surface Micromachined Accelerometers," *CICC Dig. Tech. Papers.* IEEE Custom Integrated Circuits Conference, 1995, pp. 337-344.

[2] Smith, T., Nys, O., Chevroulet, M., DeCoulon, Y., Degrauwe, M., "A 15b electromechanical sigma-delta converter for acceleration measurements," *ISSCC Digest of Technical Papers*, Feb. 1994, pp. 160-161.

[3] Burstein, A., Kaiser, W., "Mixed analog-digital highly sensitive sensor interface circuit for low cost microsensors", The 8th International Conference on Solid-State Sensors and Actuators, and Eurosensors IX, June 1995, pp. 162-165.

[4] Analog Devices, "ADXL50-monolithic accelerometer with signal conditioning," Datasheet, 1993, One Technology Way, Norwood, MA 02062.

[5] Hibbitt, Karlsson & Sorenson, Inc., *ABAQUS User's Manual*, 5.2 edition, 1080 Main Street, Pawtucket, RI, 1992.

[6] Nabors, K., Kim, S., White, J., "Fast capacitance extraction of general three-dimensional structures," *IEEE Transactions on Microwave Theory and Techniques*, July 1992, pp. 1496-506.

[7] Lu, C., Lemkin, M., Boser, B.E., "A monolithic surface micromachined accelerometer with digital output," *IEEE Journal of Solid State Circuits*, vol. 30, no. 12, December 1995. To appear.

[8] J. C. Candy and G. C. Temes. Oversampling Methods for A/D and D/A conversion. In J. C. Candy and G. C. Temes, editors, *Oversampling Delta-Sigma Data converters*, pages 1-29. IEEE Press, 1992.

[9] Degrauwe, M., Vittoz, E. Verbauwhede, I., "A micropower CMOS instrumentation amplifier," *IEEE Journal of Solid State Circuits*, SC-20(3):805-807, June 1985.

Chair: Chandu Visweswariah
Co-Chair: Karti Mayaram

Device modeling and analog simulation are increasingly challenging tasks for deep sub-micron designs. This informative and wide-ranging session addresses state-of-the-art modeling and simulation techniques.

Due to manufacturing variations, statistical modeling and simulation of IC designs are important problems. The development and extraction of device models, their robustness, and coping with transistor mismatch constitute the main topic of this session. Additional topics include delay estimation in FPGAs, optimal design of operational amplifiers for oversampled converters and the analysis of singular circuits.

Device modeling standards have been a topic of hot debate in the industry. The lead-off paper in this session examines the robustness, accuracy and computational performance of the BSIM3v3 model. While on the theme of device modeling, the second paper in the session is devoted to a compact EEPROM device model. This model is accurate during the programming, erasing and reading operations. By computing the charge and voltage of the floating gate and considering tunneling in the substrate, threshold voltage and drain current shifts are accurately modeled.

The next paper proposes a new method of statistical device parameter extraction from measured data called "Recursive Inverse Approximation." This method attempts to find an inverse mapping from the performance space to the parameter space that allows a rapid recursive parameter correction procedure. The following paper examines the impact of transistor mismatch on the trade-offs between speed, accuracy and power in analog systems. Implications of mismatch in power consumption, noise and system performance are studied.

The fifth paper of the session changes gears slightly and addresses the optimal design of operational amplifiers for sigma-delta oversampled converters. Class A op-amps are shown to be the best choice and a minimum theoretical bound on their power consumption is derived. Delay estimation of segmented channel FPGAs is the subject of the next paper. Based on the observation of inherent symmetry in routing topologies, a fast model that can be used in the inner loop of automatic timing-driven layout is proposed.

The final paper of the session addresses the analysis of singular circuits, such as filters with poles at zero or chains of pass gates that are off. The minimum least-squares solution of the singular system of equations is computed, while retaining the sparsity of the matrices involved. Thus, large circuits can be analyzed with this method.

Join us for an interesting session on device modeling and analog simulation!

An Investigation on the Robustness, Accuracy and Simulation Performance of a Physics-Based Deep-submicronmeter BSIM Model for Analog/Digital Circuit Simulation

Yuhua Cheng, Min-Chie Jeng*, Zhihong Liu, Kai Chen, Mansun Chan, Chenming Hu, Ping Keung Ko#**

Dept. of Electrical Eng. and Computer Science, Univ. of California at Berkeley, Berkeley, CA94720
* Cadence Design System
** BTA Technology Inc.,
# Hong Kong University of Science and Technology, Hong Kong

Abstract

We present an accurate and unified MOSFET model with benchmark test results for analog/digital circuit simulation. The results show that the model can pass most benchmarks suggested for a model used in circuit simulation by SEMATECH recently, and ensures good scalability and accuracy. The model has been implemented in Hspice, Spectre, SmartSpice and Spice3e2.

I. Introduction

The requirements for a MOSFET model with good robustness, accuracy (scalability) and computational efficiency have been discussed recently by both circuit designers and device model developers to meet the needs of analog and mixed analog/digital designs. Many different MOSFET models have been developed [1-4]. Some models can ensure good accuracy in specific operation regimes, but can not guarantee perfectly smooth transition of the model from subthreshold to strong inversion [1]. Some models can ensure a continuity in all operation regimes, but trail in computational efficiency due to implicit numerical iteration required [2] or lack scalability in the models based on semi-empirical approaches [3,4]. In this paper, we present a physics-based continuous BSIM (Berkeley Short-channel IGFET Model) -- BSIM3v3 with benchmark tests for its robustness, accuracy (scalability) and simulation performance according to the tests suggested by SEMATECH Compact Model Workshop recently [5,6].

II. Model Description

Unlike previous versions, BSIM3v3 introduces unified channel charge and mobility expressions. This, in effect, ensures continuity of the conductances and capacitances in all operation regimes. It accounts for the major physical effects in state-of-the art MOSFETs such as (a) Short- and narrow-channel effects, (b) Non-uniform-doping effect, (c) Mobility reduction due to vertical field, (d) Bulk-charge effect. (e)Carrier velocity saturation, (f) Drain-induced barrier lowering (DIBL) (g) Channel length modulation (CLM), (h) Substrate current induced body effect (SCBE), (I) Subthreshold conduction, (J) Source/drain parasitic resistances. A unified drain current expression in all operation regimes is developed:

$$I_{ds} = \frac{I_{dso}}{1 + \frac{R_{ds} I_{dso}}{V_{dseff}}} \left(1 + \frac{V_{ds} - V_{dseff}}{V_A}\right)\left(1 + \frac{V_{ds} - V_{dseff}}{V_{ASCBE}}\right) \quad (1)$$

where

$$I_{ds0} = \frac{W_{eff}\mu_{eff} Q_{chs0} V_{dseff}\left(1 - \frac{V_{dseff}}{2V_b}\right)}{L_{eff}\left(1 + \frac{V_{dseff}}{E_{sat}L_{eff}}\right)} \quad (2)$$

$$V_{gsteff} = \frac{2\, n\, v_t \ln\left[1 + \exp\left(\frac{V_{gs} - V_{th}}{2\, n\, v_t}\right)\right]}{1 + 2\, n\, Cox\sqrt{\frac{2\Phi_s}{q\varepsilon_{si}N_{ch}}}\exp\left(-\frac{V_{gs} - V_{th} - 2V_{off}}{2\, n\, v_t}\right)} \quad (3)$$

where R_{ds} is the drain/source parasitic series resistances, I_{dso} is the unified intrinsic *I-V* expression for R_{ds}=0, V_A is the Early voltage contributed by the CLM, DIBL effects and V_{ASCBE} is the Early voltage contributed by the SCBE effect [1], Vth is the threshold voltage, V_{gsteff} become V_{gs}-V_{th} when V_{gs}-V_{th} >>$K_B T/q$ and follows an exponential relationship with V_{gs}-V_{th} in the subthreshold region [7,8]. V_{dseff} is a function that is very close to V_{ds} when $V_{ds}<V_{dsat}$, and approaches V_{dsat} when $V_{ds}>V_{dsat}$. v_t is the thermal voltage. N_{ch} is the channel doping concentration. ϕ_s is the surface potential. V_{off} is an offset voltage in the subthreshold regime [3], *n*

0-7803-3177-6 $5.00 © 1996 IEEE

is the subthreshold swing parameter to be extracted from experiments.

Besides continuity, the accuracy of the model has also been enhanced by the following: (1) improving the threshold voltage model to simulate both the short-channel and narrow-width effects more accurately, (2) including the dependencies of R_{ds} on Vg, Vb, narrow width effects, (3) improving the body bias dependence of narrow width and DIBL effects, (4) improving the bias dependencies of bulk charge effect, (5) improving the body bias dependence of mobility, (6) introducing geometry dependence of channel length and width reduction factors dW and dL parameters, (7) including the drain bias dependence of subthreshold swing, (8) introducing the temperature dependence of R_{ds} parameter. (9) introducing a substrate current model.

III. Parameter Extraction

BSIM3v3 has 53 parameters for the I-V model, and 8 parameters for the C-V model. The parameter extraction in this paper is performed with the BSIMpro parameter extractor [9]. One large size device and three small size devices are the minimum set needed to extract parameters.

IV. Results and Discussion

The MOSFETs used in the tests are from two different 0.4um CMOS technologies with Tox of 9nm and 11 nm respectively and 3.3V power supply. The device geometry ranges from 0.4um to 6um in channel length, and from 0.6um to 20um in channel width.

(1) Qualitative tests:

A set of benchmark tests [5] has been performed and some results are shown in Figs. 1-6 along with the measured data. It can be seen from the different tests that the present model has removed the glitches and negative conductance problems of previous version of BSIM model, and can guarantee the continuities of Ids, and its derivatives as well as Gm/Id at all Vgs and Vds bias conditions.

(2) Quantitative (accuracy) tests:

The accuracy of the model down to deep-submicron can be seen in Fig. 1-6, where a 0.4um device was used. The maximum error in Ids is less than 2%. To demonstrate the model scalability further, more results for different device sizes are given in Figs. 7-12. It can be seen that the model can describe the V_{th} and I_{dsat} characteristics well, and also can match the current and G_{ds}-V_{ds} characteristics of devices of different W/L. The maximum error in Ids across different device geometries is less than 5%.

(3) Performance tests

Many different circuits have been simulated using the present model. The results of two typical circuits, a 10-bit decoder and a PLL, are shown in Table I. For comparison, the results of other SPICE3 MOSFET models are also included in the table. All simulations are performed with Cadence Spectre [10]. Eventhough the model is slower than the previous versions due to including a lot of log and exponential functions, the numbers of iterations needed in the simulation is reduced due to the model smoothness, which will lead to easier convergence in tough circuits. Also, the increase in the computation time can be compensated with the significant accuracy enhancement in the model.

IV. Summary

An investigation on the robustness, accuracy and simulation performance of a physics-based deep-submicronmeter BSIM for analog/digital circuit simulations is presented. BSIM3v3 can accurately model the MOSFET characteristics over a wide range of device geometries for various technologies. The physics-based nature of the model also makes it very suitable for statistical modeling. The present model has been implemented in Hspice, Spectre, SmartSpice and Spice3e2.

Acknowledgments

This work was supported by SRC grant BJ-324-029.

References

[1] J. H. Huang, et al, *BSIM3 Version2.0 User's Manual*, March 1994

[2] A. R. Boothroyd, et al, IEEE Trans. Computer-Adied Design, vol. 10, pp.1512-1528, 1991

[3] M. C. Jeng, *ERL memorandum ERL M90/90*, University of California, Berkeley, 1990

[4] B. J. Sheu, et al., *IEEE J. Solid-State Circuits*, vol. SC-22, pp.558-565, 1987

[5] SEMATECH, Compact Models Workshop, Sunnyvale, Aug. 1995

[6] Y. Tsividis et al., *Tech. Dig.* of CICC-93, pp14.1.1-14.1.6, 1993.

[7] Y. H. Cheng, et al., 1995 Intl Semiconductor Device Research Symposium, Charlottesville

[8] H. K. Park et al., *IEDM Tech. Dig.* Pp.485-488, 1986

[9] BSIMpro Manual, BTA Inc., 1994

[10] Cadence Spectre Manual, Cadence Design Systems, 1994

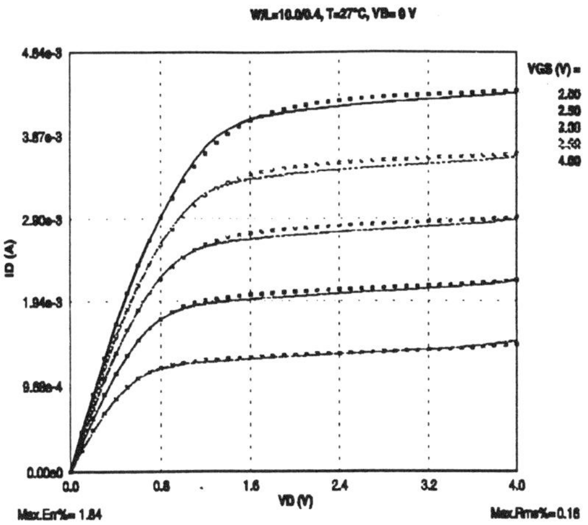

Fig.1 Measured and simulated drain current Ids versus drain bias Vds characteristics (solid lines:BSIM3v3, sympols: data).

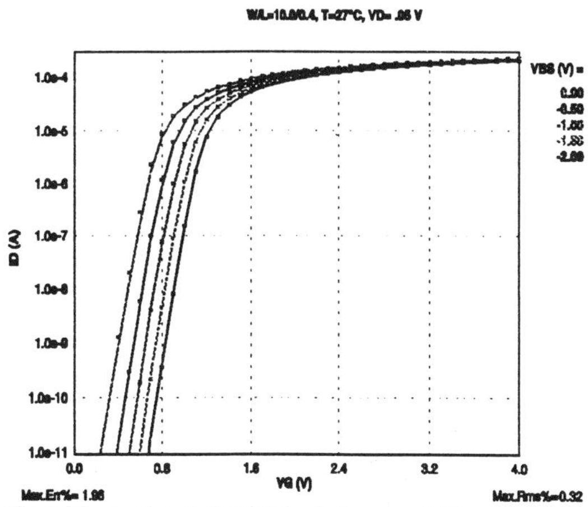

Fig.4 Measured and simulated drain current Ids versus gate bias Vgs characteristics (Log scale) (solid lines:BSIM3v3, sympols: data).

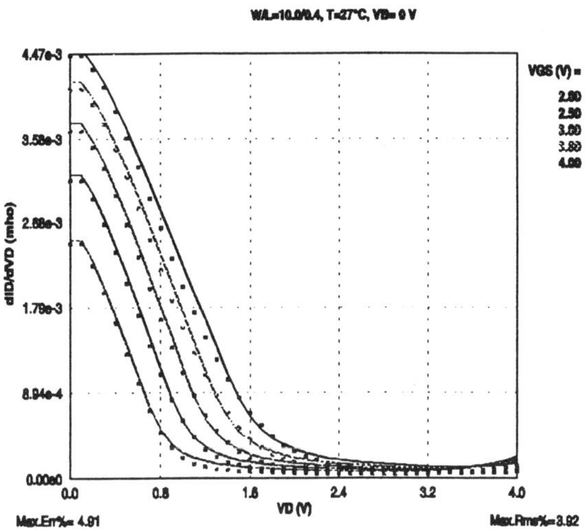

Fig.2 Measured and simulated drain current Gds versus drain bias Vds characteristics(solid lines:BSIM3v3, sympols: data).

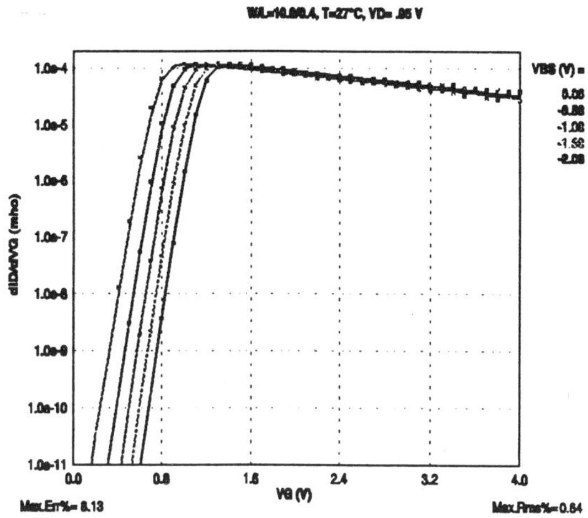

Fig.5 Measured and simulated transconductance Gm versus gate bias Vgs characteristics (Log scale) (solid lines:BSIM3v3, sympols: data).

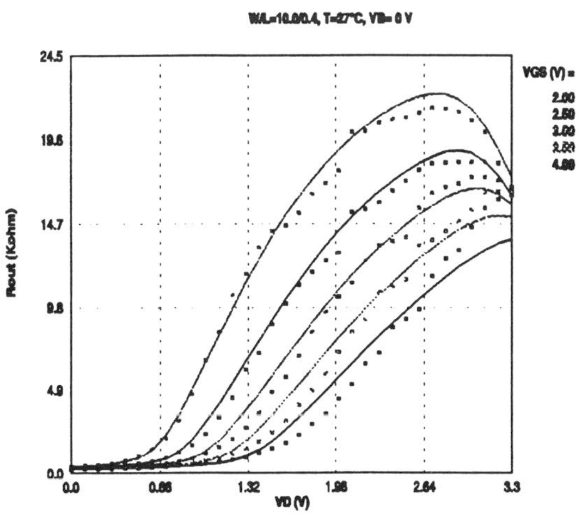

Fig.3 Measured and simulated drain current Rout versus drain bias Vds characteristics(solid lines:BSIM3v3, sympols: data).

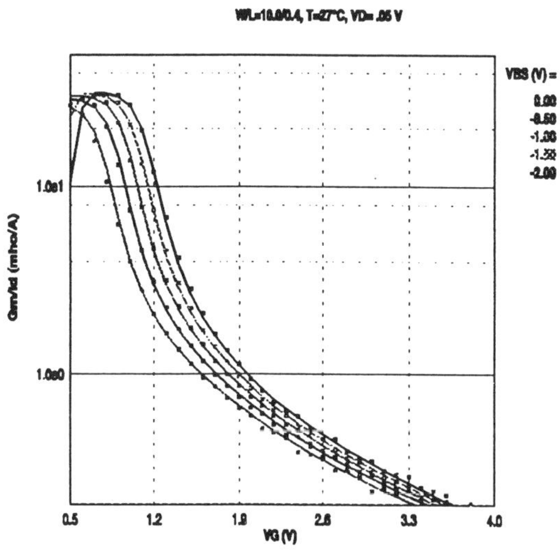

Fig.6 Measured and simulated drain current Gm/Ids versus gate bias Vgs characteristics (solid lines:model, sympols: data).

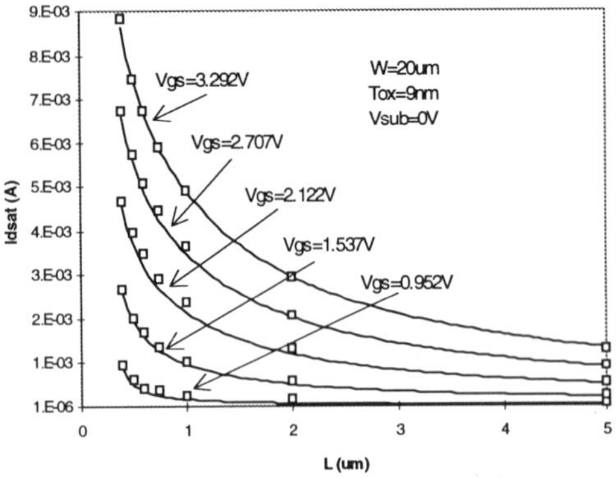

Fig.7 Measured and simulated saturation drain current Idsat versus channel characteristics(solid lines:model, sympols: data).

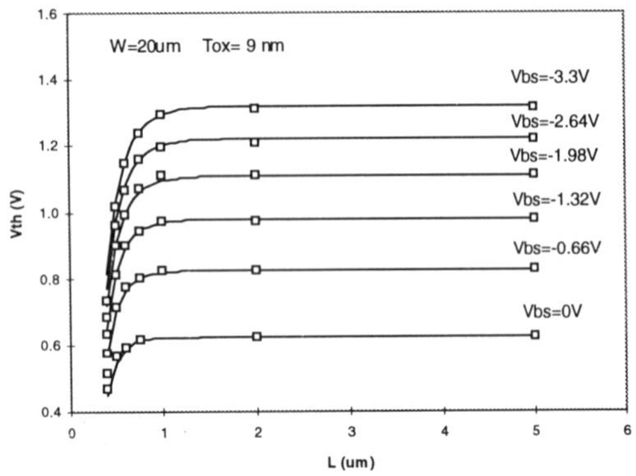

Fig.8 Measured and simulated threshold voltage Vth versus channel length L characteristics (lines: :model, sympols: data).

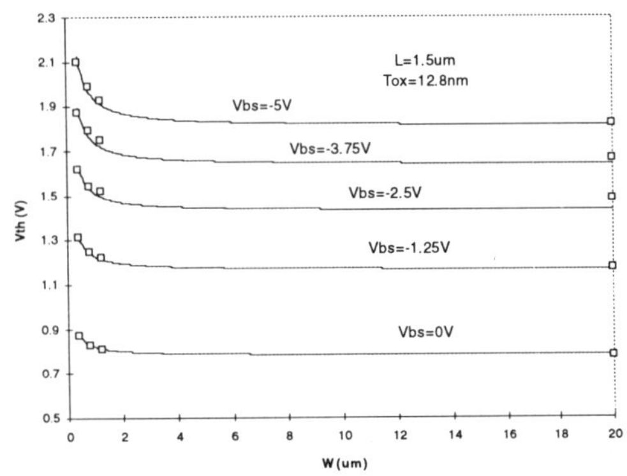

Fig.9 Measured and simulated threshold voltage Vth versus channel width W characteristics (lines: :model, sympols: data).

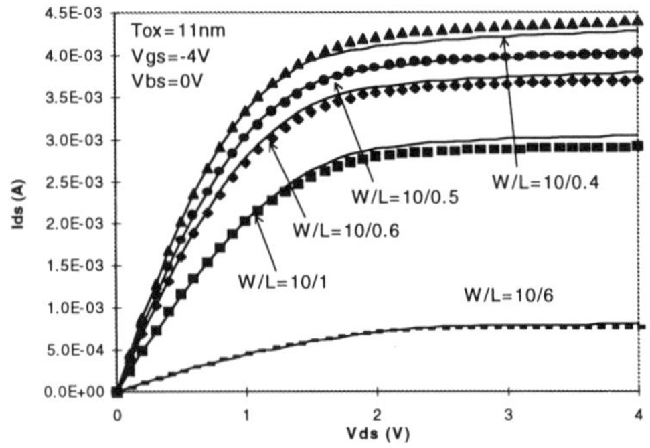

Fig. 10 Measured and simulated Ids versus Vds curves of devices of different W/L (lines: :model, sympols: data).

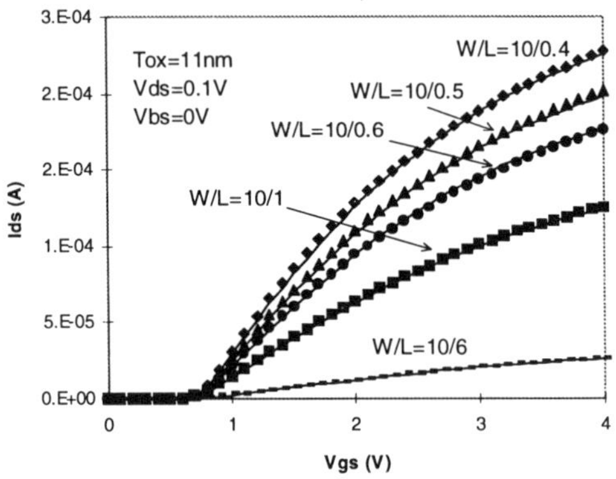

Fig. 11 Measured and simulated Gds versus Vds curves of devices of different W/L (lines: :model, sympols: data).

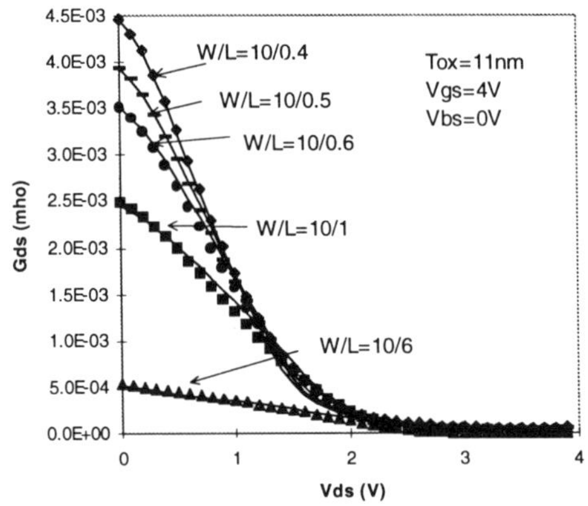

Fig. 12 Measured and simulated Ids versus Vgs curves of devices of different W/L (lines: :model, sympols: data).

Table I. Comparison of simualtion performaces of models

	Decoder			PLL		
Model	DC time (s)	Total Time (s)	Number of Iterations	DC time (s)	Total Time (s)	Number of Iterations
Level 6	2.4	684.4	16516	8.0	33.7	864
level 2	14.3	1285.7	16810	16	41.8	449
level3	1.52	575.0	13077	3.3	21.8	568
BSIM1	2.25	712.1	16330	4.9	23.4	507
BSIM2	2.75	872.0	16588	11.0	40.3	703
BSIM3v2	2.2	728.8	15016	2.15	24.0	578
BSIM3v3	6.0	1074.8	14119	18	44.7	420

An EEPROM Compact Circuit Model

P. Klein, K. Hoffmann, O. Kowarik

University of Bundeswehr Munich, Institute of Electronics ET4
Werner-Heisenberg-Weg 39, 85577 Neubiberg, Germany

Abstract — **The model allows the simulation of threshold voltage and drain current shifts as well as FN-tunnel and substrate currents caused by FN and Band-to-Band tunneling. Achieved is this by determining the floating gate charge and voltage as function of time, short channel and geometry effects during programming, erasing and reading.**

INTRODUCTION

In order to simulate the electrical behavior of EEPROM cells in a circuit a compact circuit model is necessary. Such a model has to calculate the cell threshold voltage shift and sense current and its tolerances, depending on the cells voltages, geometries, doping etc.. The model must further allow the simulation of substrate and tunnel currents as function of the programming and erasing speed. This is particularly important if an on-chip power supply has to be designed according to the consumed currents.

EEPROM MODEL

A. Determination of the floating gate potential:

In fig. 1 the concept of the developed model is shown. To calculate the tunnel current I_{tu} and the substrate current I_B the potential at the floating gate V_F must be determined. For this we use the exact charge description of [1] and [2] for the MOS-transistor drain (D), floating gate (F), source (S) and bulk (B) (sketched by thick lines in fig. 1) and its overlap regions. The stored charges are:

$$Q_D = f(V_D, V_F, V_S, V_B) \quad (1)$$

$$Q_S = f(V_D, V_F, V_S, V_B) \quad (2)$$

$$Q_B = f(V_D, V_F, V_S, V_B) \quad (3)$$

The stored charge on the control gate (C) is

$$Q_C = C_{CF} \cdot V_{CF} \quad (4)$$

with the coupling capacitance C_{CF} and the voltage $V_{CF} = V_C - V_F$ between control gate and floating gate. The total charge on the floating gate Q_F results from the charge neutrality equation

$$\begin{aligned} 0 &= Q_F + Q_D + Q_S + Q_B + Q_C \\ Q_F &= -Q_D - Q_S - Q_B - Q_C. \end{aligned} \quad (5)$$

From an initial condition e.g.

$$\begin{aligned} Q_F(t = 0\text{s}) &= 0\text{C} \quad \text{or} \\ V_F(t = 0\text{s};\ V_D, V_S, V_B, V_C = 0\text{V}) &= 0\text{V} \end{aligned} \quad (6)$$

six equations with six unknowns Q_F, Q_D, Q_S, Q_B, Q_C and the floating gate voltage V_F exist. Notice that this novel methode does not calculate the floating gate potential V_F by capacitance coupling ratios as e.g. in [3], [4], [5] but uses a correct large signal model with the charges Q_D, Q_S and Q_B. Thus all width and channel length effects are also included as the strong bias dependence of the MOS-capacitances and the overlap capacitances [1], [2]. Especially last metioned bias dependence must be taken into account, because V_F varies over the entire voltage range of e.g. -5V to about +10V during programming.

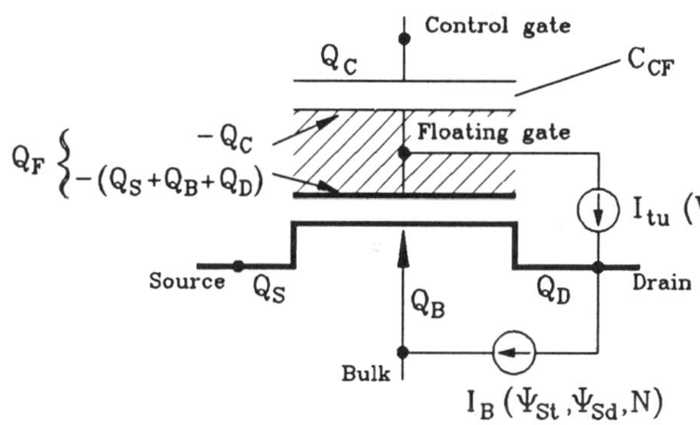

Figure 1: Charge based EEPROM model

B. Calculation of the tunnel current I_{tu}:

The tunnel current density J_{tu} depends strongly on the electrical field in the tunnel oxide $E_{tu} = V_{tu}/d_t$ [6], where V_{tu} is the tunnel voltage and d_t is the tunnel oxide thickness.

$$J_{tu} = B_1 E_{tu}^2 \exp\left(-\frac{B_2}{|E_{tu}|}\right) \quad (7)$$

B_1 and B_2 are characteristic constants [7]. Figure 2a showes how the tunnel voltage V_{tu} across the oxide is reduced by a voltage drop Ψ_{St} across the space charge region during programming. This voltage has in depletion and

0-7803-3177-6 $5.00 © 1996 IEEE

IEEE 1996 CUSTOM INTEGRATED CIRCUITS CONFERENCE

deep depletion a value of

$$\Psi_{St} = \frac{1}{4}\left(-\tilde{\gamma}_t + \sqrt{\tilde{\gamma}_t^2 - 4(V_{FD} - \tilde{V}_{FB})}\right)^2 \qquad (8)$$

with the body effect coefficient $\tilde{\gamma}_t$ of the tunnel region [1]. If the electrical field in the silicon is high enough valence band electrons tunnel through the forbidden energy gap to the conduction band leaving behind a hole (Band-to-Band tunneling, fig. 2b). Ψ_{St} approaches a fixed value of $\Psi_{St}(fix)$ e.g. 4V during voltage ramping. This occurs when an increase of Ψ_{St} is compensated by an increased Band-to-Band tunnel current density J_{BB} [8],[9].

$$J_{BB} = B_3 E_{Si} \exp\left(-\frac{B_4}{E_{Si}}\right) \qquad (9)$$

$$E_{Si} = \sqrt{\frac{2q}{\epsilon_0 \epsilon_{Si}} \Psi_{St} N_D} \qquad (10)$$

In this equilibrium case the tunnel voltage V_{tu} increases proportionally with the drain-floating gate voltage V_{DF}. This continues till the critical oxide field E_{crit} of about $E_{crit} \approx 1\text{V/nm}$ is reached (in our example $V_{tu} = 8\text{V}$ for $d_t = 8\text{nm}$). Then the tunnel current I_{tu} begins to flow. In this case the ionization rate of FN-tunnel electrons is 1.8 pairs [10]. The additional generated holes move towards the Si/SiO_2 interface and reduce the fixed $\Psi_{St}(fix)$ voltage

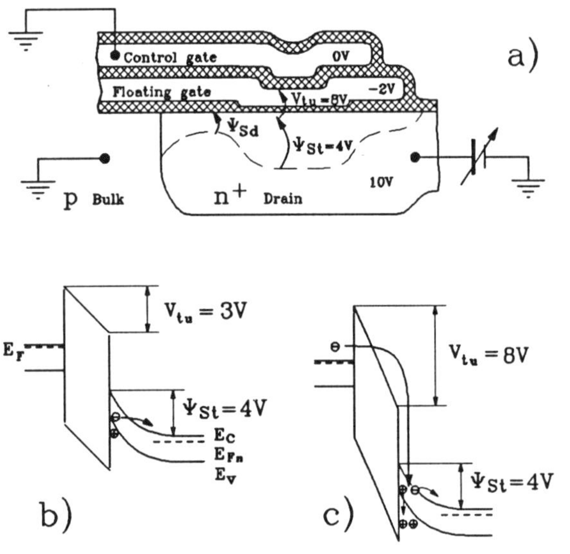

Figure 2: Voltages and surface potentials during programming

to the lower level of Ψ_{Sd} (fig. 2a). This sudden reduction of Ψ_{St} increases the tunnel voltage V_{tu} at the same time and an anomalous current peak occurs [11]. Ψ_{Sd} is calculated similar to Ψ_{St} with an appropriated value of $\tilde{\gamma}_d$.

During erasing the voltage at the control gate is increased while drain, source and bulk are grounded. In this case the drain tunnel region is accumulated and therefore the voltage drop Ψ_{St} is much lower than during programming and can be calculated similarly [12].

C. Determination of the substrat current I_B:

For calculation of the substrat current I_B during programming three special cases have to be considered. These special cases (fig. 3) depend on the doping concentration as well as on the oxide thicknesses.

1. Before FN-tunneling:

- Fig. 3a) very high doping n^{++}:
  - Surface potentials: e.g. $\Psi_{St} = 0.3V$, $\Psi_{Sd} \approx 0V$
  - Generation mechanism: none
  - Substrat current I_B: neglectable
- Fig. 3b) high doping n^+:
  - Surface potentials: e.g. $\Psi_{St} = \Psi_{St}(fix) = 4V$, $\Psi_{Sd} < \Psi_{St}$
  - Generation mechanism: Band-to-Band tunneling
  - Substrat current I_B: neglectable
- Fig. 3c) medium doping n or low oxid tickness ratio d_d/d_t (d_d is the gate oxide thickness):
  - Surface potentials: e.g. $\Psi_{Sd} \geq \Psi_{St} > \Psi_{St}(fix)$
  - Generation mechanism: Band-to-Band tunneling
  - Substrat current I_B: extremely high because $\Psi_{Sd} \geq \Psi_{St}$

The high substrat current in case c) can be calculated by

$$I_B = I_{BB} = AJ_{BB} \qquad (11)$$

where A is the effective Band-to-Band tunnel area. To avoid this excessive current a minimum oxid thickness ratio d_d/d_t versus the drain doping can be derived from our model.

2. During FN-tunneling:

When the tunnel current begins to flow 1.8 pairs as mentioned are generated. The holes move to the tunnel region surface till $\Psi_{St} = \Psi_{Sd}$ and finally flow towards the bulk as additional substrate current. This additional current caused during tunneling exists in all considered cases. However the FN-tunnel part of the substrate current is much lower than the one generated by Band-to-Band tunneling (compare fig. 3a and 3b with fig. 3c).

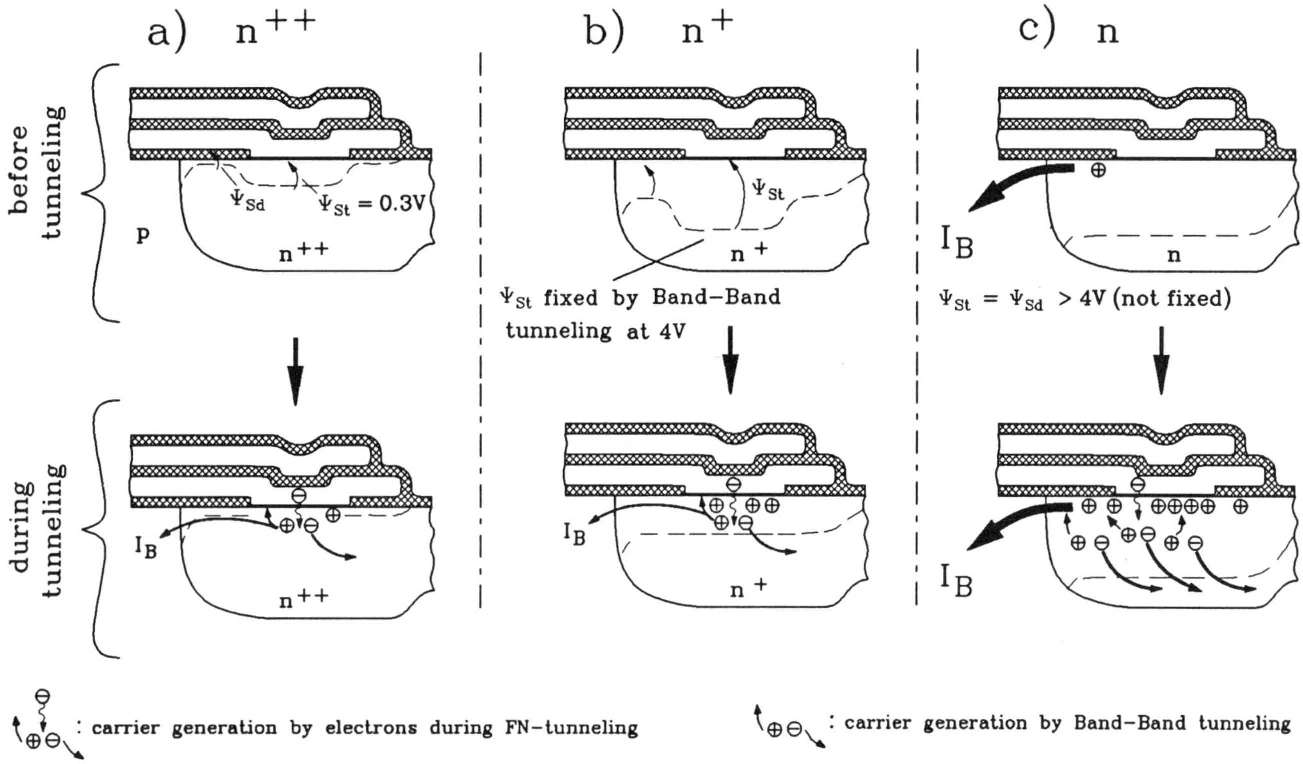

Figure 3: Sketch to explain I_B

Measurement and Simulation

In fig. 4 a comparision of a typical threshold voltage shift versus program/erase voltage is shown.

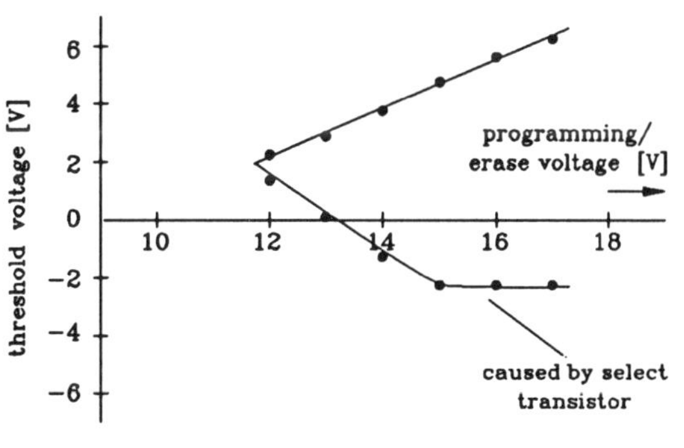

Figure 4: Threshold voltage shift versus program/erase voltage
—— simulation • • measurement

The excellent agreement between simulation and measurement indicates the correct determination of the floating-gate voltage V_F as well as the surface potential Ψ_{St}, as a basis to model I_{tu} exactly. In further simulations the influence of the substrate current I_B during programming was investigated. EEPROM cells were erased with a voltage of 16V at the control gate while drain, bulk and source were grounded. Then all cells (200x50) were programmed by a drain voltage of 16V. Hereby the control gate and the bulk are grounded over a 1 kΩ resistance to measure the currents I_C and I_B while the source is floating (compare fig. 5). The ramp speed is $8 \cdot 10^5$ V/s. Two cells with different technology have been measured and simulated.

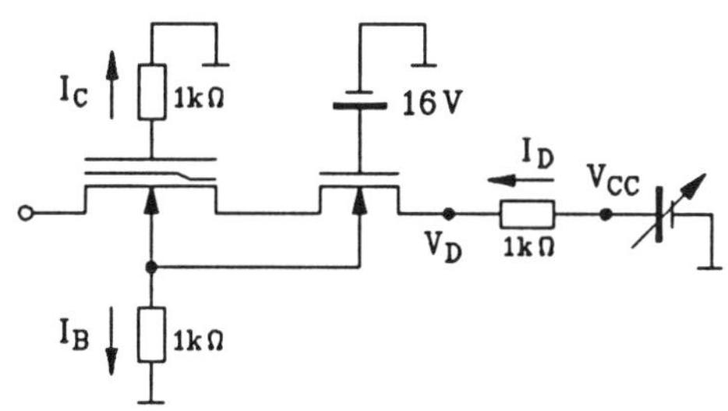

Figure 5: Measurement circuit

During the ramping of the program voltage the currents

$$I_C = -\frac{dQ_C}{dt} = \frac{d(Q_F + Q_D + Q_S + Q_B)}{dt} = -I_{tu} + \frac{d(Q_D + Q_S + Q_B)}{dt} \quad (12)$$

and

$$I_B = -\frac{dQ_B}{dt} - 1.8 I_{tu} + I_{BB} \qquad (13)$$

flow. Fig. 6 showes the measurement and simulation result of cells with a drain doping of $N_D \approx 9 \cdot 10^{18} \text{cm}^{-3}$ and an oxid thickness ratio $d_d/d_t = 3.5$. When V_{CC} reaches a value of about 8V a tunnel current $I_{tu} > 0$ flows thus I_C and I_B increase (compare (12) and (13)). The substrat current part I_{BB} caused by Band-to-Band tunneling (13) is neglectable because of a high oxide thickness ratio d_d/d_t.

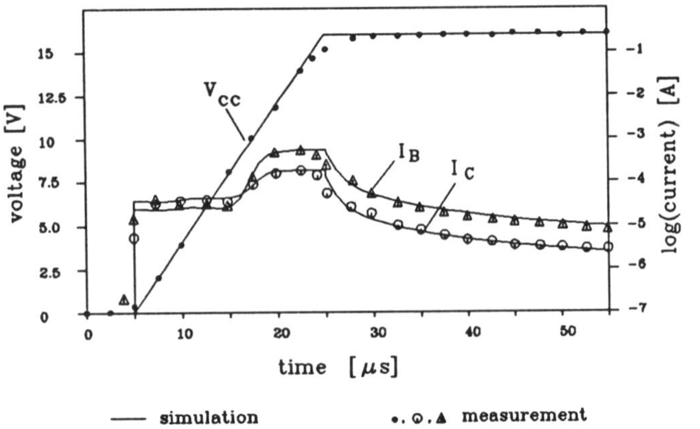

Figure 6: Measurement and simulation of I_C and I_B during programming ($d_d/d_t = 3.5$)

In fig. 7 the oxide thickness ratio is 1. Therefore a very high substrate current flows caused by Band-to-Band tunneling (I_{BB}) reducing the programming voltage dramatically. As a result the tunnel current I_{tu} respectively I_C is much lower thus the cells can not be programmed.

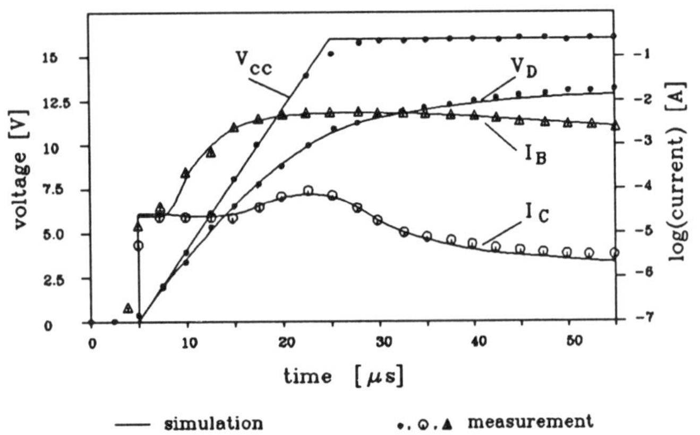

Figure 7: Measurement and simulation of I_C and I_B during programming ($d_d/d_t = 1$)

Conclusion

An EEPROM model has been implemented in the circuit simulator SABER and verified by measurements. In difference to hitherto developed models the floating gate potential is calculated by the correct charge descriptions and the charge neutrality equation. This is the base to determine all currents as function of time, geometry, doping and voltage.

Acknowledgment

This work was sponsored by the DFG (Ho 1325/2-1). The authors are grateful for the technology support by Mr. Plasa from Siemens AG Munich.

References

[1] P. Klein, K. Hoffmann, B. Lemaitre, "Description of the bias dependent overlap capacitance at LDD MOSFETs for circuit applications"; IEDM93 Washington, DC, 1993.

[2] P. Klein, K. Hoffmann, B. Lemaitre, "Submicron Charge LDD-MOSFET Model for Low Voltage Circuits"; CICC 95.

[3] A. Kolodny, S. T. K. Nieh, B. Eitan, J. Shappir, "Analysis and Modeling of Floating-Gate EEPROM Cells"; IEEE Transactions on Electron Devices, June 1986.

[4] Tat-Kwan Yu, Jack Higman, Craig Cavins, Kuo-Tung Chang and Marius Orlowski, "An EEPROM Model for Low Power Circuit Design and Simulation"; Technical Digest, IEDM 1994.

[5] L. C. Liong, P. Liu, "A Theoretical Model for the Current-Voltage Characteristics of a Floating-Gate EEPROM Cell"; IEEE Transactions on Electron Devices, January 1993.

[6] M. Lenzlinger and E. H. Snow, "Fowler-Nordheim tunneling into thermally grown SiO_2"; J. Appl. Phys., vol. 4, 1969.

[7] A. Concannon, S. Keeney, A. Mathewson, R. Bez, and C. Lombardi, "Two-Dimensional Numerical Analysis of Floating-Gate EEPROM Devices"; IEEE Transactions on Electron Devices, July 1993.

[8] S. M. Sze, Physics of Semiconductor Devices, 1981.

[9] C. Chang, J. Lien, "Corner-Field Induced Drain Leakage in Thin Oxide MOSFETs"; Technical Digest IEDM, 1987.

[10] Avinoam Kolodny, Sidney T. K. Nieh, Boaz Eitan and Joseph Shappir, "Analysis and Modeling of Floating-Gate EEPROM Cells"; IEEE Transactions on Electron Devices, June 1986.

[11] J. Suñé, M. Lanzoni, R. Bez, P. Olivo, and B. Riccò, "Transient simulation of the ERASE cycle of floating gate EEPROMs"; IEDM91.

[12] J. Suñé, P. Olivo, and B. Riccò, "Quantum-Mechanical Modeling of Accumulation Layers in MOS Structure"; IEEE Transactions on Electron Devices, July 1992.

An Efficient Approach to Device Parameter Extraction for Statistical IC Modeling*

Ming Qu † and M.A. Styblinski ‡

† National Semiconductor Corporation, M/S A2-595
2900 Semiconductor Dr., Santa Clara, CA 95052
‡ Department of Electrical Engineering
Texas A&M University, College Station, TX 77843

ABSTRACT

A technique called *Recursive Inverse Approximation* (RIA) has been developed for parameter extraction for statistical IC modeling. High accuracy and efficiency are achieved by the proposed methodology.

I. Introduction

The parameter extraction of IC device models is an essential part of statistical IC modeling. There is usually *a limited number of physical quantities* which can be measured *directly*. We refer to these measurable quantities as *IC device performances* or simply *performances*. Parameter extraction is to fit the measured performances by selecting the IC model parameters using a certain criterion. Except for very simple cases, the mapping from the measurable performances to the model parameters is very complicated and unknown. Parameter extraction is used to establish this unknown mapping through the known mapping (IC model) from the model parameters to the performances.

There are relatively few publications stressing the problems of parameter extraction for statistical IC modeling [1, 2, 3]. Most often, general purpose optimization programs were used for parameter extraction [2]. The optimization-based parameter extraction is problematic when it is used for statistical characterization and modeling. To extract the data set of model parameters corresponding to many measured performances, the optimizer has to be used for each measurement. The stopping criterion in the optimizer is very critical, since the uncertainties due to the optimizer's termination of each parameter extraction can be included in the statistical modeling. In fact, the "noise" level introduced by an optimization-based parameter extraction is sometimes very high.

Direct parameter extraction [4, 5, 3], which is based on analytical equation-solving techniques, requires the measurement of minimized data set and the analytical expressions formulated for the particular model. Whenever the analytical solutions are obtainable, direct parameter extraction is always desirable. However, obtaining the analytical solutions is not always feasible for complex models.

The special features of the proposed approach include: (1) reliable accuracy checking, (2) a composite prediction and correction procedure making the parameter extraction controllable, (3) very accurate results, and (4) high speed.

II. Problem Formulation

Let $\mathbf{p} = [p_1, p_2, ..., p_n]^T \in R^n$ denote the IC model parameter vector to be extracted, $\mathbf{y} = [y_1, y_2, ..., y_m]^T \in R^m$ denote the IC model performance vector, $\mathbf{z}^k = [z_1^k, z_2^k, ..., z_m^k]^T \in R^m$, $k = 1, 2, ..., K$ denote the measured performance vector of the kth device with K being the number of devices, $\mathbf{x} = [x_1, x_2, ..., x_k]^T \in R^k$ denote the designable IC model parameter vector, $\mathbf{v} = [v_1, v_2, ..., v_l]^T \in R^l$ denote the environmental parameter vector.

A general IC model can be written as

$$\mathbf{y}' = \phi(\mathbf{p}, \mathbf{x}, \mathbf{v}). \qquad (1)$$

Eq. (1) reflects the functional relationship between the model performance vector $\mathbf{y}'$ and the model parameter vector $\mathbf{p}$, which is parameterized by the designable parameter vector $\mathbf{x}$ and the environmental parameter vector $\mathbf{v}$. The measurements can only be made under certain selected values of $\mathbf{x}$ and $\mathbf{v}$ (discretization), so, an IC model can be simply denoted as[1]

$$\mathbf{y} = \mathbf{y}(\mathbf{p}). \qquad (2)$$

To extract a parameter vector $\mathbf{p}_{(*)}^k$ corresponding to the kth device, we want to find

$$\mathbf{p}_{(*)}^k = arg\left\{ \min_{\mathbf{p}^k \in R^n} \| \mathbf{y}(\mathbf{p}^k) - \mathbf{z}^k \| \right\} \qquad (3)$$

where, $\| \cdot \|$ denotes a suitable norm.

The weighted sum of error squares for the kth device is formed as

$$\varepsilon(\mathbf{p}^k) = \frac{1}{2}\sum_{i=1}^{m} w_i[y_i(\mathbf{p}^k) - z_i^k]^2 = \frac{1}{2}[\mathbf{y}(\mathbf{p}^k) - \mathbf{z}^k]^T \mathbf{W}[\mathbf{y}(\mathbf{p}^k) - \mathbf{z}^k] \qquad (4)$$

*Supported in part by Semiconductor Research Corporation projects No.94-DJ-229 and 95-DJ-229.

[1] Note that, in this description, $\mathbf{y}$ expresses all the measured performances for the different combinations of x_i, $i = 1, ..., k$ and v_j, $j = 1, ..., l$, so, $\mathbf{y}$'s dimension is higher than $\mathbf{y}'$'s.

0-7803-3177-6 $5.00 © 1996 IEEE
IEEE 1996 CUSTOM INTEGRATED CIRCUITS CONFERENCE

Statistical parameter extraction is defined as:

Given the known IC model mapping, Eq. (2), find the parameter vectors $\mathbf{p}_{(*)}^{k} \in R^{n}$, $k = 0, 1, 2, ..., K$ *such that*

$$\mathbf{p}_{(*)}^{k} = arg\left\{ \min_{\mathbf{p}^{k} \in R^{n}} \varepsilon(\mathbf{p}^{k}) \right\} \tag{5}$$

where, $\varepsilon(\mathbf{p}^{k})$ *is given by Eq. (4).*

It is noted that the above formulation is only conceptual, since the *exact* solution required in (5) is practically impossible for each device. Because of that, Eq. (5) is referred to as an "ideal" solution[2].

III. Recursive Inverse Approximation

In statistical IC parameter extraction, we only know the mapping from the model parameter space to the model performance space, i.e., $\mathbf{y} = \mathbf{y}(\mathbf{p})$ is known, but the *inverse mapping* from the model performance space to the model parameter space is, in general, not known. The *Recursive Inverse Approximation* (RIA) is to find an approximation to the *inverse mapping.*

The inverse mapping $\mathbf{p} = \psi(\mathbf{y})$ is referred to as the mapping from the model performance space to the model parameter space under the least-squares criterion, which can be found indirectly for the kth device, solving the problem

$$\min_{\mathbf{p}^{k} \in R^{n}} \left\{ \varepsilon(\mathbf{p}^{k}) \equiv \frac{1}{2}[\mathbf{y}(\mathbf{p}^{k}) - \mathbf{z}^{k}]^{T}\, \mathbf{W}\, [\mathbf{y}(\mathbf{p}^{k}) - \mathbf{z}^{k}] \right\}. \tag{6}$$

RIA is to find a mapping $\mathbf{p} \cong \tilde{\psi}(\mathbf{y})$ which is an approximate inverse mapping.

If we approximate the IC model around the nominal parameter vector by a linear function $\mathbf{y} \approx \mathbf{y}_{L}(\mathbf{p})$, then, under the least-squares criterion, the inverse mapping is uniquely defined if the number of measured points m for each device is no less than the number of model parameters. We denote such an inverse approximation obtained through the linear approximation of the IC model $\mathbf{y}(\mathbf{p})$ by $\hat{\mathbf{p}} \cong \tilde{\psi}_{L}(\mathbf{y})$[3]. The extracted model parameter vector for the kth device is then

$$\hat{\mathbf{p}}_{(0)}^{k} = \tilde{\psi}_{L}(\mathbf{z}^{k}) \tag{7}$$

where, the subscript "(0)" indicates the initial estimate of the parameter vector. Eq. (7) *predicts* an estimate of the model parameter vector.

In general, $\hat{\mathbf{p}}_{(i)}^{k} = \tilde{\psi}_{i}(\mathbf{z}^{k})$, $\tilde{\psi}_{i}(\cdot)$ is the ith approximation to the inverse mapping and $\tilde{\psi}_{0}(\cdot) \equiv \tilde{\psi}_{L}(\cdot)$. Iterative refinement of the estimated parameter vector for the kth device can be made using the following procedure:

$$\hat{\mathbf{p}}_{(i+1)}^{k} = \mathcal{F}_{i}(\hat{\mathbf{p}}_{(i)}^{k}) = \mathcal{F}_{i}(\tilde{\psi}_{i}(\mathbf{z}^{k})), \quad i = 0, 1, ... \tag{8}$$

subject to

$$\|\hat{\mathbf{p}}_{(i+1)}^{k} - \mathbf{p}_{(*)}^{k}\| < \|\hat{\mathbf{p}}_{(i)}^{k} - \mathbf{p}_{(*)}^{k}\| \tag{9}$$

[2]Even if a global optimization algorithm may be used to solve (5) for each device *ideally*, it is not practical, since (1) no practical implementation of a global optimization algorithm can, with 100% probability, find the true global minimum, (2) no stopping criterion in an optimizer can *really* identifies the global minimum.

[3]The hat "∧" denotes an estimate of parameter vector.

where, $\mathcal{F}_{i}(\cdot)$ is *an error correction mapping* in the $\mathbf{p}$ space, reducing the errors of extracted parameter values, $\mathbf{p}_{(*)}^{k}$ is the parameter vector satisfying Eq. (5) (an "ideal" solution).

Using this refinement, we actually establish a *recursive parameter correction* procedure as

$$\hat{\mathbf{p}}_{(i)}^{k} = \tilde{\psi}_{i}(\mathbf{z}^{k}) = \mathcal{F}_{i}(\mathcal{F}_{i-1} \cdots (\mathcal{F}_{0}(\tilde{\psi}_{L}(\mathbf{z}^{k})))). \tag{10}$$

As it is seen, the "exact" inverse mapping $\mathbf{p} = \psi(\mathbf{y})$ is recursively approximated by

$$\psi_{i}(\mathbf{y}) \cong \tilde{\psi}_{i}(\mathbf{y}) = \mathcal{F}_{i}(\mathcal{F}_{i-1} \cdots (\mathcal{F}_{0}(\tilde{\psi}_{L}(\mathbf{y})))). \tag{11}$$

In each iteration a new error correction mapping $\mathcal{F}_{j}(\cdot)$ is constructed. Eq. (10) actually corrects the "errors" of the previous estimate of the parameter vector for the kth device to obtain a more accurate parameter estimate and so is referred to as *parameter correction.* It is seen that the key point in such a procedure is to establish the transformation from the "preliminary" parameter estimate $\hat{\mathbf{p}}_{(j)}$ to the "more accurate" parameter estimate $\hat{\mathbf{p}}_{(j+1)}$.

A. Nominal Parameter Extraction

Parameter extraction for the nominal performances has been discussed extensively in literature and a number of methods have been reported [2]. One major difficulty in parameter extraction is the multi-minimum problem. The global optimization algorithm proposed in [6] is used.

B. Parameter Prediction

Once the nominal parameter vector $\mathbf{p}^{0}$ is extracted for the "nominal device", the parameter extraction for all the other devices can be performed using a linear approximation to the IC model. Since these estimates will be refined and corrected in the subsequent stages, the linear-approximation-based parameter extraction is called *Parameter Prediction.*

Suppose that $\Delta\mathbf{p} = \mathbf{p} - \mathbf{p}^{0}$ is a small change vector, by Taylor series expansion at the nominal point $\mathbf{p}^{0}$, the first order approximation of $\mathbf{y}(\mathbf{p})$ is given by

$$\mathbf{y}(\mathbf{p}) = \mathbf{y}(\mathbf{p}^{0} + \Delta\mathbf{p}) \approx \mathbf{y}(\mathbf{p}^{0}) + \mathbf{J}(\mathbf{p}^{0})\Delta\mathbf{p} \tag{12}$$

where, $\mathbf{J}(\mathbf{p}^{0})$ is the Jacobian matrix evaluated at $\mathbf{p}^{0}$. A linear least-squares problem is formed for the kth device as

$$\min_{\Delta\mathbf{p}^{k} \in R^{n}} \left\{ \varepsilon(\Delta\mathbf{p}^{k}) = \frac{1}{2}\left[\mathbf{J}(\mathbf{p}^{0})\Delta\mathbf{p}^{k} + \mathbf{y}^{0} - \mathbf{z}^{k}\right]^{T} \mathbf{W} \left[\mathbf{J}(\mathbf{p}^{0})\Delta\mathbf{p}^{k} + \mathbf{y}^{0} - \mathbf{z}^{k}\right] \right. \tag{13}$$

Since the error function in Eq. (13) is quadratic in $\Delta\mathbf{p}^{k}$, the solution is uniquely known. Let $\mathbf{H}$ represent the Hessian matrix of $\varepsilon(\Delta\mathbf{p}^{k})$. So, for the measured performance vector $\mathbf{z}^{k}$ for the kth device, an approximate estimate of the model parameter vector for the kth device is obtained from

$$\hat{\mathbf{p}}_{(0)}^{k} = \mathbf{p}^{0} + \Delta\hat{\mathbf{p}}_{(0)}^{k} \tag{14}$$

where

$$\Delta\hat{\mathbf{p}}_{(0)}^{k} = -\left[\mathbf{J}(\mathbf{p}^{0})^{T}\, \mathbf{W}\, \mathbf{J}(\mathbf{p}^{0})\right]^{-1} \mathbf{J}(\mathbf{p}^{0})^{T}\, \mathbf{W}\, (\mathbf{y}^{0} - \mathbf{z}^{k}). \tag{15}$$

Note that the solution of (15) exists only if the Jacobian matrix is of full rank. The *initial inverse approximation* is established as

$$\hat{\mathbf{p}}^k_{(0)} = \hat{\psi}_0(\mathbf{z}^k) = \mathbf{p}^0 - \left[\mathbf{J}(\mathbf{p}^0)^T \, \mathbf{W} \, \mathbf{J}(\mathbf{p}^0)\right]^{-1} \mathbf{J}(\mathbf{p}^0)^T \, \mathbf{W} \, (\mathbf{y}^0 - \mathbf{z}^k). \tag{16}$$

To calculate the Jacobian matrix, the finite difference method can be used, in which each parameter is perturbated to calculate the finite differences.

C. Accuracy Checking

Valid accuracy checking is very critical for statistical IC modeling. Our accuracy checking strategy can be considered as a *reverse verification* procedure. Suppose that we have a data set of the estimated model parameter vectors for all the measured devices. These parameter vectors can be those obtained from the parameter prediction or from the parameter correction to be discussed. We can then calculate the model performances corresponding to these *estimated parameters* (since the mapping $\mathbf{p} \in R^n \mapsto \mathbf{y} \in R^m$ defines the IC model and is known). The calculated model performances are called *simulated device performances*, denoted by

$$\tilde{\mathbf{y}}^k_{(i)} = \mathbf{y}(\hat{\mathbf{p}}^k_{(i)}), \quad k = 1, 2, ..., K \tag{17}$$

where, the subscript (i) is referred to the ith iteration. To check the accuracy of the extracted parameters, the *same parameter extraction procedure* (the one used to get $\hat{\mathbf{p}}^k_{(i)}$) is repeated using the *simulated device performances*. So, if the parameter extraction is accurate, the extracted parameters from the *simulated device performances* should be the same as the *estimated parameters*, having one-to-one correspondence. Thus, from this reverse procedure (realized through simulation), the paired parameter data set is obtained.

Let $\tilde{\mathbf{p}}^k_{(i)}$ be the extracted parameter vector from the *simulated device performances* for the kth device. It can be expressed as $\tilde{\mathbf{p}}^k_{(i)} = \tilde{\psi}_i(\tilde{\mathbf{y}}^k_{(i)})$. $\tilde{\psi}_i(\cdot)$ is actually the parameter extraction procedure whose accuracy to be checked. The *estimated parameter vectors* are given by $\hat{\mathbf{p}}^k_{(i)} = \tilde{\psi}_i(\mathbf{z}^k)$. The error difference is calculated from the paired parameter data

$$\tilde{\mathbf{d}}^k_i = \hat{\mathbf{p}}^k_{(i)} - \tilde{\mathbf{p}}^k_{(i)} = \hat{\mathbf{p}}^k_{(i)} - \tilde{\psi}_i(\mathbf{y}(\hat{\mathbf{p}}^k_{(i)})), \quad k = 1, 2, ..., K. \tag{18}$$

The paired data, one estimated by previous parameter extraction and another one extracted from the simulated device performances, are used for the accuracy checking.

D. Parameter Correction

When accuracy checking shows that the parameter extraction is not accurate enough, a procedure, called *Parameter Correction*, is performed to refine the parameter extraction. The basic idea underlying parameter correction is to correct the errors of parameter extraction based on the given IC model and the knowledge obtained from the previous stages.

Let us denote $\hat{\mathbf{p}}^k_{(i)} = \mathbf{p}^0 + \Delta\hat{\mathbf{p}}^k_{(i)}$, the extracted parameter vector for the kth device at the ith iteration. The objective of parameter correction is to find the transformation $\hat{\mathbf{p}}^k_{(i+1)} = \mathcal{F}_i(\hat{\mathbf{p}}^k_{(i)})$ such that more accurate parameter vectors can be extracted. The transformation (error correction mapping) $\mathcal{F}_i$ is selected in the form of

$$\hat{\mathbf{p}}^k_{(i+1)} = \hat{\mathbf{p}}^k_{(i)} + \mathbf{d}_i(\Delta\hat{\mathbf{p}}^k_{(i)}) \tag{19}$$

where, $\mathbf{d}_i$ is called *error correction function* and needs to be constructed.

The data set

$$\left\{\tilde{\mathbf{d}}^k_i, \, \Delta\tilde{\mathbf{p}}^k_{(i)}, \, k = 1, 2, ..., K\right\} \tag{20}$$

gives us the information relating the errors due to inaccurate parameter extraction to the extracted parameter values. A quadratic function is postulated to approximate the *error correction function* (the subscript is dropped for simplicity)

$$d_t = \sum_{j=1}^{n} \beta_j \Delta p_j + \sum_{j=1}^{n} \sum_{l=j}^{n} \beta_{jl} \Delta p_j \Delta p_l, \qquad t = 1, 2, .., n \tag{21}$$

where, $\mathbf{d} = [d_1, d_2, ..., d_n]^T$, $\Delta\mathbf{p} = [\Delta p_1, \Delta p_2, ..., \Delta p_n]^T$, β_j and β_{jl}'s are the coefficients[4].

It is noted that, in Eq (21), no intercept is included. When $\Delta\mathbf{p} = \mathbf{0}$, there is no need to correct the estimates, i.e., $\mathbf{d} = \mathbf{0}$. The coefficients can be determined by fitting Eq. (21) to the data set (20) under least-squares criterion.

Once the error correction function is established, *parameter correction* on the real measurements is performed as

$$\hat{\mathbf{p}}^k_{(i+1)} = \tilde{\psi}_{i+1}(\mathbf{z}^k) = \mathbf{p}^0 + \Delta\hat{\mathbf{p}}^k_{(i+1)}, \tag{22}$$

$$\Delta\hat{\mathbf{p}}^k_{(i+1)} = \Delta\hat{\mathbf{p}}^k_{(i)} + \mathbf{d}_i(\Delta\hat{\mathbf{p}}^k_{(i)}). \tag{23}$$

We see that

$$\tilde{\psi}_{i+1}(\mathbf{z}^k) = \mathbf{p}^0 + \Delta\hat{\mathbf{p}}^k_{(i+1)} = \mathbf{p}^0 + \mathbf{d}_i(\tilde{\psi}_i(\mathbf{z}^k) - \mathbf{p}^0) \tag{24}$$

showing the *recursive* nature.

It should be pointed out that the accuracy of parameter correction is not limited to the quadratic function, since higher order approximation is formed through iterations in the recursive procedure. The quadratic error correction functions can be replaced by other higher order polynomials when the sample size is large (that means there are enough data to estimate the coefficients). Linear functions are not suitable, since a higher order approximation cannot be realized recursively by linear functions (the linear function of linear functions is still linear).

IV. EXAMPLE

In this example, NMOS transistor parameters of SPICE level 3 model are extracted using the measured I-V data. The parameters to be extracted are: thin oxide thickness TOX, zero-bias threshold voltage VTO, body-effect parameter $GAMMA$, surface mobility UO, substrate doping $NSUB$, and lateral diffusion LD. For each I-V measurement, 200 data points are measured using selected biasing voltages of V_{DS}, V_{GS} and V_{BS} as $V_{DS} = \{1.0, 1.5, 2.0, 2.5, 3.0, 3.5, 4.0, 4.5\}$,, $V_{GS} = \{1.0, 2.0, 3.0, 4.0, 5.0\}$,, $V_{BS} = \{-4.0, -3.0, -2.0, -1.0, 0.0\}$. To check and demonstrate the proposed method, the measured devices performances (I-V data) were generated, using a statistical device model, through Monte Carlo simulation with the sample size equal to 500.

[4] These coefficients are the coefficients of the error correction function at the ith iteration. So, if totally l iterations are performed, there are l sets of these coefficients.

Table 1: Results after *Parameter Prediction*

parameter	sample μ	δ^m	sample σ	δ^s
TOX×10^8	1.4937	-0.48%	0.1135	9.87%
VTO	0.8235	-0.38%	0.06619	8.90%
GAMMA	0.6072	-0.70%	0.0690	9.81%
UO×10^{-2}	5.0829	0.23%	0.0912	41.62%
NSUB×10^{-16}	5.7518	-1.24%	0.7119	19.32%
LD×10^7	1.3090	-1.06%	0.3878	12.81%

The results obtained by *parameter prediction* are given in Table 1. The *accuracy checking* was performed on the estimated parameters obtained from *parameter prediction*. The relative errors of sample means and standard deviations, δ^m and δ^s, are defined as

$$\delta^m = \frac{mean\ of\ (extracted\,parameter\ -estimated\,parameter)}{mean\ of\ estimated\,parameter},$$

$$\delta^s = \frac{sigma\ of\ (extracted\,parameter\ -estimated\,parameter)}{sigma\ of\ estimated\,parameter}.$$

here, the *extracted parameter* is referred to as the extracted parameter from the *simulated device performances*.

It is seen, from Table 1, that: (1) all relative errors of sample means, δ^m's, are less than 1.3%; (2) relative errors of standard deviations, δ^s's, are 10 ~ 20%, except for the 42% for *UO*.

Parameter correction was run to improve the accuracy. The accuracy checking results on parameter correction are listed in Table 2. From Table 2, we see that: (1) all relative errors of sample means, δ^m's, are less than 0.04% after the 1st *parameter correction*; (2) relative errors of standard deviations, δ^s's, are improved dramatically (less than 0.90% for all), more than fifty times reduction was obtained. To see how the proposed recursive procedure can be used to increase the accuracy, a second run of *parameter correction* was performed. The accuracy is slightly improved.

Table 2: Results after the 1st *Parameter Correction*

parameter	sample μ	δ^m	sample σ	δ^s
TOX×10^8	1.4986	0.03%	0.1038	0.90%
VTO	0.8278	0.02%	0.0632	0.68%
GAMMA	0.6101	0.04%	0.0632	0.88%
UO×10^{-2}	5.0542	-0.00%	0.0593	0.80%
NSUB×10^{-16}	5.9323	0.03%	0.6239	0.52%
LD×10^7	1.3486	0.04%	0.3948	0.31%

For comparison, the optimization-based parameter extraction using the Simplex method [7] was performed. The stopping criterion in Simplex was set to $TOL = 10^{-5}$. The nominal parameter vector was also extracted using the global optimization algorithm, so that a good initial point could be used in Simplex. After a total of 45,578 function evaluations (about 3000 function evaluations in the proposed method), parameters were extracted. The results are listed in Table 3. Note that the relative errors are calculated by subtracting the true sample means from the sample means of the extracted parameters w.r.t. the true sample means. It is seen that all the standard deviations of the extracted parameters obtained from Simplex are NOT accurate. The errors are too large to be tolerable for statistical IC modeling.

Table 3: Parameters Extracted by Simplex

Parameter	Sample Mean	Relative Error (%)	Sample Sigma	Relative Error (%)
TOX×10^8	1.4920	-0.4736	0.0050	-95.1737
VTO	0.8271	-0.1208	0.0390	-38.1933
GAMMA	0.6103	-0.0328	0.0345	-45.3249
UO×10^{-2}	5.1582	2.0597	0.1148	89.7521
NSUB×10^{-16}	5.8933	-0.6507	0.1249	-80.2810
LD×10^7	1.3007	-3.4946	0.0768	-80.5373

The optimization methods with the finite-difference gradients available in IMSL MATH/Library [8] were also tried. Those methods, the quasi-Newton with BFGS update, conjugate gradient, and Levenberg-Marquardt nonlinear least-squares, failed to extract the model parameters for this example. Very often, line search was stopped because no minimum could be found in that direction.

V. Conclusion

Combining the *global optimization*, *parameter prediction*, *parameter correction*, and *accuracy checking*, a very accurate and efficient parameter extraction method (RIA) was proposed for the statistical modeling of integrated circuits. RIA fundamentally solves the accuracy problem in statistical IC parameter extraction. The proposed method is much faster than the optimization-based method. The technique was implemented in `SMIC` - a program for statistical modeling of integrated circuits. For detailed description of RIA, mathematical justification, algorithm and practical examples, the reader can refer to Chapter II of [9]. The `SMIC` implementation issues are discussed in [10].

References

[1] T. Koskinen and P. Y. K. Cheung, "Statistical and behavioural modeling of analogue integrated circuits," *IEE Proceedings, Pt.G*, vol. 140, pp. 171–176, June 1993.

[2] M. S. Sharma and N. D. Arora, "OPTIMA: a nonlinear model parameter extraction program with statistical confidence region algorithm," *IEEE Trans. on Computer-Aided Design*, vol. 12, pp. 982 – 987, July 1993.

[3] J. A. Power, B. Donnellan, A. Mathewson, and W. A. Lane, "Relating statistical MOSFET model parameter variabilities to IC manufacturing process fluctuations enabling realistic worst case design," *IEEE Trans. on Semiconductor Manufacturing*, vol. 7, pp. 306 – 318, Aug. 1994.

[4] H. P. Tuinhout, S. Swaving, and J. J. M. Joosten, "A fully analytical MOSFET model parameter extraction procedure," in *Proc. IEEE Int. Conf. on Microelectronic Test Structure*, (Long Beach, California), pp. 79 – 84, Feb. 1988.

[5] J. A. Power, A. Mathewson, and W. L. Lane, "MOSFET statistical parameter extraction using multivariate statistics," in *Proc. IEEE Int. Conf. on Microelectronic Test Structure*, (San Diego, California), pp. 209 – 214, March 1991.

[6] M. Qu and M. A. Styblinski, "A heuristic global optimization algorithm and its application to CMOS circuit variability minimization," in *Proc. ISCAS*, (Chicago), pp. 1809–1812, May 1993.

[7] J. A. Nelder and R. Mead, "A simplex method for function minimization," *Computer J.*, vol. 7, pp. 308–313, 1965.

[8] *MATH/Library User's Manual*. Houston, Texas: IMSL, Inc., 1987.

[9] M. Qu, *Statistical Modeling of Large Analog Integrated Circuits and Its Applications to Design for Quality and Manufacturability*. PhD thesis, Texas A&M University, 1995.

[10] M. Qu, "`SMIC` - Statistical Modeling for Integrated Circuits: User's Guide," Laboratory for Intelligent Design Systems, Texas A&M University, August 1995.

Impact of transistor mismatch on the speed-accuracy-power trade-off of analog CMOS circuits

Peter Kinget and Michiel Steyaert

Katholieke Universiteit Leuven, ESAT-MICAS
Kard. Mercierlaan 94, B-3001 Heverlee, Belgium

Abstract

The influence of transistor mismatch on the trade-off between speed, accuracy and power consumption of basic building blocks and analog systems is investigated. The ratio (Speed Accuracy^2)/Power for a circuit or system in CMOS is shown to be only dependent on technological constants which express the matching quality of the technology. Moreover, the minimal power consumption of high speed analog systems in CMOS imposed by transistor mismatch is two orders higher than the limit imposed by thermal noise.

Introduction

The parameters of two identically designed devices on an integrated circuit show a random variation after fabrication, which is called device mismatch. Device mismatch limits the accuracy of circuits and has important implications on the offsets of operational amplifiers or comparators, on the bit accuracy of A/D and D/A convertors, and on the power-supply-noise and common-mode rejection ratio's in differential structures. The accuracy performance can be increased in two ways: the matching of devices is improved by increasing the device size, or auto-zero and offset compensation techniques are used to reduce the effect of mismatches on the system performance. The compensation techniques require calibration phases during which the normal system operation is interrupted and the offsets of the building blocks are sampled and dynamically stored in an analog memory circuit. The maximum time between calibration phases is determined by the quality of the analog memory circuits and limits the maximal duration of a continuous operation phase. In sampled data systems, the calibration can be done in one of the clock phases but clocked circuits intrinsically have a much lower operation speed as continuous time circuits.

In many high speed analog signal processing systems the interuptions of the operation of the system cannot be tolerated or the required continous operation phases are too long. Therefor compensation techniques are not applicable and the accuracy performance of the system completely depends on the matching quality of the technology. To improve the system accuracy, large devices have to be used so that the capacitive loading of the circuit nodes increases and more power is required to attain a certain speed performance. As a result, the matching quality of the technology has a direct impact on the minimal power consumption of high precision building blocks.

I. Models for MOS transistor mismatch

The mismatch of two closely spaced, identical transistors with zero source-bulk bias, is modeled by the random variation of the difference in their threshold voltage V_T and current factor ß. A widely accepted model [1] is a normal distribution with zero mean and a variance dependent on the gate-width W and gate-length L of the devices:

$$\sigma(\Delta V_T) = \frac{A_{VT}}{\sqrt{W \cdot L}} \qquad (1)$$

$$\sigma\left(\frac{\Delta\beta}{\beta}\right) = \frac{A_\beta}{\sqrt{W \cdot L}} \qquad (2)$$

In table 1 the technology-dependent constants A_{VT} and $A_ß$ are summarized for several industrial CMOS technologies.

Using the square-law V-I relationship of a MOS transistor biased in strong inversion, the variance of the relative current difference of two identical transistors biased with the same V_{GS} is:

$$\sigma^2\left(\frac{\Delta I}{I}\right) = \sigma^2\left(\frac{\Delta\beta}{\beta}\right) + \frac{4 \cdot \sigma^2(\Delta V_T)}{(V_{GS} - V_T)^2} \qquad (3a)$$

or after substitution of (1) and (2):

$$\sigma^2\left(\frac{\Delta I}{I}\right) = \frac{1}{W \cdot L}\left(A_\beta^2 + \frac{4 \cdot A_{VT}^2}{(V_{GS} - V_T)^2}\right) \qquad (3b)$$

For a transistor pair biased with an identical drain current I_{DS}, such as in a differential pair, the variance of their gate-source voltage difference is:

$$\sigma^2(\Delta V_{GS}) = \sigma^2(\Delta V_T) + \frac{(V_{GS} - V_T)^2}{4} \cdot \sigma^2\left(\frac{\Delta\beta}{\beta}\right) \qquad (4a)$$

or

$$\sigma^2(\Delta V_{GS}) = \frac{1}{W \cdot L}\left(A_{VT}^2 + \frac{(V_{GS} - V_T)^2}{4} \cdot A_\beta^2\right) \qquad (4b)$$

From (3b) and (4b), we can conclude that the current and voltage matching depends on A_{VT} and $A_ß$ and that the relative importance of V_T and ß mismatches depends on the VGST - VGST denotes (V_{GS}-V_T). If we define a corner gate-overdrive voltage as: $(V_{GS} - V_T)_m = 2 \cdot A_{VT} / A_\beta$, the effect of the V_T mismatch is dominant over the effect of the ß mismatch in (3b) and (4b) for transistors with a VGST smaller than VGSTm . From the values of VGSTm in table 1 it is clear that the effect of V_T is indeed dominant under normal biasing conditions so that (3b) and (4b) can be simplified to:

$$\sigma^2\left(\frac{\Delta I}{I}\right) = \frac{1}{W \cdot L}\frac{4 \cdot A_{VT}^2}{(V_{GS} - V_T)^2} \qquad (5)$$

$$\sigma^2(\Delta V_{GS}) = \frac{A_{VT}^2}{W \cdot L} \qquad (6)$$

0-7803-3177-6 $5.00 © 1996 IEEE

II. Implications mismatch on the design and performance of basic circuit building blocks

The performance of a signal processing building block is characterized by three specifications: operation speed, accuracy and power consumption. The quality of a circuit is best when a high speed and accuracy performance are achieved for a low power drain. In this section closed-form expressions are derived for the relation between the power consumption and the speed and accuracy of basic circuit blocks for current or voltage signal processing. In the next section, these results are generalized towards the performance of analog systems. The same type of relationships remain valid since these systems consist of the basic blocks discussed here.

A. Current processing blocks:

The basic current processing stage is a current mirror represented in figure 1. The speed performance of this building block is determined by its bandwidth which is given by:

$$BW = \frac{g_{m1}}{2\pi \cdot (2 \cdot C_{GS})} = \frac{3 \cdot I_B}{4\pi \cdot (V_{GS} - V_T) \cdot C_{ox} \cdot W \cdot L} \quad (7)$$

The power consumption of this building block is determined by the bias current and the supply voltage:

$$P = 2 \cdot I_B \cdot V_{DD} \quad (8)$$

The relative accuracy of the current signal processing is dependent on the input offset current I_{os} and the maximal input current current signal I_{inRMS}: $Acc_{rel} = I_{inRMS} / (3 \cdot \sigma(I_{os}))$. The standard deviation of the offset current is derived from (5):

$$\sigma(I_{os}) = \frac{2 \cdot A_{VT} \cdot I_B}{(V_{GS} - V_T) \cdot \sqrt{W \cdot L}} \quad (9)$$

The maximal input signal I_{inRMS} is $I_B/(2\sqrt{2})$ for a typical bias modulation index of 1/2. The circuit specifications of (7), (8) and (9) can be combined into a quality factor for the design:

$$\frac{BW \cdot Acc_{rel}^2}{P} = \frac{(V_{GS} - V_T)}{768\pi \cdot C_{ox} \cdot A_{VT}^2 \cdot V_{DD}} \quad (10)$$

This quality factor shows that the total performance of the design is only dependent on technological constants and on the biasing of the stage. To obtain the best total performance, a current processing stage must be designed with a large VGST. It is common knowledge that to obtain a high accuracy in a current mirror a high VGST has to be used. However, (10) shows that not only a high accuracy is achieved but that the best trade-off between speed, accuracy and power performance is achieved with a high VGST for a current processing stage. The VGST for a current mirror will in practice be limited to $V_{DD}/2$ so that the optimal quality of a current mirror is given by:

$$\frac{BW \cdot Acc_{rel}^2}{P} = \frac{1}{1536\pi \cdot C_{ox} \cdot A_{VT}^2} \quad (11)$$

Equation (11) expresses a very important design conclusion: the speed, accuracy and power consumption performance of a current mirror are related to each other by technological constants only; or, once the speed and accuracy specifications for a current mirror are fixed, the lowest possible power consumption is determined by technological constants, which express the matching quality of the technology. As a result, the effect of transistor mismatch puts a boundary on the lowerst achievable power consumption.

B. Voltage processing blocks:

The basic voltage amplification block is an operational amplifier with resistive feedback as depicted in figure 2. The operational amplifier can be implemented with a one transistor amplifier as is shown in figure 2.

To obtain a well-defined amplification the $1/g_m$ of the transistor is made larger than the resistors R_1 and R_2. The voltage transfert function is then given by:

$$A(s) = -\frac{R_2}{R_1} \cdot \left(\frac{1}{1 + s/(g_m / C_{GS})} \right) \quad (12)$$

The amplication of the stage is R_2/R_1 and its speed performance and bandwidth is given by:

$$BW = \frac{3 \cdot I_B}{2\pi \cdot (V_{GS} - V_T) \cdot C_{ox} \cdot W \cdot L} \quad (13)$$

The power consumption of the circuit is:

$$P = I_B \cdot V_{DD} \quad (14)$$

The relative accuracy is determined by the maximal input signal compared to the offset voltage. The amplitude of the output signal is limited to $V_{DD}/2$ so that the maximal input signal is $V_{inRMS} = V_{DD}/(\text{Gain } 2\sqrt{2})$. The offset voltage is determined by the absolute variation of the threshold voltage of the transistor which is $\sqrt{2}$ times smaller than the value given by (6). The relative accuracy $Acc_{rel} = V_{inRMS} / (3 \cdot \sigma(V_{os}))$ is then given by:

$$Acc_{rel} = \frac{V_{DD} \cdot \sqrt{W \cdot L}}{6\sqrt{2} \cdot A_{VT} \cdot Gain} \quad (15)$$

The quality of the design can again be expressed by a combination of the specifications:

$$\frac{Gain^2 \cdot BW \cdot Acc_{rel}^2}{P} = \frac{V_{DD}}{48\pi \cdot (V_{GS} - V_T) \cdot C_{ox} \cdot A_{VT}^2} \quad (16)$$

The quality of the voltage design improves if the stage is designed with a small VGST. As far as the speed requirements allow it, the circuit should be biased with the lowest possible VGST. Again it is common knowledge that to achieve a low offset voltage in a voltage design, small VGST have to be used. However (16) shows that not only a good accuracy performance but the best combined performance when speed, accuracy and power consumption are considered, is obtained for a small VGST. The minimal power consumption of the building block is again limited by the matching quality of the technology for a given speed and accuracy. The gain of the building block plays an important role; the more gain is made the smaller the input range becomes, since the output swing of the building block is limited. This has severe implications for the accuracy performance and consequently on the power consumption as can be concluded from (16) and (17).

Implications for analog system performance

A. Multi-stage design

A general analog system consists of a cascade of different stages. In figure 3 a multi-stage voltage processing system is depicted. For current processing signal systems the same derivations can be made. The accuracy of the system will be determined by the equivalent input offset voltage which is:

$$V_{oseq} = \sqrt{(V_{os1})^2 + \left(\frac{V_{os2}}{A_1}\right)^2 + \left(\frac{V_{os3}}{A_1 \cdot A_2}\right)^2 + \cdots} \quad (17)$$

Both noise and mismatch are random quantities and put a boundary on the smallest signal that can be processed in a system. As a result, their influence is most important at the input stage where the signal levels are small. For the design of a high accuracy system the highest possible gain A_1 has to be made in the first stage and the equivalent offset voltage is then mainly determined by the offset of the first stage V_{os1}. For a first stage with an input transistor biased in strong inversion, the offset is √2 times smaller as the value in (6), and the input capacitance of the system is $C_I = 2/3 \cdot C_{ox} \cdot W \cdot L$. The offset voltage can be expressed as a function of the input capacitance:

$$V_{oseq} = V_{os1} = \sqrt{\frac{2/3 \cdot C_{ox} \cdot A_{VT}^2}{2 \cdot C_I}} \quad (18)$$

and the accuracy performance of the system is $Acc_{rel} = V_{inRMS} / (3 \cdot \sigma(V_{os}))$. For a differential input stage the offset voltage is √2 times larger but the input capacitance is 2 times smaller so that the same relationship is valid.

The maximal speed of the system is of course determined by the bandwidth of the amplifiers but is limited by the pole generated by the input capacitor and the source resistance of the signal source: $f_{\max} = 1/(2\pi \cdot R_s \cdot C_I)$. From (18) we can conclude:

$$f_{\max} \cdot Acc_{rel}^2 = \frac{V_{inRMS}^2}{6\pi \cdot R_s \cdot C_{ox} \cdot A_{VT}^2} \quad (19)$$

In a general design the speed and accuracy performance are interdependent and depend on the matching quality of the technology and the source impedance.

B. Minimal power consumption in analog systems due to transistor mismatch

In figure 3 and the derivation of the previous section the signal source has to drive the input capacitance C_I directly. For high speed and high accuracy systems this cannot be tolerated since the input pole would drastically reduce the system bandwidth. On-chip buffering is provided to increase the input impedance and reduce the loading of the signal source. The power to drive the input capacitance C_I is then delivered by the system itself and this power determines the theoretical minimal power drain for the given accuracy and speed. The best power efficiency is obtained for class B systems and the power required to drive a signal V_{inRMS} with a frequency f across a capacitor C_I is:

$$P = 8 \cdot f \cdot C_I \cdot V_{inRMS}^2 \quad (20)$$

The relative accuracy as defined in section II B is also the dynamic range (DR) of the system. After substitution of (18) in (20) we obtain:

$$P = 24 \cdot C_{ox} \cdot A_{VT}^2 \cdot f \cdot DR^2 \quad (21)$$

The matching quality of the technology puts a lower boundary on the minimal power drain of analog systems. Moreover, the important design conclusion can be drawn that also for the general system the ratio (Speed Accuracy^2)/Power is only dependent on technological constants.

C. Noise vs mismatch limits on power consumption

The effects of thermal noise on the power consumption of a circuit are derived in [3]. The minimum power for a given frequency f and a given DR is:

$$P = 8 \cdot k \cdot T \cdot f \cdot DR^2 \quad (22)$$

where k is the Boltzmann constant and T the absolute temperature. The limits on the minimal energy per cycle P/f imposed by noise [3] and by mismatch - derived in (21) - are plotted in figure 4. The limit imposed by mismatch is technology dependent and is about two orders of magnitude more important than the limit imposed by thermal noise.

In figure 4 the high dynamic range filters referenced in [3] are also represented. The dynamic range of filters is in first order not sensitive to matching. Mismatch will mainly influence the accuracy of the filter coefficients, the distortion performance and the power supply noise or common mode rejection characteristics of the filter which do not directly influence the dynamic range. So filter realizations only optimized to low power and high dynamic range can consume less than predicted by (21) from the impact of mismatch.

High speed A/D convertors, not using oversampling techniques, are a better benchmark because mismatch limits the bit accuracy directly. The limit imposed by mismatch is indeed a much better estimate for the minimal power consumption of A/D convertors. The high accuracy convertors have typically lower speeds and include digital or analog error correction circuitry, so there are only 1 order of magnitude from optimal performance. The lower bit convertors are very high speed and cannot use any error correction technique so their power consumption is about 2 to 3 orders from optimal performance. In the calculation of the minimal power consumption, only the power drain of the first stage is taken into account. In the practical realisations important amounts of power are also consumed by the subsequent stages which have also to process high speed signals. Moreover, in high accuracy designs the DR can be reduced by the effects of distortion which can have also important implications on the power consumption.

Conclusions

The impact of transistor mismatch on the performance of analog circuits is discussed. In CMOS technologies the impact of V_T mismatch on the accuracy performance is dominant. The speed, accuracy and power consumption of a circuit have a fixed relationship and the minimal power consumption of a circuit is determined by technological constants only which express the matching quality of the technology.

The limit imposed by mismatch on the minimal power consumption for a given speed and accuracy or dynamic range is shown to be several orders of magnitude higher than the thermal noise imposed limit. We can conclude that for the design of high speed, high accurate analog systems, the impact of transistor mismatch on the power consumption is most important in CMOS. Consequently whenever the system architecture or speed requirements allow it, offset compensation techniques should be used and will drastically reduce the power consumption of the system.

References

[1] M. Pelgrom et al., "Matching properties of MOS transistors", *IEEE Journal of Solid-State Circuits*, vol. 24, no. 5, pp. 1433-1439, 1989.

[2] J. Bastos, M. Steyaert, R. Roovers, P. Kinget, W. Sansen, B. Graindourze, A. Pergoot, Er. Janssens, "Mismatch characterization of small size MOS transistors", in *Proc. IEEE Int. Conf. on Microelectronic Test Structures*, vol.8, pp. 271-276, 1995

[3] E.A. Vittoz, "Future of analog in the VLSI environment", in *Proceedings ISCAS*, pp. 1372-1375, May 1990.

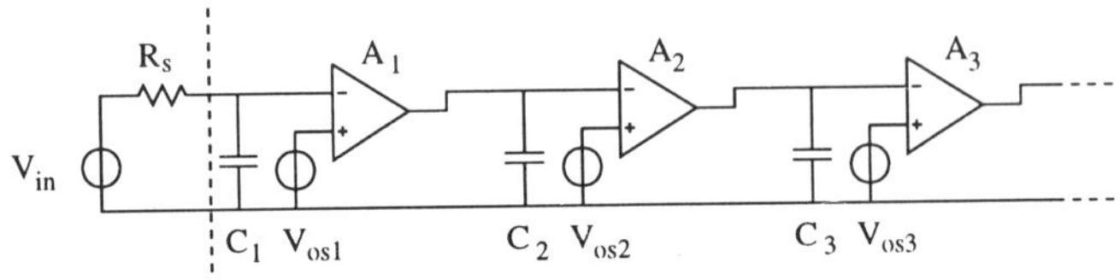

Figure 3: Multi-stage system schematic; the offset voltage, input capacitance and amplification of each stage are indicated.

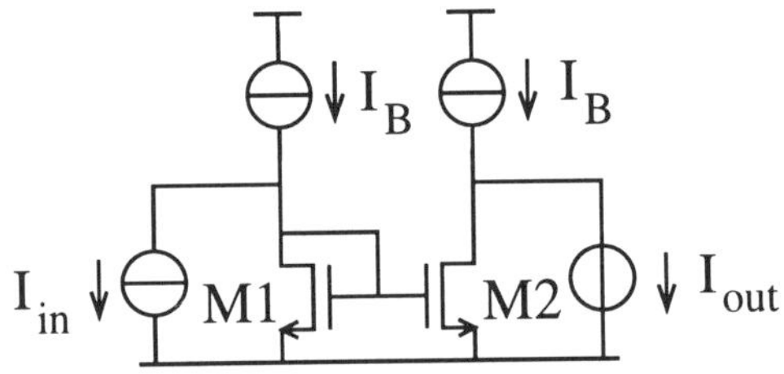

Figure 1: Current mirror schematic

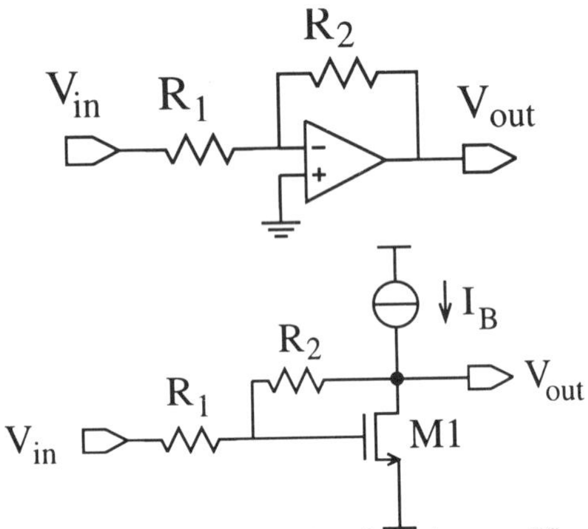

Figure 2: Top: General invering voltage amplifier schematic; Bottom: one transistor implementation

Technology	Type	A_{VT} [mV µm]	A_{β} [% µm]	$(V_{GS}-V_T)_m$ [V]
2.5 µm	nMOS	30	2.3	2.6
[1]	pMOS	35	3.2	2.2
1.2 µm	nMOS	21	1.8	2.3
[2]	pMOS	25	4.2	1.2
0.7 µm	nMOS	13	1.9	1.4
	pMOS	22	2.8	1.6

Table 1: The matching proportionality constants for threshold voltage (V_T) matching and current factor (ß) matching for different industrial CMOS processes; for transistors biased with a (V_{GS}-V_T) lower than $(V_{GS}-V_T)_m$ the V_T mismatch is dominant for the calculation of the accuracy.

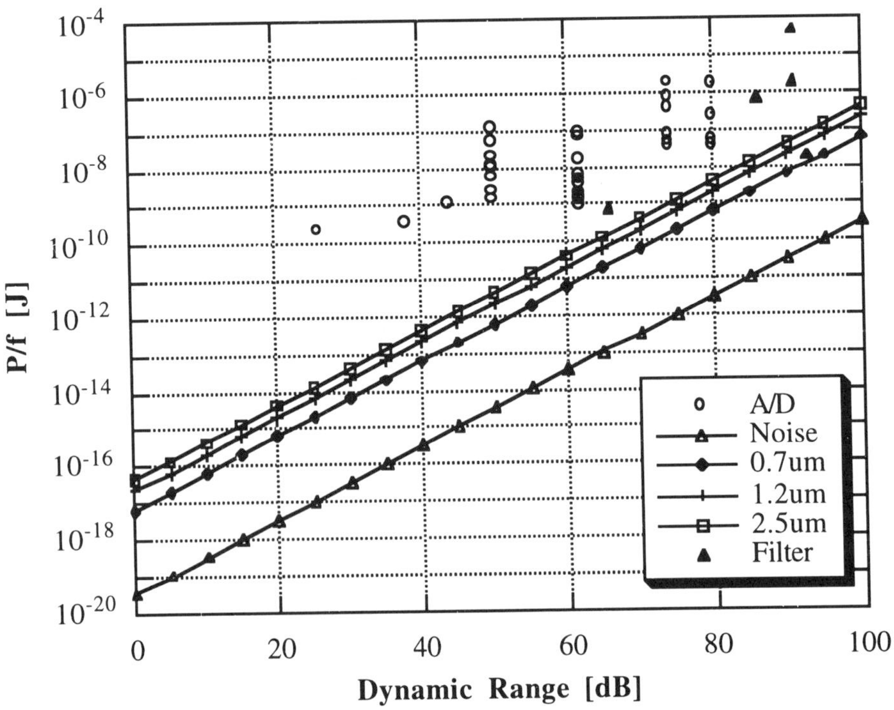

Figure 4: Limits on power consumption imposed by thermal noise and mismatch for the technologies of table 1; the power consumption of practical analog filters [3] and high speed A/D convertor designs is also indicated.

Optimal Design of Opamps for Oversampled Converters

Feng Wang and Ramesh Harjani
Department of Electrical Engineering
University of Minnesota, Minneapolis, MN 55455

Abstract: The power consumption for various opamps used in $\Sigma\Delta$ modulators are considered. We first classify the different opamp topologies used in such modulators. We then develop a generalized integrator model that allows us to compare the power consumption for all these topologies. We also derive an expression for the minimum theoretical bound on the power dissipation. We illustrate our results with two design examples: a class A and a class AB opamp. It is shown that class A opamps are the best choice for $\Sigma\Delta$ modulators.

1 Introduction

Oversampled A/D converters are widely used in instrumentation, data acquisition, telecommunication, and consumer electronics. Some advantages of oversampled A/D converters include a higher dynamic range and signal-to-noise ratio than traditional architectures. Additionally, in these converters the requirement for precisely matched components is relaxed. In this converter the operational amplifier within the modulator is the most critical component. For instance any nonidealities like finite gain, bandwidth and slew rate causes an incomplete transfer of charge in the integrator which in turn causes nonlinearities. To show this effect, in Figure 1 we plot the error voltage, (the output voltage minus the input voltage), versus the input voltage for a first order $\Sigma\Delta$ converter. The gray curve shows the ideal behavior while the dark curve shows the effect of finite settling of the opamp. The sharp peaks in both curves are the result of tones [1]. Previous research [2] has shown that some amount of harmonic distortion is inevitable. Additionally, it has been shown that the amount of distortion is input signal dependent. While a perfect opamp is not feasible, we can design an opamp that settles sufficiently for any given resolution. This accuracy in settling directly translates into a minimum bound on the power dissipation of the opamp. In [3], Castello *et al.* addressed the issue of minimum power dissipation in a general SC filter for sine wave inputs. However, their method can not be extended to $\Sigma\Delta$ modulators directly due to existence of the large feedback reference voltages in $\Sigma\Delta$ modulators. In [4], Degrauwe *et al.* developed an expression for the minimum transconductance for the opamp used in a micropower SC filter. However, in this paper the input signals are always assumed to be small such that there is no slewing. This is not true for opamps in $\Sigma\Delta$ modulators.

To cope with the large feedback signals inherent in $\Sigma\Delta$ modulators, we develop an expression for the

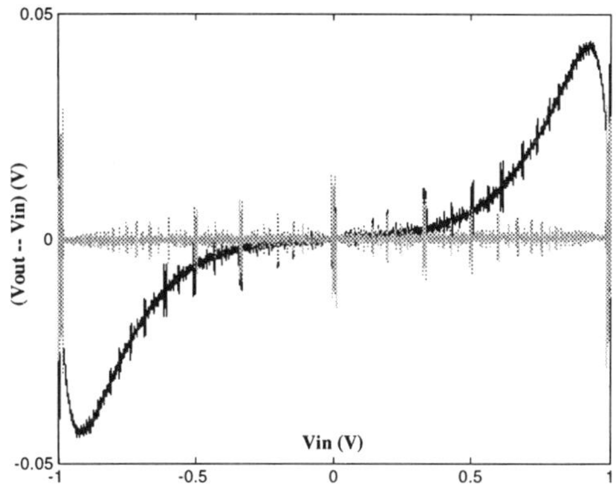

Figure 1: Nonlinearity due to finite settling

minimum achievable power dissipated by the opamps used in such modulators. We then compare this power bound with some existing opamp topologies including a design example.

2 Opamp classification

In order to make a better comparison, we shall classify some of the existing opamp topologies. Most opamps which can be used in $\Sigma\Delta$ modulators can be classified into either class A or non-class A amplifiers. For class A opamps the available output current is limited and remains fixed even when a large differential input voltage is applied. Examples of such amplifiers include the Miller-compensated two stage OTA, the symmetrical (or current gain) OTA and the folded-cascode OTA. In non-class A opamps the available output current is either not limited for large differential input signals or is simply a periodic function of time. The non-class A amplifiers can further be classified into two subcategories: signal dependent output current (SDOC) and signal independent output current (SIOC) opamps. Class AB opamps [5, 6, 7] are SDOC amplifiers, and use a low and well-controlled quiescent current which is automatically boosted when a large differential input is applied. Adaptive biased opamps [8] are also SDOC opamps. They use a conventional OTA whose tail current is boosted by monitoring the current difference in the differential input pair. In [9, 10], the output current is boosted by monitoring the differential input voltage. These too are SDOC opamp. Signal-biased dynamic opamps [11, 12] are other examples of SDOC opamps and are primarily used in switched-capacitor circuits. On the other hand, tail current-biased dynamic opamps [12, 13] are SIOC opamps. These amplifiers are based on conventional class A opamps but the tail current is periodically switched on during the start of the sampling clock. In [14], an additional booster circuit is used in parallel with the opamp to boost the output current in SC integrators. Though the booster circuit itself is not part of the opamp it behaves like a signal-dependent dynamic amplifier [13] and as such is also a SDOC opamp. Our analysis is general and covers all the topologies mentioned above.

3 A general model for SC-integrators

In this section, the simple switched-capacitor integrator used in $\Sigma\Delta$ modulators is analyzed to establish differential equations governing the time-varying behavior. For the sake of simplicity the integrator in Figure 2(top) is being selected. However, the results obtained here can be extended to other more complex configurations without much effort. In a $\Sigma\Delta$ modulator feedback is generated by the sampling of the reference voltages. For our example we shall assume a two phase non-overlapping clock scheme. During the first clock phase, Φ_1, both the input and one of reference voltages, depending on the output of the comparator in the $\Sigma\Delta$ modulator, are sampled onto the input capacitor C_1 and the reference capacitor C_2 respectively (i.e., switch S_1 and S_4 are activated). During Φ_2 the charge stored in these capacitors is then transferred to the integrating capacitor C_t (i.e., switch S_2 and S_3 are closed). For the Φ_2 clock period the integrator can be modeled by the equivalent circuit shown in Figure 2(bottom). The activation of the switches S_2 and S_3 can be transformed into step inputs whose voltage step sizes are equivalent to the input and reference voltages. In this model, V_1 is equal to the input voltage and V_2 is equal to either V_r+ or V_r- depending on the output of the comparator. The output stage of an opamp can be represented by the two current sources, $I_1(t)$ and $I_2(t)$, whose values are time-dependent. This simplified integrator schematic focuses on how node voltages and branch currents change with time without getting into the details of any particular opamp topology.

During the rising edge of V_1 and V_2 the amplifier cannot react instantaneously so the the negative input of the opamp, v_m, also makes a step jump. The magnitude of the initial step is equal to

$$V_{mi} = \frac{C_1V_1 + C_2V_2}{C_1 + C_2 + C_tC_L/C_t + C_L} \qquad (1)$$

From Kirchhoff's law of voltages and currents we can write the following set of equations

$$I_1(t) - I_2(t) + C_L\frac{dv_o}{dt} = C_t\frac{d(v_m - v_o)}{dt} \qquad (2)$$

$$C_t\frac{d(v_m - v_o)}{dt} = -(C_1 + C_2)\frac{dv_m}{dt} \qquad (3)$$

The resulting differential solution for $v_m(t)$ is given by

$$\frac{dv_m}{dt} = -\frac{I_1(t) - I_2(t)}{C_{eff}} \qquad (4)$$

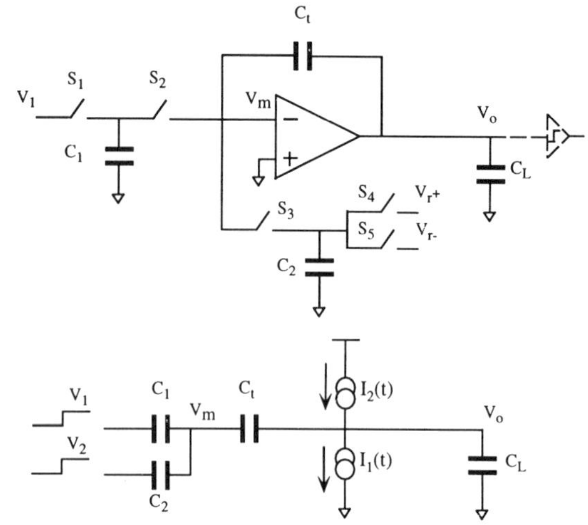

Figure 2: SC integrator

where C_{eff} is $\left[((C_L + C_t)(C_t + C_1 + C_2) - C_t^2)\right]/C_t$. Equation 4 describes the the voltage at the negative input of the opamp throughout the whole dynamic process from just after the initial voltage jump all the way to final settling. This equation is also independent of any specifics, which allows us to compare the various opamp topologies under the same framework.

Due to the large magnitude of the initial jump; V_2 is equal to $\pm V_r$ and usually of the same magnitude as the supply voltages; in $\Delta\Sigma$ converters one of the differential input transistors of the opamp will be completely off. Consequently, either $I_1(t)$ or $I_2(t)$ will be equal to zero. For example, let $I_2(t)$ be equal to zero. Now the form of $I_1(t)$ will depend on the opamp topology. For a class A opamp, $I_1(t)$ is proportional to the tail current and is a constant. For a non-class A opamp, it is a function of time. We shall refer to the time interval in which only one of current sources is present to be the large signal period. For a class A opamp this corresponds to the slewing period. When v_m becomes less than a threshold voltage ΔV, the current source $I_2(t)$ resumes conducting. This threshold voltage ΔV represents the linear region of the differential pair. Likewise, we shall refer to the time interval in which both the current sources are conducting to be the small signal period. For class A opamps this corresponds to the settling period.

4 Comparison of power consumption

The average current consumption (therefore the power consumption) can be calculated for each of the periods defined in the previous section. Suppose the time interval for the large signal and small signal period is T_{la} and T_{sm} respectively, then $T_{la}+T_{sm} = T/2$, where T is the clock period. During the large signal period, the current dissipation is represented by the output current multiplied by a constant factor which is topology-dependent. This constant factor does not vary by much and will be set to unity for simplicity. The average current during the large signal period in relation to (4) can be evaluated and is given by

$$\overline{I_{la}} = \frac{\int_0^{t_{la}} I_1(t)dt}{T_{la}} \quad (5)$$

$$= \frac{C_{eff}(V_{mi} - \Delta V)}{T_{la}} \quad (6)$$

Note that this part of current consumption is only dependent on the effective charge $C_{eff}(V_{mi} - \Delta V)$ transferred during the time T_{la}. The effect of the different opamp topologies only alters the value of T_{la}.

The current consumption during the small signal period can be represented by the current flowing through the input transistor multiplied by another constant factor. Like in the case of the large signal period, we set this factor to be unity. However, this current is evaluated in a somewhat different manner. If we assume a single-pole dominant system, valid for nominal phase margins, then the settling behavior in this period is mainly determined by the input transconductance and the effective load capacitance. For a given moderate gain-bandwidth (GBW) and load capacitance, the power is minimized when the input transistors of an opamp are operated in weak inversion [15]. For the rest of our analysis we shall assume that the input differential pair is operating in weak inversion for all the opamp topologies. The transconductance of the differential pair is then given by $g_m = I/nU_T$ [15], where I is the current flowing through the one of input transistors, n is the slope factor in weak inversion and U_T is the thermal voltage. It is important to note that the primary conclusions obtained in this paper remain valid for strong inversion as well. Only the details of the equations change somewhat. For weak inversion operation during the small signal period (4) becomes

$$\frac{dv_m}{dt} = -\frac{I(t)}{nU_T}\frac{v_m(t)}{C_{eff}} \quad (7)$$

By integrating the above equation we can derive an expression for the average current consumption flowing through the input transistor.

$$\overline{I_{sm}} = \frac{\int_{t_{sm}}^{T/2} I(t)dt}{T_{sm}} \quad (8)$$

$$= \frac{nU_T}{T_{sm}} C_{eff}\, ln\left(\frac{\Delta V}{v(T/2)}\right) \quad (9)$$

The final settling accuracy required at the output of the integrator which is related to the overall converter resolution can be transformed into a similar requirement imposed at the negative input node of the opamp. In other words, $v_m(T/2)$ should be proportional to $2^{-N}V_{REF}$, where N is the number of bits of resolution for the overall $\Sigma\Delta$ converter and V_{REF} is the magnitude of the reference voltages. We can now write the average current consumption for the entire $T/2$ period (large signal+ small signal period) as follows

$$\overline{I} = \frac{\overline{I_{la}}T_{la} + \overline{I_{sm}}T_{sm}}{T/2} \quad (10)$$

$$= \frac{Q_1 + Q_2}{T/2} \quad (11)$$

where $Q_1 = C_{eff}(V_{mi} - \Delta V)$ and $Q_2 = nU_T C_{eff} ln(2^N \Delta V/V_{REF})$. And ideal opamp would consume no power during Φ_2. Therefore, the average current consumption for an ideal opamp for the entire clock period is equal to $\overline{I}/2 = (Q_1 + Q_2)/T$. Clearly this value is only dependent on the size of the effective load capacitance C_{eff}, the maximum initial voltage jump and the resolution of the $\Sigma\Delta$ converter but is independent of the opamp topology. It sets the theoretical minimum bound on power consumption.

We can now compare the power consumption of different opamp topologies against the minimum theoretical bound established above. For each of the opamps we shall assume the most optimally designed configuration, i.e., minimum bounds for each class. For a class A opamp, the tail current is fixed and determines both the large signal and the small signal properties. Additionally, the current remains constant for both Φ_1 and Φ_2. The minimum current dissipation is thus twice as much as the theoretical value, i.e., $2(Q_1 + Q_2)/T$. For a class AB opamp and an adaptive biasing opamp, the average current dissipation is the same as the theoretical minimum value during Φ_1. But during Φ_2 they will consume the necessary quiescent current which is used to determine the settling during Φ_1. The average current for the entire clock period is $I = (Q_1 + (1 + T/(2T_{sm}))Q_2)/T$. Its minimum value is $I = (Q_1 + 2Q_2)/T$ when $T_{sm} = T/2$. This corresponds to a zero large signal period and an infinite slew rate. For signal-independent dynamic opamps, the large signal behavior and small signal behavior are well-defined by its signal-independent tail current. Hence, the opamps can also be designed to approach the minimum current value during Φ_1 and conduct a very small amount current during Φ_2. Its average current for one entire clock period is thus $(Q_1 + Q_2)/T$.

We can now conclude that purely in terms of the power dynamic opamps are the most efficient for a fixed resolution requirement. Theoretically, they can be made to approach the minimum bound. However, they suffer from poor common mode rejection ratio and are more difficult to design. On the other hand, the maximum power savings by utilizing a class AB opamp, in $\Sigma\Delta$ converters, instead of a class A opamp is always less than 50%. The primary reason is that a class AB opamp has to respond to a large feedback reference voltage in every sampling cycle. As a matter of fact, any extra circuitry used for the class AB will consume some finite amount amount of power making the saving even less significant. Additionally, it can be shown that the use of class AB opamps results in integral nonlinearity errors because of the signal dependent supply current.

5 Design example

The above analysis can be illustrated by the following design example. The integrator in the $\Sigma\Delta$ modulator is designed using two different opamp topologies, a class A and a class AB opamp. The supply current for both these designs is then compared to the theoretical bound given in (11). Both designs are generated for an 18-bit converter with a 250kHz clock. The reference voltage is selected to be 2.5V and input signal is 0.1V for a 5V power supply. The signal sampling capacitance is $5pF$, the feedback sampling capacitance is $5pF$ and the integrating and load capacitance are $10pF$ and $2pF$ respectively. For class A opamps, the fixed tail current determines both the large and small signal behaviors. Therefore, $\overline{I_{la}}$ should be equal to $\overline{I_{sm}}$, i.e.,

$$\frac{Q_1}{T_{la}} = \frac{Q_2}{T_{sm}} \qquad (T_{la} + T_{sm} = T/2) \qquad (12)$$

It can be easily shown that the current dissipation is minimized when $T_{la} = Q_1T/2(Q_1 + Q_2)$ and $I = 2(Q_1 + Q_2)/T$. This situation is shown in Figure 3. The U-shaped curve describes the total current needed vs. the fraction of time used for slewing. Following the analysis in the previous section, the ratio Q_1/Q_2 is in range of $(V_{mi} - \Delta V)/NnU_Tln2$, where N is the number of bits of resolution. A typical value for a 18-bit resolution and a step size of 2V is around 4. As long as Q_1 is larger than Q_2 which is usually true for a large range of resolutions, a larger fraction of the clock period should be assigned to the large signal period. For class A opamps, this is realized by carefully selecting the tail current. For non-class A opamps, except signal-independent dynamic ones, the duration of the large signal period is determined by the input voltage and not by the quiescent current. There is no partitioning of time. The middle curve shows the average current consumption for a class AB opamp with an infinite slew rate. The bottom curve shows the theoretical minimum bound calculated in (11).

6 Conclusions

A model which captures the dynamics of an integrator leading to a power estimate for the opamps used in $\Sigma\Delta$ modulators has been developed. We have developed an expression for the minimum achievable power dissipation. The results suggest that this lower bound of power consumption is solely dependent on the clock speed, the input signal magnitude, capacitor sizes and the resolution requirement. A carefully designed class A opamp only dissipates two times this theoretical value. On the other hand, non-class A opamps dissipate somewhere between 1X to 2X the theoretical value depending on the opamp topology used in the integrator. The optimal partition of the clock period for a class A opamp is illustrated by a design example. It shows that, to be power-efficient, we should devote more than half of the active period to slewing. The nature of the large feedback reference voltage in $\Sigma\Delta$ modulators makes class AB opamps and other slew enhanced opamps less attractive than in other environment such as biomedical and acoustics. With the result that class A opamps are the best choice for $\Sigma\Delta$ converters.

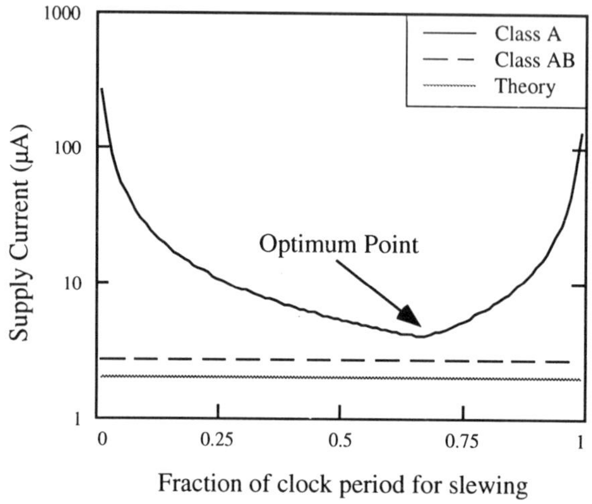

Figure 3: Power optimization

References

[1] J. C. Candy and O. J. Benjamin, "The structure of quantization noise from sigma-delta modulation," *IEEE Trans. Commun.*, vol. COM-29, pp. 1316-1323, Sept. 1981.

[2] F. Medeiro, B. Perez-Verdu, A. Rodriguez-Vazquez and J. L. Huertas, "Modeling opamp-induced harmonic distortion for switched-capacitor ΣΔ modulator design", *Proc ISCAS'94*, pp. 445-448, May 1994.

[3] R. Castello and P. R. Gray, "Performance limitations in switched-capacitor filters," *IEEE Trans. Circuits Syst.*, vol. CAS-32, pp. 865-876, Sept. 1985.

[4] M. R. Degrauwe and F. H. Salchli, "A multipurpose micropower sc filter," *IEEE J. Solid-State Circuits,* vol. sc-19, NO. 3, pp. 343-348, June. 1984.

[5] R. Castello and P. R. Gray, "A high-performance micropower switched-capacitor filter," *IEEE J. Solid-State Circuits,* vol. sc-20, No. 6, pp. 1122-1132, Dec. 1985.

[6] L. Callewaert and W. M. C. Sansen, "Class AB CMOS amplifiers with high efficiency," *IEEE J. Solid-State Circuits,* vol. sc-25, no. 3, pp. 684-691, June 1990.

[7] F. Wang and R. Harjani, "A low voltage class AB CMOS amplifier," unpublished.

[8] M. Degrauwe, J. Rijmenants, E. A. Vittoz and H. J. De Man, "Adaptive biasing CMOS amplifiers," *IEEE J. Solid-State Circuits,* No. 3, pp. 522-528, June 1982.

[9] K. Nagaraj, "CMOS amplifiers incorporating a novel slew rate enhancement technique," *Proc. IEEE 1990 Custom Integrated Circuits Conf.*, pp. 11.6.1-11.6.5, May 1990.

[10] R. Kline, B. J. Hosticka and H. J. Pfleiderer, "A very-high slew rate CMOS operational amplifier," *IEEE J. Solid-State Circuits,* vol. sc-24, pp. 744-746, June 1989.

[11] F. Krummenacher, E. Vittoz and M. Degrauwe, "Class AB CMOS amplifier for micropower sc filter," *Electron. Lett.*, vol. 17, No. 13, pp. 433-435, June 1981.

[12] B. J. Hosticka, " Dynamic CMOS amplifiers," *IEEE J. Solid-State Circuits,* vol. sc-15, No. 5, Oct. 1980.

[13] F. Wang and R. Harjani, "Dynamic amplifiers: settling, slewing and power issues," *Proc ISCAS'95,* pp. 319-326, May 1995.

[14] A. E. Stevens, G. A. Miller, "A high-slew integrator for switched-capacitor circuits," *IEEE J. Solid-State Circuits,* vol. sc-29, No. 9, pp. 1146-1149, Sept. 1994.

[15] M. R. Degrauwe and W. M. C. Sansen, "The current efficiency of MOS transconductance amplifiers," *IEEE J. Solid-State Circuits,* No. 3, pp. 349-359, June. 1984.

Fast and Accurate Delay Estimation for use in Timing Driven Layout for FPGAs with Segmented Channels

Sinan Kaptanoglu
Actel Corporation, 955 East Arques Avenue, Sunnyvale California 94086

Abstract

Timing delays for FPGAs with segmented channels can be accurately estimated by most SPICE-like circuit simulators. Such simulators however, are too slow to be of any use inside an iterative automatic timing driven layout (ATDL) engine, which may repeat the computations millions of times. Other methods such as half-perimeter approximation is very easy and fast to compute, but its accuracy is very poor for these kinds of FPGAs. The method published by Chew and Lien overcomes these problems; however, it introduces a very large number of fitted parameters, and its accuracy decreases for more complicated topologies (1). We present a method similar to their's, but with better accuracy and far fewer coefficients. This is achieved by taking advantage of the symmetry of the problem under the permutation group S_N of N objects.

Introduction

Existing fast delay estimation methods for FPGAs basically fall into three categories:

a) Very accurate (1% inaccuracy or less) simulators. These are typically so slow that the designer (or the design automation software) calls these only once or twice during the entire design cycle, usually at the very end.

b) Quite accurate (3 to 5% inaccuracy) simulators. These are 10 to 100 times faster than those of type (a). These simulators can be called many tens of times (if needed) during the design cycle. Most commonly, they are called to give feedback to the front end synthesis and technology mapping software, so that some critical paths can be re-synthesized.

c) Reasonably accurate (3 to 10% inaccuracy, but better than 5% in a great majority of cases.) These simulators are typically 10^6 times faster than those of type (a), and are used almost exclusively by iterative ATDL to improve the routability and the timing characteristics of FPGA designs. An iterative ATDL may call this kind of a simulator millions of times per FPGA design.

The subject of this paper is a simulator of type (c) that is geared specifically for FPGAs with segmented channels.

The first obvious question we must answer is why we need such a highly specialized simulator for a sub-class of FPGAs when there are methods to solve the general RC network problem (2), (3), (4), (5). These methods all suffer some inaccuracy because they model the transistors as ideal current/voltage sources. Furthermore, their application to FPGAs is not efficient enough to put them in the category (c) mentioned above. At the opposite end of the spectrum, very efficient and simple methods of delay estimation (such as the notorious half perimeter method) tend to be so inaccurate, that in some cases their prediction could be off by more than 100%. (This is due to the nature of the FPGA switching elements and the routing topologies for segmented FPGAs.) No ATDL bent on achieving the highest possible performance for an FPGA can work with such simplistic methods.

We do not intend to review here either the general routing problem for FPGAs with segmented channels, or the general problem of how to do ATDL for FPGAs. We refer the reader to the literature on both subjects (6), (7), (8), (9). We wish however, to emphasize the well-known fact that estimating delays accurately and quickly is more difficult for segmented FPGA architectures than any other one. In contrast, it is easier to do this for fully segmented architectures (most segments are of unit length) because small perturbations to the placement of logic blocks modify the existing delays incrementally. This breaks down for the segmented architectures: even the smallest perturbations to the placement of logic blocks may change some net delays very drastically. (This happens when the routing topologies end up being drastically different before and after the incremental perturbation.).

The method we shall present here is not the first one to offer a simulator of category (c) for FPGAs with segmented channels. The first such simulator was presented by Chew and Lien (1). Our method is similar, but it goes beyond their's in determining what the symmetry and the true variables of the problem are. Then, we reformulate the problem in terms of these new variables, which will turn out to be the low order Casimir invariants of the permutation group S_N. Based on this symmetry and the invariants associated with it, we will then present an approximation scheme to efficiently estimate net delays. We will also present actual numerical results from the implementation of our method.

Topologies and their Symmetries

The programmability requirements, as well as delay considerations due to the switching element resistance restrict the routing topologies for segmented FPGAs into three classes. The

0-7803-3177-6 $5.00 © 1996 IEEE

simplest class of topologies is made out of a single Steiner trunk with any number of horizontal segments driven from it as shown in Fig.1. Each horizontal segment can be of dif-

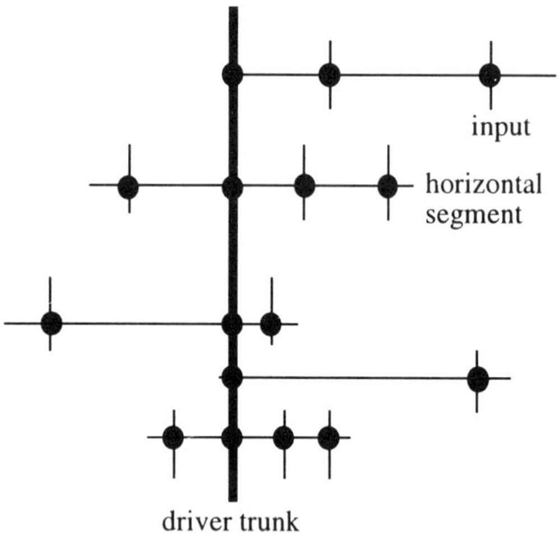

Fig. 1: Single Steiner Trunk Topology

ferent length, and each horizontal segment can drive one or more input pins on the net. All input pins of the net are driven by these horizontal segments. There are no loads (i.e. input pins) directly driven from a vertical trunk. We also consider more complicated classes of topologies with two trunks as shown in Fig. 2. Of these we distinguish two subclasses, depending on whether the trunks are driven in parallel, or in series. The first case is nothing but two single trunk topologies put together and driven in parallel. These two trunks, however, are not symmetric and not equivalent to each other for routing and delay properties. The trunks may overlap in some channels. The class of dual-series trunk topologies is the most complicated one we need to consider. In this case, the main trunk drives the secondary trunk through a connecting segment. This connecting segment itself may also drive input pins. Again, the trunks may overlap in some channels.

These three classes are sufficient to cover all practical topologies with arbitrary numbers of horizontal segments and loads on each segment. We deal with classes of topologies as opposed to the individual topologies Chew and Lien directly consider. This difference is exactly why we use vastly fewer coefficients than they do. For example, the topology class in Fig. 2 (b) covers more than 1,000 individual topologies that Chew and Lien considered. They fit the individual topologies, whereas we fit large classes of topologies.

To see how this can be done, we must understand the symmetry properties of the problem. First let us concentrate on the class of single trunk topologies. The key observation is

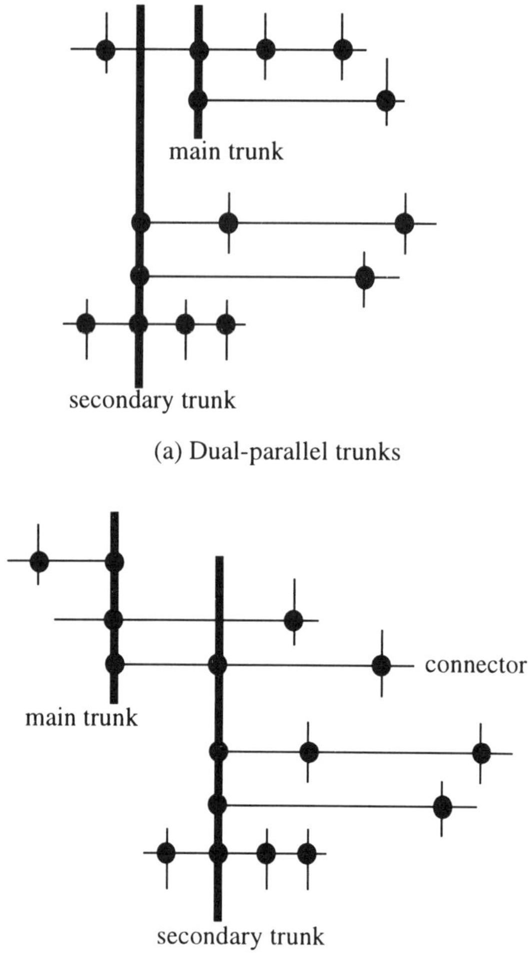

(b) Dual-series trunks with a connector

Fig. 2: Double Trunk Topologies

that the relative position of the horizontal segments on a given trunk is unimportant. If we interchange two segments with each other (along with their loads of course), the delays at the input pins do not change. This symmetry arises because the FPGA interconnection elements (whether antifuses or transistor gates) possess a much larger impedance than the metal lines of the driving trunk (which are typically extra wide metal lines to make the symmetry even better.) This symmetry obviously is not perfect, but valid to about 3% for the logic block driver size of the Actel FPGAs. This is significantly smaller than the target of better than 10% inaccuracy, therefore, we can use the symmetry as if it wasperfect. Also note that the symmetry applies to more than just the interchange of any two segments. It is valid for arbitrary permutations of all segments.

Delay Estimation and Results

Let D_i be the delay at the pins of ith horizontal segment. D_i is a function of $P_1, L_1, P_2, L_2, ..., P_N, L_N$, where P_j is the number of pins on the jth segment, and L_j is the length of the jth segment, and N is the number of segments, which we take to be very large, because we are considering a whole class of topologies. Considering the delays at all segments simultaneously, we can write $\underline{D} = \{D_1, D_2, ..., D_N\}$ like a vector. Likewise, we use a similar notation for $\underline{P}$ and $\underline{L}$ as well. With this we write $\underline{D}(\underline{P},\underline{L})$ as an N-valued function of 2N variables. We can now state the symmetry of the problem as an equation:

$$\underline{D}(\sigma\underline{P},\sigma\underline{L}) = \sigma\underline{D}(\underline{P},\underline{L}), \quad (1)$$

where σ is any element of S_N, the permutation group order N. (Among many others, σ has an N dimensional matrix representation, which has one and only one entry equal to 1 on every row and column with all other entries being 0.) Equation (1) states that $\underline{D}$ transforms as a vector under the permutation group S_N. This automatically restricts what $\underline{D}$ can be a function of. It requires that:

$$\underline{D}(\underline{P},\underline{L}) = A(\underline{P},\underline{L})\underline{\Delta} + B(\underline{P},\underline{L})\underline{P} + C(\underline{P},\underline{L})\underline{L} \quad (2)$$

where $\underline{\Delta}$ is the constant vector $\{1,1,...,1\}$, and A, B, and C are scalar (invariant) functions under S_N transformations:

$$A(\sigma\underline{P},\sigma\underline{L}) = \sigma A(\underline{P},\underline{L}) \quad (3)$$
$$B(\sigma\underline{P},\sigma\underline{L}) = \sigma B(\underline{P},\underline{L}) \quad (4)$$
$$C(\sigma\underline{P},\sigma\underline{L}) = \sigma C(\underline{P},\underline{L}) \quad (5)$$

Similar to the way we extracted the vector nature of $\underline{D}$, we must now extract the scaler nature of A, B, and C using the symmetry group S_N again. Equations (3), (4) and (5) require that A, B, and C not be arbitrary functions of $\underline{P}$, and $\underline{L}$; but only arbitrary functions of the invariants that can be formed out of $\underline{P}$, and $\underline{L}$. These are precisely the Casimir invariants of the group S_N. There are two such invariants that are of the first order in P_i, or L_i: $Q \equiv \sum_i P_i$, and $H \equiv \sum_i L_i$. There are six second order invariants, three of which can be derived from the first order ones: Q^2, H^2, and QH. Besides these, there are three others and these are none other than the scalar products of the vectors in question: $<\underline{P}|\underline{P}>$, $<\underline{P}|\underline{L}>$, and $<\underline{L}|\underline{L}>$, where for example $<\underline{P}|\underline{L}> \equiv \sum_i P_i L_i$. Of course, S_N has cubic and higher order Casimir Invariants as well, but we shall not worry about them here. We will only fit the coefficients up to (but not including) the cubic terms.

Now we can approximate A to be a linear combination of all invariants of S_N up to cubic terms:

$$A(\underline{P},\underline{L}) = a_0 + a_1Q + a_2H + a_3<\underline{P}|\underline{P}> + a_4<\underline{P}|\underline{P}> + a_5<\underline{P}|\underline{P}> + a_6Q^2 + a_7QII + a_8II^2 + O(3) \quad 6)$$

The corresponding approximation for B and C need to be done only up to (but not including) the second order, since both B and C are multiplied by vector components, which themselves are of first order:

$$B(\underline{P},\underline{L}) = b_0 + b_1Q + b_2H + O(2) \quad (7)$$

$$C(\underline{P},\underline{L}) = c_0 + c_1Q + c_2H + O(2) \quad (8)$$

These 15 coefficients in Equations (6), (7), and (8) completely determine $\underline{D}$ up to third order. If we deem this approximation to be insufficient, than we can add the third order invariants of S_N, add more coefficients for these invariants, and improve the accuracy of the delay estimation. This turns out to be unnecessary. We have determined that for $N \leq 17$, (17 happens to be the number of channels for Act1280XL) the quadratic approximation is more than sufficient to obtain the accuracy we need. The coefficients were fitted to more than 157, 000 data points calculated by a simulator of category (a) defined in the introduction.Table 1 below summarizes the results.

Table 1: Results for single trunk routing topologies:

Average inaccuracy	0.077%
Mean Square Root inaccuracy	0.729%
Worst case inaccuracy (WCI)	9.555%
Frequency of deviations worse than WCI/2	0.091%

The most important numbers are given in the third and fourth rows of Table 1. They show that the accuracy is always better than 10%, but more importantly it is better than 5% in more than 99.9% of all cases computed!

The computations for the classes of double trunk routing topologies proceed very similarly, but naturally we end up with a larger number of coefficients to fit. Due to space limitations, we will show how to compute only the most complicated class of topologies, namely the one shown in Fig. 2(b). For this case, the invariance group is not just S_N any more, but the direct product group $S_{(N-1)} \otimes S_M$, where the primary trunk has N segments (including the connector) and the secondary trunk has M segments, with $1 \leq N$, and $1 \leq M$. Note that if N=1, then $S_{(N-1)}$ is the trivial group S_0 with nothing but the identity element.

Now there will be three sperate sets of coefficients, one for the delay vector (of size N-1) of the segments driven from the main trunk, another for the delay vector (of size M) of the segments driven from the secondary trunk, and a third one for the delay of the connector (not a vector, since size is 1.) Once again we will expand everything up to third order only. Furthermore we will restrict the interaction terms between the primary and the secondary trunks to be linear with each other's invariants. It is necessary, however, to allow quadratic interaction terms for the invariants of the connector when we compute coefficients for either trunk, and quadratic terms from the main trunk when we compute coefficients for the connector.

The (N-1) dimensional vectors $\underline{D}^{(1)}$, $\underline{\Delta}^{(1)}$, $\underline{P}^{(1)}$, and $\underline{L}^{(1)}$ below refer to the segments of the main trunk. Similarly, the M dimensional vectors $\underline{D}^{(2)}$, $\underline{\Delta}^{(2)}$, $\underline{P}^{(2)}$, and $\underline{L}^{(2)}$ refer to the segments of the secondary trunk, the scalers $P^{(c)}$ and $L^{(c)}$ refer to the connector, and $\Lambda^{(c)} \equiv 1$. Then we can write:

$$\underline{D}^{(i)}(\underline{P}^{(i)},\underline{L}^{(i)}) = A^{(i)}(\underline{P}^{(i)},\underline{L}^{(i)})\underline{\Lambda}^{(i)} + B^{(i)}(\underline{P}^{(i)},\underline{L}^{(i)})\underline{P}^{(i)} + C^{(i)}(\underline{P}^{(i)},\underline{L}^{(i)})\underline{L}^{(i)}, \text{ for } i = 1,2, \text{ and } c. \quad (9)$$

$A^{(i)}$, $B^{(i)}$, and $C^{(i)}$ are then expanded in terms of the linear and the quadratic invariants for i = 1, 2, and c as follows:

$$A^{(1)} = a_0 + a_1Q^{(1)} + a_2H^{(1)} + a_3\langle\underline{P}^{(1)}|\underline{P}^{(1)}\rangle + a_4\langle\underline{P}^{(1)}|\underline{L}^{1)}\rangle + a_5\langle\underline{L}^{(1)}|\underline{L}^{(1)}\rangle + a_6(Q^{(1)})^2 + a_7Q^{(1)}H^{(1)} + a_8(H^{(1)})^2 + a_9Q^{(2)} + a_{10}H^{(2)} + a_{11}P^{(c)} + a_{12}L^{(c)} + a_{13}Q^{(1)}Q^{(2)} + a_{14}Q^{(1)}H^{(2)} + a_{15}H^{(1)}Q^{(2)} + a_{16}H^{(1)}H^{(2)} + a_{17}Q^{(1)}P^{(c)} + a_{18}Q^{(1)}L^{(c)} + a_{19}H^{(1)}P^{(c)} + a_{20}H^{(1)}L^{(c)} + a_{21}P^{(c)}Q^{(2)} + a_{22}P^{(c)}H^{(2)} + a_{23}L^{(c)}Q^{(2)} + a_{24}L^{(c)}H^{(2)} + a_{25}(P^{(c)})^2 + a_{26}P^{(c)}L^{(c)} + a_{27}(L^{(c)})^2 + O(3) \quad (10)$$

$$B^{(1)} = b_0 + b_1Q^{(1)} + b_2H^{(1)} + b_3Q^{(2)} + b_4H^{(2)} + b_5P^{(c)} + b_6L^{(c)} + O(3) \quad (11)$$

$$C^{(1)} = c_0 + c_1Q^{(1)} + c_2H^{(1)} + c_3Q^{(2)} + c_4H^{(2)} + c_5P^{(c)} + c_6L^{(c)} + O(3) \quad (12)$$

$$A^{(2)} = \bar{a}_0 + \bar{a}_1Q^{(2)} + \bar{a}_2H^{(2)} + \bar{a}_3\langle\underline{P}^{(2)}|\underline{P}^{(2)}\rangle + \bar{a}_4\langle\underline{P}^{(2)}|\underline{L}^{2)}\rangle + \bar{a}_5\langle\underline{L}^{(2)}|\underline{L}^{(2)}\rangle + \bar{a}_6(Q^{(2)})^2 + \bar{a}_7Q^{(2)}H^{(2)} + \bar{a}_8(H^{(2)})^2 + \bar{a}_9Q^{(1)} + \bar{a}_{10}H^{(1)} + \bar{a}_{11}P^{(c)} + \bar{a}_{12}L^{(c)} + \bar{a}_{13}Q^{(2)}Q^{(1)} + \bar{a}_{14}Q^{(2)}H^{(1)} + \bar{a}_{15}H^{(2)}Q^{(1)} + \bar{a}_{16}H^{(2)}H^{(1)} + \bar{a}_{17}Q^{(2)}P^{(c)} + \bar{a}_{18}Q^{(2)}L^{(c)} + \bar{a}_{19}H^{(2)}P^{(c)} + \bar{a}_{20}H^{(2)}L^{(c)} + \bar{a}_{21}P^{(c)}Q^{(1)} + \bar{a}_{22}P^{(c)}H^{(1)} + \bar{a}_{23}L^{(c)}Q^{(1)} + \bar{a}_{24}L^{(c)}H^{(1)} + \bar{a}_{25}(P^{(c)})^2 + a_{26}P^{(c)}L^{(c)} + a_{27}(L^{(c)})^2 + O(3). \quad (13)$$

$$B^{(2)} = \bar{b}_0 + \bar{b}_1Q^{(2)} + \bar{b}_2H^{(2)} + \bar{b}_3Q^{(1)} + \bar{b}_4H^{(1)} + \bar{b}_5P^{(c)} + \bar{b}_6L^{(c)} + O(3) \quad (14)$$

$$C^{(2)} = \bar{c}_0 + \bar{c}_1Q^{(2)} + \bar{c}_2H^{(2)} + \bar{c}_3Q^{(1)} + \bar{c}_4H^{(1)} + \bar{c}_5P^{(c)} + \bar{c}_6L^{(c)} + O(3) \quad (15)$$

$$A^{(c)} = x_0 + x_1P^{(c)} + x_2L^{(c)} + x_3(P^{(c)})^2 + x_4P^{(c)}L^{(c)} + x_5(L^{(c)})^2 + x_6Q^{(1)} + x_7H^{(1)} + x_8Q^{(2)} + x_9H^{(2)} + x_{10}Q^{(1)}Q^{(2)} + x_{11}Q^{(1)}H^{(2)} + x_{12}H^{(1)}Q^{(2)} + x_{13}H^{(1)}H^{(2)} + x_{14}\langle\underline{P}^{(1)}|\underline{P}^{(1)}\rangle + x_{15}\langle\underline{P}^{(1)}|\underline{L}^{1)}\rangle + x_{16}\langle\underline{L}^{(1)}|\underline{L}^{(1)}\rangle + x_{17}(Q^{(1)})^2 + x_{18}Q^{(1)}H^{(1)} + x_{19}(H^{(1)})^2 + O(3) \quad (16)$$

$$B^{(c)} = y_1Q^{(1)} + y_2H^{(1)} + y_3Q^{(2)} + y_4H^{(2)} + O(3) \quad (17)$$

$$C^{(c)} = z_1Q^{(1)} + z_2H^{(1)} + z_3Q^{(2)} + z_4H^{(2)} + O(3) \quad (18)$$

These 112 coefficients in the equations (10) through (18) completely determine $\underline{D}^{(1)}$, $\underline{D}^{(2)}$, and $\underline{D}^{(c)}$, up to third order.Table 2 below summarizes the results for Act1280XL for dual-series trunk routing topologies.

Table 2: Results for dual-series trunk routing topologies:

Average inaccuracy	0.017%
Mean Square Root inaccuracy	0.773%
Worst case inaccuracy (WCI)	3.906%
Frequency of deviations worse than WCI/2	2.646%

For this class of topologies, we actually obtained more accurate results than the class with a single trunk. Note that our accuracy is always better than 4%, and it is even better than 2% in more than 97% of all cases computed!

Conclusions

Using the (near exact) symmetry of the routing topologies for segmented FPGAs, we have developed an approximation scheme that allows very fast and accurate delay estimation for use with iterative ATDL. An actual implementation of our method proves that our results are more accurate than that of Chew and Lien, even though we use vastly fewer fitted coefficients. The improvement we get compared to their results actually gets better as the routing topologies become more complicated. This is true not only in terms of the accuracy of estimates, but also in terms of the economy of the coefficients needed.

References

(1) M. Chew, and J. C. Lien, "Fast Delay Estimation in Segmented Channel FPGAs", 2nd International ACM/SIGDA Workshop on Field-programmable Gate Arrays, February 1994.

(2) J. Rubinstein, P. Penfield, and M. Horowitz, "Signal Delay in RC Tree Networks", IEEE Transactions on Computer-Aided Design, CAD-2(3), July 1983.

(3) T. Lin, C. A. Mead, "Signal Delay in General RC Networks", IEEE Transactions on Computer-Aided Design, CAD-3(4), October 1984.

(4) P. K. Chan, K. Karplus, "Computing Signal Delay in General RC Networks by tree/link Partitioning", in 26th Design Automation Conference, pp. 485-490, 1989.

(5) L. Pillage, R.Rohrer, "Asymptotic Waveform Estimation", IEEE Transactions on Computer-Aided Design, 93 (4), October 1990.

(6) A. El-Gamal, J. Greene, S. Kaptanoglu, and V. RoyChowdhury, "Segmented Channel Routing", in 27th Design Automation Conference, pp. 567-572, 1990.

(7) J. Frankle, "Iterative and Adaptive Slack Allocation for Performance Driven Layout and FPGA routing", in 29th Design Automation Conference, pp. 536-542, 1992.

(8) S. Raman, C. L. Liu, L.G. Jones, "Timing -based Placement for an FPGA Design Environment", Oxford International Workshop on Field Programmable Logic and Applications, September 1993.

(9) A. Mathur, K. C. Chen, C. L. Liu, "Applications of Slack Neighborhood Graphs to Timing Driven Optimization Problems in FPGAs", in ACM/SIGDA International Symposium on Field Programmable Gate Arrays, pp. 118-124., 1995.

A New Technique for the Efficient Solution of Singular Circuits

Jaijeet Roychowdhury

AT&T Bell Laboratories, Murray Hill, NJ

Abstract

Singular circuits are those that have a continuum of solutions (or no solution) and are characterized by rank deficiency in the (linearized) circuit matrix. Such circuits often arise in practice - examples are filters with poles at zero, chains of transmission gates that are off, and circuits that rely on charge storage and transfer. In this paper, a technique for the efficient solution of such circuits is presented. The method is based on solving for the minimum-least-squares solution of the singular system. Unlike traditional methods for least squares solution, the new approach exploits the sparsity of the circuit matrix, making it practical for large industrial circuits. The method is best for applications with relatively small singular subspaces, as is the case in most circuits. Applications to industrial designs testify to the efficacy of the new technique; an example in which more than a week of design time would have been saved is presented.

1 Introduction

A basic operation in circuit simulation, as in many other numerical computations, is that of solving a system of linear equations $Ax = b$ for the vector x (with A a given sparse matrix and b a known vector). This operation arises, for example, in the Newton-Raphson method for nonlinear equation solution[1], as well as during the implicit integration of ordinary differential equations[2]. In most practical situations, the matrix A is assumed to be of full rank or *nonsingular*, i.e., with a nonzero determinant. This assumption makes it possible to use efficient techniques such as sparse LU factorization[3, 4] to obtain the solution x. The exploitation of sparsity is key to making large practical problems computationally tractable.

In many real-life situations, however, the nonsingularity assumption for A does not hold. For example, a circuit consisting of two isolated parts (without a common ground) except for inductive coupling through a transformer has a singular matrix. The reason for this is that one of the coils of the transformer is floating with respect to the other. Despite the singularity, the circuit is meaningful because the designer is not interested in the absolute potential but in potential differences between nodes of interest. Singularity is also encountered in the DC analysis of circuits which have nodes with no path to ground not involving a capacitor[1]. Such situations are especially common in circuits that rely on charge storage and transfer.

An important practical situation in which singular matrices occur is when real circuitry is macromodelled using idealized elements. This is common in the hierarchical design of systems, with some blocks represented in transistor-level detail and others as simple idealized macromodels. Often, the idealization is designed with a singularity – an example is a filter with a pole at zero frequency. In practical realizations, parasitics often remove exact singularities; even so, the matrix may be numerically singular owing to the finite precision to which numbers are representated in computers. Numerical singularities and near-singularities also occur in circuits with high-impedance nodes, for example inside high-gain op-amps in open loop simulation.

Previous approaches towards solving singular circuits have all been based on modifying the circuit to remove its singularities. A common method to avoid floating node problems is to insert a small conductance gmin, usually across parasitic diodes in MOS and BJT devices. Checking the circuit's topology using graph-based rules is often used to detect structures that lead to matrix singularites; the designer is required to change the circuit to remove them before proceeding with the simulation. These techniques are not always effective, even for the subclass of singular circuits that they address. For example, small values of gmin can cause the Newton-Raphson method to take very large steps, leading to non-convergence. It is impossible to detect certain singularities by graph-based rules, for example in a tank circuit excited at its resonant frequency. As shown in Section 3, the process of finding singularities can cause serious delays in design – it took more than a week for the three singularities in the circuit of Figure 5 to be found.

This work attacks the problem at the matrix level and is therefore uniformly applicable to any circuit situation that causes a singularity. As such, it does not require the explicit detection of the singularity, nor does the circuit need to be changed. Information about the location of singularities is produced as a by-product of finding the "best" possible solution of the singular circuit. Unless the singularity is the result of an error in the circuit, the solution provided is what the designer is looking for. A key feature of the method is that it uses LU factorization to exploit sparsity; hence it is applicable to large circuits, unlike other methods for dealing with singular systems[1, 5] (see Section 2.1). The method is best suited for circuits with a small number of singularities (e.g., with only a few floating nodes). For most practical circuits, this does not constitute a limitation.

The new technique is presented in Section 2. Computational considerations, crucial for the practical applicability of the method, are also addressed in Section 2. In Section 3, the method is applied to two industrial examples. In Section 4, the key features of the new technique, as well as related future work, are summarized.

2 Exploiting sparsity in singular systems

In this section, an efficient method for obtaining the "best" solution of a singular circuit is presented. The definition and properties of the minimum least-squares solution, its connection with circuits, the rôle of the singular subspace and traditional approaches for solving singular systems are the subject of Section 2.1. An outline of the new technique and its computational properties is presented in Section 2.2.

[1] It is sometimes possible in such cases to use charge conservation to eliminate the singularity using a variant of the present approach; this is a topic of current work.

0-7803-3177-6 $5.00 © 1996 IEEE

2.1 The minimum least-squares solution

Given a singular square matrix A and a vector b, the object is solve for x satisfying:

$$Ax = b \tag{1}$$

If A is nonsingular, there is always a unique x satisfying Equation 1[6]. If A is singular, however, two possibilities exist: if b is not in the *range* $R(A)$ of A, no solution x exists satisfying Equation 1; if b is in the range space of A, a continuum of solutions x satisfy the equation.

Despite the lack of a unique solution in the singular case, it can be shown[5] that a unique x^* does exist that is the "best" solution to Equation 1, whether A is singular or not. x^* is called the *minimum least-squares solution* and is defined to be the smallest possible vector (i.e., of smallest norm) that minimizes $||Ax - b||$. It is appropriate to think of x^* as the best solution because of the following properties:

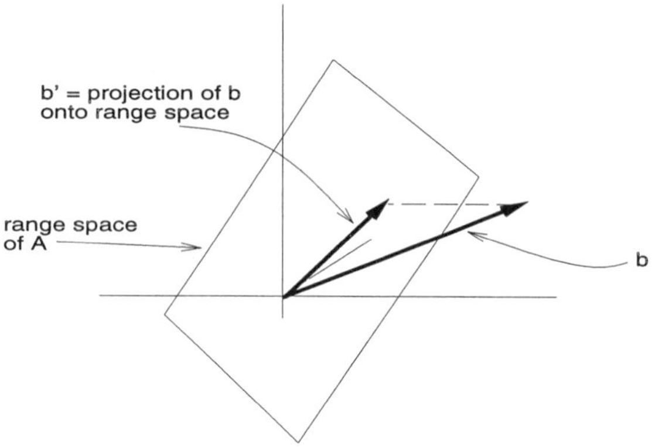

Figure 1: Orthogonal projection of b on the range of A

1. If A is nonsingular, x^* is the unique solution of Equation 1.

2. If A is singular and b is in the range of A, then x^* is the solution of minimum norm that satisfies Equation 1.

3. If A is singular and b is *not* in the range of A, then there exists a unique vector b' in the range of A that is closest to b. x^* is the solution of minimum norm satisfying $Ax = b'$.

The concept of the minimum least-squares (MLS) solution is illustrated in Figures 1 and 2.

It is useful to obtain an appreciation of how cases 2 and 3 above relate to circuits. Figures 3 and 4 show two circuits - the first is an idealization of a switched-capacitor circuit and the second a circuit segment generating a pole at zero frequency. Such structures are common in macromodels for system-level design. The circuit equations for Figure 3 can be shown to be an example of case 2, i.e., b is in the range of A and the circuit has an infinite number of solutions. The MLS solution for this case corresponds to a voltage of zero at node 2. The circuit of Figure 4 is an example of case 3 above; it does not have any solution at DC. For this case, b' corresponds to setting the current source to zero; at this value, the circuit has an continuum of solutions. The MLS solution corresponds to a voltage of zero at node 1.

The two examples above were simple linear ones. Most practical circuits are nonlinear, hence it is of interest to examine how the MLS solution generalizes to the nonlinear

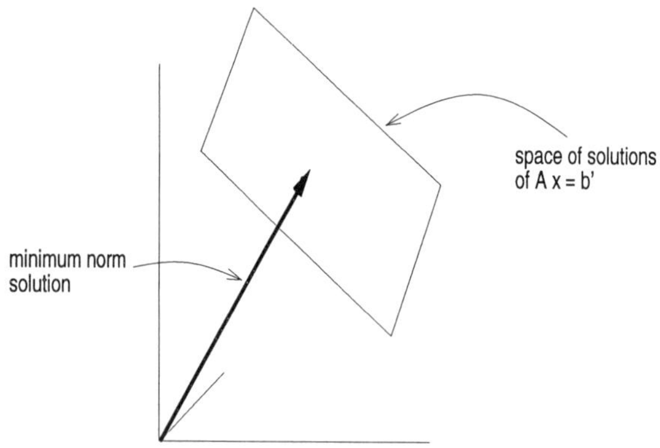

Figure 2: The MLS solution x^*

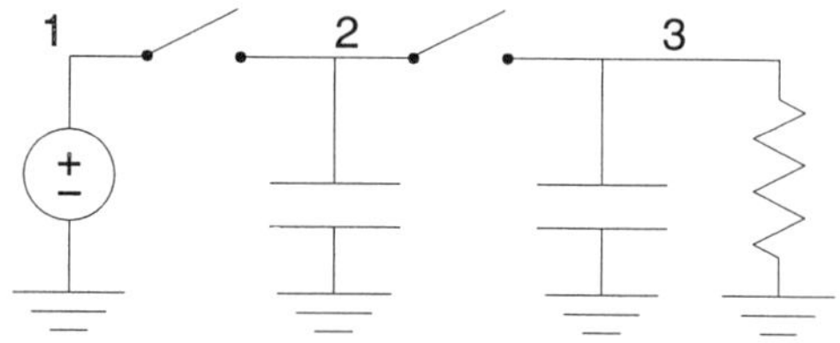

Figure 3: $b \in R(A)$: continuum of solutions

case. For an arbitrary nonlinear equation $f(x) = 0$, the analogue of the linear MLS solution is a minimum-norm x^* that minimizes $||f(x)||$ locally. Unlike the linear case, there may be several local minima, leading to multiple values for x^*, one for each local minimum of $||f(x)||$. An important connection between the linear and nonlinear cases is obtained through a generalization of the well-known Newton-Raphson method for solving nonlinear equations. The usual Newton-Raphson iteration is:

$$J\delta x = -f(x), \quad J = \frac{\partial f}{\partial x} \tag{2}$$

For J nonsingular, δx is well-defined and unique. An immediate generalization when J is singular is to use the MLS solution of Equation 2 for δx. It can be shown that this variation maintains the local quadratic convergence properties of the Newton-Raphson method; the difference is that it converges to a *local minimum* of $f(x)$[7, 8], while the nonsingular Newton-Raphson converges to a zero of $f(x)$. Hence the linear MLS solution can be used in the Newton-Raphson method to obtain nonlinear locally MLS solutions.

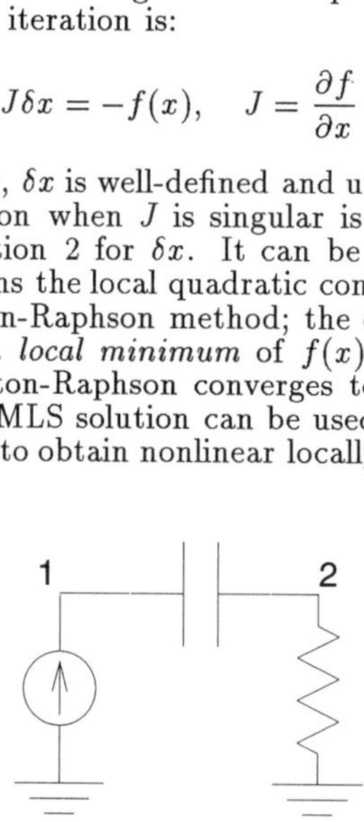

Figure 4: $b \notin R(A)$: no solutions

Let the size of A (the number of rows or columns) be denoted by n, which for practical circuits can easily be in the thousands. An important concept is the dimension of the null space $N(A)$ of A[6]. Denote this (also called the dimension of the singularity of A) by m. In circuit terms, m is the number of independent circuit segments that are not well-defined; for example, in Figures 3 and 4, m is 1. In practical circuits, m is usually much smaller than n and rarely exceeds 20 or 30. It will be seen later that this fact is key to the practical applicability of the technique of this work.

Traditional approaches for obtaining the MLS solution use the singular value decomposition (SVD) or the QR decomposition of A[1, 5], which destroy sparsity and are hence inefficient for large circuits. A property of the MLS solution that is sometimes used is the so-called *normal equation*:

$$A'Ax^* = A'b \tag{3}$$

If A is not square but of full rank with more rows than columns, it can be shown that $A'A$ is square and nonsingular. In such cases, the normal equation can be used with traditional nonsingular solution techniques to obtain the MLS This situation is not however easily applicable to circuits, which typically have square matrices; moreover, forming $A'A$ usually destroys sparsity to a considerable extent, reducing the effectiveness of sparse factorization techniques. The author is not aware of any existing approach for solving the MLS problem that exploits sparsity to the same degree as LU factorization techniques do for sparse nonsingular matrices.

2.2 Obtaining the MLS solution for sparse A

In this section, an outline of the new technique is presented.

Consider the most general case 3 above, with $b \notin R(A)$. The situation is illustrated in Figure 1. The first step in the new method is to obtain b', the orthogonal projection of b onto $R(A)$; this is the vector in $R(A)$ that is closest to b, i.e., $||b - b||'$ is the minimum least-square error achievable.

The equation $Ax = b'$ has a continuum of solutions, as illustrated in Figure 2. If x_1 and x_2 are two such solutions, then $x_1 - x_2$ in necessarily in the null space of A; hence in order to find the minimum norm solution x^*, the new technique starts with any solution x of $Ax = b'$ and subtracts from it all its components in the null space of A. In other words, x^* is obtained from any solution x by subtracting from x its projection onto $N(A)$.

In order to obtain numerical representations for the range space $R(A)$, the null space $N(A)$ and a solution x to $Ax = b'$, the LU factorization of A is used. Usually, LU factorization is performed only for nonsingular matrices; however, all square matrices, singular or otherwise, can be decomposed into LU factors. For concreteness, assume the Crout form of the LU algorithm, i.e., $L_{ii} = 1, \quad \forall i \in \{1, \ldots, n\}$. This makes L always nonsingular; U can be shown to have the same rank as A. The following example shows the LU factors of a matrix that is rank-deficient by 1 (i.e., $m = 1$):

$$A = \begin{bmatrix} 8 & 6 & 6 & 8 \\ 8 & 10 & 9 & 11 \\ 32 & 28 & 29 & 38 \\ 24 & 22 & 23 & 30 \end{bmatrix} \tag{4}$$

$$L = \begin{bmatrix} 1 & 0 & 0 & 0 \\ 1 & 1 & 0 & 0 \\ 4 & 1 & 1 & 0 \\ 3 & 1 & 1 & 1 \end{bmatrix}, \quad U = \begin{bmatrix} 8 & 6 & 6 & 8 \\ 0 & 4 & 3 & 3 \\ 0 & 0 & 2 & 3 \\ 0 & 0 & 0 & 0 \end{bmatrix} \tag{5}$$

The LU factors of a singular matrix can be obtained by small modifications to the procedure for the nonsingular case, as follows: when singularity is detected by the LU factorization routine, the remaining lower submatrices of L and U (of size m) are set to the identity and zero matrices of size m, respectively. Pivoting strategies and other heuristics used for nonsingular LU factorization carry over unchanged to this generalization.

Once the LU factors are available, they can be used to find $N(A)$, $R(A)$ and a solution to $Ax = b'$. First consider the problem of finding *a* solution x^0 of the consistent equation $LUx = b'$, assuming b' has already been computed. Since L is nonsingular, we have $Ux = L^{-1}b'$. It can be shown the last m rows of U can be chosen to be identically zero without loss of generality if A is rank-deficient by m. Therefore *a* solution is to set the last m variables of x to zero, and solve for the first $n - m$ variables using normal reverse-substitution.

A slight variant of the above can be used to produce a basis for the null space $N(A)$. To obtain m linearly independent vectors in $N(A)$, the last m variables of x are chosen as follows: $\{1, 0, \ldots, 0\}$, $\{0, 1, 0, \ldots, 0\}$, ..., $\{0, 0, \ldots, 1\}$. With each of these choices, $Ux = 0$ is solved for the top $n - m$ remaining variables using back-substitution. The resulting vectors form a basis for $N(A)$ and are linearly independent. Let the basis be denoted by $\{n^1, \ldots, n^m\}$.

The projection of x^0 onto $N(A)$ is obtained next. For this task, it does not suffice to have a basis that spans $N(A)$; it is necessary to obtain an *orthonormal basis*. Given the basis $\{n^1, \ldots, n^m\}$, the standard Gram-Schmidt orthogonalization procedure [6] is used to convert it into an orthonormal basis. Let the orthonormal basis thus obtained be denoted by $\{n^{o1}, \ldots, n^{om}\}$.

Once the orthonormal basis is available, the MLS solution x^* is easily obtained from x^0 by subtracting the projection:

$$x^* = x^0 - \sum_{i=1}^{m} \langle x^0, n^{oi} \rangle n^{oi} \tag{6}$$

where $\langle \cdot, \cdot \rangle$ denotes the dot product. Equation 6 eliminates any component of $N(A)$ from x^0, leaving the MLS solution.

The above assumed that b' had already been calculated. This is in fact the first step in finding the MLS solution, and it is achieved by first computing a basis for the *orthogonal complement of the range space* $R(A)$ of A, rather than for $R(A)$ itself. The orthogonal complement is chosen for two reasons: 1. its dimension can be shown [6] to be the same as m, hence the space is small, leading to computational efficiency, and 2. since the orthogonal complement of A is the same as the null space of A'[6], it can be computed in the same manner as the basis for $N(A)$ as described above, using $A' = U'L'$. Once an orthonormal basis for the orthogonal complement is obtained, b' is derived by subtracting from b all its components in this space, analogous to Equation 6.

The computation involved in the above procedure is dominated by the LU factorization and the Gram-Schmidt orthonormalization steps. LU factorization is almost linear in the size of the matrix n when A is sparse. The Gram-Schmidt procedure requires $O(nm^2)$ operations, i.e., it is linear in n and quadratic in m. For this reason, the method is best suited for small m, a situation that holds for most circuit applications.

3 Results

In this section, two industrial circuits with singularities are presented and the efficacy of the technique demonstrated.

The first circuit is a system-level design for a programmable line termination circuit. The block diagram of the circuit is shown in Figure 5. The purpose of the circuit is to present a programmable impedance to a communications line in order to match its characteristic impedance and minimize echoes and reflections (accurate line matching is especially important for data communications applications). The circuitry is mixed analog-digital. The line current is sensed, filtered and digitized before being input to the digital section of the circuit. A DSP-based calculation produces a digital representation of the voltage that the desired termination would generate; this voltage is then converted to analog form, filtered and placed on the line using a transducer. During system-level design, most of the blocks were macromodelled using ideal elements such as linear controlled sources.

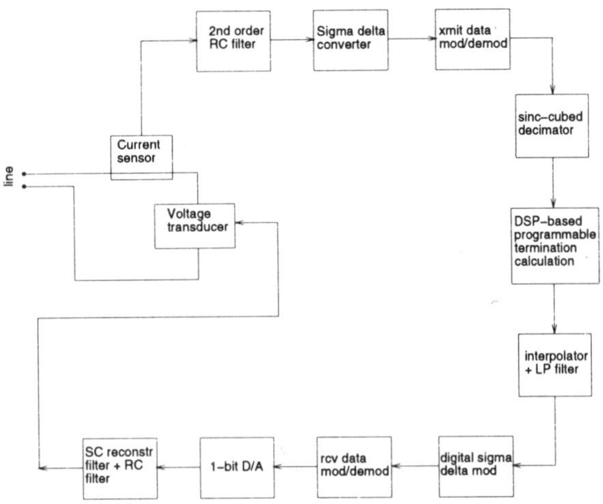

Figure 5: Programmable line termination system

The circuit as simulated by the designers contains three singularities ($m = 3$). Two of the singularities are due to deliberately designed poles at DC in the sinc^3 decimator and the interpolator/LP-filter blocks. A third singularity is the result of a benign error in the DSP macromodel, where the output of a sensing controlled current source is connected to a floating node. The circuit would not simulate in a proprietary internal simulator because the initial operating point computation would terminate with a singular matrix error. The existing topology check in the simulator having failed to locate the problems, it took the designers more than a week to find the causes of the singularities and devise workarounds.

Applying the new technique to the circuit resulted in its solution in four Newton-Raphson iterations. In addition, the degree of rank-deficiency $m = 3$ was found and information produced about the location of the singularities.

The second circuit, shown in Figure 6, consists of a chain of 32 transmission gates. The gates are alternately on and off. If the transmission gates were ideal, half the drains and sources in the chain would be floating ($m = 16$). In a well-modelled circuit, however, parasitics and leakage elements prevent exact singularities, hence the matrix is not exactly singular but nearly so. The circuit does not converge with the Newton-Raphson algorithm because the near-singular matrix leads to large steps, taking the algorithm outside its region of convergence, from which it does not recover.

When simulated with the new technique with appropriate threshold parameters, the circuit converges in 8 Newton-Raphson iterations. m is detected to be 16 and information produced about the locations of the near-singularities.

4 Conclusion

A new technique has been presented for solving circuits with singularities. The method is based on finding the minimum

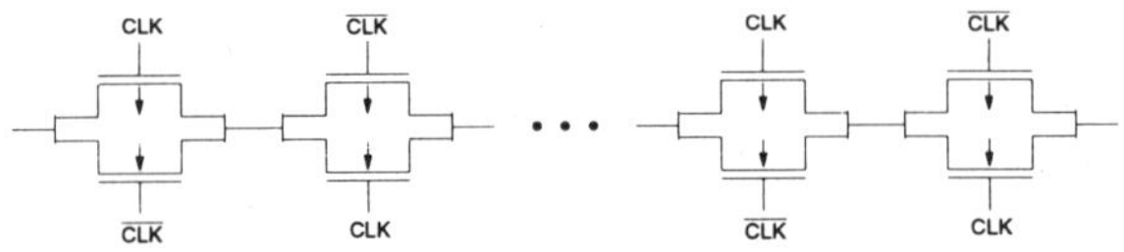

Figure 6: Chain of transmission gates

least-squares solution of $Ax = b$ efficiently when A is sparse. LU factorization and Gram-Schmidt orthogonalization are the main computational steps in the new technique.

The method is applicable to large VLSI designs, unlike traditional techniques for least-squares solution that do not take advantage of the sparsity of circuit matrices. It is best suited for circuits that have a small number of singularities because the computation involved in Gram-Schmidt orthonormalization increases as the square of the number of singularities.

The application of the method to a system-level design with singularities and to an almost-singular circuit with off transmission gates has been presented in this paper. It is relevant to a number of other areas as well – RF circuits and harmonic balance, analysis of tank circuits and oscillators, transient analysis, optimization, homotopy methods, solution using charge conservation and AHDL simulation are examples. Work is currently continuing in application to these areas.

Acknowledgments

The author is grateful to D.A. Rich for pointing out the need for the analysis of singular circuits in industrial design. He and H.S. Fetterman contributed the two examples. Thanks are due to K. Mayaram for useful discussions about singular circuits and charge-sharing and to Kishore Singhal for pointing out the rôle of the normal equations in MLS problems. R.C. Melville and M.H. Wright brought Björck's survey of least-squares methods [5] to the author's notice and P. Feldmann pointed out the application of the technique to RF CAD problems. Special thanks are due to K. Mayaram and D.A. Rich for suggesting CICC as a forum for this work and for their help in making the submission possible.

References

[1] W.H. Press, S.A. Teukolsky, W.T. Vetterling, and B.P. Flannery. *Numerical Recipes – The Art of Scientific Computing.* Cambridge University Press, 1989.

[2] C.W. Gear. *Numerical initial value problems in ordinary differential equations.* Prentice-Hall series in automatic computation. Prentice-Hall, Englewood Cliffs, N.J., 1971.

[3] K.S. Kundert and A. Sangiovanni-Vincentelli. *Sparse Version 1.3a User's Guide.* EECS Dept., University of California, Berkeley, July 1988.

[4] U. Schendel. *Sparse matrices: numerical aspects with applications for scientists and engineers.* Ellis-Horwood, 1989.

[5] Å. Björck. Least Squares Methods. Technical report, Dept. of Math., Linköping University, May 1991.

[6] K. Hoffman and R.A. Kunze. *Linear Algebra.* Prentice-Hall, 1971.

[7] E.L. Allgower and K. Georg. *Numerical Continuation Methods.* Springer-Verlag, New York, 1990.

[8] J.M. Ortega and W.C. Rheinboldt. *Iterative solution of nonlinear equations in several variables.* Academic Press, 1970.

Chair: Takayasu Sakurai
Co-Chair: Michael Thorn

Welcome to the "Video, Image and Speech Digital Signal Processing" session. This session covers two papers on motion estimation, a paper on a DCT/wavelet based CODEC, one on an image processor and two papers covering speech CODECs. All designs are with sub-micron CMOS technology and integrate about one million transistors.

In the coming years, HDTV will be a reality. The first two papers in the session deal with motion estimators for HDTV resolution. One paper describes a VLSI motion estimator for the MPEG2 MP@HL standard having a chip size of 12.2 x 12.7mm^2 using a 0.6µm triple-metal CMOS and integrating 1.2M transistors. Employing a systolic array architecture, the chip operates at 133MHz delivering 30GOPS of performance. A new macroblock clustering algorithm is proposed achieving more than ±60 pixel search range. The subsequent paper is a 200MHz, 200GOPS motion estimator also targeting HDTV resolution. Using 0.5µm CMOS technology and a novel layout approach, the chip size is 170mm^2 with 1 million transistors. The LSI is cascadable for larger search areas.

The next paper is on a video CODEC DSP that effectively handles both wavelet-based and DCT-based encoding and decoding by introduction of a special instruction. Fabricated with 0.5µm double metal CMOS, the chip size is 8.43 x 7.76 mm^2 containing 1 million transistors. Operating at 40MHz, the power consumption is 560mW. The DCT macro uses a decimation-in-frequency scheme to significantly reduce the area. The chip is targeted for the consumer markets armed with its capability for 15 frames/second H.261 video CODEC, JPEG and non-standard wavelet-based CODEC.

The fourth paper presents a 7.68GIPS image processor design with an embedded 16Mbit DRAM. Using a 0.38µm double metal CMOS process, 128 processor elements with 0.8 million transistors are integrated on a chip of 18.8 x 16.7 mm^2 with a power consumption of 992mW at 40MHz. Circuit techniques for synchronization, efficient refresh and block redundancy are shown.

The last two papers are on audio DSPs. The first describes an audio CODEC capable of handling G.711, G.722 and G.728 standards. Integrating peripherals, interfaces and memories with the processor, the LSI is used in a 4-chip implementation of a H.320 video conferencing system. The LSI area is 8.3 x 8.3 mm^2 using a 0.5µm double metal CMOS technology and operates at 36MHz. The last paper in the session presents a 1.8V, 36mW speech CODEC for PDC application. Several ideas which enable the low-power/low-voltage operation are presented. These include parallelism in the datapath, low-power RAM and ROM design, and gated clocks. The chip size of 9.8 x 10.3 mm^2 with 1.46 million transistors and an operating frequency of 22.5MHz is achieved using a 0.5µm double metal CMOS technology.

The papers in this session present ideas for DSPs with a wide range of performance, power and cost. Join us and take a look at recent achievements in the field of high-performance and low-power DSP solutions for audio and video applications.

VLSI Implementation of Hierarchical Motion Estimator for MPEG2 MP@HL

Takao ONOYE†, Gen FUJITA†, Isao SHIRAKAWA†,
Kenji MATSUMURA††, Hiromu ARIYOSHI†††, and Shuji TSUKIYAMA‡

†Dept. Information Systems Eng., Osaka University, Yamada-Oka, Suita, Osaka, 565 Japan
Phone: +81(6)879-7807, Fax: +81(6)875-5902, E-mail: {*onoe, fujita, taka2, sirakawa*}*@ise.eng.osaka-u.ac.jp*
††KCS Co., Ltd., Naka-Kosaka, Higashi-Osaka, Osaka, 577 Japan
Phone: +81(6)722-0606, Fax: +81(6)722-7546, E-mail: *matsu@k-c-s.k-c-s.co.jp*
†††Dept. Electrical and Electronics Eng., Ehime University, Bunkyo, Matsuyama, Ehime, 790 Japan
Phone: +81(899)24-7111, Fax: +81(899)23-0672, E-mail: *ariyoshi@hawk.ee.ehime-u.ac.jp*
‡Dept. Information and System Eng., Chuo University, Kasuga, Bunkyo-ku, Tokyo, 112 Japan
Phone: +81(3)3817-1871, Fax: +81(3)3817-1847, E-mail: *tsuki@kawa.elect.chuo-u.ac.jp*

Abstract

A VLSI motion estimator dedicated to MPEG2 MP@HL has been developed. Adopting a two-level hierarchical searching algorithm in detecting motion vectors, the computational labor can be reduced by 1/70 in comparison with the conventional algorithm. The proposed motion estimator is integrated in a 0.6μm triple-metal CMOS chip which contains 1,200K transistors on a 12.2 $\times$ 12.7 mm^2 die. The input clock rate can be attained up to 133MHz, which enables the real time motion estimation for MPEG2 MP@HL.

1. Introduction

The MPEG2(1) algorithm supports communication and broadcasting media as well as storage media. For a wide range of communication and broadcasting services, the real time encoding of transmitting pictures is indispensable. However, the motion estimation is the most computationally intensive among a series of encoding processes, which has been a main obstacle to the real time encoding for MPEG2 MP@HL (Main Profile at High Level).

In the process of the motion estimation, the so-called *block-matching algorithm* has been widely applied, which estimates motion vectors on the macroblock-by-macroblock basis, where a *macroblock* indicates the fundamental unit composed of 16 $\times$ 16 pels. Given a macroblock of a predicting picture and a search region of a reference picture, this algorithm is to detect the best matched macroblock in the search region.

Although in the MPEG1 the search region of a motion vector is limited to $\pm$15 in both the vertical and horizontal directions, in the MPEG2 that is not restricted to such a specific range. Moreover, the MPEG2 supports complicated prediction modes (i.e. *field prediction, 16×8 MC,* and *dual-prime*) in order to treat interlaced pictures. Thus for the real time implementation of an MPEG2 motion estimator, an extremely efficient search algorithm should be incorporated into a sophisticated VLSI architecture. Recently, a variety of approaches have been proposed for the motion estimation of MPEG2(2–7). However, they are either hard to be implemented in a single chip(2, 3), or restricted to the TV broadcasting level (MP@ML: Main Profile at Main Level)(4–7). Hence, there still remains much room for developing far more effective VLSI architectures especially for MPEG2 MP@HL pictures.

Motivated by this tendency, the present paper describes a novel architecture dedicated to MPEG2 MP@HL motion estimator, which adopts a two-level hierarchy in detecting motion vectors. Through the use of a top-down ASIC design system, *COMPASS Design Navigator*, an MPEG2 MP@HL real time motion estimator has been integrated in a 0.6μm triple-metal CMOS chip of the total die area of 12.2 $\times$ 12.7 mm^2.

2. Macroblock Clustering Algorithm

The so-called *full-search algorithm* has been employed for the block-matching, which is to detect a motion vector exhaustively within such a range as $\pm$7, $\pm$15, and $\pm$31 in both the horizontal and vertical directions by reference to MAD's (Mean Absolute Differences)(5,8). A number of architectures have been proposed for this full-search algorithm on the basis of the regulated data processing sequence. This algorithm, however, has too much computational complexity to be implemented for the real time motion estimation.

To cope with this difficulty, a *macroblock clustering algorithm* has been proposed dedicatedly for MPEG2 MP@HL(9). This algorithm adopts a two-level hierarchy in detecting motion vectors in such a way that with the use of the unit of a *cluster* ($n \times n$ macroblocks), both *cluster* and *macroblock* vectors are detected in such a different full-search fashion as summarized below.

[Motion Vector Detection]

1° Given a reference picture, construct its subsampled pattern P by subsampling pels by a factor of n^2.

2° Given a predicting picture, divide it into clusters, each composed of $n \times n$ macroblocks, i.e. $16n \times 16n$ pels (Fig. 1(a)).

3° For each cluster i, do the following:

(i) Subsample pels of cluster i by a factor of n^2, and let a set of 16×16 subsampled pels be de-

16.1.1

0-7803-3177-6 $5.00 © 1996 IEEE

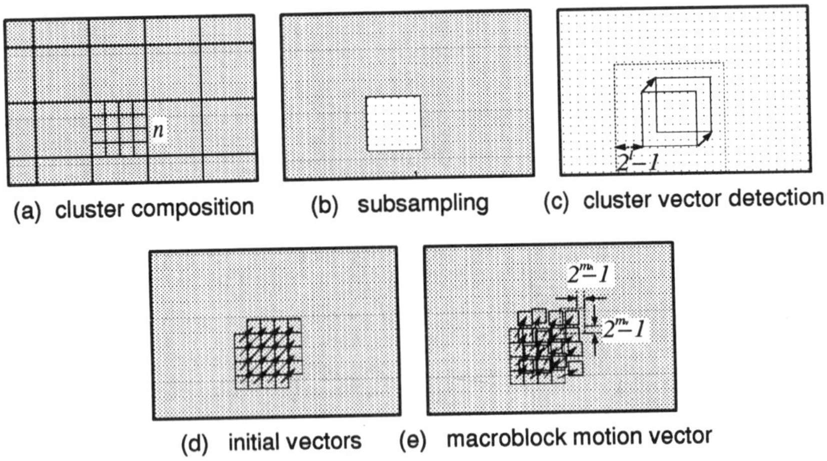

Fig. 1 Motion estimation based on macroblock clustering.

noted by a *subsampled cluster* C_i (Fig. 1(b)).

(ii) Detect a motion vector $\boldsymbol{v}_i$ of C_i on P by the full-search within the search range $\pm(2^\ell - 1)$ ($\ell = 3, 4, \ldots$) both in the horizontal and the vertical directions from the corresponding position of C_i (Fig. 1(c)).

(iii) Define an *initial vector* $\boldsymbol{u}_k$ for each macroblock k in cluster i of the predicting picture such that the direction of $\boldsymbol{u}_k$ is the same as that of $\boldsymbol{v}_i$, but the magnitude of $\boldsymbol{u}_k$ is n times as large as $\boldsymbol{v}_i$ (Fig. 1(d)).

(iv) Seek a motion vector $\boldsymbol{w}_k$ for each macroblock k of the predicting picture by the full-search within the search range of $\pm(2^{m_v - 1})$ and $\pm(2^{m_h - 1})$ ($m_v, m_h = 1, 2, \ldots$) from the destination of the initial vector $\boldsymbol{u}_k$ in the horizontal and the vertical directions, respectively (Fig. 1(e)).

Fig. 1 illustrates a clustering example in the case of $n = 4, \ell = 3, m_v, m_h = 3$.

This approach can reduce the computation labor by 1/10–1/100 according to the values of n, ℓ, m_v, and m_h, in comparison with the original full-search algorithm. According to the experimental results, these parameter are fixed in our motion estimator such that $n = 4$, $\ell = 4$, $m_v = 2$, and $m_h = 3$ which may facilitate real time processing of MP@HL pictures without degrading the quality of detected vectors.

3. VLSI Architecture

Henceforth, the detailed architecture is outlined, which can admit the high speed execution of the clustering algorithm.

To implement this new full-search algorithm, the so-called *systolic array architecture* is employed, as has been done in (4–8), where a PE (Processing Element) is used for calculating an absolute difference between pixel data of a reference picture and those of a current predicting picture, and then for adding the difference to the *intermediate sum data* output from the preceding PE. To achieve the real time encoding of MPEG2 MP@HL, a sophisticated scheme of reusing reference pixel data should be introduced into the full-search procedure especially in terms of lightening the memory access bottleneck.

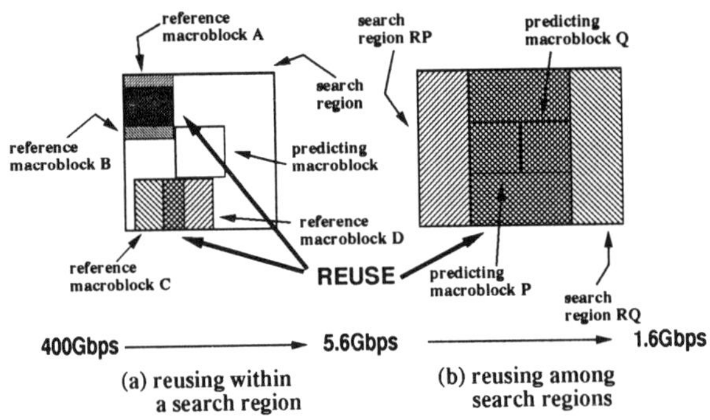

Fig. 2 Reusing mechanism of reference picture.

An effective memory architecture has been introduced to achieve high performance of PE's(8), in which an approach is attempted for reusing reference pixel data in the process of detecting a motion vector for a macroblock in a predicting picture as exemplified in Fig. 2(a). Through this reusing mechanism in executing the clustering algorithm, the required bandwidth of frame memory interface can be reduced from 400Gbps to 5.6Gbps. However, there still remains a disadvantage in this approach that the reference pixel data once used for detecting a motion vector for a macroblock will not be reused for detecting another motion vector even for a neighbor of the macroblock.

It should be noticed here that the block-matching is executed essentially macroblock by macroblock, and that there is considerable overlapping among search regions for adjacent macroblocks. Thus in order to exploit this memory architecture for the macroblock clustering algorithm, a far more effective reusing mechanism of reference pixel data should be constructed so as to be applied to detecting motion vectors for adjacent macroblocks.

A novel architecture is devised in our motion estimator, which attempts the maximum possible reuse of reference pixel data between both pairwise adjacent clusters and macroblocks. Fig. 2 exemplifies the key concepts of our reusing mechanism, where the reusing is attempted not only within a search region (Fig. 2(a)) but also between pairwise adjacent search regions (Fig. 2(b)). Consequently, the bandwidth of frame memory interface can be successfully reduced to 1.6Gbps.

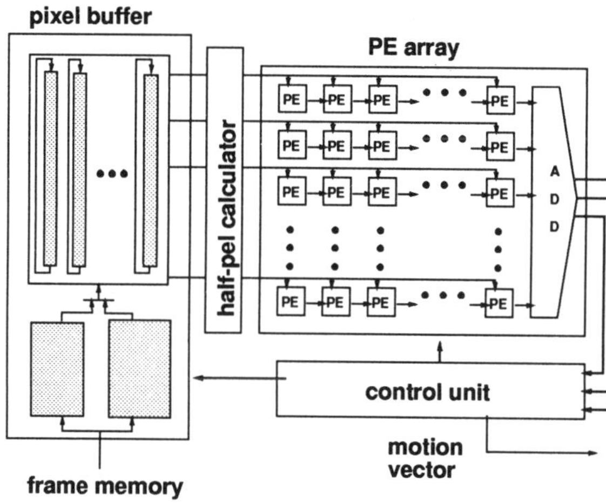

Fig. 3 Overview of motion estimator.

As illustrated in Fig. 3, our motion estimator consists of a PE array, a pixel buffer, a half-pel calculator, and a control unit, which are capable of executing the macroblock clustering algorithm. The PE array is of a *broadcasting type*, and is used both for the cluster vector search and for the macroblock vector search.

In should be added that our motion estimator is intended to deal with the half-pel and dual-prime prediction for P-pictures *(predictive pictures)*. These facilites were once attempted by using a specific unit(5, 6), which occupied very large area. On the contrary, our motion estimator adopts the PE array for seeking half-pel and dual-prime motion vectors, which can be shared with both the cluster vector search and the macroblock vector search.

[PE Array]

As shown in Fig. 4, each PE adopts a two-stage pipeline, where high-speed 8- and 12-bit datapaths are employed in the first and second stages, respectively. Final adders collect the outputs of all rows so as to generate a *frame vector*, two field vectors, and two 16×8 MC vectors simultaneously.

[Pixel Buffer]

The pixel buffer is constructed of an *active buffer* and a *preserve buffer* (see Fig. 5). The active buffer composed of a 64-bit tri-port register file feeds reference pixel data to the PE array. The half-pel prediction can be implemented with the use of this register file and a

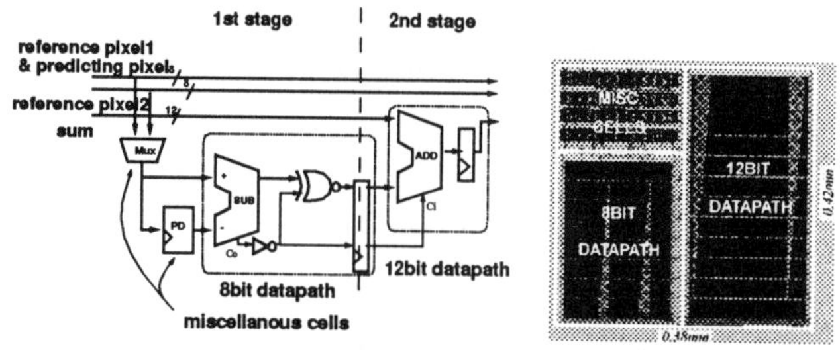

Fig. 4 High-speed datapath for each PE.

half-pel calculator.

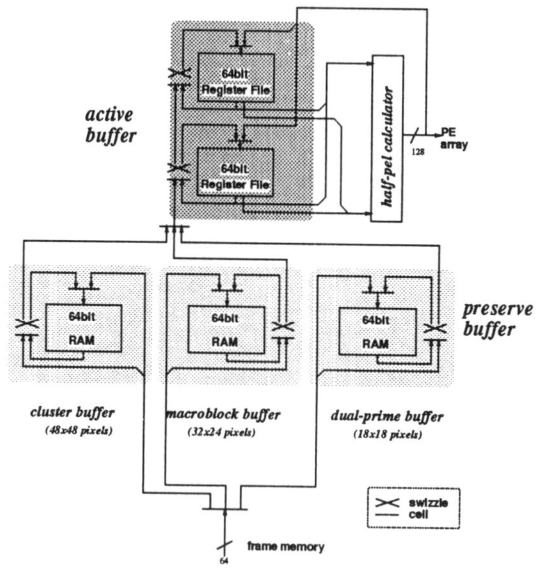

Fig. 5 Block diagram of pixel buffer.

To reduce the amount of control circuits, each register is made to contain 8 pixel data in cascade. In the active buffer, the shifting of pixel data to the upper row is executed by means of a bit swizzle cell as illustrated in Fig. 6. The preserve buffer consists of a *cluster buffer*, a *macroblock buffer*, and a *dual-prime buffer*. Each of the cluster buffer and the macroblock buffer can store all pixel data of the search region (48 × 48 for a cluster; 32 × 24 for a macroblock) so as to facilitate the reusing mechanism of reference pixel data. The dual-prime buffer is made to contain pixel data of the search range for a dual-prime vector. In contrast with the active buffer, the preserve buffer, which need not feed pixel data to the PE array, is implemented by a 64-bit RAM.

[Control Unit]

The control unit supervises the overall status of the motion estimator. The main tasks of this unit include controlling the PE array, outputting the final motion vectors, and generating frame memory addresses for reference and current predicting pictures. Fig 7 illustrates the input timing of reference picture to the PE array. To store both the reference and predicting pictures, a set of three frame memories is empolyed for our motion estimator, each realized by two 16M SDRAM's.

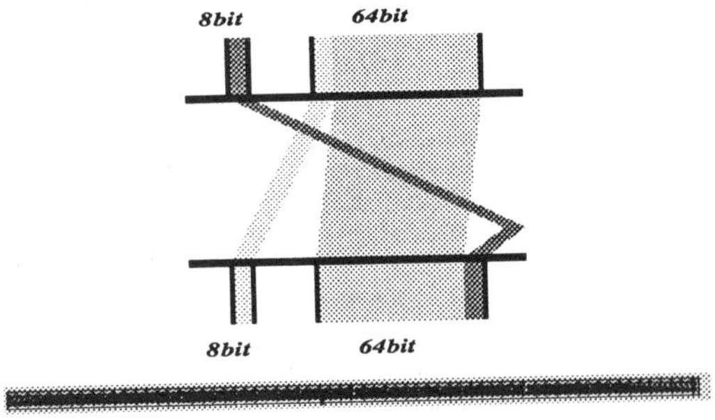

Fig. 6 Swizzle cell for shifting reference pixel data.

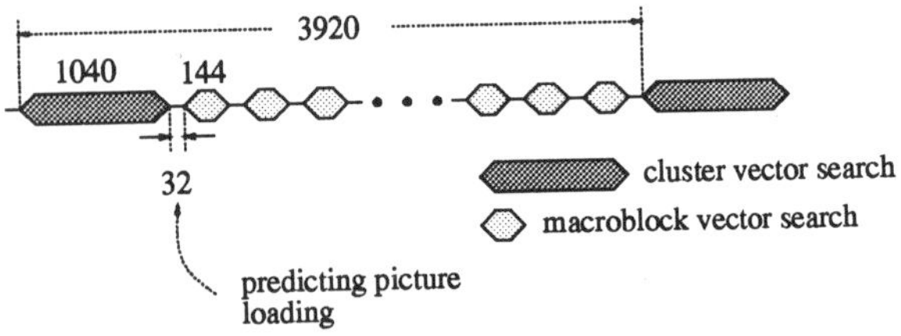

Fig. 7 Input timing to PE array.

Table 1 Main features of the motion estimator.

Technology	0.6 μm CMOS triple-level Al
Chip size	12.2 mm × 12.7 mm
Transistors	1,177,760
Clock frequency	133MHz
Support image	MPEG2 MP@HL
Search range	$-72 \sim +67$ (vertical) $-68 \sim +63$ (horizontal)

4. Implementation Results

The proposed motion estimator is distinctive in that by means of the maximum possible reuse of reference pixel data, the motion estimation can be accomplished for all macroblocks in a cluster within 3,920 cycles, as can be observed from Fig. 7. This motion estimator has been integrated in a single chip through the use of a top-down ASIC design system, *COMPASS Design Navigator*. Table 1 indicates the main chip features of the motion estimator, and Fig. 8 shows its layout patterns obtained by using 0.6 μm CMOS standard-cell and datapath libraries. It can be readily verified that the input clock rate can be attained up to 133MHz, which accomplishes MP@HL real time block-matching (16,200 clusters per second). The peak computational power is more than 30GOPS (Giga Operations Per Second), for which 2.1TOPS (Tera Operations Per Second) might be necessary if the original full-search algorithm were employed, mainly because the search range grows to the extent of $\pm(2^6 - 1)$ in both the horizontal and vertical directions.

5. Conclusion

This paper has described a VLSI motion estimator dedicated to MPEG2 MP@HL. The two-level hierarchical searching algorithm has been synthesized in a single chip by means of the multi-bit high-speed datapath in the PE array and the effective memory architecture in the pixel buffer intended for the maximum possible reuse of reference pixel data. In addition, the designed motion estimator can yield the half-pel and the dual-prime vectors. With the use of a 0.6μm triple-metal CMOS process, the motion estimator can operate at a clock rate of 133MHz which enables the MPEG2 MP@HL real time motion estimation.

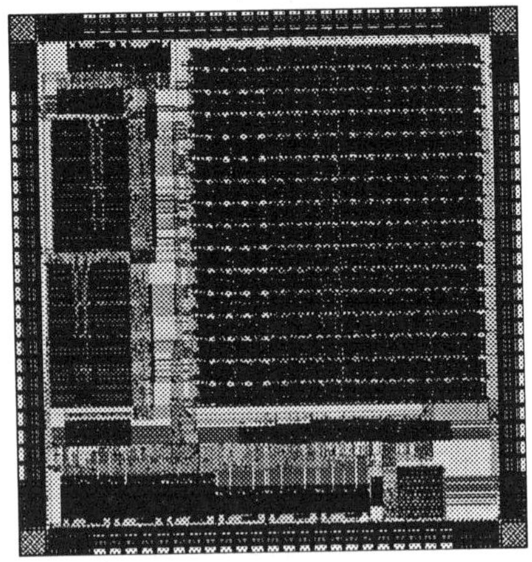

Fig. 8 Final layout patterns of single chip motion estimator.

References

(1) ITU-T Rec. H.262, ISO/IEC 13818-2: "Generic coding of moving pictures and associated audio", Draft International Standard, Oct. 1994.

(2) Y.-L. Chan and W.-C. Siu: "A new block motion vector estimation using adaptive pixel decimation", in *Proc. IEEE Int'l Conf. Acoustics, Speech, and Signal Processing*, pp. 2257-2260, May 1995.

(3) M. C. Chen and A. N. Willson Jr.: "A high accuracy predictive logarithmic motion estimation algorithm for video coding", in *Proc. IEEE Int'l Symp. Circuits and Systems*, pp. 617-620, May 1995.

(4) Y. Kim, C.S. Rim, and B. Min: "A block matching algorithm with 16:1 subsampling and its hardware design", in *Proc. IEEE Int'l Symp. Circuits and Systems*, pp. 613-616, May 1995.

(5) K. Ishihara, S. Masuda, S. Hattori, H. Nishikawa, Y. Ajioka, T. Yamada, H. Amishiro, and M. Yoshimoto: "A half-pel precision MPEG2 motion-estimation processor with concurrent three-vector search", in *IEEE ISSCC Digest of Technical Papers*, pp. 288–289, Feb. 1995.

(6) A. Otani, Y. Matsumoto, M. Gion, H. Yoshida, T. Araki, A. Ubukata, M. Serizawa, K. Aoki, A. Sota, A. Nagata, and K. Aono: "A motion estimation processor for MPEG2 video real time encoding at wide search range", in *Proc. IEEE Custom Integrated Circuits Conference*, pp. 405–408, May 1995.

(7) Y. Senda, H. Harasaki, and M. Yano: "A simplified motion estimation using an approximation for the MPEG-2 real-time encoder", in *Proc. IEEE Int'l Conf. Acoustics, Speech, and Signal Processing*, pp. 2273–2276, May 1995.

(8) E. G. Tzeng and C.-Y. Lee: "An efficient memory architecture for motion estimation processor design", in *Proc. IEEE Int'l Symp. Circuits and Systems*, pp. 712-715, May 1995.

(9) T. Onoye, M. Takatsu, G. Fujita, and I Shirakawa: "A VLSI-suited motion estimation algorithm based on macroblock clustering", *IEICE Tech. Report*, CAS95-43, Jun. 1995 (in Japanese).

A Cascadable 200 GOPS Motion Estimation Chip for HDTV Applications

Jan Peter Berns, Tobias G. Noll

Chair of Electrical Engineering and Computer Systems,
University of Technology RWTH Aachen, Germany

Abstract

A flexible block matching motion estimation chip is described with variable sized blocks between 8 × 8 and 32 × 32 pixels. Each chip performs block matching with a search area of ± 15 vertically and horizontally for a block size of 32 × 32. For larger search areas devices can be cascaded. Besides full search, fast algorithms can be emulated. Sub-pel precision motion vectors can be calculated using a smaller search area or cascading devices. The chip will have a computational power of more than 200 GOPS and a die size of 170 mm^2 in an 0.5-μm CMOS technology.

Introduction

Methods of low bit-rate digital coding of image sequences have been subject of increasing interest. Most of the temporal correlation between frames of an image sequence can be removed by using motion compensation to predict each frame and coding only the prediction error and motion field [1, 2]. Beside applications in the field of motion-compensated predictive coding (H.26X, MPEG, multimedia applications) motion-compensated interpolation and filtering is of broad interest (MPEG, transcoding applications).

Block matching is a pragmatic solution for the problem of motion estimation using highly regular structures. The current picture is divided into blocks of typically square shape. Each of these blocks is then placed at different positions within the corresponding search area of the previous frame. The matching criterion is the L1-norm given as the sum over all absolute pixel differences, also known as the displaced frame difference (DFD).

$$DFD = \sum_{k=1}^{n_x} \sum_{l=1}^{n_y} |s_T(k,l) - s_{T-1}(k+\Delta i, l+\Delta j)|$$

The best match is the minimum of the DFDs and provides the resulting motion vector (Fig. 1).

Parameters of the algorithm are the block size, search area, image format and the frame rate. These parameters typically extend over considerable ranges for different applications. Moreover, various search methods have been proposed in the literature, ranging from a full search to hierarchical search strategies and even to the investigation of single vector candidates. For numerous applications optimal block matching parameters have not been specified yet. Search areas of up to ± 128 horizontally and ± 32 vertically for a picture format of 1920 × 1080 with 25 frames per second are discussed for HDTV applications [3]. This results in a required computational power of more than 800 GOPS for full pel and full search and more than 3200 GOPS (3.2 TOPS) for half-pel precision motion vectors with an operation being defined as an 8bit accumulation of absolute differences. A simulation of these algorithms for longer video-sequences with standard RISC or DSP processors is not possible in acceptable time. Therefore there seems to be a need for a real-time emulation environment to investigate different parameters as well as different search strategies for research and development purposes. In addition to standard block matching the system must be capable of emulating new object-based algorithms. An adaptively refined block size and also a variable shape is necessary to implement region oriented algorithms. Inherent disadvantages of standard block matching like the detection of only translational motion can be avoided by rotating and zooming blocks in a preprocessing unit.

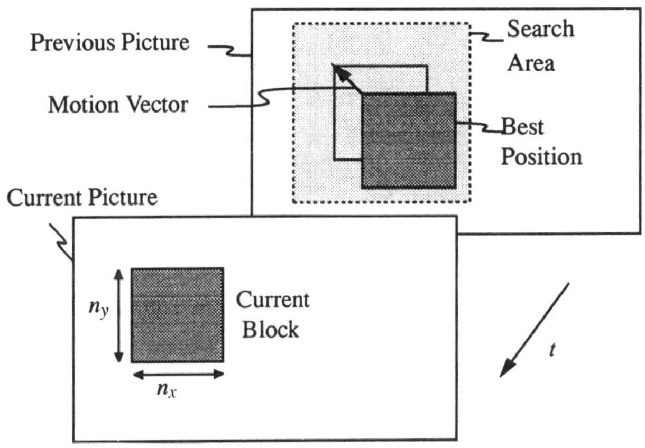

Fig. 1: Block Matching

For coding purposes standard block matching with full search has an acceptable performance, but for applications like motion-compensated interpolation the true motion vectors are essential. The detection of rotation and zoom especially leads to better pixel-based true motion vectors.

Requirements of the Target Architecture

Block matching parameters for HDTV applications are:

- block sizes between 8 × 8 and 32 × 32, typically 16 × 16 pixels,

16.2.1

0-7803-3177-6 $5.00 © 1996 IEEE

- a search area of up to ± 128 horizontally and ± 32 vertically,
- full pel, half pel or quarter pel precision, and
- variable picture formats and frame rates of up to 1920 × 1080 pixels and 50 frames per second progressiv.

In addition to full search block matching algorithms, the execution of fast algorithms with reduced computational complexity must be implementable in a flexible and programmable architecture. Key features of some of these algorithms are

- subsampling in the motion field, possibly combined with a hierarchical search method,
- the investigation of candidate vectors or displacement fields around candidate vectors, and
- pixel decimation for data of the current and/or the previous field/frame.

For future standards it also must be possible to mask pixels and to emulate object based algorithms.

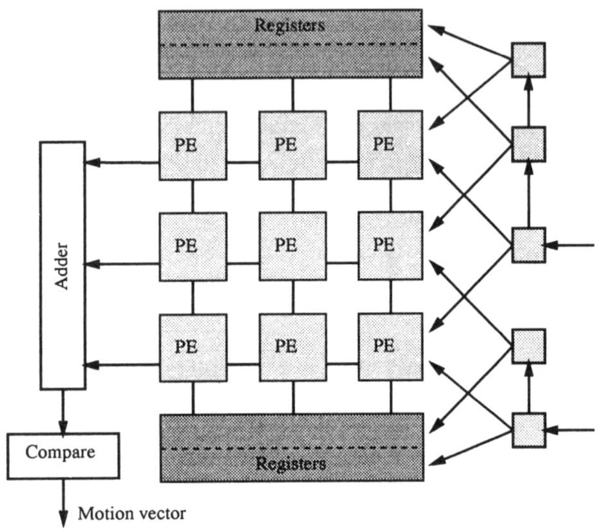

Fig. 2: 2-D Systolic Array for Block Matching

Because the full search block matching determines the upper limit of the required performance the architecture has to be optimized on that algorithm. The number of operations P_{min} which have to be calculated in parallel for the full search depends only on the number of different search positions, m_x horizontally and m_y vertically, the pixel-rate f_s, and the clock frequency f_c

$$P_{min} = \frac{f_s}{f_c} \cdot (m_x + 1) \cdot (m_y + 1).$$

Given the worst case parameters for HDTV and a clock frequency of 100 MHz, the number of operations per period (assuming a utilization of the PE array of 100%) is approximately 16000. A two-dimensional array of processing elements (PE) is therefore necessary, each PE calculating the basic operation. Different solutions to the mapping problem have been proposed, e.g. in [4]. One possibility is to assign each PE to one pixel of the current block. The pixels of the search area are moved over the array in a meander-like data flow in order to have only local communication between the PEs and no global memory access for each PE. The results of all PEs are accumulated to the resulting DFDs which are compared successively to get the final motion vector. Beside the array of PEs additional registers above and below the PE array and an input data-flow-converter (DFC) are necessary to realize the data-flow and to keep the I/O bandwidth low. This architecture is shown in Fig. 2.

Another possibiltity is to assign each PE to one pixel of the search area. It can be shown that this architecture is dual to the above block oriented structure [5].

Programmable Algorithms

A. Object-Based Algorithms

The proposed architecture is prepared for the implementation of object-based algorithms. A variable shape of the matched region is achieved by using an additional mask bit in the current block which controls a multiplexer in the accumulation path of each PE. If a pixel is masked its corresponding absolute difference is not added to the DFD. Furthermore rotation and zoom effects can be estimated. This leads to more precise motion vector fields and better results especially for interpolation applications. The details of this technique are beyond the scope of this paper.

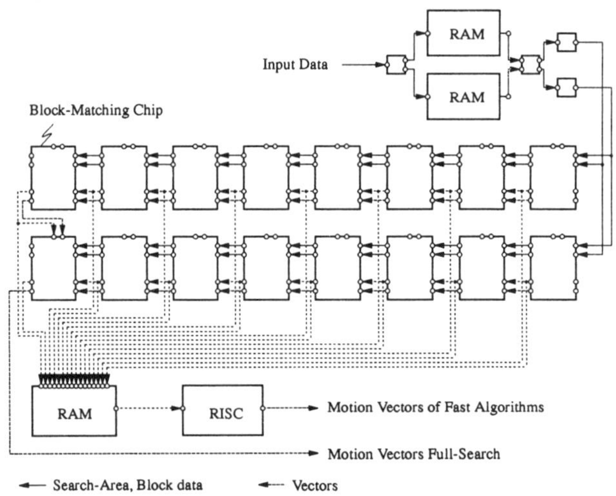

Fig. 3: Overall System Architecture using Cascaded Chips

B. Reduced Computational Complexity Algorithms

All DFDs have to be calculated for the full-search algorithm. To implement reduced complexity algorithms fewer DFDs have to be evaluated. Therefore the DFDs are written into a SRAM. An external RISC processor (Fig. 3) selects the motion vectors out of the set of DFDs. Masking of the input data is necessary to handle different word lengths for pixel decimation algorithms. The additional control overhead is negligible in comparison to the PE array.

C. Variable Block Size

The block matching algorithms are based on the assumption that the motion can be considered constant in each given image block. This assumption, however, does not hold for large blocks, especially if motion boundaries are encountered. In this case it is not possible to find the exact match of the block in the previous frame. Although using smaller blocks would often lead to more accurate motion estimates, small blocks contain less discriminative features and a higher bandwith is needed for transmitting the motion vectors.

With a class of algorithms known as variable size block matching (VSBM) a good compromise between these contrary effects could be found [6, 7] which leads to a better stability of block matching. Depending on the quality of the matching for the largest block size it is iteratively refined and the complete search is repeated using smaller blocks (Fig. 4).

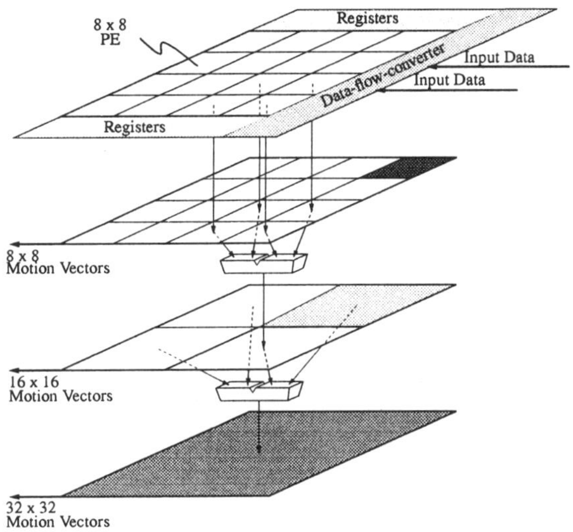

Fig. 4: Structure of the Implementation of VSBM

Since this algorithm works iteratively a straightforward VLSI implementation seems to be impossible. Using the basic architecture of Fig. 2 the array can be divided into subarrays with separate add and compare units. The resulting local DFDs of each subarray are then added and compared to get the global DFDs and motion vectors for the larger blocks. Here an array of 32 × 32 PEs can be divided into two steps down to a block size of 8 × 8.

This new massive parallel approach leads to an area overhead for VSBM of only about 10%.

Realization

A. Cascadability

The computational power required for HDTV parameters can not be implemented on a single chip ASIC today. It is advantageous to cascade identical devices in order to accomplish the desired computational power. In the proposed architecture an identical current block of 32 × 32 pixels is matched with a horizontally and vertically overlapping part of the search area on each chip. By cascading, each chip has only local communication with its horizontal neighbours, thus the external memory access can be minimized (Fig. 3). Half-pel and quarter-pel motion vectors can be calculated using a smaller search area and an interpolation filter at the input data line.

B. New Data Flow Converter

Because of the meander-like data-flow, only every n_y clock-cycles a new column of pixels of the search area has to be loaded into the array of PEs. In order to keep the I/O bandwidth low a DFC is necessary that intermediately stores the pixels which are loaded serially onto the chip. Depending on the status of the meander scan the data is multiplexed into the top or bottom part of the PE array and the attached registers. The solution proposed in [4] applies a chain of registers as DFC with attached multiplexers which controls the data flow into the two different directions. Caused by the intersections (Fig. 2) m_y busses, each 8-bit wide, have to lay next to each other over the full height of the PE array. For a search area of $m_y = 64(\pm 32)$ the width of this structure is 841 μm. Assuming a height of the array of $15 mm$ the area for the DFC adds up to approximately 13 mm^2 which is about 10% of the typical chip-area.

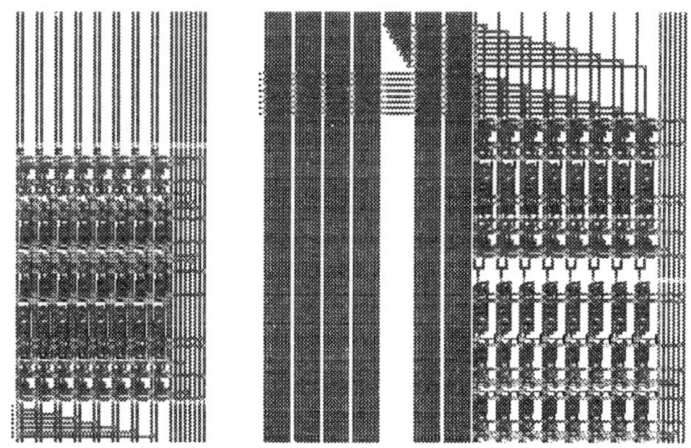

(a) (b)

Fig. 5: Layout of the New (a) and the Old (b) DFC for an Unrealistic Small Search Area of ± 4

To avoid this overhead in area, speed and especially power consumption a new DFC is proposed which connects each PE and register of the first column of the array to a register of the DFC. In this solution the meander scan is realized also in the DFC. Data can be shifted either up or down through the chain of registers. Therefore the registers turn out to be more complex than the old ones but the multiplexers are not needed anymore. Layouts of these two DFCs are shown in Fig. 5, the exact dimensions are listed in Table 1. To distinguish the layouts in Fig. 5 an unrealistic small search area of ± 4 has been chosen.

Width	New DFC	Old DFC
Cell	$68\mu m$	$73\mu m$
Wiring	0	$m_y \cdot 12\mu m$
$\Sigma(m=64)$	$68\mu m$	$841\mu m$

Table 1: Width-Difference of the Old and New DFC

C. Basic Cell

The layout of the basic cell for the chip is shown in Fig. 6. The basic cell is the most important part of the design. This cell is used 1024 times on each chip with a PE-array size of 32×32. As subtractor a fast carry-ripple adder is used. The pixel DFDs are accumulated in a pipelined cascade of carry-save adder stages. The absolute value can be easily realized by XOR gates. Dynamic register structures like transmission gates and C^2MOS latches are used for synchronisation. The features derived from the first simulations of the basic cell are shown in Table 2. Fabrication of the chips will be in 1996.

Conclusion

A flexible block matching motion estimation chip as a component of an emulator, required for research and development purposes for advanced video coding techniques, has been described. Different search algorithms as well as variable sized blocks between 8×8 and 32×32 are possible. A large variety of search areas can be implemented by cascading identical chips. The chip is prepared for object based algorithms. Each chip has a computational power of more then 200 GOPS.

References

[1] H. G. Musmann, P. Pirsch, H. J. Grallert "Advances in Picture Coding," *Proceedings of the IEEE*, vol. 73, no. 4, pp. 523–548, April 1985

[2] R. Srinivasan, K. R. Rao "Predictive Coding Based on Efficient Motion Estimation," *IEEE Transactions on Communications*, vol. 33, no. 8, pp. 888–896, August 1985

[3] *Grand Alliance HDTV System Specification*, Version 2.0, December 7th, 1994

[4] L. De Vos, M. Stegherr "Parameterizable VLSI Architectures for the Full-Search Block-Matching Algorithm," *IEEE Transactions on Circuits and Systems*, vol. 36, no. 10, pp. 1309–1316, October 1989

[5] J. P. Berns, A. Schuur, T.G. Noll "Verfahren zur Kaskadierung und Parallelisierung eines HDTV-Block-Matching-Prozessors," *DSP Deutschland '95*, pp. 186–196, September 1995

[6] M. H. Chan, Y. B. Yu, A. G. Constantinides "Variable size block matching motion compensation with applications to video coding," *IEE Proceedings*, vol. 137, Pt. I, no. 4, August 1990

[7] Y. Huang, X. Zhuang "An Adaptively Refined Block Matching Algorithm for Motion Compensated Video Coding," *IEEE Transactions on Circuits and Systems for Video Technology*, vol. 5, no. 1, pp. 56–59, February 1995

Parameter	PE
Area	0.1 mm^2
f_C (worst case)	200 MHz
P_V	15 mW @200 MHz
Transistors	1234
Techn.	0.5-μm CMOS

Table 2: Technical Data

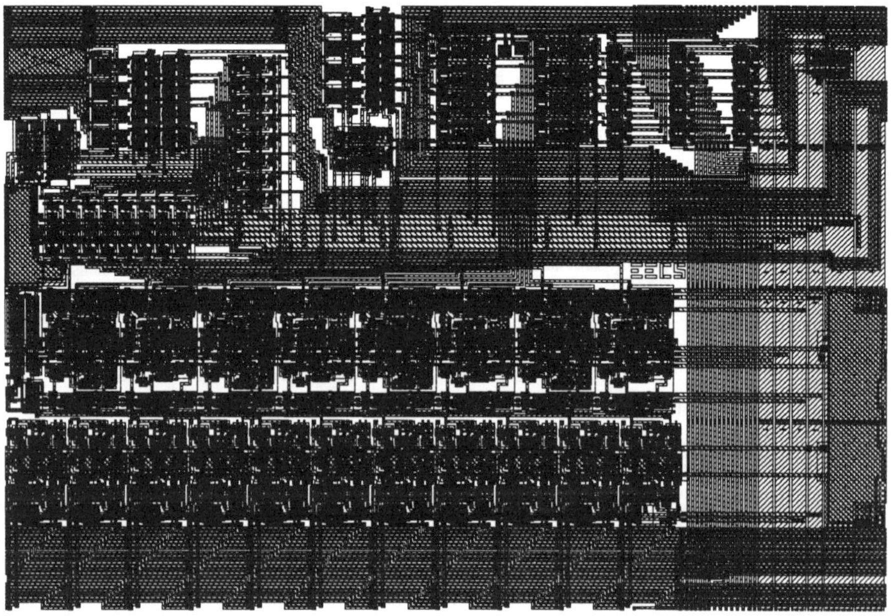

Fig. 6: Layout of the Basic Cell

A DSP for DCT-Based and Wavelet-Based Video CODEC's for Consumer Applications

K.Okamoto*, T.Jinbo*, T.Araki*, Y.Iizuka**, H.Nakajima**, M.Takahata**, H.Inoue***, S.Kurohmaru***, T.Yonezawa***, and K.Aono*

*Semiconductor Research Center, Matsushita Electric Industrial Co.,Ltd.
**LSI Development Center, Matsushita Electronics Corporation
***Corporate Research Division, Matsushita Electric Industrial Co.,Ltd.

3-1-1 Yagumo-Nakamachi Moriguchi, Osaka 570 Japan

Abstract

We have developed a video DSP which performs real-time encoding and decoding for DCT-based algorithms, such as ITU-T H.261, and wavelet-based subband encoding algorithms. This LSI is suitable for consumer applications as it was implemented using 0.5μm-CMOS process technology to realize compactness (1 million transistors on 65mm^2) and low power (560mW). It features a processing unit which performs wavelet filters at high speeds, a compact DCT circuit, and a fast, flexible DRAM interface for low-cost systems. At 40MHz, a single chip is capable of processing QCIF size pictures at a rate of over 15 frames/second.

1. Introduction

With the expansion of markets for PC-based videoconferencing and image database applications, the need for low-cost, low-power LSI's suitable for video compression is growing. Software implementations pose a great burden on the CPU and generally do not have high throughput.

Algorithms suitable for such applications include DCT-based algorithms , such as ITU-T H.261 and JPEG, and non-standard wavelet-based subband encoding algorithms. DCT-based algorithms offer interoperability through standardization. Wavelet-based algorithms, though not standardized, do not suffer from DCT-related block noise, and can be used in closed systems requiring higher picture quality.

The DCT consists of many multiplications and additions and usually requires a large, dedicated circuit to achieve enough throughput. Wavelet-based algorithms consist of applying different types of filters over the entire picture frame many times and require flexible, high performance hardware. Incorporating the two functions into a single LSI has been difficult due to size and performance requirements.

This LSI was based on a DSP(160mm^2,2.4W) for DCT based algorithms(1), which was previously developed using 0.8μm-CMOS technology. Compactness and low power were realized by a change in the process rule to 0.5μm. The scope of applications was broadened to cover wavelet-based algorithms with the addition of a new type of instruction which performs filters efficiently. The DCT circuit size was decreased, and the memory interface was changed to support DRAM's instead of SRAM's to allow low-cost systems.

2. Architecture

A. Block Diagram

A block diagram of this LSI is shown in Fig.1. This LSI contains many of the features of the previous LSI, including a vector-pipeline(VP) architecture in which 2-dimensional image vector data is read from memory by a Space Address Generator(SAG), operated by processing units, and written into memory. The SAG contains parametric registers which define 2-dimensional access regions. These parameters are the start address(SA), the horizontal and vertical step sizes(DX, DY), the number of horizontal and vertical data(NX, NY). As general processing units, it contains a 16-bit multiplier(MPY), a 24-bit accumulator(ACC) and a 16-bit enhanced ALU(EALU) capable of such functions as absolute difference, clipping and maximum/minimum value computation, as well as conventional arithmetic and logical functions. As special processing units, it contains dedicated hardware for DCT and inner-loop filtering(FLT). There are 3 data memories(BM1,BM2,DM), each with a capacity of 1kwd. There are 2 external memory ports(D1, D2), each of which has a data width of 16 bits, as well as serial(SIO) and parallel(PIO) ports.

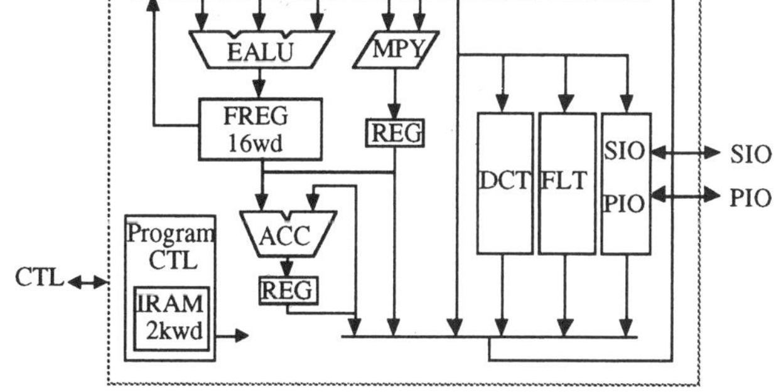

Fig.1. Block diagram of the new video DSP

0-7803-3177-6 $5.00 © 1996 IEEE

IEEE 1996 CUSTOM INTEGRATED CIRCUITS CONFERENCE

The VP architecture supports vector-type(P instruction) and convolution-type(Q instruction) pipeline operations suitable for image processing, as shown in (1) and (2).

$$Yi = Ai \# Bi \quad (i = 1,2,...,n \; ; \; \# \text{ - operator}) \qquad (1)$$
$$Y = \sum_i Ai \# Bi \qquad (2)$$

In addition to the modifications which will be explained in the following subsections, the instruction memory(IRAM) was increased from 1kwd to 2kwd, in order to accomodate both encoding and decoding programs for CODEC applications, and the number of general purpose registers were increased from 8 to 16 to eliminate the use of data memories for parameter storage, and to greatly facilitate the management of memories and registers.

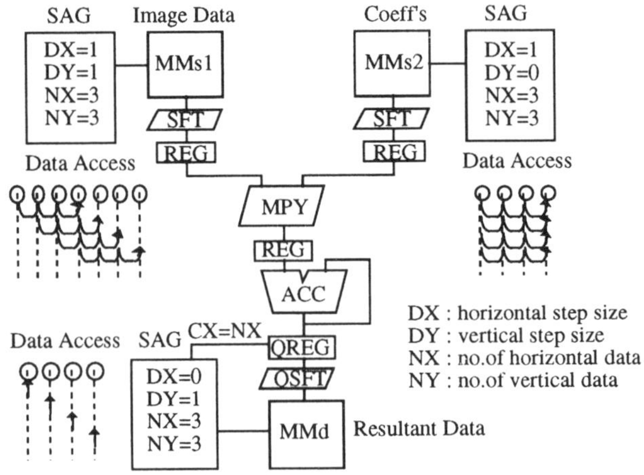

Fig.2. Performing a 4-tap filter with the QP-instruction

B. A Fast, Flexible Instruction for Wavelet filters

Wavelet filters are a class of band-pass polyphase filters which can be used for subband encoding algorithms. Usually, band regions are formed through the iterative use of low and high-pass filters, and subsampling. In our previous architecture, wavelet filtering can be performed by repetitively using a type of Q instruction, known as the QMA instruction. The QMA instruction performs one convolution, i.e. several multiplications and one accumulation. This, however, resulted in overhead instructions with each convolution such as the setting of the SA of the SAG, data transfer, and the initialization of the accumulation register. Since one convolution is performed for each pixel of the frame, the overhead cycle time was large.

In our new DSP, we created a new type of instruction(QP-instruction), which is a hybrid of the Q and P instructions and performs many convolutions at a time, as shown in (3).

$$Yi = \sum_j Aij \# Bij \qquad (i, j = 1,2,...,n) \qquad (3)$$

The data flow for a 4-tap filter using the QP-instruction is shown in Fig.2. The necessary data access can be performed with the SAG. The 1st source memory(MMs1) contains the pixel data and is accessed in a sliding horizontal window of 4 pixels. The 2nd source memory(MMs2) contains the 4 filter coefficients and is accessed repetitively. The resultant data is written into the destination memory(MMd). The initialization of the accumulation register and the write operation to MMd is executed by the SAG of the MMd, and is set to occur every 4 cycles.

The filter tap size and frame size can be changed by setting the SAG parameter appropriately. The filter coefficients can be changed by rewriting the proper memory. The time required to perform the wavelet filtering in our original algorithm, which will be explained in the next section, was reduced to approximately 1/2 of that of the previous DSP.

Furthermore, the QP-instruction can perform other operations, such as the block matching operation for motion estimation, by reconfiguring the pipeline to include the EALU instead of the MPY.

C. A Compact DCT Circuit

In the previous DSP, the DCT circuit , which performs both forward and inverse DCT calculations, occupied approximately 1/6 of the total circuit area (not including the pad area), and it was the largest functional block of the LSI. The Distributed Arithmetic(2) algorithm was used for implementing the previous DCT circuit, and at the time, it was significantly smaller than its predecessor which used multiply-accumulate units(3). It required, however, a ROM-accumulation(RAC) stage with 16 ROM's as coefficient look-up-tables(LUT) each with a capacity of 512words x 16bits, and the LUT's occupied roughly half the circuit area.

In this DCT circuit, we employed a decimation-in-frequency scheme, in addition to the distributed arithmetic algorithm, to reduce the required capacity of each LUT to 32words x 16bits, which is 1/16 that of the previous. Furthermore, since the required capacity of the LUT's were very small, a combinational logic equivalent offered a more area-efficient implementation, and this resulted in a further reduction of size. The DCT block diagram is shown in Fig.3.

The decimation-in-frequency for the inverse DCT, however, is performed after the RAC stage and truncation of the result is required to take place after the decimation-in-frequency stage, not the RAC stage, in order to satisfy precision requirements. In this DSP, we created a circuit which keeps track of the carry or borrow generated from the LSB's of thepartial products of two RAC's which would be paired off in a decimation-in-frequency calculation. This circuit can be created with just a few gates and a flip-flop. The final carry or borrow bits and the truncated results of the two RAC's are added/subtracted to perform the decimation-in-frequency.

With the new approach, the 8 registers for the RAC stage, 8 holding registers, and the adder/subtracter for the decimation-in-frequency stage only needs to be 21 bits, not 34 bits, resulting in significant savings, as shown in Fig.4.

Other circuit reduction techniques lead to an overall DCT circuit size of 4.5 mm^2. Table 1 shows a comparison of the new and previous DCT circuits. This DCT circuit is also smaller than other circuits which have been previously reported.(4) (5)

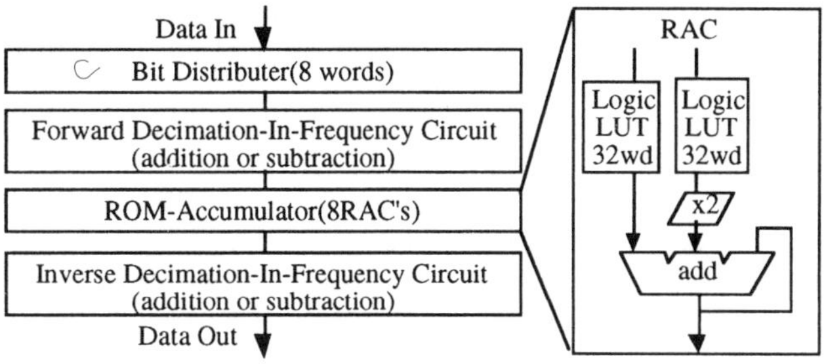

Fig.3. DCT Block

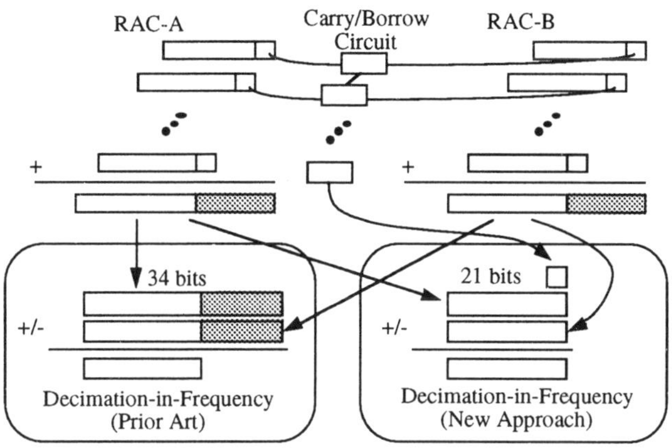

Fig.4. Carry/Borrow operation

Table.1. DCT circuit comparison

	Prior DCT	New DCT
Algorithm	Distrib.Arith(DA)	DA & Fast Algorithm
Process	0.8μm-CMOS	0.5μm-CMOS
Size	$21.0mm^2$	$4.5mm^2$
SizeRatio	100[100*]	21[41*]
Power	300mW	80mW
Power Ratio	100[100*]	27[80*]

* After adjustments for process rule and supply voltage.

D. DRAM interface

In this DSP, the memory interface was modified to support DRAM's instead of SRAM's to allow low system cost. A mode for EDO-DRAM's was employed, in addition to one for conventional DRAM's, in order to realize access rates similar to SRAM's. In the EDO-DRAM mode, a maximum access rate of 50ns/word can be achieved, as opposed to the maximum access rate of 75ns/word of the conventional DRAM mode. Fast-page mode is used for data access, and CAS-before-RAS(CBR) mode is used for refresh, and refresh can be initiated via external pins or by an internal timer. The page size can be changed to support DRAM's with capacities from 1Mb, which is enough to hold a QCIF size picture, to 16Mb, which is enough to hold high resolution pictures for JPEG applications.

3. Performance Evaluation

A. Wavelet-Subband Encoding/Decoding

As part of our work, we developed our own wavelet-subband algorithm. First, the frame is divided into many 2-dimensional subband regions. This is done through the successive use of a low-pass filter(LPF), a high-pass filter(HPF) and subsampling, as shown in Fig.5. The resultant data were encoded with the basic processing unit being a 22 pixel x 18 pixel block(MBK) which is the size of the lowest frequency (LL3) region. The MBK are analogous to the macroblock(MB), and we used the same code structure, down to the picture level, as that of ITU-T H.261. At the MBK level and below, we implemented our original code structure.

With a target performance of QCIF(176x144) frame size at 15 frames/second, we evaluated this DSP at 40Mhz when encoding and decoding according to our wavelet-subband algorithm. The number of cycles was measured using a software simulator which was developed concurrently. Using the Q instruction and the conventional DRAM mode with a maximum rate of 75ns/word, the processing time required for encoding 1 frame is 68.15ms, which is equivalent to only 14 frames/second. Using the QP instruction and the EDO-DRAM mode with a maximum rate of 50ns/word, the processing time for 1 frame can be reduced to 40.32ms, which is equivalent to 24 frames/second. A graph of the processing time for encoding 1 frame is shown in Fig.6. When decoding, only 13 frames/second is possible with the previous approach, whereas 23 frames/second is possible using the new approach.

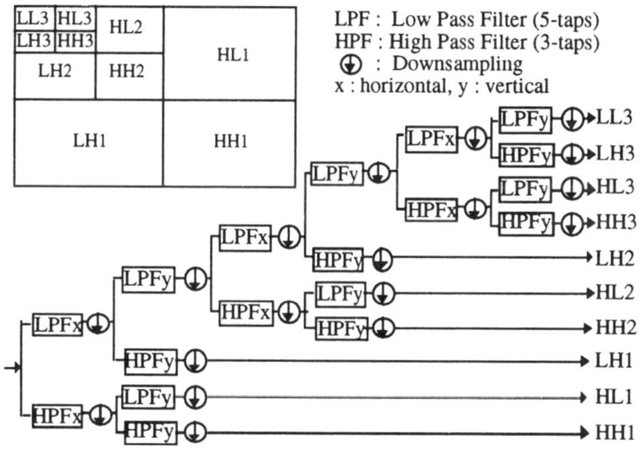

Fig.5. Data flow for wavelet filtering

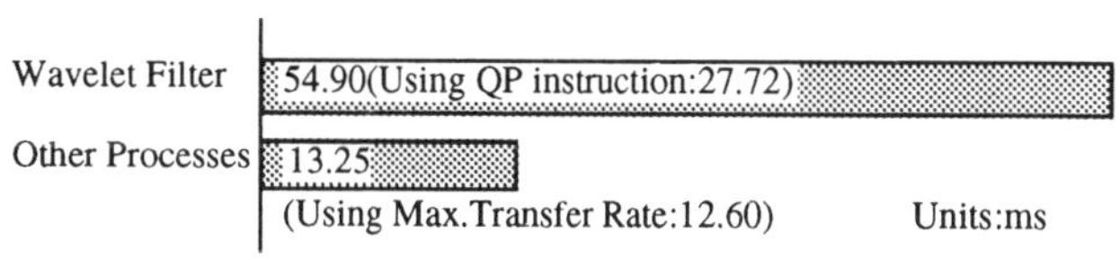

	Normal Condition	Using QP & Max.Rate
Time for 1Frame	68.15ms	40.32ms
Frames/Second	14	24

(Estimate for Decoding:23frames/second)

Fig.6. Processing time for 1 frame for wavelet-subband encoding

B. ITU-T H.261

A graph of the processing time for encoding 1 macroblock(MB) is shown in Fig.7. Using the conventional DRAM mode, the processing time required by this DSP when encoding 1 MB is 485.75μs, which is equivalent to a frame rate of 20 frames/second . Using the EDO-DRAM mode, the processing time for 1 MB can be reduced to 429.27μs, which is equivalent to 23 frames/second. When decoding, a frame rate of 45 frames/second was confirmed using the conventional DRAM mode. Moreover, a real-time CODEC can be realized, as both encoding and decoding can be performed in 653.71μs, which is equivalent to 15 frames/second.

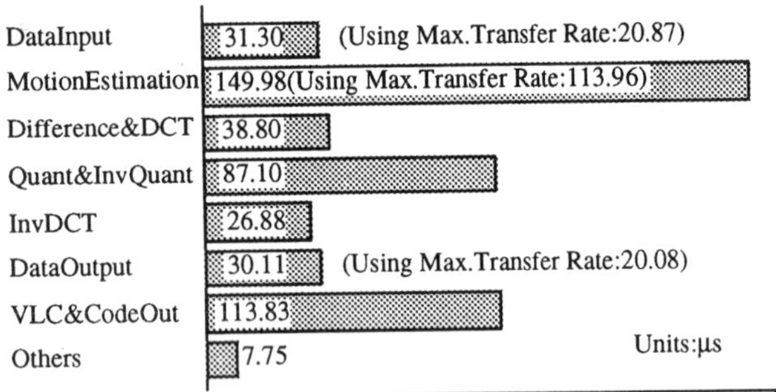

	Normal Condition	Using Max.Transfer Rate
Time for 1MB	485.75μs	429.27μs
Frames/Second	20	23

(Estimate for Decoding:45 frames/second, CODEC:15frames/second)

Fig.7. Processing time for 1 MB for ITU-T H.261 encoding

4. Implementation

The decrease in total circuit size from 160mm^2 of the previous DSP to 65mm^2 of the new DSP, shown in Table.2, resulted in great reduction of cost. The change in the process rule from 0.8μm to 0.5μm lead to a 61% decrease in size. The increase in the number of general registers(FREG) , the capacity of instruction memory(IRAM) , and the change in the memory interface, lead to an increase of 12%, but was mostly offset by the decrease in the DCT circuit, which was 10%.

The maximum power of this DSP was measured to be 560mW (3.3V, 40Mhz), as opposed to 2.4W (5V, 60MHz) of the previous DSP. The low power consumption, mainly attributable to a reduction in supply voltage, frequency, and total capacitance, allowed the use of a plastic QFP package for further cost reduction. The LSI features are shown in Table.3.

Furthermore, all image signal processing functions can be accomplished with a low-cost system consisting only of this DSP, DRAM's for the data buffer and the frame memory, and FIFO's for the bitstream. A chip microphotograph is shown in Fig.8.

Table 2. Reduction in DSP circuit size

Prior DSP	(160mm^2)	120mm^2 *
Process Change (0.8μm->0.5μm)		-61%
Circuit Increase (IRAM, FREG, DRAM-IF)		+12%
Circuit Decrease (DCT)		-10%
New DSP	(65mm^2)	48mm^2 *

* Not including pad area

5. Conclusion

We have developed a compact (65mm^2), low power (560mW), plastic package, video CODEC DSP for consumer applications, using 0.5μm-CMOS process technology. By creating a new instruction to perform wavelet filters at high speeds, in addition to a compact DCT circuit, we were able to support wavelet-based algorithms, as well as DCT-based algorithms. A DRAM interface was added which flexibly supports different speeds and capacities to accomodate various low-cost sytems. A real-time(QCIF, 15 frames/second or more) CODEC for ITU-T H.261, and an encoder or decoder for our wavelet-based algorithm can be realized using one chip. This LSI can also be used for JPEG and ITU-T H.263.

Table.3. LSI features

Data Format	16bit Fixed Point
Instruction Cycle	25ns (Max.12.5ns)
Internal Memory	Data : 1kW x 16b x 3 Instruction : 2kW x 32b
External Memory	Max : 4MW x 16bit x 2 (DRAM) Max.Transfer Rate : 50ns
Processing Units	Enhanced ALU, MPY, Accumulator DCT, Loop Filter
Power Consumption	560mW (3.3V)
Process Technology	0.5μm-CMOS (2-Metal)
Size	8.43mm x 7.76mm = 65mm^2
Package	128 pin PQFP

Acknowledgements

We wish to thank Y.Kakinuma, K.Aruga, A.Teranishi, J.Michiyama, T.Katsura, and K.Hasegawa for their support.

Reference

(1) K.Aono et al,"A video digital signal processor with a vector-pipeline architecture",IEEE Journ. Solid-State Circ.,vol.27,no.12,pp.1886-94,Dec.1992

(2) A.Peled and B.Liu,"A new hardware realization of digital filters",IEEE Trans. Acoust.,Speech,Sig.Proc.,vol.22,no.6,pp.456-462,Dec.1976

(3) K.Aono et al,"A 30ns (600MOPS) image signal processor with a reconfigurable pipeline architecture",Proc.IEEE 1989 CICC,pp.24.4.1-24.4.4,May 1989

(4) S.Uramoto et al,"A 100-MHz 2-D discrete cosine transform core processor",IEEE Journal Solid-State Circuits.,vol.27,no.4,pp.492-499,Apr.1992

(5) M.Matsui et al, "A 200-MHz 13mm2 2-D DCT macrocell using sense-amplifying pipeline flip-flop scheme",IEEE Journal Solid-State Circuits.,vol.29,no.12,pp.1482-1490,Dec.1994

Fig.8. Chip microphotograph

Control and Synchronization Scheme for Parallel Image Processing RAM with 128 Processor Elements and 16-Mb DRAM

Yoshikazu YABE, Tohru KIMURA, Yoshiharu AIMOTO, Hideki HEIUCHI, Youetsu NAKAZAWA, Takuya KOGA, Yoshihiro FUJITA, Masayuki HAMADA, Takaho TANIGAWA, Hajime NOBUSAWA, and Kuniaki KOYAMA

NEC Corporation
1120 Simokuzawa, Sagamihara, Kanagawa, 229, Japan

ABSTRACT

A newly developed parallel image processing RAM (PIP-RAM) integrates 128 processor elements and a 16-Mb DRAM on a single chip. The paper presents three novel circuit design techniques: a data path control and synchronization scheme between processors and memory; a refresh scheme which enable refresh operations in parallel with arithmetic-logical operations; and a special block redundancy scheme.

1. INTRODUCTION

As image processing applications come to require increasingly large-capacity memories and high-performance processors, data transfer between processor and memory becomes a more highly pronounced bottleneck [1][2]. To overcome this problem, we have developed a Parallel Image Processing RAM (PIP-RAM) which integrates 128 processor elements (PEs) and a 16-Mb DRAM on a single chip [3]. Parallel processing is particularly well-suited to image-processing applications because of the large number of simple parallel calculations that they require. The integration of a large number of processors and a large capacity memory on a single chip helps the PIP-RAM achieve a wide memory bandwidth, high performance, and low-power consumption.

Integrating processors and a DRAM on a single chip may lead, however, to the following problems: 1) decrease in the hold time of data due to high power dissipation of PEs and increase in signal noise from logic circuits; 2) increase in both a signal distribution delay and signal skew, due to increased chip size; and difficulty in designing appropriate control circuits for PEs and for DRAM, such as 3) internal interface, 4) external interface, 5) refresh control, and 6) redundancy control.

In the design of the PIP-RAM, our response to these problems has included: 1) use of a triple well process to isolate the wells of processors from those of the DRAM, as well as to isolate their respective supply lines (which also helps to suppress logic circuit noise and to reduce processor power dissipation); 2) a new data distribution and control scheme which excludes both decoding delay and interconnect delay from cycle times; 3) a new interface for synchronizing the processors and the DRAM (in the scheme, each processor controls its own local DRAM element); 4) a high speed external interface (143MHz, 8-b, 2-port serial I/O) and 8-b memory bus; 5) a new scheme for executing refresh operations in parallel with logical-arithmetic operations; and 6) a new block

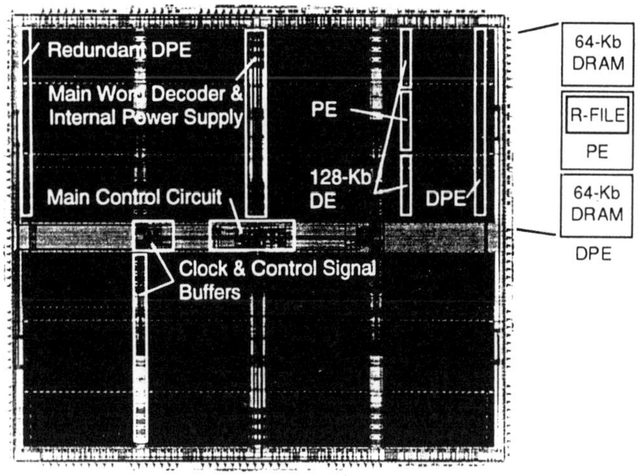

Figure 1. Microphotograph of the PIP-RAM.

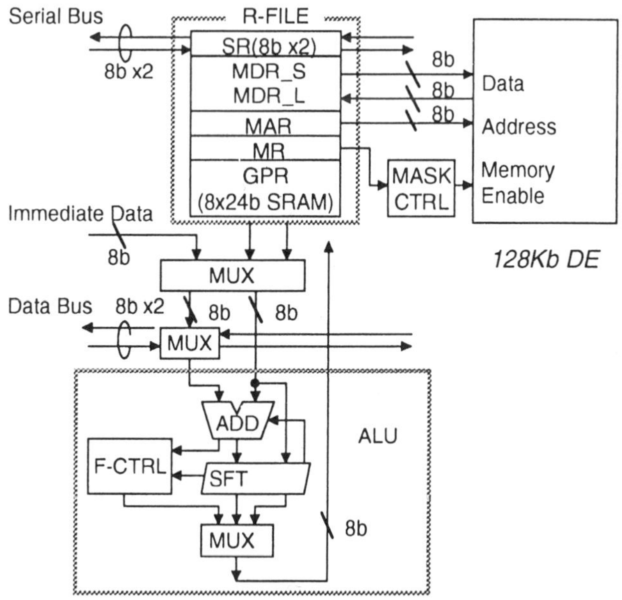

Figure 2. Block diagram of a processor element (PE).

16.4.1

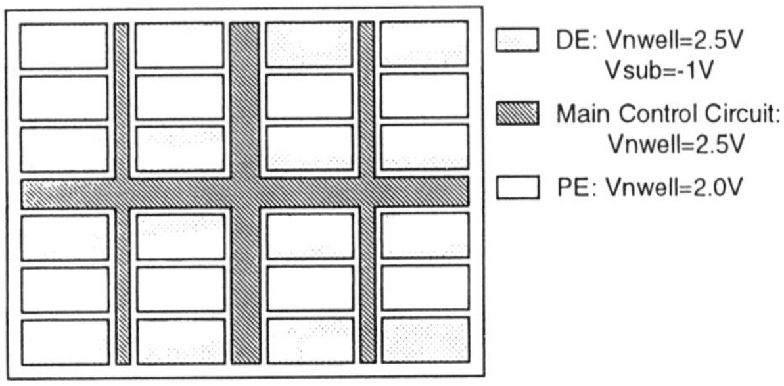

Figure 3. Wells & supply line layout.

redundancy scheme which can compensate for defects in both processors and DRAM elements.

A fabricated model of our PIP-RAM has achieved 7.68-GIPS processing performance with a 3.84-GB/s memory bandwidth in 30-MHz operations.

2. CHIP ARCHITECTURE

Figure 1 shows a microphotograph of the PIP-RAM, which contains 2 rows of 64 DRAM processors (DPEs) each, with each DPE consisting, as illustrated, of one processor element (PE) and two 64Kb DRAM elements (DEs). The PIP-RAM employs a single instruction stream multiple data stream (SIMD) architecture in which a single external instruction controls 128 PEs. With the exception of two bus lines used for transferring data serially, the entire external I/O interface is synchronized with the system clock (@30MHz).

Figure 2 is a block diagram of a PE, which consists of a 8-bit adder/subtractor (ADD), a 16-bit shifter (SFT), a flag controller (F-CTRL), and a register file (R-FILE). The F-CTRL is employed to extend data from 8-b to 16-b. Both logical-arithmetic operations and memory access operations are handled within a single cycle. Multiply operations are 8-b and are conducted by repeating ADD and SFT operations over a period of 8 cycles.

The R-FILE consists of three memory control registers, two serial data transmission registers (SRs), a mask register (MR), and 24 general purpose registers (GPRs). The MR is 1-b and is used for masking load/store operations of individual PEs. All the other registers are 8-b width. Each memory control register consists of a memory address register (MAR) and two memory data registers (MDR), one for load (MDR_L) and the other for store (MDR_S). Since MDR_S behaves as a store buffer, ALU and memory access operations can be executed at the same time. Each MAR holds an index address corresponding to an individual data item stored in a coefficient table in the memory. SRs are used for serial transfer of picture data to and from the external interface. The GPR is composed of a 3-port (2-read + 1-write) 24-word 8-bit SRAM. This GPR was designed this large so that a PE can continue executing ALU operations even when a processor's accessing of a DE has been stalled by a segment miss [3].

In order to help suppress signal noise, we have completely isolated (by means of a triple well process) the wells and power

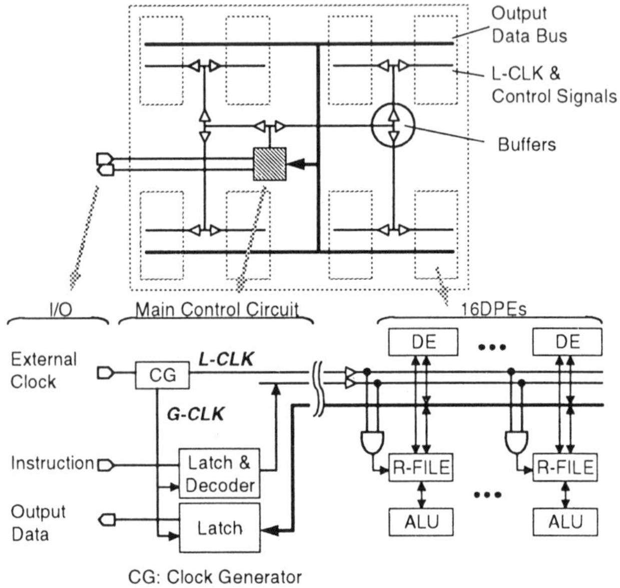

Figure 4. Distribution of clock & signals.

supply lines of the PEs from those of the DEs (see Fig. 3). Isolation also allows us to set operational voltages for PEs and DEs separately, each to the respective value for which power consumption will be reduced as much as possible; the PEs operate at 2.0V, the DEs at 2.5V. With these voltage levels, we are able to achieve same-speed operations of the PEs and DRAMs, while reducing the PE power dissipation to 64% of that of 2.5V operation.

3. SYNCHRONIZATION

Integration of such a large number of processors and such a large capacity memory results in clock skew and in delays in the distribution of instructions. Here, we use the term "delay" to include decoding delay, buffering delay, and interconnect delay (2.5ns, 2.2ns, 0.3ns, respectively). While it is important to try to avoid such delay's being added to the cycle time, we cannot for this purpose simply provide pipeline registers to each PE since this would significantly increase chip size[4]. What we have done here is to develop a new synchronization scheme which we refer to as Local-Synchronization Architecture.

Figure 4 shows the distribution of clocks and signals. The main control circuit consists of a clock generator (CG), instruction and output data latches, and an instruction decoder. The CG generates the local-clocks (L-CLKs) to be distributed to each DPE, as well as the global-clocks (G-CLKs) provided to instruction and output data latches in the main control circuit. L-CLK and register control signal decoding instructions are distributed to each DPE by means of tree-structure drivers.

Figure 5 is a timing chart of the synchronization scheme. So that the R-FILE in each DPE is able to hold data on rising edges of the L-CLK, the L-CLK has been designed slower than the G-

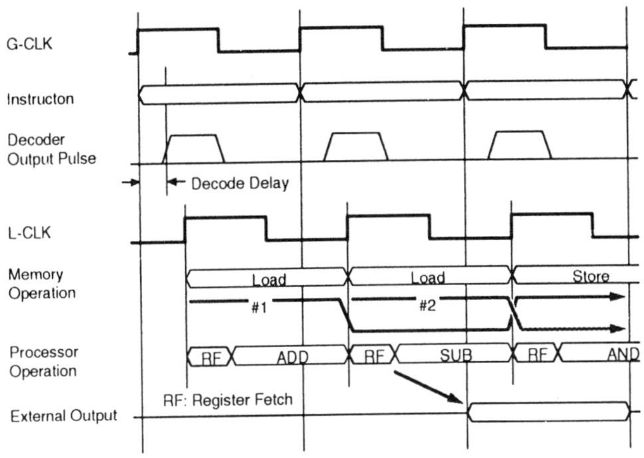

Figure 5. Timing chart of local synchronization.

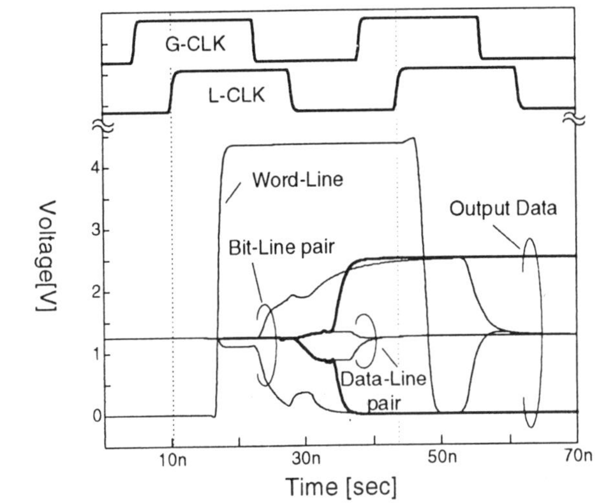

Figure 6. Simulated waveforms of data path control scheme.

CLK by the distribution delay time. Since external output data from R-FILE in selected PE are faster than one cycle time, external I/O interfaces can be synchronized by G-CLK. Consequently, the PIP-RAM successfully excludes distribution delay (5.0ns) from the cycle time, and PIP-RAM achieves a 30-MHz operation.

Control of the data path between PEs and DEs is important because a wrong sequence of memory control signals could cause data in DRAM cells to be destroyed. In our system, memory control signals with the same delay are generated from the R-FILE in each PE, synchronized by a single clock. Memory control signals include address, data, read/write select, and a memory operation enable signal. Figure 6 shows simulated waveforms of data path control scheme between a PE and a DE. Since all critical memory control signals are synchronized with PE operations, all wrong memory control signal sequences are completely eliminated.

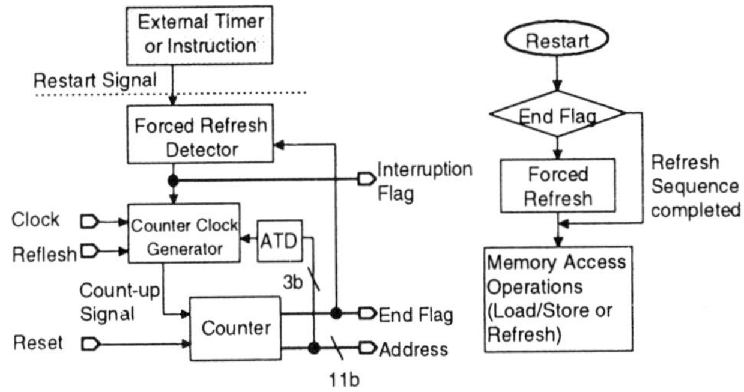

Figure 7. Block diagram of refresh circuit.

4. REFRESH CIRCUIT

The handling of refresh operations is important in logic integrated DRAMs because refresh operations are likely to interrupt both instruction streams and data streams. Figure 7 shows a block diagram of our refresh circuit. A counter block is controlled by a count-up clock generator (CCG), and it issues DRAM address. An address transition detector (ATD) is equipped to handle with both word address transitions (2-cycle) and column-segment address transitions (1-cycle). The end flag signal, which is generated by the counter, signifies the end of a refresh sequence. In order to avoid an incomplete refresh sequence, a detector determines whether or not a forced refresh is required, and when it is, it issues a forced-refresh instruction and temporarily blocks the reception of other instructions. Refresh operations require only 2,048-cycles/8ms. So that percentage of refresh operations to all memory access operations is 0.85%. Refresh operations, as well as load/store operations, are executed in parallel with arithmetical-logical operations. In our image processing applications, the maximum percentage of load/store instructions to all instructions is 30%. So that almost all refresh operations can be executed without interrupting other instructions.

5. BLOCK REDUNDANCY

Achieving a high chip yield is an important problem for large-size LSIs. We have developed a new redundant scheme which compensates for both PE and DE defects. The PIP-RAM has 4 redundant DPEs in a die, two for each row. It is necessary to reassign the PE address and reconnect the serial data bus in order to replace defective DPEs. We have implemented the novel redundant control circuit shown in Figure 8. If defects are found in a DPE, a fuse in the redundant switch block is cut by a laser. No PE address is assigned to a defective DPE, and the serial data bus bypasses it. This redundant circuit scheme takes up only 0.9% of chip area.

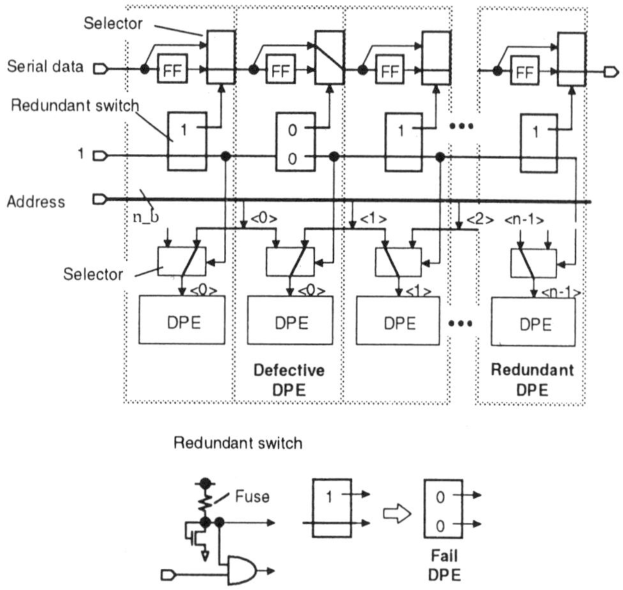

Figure 8. Block diagram of block redundancy.

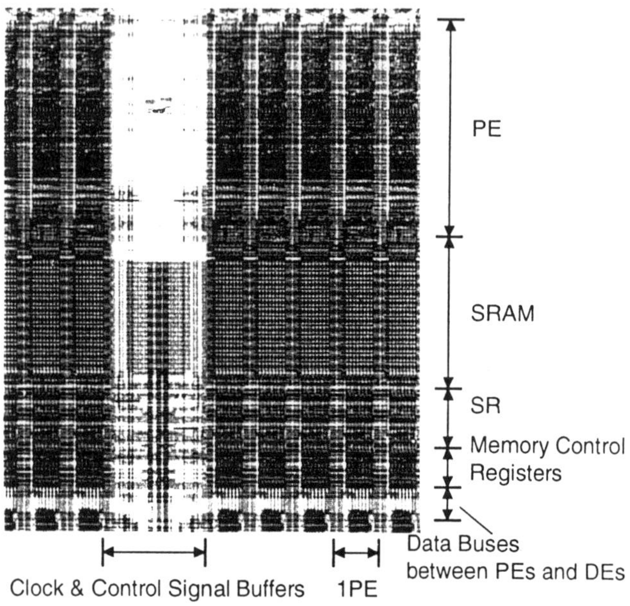

Figure 9. Microphotograph of the PE.

6. CHIP IMPLEMENTATION

External interfaces of the PIP-RAM are summarized in Table 1. The instruction field is 34-b, and the data field is 24-b. Table 2 shows PIP-RAM characteristics. Fabrication technology was a standard 0.38-μm CMOS Triple-well 2-level metal DRAM process. The logic circuit contains 801,064 transistors and takes up 32.8% of the total chip area. The logic circuit needs a large area for wiring (Figure 9). If the PIP-RAM were to be fabricated by means of a 3-level metal process, its chip area might be reduced by 45%.

Table 1. External interface of the PIP-RAM.

I/O		bit
Instructions	Fnction field	6
	Operand fields	5 x 3
	Immediate data field	8
	Memory access field	3
	Data bus select field	2
Special data bus	Serial data I/O	8 x 2
	Memory bus	8
Mode select for multi-chip	Chip select	1
	Mask select	1

Table 2. PIP-RAM characteristics.

Fabrication technology	0.38-μm CMOS, P-sub Triple-Well Two metal layers
Die size	18.8 x 16.7 mm^2
Number of PEs	128
General purpose registers	24 x 8 b/PE (three-port SRAM)
Number of transistors	801,064
DRAM Cell size	1.69 x 0.85 mm^2
DRAM capacity	16-Mb
Clock frequency	30-MHz
Memory bandwidth	3.84-GB/s
Performance	7.68-GIPS
Supply voltage	DE: 2.5-V, PE: 2.0-V
Power dissipation	992mW at 30-MHz

7. CONCLUSION

We have presented here new schemes for synchronization, memory access, refresh, and block redundancy. Additionally, a triple well process has helped us reduce noise interference and power consumption in the logic circuit. The synchronization scheme, which eliminates the problem of decoding delay and wiring delay, appears particularly promising for use in large size LSIs. Instructions for both arithmetical-logical operations and memory access operations are suitable to hide a refresh cycle. A novel redundancy scheme helps increase a chip yield. Our newly developed circuit techniques for the PIP-RAM also appear promising for use in future parallel processor integrated DRAMs.

REFERENCE

[1] N. Yamashita et al, "A 3.84 GIPS Integrated Memory Array Processor with 64 Processing Elements and a 2-Mb SRAM", *IEEE J. of Solid-State Circuits*, pp. 1336-1343, Nov. 1994.

[2] T. Kimura et al, "Design of 1.28-GB/s High Bandwidth 2-Mb SRAM for Integrated Memory Array Processor Applications," *IEEE J. of Solid-State Circuits*, pp. 637-643, June 1995.

[3] Y. Aimoto, T. Kimura, Y. Yabe et al, "A 7.68-GIPS 3.84-GB/s 1-W Parallel Image Processing RAM Integrating a 16-Mb DRAM and 128 Processor Elements," *ISSCC Digest of Technical Papers*, 1996 (to be published).

[4] K. Nakamura et al., "A 220MHz Pipelined 16Mb BiCMOS SRAM with PLL Proportional Self-Timing Generator," *ISSCC Digest of Technical Papers*, pp. 258-259, 1994.

A 36 MHz CMOS Fixed-point G.728 Low Delay CELP Integrated Decoder-Encoder

William Bong Hock Soon, Sameer Goyal, Liu Chong Hsu, and Yeo Lik Wee

Siemens Components Pte Ltd, Microelectronics Design Center,
166 Kallang Way, Singapore 349249
Republic of Singapore

ABSTRACT

The highly integrated JADE (Joint Audio Decoder-Encoder) device implements audio compression using either G.711, G.722 or G.728 compression algorithms. The integration of peripherals, interfaces and memories with a fixed point DSP (digital signal processor) to implement complex speech algorithms such as G.728, opens up the market for low cost digital telecommunications applications. Using the G.728 Low Delay Code Excited Linear Prediction (LD-CELP) algorithm, the device compresses / decompresses a digitized PCM (64 kbps) or linear (128 kbps) voice signal to / from a 16 kbps bit stream. The device has been fabricated in a 0.5 um n-well CMOS process.

1. INTRODUCTION

In May 1992, the first worldwide standardized parametric speech codec was published under the ITU recommendation G.728 (1), also known as LD-CELP (Low Delay Code Excited Linear Prediction). This codec uses a 2 bit coding at 8 kHz and a short delay of 5 sample periods (0.625 ms) without degrading speech quality but increasing hardware processing complexity (2). Conventionally, the G.728 algorithm requires a very high performance DSP and is offered as a software library which must be integrated by the user into his DSP environment. But in JADE, the G.728 algorithm is implemented using an embedded high performance DSP and can be plugged into any application. The G.728 algorithm is implemented in JADE in 16 bit arithmetic and complies with the 16 kbps fixed point specification (3).

JADE supports full duplex communication. In addition to the aforementioned algorithms, within G.711 (64 kbps Pulse Code Modulation for 3.4 kHz audio data), automatic A-law and u-law recognition (as in G.725) is also performed.

To ease integration into diverse applications, JADE communicates through many flexible interfaces. The IOM-2 (ISDN Oriented Modular interface standard) (4) enables communication to other ISDN (Integrated Services Digital Network) integrated circuits. A host (e.g. a microcontroller or a PC) can communicate with JADE through the parallel host interface. Additional algorithms can be loaded by the host into off-chip SRAM over the External Memory Interface and JADE can execute these programs. A programmable Serial Audio Interface connects to a VCP (Video Codec Processor) and 2 universal serial HDLC (High-level Data Link Control) /transparent data controllers enable reception and transmission, as done for example within the ISDN D-Channel Protocol. The device can be entirely controlled via the IOM-2 Monitor channel, enabling standalone applications to be implemented.

2. CHIP DESCRIPTION

A block diagram of the chip is shown in Fig. 1. The basic version of JADE is described.

The major functional units are an embedded high-end 16 bit fixed-point digital signal processor, called Signal Processor Core (SPC), program ROM (16 kwords) (each word is 16 bits), data RAM (4 kwords) organized as a 3 kword XRAM and a 1 kword YRAM, data ROM (3 kwords), mailbox RAM (256 bytes for SPC read/host write, 256 bytes for SPC write/host read), a programmable PLL (phase locked loop) based clock generator and the interfaces mentioned above. The program ROM size allows the integration of G.728 and G.711 or G.728 and G.722 SB-ADPCM (Sub-Band Adaptive Differential Pulse Code Modulation) on chip.

Bus Interface Modules are used extensively in the design to reduce the capacitive loading on the main data bus by dynamically segmenting the bus. Full custom layout is used in the PLL unit and FIFOs, the RAMs and ROMs are module generated and the SPC has regular datapaths. The rest of JADE is comprised of standard cells, using automatic place-and-route tools to reduce layout time.

Fig. 2 shows a photograph of the die. Table 1 summarizes the device features.

2.1 Signal Processor Core

SPC has a 3 stage pipelined Harvard architecture, with 16 bit program address bus, 16 bit

0-7803-3177-6 $5.00

instruction bus, 16 bit data address bus and 16 bit data bus. SPC comprises the following units:

- Program Control Unit (PCU)
  - Program decode and control logic
  - Exception handling logic
  - Program counter
  - Repeat and block repeat counter

- Data Address Arithmetic Unit (DAAU)
  - Registers and logic for using direct, pointer and modulo addressing

- Computation Unit (CU)
  - 16 x 16 multiplier with 32 bit output
  - 36 bit ALU
  - Two 36 bit accumulators

- Hardware stack
  - 17 levels deep

2.2 Mailbox

System functions and communication between the embedded SPC and an external controller is supported by interrupts, via a full duplex communication memory mailbox. The mailbox is implemented as 2 physically separate 256-byte memory blocks with auto-increment capability. One is read-only by the SPC and write-only by the host while the other is write-only by the SPC and read-only by the host.

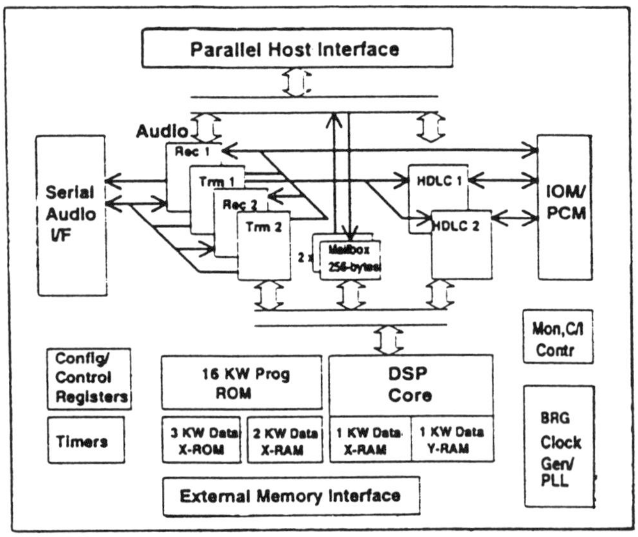

Fig. 1 Block diagram of chip

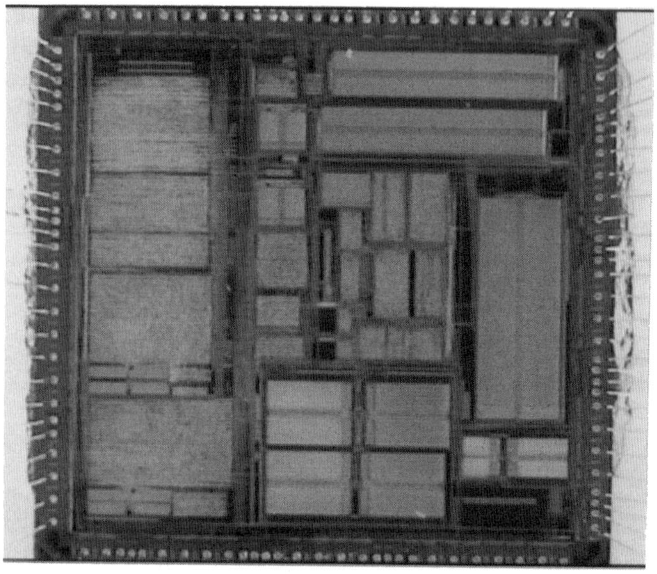

Fig. 2 Photograph of the die.

2.3 Serial Audio Interface

The Serial Audio Unit is a generic five line interface consisting of two independent transceivers. This interface can be used as a general Time-Division Multiplex highway with timeslots of programmable lengths and locations for audio and data. It is also used typically to transfer messages or blocks of compressed or uncompressed data, preceded by a header of control information pertaining to the transferred data block and synchronous to it. The blocks can be received and transmitted using one of the HDLC controllers in the transparent mode. An application of this mode of operation is the synchronous transfer of H.221 oriented data between JADE and the attached VCP. The interface to each of the audio channels is a 32-bit wide buffer shift register. The buffering up to 32 bits reduces the reaction time of the DSP software.

2.4 External Memory Interface

The external memory interface allows the connection up to 64 kwords of program memory, eg. SRAM or EPROM, with programmable wait states. Using external SRAM eases the testing and debugging of other algorithms. The host can download programs into the SRAM. The interface can be deactivated to reduce power dissipation.

2.5 IOM-2 Interface

The IOM-2 functions supported include (4):

- Layer 1 functions in terms of the frame structure supporting any number n of 4 byte multiplexes (n=1,...,16)
- One monitor channel of programmable location
- Two Command/Indicate channels

2.6 Clock Generator

The clock generator includes a PLL which provides the internal master clock for the SPC. The PLL is designed to take in an input frequency of 7.68 MHz. This solution is found to be most cost effective because of the availability of a 7.68 MHz clock in ISDN applications. In addition, there are different clock modes in the PLL to cater to 4 different input clock frequencies.

A programmable baud rate generator with factor 1,2,3,..,2^{19} is derived from the SPC clock. The wide range of division factors for the baud rate generator output allows it to be used as a time marker to synchronize other devices to the chip time base. The output clock can also be used as a system clock for other devices when the SPC clock is disabled. In addition, there are three timers with programmable division factors to generate interrupt status and maskable interrupts.

2.7 HDLC/Transparent Data Controllers

The two HDLC controllers can be independently serviced either by the SPC or via the parallel host interface. Each HDLC controller can be operated either in HDLC mode or transparent mode. In the latter mode, data is received and transmitted fully transparently without HDLC framing. Each controller has a receive FIFO of 2 x 32 bytes and a transmit FIFO of 2 x 32 bytes. One half of each receive/transmit FIFO is connected to the receive/transmit shift register while the second half is accessible from the controlling software.

In HDLC mode, the controller supports the following bit functions:

- Flag generation/detection
- Zero bit insertion/deletion
- CRC (cyclic redundancy check) generation/check according to CCITT polynomial of order 16 or 32
- Abort generation
- Inter-frame time fill generation

Reception of back-to-back frames, consecutive frames with a shared flag and flags with shared "0"s is implemented.

The mailbox allows efficient full duplex data transfer between an external application software and the JADE HDLC controllers, for implementing high speed data link applications; the data units may be frames or "raw data". For non-HDLC serial protocols, the transparent mode can be used.

3. APPLICATIONS

Applications of the device include ISDN videophones; video conference systems; corporate network voice concentrators, multiplexers and gateways; data-over-voice and voice-over-data terminals; networks for packetized voice; digital added main-line (DAML) and digital circuit multiplication equipment (DCME); and voice storage in PC based applications.

The integrated HDLC controllers allow the cost effective realizations of applications such as:

- Data-over-voice and voice-over-data terminal equipment, where the HDLC controller shares one B-channel with an audio channel for simultaneous voice and data transmission (e.g. 16 kbps G.728 plus 48 kbps data).
- D-channel handling in point-to-point configurations e.g. on DCME.
- Packetized voice with G.728.
- Conversion of analog modem data into an ISDN 64 kbps data stream using rate adaptation e.g. V.120.
- One-chip V.34 (V.fast) modem. The HDLC controller formats the data and the output result is modulated.

Applications of the device are elaborated in (5).

4. RESULTS

The system performance of the device has been verified in a H.320 videophone reference board with PCI (Peripheral Component Interconnect) local bus connection for the PC with Windows software. The complete system includes a videocamera and a handset speakerphone (Fig. 3). The other principal chips used on the reference board are ARCOFI-SP (Audio Ringing Codec Filter with Speakerphone), a hands-free codec that performs detection and elaborate balancing of audio signals to suppress undesirable effects due to acoustical feedback, and ISAC-S TE (ISDN Subscriber Access Controller for Terminals), the interface between IOM-2 and the ISDN. The VCP is responsible for video compression and multiplexing of audio/video data.

In enhanced versions, other compression methods e.g. G.722 resulting in 48, 56 or 64 kbps for wideband 7 kHz audio, will be implemented in the firmware.

5. CONCLUSION

JADE has enabled the realization of an extremely cost effective H.320 video conference system comprising 4 main ICs on one PC daughterboard. The audio signal is compressed to a bit stream of 16 kbps without degrading the acoustic speech quality, enabling simultaneous transmission of audio and video via one B channel. The optional use of a second B channel for the video signal almost doubles the bandwidth (112 kbps) which results in the best video and audio quality.

Table 1. Integrated Decoder/Encoder Features

Technology	0.5 um 2 metal n-well CMOS
Die Size	8.3 mm x 8.3 mm
Package	P-TQFP 100
Transistor Count	1.22 million (approx.)
Operating Frequency	36 MHz (internal)
Supply Voltage	3.0 to 3.6 V Additional 4.5 to 5.5 V for connection to 5 V systems.
Boundary Scan	IEEE Std. 1149.1

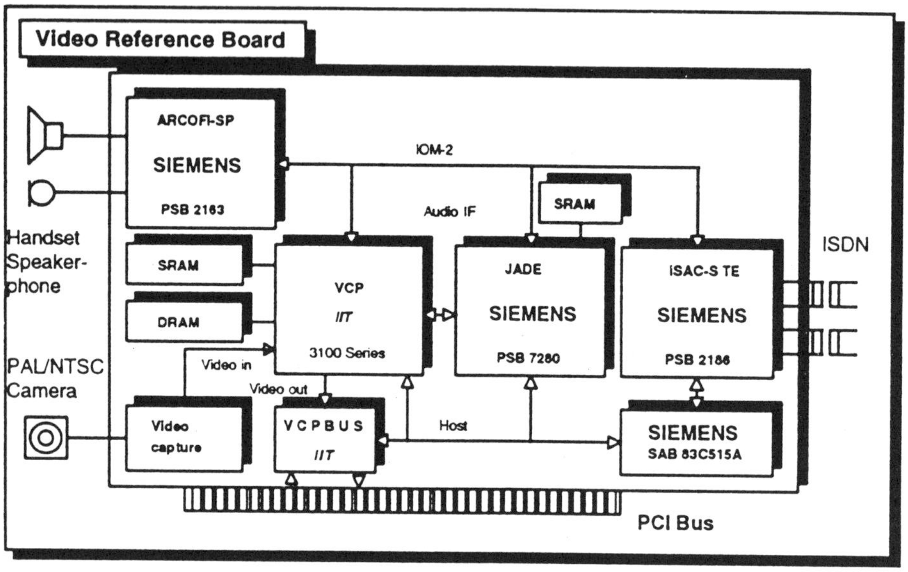

Fig. 3 System application of JADE in PC-based videoconferencing.

References

(1) ITU Recommendation G.728 - Coding of speech at 16 kbit/s using low-delay code excited linear prediction.

(2) J. H. Chen, R. V. Cox, Y. C. Lin, N. Jayant, and M. J. Melchner, "A low-delay CELP coder for the CCITT 16 kb/s speech coding standard," *IEEE J. on Selected Areas in Communication*, vol. 10, no. 5, June 1992.

(3) ITU Annex G to Recommendation G.728 - Coding of speech at 16 kbit/s using low-delay code excited linear prediction: 16 kbit/s fixed point specification.

(4) Siemens Semiconductor, *"ICs for Communications, IOM-2 Interface Reference Guide"*.

(5) Paul M. Gerber and Charles Bry, "Siemens PSB7280 - A low-cost single chip solution of an LD - CELP Codec for 16 kbit/s speech coding standard (G.728)", *DSPs in der Praxis: Architekturen, Werkzeuge, Software, Applikationen*, Sept 1994, Munich, Germany, pp.129 - 134

(6) Siemens Semiconductor, *"ICs for Communications, PSB 7280 Joint Audio Decoder-Encoder"*.

A 1.8V 36mW DSP for the Half-Rate Speech CODEC

Taketora Shiraishi, Koji Kawamoto, Kazuyuki Ishikawa, Hisakazu Sato, Fumiyasu Asai, Eiichi Teraoka, Toru Kengaku, Hidehiro Takata, Takeshi Tokuda, Kouichi Nishida, and Kazunori Saitoh
System LSI Laboratory, Mitsubishi Electric Corporation
4-1 Mizuhara, Itami, Hyogo, 664 Japan

Abstract

A low-power 16-bit DSP has been developed to realize a low bit-rate speech CODEC. A dual datapath architecture and low-power circuit design techniques are employed to reduce power consumption. The PDC half-rate speech CODEC is implemented in the DSP with 36 mW at 1.8V.

Introduction

The PDC half-rate system with both the PSI-CELP (Pitch Synchronous Innovation Code Excited Linear Prediction) algorithm and the VSELP (Vector Sum Excited Linear Prediction) algorithm has gone into operation commercially in Japan. To reduce the bit-rate to half of the full-rate VSELP (6.7kbps to 3.45kbps), the PSI-CELP requires higher performance (approximately 23.4MOPS) than the VSELP (approximately 7.8MOPS) [1]. DSPs for the PSI-CELP CODEC have already been developed [2][3]; however, power consumption of these DSPs (100~200mW) is several times higher than the VSELP CODEC. Reduction of power consumption is achieved by combining the throughput enhancement and supply voltage reduction [4]. Because over 50% of instruction cycles are multiply-accumulate (MAC) operations in the PSI-CELP, parallelism in the datapath, such as parallel datapath [2] or pipelined datapath [3][5], is effective for throughput enhancement. We have developed a new low-power 16-bit DSP employing a dual datapath architecture and low-power circuit design techniques. This paper describes the low-power design and evaluation results of the DSP.

Architecture Overview

Fig. 1 shows the block diagram of the DSP. The DSP core is constructed of a program control unit (PCU), an address arithmetic unit (AAU), and a data arithmetic & logical operation unit (DALU).

The DALU has a dual datapath which includes four 16-bit input registers, two 17-bit × 17-bit multipliers, a 32-bit ALU, a 32-bit add-subtract unit, and four 32-bit accumulators.

The program memories comprise 16K words of ROM and 128 words of RAM. The data memories are organized into 6 banks: four 1K word RAMs, one 6K word ROM, and one 12K word ROM.

The DSP has a 32-bit instruction width. The instruction set includes sequence, move, load, bit-manipulation, and operation instructions. With orthogonally encoded fields of the operation instructions, it is possible to specify memory access, data operation, data transfer, and sequence control.

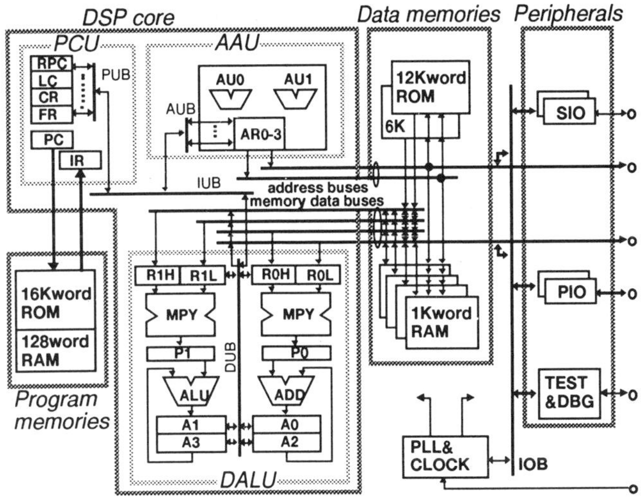

Fig.1 Block Diagram of DSP

Dual Datapath Architecture

The dual datapath architecture allows the DSP to reduce the power supply voltage while maintaining the high throughput.

In this architecture an arbitrary bank of data memory can be connected to any of four 16-bit memory data buses through the memory access field of the operation instructions. Four input registers (R1H, R1L, R0H, R0L), which are directly connected to these four memory data buses, can supply operands to any input of either multipliers, ALU, and add-subtract unit according to the data operation field. Using the inside data bus (DUB) of the DALU, data transfer among four input registers and four accumulators is controlled by the data transfer field. With this architecture, high throughput dual MAC operations can be achieved for single precision, double precision, and mixed precision operations which are necessary to obtain the required precision of the PSI-CELP algorithm in a 16-bit DSP implementation.

Typical examples for MAC operations for FIR filtering using the dual datapath are shown in Fig. 2. Fig. 2(a) shows the data flow for a single precision operation of FIR filtering with coefficients h(i), input x(n), and output y(n). In the operation, the instruction fields of memory access, data operation, and register transfer specify data flow as follows: (i) data load from two memory banks to R1H and R0H; (ii) two multiply and accumulate operations (R1H × R1L and R1H × R0H); and (iii) data transfer from R0H to R1L. With this organization of the instruction fields, one datapath calculates y(n) (see Eq. 1), and the other

0-7803-3177-6 $5.00 © 1996 IEEE

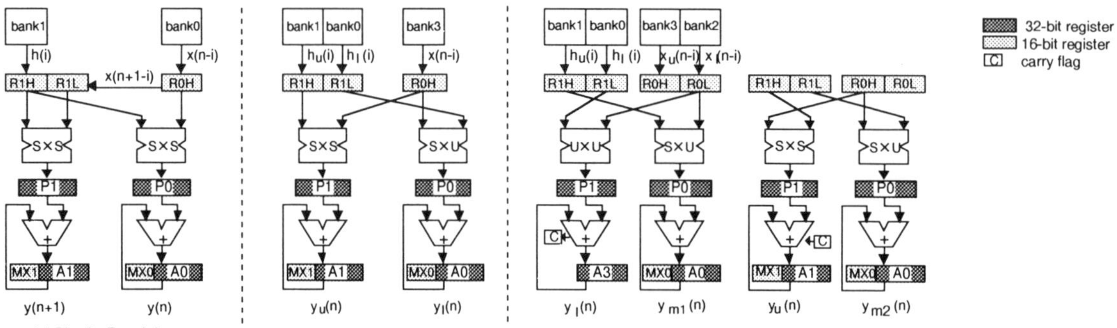

(a)Single Precision (b)Mixed Precision (c)Double Precision

Fig. 2 Multiply-Accumulate Data Flow

calculates y(n+1) (see Eq. 2). Thus two single precision FIR filter operations are realized in N clock cycles.

$$y(n) = \sum_{i=0}^{N-1} h(i) \cdot x(n-i) \quad (1)$$

$$y(n+1) = \sum_{i=0}^{N-1} h(i) \cdot x(n+1-i) \quad (2)$$

where N is the number of taps.

Fig. 2(b) shows a mixed precision operation. In this operation, the DSP executes: (i) data load from three memory banks to R1H, R1L, and R0H; and (ii) two multiply and accumulate operations (R1H × R0H and R0H × R1L). Thus, one datapath calculates $y_u(n)$ (see Eq. 3), and the other calculates $y_l(n)$ (see Eq. 4). Then an add operation is executed to maintain alignment of the binary point. Thus a mixed precision FIR filter operation is realized in N clock cycles.

$$y_u(n) = \sum_{i=0}^{N-1} h_u(i) \cdot x(n-i) \quad (3)$$

$$y_l(n) = \sum_{i=0}^{N-1} h_l(i) \cdot x(n-i) \quad (4)$$

where h_u holds the upper 16 bits of the coefficient, and h_l the lower 16 bits of the coefficient.

Fig. 2(c) shows a double precision operation. In this operation, two sets of instructions are executed. First, the DSP performs: (i) data load from four memory banks to R1H, R1L, R0H, and R0L; and (ii) two multiply and accumulate operations (R1L × R0L and R1H × R0L). Then, the DSP executes two multiply and accumulate operations (R0H × R1H and R0H × R1L). In this way, $y_l(n)$ (see Eq. 5), $y_{m1}(n)$ and $y_{m2}(n)$ (see Eq. 6), and $y_u(n)$ (see Eq. 7) are calculated. After that, an add operation is executed to maintain alignment of the binary point. Thus a double precision FIR filter operation is realized in 2N clock cycles.

$$y_l(n) = \sum_{i=0}^{N-1} h_l(i) \cdot x_l(n-i) \quad (5)$$

$$y_m(n) = y_{m1}(n) + y_{m2}(n) = \sum_{i=0}^{N-1} h_u(i) \cdot x_l(n-i) + \sum_{i=0}^{N-1} h_l(i) \cdot x_u(n-i) \quad (6)$$

$$y_u(n) = \sum_{i=0}^{N-1} h_u(i) \cdot x_u(n-i) \quad (7)$$

where h_u holds the upper 16 bits of the coefficient, h_l the lower 16 bits of the coefficient, x_u the upper 16 bits of the data, and x_l the lower 16 bits of the data.

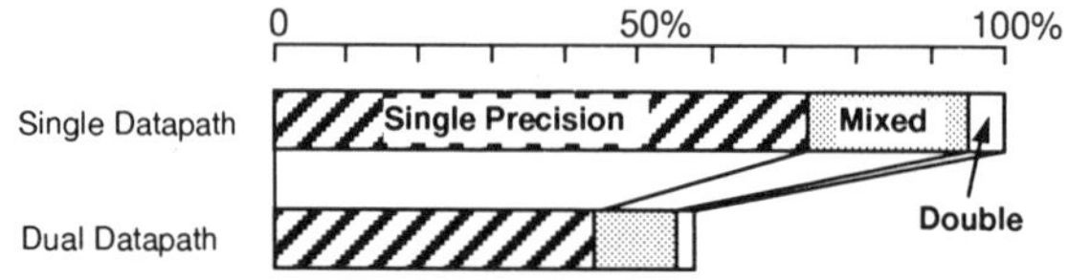

Fig. 3 Reduction of Instruction Cycles for MAC in PSI-CELP Algorithm

The reduction of instruction cycles with dual MAC operation is shown in Fig. 3. 86.3% of the total MAC operations of the PSI-CELP algorithm are executed in parallel in this dual datapath architecture, thus reducing the total instruction cycles of MAC operations to 56.9% of that for a single datapath.

Furthermore, there are two other operations which are executed efficiently in a dual datapath. One is square distance operation for the codebook search: a subtract operation and a square-accumulate operation are executed in parallel. So only one instruction cycle is required for execution of square distance operation. The other operation is the add compare select operation for the Viterbi decoding. Owing to parallel execution of an addition and a subtraction in the dual datapath, six instruction cycles are required for add compare select operation. These cycles are almost the same as those of dedicated hardware for add compare select operation [6].

Owing to the dual datapath with a powerful instruction set, the PSI-CELP CODEC is performed at 22.5 MHz. The area overhead for the dual datapath is only 2.4 mm^2, 2.4% of the total area.

Low-Power Circuit Design

Analysis of power consumption of a 24-bit floating point DSP for full-rate speech CODEC [7][8], shows that 50% of the total power is used for memory and clock circuits. So, low-power circuit design has been focused on memory design and clock distribution.

A. Low-Power RAM

The DSP has four 1K word RAMs. Each RAM is divided into 4 blocks of 256 words to reduce power consumption by decreasing the activated portion. The circuit diagram of each RAM is shown in Fig. 4. Because sequential memory accesses comprise more than 90% of all data memory access in the PSI-CELP CODEC, we have

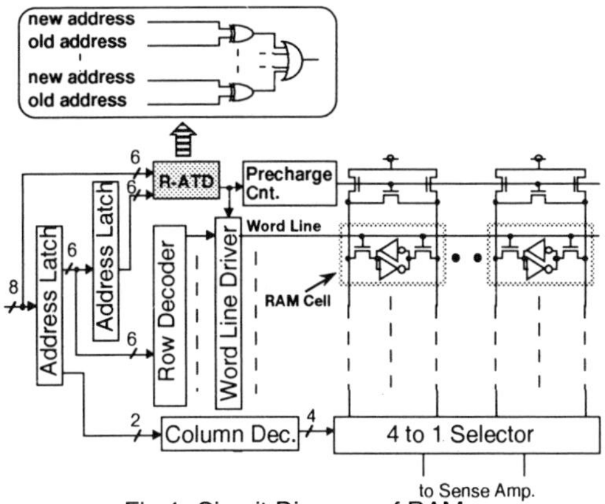

Fig.4 Circuit Diagram of RAM

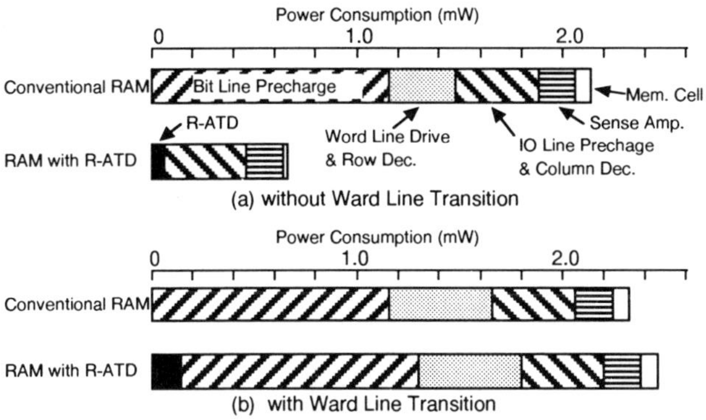

Fig. 5 Simulated Power Consumption for 256×16-bit RAM

focused on sequential address accessing. With the proposed row-address data transition detector (R-ATD), bit lines are precharged and the active word line is changed only when the row decoder address data changes. Higher address data (6 bits) drives the row decoder, and lower address data (2 bits) the column decoder. Thus, the number of times that bit lines are precharged and the active word line is changed are reduced to 1/4 in the sequential address case.

The result of the circuit simulation is shown in Fig. 5. By using R-ATD, power consumption is reduced to 29.8% when memory is accessed without word line transition, while power consumption increases by 6.2% when memory is accessed with word line transition. Consequently, a power reduction of 49.8% is achieved in the sequential address case.

B. Low-Power ROM

Similarly to the RAM, every ROM is divided into 2K word blocks. Thus, the program ROM is divided into 8 blocks and the two data ROMs are divided into 3 and 6 blocks, respectively. The circuit diagram of each ROM is shown in Fig. 6. As only bit lines selected by the column selector are precharged through the IO lines, the number of bit lines precharged is reduced from 256 lines to 16 lines. And with feedback to the precharge circuit from the output of the sense amplifier, the voltage of a bit line is limited to the threshold voltage of the inverter which is used as a

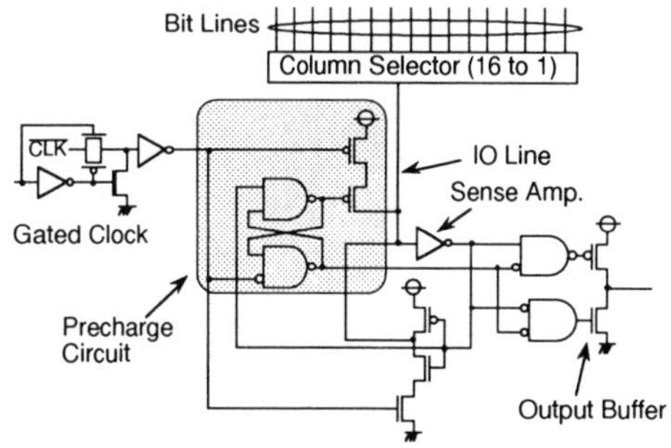

Fig.6 Circuit Diagram of ROM

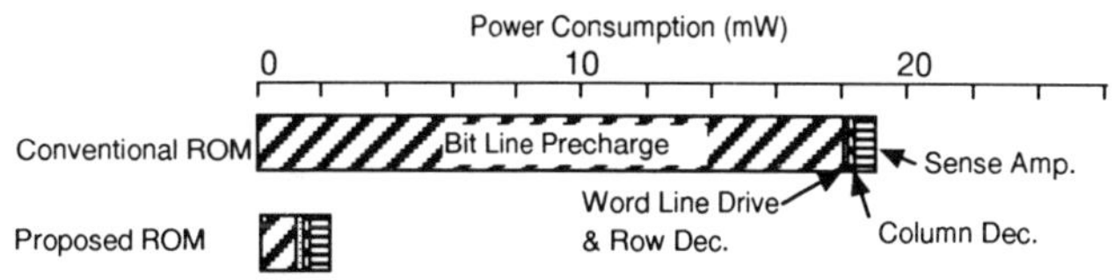

Fig. 7 Simulated Power Consumption for 2K×16-bit ROM

sense amplifier, so that current for extra precharging is reduced. The result of the circuit simulation is shown in Fig. 7. A power reduction of 89.3% is achieved.

C. Gated Clock

The circuit diagram of a gated clock is shown in Fig. 8. A gated clock is generally used in order to reduce clock lines capacitance. But the effect of low-power consumption by the gated clock depends on activation rate and the number of latches to which the gated clock are applied. To estimate power consumption, we have used a circuit simulation model of a gated clock which considers clock line lengths and the layout. The result of the simulation, the relation between power reduction rate and activation rate, is shown in Fig. 9. When Pgate/P0 (Pgate = power consumption with gated clock; P0 = power consumption without gated clock) is less than 1, a gated clock is useful for power reduction. By estimating activation rate of latches in the PSI-CELP CODEC, a gated clock is applied when Pgate/P0 is less than 1.

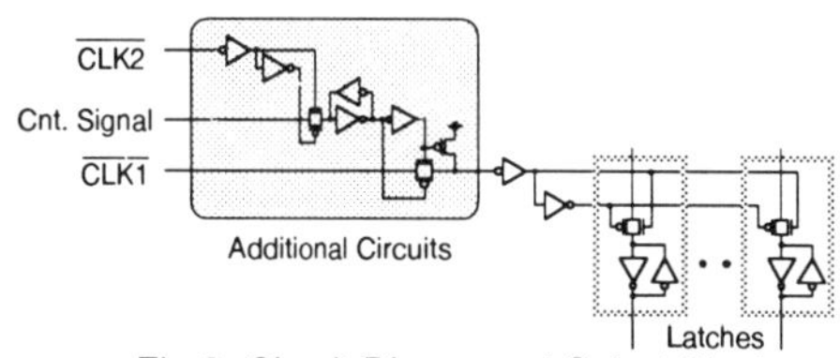

Fig.8 Circuit Diagram of Gated Clock

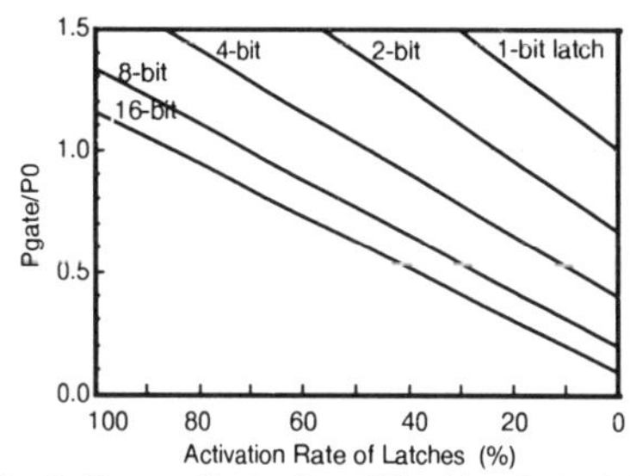

Fig. 9. Power Reduction Effect by Gated clock

Pgate: Power Consumption with gated clock

P0 : Power Consumption without gated clock

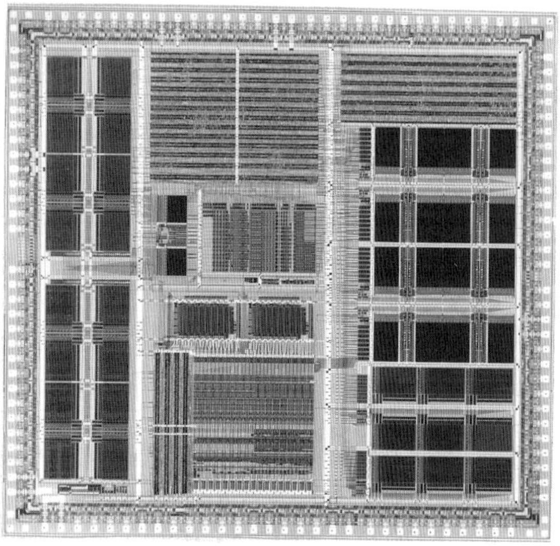

Fig. 10. Photomicrograph of the DSP

Table1 Features

Fabrication technology	0.5 μm double-metal CMOS Vtn=0.55V Vtp=-0.55V
Power Dissipation	36mW at 1.8V VDD
Die Size	9.8mm × 10.3mm
Number of Transistors	1.46M (Logic 134k, RAM 455k, ROM 868k)
Package	100-pin QFP

Evaluation Results

The DSP is fabricated using 0.5 μm single-poly, double-metal CMOS technology. The photomicrograph of the DSP is shown in Fig. 10. The LSI characteristics of the DSP are summarized in Table 1. It integrates 1.46 million transistors (134k transistors for logic, 455k transistors for RAM and 868k transistors for ROM) in a 9.8mm × 10.3mm die size, and is packaged in a 100-pin quad-flat-package (QFP). PSI-CELP CODEC and VSELP CODEC are implemented in the program and data ROMs. A shmoo plot, which shows the instruction cycle time vs. supply voltage VDD at room temperature is shown in Fig. 11. Fig. 12 shows power consumption when PSI-CELP CODEC (including noise canceling and error correct coding/ decoding) is executed at 22.5MHz. At 1.8V VDD, power consumption is 36mW.

Conclusion

A low-power consumption 16-bit DSP has been developed to implement the PDC half-rate CODEC. By employing a dual datapath architecture, throughput of a MAC operation is enhanced and power consumption is reduced by decreasing power supply voltage. Based on an analysis of the speech CODEC algorithm, power reducing techniques focused on memory design and clock distribution are applied extensively in the circuit design. The DSP, fabricated in 0.5 μm single-poly, double-metal CMOS technology, integrates 1.46million transistors in a 9.8mm × 10.3mm die size. The PSI-CELP algorithm for the PDC half-rate CODEC can operate perfectly at 22.5MHz, resulting in a low-power consumption of 36mW with a 1.8V power supply.

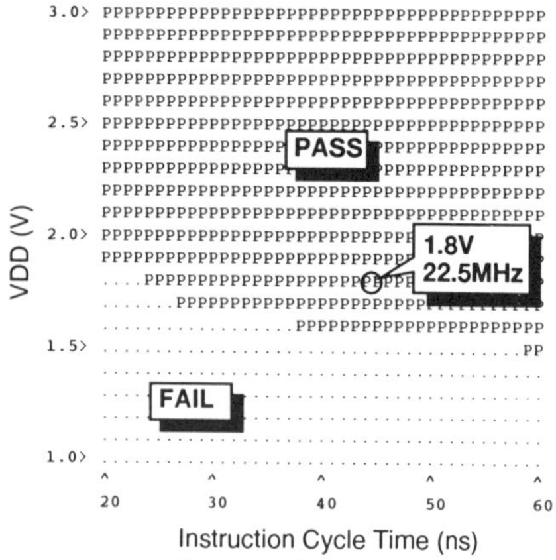

Fig. 11. Shmoo Plot

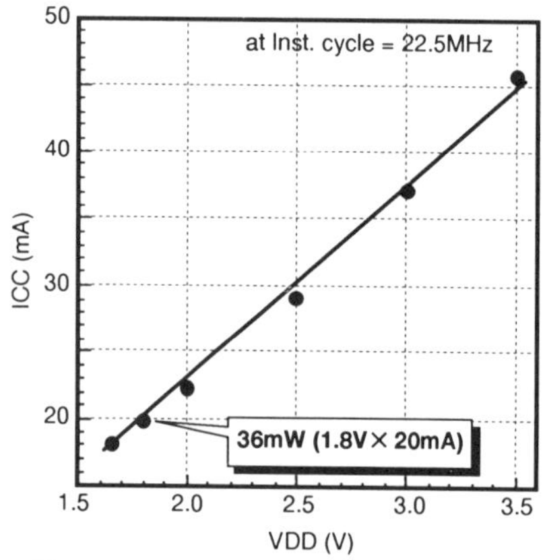

Fig.12 ICC vs. VDD for PSI-CELP CODEC

Acknowledgments

The authors would like to acknowledge the technical contributions of Y. Seguchi, H. Wakada, Y. Arima, T. Asai, K. Aota, S. Yamazaki, T. Higuchi, K. Yamazaki, S. Furubeppu, M. Nakajima, M. Yamaguchi, H. Momono, T. Inada, M. Nozaki, and Y. Shimazu. We would also like to thank Dr. Y. Horiba for their encouragement.

References

[1]F. Masui and M.Oguchi, "Activity of the Half Rate Speech Codec algorithm Selection for the Personal Digital Cellular System", Technical Report of IEICE, RCS93-77, pp. 55-62, 1993.

[2]Y. Okumura, T. Ohya, Y. Miki, and T, Miki, "A Study on DSP Circuits Applied to Speech CODEC for Digital Mobile Communications", Proceedings of the IEICE fall conference, p. 2-294, 1993.

[3]H. Kabuo, *et al.*, "A 16bit Low-Power-Consumption Digital Signal Processor Using an 80 MOPS Redundant Binary MAC", IEEE Digest of Technical Papers Symposium on VLSI Circuits, pp.63-64, 1995.

[4]A. P. Chandrakasan, S. Sheng, and R. W. Brodersen, "Low-Power CMOS Digital Design", IEEE Journal of Solid-State Circuits. Vol. 27, NO.4, April 1992.

[5]K. Ueda, *et al.*, "A 16b Low-Power-Consumption Digital Signal Processor", IEEE Digest of Technical Papers ISSCC, pp. 28-29, 1993.

[6]Texas Instruments, "TMS320C540 User's Guide"

[7]T. Kengaku, *et al.*, "A Single-Chip DSP for Mobile Telecommunication Applications", Proceedings of IEEE CICC, 1991.

[8]T. Tokuda, *et al.*, "A Mixed-Signal Digital Signal Processor for Single-Chip Speech Codec", IEICE Trans. Electron., Vol.E75-C, No.10 October 1992.

Chair: Paul Billig
Co-Chair: Neill Mullinger

The complexity of deep sub-micron design forces designers in the late 1990s to deal with many new issues on the road to a successful design. This session looks at ways to meet a number of these challenges: logical verification of very large designs, analysis of cyclic combinational circuits, modeling of noise and the problem of power analysis and minimization.

Current designs are becoming too large to allow good functional coverage using vectors and simulation. Formal verification methods are relatively recent techniques for achieving 100% functional verification coverage of, even, very large designs. The first paper is an introductory tutorial on formal methods.

The complete inability of today's logical and timing analysis tools to handle cyclic combinational circuits significantly impacts the design of large ICs, especially those with datapaths. The second paper details a method for decomposing cyclic combinational circuits into multiplexed acyclic circuits. These acyclic circuits may then be used with existing logical and timing analysis tools.

Another common issue is noise immunity in today's high density, high performance designs. The third paper shows a method for efficiently modeling noise in mixed-signal ICs.

A significant problem in today's deep submicron designs is the analysis and minimization of power, especially in portable, handheld and other battery applications. The fourth paper details a methodology for analyzing power distribution early in the design cycle when the highest leverage exists for minimizing power. The technique utilizes power analysis at the gate-level and is annotated on to the floorplan for analysis. Thus, optimization for power may be done prior to the actual detailed placement and routing. The fifth paper details a very fast technique for probabalistically analyzing power in combinational circuits. The sixth paper shows a novel technique for minimizing power in embedded systems using network flow to minimize on-chip/off-chip memory accesses. This is important due to the significant amount of power used by memories in today's embedded systems. The final paper shows a new high-level synthesis technique for implementing low-power, real-time applications. Current high-level synthesis programs typically minimize area by utilizing hardware sharing. This paper shows how power can be optimized utilizing reduced hardware sharing, while minimally impacting area.

There are many issues facing deep sub-micron IC designers today and more to come in the next few years. Join us at this session to learn some innovative techniques for attacking these problems.

A Brief Introduction to Formal Methods *

Paul E. Black Kelly M. Hall Michael D. Jones Trent N. Larson Phillip J. Windley

Laboratory for Applied Logic
Brigham Young University

info-lal@lal.cs.byu.edu

Abstract

As hardware designs grow in size and complexity, current design methods are proving less adequate. Current methods for specification, design, and test are typically empirical or informal, that is, they are based on experience and argument. Formal methods are solidly based on mathematical logic systems and precise rules of inference. Formal methods offer a discipline which complements current methods so designers can successfully meet the demand for high performance systems.

Formal methods covers a broad and diverse set of techniques aimed at improving computer correctness. This paper explains the role of specifications and implementation models in formal methods, and different approaches to proving their correspondence. We refer to excellent overview papers and cite some recent successful examples of using formal methods in hardware design.

Introduction

Current hardware and software designs are orders of magnitude larger and more complex than they have been. It is therefore more difficult to design correct systems using only informal techniques and practices. The term *formal methods* includes a set of techniques based on mathematical foundations and analysis. Formal methods [10] improve computer design by reducing design errors when used as a complement to empirical techniques currently used. This paper provides a brief introduction to formal methods for hardware design.[1] We discuss what they are, describe different methodologies grouped under the heading *formal methods*, and suggest where they can be used successfully. Due to space considerations, the bibliography is not extensive, but it was carefully chosen to provide a good starting place for further exploration.

*This work was sponsored by the National Science Foundation under NSF grant MIP–9412581 and the Department of Defense under contract MDA904-94-C-6115.

[1] This paper focuses exclusively on formal methods in hardware design. Formal methods can also be used in broader system design, including software, but such discussion is beyond the scope of this paper.

What are Formal Methods?

Formal methods are an analytical approach relying on mathematical models for excluding design errors in hardware. Other approaches to design fault exclusion, such as simulation, are empirical in nature. The chief benefit of analytical techniques is that they offer 100% coverage of the design space. That is, with a precise mathematical model, one can reason about all possible cases. The chief drawback is the difficulty of building models and conducting analysis. The precise nature of formal methods precludes informal hand waving to dismiss difficult, extreme cases.

All formal methods involve one or more of the following three parts:

1. a mathematical model of the design's intended behavior or properties, called the *specification,*
2. a mathematical model of the design's structure, called the *implementation model,* or more briefly, the implementation, and
3. mathematical expressions stating relationships between the models established using analysis (proof) to demonstrate that the relations hold.

Formal methods begin with a specification, an implementation model, and a mathematical expression stating the relationship between them. They finish by demonstrating the relationship via precisely defined rules. However, formal methods need not include all three aspects. Benefits accrue from simply writing a formal specification which then serves as an unambiguous reference for implementation, simulation, and testing.

A. Simple Example of Formal Methods

As an example of the activities and models discussed above, we might specify the behavior of an exclusive–or gate with the following mathematical formula:

$\vdash_{def}$ `xor2_spec a b out = (out = ¬(a = b))`

17.1.1

0-7803-3177-6 $5.00 © 1996 IEEE IEEE 1996 CUSTOM INTEGRATED CIRCUITS CONFERENCE

That is, the behavior of an exclusive–or relates inputs, a and b, and the output, out. Note that the above formula can easily be assigned a rigorously defined meaning. The implementation model could be described using a netlist:

```
MODULE out .xor2_imp a b;
       BEGIN
       p    .nand2  a b ;
       q    .nand2  a p ;
       r    .nand2  p b ;
       out  .nand2  q r ;
       END;
```

This implementation is four interconnected **NAND** gates. In addition, we must have rigorous definitions of the meaning of **MODULE**, **.nand2**, **BEGIN**, etc., so that the above implementation model also has an unambiguous meaning.

We wish to show that the implementation satisfies the specification. We can express this with the mathematical relation *implies* and express rigorous definitions of the netlist (not given here for brevity) as a function **INTERP** in the following manner :

```
⊢ ∀ a b out. INTERP(out .xor2_imp a b) ⇒
             xor2_spec a b out
```

One can also read the formula as, for all `a`, `b` and `out`, the interpretation of an **XOR2** implementation (as defined above) on `a`, `b` and `out` implies the **XOR2** specification (also defined above) on `a`, `b` and `out`. Using mathematical analysis and the definitions of `.xor2_imp`, `xor2_spec`, and **INTERP**, we can prove that the implementation satisfies the specification.

Notice that the relationship covers *all* values of the inputs and output (`a`, `b`, and `out`), not just some test values. Of course, in this simple example an exhaustive simulation is trivial, but many formal methods can be applied to circuits with 10^{100} states or more and still show that the relationship holds for all possibilities.

How Do I Put Formal Methods to Work?

Various formalisms and techniques are applicable to each part of the process described in the previous section. To write a formal specification, one must make choices about which formalism to use (first order logic, higher order logic, temporal logic, state machines, automata, trace specifications, etc.) and the kinds of criteria to specify (functional correctness, liveness, safety, timing, and so on). To model a circuit, one must decide which level of abstraction (gate level, switch level, circuit level, register-transfer level, or higher) is appropriate as well as which formalism to use (first order logic, higher order logic, automata, etc.). The relationship of implementation and specification may be equivalence, implication, etc. How one handles each of the three parts forms a taxonomy of formal methods tools and techniques [3].

A. The Specification

Writing a specification for a design is perhaps the most difficult aspect of the formal methods process. Formal specifications require the designer to clearly, concisely, and unambiguously state what a circuit must do. To be of any benefit, the specification must be an abstract representation of the implementation. That is, it should state *what* a circuit must do, not *how*. The abstractions may be any combination of structural (an ALU instead of gates), data (numbers instead of bit vectors), temporal (instruction cycles instead of clock cycles), or behavioral (a page from memory is saved to disk instead of which page is saved to which cylinder). Specifications may be quite comprehensive, or they may include relatively few fundamental requirements such as a request is eventually granted or two communicating devices never deadlock. Specifications can also indicate timing properties, load characteristics, and other properties of the device.

The idea of formal methods is to show that the implementation meets the specification; but how does one ensure that the *specification* is correct? Ultimately it must be *validated* by the designers: they must examine the specification and decide that it expresses what they want. Higher level abstractions help by making it easier to state desired properties and behaviors. More powerful representations can more easily and concisely express the designer's desires. A specification of a few fundamental properties may be easy to judge correct, but leaves other important properties only informally specified. Some representations are executable, allowing designers to validate the specification by simulation in addition to review.

One of the most important choices to make is the level of abstraction in the specification. Higher level abstractions tend to allow more concise specifications, since less detail is included. Abstraction causes the specification to be more easily modified and validated. On the other hand, an abstract specification is more difficult to relate to the implementation. Multi-level verification treats the one level's specification as the next higher level's implementation. Thus several simple abstractions can be independently verified to yield the overall proof.

Related to the level of abstraction is the expressiveness of the language. A simple language, such as state machines or first order logic, is easy to reason about — in fact, many simple languages are decidable: they have completely automatic algorithms for calculating the correctness of statements. More expressive languages, such as higher-order logic, can more concisely express a wide range of specifications, but they are more difficult to reason about.

More abstract and expressive languages are more powerful in the long run, but tend to require more initial investment since they are more mathematical and less like representations with which designers are familiar.

B. The Implementation Model

Creating a model of the implementation is a standard task for hardware designers. Implementation models are similar to simulation models in use by designers now. An implementation model may be extracted from the simulation model, or, potentially, the same model may be used for both verification and simulation. The implementation model *must* have a well defined interpretation or meaning. A model with simple primitives is easier to reason about, but is a poor representation of the circuit. A more detailed model is a better representation of the circuit, but it is more difficult to use in a verification.

How does one ensure that the implementation model actually represents the physical device? As with the specification, validating the implementation can not be done by machine. Since the model only represents certain characteristics of the device, the final design must be checked to ensure that it has those characteristics.

Few formal methods tools accept models written for standard simulation tools without significant syntactic changes to the model. Most tools require a completely new model expressed in a different modeling framework. Thus, a designer often must construct multiple models of their circuit, one for each design tool (simulator, formal methods tool, etc.). Multiple versions raise the cost of design maintenance and can lead to version skew problems. Current research in formal methods is aimed at using standard HDLs for implementation modeling and providing increased simulation capability.

C. Relating the Implementation and the Specification

There are several methods currently being used for relating implementations to specifications. These include theorem proving, model checking, equivalence checking, and language containment. Among these, the most commonly used are model checking and theorem proving.

In the model-checking approach, the specification is expressed as a formula in temporal logic. Such logics make statements about a world that changes through time, and they allow reasoning about dynamically changing situations. Implementation models are usually in the form of state transition graphs. They can be compared with the specification automatically. The system may verify that the models are valid, or it may provide counter-examples for any specifications falsified by the implementation.

Since model checking typically uses state-based hardware descriptions, it is better suited for checking control structures, as it expresses things such as concurrency and synchronization. Also, by modeling a particular kind of logic and world, model checkers aren't excessively complex. Of course, this means that they are not readily used on other types of problems.

As an example, the well known model checker SMV [6] uses specifications written in the temporal logic CTL and implementation models written in a VHDL-like language for creating state machines.

Theorem proving, in contrast, is a more interactive technique. When one uses a theorem prover to verify hardware the usual process is to design a specification and implementation as logic descriptions first-order predicate logic, higher-order logic, etc. The designer then guides the proof assistant tool through rigorous proof steps showing that the implementation model satisfies the logical specification. The level of interaction required of the user varies widely between theorem prover tools: some tools demand much detail but offer great flexibility (e.g., HOL); other tools are more automatic at the expense of flexibility (e.g., PVS, NQTHM).

Theorem provers ultimately rely on the designer to create an appropriate model of the hardware, and even to guide the system (sometimes explicitly) along the path to a proof. This can be a complex process, but theorem provers are very general and can be employed in a wide range of applications. Theorem provers are not as useful for reasoning about temporal aspects of hardware. But theorem provers are well suited for hierarchical methods of development (due to their abstraction mechanisms) as well as reasoning about functional specifications and parameterized descriptions.

The NQTHM theorem prover uses Boyer-Moore quantifier-free first order logic (with equality) to represent both specifications and implementation models [4]. The HOL theorem prover [9] uses higher-order logic to produce a flexible, but demanding environment for creating specification and implementation models.

What Will Formal Methods Do?

Formal methods will ensure that an implementation meets a specification, but they will not guarantee that the final product will always operate perfectly. What formal methods can do is limited by the philosophical limits on what can be proven, informally defined description languages and extra-logical factors. These limitations ought to be kept in mind when describing and discussing the results of a formal verification project. A more complete discussion of what formal methods will and will not do can be found in [2].

There are aspects of the design process to which the notions of specification and implementation simply do not apply. Specifications can not convey design intentions and implementation models can not describe physical properties. Consequently, the verifier is forced to choose a level of abstraction for the specification and a sufficiently concrete level for the implementation model. These decisions are part of the process, but should be explicitly stated when making a claim about verification. For example, claiming that a microprocessor has been verified at the gate-level implementation to meet a functional block specification more completely conveys the scope of the verification than simply stating that the processor has been verified.

In practice, the designer, verifier and each manufacturer use different, informally defined languages to describe the design. Often, the translations between these languages requires a combination of experience, intuition and luck. This is especially true of low-level design descriptions which may consist of nothing more than annotated diagrams and a few paragraphs of text. Without mathematically precise definitions of the design-description language, it is impossible to know if the verifier's interpretation of the design is the same as the designer and if the design manufactured is the same as the one described by the verifier. These gaps can be bridged by using a common, formally defined language at all three levels. Work on these types of languages is underway at Computational Logic, Inc. and Brigham Young University [11].

Claims about verified devices, especially in safety or security critical applications, should be strictly limited to factors covered by the logic. Faulty communication, social hierarchies, political climates, and so forth are usually not covered in the verification process. For example, verification demonstrating that the low-level model of a chip prevents unauthorized users from accessing data does not guarantee that a passer-by could not read sensitive data on a monitor. Consideration of these factors ought to temper broad guarantees about verified devices.

Formal methods have been successfully used in commercial and academic designs. We mention a few to suggest the wide applicability and utility of formal methods.

Johnson, Miner, and Camilleri[5] compare several formal tools by implementing a simple circuit in each of them. They "contrast how the underlying formalisms influence one's perspective on design and verification."

Windley and Coe[8] verified the correctness of a simple pipelined microprocessor using HOL, and Srivas and Miller[7] report the formal verification of a commercial microprocessor. Bainbridge, Camilleri, and Fleming[1] relate verifying memory protocols in an industrial setting. They point out formal methods can be used even with a short *time to market.*

Conclusion

Formal methods are a useful addition to the hardware design process. To make use of formal methods, first a specification must be written which expresses the design criteria, then an implementation model must be written or captured which represents the design. Formal verification demonstrates that an implementation model meets a specification for *all* cases. However formal methods are not a *silver bullet* which prevents all errors. Rather formal methods are a complement to good design methodology and testing. The chief benefit of formal methods is not the final *true* from the tool, but rather the process that is required to get the final result.

While the wide variety of specification and implementation modeling techniques and the many ways of relating them can make the choice of how and when to apply formal methods sound daunting, in fact there are relatively few tools to choose from. Of these tools, the best criteria for choosing is how they handle the task of relating the implementation model and the specification. Once a particular tool is chosen, many of the options for creating the implementation model and specification are eliminated.

References

[1] Simon Bainbridge, Albert Camilleri, and Roger Fleming, *Theorem Proving as an Industrial Tool for System Level Design*, in V. Stavridou, T. F. Melham, and R. T. Boute (eds.) *Theorem Provers in Circuit Design*, North-Holland, 1992.

[2] Avra Cohn, *The Notion of Proof in Hardware Verification*, Journal of Automated Reasoning, Vol 5, No 4, pp. 127–139, 1989.

[3] Aarti Gupta, *Formal Hardware Verification Methods: A Survey*, Formal Methods in System Design, Vol. 1, 1992, pp. 151–238, Kluwer Academic Publishers.

[4] William A. Hunt, Jr., *Microprocessor Design Verification*, Journal of Automated Reasoning, Vol 5, No 4, pp. 429–460, 1989.

[5] Steven D. Johnson, Paul S. Miner, and Albert Camilleri *Studies of the Single Pulser in Various Reasoning Systems*, in R. Kumar and T. Kropf (eds.), *Theorem Provers in Circuit Design*, Springer, 1994.

[6] Kenneth L. McMillan, *Symbolic Model Checking, An Approach to the State Explosion Problem*, Ph.D. thesis, School of Computer Science, Carnegie Mellon University, Pittsburgh, PA, 1992.

[7] Mandayam Srivas and Steven P. Miller, *Applying Formal Verification to a Commercial Microprocessor, Proceedings of the 1995 IFIP International Conference on Computer Hardware Description Languages.*

[8] Phillip J. Windley and Michael Coe, *A Correctness Model for Pipelined Microprocessors*, in T. Kropf and R. Kumar (eds.), *Proceedings of the 1994 Conference on Theorem Provers in Circuit Design.*

[9] *Description of the HOL Theorem Proving System*, (URL: http://lal.cs.byu.edu/lal/hol-desc.html)

[10] *Other Formal Methods Sites page*, (URL: http://lal.cs.byu.edu/otherFMSites.html)

[11] *The Laboratory for Applied Logic homepage*, (URL: http://lal.cs.byu.edu/)

Practical Analysis of Cyclic Combinational Circuits

Arvind Srinivasan
Mentor Graphics Corp., San Jose, CA *

Sharad Malik
Princeton University, Princeton, NJ †

Abstract

Cycles arise in combinational logic circuits under a variety of different circumstances. Typically a cyclic combinational circuit consists of a set of acyclic circuits multiplexed to share resources in a cyclic topology. Recently Malik [1] provided an algorithm for deciding if a given circuit with cycles is combinational and also an algorithm for static timing analysis of such circuits. The logical analysis technique was based on OBDDs and the timing analysis technique was based on solving the false path problem for a large and complex circuit. As a result that method does not scale well to very large circuits. Prior to the present work, the only option left for designers was to do a case analysis by manually identifying the different acyclic circuits multiplexed in the cyclic structure, and then performing logical and timing analysis on this set of acyclic circuits. In this paper, we provide an analysis technique that automatically identifies the set of acyclic circuits constituting the cyclic circuit. This enables designers to directly use existing logical and timing analysis tools on this set of acyclic circuits. In addition to providing a more intuitive analysis, the proposed algorithm also scales easily to handle the very large circuits encountered in practice, with runtimes of a few hundred seconds on circuits with over 15000 gates.

1 Introduction

A cycle is combinational if the circuit outputs do not depend on the initial state of the cycle. After the input stimulus is applied, the circuit outputs may be unstable for a few iterations, but they must eventually settle to a stable value that is independent of the internal state of the cycle. The necessary condition for a cycle to be false is derived in [1]. Essentially, the condition states that every non-simple path (a path that intersects itself at least once) from the primary inputs to the primary outputs is blocked for all valid input assignments. Figure 1 illustrates this concept. In the figure, either the side input of g_1 is 0 or or the side input of g_2 is 1 for all input assignments. This blocking side input determines the value on the path independent of the state in the loop. A practical implementation of the test for false cycles requires testing for tautology – an operation that is usually performed using OBDDs or ATPG techniques. Consider the following function:

$$\textbf{if } (c) \textbf{ then } z = (a+b) \times d \textbf{ else } z = a \times d + b$$

and its implementation shown in Figure 2. An area minimal implementation of this design contains a false cycle and is only slightly slower (two additional multiplexer delays) than the cycle-free implementation which requires duplication of operators. For the purposes of timing analysis, breaking a cycle at an arbitrary net may lead to an underestimate of delay. For example, if the cycle in the above example is broken at X the delay from a to z through the adder followed by the shifter is considered during timing analysis. The delay of the other path through the circuit from a to z through the shifter followed by the adder is not considered. These two delays are equal only if the operators are symmetric logic functions. Such cycles may arise when circuits are synthesized from behavioral descriptions Generalizing the above idea, let $\{g_1, g_2, \ldots, g_m\}$ be a set of m distinct Boolean functions of n variables each. A cyclic implementation of the $m(m-1)$ Boolean functions $\{g_1(g_2), g_1(g_3), \ldots, g_m(g_1), \ldots, g_m(g_{m-1})\}$ uses one implementation each of $\{g_1, g_2, \ldots, g_m\}$ and $O(m)$ selectors with $O(\log(m(m-1)))$ lines. Such a design may be significantly smaller than and almost

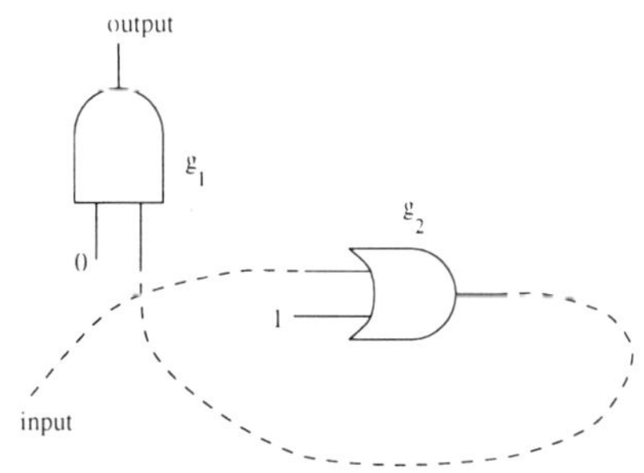

Figure 1: Boolean conditions for a cycle to be false

*arvind_srinivasan@mentorg.com
†sharad@ee.princeton.edu

0-7803-3177-6 $5.00 © 1996 IEEE

as fast as a cycle-free implementation. Logical multiplexing of several acyclic circuits is at the heart of all cyclic combinational circuits.

Cycles pose a problem during logic design for several reasons: (1) Circuit delays cannot be determined by a block-oriented approach for primary outputs and latch elements which depend on the cycles because of cyclic dependencies in the static analysis timing graph. (2) Many logic optimization and test-pattern generation techniques cannot deal with cyclic equations.
Estimating the delay involves determining the longest true simple path, a NP-hard problem for which efficient heuristics are not known. Malik [1] described a technique for approximating the delay by unrolling the cyclic circuit to obtain an equivalent acyclic circuit and then estimating the longest true path in the corresponding acyclic circuit. The technique relied on OBDD-based representations of circuit Boolean functions, which can be exponential in size for some commonly encountered functions.

In this paper, we combine structural analysis of feedback vertex sets with logical analysis to efficiently determine whether a cyclic circuit is combinational. We also obtain a concise set of Boolean conditions such that when each condition is true, all cycles are broken. These conditions help identify the set of acyclic circuits being multiplexed in the cyclic topology – each condition corresponds to one such circuit. For example, for the circuit in Figure 2, the conditions $c = 1$ results in the acyclic circuit with the adder followed by the multiplier and $c = 0$ results in the acyclic circuit where the multiplier is followed by the adder. Between them, these two conditions cover the entire Boolean space of inputs, i.e. $c + \bar{c} = 1$ and thus take care of all possible cases. These acyclic circuits can then be used for timing analysis or logic optimization. Between them they contain all possible circuit behaviors. Our techniques are shown to be practical for large designs, including some for which a compact OBDD representation does not exist.

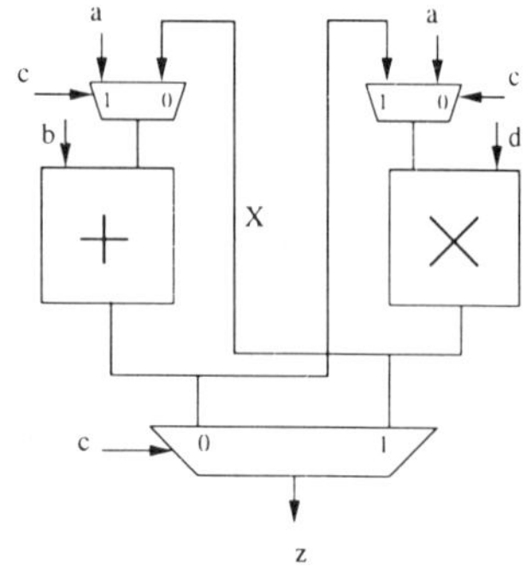

Figure 2: Logic circuit with a false cycle

2 Problem Definition

A combinational logic circuit C consists of a set of gates and connections between them. To simplify the notation, we assume that the circuit is composed of AND, OR and NOT gates only. The *controlling value* of a gate input is one that determines the output of the gate independent of the other inputs. For example, 1 is a controlling value for an OR gate. A path is an alternating set of gates and connections $\{g_0, c_0, \ldots, g_{n-1}, c_n, g_n\}$, where connection $c_i, 0 \leq i \leq n$ connects the output of g_i to g_{i+1}. A path is said to be *simple* if each gate in the path appears only once along that path. A *cycle* is a path such that the first and last gates are the same. A side input to a path is an input to some g_i besides c_{i-1}, i.e. an input that does not lie on the path. A path is said to be blocked for some input vector if some side input to that path has a controlling value. A *feedback gate set* is a subset S of gates such that removal of all the gates in S eliminates all cycles in the circuit. We assume that in a practical design non-simple paths are blocked in the cycle rather than the sub-path from the cycle to the primary outputs. This is a stricter condition than that imposed by Malik [1], and one that we impose only to make our solution more practical.

Let B be the valid Boolean input subspace for the circuit (i.e. one that excludes the don't care conditions). Define $B(g)$ to be the Boolean subspace such that some input connection of g that is not on any cycle is controlling. For a gate g on a cycle, $B(g)$ is the condition under which all cycles containing g are blocked by g independent of the other gates in the circuit.

Definition 1 *The cost of a feedback gate set, $w(S)$ is defined to be the number of literals in the minimum sum-of-products (SOP) form of*

$$B(S) = \bigcap_{g \in S} B(g).$$

Definition 2 *A blocking set is a set $\tilde{S}$ of feedback gate sets such that:*

$$B \subseteq \bigcup_{S_i \in \tilde{S}} \left(\bigcap_{g \in S_i} B(g) \right) \quad (1)$$

Thus, for a set $\tilde{S}$ of feedback sets to be *blocking*, the union of the conditions $B(S_i), S_i \in \tilde{S}$ must cover the Boolean input subspace of the circuit. Thus, each element of the blocking set is a logical condition that results in all cycles in the circuit being blocked at some gate. The removal of these gates results in an acyclic circuit. Since the union of the conditions $B(S_i)$ is equal to B, all possible input conditions are contained in some acyclic circuit derived from a feedback set. Thus, the set of feedback sets provides a direct mechanism for obtaining the set of acyclic circuits needed for subsequent analysis. The following problem defines an optimality condition that is desirable for the blocking sets.

Problem 1 *The minimum cost blocking set problem is to find a blocking set such that* $\sum_{\bar{S}} w(S_i)$ *is minimum.*

Theorem 1 *Problem 1 is NP-hard* [1]

3 Heuristic Solution

3.1 Notation

While it is desirable to find a minimum cost blocking set, it is not always necessary or practical from a computational standpoint to search for one. Rather, a useful solution to the designer is one containing a small number of terms that can be enumerated in a practical design flow. The heuristic we use to "solve" the blocking set problem is an adaptation of the minimum feedback vertex set algorithm by Smith and Walford [2]. We follow their terminology:

The Boolean function of primary inputs represented by the output of a gate g is denoted by $f(g)$. We also write $g = 0(1)$ when we mean $f(g) = 0(1)$ and $\bar{g}$ when we mean $\bar{f}(g)$. A gate g is said to break a cycle if g is not on the cycle, the output of g is an input to a gate g' on the cycle, and there is an input assignment such that g is controlling for g'. Since we allow only simple gates, either 0 or 1 can be a controlling value, but not both. A *partitioned circuit* is a circuit which is split into two or more mutually exclusive sub-circuits, each of which is said to be a partition. A *maximal strongly connected* component of a circuit $\mathcal{C}$ denoted by $MSC(\mathcal{C})$. A gate g is *covered* by g' if g' breaks all cycles broken by g. A set of controlling output values $\{v_1, v_2, \ldots, v_n\}$ of gates $\{g_1, g_2, \ldots, g_n\}$ is *consistent* if the Boolean condition

$$(g_1 = v_1) \wedge (g_2 = v_2) \wedge \ldots \wedge (g_n = v_n)$$

is satisfiable. A consistent set of controlling values, denoted by a vector F (an F set), induces partitions $\mathcal{C}_F$ and $\mathcal{C}_R$ on $\mathcal{C}$ such that each gate of $\mathcal{C}_F$ is contained in some cycle(s) of $\mathcal{C}$ and only in cycles also broken by an element in F. The remaining gates are in $\mathcal{C}_R$. The set of gates on cycles that are controlled by the F set is denoted by S_F. If each cycle in the circuit is contains at least one element of S_F, it is said to be a *complete set.* A complete set of minimum cardinality is known as an *optimum set.* The *index*, $I(\mathcal{C})$ of a circuit is the cardinality of an optimum set of $\mathcal{C}$. An *essential gate* is an element of at least one optimum set.

[1] The interested reader may contact one of the authors for an extended manuscript containing proofs of the theorems stated in this paper

Theorem 2 (Smith and Walford) *If $S_1, S_2, \ldots, S_K$ are optimum sets for partitions $\mathcal{C}_1, \mathcal{C}_2, \ldots, \mathcal{C}_K$ of $\mathcal{C}$ induced by F sets $F_1, F_2, \ldots, F_K$ and if the set*

$$S = \bigcup_K S_k$$

is complete for $\mathcal{C}$, it is also optimum for $\mathcal{C}$.

Theorem 3 (Smith and Walford) *If $I(\mathcal{C}_F) = |S_F|$, then S_F is a subset of an optimum set of $\mathcal{C}$.*

3.2 Algorithm

Since Problem 1 is NP-hard, we attempt to solve it using a heuristic approach that searches for an optimum set rather than a minimum cost blocking set. Trial F sets of increasing size are generated upto a user defined limit K, starting with trial sets containing a single gate. Upon generation of a trial set that satisfies the condition of Theorem 3, we partition $\mathcal{C}$ into $\mathcal{C}_F$ and $\mathcal{C}_R$ and recursively generate an optimum set for $\mathcal{C}_R$. Trial F sets are generated as long as Equation 1 is not satisfied or a user defined limit on the maximum trial set size is not exceeded. The runtime of the algorithm depends on the number of trial sets generated.

Each candidate F set is subjected to three *essentiality tests*:

1. The F set must not be in another previously generated set.
2. The gates in S_F must be in a cycle with all the other elements removed from the F set. If this test is not satisfied, the F set cannot be optimum and it is rejected.
3. The F set must satisfy the condition of Theorem 3. This can be checked by generating all possible trial sets F' for $MSC(\mathcal{C}_F)$ such that $|S_{F'}| = |S_F| - 1$ and checking whether one of the subsets breaks all cycles in $MSC(\mathcal{C}_F)$. If all

cycles can be broken by any one of those trial sets, $I(\mathcal{C}_F) < |F|$ and the F set is not optimum for $\mathcal{C}_F$.

Algorithm 1

```
Blocking set algorithm:
begin
   funct blocking_set(C, B) ≡
      ⌈B_S ← ∅
      while (B ⊄ B_S) do
            S ← min_feedback(C)
            B_S ← B_S ∪ S
            C : Mark elements in S as considered
            mark_elements(S)
      od
      B_S ← reduce(B_S)
      B_S⌋.
end
```

```
1  funct min_feedback(C) ≡
2     ⌈S ← ∅
3     for MSC ∈ C do
4        k ← 1
5        while (k ≤ K) do
6              for each trial F set of size k  do
7                  calculate C_F and C_R
8                  if F is essential  then  go to 12 fi
9              od
10             k ← k + 1
11       od
12       S ← S ∪ S_F ∪ min_feedback(C_R)
13    od
14    S ← reduce(S)
15    S⌋.
```

```
funct reduce(S) ≡
   ⌈S' ← S
   for s ∈ S' do
      if S' ⊆ S' − s then S' ← S' − s fi
   od
   S'⌋.
```

A desirable feature of the algorithm is that it can provide the designer with useful information even if it does not return a blocking set (this can occur if the user defined limit on the size of trial sets is exceeded).

4 Experimental Results

We have implemented the blocking set heuristic in the Mentor Graphics AutologicII Logic Synthesis product. The implementation uses OBDDs to perform all necessary Boolean operations. Table 1 shows the results of running the heuristic on a set of cyclic combinational circuits on a Hewlett-Packard HP9000/735 workstation. All the results were obtained on flat designs containing simple gates. The *chain16* and *chain32* designs are portions of a complex 32-bit datapath design containing multiple nested loops. The *addmult* design consists of an 32-bit adder and a 32-bit multiplier in a cycle, similar to Figure 2 and *multdiv* contains a 32-bit datapath composed of a multiplier and a divider. *reconv* is a 16-bit arithmetic circuit with deeply nested loops, capable of producing 64 different output functions using 8 functional elements.

Table 1: Experimental Results

Example	Gates	CPU seconds
chain16	2246	39
chain32	6696	70
addmult	9110	43
multdiv	13044	188
reconv	17740	294

5 Conclusions

Cyclic combinational circuits consist of topologically multiplexed acyclic circuits. This paper described an algorithm for automatically deriving this set of acyclic circuits, providing for an easy and intuitive logical and timing analysis, since all existing logical and timing analysis techniques can be used on the set of acyclic circuits contained in the cyclic one. A practical implementation with small runtimes for large practical circuits demonstrates the efficacy of this algorithm.

References

[1] Sharad Malik. Analysis of cyclic combinational circuits. In *International Conference on Computer-Aided Design*, November 1993.

[2] George W. Smith and Robert B. Walford. The identification of a minimal feedback vertex set of a directed graph. *IEEE Transactions on Circuits and Systems*, CAS-22(1), January 1975.

SUBWAVE: a Methodology for Modeling Digital Substrate Noise Injection in Mixed-Signal ICs

Paolo Miliozzi, Luca Carloni, Edoardo Charbon†, and Alberto Sangiovanni-Vincentelli

Department of Electrical Engineering and Computer Sciences,
University of California, Berkeley, CA

† Cadence Design Systems, Santa Clara, CA

Abstract

A methodology is presented for generating compact models of substrate noise injection in complex logic networks. For a given gate library, the injection patterns associated with a gate and an input transition scheme are accurately evaluated using device-level simulation. Assuming spatial independence of all noise generating devices, the cumulative switching noise resulting from all injection patterns is efficiently computed using a gate-level event-driven simulator. The resulting injected signal is then sampled and translated into an energy spectrum which accounts for fundamental frequencies as well as glitch energy. Preliminary results demonstrate the validity of the assumptions and the accuracy of the approach on a set of standard benchmark circuits.

1 Introduction

With the continuous increase of chip complexity, device density and circuit speed, noise immunity has become a major source of concern in the design of reliable VLSI ICs. Due to the decline of the distance between high-swing high-frequency noise sources and sensitive devices, the substrate has become a major carrier of spurious signals. In mixed-signal circuits in particular, coupling of analog, relatively slow signals with quickly switching digital ones is often disastrous. To alleviate the problem, heavily over-designed structures are generally adopted, thus seriously limiting the advantages of innovative technologies. For this reason substrate modeling has received attention from mixed-signal circuit designers, attempting to integrate RF analog and baseband digital circuitry on the same chip.

In digital circuits a large number of gates undergoes a transition periodically. When a transition occurs a spike of current is absorbed from a power bus and used to charge a load in the signal path. In general, a significant portion of this current is discharged to a ground bus through direct feedthrough, or it is injected directly into the substrate through various mechanisms [1]. The cumulative contribution of currents injected by switching gates is felt in sensitive circuits in a form of a spurious signal, referred to as *switching noise*.

The spatial location and rate of occurrence of transitions in digital circuits are generally time-variant and difficult to characterize exactly. In addition, due to the complexity of the circuits, an exact waveform characterization of switching noise is impractical. For this reason, switching noise is often modeled based on its macroscopic appearance. For example, if the number of switching gates is large enough and the global switching activity of the digital circuit is uniformly distributed over a large section of the spectrum, the switching noise can be modeled as a single Gaussian white or pink noise source. An alternative modeling technique, proposed in [2, 3], consists of representing switching noise as a capacitively and/or resistively coupled signal derived from the global clock of the circuit.

Switching noise models in this compact form are useful to efficiently estimate the impact of an architecture or a floorplan to system performance, and they can be used to drive a number of global optimization tools. However, the approximations for injected switching noise using such a method often capture only a relative small portion of the entire noise energy spectrum. Thus, potentially detrimental noise components may be underestimated or ignored. Moreover, with the emergence of submicron technologies, the problem of high-speed substrate noise interference may compromise performance at the nominal frequencies for which the circuits are designed.

In this paper a novel method is proposed for accurate and efficient characterization of switching noise generated by the logic components of mixed-signal ICs. The method is structured as follows. First, the noise injected in the substrate from a certain gate with a given input transition pattern is accurately simulated using detailed extraction. Call *noise signature* such a signal. Second, an event-driven simulator computes the rate of occurrence for each noise signature cumulatively over all the gates present in the circuit. Third, the noise signal associated with each signature is evaluated over the entire simulation time span. Fourth, the energy spectrum of the resulting signal is computed using the Fast Fourier Transform (FFT). Figure 1 shows all the phases of SUB-

0-7803-3177-6 $5.00 © 1996 IEEE

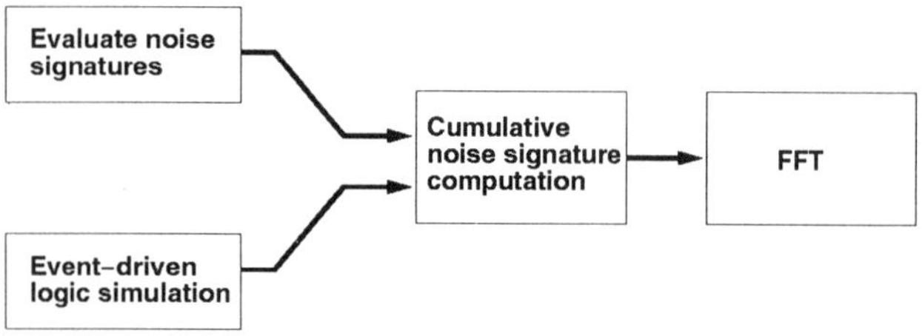

Figure 1: Flow diagram of SubWave

WAVE.

There are several advantages to this procedure. First, the substrate noise model is compact and, although accurate, it can be translated into a single noise signature which is unique for a given digital component and hence it can be re-used in future re-design loops. Second, the energy spectrum captures the energy associated with glitches and other high-speed phenomena. Finally, due to the efficiency of the event-driven evaluation techniques and of the FFT, a rapid characterization of injected noise can be obtained and possibly drive the synthesis of *silent* digital circuits.

The paper is organized as follows. Section 2 describes the techniques for the generation of the noise signatures associated with the gates and corresponding input transition schemes. In section 3 the event-driven simulator and the cumulative noise computation are described in detail. Finally, in section 4 a number of experiments on a number of benchmarks is presented and discussed.

2 Modeling Substrate Injection

Substrate noise is injected mainly through devices and capacitive coupling from interconnects.

Device noise injection is caused by impact ionization and drain/source-substrate junctions. The effects of inverse biased n/p junctions are effectively modeled through non-linear capacitances. Impact ionization on the contrary, is a highly localized phenomenon due to high electric fields experienced in the drain-channel-substrate interface. Impact ionization is particularly important in sub-micron technologies, where small channel size and a reduced oxide thickness cause electric fields to exceed the critical value of $4 \cdot 10^4 V/cm$, thus resulting in large electron-hole pair generation. Impact ionization currents can be expressed through the following equation [4]

$$I_{impact} = C_1(V_{ds} - V_{dsat}) I_d exp\left(-\frac{C_2 t_{ox}^{1/3} x_j^{1/2}}{V_{ds} - V_{dsat}}\right)$$

Device level simulations have been performed to extract parameters C_1, C_2 using the 2-D device level simulator PISCES [5]. These parameters have been then used in standard SPICE models.

There exist two main classes of substrates: one referred to as *low-resistivity substrate* which consists of a thick, high-resistivity epitaxial layer (d $\simeq 10\mu m$, $\rho \simeq 10 \div 15\ \Omega cm$) and a low-resistivity bulk($\rho \simeq 1m\ \Omega cm$). A second class, referred to as *high-resistivity substrate*, is composed of a uniformly doped layer with a resistivity coefficient of $20 \div 50\ \Omega cm$. Recently, low-resistivity substrates have been widely adopted for desirable latch-up suppression properties [4]. In general, it has been found that at low and medium frequencies, typically less than $5GHz$, substrate shows a resistive behavior. At higher frequencies on the contrary, the transport patterns are too complex to be accurately modeled using RC meshes [1].

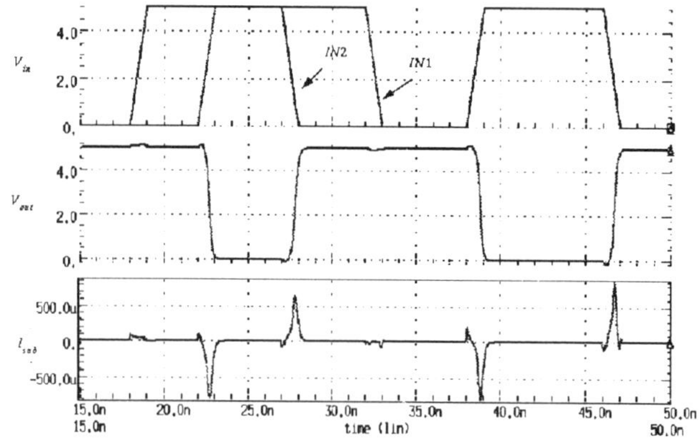

Figure 2: Substrate noise signature of the nand gate

The gate count of realistic digital circuits is typically in the millions. A complete extraction and detailed simulation of each individual substrate noise injector is impractical. If the substrate underlying the circuit can be approximated to be equipotential, then simultaneous substrate currents injected in different locations contribute in a cumulative fashion to the spurious potential sensed remotely. Each transition in the input pattern of each gate causes it to inject a unique noise signature. Hence, a set of S_C *signature classes* can be identified for a circuit, with $S_C \leq \sum_{c=1}^{N_L} s_c$, where N_L is the size of the technology library and s_c is the number of unique signatures per gate. Assuming that the activity of each signature class is known, then the resulting injection patterns due to all gates can be cumulatively evaluated and propagated to the sensing point.

As shown in Figure 2, when a circuit is realized with basic logic gates (nand, nor, not,..), two classes are necessary to capture the behavior of each gate with respect to substrate injection. In fact, it can be seen that the injection depends only on the output transition (rise and fall).

For a given technology a table (*Wave_Table*) is generated which stores, for each class, a waveform obtained from a SPICE simulation.

```
SIMULATE(Mapped_Network, Input_Vectors)
  foreach vector k ∈ Input_Vectors
    foreach node p ∈ PI
      SetNode(p, t_0^k, input_value(p))
      foreach node n ∈ FO(p)
        Mark(n, (t_0^k + wire_delay(p, n)))
    foreach instant t ∈ T^k
      foreach node n ∈ (DFO − PI)
        if IsMarked(n, t))
          SimulateNode(n, t)
          if (Event(n, t) = Rise or Event(n, t) = Fall)
            g_n ← gate_of(n)
            UpdateGateTable(g_n, Event(n, t), t)
            foreach fanout n' ∈ FO(n)
              Mark(n', t + wire_delay(n, n'))
          else
            SetNode(n, t, value_at_time(t − 1))
        else
          SetNode(n, t, value_at_time(t − 1))
  return Gate_Table
```

Figure 3: Event-driven logic simulator

3 Event-Driven Logic Simulation and Cumulative Noise Signal Computation

The simulator is described in Figure 3. The technology mapping of the Boolean network associated with the circuit (*Mapped_Network*) and a sequence of inputs (*Input_Vectors*) are passed to the simulator. The simulator produces the switching activity of each signature class as an output *Gate_Table*. The notation used in Figure 3 is the same as the one used in [6].

Let $\mathcal{N}$ be the set of nodes of the Boolean network, f_n the logic function associated with node n, and y_n the logic variable associated with n. Let us define $\mathcal{PI}$ as the primary input set, $\mathcal{FO}(n)$ as the set of fan-out of a node n, and $\mathcal{TFI}(n)$ as the set of transitive fan-in of a node n. Moreover, it is useful to introduce the ordered set $\mathcal{DFO}$, which contains all nodes in *Depth-First Order* from the outputs.

Let assume that every node n has been bound to a basic logic gate g_n by the technology mapping procedure. Hence, every arc in the graph is individually associated with a wire connecting two different gates of the circuit. Furthermore, for each gate g_n two signature classes, one for the rise and one for the fall transition are necessary.

SIMULATE performs internally an event-driven gate level simulation based on a *pure bounded wire delay model* [7]: in event-driven simulation each change in value of a wire in the circuit is regarded as an event, while a pure bounded wire delay model implies that there is exactly one delay element per gate input. Therefore, for a given sequence of input vectors, SIMULATE determines the output vector for each input vector, by evaluating for each node n the Boolean equation f_n, after considering the delay elements associated with g_n. This model permits to consider implicitly and accurately every possible spurious transition (*glitch*) in the circuit.

SIMULATE introduces a discretization of time based on the delay informations available from the gate library and the network topology. Let $t_0 = 0$ be the instant when the simulation begins, t_{max} the delay of the longest path in the network, and K is the number of input vectors, then the simulation interval is $\mathcal{T} = [t_0, t_{end}]$, with $t_{end} = K \times t_{max}$. Moreover, while processing the kth vector, just the sub-interval $\mathcal{T}^k = [t_0^k, t_{end}^k]$, where $t_0^k = (k-1) \times t_{max}$ and $t_{end}^k = k \times t_{max}$, has to be considered.

Suppose node m changes its output at t, each fan-out n of m is marked by **Mark**(n, t'), with $t' =$ *wire_delay*(m, n)), so that when SIMULATE processes the instant t', **IsMarked**(n, t') returns a true value and function f_n is evaluated by **SimulateNode**(n, t'). Hence, **SimulateNode**(n, t') computes the output value for node n at instant t', detecting if an event, i.e. a logic transition, has occurred, while **Event**(n, t') returns its type. When an event is detected for a node n at the instant t', SIMULATE calls **UpdateGateTable** to update the data associated to g_n within the table *Gate_Table*.

Gate_Table contains a complete trace of the switching activity of each signature class c within the digital circuit, for the overall simulation time interval $\mathcal{T}$. Let us call $tr_c(t)$ such a trace. Each signature class is also associated with a unique substrate injection waveform $w_c(t)$ stored in *Wave_Table*. The noise cumulative injection signal $i_c(t)$ for signature c is computed as the convolution

$$i_c(t) = \sum_{t'=t_0}^{t_{end}} tr_c(t - t') w_c(t')$$

The global injection noise $i(t)$ is obtained by accumulating the noise signals associated with all S_C signature classes as

$$i(t) = \sum_{c=1}^{S_C} i_c(t)$$

As a final step, the FFT $i(t)$ yields the power spectrum density $I(f)$.

4 Results

The methodology proposed in this paper is supported by several tools which implement the various functions described in Figure 1. The tools are implemented in C/C++ running under the UNIX operating system. Figure 4 shows the flow adopted for the testing of the methodology. A Boolean network is mapped onto a given technology using the synthesis tool SIS [8] and compiled in standard cell style within the OCTTOOLS environment. Event-driven simulation is performed on the mapped network and the resulting signal is convolved and transformed in the frequency domain. As a verification step, the layout and underlying substrate is fully extracted

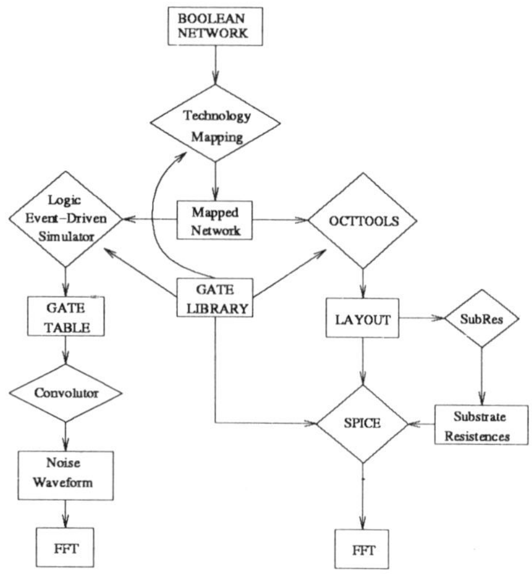

Figure 4: Testing SubWave methodology

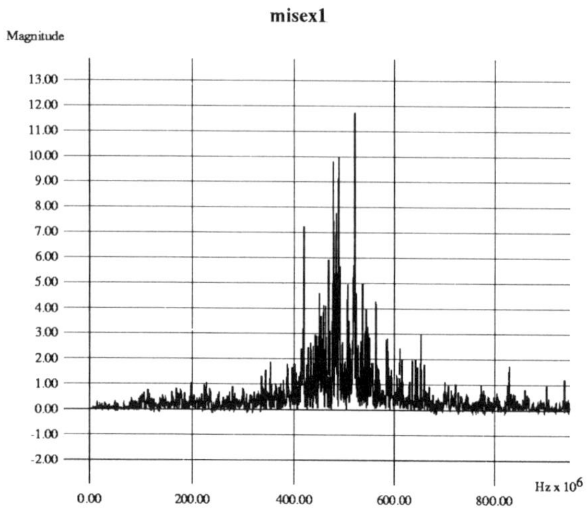

Figure 5: Substrate noise spectrum of **misex1**

with SUBRES [1] and the injected currents are evaluated at a common sensing point using SPICE.

Several circuits from the MCNC91 benchmark suite have been tested using one thousand randomly generated input vectors. In Table 1 the CPU times for the generation of the models and for the verification step are reported for a DEC AlphaServer 2100 5/250. The model for injected switching noise obtained through full simulation was in general two order of magnitudes slower than that obtained with SUBWAVE. A direct comparison of the waveforms obtained with the two approaches shows a contained error estimated to be less than 10% for all the benchmarks. The largest circuits, such as the 16 bit multiplier, could not be simulated using SPICE due to their extreme complexity. Figure 5 shows the spectrum generated for the `misex1` circuit. It results that the most of the spectrum is centered around the 500 MHz frequency. The same occurred for all the other benchmark mapped on the same library based on gates with a common 1 ns delay. This suggests that the substrate noise spectrum is mainly dependent on the technology rather than other

Mapped_Network				SubWave	SubRes	Spice
Circuit	function	I/O	Gates	*sec*	*sec*	
misex1	pla	8/7	70	75	257	395
z4ml	2b add	5/16	75	54	69	509
mux	mux	21/1	115	75	129	9139
my_adder	16b add	33/17	242	134	3383	-
alu2_cl	alu	10/6	506	430	10105	-
C1355	ecc	41/32	1215	202	-	-
C6288	16b mlt	32/32	2731	1426	-	-

Table 1: Results on MCNC91 Benchmarks

design parameters (clock, topology). Hence, noise models based on clock rate may account in general for less than 10-20% of the energy injected in the substrate.

5 Conclusions

A novel methodology has been proposed for the synthesis of compact injection noise models for large logic circuits. The injection signatures associated with given gates and all possible input transitions are evaluated and stored. All noise generating devices are cumulatively accounted for and the resulting switching noise is efficiently derived using a gate-level event-driven simulator. Finally, the FFT of the switching noise is performed. A set of standard benchmarks has been successfully used to test the accuracy of the models by comparing them against full extraction of the circuit and simulation. A strong dependency between the noise spectrum and the gate delay has been pointed out.

Acknowledgments The authors would like to thank Ranjit Gharpurey and Alex Saldanha for their support and useful discussions. A special thank goes to Alberto Ferrari for his invaluable technical support.

References

[1] R. Gharpurey and R. G. Meyer, "Modeling and Analysis of Substrate Coupling in ICs", in *Proc. IEEE CICC*, pp. 125–128, May 1995.

[2] B. R. Stanisic, N. K. Verghese, D. J. Allstot, R. A. Rutenbar and L. R. Carley, "Addressing Substrate Coupling in Mixed-Mode ICs: Simulation and Power Distribution Synthesis", *IEEE Journal of Solid State Circuits*, vol. SC-29, n. 3, pp. 226–237, March 1994.

[3] S. Mitra, R. A. Rutenbar, L. R. Carley and D. J. Allstot, "Substrate-Aware Mixed-Signal Macro-Cell Placement in WRIGHT", in *Proc. IEEE CICC*, pp. 529–532, May 1994.

[4] C. Hu, *VLSI Electronics: Microstructure Science*, volume 18, Academic Press, New York, 1981.

[5] M. R. Pinto, C. S. Rafferty, H. R. Yeager and R. W. Dutton, *PISCES-IIB*, Stanford University, Stanford, CA, 1985.

[6] R. K. Brayton, G. D. Hachtel and A. L. Sangiovanni-Vincentelli, "Multilevel Logic Synthesis", *Proceedings of the IEEE*, vol. 78, n. 2, pp. 264–300, February 1990.

[7] L. Lavagno and A. Sangiovanni-Vincentelli, *Algorithms for synthesis and testing of asynchronous circuits*, Kluwer Academic Publishers, 1993.

[8] E. M. Sentovich, K. J. Singh, C. Moon, H. Savoj, R. K. Brayton and A. L. Sangiovanni-Vincentelli, "Sequential Circuit Design Using Synthesis and Optimization", in *Proc. Intl. Conf. on Computer Design*, pp. 328–333, October 1992.

Forward Power Annotation on Physical Layout Floor-plan

Roberto Zafalon, Carlo Guardiani, Marco Casale Rossi, Roberto Rambaldi[+]

SGS-THOMSON Microelectronics,
v. C.Olivetti, 2 ,
20041 Agrate (MI), ITALY

[+]Università di Bologna,
Dipartimento di Elettronica Informatica e Sistemistica, Viale Risorgimento 2, Bologna, ITALY

Abstract

A design methodology that provides the information about the distribution of power on the physical layout is presented in this paper. Unlike previous work, this information can be used to detect possible power distribution problems early in the design cycle. The methodology consists in creating the required links between a gate-level, probabilistic power estimation tool and a floor-planner. In this way the power consumption data can be properly localized with respect to the regioning generated on the physical layout before the actual detailed placement and routing occurs. The application of this technique to the design of a low-power IDCT circuit [22] on a 0.5μm sea-of-gate CMOS technology is presented in this paper, showing the considerable advantages of the proposed method.

Introduction

The application of the top-down design concept to digital VLSI circuits and systems has been the foundation of the outstanding progress of microelectronics during the last decade. A key factor for the success of top-down design methodology is the capability of resolving most of the design issues at higher level of abstraction [1][16]. In this way the most critical design performance constraints can be met without costly loops across the entire design cycle. In this paper we present a design methodology implementing an effective predictor of power consumption distribution across the physical layout of the circuit before the actual detailed placement and routing information is available. This predictive capability is of great importance to detect and avoid such insidious problems as incorrect power rail sizing, supply pads distribution and hot spots early in the design cycle [9]. Therefore the proposed methodology fits snugly in a "power conscious" design flow, spanning from process-level technology optimization up to high-level low power system design techniques [2]. The approach proposed in this paper consists in generating the gate level power consumption estimates by using a probabilistic power simulation tool [21]. The power data thus generated are then automatically annotated on the optimized physical regioning generated by a floorplanner. The resulting power consumption map can then be graphically represented and optimized either manually or automatically.

Early design exploration

As was already mentioned the above described forward power annotation (FPA) methodology allows a qualitative exploration of the design space in the initial phase of the design process. This capability is essential to shorten the design cycle by dramatically reducing the number of iterations that are required to achieve a consistent front to back estimation of the circuit performance. In particular FPA does not reduce the need for a final detailed power verification, based on accurate, post-layout simulation (e.g. to check for the correct arrangement of the power budget, internal VDD busing and electromigration constraints [7]), but it rather allows to initially select the optimal architecture and circuit partitioning, thus decreasing the need for a complex power optimization at gate or transistor level. In fact it has been recently observed that circuit and gate-level optimization techniques typically have less than a 2x impact on power [3][4], while published results on architecture and system-level strategies offer savings of 10-100x or more [2][5][6].

Front-end floorplanning

The proposed design flow is based upon Front-End Floor-Planning (FEFP) technology, which is emerging as a fundamental technique filling the gap between front-end and back-end of the design cycle. In particular the application of FEFP techniques in the FPA design flow was essential to provide:

1. Accurate RC information for the area/timing driven synthesis tool
2. Distribute the net toggling activity across the floorplan blocks and display power partitioning information on the back-end regioning.
3. Display power distribution map on the physical layout after detailed placement.

Power consumption model

The approach described in [13] (UniPower) has been used to estimate the power consumption in CMOS VLSI circuits, at both gate and macro level. This approach is based on a power analysis tool that generates a model of the energy consumption associated with each transition event. This model incorporates the effect of output capacitive loading (including the

0-7803-3177-6 $5.00 © 1996 IEEE

IEEE 1996 CUSTOM INTEGRATED CIRCUITS CONFERENCE

correlation between two different outputs) as well as input slew sensitivity. As an example the equation representing the energy consumption associated with a rising edge event on the clock input (CP) of a D-FF, with the D input high that is obtained after characterization and fitting is as follows (energy in [pJ]):
1) "CPr_Qlh" 6.30+2.85*rise(CP)+24.74*c(Q)+0.0 *c(QN)
where CP is the clock input, Q and QN, respectively, the direct and complemented output.

Link to the floor-planner

A key point of the proposed methodology consists in creating the required links between the gate-level, probabilistic power estimation tool [8][11][12] and the floor-planner [27]. In this way the power consumption data can be properly localized with respect to the regioning generated on the physical layout before the actual detailed placement and routing occur.
Unlike previous work, this information can be used to detect possible power distribution problems early in the design cycle. The main data entries in the flow come from the logic design on one side, and from the cell library which provides the suitable power characterized model on the other side. More specifically:
A) design: the logic netlist (edif/verilog/VHDL) goes through the front-end floor-planner. Two kind of information can be obtained:

1. the partitioning of the netlist into regions, one for each functional block (hierarchical depth can be specified)
2. the detailed placement of all the components together with the interconnect RC information.

B) library: the power-model for the cells goes through an analysis tool in conjunction with the interconnect RC information, in order to generate an instance-based energy dissipation report which may include:
- the average power dissipation for each cell, assuming a given load estimation and an activity rate of 1 MHz at 1ns input slew (in our case).
- the power dissipation for each cell, assuming a given load and the resulting activity from a logic co-simulation (UniPower or PowerSim [20]) or probabilistic [9][10] estimation (e.g.: Design Power [21]).
The results of A.1+B are then processed by a specific graphic interface (PowerDisplay), which assigns the proper power dissipation to each region (i.e. functional block). PowerDisplay also allows the user to specify a toggling activity percentage and an operating frequency for each region as shown in FIG.1.In alternative, PowerDisplay can be fed by A.2 followed by B, assigning the proper power dissipation to the relevant grid-based binding of instances (i.e.: user specified BIN). The user can still specify a toggling activity percentage and an operating frequency but for the overall design as shown in FIG.2.It is worth to note that PowerDisplay only needs to process A.1 or A.2 in addition to the results of a power analysis tool, reported on either instance or region - based. By estimating the gate complexity and the power dissipation (in terms of nand2 equivalent gates) for each region, i.e. functional block, only the nand2 power characterization and a desk-calculator would be necessary.While, by estimating the power dissipation as:

```
foreach instance
    foreach event
        sumof power
    endforeach
    divided by number of events
endforeach
```

a complete power characterization of the cell library is required in order to provide accurate data for the described methodology.
Of course as the design cycle is transversed and more details about the physical domain become available the power distribution information can be correspondingly refined [16][19] thus incrementally improving the accuracy of the results.

The IDCT design as an application case

An application of the FPA technique to the design of a low-power IDCT circuit on a 0.5μm sea-of-gate CMOS technology has been used to demonstrate the advantages of the method.The discrete cosine transform, and its inverse transform, are well known algorithms used by both the CCITT H.261 and MPEG standards for video transmission to exploit the spatial redundancy to compress images [24][25][26].Both of them treats the pixel data by a unit block which is a matrix of 8 rows X 8 columns. An IDCT processor is a good test for a power analysis because of the huge amount of computational power required at the different targets that the video compression can aim to: low-power for very low-bit rates transmission lines or high performance for HDTV quality. If we consider the CCITT H.261 recommendation which best suites the transmission at low-bit rates over very slow bit-lines, a 8X8 block of the inverse-transform, is the same as 64X64 = 4096 floating point additions and multiplications, or in general 8KFLOP. At a speed rate of 7 frame per second the computational power required is equivalent to 27 MFLOP/s.
A careful architecture should be developed to achieve this performance at extremely low-voltage power supply [15][18] (typically ranging from 0.8 volt to 1.1 volt). Computational power increases as the frame rate and frame dimensions increase up to the 1920 X 1152 images for HDTV. In Table 1 the results of the low power version of the IDCT processor are compared with those obtained after the completion of the full back-end flow, showing a good agreement between the values predicted with the proposed methodology and the reference ones. The hazard scaling is a glitch shape reducing factor, aimed to give the glitch peak value as a typical ratio of VDD. Hazard=0.5 means how the glitch peak is bounded to VDD/2.

Table 1: Total Power consumption results: @ VDD=3.3V, Freq=1MHz

Predicted (PowerDisplay)	Measured (full layout)
Hazard Scaling 0.5	Hazard Scaling 0.5
Power ....3.3 mW	Power .... 3.252 mW

Conclusion

A novel methodology (Forward Power Annotation), allowing to quickly predict with good accuracy the distribution of the power consumption across the physical layout of complex ICs, has been presented in this paper. This methodology can be used to sensibly reduce the design cycle time by dramatically decreasing the inconsistency between high level power estimation and circuit level design validation. The application of the FPA methodology to the power conscious design of an IDCT circuit has been presented, showing the considerable advantages of this technique.

References

[1] M.Pedram, J.Rabey et al., "Power Conscious CAD Tools and Methodologies: A Perspective", survey of IEEE Proceedings 1995.

[2] Jan Rabaey et al., "An Integrated CAD environment for Low-Power Design", University of California, Berkeley, 1994.

[3] A. Chandrakasan, *"Low-power CMOS Design"*, PhD Thesis, University of California, Berkeley, Aug 1994.

[4] F. Somenzi et al., An Application of ADD-based Timing Analysis to Combinational Low Power Re-Synthesis", 1994 Int. Workshop on Low Power Design, April 1994.

[5] B. Ricco' et al., "Analysis of Hazard Contributions to Power Dissipation in CMOS IC's", 1994 Int. Workshop on Low Power Design, April 1994.

[6] A. Chandrakasan, R. Brodersen et al., "Design of Portable Systems", Proceedings of CICC '94, San Diego. May 1994.

[7] Harry J.M. Veendrick, "Short-Circuit Dissipation of Static CMOS Circuitry and Its Impact on the Design of Buffer Circuits", IEEE Journal of Solid State Circuits, August 1984.

[8] M. Dana et al, "Probabilistic power consumption estimation in digital circuits", Workshop on Power, Timing, Modelling (Paris Sep. 1992), TELECOM Paris University.

[9] F.N.Najm et al, "Probabilistic simulation for reliability analysis of CMOS VLSI circuits", IEEE Trans. on CAD, Vol. 9, No. 4, Apr 1990.

[10] Srinivas Devdas, Kurt Keutzer and Jacob White, "Estimation of Power Dissipation in CMOS Combinational Circuits Using Boolean Function Manipulation", IEEE Trans. on CAD, Vol. 11, No. 3, March 1992.

[11] R. Burch, F. Najm, P. Yang and D. Hocevar, "Pattern-independent current estimation for reliability analysis of CMOS circuits" in Proc. 25th Design Automation Conf., June 88, pp. 294-299.

[12] S. Chowdhury and J.S. Barkatullah, "Estimation of maximum currents in MOS IC logic circuits" IEEE Trans. on CAD, vol. 9 June 90, pp. 642-654.

[13] Pawan Kulshreshtha, Roberto Zafalon, Milena Magni, "UNIPOWER: A Power Analyser for CMOS VLSI Circuits", Workshop on Power, Timing, Modelling (Paris Sep. 1992), ST Central R&D - Agrate.

[14] A. Chandrakasan, T. Burd, A. Burstein, S. Narayanaswamy, S. Sheng, R. Brodersen, "System Design of a Multimedia I/O Terminal", VLSI Signal Processing, IV, IEEE Special Publications, 1993.

[15] E. Tsern et al.,"Video Compression for Portable Communication Using Pyramid Vector Quantization of Subband Coefficients", Proc. of IEEE workshop on VLSI sign. proc., Veldhoven, Oct 1993.

[16] P. Landman and J. Rabaey, "Power Estimation for High-Level Synthesis," European Design Automation Conference, Paris, France, February 1993.

[17] A. Dharchoudhury, "Worst Case Analysis and Optimization of VLSI Circuit Performances", IEEE Transaction on CAD, vol. 14, n 4, April 1995.

[18] S. R. Powell et al, "Estimating Power Dissipation of VLSI Signal Processing Chips: The PFA Technique," *VLSI Signal Processing IV*, pp. 250-259, 1990.

[19] R. Mehra, J. Rabaey, "Behavioral Level Power Estimation and Exploration", Int. Workshop on Low Power Design, 1994.

[20] System Science Inc., Palo Alto (CA), "PowerSim: Power Simulation System User's Manual", v 2.002, June 1995.

[21] Synopsys, "DesignPower Reference Manual", v 3.2b February 1995.

[22] K.R. Rao, P.Yip, "Discrete Cosine Transform: Algorithms, Advantages,Applications" Academic Press, San Diego, 1990.

[23] G.Proakis, D. G.Manolakis, "Digital Signal Processing. Principles, Algorithms, and Applications", 2nd Edition, Macmillan, New York, 1992

[24] B. G. Lee, "A New Algorithm to Compute the Discrete Cosine Transform", IEEE Transactions on Acoustic, Speech and Signal Processing, Vol. ASSP-32, No.6, pp. 1243-1245, December 1984

[25] C. Loeffler, A. Ligtenberg, G. Moschytz, "Pratical, Fast 1-D DCT Algorithm with 11 Multiplications", Proceedings of ICASSP-89, Vol.2, pp. 988-991

[26] T.Masaki et al., "VLSI Implementation of IDCT amd Motion Compensation for MPEG2 HDTV Video Decoding", IEEE Transaction on Circuits And Systems for Video Technology, Vol. 5, No. 5, October 1955, pp. 387-395

[27] Compass Design Automation, "ChipPlanner, Reference Manual", v 8r4.8, June 1995.

FIG.1 **IDCT Power dissipation Map (region based)**

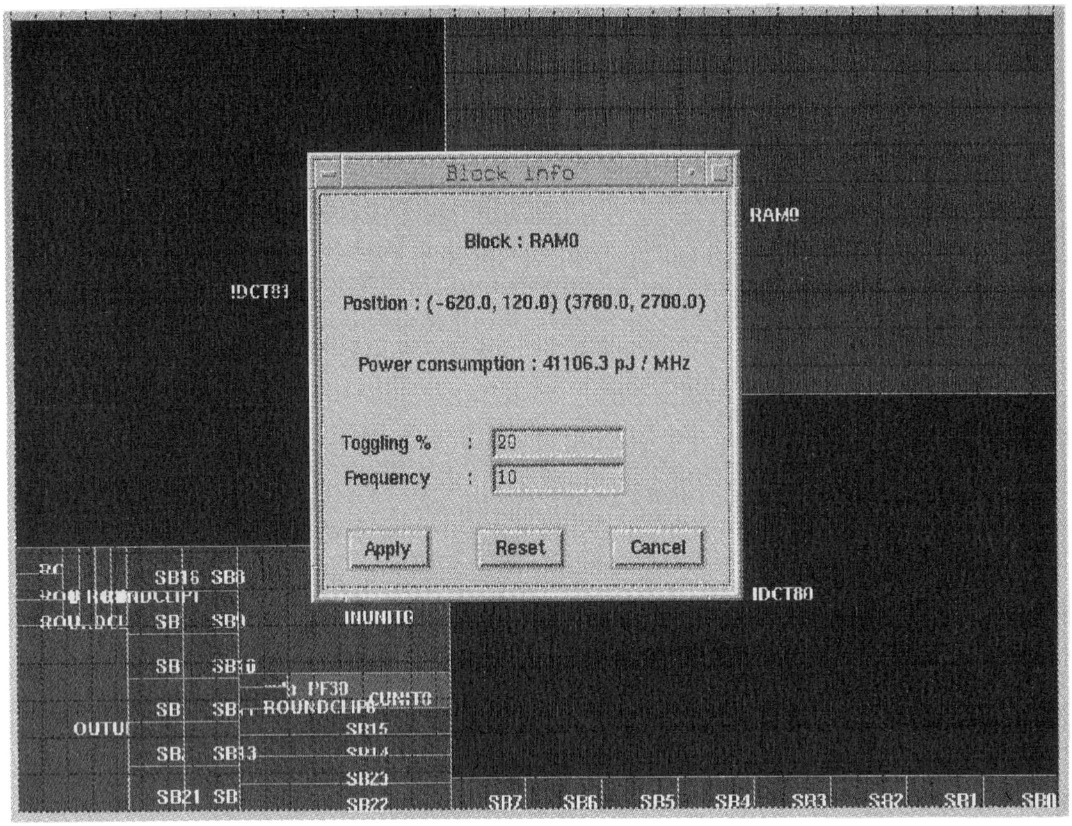

FIG.2 **IDCT Power dissipation Map (grid placement based)**

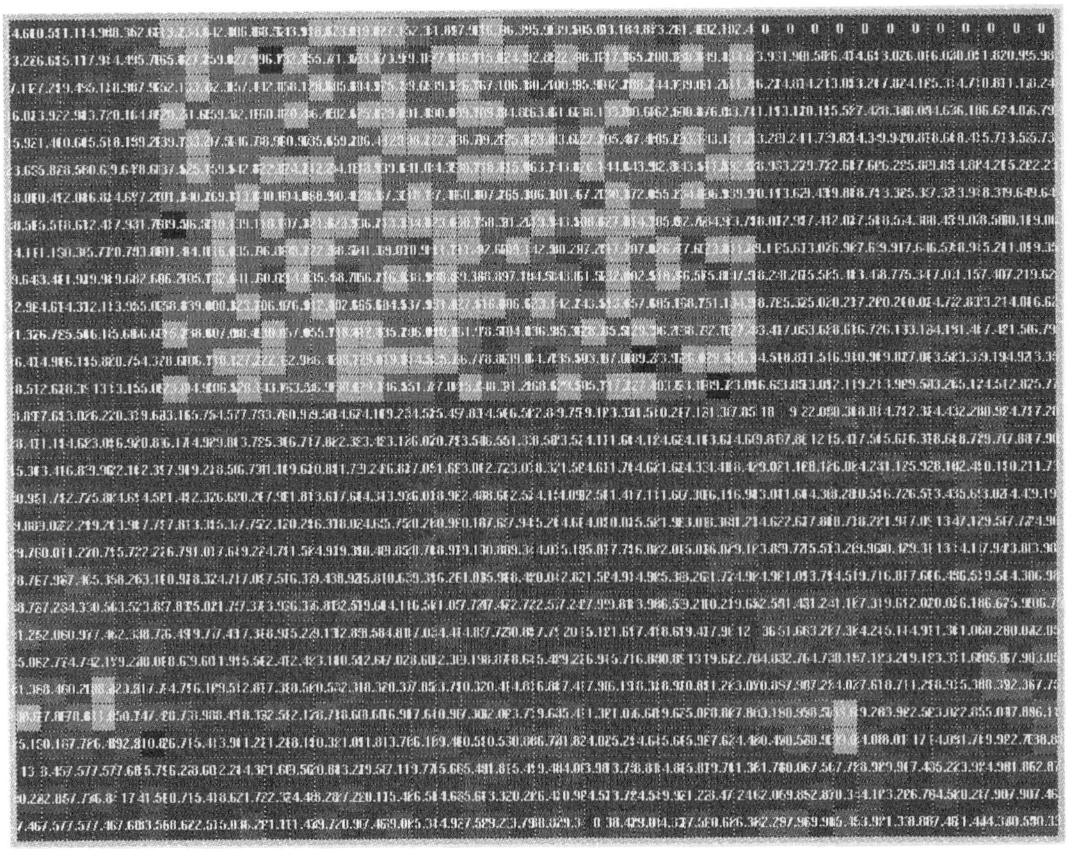

17.4.4

Efficient power analysis of combinational circuits

Shankar Krishnamoorthy and Adel Khouja
Synopsys Inc.,
700, E. Middlefield Road,
Mountain View, CA 94043-4033

Abstract

Designers of low power chips are finding it increasingly useful to obtain estimates of average power consumption early in the design phase. They need a fast power estimation capability which gives reasonably accurate power estimates. In this paper, we present some ideas to obtain an efficient implementation of probabilistic power analysis for combinational circuits. These ideas are based on the observation that in order to obtain a fast implementation, it is very important to control the sizes of the Reduced Ordered Binary Decision Diagrams (ROBDDs) that are created during analysis. However, controlling sizes of ROBDDs could impact the accuracy of the result. We explore the consequences of making the trade-off between performance of the implementation and accuracy. Our experimental results illustrate that our implementation can provide estimates that are within 5% of a commonly used probabilistic method and runs 10 times faster, on average. Our method also scales very well for large circuits.

1 Introduction

The growing popularity of portable devices has contributed significantly to research and development in the area of low power circuit design and CAD tools that enable low power design. Components that dissipate less power significantly increase the "battery life" of portable applications, which is a major competitive advantage in the domain of portable and hand-held devices. Several techniques for power analysis have been proposed using circuit simulation and probabilistic approaches. Power analysis using probabilistic methods can be considerably faster than simulation-based analysis. However, there is an accuracy issue that needs to be dealt with. Innovative methods have been proposed to improve the accuracy of probabilistic analysis by accounting for different forms of correlation amongst signals in the circuit.
Designers who use a high-level design methodology are finding it increasingly tedious to design low power chips. Presently in the design exploration phase, they have to use gate-level or transistor-level simulation tools in order to evaluate an architecture or algorithm. Simulation-based tools are accurate but time-consuming. The need for a fast power analysis capability is obvious at the exploration stage, even if it comes at an accuracy cost. We shall refer to this observation as the *speed-accuracy trade-off* in power estimation.
In this paper, we shall describe an efficient implementation which achieves a significant improvement in analysis speed over a conventional power analysis implementation. For most circuits, this improvement is accompanied only by a marginal loss of accuracy. The rest of the paper is organized as follows: In section 2, we present a classification of existing power analysis methods and review techniques proposed for speeding up analysis. In section 3, elements of the proposed implementation will be presented and experimental results will be reviewed in section 4.

2 Previous work

The switching power for a circuit is given by the following equation:

$$P_{sw} = 0.5 \times Vdd^2 \times \sum_{i \in nets} C_i \times f_i \qquad (1)$$

Since C_i - the capacitance seen at a gate output is known, all power estimation methods attempt to accurately determine the value of f_i - the switching activity for every gate output in the circuit. Power estimation methods can be classified into simulation-based and probabilistic methods. A comprehensive survey of these methods as well as definitions of basic concepts in power analysis is provided in [1]. Since our focus is on efficient power analysis, we shall review existing probabilistic methods with respect to efficiency.
Power analysis methods can be broadly classified into two categories: *local propagation* and *global propagation*. These categories reflect the computation methods employed in the power analysis techniques. Local propagation techniques rely on computing the switching information of a gate in the logic circuit, by looking only at the immediate fanins of the gate. A single pass is made over the entire circuit from primary inputs to primary outputs and the switching information is computed at the output of every gate in the circuit. Examples of local propagation methods are described in [2], [3] and [4]. The techniques described in [3] and [4] store additional information at every gate (spatio-temporal correlation coefficients and probability waveforms respectively) in order to improve the accuracy of local propagation, while incurring a performance and memory overhead. (In this paper, the word 'performance' will be used to signify CPU speed or run time of the power analysis implementation.)
The second class of methods can be referred to as global propagation methods. In these techniques, the logic function of every gate output is constructed with respect to the primary inputs that transitively fanin to the gate. By considering the logic function with respect to the primary inputs, any spatial correlation that exists at the immediate fanins of the gate is included in the computation. These 'global' functions are then manipulated to obtain the switching information for the output of each gate. Examples of power estimation algorithms based on global propagation, can be found in [2], [5] and [6].
The global propagation methods described in the literature use Reduced Ordered Binary Decision Diagrams (ROBDDs) in order to represent and manipulate logic functions. ROBDDs have an exponential space complexity with respect to number of variables. This implies that for many cases, ROBDDs can grow very quickly and this could affect the performance of the power estimation method. It is also very likely that ROBDDs cannot be built for some gates in large combinational circuits, due to space limitations. The performance of all the global propagation techniques for power estimation is influenced by two factors: the cost of building ROBDDs and the cost of processing them. Therefore, in order to obtain an efficient implementation, it is very important to be able to control and manage the number of nodes in the ROBDDs that get created.
It is difficult to determine whether or not local propagation methods outperform global propagation methods or vice-

0-7803-3177-6 $5.00 © 1996 IEEE
IEEE 1996 CUSTOM INTEGRATED CIRCUITS CONFERENCE

versa. The performance gains that can be obtained by eliminating the need to build large functions with respect to primary inputs (in the global propagation methods) seems to be offset by the volume of information that needs to be computed and maintained at each gate in the circuit (in the local propagation methods). In this paper, we present a framework for an efficient implementation of probabilistic power analysis (using global propagation) by controlling am managing the memory usage of the ROBDD package.

3 Efficient implementation of power analysis

In this section, we review the details of our implementation of power analysis. The algorithm that is proposed is not bound to any one power estimation formulation. It is a generic method that can be used for all global propagation techniques. In section 3.1, we introduce the power estimation model that is used to obtain average power consumption. Section 3.2 will illustrate the procedure of controlling the memory usage of ROBDDs in order to obtain an efficient implementation. Section 3.3 describes the actions that are taken when memory overflow occurs during the analysis.

3.1 Power estimation model

We use the power estimation model proposed in [2] for our system. We introduce two terms in order to review the ideas presented in [2].

Signal probability: For a node *f* in a circuit, the signal probability of *f* (denoted by P(*f*)) reflects the fraction of the time that the node takes the value 1 as opposed to the value 0. If node *f* implements a function F, then P(*f*) can be computed by the recursive formulation:

$$P(f) = P(x) \times P(F_x) + P(\bar{x}) \times P(F_{\bar{x}}) \qquad (2)$$

Transition density: For a node *f* in a circuit, the transition density of the node (denoted by D(*f*)) is the average number of transitions observed at the output of *f* (0 to 1 and 1 to 0) per unit time. If the node implements the function F, then D(*f*) is given by:

$$D(f) = \sum_{x_i \in INPUTS(F)} P\left(\frac{\partial F}{\partial x_i}\right) \times D\left(x_i\right) \quad \text{where,}$$

$$\frac{\partial F}{\partial x} = F_x \oplus F_{\bar{x}} \qquad \text{(boolean difference)} \qquad (3)$$

The transition density of a node is exactly its switching activity. This is multiplied with the capacitance seen at the output of the node to obtain the average power dissipation at that node in the circuit.

In our system, we adopt the global propagation extension of [2] in order to account for the spatial correlation between intermediate nodes in the circuit. Hence the function F for the node *f* described above is expressed in terms of the primary inputs that transitively fanin to *f*. Also, a zero delay model is assumed for the analysis in order to enable fast turnaround time for the power estimation. As a result, the power due to glitching is ignored (glitching averages about 15-20% of the total power dissipation). Compared to other global propagation methods, the transition density formulation is amenable to a fast implementation. The computations of signal probability and boolean difference can be implemented efficiently using ROBDDs. However, the method does not account for primary inputs switching simultaneously and this impacts the accuracy of the power estimation[6].

3.2 Controlling memory usage of ROBDDs

The efficiency of the ROBDD package has a big impact on the efficiency of the power analysis implementation.We used an ROBDD package similar to the one described in [7] for our implementation. The package provides the primitive functions required to build ROBDDs like AND, OR, EXOR, NOT etc. The package also allows the users to specify a capacity or an upper bound (in terms of ROBDD nodes) for memory usage. This means that while performing an ROBDD operation (such as AND), if the memory limit set by the user is exceeded, then the operation is aborted and the user is informed about the memory overflow. All the ROBDDs that existed prior to execution of the failed operation continue to exist after the operation has been aborted.

In our power analysis implementation, we place an upper bound on the memory usage of the ROBDD package instead of trying to control the size of each individual ROBDD. This is because for any combinational circuit, ROBDDs of the gates have varying sizes. ROBDDs for some gates are small while others are large. Setting a memory limit on the package as opposed to each individual ROBDD has the advantage that ROBDDs of different sizes can be built and processed (as long as it is within the memory limit of the package), resulting in no loss in accuracy. For those circuits where memory overflow occurs, a memory recovery strategy must be devised in order to continue the analysis with some loss in accuracy. In order to describe the bound used in our implementation, we first describe the nodes-per-gate metric.

Nodes-per-gate (μ) reflects the memory (expressed in terms of ROBDD nodes) that is allocated per gate in the circuit. This is an invariant for the power estimation model and is determined by performing an empirical analysis over several combinational circuits of different sizes.

The bound placed on the memory usage (B) for a circuit is given by the equation:

$$B = \mu \times C \qquad (4)$$

C is the number of gates in the circuit. This formulation helps us to tune the memory limit based on the size of the circuit. The value of μ must be chosen carefully. If the value is too low (or too high), similar symptoms to a low (or a high) static bound are exhibited. The experimental results section illustrates how the accuracy and runtime are impacted by different values of μ.

3.3 Memory recovery strategies

In this section, the recovery technique that is employed during memory overflow will be described. The requirements of such a method with respect to fast power analysis will be elaborated and an algorithm will be presented that attempts to meet these requirements.

3.3.1 Management of ROBDDs

Since a bound has been placed on the memory usage, it is important to prevent any wasteful usage of ROBDDs. This is achieved by enforcing the paradigm that ROBDDs are created only on-demand and are freed as soon as they are not needed. If the ROBDD for a gate in the circuit is created, then it is freed as soon as all the fanouts of that gate have been processed. This ensures that memory overflow occurs only when one of the following cases occurs:

- The memory usage of the ROBDD package was well below the limit and the creation of the next ROBDD caused the overflow. We shall refer to this case as "explosive ROBDD growth".

- The memory usage of the package was nearing the upper bound and the overflow was caused by the next ROBDD created (small or large). We shall refer to this case as "gradual ROBDD growth".

Both the aforementioned scenarios occur very frequently in a global propagation method. The memory recovery strategy that is devised must not only be efficient (both in the number of times it is called as well as time spent in each invocation) but should also have minimal impact on accuracy of the computation.

3.3.2 Terminology

A **frontier** is defined as a set of gates in the circuit which have one or more unprocessed fanout nodes. As gates are processed from primary inputs to primary outputs, the frontier is continuously updated. Whenever an ROBDD is created(deleted) for a gate, it is added to (removed from) the frontier.

A **reduction factor** is a metric that is used to indicate a target memory level for the recovery strategy. It takes a value between 0% and 100%. ROBDDs must be freed until the memory usage of the ROBDD package falls below the target memory level.

A **startpoint** is defined as a gate whose ROBDD is freed during the memory recovery process. This gate is then treated as a primary input for the purposes of analysis of the unprocessed gates in the circuit. Spatial correlation between the output of the gate and other signals in the circuit is ignored, resulting in some loss of accuracy.

3.3.3 Description of memory recovery method

The algorithm `blowup`, which is used for memory recovery is described in Figure 1. The function `mem_usage` when applied to the ROBDD manager returns the memory consumption[1] of the ROBDD directed acyclic graph (DAG). As the function is successively called, the memory usage is monitored until it falls below the target memory level. The complexity of the recovery strategy is O(N) where N is the number of gates in the circuit. In reality, the size of the frontier is usually a small percentage of the total number of gates in the circuit[2]. The `blowup` algorithm is greedy with respect to memory recovery. However, the accuracy loss is limited by the reduction factor. The value of the reduction factor is empirically determined from analysis of several combinational circuits. If it is set to a high percentage (say 80%), then it may result in marginal memory reduction. If it is set too low (say 10%), then too many startpoints may be created resulting in significant loss in accuracy.
For the "explosive ROBDD growth" case (described in 3.3.1), the algorithm only frees the fanin ROBDDs of the node where the overflow occurred. This is usually sufficient to reduce the memory usage below the target memory level. For the "gradual ROBDD growth" case, `blowup` may have to create more startpoints in order to get the desired memory reduction. For most circuits, we have seen that the percentage of ROBDDs freed with respect to ROBDDs at the frontier varies between 1% (for the "explosive growth" case) and 20% (for the "gradual growth" case). Both these cases can appear during analysis of the same combinational circuit.
Figure 2 illustrates the profile of memory usage of the ROBDD package before and after the recovery method. The memory overhead of the recovery technique proposed in `blowup` is that some additional ROBDD variables are added corresponding to the startpoints. This is represented by item D in the profile. The area between B and C is the recovered memory as a result of application of `blowup`. As described in section 3.3.2, the memory limit of the package (item A) is variable depending on the size of the circuit.
The analysis algorithm is a typical implementation of a global propagation probabilistic technique. For brevity, we have omitted the description of the overall power analysis algorithm. It is explained in detail in [8].

```
algorithm blowup(C, M, n, r) {
/* C is the circuit */
/* n is the node at which the overflow
      occurred */
/* M is the ROBDD package manager */
/* r is the reduction factor */
F = get_frontier(C);
Sort F in decreasing order of ROBDD sizes;
K = K0 = mem_usage(M);

while (K > r*K0) {
      s = get_next_node(F);
      B = BDD(s);
      free(B);
      create_startpoint(s);
      K = mem_usage(M);
}
Build ROBDD for n
return;
}
```

Figure 1: Algorithm for memory recovery

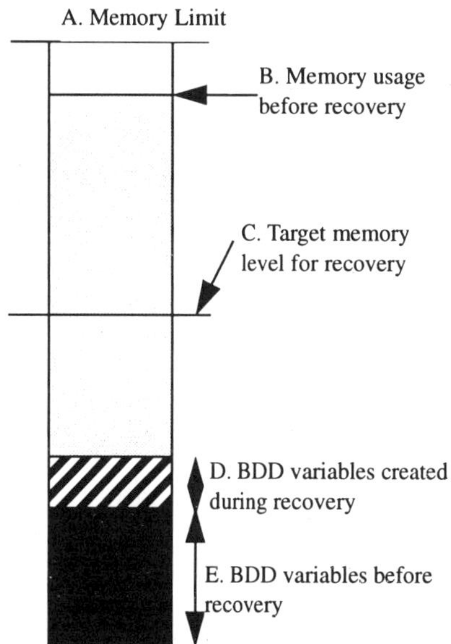

Figure 2: Profile of BDD memory usage

1. number of ROBDD nodes
2. 15% on average for our experiments

4 Experimental results

In this section, we report the results of the experiments that were conducted. Our hypothesis was that the `blowup` algorithm could be incorporated within a power analysis implementation to obtain power estimates considerably faster than existing global propagation techniques with a minimal loss in accuracy. To verify this, the nodes-per-gate (μ) parameter was iteratively assigned a value from the range (1, 5, 50, 500, and 1250).

Setting($\mu = 1$) represents a purely local propagation through the circuit where spatial correlation is completely ignored. This point corresponds to the local propagation method described in [2], and it provides a lower bound on the runtime of the power analysis implementation. Setting ($\mu = 1250$) represents a near-infinite memory limit where ROBDD sizes are "unbounded". Theoretically, this point should correspond to the global propagation method described in [2] as both techniques account for all spatial correlation between internal signals of the circuit. Therefore, the power estimates obtained from this data-point are considered the "accurate" power values for comparison against results for other values of μ.

For the experiments, we used ISCAS and IWLS benchmark circuits. The circuits were mapped to the lsi_10k library to produce gate-level implementations. The primary inputs were assumed to be pseudo-random signals. Each input was assigned a static-probability of 0.5 and a transition density of 0.025(transitions per ns). The tests were run on SPARC-10 class machines.

As expected, for all the tested circuits, higher values of μ resulted in longer runtimes and more accurate average power results. Furthermore, increasing μ had a diminishing benefit on the accuracy of the power results. In fact, there was usually a certain point beyond which the power results did not appear to change, implying that higher values of μ (beyond that point) were only producing higher runtimes without a noticeable improvement in the accuracy of the results. Empirical evidence showed that at ($\mu = 50$) the power values had usually converged to the "accurate" value" (i.e. power errors were small).

Table 1: Runtime/accuracy results

Circuit	Gate count	$\mu = 50$		$\mu = 1$ (Local propagation)	
		Runtime speedup[1]	Power error[1]	Runtime speedup[1]	Power error[1]
C1355	259	14.3x	4.2%	93.0x	36.8%
C1908	208	5.3x	1.4%	22.4x	10.2%
C3540	527	13.3x	12.4%	146.8x	31.6%
C499	195	13.9x	4.5%	101.0x	3.7%
C5315	706	1.9x	0.0%	17.3x	5.6%
C6288	1654	19.9x	291.1%	2145.7x	316.3%
C7552	889	19.6x	0.1%	169.9x	11.6%
C880	181	2.8x	0.2%	8.3x	0.9%
des	1629	1.0x	0.0%	1.6x	18.0%

1. Versus runtime and power results for μ=1250 (global propagation with "unbounded" ROBDDs)

Table 1 shows the runtime speedup and power estimate errors for two values of μ (1 and 50) for some circuits that we tested. A complete set of benchmark results is included in [8]. For each value of μ, the runtime and power values are compared with the corresponding results for ($\mu = 1250$) which represents the "accurate" global propagation method. The runtime speedup values denote the number of times our implementation is faster than the "accurate" method. The power error values denote the relative error in the total average power value compared to the "accurate" method.

Except for C6288 and C3540, the error values are less than 5% with most being almost 0%. The average runtime improvement is 10x with a maximum improvement of up to 20x. C6288 is a particularly interesting design. It contains a 16x16 multiplier which poses known difficulties for ROBDD implementations. Even at ($\mu = 1250$), we found that the power analysis tool still hit the memory limit while trying to construct ROBDDs. The excessive number of blowups coupled with the large degree of internal reconvergent paths in the multiplier (which lead to a high degree of internal spatial correlation) contribute to the high errors for C6288.

5 Conclusion

We have described a general method to improve the efficiency of global propagation power analysis techniques. We illustrated the importance of placing a bound on the ROBDD memory usage. We also described a memory recovery strategy that can be used when the memory bound is exceeded. By controlling memory in this way, our technique achieves a better speed-accuracy trade-off than existing methods.

Significant runtime improvements (ranging up to 20x) were achieved with our implementation with minimal accuracy impact. We would like to better understand the results that were observed on C6288 with the hope of incorporating some extensions to the memory recovery algorithm that might improve handling of such circuits which are inherently difficult to represent using ROBDDs.

As designers move upwards in levels of abstraction, they will continue to require faster power analysis tools to allow evaluation of architectures at a high-level. By reducing the turnaround times for power analysis, designers can more exhaustively explore architectural and gate-level alternatives for low power.

6 References

[1] F. Najm. "A Survey of Power Estimation Techniques in VLSI circuits" In IEEE *Transactions on VLSI Systems*, 2(4): pp. 446-455, December 1994.

[2] F. Najm. "Transition Density, A Stochastic Measure of Activity in Digital Circuits" In *Proceedings of the 28th Design Automation Conference*, pp. 644-649, June 1991.

[3] R. Marculescu, D. Marculescu, M. Pedram. "Switching activity analysis considering spatiotemporal correlations" In *Proceedings of the International Conference on Computer-Aided Design*, pp. 294-299, November 1994.

[4] C-Y. Tsui, M. Pedram, A.M. Despain. "Efficient estimation of dynamic power consumption under a real delay model" In *Proceedings of the International Conference on Computer-Aided Design*, pp. 224-228, November 1993.

[5] A. Ghosh, S. Devadas, K. Keutzer, J. White. "Estimation of average switching activity in combinational and sequential circuits" In P*roceedings of the 29th Design Automation Conference*, pp. 253-259, June 1992.

[6] T. L. Chou, K. Roy, S. Prasad. "Estimation of circuit activity considering signal correlations and simultaneous switching" In *Proceedings of the International Conference on Computer-Aided Design*, pp. 300-303, November 1994.

[7] K. Brace, R. Rudell, R. Bryant. "Efficient Implementation of a BDD package" In *Proceedings of the 27th Design Automation Conference*, pp. 40-45, June 1990.

[8] S. Krishnamoorthy, A. Khouja. "Efficient power analysis of combinational circuits" *Synopsys Inc. Internal Technical Report*, April 1995.

Low Energy Memory Component Design for Cost-Sensitive High Performance Embedded Systems

Catherine H. Gebotys
Department of Electrical and Computer Engineering
University of Waterloo,
Waterloo, Ontario. Canada

Although memory has been shown to be a crucial component of the total system energy dissipation, few low energy memory design techniques exist. This paper presents for the first time a network flow approach to minimizing the energy dissipation of memory components during systems synthesis. The number of external and internal memory accesses and the number of extra computations (or data regeneration) for each task is optimized to minimize the total estimated system energy dissipation. This is unlike previous research which has only discussed adhoc suggestions for this problem. Results for a large complex real industrial application, audio compression, donated by Motorola, show that estimated energy savings of 2 to 10 times per task contributed 2.7 times energy savings for the embedded system. This research is important for industry since energy dissipation consideration at the early stages of design is crucial for mapping high performance applications into cost-efficient and reliable systems.

1.Introduction

System design for low power is very important for industries such as portable electronics, automotive, telecommunications, information technology, etc... This is due to the fact that low power designs can offer significant reductions in system packaging costs and improvements in system reliability[1]. Tools for low energy design need to be developed to support the mapping of applications into low energy system architectures. Although it is important to explore energy versus cost tradeoffs, for many cost sensitive markets, such as multimedia[17], new techniques need to be developed which provide low energy without an increase in cost or decrease in performance. For example in multimedia applications low power is believed to be crucial[2]. This includes audio and video algorithms which process large amounts of data, performing computations in real time. High level design issues, such as determining the number of external/internal memory accesses, are seen as having a significant impact on the final system's cost, performance and power dissipation. Clearly there is a need to study high level techniques for designing low power, high performance systems.

It is critical to consider memory requirements during low power systems design, since it is well known that external memory accesses will dissipate significantly more energy than using on-chip memory or cache which in turn may dissipate more energy than performing some computation (such as addition) on the chip. In fact a minimum energy solution may be to perform an on-chip memory access plus a on-chip computation in place of performing one off-chip external memory access. This has been referred to as data regeneration[1]. However no known techniques exist for determining how much data regeneration if any is necessary[1]. It is known that these early design decisions have a significant impact on the final embedded system. The low energy system synthesis problem is most likely NP-hard ,since many of its subproblems , such as simultaneous scheduling and allocation, have been defined as NP-complete[4].

2.Problem Description and Previous Research

The following problem, problem 1 given below, is an important part of the low energy system design problem that will be studied in this paper. For simplicity let us assume that an algorithm to implement the application has already been assigned based upon accuracy required, low power implementation, etc.. The algorithm is given as a partially ordered list of tasks (where each task is represented by a partially ordered list of code operations). For the problem definition below we assume that there is one chip with on-chip cache and external main memory.

Problem 1. Assume we are given an initial ordering of tasks, represented by $t_1, t_2, ...t_N$, where the subscript refers to the *i'th* task being executed. For each task, t_i, determine if the data, d_i, accessed by the task was computed earlier. If it was then see if it was recently used by the previous task, t_{i-1} (and therefore is still in the on-chip cache). If the data required as input to the task was not recently used by the previous task but generated earlier in the application by another task, t_j where $j \ll i$, we assume that it would not be located in the on-chip cache. Therefore the problem is to determine what actions below will be taken to obtain d_i for task t_i, such that the estimated energy dissipation is minimized:

- access the data or an intermediate form of the data from external memory and store it in on-chip cache.
- access an intermediate form of the data from on-chip

0-7803-3177-6 $5.00 © 1996 IEEE

cache memory.

• regenerate the data needed by performing a number of computations on data accessed from on-chip cache (which may have been just accessed from external memory or used recently by previous task).

The objective is to minimize the estimated energy dissipation of the task given the average switched capacitances for off-chip word transfers, on-chip computations (multiplications, additions, etc), and on/off-chip memory accesses. (The solution of this problem also determines what data should be stored in external and internal memory.)

Although this problem is defined above for single chip systems where on-chip cache is available along with external memory, it can also be extended to multichip systems.

The total energy per computation, for example in an adder (based upon CMOS VLSI technology [3]), is given as $=P_{total}/f_{clk}=C_{effective}V_{dd}^2$ where $C_{effective}$ is the average effective capacitance being switched per clock cycle (clock cycle = $1/(f_{clk})$). The term P_{total} is the total energy dissipated in the system over the interval of time being one clock cycle. For applications with fixed throughput requirements (or with a fixed number of data samples which must be processed per second), it is generally agreed that energy should be used to compare design techniques and VLSI architectures as opposed to power[6]. From now on we will refer to energy savings. In order to reduce energy one can reduce the average capacitance of the system being switched. Some researchers have studied different choices of algorithms[7], or modifications of the algorithm[8,3]. A very popular technique, called voltage scaling[9,1] is to increase the parallelism in the system so that the time to perform the computations is very short. Then by lowering the supply voltage, the linear increase in capacitance (from added parallelism) is offset by the quadratic decrease in voltage, thus reducing the energy dissipated, at expense of increase in cost (or chip area). Other studies identified that the energy dissipation of memory accesses and off-chip transfers can be very significant[8,11,2]. For example a 16 bit multiplication, on-chip memory read, on-chip memory write, and off-chip transfer required 4, 5, 10, and 11 times more energy dissipation respectively than a 16 bit addition operation in a CMOS library optimized for low energy[11]. Some recent algorithms developed for memory allocation either have not addressed low energy [18], or are very application-specific[13] or only deal with on-chip memory where allocation of variables to proper memory banks is performed to maximize simultaneous data accesses[12].

Although researchers have discussed data regeneration as a method for saving energy, few techniques have been developed for determining the number of computations, and the number of on-chip/off-chip memory accesses which will minimize the overall estimated energy. In this manuscript a new approach is presented to solve problem 1, low energy data-regeneration/memory design for VLSI embedded systems synthesis. Unlike previous research, a network flow approach to finding minimum energy solutions is introduced. A globally optimal solution can be obtained in polynomial time using very efficient algorithms[14,16].

3.Assumptions and Notation

The following terminology will be used in this paper: $Energy_{system}$= energy dissipated by the system. E_f= energy dissipated by functional units (ie. multiplier-accumulators, alus,.etc) on each chip. E_m= energy dissipated by the registers, register files, RAMs and ROMs on each chip. E_{off}= energy dissipated by transfers on and off of each chip. $Vsc(T)$= the value of the scaled voltage where T is the execution time or latency of the scheduled task flow graph. $Cap(x)$= the average switched capacitance of module x (taken from [11] for results of this paper), where $x=f,m,$ *or* off, representing computations, memory accesses or off-chip transfers. The embedded system energy model is similar to [9] but includes voltage scaling and is given as: $Energy_{on-chip}= E_f+E_m+E_{off} =[Cap(f)+Cap(m)+Cap(off)]Vsc(T)^2$ The energy model presented above will be used to estimate the relative power savings in the rest of this paper. The next section will provide an introduction to network flow. The following section will show how it is used to solve problem P1, by looking at two examples.

4. Introduction to Network Flow

Given a directed acyclic graph, G, composed of nodes and arcs, each arc, (i,j), has a real valued quantity associated with it, called capacity, $c_{i,j}$. There are two special nodes in this graph called node s and node t. If we partition the nodes of G into two sets, S and T, where $s \varepsilon S$ and $t \varepsilon T$, then we can define a cutset. A cutset is the set of arcs, C, such that C contain directed arcs from set S to set T. We can also define a capacity of a cutset as follows : $C(S,T)=\sum_{i \varepsilon S}\sum_{j \varepsilon T} c_{i,j}$, which is the sum of the capacities of all arcs directed from S to T. The min-cut problem is to find the cutset whose capacity is minimum over all possible cutsets in the graph. This optimal value is equivalent to the value of maximum flow in the graph and is fundamental to the famous maxflow/mincut theorem [16]. This problem can be solved in polynomial time with very efficient algorithms[14,16]. The bold arcs in figure 1 shows an example of a cutset and the the dashed line represents the partition of nodes into sets S and T. Previous research has applied this to many problems including the problem of mapping algorithms sequentially onto multiprocessors with interprocessor delays[15].

5. Methodology and Modeling

This section will briefly describe the methodology for low energy systems synthesis followed by a description of how network flow is used to solve problem P1. First an algorithm is selected for the application (based upon cost, performance and energy dissipation requirements), each task is placed in an ordered list, and low energy transformations are performed within each task (such as loop tiling, precomputation, etc..). Finally the network flow approach is applied to obtain further reductions in energy dissipation. After this stage is completed, detailed scheduling and allocation for embedded system synthesis is completed by mapping into specific instructions of an embedded processor or into a custom VLSI architecture chip. The two sections below will describe in detail how network flow is used to solve problem P1.

5.1. Min-Cut Modeling Example #1

One approach to reducing the number of computations performed are to precompute as many values where ever possible. The example below is derived from a real application, the audio compression algorithm[10]. The values V_i, $i=a,b,..f$, in equation (1), are computed once and the algorithm must then find the optimal set of two values $g0,g1$ out of 256 sets which minimizes the function (only part of the actual application is shown below for illustration purposes):

$$\sqrt{g0g1}\, V_a + g0\sqrt{g1}\, V_b + (\sqrt{g0g1} - g0\sqrt{g1})\, V_c + \qquad (1)$$

$$g1g0\sqrt{g0}\, V_d + (\sqrt{g0g1} - g0\sqrt{g1})/8\, V_e + 8g1g0\sqrt{g0}\, V_f$$

Determining what to store and what if any computations to perform can be modeled as a network flow problem as we will see below. To model this problem as a network flow problem, we first have to develop a graph. The computations of the coefficients of equation (1) and the data transfers are represented as arcs as shown in figure 1, where arc i represents the computation of the coefficient of V_i, $i=a,...f$, and arc $i \rightarrow j$ represents the data transfer arc of coefficient of V_i to computation of coefficient of V_j (arc is defined as directed from tail of arc i to head of arc j). Then we add the input values $g0,g1$ to the graph as edges i_i,$i=g0,g1$, where both edges have tails equal to node s. All inputs of the computations are connected to s. All outputs of the computations are connected to node t, as shown in figure 1. Finally computation arcs are added, shown as arcs (not arrows) in figure 1. Now we assign capacities to each arc in the graph as follows, the capacity of data-transfer arcs is the average switched capacitance of accessing one word from memory (using capacitance model in [5] for example) and the capacity of arc c_i = average switched capacitance of performing computation i. Now the solution to problem P1 in this example is equivalent to finding the min-cut of this graph. In figure 1 one possible solution of the mincut problem is the cutset shown in bold, where coefficients a,b, and d are accessed from memory and coefficients c,e,f are computed to obtain minimum energy dissipation. Note that in this example we could use this analysis to model data regeneration for all on-chip memory or off-chip memory.

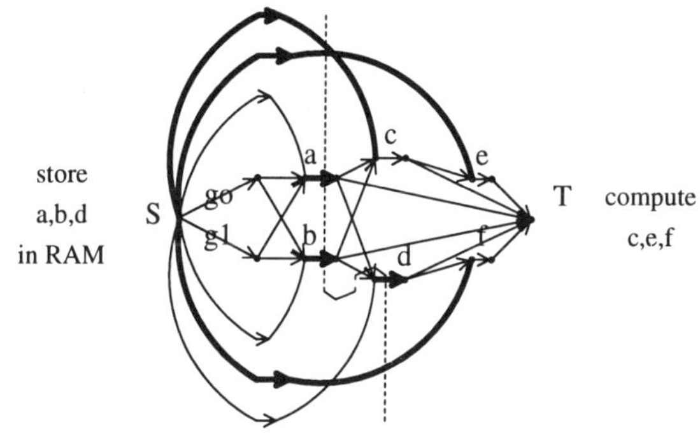

Figure 1. The graph represents network flow problem for example #1, where cutset is shown in bold and partition of nodes shown by dashed line.

5.2. Min-Cut Modeling Example #2

A second example illustrating the use of the network flow is to consider the task flow graphs in figure 2 b) and c). Task T1 produces a large data structure, $f_i(n)$, that is required later in the application by task T4. In figure 2c) this large data structure could be accessed from external memory. Alternatively it could be regenerated, using task *reg*, (where $f_{i*}(n) = G(u_{i*}(n),h(n))$), as in figure 2b), from data which is currently in the cache (because it was used during the last task, T3, $u_i(n)$ and $h(n)$), along with a small amount of data from task T1 ($i*$). This partial example is taken directly from the VSELP application[10]. In this example the data regeneration calculation is obtained by simple transformations. The problem modeled as a network flow problem, is shown in figure 2a). Cut 1 solution includes energy dissipated by computations to perform G in task *reg* and energy dissipation of on-chip cache access to obtain $h(n),u_{i*}(n)$. Cut 2 solution illustrates a solution that involves accessing $f_{i*}(n)$ from external main memory. In this example we use capacities of the network flow problem to represent the average switched capacitance of on-chip memory access (arrows from s to nodes in figure 2a), on-chip computations (arcs from s to nodes in figure 2a)) and off-chip memory accesses (arrows from nodes to node t in figure 2a)).

6. Experimental Results

An audio compression algorithm is used to illustrate the low energy methodology. A large real complex industrial audio compression algorithm[10], VSELP, donated by

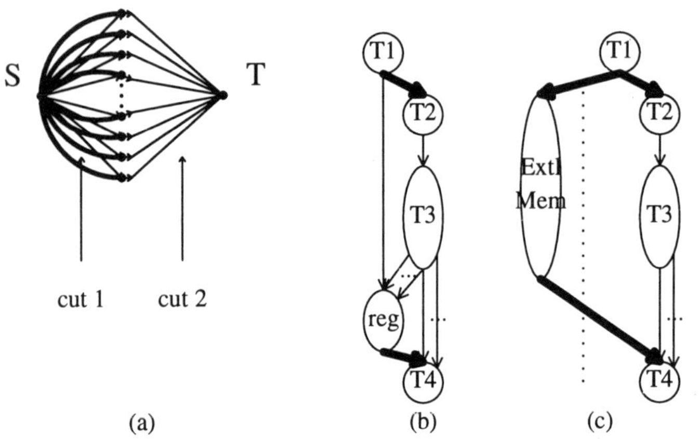

Figure 2. Min-cut model in (a) where cut 1 is illustrated in (b) and cut 2 is illustrated in (c). Bold arrows in (b) and (c) indicate large amount of data to be stored between tasks.

Motorola, is composed of over 123,000 multiply accumulate operations which dominated the application. Unfortunately there is limited research on energy savings in systems so it is not possible to make a comparison with previous research.

The network flow approach was used for several tasks in the VSELP application. The code book search, similar to example #1 from section 5.1, was modeled as network flow. The optimal solution in this case provided 1.7 times improvement in estimated energy dissipation for the off-chip memory access analysis. Three other tasks used the network flow technique similar to example #2 of section 5.2. Using a on-chip cache, with 4 words to a block, the savings in estimated energy dissipation was up to 10 times for these tasks. Other tasks in this application also benefited from the network flow technique and together contributed to 2.7 times savings in estimated energy for the entire embedded system.

7. Discussions and Conclusions

In summary energy savings due to the network flow approach was found to provide 2.7 times savings in energy. In contrast to previous research[13,12,2] which examined memory design for low energy systems but did not suggest any techniques for data regeneration, we have introduced the application of network flow to this problem, which has been shown here for the first time to be beneficial in reducing estimated energy dissipation at negligible cost. The network flow technique could also be extended to answer the question, "what is the optimal width of memory that should be used?". Since large network flow problems have been solved with very efficient algorithms [14], the application to even larger examples should also be very practical. This network flow technique has also been applied to multichip systems and different applications.

We have introduced a methodology for energy minimization in large realtime compute intensive applications for embedded systems. The methodology incorporates a new approach to low energy memory design. For the first time network flow is applied to the problem of determining the amount of off-chip/on-chip memory accesses and data regeneration that should be employed to minimize estimated energy dissipation. The network flow approach finds globally optimal solutions in polynomial time. Results showed that up to 10 times improvement in energy dissipation for tasks is attainable with a large real industrial audio compression application. This contributes to 2.7 times improvement in the estimated energy dissipation of the embedded system. It is also interesting to note that this approach comes at no (or negligible) decrease in performance or increase in cost, unlike other techniques where cost may be traded for energy dissipation. Finally the audio compression application which is quite complex and in general difficult to optimize for energy, is targeted for portable digital cellular phones which is a very cost-sensitive market, so a low energy implementation should provide a significant decrease in battery size, packaging costs and perhaps reliability. This research is important for industry since energy dissipation consideration during these early stages of design are critical to ensuring that the final product will be cost effective, competitive, and meet high performance requirements.

References

[1] K.Keutzer, "The Impact of CAD on the Design of Low Power Digital Circuits", IEEE Symposium on Low Power Electronics, 1994, p42-45.

[2] S.Wuytack,F.Catthoor,F.Franssen,L.Nachtergaele,H.DeMan, "Global Communication and Memory Optimizing Transformations For Low Power Systems", International Workshop on Low Power Design, 1994, p 203-208.

[3] A.Chandrakasan,et.al."Optimizing Power Using Transformations", IEEE Transactions on CAD, Jan 1995,Vol14,No.1, p12-31.

[4] Garey and Johnson,**Computers and Intractability** N.Y.Freeman & Co 1979.

[6] M.Horowitz,T.Indermaur,R.Gonzalez, "Low-Power Digital Design" IEEE Symposium on Low Power Electronics, 1994, p8-11.

[7] V.Tiwari, S.Malik,A.Wolfe, "Power Analysis of Embedded Software ;A First Step Towards Software Power Minimization, IEEE Trans on VLSI, Vol. 2, No. 4, Dec 1994, p437-445,

[8] A.Chandrakasan, A.Burstein R.Brodersen," A Low-Power Chipset for a Portable Multimedia I/O Terminal", IEEE Journal of Solid State Circuits, Dec 1994 p 1415-1428.

[9] S.Powell,P.Chau, " A Model for Estimating Power Dissipation in a Class of DSP VLSI Chips", IEEE Trans on Circuits and Systems vol38,no6,June1991.

[10] M.Sunwoo,S.Park,"Principles of VSELP Speechcoder and its implementation on the DSP56156", Motorola Inc, DSP Operations, Austin Texas.

[11] P.A.Beerel, USC, Panel Presentation on "Where Does the Power Go?" at International Symposium on Low Power Design, April 1995.

[12] Lee,Tiwari, "A Memory Allocation Technique for Low-Energy Embedded DSP Software", Symposium on Low Power Electronics, Oct, 1995, p24-5.

[13] Lidsky,Rabaey, "Low-Power Design of Memory Intensive Functions", IEEE Symposium on Low Power Electronics, 1994, p16-7.

[14] Nemhauser, Wolsey, **Integer and Combinatorial Optimization**, New York Wiley Interscience, 1988.

[15] H.S.Stone, "Multiprocessor Scheduling with the Aid of Network flow Algorithms", IEEE Trans on Software Engineering, Vol.SE-3,1977.

[16] E.Lawler **Combinatorial Optimization Networks and Matroids**, Holt, Rinehart and Winston, 1976.

[17] Panel on "How do we reduce Digital System Power at the System Level?", Int'l Symp on Systems Synthesis, 1995.

[18] H.Schmitt,D.Thomas, "Array Mapping in Behavioral Synthesis ", Int'l Symp on Systems Synthesis, 1995.

Exploiting Locality for Low-Power Design

Renu Mehra, Lisa Guerra, and Jan Rabaey

Department of Electrical Engineering and Computer Sciences
University of California at Berkeley
Berkeley, CA 94720

Abstract

We propose a new high-level synthesis technique for the low-power implementation of real-time applications. The technique uses algorithm partitioning to preserve locality in the assignment of operations to hardware units. This results in reduced usage of long high-capacitance buses, fewer accesses to multiplexors and buffers, and more compact layouts. Experimental results show average reductions in bus and multiplexor power of 62.9% and 38.5%, respectively, resulting in an average reduction of 18.5% in total power.

1. Introduction

High-level synthesis is steadily making an inroad into the digital design community. Most work to date has focused on techniques for area and speed optimization [1]. Recently, there has been significant interest in techniques and tools for power optimization. While for area optimization, high resource utilization through hardware sharing is one of the main goals, for power optimization, reduced hardware sharing often gives better results.

Consider Wu's comparison of an automatically-generated maximally time-shared and a manually-generated fully-parallel implementation of a QMF sub-band coder filter [2]. In the manual design, a number of optimizations were used to obtain power savings in the various components. The power consumption of both versions is documented in Table 1. For the same supply voltage, an improvement of a factor of 10.5 was obtained at the expense of a 20% increase in area.

Note that the interconnect elements (buses, multiplexors, and buffers) consume 43% and 28% of the total power in the time-shared and parallel versions, respectively. Further, these elements contribute the most to the power reduction achieved in the parallel version. Power improvement factors of 16.9, 15.1, and 12.5 were obtained for buses, multiplexors, and buffers, respectively, mainly due to dedicated communication and reduced usage of multiplexors and buffers. This points to the large opportunity available for interconnect power reduction and highlights its significance.

While in this example, the fully-parallel implementation resulted in large power gains with low area overhead, this may not always be the case. Parallel implementations may be too area intensive and may not necessarily result in reduced interconnect power. If the area overhead is too high, the increase in the required bus lengths may offset the power gains due to other factors.

Table 1. Power consumption (mW) in the maximally time-shared and fully-parallel versions of the QMF sub-band coder filter.

Component	Time-shared	Fully-parallel	Improvement factor
Functional units	8.52	1.03	8.3
Registers	9.76	1.08	9.0
Buses	23.69	1.40	16.9
Multiplexors	3.77	0.25	15.1
Buffers	4.36	0.35	12.5
Others	23.99	2.92	8.2
Total	74.09	7.03	10.5

In this work, techniques are presented to achieve low-power designs by reducing the interconnect power while incurring low area overhead. The approach aims to capture some of the optimizations of the above example in an automated way while maintaining a balance between the maximally time-shared and the fully-parallel implementations. The next section illustrates the main idea behind our proposed low-power synthesis technique.

2. The impact of exploiting locality

The main idea behind our approach is to synthesize designs with localized communications. We achieve this by dividing the algorithm into *spatially local clusters* and performing a *spatially local assignment.* A spatially local cluster is a group of algorithm operations that are close to each other in the flowgraph representation. A spatially local assignment is a mapping of the algorithm operations to specific hardware units such that no operations in different clusters share the same hardware. Partitioning the algorithm into spatially local clusters ensures that the majority of the data transfers take place within clusters and relatively few occur between clusters. The spatially local assignment restricts intra-cluster data transfers to buses that are local to a subset of the hardware (local buses); thus only inter-cluster data transfers use buses that are shared by all resources (global buses). The combined result is that local buses which are

0-7803-3177-6 $5.00 © 1996 IEEE

IEEE 1996 CUSTOM INTEGRATED CIRCUITS CONFERENCE

shorter are used more frequently than longer highly-capacitive global buses.

Consider the two different assignments for maximum throughput implementations of the fourth-order parallel-form IIR filter shown in Fig. 1. Indicated beside each operation is the hardware resource that it is assigned to. (A_i are adders and M_i are multipliers). In Fig. 1a, the graph is divided into two spatially local clusters and the operations in each cluster are mapped to mutually exclusive sets of hardware resources (A_1, A_2, and M_1 are used for operations in cluster 1 and A_3, A_4, and M_2 are used for those in cluster 2). As a result, a large number of the communications are restricted to only a subset of the hardware. In Fig. 1b, however, the hardware is not partitioned and all communications are global. The number of global data transfers (shown with solid lines in both cases) for the local and the non-local assignments are 2 and 20, respectively.

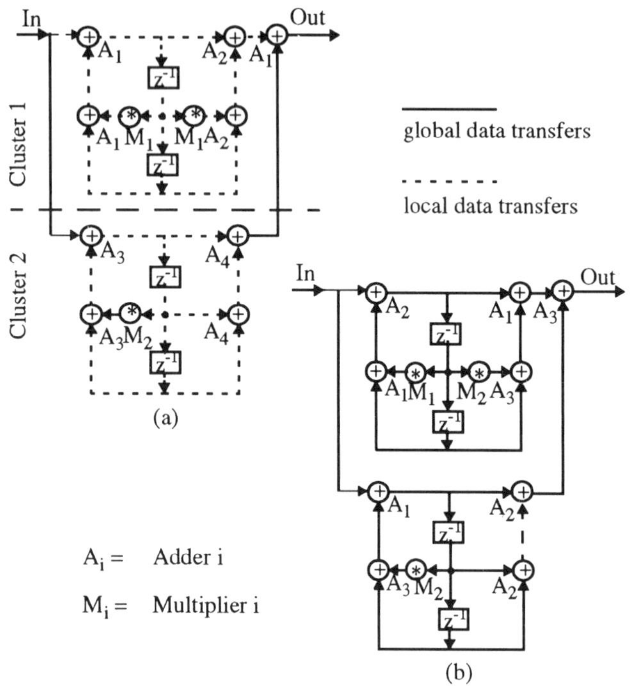

Fig. 1. A fourth-order parallel-form IIR filter: (a) Local assignment, (b) Non-local assignment.

Notice that the local version needs 4 adders and 2 multipliers whereas the non-local assignment requires just 3 adders and 2 multipliers. This increase in the number of functional units does not necessarily translate into a corresponding increase in the overall area since localization of interconnect makes the design more conducive to compact layout. Furthermore, reduced hardware sharing results in additional power savings due to fewer accesses to multiplexors and buffers.

Varying the number of clusters trades off local and global bus power. This is because, as the number of clusters is increased, the number of inter-cluster communications increases while the local bus lengths decrease.

3. Low-power synthesis system

The high-level synthesis process generates an architectural level netlist from a behavioral description and a set of performance constraints. In this section we present a new high-level synthesis strategy based on exploiting the locality of algorithm operations. The core of the approach is a partitioning and assignment scheme. The techniques have been integrated into the *Hyper-LP* system [3].

While the basic synthesis flow of the *Hyper-LP* system is the same as that of the *Hyper* system [4], a new partitioning step is added preceding the assignment phase and the assignment algorithm itself is modified to exploit spatial locality.

3.1 Partitioning methodology

Previous works in partitioning for high-level synthesis have targeted area minimization, with a significant portion of the gains resulting from interconnect reduction [5, 6]. For power minimization, however, it is better to have two global buses each accessed twice rather than one bus accessed six times. The goal for reducing interconnect power, therefore, is to minimize the *number of accesses* to long global buses.

Our partitioning methodology consists of two phases — the first phase generates several candidate partitioning solutions and the second phase evaluates them and selects the best one.

The generation of candidate partitions is based on a spectral partitioning technique used in a number of partitioning algorithms [7, 8, 9]. The technique was introduced by Hall [7] who proved that the second smallest eigenvector of the Laplacian of a graph gives a one-dimensional placement of graph nodes such that the sum of squares of edge lengths is minimized. Large gaps in this ordering are used to delimit the clusters. For example, Fig. 2 shows an eighth order cascade filter and the corresponding eigenvector placement. The spacing between nodes in the placement clearly indicates four distinct clusters which are also evident from the structure. We use $m+3\sigma$ as the threshold for detecting these gaps, where m is the mean of all the distances between the nodes and σ is the standard deviation of the distances. In the cascade example of Fig. 2, this threshold delimits the expected four clusters.

In our scheme, several different candidate partitions are generated by varying the targeted number of clusters. For example, in the cascade filter of Fig. 2, in addition to the 4-cluster partition, a 2-cluster partition may also be proposed. As discussed before, varying the number of clusters trades off between global and local bus power.

In the second phase a rough estimate of the total bus power is used to evaluate and compare the candidate partitions. The metric used as a measure of the global bus power is the number of global data transfers times the estimated global bus

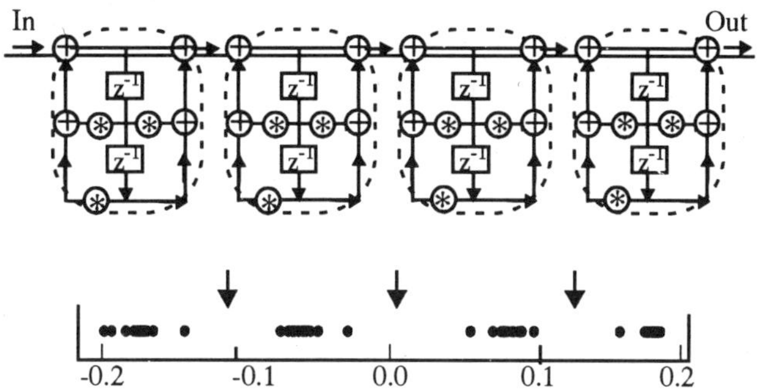

Fig. 2. An eighth-order cascade-form IIR filter and the corresponding eigenvector placement.

length. Similarly, the measure used for the local bus power of each cluster is the number of data transfers local to it times the cluster's estimated bus length.

Since the lengths of the buses have been shown to be proportional to the square root of the area, the cluster and total chip area estimates are used as measures of the local and global bus lengths, respectively. Estimates of these areas are in turn given by the maximum value of the weighted concurrency distribution graph [10]. The distribution graph gives the amount of concurrent hardware needed by the computation in each time slot.

At the culmination of the partitioning phase, the single most promising candidate partition is applied to the algorithm.

3.2 Assignment methodology

Once the partitioning is complete, all operations have an associated cluster number. Also, each data transfer is classified as either an inter- or intra-cluster transfer. The functional unit assignment performs the random assignment with iterative improvement approach of the *Hyper* system [4] and uses the clustering information to ensure that each cluster is assigned to mutually exclusive hardware.

Graph coloring is a commonly used technique to assign data transfers to shared buses such that there are no timing conflicts. In our scheme, we assign data transfers to buses avoiding not only timing but also clustering conflicts. Local buses are shared only among transfers within a cluster. Global buses are used only by inter-cluster data transfers. This ensures that intra-cluster data transfers occur on short local buses and only inter-cluster ones use the long highly-capacitive global buses.

4. Results

In this section we present the results of our partitioning-based synthesis scheme. Implementations generated using the *Hyper-LP* and the *Hyper* systems are compared. The SPA architectural power estimation tool [11] is used for power estimations. Estimates of the total chip area and bus lengths are obtained using models presented in [12]. The bus length model was enhanced to estimate the local and global bus lengths separately.

4.1 Cascade filter

The first result compares the *Hyper-LP* and *Hyper* implementations of the eighth-order cascade filter (Fig. 2). Given a throughput constraint of 21 clock cycles, the *Hyper* implementation uses 4 adders and 3 shifters while the *Hyper-LP* implementation uses one adder and one shifter for each cluster resulting in a total of eight units. Layouts of the two implementations are shown in Fig. 3. In the *Hyper* implementation, 2 of the 7 functional units are merged by the layout tool. In the *Hyper-LP* implementation the 4 datapaths pictured correspond to each of the 4 clusters.

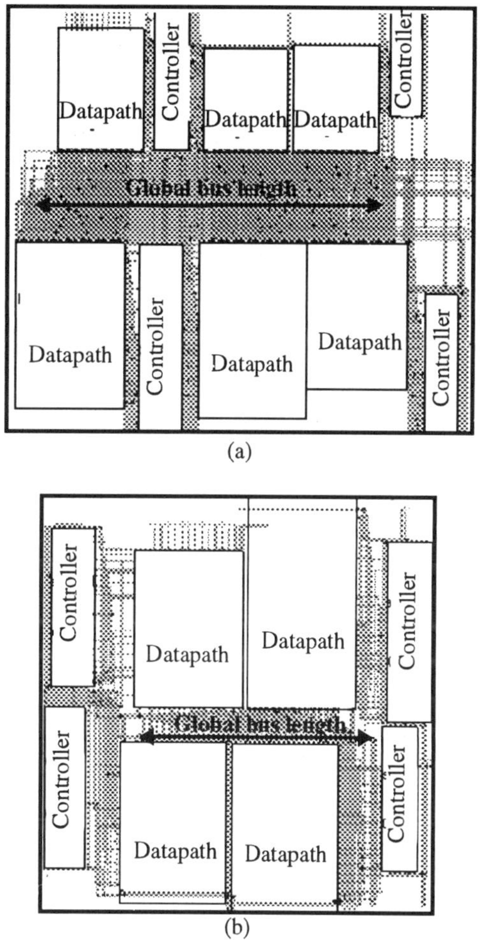

Fig. 3. Cascade filter layouts: (a) Non-local implementation from *Hyper*, (b) Local implementation from *Hyper-LP*.

Table 2 compares the power dissipated in the two implementations. An overall reduction of 34% in the power consumption was realized by the *Hyper-LP* approach. The average length of the global buses reduced by approximately 55%, from 2100 to 950 microns and the bus power reduced 5-fold, from 2 mW to only 0.4 mW. The multiplexor power reduced by 60% as the reduced time-sharing of units results in lower usage of multiplexors. Notice that the contribution of interconnect (buses, multiplexors, and buffers) to the total power dissipation was reduced from 30% to 17%.

Table 2. Comparison of power consumption (mW) in the *Hyper* and *Hyper-LP* implementations of the cascade filter.

Component	*Hyper*	*Hyper-LP*	Percentage reduction
Buses	2.0	0.4	80.0
Multiplexors	3.7	1.5	59.5
Buffers	1.0	0.9	10.0
Others	8.7	7.4	14.9
Total	15.4	10.2	33.77

4.2 *Other examples*

This section summarizes our experimental results for the cascade and several other DSP filter and transform examples. Some are in their original form (DCT, FFT, and parallel-form IIR) and others are transformed using either constant multiplication expansion (cascade-form IIR, direct-form IIR, and wavelet) or retiming (wave digital filter).

Table 3. Comparison of power consumption (mW) in the *Hyper* and *Hyper-LP* implementations.

Design	*Hyper*			*Hyper-LP*			
	Bus	Mux	Total	Clusters	Bus	Mux	Total
Cascade	2.0	3.2	21.3	4	0.4	1.5	16.3
Direct form	29.8	38	144.6	3	10.3	21.1	110.4
Wave digital	1.5	3.2	20.4	2	0.5	1.7	18.0
DCT	9.0	3.8	41.5	2	4.5	2.3	37.2
FFT	17.6	4.7	48.6	2	5.2	3.8	36.8
Parallel IIR	15.1	2.8	57.5	2	3.2	2.2	48.8

Table 3 shows the bus, multiplexor, and the overall power dissipation for both implementations of each example. Table 4 summarizes the percentage power improvements. The *Hyper-LP* implementations uniformly dissipate less power than the *Hyper* implementations. Power consumed by buses is reduced drastically in all examples (up to 80%). Up to 60% reduction in multiplexor power is seen due to reduced and more localized hardware sharing. The average reduction in bus, multiplexor, and total power is 62.9%, 38.5%, and 18.5%, respectively.

The power reduction comes at the cost of an increase in the number of units. However, since the communications are localized, the designs are more conducive to compact layout. Further, overhead elements such as multiplexors and buffers are reduced. Table 4 shows the estimated area penalty obtained in the *Hyper-LP* designs.

Table 4. Overall power reduction and area overhead.

Design	Percentage power reduction			Percentage change in area
	Bus	Mux	Total	
Cascade	80.0	59.6	25.9	-15.1
Direct form	65.4	44.5	23.7	+2.1
Wave digital	33.3	46.9	11.8	+7.2
DCT	50.0	39.5	10.4	-23.4
FFT	69.7	19.1	24.3	+0.8
Parallel IIR	78.7	21.4	15.1	+22.1
Average	62.9	38.5	18.5	-1.05

5. Conclusions

We have presented a technique for power reduction based on exploiting the locality in a given application. At the core of the approach is a partitioning and assignment strategy. It was seen that the proposed scheme improves the implementation in a variety of ways. The predominant effect is the reduction of accesses to highly-capacitive global buses. Our results showed average bus, multiplexor, and overall power reductions of 62.9%, 38.5%, and 18.5%, respectively, and low associated area overheads. The partitioning and assignment techniques have been integrated into the *Hyper-LP* system.

6. References

1. D.D. Gajski, *High-Level Synthesis: Introduction to Chip and System Design*, Boston, Kluwer Academic, 1992.
2. S. Wu, "A Hardware Library Representation for the Hyper Synthesis System," *Masters' Thesis*, University of California, Berkeley, Memorandum No. UCB/ERL M94/47, June 1994.
3. A. Chandrakasan, M. Potkonjak, R. Mehra, J. Rabaey, and R. W. Brodersen, "Optimizing Power Using Transformations," *IEEE Trans. on CAD*, Vol. 14, No. 1, Jan. 1995, pp. 12-31.
4. J. M. Rabaey, C. Chu, P. Hoang, and M. Potkonjak, "Fast Prototyping of Datapath-Intensive Architectures," *IEEE Design & Test of Computers*, June 1991, pp. 40-51.
5. M.C. McFarland and T.J. Kowalski, "Incorporating Bottom-up Design into Hardware Synthesis," *IEEE Trans. on CAD*, Vol. 9, No. 9, Sept. 1990, pp. 938-949.
6. E.D. Lagnese and D.E. Thomas, "Architectural Partitioning for System Level Synthesis of Integrated Circuits," *IEEE Trans. on CAD*, Vol. 10, No. 7, July 1991, pp. 847-860.
7. K. M. Hall, "An r-Dimensional Quadratic Placement Algorithm," *Management Science*, Vol. 17, No. 3, Nov. 1970, pp. 219-229.
8. L. Hagen and A. B. Kahng, "New Spectral Methods for Ratio Cut Partitioning and Clustering," *IEEE Trans. on CAD*, Vol. 11, No. 9, Sept. 1992, pp. 1074-1085.
9. B. Hendrickson and R. Leland, "The Chaco User's Guide, V. 1.0," *Tech Report SAND93-2339, Sandia National Lab*, Oct. 1993.
10. P. G. Paulin and J.P. Knight, "Force-Directed Scheduling for Behavioral Synthesis of ASIC's," *IEEE Trans. on CAD*, Vol. 8, No. 6, June 1989, pp. 661-679.
11. P. E. Landman and J. M. Rabaey, "Architectural Power Analysis: The Dual Bit Type Method," *IEEE Trans. on VLSI Systems*, Vol.3, No.2, June 1995, pp. 173-87.
12. R. Mehra and J. M. Rabaey, "Behavioral Level Power Estimation and Exploration," *Proc. of the Int'l. Workshop on Low-Power Design*, April 1994, pp. 197-202.

Deep Sub Micron Chip Design - What Business Issues will Dominate?

ORGANIZERS:
Rakesh Kumar and Michael Beunder

MODERATOR:
Rakesh Kumar
Cadence Spectrum Design

PANELISTS:

Don Ciffone
VP & Gen. Mgr.
VLSI Technology

Robert Pepper
President & CEO
Level One Communications

Conrad Dell Oca
President
ASPEC

J. Robinson
Vice President
Sun Microsystems

John Luke
President
TSMC

Ed Ross
President WWM
Cirrus

While the technical challenges of Deep Sub Micron (DSM) design have had much attention in the industry, it is the business issues that may dominate product shipment on schedule and the meeting of any company's business objectives.

In order to more effectively control chip costs, Intellectual Property (IP) and DSM technical solutions, more and more companies are choosing to use the Customer Owned Tooling (COT) approach to design their chips. This involves setting up the infrastructure to manage foundry relations and Wafer capacity allocations. The recent Wafer supply shortages from both ASIC suppliers and foundries are projected to continue well into the latter part of this decade. Even though many more fabs are projected, the increased amount of fab investment has prompted financial linkups between the larger design house and the foundries. Will the smaller, emerging fabless design houses lose the ability to get silicon?

The increased investment required to build Wafer fabs and to develop process technologies raises issues of profitability and affordability. Will only the large, merchant suppliers be able to afford leading edge technology? And will this affect the ability of ASIC suppliers and foundries to offer leading ASIC edge technology?

Key Questions for the panel are:

1. Is the COT approach really better for DSM technical solutions?
2. Where and how will the smaller, emerging Fabless Design Houses get silicon?
3. How will ASIC suppliers and foundries be able to support the latest technology?

Technical Challenges in One-Million-Gate Chip Designs

ORGANIZERS:
Jim Lipman, Neill Mullinger and Paul Billig

MODERATOR:
Jim Lipman
EDN Magazine

PANELISTS:

Jean-Louis Bories
VP ASIC Design
LSI Logic

Glenn House
VP/GM, Silicon Systems Div.
Mentor Graphics

Jeff Lewis
VP Marketing
COMPASS Design Automation

Naeem Zafar
VP Marketing
QuickTurn Solutions

You're ready to design your system-on-a-chip. You've chosen the appropriate deep submicron CMOS technology, identified the core and other blocks to go on the chip, along with some custom logic, and selected the EDA tools you will need to put all the pieces together. If you believe the EDA and ASIC vendors, after about six months to a year of effort in designing and verifying your chip, it will be ready to drop into your waiting system. Not so fast!

Million-plus-gate chip designs, while possible, still remain difficult to do and full of unpleasant surprises to the unwary. As a designer, you need to work closely with software and hardware vendors to successfully complete the design and meet your system's "window of opportunity". Our Technical Challenges in One-Million-Gate-Chip Designs panel will give you the opportunity to quiz representatives from leading ASIC, EDA, computer chip and emulation companies in the deep submicron development arena to determine where the "rough spots" exist in system-on-a-chip design.

Chair: Yusuf Haque
Co-Chair: Masao Hotta

Oversampled delta sigma converters have been known to be prone to instability caused by either high signal levels or by input overload conditions. The first two papers present use of local feedback techniques to stabilize the modulator used in these ADCs. The first paper demonstrates a 3rd order modulator which achieves 12 bits of resolution with 64 times oversampling ratio. Low voltage and low power techniques are used to achieve sub 2V operation with only 0.34mW dissipation in a 1.2μ CMOS process. The second paper discusses a fourth order delta sigma ADC that achieves 24 bits of resolution with 102 dB peak S/N+D in a 0.7μ CMOS process.

Digital communication and ultrasound imaging have been driving the need for ADCs with 8 or more bits of resolution at conversion rates above 40MHz with low power dissipation. The third paper is the first CMOS device to report 100MHz conversion rates with 10 bits of resolution in a 1μm process. The next paper shows how to build an 8 bit pipelined ADC with just 3 opamps instead of 7 required in a conventional architecture. Using two such pipelines in parallel, a 50Ms/s device dissipating only 250mW from a 5V source is demonstrated in a 0.9μm CMOS process.

The next two papers focus on achieving differential nonlinearity (DNL) and integral nonlinearity (INL) specifications without requiring trim or calibration. The fifth paper presents an ADC design technique suitable for achieving 12 bits of DNL performance with 7 bits of capacitor matching and without requiring calibration. The last paper describes the design of a 12 bit current steering DAC built in a 0.7μ standard CMOS process that achieves better than 1 lsb INL without using trimming or calibration.

A 1.95V, 0.34mW 12-bit Sigma-Delta Modulator Stabilized by Local Feedback Loops

Stephen Au and Bosco Leung

Department of Electrical & Computer Engineering,
University of Waterloo,
Waterloo, Ontario, Canada N2L 3G1

Abstract

The design of a low power, low voltage, 12-bit 8-kHz bandwidth $\Sigma\Delta$ modulator that achieves 0.34mW power consumption at 1.95V supply is described. The modulator employs a novel architecture in which a third-order modulator is stabilized by a local feedback loop around each integrator. Unlike multi-stage $\Sigma\Delta$ modulators, this architecture is very tolerant to the modest d.c. gain of low voltage opamps. The architecture, together with special circuit techniques, permits a low voltage switched capacitor implementation at 1.95V-3.3V supply using standard 1.2μm CMOS technology.

1. Introduction

Oversampled A/D converter has become the most popular choice for high resolution, low-to-medium bandwidth applications. Because of the often high sampling frequency needed the power consumption tends to be excessive for many portable systems. The power consumption of both the analog $\Sigma\Delta$ modulator and the digital decimation filter can be substantially reduced by lowering the supply voltage. In addition, the significant power consumed by the modulator will decrease if a high order (> 2) architecture is used to reduce the sampling frequency. The lowering of both the supply voltage and the clock frequency drastically reduces the overall power consumption.

In stabilizing typical high order single-stage, single-bit modulators by decreasing integrator gains, their dynamic range is reduced [1]. The stable multi-stage (MASH) architecture with first or second order individual stages requires high gain opamps to achieve the necessary quantization noise cancellation [2]. High gain cascode opamps are not attractive choices at low supply because of the further loss in output swing. Although high gain opamps in low supply voltage can be realized with cascaded stages, they tend to consume more power due to the larger number of current branches [3]. Therefore a new single-stage, single-bit modulator architecture that is stabilized by local feedback loops is adopted. This architecture can tolerate low gain opamps which are readily realized at low supply and hence achieve low power dissipation.

2. Modulator Architecture

Fig. 1 shows the modulator scheme [4]. Apart from the local feedback loops which include the overload detectors OVLs, local DACs and l scale factors, and the digital compensators H_1, H_2 and H_3, the basic circuit is a third-order single-stage modulator. When the integrators operate within their normal range ($-V_{ovl}$, V_{ovl}) the OVLs output 0 (Fig. 1) and therefore both the local feedback loops and the digital compensators are not activated. In essence, the modulator operates just like an ordinary third-order single-stage modulator which is very tolerant to modest opamp gain and component mismatch, but is prone to instabilty, particularly at higher input level. If an exceedingly large integrator level is detected by any of the OVLs, the corresponding integrator is forced back to the normal operating range ($-V_{ovl}$, V_{ovl}) via the tri-level DAC. The associated digital compensator, whose transfer function matches that of the corresponding local feedback signal, cancels it digitally at the final output to restore the original SNR of the basic modulator.

As will be shown in the next section, a tri-level DAC can be realized with good linearity in a differential architecture. The digital compensators can also be approximated in the baseband by the simple FIR filters:

$$H_1(z) = \frac{l_1}{j_1} z^{-3} \quad (1)$$

$$H_2(z) = \frac{l_2}{j_1 k_2} z^{-2}(1 - z^{-1}) \quad (2)$$

$$H_3(z) = \frac{l_3}{j_1 k_2 k_3} z^{-1}(1 - z^{-1})^2 \quad (3)$$

The accuracy of this digital compensation does depend on the matching between the analog modulator and the

0-7803-3177-6 $5.00 © 1996 IEEE

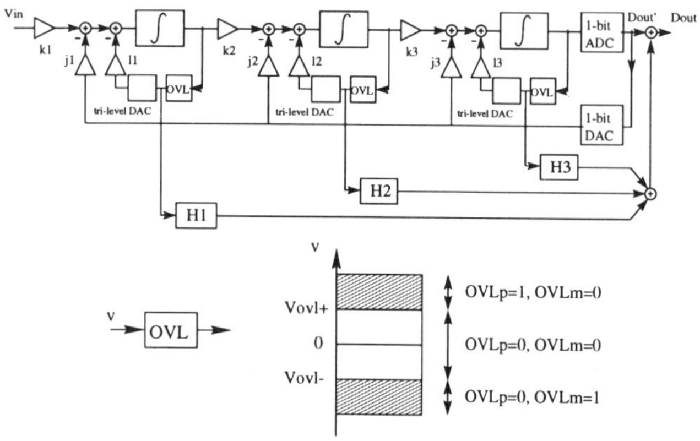

Figure 1: Single-stage third-order modulator stabilized with local feedback loops.

digital compensators like that in a multi-stage architecture. However, since in reality the OVLs are invoked relatively infrequently, the requirement on the matching is much relaxed. Therefore, simple opamp without cascode devices or cascaded stages having only modest d.c. gain, but is readily realized even at low supply voltage, can be used as in an ordinary single-stage modulator without its inherent stability problem.

In general the modulator k, j and l coefficients have to satisfy certain relationships which are sufficient (but not necessary) conditions for stability [4]. Practically, however, some of these criteria can be relaxed to accomodate easier implementation and better SNR performance without jeopardizing the modulator stability. One of the key issues is to select the coefficients such that most of the overloadings for high input level would occur at OVL2 and OVL3 rather than at OVL1 (see Fig. 7).

Simulation also shows that the same $SINC^4$ comb decimator for a conventional third-order modulator can be used for this modulator. This architecture can also be extended easily to stabilize modulator of even higher order.

3. Circuit Implementation

The system level schematic is shown in Fig. 2. Each overload detector is implemented with two comparators, one to detect positive overloading and the other negative. The local tri-level DAC and the global single-bit (bi-level) DAC is incorporated in the capacitor arrays (Fig. 3). The size of the input capacitors for the first integrator is decided by thermal noise (kT/C noise) considerations, while those for the second and third integrator are half-sized because the higher kT/C noise would be subject to first and second order noise shaping respectively. This allows the second and third opamp to operate at lower currents. The clock x_H is a boosted version of the clock x. Its complement is not boosted (to voltage below V_{ss}). The other switch control signals $Dout'$, $\overline{Dout'}$, $OVLp$ and $OVLm$ are not boosted because the fixed d.c. level V_{ref+} or V_{ref-} (1.35V and 0.60V respectively for $V_{dd} = 1.95V$) is handled easily by either the NMOS or the PMOS transistors. On the other hand, the switches controlled by x_H have to handle either varying voltage level or the analog GND level, at which the conductance of an unboosted CMOS switch is lowest.

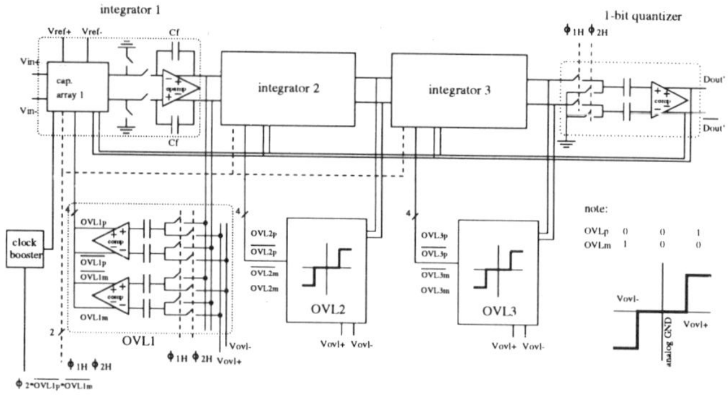

Figure 2: System level schematic. Digital compensators are not shown.

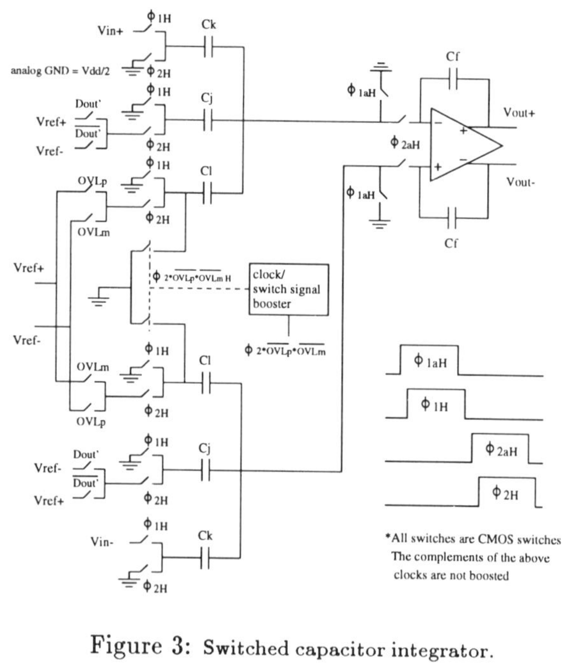

Figure 3: Switched capacitor integrator.

The clockbooster is shown in Fig. 4. When the input ϕ_1 is low, M2 is off and M3 is on, discharging C_{load} so that ϕ_{1H} is at 0V. Meanwhile the top plate of C_1 is charged to $V_{dd} - V_t$ through M1. When ϕ_1 goes from 0V to V_{dd}, M2 is on and M3 is off. As the bottom plate of C_1 jumps from 0V to V_{dd}, the top plate voltage will jump to $(2V_{dd} - V_t) * C_1/(C_1 + C_{load})$. The small capacitor C_{m2} is used to store the high voltage level required by the substrate of the PMOS transistor M2. Identical clockbooters are also used to boost the other clocks ϕ_2, ϕ_{1a} and ϕ_{2a}. Clockbooster that generates higher clock level [5] has not been used because of the extra complexity and a high internal node voltage ($2V_{dd}$), which means higher stress for the thin oxide in fine-line CMOS processes. The PMOS in a switch also lowers the HIGH clock level needed for the NMOS.

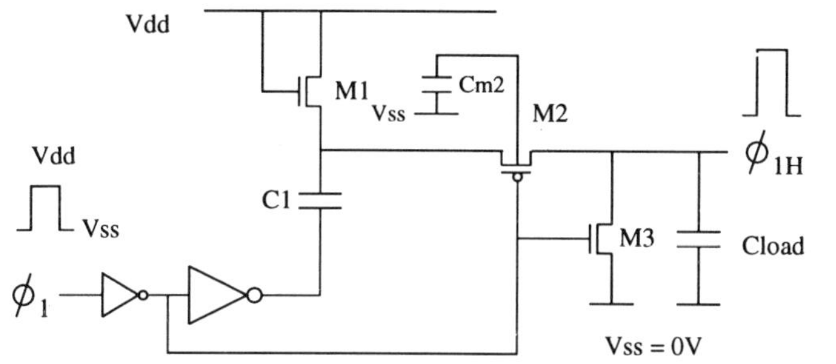

Figure 4: Clockbooster circuit schematic.

Scaled-down clockbooster circuit is also used to boost the switch control signal $\overline{OVLp} * \overline{OVLm}$. Since an integrator may not overload for many clock cycles, its $\overline{OVLp} * \overline{OVLm}$ signal would be HIGH for the same number of clock cycles, and therefore $\overline{OVLp} * \overline{OVLm}$ has to be ANDed with ϕ_2 before being fed into a clockbooster (Fig. 3).

As explained previously because an opamp with only modest gain is needed, a fully differential current mirror opamp with no cascode device (Fig. 5) is used to maximize the output range. A current mirror ratio of 2 and longer channel output devices increase the d.c. gain to around 300, more than adequate for the modulator. The opamp employs a dynamic common mode feedback circuit and has a simulated unity gain-bandwidth of 10MHz. The d.c. gain of an opamp is usually smaller close to its output swing limits. But since the opamp outputs are kept within ($-V_{ovl}$, V_{ovl}) most of the time, the nonlinearity due to the above effect is negligible.

A comparator design with a simple differential pair preamplifier with low bias current followed by a regenerative latch is used. A total of 7 clockboosters are needed, 4 larger ones (ϕ_1, ϕ_2, ϕ_{1a} and ϕ_{2a}) and 3 small ones.

4. Experimental Results

The microphotograph is shown in Fig. 6. The clockboosters occupy minimal area. The testing results are depicted in Fig. 7-9. As shown in Fig. 7 and 8, the modulator achieves over 12-bit dynamic range at V_{dd} down to 1.95V at 1MHz clock frequency, at which point the power consumption is 0.34mW. Better resolution is attained at higher V_{dd} but at the expense of higher power consumption. As shown in Fig. 7, the activations of the local feed-

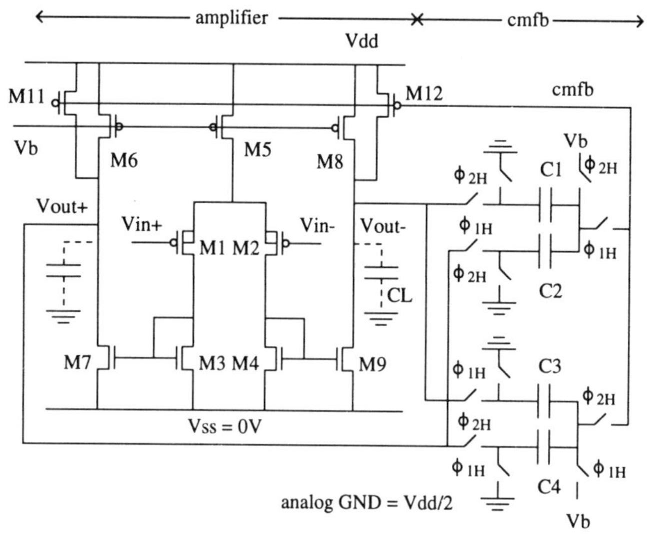

Figure 5: Opamp circuit schematic.

back loops due to overloadings at higher input level stabilize the modulator with almost no loss in S/(N+THD), despite the modest opamp gain and the approximate nature of the digital compensators. Thus the robustness of a single-stage ΣΔ modulator and the stability of a multi-stage ΣΔ modulator are achieved simultaneously. The dynamic range, which is defined as the difference in input level (in dB) between the point when S/(N+THD) = 0 and the point at which S/(N+THD) drops significantly, is also increased due to the stabilization effects of the local feedback loops. Fig. 9 shows the measured output spectrum at high input level at 1.95V supply. Table I summarizes the characteristics of the chip.

5. Conclusions

A LV/LP ΣΔ modulator implemented using a special high order single-stage architecture stabilized by local feedback loops has been presented. The difficulties involved in the analog circuit design at low V_{dd} using a standard CMOS process, namely insufficient gate drive for CMOS switches, modest d.c. gain of a simple LV opamp, are solved by a combination of novel modulator architecture and circuit approaches. The lower V_{dd}, together with a lower sampling frequency made possible by a stable high order modulator, reduces the power consumption of the analog modulator as well as the entire oversampled A/D converter without loss in resolution.

Acknowledgements

The authors would like to thank their colleague F. Chen for his many technical advices and help. The chip is fabricated by the Canadian Microelectronic Corporation. This work was supported by NSERC.

References

[1] S. H. Ardalan, J. J. Paulos, "An analysis of nonlinear behaviour in delta-sigma modulators," *IEEE Trans. Circuits Syst.*, vol. CAS-34, pp. 593-603, Jun. 1987.

[2] L. A. Williams, B. A. Wooley, "Third-order cascaded sigma-delta modulators," *IEEE Trans. Circuits Syst.*, vol. CAS-38, pp. 489-498, May. 1991.

[3] R. G. H. Eschauzier, et al., "A programmable 1.5V CMOS class-AB operational amplifier with hybrid nested Miller compensation for 120dB gain and 6MHz UGF," *ISSCC Digest of Technical Papers*, pp. 246-247, 1994.

[4] S. M. Moussavi, B. H. Leung, "High-order single-stage single-bit oversampling A/D converter stabilized with local feedback loops," *IEEE Trans. Circuits Syst.*, vol. 41, pp. 19-25, Jan. 1994.

[5] T. Cho, P. Gray, "A 10-bit, 20-Ms/s, 35-mW Pipeline A/D Converter," *Proceedings of the IEEE CICC*, pp.499-502, 1994.

Figure 6: Micro-photograph of the third-order modulator stabilized with local feedback loops. The digital compensators are also shown.

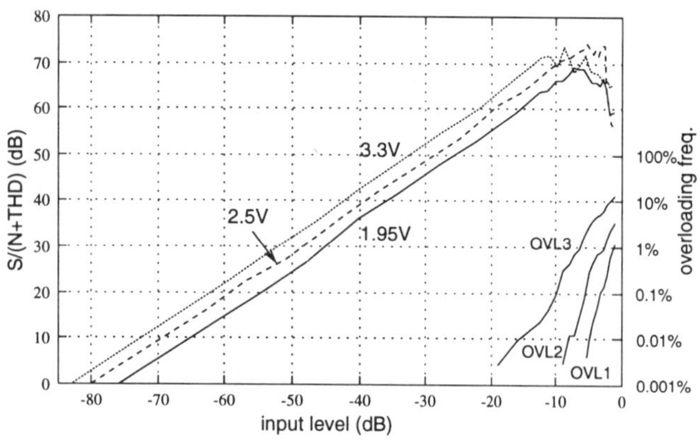

Figure 7: Measured S/(N+THD) versus input level at various supply voltages. The measured overloading frequencies of the integrators at Vdd = 1.95V are also shown.

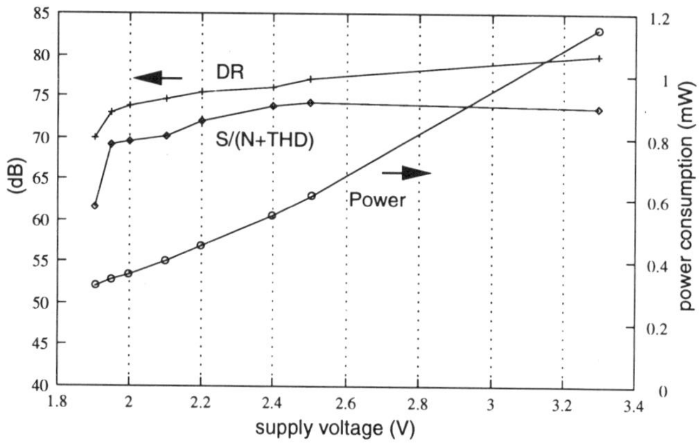

Figure 8: Measured peak S/(N+THD), dynamic range and power consumption as functions of supply voltage.

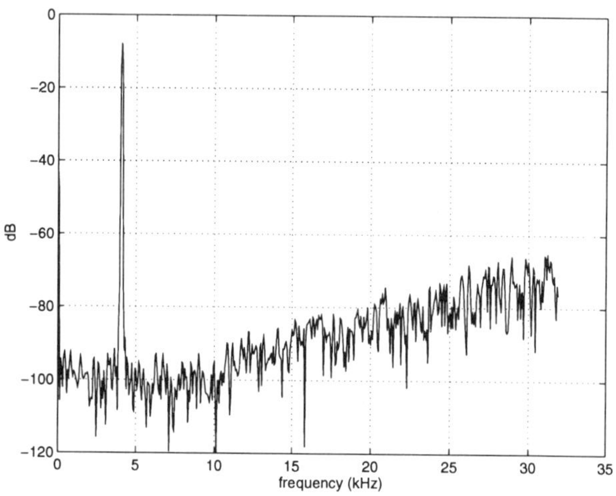

Figure 9: 16k-point FFT plot of modulator output at input = -6dB, Vdd = 1.95V.

Table I: Measured chip performance

technology	1.2-μm CMOS ($V_{tn} = 0.77V$, $V_{tp} = -0.79V$)
area	1.2mm x 2.8mm core
bandwidth	8-kHz
sampling freq.	1.024MHz (oversampling ratio = 64)
power	0.34mW@1.95V, 0.61mW@2.5V
peak S/(N+THD)	69dB@1.95V, 74dB@2.5V
dynamic range	73dB@1.95V, 78dB@2.5V

A 5v Single-Chip Delta-Sigma Audio A/D Converter with 111dB Dynamic-Range

Ichiro Fujimori, Kazuo Koyama
Asahi-Kasei Microsystems Atsugi, Kanagawa, Japan
David Trager, Fred Tam, Lorenzo Longo
Oasis Design Inc. Austin, Texas

Abstract-- *A 5v 24bit audio delta-sigma A/D converter has been developed. The single chip integrates stereo delta-sigma modulators, voltage reference and decimation filter. A 4th-order cascaded delta-sigma with use of a local feedback was employed to improve noise performance. A new multi-stage merged comb filter was used for reduction of digital noise. The 25.8mm² chip was fabricated in 0.7u CMOS. Measured results show 111dB dynamic-range and 102dB peak S/N+D.*

Introduction

In the past decade, oversampled delta-sigma modulation has been the technology of choice for audio converters, due to the insensitivity to component matching and the cost merit when using a fine-line CMOS process. Recent development of such converters can be divided into two categories. One is low-end applications like multi-media audio that require 16bit resolution or less. The other is high-end performance driven applications that require 18bits or more. Several high-end A/D converters with 18bit resolution have been previously reported [1][2], but all require +/-5v dual-supplies that limit the application to professional use. For consumer audio, operation with a 5v single-supply is mandatory. Also, these converters are in 2-chip configurations. A more cost effective single-chip solution is required for consumer applications. Until now, no 5v single-supply single-chip A/D converter with true 18bit performance has been reported.

This paper describes the development of a 5v 24bit audio A/D converter for high-end consumer applications. The single chip integrates stereo delta-sigma modulators (DSMs), voltage reference and decimation filter. The target dynamic-range is 110dB which is true 18bit performance. Two major challenges were overcome to accomplish this goal. One was the achievement of analog performance with reduced signal swing and head room at 5v operation. The other was the reduction of the effect of digital noise coupled onto the substrate. Detailed design regarding these issues will be described in the following sections.

Delta-Sigma Modulator Architecture

The key concern for the DSM architecture design is to allocate maximum noise budget to analog noise (thermal and 1/f noise), since analog noise determines power dissipation and analog circuit size. With a reduced overall budget due to small signal swing, the analog budget must be as large as possible for a practical realization. One direct approach is to reduce the DSM inband quantization noise so that its contribution is negligible within overall performance. Fourth-order noise shaping at 128 times over-sampling was chosen to meet this criteria.

An additional key design issue is to avoid instability and overload for large level input signals near full-scale (FS). One common solution is input gain scaling [3]. The concept of input gain scaling is to attenuate the signal level prior to the DSM and compensate the lost gain at the decimation filter afterwards. This allows the DSM to operate in a linear region where 1bit data duty cycle is closer to 50%. The disadvantage of input gain scaling is that analog and quantization noise generated in the DSM will also be increased by the gain compensation when referred to the A/D converter output. For this reason, a new DSM architecture that does not require input gain scaling was developed.

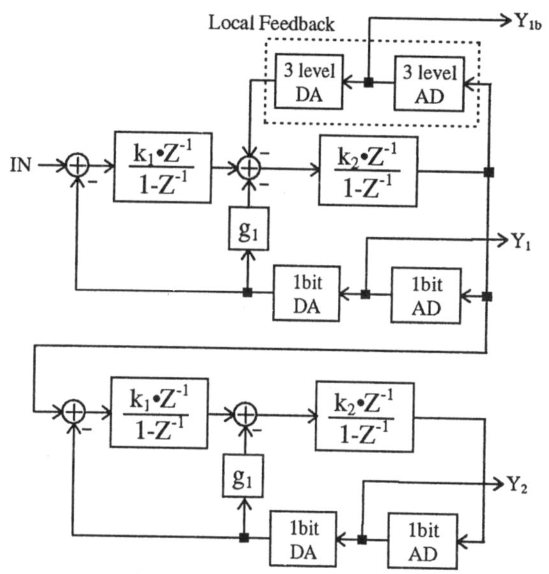

Fig.1 Delta-Sigma Modulator

The DSM system architecture is shown in fig.1. A 2-2 cascade DSM was chosen to implement 4th-order noise shaping. A cascade of inherently stable 2nd-order loops will maintain stability without input gain scaling. Quantization noise leakage caused by coefficient error of 1st stage integrator gains k1 and k2 will be suppressed by 2nd-order noise shaping [4]. Even though stability is ensured, a cascaded DSM will still encounter rapid increase of quantization noise with a large level input. This is usually referred to as overload and is caused by the non-linear nature of the 1bit AD. Since the 1bit DA can only feedback a limited and discrete level, the loop will enter a non-linear state when the 1bit AD input exceeds a certain

0-7803-3177-6 $5.00 © 1996 IEEE

IEEE 1996 CUSTOM INTEGRATED CIRCUITS CONFERENCE

level. In this design, instead of input gain scaling, the local feedback (LFB) technique [5] is used to prevent overload. In fig.1, a LFB is added to the 2nd integrator of the 1st stage (2nd LFB). The 2nd LFB will continuously monitor the output of the 2nd integrator and will control the swing by means of negative feedback. The AD and DA are 3-level so will only be active for large level inputs when overload occurs. The quantization noise of the output Y1b will be canceled by noise cancellation logic the same way as quantizaton noise of the 1st stage output Y1 is canceled in conventional cascaded DSM.

This LFB technique was introduced by Moussavi [5] in order to stabilize a 3rd-order single-loop. One drawback of using this technique in a single-loop is the necessity of a LFB around the 1st integrator (1st LFB) to prevent overload. The presence of the 1st LFB causes direct and severe quantization noise leakage with coefficient mismatch between the 1st LFB gain and noise cancellation logic. Also, the 1st LFB will add an analog noise source that is directly referred to the DSM input. The advantage of using the LFB technique in the cascaded arrangement is the omission of this 1st LFB. A cascaded DSM has a particular nature that unless the 2nd stage is in overload, even if the 1st stage is, no noise degradation will appear at the noise cancellation output. This is because 1st stage quantization noise is canceled. To prevent overload of only the 2nd stage, the 2nd stage input only needs to be controlled not to exceed a certain boundary regardless of whether the 1st stage is in overload state or not. For this task, the 2nd LFB alone is good enough. Coefficient error of the 2nd LFB gain will only cause 1st-order noise-shaped quantization noise leakage that is 60dB less severe than direct leakage at 128 times oversampling. A 2-2 cascade with 2nd LFB will prevent overload with a FS input even without input gain scaling.

Delta-Sigma Modulator Implementation

The DSM samples the analog input with a 6.144MHz rate that is 128 times the 48kHz A/D converter output rate (fs). A fully-differential SC implementation was chosen because of the better common-mode rejection of digital noise injected to the substrate and to the power supply.

Fig. 2 shows the 1st integrator of the 1st stage with the 1bit DA. The left side of capacitor Cs1 is switched in a cross-coupled fashion from differential input $V_{IN}+$ to input $V_{IN}-$, or vice versa. This will double the effective signal swing across Cs1 to improve the S/N ratio set by kT/C noise [1]. On the contrary, sampling by Cr1 in the 1bit DA is done in a single-ended manner between the reference voltages $V_{ref}+$, $V_{ref}-$ and analog common V_{CM}. This is to prevent Y1 data dependent loading on the reference during the integrating phase (ϕ_2). During ϕ_2, the right side of Cr1 will be connected to either positive or negative input of the opamp depending on Y1. Since the opamp has input offset, voltage across Cr1 will be data dependent. Therefore, if Cr1 samples in a cross-coupled fashion, charge transferred from reference during ϕ_2 will become data dependent. Any Y1 data dependency on the reference will modulate high frequency quantization noise to inband which degrades noise performance severely.

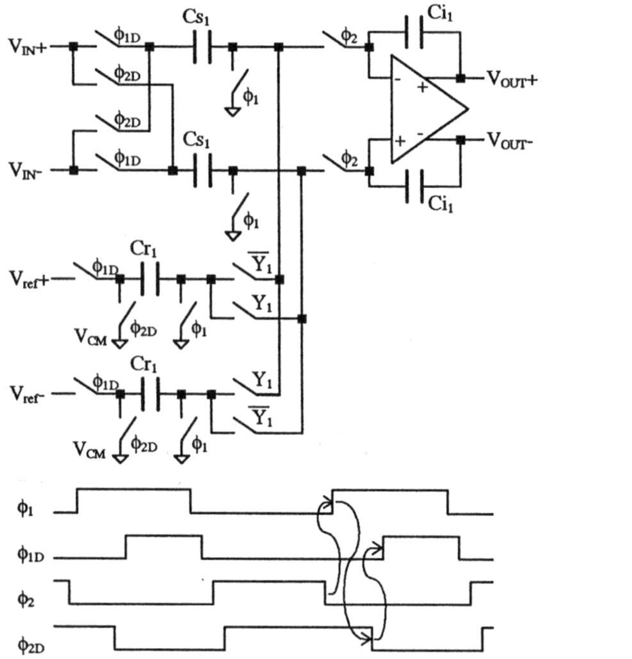

Fig.2 1st SC Integrator with 1bit DA

Even with single-ended sampling, the reference will still see data dependent loading during the sampling phase (ϕ_1). Since voltage dependency across Cr1 still remains after ϕ_2, the reference will see a data dependent load when recharging Cr1 during ϕ_1. To avoid this mechanism, a new 4-phase non-overlap clock scheme was developed. The difference from the conventional 4-phase clock [6] is the clock sequence which ϕ_1 goes high prior to ϕ_{2D} going low. This will make a short period where both terminals of Cr1 are connected to V_{CM}. This period will deload any data dependency across Cr1 before charging of the reference voltage during the sampling phase.

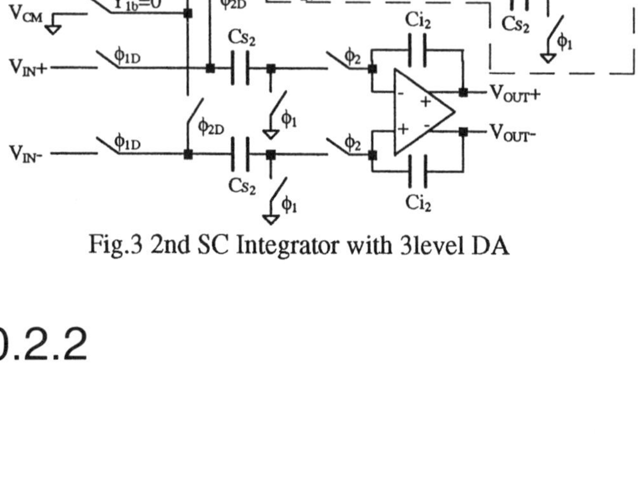

Fig.3 2nd SC Integrator with 3level DA

20.2.2

The 2nd integrator of the 1st stage and the 3-level DA of the 2nd LFB are shown in fig.3. A shared-capacitor switching scheme is used to minimize quantization noise leakage of the 2nd LFB. Capacitor Cs2 is shared by both input sampling and 3-level DA. A 3-level DA is implemented by connecting the left side of Cs2 to either Vref+, Vref- or Vcm depending on Y1b data. Sharing Cs2 will eliminate coefficient error of the LFB gain, thus minimizing quantization noise leakage. The drawback of shared capacitor implementation is the Y1b and Vin+/- dependent load on the voltage reference. This effect is equalized by a dummy SC network which samples Vin+ or Vin- and then discharges to the potential opposite from the shared capacitor network.

The SC opamp is shown in fig.4. The fully-differential folded-cascode structure achieves high bandwidth with capacitive loads. A SC divider is used in the common-mode feedback. Use of a pMOS differential pair eliminates need of chopper stabilization due to 30 times less 1/f noise compared to nMOS in this process. Fully-differential gain enhancement [7] is employed to increase output impedance of the cascode loads, which enhances both DC gain and PSRR. Insufficient DC gain will cause integrator gain error and will increase quantization noise leakage. For 5v operation, cascode loads must be kept in high-swing condition [8]. This condition can be met by using the high swing cascode bias potentials bpc, bnc as common mode references of the auxiliary opamps. The opamp gain bandwidth is 90MHz and the DC gain is 105dB.

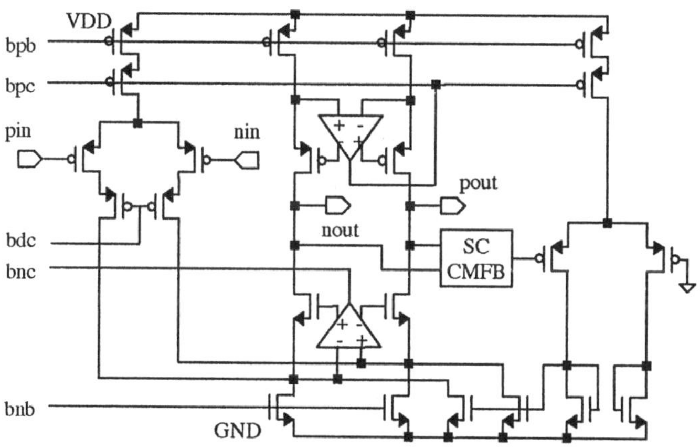

Fig.4 OPAMP Diagram

Decimation Filter

The digital decimation filter provides decimation by 128 and greater than 110 dB anti-aliasing protection, with less than 0.001dB passband ripple. The passband is specified from DC to 0.4535fs, and the stopband from 0.5465fs to 127.4535fs. In addition, the decimation filter is designed with sufficient performance to limit the total modulator quantization noise and numerical quantization noise to a negligible level. This practically ensures that the digital filter numerical performance has no effect on the overall A/D converter performance.

A major challenge in the design of the decimation filter was to develop and implement techniques to minimize the effects of digital switching noise on analog performance. The first technique was to minimize the power dissipation of the digital circuitry. This was accomplished primarily by reducing the operating voltage to 3.3v, reducing operating frequencies, and sizing transistors to minimize short-circuit current. A second technique was to develop filter architectures that have short critical paths, and operate at or below the analog sampling rate. This allows maximum time for the substrate to settle before the critical analog sampling events. A final technique used was to rigorously avoid operating digital circuits at 1/2 the analog sampling rate to avoid modulating quantization noise into the passband. To efficiently perform the decimation by 128, a multi-stage decimation filter was used. A front-end filter decimates by 64, and a back-end filter does the final decimation by 2.

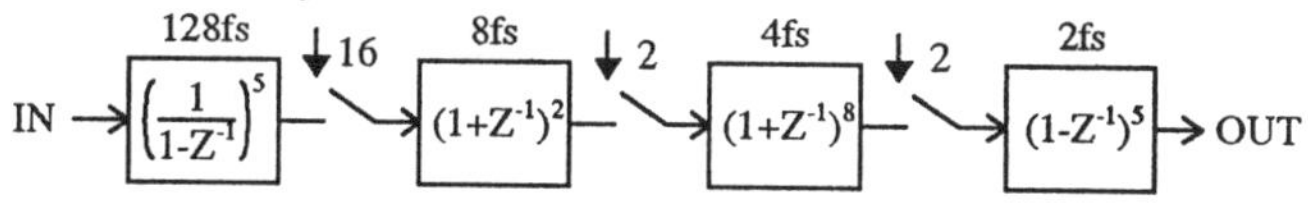

Fig.5 New Merged Multi-Stage Comb Filter

The front-end filter accepts input data at a high sample rate, and performs the bulk of decimation. Therefore, a simple structure must be used to limit implementation size. Comb filters, are an excellent choice for the front-end due to their simple implementation, and good frequency response characteristics [9]. Directly implementing the decimate by 64 front-end filter using a conventional comb filter would be prohibitively expensive due to the high order needed which requires a large input word length. Multi-stage comb filters [10] can be used to limit the maximum word length, but can result in additional add and delay elements compared to the conventional approach. For this project, a new merged multi-stage comb filter was developed which minimizes both the maximum word length and number of add and delay elements (fig.5). By implementing some comb filter orders as $1+z^{-1}$ in the final stages, the total add and delay elements can be minimized, while maintaining reduced word length. The new merged multi-stage comb filter can be implemented in 45% the area required for the conventional comb filter. This is important since separate structures are used for each channel to allow operation at analog sampling rate.

The back-end filter performs the final decimation by 2 and compensates for the passband droop introduced by the front-end filter. A 135-tap FIR filter is implemented as a polyphase filter, and common hardware is used for both channels. A bit-serial architecture is used which has several beneficial properties. First, the bit-serial network allows the maximum logic depth to be only 8 full-adders. Second, the bit-serial network is a hard-wired implementation of the filter coefficient multiplication and summation operations. This eliminates the need for a ROM which can inject a large amount of noise into the substrate. Finally, the bit-serial implementation can be operated at a low clock frequency, allowing a RAM timing scheme to be used which has no asynchronous switching activity. This ensures that all digital noise occurs at well-defined times.

Measured results

Measurements were taken with analog and digital power supplies of 5v and 3.3v, respectively. The output rate fs is 48kHz with a passband of DC to 22kHz. FS input range is +/-2.6v differential. Fig.6 shows a 2048 point FFT spectrum for a 1.5kHz sine-wave at -20dBFS. The spectrum has a flat noise floor without any spurious components. Measured S/(N+D) at -20dBFS is 91dB, yielding 111dB dynamic-range. Fig.7 shows S/(N+D) vs. input level for a 1.5kHz signal. Peak S/(N+D) is 102dB at -4dBFS. The A/D converter performance is summarized in table1. Fig.8 shows the chip micrograph. A 0.7u double-metal, double-poly Nwell CMOS process with low threshold devices was used. The die is 4.1mm x 6.3mm. Wide power supply lines are placed between analog and digital circuitry for physical isolation. The substrate ground is connected by 6 bonding wires for fast settling of digital noise coupled onto the substrate.

Table 1 A/D Converter Performance

Resolution	24bits
Full-scale input range	+/-2.6v (differential)
Dynamic range	111dB
Peak S/(N+D)	102dB
Channel separation	120dB
Bipolar zero offset	+/-10 LSB
Power dissipation: analog	540mW
digital	20mW
Chip size	4.1mm x 6.3mm
Package	28pin SOP

References

[1] I. Fujimori et al., "A fifth-order delta-sigma modulator with 110dB audio-band dynamic range," 93rd Audio Engineering Society Conv., San Francisco, Oct. 1992, preprint 3415

[2] P. Ferguson et al., "An 18b 20kHz dual delta-sigma A/D converter," in Proc. 1991 ISSCC., Feb. 1991, pp.68-69.

[3] N. Sooch, "Gain scaling of oversampled ADC," USP 4851841

[4] L. Longo and M. Copeland, "A 13 bit ISDN-band oversampling ADC using two-stage third order noise shaping," in Proc. 1988 CICC, May 1988, pp.21.2.1-4

[5] S. Moussavi and B. Leung, "High-order single-stage single-bit oversampling A/D converter stabilized with local feedback loops," IEEE Trans. on Circuits and Systems-II: Analog and Digital Signal Processing, Vol.41, No.1, Jan. 1994, pp.19-25

[6] D. Haigh and B. Singh, "A switching scheme for switch capacitor filters which reduces the effect of parasitic capacitances associated with switch control terminals," in Proc. 1983 IEEE Int. Symp. Circ.and Sys., May 1983, pp.586-589.

[7] J. Lloyd and H.S. Lee, "A CMOS OpAmp with fully-differential gain-enhancement," IEEE Trans. on Circ. and Sys.-II: Analog and Digital Signal Processing, vol.41,No.3, Mar. 1994.

[8] R. Gregorian and G.C. Temes, Analog MOS integrated circuits for signal processing, Wiley Interscience, 1986, pp. 247

[9] J. Candy and G. Temes, "Oversampling methods for A/D and D/A conversion," in Oversampling delta-sigma data converters. Theory, design and simulation, IEEE press, 1992

[10] S. Chu and C. Burrus, " Multirate filter designs using comb filters," IEEE Trans. Circuits and Sys., vol.CAS-31 pp.913-924, Nov. 1984.

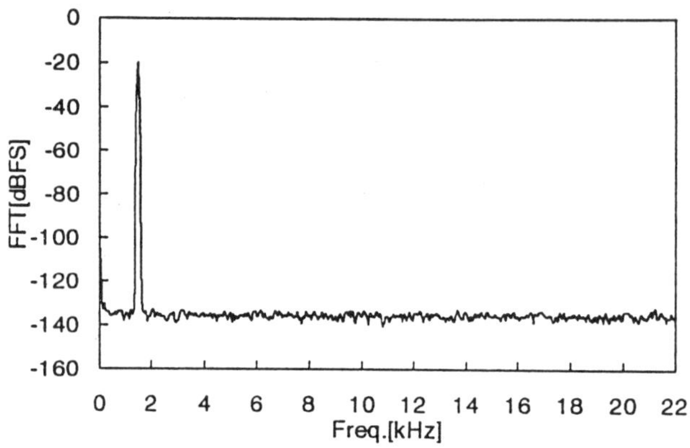

Fig. 6 FFT of -20dBFS 1.5kHz input

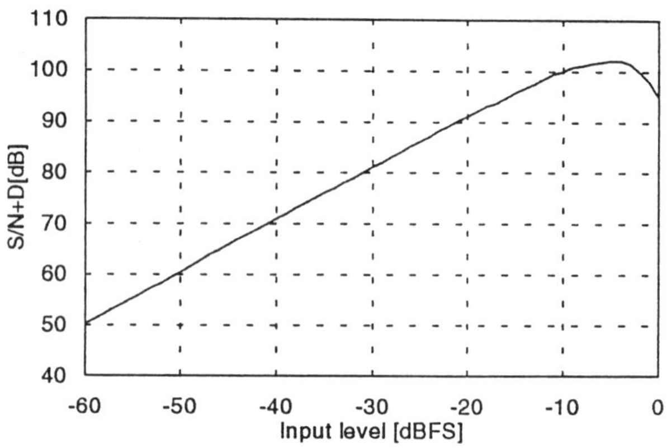

Fig. 7 S/(N+D) vs. Input level, 1.5kHz input

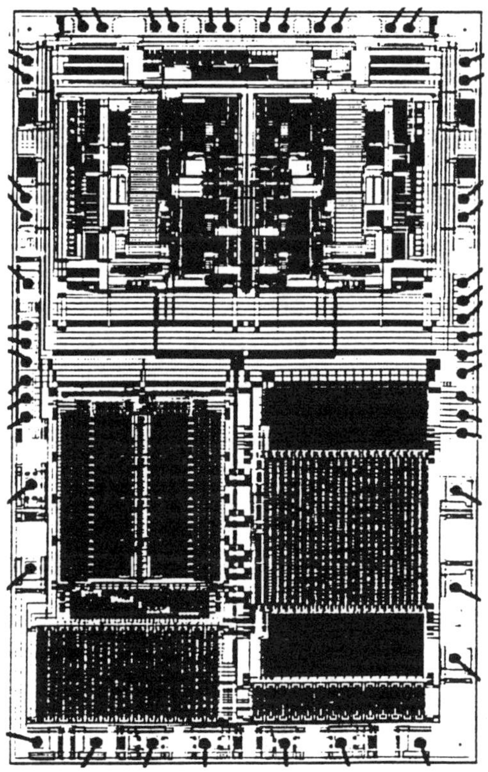

Fig. 8 Chip Micrograph

A 10-bit, 100 MS/s CMOS A/D Converter

K. Y. Kim, N. Kusayanagi, and A. A. Abidi

Integrated Circuits & Systems Laboratory
Electrical Engineering Department
University of California
Los Angeles, CA 90095-1594

Introduction

Applications such as high-end video signal processing, high-performance digital communications, and medical imaging require A/D converters with sample rates approaching 100 MS/s and a dynamic range at the Nyquist bandwidth close to 60 dB. In response to these needs, there is a continued search for architectures and circuit techniques enabling a monolithic A/D converter to meet these specifications with a reasonable chip area and power dissipation. It is of particular interest if such an A/D converter is fabricated in a standard CMOS technology.

Most published A/D converters which come close to these specifications are bipolar ICs [1-5]. Their high speed and wide dynamic range owes to the use of *open-loop* precise building blocks, including *low-offset* comparators. A recent A/D converter attains up to a 70 dB dynamic range with this approach [5]. CMOS A/D converters, by contrast, tend to use closed-loop, op amp-based circuits for precise analog signal processing, and closed-loop autozeroed comparators in quantizers resolving 4b or more. It is therefore difficult for these circuits to attain both the dynamic range and the speed of a bipolar ADC built with devices of comparable f_T. For instance, an open-loop folding and interpolation ADC in 0.8µm CMOS clocks at up to 75 MHz, but is limited to about 45 dB in dynamic range [6]. Device layout and other averaging techniques yield limited improvements in the accuracy of the intrinsically poorly-matched MOSFET. Parallelism is one reliable way to obtain a high throughput while retaining wide dynamic range. For instance, an array of identical high-resolution but low-speed ADCs may be *time-interleaved* in parallel to increase overall throughput, with proportional increase in hardware complexity and power dissipation [7].

This work addresses some of the known problems inherent in time-interleaved, or parallel, ADCs with a new architecture. A prototype of this architecture demonstrates, for the first time, 10-bit operation at 100 MS/s in a 1-µm CMOS technology.

The Parallel ADC Architecture

A 10-bit A/D converter must use some form of multi-stage quantization, with intermediate stages between each sub-quantizer to form and amplify the analog residue (Fig.1). It is straightforward in CMOS to make the residue amplifiers into sample-and-hold circuits, and thus to pipeline each sub-quantizer. The overall conversion rate may be increased *m*-fold in a multiplicity (*m*) of such pipeline ADCs in parallel, their clocks uniformly interleaved in time (Fig.2). The input waveform is sampled at a rate f_0 in round-robin sequence by the input track-and-hold of each pipeline ADC, so the system as a whole acquires mf_0 samples of the input per second. After the inherent latency, digitized samples representing the input waveform appear interleaved in time at the outputs of the *m* channels.

Errors in such an ADC arise from three sources. *Timing skews* in the instants when the various channels sample the dynamic analog input cause errors in the digitized samples proportional to the instantaneous slope of the input signal, which leads to distortion in the reconstructed digitized signal. *Gain dispersion* produces spurious in-band tones in the digitized spectrum [8]. *Dispersion in input offsets* among the channels results in fixed-pattern noise, which contributes tones at 1/*m* times the clock frequency and its integer multiples across the Nyquist bandwidth. Gain and offset dispersion, in principle, may be digitally calibrated and compensated, but it is very difficult to automatically eliminate timing skews between the strobes to each front-end T/H in the various channels.

There are four significant pieces of prior art relating to the parallel ADC architecture. Black & Hodges first demonstrated time-interleaved multiple flash A/D converters on the same IC [7]. This concept reappeared in a *m=4* flash A/D converter (not monolithic) that Hewlett-Packard uses in their high-speed digitizing oscilloscopes, where a *single full-speed T/H* acquires samples of the input, and each channel then reacquires the non-dynamic *held* sample for digitization [9]. Lewis & Gray showed that CMOS ADCs are well-suited to the pipeline architecture [10]. In later work, by combining four channels in parallel, Conroy, *et al.* [11] attained 8 bits resolution at 80 MHz in a 1-µm CMOS implementation. At the same time Yotsuyanagi *et al.* at Sony applied a full-speed track-and-hold in front of a 50 MHz 0.8-µm CMOS ADC with *m=2* [12], much as in the earlier work by H-P.

This work focuses on how to reconfigure a parallel, pipelined ADC to solve all three of the known limitations. Only the simplest calibration methods are used when necessary. Key features and the underlying rationale of this architecture (Fig.3) are now listed.

1 A *full-speed* (100 MHz) *single track-and-hold* at the front-end eliminates timing skews. The T/H output does not reset between samples, giving subsequent circuits the full clock period to operate on the held sample.

2 A *full-speed* (100 MHz) *4-bit quantizer and reconstruction DAC* following the track-and-hold forms the *residue*, and distributes it unamplified to time-interleaved channels. Subsequently, the residue needs only be quantized to a 7-bits, which means that typical mismatches in channel gains (~1 part in 2^9) and in their input-referred offsets (~0.5 mV rms) may be tolerated with no adverse effect on accuracy. Programmable inter-channel offset correction is added to a later stage of the pipeline.

3 The residue is distributed to *two channels (m=2),* each with a throughput rate of 50 MHz.

4 Each channel consists of a *cascade of two 4-bit quantizers,* which are scaled replicas of the first quantizer.

Research supported by Analog Devices, Burr-Brown, Signal Processing Technologies, Raytheon Semiconductor, and the State of California MICRO Program. N. Kusayanagi was on leave from Yokogawa Electric Co., Tokyo, Japan.

0-7803-3177-6 $5.00 © 1996 IEEE

IEEE 1996 CUSTOM INTEGRATED CIRCUITS CONFERENCE

5 The *residue is amplified* 8× between quantizers. To obtain settling to better than 7b accuracy, the first residue amplifier consists of a closed-loop 50 MHz gain-of-2 amplifier, which distributes its output to two time-interleaved 25 MHz gain-of-4 amplifiers. Each feedback amplifier thus operates at the same *peak* gain-bandwidth product. Prior to the third quantizer, however, as the residue need only be specified to 4b accuracy, a single 50 MHz gain-of-8 amplifier suffices.

The operation of the A/D converter is now explained in reference to the timing diagram (Fig.4). A capacitor and switch acquires a sample of the balanced input voltage, and at $t=0$ the capacitor connects in feedback around the T/H op amp. The first 4-bit quantizer follows the T/H output from $t=0$ to $5ns$, and then it is strobed. At $t=6ns$, the quantizer output selects one tap from a 4b resistor-ladder DAC to reconstruct the digital code as an analog threshold. Meanwhile over $t=0$ to $10ns$, the Channel 1 gain-of-2 amplifier, say, is in reset mode as its input capacitor follows the T/H output. At $t=10ns$ this capacitor is switched to the reconstruction DAC output, and the op amp is released from reset to amplify the difference charge which constitutes the residue. Meanwhile, the corresponding amplifier in Channel 2 settles to 2× the previous sample's residue, and now starts to follow the T/H output to acquire the next sample.

Overall 10b linearity requires that the input T/H be linear to better than 10b, and that the 15 thresholds of the first reconstruction DAC should be 10b accurate. Although neither gain inaccuracy nor input offset at the input T/H affect overall linearity, the residue amplifier gains following the input T/H must be 7b accurate. Their input-referred offset may also be any reasonable value, but the *difference* in these offsets between the two channels must be less than ½ LSB at the 10b full-scale to avoid a noticeable fixed-pattern at the output. An inter-channel offset difference may be corrected with a digital word at the input of the last residue amplifier.

Circuit Description

Input Track-and-Hold Circuit

The input track-and-hold provides a balanced non-reset output, while acquiring a new sample every 10 ns. A capacitor tracks the input over 5 ns, and then transfers the charge to a non-reset track-and-hold circuit, which holds the sample over a full 10 ns, allowing the quantizer to settle, and the residue to be formed. This contributes less thermal noise than two reset track-and-hold circuits in cascade to accomplish the same function.

A passive capacitor-switch circuit tracks the input with a 300 MHz wide bandwidth (Fig.5(a)). The turnoff time of the switch connected to a fixed common-mode voltage defines the sampling instant, and is not influenced by the instantaneous value of the input voltage. The non-reset buffer (Fig.5(b)) [13] makes use of the balanced op amp outputs to remove memory of the previously held sample, and to charge the hold capacitor to exactly the voltage on the sampling capacitor. One disadvantage of this approach is that any negative feedback action through one hold capacitor is exactly nulled by positive feedback through the other capacitor, so the hold capacitor pair on each side of the op amp appears as an input load without aiding the feedback factor.

The track-and-hold, and all residue amplifiers, use a high-speed, gain-boosted super-cascode op amp [14] capable of 2V ptp output voltage swing. The T/H op amp, which drains a 28 mA bias current, is sized to drive a total load capacitance of 9.6 pF at 100 MHz. When sampling a 2V ptp, 50.001 MHz input sinewave at 100 MHz, the simulated 3rd harmonic distortion at the T/H output is –66 dBc.

4b Flash Quantizers

The fast, low-latency differential flash quantizer used in all blocks comprises comparators to amplify the difference between the balanced analog input and differential taps taken from a resistor ladder (Fig.6). The ladder folded back on itself at midpoint conveniently illustrates the concept, but on the IC it is laid out as a straight piece of polysilicon to avoid corner discontinuities in current flow which detract from 10b accuracy.

The high-speed comparator with two differential input ports consists of a two-stage preamplifier followed by a latch (Fig.7). The preamplifier senses the difference between two balanced inputs by subtracting the output currents of two differential pairs, and amplifies it 5× prior to the latch input. Due to this large gain, and with careful latch design so there is no large common-mode jump at strobing, only mismatch in the input differential pair FETs determines the comparator offset. In a first prototype of the ADC, the measured offset in comparators across the array was 9 mV rms, which is small enough for satisfactory operation with the 62.5 mV LSB in this ADC. The bandwidth of the two-stage cascaded preamplifier is 250 MHz, sufficient to track a full-scale Nyquist frequency input.

Each differential pair in the first preamplifier stage compares one side of the balanced input with one side of the balanced reference tap voltage. The output common-mode of the T/H is slaved to the common-mode voltage set by the center tap of the resistor ladder. Thus, the differential pairs in the critical comparator detecting the closest threshold to the input are always biased at maximum gain. Any error or noise in the input or threshold common-mode values is subtracted at the comparator output. However, the input common-mode range of the preamp first stage must be equal to the full-scale voltage.

The latch samples the preamplified error voltage through switches onto two common-source PFETs, which are the loads for a cross-coupled pair of NFETs. Bootstrap capacitors speed regeneration. The simulated regeneration time constant of the loaded latch is 0.27ns. When combined with the dynamics of the following flip-flop, a bit-error rate less than 10^{-10} is projected at 100 MHz.

NOR gates decode the thermometer code at the flip-flop outputs to turn on one of the 15 switches connected to a second resistor ladder, and the selected tap on the ladder reconstructs the quantizer output code as an analog threshold. The physical layout (Fig.8) illustrates how the signal flows unidirectionally from the quantizer, through the encoder, into the reconstruction DAC. Routing back the encoder output to select a tap from the quantizer reference ladder will lead to a prohibitively large delay due to the wiring capacitance. Also, the kickback charge injected by the regenerating comparators on the quantizer reference ladder will cause the reconstructed output to glitch. The separate ladder for the DAC eliminates both these problems. To attain 10b accuracy in its tap voltages, this ladder is made wider than the quantizer ladder, resulting in a total polysilicon resistance of 150Ω. The first two taps at each end are discarded to eliminate edge effects. Measurements across 20 ICs of a first prototype indicate a worst-case 0.6 LSB INL across this ladder; the average INL is much lower.

The preamps in the 2nd and 3rd quantizer constitute a large fraction of the capacitive load on the residue amplifiers which must drive them. In each case, the input capacitance of the quantizer is halved by attaching the first preamplifier only on *every other tap* of the resistor ladder, and interpolating their outputs with the second stage (Fig.9). The slightly longer propagation delay through this arrangement is tolerable at the 50 MHz quantization rate.

Residue Amplifiers

The gain of the first bank of residue amplifiers is sufficiently accurate to align its full-scale output voltage with the reference of the subsequent quantizer within ½LSB at 7b. Using op amps with a large DC gain and a short settling time, the overall gain accuracy of these SC amplifiers is set by capacitor ratio, which well exceeds the 7b requirement when a minimum capacitor size of about 1 pF is used. The input offset of these amplifiers is autozeroed so as not to cause any loss in dynamic range, but final adjustment of offset differences is made later.

A folded cascode auxiliary amplifier in the gain-boosted op amps guarantees a DC gain greater than 70 dB. In autozero-mode, the input is shorted to the output common-mode voltage. However, for maximum swing in amplify-mode, this required common-mode voltage lies outside the input common-mode range. Therefore, the reference in the common-mode feedback loop switches between 2.1V in autozero-mode and 2.7V in amplify-mode. For acceptably good settling, the common-mode feedback loop bandwidth is almost as wide as that of the signal feedback loop around the op amp.

The 4× residue amplifiers operate at 25 MHz, but drive a larger load capacitance than the 2× amplifiers, and thus they become the bottleneck to overall speed. A shorter settling time is obtained than in a conventional implementation with feedback factor of 1/4 by sampling the input voltage on both the input and feedback capacitors during autozero, such that a capacitor ratio of 3, with feedback factor of 1/3, yields a gain of 4× [10]. The multiplexed output of the two 4× residue amplifiers at 50 MHz is a non-reset waveform.

The second bank of residue amplifiers consists of a 2× amplifier cascaded with a 4× amplifier, both operating at 50 MHz. Incomplete settling is no longer a source of inaccuracy for the final 4b quantizer. A binary-weighted array of small capacitors injects a programmable correction charge derived from a DC reference into the inputs of the 2× amplifiers for correction of inter-channel offset.

Digital Circuits

Each quantizer encodes its output as Gray code, and one bit of the output is dedicated to error-correction between quantizers. All error-correction logic and pipeline delay registers are on-board. Two non-overlapping phases are generated on-chip at the externally-supplied clock frequency, as well as at one-half and one-fourth this frequency. Differential output buffers drive ECL logic levels into a remotely terminated logic analyzer without producing large data-dependent current glitches on the chip power supplies.

Experimental Results

A prototype of this pipeline ADC, implemented in a 1-μm CMOS process through MOSIS, occupies an active area of 50 sq mm (Fig.10), and dissipates 1100 mW at 100 MHz from a 5V supply. The active area may be reduced by more aggressively scaling down some of the later quantizers, or better yet, by using a more efficient implementation of a single-stage 50 MHz, 7b ADC in the two channels following the full-speed front-end. However, the primary goal of this work was to demonstrate the viability of an architecture, and therefore the high-performance, but large, front-end blocks are re-used in the subsequent stages.

The ADC performance is evaluated at various conversion rates, with various input frequencies. A one-time test at a low conversion rate initially digitizes the inter-channel offset mismatch, and on this basis offset-correction words are programmed at the third residue amplifier. The measurements are summarized in Figs. 11 and 12. When combined with a peak measured DNL of 1 LSB at 90 MHz, these results indicate true 10-bit performance with 60 dB signal/noise+distortion at low conversion rates, and a spurious-free dynamic range of 75 dB. At a fixed low input frequency, there is a loss of only 3 dB in SNDR at a conversion rate of 80 MHz. However, slew-rate limit induced distortion appears to in the T/H at high analog input frequencies, possibly because the capacitive load was inadequately modelled during ADC design.

Three representative reconstructed spectra (Fig. 13 (a), (b), and (c)) illustrate the very low levels of spurious tones, fixed-pattern noise, and tones induced by channel-mismatch in this ADC, which verify that this architecture accomplishes its intended purpose. The dynamic performance is as yet unattained by any other CMOS ADC.

References

1 R. Petschacher, B. Zojer, B. Astegher, H. Jessner, and A. Lechner, "A 10b 75 MSPS Subranging A/D Converter with Integrated Sample and Hold," *IEEE J. of Solid State Circuits*, vol. 25, no. 6, pp. 1339-1346, 1990.

2 P. Vorenkamp and J. P. M. Verdaasdonk, "A 10b 50MS/s Pipelined ADC," presented at *Int'l Solid State Circuits Conf.*, San Francisco, pp. 32-33, 1992.

3 W. T. Colleran and A. A. Abidi, "A 10-b, 75-MHz Two-Stage Pipelined Bipolar A/D Converter," *IEEE J. of Solid-State Circuits*, vol. 28, no. 12, pp. 1187-1199, 1993.

4 K. Sone, Y. Nishida, and N. Nakadai, "A 10-b 100-Msample/s Pipelined Subranging BiCMOS ADC," *IEEE J. of Solid-State Circuits*, vol. 28, no. 12, pp. 1180-1186, 1993.

5 F. Murden and R. Gosser, "12b 40 MSample/s Two-Stage A/D Converter," presented at *Int'l Solid State Circuits Conf.*, San Francisco, pp. 216-217, 1995.

6 B. Nauta and A. G. W. Venes, "A 70 MSample/s 110 mW 8b CMOS Folding Interpolating A/D Converter," presented at *Int'l Solid State Circuits Conf.*, San Francisco, pp. 214-215, 1995.

7 W. C. Black, Jr and D. A. Hodges, "Time Interleaved Converter Arrays," *IEEE J. of Solid State Circuits*, vol. SC-15, no. 6, pp. 1022-1029, 1980.

8 Y.-C. Jenq, "Digital Spectra of Nonuniformly Sampled Signals: Fundamentals and High-Speed Waveform Digitizers," *IEEE Trans. on Instrumentation & Measurement*, vol. 37, no. 6, pp. 245-251, 1988.

9 K. Poulton, J. J. Corcoran, and T. Hornak, "A 1 GHz 6 bit ADC System," *IEEE J. of Solid State Circuits*, vol. SC-22, no. 6, pp. 962-970, 1987.

10 S. H. Lewis and P. R. Gray, "A Pipelined 5 MS/s 9-bit Analog to Digital Converter," *IEEE J. of Solid State Circuits*, vol. SC-22, no. 6, pp. 954-961, 1987.

11 C. S. G. Conroy, D. W. Cline, and P. R. Gray, "An 8-b 85-MS/s Parallel Pipeline A/D Converter in 1-μm CMOS," *IEEE J. of Solid State Circuits*, vol. 28, no. 4, pp. 447-454, 1993.

12 M. Yotsuyanagi, T. Etoh, and K. Hirata, "A 10-b 50-MHz Pipelined CMOS A/D Converter with S/H," *IEEE J. of Solid State Circuits*, vol. 28, no. 3, pp. 292-300, 1993.

13 G. Nicollini, P. Confalonieri, and D. Senderowicz, "A Fully Differential Sample-and-Hold Circuit for High-Speed Applications," *IEEE J. of Solid State Circuits*, vol. 24, no. 5, pp. 1461-1465, 1989.

14 K. Bult and G. Geelen, "A Fast Settling CMOS Op Amp for SC-Circuits with 90 dB DC-Gain," *IEEE J. of Solid State Circuits*, vol. 25, no. 6, pp. 1379-1384, 1990.

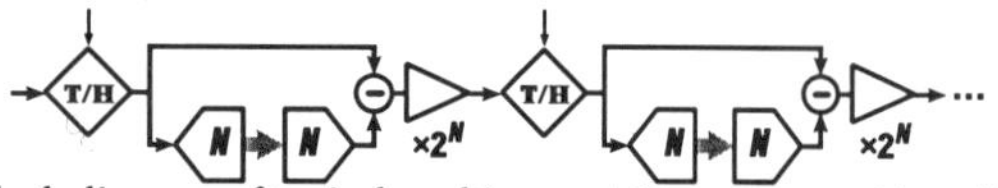

Fig.1 Block diagram of typical, multi-stage A/D converter, with residue amplification and feedforward.

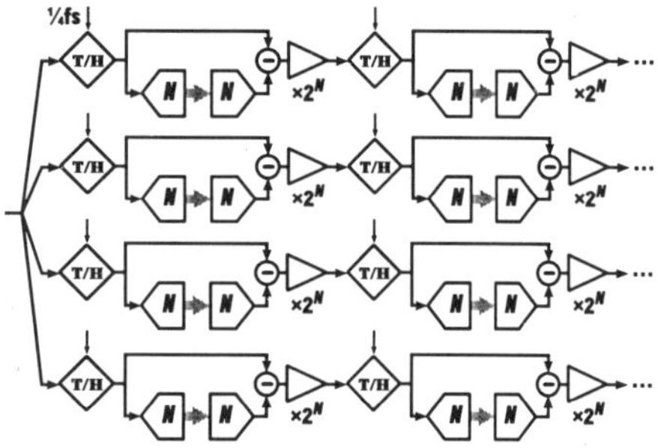

Fig.2 Throughput rate of ADC may be increased by time-interleaving the operation of multiple channels in parallel.

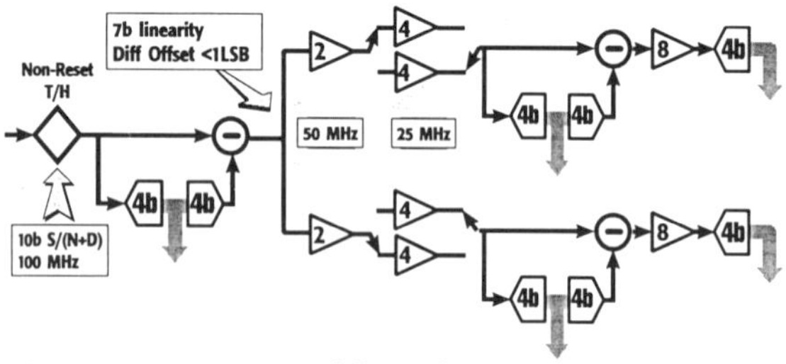

Fig.3 Our A/D converter uses a full-speed single-channel T/H, followed by a single full-speed quantizer. The *residue* is distributed to two channels for amplification and subsequent quantization.

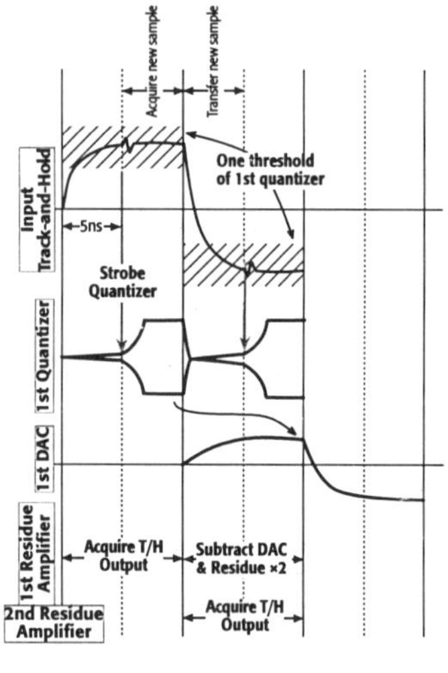

Fig.4 Timing diagram of our ADC front-end.

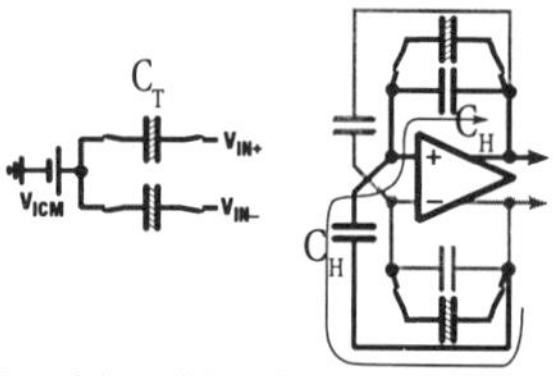

Fig.5 (a) Wideband passive switch/capacitor network tracks balanced analog input. (b) Capacitor transfers sample to non-reset amplifier.

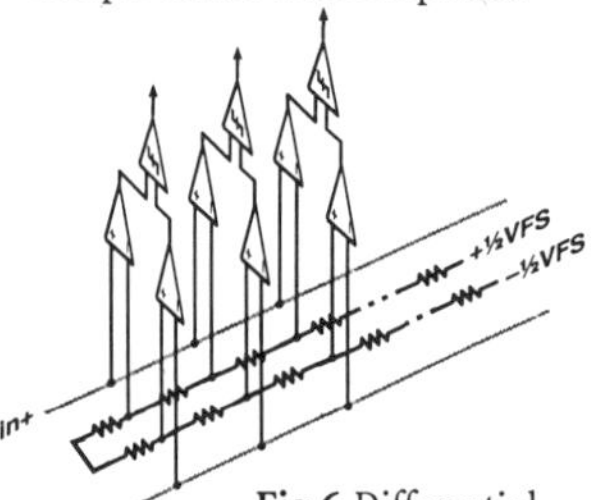

Fig.6 Differential flash quantizer.

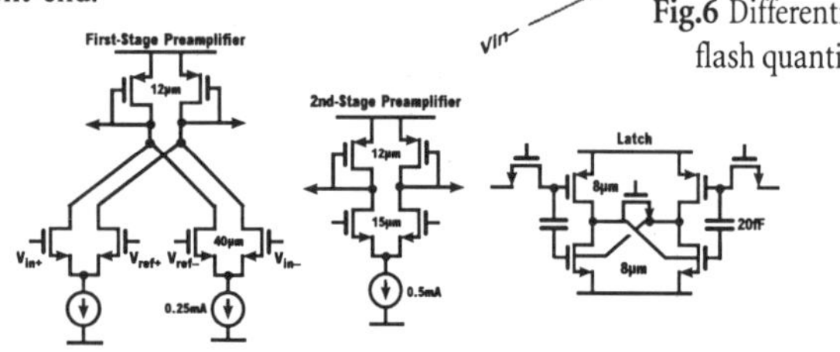

Fig.7 Each comparator consists of a differential differencing amplifier, a second stage amplifier, and a high-speed latch.

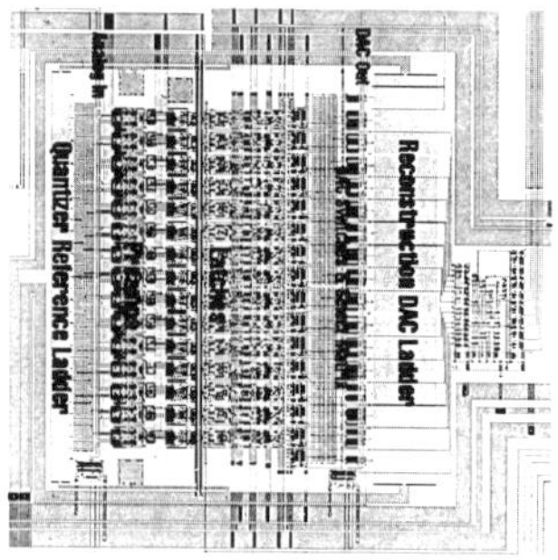

Fig.8 Layout of first 4-bit quantizer and reconstruction DAC.

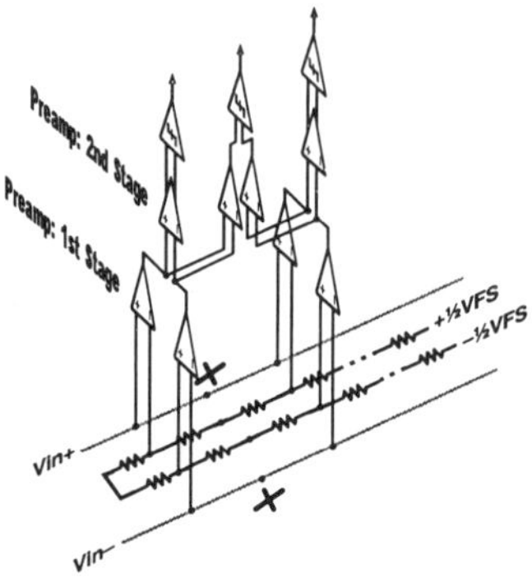

Fig.9 Preamp interpolation in 2nd and 3rd stage quantizers halves the input capacitance.

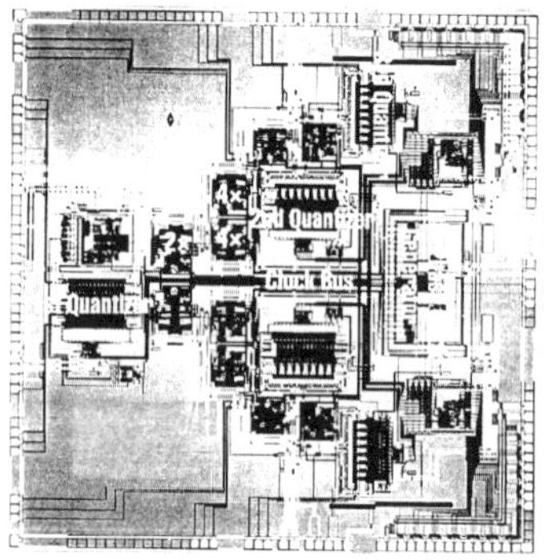

Fig.10 ADC chip layout, showing the major blocks.

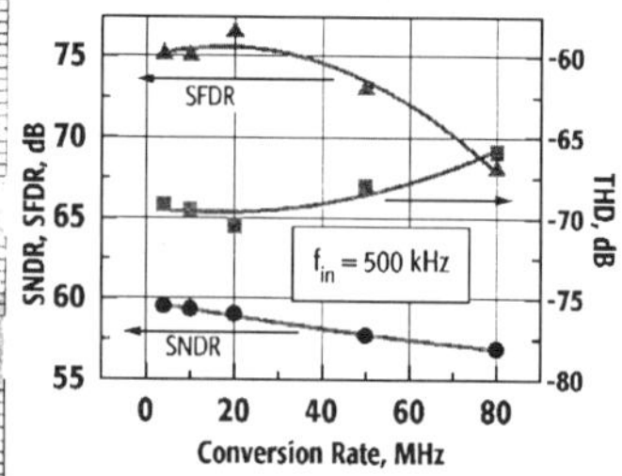

Fig.11 Measured SNDR, SFDR, and THD as a function of ADC clock rate.

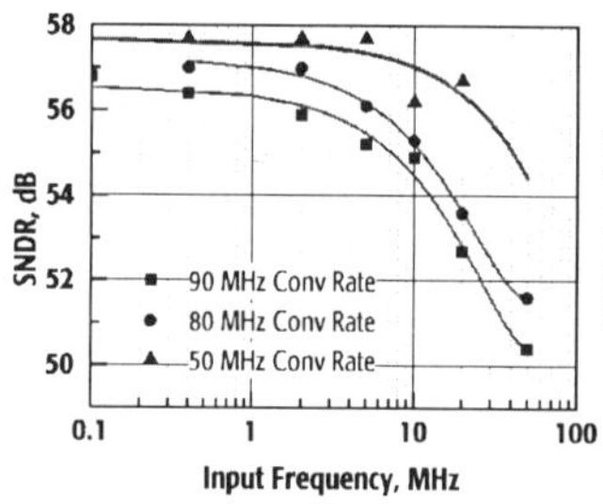

Fig.12 Measured dynamic performance of ADC at clock rates of 50, 80, and 90 MHz.

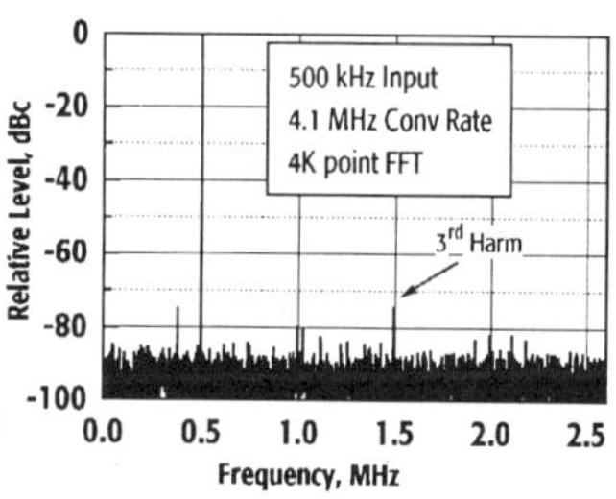

Fig.13 (a) Reconstructed spectrum at low conversion rates.

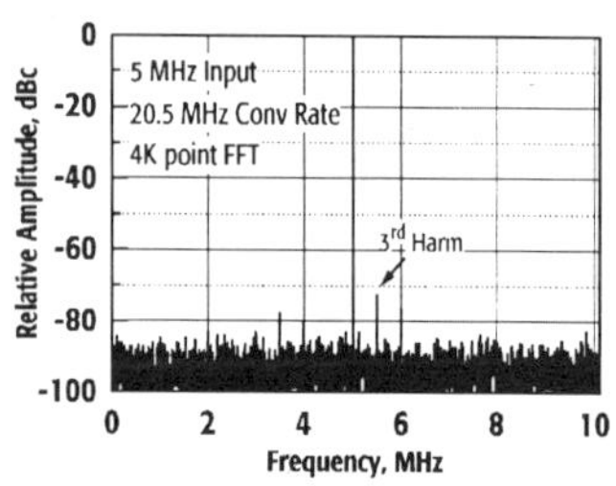

Fig.13 (b) Reconstructed spectrum at medium conversion rates.

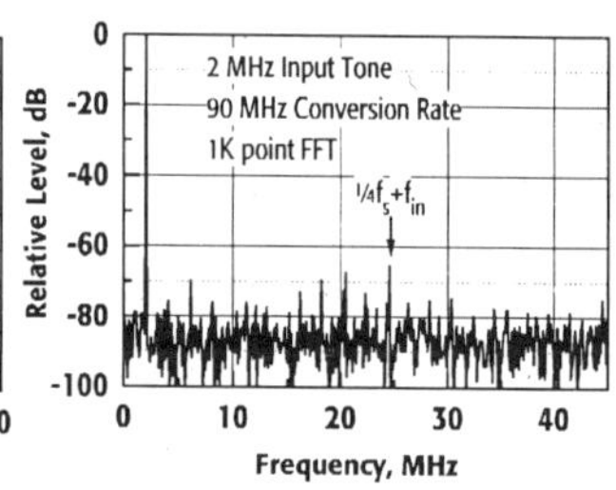

Fig.13 (c) Reconstructed spectrum at high conversion rates.

An 8-Bit 50+ Msamples/s Pipelined A/D Converter With An Area And Power Efficient Architecture

K. Nagaraj, H.S. Fetterman, R.S. Shariatdoust, J. Anidjar
S.H. Lewis*, J. Alsayegh and R.G. Renninger

AT&T Bell Laboratories, Allentown, PA
* Dept. of EECS, University of California, Davis, CA

Abstract

An efficient architecture for a pipelined A/D converter is described. By sharing amplifiers along the pipeline and also completely eliminating the amplifier from the last stage, an 8-bit converter is realized using just 3 amplifiers (instead of 7 amplifiers with a conventional pipeline architecture). By using two such pipelines in parallel, a sampling rate of over 50 Msamples/s has been achieved in a 0.9-μm CMOS technology.

Introduction

High-speed, medium-resolution A/D conversion is important in a wide variety of commercial applications, including digital video, data communications and storage, and medical imaging. Conversion rates of at least 50 Msamples/s with about 8-bit resolution are often required. A pipelined architecture is attractive for such applications. However, straightforward pipelined converters are not generally fast enough in 1-μm CMOS technologies to reach conversion rates of 50 Msamples/s. To overcome this problem, more than one pipelined A/D converter can be time interleaved [1-3]. However, this increases the required power dissipation. One approach to reduce power dissipation is to reduce capacitor sizes as the stage number increases [4]. This approach has the disadvantage that it requires a new design and layout for each stage. A possible low-power approach that has not received much attention is to reduce the number of amplifiers in a pipelined converter. In a parallel pipelined A/D converter, this can be accomplished by sharing amplifiers across channels using double sampling [5]. A key disadvantage of this technique is that the required layout is difficult. Also, this architecture does not possess the desirable modularity that would allow a single pipeline to be used in other configurations. This paper presents a prototype pipelined A/D converter with parallelism that shares amplifiers along each pipeline (instead of across two pipelines) to reduce the power dissipation and area.

Review of the Basic Pipelined A/D Converter

Fig. 1 shows the circuit schematic of one stage of a typical '1 bit-per-stage' pipelined converter [6]. N-1 such stages are required for N bits of resolution. Each stage consists of two parts: an analog arithmetic unit built around an operational amplifier and two identical capacitors C_1 and C_2, and a decision circuit made of two comparators. Its operation is as follows. During clock phase 1, both the capacitors sample the analog residue $V_{RES}(n-1)$ from the previous stage. During clock phase 2, C_2 is connected around the amplifier whereas the bottom plate of C_1 is connected to $+V_R$, $-V_R$ or 'zero', depending on the output of the decision circuit in the previous stage (represented by D_{n-1} in Fig. 1). This operation results in a new analog residue $V_{RES}(n)$ at the amplifier output. This is compared with the voltages $\frac{1}{4}V_R$ and $-\frac{1}{4}V_R$ in the decision circuit. The comparator outputs are encoded into a two-bit binary word D_n which is passed on to the next stage for controlling its arithmetic operation. The outputs of the decision circuits from all the stages in the pipeline are processed by a digital-error-correction logic (not shown in Fig. 1) that eliminates the redundancy. Since the required closed-loop gain of the amplifier is only two here, the resulting bandwidth of the amplifier is large. Also, the built-in redundancy provides a large tolerance to decision errors.

Amplifier Sharing

In Fig. 1, no attempt is made to sample the amplifier offset when the input is sampled (phase 1), unlike in conventional high-precision switched-capacitor circuits. Eliminating the need for offset cancellation is an important property that results from the use of redundancy. After the redundancy is eliminated by digital error correction, amplifier offsets no longer cause linearity errors in the A/D converter. Instead, they simply cause an input-referred offset voltage. As a result, the amplifier is idle during phase 1. Taking advantage of this, we can time-share the amplifier with the adjacent stage in the pipeline [7,8], as shown in Fig. 2. Here, the capacitors C_1 and C_2 are part of the nth stage whereas C_3 and C_4 are part of the the $(n+1)^{th}$ stage. The two switched-capacitor networks operate on opposite clock phases, with the amplifier alternating between them. The operations occurring when a sample passes through this composite stage are as follows. During phase 1, C_1 and C_2 sample $V_{RES}(n-1)$. During phase 2, the residue $V_{RES}(n)$ is developed by the combination of C_1, C_2 and A_1 operating under the control

0-7803-3177-6 $5.00 © 1996 IEEE

of the digital input D_{n-1}. This is sampled by C_3 and C_4 and is simultaneously processed by the decision circuit 1, to generate the digital signal D_n. During phase 1 of the next clock interval, C_3, C_4 and A_1 combine to generate the analog residue $V_{RES}(n+1)$ under the control of D_n. This is processed by the decision circuit 2, to generate the digital output D_{n+1}.

Effects of Amplifier Sharing :

The amplifier sharing introduces several effects. Firstly, the additional switches that are used to implement amplifier sharing, introduce series resistances which in combination with the amplifier input capacitance, affect the settling behavior of the stage. The switch resistances can be reduced by using large switches at the expense of a potential increase in offsets due to charge feedthrough. The amplifier input capacitance can be reduced by using source followers [9]. In the present implementation however, simulations show that this effect can be kept insignificantly small without the need for unreasonable switch sizes or source followers at the amplifier inputs.

Another result of amplifier sharing is that the non-zero input voltage of the amplifier (resulting from finite amplifier gain) is never reset. Thus, every input sample is affected by the finite-gain error component from the previous sample. In conventional 'double sampled' switched-capacitor circuits, the finite-gain error is a linear function of the previous sample. This simply results in a low pass filtering effect. In the present situation, the finite-gain error is a non-linear function of the previous input sample, thus producing a more complex effect on the conversion. However, simulations show that with an amplifier gain of 60 dB, the magnitude of this error is insignificant at the 8 bit level.

Further Reduction of Amplifiers

In a '1 bit-per-stage' pipeline, every stage except the first, yields one data bit and a carry bit which is used to correct the result from the previous stage. The first stage yields two data bits. Thus, an 8-bit converter would employ 7 stages. With the amplifier sharing described above, the first 6 stages would use 3 amplifiers, but the 7th stage alone would need a fourth amplifier. The fourth amplifier here is not used efficiently. This is avoided in the present implementation by converting the last stage into a successive-approximation stage that uses only a decision circuit and no amplifier. With this, an 8-bit pipeline only needs 3 amplifiers. The particular scheme employed here has the desirable feature that the last stage uses exactly the same type of decision circuit that is used in the other stages and produces exactly the same output format. To understand the operation of this, consider the case where this function is realized in the conventional algorithmic manner, as in the previous stages. The stage would have to generate a residue $V_{RES}(7) = 2V_{RES}(6) + D_6 V_R$, and then compare it with $\frac{1}{4}V_R$ and $-\frac{1}{4}V_R$ to determine D_7. Using some simple arithmetic manipulations, we can show that this operation is exactly equivalent to the following.

If $D_6 = 1$ compare $V_{RES}(6)$ with $-\frac{3}{8}V_R$ and $-\frac{5}{8}V_R$.

If $D_6 = -1$ compare $V_{RES}(6)$ with $\frac{5}{8}V_R$ and $\frac{3}{8}V_R$.

If $D_6 = 0$, compare $V_{RES}(6)$ with $\frac{1}{8}V_R$ and $-\frac{1}{8}V_R$.

If the voltages $\pm\frac{1}{8}V_R$, $\pm\frac{3}{8}V_R$ and $\pm\frac{5}{8}V_R$ are available, the above operations can be performed without the need for generating $V_{RES}(7)$, as shown in Fig. 3. These additional reference voltages can simply be tapped off from the same resistor string that generates the references for the other stages, as shown in Fig. 4. Note that the references $\pm\frac{3}{8}V_R$, $\pm\frac{1}{8}V_R$ and $\pm\frac{5}{8}V_R$ are slightly asymmetrical around V_{CM}. With fully differential, capacitively coupled comparators, this variation in common mode voltage is converted into a differential error signal by any unbalance between the two signal paths. However, this error is insignificant in practice because of the large tolerance of this scheme to decision errors.

Overall Architecture of the A/D Converter

With the goal of achieving sampling rates in excess of 50 Msamples/s, a dual pipeline architecture has been employed, as shown in Fig. 5. A critical design consideration here is the matching between the two pipelines [2]. Great care has been taken in the design and layout of the clock generator to minimize the difference in sampling instants for the two pipelines. Also, to minimize the gain error between the two pipelines, they share the same reference voltage generator.

Experimental Results

Fig. 6 shows the photomicrograph of the prototype A/D converter, which has been fabricated in a 0.9 micron CMOS technology. The design uses telescopic-cascode amplifiers and simple comparators [6]. The measured output spectrum for a 6-MHz sinusoidal input and a sampling rate of 52 Msamples/s is shown in Fig. 7. The signal to noise + distortion ratio (SNDR) is about 48 dB. Fig. 8 shows the variation of the SNDR as a function of the input frequency. The degradation at higher input frequencies is partly due to the mismatches in the sampling instants between the two pipelines and partly due to the non-linear settling behavior of the input sample-and-hold circuits. The measured differential non-linearity (DNL) is less than 0.25 LSB and the offset voltage is less than 2 LSB. The component at the Nyquist frequency (due to offset mismatches between the two pipelines) is 40 dB below full scale, which is negligible in the telecommunication applications for which this converter is intended. The power consumption of the complete A/D converter (including the core, the band-gap reference and biasing circuits, the clock drivers, and the output buffers driving external loads) is 250 mW from a 5 V supply. In comparison, a version of the same A/D converter without amplifier sharing, consumes 350 mW.

References :

1) M. Yotsuyanagi et al., "A 10-b 50-MHz Pipelined CMOS A/D Converter with S/H," IEEE J. of Solid State Circuits, Vol.28, pp 292-300, March 1993.

2) C. Conroy, et al., "An 8-b 85 MS/s Parallel Pipeline A/D Converter in 1-μm CMOS," IEEE J. of Solid State Circuits, Vol.28, pp 447-454, April 1993.

3) K. Nakamura et al., "An 85 mW, 10 b, 40 Msample/s CMOS Parallel-Pipelined ADC", IEEE J. of Solid State Circuits, Vol.30, pp 173-183, March 1995.

4) T. B. Cho, et al., "A 10-b, 20-Msample/s, 35 mW Pipeline A/D Converter," IEEE J. of Solid-State Circuits, Vol. 30, pp. 166-172, March 1995.

5) T. C. Choi, et al., "Considerations for High-Frequency Switched-Capacitor Filters," IEEE Trans. on Circuits and Systems, Vol. CAS-27, pp. 545-551, June 1980.

6) S. H. Lewis, et al., "A 10-b 20-Msample/s Analog-to-Digital Converter," IEEE J. of Solid State Circuits, Vol.27, pp 351-358, March 1992.

7) K. Nagaraj, "Efficient circuit configurations for algorithmic analog to digital converters", IEEE Trans. on Circuits and Systems, Part II, Vol.40, pp 777-785, December 1993.

8) R. Ginetti and P.G.A. Jespers, "A 1.5 Ms/s 8-bit pipeline RSD A/D converter", ESSCIRC Dig. of Tech. Papers, September. 1990.

9) W-C. Song, et al., "A 10 b 20-Msamples/s low-power CMPS ADC", IEEE J. of Solid State Circuits, Vol.30, pp 514-521, March 1995.

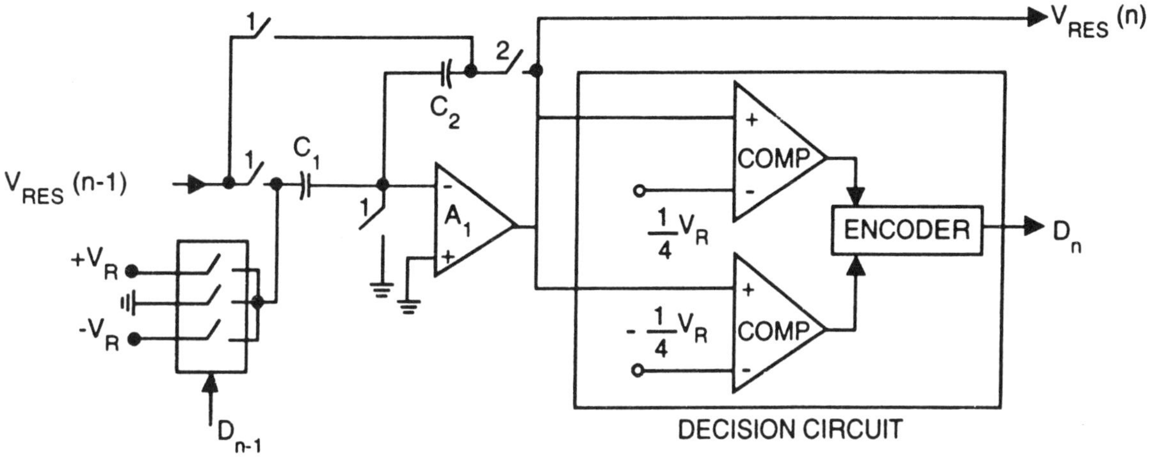

Figure 1: Circuit schematic of one stage of a typical pipelined A/D converter

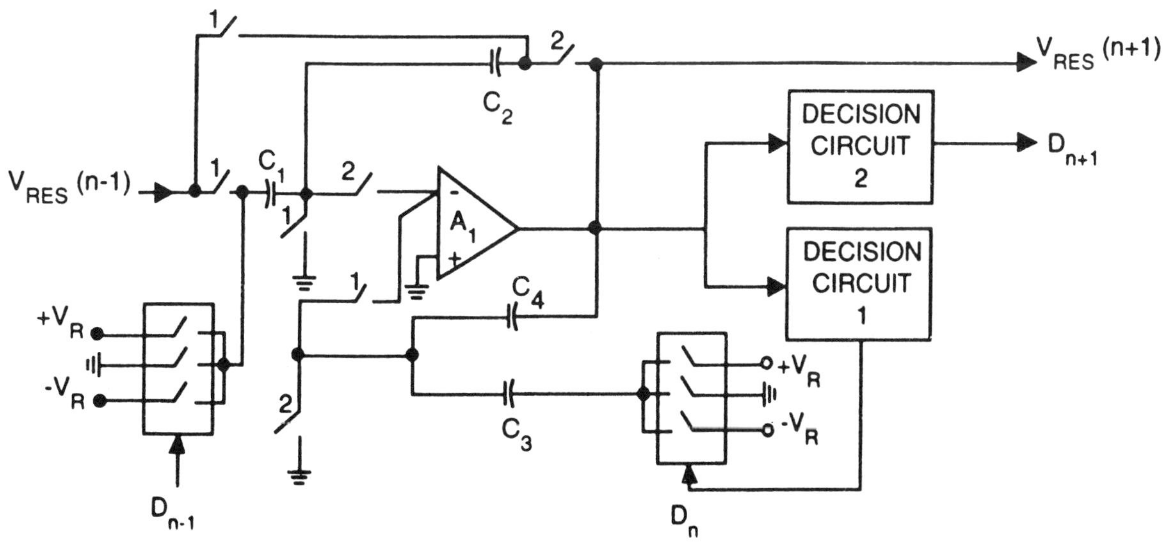

Figure 2: Circuit schematic showing the sharing of a single amplifier by two consecutive stages

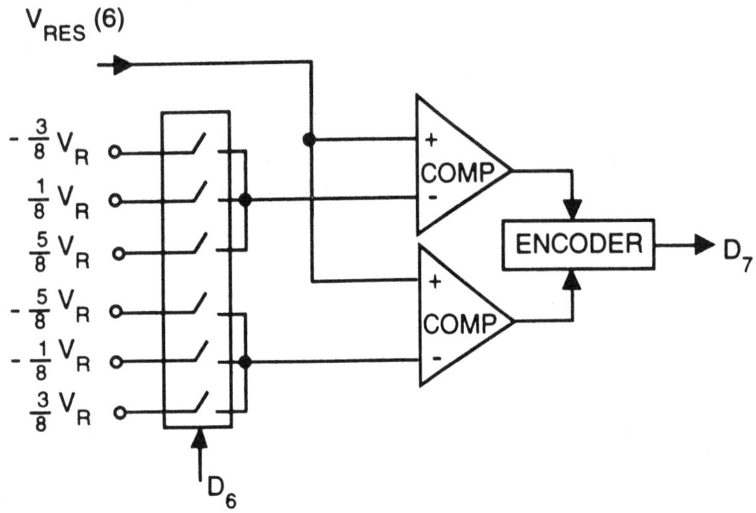

Figure 3: Circuit schematic of the last stage

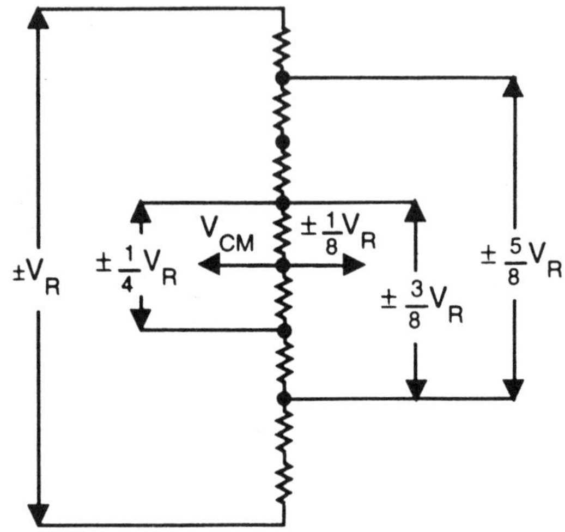

Figure 4: Generation of the reference voltages

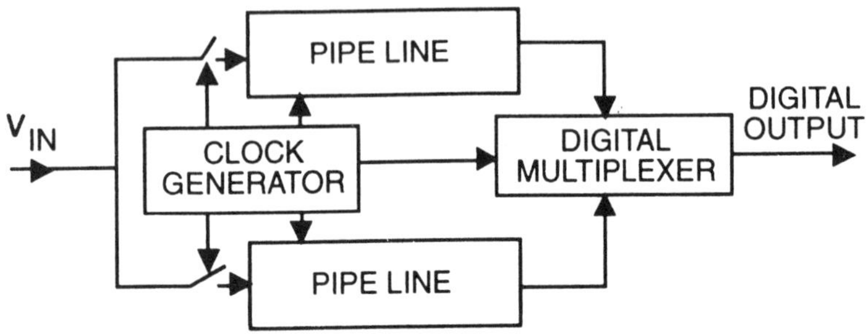

Figure 5: Overall architecture of the A/D converter

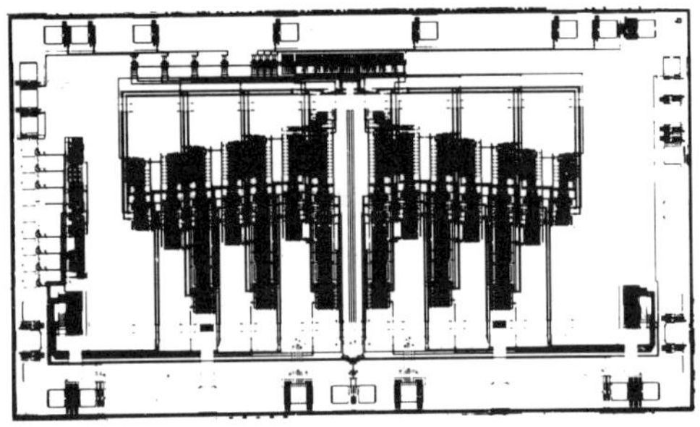

Figure 6: Photomicrograph of the device

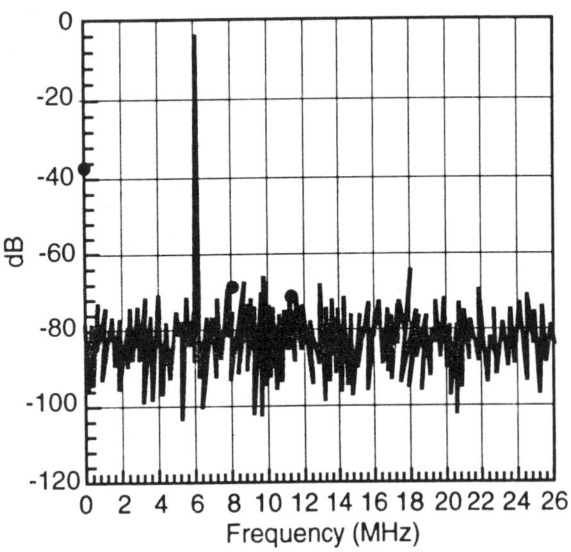

Figure 7: Output spectrum for a 6 MHz input at 52 Msamples/s

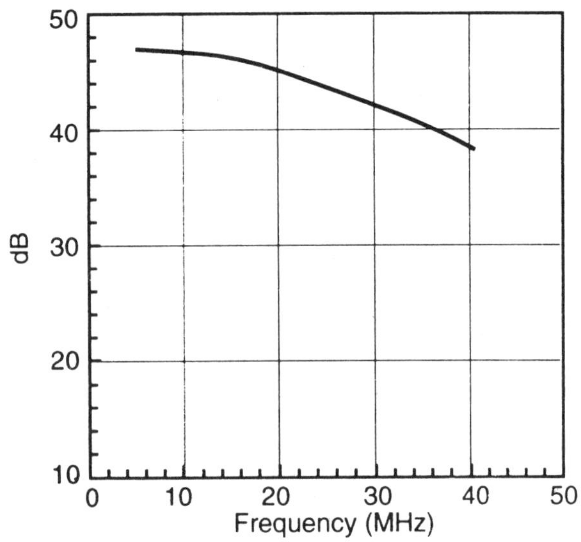

Figure 8: Variation of SNDR with input frequency

A CMOS 12-bit 4MHz Pipelined A/D Converter with Commutative Feedback Capacitor

Jungwook Yang and Hae-Seung Lee

Department of Electrical Engineering and Computer Science
Massachusetts Institute of Technology
Cambridge, Massachusetts 02139

Abstract

A 2-bit/stage CMOS 12-bit 4MHz pipelined A/D converter (ADC) using commutated feedback capacitor switching scheme is presented. The technique improves the DNL without using complicated calibration circuitry, nor requiring extra calibration cycle. Approximately, only 7-bit matched capacitors are required for 12-bit ADC's. Very high output swing is achieved in a gain enhanced folded cascode amplifier by operating output stage transistors in the triode region. This 12-bit pipelined ADC is integrated in a standard 0.8μm single poly, triple metal CMOS process, and dissipates 45mW.

Introduction

Capacitor matching determines the accuracy of CMOS pipelined A/D converters. For good capacitor matching, large capacitors are needed. Consequently, the conversion speed goes down. A number of calibration techniques have been developed to improve the accuracy of ADC's [1]-[3], but they usually require complicated circuitry and extra calibration cycles.

The A/D converter presented in this paper is based on the commutative feedback capacitor switching [4] to significantly relax matching requirements for capacitors. Although any number of bits per stage can be employed, we have chosen 2-bits/stage to demonstrate the concept.

In a conventional ADC, the feedback capacitor C_f is used to amplify the sampled input as in Figure 1-(a) shown without standard digital error correction for clarity. The voltage drops δ_i at the transition points in the residue characteristic in Figure 1-(b) should be exactly V_{ref} in an ideal ADC. With capacitor mismatch, however, the resulting voltage drop is approximately $V_{ref}(1+\epsilon)$, where ϵ is the percentage capacitor mismatch error. This causes DNL errors at the transition points as shown in Figure 2.

Commutative Feedback Capacitor for Multi-bit/stage

In this design, one of the sampling capacitors are selected as the feedback capacitor during the amplification period according to the input signal range determined by the comparator outputs as shown in Figure 3. The voltage δ_i at the transition point is shown to be approximately $V_{ref}(1+\epsilon^2)$, which gives smooth code transitions as shown in Figure 4. It can be shown that INL is also improved by a factor of 2 compared with conventional techniques. 12-bit monotonicity is feasible with 7-bit capacitor matching. This technique is applicable to any number of bits per stage. Since the capacitor matching does not have to be accurate, the capacitance can be reduced to that limited by $\frac{kT}{C}$. Hence, the operating speed and power consumption are improved.

For a 2-bit/stage ADC, four sampling capacitors whose capacitance is 0.1pF. To reduce the area, all three metal layers are used. The first and third metal layers are used for the bottom plate. The shielded second-level metal is used for the top plate to reduce noise coupling to it. A digital error correction is used to correct the comparator errors. Two more comparators at the voltage of V_{ref} and GND are added as shown in Figure 5-(a) to fold the residue output, when the output of the previous stage is greater than V_{ref} or smaller than GND without attenuating the gain reduced to half [4]. The additional information from the comparators at V_{ref} and GND are used to correct the error in the previous stage. Consequently, two additional sampling capacitors are added. To maintain the gain of 2^{n_s}, where n_s is the number of bits per stage, the extra two capacitors sample V_{ref} and GND, respectively. During the amplification period, one of the sampling capacitors are sequentially selected as shown in Table 1.

ADC Architecture

The prototype ADC has five stages, with a 2-bit first stage. The second through fourth stages give two and a half bits each, which is used for correcting the digital code error in the previous stage. The fifth stage is a 5-bit flash ADC, whose MSB is also used to correct the digital error in the fourth stage. The standard error correction with gain reduced by half is employed between the fourth and the fifth stages.

A folded cascode opamp with fully differential active cascode gain enhancement [5] is used as shown in Fig-

0-7803-3177-6 $5.00 © 1996 IEEE

IEEE 1996 CUSTOM INTEGRATED CIRCUITS CONFERENCE

ure 6 to obtain gain of more than 80dB and phase margin of 60°. The drawback of the conventional folded cascode opamp at a low voltage supply is that the output swing is significantly reduced. The output swing has been increased by reducing the drain-source voltages of M5, M6, M11, and M12, which makes these transistors operate in the triode region. However, the overall gain is still large enough for a 12-bit ADC because the reduced output resistance is compensated by the gain enhancement amplifiers. Figure 7 shows the simulated opamp gain versus the drain-source voltage of M5-6 and M11-12. The gain of 80dB is still obtained while maintaining only 0.2V for the drain-source voltage of M5-6 and M11-12.

The configuration of the gain enhancement amplifiers are also fully-differential folded cascode amplifiers[5]. Op1 at the top side in Figure 6 has nMOS input stage as shown in Figure 8, and Op2 at the bottom side has pMOS input stage to have better common mode input range. They are working in unity gain amp configuration, so the output follows the input voltage. The gate voltage of M0 in Figure 8 sets up the input common mode voltage, which also sets the common-mode output voltage controlling V_{DS} of M5-6 and M11-12 in Figure 6.

Experimental Results

The ADC has been fabricated in a single poly, triple metal 0.8μm CMOS process provided by MOSIS. Figure 9 shows the DNL and INL. Most of the DNL peaks in Figure 9 are caused by large offsets of about ±0.5LSB's in the comparators of the last stage flash ADC. Note that INL is smooth as predicted in Figure 4 despite large capacitor mismatches more than 0.2% in this chip. The frequency spectrum is shown in Figure 10 with the conversion rate of 1MHz with a 50kHz input frequency. The SNR at this frequency is 71.0dB while SDR is 62.4dB. Since the commutative feedback improves the INL only by a factor of 2, harmonic distortion caused by capacitor mismatch is still present in Figure 10. However, INL is not critical for many applications. Figure 11 shows the SNR versus the conversion rate. Table 2 summarizes the measured performance of the ADC.

Conclusion

A microphotograph of the chip is shown in Figure 12. Since the commutative feedback makes it possible to use small capacitors for high resolution ADC's, it is suitable for high speed and low power video applications.

Acknowledgements

This research was partially funded by NSF and ARPA under Grant 9117724-MIP. The fabrication was provided by MOSIS. The authors acknowledge Paul Yu, Jennifer Lloyd, and Andrew Karanicolas for helpful discussions.

References

[1] H. S. Lee, et. al., "A Self-Calibrating 15-bit CMOS A/D Converter," *IEEE J. Solid-State Circuits*, vol. SC-19, pp. 813-819, Dec. 1984.

[2] S. H. Lee, et. al., "Digital-Domain Calibration of Multistep Analog-to-Digital Converters," *IEEE J. Solid-State Circuits*, vol. 27, No. 12, pp. 1679-1992, Dec. 1992.

[3] A. N. Karanicolas, et. al., "A 15-b 1-Msample/s Digitally Self-Calibrated Pipeline ADC," *IEEE J. Solid-State Circuits*, vol. 28, No. 12, pp. 1207-1215, Dec. 1993.

[4] Paul C. Yu, et. al., "A Pipelined A/D Conversion Technique with Near-Inherent Monotonicity", *IEEE Tr. on Circuit and Systems-II*, Vol. 42, No. 7, Jul. 1995.

[5] Jennifer Lloyd, et. al, "A CMOS Op Amp with Fully-Differential Gain-Enhancement," *IEEE Tr. on Circuits and Systems*, Vol. 41, No. 3, pp. 241-243, Mar. 1994.

Table 1. Capacitor configurations during sampling and amplification period

Sampling Period						
Input range	C0	C1	C2	C3	C4	C5
	V_{ref}	V_{in}	V_{in}	V_{in}	V_{in}	G
amplification Period						
Input range	C0	C1	C2	C3	C4	C5
$V_{in} < 0$	FB	G	G	G	G	G
$0 < V_{in} < \frac{V_{ref}}{4}$	V_{ref}	FB	G	G	G	G
$\frac{V_{ref}}{4} < V_{in} < \frac{V_{ref}}{2}$	V_{ref}	V_{ref}	FB	G	G	G
$\frac{V_{ref}}{2} < V_{in} < \frac{3V_{ref}}{4}$	V_{ref}	V_{ref}	V_{ref}	FB	G	G
$\frac{3V_{ref}}{4} < V_{in} < V_{ref}$	V_{ref}	V_{ref}	V_{ref}	V_{ref}	FB	G
$V_{ref} < V_{in}$	V_{ref}	V_{ref}	V_{ref}	V_{ref}	V_{ref}	FB

Table 2. Measured Performance of the ADC

Resolution	12b
DNL	+0.6/-0.6LSB
INL	+2.5/-5.5LSB
Maximum Conversion rate	4MHz
Input range	±1.25V
peak SNR at 1MHz clock	71.0dB
peak SDR at 1MHz clock	62.4dB
RMS Noise when V_{in}=0V	0.304LSB
Supply Voltage	±2V
Analog Power Consumption	45mW
Input Capacitance	0.6pF
Area	3.3 × 3.3mm^2
Technology	0.8μm 1-poly 3-metal CMOS

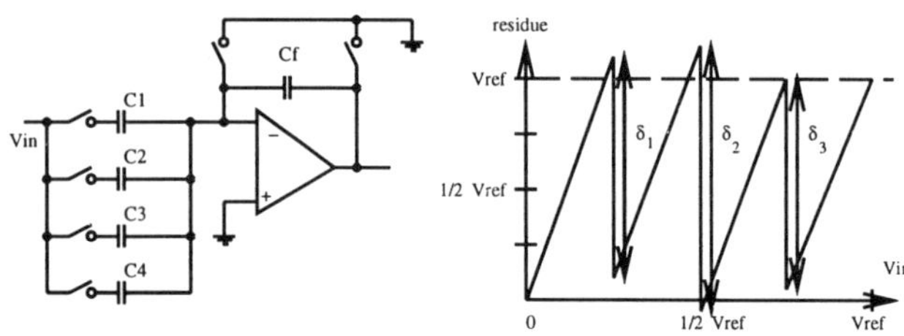

(a) Schematic of a conventional sub–ADC stage (b) Residue Plot

Figure 1: Conventional ADC

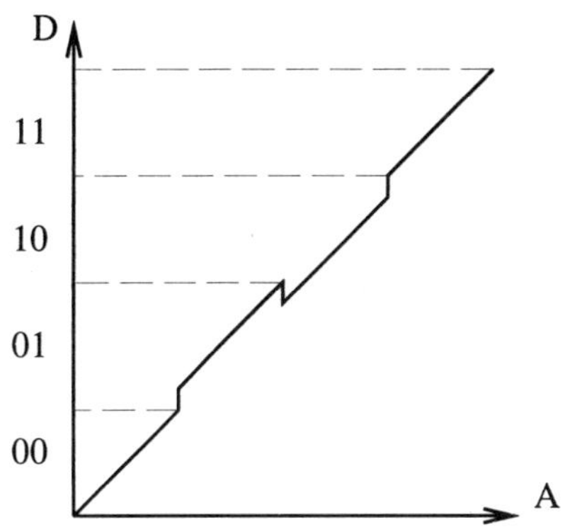

Figure 2: Characteristic of conventional pipelined ADC

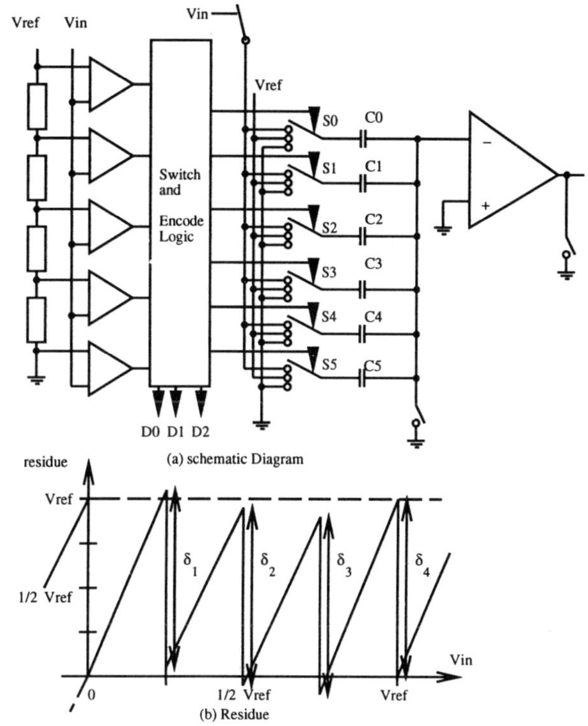

Figure 5: 2.5-bit Converter Stage for Error Correction

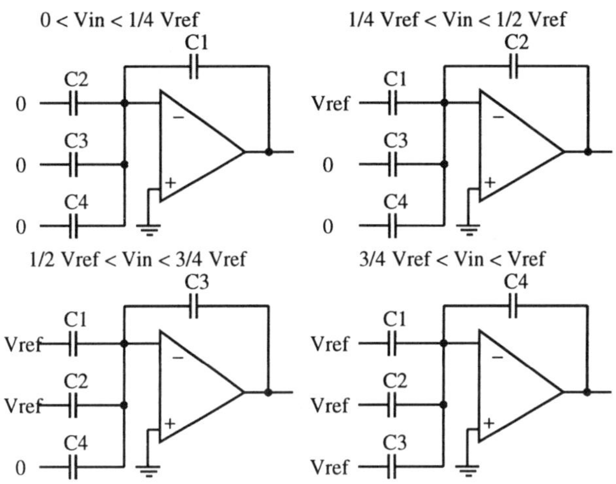

Figure 3: Commutative feedback capacitor switching

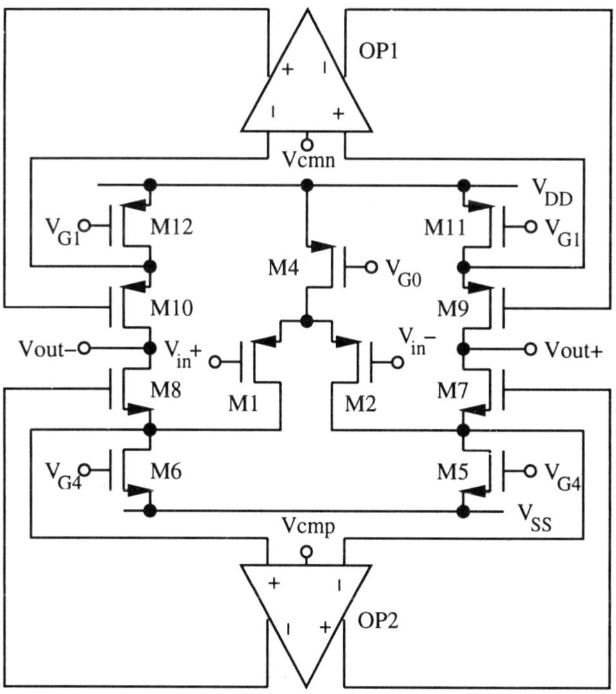

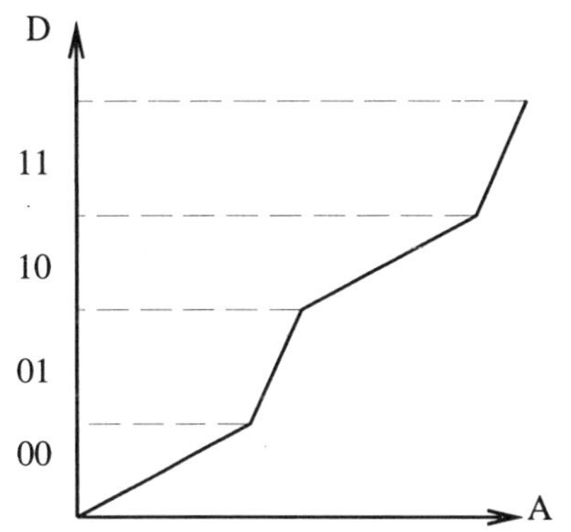

Figure 4: Characteristic of ADC with commutative feedback

Figure 6: Folded Cascode Opamp with Gain Enhancement Amplifiers

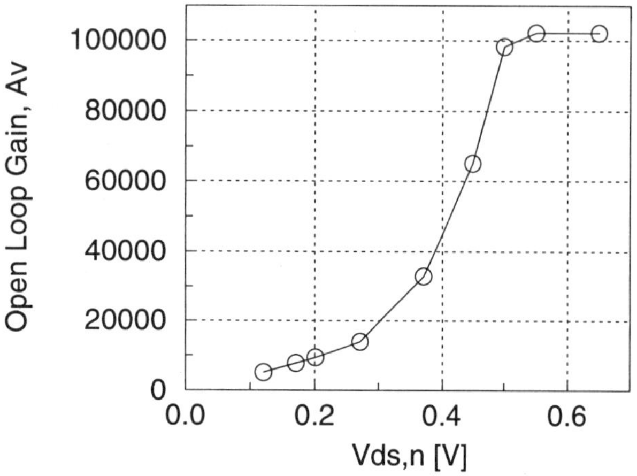

Figure 7: Simulated Gain vs. V_{DS} of M5-6 and M11-12

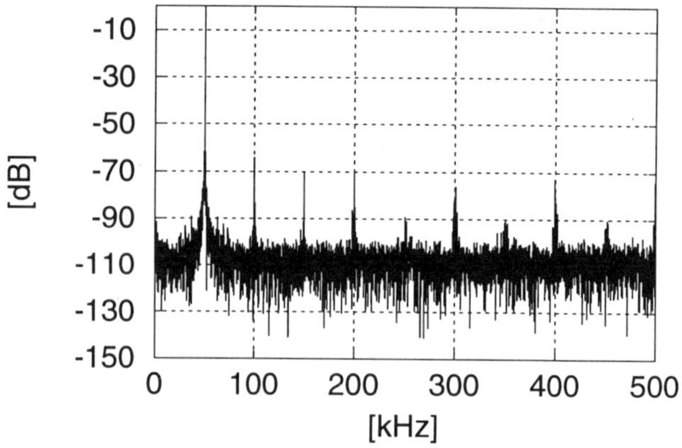

Figure 10: FFT plot for 50kHz input and 1MHz conversion

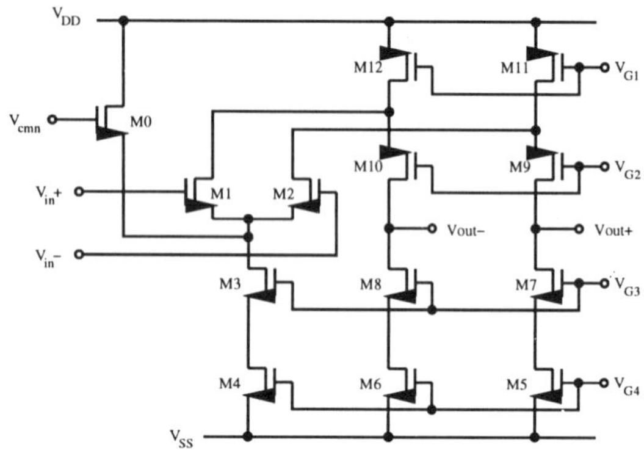

Figure 8: Gain Enhancement Amplifier

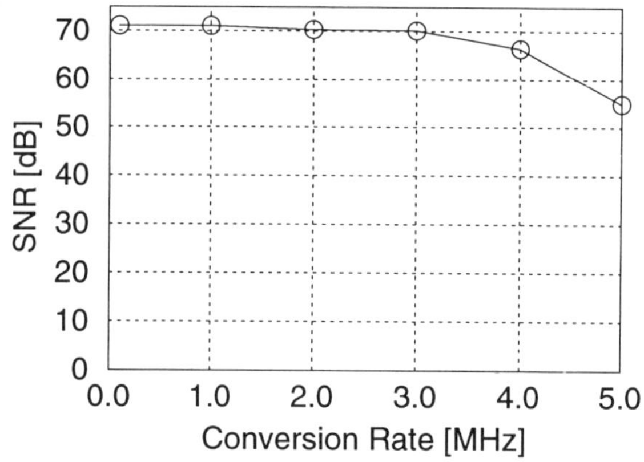

Figure 11: SNR versus clock frequency

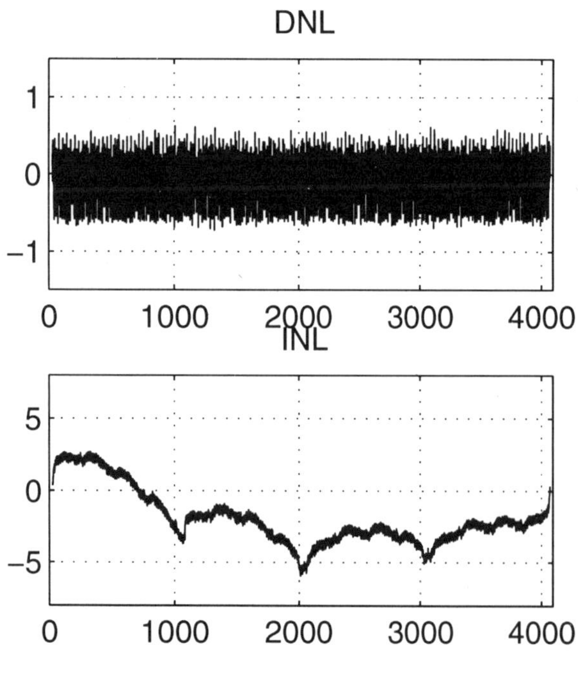

Figure 9: Differential Nonlinearity and Integral Nonlinearity

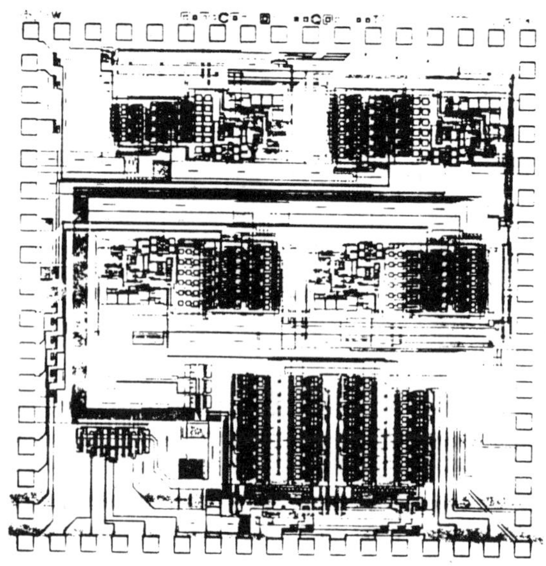

Figure 12: Microphotograph of the chip

A High Yield 12-bit 250-MS/s CMOS D/A Converter

Jose Bastos, Michiel Steyaert, and Willy Sansen

Katholieke Universiteit Leuven, ESAT-MICAS
Kard. Mercierlaan 94, B-3001 Heverlee, Belgium

Abstract

A 12-bit linearity binary-weighted all MOS transistor D/A converter is presented. Experimental results demonstrate the feasibility of fabricating with high yield such a converter in a standard CMOS 0.7μm technology. The output drives a doubly terminated 50Ω coaxial cable. The full scale 10-90% rise/fall time is 4 ns. The active chip area is $1mm^2$.

Introduction

Recent interest in using digital signal processing in such fields as wireless personal communications and video signal processing has created a demand for fast D/A converters with accuracy higher than 10 bits. The accuracy of D/A converters that rely on static mirrors is limited by the achievable transistor matching in a given technology. In CMOS VLSI technologies the maximum accuracy is typically 10 bits [1]-[2]. In this paper we present a 12 bit linearity 250MS/s all MOS transistor D/A converter, which demonstrates the feasibility of fabricating with high yield such a converter in a CMOS 0.7μm digital standard technology.

Chip Architecture

In order to achieve high speed current steering is the preferred architecture. Current steering D/A converters are based on current cells which are binary weighted, or monotonic decoded unit cells. The two architectures trade a reduction in layout area and complexity for increased differential non linearity (DNL) and glitch energy. Independent of the architecture, the most important static specification, the integral non linearity (INL) has the same difficulty requirement. As our most important goal is to achieve 12 bit integral linearity, we have chosen a binary weighted architecture because it has minimum layout area and it offers considerable advantages in eliminating systematic errors introduced by process gradients. A block diagram of the DAC architecture is shown in Fig. 1.

Design for Linearity

A. Random Errors

Due to the matching fluctuation of the current sources, the INL specification varies randomly in a set of several DACs realized in the same process technology. Hence it is of primary importance to establish the circuit yield as a function of the matching accuracy in the current sources.

Theoretical yield estimations are found in the literature [3]-[4]. However, a correct yield estimation can only be obtained by Monte-Carlo simulations. The result of such a simulation is shown in graphical form in Fig. 2. We can see that, with a unit current source relative standard deviation (σ_I/I) of 0.3%, we can guarantee a 99% yield.

This result allows to estimate the minimum channel area of the unit current source. The relative standard deviation of matching between two unit current sources as function of the bias point and the channel dimensions is approximately given by [5]

$$\begin{aligned}
&\sigma^2\left(\frac{\Delta I}{I}\right) = 2\frac{\sigma_I}{I} \\
&\sigma^2\left(\frac{\Delta I}{I}\right) \cong \sigma^2\left(\frac{\Delta \beta}{\beta}\right) + \frac{4\,\sigma^2(\Delta V_T)}{(V_{GS}-V_T)^2} \\
&\sigma^2\left(\frac{\Delta \beta}{\beta}\right) = \frac{A_\beta}{WL} \\
&\sigma^2(\Delta V_T) = \frac{A_{V_T}}{WL}
\end{aligned} \qquad (1)$$

The parameters A_{VT} and A_β are technology constants previously extracted for the 0.7μm CMOS technology, and are shown in Table I [6]. Defining a bias point of $V_{GS}=2.0V$, a minimum channel area of the unit current source is obtained. This value is $(WL)_{min}=50\mu m^2$. This is the first and most important constraint of the design.

A second constraint arrives from the total drain current specification I_{total} of 20mA, for a 0.5V voltage output over a 25Ω load. This constraint imposes a maximum W/L ratio

0-7803-3177-6 $5.00 © 1996 IEEE

IEEE 1996 CUSTOM INTEGRATED CIRCUITS CONFERENCE

of 1/12.5. The unit current source transistor has thus channel dimensions $W/L=2/25\ \mu m/\mu m$.

B. Systematic Errors

The unit current source finite output impedance and the voltage drop along the ground line can introduce significant linearity errors [7]. Due to the large L of the unit current source, the output resistance of the cascode configuration of the unit current source and the current switch transistor is sufficiently high. Maximum resistance for the column ground lines and the backbone ground line is not exceeded.

Parameter gradients over the die can represent the limiting accuracy obtainable for large dies. For this reason, the distance influence on matching has been studied for the applied technology. Table I shows the process parameters S_{VT} and S_{β} which model the matching degradation with distance for the threshold voltage and the current factor. Although the wafer gradients introduce a deterministic matching error, we do not know the position of the die in the wafer, which means that the parameter gradient can have any direction in the die. This problem has been tackled in the linear architecture by introducing a special wiring method or switching sequence [1]-[2], which increases the wiring complexity and silicon area. We have defined an algorithm, which builds the binary weighted current sources by connecting the outputs of **centroid** distributed unit current sources over the array area. In this way all the two dimensional systematic error sources whose period is a rational fraction of the array length, do not contribute to the degradation of the integral linearity. Fig. 3 shows this distribution for the four most significant bits.

Layout

A straightforward approach to the layout of a matrix with 4096 unit current sources spends a considerable amount of layout area. For this technology, if we route 12 bit lines in a 64x64 square matrix, the total area occupied by the routing is 2.4mm^2, roughly 10 times more area than the gate area of the transistors! A straightforward layout is thus not acceptable.

The solution is to reduce the array size and to route less bit lines. We define a 16x16 matrix of "new" unit current sources, each having a value of 16 "old" unit current sources, for the 8 most significant bits. The channel dimensions of the current source are now 32/25 μm/μm, and the routing of the 8 bit lines is done **over** the transistor. In this approach no silicon area is used in the routing, and a most compact layout for the array is obtained. The four least significant bits are constructed by connecting **in series** the necessary number of "new" unit current sources. These current sources are layed out around the array, and also serve to provide identical surroundings to all the array elements, thus avoiding micro loading effects. The layout of the unit current source is shown in Fig. 4. The sources are connected by the vertical running metal 1, over the gate, and thus occupies no extra area. The drains are connected to the corresponding bit line (running parallel in metal 2), also over the gate. The complete layout of the unit current source array occupies only650x650μm^2.

Design for Speed

The conventional D/A has three main dynamic error sources: imperfect synchronization of the inputs, digital signal feedthrough, and current variation due to drain voltage variation of the current sources.

The transistors used in the differential switches have the same (minimum) length but a width proportional to the current they are switching. For the most significant bits, the width is very large, and thus their gate and drain capacitance are also large, and significant delays are introduced. To synchronize the inputs a 4-stage tapered driver stage with a size factor of 2 has been used to buffer the bit lines.

Charge feedthrough of digital input signals to the output can, due to the large gate-drain capacitance of the MSBs, also affects negatively the settling time. This source of error is significantly reduced by using cascode switches.

The cross point of rising/falling waveforms in the input drivers is also very important. In a symmetrical switching circuit the cross point is too low and there is a significant voltage swing at the drain of the current sources. We use an input driver scheme which combines the advantages of decreased gate voltage swing and delayed switching. The scheme is shown in Fig. 5 In this circuit, the cross coupled NMOS transistors work as delay elements, so that the on-off transition is slightly delayed. The delay can be fine tuned by changing the voltage V_S at the source of the transistors. Changing the voltage V_S has the extra advantage that, by decreasing the voltage swing at the gates of the switching transistors, there is less digital signal feedthrough.

Measurement Results

Figure 6 shows a microphotograph of the DAC. The active area occupied by the circuit is 1mm^2.

The INL error measured in 30 out of 30 functionally working samples is less than 1/2 LSB. It can be stated that the design yield is greater than 90%, at a 95% confidence level. Figure 7 shows a typical integral linearity plot. The average INL is 0.39 LSB. In most of the cases, the maximum INL occurs at half the scale, and coincides with the DNL at the MSB transition, as expected. This fact demonstrates that the centroid layout has successfully minimized the effect of the die gradients.

Figure 8 shows a full scale positive and negative output step (50Ω doubly terminated coaxial cable, 20pF load). The rise/fall time is less than 4ns. The dies have been bonded to a 28pin DIP. The bond wire inductance is the remaining source of error, causing the extra ringing at the output for a full scale voltage swing.

Conclusion

A 12 bit linearity current steering CMOS D/A converter capable of video speed has been presented. Taking advantage of the matching properties of a modern CMOS 0.7μm technology and using a very compact layout, a design yield greater than 90% is obtained.

Acknowledgment

J. Bastos is supported by a grant from the Portuguese National Research Board (JNICT).

References

[1] Y. Nakamura et al., "A 10-bit 70MS/s CMOS D/A Converter", IEEE J. Solid-State Circuits, vol. 26, pp. 637-642, April 1991.
[2] T. Wu, C. Jih, J. Chen, C. Wu, "A Low Glitch 10-bit 75-MHz CMOS Video D/A Converter", IEEE J. Solid-State Circuits, vol. 30, pp. 68-72, January 1995.
[3] K. Lakshmikumar, R. Hadaway, and M. Copeland, "Characterization and Modeling of Mismatch in MOS Transistors for Precision Analog Design", IEEE J. Solid-State Circuits, vol. SC-21, no. 6, pp. 1057-1066, December 1986.
[4] K. Lakshmikumar, M. Copeland, and R. Hadaway, "Reply to 'A Comment on 'Characterization and Modeling of Mismatch in MOS Transistors for Precision Analog Design''", IEEE J. Solid-State Circuits, vol. 23, pp. 296, February 1988.
[5] M. Pelgrom, A. Duinmaijer, and A. WelbersMatching Properties of MOS Transistors", IEEE J. Solid-State Circuits, vol. SC-24, pp. 294-296, February 1988.
[6] J. Bastos et al., "Matching of MOS Transistors with Different Layout Styles", IEEE International Conference on Microelectronic Test Structures, Trento, 1996, in press.
[7] T. Miki et al., "An 80 MHz 8-bit CMOS D/A Converter", IEEE J. Solid-State Circuits, vol. SC-21, pp. 983-988, December 1986.

Table I

Mismatch Parameters

Parameter	*n-channel*	*p-channel*	*unit*
A_{VT}	13	22	$mV\ \mu m$
A_{β}	1.9	2.8	$\%\ \mu m$
S_{VT}	0.4	0	mV/mm
S_{β}	0.2	0.3	$\%/mm$

Table II

Characteristics of the DAC

Resolution	12 bit
Differential Nonlinearity	<0.5 LSB
Integral Nonlinearity	<0.5LSB
Conversion Rate	250MS/s
Rise/Fall Time (10-90%)	<4ns
F.S. Output Current	20mA
Supply Voltage	3.3V
Process	0.7μm CMOS
Chip Size (without pads)	1mm²

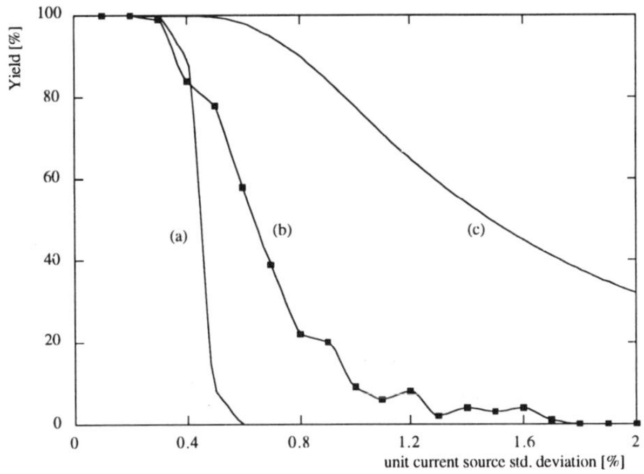

Fig. 2. DAC yield as a function of the unit current source accuracy, (a) as predicted by [1], (b) as obtained by Monte-Carlo simulations, and (c) as predicted by [2].

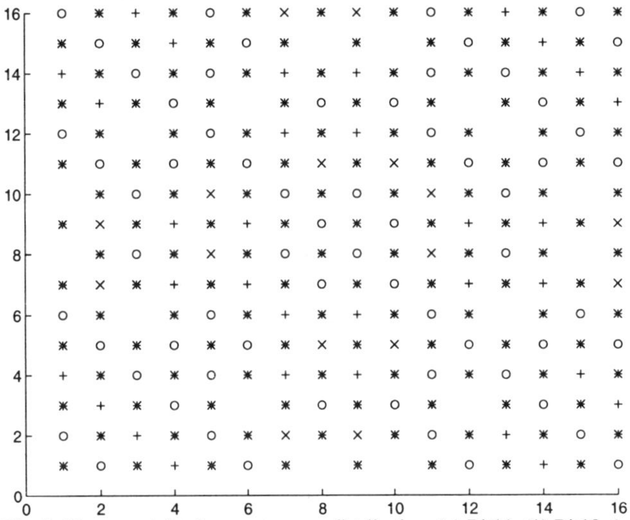

Fig. 3: Binary weighted current source distribution. (o) Bit11, (*) Bit10, (+) Bit9, (x) Bit8.

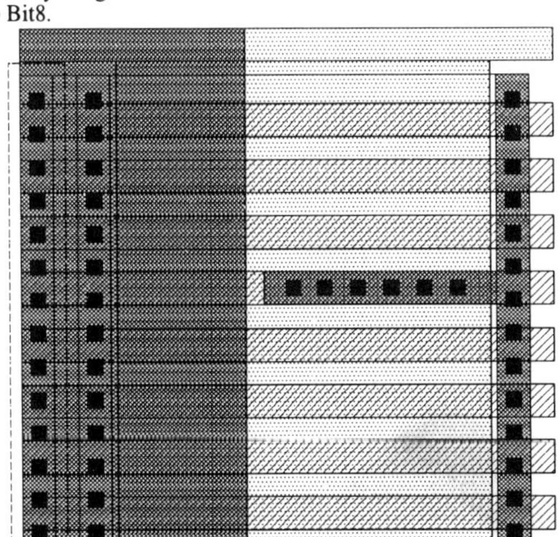

Fig. 4. Unit current source layout.

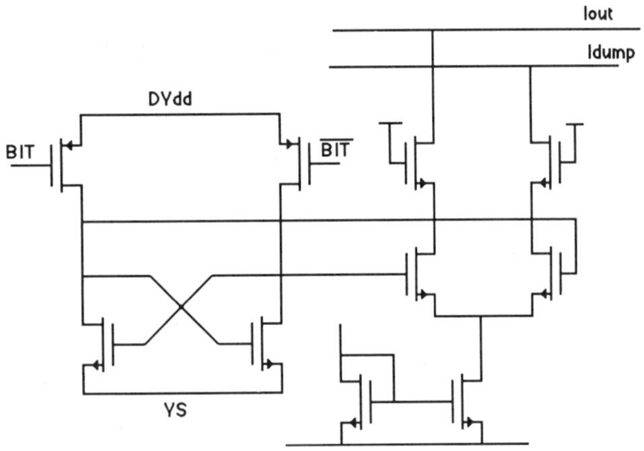

Fig. 5. Basic current cell.

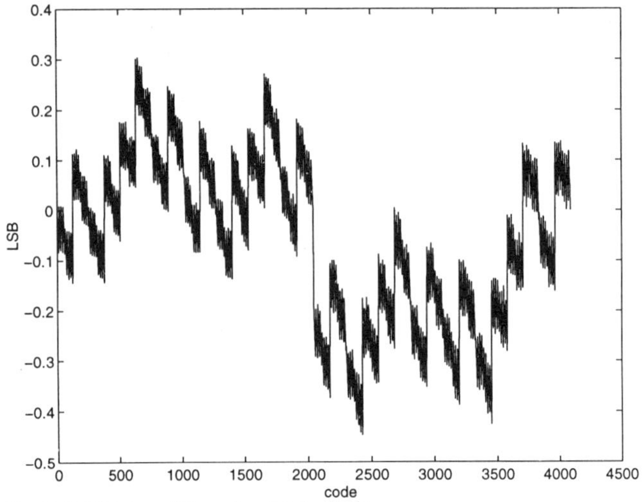

Fig. 7. DAC integral linearity plot (typical).

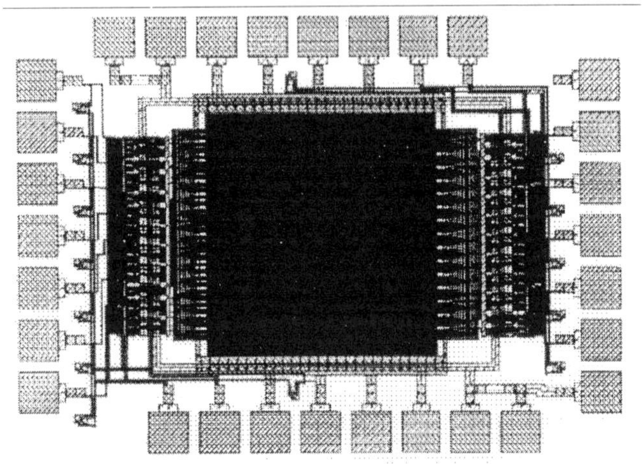

Fig. 6. Die microphotograph.

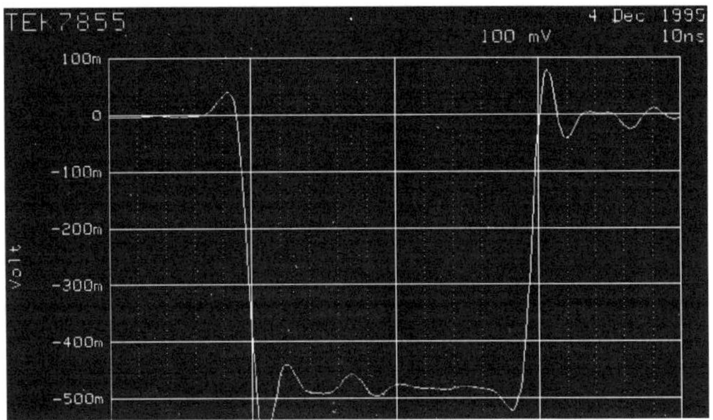

Fig. 8. Full scale output transitions.

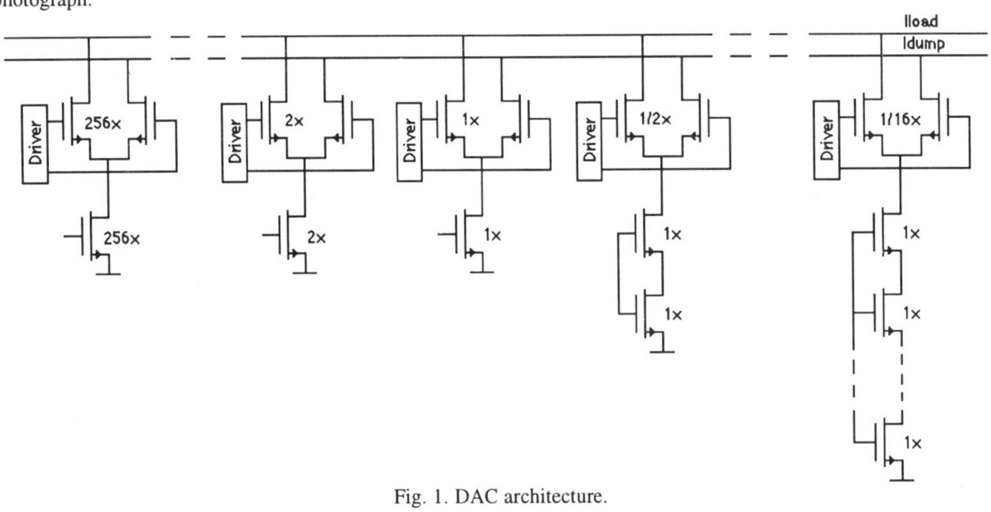

Fig. 1. DAC architecture.

Chair: Brian Antao
Co-Chair: George Stamoulis

RF integrated circuit sections feature prominently in the design of modern integrated circuit and system design that include analog, digital and communication sections on chip. The design of integrated systems in emergent technologies requires efficient verification of these various sections, and the integrated verification of the entire chip. The simulation of analog-digital systems has received considerable attention, with a number of simulation techniques and methodologies being developed. The simulation of integrated RF sections and circuits is however a new field, as most of the RF designs were largely discrete circuit designs. This session on modeling and simulation of RF designs includes a number of new papers that present new techniques for simulation of RF integrated circuits.

The first paper is an invited tutorial on "Fast Simulation Algorithms for RF Circuits" that describes algorithms for simulation of RF and high-frequency circuits in the time and frequency domain as well as mixed frequency-time domain methods. The second paper in this session is on the "Parasitic Characterization of RF Circuits Using Mixed-Mode Simulation", combining device and circuit-level models. The third paper which is on "Receiver Characterization Using Periodic Small-Signal Analysis", describes methods for the periodic small-signal analysis including the characterization of the transfer function, noise and distortion of communication circuits. Simulation of high-frequency noise effects is an important problem in the analysis of RF circuits and the paper, "Simulation and Modeling of Phase Noise in Open-Loop Oscillators", presents an approach for the simulation and modeling of phase noise in oscillators. Oscillators are widely used component in RF circuits. The next paper is on "Dispersive Coupled Transmission Line Simulation Using an Adaptive Block Lanczos Algorithm", where the authors present a novel block Lanczos algorithm for reduction of transmission line models. Transmission lines models can also be used in the simulation of IC interconnect effects in RF circuits. The last paper in this session is "Efficient Frequency Domain Analysis of Large Nonlinear Analog Circuits", which presents a new implementation of the harmonic balance method, that can be applied to circuits that are 2-3 orders of magnitude larger than conventional circuits.

Fast Simulation Algorithms for RF Circuits

R. Telichevesky, K. Kundert
Cadence Design Systems
San Jose, California

I. Elfadel, J. White
Massachusetts Institute of Technology
Cambridge, Massachusetts

Abstract

RF integrated circuit designers make extensive use of simulation tools which perform nonlinear periodic steady-state analysis and its extensions. However, the computational costs of these simulation tools have restricted users from examining the detailed behavior of complete RF subsystems. Recent algorithmic developments, based on matrix-implicit iterative methods, is rapidly changing this situation and providing new faster tools which can easily analyze circuits with hundreds of devices. In this paper we present these new methods by describing how they can be used to accelerate finite-difference, shooting-Newton, and harmonic-balance based algorithms for periodic steady-state analysis.

1 Introduction

The intensifying demand for very high performance portable communication systems has greatly expanded the need for simulation algorithms that can be used to efficiently and accurately analyze frequency response, distortion, and noise of RF communication circuits such as mixers, switched-capacitor filters and narrow-band amplifiers. Although methods like multitone harmonic balance, linear time-varying, and mixed frequency-time techniques [6, 3, 1, 4] can perform these analyses, the traditional implementations of these techniques grow so rapidly with increasing circuit size that they have been too computationally expensive to use for more complex circuits.

Recent algorithmic developments, based on preconditioned matrix-implicit Krylov-subspace iterative methods, is rapidly changing the situation and providing new tools which can easily analyze circuits with hundreds of devices. Preconditioned iterative techniques have been applied to accelerating periodic steady-state analysis using both harmonic balance methods [2, 5] and time-domain shooting methods [9]. Additional results for more general analyses are also under investigation.

In this tutorial paper, we try to present these accelerated methods for periodic steady-state analysis in a cohesive fashion, and perhaps make the more specialized papers that have or will shortly appear a little more approachable. We start in the next section by presenting the time domain finite-difference-Newton and shooting-Newton methods, give a brief idea of their computational costs using standard factorization, and give the faster matrix-implicit algorithm. In section three we present, using as much common notation as possible, the same aspects of the harmonic-balance approach. Where possible, we give some computational results so as to give the reader a feel for the cost reductions. Also, we have tried to pay particular attention to providing approachable references in the mathematical literature. Finally, conclusions and acknowledgements are given in Section 4.

2 Time Domain Methods

Finding the periodic steady-state solution of a circuit means finding the initial condition for the circuit's associated system of differential equations such that the solution at the end of the period matches the initial condition. More precisely, the steady-state solution must be a particular solution to the circuit equations, as in

$$f(v(t),t) = \frac{d}{dt}q(v(t)) + i(v(t)) + u(t) = 0, \qquad (1)$$

where $u(t) \in \Re^N$ is the vector of input sources, $v(t) \in \Re^N$ is the vector of node voltages, and $i(v(t)), q(v(t)) \in \Re^N$ are the vectors of resistive node currents and node charges (or fluxes), respectively. In addition to satisfying (1), the periodic steady solution must also satisfy the two-point constraint

$$v(T) - v(0) = 0. \qquad (2)$$

A sampling discretization is usually introduced to numerically solve the combination of (1) and (2), by

*This work was supported by a grants from Motorola and Cadence Design Systems

0-7803-3177-6 $5.00 © 1996 IEEE

IEEE 1996 CUSTOM INTEGRATED CIRCUITS CONFERENCE

which we mean that $v(t)$ is represented over the period T by a sequence of M numerically computed discrete points,

$$\begin{bmatrix} v(t_1) \\ v(t_2) \\ \vdots \\ v(t_M) \end{bmatrix} \approx \begin{bmatrix} \hat{v}(t_1) \\ \hat{v}(t_2) \\ \vdots \\ \hat{v}(t_M) \end{bmatrix} \equiv \hat{\boldsymbol{v}}, \tag{3}$$

where $t_M = T$, the hat is used to denote numerical approximation, and $\hat{\boldsymbol{v}} \in \Re^{MN}$ is introduced for notational convenience.

A variety of methods can be used to derive a system of equations from which to compute $\hat{\boldsymbol{v}}$. For example, if the backward-Euler method is used to approximate the derivative in (1), then $\hat{\boldsymbol{v}}$ must satisfy

$$F_1(\hat{\boldsymbol{v}}, \hat{v}(t_0)) \equiv \frac{q(\hat{v}(t_1)) - q(\hat{v}(t_0))}{h_1} + i(v(t_1)) + u(t_1) = 0 \tag{4}$$

and for $j \in \{2, ..., M\}$

$$F_j(\hat{\boldsymbol{v}}) \equiv \frac{q(\hat{v}(t_j)) - q(\hat{v}(t_{j-1}))}{h_j} + i(v(t_j)) + u(t_j) = 0, \tag{5}$$

where $h_j = t_j - t_{j-1}$ and $\hat{v}(t_0)$ is the unknown differential equation initial condition which generates the periodic steady-state. Combining (2) with (5) yields a system of nonlinear equations for $\hat{\boldsymbol{v}}$ and $\hat{v}(t_0)$.

Two variants of Newton's method are commonly used to solve the combination of (2), (4) and (5): the finite-difference-Newton and the shooting-Newton methods [10, 11]. The finite-difference method is a direct Newton method but the shooting-Newton method uses an additional inner Newton iteration. The use of an inner iteration implies that the shooting-Newton method is a multi-level Newton method, and therefore has typically better global convergence properties for stable problems. Despite this convergence property difference, computing the Newton iterates in these two methods leads to surprisingly similar systems of linear equations, and both these linear systems are easily solved using iterative methods like preconditioned GMRES [7]. Below we will present the two methods, and then describe how to apply the iterative matrix-implicit solution techniques.

2.1 Finite-Difference-Newton Method

The finite-difference-Newton method is precisely Newton's method applied directly to (5), where (2) is included by the identification $\hat{v}(t_0) = \hat{v}(t_M)$. Before giving the finite-difference-Newton iteration equation, we establish some notation. Let L be the block lower bidiagonal matrix given by

$$L \equiv \begin{bmatrix} \frac{C_1}{h_1} + G_1 & & & \\ -\frac{C_1}{h_2} & \frac{C_2}{h_2} + G_2 & & \\ & \ddots & \ddots & \\ & & -\frac{C_{M-1}}{h_M} & \frac{C_M}{h_M} + G_M \end{bmatrix}, \tag{6}$$

and define B as

$$B \equiv \begin{bmatrix} 0 & \cdots & 0 & -\frac{C_M}{h_1} \\ & 0 & & \\ & \ddots & \ddots & \\ & & 0 & 0 \end{bmatrix}, \tag{7}$$

where $C_j, G_j \in \Re^{N \times N}$ denote $dq(\hat{v}(t_j))/dv$ and $di(\hat{v}(t_j))/dv$, respectively. For most practical circuits, the connectivity is very sparse. This implies that for typical problems, there are fewer than ten entries per row in C_i and G_i, regardless of the circuit size.

Using the above notation, the finite-difference-Newton iteration equation is given by

$$(L + B)\left(\hat{\boldsymbol{v}}^{k+1} - \hat{\boldsymbol{v}}^k\right) = -\boldsymbol{F}_{fd}(\hat{\boldsymbol{v}}^k), \tag{8}$$

where k is the Newton iteration index and

$$-\boldsymbol{F}_{fd}(\hat{\boldsymbol{v}}^k) \equiv \begin{bmatrix} -F_1(\hat{\boldsymbol{v}}^k, \hat{v}^k(t_M)) \\ -F_2(\hat{\boldsymbol{v}}^k) \\ \vdots \\ -F_M(\hat{\boldsymbol{v}}^k) \end{bmatrix}. \tag{9}$$

2.2 Shooting-Newton Method

Another approach to solving (5) is to exploit the fact that if $\hat{v}(t_0)$ is given, then the nonlinear equation

$$\frac{q(\hat{v}(t_1)) - q(\hat{v}(t_0))}{h_1} + i(\hat{v}(t_1)) + u(t_1) = 0 \tag{10}$$

can be solved, presumably using Newton's method, for $\hat{v}(t_1)$. Then, $\hat{v}(t_1)$ can be used to solve

$$\frac{q(\hat{v}(t_2)) - q(\hat{v}(t_1))}{h_2} + i(\hat{v}(t_2)) + u(t_2) = 0 \tag{11}$$

for $\hat{v}(t_2)$. This procedure can be continued, effectively integrating the differential equation one timestep at a time in the standard fashion, using the initial condition $\hat{v}(t_0)$. And since the nonlinear equations are solved at each timestep, $\hat{v}(t_j)$ is an implicitly defined function of $\hat{v}(t_0)$. This function, computed by the numerical integration, is referred to as the state-transition function of the discretized system and is denoted as

$$\phi(\hat{v}(t_0), t_0, t_j). \tag{12}$$

Note that if $\hat{v}(t_j)$ is computed by solving the nonlinear equation F_j in (4) or (5) at each timestep, or equivalently if $\hat{v}(t_j)$ is computed by evaluating $\phi(\hat{v}(t_0), t_0, t_j)$, then (4) and (5) will always be satisfied, regardless of the choice of $\hat{v}(t_0)$. If $\hat{v}(t_0)$ equals $\hat{v}(t_M)$, then (2) is also satisfied and a periodic steady-state solution has been computed. This statement is mathematically equivalent to the condition

$$F_{sh}(\hat{v}(t_0)) \equiv \phi(\hat{v}(t_0), t_0, t_M) - \hat{v}(t_0) = 0. \tag{13}$$

The shooting-Newton method then computes $\hat{v}(t_0)$ for the periodic steady-state by applying Newton's method to solving (13), resulting in the iteration equation

$$[J_\phi(\hat{v}^k(t_0), 0, T) - I][\hat{v}^{k+1}(t_0) - \hat{v}^k(t_0)] = -F_{sh}(\hat{v}^k(t_0)), \tag{14}$$

where k is the Newton iteration index, I is the identity matrix, and

$$J_\phi(v, 0, T) \equiv \frac{d}{dv}\phi(v, 0, T) \tag{15}$$

is referred to as the sensitivity matrix.

To complete the description of the shooting-Newton method, and to see why we refer to it as a multilevel-Newton method, it is necessary to present the procedure for computing $\phi(v, 0, T)$ and $J_\phi(v, 0, T)$. As mentioned above, computing the state transition function is equivalent to solving the backward-Euler equations in (4) and (5) one timestep at a time. Solving the backward-Euler equations is usually accomplished using an inner Newton iteration, as in

$$\left[\frac{C_{jkl}}{h_j} + G_{jkl}\right] (\hat{v}^{k,(l+1)}(t_j) - \hat{v}^{k,l}(t_j)) =$$
$$-\frac{1}{h_j}\left(q(\hat{v}^{k,l}(t_j)) - q(\hat{v}^{k,l}(t_j))\right) - i(\hat{v}^{k,l}(t_j)) - u(t_j) \tag{16}$$

where j is the timestep index, k is the shooting-Newton iteration index, l is the inner Newton iteration index, $C_{jkl} = \frac{dq(\hat{v}^{k,l}(t_j))}{dv}$ and $G_{jkl} = \frac{di(\hat{v}^{k,l}(t_j))}{dv}$. Sometimes, there are just too many indices.

To see how to compute $J_\phi(v, 0, T)$ as a by-product of the Newton iteration in (16), let $l = *$ denote the inner Newton iteration which achieves sufficent convergence, and let $\hat{v}^{k,*}(t_j)$ denote the associated converged solution. Using this notation,

$$\frac{q(\hat{v}^{k,*}(t_j)) - q(\hat{v}^{k,*}(t_{j-1}))}{h_j} + i(\hat{v}^{k,*}(t_j)) + u(t_j) = 0 \tag{17}$$

to within the Newton iteration convergence tolerence. Implicitly differentiating (17) with respect to $\hat{v}^k(t_0)$ results in

$$\left[\frac{C_{jk*}}{h_j} + G_{jk*}\right] \frac{d\hat{v}^{k,(*+1)}(t_j)}{d\hat{v}^k(t_0)} = \frac{C_{(j-1)k*}}{h_j} \frac{d\hat{v}^{k,*}(t_{j-1})}{d\hat{v}^k(t_0)} \tag{18}$$

By recursively applying the above equation, one can derive that

$$J_\phi(\hat{v}^k(t_0), t_0, t_M) = \prod_{j=1}^{M} \left[\frac{C_{jk*}}{h_j} + G_{jk*}\right]^{-1} \frac{C_{(j-1)k*}}{h_j}, \tag{19}$$

where the notation $\prod_{j=1}^{M}$ indicates a product rather than a sum.

2.3 Direct Matrix Solution

In comparing the shooting-Newton iteration, (14), to the finite-difference-Newton iteration, (8), there *appears* to be an advantage for the shooting-Newton method. The shooting-Newton method is being used to solve a system of N nonlinear equations, whereas the finite-difference-Newton method is being used to solve an NM system of nonlinear equations. This advantage is not as significant as it seems, primarily because computing the sensitivity matrix according to (19) is more expensive than computing the finite-difference Jacobian. In this section we will show that using direct factorization to solve either (14) or (8) leads to nearly equivalent computations.

To start, consider that L defined in (6) is block lower bidiagonal, where the diagonal blocks have the same structure as the single timestep Jacobian in (16). It then follows that the cost of applying L^{-1} is no more than computing one Newton iteration at each of M timesteps. One simply factors the diagonal blocks of L and backsolves. Formally, the result can be written as

$$(I_{NM} + L^{-1}B)\left(\tilde{v}^{k+1} - \tilde{v}^k\right) = -L^{-1}F(\tilde{v}^k), \tag{20}$$

though L^{-1} would never be explicitly computed. Here, we have denoted the identity matrix in $\Re^{NM \times NM}$ by I_{NM}.

Examining (20) reveals an important feature, that $L^{-1}B \in \Re^{NM \times NM}$ has nonzero entries *only in the last N columns.* Specifically,

$$(I_{NM} + L^{-1}B) = \begin{bmatrix} I_N & \dots & 0 & P_1 \\ & I_N & & \\ & \ddots & \ddots & \\ & & 0 & I_N + P_M \end{bmatrix} \tag{21}$$

where $P_i \in \Re^{N \times N}$ is the $((i-1)*N)+1$ through $i*N$ rows of the last N columns of $L^{-1}B$. This bordered-block diagonal form implies that $\tilde{v}^{k+1} - \tilde{v}^k$ in (20) can be computed in three steps. The first step is to compute P_M. This can be accomplished by forming the N products Be_i, where the $e_i's$ are the first N unit vectors in $\Re^{NM}$, and then backsolving N times with L. The second step is to use the computed P_M to determine the last N entries in $\tilde{v}^{k+1} - \tilde{v}^k$. This second task can be performed by directly factoring $I + P_M$, and using it to solve solving the last N equations in (20). The last step in solving (20) is to compute the rest of $\tilde{v}^{k+1} - \tilde{v}^k$ by backsolving again with L.

The close relation between solving (20) and (14) can now be easily established. If L and B are formed using C_{jk*} and G_{jk*} as defined in (18), then by explicitly computing $L^{-1}B$ it can be shown that $J_\phi(\hat{v}^k(t_0), t_0, t_M) = P_M$. The importance of this observation is that solving the shooting-Newton update equation is nearly computationally equivalent to solving the finite-difference-Newton update equation. Specifically, (14) can be solved by following the first two steps of the above procedure, except the factorization of $P_M - I_N$ replaces the factorization of $I_N + P_M$.

The strong connection between the computational steps in solving the shooting-Newton and finite-difference-Newton update equations implies their computational costs are very similar. For both methods, the dominant costs are the more than order MN^2 operations required to form P_M and the N^3 operations required to factor the dense $I_N + P_M$ or $P_M - I_N$ matrix. The reason forming P_M is so expensive is that it requires N backsolves with L, and each backsolve with L requires *at least* order NM operations. We say *at least* because the cost of backsolving with L depends critically on the amount of fill-in produced by factoring the $N \times N$ sparse diagonal blocks $\frac{C_j}{h_j} + G_j$. Since these blocks are generated by the linearization of a circuit, their structure is problem dependent. For example, if the circuit happens to be a tree structure, then $\frac{C_j}{h_j} + G_j$ will have order N elements, the factorization will produce no fill, and backsolving with a factored L will require only order NM operations. However, if the circuit is a three-dimensional mesh, then factoring $\frac{C_j}{h_j} + G_j$ will produce substantial fill-in, and backsolving with a factored L will cost order $N^{4/3}M$ operations.

2.4 Matrix-Implicit Iterative Methods

As described above, applying direct factorization to solving (14) or (20) results in an algorithm whose computational complexity grows faster than order $MN^2 + N^3$. This rapid growth of computation time with problem size severely limits the size of circuit which can be analyzed, and in this section we examine how to reduce matrix solution cost using matrix-implicit iterative methods. To begin, consider solving the linear system in (8) using an iterative method like the Krylov-subspace based GMRES algorithm [7]. A simplified version of GMRES is given below.

GMRES algorithm for solving $Ax = b$

Guess at a solution, x^0.
Initialize the search direction $p^0 = b - Ax^0$.
Set $k = 1$.
do {
  Compute the new search direction, $p^k = Ap^{k-1}$.
  Orthogonalize, $p^k = p^k - \sum_{j=0}^{k-1} \beta_{k,j} p^j$.
  Choose α_k in
    $x^k = x^{k-1} + \alpha_k p^k$
  to minimize $\|r^k\| = \|b - Ax^k\|$.

  If $\|r^k\| < tolerance_{\text{gmres}}$, return v^k as the solution.
  else Set $k = k + 1$.
}

When the GMRES algorithm is applied to solving (8), if the number of iterations required to achieve convergence is bounded by a constant independent of problem size, then the resulting solution algorithm is much faster than direct factorization. The cost is reduced from order $MN^2 + N^3$ to order NM operations. To see this, consider that forming $(L+B)p^k$ costs order NM operations because of the circuit matrix sparsity, and we have assumed a constant number of iterations. Unfortunately, for typical circuit problems, $(L + B)$ in (8) is much too ill-conditioned for GMRES to converge rapidly. Instead, the iterative method should be preconditioned, or premultiplied, with L^{-1}, and this preconditioner insures rapid convergence [8].

Preconditioning using L^{-1} corresponds to applying the GMRES algorithm to solving (20) instead of (8). It is crucial to observe that $I_{NM} + L^{-1}B$ is not explicitly required in the GMRES algorithm, it is only necessary to be able to compute the matrix-vector products $(I_{NM} + L^{-1}B)p^k$. This observation implies that each preconditioned GMRES iteration can be performed in nearly order NM operations by multiplying p^k by B, then backsolving with a factored L, and finally adding p^k. Therefore, the cost per iteration is somewhat more than order NM operations. Also, since the L^{-1} preconditioning has insured rapid convergence independent of problem size, the entire solution algorithm is nearly order NM operations. Finally, an almost identical algorithm can be used to solve (14). Consider that computing the matrix-vector product $(J_\phi - I_N)p^k$, formed when GMRES is applied to solving (14), is equivalent to padding the N-length p^k with zeros to make it an NM-length vector, and then computing $(L^{-1}B - I_{NM})p^k$. If such an approach is used, $P_M - I_N$ is represented "quite" implicitly and the so derived shooting-Newton-GMRES method has the same nearly order NM operation cost as finite-difference methods [9]. Be reminded, however, that shooting-Newton methods still have superior global convergence properties for stable problems.

2.5 Results

In this section we experimentally examine the performance of three shooting-Newton schemes: direct factorization or Gaussian elimination, forming GMRES, and matrix-implicit GMRES.

Table 1 compares the performance of the various Newton-Raphson shooting method approaches implemented experimentally into the Spectre circuit simulator. The test suite includes *xtal*, a crystal filter; *mixer* is a small GaAs mixer; *dbmixer* is a double balanced mixer; *lmixer* is a large bipolar mixer; *cheby* is an active filter; and *scf* is a relatively large switched capacitor filter. The second column in Table 1 lists the number of equations in each circuit. The third column represents the number of one-period transient analyses that were necessary to achieve steady-state using the shooting-Newton method. The fourth, fifth, and sixth columns represent, respectively, the time in seconds to achieve steady-state using Gaussian elimination, explicit GMRES, and the matrix-implicit form. All the results were obtained on a HP712/80 workstation. The sixth column demonstrates the effectiveness of the matrix-implicit approach, listing the speedup obtained in respect to the Gaussian-elimination method. Note that the speed-up over the explicit GMRES algorithm would be similar for the size examples examined.

3 Frequency-Domain Methods

Any square integrable T-periodic waveform, $x(t)$, can be represented as a Fourier series,

$$x(t) = \sum_{k=-\infty}^{k=\infty} X[k] e^{j2\pi f_k t} \tag{22}$$

where $f_k = \frac{kt}{T}$ and

$$X[k] = \frac{1}{T} \int_{-T/2}^{T/2} x(t) e^{-j2\pi f_k t} dt. \tag{23}$$

circuit	eqns	it	GE	GMRES	MI	GE/MI
xtal	29	3	0.50	0.50	0.39	1.28
mixer	24	4	1.85	1.74	1.20	1.54
dbmixer	100	4	4.15	4.07	1.34	3.09
lmixer	126	3	3.72	3.63	1.03	3.61
cheby	237	4	23.39	21.97	3.01	7.96
scf	377	6	2962	2954	281.4	10.52

Table 1: Comparison of different shooting method schemes

If in addition to being periodic, $x(t)$ is sufficiently smooth, formally infinitely continuously differentiable, then the $X[k]$'s vanish exponentially fast with increasing k. This implies $x(t)$ can be accurately represented with a truncated Fourier series, that is $\hat{x}(t) \approx x(t)$ where $\hat{x}(t)$ is given by the truncated Fourier series,

$$\hat{x}(t) = \sum_{k=-K}^{k=K} \hat{X}[k] e^{j2\pi f_k t} \tag{24}$$

where the number of harmonics, K, is typically fewer than 15. Note that the time derivative of $\hat{x}(t)$ is given by

$$\frac{d}{dt}\hat{x}(t) = \sum_{k=-K}^{k=K} \hat{X}[k] j2\pi f_k e^{j2\pi f_k t}. \tag{25}$$

Representing a periodic waveform as a truncated Fourier series can be used to approximately solve (1) and (2), in which case the resulting methods are referred to as harmonic balance methods [3], or Fourier spectral methods [13]. Of particularly wide use in circuit simulation are the spectral-collocation-on-charge variants of these methods [3]. To derive the collocation method requires a little notation. If $q(v)$ in (1) is globally invertible, then we can denote $\tilde{q}(t) \equiv q(v(t))$ and $\tilde{i}(\tilde{q}(t)) \equiv i(v(t))$. Using this notation, (1) can be rewritten as a differential equation in normal form in terms of q,

$$\frac{d}{dt}\tilde{q}(t) + \tilde{i}(\tilde{q}(t)) + u(t) = 0. \tag{26}$$

As an aside, standard integration methods are typically applied to the above form of (1) to insure charge conservation.

Introducing the truncated Fourier series approximation for $\tilde{q}(t)$,

$$\hat{\tilde{q}}(t) = \sum_{k=-K}^{k=K} \hat{Q}[k] e^{j2\pi f_k t}, \tag{27}$$

into (26) results in

$$\sum_{k=-K}^{k=K} \hat{Q}[k] j2\pi f_k e^{j2\pi f_k t} + \tag{28}$$

$$\tilde{i}\left(\sum_{k=-K}^{k=K} \hat{Q}[k] e^{j2\pi f_k t}\right) + u(t) = 0.$$

Of course, it would be best if (28) holds for all t, but there are only $M = 2K + 1$ degrees of freedom, the $\hat{Q}[k]'s$. Note, each $\hat{Q}[k]$ is an $N-$length vector, so we are using "degrees of freedom" in a very loose sense. In order to derive M systems of equations from which to determine the $\hat{Q}[k]$'s, (28) is enforced only at selected collocation points $t_1, t_2, ..., t_M = T$, as in

$$\sum_{k=-K}^{k=K} \hat{Q}[k] j2\pi f_k e^{j2\pi f_k t_j} + \tag{29}$$

$$\tilde{i}\left(\sum_{k=-K}^{k=K} \hat{Q[k]} e^{j2\pi f_k t_j}\right) + u(t_j) = 0.$$

Convergence analysis of spectral-collocation methods can be found in [13, 14].

Note that (29) represents NM equations in NM unknowns, but its form is not so readily identified with the original functions $q(v)$ and $i(v)$. The relation is easily unraveled, however. Let $\Gamma^{-1} \in C^{NM \times NM}$ denote the discrete Fourier transform matrix which represents the relationship between the $\hat{Q}[k]'s$ and $\hat{\tilde{q}}(t_j)'s$, as in

$$\Gamma^{-1} \begin{bmatrix} \hat{Q}[K]) \\ \hat{Q}[(K-1)] \\ \vdots \\ \hat{Q}[-K] \end{bmatrix} = \begin{bmatrix} \hat{\tilde{q}}(t_1) \\ \hat{\tilde{q}}(t_2) \\ \vdots \\ \hat{\tilde{q}}(t_M) \end{bmatrix}. \tag{30}$$

Using the definitions of $\tilde{q}$ and $\tilde{i}$, (29) can be simplified as

$$\boldsymbol{F}_{time}(\hat{\boldsymbol{v}}) = \Gamma^{-1}\Omega\Gamma\boldsymbol{q}(\hat{\boldsymbol{v}}) + \boldsymbol{i}(\hat{\boldsymbol{v}}) + \boldsymbol{u} = 0 \tag{31}$$

where

$$\boldsymbol{q}(\hat{\boldsymbol{v}}) \equiv \begin{bmatrix} q(\hat{v}(t_1)) \\ q(\hat{v}(t_2)) \\ \vdots \\ q(\hat{v}(t_M)) \end{bmatrix}, \quad \boldsymbol{i}(\hat{\boldsymbol{v}}) \equiv \begin{bmatrix} i(\hat{v}(t_1)) \\ i(\hat{v}(t_2)) \\ \vdots \\ i(\hat{v}(t_M)) \end{bmatrix}, \tag{32}$$

$$\boldsymbol{u} \equiv \begin{bmatrix} u(t_1) \\ u(t_2) \\ \vdots \\ u(t_M)) \end{bmatrix}, \tag{33}$$

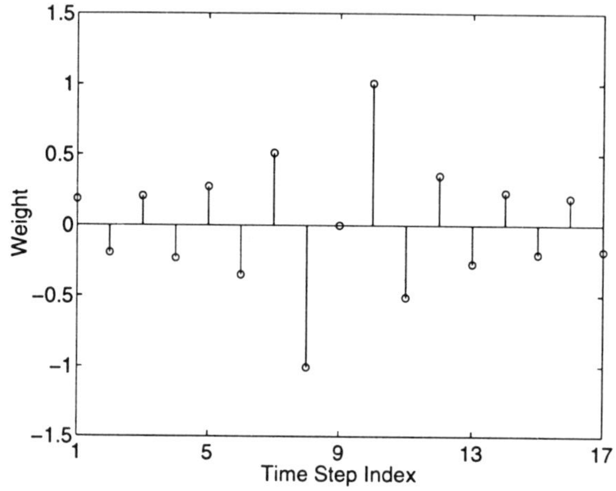

Figure 1: The harmonic balance discretization weights for t_9 where $T = 17$ and $M = 17$.

and Ω is the diagonal matrix given by

$$\Omega \equiv \begin{bmatrix} j2\pi f_K I_N & & & \\ & j2\pi f_{K-1} I_N & & \\ & & \ddots \quad \ddots & \\ & & & j2\pi f_{-K} I_N \end{bmatrix}. \tag{34}$$

Using a change of variables, $\hat{\boldsymbol{V}} = \Gamma\hat{\boldsymbol{v}}$, (31) can be written in the form

$$\boldsymbol{F}_{freq}(\hat{\boldsymbol{V}}) = \Omega\Gamma\boldsymbol{q}(\Gamma^{-1}\hat{\boldsymbol{V}}) + \Gamma\boldsymbol{i}(\Gamma^{-1}\hat{\boldsymbol{V}}) + \Gamma\boldsymbol{u} = 0. \tag{35}$$

The form in (35) is often referred to as the harmonic balance form because the unknowns are the Fourier coefficients and the equations are in the frequency domain.

From the formulation in (31), it is clear that $\Gamma^{-1}\Omega\Gamma$ is an approximate differentiation operator. The weights for this spectral differentiation operator for the case of $T = 17$, $M = 17$, and at timepoint $t_9 = 9$ are plotted in Figure (1). Note that the weights at t_8 and t_{10} are approximately -1 and 1 respectively, so spectral differentiation is somewhat similar to a central-difference scheme in which

$$\frac{d}{dt}x(t_9) \approx \frac{x(t_{10}) - x(t_8)}{t_{10} - t_8} \tag{36}$$

The connection between spectral differentiation and standard differencing schemes can be exploited when developing preconditioners, a point we will return to subsequently.

3.1 Applying Newton's Method

Newton's method applied to (31) or (35) yield the iteration equations

$$\left(\Gamma^{-1}\Omega\Gamma \boldsymbol{C} + \boldsymbol{G}\right)\left(\hat{\boldsymbol{v}}^{k+1} - \hat{\boldsymbol{v}}^{k}\right) = -\boldsymbol{F}_{time}(\hat{\boldsymbol{v}}) \quad (37)$$

or

$$\left(\Omega\Gamma \boldsymbol{C}\Gamma^{-1} + \Gamma \boldsymbol{G}\Gamma^{-1}\right)\left(\hat{\boldsymbol{V}}^{k+1} - \hat{\boldsymbol{V}}^{k}\right) = -\boldsymbol{F}_{freq}(\hat{\boldsymbol{v}}) \quad (38)$$

where k is the Newton iteration index, $\boldsymbol{F}_{time}$ and $\boldsymbol{F}_{freq}$ are given in (31) and (35), and $C, \mathrm{G} \in \Re^{\mathrm{NM}\times\mathrm{NM}}$ are block diagonal matrices given by

$$\boldsymbol{C} \equiv \begin{bmatrix} C_1 & & & \\ & C_2 & & \\ & & \ddots & \\ & & & C_M \end{bmatrix}, \quad (39)$$

$$\boldsymbol{G} \equiv \begin{bmatrix} G_1 & & & \\ & G_2 & & \\ & & \ddots & \\ & & & G_M \end{bmatrix}, \quad (40)$$

where C_j denotes $dq(v(t_j))/dv$ and G_j denotes $di(v(t_j))/dv$.

The Jacobian in (37) has a structure that is similar, but much denser, than the Jacobian associated with the backward-Euler based finite-difference-Newton Method. This is clear from the explicit representation of $\left(\Gamma^{-1}\Omega\Gamma \boldsymbol{C} + \boldsymbol{G}\right)$,

$$\begin{bmatrix} G_1 & \alpha_1 C_1 & \alpha_2 C_1 & ,\ldots, & \alpha_{-1} C_1 \\ \alpha_{-1} C_2 & G_2 & \alpha_1 C_2 & ,\ldots, & \alpha_{-2} C_2 \\ & \ddots & \ddots & \ddots & \\ \alpha_1 C_M & \alpha_2 C_M & ,\ldots, & \alpha_{-1} C_M & G_M \end{bmatrix} \quad (41)$$

where the α_i's are the weights for the spectral differentiation. Note that the block diagonal terms depend only on the G's (incremental conductances). Also, for linear time-invariant problems, where $C_j = C$ and $G_j = G$ for all j, the matrix density is basically unchanged.

The explicit representation for $\Omega\Gamma \boldsymbol{C}\Gamma^{-1} + \Gamma \boldsymbol{G}\Gamma^{-1}$ is, in general, similarly dense. However, one of the advantages of the harmonic balance form of the equations is that if the problem is linear and time-invariant the matrix is block diagonal and is given by

$$\begin{bmatrix} \frac{j2\pi K}{T}C + G & & & \\ & \frac{j2\pi(K-1)}{T}C + G & & \\ & & \ddots \quad \ddots & \\ & & & \frac{j2\pi(-K)}{T}C + G \end{bmatrix} \quad (42)$$

where $C = C_j$ and $G = G_j$ for all j. In this case, of course, the matrix just represents phasor analysis at a collection of isolated frequencies.

3.2 Iterative Matrix Solution

As mentioned in Section 2.4, if GMRES is applied to solving a linear system of equations, then the matrix representing the system is not needed explicitly. It is only necessary to be able to compute matrix-vector products. This implies that if GMRES is applied to solving (37) or (38), then the iterations can be performed in order $NMlogM$ operations by employing the fast Fourier transform [12]. This is easily shown, consider computing

$$\left(\Gamma^{-1}\Omega\Gamma \boldsymbol{C} + \boldsymbol{G}\right) p^k \;=\; \Gamma^{-1}\Omega\Gamma \boldsymbol{C} p^k + \boldsymbol{G} p^k. \quad (43)$$

Forming $\boldsymbol{G}p^k$ and $y = \boldsymbol{C}p^k$ require only order NM operations, due to the sparsity of C_i and G_i. Forming $w = \Gamma y$ can be accomplished in $NMlogM$ operations using N fast Fourier Transforms. Multiplying w by Ω is order NM operations because Ω is diagonal. And finally, forming $\Gamma^{-1}\Omega w$ requires order $NMlogM$ operations using N inverse fast Fourier transforms. A similar argument demonstrates that

$$\left(\Omega\Gamma \boldsymbol{C}\Gamma^{-1} + \Gamma \boldsymbol{G}\Gamma^{-1}\right) p^k \quad (44)$$

can be computed in order $MNlogM$ operations.

GMRES applied directly to solving (37) or (38) converges too slowly to be practical. And, since the matrices in (37) or (38) are related by an orthonormal similarity transform, GMRES will converge identically slowly in either case. However, when one considers preconditioners, there is a reason for choosing between the formulations. For the harmonic-balance form, a block-diagonal, sometimes called a block Gauss-Jacobi, preconditioner can be extremely effective when the problem is nearly linear and time-invariant. This is because the off-diagonal blocks of the matrix in (38) are produced by frequency translations which are minimal for the nearly linear and time-invariant case. Variations on this idea include adding in some of the off-diagonal blocks, using bands of blocks and incomplete factorization, or using the block lower triangular matrix [2, 5, 3, 15].

For rapidly time-varying or nonlinear problems, frequency coupling is more severe, and preconditioners based on discarding off-diagonal blocks of the matrix in (38) become ineffective. A different approach, which involves preconditioning (37), uses matrices associated with finite-difference methods as preconditioners, like the L matrix defined in (6) [14]. The key concept is that both matrices have an approximate differentiation operator, but, for example, the backward-Euler or a central-difference discretization use much sparser operators than spectral discretization. It is worth noting that backward-difference methods have a computational advantage over central-differencing schemes in that they are more lower triangular. Finally, the L matrix given in (6) is not a very effective preconditioner because the backward-Euler operator is not sufficiently accurate to properly precondition the very high order spectral differentiation operator. Matrices associated with higher order differencing schemes are much more effective.

4 Future Work and Acknowledgements

Although in this tutorial paper we have focussed on periodic steady-state analysis, most RF communication circuit problems require multitone analysis. For example, mixers used for down conversion generate sum and difference frequencies that can be seperated by several orders of magnitude. For this reason, the newest work in this area is applying matrix-implicit iterative techniques to accelerating multitone problems using multitone harmonic balance [5], linear time-varying noise analysis [8], and mixed frequency-time techniques [3]. The authors would like to thank Robert Melville, Peter Feldmann, Jaijeet Roychowdhury, Michael Steer, and David Sharrit for many valuable discussions over the years.

References

[1] A. Ushida, L. Chua, and T. Sugawara, "A Substitution Algorithm for solving nonlinear circuits with multifrequency components. *International Journal on Circuit Theory and Applications*, vol. 15, 1987.

[2] P. Heikkilä. *Object-Oriented Approach to Numerical Circuit Analysis.* Ph. D. dissertation, Helsinki University of Technology, January 1992.

[3] K. Kundert, J. White and A. Sangiovanni-Vincentelli. *Steady-State Methods for Simulating Analog And Microwave Circuits.* Kluwer Academic Publishers, Boston 1990.

[4] R. Gilmore and M. Steer, "Nonlinear circuit analysis using the method of harmonic balance - a review of the art. Part I - Introductory Concepts." *Int. J. on Microwave and Millimeter Wave Computer Aided Engineering*, Vol. 1, No. 1, 1991.

[5] R. Melville, P. Feldmann, and J. Roychowdhury. "Efficient multi-tone distortion analysis of analog integrated circuits." *Proceedings of the 1995 IEEE Custom Integrated Circuits Conference*, May 1995.

[6] M. Okumura, H. Tanimoto, T. Itakura, and T. Sugawara. "Numerical Noise Analysis for Nonlinear Circuits with a Periodic Large Signal Excitation Including Cyclostationary Noise Sources." *IEEE Transactions On Circuits and Systems - I Fundamental Theory and Applications.*, vol. 40, no. 9, pp. 581–590, September 1993.

[7] Y. Saad and M. H. Schultz. "GMRES: A generalized minimal residual algorithm for solving nonsymmetric linear systems." *SIAM Journal on Scientific and Statistical Computing*, vol. 7, pp. 856–869, July 1986.

[8] R. Telichevesky, K. Kundert, and J. White. "Matrix-Implicit Iterative Techniques For RF Circuit Analysis," *in preparation.*

[9] R. Telichevesky, Kenneth S. Kundert, Jacob K. White. "Efficient Steady-State Analysis based on Matrix-Free Krylov-Subspace Methods." *Proc. Design Automation Conference*, Santa Clara, California, June 1995.

[10] H. Keller. *Numerical Solution of Two Point Boundary-Value Problems*, SIAM, 1976.

[11] Thomas J. Aprille and Timothy N. Trick. "Steady-state analysis of nonlinear circuits with periodic inputs." *Proceedings of the IEEE*, vol. 60, no. 1, pp. 108–114, January 1972.

[12] J. P. Boyd, *Chebyshev and Fourier Spectral Methods*, Springer-Verlag, New York, 1989.

[13] D. Gottlieb and S. Orszag, *Numerical Analysis of Spectral Methods: Theory and Applications*, SIAM, 1977.

[14] C. Canuto, M. Y. Hussaini, A. Quarteroni and T. A. Zang, *Spectral Methods in Fluid Mechanics*, Springer-Verlag, New York, 1987.

[15] B. Troyanovsky, Z. Yu, L. So. and R. Dutton, "Relaxation-Based Harmonic Balance Technique for Semiconductor Device Simulation," *Proc. International Conference on Computer-Aided Design*, Santa Clara, California, November 1995.

Parasitic Characterization of Radio-Frequency (RF) Circuits Using Mixed-Mode Simulation

Jaejune Jang, Edwin C. Kan, Lydia So*, Robert W. Dutton
AEL 201, Integrated Circuit Lab, Stanford University, Stanford, CA 94305, U.S.A.
Ted Johansson and Torkel Arnborg
Ericsson Microsistems, Sweden

Abstract

Accurate estimation of the parasitics in high-speed circuits is critical in optimizing circuit performance. A new method for parasitic characterization of high-speed circuits employing mixed-mode circuit and device simulation is proposed. The intrinsic characteristics are captured using a device simulator and the equivalent circuit for passive extrinsic elements are evaluated in a straightforward manner from impedance and admittance representation of measured S-parameters. This decoupling enables total performance optimization based on separate tuning of layout and the fabrication process recipe.

I. Introduction

For monolithic microwave and RF integrated circuit applications, accurate modeling of parasitic element is critical for the device performance analyses such as power gain, bandwidth and stability analyses. Usually, the small-signal equivalent circuit is obtained at many bias settings using S-parameter (Scattering parameter) measurements over the frequency range of interest, either by analytical methods or using optimization techniques. The disadvantage of the optimization approach is that the results can vary depending upon the fitting method and the starting guess. In the case of analytical parameter fitting, additional test structures are usually required to consider de-embedding the S-parameters from the parasitic environment, and other steps (cold measurements [1]) are necessary to distinguish between bias dependent and independent elements.

In this paper, we use a mixed-mode (circuit and device) simulation approach for characterization of device-level behavior modeling. The circuit is separated into the intrinsic and extrinsic parts. The active intrinsic device characteristics are modeled using device simulation [3, 4] instead of a fixed equivalent circuit model such as the Gummel-Poon representation. The extrinsic lumped elements are extracted analytically by comparing simulated S-parameters of the intrinsic device and experimental S-parameters extracted from the entire circuit. This method decouples the circuit element dependencies between layout and doping profiles; hence the distributive aspects of the intrinsic device can be taken into consideration directly. Due to inefficiencies in simulating the entire multi-finger structures, only a 2-D cut of a single emitter and base finger structure is considered in device simulation. The full structure is analytically combined to represent the intrinsic part. More accurate simulation is possible using 3-D simulations, at the cost of more computational power.

*L. So is currently with Siliconix Inc., CA.

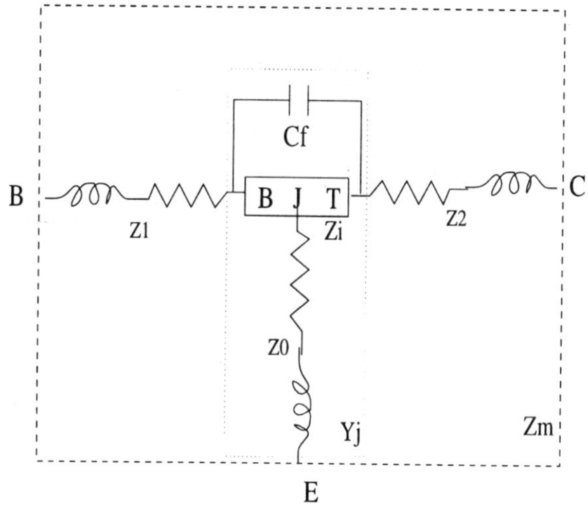

Figure 1: A T-like network for characterization of monolithic RF BJT devices.

II. Methodology

The basic tool for parasitic extraction is the mixed-mode circuit and device simulator [2] which interfaces Stanford's device simulator PISCES-2ET [4] and UC Berkeley's circuit simulator SPICE-3f2. The characterization of device performance at high frequencies using S-parameter extraction is widely accepted. For devices which are fabricated on monolithic substrate, the circuit often exhibits parasitic behavior due to elements that can be represented by a two-port network configuration as shown in Fig. 1.

This T-like topology has been used successfully by many authors [5, 6] . For this network it is assumed

IEEE 1996 CUSTOM INTEGRATED CIRCUITS CONFERENCE

that over the frequency range of concern (from 300 MHz to 3 GHz) the equivalent circuit composed of lumped RLC elements, can be described using general interconnect parasitics. In reality, at high frequencies the interconnect circuit parameters such as resistance may vary with frequency, due to skin and other distributive effects. However, for the frequency range considered here, frequency-independent modeling parameters are usually sufficient.

Accurate physical modeling of the intrinsic device, excluding any other parasitics such as interconnect, bond wire, and contacts is performed using the device simulator. The input doping profiles are extracted from SUPREM 1D simulation, calibrated with measurements. For electrical characteristics, the Gummel plot and the cut-off frequency f_T are fitted to measurements. Standard bipolar models such as Auger, concentration dependent SRH, and concentration dependent mobility are used and parameters such as lifetimes and ambient temperature are adjusted slightly. Also adaptive gridding is essential for accurate simulation at low current levels. This device is then simulated using mixed-mode approach at specific bias conditions; Y-parameters are thus obtained. Using this procedure, V_{be} of the intrinsic device can be determined at the same current levels as thosed used in measurements, since Gummel plots are used for calibration. The effect of small variations of V_{ce} on S-parameters are negligible. Hence, V_{ce} of the intrinsic device can be chosen by comparing with measured external V_{ce}. Another method for determining bias conditions of the intrinsic device is to first perform extraction at a small V_{be} level since the effects of resistive parasitics plays a negligible role there.

As shown in Fig. 1, construction of the two-port matrices from the inner shell $[Z_i]$ to the outer one $[Z_m]$ or vice versa is straightforward. The analytical expression for Z_m, obtained by expanding from the intrinsic matrix Z_i, is shown below.

The intrinsic device is characterized using a two port network.

$$Z_i = \begin{vmatrix} Z_{11} & Z_{12} \\ Z_{21} & Z_{22} \end{vmatrix}$$

Inclusion of Z_0 and C_f results in,

$$Y_j = \frac{1}{\Delta Z}\begin{vmatrix} Z_{22}+Z_0 & -Z_{12}-Z_0 \\ -Z_{21}-Z_0 & Z_{11}+Z_0 \end{vmatrix} + jwC_f\begin{vmatrix} 1 & -1 \\ -1 & 1 \end{vmatrix}$$

$$\Delta Z = Z_{22}Z_{11} - Z_{12}Z_{21} + Z_0(Z_{22}+Z_{11}-Z_{12}-Z_{21})$$

$$Z_0 = R_0 + jwL_0$$

Inclusion of Z_1 and Z_2 at the input and output ports gives,

$$Z_m = \frac{1}{det(Z_j)}\begin{vmatrix} Z_{j11} & Z_{j12} \\ Z_{j21} & Z_{j22} \end{vmatrix} + \begin{vmatrix} Z_1 & 0 \\ 0 & Z_2 \end{vmatrix}$$

$$det(Z_j) = 1 + jwC_f(Z_{22}+Z_{11}-Z_{12}-Z_{21})$$

$$Z_{j11} = Z_{11} + Z_0 + jwC_f\Delta Z$$

$$Z_{j12} = Z_{12} + Z_0 + jwC_f\Delta Z$$

$$Z_{j21} = Z_{21} + Z_0 + jwC_f\Delta Z$$

$$Z_{j22} = Z_{22} + Z_0 + jwC_f\Delta Z$$

$$Z_1 = R_1 + jwL_1$$

$$Z_2 = R_2 + jwL_2$$

The first step is to obtain C_f from

$$Z_{m12} - Z_{m21} = \frac{Z_{12}-Z_{21}}{det(Z_j)}$$

By inserting the extracted C_f in Z_{m12}, Z_{m11}, and Z_{m22}, all other parasitics Z_0, Z_1, and Z_2 can be found.

$$jwC_f = \frac{Z_{12}-Z_{21}+Z_{m21}-Z_{m12}}{(Z_{m12}-Z_{m21})(Z_{22}+Z_{11}-Z_{12}-Z_{21})} \quad (1)$$

$$Z_0 = \frac{Z_{m12}det(Z_j) - Z_{12} - jwC_f(Z_{11}Z_{22}-Z_{12}Z_{21})}{1+jwC_f(Z_{11}+Z_{22}-Z_{12}-Z_{21})} \quad (2)$$

$$Z_1 = Z_{m11} - \frac{Z_{11}+Z_0+jwC_f\Delta Z}{1+jwC_f(Z_{11}+Z_{22}-Z_{12}-Z_{21})} \quad (3)$$

$$Z_2 = Z_{m22} - \frac{Z_{22}+Z_0+jwC_f\Delta Z}{1+jwC_f(Z_{11}+Z_{22}-Z_{12}-Z_{21})} \quad (4)$$

III. Simulation Results and Discussion

It is critical to model the intrinsic device as accurately as possible. The cross sectional view of the modeled intrinsic device and the fitted Gummel plot is shown in Figs. 2 and 3. The most sensitive parameters in fitting the Gummel plot are carrier lifetimes and doping profiles in the active transistor region. Mobility is a relatively insensitive parameter in fitting the Gummel plot. For f_T, the simulated value of 3.5 GHz is achieved in comparison with a measured value of 3.15 GHz.

Using the parasitic extraction method previously described, each component is calculated over the frequency range of 300 MHz to 3 GHz, which is plotted in Figs. 4 and 5. The measured S-parameters have included heuristic system and probe error compensation. The C_f, L_0,and L_1 are almost constant

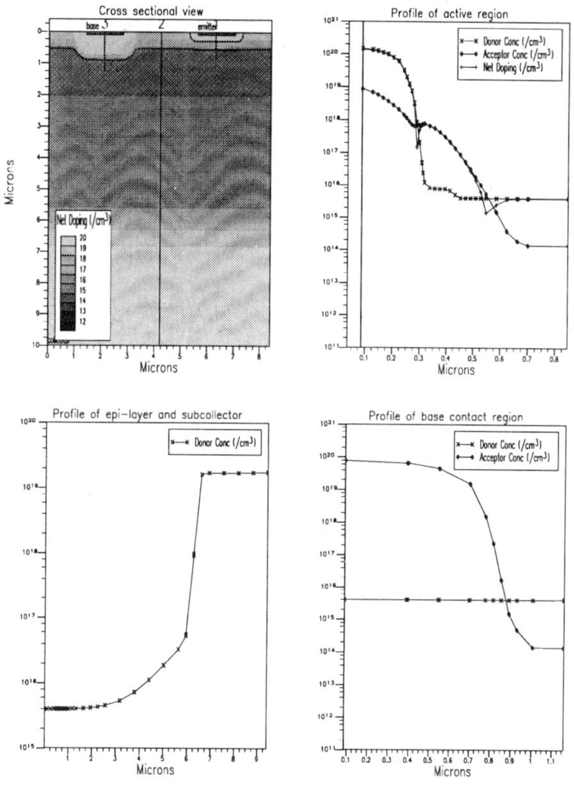

Figure 2: Cross section and doping profiles used for modeling the intrinsic device.

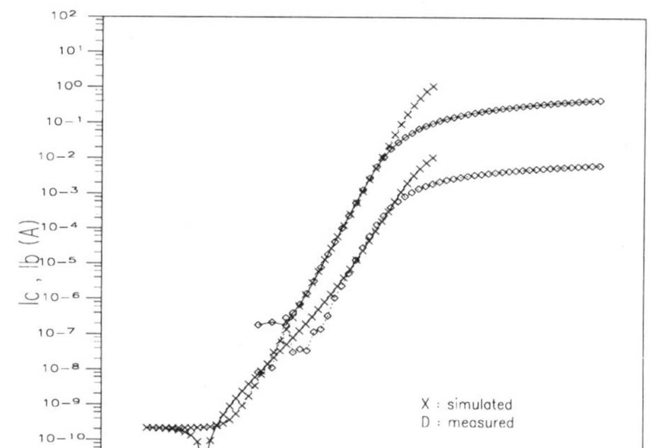

Figure 3: The cross symbols are the Gummel plot at V_{ce} =5V for the intrinsic device. Simulation data and direct measurements on the intrinsic device (a single finger without packaging) have been matched completely for $V_{be} > 0.5V$. The diamond symbols are measured data from the entire circuit with multiple fingers and packaging. The currents of the intrinsic device is multiplied by the number of fingers for easy comparison.

throughout the frequency range. However, the resistance noise and the fact that L_2 is negative can be explained in two ways. First, those extracted values can be comparable to the noise included in measurements. Second, presently the intrinsic device is calibrated *only* against the Gummel plot and f_T. The device behavior in the collector contact region (in this case, an epi-layer) is not accurately accounted for during the calibration, as is manifested in the fact that both the Gummel plot and f_T will have negligible change for large variations of V_{ce}. More detailed calibration with the test structure including only the base-collector junction can remove this error. Also, modeling parameters will affect some of the parasitic values. Mobility has a significant influence on subcollector resistance. If higher mobilities are used, R_2 will become larger due to the reduced resistance of the intrinsic device. The extracted values of R_0, R_2, C_f, L_0,and L_2 are very sensitive to the V_{be} hence an accurate estimation of the bias conditions for the intrinsic device are necessary. On the other hand, for small variations of V_{ce}, only slight variations of C_f are observed. Presently the final resistance values are determined as below. R_1 is first selected by comparing the resistive component of S_{11}, since R_1 dominates the resistive part of S_{11}. By satisfying the bias conditions and the current constraints, R_0 can be determined. R_2 is selected in the same manner as for selecting R_1. These selected resistance values are actually in the range of extracted data from (2), (3), and (4).

Using the average values of extracted data for L_0, L_1, and C_f over the frequency range, and using resistance values from the above procedure, the full circuit is simulated and compared with experimental data in Fig. 6. It can be seen that the high-frequency circuit behavior is well captured by the mixed-mode simulation with frequency-independent parasitic elements. As mentioned before, the extraction procedure can be further improved by minor mobility adjustments (distinguishing majority and minority mobilities) and inverted-mode measurements of the intrinsic device.

After the parasitic characterization procedure, a mixed-mode representation of the RF circuit is obtained. High-frequency performance optimization can be obtained from separate adjustments of the mask layout and the doping profiles. Also, other virtual instrument techniques such as the harmonic balance analyzer [7] can be applied for detailed analyses of the analog circuit behavior.

Acknowledgment

This work is supported through ARPA Contract # DABT63-94-C-0055 (Computational Prototyping of 21st Century Semiconductors) and Ericsson Microsystems. The authors wish to thank Professor

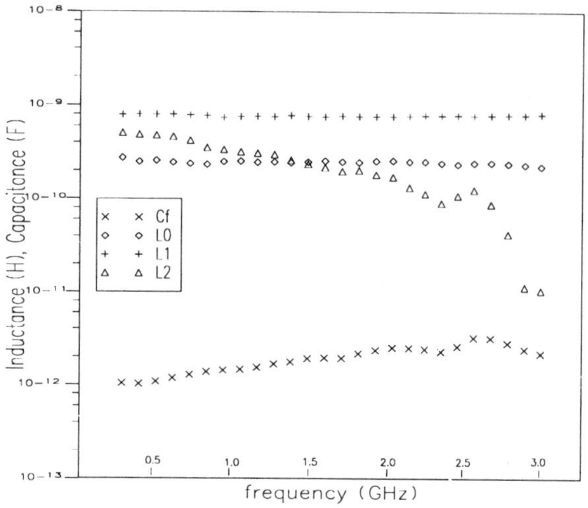

Figure 4: The extracted inductances and capacitance using (1) to (4) (absolute values of L2 are used).

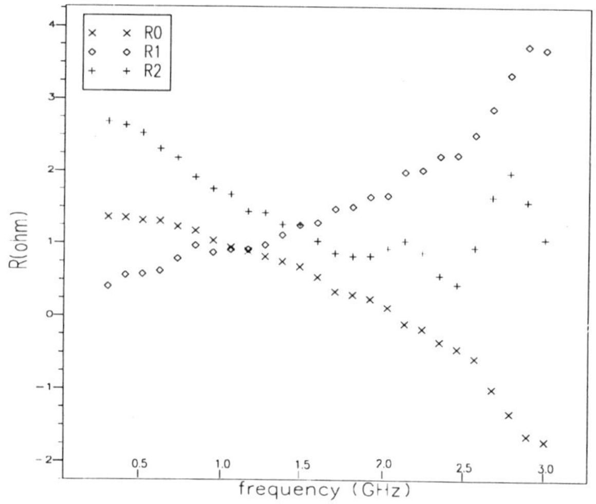

Figure 5: Extracted parasitic resistances using (2) to (4).

Thomas H. Lee of Stanford University on helpful discussion of RF device characterization.

References

[1] G. Dambrine, et. al., "A new method for determining the FET small-signal equivalent circuit,"*IEEE Trans. Microwave Theory Tech.*, vol. MTT–36, pp. 1151–1159, 1988.

[2] F. Rotella, Z. Yu and R. W. Dutton, "Mixed-mode device/circuit simulator: PISCES-2ET and its application subsystems," Stanford University, CA.

[3] *ATLAS - Framework for two dimensional device simulation*, SILVACO International, 1994.

[4] Z. Yu, D. Chen, L. So, and R. W. Dutton, "PISCES-2ET— two dimensional device simulation for silicon and heterostructures," Stanford University, CA.

[5] R. Ramachandran, et. al., "A high efficiency HBT MMIC power amplifier," *GaAs IC Symp.*, vol. 22, pp. 357–360, 1990.

[6] U. Schaper and B. Holzapfl, "Analytical parameter extraction of the HBT equivalent circuit with T-like topology from measured S-parameters," *IEEE Trans. Microwave Theory Tech.*, vol. 43, pp. 493–498, 1995.

[7] R. W. Dutton, et. al., "Virtual instruments for development of high performance circuit technologies," *IEEE CICC.*, pp. 225–228, 1995.

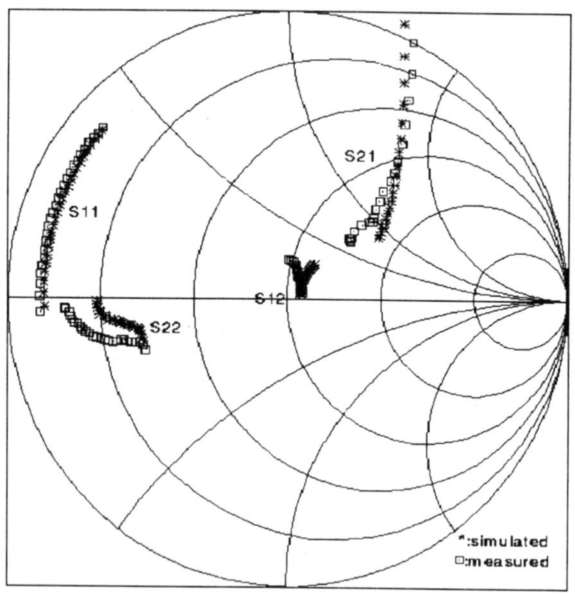

Figure 6: Simulated and measured S-parameters at V_{be}=0.82 V and V_{ce}=5V.

Receiver Characterization using Periodic Small-Signal Analysis

Ricardo Telichevesky and Ken Kundert
Cadence Design Systems, San Jose CA

Jacob White
Massachusetts Institute of Technology, Cambridge MA

***Abstract*— Periodic small-signal analysis is a recently developed approach for AC and noise analysis of linear periodically-varying circuits. This paper presents several new approaches to characterizing the transfer functions, noise, and distortion of communication circuits using this new technique. A receiver is used as an example, but the techniques presented are applicable to a wide variety of circuits, including mixers, samplers, sample-and-holds, chopper-stabilized amplifiers, multipliers and switched-capacitor filters.**

I. Introduction

The increasing demand for low-cost mobile communication systems has greatly expanded the need for simulation algorithms that are both efficient and accurate when applied to RF communication circuits such as mixers. The difficulty is that RF circuits often process signals that contain closely-spaced large frequencies. For example, a typical cellular phone service consists of 10 kHz voiceband channels near 900 MHz. A SPICE transient analysis is extremely expensive for circuits that contain such signals because it needs to take very small time steps to follow the 900 MHz carrier, and simulate for a long time to adequately resolve the voiceband modulation. However, circuits such as amplifiers can be linearized about their DC operating point and a small-signal analysis such as AC analysis used to efficiently determine the sinusoidal steady-state response of the circuit over an arbitrarily range of frequencies. The same is not true of circuits such as mixers or switched-capacitor filters that must be driven with an additional large periodic stimulus in order to operate properly.

II. Periodic Small-Signal Analysis

Periodic small-signal analyses address periodically-driven circuits using a two-step process. First, the *periodic operating point* is found by performing a *periodic steady-state (PSS) analysis* with only the large periodic stimulus applied. For the case of a mixer, only the LO is applied; with switched-capacitor filters, only the clock is applied. The circuit is then linearized about the periodic operating point, and the steady-state response of the periodically-varying linear circuit is computed assuming that it is driven with a small sinusoid at an arbitrary frequency. The time required to compute the response of the linearized circuit is about the same as needed to compute the periodic steady-state response to the large stimulus alone, regardless of the frequency of the sinusoid.

An important aspect of periodically-driven circuits such as mixers and switched-capacitor filters is that they exhibit frequency translation. Conventional small signal analyses (AC and noise) linearizes about the DC operating point to create a linear time-invariant representation of the circuit, which cannot exhibit frequency conversion effects. With the periodic small-signal analyses, the circuits are linearized about a time-varying operating point and so can exhibit frequency conversion effects.

The original approach to periodic small-signal analysis used Gaussian elimination to factor and solve the set of N dense linear equations that describe the periodically-varying linearized circuit at a particular frequency where N is the number of nodes [1], [2]. The time required to factor a dense $N \times N$ matrix is on the order of N^3, which limits this approach to circuits containing at most a few hundred nodes. However, by applying matrix-free Krylov-subspace techniques, the time required to solve the system of equations can be reduced to the order of N, allowing circuits with up to 5000 nodes to be simulated over a broad range of frequencies in a few hours [3], [4].

A. Periodic AC Analysis

The periodic AC (PAC) analysis computes the responses of a periodically driven circuit to a small sinusoidal input at an arbitrary frequency.

When applying a small sinusoid to a linear time-invariant circuit, the steady-state response is a sinusoid at the same frequency. However, when applying a small sinusoid to a linear periodically time-varying circuit, the circuit responds with sinusoids at many frequencies, as shown in Figure 1. Thus, in a sense, PAC computes a series of transfer functions, one for each frequency. However, with these transfer functions the input and output frequencies are offset by the harmonics of the LO.

The small signal frequency can be arbitrarily close, or

Manuscript received December 8, 1995.
E-mail addresses: ricardo@cadence.com, kundert@cadence.com, and white@rle-vlsi.mit.edu.

0-7803-3177-6 $5.00 © 1996 IEEE

IEEE 1996 CUSTOM INTEGRATED CIRCUITS CONFERENCE

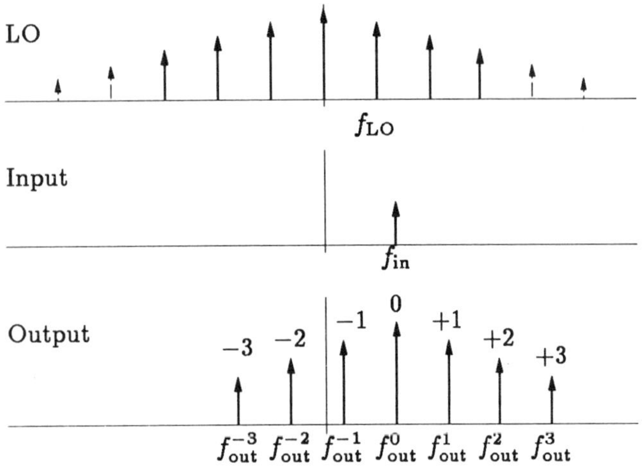

Fig. 1. The small-signal response of a mixer as computed by PAC analysis. The input signal is replicated and translated by each harmonic of the LO. In down-conversion mixers, the −1 sideband usually represents the IF output.

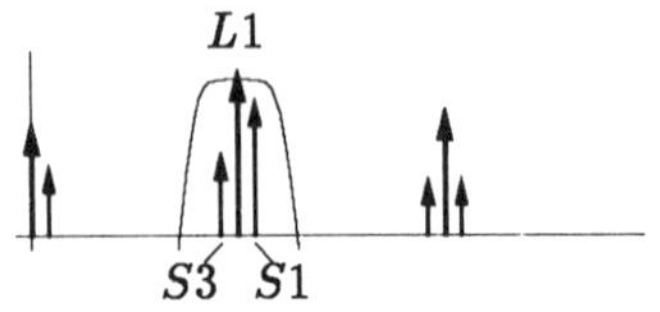

Fig. 2. Intermodulation distortion is efficiently measured by applying one large tone ($L1$), performing a PSS analysis, and then applying the second small tone ($S1$) with a PAC analysis. In this figure, V_{L1} is the fundamental of the response due to the large input tone, V_{S1} is the fundamental of the response due to the small input tone and is the upper sideband of V_{L1}. V_{S3} is the lower sideband of V_{L1}, and represents the intermodulation distortion.

even equal to, the LO frequency with no performance penalty.

B. Measuring Intermodulation Distortion using PAC

A important use of PAC analysis is to measure the intermodulation distortion of amplifiers and mixers. The traditional approach to measuring intermodulation distortion is to apply two large closely spaced tones to the input and measure the third-order intermodulation products. Applying this approach with a circuit simulator is expensive because two tones that are closely spaced in frequency require a long analysis interval. The approach used with PSS/PAC is slightly different. Here only one large tone is applied. Then PAC is used to apply the second tone close to the first. If the small input signal is considered to be one sideband of the large input signal, then the response at the other sideband is the third-order intermodulation distortion, as shown in Figure 2. In this case, the first tone (the tone applied in PSS analysis) drives the circuit hard enough to cause distortion and the second tone (the one applied in PAC) is a test tone that is only used to measure the intermodulation distortion. Once V_{L1}, V_{S1}, and V_{S3} are measured in decibels, the third-order intercept point is

$$IP_3 = V_{L1} - \frac{V_{S3} - V_{S1}}{2}. \tag{1}$$

The intermodulation distortion of a mixer is measured in a similar manner, except the PSS analysis must include both the LO and the one large tone.

This approach gives slightly different results than the approach typically used when measuring intermodulation distortion on the bench. There, two large tones of equal amplitude are applied, both of which drive the circuit to behave nonlinearly. In this new method, only one tone is large enough to drive the circuit to behave nonlinearly. To mimic the PSS/PAC approach on the bench, simply reduce the amplitude of one of the tones. Both approaches are measuring intermodulation distortion; they simply use different ways of quantifying the results. Typically, when distortion levels are small, which they must be when computing IP_3 using (1), then the two methods produce results that are within one or two decibels.

Besides dramatically reducing the simulation time and memory, the PSS/PAC approach has other intesting benefits. If intermodulation distortion is measured using two large tones, then two intermodulation tones are produced and they are usually slightly different. A difference in amplitude in the intermodulation products usually indicates that the amplitude response of the circuit as a function of frequency is not flat at the measurement frequency. (It might also indicate that the two tones are not precisely the same size). On the bench, one can make the two test tones closer in frequency to reduce this effect, but this approach dramatically increases the time required for simulation. With the PSS/PAC approach, the separation between the tones can be arbitrarily small or they can be even at the same frequency. In this case, the ambiguity is resolved by accessing the signals by sideband index, rather than frequency, enabling accurate measurement of intermodulation distortion even where the frequency response is not flat, such as along the skirt of a filter.

C. Periodic Transfer Function Analysis

A periodic transfer function (PXF) analysis computes the transfer function from every source in the circuit at any frequency to a single output at a single frequency. It differs from a PAC analysis in that the PAC analysis computes the response from a single stimulus at a single frequency to every node in the circuit at every frequency. This is illustrated in Figure 3. Generally, PAC is used when several transfer functions are needed that share the same input, and PXF is used when several transfer functions are needed that share the same output. The PXF analysis can simultaneously compute such useful quantities as conversion efficiency (transfer function from input to output at desired frequency), image and sideband rejection (input to output at undesired frequency), and LO feed-through and

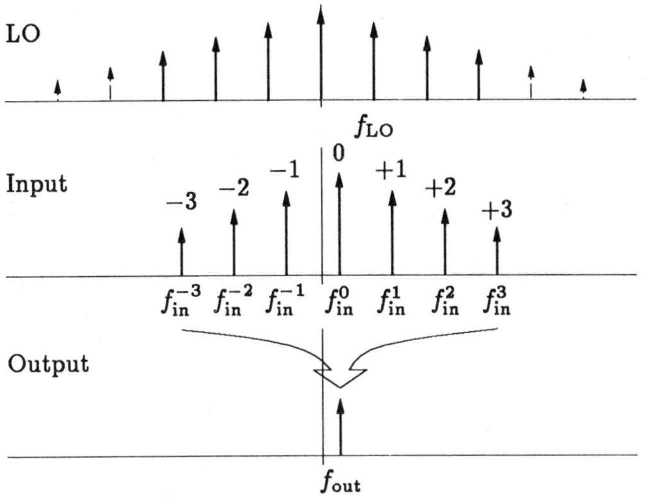

Fig. 3. The transfer functions computed by PXF analysis. The output is sensitive to signals at many frequencies at the input of the mixer. The input signals are replicated and translated by each harmonic of the LO. The ones shown are the ones that end up at the output frequency of interest.

power supply rejection (undesired input to output at all frequencies). Frequency translation is taken into account in these quantities, so it is possible to determine how sensitive the output is to noise either up or down-converted from the supplies or the LO.

D. Periodic Noise Analysis

The periodic noise (PNoise) analysis is similar to the conventional noise analysis, except that it includes frequency conversion effects. Hence is it useful for computing the noise behavior of mixers, switched-capacitor filters, chopper-stabilized amplifiers, and the like.

PNoise is a two step process where PSS is used to compute the response to a large periodic signal such as a clock or LO. Results are labeled LO and shown in Figure 4. In the second step, which is the actual PNoise analysis, the resulting noise performance is computed. With periodic systems, there are two effects that act to translate noise in frequency. First, for noise sources that are bias dependent, such as shot noise sources, the time-varying operating point acts to modulate the noise sources. Second, the transfer function from the noise source to the output is also periodically time-varying and so acts to modulate the contribution of the noise source to the output. Modulation is a multiplicative effect in the time-domain and so in the frequency-domain the spectrum of the noise source is convolved with the spectrum of the transfer function. The transfer function is periodic and so has a discrete line spectrum. Convolution with a discrete spectrum involves a countable number of scale, translate, and sum operations, in a process known as *noise folding*, as shown in the rest of Figure 4. The final result of the analysis is the sum of the noise contributions both up-converted (negative labels)

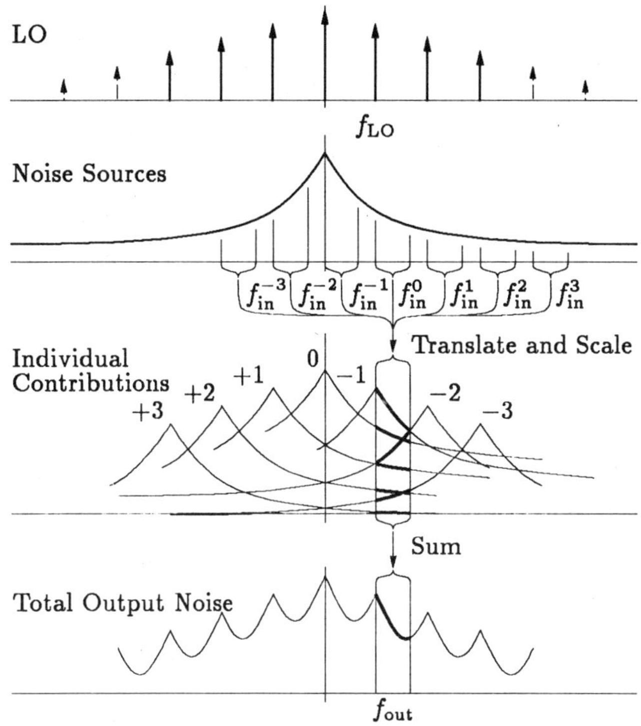

Fig. 4. How noise is moved around by a mixer. The noise is replicated and translated by each harmonic of the LO. The PNoise analysis computes the total noise including the modulation effects caused by the LO.

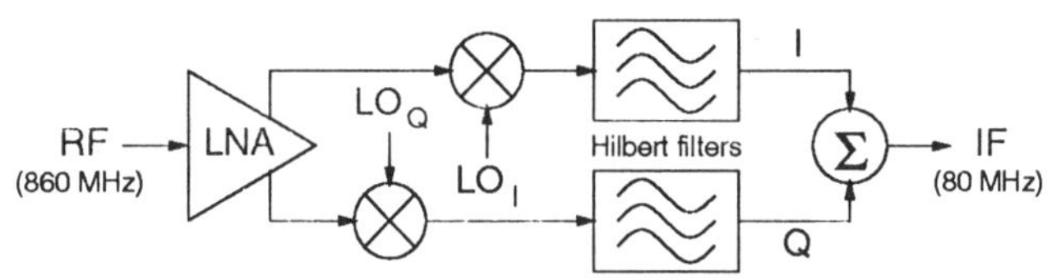

Fig. 5. The block diagram of the high-performance image-rejection receiver being characterized.

and down-converted (positive labels) to the desired output frequency.

III. Characterizing the Receiver

The receiver being characterized is a rather large RF circuit designed by Tomas Melander of Ericsson Radio, and is shown in Figure 5. It is an image rejection receiver that consists of a low-noise amplifier, a splitting network, two double-balanced mixers, and broad-band Hilbert transform output filter combined with a summing network that is used to suppress the undesired sideband. It contains 167 bipolar transistors and uses 378 nodes. It is being simulated at the circuit level using the Gummel-Poon model for each transistor. It generated 986 equations (due to the large number of inductors and the BJTs parasitic resistors).

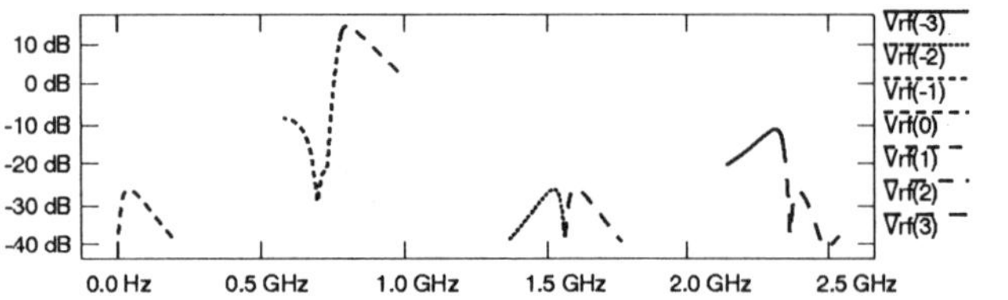

Fig. 6. The transfer functions from the V_{rf} input to V_{out}.

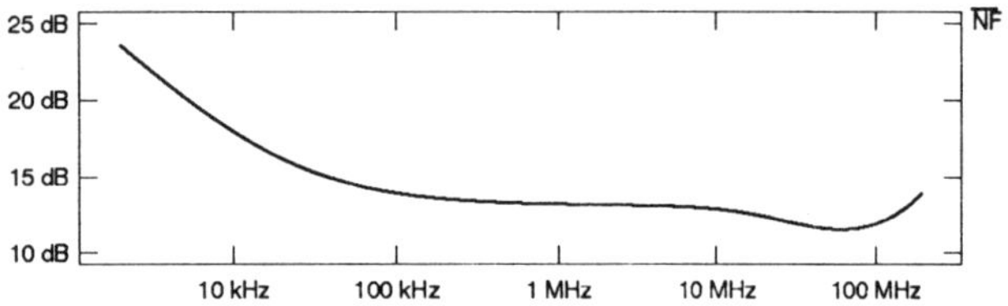

Fig. 7. The noise figure from *Vrf*, assuming 50 Ω source resistance.

A. Conversion Gain

PXF analysis is performed to compute the conversion gain and the results are shown in Figure 6. Several transfer functions are shown. All are transfer functions from the RF input to the IF output, however each represents a different sideband. In Figure 6 we chose to label the X-axis with the input frequency. The output frequency range for all of the transfer functions shown is baseband, which is being swept from 20 kHz to 200 MHz. For example, $V_{RF}(1)$ is the transfer function from the first upper sideband (780.02 MHz–980 MHz) to baseband (20 kHz–200 MHz), and $V_{RF}(-1)$ is the transfer function from the first lower sideband (580 MHz–797.98 MHz) to baseband.

Less than 4 minutes were required on an HP735/125 for this analysis, which contained 50 frequency points.

B. Noise Figure

The PNoise analysis is similar to the traditional noise analysis, except it includes the effect of noise moving from one frequency to another as it mixes with the LO and its harmonics. The noise measured at the output includes contributions from noise generated at frequencies $f_{\text{out}} \pm k f_{\text{LO}}$, where $|k| \leq 7$, and 7 is an user setable parameter.

The noise figure as computed by PNoise is shown in Figure 7. Less than 4 minutes were required on an HP735/125 for 50 frequency points.

C. Intermodulation Distortion

The first step in computing intermodulation distortion is to configure PSS analysis to measure harmonic distortion. Thus, the LO and one large input tone are applied. Since the resulting signals must be periodic, the applied signals must be commensurate and have the smallest possible common period. In this circuit the LO is given as 780 MHz and the input tone is chosen to be 840 MHz, which gives a 16.67 ns common period. The PSS analysis takes 2.5 minutes and 93 MB to compute 787 time-points. So far, only enough information is available to compute the receiver's harmonic distortion. As a second step, a small 850 MHz signal is applied and PAC analysis is run. Since only one frequency point is needed from the PAC analysis, it is very efficient, requiring only 47 seconds to compute the perturbation to the PSS solution due to the small 850 MHz RF test signal.

IP_3 is computed using V_{L1}, V_{S1}, and V_{S3}, where $V_{L1} = -25.7$ dBV is the magnitude at the output of first fundamental signal at 60 MHz, $V_{S1} = 2.6$ dBV is the magnitude at the output of the second fundamental signal at 70 MHz, and $V_{S3} = -49.7$ dBV is the IM_3 term at the output at 50 MHz.The third-order intercept point computed using (1) is 0.45 dBV.

To confirm this result, a 3-tone PSS analysis was performed. The results were within 1 dB, with the PSS/PAC approach requiring 6 times less time and memory. This ratio would be much greater if the receiver were followed by a narrow-band filter, and so required closer spacing of the intermod tones.

IV. Conclusion

Conventional small-signal analyses, such as AC and noise analyses, are widely considered essential by those engineers involved in designing amplifiers and filters. Periodic small-signal analyses provide these same measurement capabilities for periodically-driven circuits such as mixers, detectors, samplers, frequency multipliers, phase-locked loops, oscillators, parametric oscillators, switched-capacitor and switched-current filters, chopper-stabilized and parametric amplifiers, sample-and-holds, and etc. In addition, it is shown how to use a periodic small-signal analysis to efficiently measure intermodulation distortion. It is felt that these measurement capabilities will soon be considered as essential for periodically-driven circuits as AC and noise analysis are for amplifiers and filters.

References

[1] M. Okumura, T. Sugawara, and H. Tanimoto. "An efficient small signal frequency analysis method for nonlinear circuits with two frequency excitations." *IEEE Transactions of Computer-Aided Design of Integrated Circuits and Systems*, vol. 9, no. 3, pp. 225–235, March 1990.

[2] M. Okumura, H. Tanimoto, T. Itakura, and T. Sugawara. "Numerical Noise Analysis for Nonlinear Circuits with a Periodic Large Signal Excitation Including Cyclostationary Noise Sources." *IEEE Transactions On Circuits and Systems — I Fundamental Theory and Applications.*, vol. 40, no. 9, pp. 581–590, September 1993.

[3] Ricardo Telichevesky, Kenneth S. Kundert, Jacob K. White. "Efficient Steady-State Analysis based on Matrix-Free Krylov-Subspace Methods." 1995 Design Automation Conference.

[4] Ricardo Telichevesky, Kenneth S. Kundert, Jacob K. White. "Efficient AC and Noise Analysis of Two-Tone RF Circuits." Submitted to the 1996 Design Automation Conference.

Simulation and Modeling of Phase Noise in Open-Loop Oscillators

Alper Demir   Alberto L. Sangiovanni-Vincentelli

Department of Electrical Engineering & Computer Sciences
University of California, Berkeley, CA 94720

Abstract*

This paper presents

- a discussion about the definition of *phase noise* for general oscillation waveforms,
- a numerical method for transistor-level simulation and characterization of phase noise in open-loop (free running) oscillators,
- application of the phase noise simulation method to harmonic, relaxation and ring oscillators and a technical discussion of the phase noise characterization results,
- comparisons of the results of phase noise simulations for harmonic, relaxation and ring oscillators with the results of related work in literature.

1 Introduction

The market for portable, personal RF communications devices has exploded in the past few years. Considerations of power dissipation, form factor and cost push the RF/IF sections of these devices to higher and higher levels of integration [1]. Higher levels of integration create new challenges in the design of the RF and the analog signal processing subsystems of these mobile communications devices. Any modeling and simulation technique which can help designers come up with and evaluate new architectures for highly integrable RF transceiver circuits is bound to be relevant.

VCOs in conjunction with a crystal reference in a feedback loop are used for frequency synthesis of the LO signal in RF transceivers. The "phase noise" in the LO signal generated by the frequency synthesizer is of major concern in RF transceiver design, because it contributes directly to "phase noise" on the IF or baseband signal after frequency conversion, it mixes with adjacent channel signal energy resulting in an overall receiver SNR degradation, and in the transmitter, it contributes noise energy outside the band of the channel being transmitted [1]. Low "phase noise" frequency synthesizers are probably the most challenging (in terms of both design and CAD) subblock of an RF transceiver design.

In most applications, the synthesizer phase noise in the frequencies of interest is dominated by the inherent VCO phase noise [1]. Thus, characterization of the phase noise performance of open-loop VCOs is necessary. There are numerous publications in the literature on this topic. [2][3][4][5][6] approach the problem by concentrating on a simplified version of a specific oscillator circuit and use approximate analytical techniques to predict phase noise/timing jitter. [4][6] use a simplified linear time-invariant model (LTI) of the oscillator circuit for noise analysis. [2][3][5] also consider the effects of nonlinearities together with noise. These approaches are all based on simplified circuit models and make simplifying approximations in the analysis, because the problem of analyzing a general nonlinear system in the presence of noise is not analytically tractable. [3][5] use Monte Carlo transient noise analysis to verify the results of their analytical analysis. They obtain good qualitative and quantitative results for their specific oscillator circuits, and offer intuitive guidelines for low phase noise oscillator design.

In this paper, we discuss the definition of "phase noise" for a general oscillation waveform, which characterizes the noise in the "zero-crossing" times of the waveform. A numerical method for transistor-level phase noise analysis of open-loop oscillators is presented. This technique is based on our time-domain non-Monte Carlo noise simulation algorithm [7] for nonlinear dynamic circuits, and it is applicable to any type of oscillator circuit without restrictions or simplifications on the circuit complexity and device models. We use this method for phase noise characterization of harmonic, relaxation and ring-oscillators. A technical discussion of the simulation results follows with comparisons to the results obtained in [2][3][5].

* This work was supported by the Semiconductor Research Corporation under contract 95-DC-324 and by Motorola Inc.

2 Time-domain Noise Simulation for Nonlinear Dynamic Circuits

We have recently proposed a time domain numerical method for simulation of noise in nonlinear dynamic circuits [7]. Let us assume that the circuit equations for the oscillator circuit without noise sources can be written in the form

$$\dot{y} = F(y) \qquad y \in \Re^d \tag{1}$$

where F is a nonlinear map and y is a d-dimensional vector of the state variables for the oscillator circuit. (State equation formulation is chosen for notational simplicity only. In [7], circuit equations are formulated using modified nodal analysis.) Let the *steady-state* solution of the system of first-order differential equations in (1) for the autonomous oscillator circuit be $y_s(t)$. At this point, we assume that y_s corresponds to a stable limit cycle in the d-dimensional space. Hence

$$y_s(t) = y_s(t+kT) \qquad t \in \Re, k \in Z \tag{2}$$

where T is the period of oscillation. If we include the noise sources, the circuit equations become

$$\dot{y} = F(y) + B(y)\, v \qquad y \in \Re^d, v \in \Re^p \tag{3}$$

where v is a vector of uncorrelated, zero mean, *white Gaussian* stochastic processes that represents the shot and thermal noise sources in the oscillator circuit [7]. In this paper, we concentrate on the shot and thermal noise sources only. $1/f$ noise is usually not of practical importance, because we assume that the oscillator will be used as the VCO in a PLL, and the loop tracks out the low frequency fluctuations due to thermal drift and the ones caused by $1/f$ noise [3][5]. $B(y)$ in (3) is a $d \times p$ matrix, the entries of which are nonlinear functions of y, representing the feedback of the state variables on the intensity of the noise sources. Now, we assume that the solution of the system of stochastic differential equations in (3) can be approximated in the form

$$y_{sn}(t) = y_s(t) + y_n(t) \tag{4}$$

where $y_s(t)$ is the deterministic solution that corresponds to the stable limit cycle and $y_n(t)$ is a vector of stochastic processes which represents the effect of noise. Using a first-order expansion of the nonlinear map $F(y)$ around the limit cycle $y_s(t)$, one can derive a system of linear stochastic differential equations for $y_n(t)$ with periodically time-varying coefficient matrices [7]:

$$\dot{y}_n = H(t)\, y_n + G(t)\, v \tag{5}$$

where

$$H(t) = \frac{\partial}{\partial y} F(y) \Big|_{y = y_s(t)} \qquad G(t) = B(y_s(t))$$
$$H(t) = H(t+kT) \qquad G(t) = G(t+kT) \qquad t \in \Re, k \in Z \tag{6}$$

Then, one can proceed as in [7] using some results from the theory of stochastic differential equations and derive the following system of *linear* ordinary differential equations with time periodic coefficient matrices for the autocorrelation function of $y_n(t)$:

$$\dot{K}(t) = H(t) K(t) + K(t) H(t)^T + G(t) G(t)^T$$
$$\frac{\partial}{\partial t_2} K(t_1, t_2) = K(t_1, t_2) H(t_2)^T \qquad K(t_1, t_2) \in \Re^{d \times d} \tag{7}$$
$$K(t_1, t_2) \equiv E[y_n(t_1)\, y_n(t_2)^T] \qquad K(t) \equiv K(t, t)$$

The numerical solution of these differential equations, which was implemented in SPICE coupled with transient analysis [8], provides us with the second-order probabilistic characteristics of the vector of stochastic processes $y_n(t)$.

0-7803-3177-6 $5.00 © 1996 IEEE

IEEE 1996 CUSTOM INTEGRATED CIRCUITS CONFERENCE

3 Extraction of Phase Noise Statistics

The notions of "phase" and "amplitude" are clearly defined for a sinusoidal waveform. Similarly, the concepts of "phase noise" and "amplitude noise" were first introduced as noise performance measures for harmonic (LC/resonant) oscillators with sinusoidal waveforms. The waveforms produced by a practical oscillator (harmonic, relaxation or ring oscillator) are, in general, not sinusoidal as a result of the nonlinear behavior. In the context of oscillators which generate "square-wave like" waveforms, the concept of "timing jitter" is used as the noise performance measure [2][3]. Actually, "phase noise" for sinusoidal waveforms and "timing jitter" for square-wave like waveforms, both characterize the same phenomenon: the noise in the "zero-crossing" times of an oscillation waveform. In the below discussion, we first present a formal definition of "phase noise" in general oscillation waveforms. Even though "phase" and "phase noise" are not defined for general oscillation waveforms, we retain the terminology "phase noise" (to characterize the noise in the zero-crossing times) for historical reasons. Using the noise simulation method outlined in the previous section, we are able to numerically calculate the noise autocorrelation function of circuit variables (i.e. node voltages) for an oscillator circuit described at the transistor-level. Below, we also discuss how to extract the autocorrelation function for the "phase noise" of the oscillator from the noise autocorrelations of circuit variables. Then, we consider a "noisy" sinusoidal waveform (for which the definitions of phase and phase noise are clear) to motivate for our definition of "phase noise" in general oscillation waveforms.

Let $x_s(t)$ be a component of $y_s(t)$ and $x_n(t)$ be the corresponding component of $y_n(t)$:

$$x_s(t) = (y_s(t))_i \qquad x_n(t) = (y_n(t))_i \qquad i \in \{1, 2, \dots, d\} \tag{8}$$

where $(z)_i$ denotes the ith component of the vector z, $y_s(t)$ and $y_n(t)$ were defined in Section 2, and d is the dimension of y in (1).

Now, we define

$$S(t) = \frac{d}{dt}x_s(t) \qquad \Gamma = \{\tau \geq 0 : S(\tau) = \max_t S(t)\} \tag{9}$$

Note that

$$\Gamma = \{\tau_1, \tau_2, \tau_3, \dots\} \qquad \tau_k - \tau_{k-1} = T = \frac{1}{f_c}$$
$$S(\tau_l) = S(\tau_m) \qquad \tau_l, \tau_m \in \Gamma \tag{10}$$

$\tau_k \in \Gamma$ are the low-to-high "transition" (zero-crossing) times of the noiseless oscillator waveform $x_s(t)$, and $S(\tau_k)$ denote the "slew rates" of the waveform at these transition times. f_c is the oscillation frequency.

Now, we formally *define* the sampled (at the zero-crossings) "phase noise" for a general oscillator waveform with

$$\phi(\tau_k) = \left(\frac{2\pi f_c}{S(\tau_k)}\right) x_n(\tau_k) \qquad \tau_k \in \Gamma \quad k \geq 0 \tag{11}$$

One problem with this definition of phase noise for a general oscillator waveform is that low-to-high "transition" times described by (9) and (10) may not be well-defined for all $i \in \{1, 2, \dots, d\}$, but we can always choose i in (8) (e.g. choose a particular node in the oscillator circuit) such that the low-to-high transition times, designated by the time instants with the largest slope of the noiseless waveform, are well-defined. All choices of i among the ones which satisfy this condition give the same phase noise characterization result, which will be discussed in Section 4.3. In the above definition of phase noise, we have arbitrarily chosen to use the "low-to-high" transitions of the waveform. This choice does not affect the phase noise characterization results, which will also be discussed in Section 4.3.

Now, we can obtain the autocorrelation function for the sampled phase noise process defined in (11) by sampling the autocorrelation function of $y_n(t)$ which can be calculated by our noise simulation method described in the previous section:

$$E[\phi_k \phi_{k+m}] = \mu^2 E[x_n^k x_n^{k+m}] = \mu^2 (K(\tau_k, \tau_{k+m}))_{i,i}$$
$$E[\phi_k^2] = \mu^2 E[(x_n^k)^2] = \mu^2 (K(\tau_k))_{i,i} \tag{12}$$

where

$$\phi_k \equiv \phi(\tau_k) \qquad x_n^k \equiv (y_n(\tau_k))_i \qquad \mu \equiv \left(\frac{2\pi f_c}{\max_t S(t)}\right) \tag{13}$$

and $K(t_1, t_2)$, $K(t)$ are defined by (7). $E[\phi_k^2]$ and $E[\phi_k \phi_{k+m}]$ in (12) are in rad².

Now, we consider a sinusoidal waveform to motivate for our definition of "phase noise" for general oscillation waveforms,

$$x_s(t) = A\cos(2\pi f_c t) \qquad A > 0 \tag{14}$$

where the amplitude A and the frequency f_c are constants. Then, we represent a *noisy* sinusoidal oscillator waveform with

$$x_{sn}(t) = (A + a(t))\cos(2\pi f_c t + \phi(t)) \tag{15}$$

where $a(t)$ and $\phi(t)$ are, possibly *nonstationary,* zero mean stochastic processes representing the "amplitude" and "phase" noise of the oscillator waveform respectively [9]. Assuming that $|a(t)| \ll A$ and $|\phi(t)| \ll 1$ (small angle approximation [3]), $x_{sn}(t)$ can be approximated by

$$x_{sn}(t) \cong A\cos(2\pi f_c t) + a(t)\cos(2\pi f_c t) - A\phi(t)\sin(2\pi f_c t)$$
$$x_{sn}(t) \cong x_s(t) + x_n(t) \tag{16}$$
$$x_n(t) = a(t)\cos(2\pi f_c t) - A\phi(t)\sin(2\pi f_c t)$$

In (16), the noisy oscillator waveform is expressed as the summation of the noiseless deterministic oscillator waveform $x_s(t)$ and a noise term $x_n(t)$ which is a combination of two terms resulting from amplitude and phase noise. This is similar to (4). Motivated by the sinusoidal oscillator waveform case, we assume that the components of $y_n(t)$ represent a mixture of "amplitude" and "phase noise" for the waveforms of a general oscillator. The numerical noise simulation method described in the previous section gives us the autocorrelation function for the components of $y_n(t)$. Following the assumption stated above, these autocorrelation functions are concluded to be a combination of "amplitude" and "phase noise" autocorrelations as well as the cross-correlation between "amplitude" and "phase noise". We are not interested in calculating the probabilistic characteristics of amplitude noise, because it is not important from a design point of view and can be eliminated using a limiter. We would like to extract the autocorrelation function of phase noise from the autocorrelation function of $y_n(t)$ rejecting the portions due to the amplitude noise autocorrelation and the cross-correlation between amplitude and phase noise.

If we sample the noise term $x_n(t)$ in (16) at the time instants in Γ (low-to-high zero crossing times for $x_s(t)$ in (14)) and apply the formula in (11), we *reject* the portion of $x_n(t)$ due to amplitude noise and obtain a sample of the phase noise process $\phi(t)$ only for the noisy sinusoidal oscillator waveform.

4 Experimental Results with Phase Noise Simulation for Open-Loop VCOs

In this section, we use the numerical method described in the previous sections to characterize the phase noise performance for three different kinds of oscillators, namely harmonic, relaxation and ring oscillators.

4.1 Ring-Oscillator

The ring-oscillator circuit we have chosen is a three stage oscillator with fully differential ECL buffer delay cells (differential pair followed by emitter followers). This circuit is from [3] (p. 116). For a ring-oscillator, the natural choice for the circuit variable (choice of i in (8)) to be observed for phase noise characterization is any tap voltage in the ring (in this case, difference of two node voltages). Fig. 1 shows the portions of the autocorrelation function of the noise content $x_n(t)$ for the differential tap voltage.

The waveform obtained for the autocovariance $E[x_n(t)^2]$ in Fig. 1 (top plot), is an oscillatory waveform with a *linear ramp* envelope, with the initial condition $E[x_n(0)^2] = 0$. The peaks of this oscillatory waveform coincide with the low-to-high and high-to-low transition times (as defined by (9) and (10)) of the deterministic large-signal oscillation $x_s(t)$. Hence, if the waveform for $E[x_n(t)^2]$ is sampled to calculate the autocovariance function for the sampled phase noise process using (12) and (13), we obtain

$$E[\phi_k^2] = \mu^2 \alpha k T \qquad k \geq 0 \tag{17}$$

where α (in v^2/sec) is the slope of the linear ramp envelope for $E[x_n(t)^2]$. $E[\phi_k^2] = 0$ at $k = 0$ is a result of the initial condition $E[x_n(0)^2] = 0$ at the start of nonlinear noise simulation.

The waveform obtained for a portion of the autocorrelation $E[x_n(t_1)(x_n(t_2))]$ in Fig. 1 (bottom plot), is a periodic waveform with the same period as the oscillator. For the simulation shown, t_1 is fixed and was chosen to coincide with one of the peaks in $E[x_n(t)^2]$. The initial condition for the calculation of $E[x_n(t_1)(x_n(t_2))]$ is chosen as $E[x_n(t_1)^2]$ [7]. Similarly, the peaks in $E[x_n(t_1)(x_n(t_2))]$ coincide with the transition times of $x_s(t)$. If we sample $E[x_n(t_1)(x_n(t_2))]$ to obtain a portion of the autocorrelation for the sampled phase noise process, we arrive at

$$E[\phi_k \phi_{k+m}] = E[\phi_k^2] = \mu^2 \alpha k T \qquad m \geq 0, t_1 = kT \qquad (18)$$

If we combine (17) and (18),

$$E[\phi_k \phi_m] = \mu^2 \alpha \min\{k, m\} T \qquad k, m \geq 0 \qquad (19)$$

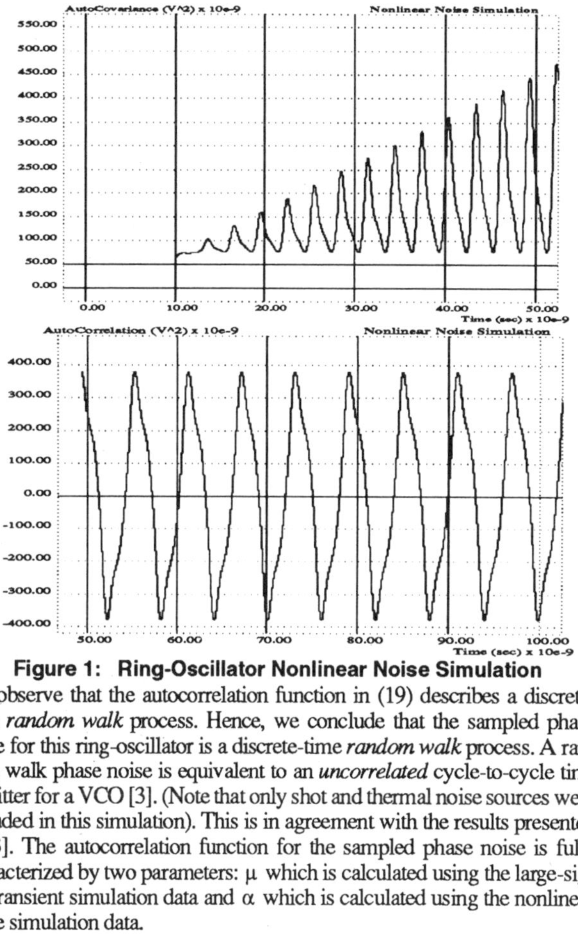

Figure 1: Ring-Oscillator Nonlinear Noise Simulation

We observe that the autocorrelation function in (19) describes a discrete-time *random walk* process. Hence, we conclude that the sampled phase noise for this ring-oscillator is a discrete-time *random walk* process. A random walk phase noise is equivalent to an *uncorrelated* cycle-to-cycle timing jitter for a VCO [3]. (Note that only shot and thermal noise sources were included in this simulation). This is in agreement with the results presented in [3]. The autocorrelation function for the sampled phase noise is fully characterized by two parameters: μ which is calculated using the large-signal transient simulation data and α which is calculated using the nonlinear noise simulation data.

In practice, phase noise is usually characterized in the frequency domain. The autocorrelation function defined by (19) fully characterizes the *second-order* statistics of the sampled phase noise process. The corresponding quantity in the frequency domain, to characterize the second-order statistics of a stochastic process, is the *spectral density*. Spectral density is defined as the Fourier transform of the autocorrelation function for a *wide-sense stationary* stochastic process. There are some generalized definitions of spectral density for nonstationary processes [10], but they are seldom used. Obviously, the sampled phase noise process is a *nonstationary* process, since the autocorrelation function in (19) can not be expressed just as a function of $k - m$. Actually, the variance of ϕ_k increases linearly with k. Thus, spectral density (as defined for wide-sense stationary processes), in the strict probabilistic sense, does not exist for a random walk phase noise process. But, for practical purposes, a spectral density expression can be derived by *treating* it as a wide-sense stationary process. For the particular phase noise process we are considering, it takes the form

$$S_\phi(f) = \mu^2 \alpha \frac{1}{(2\pi f)^2} \qquad rad^2/Hz \qquad (20)$$

(20) is obtained by considering the continuous-time equivalent of a random walk process, namely a Wiener process or Brownian motion, and *formally* expressing it as the *integral* of a stationary white noise process. This spectral density expression is similar to the results presented in [3][5] for ring-oscillators. It coincides with the sideband phase noise that can be measured using a spectrum analyzer, when it is expressed in dBc/Hz and the frequency f in (20) is considered to be the offset frequency from the carrier.

$$S_\phi(f_o) = 10\log\left(\mu^2 \alpha \frac{1}{(2\pi f_o)^2}\right) \qquad \text{dBc/Hz} \qquad (21)$$

Note that a spectrum analyzer does not distinguish amplitude noise from phase noise. For accurate phase noise measurements using a spectrum analyzer, either the amplitude noise for the oscillator signal should be significantly below its phase noise or certain special measurement techniques must be used [11].

Simulation No:	Ω R_c	Ω r_b	μA I_{ee}	$v/\mu sec$ S	v^2/sec α	MHz f_c	$fsec$ α/S^2
1	500	58	331	186	9.3	167.7	0.269
2	2000	58	331	287	12.2	74	0.149
3	500	1650	331	159	17.38	94.6	0.686
5	500	58	450	259	12.32	169.5	0.182
6	500	58	600	330	16.40	169.7	0.151
7	500	58	715	373	19.77	167.7	0.142

Table 1: Ring-Oscillator Phase Noise Simulation Results

For further comparison of phase noise simulation results obtained by our numerical method and the ones presented in [3], we performed several other phase noise simulations for the particular ring-oscillator we are considering. The results are shown in Table 1, where R_c denotes the collector load resistance for the differential pair (DP) in the delay cell, r_b is the zero bias base resistance for the BJTs in the DP, I_{ee} is the tail bias current for the DP, and f_c is the oscillation frequency for the three stage ring. In Table 1,

$$S = \max_t S(t) \qquad (22)$$

$S(t)$ was defined in (9). In all the simulations, $E[x_n(t)^2]$ had a linear ramp envelope, so the random walk phase noise model is valid for all the cases. Hence, the phase noise performance can be characterized by the slope of this ramp envelope, α, together with $\mu = 2\pi f_c / S$. Note that changes in R_c and r_b affect the oscillation frequency, unlike the changes in I_{ee}.

Fig. 2 shows a plot of $\mu^2 \alpha$ vs I_{ee} using the data from Table 1. This prediction of the dependence of phase noise performance on the bias current is in agreement with the results presented in [3][5] for ring-oscillators with ECL type delay cells.

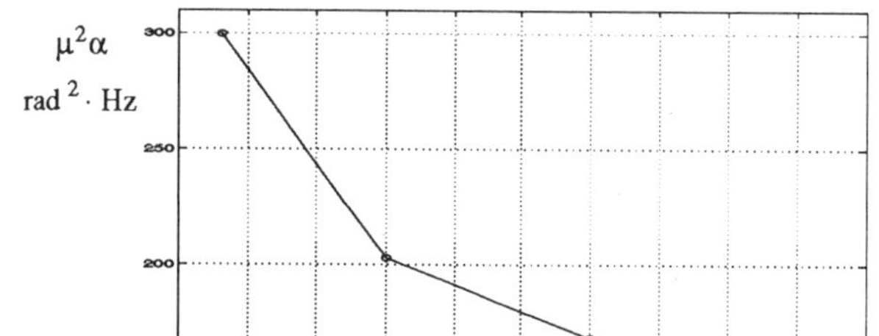

Figure 2: Phase Noise Performance vs I_{ee}

4.2 Relaxation Oscillator

The emitter-coupled multivibrator with a floating timing capacitor [12] (p. 629) was chosen as the relaxation oscillator example. [2] analyzes the process of jitter production for this circuit by describing the circuit behavior with a single first-order stochastic differential equation based on a simplified version of the circuit, and lumping all the noise sources into a single stationary current noise source. [2] arrives at intuitive qualitative results for low jitter relaxation oscillator design. Obviously, a relaxation oscillator operates in a highly nonlinear fashion. The analysis of the process of jitter production is not analytically tractable without reverting to simplifications. With our general phase noise simulation methodology, we can simulate the process of jitter production in a relaxation oscillator numerically without simplifying the circuit. Phase noise simulation for relaxation oscillators using a harmonic balance based method would be too inefficient and perhaps impossible because of the highly nonlinear behavior arising from the regenerative switchings.

Fig. 3 shows the noise variance waveform for the voltage across the timing capacitor. The envelope for the variance waveform is also a *linear ramp* as it was the case for the ring-oscillator. The spikes in the variance waveform coincide with the regenerative switchings of the multivibrator. We conclude that the random walk phase noise model is also valid for this relaxation oscillator (considering thermal and shot noise sources only). The large signal waveform for the timing capacitor voltage is triangular. Hence, switching times characterized by largest slope are not well defined. We use the output of the multivibrator for phase noise characterization. The phase noise spectrum is simulated to be given by (20) where $f_c = 0.88\,\mathrm{MHz}$, $\alpha = 1.3\times10^5\,v^2/\mathrm{sec}$ and $S = 3.29\times10^9\,v/\mathrm{sec}$, resulting in $\mu^2\alpha = 0.37\,\mathrm{rad}^2\cdot\mathrm{Hz}$. This corresponds to 103ppm timing jitter, or $-120.3\,\mathrm{dBc/Hz}$ phase noise 100KHz away from the carrier.

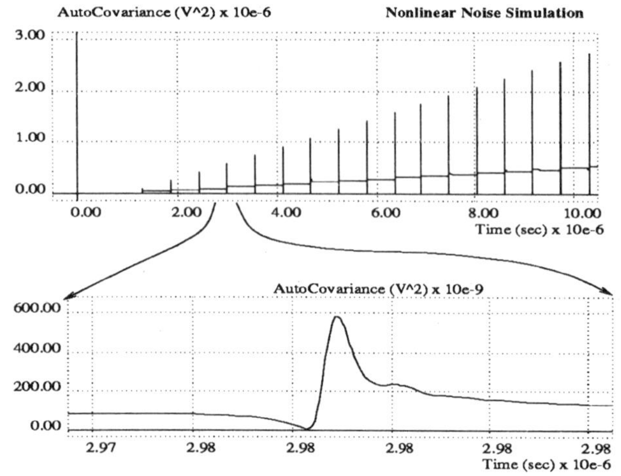

Figure 3: Multivibrator Nonlinear Noise Simulation

4.3 Harmonic Oscillator

The harmonic oscillator we have chosen uses an LC tank circuit and a single BJT with a Colpitts feedback circuit around it. The oscillation frequency is $f_c = 773.2\,\mathrm{MHz}$. Nonlinear noise simulation results showed that random walk phase noise model is also valid for this oscillator. Hence, phase noise performance can be summarized with the parameters α and μ defined before. We have calculated $\mu^2\alpha$ values for three different nodes in the circuit to see the effect of the choice of i in (8) on the phase noise characterization results (Table 2). Note that $\mu = 2\pi f_c/S$. We can observe in Table 2 that even though the large signal slew rate at the transitions and the slope of the linear ramp envelope for the noise variance waveform have different values, the calculated value of $\mu^2\alpha$ is equal for the three nodes considered. Hence, the same phase noise spectrum is calculated independent of the node that was observed. This is actually what we expect. Phase noise is equivalent to time synchronization error and should be equal in all the waveforms of an oscillator circuit. The $\mu^2\alpha$ value in Table 2 corresponds to $-135\,\mathrm{dBc/Hz}$ phase noise 1MHz away from the carrier.

Node	v/nsec S	v^2/sec α	$\mathrm{rad}^2\cdot\mathrm{Hz}$ $\mu^2\alpha$
1	1.53	0.127	1.28
2	1.104	0.063	1.22
3	1.69	0.151	1.25

Table 2: Harmonic Osc. Phase Noise with Different Nodes

The ring-oscillator and the relaxation oscillator we have considered in Section 4.1 and Section 4.2 are differential circuits. Hence, the low-to-high and high-to-low transitions of the large signal waveforms were symmetric. Thus, sampling the phase at either kind of transition is equivalent. On the other hand, the harmonic oscillator we are considering in this section is not a differential circuit, and the low-to-high and high-to-low transitions are not symmetric (i.e. the largest slew rate value is different). But the phase noise spectrum calculation done by sampling the phase noise at either kind of transition for the same node gave the same result, similar to the case where results were compared by calculating phase noise spectrum observing different nodes.

4.4 Comments on the Results of Phase Noise Simulations

We have characterized the phase noise performance of three different kinds of oscillators by calculating the autocorrelation function of the sampled phase noise process using our numerical phase noise characterization methodology. We have found out that, in all three cases, the sampled phase noise is a *random walk* process, hence has a "spectral density" with a $1/f^2$ frequency dependence. In these characterizations, only the shot and thermal noise sources were considered.

Even though it was the case for all the oscillator circuits we have considered, it is not always true that the sampled phase is a random walk process for an oscillatory system that can be described by a system of equations of the form given in (3) where the noise sources in the system are *white*. An oscillatory system with large "time constants" will not have a random walk phase noise. For instance, a circuit with $1/f$ noise sources falls into this category, if we consider that $1/f$ noise sources can be synthesized by passing a white noise source through an LTI network with some large time constants [8]. Our numerical method for phase noise characterization can still be used for such an oscillatory system. The deviation from the random walk phase noise model will cause the envelope for the oscillatory noise variance waveform (i.e. $E[x_n(t)^2]$) for a state variable of the system not to be a linear ramp.

5 Conclusions

We have presented a numerical methodology for transistor-level phase noise characterization for open-loop oscillators. One we know how to characterize phase noise for open-loop oscillators, which are the main phase noise contributors to the LO signal in a transceiver, we would like to simulate and characterize the phase noise and the spurious tones of the LO signal produced by a phase-locked loop frequency synthesizer. We approach this problem with a "behavioral level" modeling and simulation strategy, which will be presented in another paper [13].

Acknowledgments

The authors would like to thank Iasson Vassiliou and Todd Weigandt of U.C. Berkeley, Branko Avanic, Joe Heck, Brian Mulvaney and members of the MCSPICE group of Motorola Inc. for many helpful discussions. This work was supported by the Semiconductor Research Corporation under contract 95-DC-324, and by Motorola Inc. in a summer internship.

References

[1] P.R. Gray and R.G. Meyer, "Future Directions in Silicon ICs for RF Personal Communications", *Proc. IEEE CICC*, May 1995.
[2] A.A. Abidi and R.G. Meyer, "Noise in Relaxation Oscillators", *IEEE JSSC*, December 1983.
[3] J.A. McNeill, *Jitter in Ring Oscillators*, Ph.D. Thesis, Boston University, 1994.
[4] D.B. Leeson, "A Simple Model of Feedback Oscillator Noise Spectrum", *Proc. of the IEEE*, February 1966.
[5] T.C. Weigandt, B. Kim and P.R. Gray, "Analysis of Timing Jitter in CMOS Ring Oscillators", *Proc. IEEE ISCAS*, June 1994.
[6] B. Razavi, "Analysis, Modeling and Simulation of Phase Noise in Monolithic Voltage-Controlled Oscillators", *Proc. IEEE CICC*, May 1995.
[7] A. Demir, E.W.Y. Liu and A.L. Sangiovanni-Vincentelli, "Time-Domain non-Monte Carlo Noise Simulation for Nonlinear Dynamic Circuits with Arbitrary Excitations," *Proc. ACM/IEEE ICCAD*, November 1994.
[8] A. Demir, E.W.Y. Liu and A.L. Sangiovanni-Vincentelli, "Time-Domain non-Monte Carlo Noise Simulation for Nonlinear Dynamic Circuits with Arbitrary Excitations," to appear in the *IEEE Trans. on CAD of Integrated Circuits and Systems*.
[9] E. Hafner, "The Effects of Noise in Oscillators", *Proc. of the IEEE*, February 1966.
[10] W.A. Gardner, *Introduction to Random Processes with Applications to Signals & Systems*, Second Edition, McGraw-Hill, 1990.
[11] U. L. Rohde, *Digital PLL Frequency Synthesizers: Theory and Design*", Prentice-Hall, 1983.
[12] P.R. Gray and R.G. Meyer, *Analysis and Design of Analog Integrated Circuits*, Second Edition, John Wiley & Sons, 1984.
[13] A. Demir and A.L. Sangiovanni-Vincentelli, "Simulation and Modeling of Phase Noise and Spurious Tones in Phase-Locked Loop Frequency Synthesizers", in preparation.

Dispersive Coupled Transmission Line Simulation using an Adaptive Block Lanczos Algorithm

Tuyen V. Nguyen*, Jing Li*, and Zhaojun Bai**

* Advanced Design Technology, Motorola Inc., Austin, TX 78730

** Dept. of Mathematics, University of Kentucky, Lexington, KY 40506

Abstract

In this paper, an effective method for the time domain simulation of dispersive coupled transmission lines with measured or tabulated frequency data is presented. Accurate high order approximations of the line parameters are obtained using a two phase least square fit procedure with frequency partitioning [1][2]. Reduced order matrix Padé approximations of the fitted high order models are then computed using a block Lanczos algorithm with a reliable adaptive scheme for treating breakdowns [3].

1.0 Introduction

The need for accurate and efficient simulation of high speed packages and interconnects with frequency dependent effects is well known. The best approach for an effective verification of such circuits is to have the extractor directly produce accurate reduced order models of the interconnects and/or packages in terms of rational function representation [4]. These reduced order models can then be incorporated into a circuit simulator using recursive convolution [5] or state space simulation [1]. However, until the above extraction approach finds its way into production usage, the practice in vogue is to use measurement or packaging analysis tools to generate a table of interconnect/package parameter data at a set of discrete frequencies. A high order rational function is fitted to this parameter data, and this high order model is then reduced using model reduction techniques to make it more tractable for simulation.

In this paper, a two phase least square fitting procedure with frequency partitioning [1][2] is used for obtaining accurate high order approximations of dispersive coupled transmission line parameters. An adaptive block Lanczos algorithm with a reliable look-ahead scheme for dealing with breakdowns [3] is employed to reduce the fitted model. In [2], the model reduction is performed using truncated balanced realization [6]. However, this approach is considerably more expensive than computing the Padé approximation.

The application of Padé approximation to linear circuit analysis has been made popular by the well known Asymptotic Waveform Evaluation (AWE) technique [7]. For lumped circuit models of interconnects and distributed transmission line models with constant (frequency independent) parameters, AWE has been proven quite successful [7][8]. In order to address the inherent numerical instability of AWE in computing high order approximations, Complex Frequency Hopping [9] and multi-point Padé approximation [10] techniques have been developed. However, these methods suffer from high computational cost. Recently, the Padé -via-Lanczos (PVL) method [11] has elegantly addressed the numerical stability problem in AWE and has promoted Krylov subspace methods as the method of choice for the reduced order modeling of large lumped linear circuits.

For distributed transmission line models, an effective approach is to extract the mode delays to capture most of the distributed nature of the line, then approximate the rest of the propagation function by a high order rational function approximation. In formulating the line equations, the matrix exponential method is used in [9] to ensure the numerical accuracy of higher order moments. The drawback of this approach is that very high order lumped models must be used due to the difficulty in extracting mode delays. On the other hand, the generalized method of characteristics [12] allows the extraction of mode delays, but requires matching eigenvectors at different frequencies for frequency dependent coupled lines. In this paper, a simple procedure that avoids the eigenvector matching is used to extract the mode delays. This technique together with the formulation of the line equations is discussed in the next section.

2.0 Formulation of Dispersive Coupled Transmission Lines

The differential equations for a lossy $(n+1)$ conductor transmission line (n signal lines plus a reference conductor) can be expressed in matrix form as

$$\begin{aligned} \frac{\partial}{\partial x}V(x,s) &= -Z(s)\,I(x,s) \\ \frac{\partial}{\partial x}I(x,s) &= -Y(s)\,V(x,s) \end{aligned} \qquad (1)$$

where V is an n-vector of line voltages with respect to the reference conductor and I is an n-vector of line currents. Here $Z(s) = R(s) + sL(s)$ and $Y(s) = G(s) + sC(s)$. $R, L, C,$ and G are $n \times n$ real symmetric matrices of per unit length line resistances, inductances, capacitances, and conductances respectively. $R, L, C,$ and G may be functions of frequency, but they are independent of x. Let d be the line length. In the frequency domain, the solution of the above differential equations is given by

$$\begin{aligned} I_1(s) &= Y_0(s)\,V_1(s) + J_1(s) \\ I_2(s) &= Y_0(s)\,V_2(s) + J_2(s) \end{aligned} \qquad (2)$$

where

$$\begin{aligned} J_1(s) &= -e^{-dQ(s)}\,[Y_0(s)\,V_2(s) + I_2(s)] \\ J_2(s) &= -e^{-dQ(s)}\,[Y_0(s)\,V_1(s) + I_1(s)] \end{aligned} \qquad (3)$$

$$\begin{aligned} V_1(s) &= V(0,s) \qquad & V_2(s) &= V(d,s) \\ I_1(s) &= I(0,s) \qquad & I_2(s) &= -I(d,s) \end{aligned} \qquad (4)$$

with $Q^2(s) = Y(s)\,Z(s)$, and $Y_0(s) = Q^{-1}(s)\,Y(s)$. In the method of characteristics, the modal decomposition $Q(s) = B^{-1}(s)\,\Gamma(s)\,B(s)$ obviates the need to match the eigenvector matrix $B(s)$ at different frequencies. In order to avoid this matching and at the same time to allow the extraction of mode delays, the exponential propagation matrix is rewritten as

$$e^{-dQ(s)} = e^{-sdQ(\infty)}E(s) \quad \text{with} \quad E(s) = e^{sdQ(\infty)}e^{-dQ(s)} \qquad (5)$$

where $Q(\infty)$ is the value of $Q(s)$ at infinite frequency. If this value is not available, the value at the highest possible frequency is used. Since the time domain simulation of $e^{-sdQ(\infty)}$ can be performed using a frequency independent modal decomposition of $Q(\infty)$, the task now is to compute the rational function approxima-

0-7803-3177-6 $5.00 © 1996 IEEE

IEEE 1996 CUSTOM INTEGRATED CIRCUITS CONFERENCE

tions of the characteristic admittance matrix $Y_0(s)$ and the propagation matrix $E(s)$. The procedure to generate these approximations is discussed in the next section.

3.0 Least Square Fitting of Line Parameters

The least square fitting procedure consists of two phases. In the first phase, frequency partitioning (or sectioning) is used to reduce the dynamic range of a wide band of frequency. A local rational function approximation is computed for each partitioned frequency interval. This partitioning reduces the ill-conditioning of the problem. It also improves the convergence in the case of nonlinear least square fit [1]. In the second phase, a final optimization is performed to compute the global fit over the entire frequency range.

If the rational function is of the form of a ratio of two polynomials, a linear weighted least square fit procedure can be used to compute the polynomial coefficients [2]. The drawback of this procedure is that the poles are not guaranteed to be in the left half plane. However, due to partitioning, low order approximations are usually sufficient in each partitioned frequency interval, thus reducing the occurrence of positive poles. Moreover, these positive poles can usually be discarded [2]. In practice, this procedure was found to work well in most cases, especially for the admittance matrix. For the propagation matrix, sometimes this linear procedure fails to produce adequate approximations, and it is necessary to enforce the Lyapunov stability conditions (poles in the left half of the complex plane) using nonlinear least square fit. One advantage of nonlinear least square fit is that constraints can be included more easily. For example, the minimum phase condition (both poles and zeros are in the left half plane) can be enforced naturally in this procedure [1], which may help to ensure the passivity or realizability of the rational function approximation. After the local approximations in all the partitioned frequency intervals have been computed, an accurate set of poles is obtained. However, it is necessary to perform a final global fitting to adjust the residues. This is a fairly well conditioned linear constrained least square fit. The constraint here is usually to enforce a match at zero frequency for correct steady state computation. A nonlinear least square fit can also be performed at this stage if a global optimization of both the poles and residues is necessary.

Once the rational function approximations of all the entries have been obtained, the approximated admittance matrix or propagation matrix can be represented by a multiple input, multiple output system as

$$\dot{x} = Fx + Bu \quad \text{and} \quad y = C^T x \tag{6}$$

where x is the state vector of size n, u is the input vector (the port voltages in this case), y is the output vector (the port currents in this case) of size p. F is an $n \times n$ square matrix, where n is the total system order. B and C are matrices of size $n \times p$, where p is the number of ports, which is equal to the number of coupled lines in this case. The matrix transfer function of system (6) can be written as

$$H(s) = C^T(sI - F)^{-1}B = L^T(I - \sigma A)^{-1}R \tag{7}$$

where $s = s_0 + \sigma$ (s_0 is a real frequency shift), $A = (F - s_0 I)^{-1}$, $L = C$, and $R = -AB$. In order to reduce (6), the matrix Padé approximation to (7) is computed. The next section discusses an adaptive block Lanczos method, which is applied to the matrix A for this purpose.

4.0 Adaptive Block Lanczos Method

The adaptive block Lanczos method ABLE [3] is developed to solve large scale non-Hermitian eigenvalue problems. Its basic steps are:

1. Choose initial $n \times p$ matrices P_1 and Q_1 so that $P_1^T Q_1 = I$
2. Compute $R_1 = (P_1^T A)^T$ and $S_1 = AQ_1$
3. For $j = 1, 2, \ldots$ until convergence
   - 3.1. Compute $A_j = P_j^T S_j$
   - 3.2. Set $R_j := R_j - P_j A_j^T$ and $S_j := S_j - Q_j A_j$
   - 3.3. QR decompositions: $R_j = P_{j+1} B_{j+1}^T$ and $S_j = Q_{j+1} C_{j+1}$
   - 3.4. Singular Value Decomposition: $P_{j+1}^T Q_{j+1} = U_j \Sigma_j V_j^H$
   - 3.5. Set $B_{j+1} := B_{j+1} U_j \Sigma_j^{1/2}$ and $C_{j+1} := \Sigma_j^{1/2} V_j^H C_{j+1}$
   - 3.6. Set $P_{j+1} := P_{j+1} \overline{U}_j \Sigma_j^{-1/2}$ and $Q_{j+1} := Q_{j+1} V_j \Sigma_j^{-1/2}$
   - 3.7. Compute $R_{j+1} = (P_{j+1}^T A - C_{j+1} P_j^T)^T$
   - 3.8. Compute $S_{j+1} = AQ_{j+1} - Q_j B_{j+1}$

The matrices generated above satisfy the three term recurrent relation

$$\begin{aligned} B_{j+1} P_{j+1}^T &= P_j^T A - A_j P_j^T - C_j P_{j-1}^T \\ Q_{j+1} C_{j+1} &= AQ_j - Q_j A_j - Q_{j-1} B_j \end{aligned} \quad \text{for } j = 1, 2, \ldots \tag{8}$$

Let $\hat{P}_j = [P_1 \ldots P_j]$ and $\hat{Q}_j = [Q_1 \ldots Q_j]$, then

$$\begin{aligned} \hat{P}_j^T A &= T_j \hat{P}_j^T + E_j B_{j+1} P_{j+1}^T \\ A\hat{Q}_j &= \hat{Q}_j T_j + Q_{j+1} C_{j+1} E_j^T \end{aligned} \quad \text{with } T_j = \begin{bmatrix} A_1 & B_2 & & \\ C_2 & A_2 & & \\ & & \ldots & B_j \\ & & C_j & A_j \end{bmatrix} \tag{9}$$

Here E_i is an $jp \times p$ tall thin matrix with an identity matrix of dimension p at the ith block and zeros elsewhere, where $1 \le i \le j$.

The initial block vectors P_1 and Q_1 in this application are generated by a two sided Gram-Schmidt process such that

$$L = P_1 B_1^T, \quad R = Q_1 C_1, \text{and} \quad P_1^T Q_1 = I \tag{10}$$

The block vector R_j can be shown to be orthogonal to Q_s and S_j to P_s for $1 \le s \le j$ after step 3.2. The QR and the Singular Value Decomposition (SVD) results in steps 3.3 and 3.4 are used to ensure $P_{j+1}^T Q_{j+1} = I$ as well as to maintain $R_j = P_{j+1} B_{j+1}^T$ and $S_j = Q_{j+1} C_{j+1}$. In addition, the singular values σ_{jk} in $\Sigma_j = diag(\sigma_{jk})$, $1 \le k \le p$, can be used to monitor the condition of the Lanczos vectors P_j and Q_j. In fact, it can be proven that

$$cond(\hat{Q}_j) = \|\hat{Q}_j\|_2 \|\hat{Q}_j^+\|_2 \le \sum_{i=1}^{j} [1/(min(\sigma_{ik}))] \tag{11}$$

where $\hat{Q}_j^+$ is the pseudo-inverse of $\hat{Q}_j$. The same upper bound also holds for $\hat{P}_j$. The matrix sequences $\{P_j\}$ and $\{Q_j\}$ satisfy the duality (or biorthonormality) condition:

$$P_i^T Q_j = \delta_{ij} = \begin{cases} I & \text{if } i = j \\ \emptyset & \text{if } i \ne j \end{cases} \tag{12}$$

i.e. $\hat{P}_j^T \hat{Q}_j = I$. These matrices span the two Krylov subspaces

$$\begin{aligned} K_j(P_1^T, A) &= span(P_1^T, P_1^T A \ldots P_1^T A^{j-1}) = span(P_1^T \ldots P_j^T) \\ K_j(Q_1, A) &= span(Q_1, AQ_1 \ldots A^{j-1} Q_1) = span(Q_1 \ldots Q_j) \end{aligned} \tag{13}$$

More importantly, the singular value decomposition is employed to detect breakdowns, which occur when $R_j^T S_j$ is singular. The breakdowns are considered easy when either R_j or S_j or both are rank deficient. Serious breakdowns occur when both R_j and S_j are not rank deficient. For easy breakdowns, if $S_j = \varnothing$, ABLE restarts the iteration with a new $n \times p$ matrix Q_{j+1} such that $Q_{j+1}^T P_s = \varnothing$ for $s \leq j$. If S_j is rank deficient, ABLE dualizes Q_{j+1} with respect to all previous P_s. For serious breakdowns, ABLE augments the current P_{j+1} and Q_{j+1} by first generating two random $n \times (p-r)$ matrices $\hat{P}$ and $\hat{Q}$ and then dualizing them against the previous Lanczos vectors. Here r is the number of singular values of $P_{j+1}^T Q_{j+1}$ above a certain threshold. It is found that the adaptive scheme in ABLE is robust and effective. ABLE also enjoys the advantage of using a fixed instead of a variable block size to treat breakdowns. For a detailed discussion of breakdown treatments, refer to [3]. The relation between a block Lanczos process and a matrix Padé approximation is presented in the next section. Note that the three term recurrence relation (9), which is the key for proving the Padé connection, still holds after treatments of breakdowns. The only difference is that the block size may be greater than p after a breakdown treatment.

5.0 Matrix Padé Approximation

The reduction of an eigenvalue problem of order n to order j by a Lanczos process is closely related to the Padé approximation [13]. The application of this relationship to model reduction is discussed in [11]. It will be shown that the Padé connection still exists for a block Lanczos process.

Using the series expansion of the inverse operator, the matrix transfer function (7) can be written as

$$H(\sigma) = \sum_{k=1}^{\infty} M_k \sigma^k \qquad (14)$$

where $M_k = L^T A^k R$ is the kth matrix moment. The jth matrix Padé approximation to $H(\sigma)$ can be obtained from the jth block tridiagonal matrix T_j. Rewrite the subexpressions of the matrix moment as

$$\begin{aligned} A^s R &= A^s Q_1 C_1 = A^s \widehat{Q}_j E_1 C_1 \\ L^T A^s &= B_1 P_1^T A^s = B_1 E_1^T \widehat{P}_j^T A^s \end{aligned} \qquad (15)$$

Using the recurrence relation (9) and the fact that T_j^s has s upper and lower subdiagonal blocks, it can be shown by induction that

$$A^s \widehat{Q}_j E_1 = \widehat{Q}_j T_j^s E_1 \quad \text{and} \quad E_1^T \widehat{P}_j^T A^s = E_1^T T_j^s \widehat{P}_j^T \qquad (16)$$

for $1 \leq s \leq j-1$. Hence, for $0 \leq k \leq 2j-2$,

$$M_k = L^T A^k R = (E_1 B_1^T)^T \; T_j^k \; (E_1 C_1) \qquad (17)$$

Again using (9), (16) and the duality condition (12), one can show that (17) still holds for $k = 2j-1$. Therefore, the jth order matrix Padé approximation to $H(\sigma)$ is given by

$$\hat{H}(\sigma) = (E_1 B_1^T)^T \; (I - \sigma T_j)^{-1} \; (E_1 C_1) \qquad (18)$$

For time domain simulation, note that the above transfer function corresponds to the following reduced system, assuming $\sigma = s$,

$$T_j \dot{z} = z + (E_1 C_1) u \quad \text{and} \quad y = -(E_1 B_1^T)^T z \qquad (19)$$

6.0 Example

The following simple example was adopted from [14]. It was chosen for easy computation of the exact solution using FFT. The schematic of the circuit is shown in Fig. 1. The line length is 10cm and the line parameters are given below

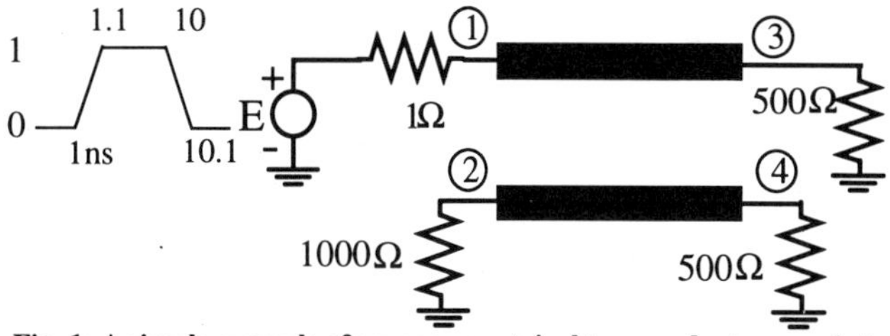

Fig. 1: A simple example of an unsymmetrical two conductor coupled line with frequency dependent parameters.

$$R = \begin{bmatrix} 0.5 & 0 \\ 0 & 0.2 \end{bmatrix} \frac{\Omega}{cm} \quad L = \begin{bmatrix} 9 & 3 \\ 3 & 7 \end{bmatrix} \frac{nH}{cm} \quad C = \begin{bmatrix} 1.2 & -0.6 \\ -0.6 & 1.4 \end{bmatrix} \frac{pF}{cm}$$

$$G = \begin{bmatrix} 5\,(10^{-5}) & 0 \\ 0 & 10^{-4} \end{bmatrix} \frac{1}{\Omega cm}, \; K = \begin{bmatrix} 2.8\,(10^{-4}) & 0 \\ 0 & 1.2\,(10^{-4}) \end{bmatrix} \frac{\Omega\sqrt{s}}{cm}$$

Note that the line is unsymmetrical, so the eigenvectors will not be independent of frequency. Note also that the function $A + B\sqrt{s}$ has been used to simulate the skin effect. Thus, the impedance matrix is given by $Z(s) = R + sL + K\sqrt{s}$. In this example, the table of data for the admittance and the propagation matrices are generated over a wide frequency range from 1KHz to 1000GHz with 3 points per decade. Then each entry of these matrices is fitted individually. Finally, the state space models for these matrices of rational function approximations are reduced to lower order systems using ABLE.

For the admittance matrix, the least square fitting produces 29 poles each for Y11 and Y22, and 23 poles each for Y12 and Y21. Therefore, the total order of the state space model for $Y_0(s)$ is 104. This model is reduced to a model of order 32. Similarly, the initial state space model for the propagation matrix $E(s)$ has 94 states, and the reduced model contains 38 states. All the matrix Padé approximations are obtained about s=0 for convenient incorporation into time domain simulation. Note that in the original approximation each entry of the matrix has its own set of poles. In the reduced model, all the entries share the same set of poles.

The real and imaginary parts of an off diagonal entry of $E(s)$ from the generated data, the original fitted model, and the reduced models are shown in Fig. 2 and Fig. 3. This entry is chosen because it

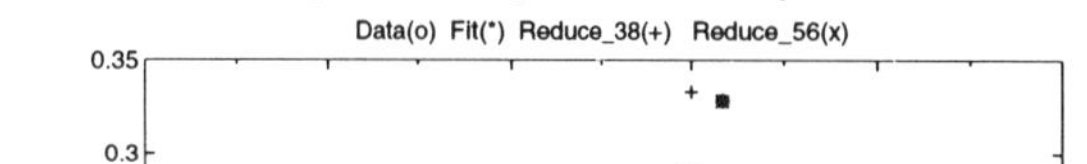

Fig. 2: The real part of E21 from the data, fitted model (order 94), and reduced models (orders 38 and 56)

represents the worst fit among all approximations. The original model fits the data almost perfectly. There are some discrepancies

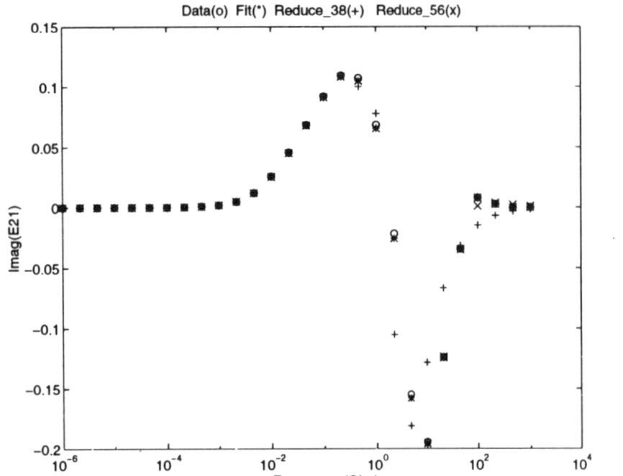

Fig. 3: The imaginary part of E21 from the data, fitted model (order 94), and reduced models (orders 38 and 56)

between the reduced model of order 38 and the data at high frequencies. This is due to the expansion about s=0. As the order of the reduced model is increased to 56, a good fit at these high frequencies is achieved as well. For this example, the reduced models of order 32 for $Y_0(s)$ and order 38 for $E(s)$ are sufficiently accurate to produce excellent time domain waveforms. Fig. 4 and Fig. 5 show the waveforms at the near ends (nodes 1 and 2) and far ends (nodes 3 and 4) of the line for both the exact solutions and the approximate results using the reduced models. The exact solutions

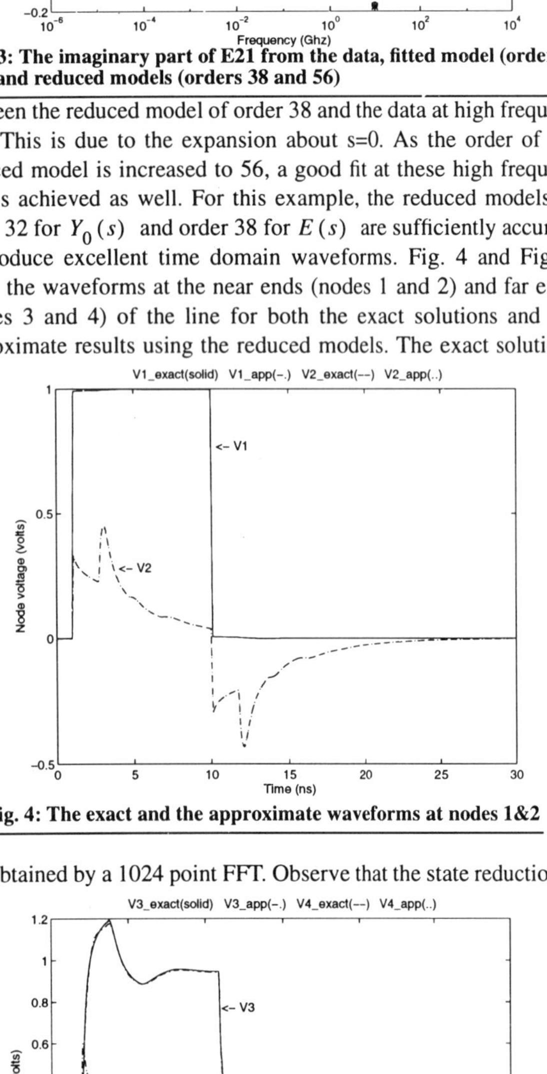

Fig. 4: The exact and the approximate waveforms at nodes 1&2

are obtained by a 1024 point FFT. Observe that the state reduction is

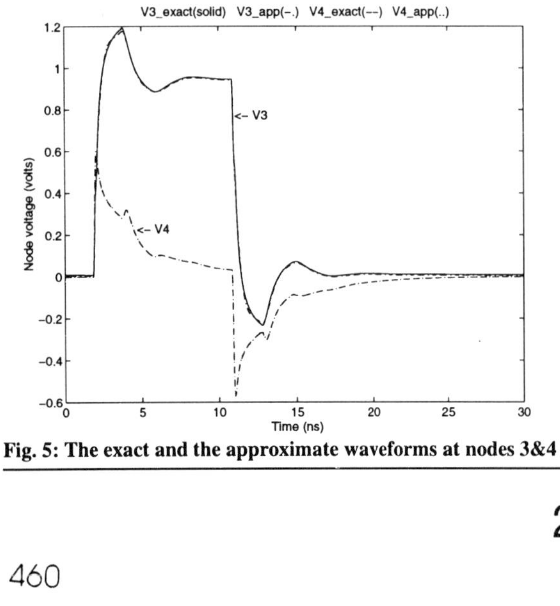

Fig. 5: The exact and the approximate waveforms at nodes 3&4

fairly significant even for this small example where the coupling between the two lines is strong and the system size is relatively small. For larger number of coupled lines, where the coupling among non-adjacent lines is weaker and the system size is larger, further reductions can be expected.

7.0 Summary and Acknowledgments

This paper presents the application of a robust adaptive block Lanczos algorithm to the reduced order modeling of dispersive coupled transmission lines. A simple procedure that avoids matching eigenvectors at different frequencies is used to extract the mode delays. Accurate high order rational approximations of all the entries of the remaining propagation matrix and the characteristic admittance matrix are then obtained using a least square fitting procedure with frequency partitioning. The effectiveness of the proposed approach is demonstrated using an example of two unsymmetrical coupled lines with skin effect.

The first two authors would like to thank Dr. Brian Mulvaney, the manager of the RF & Analog Simulation group at Motorola, for his constant support during the course of this work. Other members of the group are also acknowledged for their cordial friendship and helpful discussion.

References

[1] T. V. Nguyen, "Efficient Simulation of Lossy and Dispersive Transmission Lines," *Proc. ACM/IEEE DAC*, pp. 622-627, June 1994.

[2] L. M. Silveira, I. M. Elfadel, J. K. White, M. Chilukuri, and K. S. Kundert, "An Efficient Approach to Transmission Line Simulation using Measured or Tabulated S-parameter Data," *Proc. ACM/IEEE DAC*, pp. 634-639, June 1994.

[3] Z. Bai, D. Day, and Q. Ye, "ABLE: an Adaptive Block Lanczos Method for Non-Hermitian Eigenvalue Problems," Research Report 95-04, Dept. of Mathematics, Univ. of Kentucky, May 1995.

[4] L. M. Silveira, M. Kamon, J. K. White, "Direct Computation of Reduced-Order Models for Circuit Simulation of 3-D Interconnect Structures," *3rd Topical Meeting on Electrical Performance of Electronic Packaging*, pp. 246-248, Nov. 1994.

[5] T. V. Nguyen, "Recursive Convolution and Time Domain Simulation of Lossy Coupled Transmission Lines," *IEEE Trans. Computer-Aided Design*, vol. 13, pp. 1301-1305, Oct. 1994.

[6] B. C. Moore, "Principal Component Analysis in Linear Systems: Controllability, Observability, and Model Reduction," *IEEE Trans. Automatic Control*, vol. 26, pp.17-31, Feb. 1981.

[7] L. T. Pillage and R. A. Rohrer, "Asymptotic Waveform Evaluation for Timing Analysis," *IEEE Trans. Computer-Aided Design*, vol. 9, pp. 352-366, Apr. 1990.

[8] J. E. Bracken, V. Raghavan, and R. A. Rohrer, "Interconnect Simulation with Asymptotic Waveform Evaluation," *IEEE Trans. Circuits Syst.*, vol. 39, pp. 869-878, Nov. 1992.

[9] E. Chiprout and M. Nakhla, "Analysis of Interconnect Networks using Complex Frequency Hopping," *IEEE Trans. Computer-Aided Design*, vol.14, pp. 186-200, Feb. 1995.

[10] M. Celik, O. Ocali, M. Tan, and A. Atalar, "Pole-Zero Computation in Microwave Circuits using Multipoint Padé Approximation," *IEEE Trans. Circuits Syst. I: Fundamental Theory and Applications*, vol. 42, pp. 6-13, Jan. 1995.

[11] P. Feldmann and R. W. Freund, "Efficient Linear Circuit Analysis by Padé Approximation via the Lanczos Process," *IEEE Trans. Computer-Aided Design*, vol. 14, pp. 639-649, May 1995.

[12] F. Y. Chang, "The Generalized Method of Characteristics for Waveform Relaxation Analysis of Lossy Coupled Transmission Lines," *IEEE Trans. Microwave Theory Tech.*, vol. 37, pp. 2028-2038, Dec. 1989.

[13] W. Gragg, "Matrix Interpretation and Application of the Continued Fraction Algorithm," *Rocky Mountain Journal of Mathematics*, vol. 4, pp. 213-225, 1974.

[14] F. Y. Chang and O. Wing, "Transient Simulation of Lossy Coupled Transmission Lines Characterized with Frequency-Dependent Parameters," *Int. J. High Speed Electronics*, vol.2, pp. 319-353, 1991.

Efficient Frequency Domain Analysis of Large Nonlinear Analog Circuits

Peter Feldmann Bob Melville David Long
AT&T Bell Laboratories
Murray Hill, NJ 07974

Abstract—**In this paper, we present a new implementation of the harmonic balance method which extends its applicability to circuits 2-3 orders of magnitude larger than was previously practical. The results reported here extend our previous work [1] which only considered large circuits operating in a mildly nonlinear regime. The new implementation is based on quadratically convergent Newton methods and is able to simulate general nonlinear circuits. The significant efficiency improvement is achieved by use of Krylov subspace methods and a problem-specific preconditioner for inverting the harmonic balance Jacobian matrix. The analysis of radio-frequency mixers, implemented in integrated circuit technology, is an important application of our new method. We describe the theory behind the method, then report performance results on a complete receiver design using detailed transistor models.**

I. Introduction

The method of *Harmonic Balance* is well established for fast and accurate steady-state analysis of nonlinear circuits, especially when driven by two or more periodic inputs at widely separated frequencies [2,3,4]. Unlike conventional transient analysis, the computer time required for a harmonic balance analysis is almost insensitive to the numeric values of the stimuli frequencies. The gist of the harmonic balance method is to express each of the n circuit waveforms as a Fourier series truncated to N coefficients, then replace the circuit's differential equations by a system of nonlinear algebraic equations involving the Fourier coefficients. This is possible because the derivative with respect to time of a Fourier series is just an algebraic operation. A numerical technique, such as Newton's method, is then employed to solve the resulting system of nonlinear equations. This approach is especially intuitive for radio-frequency designers because of the natural connection to spectrum analyzer measurements.

As stated above, we use n to denote the number of waveforms which must be stored to capture the behavior of the circuit. Even a medium-sized IC can have hundreds of nodes, especially if a sophisticated transistor model is used. The transform dimension N can also become large in the case of a multi-tone analysis. For the simulation of a mixer, for example, the designer might request 16 harmonics of the local oscillator and 8 harmonics of the signal frequency. Because two real numbers are needed to capture the amplitude and phase of each spectral component, the final transform dimension is on the order of $N = 2 \times 8 \times 16 = 256$. Note that the overall dimesion of the system of harmonic balance equations is nN, which can become quite large in practical examples (over 800,000 for the circuit example in the results section). Letting X denote the aggregrate vector of all unknown Fourier coefficients, the harmonic balance equations are written as a system $H(X) = 0$ of size nN.

The preferred methods for solving harmonic balance equations are Newton-based techniques [13] because of their fast convergence and high accuracy. However, most implementations of this approach require the *Jacobian matrix* $\partial H/\partial X$ which is of dimension nN and much more dense than conventional matrices which arise in numerical circuit simulation. Hence, inversion of this matrix becomes a computational bottleneck and is largely responsible for the limitation of harmonic balance techniques to relatively small examples.

The technical contribution of our paper is a method for fast exact inversion of the harmonic balance Jacobian matrix using preconditioned Krylov subspace algorithms. Before going into details, it might be worthwhile to motivate the need for exact inversion, especially since methods have been proposed for solving the harmonic balance equations using approximate inversions of the Jacobian matrix[6], or even not using this matrix at all[5]:

- Newton's method with accurate Jacobian matrix inversion exhibits local *quadratic convergence*, i.e, a solution to the harmonic balance system can be obtained to machine precision in a fixed, small number of iterations. Thus, the overall solution time would be small if the inversion could be done efficiently.
- Widely convergent continuation methods[7] for solving the harmonic balance equations also require solving linear systems involving $\partial H/\partial X$ as a sub-matrix.
- The inverted matrix can be interpreted as a linearization of the circuit in a small-signal sense, which can then be used for sensitivity computation [9] and noise analysis [8].

Hence, we see fast accurate inversion of the Jacobian matrix as a key enabling technology for progress in this area. Recent work by Rizzoli at al. [10] also emphasizes the need for accurate Jacobian matrix inversion and attacks the problem using hierarchical decomposition.

II. Theoretical considerations

The most general form of circuit equations, resulting from one of the known formulations, e.g., MNA [12], is

$$\int_{-\infty}^{t} y(t-\tau)x(\tau)d\tau + \frac{d}{dt}q(x(t)) + f(x(t)) - b(t) = 0 \quad (1)$$

0-7803-3177-6 $5.00 © 1996 IEEE

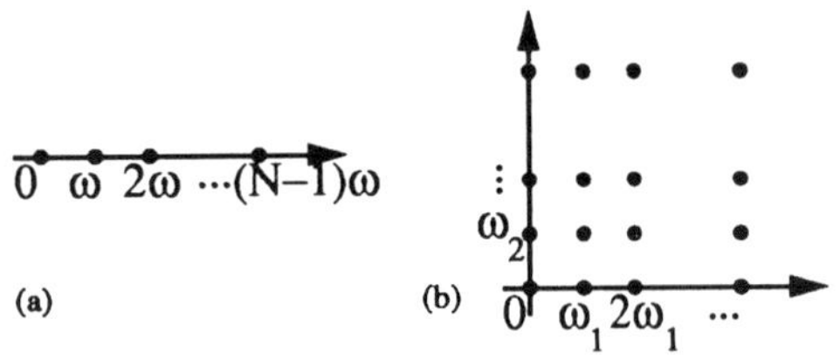

Fig. 1. Significant frequencies: (a) one-tone, (b) two-tone

This is a system of n equations where $x(t)$ is an n-vector of circuit variables (currents, voltages, charges, and fluxes). The first term represents the contribution of distributed elements expressed using Green functions; $\frac{d}{dt}q(x(t))$ represents the contribution of reactive circuit components, such as capacitors, inductors, etc.; and $f(x(t))$ represents the contribution of non-reactive circuit components, such as resistors. Semiconductor devices such as diodes and transistors contribute to both of the last two terms. Finally, $b(t)$ is the excitation term due to independent sources, which, in our case, are sinusoidal of one or more incommensurate frequencies (tones)

$$b(t) = \sum_{i=1}^{p} b_k e^{j\omega_i t}. \qquad (2)$$

In the absence of chaos, the response of most practical circuits has a discrete spectrum with components at integer combinations of the excitation frequencies,

$$x(t) = \sum_{k_1,\ldots,k_p \in Z} x_{k_1,\ldots,k_p} e^{j(k_1\omega_1+\ldots+k_p\omega_p)t}, \qquad (3)$$

Circuits that oscillate will generate, in addition, components at other frequencies. The methods presented in this paper assume response spectra at combinations of the excitation frequencies but can be extended to handle oscillation. Due to the attenuation of the high frequencies in practical circuits we can further assume that only a finite subset of the frequency components is significant

$$\{\omega_0, \ldots \omega_{N-1}\} \subset \{k_1\omega_1 + \ldots + k_p\omega_p\}_{k_1,\ldots,k_p \in Z}. \qquad (4)$$

For the case of a one-tone analysis the subset of significant frequencies consists of a DC component, the fundamental frequency and a number of harmonics. For a two-tone analysis the subset contains the DC component, the two fundamental frequencies, and a number of harmonics and intermodulation products, as illustrated in Figure 1. Subsets for three or more tones are similar. Thus, the response of the circuit is defined by the N coefficients corresponding to each of the n circuit variables

$$x(t) = \sum_{k=0}^{N-1} X_k e^{j\omega_k t} \qquad (5)$$

We apply the Fourier transform to the circuit equations (1). The time-domain convolution is mapped into a frequency-domain product, and we obtain

$$Y(\omega)X(\omega) + j\omega Q(\omega) + F(\omega) - B(\omega) = 0 \qquad (6)$$

where $Y(\omega)$, $X(\omega)$, $F(\omega)$, $Q(\omega)$, and $B(\omega)$ are Fourier transforms of the time domain functions $y(t)$, $x(t)$, $f(x(t))$, $q(x(t))$, and $b(t)$ respectively. The nN equations, necessary to determine nN coefficients, are obtained by forcing the system of n functional equations (6) to hold at the N frequency points of interest

$$\begin{aligned} Y(\omega_0)X_0 + j\omega_0 Q(\omega_0) + F(\omega_0) - B(\omega_0) &= 0 \qquad (7)\\ Y(\omega_1)X_1 + j\omega_1 Q(\omega_1) + F(\omega_1) - B(\omega_1) &= 0 \\ \ldots \end{aligned}$$

For notational simplicity, we assume that (1) consists of a single equation. The nonlinear functions F and Q depend on the frequency coefficients $[X_0, \ldots, X_{N-1}]^T$. We illustrate the dependence for F. Starting from the frequency coefficients $[X_0, \ldots, X_{N-1}]^T$, we transform to the time domain and obtain a sample of circuit variables at different time points $[x(t_1), \ldots, x(t_{N-1})]^T$. We evaluate the function f for each time point $[f(x(t_1)), \ldots, f(x(t_{N-1}))]^T$ and transform the results back to frequency domain. In matrix notation we have

$$\begin{bmatrix} F(\omega_0) \\ \vdots \\ F(\omega_{N-1}) \end{bmatrix} = \Gamma \begin{bmatrix} f(\cdot) & & \\ & \ddots & \\ & & f(\cdot) \end{bmatrix} \Gamma^{-1} \begin{bmatrix} X_0 \\ \vdots \\ X_{N-1} \end{bmatrix} \qquad (8)$$

where Γ represents the time-to-frequency-domain transform and the diagonal "matrix" denotes the application of a nonlinear operator. For the one-tone case, both the frequency-domain and time-domain samples are equally spaced and time-to-frequency and frequency-to-time transforms are just the DFT and IDFT computed using the efficient FFT algorithm. The use of the FFT can be extended to the multitone case (unevenly spaced frequencies), through appropriate selection of time-points [2].

The discretized system of circuit equations (7) becomes

$$\begin{bmatrix} Y(\omega_0) & & \\ & \ddots & \\ & & Y(\omega_{N-1}) \end{bmatrix} \begin{bmatrix} X_0 \\ \vdots \\ X_{N-1} \end{bmatrix} + \qquad (9)$$
$$\begin{bmatrix} j\omega_0 & & \\ & \ddots & \\ & & j\omega_{N-1} \end{bmatrix} \Gamma \begin{bmatrix} q(\cdot) & & \\ & \ddots & \\ & & q(\cdot) \end{bmatrix} \Gamma^{-1} \begin{bmatrix} X_0 \\ \vdots \\ X_{N-1} \end{bmatrix} +$$
$$\Gamma \begin{bmatrix} f(\cdot) & & \\ & \ddots & \\ & & f(\cdot) \end{bmatrix} \Gamma^{-1} \begin{bmatrix} X_0 \\ \vdots \\ X_{N-1} \end{bmatrix} - \begin{bmatrix} B(\omega_0) \\ \vdots \\ B(\omega_{N-1}) \end{bmatrix} = 0$$

or more compactly

$$YX + \Omega\Gamma q(\cdot)\Gamma^{-1}X + \Gamma f(\cdot)\Gamma^{-1}X - B = 0. \qquad (10)$$

Here the diagonal matrix Ω represents the frequency-domain differentiation operator. We have just obtained the *harmonic balance* system of equations $H(X) = 0$. For a system of n circuit equations, additional permutation operators need to be inserted to insure the required order of the vectors for the various operators.

Using Newton's method to solve $H(X) = 0$ requires a linear solve with the Jacobian matrix $J = \partial H / \partial X$ at each iteration. The Jacobian matrix of (9) is

$$J = Y + \Omega\Gamma C\Gamma^{-1} + \Gamma G\Gamma^{-1} \qquad (11)$$

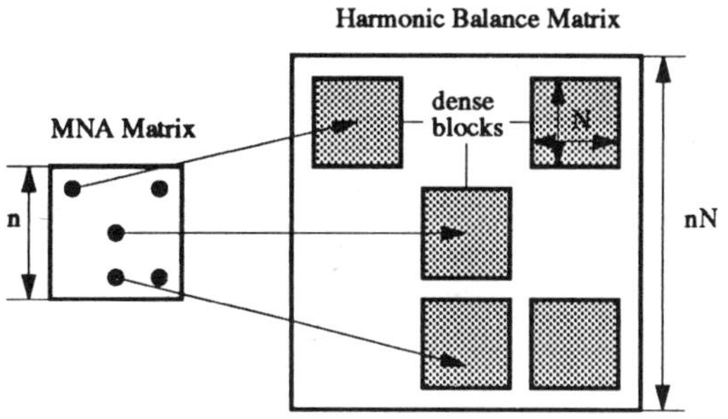

Fig. 2. Structure of the Jacobian matrix

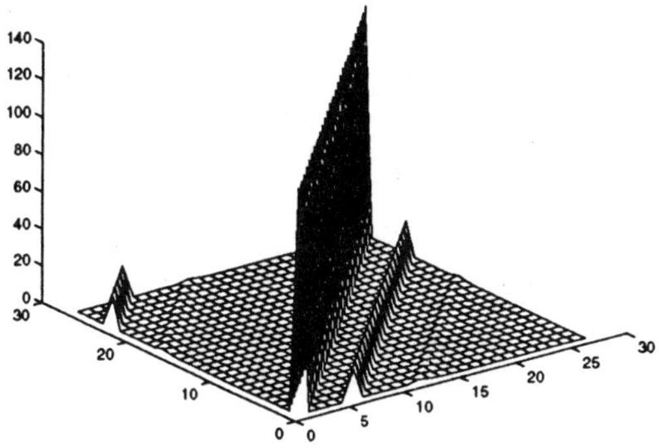

Fig. 3. Magnitudes of circulant matrix entries

where C and G are diagonal (block-diagonal in general) matrices, with the diagonal elements representing circuit linearizations at the sampled time points

$$G = \begin{bmatrix} \left.\frac{\partial f}{\partial x}\right|_{t_1} & & \\ & \ddots & \\ & & \left.\frac{\partial f}{\partial x}\right|_{t_N} \end{bmatrix}, \quad C = \begin{bmatrix} \left.\frac{\partial q}{\partial x}\right|_{t_1} & & \\ & \ddots & \\ & & \left.\frac{\partial q}{\partial x}\right|_{t_N} \end{bmatrix} \tag{12}$$

Naive Jacobian matrix evaluation results in a large, $(nN \times nN)$, and dense matrix, which is illustrated in Figure 2. Direct factorization or even storage of such a matrix would overwhelm any computer even for medium sized circuit.

Iterative methods are the only hope to solve such linear systems. Krylov subspace based iterative methods such as QMR [11] are particularly suited for this problem since they derive all matrix information solely from products of the matrix with vectors. It is easy to see from the expression

$$JZ = YZ + \Omega\Gamma C\Gamma^{-1}Z + \Gamma G\Gamma^{-1}Z \tag{13}$$

that such matrix-vector products can be computed very efficiently, without ever forming the matrix — only through multiplications with sparse matrices and FFTs.

Unfortunately iterative linear solvers and Krylov subspace methods in particular, do not converge reliably. Convergence is achieved through *preconditioning*, i.e., system $JZ = W$ is replaced with the *preconditioned* system $\tilde{J}^{-1}JZ = \tilde{J}^{-1}W$, which has, the same solution. For robust and efficient convergence, the preconditioner matrix $\tilde{J}$ should be, in some sense, a good approximation of J, and also "easy" to invert. In the remainder of this section we describe a preconditioning method specific to the structure of the harmonic balance Jacobian matrix. Being problem specific, it is superior to any general purpose preconditioning technique.

The terms that introduce dense blocks in the Jacobian matrix are $\Gamma C\Gamma^{-1}$ and $\Gamma G\Gamma^{-1}$. They result in circulant (block-circulant) matrices [14]. We analyze the first one

$$\Gamma \begin{bmatrix} c_1 & & \\ & \ddots & \\ & & c_N \end{bmatrix} \Gamma^{-1} = \begin{bmatrix} \mathcal{C}_1 & \mathcal{C}_2 & \dots & \mathcal{C}_N \\ \mathcal{C}_N & \mathcal{C}_1 & \ddots & \vdots \\ \vdots & \ddots & \ddots & \mathcal{C}_2 \\ \mathcal{C}_2 & \dots & \mathcal{C}_N & \mathcal{C}_1 \end{bmatrix} \tag{14}$$
$$= \text{circulant}\,(\mathcal{C}_1, \dots, \mathcal{C}_N)$$

where $[\mathcal{C}_1 \; \dots \; \mathcal{C}_N]^T$ is the DFT of $[c_1 \; \dots \; c_N]^T$. If the circuit is linear, the elements $c_1, \dots, c_N$ on the diagonal of C are all identical — the linearization of a linear circuit does not vary. In this case only $\mathcal{C}_1$ is nonzero and equal to the diagonal elements. The analysis of the term containing G is similar. The Jacobian matrix for a linear circuit becomes diagonal (block-diagonal)

$$J_{\text{lin}} = \begin{bmatrix} Y(\omega_0) + \mathcal{G} + j\omega_0\mathcal{C} & & \\ & \ddots & \\ & & Y(\omega_{N-1}) + \mathcal{G} + j\omega_{N-1}\mathcal{C} \end{bmatrix} \tag{15}$$

and, therefore, easy to invert.

When the circuit is nonlinear the circulant matrix fills up. However, as shown in Figure 3, the strongest entries in the matrix (in this case comming from a simple mixer) are concentrated on the diagonal. We can still set elements $\mathcal{C}_2, \dots \mathcal{C}_{N-1}$ and $\mathcal{G}_2, \dots \mathcal{G}_{N-1}$ to zero and use the resulting diagonal matrix as a preconditioner. In this case the diagonal elements $\mathcal{C}_1$ and $\mathcal{G}_1$ represent the average of the circuit linearizations at all time-points. If the circuit is "mildly" nonlinear this diagonal matrix is a good approximation of the Jacobian matrix and serves as an effective preconditioner. When stronger nonlinearities exist in the circuit, some off-diagonal elements of the circulant matrices must also be included in the preconditioner. In this case, the Jacobian matrix approximation loses its diagonal structure but remains much sparser than the original Jacobian matrix and, therefore, within the capabilities of direct factorization algorithms.

III. Results

As an example, we describe a variety of simulations on the front-end for a 900 MHz integrated direct conversion receiver. The circuit consists of a low-noise amplifier (LNA) driving I and Q mixers. It has 27 bipolar transistors, plus a variety of linear devices. The steady-state solution is described by just over 300 waveforms, many of which are for internal nodes of the detailed transistor model that was used. One-tone simulations of the LNA alone were used to measure gain and the 1dB compression point. Local oscillator (LO) feedthrough from the mixers back into the LNA

TABLE I

Summary of simulations

Circuit	n	Transform	Unknowns	Purpose	Time
LNA	223	1T, $N = 3$	669	LNA gain	10s
LNA	223	1T, $N = 35$	7,805	Compression point	1m30s
LNA	223	2T, $N = 175$	39,025	Intermodulation	1m30s
LNA+mixer	317	1T, $N = 63$	19,971	LO feed-through	15m/2m30s
LNA+mixer	317	2T, $N = 175$	55,475	Conversion gain	20m/5m
LNA+mixer	317	3T, $N = 2,025$	641,925	Desensitization	3h
LNA+mixer	317	3T, $N = 2,625$	832,125	Intermodulation	1h40m

was also measured with one-tone simulations. For the latter, we raised the LO amplitude in steps in order to keep the Newton loop under control. Once the LO was being driven with the desired amplitude, we captured the solution and used it as an starting point for subsequent runs. This allowed us to change component values and quickly resimulate in order to explore design alternatives. Information from the LO feedthrough simulations was used in designing the single-ended to differential conversion circuitry between the LNA and the mixers.

Two-tone simulations were used to measure intermodulation in the LNA alone, and conversion gain in the full circuit. As in the LO feedthrough simulation, for the conversion gain tests we captured the solution from the initial run and used it to speed subsequent tests for a number of slightly different circuits.

The most demanding simulation was a blocking test. The LNA input network was driven with an RF signal at -98dBm and with a blocking signal 3MHz away at -23dBm. A third frequency was needed for the LO, bringing the total transform size to over 2,000 and the total number of unknowns to over 600,000! A transform of this size was needed because the strong blocker and LO signals required many harmonics to prevent aliasing. The simulation required three hours on an SGI system (150MHz R4400). Figure 4 shows one of the outputs of the front-end. The designers were very interested in seeing how much the conversion gain was reduced in the presence of the blocker. Because of the large range of frequencies, it was estimated that getting this information using transient analysis would have required about one week of CPU time. Table I summarizes performance data for these simulations.

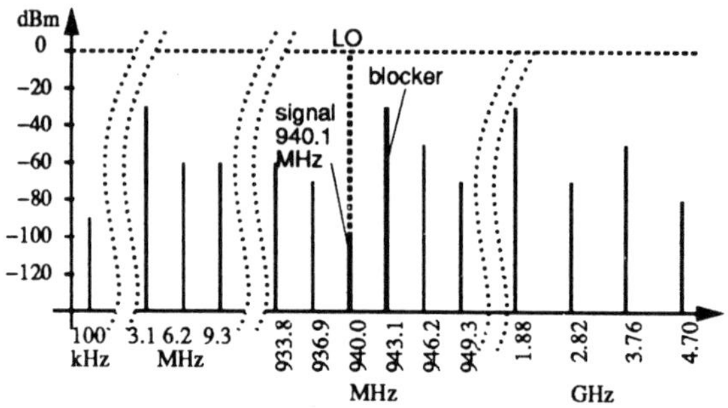

Fig. 4. Receiver front-end output spectrum

IV. Conclusion

This paper presents a new harmonic balance algorithm capable of solving problems 2-3 orders of magnitude larger than its predecessors. The efficiency gain is due to (1) the decomposition of the large and dense Jacobian matrix into a number of inexpensive-to-apply linear operators, and (2) use of an iterative linear solver which exploits the decomposition and avoids the storage and factorization of the Jacobian matrix. The convergence of the linear solver is ensured by a problem specific preconditioner rooted in engineering intuition.

References

[1] Melville, R.C., P. Feldmann, and J. Roychowdhury, "Efficient multi-tone distortion analysis of analog integrated circuits," in *Proc. of the Custom Integrated Circuits Conf.*, May 1995.

[2] Kundert, K.S., White, J. K., Sangiovanni-Vincentelli, A., "Steady-state Methods for simulating analog and microwave circuits", Kluwer, Boston, MA., 1990.

[3] Gilmore, R.J, and Steer, M.B., "Nonlinear circuit analysis using the method of harmonic balance — a review of the art. Part I. Introductory concepts," Int. J. on Microwave and Millimeter Wave Computer Aided Engineering, Vol. 1, No. 1, 1991.

[4] Rizzoli, V., Cecchetti, C., Lipparini, A., Fastri, F., "General-purpose harmonic balance analysis of nonlinear microwave circuits under multitone excitation," IEEE Trans. Microwave Theory and Tech., vol. MTT-36, pp 1650-1660, Dec. 1988.

[5] Maas, S. A., *Nonlinear Microwave Circuits,* Artech House, 1988.

[6] Rizzoli, V., Mastri, F., Sgallari, F., Frontini, V., "The exploitation of sparse matrix techniques in conjunction with the piecewise harmonic balance method for nonlinear microwave circuit analysis," IEEE MTT-S Int. Microwave Symp. Digest, Dallas, pp. 1295-1298 May 1990.

[7] L. Watson, "Globally convergent homotopy methods: a tutorial," Appl. Math. and Comp., vol. 31BK, pp. 369-396, 1989.

[8] Rohrer, R.A., Nagel L., Meyer, R. and Weber, L., "Computationally efficient electronic-circuit noise calculations," IEEE J. Solid State Circuits, vol. SC-6, pp. 204-213, 1971.

[9] Director, S.W. and Rohrer, R.A., "Automated network design — the frequency domain case," IEEE Trans. Circuit Theory, Vol. CT-16, pp. 370-337, 1969.

[10] Rizzoli, V., F. Mastri, D. Masotti, "A hierarchical harmonic-balance technique for the efficient simulation of large size microwave circuits," 25th European Microwave Conference, Bologna, Italy, Sept. 1995, pp. 615-619.

[11] Freund, R., G.H. Golub, N.M. Nachtigal, "Iterative solution of linear systems", Acta Numerica, pp. 57-100, 1991.

[12] J. Vlach and K. Singhal, *Computer Methods for Circuit Analysis and Design*. New York, N.Y.: Van Nostrand Reinhold, 1983.

[13] Ortega, J.M. and Rheinboldt, W.C., *Iterative Solutions of Nonlinear Equations in Several Variables*. Academic Press, New York, 1969.

[14] Roychowdhury, J. S. "Structure in the Jacobian matrices of nonlinear circuits in steady state,", AT&T internal document.

Chair: Linda Milor
Co-Chair: Robert Aitken

As technology scales, test and design engineers face new challenges in test set design, avoiding reliability hazards in designs, and knowing how much design-for-test circuitry to use. This session addresses a wide scope of these concerns for digital, analog, and mixed-signal circuits, and for DRAM memory.

One impact of technology scaling is the increased density of devices, which can lead to a significant increase in chip temperature, which in turn may induce many failure mechanisms, like electromigration and hot carrier effects. Y-K. Cheng and S-M. Kang present an algorithm to find the steady-state on-chip temperature, hot spots, and power consumption efficiently. Their tool can help find not just temperature-dependent reliability problems, but also temperature-dependent timing problems.

Problems with chip temperature are partially addressed by lowering the power supply voltage. C. Tong, M. Wen, and S. Su consider the question of whether Iddq tests can be effectively used for low power circuits, a technique currently gaining acceptance for its effectiveness in screening circuits for bridging faults.

Another impact of technology scaling is that it becomes feasible to design more complex circuits, which are also consequently more difficult to test. For sequential circuits, scan paths help to simplify the test set design problem. However, traditional approaches to scan design may require significant chip area and delay overhead. C. Lin, M. Marek-Sadowska, and K. Cheng show how the area overhead in scan design can be reduced through exploiting the controllability of primary inputs to obtain cost-free scan paths and through test-mode point insertion. Moreover, they propose an algorithm for test-mode point insertion.

A. Chaven et al. demonstrate the use of scan design for testing their mixed-signal chip. Specifically, they propose a test architecture consisting of multiple selectable sub-scan chains which allow the test engineer to design test sets with simple modular tests. In their design, boundary scan cells are used at the analog-digital boundary to help test the analog section. In fact, testing the entire chip was done using a digital tester!

Nevertheless, techniques to evaluate the quality of analog test sets are still at their infancy due to the lack of analog fault models. M. Ohletz presents a realistic analog fault model which can be derived from a circuit schematic and used to build up a dictionary for fault simulation.

Finally, M. Hashimoto, I. Uchida, and S. Hatakoshi attack the problem of reducing the testing time of the data retention test for DRAMs. They propose a charge offset circuit and demonstrate accurate results, low area overhead, and a significant reduction in testing time, without the need for extreme test conditions, such as a very low operating voltage or very high temperature!

Together, these papers span a wide range of topics in test and reliability and provide new insight into many of today's toughest problems. Come and learn the solutions at the test and reliability session.

IDDQ Testing in Low Power Supply CMOS Circuits

Carol Q. Tong [1] Mindy Wen and Shyang-Tai Su†

Dept. of Electrical Engineering, Colorado State University, Fort Collins, CO 80523
†Hewlett-Packard Company, 3404 E. Harmony Rd., Fort Collins, CO 80525

Abstract

The I_{DDQ} current testing for low power circuits is investigated. Two examples of low power circuits are considered. One involves strictly a decrease in supply voltage, the other reduces supply voltage simultaneously with a scaling down of transistor parameters. Based on the theoretical and simulation results, the gap between faulty Iddq and fault-free Iddq can be seen to decrease in low power supply circuits. For I_{DDQ} testing to be effective, the gap should be discriminative. A method of partitioning is provided to solve this problem.

1. Introduction

With the increasing complexity of circuits, new techniques are required to ensure IC quality and reliability. Conventional logic testing using the stuck-at fault model is not able to produce high quality tests since the tests may not detect all defects which cause shorts and opens in CMOS circuits. One technique which is gaining acceptance is monitoring the quiescent power supply current (Iddq). If a current sensor can differentiate the Iddq of a faulty circuit from that of a defect-free circuit, the defect is detected by I_{DDQ} testing[1]-[3]. It has been shown that I_{DDQ} testing has high fault coverage and is very effective for CMOS VLSI.

The purpose of this paper is to analyze the effectiveness of I_{DDQ} testing in low power circuits. The paper is organized as follows: Section 2 discusses leakage current in conventional and lower power circuits. Section 3 introduces the I_{DDQ} method in low power circuit. Section 4 gives experimental results about the gap between fault-free and faulty currents. Section 5 reviews the results and proposes a method to solve the problem. And Section 6 concludes the paper.

2. Low Power Circuit Design

2.1 Why low power?

One of the major advantages of CMOS IC is the low power consumption. However, as the density of CMOS circuit increases, the power consumption becomes an important issue. The high dissipated power will increase the temperature of the IC chip, which will reduce the switching speed and may cause some malfunctions. At the worst case, it might burn up the chip. For high density IC chips, the difficulty in providing adequate cooling might either add significant cost to the system or provide a limit on the amount of functionality that can be provided. Thus, it is evident that the design of high performance, low power system is needed [5].

2.2 Design of low power circuits

Scaling of device feature sizes, along with the development of high-density, low-parasitic packaging, such as multichip modules, will alleviate the overriding concern with the numbers of transistors being used. When MOS technology has scaled to 0.2 μm minimum feature size it will be possible to place from 1 to 10 $\times 10^9$ transistors in an area of 8 in x 10 in if a high-density packaging technology is used. The advantages of scaling down device feature size include: 1) to increase circuit speed, 2) to keep the circuit's physical integrity, 3) to improve short channel effects, and 4) to improve power supply dissipation. However, while scaling down the transistor threshold voltage, the leakage current is increased. The reason will be addressed later.

We will investigate two methods used in the design of low power circuits: constant field scaling and reduction of Vdd.

3. I_{DDQ} Testing in Low Power CMOS Circuits

Basically, Iddq is the leakage current for CMOS circuit when it is in stable state(quiescent state). A current sensor, built-in (BIC) or off-chip, is usually used to monitor the current from the supply power to the ground of a CMOS circuit[3]. If it can differentiate the Iddq of a faulty circuits from that of a defect-free circuit, the fault is detected. Therefore, to investigate the possibility of using I_{DDQ} testing in low power circuits, the leakage current for normal and low power CMOS ICs need to be analyzed in details.

3.1 Leakage current in CMOS circuit

A MOS transistor has three different working regions: linear, saturation and cut-off. The drain-source current, I_{ds}, in cut-off region is the so-called leakage current and it is approximated as [6]:

$$I_{ds} = \frac{\mu C_{ox} W}{L} \times (\eta - 1) \times {V_t}^2 \times e^{\frac{V_{gs}-V_{th}}{\eta V_t}} \times (1 - e^{\frac{-V_{ds}}{V_t}}) \quad (1)$$

where μ $\rightarrow$ the effective surface mobility of the carriers

C_{ox} $\rightarrow$ the gate-oxide capacitance

W $\rightarrow$ the gate width

L $\rightarrow$ the gate length

V_t $\rightarrow$ the thermal voltage:

$V_t = \frac{KT}{q}$

K=1.38×10^{-23}J/K:Boltzmann's constant

[1] Contact author

0-7803-3177-6 $5.00 © 1996 IEEE

IEEE 1996 CUSTOM INTEGRATED CIRCUITS CONFERENCE

T-temperature in Kelvin system

q=1.6×10^{-19}:the charge of an electron

η $\rightarrow$ a factor (1~3),signifies capacitive coupling between the gate and silicon surface

V_{th} $\rightarrow$ the gate threshold voltage

To find the relationship between leakage current and V_{ds}, we have to theoretically analyze the gate threshold voltage V_{th}. V_{th} is defined as the voltage applied between the gate and the source of an MOS device below which the drain-source current I_{ds} effectively drops to zero. The word "effectively" is used because the drain current never really is zero but drops to a very small value, i.e., the leakage current. In general, the threshold voltage is a function of a number of parameters including the following [5]:gate conductor material, gate insulator material, gate insulator thickness-channel doping, impurities at the silicon-insulator interface and voltage between the source and the substrate, V_{sb}(it is called body effect).

In addition, for long channel devices, the surface potential is constant over most of the channel length. As the channel length is reduced, keeping all other parameters constant, the peak of the surface potential is reduced and is constant only over a small fraction of the channel length. Since the peak potential is reduced, the current will increase. If V_{ds} is reduced, the peak is further reduced and the region of constant potential is also reduced. Therefore, the short-channel effect has been attributed to the penetration of the junction eletric fields into the channel region, causing barrier lowering, which in turn leads to V_{th} reduction. This reduction in V_{th} depends linearly on V_{ds}, i.e.

$$V_{th}(V_{ds}) \equiv V_{thd} = V_{th}{}^{*} - \sigma V_{ds} \qquad (2)$$

where $V_{th}{}^{*}$ is threshold voltage at low $V_{ds}(< 0.1v)$ and σ is called the DIBL(Drain Induced Barrier Lowing)parameter. This effect is also shown in Figure 1. The body effect($V_{sb} \neq 0$) is considered in Figure 1b[6].

3.2 Leakage current in low power circuit

The low power circuits can be formed by constant field scaling and reducing V_{dd}. The leakage current in these two methods are considered below.

A. Constant field scaling: the scaled device is obtained by applying a dimensionless factor α to all dimensions including those vertical to the surface, device voltage and the concentration densities.

If device dimensions(which include channel length, L; channel width, W; oxide thickness, t_{ox}; applied voltages; and substrate concentration density) are scaled by the constant parameter α, then the depletion layer thickness, the threshold voltage and drain-source current are also scaled [5](see Table 1). One of the important factors to be noted is that since the voltage is scaled, electrical field, E, in the device remains constant. This has the desirable effect that many nonlinear factors essentially remain unaffected.

Consider the leakage current described in equation (1). Since V_{ds} is larger than $4V_t$(V_t = 25mv at room temperature) in most cases, I_{ds} can be simplified as:

$$I_{ds} = \frac{\mu C_{ox} W}{L} \times (\eta - 1) \times V_t{}^2 \times e^{\frac{V_{gs}-V_{th}}{\eta V_t}} \qquad (3)$$

Table 1. Influence of scaling on MOS-Device Characteristics

PARAMETER	SCALING MODEL
	Constant field
Length(L)	$1/\alpha$
Width(W)	$1/\alpha$
Supply voltage(V)	$1/\alpha$
Gate-oxide thickness(t_{ox})	$1/\alpha$
Current(I=(W/L)(1/t_{ox})V^2	$1/\alpha$
Transconductance(g_m)	1
Junction depth(X_j)	$1/\alpha$
Substrate doping(N_A)	α
Electric Field across gate oxide(E)	1
Depletion layer thickness(d)	$1/\alpha$
Load Capacitance(C=WL/t_{ox})	$1/\alpha$
Gate Delay(VC/I)	$1/\alpha$
	RESULTANT INFLUENCE
DC power dissipation(P_s)	$1/\alpha^2$
Dynamic power dissipation(P_d)	$1/\alpha^2$
Power-delay product	$1/\alpha^3$
Gate Area(A=WL)	$1/\alpha^2$
Power Density(VI/A)	1
Current Density	α

The threshold voltage V_{th} is reduced by factor α, the gate-oxide capacitance $C_{ox}(C_{ox} = \epsilon \cdot W \cdot L/t_{ox})$ decreases by factor α. Therefore I_{ds} is increased.

B. Reducing supply voltage V_{dd} only: if supply voltage V_{dd} is reduced, the drain-source voltage will decrease. According to equation(2) and Figure 1, it can be seen that the threshold voltage V_{th} will increase. Since we only reduce the supply voltage, the parameters which may change due to reduced V_{dd} are V_{ds} and V_{th} in equation(1). Therefore, the leakage current will decrease in the circuits where only the supply voltage is reduced.

3.3 I_{DDQ} method in low power supply circuits

Iddq is the current when CMOS circuit is in the stable state(quiescent state). If we assume V_{ds} is greater than $4V_t$, Iddq current, I, in fault-free circuits can be estimated by equation(3) as follows:

$$V_{gs} = 0 \qquad I = \frac{\mu C_{ox} W}{L} \times (\eta - 1) \times V_t{}^2 \times e^{\frac{-V_{th}}{\eta V_t}} \qquad (4)$$

In faulty cases (there are defects which cause conduction paths between V_{dd} and ground), the Iddq current, I_f, can be approximated as the saturation current. In order for I_{DDQ} testing to work, the I_f to I ratio, R, should be around $10^3 - 10^5$ [6].

A. Constant field scaling circuits (α: scaling factor):

$$I_f{}' = \frac{1}{\alpha} I_f \qquad I' = \frac{\mu C_{ox} W}{\alpha L} \times (\eta - 1) \times V_t{}^2 \times e^{-\frac{V_{th}}{\alpha \eta V_t}} \quad (5)$$

Therefore, faulty current $I_f{}'$ is decreased and fault-free current I' is increased. The ratio R is reduced. In turn, this will reduce the gap between the faulty IDDQ value and fault-free Iddq value. It implies that it is difficult to distinguish between the fault free and faulty Iddq values with a low power supply.

B. Reducing Vdd only

In this case, both the faulty and fault free Iddq will be reduced. If the rate of faulty current reduction is less than that of fault-free current change, the I_{DDQ} test method will still work in this kind of low power circuits. We will give the values in Section 4.

4. Simulation Results

The Accusim simulation (Mentor Graphics) was performed on a 3-8 decoder static CMOS circuit which con-

Table 2. GAP between faulty and fault-free current when reducing V_{dd} only

3-to-8 decoder:	CMOSN12			CMOSN08		
V	$I_f(\mu A)$	$I(\mu A)$	GAP(μA)	$I_f(\mu A)$	$I(\mu A)$	GAP(μA)
5	1852.02	4.86	1847.16	2114.72	33.94	2080.78
4	1175.70	4.50	1171.20	1373.38	32.55	1340.83
3	610.84	4.21	606.63	751.68	30.326	721.35

Benchmark C432:	CMOSN12			CMOSN08		
V	$I_f(\mu A)$	$I(\mu A)$	GAP(μA)	$I_f(\mu A)$	$I(\mu A)$	GAP(μA)
5	3768.23	1822.42	1945.81	4376.43	2223.93	2152.50
4	2360.76	1183.80	1166.96	2801.05	1449.97	1351.08
3	1248.00	638.38	609.62	1522.13	792.53	729.60

Table 3. GAP between faulty and fault-free currents in constant field scaling

3-to-8 decoder					
$t_{ox}(\mu m)$	W/L(μm)	Volts(v)	I_f (μA)	$I(\mu A)$	GAP (μA)
2.12	3.6/1.00	5	1852.02	4.8613	1847.158
1.73	2.9/0.80	4	1679.50	55.109	1624.39

Benchmark C432					
$t_{ox}(\mu m)$	W/L(μm)	Volts(v)	I_f (μA)	$I(\mu A)$	GAP (μA)
2.12	3.6/1.00	5	3768.23	1822.42	1945.81
1.73	2.9/0.80	4	3617.84	1889.21	1728.63

sists of 76 transistors, and a Benchmark circuit C432 which has 160 gates.

The effects of power supply reduction were observed for normal and faulty(has one bridging fault injected) circuits. Simulation was performed with three different supply voltages: 5v, 4v, and 3v. The faulty and fault-free circuits were simulated one at a time. Table 2 shows the Iddq value in the case of only reducing supply power. The models used in simulation are named CMOSN12 and CMOSN08. These are two SPICE models for low power MOS transistors. Then, a constant field scaling was simulated in faulty and normal circuits. The size and oxide thickness of transistor and supply voltage were scaled by α=1.22. The simulation results are summarized in Table 3.

In Figure 2 and 3, the trends in leakage current change are shown. The accuracy of the current values is limited by the simulator and the models used. These figures show that when reducing V_{dd} only the fault-free current I is decreased, but the faulty current I_f is also decreased. For the small circuit, ratio R is more than three orders of magnitude, suggesting that I_{DDQ} testing will work. But if we look at a big circuit(the darker traces) with more transistors, the ratio will be reduced. In constant field scaling the fault-free current increases. In addition, if we approximate the faulty Iddq to the saturation current it will be reduced. Therefore, the gap between faulty and fault-free Iddq values are decreased in constant field scaling.

The distinction between the normal Iddq value, I, and the faulty Iddq value, I_f, is easy to see if the two distributions are well separated[7]. For a large low power supply circuit, the two distributions will be overlapping. Hence the BIC sensor or off-chip current monitor becomes unable to distinguish the difference between these two current distributions.

5. How to Solve the Problem ?

As discussed above, for low power supply devices, the distinction between the normal and faulty current values may be hard to make. In addition to this, increasing the number of gates (transistors) in a single chip also pushes the defect-free Iddq distribution closer to the defective Iddq distribution. The way to solve this problem is by either increasing the defective Iddq distribution or decreasing the defect-free Iddq distribution. To increase the defective Iddq distribution is not feasible because this involves the characteristics of defects. Therefore, decreasing the defect-free Iddq is the only one way for increasing the Iddq observability. This can be done by monitoring individual contributions of circuit partitions to the overall Iddq. The size of these partitions is determined such that the defect-free Iddq in each is distinguishable from the defective Iddq.

The idea of partitioning circuits for I_{DDQ} test has been proposed in articles [3],[7],[8]. Based on the idea of partitioning a large circuit for built-in-current-testing, a method to determine the size of each circuit partition is proposed in [7]. However, for a circuit with a large number of partitions, the overhead of testing pins will not be tolerable. [8] investigates a partitioning of circuits utilizing the dynamic power supply current, i_{DDT}, as a window of observability into the switching behavior. A design for I_{DDQ} test scheme with a low overhead of testing circuit and testing pins becomes very important for implementing the I_{DDQ} test to the low power supply circuit.

Figure 4 shows one pseudo-partition module that only requires an overhead of a shared BICS and two switching transistors and one latch for each partition. Two Vdd traces are provided;"NormalV_{dd}" is for normal operation and"AlternateV_{dd}" is for observing Iddq. Each partition can be connected either to NormalV_{dd} or AlternateV_{dd} depending on the logic value stored in the latches on top of each partition. Only one partition is allowed to be connected to the AlternateV_{dd} at a given time in the test mode, hence the Iddq of each partition can be monitored individually. Once the Iddq measurement for the first partition is finished, the Iddq measurement for the second partition can be performed by switching the first partition back to NormalV_{dd} and switching the second partition to AlternateV_{dd}. The Iddq from the rest of partitions can be measured in the same manner. The same scheme can also be modified for off-chip current testing simply by removing the BICS and adding a test pin to the end of the trace of AlternateV_{dd}.

5.1 Testing time

When the logic "1" is in latch L_i, it means partition P_i is under monitoring of the shared BICS. This scheme requires extra $n * \Delta t$ testing time where n is the total number of partitions in the circuit and Δt is time for switching between partitions and measuring Iddq. In general, the I_{DDQ} testing has to be performed in a speed much slower than its normal operating speed because it takes a very long time for power supply current to settle down. Compare this settling down time with Δt, Δt becomes very small because only two transistors are switched on and off. Therefore, the extra testing time overhead is ignorable.

5.2 Equillibium of performance and area overhead

The number of partitions will increase if the package density is growing while the power supply is getting lower. In this situation, the problem of balancing performance and area overhead is important. How big each partition should be and appropriate sizes for the two switching transistors are problems to be addressed. To pursue good performance, we need big size transistors, but this increases area overhead.

Control transistor size can be chosen by analyzing the influence of the size on the delay time along critical paths inside partitions. The simulation results are shown in Figure 5. From this we can identify an optimal size for the control transistor that will take up the least amount of chip area.

6. Conclusions

We have discussed the problems that exist for the I_{DDQ} testing of low power supply circuits. A method of partitioning has been introduced that can avoid these problems.

7. Acknowledgment

The authors would like to thank Michael Porter for his work on the transistor-level layout of the Benchmark circuit.

References

[1] C.F. Hawkins, J.M. Soden, R.R. Fritzemeier and L.K. Horning, "Quiescent power supply current measurement for CMOS IC defect detection," in *IEEE Trans. Industrial Electronics,* May 1989.

[2] P. Maxwell, R. Aitken, V. Johansen, and I. Chiang, "The Effectiveness of Iddq, Functional and Scan Tests: How Many Fault Coverages Do We Need?", *IEEE Proc. of Int. Test Conf.,* pp. 168-177, 1992.

[3] W. Maly et al., "Built-in Current Testing - Feasibility Study", *Research Report No. CMUCAD-88-27,* pp. 1-23, June 1988.

[4] N.H.E.Weste and K. Eshraghian, *Principles of CMOS VLSI Design,* Addison-Wesley Publishing Co., 1993

[5] A.P. Chandrakasan, S. Sheng, and R.W. Brodersen, "Low-Power CMOS Digital Design," *IEEE Jour. Solid-State Circuits,* Vol.27, No.4, pp.473-484, April 1992.

[6] N. Arora, *MOSFET Models for VLSI Circuit Simulation-Theory and Practice,* Springer-Verlag Wien, New York, 1993.

[7] Y. Malaiya, A. Jayasumana, C.Q. Tong, and S. Menon, "Enhancement of Resolution in Supply Current Based Testing for Large ICs", *Proc. IEEE VLSI Test Symp.,* Atlantic City, NJ, April 1991.

[8] S.T. Su, R.Z. Makki, T. Nagle, "Transient power supply current monitoring-A new test method for CMOS VLSI circuits", *Journal of Electronic Testing: Theory and Applications,* vol. 6, pp. 23-43, February 1995.

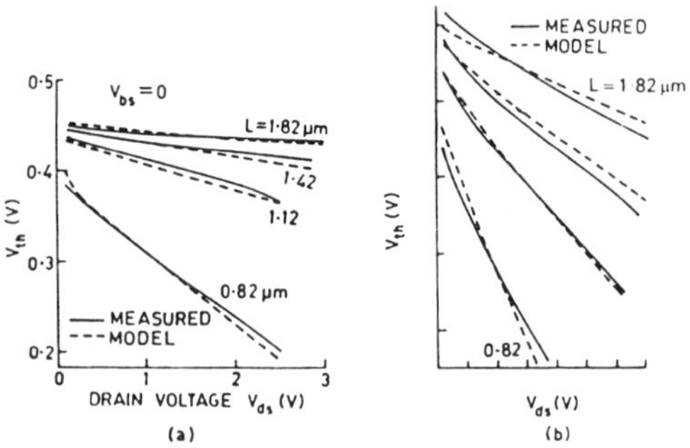

Fig.1 V_{th} versus V_{ds} (a) V_{sb}=0 and (b) V_{sb}=-2v

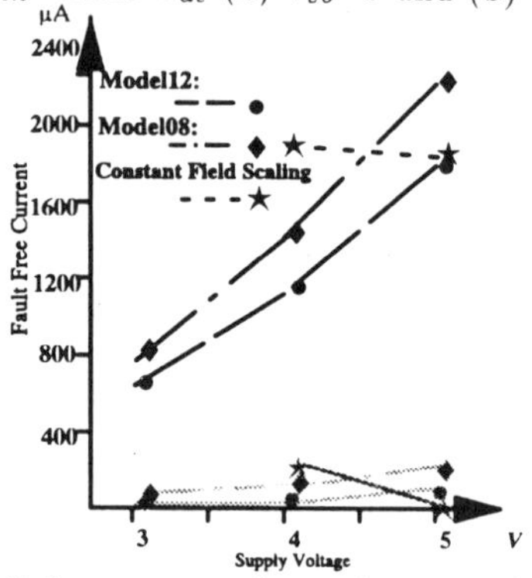

Fig.2 Fault-free current for various supply voltages

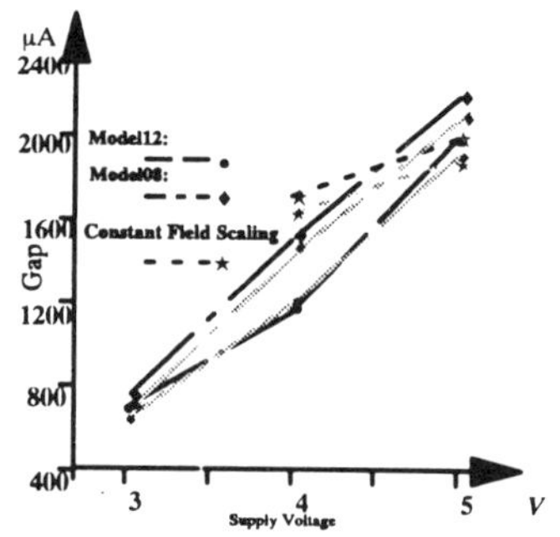

Fig.3 Gap for various supply voltages

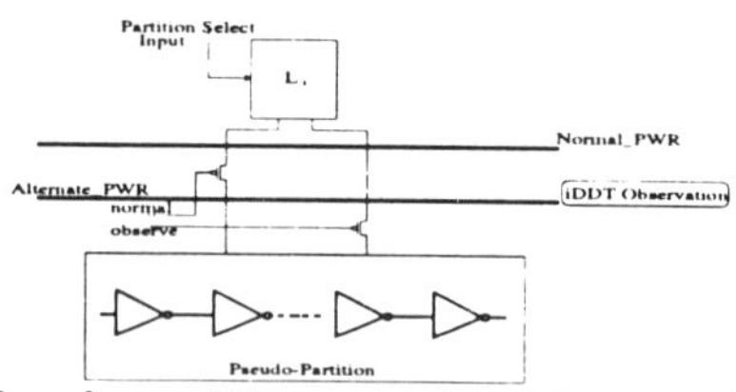

Figure 4. Pseudo-partition module with control transistors

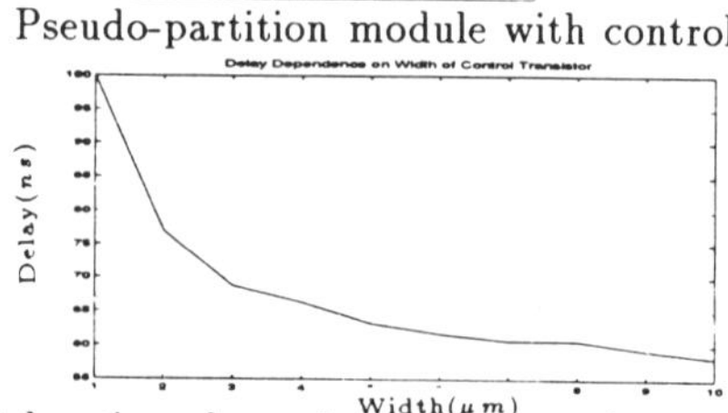

Figure 5. Delay time dependence on size of control transisitors

A CMOS Mixed-Signal Integrated Circuit Having a Reduced Vector Set and Closed-Loop Analog Test Architecture

A.V. Chavan[1], D.W. Stringfellow, S.R. Mallarapu, and R. Ardeishar

Delco Electronics Corporation
Kokomo, Indiana 46904

Abstract

This paper describes a mixed-signal IC that employs a test architecture consisting of multiple selectable sub-scan chains (SSC's), which the authors have termed "pseudo full scan". This implementation reduces test pattern length and improves fault grading. An additional scan chain of boundary shift register latches (BSRL's) is used to sensitize the analog section of the IC to permit closed-loop testing. A mixed-signal IC using this architecture can be tested with a less expensive Digital IC tester. Modeling for standard automatic test pattern generation (ATPG) tools is also presented.

1. Introduction

Previous research in scan-testable CMOS circuits [1, 2, 3] has focused on classical line stuck-at fault models when applied to pure digital circuits. In this paper, we present an extension of this previous work to the emerging mixed-signal problem. Specifically, we describe a test architecture called "pseudo full scan" which uses a combination of full scan and boundary scan. We depart from the classical scan techniques in that we combine normal SRL's and Boundary SRL's (BSRL's) on the same scan path. Further, the scan chain is split into multiple sub-scan-chains (SSC's), with the BSRL's embedded as I/O at the analog-digital boundary. These features reduce the test development time required to achieve the mandatory 99% digital fault coverage stipulated by Delco Electronics.

We had two seemingly contradictory requirements: a short test development cycle and a high fault coverage vector set. The design team also had to consider tester utilization and memory constraints. To this end, we chose an approach which allows the IC to be tested using short, modular tests and permits closed-loop testing of analog functions. A significant result is that a mixed-signal IC using this architecture can be tested with a less expensive Digital IC tester.

The IC under discussion is not fully 1149.1 compatible; we did not have enough external pins available to implement the connections to the standard JTAG TAP. The clock generator circuit also used certain "untestable" logic topologies, so we had to use functionally equivalent models and modify the gate level netlist to make it suitable for our ATPG tool.

In the remainder of this paper, we will describe the basic test architecture, briefly discuss the mixed-signal IC, and list the modeling and test generation procedures. Lastly, we will describe the limitations of the test architecture and discuss possible future work for our "pseudo full scan" approach.

1. Also with The Department of Electrical Engineering and Computer Science, University of Michigan, Ann Arbor, Michigan, USA.

2. IC Description

The IC under discussion is used in conjunction with a host microprocessor, hall effect sensors, and high-current FET's to drive a multi-phase brushless motor. It can be used in automotive electronics applications such as electric steering assist, electric/hydraulic steering assist, and four wheel steering. It can dynamically adjust between current mode and voltage mode operation, contains a digital slope compensation circuit which improves current-mode accuracy at high duty cycles, and provides open and short fault detection.

The IC is fabricated using a 1.0 μm, dual-metal modular high-voltage CMOS process. The die contains 12,605 transistors (2,211 analog and 10,394 digital) and has a die size of 162 mil x 158 mil (4.11 mm x 4.01 mm).

Fig. 1. Photograph of the Brushless Motor Driver IC

3. Background

Morley and Marlett [4] present a very efficient scan test architecture which provides test program length reductions of at least 60% and as high as 75%. Their architecture uses prediction techniques combined with probability theory to parti-

0-7803-3177-6 $5.00 © 1996 IEEE

IEEE 1996 CUSTOM INTEGRATED CIRCUITS CONFERENCE

tion the scan path into two pieces. We chose to partition the scan chain into multiple parts corresponding to the on-chip functions. In contrast to Morley, our partitions are dynamically reconfigurable and can simplify the work of test developers and test engineers, especially during design debug.

Several architectures have been presented for mixed-signal testing [6,7,8,9]. While signal distortion and loading effects have been substantially reduced in approaches described in [6,7,8], we could not afford the required control lines and area overhead. Thus we devised a hybrid scheme using both SRL's and BSRL's on the same scan path. Since the BSRL's can participate in scan operations, inputs coming from the analog side are treated as primary inputs, and a standard ATPG tool can be used to produce a high fault coverage for the digital portion of the IC.

4. Test Architecture

A. Overview

There are a total of 298 SRL's in the IC; 28 are BSRL's, and the remaining 270 SRL's make up the "data" and "control" scan chains. The "data" scan chain has 261 SRL's and the "control" scan chain has 9 "modified" flip-flops which are called "test port" (TP) logic cells. Each sub-scan chain has a single TP bit, which is used to either select or bypass it's corresponding sub-scan chain. The TP cell for an unselected sub-scan chain acts as a place holder bit in the "data" scan chain and must be accounted for in the test patterns. The following test modes were implemented: scan "control", scan "data", hold boundary SRL, latch boundary SRL, and normal functional. Using the analysis of [4], it can be shown by using a test architecture which permits combining smaller sub-scan chains to form a single, larger data scan chain, test vector length can be reduced by a minimum of 50%.

B. Block Diagram of Test Architecture

The Test Access Port (TAP) configures the clock generator to operate in one of the test modes. The clock generator's P1 and P2 outputs supply a two-phase non-overlapping clock scheme appropriate for master/slave flip-flops or SRL's [3]. The DSC signal clocks the master portion of a master-slave flip-flop when scan "data" mode is enabled. In this mode, P1 is negated. The CSC signal clocks the master portion of a master-slave flip-flop when scan "control" mode is enabled. In this mode, P1 is also negated.

The IC under discussion uses a single test input pin, TP1, and a single test output pin, TP2. When active, TP1 configures certain external pins to function as the standard TAP inputs C1 and C0 and output SO shown in Fig. 3.a). When TP1 is active, the SI pin serves as the IC's scan data input.

The SenseA and SenseB input pins are "synchronized" to the IC's internal slave clock, producing C1D and C0D, respectively. The local scan chain's master clock, LDSC, is enabled when the SE signal from the TP bit is set to logic "1".

C. Test Port (TP) Bit

The TP bit used in the TAP serves as the connecting point from the last sub-scan bit in the "data" scan chain to the TP2 test output pin. At power-up, SE=0, and the "default path" of the 2:1 mux is selected, bypassing the sub-scan chain associated with a given TP bit. When SE=1, the corresponding sub-scan chain output is selected by the mux and passed through the 1/2-latches, allowing the corresponding sub-scan chain to be combined with other enabled sub-scan chains to form a "data" chain for use in scan test operations. The TAP contains a 3 bit test register (bits MT1, MT2, and MT3) which controls the BSRL's (see Fig. 3.a). The BSRL Control Logic shown in Fig. 2. selects a test mode according to Table 1.

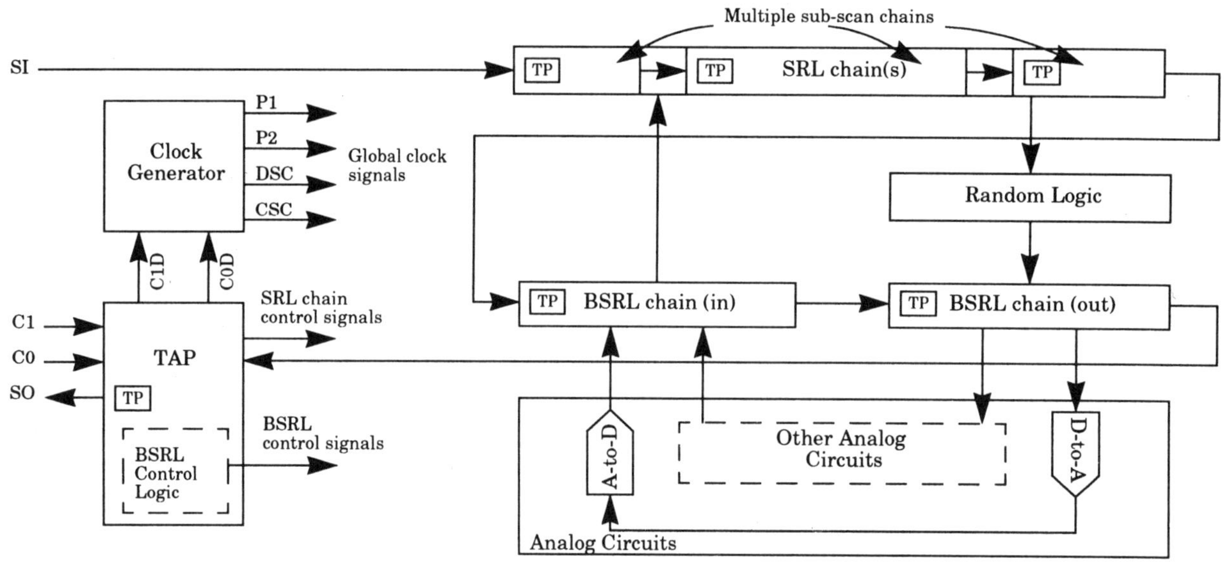

Fig. 2. Scan Test Architecture

D. Scan Chain Organization

The "data" scan chain is composed of smaller sub-scan chains that are dynamically reconfigurable using the "control" scan chain during IC testing. Each functional block of the IC has a corresponding sub-scan chain. The typical sequence for scan testing consists of a scan-in operation, a functional clocking operation, and a scan-out operation. The exact clock timing is described in Fig. 4.

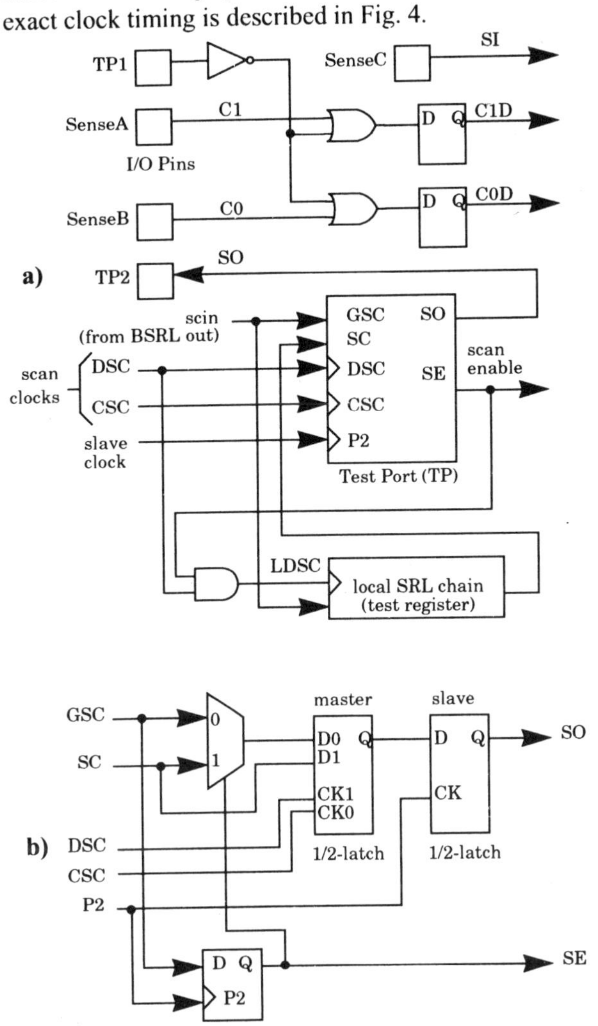

Fig. 3. a) Test logic in the TAP, b) TP (test port) bit

Each sub-scan is selected for scan operations by setting it's TP bit's SE output to logic "1". DSC is enabled when the IC is in scan "data" mode (C1=0, C0=0). CSC is enabled when the IC is in scan "control" mode (C1=0, C0=1). Only one scan "control" operation is needed for multiple scan-in/run functional/scan-out operations. The input clock can free-run when switching between different scan modes.

E. Boundary SRL's

To test the analog circuits, values can be set up and held in the output BSRL's, and then normal operation (functional mode) can proceed. Next, the IC can be configured to scan out the analog circuit's response. Hence, additional pins or multiplexed nodes on the analog side are not required.

Table 1.
BSRL Test Register bits contained in TAP

MT1 (misc reg 7)	*MT2 (misc reg 5)*	*MT3 (misc reg 6)*	*Mode*
0	0	0	Bypass, No Hold (Normal Operation)
0	0	1	Bypass, BSRL out hold
0	1	0	Bypass, BSRL in hold
0	1	1	Bypass, Both (BSRL in/out) hold
1	0	0	Both (BSRL in/out) latched, No hold
1	0	1	Latched BSRL in, Hold BSRL out
1	1	0	Latched BSRL out, Hold BSRL in
1	1	1	Both Latched, Both Hold mode

Our IC has a D-to-A which, in test mode, has its output is routed to an analog pin by an analog mux (not shown in Fig. 2.). All the analog signals driving the logic must be level-shifted to be compatible with the digital circuitry.

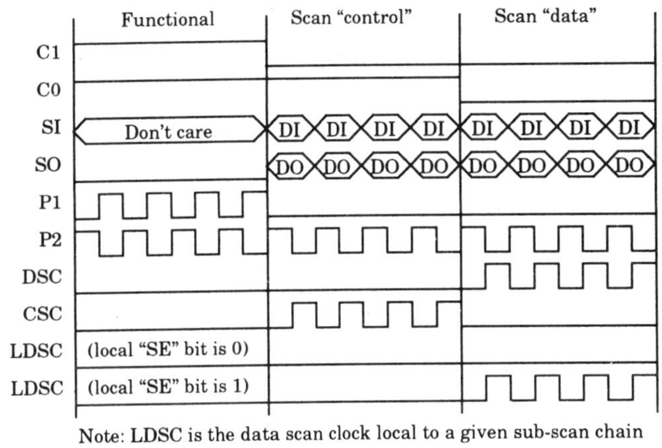

Fig. 4. Scan mode timing diagram

In test mode, the hold mode control inputs coming from the TAP are enabled and keep data from changing as scanning takes place at the inputs of the n-bit D-to-A converter. In this way, values can be scanned into the BSRL output registers to replace signals from the digital circuitry. During normal operation, these BSRL's are "bypassed" so that they do not add any extra delay to the signals flowing in both directions. During test, we need to capture and hold the values from the logic and analog circuitry using the BSRL's. The following three test scenarios are thus supported.

Scenario 1: Normal scan testing of digital logic. In the actual IC, the tester does not have direct access to these pins. On power-up at the tester, the IC's analog inputs are initialized randomly from one IC to the next. Thus, IC test results do not necessarily match the ATPG results. With our scheme, IC test and fault grading results can be guaranteed equal.

Scenario 2: Testing of the D-to-A and other analog functions. In this mode it is necessary to scan values into the BSRL outputs and hold these inputs to the D-to-A at fixed values.

Scenario 3: Prevent digital outputs from changing analog circuit inputs during scan operations. We chose to not expose the analog circuits to this situation in which analog circuits may potentially inject current into the substrate.

5. IC Testing

The design team used ATTEST's TDX software for the test generation. This tool extracted 18,086 faulted nodes and required about 30 minutes on an HP-715/100 to produce a high quality 95%+ fault graded pattern set.

To minimize area overhead, we used the following logic constructs: 1) gated clocks, 2) clocks which are outputs of other state elements, and 3) clocks which are not primary inputs (internally generated). These techniques are normally considered incompatible with classical scan testing, so special ATPG models were required.

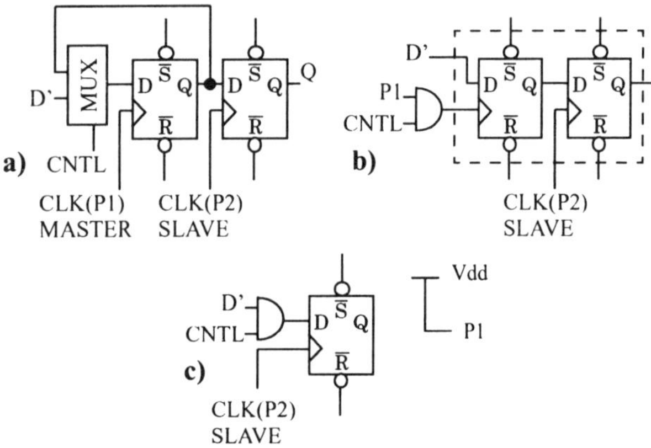

Fig. 5. a) conventional scan element, b) scan element for this IC, c) ATPG model of scan element

A conventional state element implementation, which uses a "data hold" feature, is shown in Fig. 5.a). This type of state element is normally required when the test architecture is LSSD-based [3]. This construct becomes very expensive for a design having registers and counters, because each flip-flop requires a 2:1 mux. Since our design had several register and counter building blocks, we chose the more efficient construct shown in Fig. 5.b), which replaces multiple muxes with a single gate. The ATPG software, however, expected the logic construct shown in Fig. 5.a)., so in order to generate valid fault vectors, the gated clocks were modeled as shown Fig. 5.c). Note here that P1 is connected to a logic "1", or Vdd, so that the master element's clock is always enabled.

6. Conclusions

We can conclude from this particular design that our "pseudo full scan" architecture is very suitable for mixed-signal designs which have short design cycles and that are cost-driven. When this architecture is used, simple modular tests can be developed. Analog testing can also be done using a less expensive digital IC tester. Significant results include:

a. Area overhead due to special circuits, FF's and routing was < 15%
b. Total test vector length to achieve 95% fault coverage was approx 230,000 vectors
c. Testing of critical analog functions possible on a lower-cost Teradyne A580 (digital) tester
d. Total time expended for 95% fault coverage was roughly 2 person-months
e. TDX produced very small fault dictionaries due to smaller SSC's.

One limitation is that two slightly different flip-flop models are required: one for vector generation and one for functional simulation. Thus, two passes are needed: one to generate vectors using a simple model and a second to obtain an accurate fault grade percentage using a fault simulator.

We are considering the following enhancements: 1) Achieve full IEEE 1149.1 compatibility (expand this chain across multiple IC's on a PCB to reduce board test time), and 2) Address analog fault grading issues.

Acknowledgments

We would like to acknowledge the inputs received from Wayne Chen of Texas Instruments. We would also like to thank Tony Frisch and Paul Ainslie for their encouragement.

References

[1] D.K. Liu and E.J. McCluskey, "Design of CMOS VLSI Circuits for Testability", *IEEE 1986 Custom Integrated Circuits Conference*, pp. 421-424.

[2] B-H Park and P.R. Menon, "Design of Scan-Testable CMOS Sequential Circuits", *IEEE Int. Test Conf. Proc.*, 1990, pp. 369-376.

[3] E.B. Eichelberger and T.W. Williams, "A Logic Design Structure for LSI Testability", *Proc. 14th Des. Autom. Conf.*, pp. 462-468, June 1977.

[4] S.P. Morley and R.A. Marlett, "Selectable Length Partial Scan: A Method to Reduce Test Vector Length", *IEEE Int. Test Conf. Proc.*, 1991, pp. 385-392.

[5] IEEE Standard 1149.1-199x, "Standard Test Access Port and Boundary-Scan Architecture", *IEEE Standards Board*, New York, 199x.

[6] A.H. Bratt, A.M.D. Richardson, R.J.A. Harvey, and A.P. Dorey "A Design For Test Structure for Optimising Analogue and Mixed-Signal IC Test", *Europ. Test Conf. Proc.* 1995, pp. 24-33.

[7] J.S. Matos, J.C. Ferreira, A.C. Leao, J.M. Silva, "Architecture of Test Support IC's for Mixed-Signal Testing", *Europ. Test Conf. Proc.* 1994, pp. 240-246.

[8] B.R. Wilkins, S. Oresjo, and B.S. Suparjo, "Towards Mixed-signal Testability Bus Standard P1149.4", *IEEE Int. Test Conf. Proc.*, 1993, pp. 58-65.

[9] S. Patel, "Brushless Motor Driver IC Specification", Delco Electronics Corp. SK-4292, Revision D, 2/14/94.

L^2RFM - Local Layout Realistic Faults Mapping Scheme for Analogue Integrated Circuits

Michael J. Ohletz

Institut für Theoretische Elektrotechnik
Universität Hannover - Germany
email: mo@tet.uni-hannover.de

Abstract

A new fault modelling scheme for analogue ICs called *Local Layout Realistic Fault Mapping*[1] is introduced. It is aimed at realistic fault assumptions prior to the final layout. Defects are assumed and their electrical failure modes are evaluated. It turned out that some faults at schematic level are unrealistic, new types of fault emerge and the distribution of faults chances. For a CMOS operational amplifier the number of faults dropped from 45 to 27.

Introduction

Traditionally, analogue testing (1, 2) is understood as testing against specifications. But, an out-of-specification response can be due to any defect or failure mechanism. This allows for a defect oriented view of testing. The variety of possible failure modes makes analogue fault modelling a difficult task. An appropriate fault model is always a trade-off: a) it shall be based upon realistic data; b) the fault model must be manageable within an automatic fault simulator in terms of effort of implementation and simulation time (3), (4), (5). Fault modelling for integrated analogue circuits is a relatively recent and still controversal issue. Nevertheless fault models have been used for more than two decades. It is common use (1, 2, 6, 7) to classify faults in analogue circuits into two categories:

- *Catastrophic Faults or Hard Faults*
- *Deviation Faults or Soft Faults*

Hard Faults are used to describe random defects that can cause short and open circuits or large scale deviations of design parameters (4). Soft faults represent out-of-tolerance parameter deviations e.g. caused by statistical fluctuations in the manufacturing environment (4). From reliability investigations various failures are known (8, 9, 10). Many electrical failure modes (EFM) are short and open circuits or leaky junctions. In (11, 12) it is reported that 54% of all faults observed on the I/O pins were shorts and opens. An amount of 28% failures was attributed as "mal-functions" which potentially has been caused by "internal" shorts or opens. Others report that 75% (13) and 83.5% (14), respectively, of all failures observed were catastrophic. In general the single faults are assumed. For example for a transistor six hard faults are assumed at the schematic level irrespective whether or not this assumption is realistic:

Shorts: Gate-Drain, Gate-Source, Drain-Source
Opens: Gate, Drain, Source

thus, it appears that at least hard faults should be taken into account. In order to prove whether or not assumed faults are realistic the layout of the circuit must be investigated. In the context of this paper the term "realistic" only refers to the possibility of the occurence of a fault and not to its likelihood. A successful layout based approach known as "Inductive Fault Analysis (IFA)" has been proposed for digital circuits (15) and for an analogue CMOS circuit (16). But some restrictions apply for IFA: First, the fault list can only be obtained after the complete layout. Consequently, fault simulations can only be performed afterwards. Alterations of the layout directly effect the fault list which thus, has to be updated by re-running the IFA. Secondly, the effort to perform a complete IFA can be time consuming. An alternative to assume realistic faults prior to the final layout could be the L^2RFM scheme.

Local Layout Realistic Faults Mapping Scheme

The scheme of *Local Layout Realistic Faults Mapping, L^2RFM,* is based on the fact that analogue ICs are composed of only a few frequently occuring building blocks such as differential pairs, current mirrors etc.. These blocks can be easily identified from the schematic. Further it turns out that there are local structures which do not vary too much e.g. due to matching considerations. Investigating such typical structures with respect to the impact of characteristic defects opens a door to draw some conclusions about realistic or unrealistic fault assuptions which then could be applied to other analogue ICs prior to the precise knowledge of the actual final layout. The only information required are the schematic (netlist) and some technology data available anyway in the early design process. After this investigation the obtained realistic and unrealistic faults can be used to build up a dictionary for further fault simulations (17). In turn assumed hard faults can be verified at the layout level. This process aims at a reduction of the later simulation time by excluding non-realistic faults from the fault list

[1] This research was performed within the ESPRIT III project ARCHIMEDES and supported by the European Union under grant 7107.

0-7803-3177-6 $5.00 © 1996 IEEE

IEEE 1996 CUSTOM INTEGRATED CIRCUITS CONFERENCE

and at enhancing the initial fault list by realistic faults. In particular this scheme seems to be advantages for analogue or mixed-signal ASICs.

The initial fault list is the complete set of single hard faults possible on each element of the circuit. Defects impacting devices topologically close but electrically not connected to each other directly are excluded. This is done since these faults are layout dependent and therefore can not be generalised. The following chapters will describe the scheme for an operational amplifier.

L^2RFM for a CMOS Operational Amplifier

The example shown in Fig. 1, is a low current CMOS operational amplifier (18) in n-well CMOS technology comprising of eight enhancement mode transistors (n- and p-channel) and one poly-Si/Al capacitor for the internal frequency compensation. Fig. 2 depicts the layout. The biasing of the input and output stage can be adjusted by an external biasing cell e.g. from an analogue cell library. For the L^2RFM scheme (19) the following defects will be addressed:

- Contact Failures
- Polysilicon and Metal Cracks
- Pin Holes and other Failures

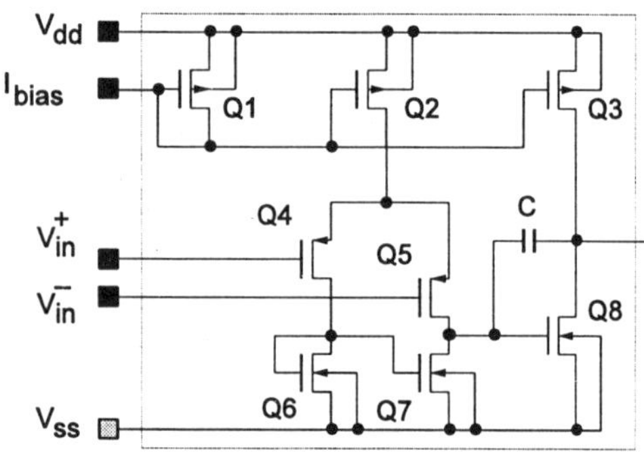

Fig. 1 CMOS operational amplifier with external biasing

Looking at the schematic it appears that the complete initial list of "schematic faults" comprises 25 possible open and 22 possible short faults.

A. Contact Failures

The first class of failures are defects causing "missing contacts". Fig. 2 shows the different failure locations within the layout. Each failure is denoted by a letter followed by a number representing the number of contacts associated. There are four different types:

- *gate open (floating gate)*
- *drain open*
- *source open*
- *floating substrate/n-well*

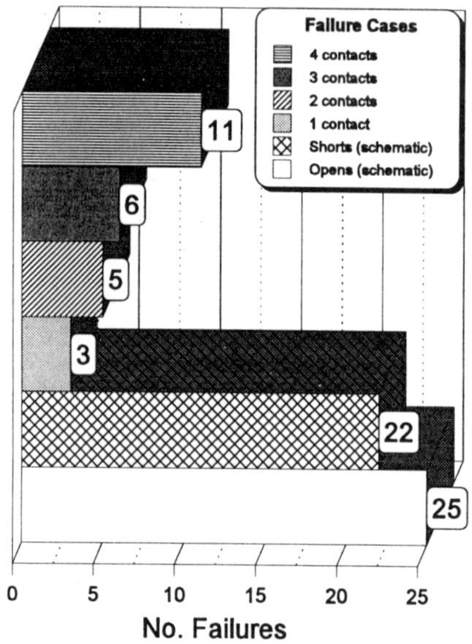

Fig. 2 Distribution of faults versus different number of contact failures (Bottom bars are faults from the schematic)

Some missing contact cause floating gates, e.g. B1 and C1. But there are also instances where multiple open gate faults are caused. If the contact A1 is not in place all gates of the transistors Q1-Q3 are floating. But due to the layout of the gate connection their gates remain connected to each other. This layout configuration is a typical structure which can be encountered very often. Since only single faults rather than single defects were assumed originally, this type of fault did not emerge in the initial fault list. Now this fault situation can be identified from the schematic easily. D2 causes a multiple open gate, while E2 causes a DC-open gate fault. DC-open refers to the fact that due to the capacitor configuration the Q8 gate is capacitively connected to the output. There are other contact failures which only occur if more than one contact is absent. Therefore it seems reasonable to assume that these cases are less likely compared with contact failures of lower contact count. Transistors with sources connected to the power rails are "nailed" by chains of many contacts (29 -31 !!). to the power rails. Due to this implementation open faults caused by missing contacts appear to be extremely unlikely. From that circumstance the following rule for the mapping could be stated: *Open faults on transistor sources connected to the power rails are unrealistic and can therefore be discarded from the initial fault list.* If contact failures do not impact all contacts which are necessary to yield an open fault this failure potentially can produce a soft fault. This type of fault diminishes the overall dynamic performance of the circuit. If the characteristic of the particular contact is known this type of soft fault can be included in the fault list as a realistic soft fault.

Consequently, the number of open faults which must be addressed has decreased as can be seen from fig. 3. Only 3 faults must be taken into account if the absence of only one contact is assumed.

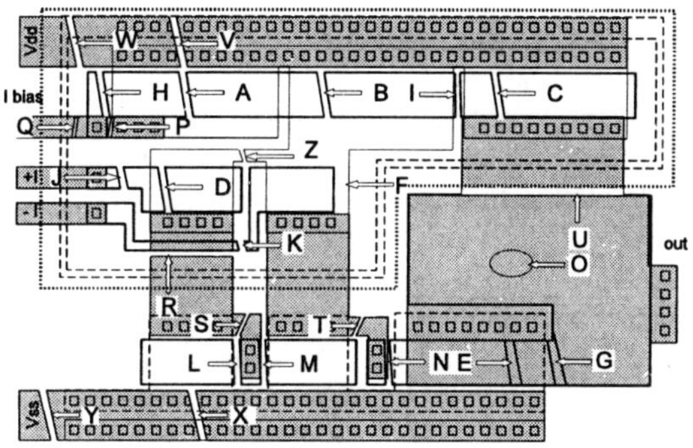

Fig. 3 Assumed cracks in the polysilicon and metal layer

B. Cracks in the Polysilicon and Metal Leads

Fig. 4 shows the layout along with assumed cracks. Cracks within the diffusion are not considered to the same extend since they seem to be very unlikely. Again there are different types of failures depending on where the failure is located:

- ❑ *polysilicon cracks over gate oxide area*
- ❑ *polysilicon cracks over field oxide area*
- ❑ *metal cracks*

C. Cracks within the Polysilicon Layer

At first *polysilicon cracks within the active area* (gate) of a transistor are considered. Such failures cause source drain shorts and a conducting channel. Thus, the transistor exhibits a failures which could either be represented as a short fault or as a soft fault. Depending on the location and the width of this gap the transistor operation might only be diminished more or less. If the resistance of the created channel is low enough the transistor does not operate anymore at all and it is realistic to assume a single short fault. But, if the gap occurs more closely to the site where the gate is connected to other components (gate input), it is additionally cut off and the short fault is accompanied by a gate open on the same transistor. As a result a double fault situation on the particular transistor occurs which comprises of a drain-source short and a gate open fault.The same fault constellation is established on all other transistors, e.g. like in case of failure D. The only difference is the number of additionally generated open faults on other transistors which depends on the location of this failure. In case of failure A on Q1 it can be easily recognised from fig. 4 that in addition to the failure situation on transistor Q1 the gates of Q2 and Q3 are also effected. Depending on the location of the short either a triple fault constellation emerges which comprises a soft fault (leaky Q1 transistor) along with two opens (Q2, Q3), or a quadruple fault situation is established. A similiar fault situation is caused by the failure B on transistor Q2. These types of faults did not occur in the initial fault list. It is the special gate configuration often encountered along with biasing transistors which causes these new types of faults. Another type of failure which causes a similiar fault situation, denoted as failure F, is shown in fig. 4. Here the assumed failure e.g. due to any lithographic defect provokes that the width of the respective transistor gate becomes too small. As a result the respective transistor is shunted by this failure. Depending on the amount of misalignment the transistor either behaves like a leaky transistor which resembles a soft fault or like a short with a neglegible transistor operation. From the previous consideration it can be summarised: *Crack failures within the gate area of a transistor yield a multiple fault constellation consisting of a drain-source short and one or more gate opens.*

Eventhough a single open fault on the capacitor can be easily assumed it turns out from the layout that an open fault on the capacitor seems to be very unlikely. The capacitor´s one electrode is a metal plate whereas the other one is made from polysilicon directly "integrated" into the gate of transistor Q8. An open due to a polysilicon crack only is possible if the crack is located in the small area right at the end of the gate of transistor Q8 (failure G) or if the crack is large. In consequence the likelihood for such a failure could be considered to be not too high. An example for a realistic soft fault are pin holes in the polysilicon layer within the area of a capacitor (failure O) or a transistor causing a degradation of the dynamic performance.

Polysilicon cracks outside the gate areas normally cause open or floating gates. Such failures could cause single open faults like the failures I - M. From these the failure cases J and K seem to be the most likely ones. For the failures I and L-M the area to establish a single open is very small. Therefore it could be considered as less likely. A situation described already before occurs with the crack N which again produces a DC-open fault on Q8. That type of fault was neither assumed in the initial fault list, since faults were attributed only to elements rather than to nodes. Also multiple open constellations can be created.

D. Cracks within the Metal Layer

Beside the polysilicon metal is the most sensitive layer with respect to opens. Of course both layers are not only subject to opens but also to shorts. Since shorts are much more depending on the global layout they are not further considered here. Metal cracks cause single as well as multiple open faults. There two types of enhancement transistors. Connections between transistors of the same type are normally made through diffusion or polysilicon whereas connection between transistors of different type are made by metal. Due to the associated contacts and steps the chance for defects such as R (insufficient step coverage) is higher compared with an open e.g. Z of a "simple" diffusion line. For wide transistors the probability for an open fault may be less than for narrow transistors. The impact of cracks in power supply rails strongly depends on its actual location. If the power supply rail is broken outside the contact row area (failure W for V_{dd} or Y for V_{ss}) all sources of the respective transistors are disconnected from the power supply. Thus, the whole circuit does not operate properly. Eventhough this type of failure might occur it does not need to be considered since the malfunction of the circuit can be easily detected. If the defect occurs within the contact rows (failures V and X) then the location of the crack again effects the circuit with respect to its performance. Such defects do not cause source open faults on the transistors directly connected to the power rails because the sources are also connected together through diffusion. In general the defect diminishes the overall dynamic performance.

E. Fault Distribution

Fig. 5 depicts the fault distribution. It appears that the number of single gate open faults dropped from eight to five realistic and finally to only two realistic and likely single gate opens. Also dramatic is the drop of single drain/source open faults by 69% from 16 to five. All of them are likely. The capacitor open at schematic level disappears since even it is "realistic" it is not very likely. Not included in the schematic initial fault list were multiple gate open constellations. But there are four cases of such fault situations which remain and likely. Considering short faults it immediately is conspicuous that non of the eight initially assumed single drain-source shorts are realistic. This type of short always is accompanied by at least one gate open. Thus, all eight schematic single drain-source shorts are transformed into eight open-short comibnations. There are eigth gate to channel short. Initially 14 schematic short fault between gate drain/source were assumed. All of them are realistic. Due to the very small area available to establish the respective failure they are assumed to be realistic but unlikely for this investigation and thus, are discarded. But due to the likelihood of occurence within the "global" it could be reasonable to included them in the fault list. Comparing the three different distributions of faults with respect to the overall number of faults it turns out that in the initial schematic fault list 47 different faults were assumed. From that the number dropped to overall 45 realistic and finally to 27 if the likelihood of the realistic faults is introduced.

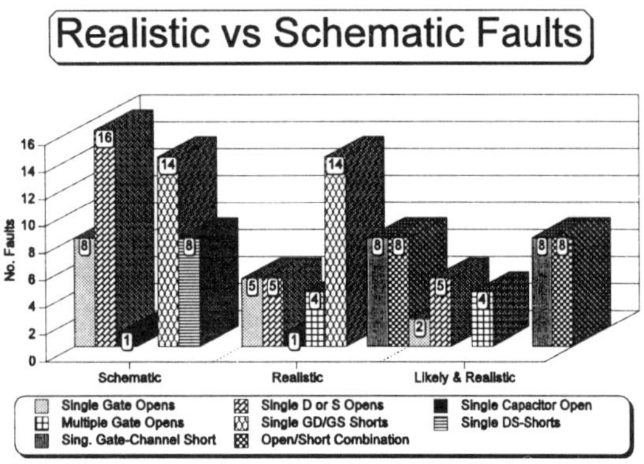

Fig. 4 Comparision between initially assumed schematic faults and likely realistic faults

Conclusion

A new scheme to obtain realistic fault assuptions for the fault simualtion of analogue circuits has been introduced. It allows to make the fault assumption prior to the final layout and thus more effective test development is possible. It has been shown that some fault assuptions at schematic level are not realistic while other important faults are not covered. Once enough layout structures have been investigated a dictionary of realistic fault can be implemented in an analogue fault simulator. For an example of a simple operational amplifier the L²RFM scheme has been demonstrated. It was shown that the number of faults could be reduced from 45 to 27.

References

(1) P. Duhamel, J.-C. Rault, "Automatic Test Generation Techniques for Analog Circuits and Systems: A Review", IEEE Trans. Circuit and Systems, Vol. CAS-26, No. 7, 1979.

(2) J. W. Bandler, A. E. Salama, "Fault Diagnosis of Analog Circuits", Proc. IEEE, Vol. 73, No.8, 1985

(3) H. Lipinski, " Untersuchungen zum Single Hard Fault Model", Report, University of Hannover, July 21, 1988.

(4) Linda Milor, V. Visvanathan, "Detection of Catastrophic Faults in Analog Integrated Circuits", Trans. on CAD, Vol. 8, No. 2, 1989.

(5) M. J. Ohletz, "Selbsttest monolithischer analog-digitaler Schaltungen", Dissertation, University of Hannover, Germany, November 1989.

(6) W. A. Plice, "Overview of Current Automated Analog Test Design", ITC 1979.

(7) M. J. Ohletz, "Hybrid Built-In Self-Test (HBIST) for Mixed Analogue/Digital Integrated Circuits", ETC91, pp.307-316.

(8) F. Fantini, C. Morandi, " Failure Modes and Mechanisms for VLSI ICs - A Review.", IEE Proc., Vol. 132, Pt. G, No. 3, pp. 74-81, Juni 1985.

(9) F. Fantini, M. Vanzi, " VLSI Failure Mechanisms", Proc. CompEuro 87, pp. 937-943, Hamburg, Mai 1987.

(10) E. A. Amerasekera, D. S. Campbell, " Failure Mechanisms in Semiconductor Devices", John Wiley & Sons, 1987

(11) Integrated Circuit Engineering Corporation, " Microcircuit Manufacturing Control Handbook - A Guide to Failure Analysis and Process Control", 2nd ed., Scottsdale, AZ, 1977.

(12) H. K. Dicken, " What Can Happen to an IC in Your System? Evaluation Engineering", pp. 36-40, July/August 1977 and pp.46-47, November/December 1978.

(13) J. Galiay, Y. Crouzet, M. Verginault, " Physical Versus Logical Fault Models for MOS LSI Circuits: Impact of their Testability", IEEE Trans. on Computers, Vol. C-29, pp. 527-531, Juni 1980.

(14) C. H. Stapper, F. M. Amstrong, K. Saji, " Integrated Circuit Yield Statistics", Proc. IEEE, Vol. 71, No. 4, pp. 453-470, April 1983.

(15) John P. Shen, W. Maly, F. Joel Ferguson, " Inductive Fault Analysis of MOS Integrated Circuits", IEEE Design & Test, Dec. 1985.

(16) D.B. Feltham, W. Maly, " Physically Realistic Fault Models for Analog CMOS Neural Networks", IEEE J. of Solid State Circuits, vol. 26, No. 9, pp. 1223-1229, Sept., 1991.

(17) Chr. Sebeke, J.P. Teixeira, M.J. Ohletz, "Automatic Fault Extraction and Simulation of Layout Realistic Faults for Integrated Analogue Circuits", Proc. EDTC 1995, pp. 464-468

(18) Malcom R. Haskard, Ian C. May, " Analog VLSI Design", Prentice Hall, ISBN 0 7248 0027 1, 1988.

(19) Michael J. Ohletz, " Realistic Faults for Analogue ICs Based upon Typical Layout Structures", ESPRIT III, 1st open workshop "Mixed-Signal and Analogue Testing", Hannover, Germany, June 8-9, 1993.

Fast Thermal Analysis for CMOS VLSIC Reliability *

Yi-Kan Cheng and Sung-Mo (Steve) Kang
Department of Electrical and Computer Engineering
Coordinated Science Laboratory
University of Illinois at Urbana-Champaign
1308 W. Main St., Urbana, IL 61801

Abstract

In this paper, we present a fast thermal reliability diagnosis tool (iTHREAD) for CMOS VLSI chips. Unlike existing circuit-level electrothermal simulators, it finds the steady-state temperature distribution, the hot spots, as well as the resulting power consumptions in an extremely efficient way. By using a fast timing simulator with accurate temperature-dependent device models and a novel 3-D analytical thermal simulator, it can easily handle very large circuits. With iTHREAD, temperature-dependent reliability and timing problems of VLSICs can be accurately and quickly predicted. It can be further applied to guide the module placement, timing verification, and many other IC reliability diagnoses.

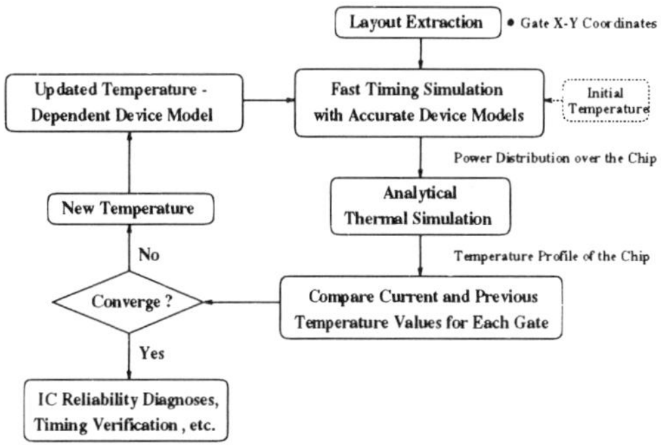

Figure 1: Flowchart of iTHREAD electrothermal simulation.

1. Introduction

In state-of-the-art VLSI technologies, the increasing device packing density leads to dramatic temperature rise in a VLSI chip. Since many known IC failure mechanisms are thermally activated or related (i.e., electromigration, hot carrier, etc), on-chip temperature distribution must be predicted prior to any reliability diagnoses. In addition, temperature profile plays an important role in the applications of thermal stress evaluation and packaging design at the chip or print-circuit-board level.

Although some research has been conducted for the circuit-level temperature profile estimation [1, 2], there still exists many bottleneck problems: Firstly, existing *coupled* electrical and thermal (i.e., electrothermal) simulations are inefficient in nature. In those approaches, time frame is divided into many small intervals and the electrical and thermal simulations are coupled to find the transient or steady-state temperatures for each interval. Secondly, SPICE-like programs were often utilized for the electrical power calculation. This results in low simulation speed which is not acceptable for VLSI chips containing thousands or millions of transistors. Thirdly, customary circuit-level thermal simulators employ the numerical finite-difference method to solve the 3-D heat equation. In this method, each device is treated as a node in a 3-D RC network. The temperatures are then determined by solving the admittance matrix of the network. Since the matrix solving routine has the time complexity of $O(n^3)$ (i.e., n is the number of nodes), the numerical approach becomes extremely expensive as the circuit size grows. Although alternative method [2] was proposed recently to solve the heat equation, the infinite Fourier-series summation still requires a large amount of computational time. Fourthly, power calculation in existing electrothermal simulators relies heavily on well-established temperature-dependent device models. This restricts their usefulness when the device models are not fully developed or characterized.

iTHREAD is developed to resolve the above problems. It aims at finding the on-chip steady-state temperature distribution by using a sequence of procedures: layout extraction with x-y coordinates for each logic gate, fast and accurate power calculation using highly adaptive temperature-dependent device models, efficient *analytical* thermal simulation, followed by an *uncoupled* electrothermal simulation until the temperature is converged. The block diagram of iTHREAD simulation procedures is shown in Fig. 1. Note that iTHREAD calculates the *average* instead of transient power for each gate, thus uncouples the electrical and thermal simulations. This procedure greatly increases the simulation speed and is justified by the fact that, the time for temperature to reach steady state in silicon (i.e., milliseconds) is much longer than the period of timing simulations [3].

2. Temperature Dependence Device Modeling

When temperature rises in a VLSI circuit, the drain current I_{ds} in MOS devices tends to decrease due to the lowered carrier mobility. The reduction on I_{ds} also decreases the short-circuit power dissipation, which forces the temperature to redistribute by heat diffusion. Therefore it is essential to develop a simple yet accurate temperature modeling method for CMOS devices. The fast timing simulator employed in iTHREAD uses the regionwise quadratic (RWQ) modeling technique [4] for accurate submicron device modeling. In current work, we further develop this technique to incorporate the temperature dependence of MOS devices.

The RWQ modeling method is a highly accurate fitting technique for generating MOS I-V data in the fast timing

*This research was supported in part by Intel Corporation, Rome Laboratory (F30602-95), JSEP (N00014-94-J-1270), and Semiconductor Research Corp. (SRC94-DP-109).

0-7803-3177-6 $5.00 © 1996 IEEE

IEEE 1996 CUSTOM INTEGRATED CIRCUITS CONFERENCE

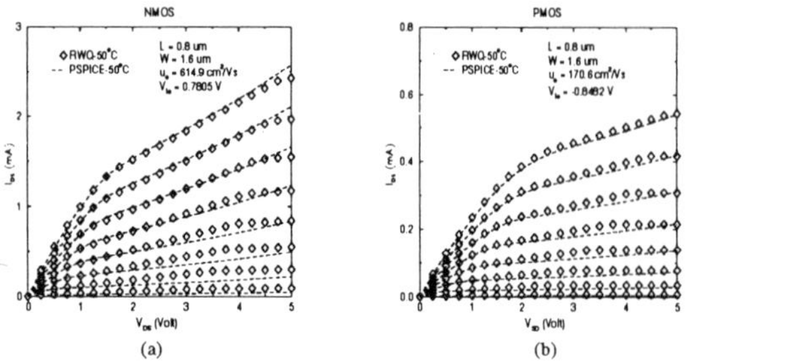

Figure 2: Device modeling at 50 $^{\circ}C$.

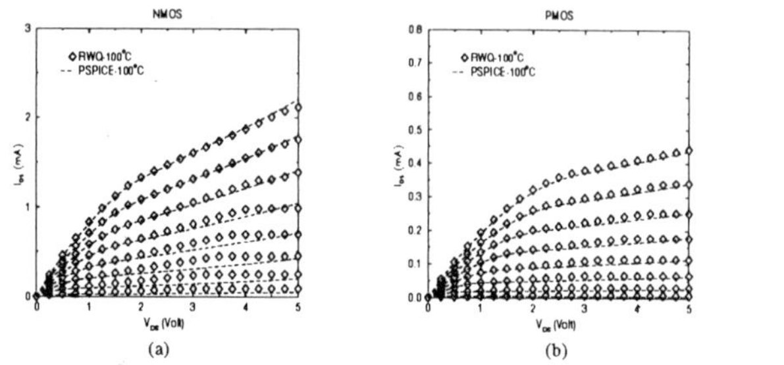

Figure 3: Device modeling at 100 $^{\circ}C$.

simulation. Once the measured data points of (V_{ds}, V_{gse}, I_{ds}) are given, (i.e., $V_{gse} = V_{gs} - V_{t0}$ and V_{t0} is the zero-bias threshold voltage), we optimally partition the (V_{ds}, V_{gse}) plane and numerically fit a quadratic I_{ds} model in each region. For example, we have in the kth region,

$$\frac{I_{ds}}{\beta} = (\alpha_0^{(k)} + \alpha_1^{(k)} V_{gse} + \alpha_2^{(k)} V_{ds} + \alpha_3^{(k)} V_{gse}^2 + \alpha_4^{(k)} V_{gse} V_{ds} + \alpha_5^{(k)} V_{ds}^2), \quad k = 1, \ldots, n_r \tag{1}$$

where $\beta = \frac{1}{2}\mu_0 C_{ox} \frac{W}{L}$, n_r is the number of regions chosen for best fitting, and α's are fitting parameters in the kth region.

In (1), there are two strongly temperature-dependent physical parameters: μ_0 and V_{t0}. For comparison purpose, we use the same temperature dependence equations of μ_0 and V_{t0} as used in SPICE Level 3 model. We further assume that: (i) μ_0 and V_{t0} in (1) vary with temperatures; (ii) the regionwise partition of the RWQ model and the coefficients ($\alpha_0 \sim \alpha_5$) in each region remain unchanged for *certain interval* of temperatures. By default, users need to provide three sets of measured I-V data at T_1, T_2 and T_3 (T_1, T_2, T_3 are any user-specified temperatures). These data are then fitted by using the RWQ technique. We denote the RWQ-fitted models at temperature T by RWQ(T). In other words, each device is assigned with three RWQ models - RWQ(T_1), RWQ(T_2) and RWQ(T_3). Whenever the localized temperature T of a device satisfies $T_1 \leq T < T_2$ during the timing simulation, RWQ(T_1) is used and only μ_0 and V_{t0} are updated subject to T. Similarly, RWQ(T_2) or RWQ(T_3) is used if $T_2 \leq T < T_3$ or $T_3 \leq T$ is satisfied, respectively. The above procedure implicitly takes into account the suppressed degree of freedom of V_{t0} in the RWQ fitting process. It assures the modeling accuracy while still remaining the simple expression in (1).

To demonstrate the accuracy of temperature-dependent device modeling in iTHREAD, we use PSPICE to generate I-V data at 27 $^{\circ}C$, 77 $^{\circ}C$, and 127 $^{\circ}C$. The data are then RWQ-fitted. To find I_{ds} at, say, 50 $^{\circ}C$ and 100 $^{\circ}C$, RWQ(27) and RWQ(77) are used as discussed above. The modeling results are compared with PSPICE as shown in Fig. 2 and Fig. 3. Note that the above modeling procedure makes iTHREAD highly adaptive to advanced device technologies, especially for which only measured data are available and the device models are not well-built.

3. Analytical Thermal Simulation

The heat diffusion equation for an isotropic material is [5]

$$\nabla^2 T(\mathbf{r}, t) + \frac{g(\mathbf{r}, t)}{k} = \frac{1}{\alpha}\frac{\partial T(\mathbf{r}, t)}{\partial t} \tag{2}$$

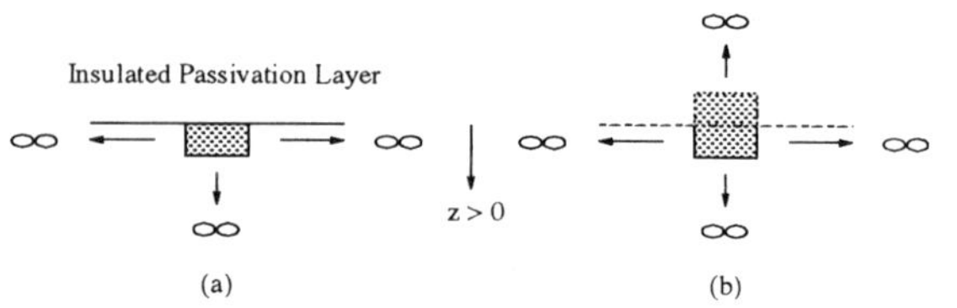

Figure 4: Method of images.

subject to the general boundary condition (BC):

$$k_i \frac{\partial T(\mathbf{r}, t)}{\partial n_i} + h_i T(\mathbf{r}, t) = f_i(\mathbf{r}, t) \tag{3}$$

where $T(\mathbf{r}, t)$ is the temperature, $g(\mathbf{r}, t)$ is the power density of the source(s) (W/m^3), k_i is the thermal conductivity ($W/m\ ^{\circ}C$) of the bulk, α is the thermal diffusivity (m^2/sec), h_i is the heat transfer coefficient ($W/m^2\ ^{\circ}C$), and n_i is the outward direction normal to the surface i. For steady-state case, the $\frac{\partial T(\mathbf{r},t)}{\partial t}$ term is zero.

To solve (2) exactly, both finite-difference numerical approach or infinite-series analytical approach can be applied. Due to the large number of heat sources (i.e., gates) in a VLSI chip, however, these methods are limited by their simulation efficiency. Here we propose to use the *closed-form* Green's function (GF) [6] method to approximate the on-chip temperatures. This approach is in light of the fact that the dimensions of the heat sources are small compared to the dimensions of the chip. In other words, all heat sources are considered to be located in an *infinite* body and the original boundary condition problems are greatly simplified.

Consider a heat source residing in a chip as shown in Fig. 4(a). Since ICs have a passivation layer, the top of the chip is assumed to be insulated in this case. We therefore have a boundary value problem with infinite body in the x-y plane while with semi-infinite body in the z direction. Moreover, the BC at $z = 0$ is $\frac{\partial T(\mathbf{r},t)}{\partial z} = 0$. To find the temperature subject to this specific geometry and BCs, we employ the method of images as in the analogous electromagnetics problems: We add an identical heat source symmetric with respect to $z = 0$ and remove the insulating boundary. Now the problem in Fig. 4(a) is transformed to that in Fig. 4(b).

The GF $G(\mathbf{r}, t|\mathbf{r}', \tau)$ for a point source in an infinite medium can be derived as $G(\mathbf{r}, t|\mathbf{r}', \tau) = G_x \cdot G_y \cdot G_z$, where

$$G_x = \frac{1}{[4\pi\alpha(t-\tau)]^{1/2}} \exp[-\frac{(x-x')^2}{4\alpha(t-\tau)}] \tag{4}$$

G_y and G_z can be obtained similarly. We therefore write the resulting temperature for a parallelepiped heat source with dimensions $a \times b \times c$ as

$$T(\mathbf{r},t) = T_a + \frac{\alpha P_0}{k(abc)} \int_{\tau=0}^{t} \int_{-c}^{c} \int_{-b/2}^{b/2} \int_{-a/2}^{a/2} G(\mathbf{r},t|\mathbf{r}',\tau) dv' d\tau \tag{5}$$

where the origin is set to be at the center of the source, T_a is the ambient temperature, and P_0 is the source power. We then further derive the temperature at $\mathbf{r} = \mathbf{0}$ by using error functions:

$$\begin{aligned} T(\mathbf{0},t) &= T_a + \frac{\alpha P_0}{k(abc)} \int_0^t \mathrm{erf}(\frac{a}{4\sqrt{\alpha\tau}}) \mathrm{erf}(\frac{b}{4\sqrt{\alpha\tau}}) \mathrm{erf}(\frac{c}{2\sqrt{\alpha\tau}}) d\tau \\ &= T_a + \frac{\alpha P_0}{k(abc)} \int_0^t \mathcal{G}(\tau) d\tau \end{aligned} \tag{6}$$

In order to perform the integration in (6) analytically, the error functions are approximated by $\mathrm{erf}(x) \approx 2x/\sqrt{\pi}$ if $x \leq \sqrt{\pi}/2$ and $\mathrm{erf}(x) \approx 1$ if $x \geq \sqrt{\pi}/2$ from the math table. As a consequence we obtain, for example,

$$\begin{aligned} \mathrm{erf}(a/4(\alpha\tau)^{1/2}) &\approx a/(2\sqrt{\alpha\pi\tau}) \\ &= (t_a/\tau)^{1/2}; \quad \tau \geq t_a \\ &\approx 1; \quad \tau \leq t_a \end{aligned} \tag{7}$$

where t_a has been defined as $a^2/4\pi\alpha$. Similarly, we define $t_b = b^2/4\pi\alpha$ and $t_c = c^2/\pi\alpha$ in order to facilitate the error function integration in (6).

By assuming $2c \leq b \leq a$ (i.e., $t_c \leq t_b \leq t_a$) without loss of generality, the steady-state temperature is carried out explicitly:

$$\begin{aligned} T(\mathbf{0},\infty) &= T_a + \frac{\alpha P_0}{k(abc)} (\int_0^{t_c} + \int_{t_c}^{t_b} + \int_{t_b}^{t_a} + \int_{t_a}^{\infty}) \mathcal{G}(\tau) d\tau \\ &= T_a + \frac{P_0}{\pi k a} \{2 + \ln(\frac{a}{b}) - \frac{c}{b}\} \end{aligned} \tag{8}$$

Equation (8) is the temperature at the center of the source due to the heat dissipation from *itself* only. For a structure with n heat sources as in real VLSI chips, the temperature at the center of source p should be obtained by superposing all single source solutions in the chip:

$$T_p = T_p(\mathbf{0},\infty) + \sum_{k=1}^{n-1} T_k(\mathbf{r}_k,\infty) - (n-1)T_a \tag{9}$$

where T_k is the temperature contribution from source k.

$T_p(\mathbf{0},\infty)$ in (9) can be found from (8). In order to find $T_k(\mathbf{r}_k,\infty)$, we propose the following approach. For a unit spherical-surface source of radius ρ' in an infinite medium, the steady-state GF is

$$\begin{aligned} G(\rho|\rho') &= \frac{1}{4\pi\rho'}; \quad \rho < \rho' \\ &= \frac{1}{4\pi\rho}; \quad \rho > \rho' \end{aligned} \tag{10}$$

By using (10), we find the temperature due to a spherical heat source i with radius $\mathcal{R}_i$ and power P_i to be

$$\begin{aligned} T_i(\rho_i) &= T_a + \frac{P_i}{4\pi k}\rho_i^{-1}; \quad \rho_i \geq \mathcal{R}_i \\ &= T_a + \frac{P_i}{8\pi k}(3\mathcal{R}_i^{-1} - \rho_i^2\mathcal{R}_i^{-3}); \; 0 \leq \rho_i \leq \mathcal{R}_i \end{aligned} \tag{11}$$

For a parallelepiped source shown in Fig. 5(a), we can approximate it by s spherical sources as shown in Fig. 5(b). The

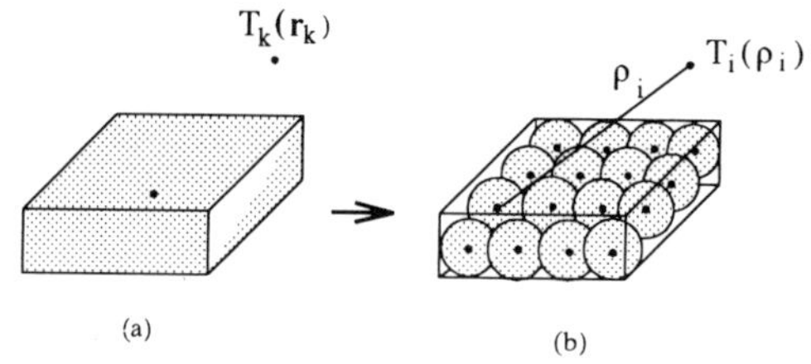

Figure 5: Approximate problem at near field.

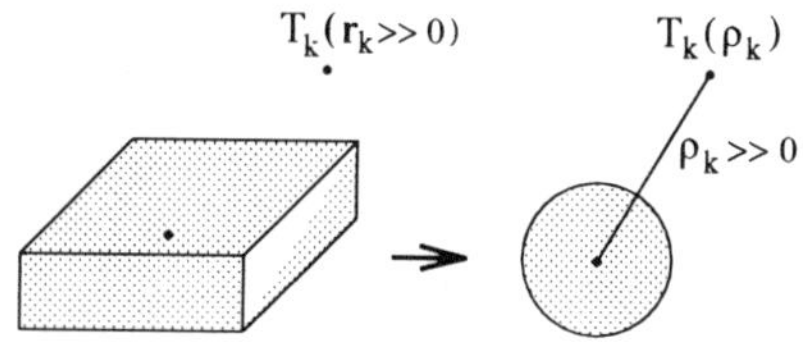

Figure 6: Approximate problem at far field.

temperature at $\mathbf{r}_k$ due to source k can be therefore determined by superposing the solution from the comprised spheres:

$$T_k(\mathbf{r}_k,\infty) = \sum_{i=1}^{s} T_i(\rho_i) - (s-1)T_a \tag{12}$$

For the position far from the source as shown in Fig. 6, we can further simplify the problem by treating the parallelepiped source as a single spherical one. By doing this, the simulation efficiency is even more enhanced by sacrificing little accuracy. With (8), (11) and (12), (9) now can be evaluated for every on-chip source p, $p \in \{1 \sim n\}$.

4. Experimental Results

In order to facilitate the testing of on-chip temperatures, a CMOS tester chip design has been sent to MOSIS for fabrication and packaging. Figure 7 shows the layout of the chip. Three shorter blocks are high frequency 3-stage ring oscillators which are designed in a standard super-buffer configuration to drive an off-chip load capacitance of 50 pF; two longer blocks are 149-stages ring oscillators; white dots are diodes. Here we call the 3-stage oscillators "Rosc3"s, and the 149-stage oscillators "Rosc149"s. Each ring oscillator has an enable signal which can activate or deactivate the oscillator. Since the operating frequency and the load capacitance of Rosc149s are smaller than those of Rosc3s, power dissipation due to Rosc149s is negligible in comparison to Rosc3s. We therefore only treat Rosc3s as heat sources. The elevated on-chip temperature can be determined either by measuring the voltage drop across the forward-biased diodes (i.e., according to $V_d = (kT/q)\ln(I/I_s + 1)$), or by measuring the frequency shift of Rosc149s (i.e., due to the increased voltage rise time/fall time at higher temperatures).

Table 1 shows the iTHREAD simulation results of the tester chip. Each block is labeled from I to V as shown in Fig. 7. In the experiments, we always keep Rosc149s activated while changing the power consumption of the chip by activating different Rosc3s. In Exp.1, only block III is activated. In Exp.2, block I and block III are activated. In Exp.3, all Rosc3s are activated. By doing so, the number of

Tester Chip	Chip size	n_{hsrc}	P^1_{avg}	T_{chip}	T_{blk2}	Freq. shift iTHREAD	Freq. shift SPICE	n^2_{run}	Timing sim. CPU time[3]	Thermal sim. CPU time[3]
Units	μm^2	-	Watt	^oC	^oC	MHz	MHz	-	sec	sec
Exp. 1	600 × 600	3	0.67	29.09	31.25	8.27 → 8.09	8.28 → 8.12	2	28.19	9.38
Exp. 2	600 × 600	6	1.34	33.08	38.71	8.27 → 7.85	8.28 → 7.97	3	42.57	28.11
Exp. 3	600 × 600	9	2.02	39.27	38.96	8.27 → 7.85	8.28 → 7.85	3	44.12	42.93

[1]Under steady-state temperature distribution. [2]Number of iterations under the convergence criterion: $\Delta T < 0.1\ ^oC$. [3]On SUN SPARC 10.

Table 1: iTHREAD simulation results of tester chip

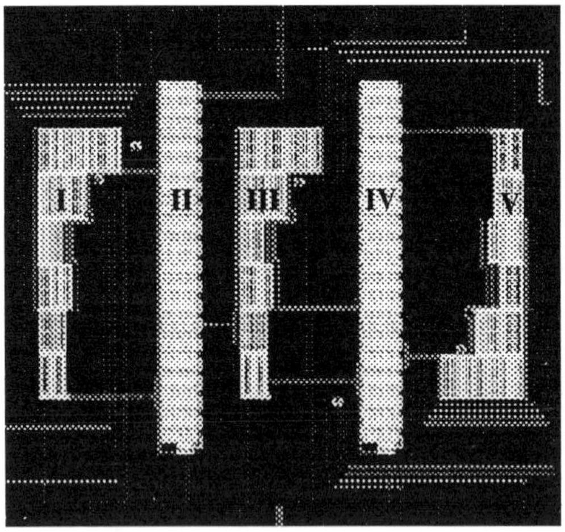

Figure 7: Layout of tester chip. Long blocks are Rosc149s and short blocks are Rosc3s. White dots are diodes.

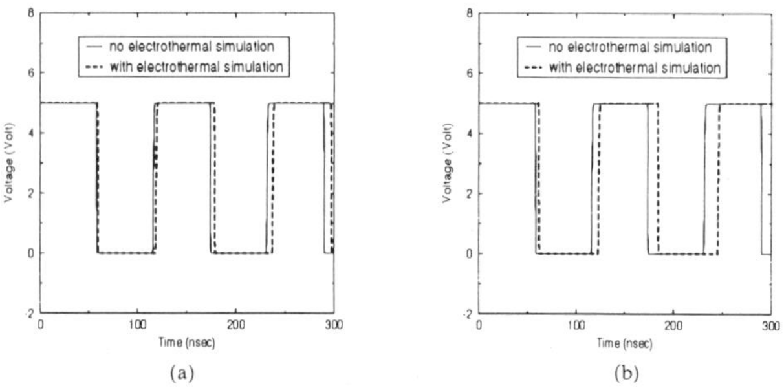

Figure 8: Frequency shift: (a) Exp. 1 (b) Exp. 3.

Tester Chip	Exp. 1	Exp. 2	Exp. 3
Grid Number (FD)	1250	1875	2625
CPU Time Saved	73.29%	76.48%	81.91%

Table 2: CPU time saved in thermal simulations.

heat sources (n_{hsrc}) and the average power (P_{avg}) of the chip vary consequently. In Table 1, T_{chip} and T_{blk2} represent the average temperature of the chip and block II, respectively. By comparing Exp.2 and Exp.3, it shows that T_{blk2} is hardly affected by the activation of block V. This verifies the concept of the heat diffusion length [3], beyond which the temperature rise due to the presence of a heat source is small. The frequency shifts of block II simulated by both iTHREAD and PSPICE are demonstrated in Table 1. In PSPICE simulation, block II is given the *constant* temperature T_{chip} by specifying the .TEMP statement in the input deck (i.e., the temperature gradient across the chip is ignored). This causes inaccurate simulation result compared to iTHREAD as expected, especially in Exp.2. The plots of frequency shift of block II for Exp.1 and Exp.3, both simulated by iTHREAD, are shown in Fig. 8. The last two columns of Table 1 demonstrate the superior CPU time usage of iTHREAD simulations.

Table 2 compares the CPU time of thermal simulations between current work and [1], where the finite-difference (FD) method was used. The assigned number of variable grids in the FD method increases with the number of heat sources. The maximum percentage error of iTHREAD-simulated temperatures with respect to those from the FD method (i.e., temperatures were compared all over the chip and the maximum difference was recorded) is within 5% in all cases, whereas the average percentage error is less than 2.5%. Table 2 shows that the CPU times saved by our approach are substantial. If the chip size or the source density increases, the expected CPU time saving will be even more.

5. Conclusions

In this paper, we presented a new fast thermal diagnosis tool for CMOS VLSIC reliabilities. Unlike existing simulators, iTHREAD is computationally very efficient. To achieve this goal, accurate temperature-dependent device models were developed for the fast power calculation. Moreover, closed-form analytical solutions of the heat equation were derived and implemented for the fast thermal simulation. Finally, steady-state on-chip temperature profile was found efficiently by using the uncoupled electrothermal simulation. With the information provided by iTHREAD, issues such as temperature-driven module placement and many temperature-dependent VLSIC reliabilities, can be addressed.

REFERENCES

[1] S. S. Lee and D. J. Allstot, "Electrothermal simulation of integrated circuits," *IEEE J. Solid-State Circuits*, pp. 1283–1293, Dec. 1993.

[2] V. Koval, I. W. Farmaga, A. J. Strojwas, and S. W. Director, "MONSTR: a complete thermal simulator of electronic systems," in *Proc. Design Automation Conf.*, pp. 570–575, 1994.

[3] Y. K. Cheng, P. Raha, E. Rosenbaum, and S. M. Kang, "Full-Chip Electrothermal Analysis," Nov. 1994. Research Report No. UI-94-11 submitted to Intel Corporation.

[4] A. Dharchoudhury, K. H. Kim, S. H. Lee, and S. M. Kang, "Fast and accurate timing simulation with RWQ models of MOS I-V characteristics," in *Proc. IEEE Int. Conf. Computer-Aided Design*, pp. 208–211, Nov. 1994.

[5] S. Kakac and Y. Yener, *Heat Conduction*. Taylor & Francis, 1993.

[6] G. F. Roach, *Green's Function*. Cambridge, 1982.

Over an Order of Magnitude DRAM Test Time Reduction

by Charge Offset

Masashi Hashimoto[1], Isao Uchida[2], and Seishi Hatakoshi[3]

Texas Instruments Japan Ltd.
[1] Design Center, Miho, Ibaraki (visiting Dallas)
[2] MOS Memory Product Engineering, Hiji, Oita
[3] MOS Memory Product Engineering, Miho, Ibaraki

Abstract

A circuit technique of reducing DRAM data retention test time by over an order of magnitude is described. The circuit requires almost zero area overhead. Data retention test time reduction is achieved by using a charge offset realized by manipulating a dummy cell reference voltage. The retention time parallel shift in logarithmic time domain with respect to charge offset was confirmed experimentally. Overall DRAM test time can be reduced by 2/3 since this amount of time is being utilized to perform the data retention test.

Introduction

A small capacitor fabricated on a silicon surface is used as a memory element in DRAMs. Existence or non-existence of charge in the capacitor determines a logical 1 or 0 level. As long as the charge in the capacitor is retained, a written logic level is read out properly. Unfortunately, the charge in the capacitor may be dissipated by several leakage mechanisms such as PN junction leakage, MOS capacitor oxide tunneling and/or memory cell isolation transistor (or pass gate) subthreshold leakage [1].

Charge loss beyond a critical level, caused by process parameter variation plus sense amp characteristics coupled with memory array circuit design and layout, causes an erroneous read operation or failure. The period of time to fail defines data retention time. In order to maintain the data in memory cell beyond its data retention time, DRAM requires a "refresh" operation.

In the refresh operation, reading data from the memory cell by the sense amp causes the memory cell charge to be replenished by feeding the sensed data back to the cell.

TABLE 1
DRAM by-generation refresh overhead.

DRAM Size	Refresh Time	Number of Rows to be Refreshed
16K	2ms	128
64K	4ms	256
256K	4ms	256
1M	8ms	512
4M	16ms	1024
16M	64ms	4096
64M	64ms	8192

Since only one row can be refreshed by one refresh cycle, the overhead of refreshing tends to get worse as memory size increases. Therefore a longer refresh cycle time, the time needed between refreshing the charge in a memory cell, is desired in terms of DRAM utilization efficiency in a system.

Refresh is a major component of DRAM stand-by power. Longer refresh cycle time is also desirable to achieve reduction in stand-by power.

Table 1 shows DRAM generation to generation refresh time together with the number of rows to be refreshed within the refresh period. The cell data retention time must be long enough to satisfy the refresh cycle time requirement. The refresh time has either improved or stayed constant with each succeeding DRAM generation, hence the data retention time has improved by generation. This improvement, however, is causing DRAM test time to increase (note that about 2/3 of the test time is being occupied by the data retention test). In addition, DRAM density is growing at a steady pace. DRAM test time is therefore becoming a crucial concern.

In this paper, a circuit technique for reducing DRAM data retention test time is introduced. Next, a conventional test method of data retention time is

0-7803-3177-6 $5.00 © 1996 IEEE

IEEE 1996 CUSTOM INTEGRATED CIRCUITS CONFERENCE

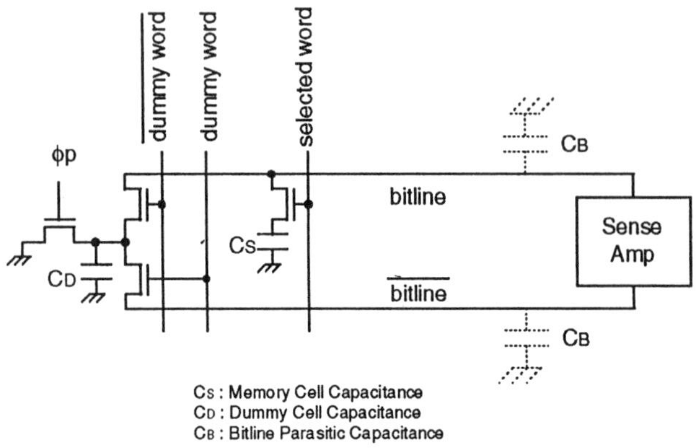

Figure 1: Dummy Cell Associated Memory Array

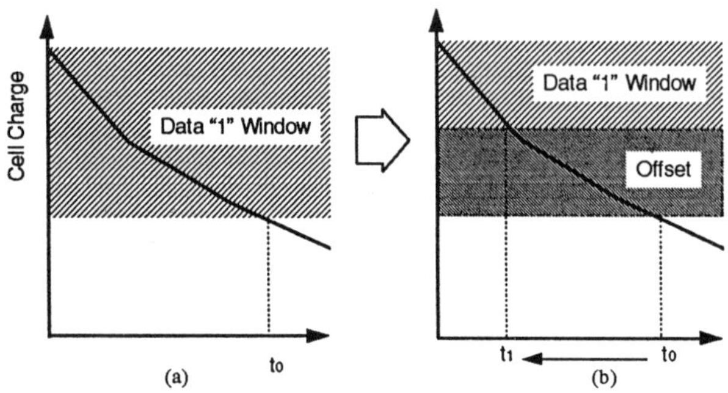

Figure 2. Memory Cell Data Retention Time Conceptual Figure without offset (a) and with offset (b).

described followed by the circuit technique for achieving data retention test time reduction. Some experimental results will be shown in the last section.

DRAM Data Retention Test Flow

In order to ensure the DRAM refresh cycle time specification, two tests are usually performed. One is "Simple pause time test", the other is "Pause test with disturbance pattern".

DRAM memory cells are first examined by a simple pause time test. In this test, a topological checkerboard of 0/1 data written at the memory cell layout level is often employed. After completion of writing the checkerboard pattern, the device is left alone for a predetermined time interval that should be long enough to assure the refresh period. Then, data in the memory cells are read out for verification. Usually, a memory cell having a relatively high level of leakage is detected by this test. Since all cells are examined at one time, required test time is in a practical range.

The second test is the most time consuming in DRAM testing. After writing data to the memory cell under test, the most effective disturb conditions (determined by design parameters and prior DRAM testing/usage knowledge) to induce cell leakage are applied to that cell. The disturbances can be simultaneously applied to limited number of cells only. Therefore, with increasing memory size, the pause time test with disturbance pattern gets long.

Charge Offset Circuit

A differential amplifier is used in DRAMs to sense a small voltage on a bitline. By using a double ended bitline configuration, one of the bitline pairs is used as a data line and the other as a reference. Because the cell leakage for the condition of holding data "1" differs from that of holding data "0", data read margin equalization is achieved by adjusting a reference level.

In general, cell leakage happens when holding data at a "1" in a memory cell formed on P-substrate. For that case, the reference level is shifted down by a dummy cell to improve the data read margin of the logic level "1". Fig.1 shows an example of a circuit that can be used to achieve the improved "1" read margin.

Since the concept of the charge offset is an extension of the concept of the data read margin equalization, the reference level adjustment basic operation will be briefly described.

Prior to a read operation, the bitline pair is precharged to half of the power supply operation voltage, $V_{DD}/2$, and the dummy cell is discharged to ground level. The resultant voltage difference on the bitline pair after activating the selected word and the dummy word can be calculated from a simple charge conservation law. The final voltage differences for the conditions of holding a data "1" and a data "0" are given by;

$$\delta V_H = \frac{V_{DD}}{2} \times \left(\frac{C_S}{C_B + C_S} + \frac{C_D}{C_B + C_D} \right) \quad \cdots\cdots\cdots \text{eq.1}$$

$$\delta V_L = \frac{V_{DD}}{2} \times \left(\frac{-C_S}{C_B + C_S} + \frac{C_D}{C_B + C_D} \right) \quad \cdots\cdots\cdots \text{eq.2}$$

The value of C_D is determined from eq.2. For the condition of no cell leakage at the data "0" level (again, this is the case in general), a safe minimum voltage to use for sensing data "0" by the sense amp is δV_L. The safe minimum voltage of course depends upon process parameter deviation [2], sense amp architecture [3] [4],

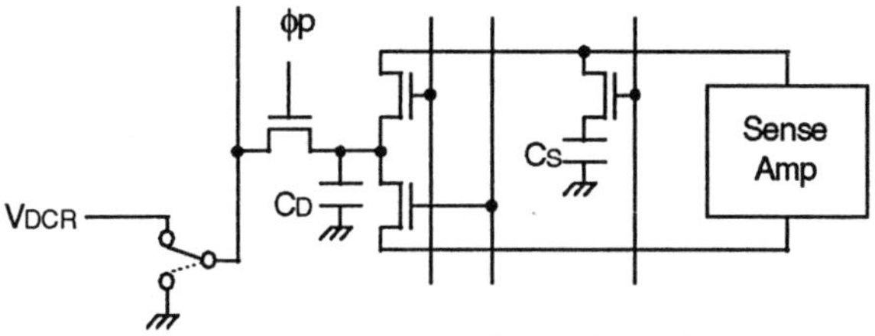

Figure 3. Memory Array with Charge Offset Concept

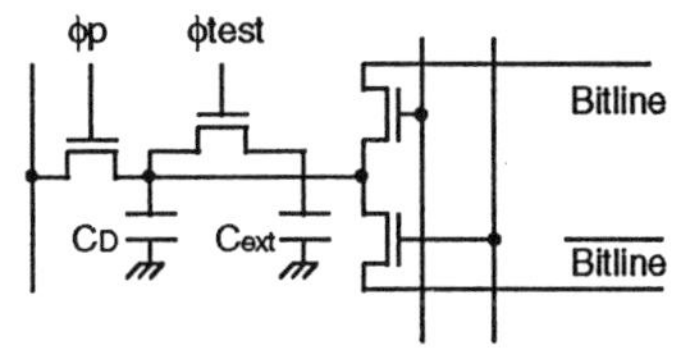

Figure 4. Modified Dummy Cell

topological asymmetry, etc. From the results of the calculation , the capacitance value C_D is typically 30% to 50% of C_S.

As is shown in Figure 2 (a), charge in the memory cell leaks with time (conceptual figure). As long as cell charge is in the data "1" window range, the data can be read out correctly. A pause time of at least t_0 is required before an erroneous read will occur.

A narrower data "1" window results in a shorter pause time. The charge offset idea is to narrow the data "1" window only during the data retention time test. Figure 2 (b) illustrates the idea. An erroneous read occurs at time t_1. Since a cell having data retention time shorter than t_0 may show a read "1" fail before time t_1 using the charge offset condition (and this is the real situation as will be discussed in the next section), the time t_1 can be used for screening cells with the condition of data retention time t_0.

The charge offset is achieved by supplying V_{DCR} to the dummy cell during dummy cell precharging. Figure 3 shows a schematic of the circuit. Resulting voltage differences on the bitline pair for data "1" and data "0" are given by;

$$\delta V_{Htest} = \delta V_H - V_{DCR} \times \frac{C_D}{C_B + C_D} \quad \text{........ eq.3}$$

$$\delta V_{Ltest} = \delta V_L + V_{DCR} \times \frac{C_D}{C_B + C_D} \quad \text{........ eq.4}$$

The second term in the above two equations represents the charge offset. An arbitrary offset can be made by changing V_{DCR} externally. This avoids design iteration work for the offset circuit.

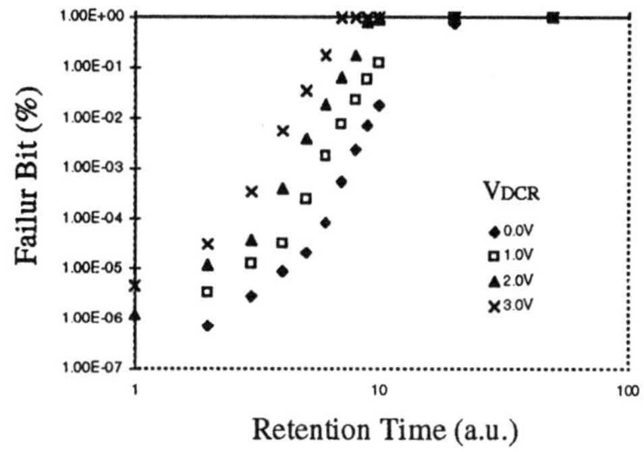

Figure 5. Data Retention Time VDCR Dependency

Although the dummy cell size is pre-determined at chip design, the offset value is adjusted by using the V_{DCR} level. An extra capacitance can be added to C_D during test mode by making a simple logic change. Figure 4 shows an example of the circuit. A steep offset can be achieved by adopting this dummy cell scheme.

Note that V_{DCR} and ϕ_{test} will be activated only during the data retention test. Therefore, use of the offset circuit will not disturb normal DRAM operation. The major elements required to realize charge offset are already embedded in the conventional design. The logic implementation overhead is minor.

Experimental Results

The charge offset concept was examined by using 4Mb BRAM (Block Random Access Memory), one of TI's ASM (Application Specific Memory) family.

Figure 5 shows the failure bit data retention time dependency under different V_{DCR} conditions. As shown, data retention time parallel shift in logarithmic time domain with respect to V_{DCR} was observed. The condition at V_{DCR} = 0V corresponds to a performance of data retention of the device under the normal operation.

It is known that the time dependent fail bit data retention curve is composed of two sections, the steep main section and the relatively gentle tail slope section [5]. The memory cells corresponding to the tail section have a somewhat different leakage mechanism than the cells in the main section [6]. Ideally, devices containing the tail section of the curve are rejected at device test. However, from a practical perspective, as long as memory cells in the tail section meet the data retention spec, they are shipped as good bits. For this very reason, data retention time of DRAMs is governed by the bits in the tail section. Considering this, the data retention time marginal bits, those in the curve tail section, were carefully examined. The worst ten bits' retention time VDCR dependency were traced for several devices. All these bits followed the same VDCR dependency as that observed for good bits with a statistical scattering error also similar to that of the good bits.

It should be noted that the measurement was performed at a voltage far below that of normal device operation. The amount of charge held in the memory cell is proportional to the applied voltage. Lower operating voltage makes initial charge in the memory cell less. This leads to a shorter data retention time. This is a commonly employed scheme to shorten the duration of memory cell data retention time test. This scheme, however, raises a concern about improper response of sense amps and peripheral circuitry. There is a possibility of widening or narrowing the data "1" readable window not because of cell charge retention but because of the circuit response. A positive aspect about the charge offset based data retention test time reduction is that the test time reduction is achieved without the need to change the operating voltage.

In our measurements, lowering the operating voltage was not for the purpose of test time reduction but instead was a means to study offset charge limitations since the circuit for adding extra capacitance (as shown in Figure 4) was not implemented in the 4M BRAM. The charge offset was limited to the pre-determined dummy cell capacitance. In order to achieve a further charge offset, an extreme operating voltage condition was employed as the second best approach.

The data retention time at VDCR = 3V became 1/3 of that at VDCR = 0V. Theoretically, the retention time can be reduced to an infinitesimal value by increasing the offset. By taking into account the uniform consistency of the tail bits' retention time VDCR dependency for the extremely low operating voltage (low charge in the memory cell), over an order of magnitude retention test time reduction should be achieved .

Conclusion

It was shown that the charge offset concept was a very effective way of reducing DRAM memory cell data retention test time. The experimental results indicate over an order of magnitude test time reduction. Since the data retention test time consumes 2/3 of total DRAM testing time, overall DRAM test time is expected to be reduced by this amount.

It was confirmed that there was a good consistency in the worst ten bits' data retention time charge offset dependency with other bits. The data retention time was shifted in parallel in logarithmic time domain with respect to charge offset.

The charge offset concept requires no extreme test conditions such as very low operating voltage or very high temperature. Therefore, concerns about sense amp or other circuitry related data retention time degradation by using extreme conditions is negated. Yield overkill or test escape problems are fully avoided.

Since most of the elements that are required in this charge offset circuit already exist in current DRAM, very little overhead is required to implement the idea.

This charge offset can be used as a tool to determine sense amp sensitivity, optimize cell capacitance and obtain other information required during the initial stages of device development. Expensive design iterations can be avoided.

Acknowledgment

The authors wish to thank M.Nomura and his design team for the circuit implementation, T.Yoshii for device measurement support, S.Kim for a helpful discussion, B.Strong for reviewing and proofreading the paper, N.Kitagawa and H.Nakano for their strong support on this study.

References

[1] W.F.Richardson "Trench Capacitor Leakage, Mechanisms and Measurements on 16Mb DRAMs" *TI Technical Report,* No. 03-91-02.

[2] H.Geib, W.Weber, E.Wohlrab, and L.Risch "Experimental Investigation of the Minimum Signal for Reliable Operation for DRAM Sense Amplifiers", *IEEE J. Solid-State Circuits*, vol. 27, pp.1028-1034, Jul. 1992.

[3] R.Kraus, and K.Hoffmann "Optimizaed Sensing Scheme of DRAM's", *IEEE J. Solid-State Circuits*, vol. 24, pp.895-899, Aug. 1989.

[4] R.Sarpeshkar, J.L.Wyatt Jr., N.Lu, and P.Gerber "Mismatch Sensitivity of a Simultaneously Latched CMOS Sense Amplifier", *IEEE J. Solid-State Circuits*, vol. 26, pp.1413-1422, Oct. 1991.

[5] S.S.Kim and W.Wijaranakula "The Effect of the Crystal Grown-in Defects on the Pause Tail Characteristics of Megabit Dynamic Random Access Memory Devices", *J. Electrochem. Soc.*, vol. 141, pp.1872-1878, Jul. 1994.

[6] M.Ogasawara et al., "Physical Model of Bit-to-Bit Variation in Data Retention Time of DRAMs", *53rd Annual Dev. Res. Conf. Digest*, pp.164-165, Jun. 1995.

Scan Paths through Functional Logic

Chih-chang Lin, Malgorzata Marek-Sadowska, Kwang-Ting Cheng, and Mike Tien-Chien Lee*

Electrical and Computer Engineering
Univ. of California
Santa Barbara, CA 93106

Fujitsu Laboratories of America*
3350 Scott Blvd., Bldg #34
Santa Clara, CA 95054

ABSTRACT

Conventional scan design imposes considerable area and delay overhead due to the use of larger flip-flops and additional connections between flip-flops. We propose a low-overhead scan design methodology which exploits the possibility of utilizing input vectors and the test-mode point insertion technique to establish scan paths through the combinational logic. The technique re-uses the existing functional logic; as a result, the DFT overhead can be reduced.

I. Introduction

Automatic test pattern generation (ATPG) for sequential circuits is a difficult problem [1, 2] because of the lack of direct controllability of the present state lines and direct observability of the next state lines. Design-for-Testability (DFT) techniques aiming at improving controllability and observability of the state lines have been proposed, such as full scan [3, 4] and partial scan [5, 6, 7, 8, 9]. However, the area and delay overheads imposed by conventional scan can be significant, due to the extra scan multiplexers (MUXs) in the scan flip-flop (assuming the MUXed D flip-flops are used) and routing area for the scan chains.

To alleviate the above DFT penalty, we proposed low-overhead scan design methods which exploit the controllability of primary inputs and test-mode point insertion technique to establish scan paths through the existing combinational logic. A scan path is defined as a physical path between two flip-flops that can be fully sensitized in the test mode. The scan design methodology called, *free-scan* [10], is based on an observation that by setting appropriate values at primary inputs during the test mode, some combinational paths between flip-flops can be sensitized and thus a portion of the scan chain can be established without any DFT overhead. In [11] we proposed a test-mode point insertion technique which further extends the free-scan technique and utilizes more existing combinational logic for establishing the scan paths. In this paper, we study the relation of finite state machine (FSM) encodings with the free-scan design and show an algorithm which integrates the free-scan and test-mode point insertion methods together.

Our essential ideas of designing scan paths through combinational logic can be illustrated in Figure 1 and Figure 2. In Figure 1, a scan path from $F2$ to $F1$ is formed through the combinational logic (indicated by the dashed line) by assigning 0 and 1 at the inputs $x1$ and $x2$, respectively. Since $F1$ costs no scan overhead for the scan order from $F2$ to $F1$, $F1$ is called a free-scan flip-flop. Figure 2 shows an example of a test-mode point insertion. The test-mode points are established by ap-

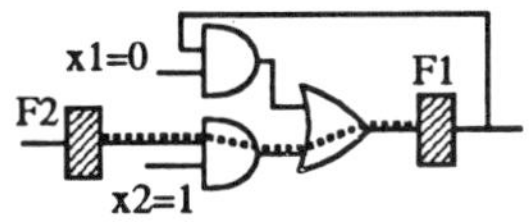

Fig. 1. Example of a sequential circuit. A free-scan path (in dashed line) is established from $F2$ to $F1$ by $(x1x2) = 01$.

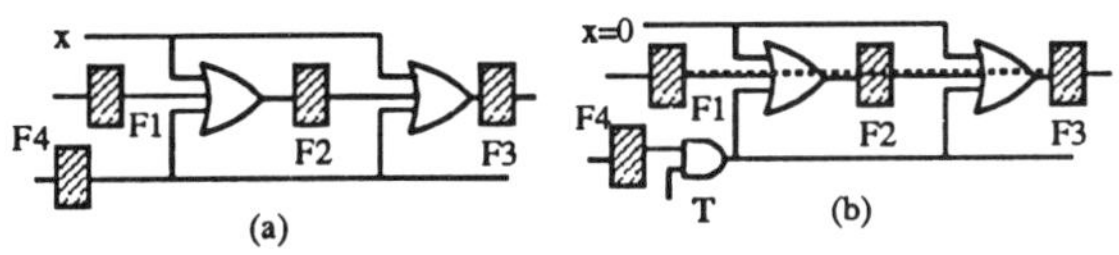

Fig. 2. Example of a test-mode point insertion.

propriately inserting two-input AND or OR gates with a common test-mode input. Figure 2(a) shows a portion of a sequential circuit. By inserting a test-mode point at the output of $F4$ and setting the primary input x to 0 during the test mode, a scan chain $F1 \rightarrow F2 \rightarrow F3$ can be formed through the combinational logic, as shown in the dotted line of Figure 2(b).

In the following, we review some related previous works. The works in [12, 13] presented algorithms to reduce scan overhead by attempting to merge scan MUXs into the combinational logic during logic synthesis. In [13], the concept of *test synthesis constraints* was proposed, where a mechanism was used to describe the conditions which must be met by the synthesized circuit in order to be fully testable. The concept of embedded scan was also proposed where attempts were made to embed the multiplexers for scan into the logic immediately preceding the scan flip-flops. A technique of testpoint insertion to improve the sequential ATPG fault coverage was proposed in [14]. This technique is different from the scan method. The idea was to insert a set of test-cells into a circuit to improve the observability and controllability of some internal signals. The size of a test-cell could be large and the compound effect of adding such cells may result in significant area overhead (a test-cell requires at least one flip-flop and two multiplexers). On the other hand, the test-mode point used in [11] is simply a two-input AND or a two-input OR gate and the purpose of inserting test-mode points is to establish scan paths by utilizing existing logic, which in turn makes scanned flip-flops fully observable and controllable.

II. Free-scan design

To utilize the concept of free-scan, we have to analyze the entire circuit to determine all the free-scan flip-flops and the associated input assignments, and then select the best input assignment to maximize the number of free-scan flip-flops for constructing the scan chain. In the following, we assume a sequential circuit has primary inputs

0-7803-3177-6 $5.00 © 1996 IEEE

X and n flip-flops, $F_1, \cdots, F_n$. The input and output of flip-flop F_i are denoted as z_i and y_i, respectively.

A. Enabling vectors

Given a pair of flip-flops F_i and F_j, the set of input vectors whose assignments at the primary inputs X will make z_i depend only on y_j, regardless of the values of other present state lines is the on-set of the following Boolean function [10]:

$$K^{j,i}(X) = \mathcal{C}_Y(f(X,Y) \equiv y_j) + \mathcal{C}_Y(f(X,Y) \equiv \overline{y_j}),$$

where $Y = \{y_1, \cdots, y_n\}$ is the set of the present state lines and $f(X,Y)$ represents the Boolean function of z_i. The first (second) term represents all the minterms whose assignments make z_i equal to y_j ($\overline{y_j}$). The symbol $\mathcal{C}_Y$ denotes Boolean consensus operator. Notice that we only need to compute $K^{j,i}(X)$ for flip-flops where there exists combinational paths between them.

B. The best enabling vector X^* selection

To find the best enabling vector X^*, [10] attempts to find the one that maximizes the number of 1's in $K^{j,i}(X)'s$. However, because a flip-flop can at most have one successor in the scan chain, it is essential to merge the set of input patterns $K^{j,i}(X)'s$ which make some flip-flops F_i's depend only on flip-flop F_j, i.e., $K^j(X) = \bigvee_{i=1}^{n} K^{j,i}(X)$. As a result, finding an enabling vector to maximize the number of 1's in $K^j(X)'s$ models more closely the number of free-scan paths in the final scan chain. Besides, the BDD computation is simpler also.

We can use ROBDD [15] technique to implicitly enumerate all the input patterns. An auxiliary Boolean variable q_j is assigned to represent $K^j(X)$ and the characteristic function, $\mathcal{X}(X, q_1, \cdots, q_n) = \prod_{j=1}^{n}(q_j \equiv K^j(X))$, captures the relation of X and $K^j(X)$. A depth-first traversal of $\mathcal{X}$'s BDD graph can determine the best enabling vector.

C. Cycle-free scan paths construction

After finding the best enabling vector X^*, we check whether free-scan path between any pair of flip-flops F_i and F_j is establishable and construct a graph $G(X^*)$ where each vertex corresponds to a flip-flop and an edge to a free-scan path. If under X^*, a free-scan path can be established (i.e., $K^{j,i}(X^X)$ is 1); an edge from vertex F_j to vertex F_i is added. The graph $G(X^*)$ constructed in this manner may have cycles, but the scan chain must be acyclic. Therefore, to break all the cycles in $G(X)$, some flip-flops need to be converted into MUXed scan flip-flops. The problem of finding such a minimum set is referred to as the *minimum feedback vertex set* (MFVS) problem, and in general is an NP-hard problem [8]. Fortunately, $G(X^*)$ has the following properties by which a simple linear-time algorithm can be developed to break all the cycles optimally.

Theorem 1 *[10] The in-degree of a node in $G(X)$ must be at most 1. Any disjoint component of $G(X)$ can have at most 1 cycle.*

III. State encoding and free-scan design

In this section, we discuss the relation between FSM state encodings and free-scan design. Given a finite state machine FSM (represented by its state transition graph), if dense encoding techniques like Jedi or Nova [16] are used, intuitively, the obtained sequential circuit will have very little chance for finding any free-scan path. This is because the states are encoded in such a way that state variables strongly depend on each other. On the other hand, if we employ one hot encoding, the potential gain for free-scan design is high, since the original state transition graph reflects the flip-flop dependence directly. For a one-hot encoded circuit, we found that, under an particular input assignment, a large number of flip-flops depend on exactly one other flip-flop. For example, Figure 3 shows the resulting state transitions under the input assignment 0000000 for the MCNC benchmark *keyb* FSM. 15 out of 19 flip-flops are free-scannable. Notice that, *st0* depends on two states *st17* and *st18*, thus *st0* is not free-scannable. Similarly, *st14* neither.

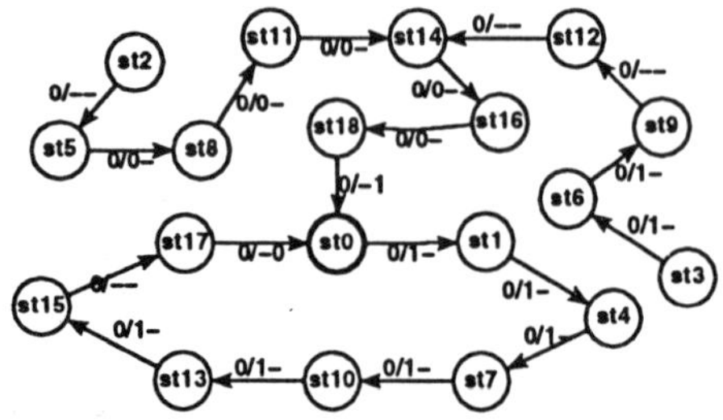

Fig. 3. State transition graph of MCNC 91 benchmark *keyb* under the input assignment 0000000 (we use 0 to denote 0000000). Under one-hot encoding, each state corresponds to a flip-flop. Except *s11*, *s12*, *s17* and *s18*, all other flip-flops are free-scannable. Thus, the number of free-scan flip-flops is 15.

TABLE I
Experimental results for MCNC91 benchmark.

circuit	#FF/B	#FF/A	#free	saving
dk16	27	27	7	26%
planet	48	48	37	77%
pma	24	24	10	42%
s1488	48	48	37	77%
s1494	48	48	37	77%
s298	218	135	63	47%
s510	47	47	46	98%
s820	25	24	15	62%
s832	25	24	15	62%
sand	32	32	23	72%
scf	122	97	64	66%
styr	30	30	13	43%
total	694	584	367	63%

We tested the free-scan method on all MCNC91 FSM benchmarks using one-hot encoding. The results on FSMs with more than 20 states are shown in Table I. The second (#FF/B) and third (#FF/A) columns show the number of states before and after state minimization by SIS-1.2 [16], while the fourth column (#free) shows the number of free-scan flip-flops. On the average, 63% of the flip-flops are found free-scannable.

IV. Test-mode point insertion

In Section II, we discussed how to find an input assignment to maximize the number of free-scan flip-flops. In this section, we show a new technique, called test-mode point insertion, to further utilize the combinational logic to increase the number of scan paths without direct scan flip-flop substitution.

For a pair of flip-flops connected by a combinational path, to include this path into a scan chain, all the side-inputs along this path have to be disabled (i.e., to be assigned to non-controlling values). If a value *zero* is desired at a connection c for disabling the connection during the test mode, we can insert a two-input AND gate at c with the test-mode input T as one of its inputs. The value of T is assumed to be 1 in the normal mode and 0 in the test mode. On the other hand, if a value *one* is desired, we can insert a two-input OR gate with T' as one of its inputs, where T' is the negation of T.

In general, assigning a constant value at a connection (by inserting a test-mode point) may potentially disable more than one side-input because the connection may have multiple fanouts. The circuit's topology has to be analyzed to determine the global effect of inserting a particular test-mode point, so that we can establish as many scan paths with as few test-mode points as possible.

A. Path matrix construction

Given a sequential circuit, we build a sparse matrix A, where the entry A_{ij} represents a set of combinational paths from flip-flop F_i to flip-flop F_j. Since there might exist a huge number of paths in the circuit and in general it is more costly to establish a scan path using a combinational path with a large number of side-inputs, we heuristically limit the number considered paths and record only those paths with a number of side-inputs smaller than a user specified upper bound K_{bound} to save the computation time.

Given a combinational path p_k in A_{ij}, let $|p_k|$ denote the number of side-inputs along this path. During the iteration of test-mode point insertion and the forward implication of the assigned constants, side-inputs of p_k might have sensitizing, controlling or unknown values. If there exists a side-input having a controlling value, it will be impossible to build a scan path through it. We call such a path a *nullified* path and remove it from A_{ij}. On the other hand, if there is no side-input with a controlling value, we use w_k to denote the number of side-inputs with an unknown value. To make a path become a scan path, all the side-inputs must have sensitizing values. The gain of setting one of the side-inputs to a sensitizing value is $1/w_k$. Notice that, for each path p_k, the number $|p_k|$ does not change while w_k is decreasing during the process. When w_k is reduced to zero, the path p_k successfully becomes a scan path.

B. Effect analysis

Given a connection c, if we insert a test-mode point with value v at c, forward implication of v at c may imply new values v_i's at some connections c_i's in c's fanout cone. We denote them as $\{(c_1, v_1), \cdots, (c_h, v_h)\}$. With these new constant values, paths passing through $c, c_1, \cdots, c_{h-1}$, or c_h will be nullified. Also, paths with c or c_i's as side-inputs and v or v_i's being controlling values will be nullified. On the other hand, paths with c or c_i's as side-inputs and v or v_i's being sensitizing values will have their w_k's reduced. We denote the set of such paths as S_c. Thus the gain of inserting a test-mode point with value v on c is as follows:

$$\sum_{j=1}^{n}(\text{MAX}_{i=1}^{n}(\text{MAX}_{p_k \in A_{ij}, p_k \in S_c} \frac{1}{w_k})), \qquad (1)$$

where n is the number of flip-flops. We sum the contribution of making flip-flops F_j's as a part of the scan chain. Among all the paths in A_{ij}'s ending at a flip-flop F_j, we choose the maximal contribution instead of their summation because our objective is to establish exactly one path from a flip-flop to flip-flop F_j in the scan chain.

C. Greedy insertion

Based upon the cost function in Equation 1, we can iteratively choose a connection and a value (c, v) with the highest gain as a test-mode point and update the entries A_{ij}'s in the matrix A by removing the nullified paths. During the iteration of the greedy insertion process, if the scan path $F_i \rightarrow F_j$ is established, we record this path as part of the final scan chain. Since the scan chain has to be acyclic, we also remove some entries A_{pq}'s if adding the path $F_p \rightarrow F_q$ to the scan chain would result in a cycle.

V. Overall algorithm: TPGREED2

In this section, we discuss an algorithm TPGREED2 which combines the free-scan input assignment method and test-mode point insertion technique to establish the scan paths through the existing combinational paths. The overall algorithm can be divided as following steps:

1) free-scan input assignment.
2) support reduction input assignment.
3) test-mode point insertion:
   - 3.1) construct path matrix with bound K_{bound}.
   - 3.2) analyze effects and choose a test-mode point with the highest gain.
   - 3.3) if gain is greater than lower bound gain$_{bound}$ update path matrix goto step 3.2
   - 3.4) remove redundant test-mode points.
4) post-input assignment for test-mode point reduction.

The first step performs input assignment to maximize the number of free-scan flip-flops (see Section II). The established free-scan paths and input assignment are recorded. Next, before going to test-mode point insertion, we use the remaining un-assigned inputs to reduce the dependencies between flip-flops. The goal here is to let a flip-flop depend on at most two other flip-flops. The detail procedure is similar to the one used by free-scan design where a flip-flop depends on at most one other flip-flop.

The third step performs greedy insertion of test-mode points. This procedure requires two parameters K_{bound} and gain$_{bound}$. K_{bound} is used to limit the number of side-inputs for paths considered for establishing scan paths, while gain$_{bound}$ is used to terminate the algorithm when the highest gain computed by Equation 1 for all candidate connections is smaller than gain$_{bound}$. During the iteration, some scan paths may be established. Besides adding them as a portion of the scan chain, we also have to make sure that the subsequent insertions will not destroy the previously established scan paths. Since the insertion is iterative, the exact effect of inserting a test-mode point cannot be computed until the end of insertions. As a result, there might exist redundant test-mode points. In step 3.4, we perform redundant test-mode points removal to eliminate those unnecessary test-mode points.

In the fourth step, after performing the greedy insertion, we know exactly at which connections (i.e., $c_1, \cdots, c_m$) test-mode points should be inserted and also what values (i.e., $v_1, \cdots, v_m$) they should have. Before physically inserting AND (for value 0) or OR (for value 1) gates, we make attempts to set up as many the values v_i's as possible at the connections c_i's by assigning appropriate values at those un-assigned inputs to avoid inserting of unnecessary test-mode points.

TABLE II
EXPERIMENT RESULTS FOR ISCAS89 AND MCNC91 CIRCUITS.

		free-scan			free-scan + test-mode point			
circuit	#FF	#free	save	CPU	#T	#scan	save	CPU
s1423	74	5	7%	665s	22	20	12%	150s
s5378	152	48	32%	142s	22	65	36%	321s
s9234	135	30	22%	229s	15	60	39%	268s
s13207	453	96	21%	509s	99	218	37%	1412s
s15850	540	114	21%	3764s	40	214	36%	1016s
s35932	1728	1431	83%	1592s	0	1431	83%	328s
s38417	1636	324	20%	2376s	132	519	33%	10519s
s38584	1294	1134	88%	3577s	0	1134	88%	3841s
bigkey	224	0	0%	168s	112	112	25%	534s
clma	33	0	0%	1236s	0	0	0%	3754s
dsip	224	220	98%	146s	0	220	98%	283s
mult32a	32	31	97%	961s	0	31	97%	992s
mult32b	61	1	2%	13s	30	31	26%	1278s
sbc	27	3	11%	11s	7	8	17%	59s

VI. Experimental results

We tested the proposed algorithm TPGREED2 on a number of ISCAS89 and MCNC91 sequential benchmarks. All circuits are optimized by SIS *script.algebraic* script and mapped using technology libraries *nand-nor.genlib* and *mcnc_latch.genlib* for minimal area. In the current implementation, we can only handle primitive gates, including AND, OR, NAND, and NOR gates.

The results of test-mode point insertion are shown in Table II. The CPU time (second) is measured on a SUN SPARC 5 with 128 Megabyte memory. In our experiments, the parameters K_{bound} and gain_{bound} are set to 10 and 0.5, respectively. The columns #FF and #free show the number of flip-flops in the circuits and the number of free-scannable flip-flops, respectively. The column #T indicates the number of test-mode points inserted. For example, the circuit *s15850* has 540 flip-flops. Among them, 114 flip-flops are free-scannable. We inserted 40 test-mode points during the iteration to establish totally 214 scan paths.

The cost of inserting a test-mode point is one AND (OR) gate and a connection from the test-mode input T, while converting a flip-flop into a MUXed scan flip-flop requires a multiplexer, a connection from another flip-flop, and a connection from the test-mode input T. Inserting test-mode points is advantageous in terms of area. Assuming that the area costs of inserting a multiplexer and a test-mode point are 2 and 1, respectively, the reduction of area overhead will be

$$1 - \frac{2(\#\text{FF} - \#\text{scan}) + \#\text{T}}{2\ \#\text{FF}}.$$

The term (#FF-#scan) represents the number of remaining flip-flops which requires a multiplexer for each of them. The amount of saving depends on a circuit's structure, the logic synthesis algorithm, and our algorithm. In the case of *s35932*, as much as 83% in saving of the area overhead can be achieved by free-scan design alone. In the case of *bigkey*, there are no free-scannable flip-flops. Using test-mode point insertion, we obtain a 25% saving in the area overhead.

VII. Conclusion

We propose low-overhead scan design methods which exploit the use of a primary input vector and employ the test-mode point insertion technique to establish scan paths through the combinational logic. Applications for reducing area overhead are addressed and the experimental results demonstrate its usefulness.

Acknowledgements

This work was supported under NSF grant MIP9419119 and MICRO/Fujitsu.

References

[1] F. C. Hennie, "Fault detecting experiments for sequential circuits," *Proc. 5th Ann. Symp. Switching Circuit Theory & Logic Design*, pp. 95–110, 1964.

[2] A. Miczo, *Digital Logic Testing and Simulation*. New York: Harper and Row, 1986.

[3] M. J. Y. Williams and J. B. Angell, "Enhancing testability of large scale integrated circuits via test points and additional logic," *IEEE Trans. on Computers*, vol. C-22, pp. 46–60, Jan. 1973.

[4] V.D. Agrawal, S.K. Jain and D.M. Singer, "Automation in design for testability," *IEEE Custom Integrated Circuits Conference*, pp. 159–163, 1984.

[5] E. Trischler, "Incomplete scan path with an automatic test generation approach," *Int. Test Conf.*, pp. 153–162, 1980.

[6] V.D. Agrawal, K.T. Cheng, D.D. Johnson and T. Lin, "Designing circuits with partial scan," *IEEE Design and Test of Computers*, pp. 8–15, Apr. 1988.

[7] K.T. Cheng and V.D. Agrawal, "A partial scan method for sequential circuits with feedback," *IEEE Trans. on Computers*, vol. 39, pp. 544–548, Apr. 1990.

[8] D. H. Lee and S. M. Reddy, "On determining scan flip-flops in partial-scan designs," *Proc. ICCAD*, pp. 322–325, 1990.

[9] S.T. Chakradhar and A. Balakrishnan and V.D. Agrawal, "An exact algorithm for selecting partial scan flip-flops," *Proc. ACM/IEEE Design Automation Conf.*, pp. 81–86, 1994.

[10] C.C. Lin, Mike T.C. Lee, M. Marek-Sadowska and K.C. Chen, "Cost-free scan: A low-overhead scan path design methodology," *Proc. ICCAD*, 1995.

[11] C.C. Lin, M. Marek-Sadowska, K.T. Cheng and Mike T.C. Lee, "Test-mode point insertion: Scan paths through combinational logic," Report UCSB-ECE-CAD-95-10, University of California, Santa Barbara, 1995.

[12] B. Vinnakota and N. K. Jha, "Synthesis of sequential circuits for parallel scan," *Proc. Int. European Conf. Design Automation*, pp. 289–293, 1992.

[13] H. Cox, "On synthesizing circuits with implicit testability constraints," *Int. Test Conf.*, pp. 989–998, 1994.

[14] H.H.S. Gundlach and K.D. Muller-Glaser, "On automatic testpoint insertion in sequential circuits," *Int. Test Conf.*, pp. 1072–1079, 1990.

[15] R. E. Bryant, "Graph-based algorithms for Boolean function manipulation," *IEEE Trans. on Computers*, vol. C-35, pp. 667–691, 1986.

[16] "SIS: A system for sequential circuit synthesis," Report M92/41, University of California, Berkeley, 1992.

Chair: Jim Lipman
Co-Chair: Hidetoshi Onodera

Design reuse has become a real "buzz" phrase as deep submicron designs have entered into the mainstream of both ASIC- and ASSP-based systems. Increasing gate densities and system speed have emphasized the need for efficiently designed cell libraries and application-specific functional cores. The papers in this session show some of the advanced work you can use to help you do your chip designs.

The first four papers deal with CMOS standard cell library generation. The first, from Cascade Design Automation, discusses a methodology for optimizing standard cell libraries for either high speed, low power or minimum area. With different libraries within the same process, you can choose the one that best meets application-specific design requirements, such as low power library for portable systems. The authors describe the three algorithms, used on 80 different processes from 25 silicon foundries, and show some of the design results they have obtained. Next is Mentor's paper describing their approach to process-portable standard cell generators. Employing a symbolic layout system, ICGen, that uses a designer's expertise to guide placement and routing, the authors were able to create a qualified 0.35-micron CMOS standard cell library, only one month after process freeze, with density as good as that obtained with custom layout.

From the University of Washington we have a paper that describes a parallel strapping layout style for standard cells fabricated in a process requiring diffusion strapping. The authors present an algorithm for generating static CMOS cells with a minimum number of diffusion breaks, and hence minimum width. Their results show significant area savings compared to cells generated with traditional full diffusion strapping. Addressing the design of dynamic CMOS leaf cells, Carnegie Mellon's paper describes a transistor ordering technique for minimizing cell area. Their algorithm, which employs a Eulerian trail method and iterative improvement strategy, guarantees optimum cell width while reducing cell height. The authors present results they've obtained on a number of benchmark circuits.

Using local interconnect technology, which is suitable for intracell routing but too resistive for global routing, IBM Microelectronics has created a 0.4-micron gate array library with higher density than a similar library without local interconnect. The use of this technology with a new intracell routing scheme makes more efficient use of global routing resources. Benchmark results on a 520K gate ASIC design show the advantages of a local interconnect metal layer.

Ending the session, we have another paper from Carnegie Mellon, this one discussing a comparison of Wallace tree and array multipliers. The comparison includes physical layout effects, not just logical descriptions. The results, obtained on multipliers ranging from 8 to 24 bits, indicate that Wallace tree multipliers take more area, dissipate less power and have slightly less delay compared to array multipliers.

Optimization of Standard Cell Libraries for Low Power, High Speed, or Minimal Area Designs

Cameron Fisher, Rob Blankenship, John Jensen, Tom Rossman, and Kevin Svilich
Cascade Design Automation
3650 131st Avenue SE
Bellevue, WA 98006

Abstract

A methodology for optimization of device sizes in CMOS standard cell libraries is described. The libraries are generated by compilers that allow for design rule and device size parameterization. Device sizes are optimized to achieve a low power, high speed, or minimal layout area cell library. The device sizing algorithms have been exercised on more than 80 different processes from 25 silicon foundries. This paper describes the three sizing algorithms and presents cell performance data and results from standard cell designs implemented with the three cell libraries.

Introduction

Digital standard cells are used in a variety of integrated systems. These systems have a wide variation in design requirements. In some systems, power consumption is critical, in others minimal silicon area is driven by cost, and in some systems operating speed is the dominant consideration. Given the varying requirements of system designs and the widespread usage of digital standard cells, it is desirable to have standard cell libraries which are tuned to these specific needs.

Previous standard cell optimizations have focused on specific circuits and implementations. Typical optimization has been limited to the design of a buffer or inverter chain, optimized to minimize power dissipation or delay. The delay optimization efforts have been primarily targeting the ratio of successive stages rather than the NMOS to PMOS ratio. See (1) for an example of low power optimization and (2), (3) for examples of minimal delay buffer optimization.

The standard cell compilers all use one of three sets of transistor sizes. A routine was developed which returns the standard cell height as a function of the transistor sizes. This provides a means of determining an area cost when performing device size optimization. The three sets of transistor sizes are optimized for low power, high speed, or minimal area designs resulting in standard cell libraries that are optimized accordingly.

Transistor Size Parameterization

Each set of transistor sizes consists of four NMOS and four PMOS sizes. The four sizes are selected based on the number of transistors connected in series. For example, a two input nand gate has two NMOS devices in series, therefore it would use the transistor size selected for two NMOS transistors in series. The notation used for this transistor size is N_2s. The same two input nand gate has one PMOS device in series, therefore it would use the transistor size selected for a single PMOS transistor in series, P_1s.

The goal of transistor size optimization is thus to determine the four NMOS and four PMOS transistor sizes which will optimize the standard cell library to achieve minimal power dissipation, maximum operating speed, or minimal silicon area. Details of the three optimization procedures follow.

Low Power Optimization

Power dissipation is minimized by reducing capacitance as much as is practical. This is implemented with minimal device sizes, subject to a lower limit imposed by setting a minimum noise margin for the gates. The noise margin limit insures that the standard cells will be sufficiently immune to power supply noise. Noise margin is defined as the difference between the logic threshold of a gate and the power supply. The gate's logic threshold is evaluated with the output of the inverting gate shorted to all inputs. Logic thresholds are evaluated for all gates in the standard cell library across temperature, power supply, and process variations. Transistor sizes are adjusted if a gate fails to meet a noise margin requirement and the process is restarted. This is repeated until all gates pass the noise margin requirement. A preference for reducing transistor sizes is included in the process, however, transistor sizes will be increased if required to meet the noise margin requirements. In addition, the output rise and fall times for the gates are limited, by limiting the maximum capacitance that a gate can drive. The maximum capacitance limits the feed through current through the PMOS and NMOS devices from the power supply to ground. This algorithm is shown in Figure 1, "Low Power Device Sizing".

High Speed Optimization

The goal of high speed optimization is to achieve minimal gate delays without excessive area increase. The area increase is determined by the cell height, as the standard cell widths do not vary significantly with the device sizes.

To accomplish this, a minimal standard cell height is computed, based on minimum sized NMOS and one and one half (1.5) times minimum PMOS. The standard cell height is allowed to grow nominally twenty five percent (25%) over this minimum. Beyond this 25%, the standard cell height is allowed to increase further only when the increase in speed exceeds the square of the height increase.

0-7803-3177-6 $5.00 © 1996 IEEE

Figure 1, Low Power Device Sizing

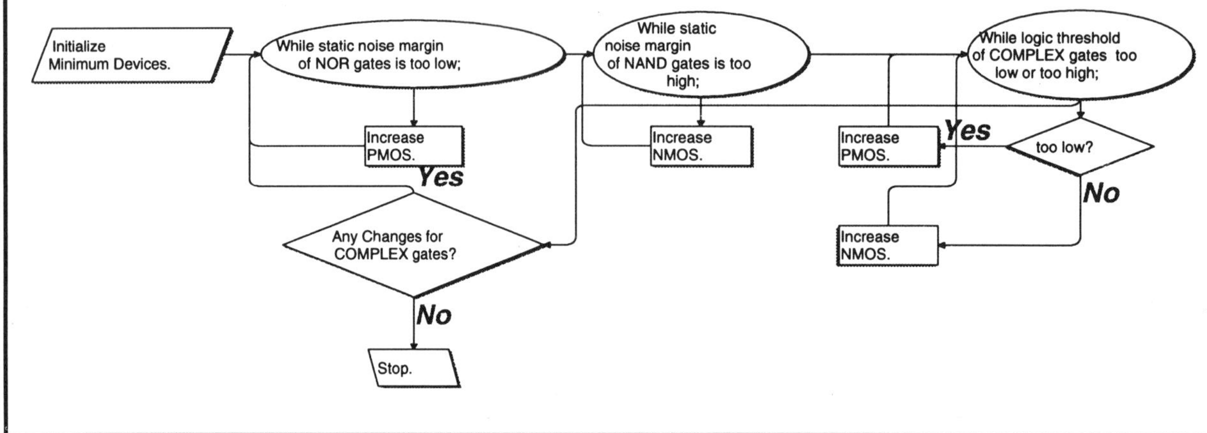

The device sizes are calculated to fit into this standard cell height using iterative SPICE simulations which select the optimum NMOS and PMOS transistor sizes to minimize gate delays. The gate delay is measured as the average of the input rising and input falling delays for three gates in series. Unused inputs are connected high for nand gates or low for nor gates. The load on each gate consists of the following gate plus a standard load. A standard load is based on the input capacitance of four clocked inverters plus a routing capacitance which is scaled by the standard cell height. This load changes as the device sizes are optimized, reflecting the effect of load capacitance as device sizes change. Optimization is done on a set of nand and nor gates as identified in Table 1, "High Speed Gate Optimaztion Steps" determining the transistor sizes for that gate. SPICE based optimization establishes the one, two, and three in series NMOS and PMOS transistor sizes required to optimize the standard cell library for high speed operation. The four in series NMOS and PMOS transistor sizes are set such that four in series devices will provide the same pull up or pull down current as the three in series transistor sizes. However, the four in series devices sizes are limited to the maximum NMOS or PMOS device width as computed by the standard cell height routine in this step. A final static noise margin check, as described for the Low Power Optimization, insures that the high speed standard cell library meets the same noise margin requirements as the low power optimized cells. This process is detailed in the Figure 2, "High Speed Device Sizing".

Table 1: High Speed Gate Optimization Steps

Step	Gate Optimized	NMOS Devices Sized	PMOS Devices Sized
1	2 input nand	2 in series	1 in series
2	2 input nor	1 in series	2 in series
3	3 input nand	3 in series	
4	3 input nor		3 in series

Figure 2, High Speed Device Sizing

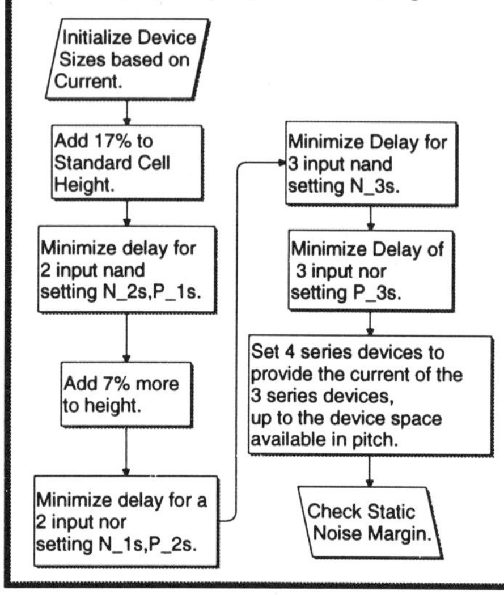

Minimal Area Optimization

Minimal area optimization is designed achieve the fastest possible cells in the minimum area. The devices are sized to fill up the minimum standard cell height. In addition to the check for minimum noise margin, the ratio of gate output rise to output fall times are controlled. To achieve this goal, an initial set of transistor sizes is selected taking in to considerations the standard cell height requirements. This initial set of transistor sizes is further optimized using iterative SPICE simulations to minimize gate delays without increasing the standard cell height. In this process, maximum speed gates are obtained within a minimal layout area. The initial set of transistor sizes approximately equalizes the rise and fall times of the various gates, using the current ratios specified in Table 2, "Initial Current Ratios for Area Optimization". These sets of currents were empirically found to provide a good tradeoff between area and performance.

Table 2: Initial Current Ratios For Area Optimization		
	NMOS	PMOS
1 Series	1 (Minimum)	5/6
2 Series	2/3	5/6
3 Series	5/6	5/6
4 Series	5/6 (Limited)	5/6 (Limited)

These sizes are then used to calculate a standard cell height. This initial device set is then further optimized using a sequence similar to the high speed optimization, with the exception that the initial standard cell height is not allowed to increase. Optimization is done on a set of inverter, nand and nor gates as identified in Table 3, "Minimal Area Gate Optimization Steps. This process establishes the set of transistor sizes required to optimize the standard cell library for minimal area. The optimization determines the one, two, and three in series NMOS and PMOS transistor sizes. The four in series NMOS and PMOS transistor sizes are set such that four in series devices will provide the same pull up or pull down current as the three in series transistor sizes. However, the four in series devices sizes are limited to the maximum NMOS or PMOS device width as computed by the standard cell height routine in this step.

Next the ratio of pull up to pull down current is checked for the inverter, nand and nor gates. This ratio of pull up to pull down current is limited. The NMOS or PMOS transistor width is reduced if possible, otherwise the opposite (PMOS/NMOS) transistor size is increased, as required.
A final static noise margin check, as described for the Low Power Optimization, insures that the resulting standard cell library meets the same noise margin requirements as the low power optimized cells. This process is detailed in the flow diagram Figure 3, "Area Optimized Device Sizing".

Figure 3, Area Optimized Device Sizing

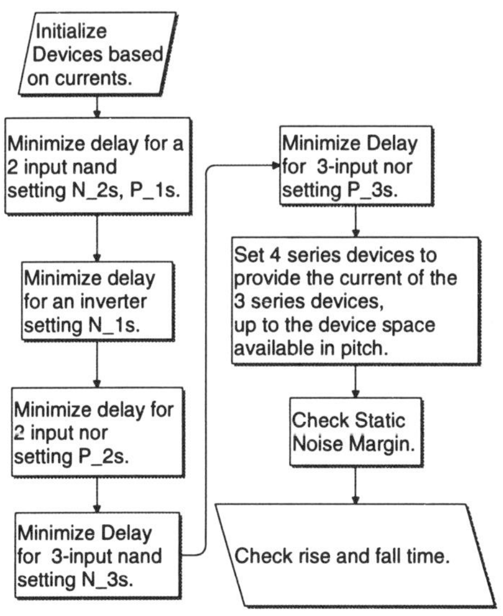

Table 3: Minimal Area Gate Optimization Steps			
Step	Gate Optimized	NMOS Devices Sized	PMOS Devices Sized
1	2 input nand	2 in series	1 in series
2	inverter	1 in series	
3	2 input nor		2 in series
4	3 input nand	3 in series	
5	3 input nor		3 in series

Standard Routed Cell Test Cases

To test the relative performance of the three standard cell libraries, 12 standard cell test cases were synthesized using Synopsys and routed. A three layer metal, 0.5 micron process was used. The "standard" test case reflects the minimal area optimized device sizes. Figures 4, 5, and 6 compare these test cases.

Figure 4, Area Comparison

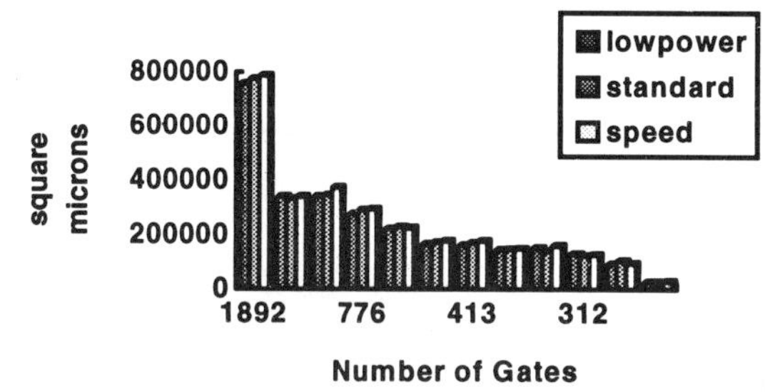

Figure 5, Power Comparison

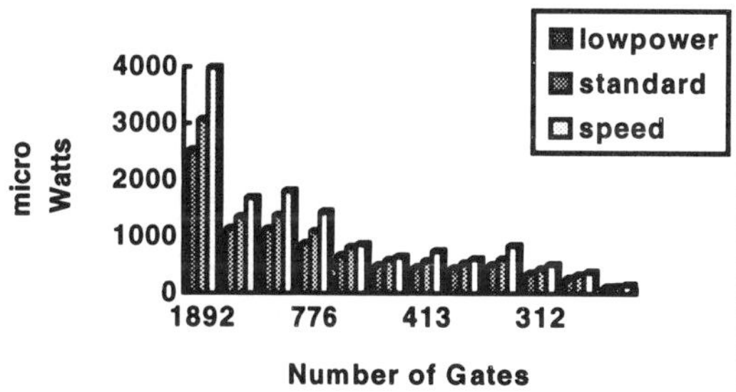

Figure 6, Critical Path Comparison

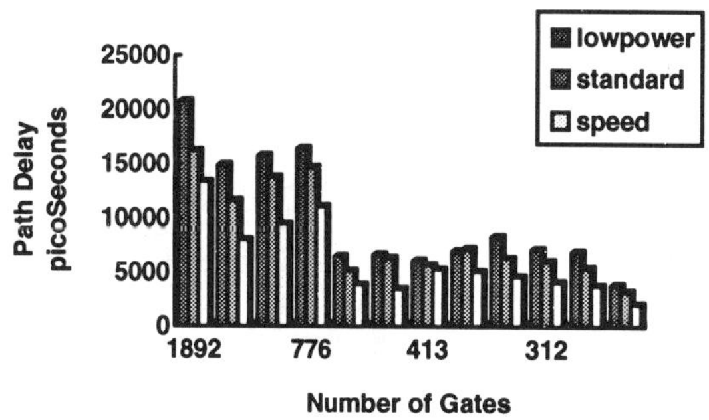

Test Die

As part of qualification for this process, a test die was generated, simulated, produced, and tested. This test die has three standard cell blocks, one optimized using each algorithm previously described. This test die is pictured in the Test Die Photo below. In this photo, the upper left corner is a group of "speed" cells, the upper right is a group of "lowpower" and the lower right is a group of "standard" cells. The average input rising and input falling delays for strings of gates, measured for this 3V 0.8 micron process is shown in Figure 7, "Measured Delay Comparison" for several gates.

Conclusion

The optimization procedures presented allow a tradeoff between power, speed, and silicon area. Standard cell designs can better meet diverse system requirements with standard cell libraies specifically optimized for low power, high speed, or minimal area.

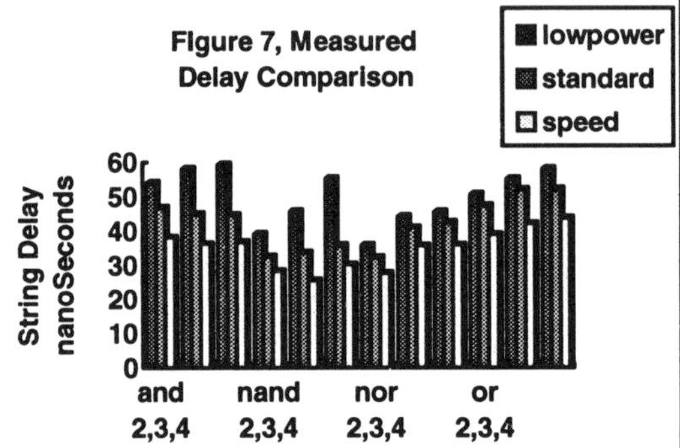

References

(1) Sha Ma and Paul Franzon, "Energy Control and Accurate Delay Estimation in the Design of CMOS Buffers," IEEE Journal of Solid-State Circuits, vol. 29, no. 9, pp. 1150-1153, September, 1994.

(2) S. R. Vemura and A.R. Thorbjornsen, "Variable-taper CMOS buffer," IEEE Journal of Solid-State Circuits, vol. 26, no. 9, pp. 1265-1269, September, 1991.

(3) Craig Prunty and Laszlo Gal, "Optimum Tapered Buffer," IEEE Journal of Solid-State Circuits, vol. 27, no. 1, pp. 118-119, January, 1992.

Test Die Photo

Efficiently Embedding Expertise in High-Density Process-Portable Standard Cell Generators

J. Duh and T. G. Matheson

Mentor Graphics Corporation

Ed Hepler

VLSI Concepts Inc. and Villanova University

Abstract

A new approach to the creation of process-portable standard cell generators is described. The system efficiently captures designer expertise while automating the details of the cell layout process. The system takes advantage of the extensibility of the underlying symbolic layout system to provide process portability without sacrificing layout density. The system includes novel transistor-bending and mask-processing automation.

Introduction

As more and more ASIC designs move to deep submicron (DSM) technologies, cell-based and library-based design methodologies are becoming increasingly important. Rising time-to-market pressures and shortening fabrication-process lifetimes demand that the layout design be automated. Automation is needed not only to shorten design times but also to enable concurrent development of the cell library and the fabrication process; process-portable layout automation makes it possible to begin cell development while the fabrication process is still unstable. The layout automation can automatically track the process when design rule changes require layout modification or shifting electrical models require transistor resizing. Concurrent development of the cells and the process can dramatically shorten the time it takes to produce a qualified library after the process freezes.

Simple economics based on the value of silicon real estate constrains the automation by demanding that the automation not sacrifice any layout density. Manual layout is still widely employed because cell density remains highly leveraged; cells often appear in hundreds of different designs and their density is directly reflected in the density of the resulting designs. While manual layout sets the standard for density, manual techniques make it nearly impossible to concurrently develop cell libraries and fabrication processes.

New approaches to cell layout and creation are needed that address process portability and the shortening design cycle while providing layout densities as good or better than those achieved by hand. This paper presents such a technique.

Existing Techniques

A spectrum of techniques has been applied to the creation of libraries of standard cell layouts. The three most dominant have been layout synthesis (1), symbolic layout with compaction (2) (3), and procedural module generation (4) (5). While all of these automation techniques speed the creation of cell layouts, they track process changes with varying degrees of success and usually sacrifice some layout density in exchange for the automation.

Layout synthesis starts with a net list and produces a complete cell layout, typically using a variety of graph-theoretical techniques and heuristics for transistor chaining, clustering, and placement, followed by routing and possibly compaction. Because it is completely automatic, layout synthesis tracks process changes reasonably well. While it is an active area of research and new commercial products have been introduced, work remains to develop techniques that consistently approach hand density. Designers can often optimize placement and routing better than the layout synthesis heuristics.

Symbolic layout with compaction starts with a topology created by a designer and then applies virtual-grid or constraint-graph compaction to minimize the cell area for a given set of design rules. Symbolic layout with compaction can approach hand density, but usually only when designers direct the compactor by hand-annotating constraints. When the process changes, existing constraints often interact with the changes in unpredictable ways, making it difficult to keep the layout optimal. The result is often a loss of layout density.

Procedural module generation encodes the procedure for assembling a cell in an executable program that, when run, produces layout tailored to the design rules. Procedural layout can also approach hand density and usually tracks small changes in layout rules well. The procedural programs become difficult to modify, however, in the same way that compactor-based systems do. The programs effectively capture constraints in procedural code that often interact with

0-7803-3177-6 $5.00 © 1996 IEEE

IEEE 1996 CUSTOM INTEGRATED CIRCUITS CONFERENCE

process changes in unexpected and non-obvious ways, resulting in a loss of layout density.

Design Goals and Approach

Our design goals were to:

1. Produce an automation system specific to the domain of standard cell layout that could meet or exceed manual layout density.
2. Develop a system that made it easy to modify the layout whenever necessary.
3. Produce a system that could track process changes, including physical and electrical changes.

To accomplish the first goal we chose to not take a fully automated layout synthesis approach because the heuristic-driven placement and routing that is a part of layout synthesis systems can often "get in the way" of layout designers by preventing them from using their expertise in layout design. Instead we chose to design a system that could leverage designers' expertise by allowing them to direct the placement and routing while still automating much of the details of the layout process.

The way we addressed the second goal is closely related to how we approached the first goal. By only presenting facets of the layout to designers where they could add value we kept the capture of the layout simple making it much easier to understand and use. Choosing to take a domain-specific approach to the automation simplified the problem.

We addressed the third design goal by building the system on top of a symbolic module generation system. The next section describes this in more detail.

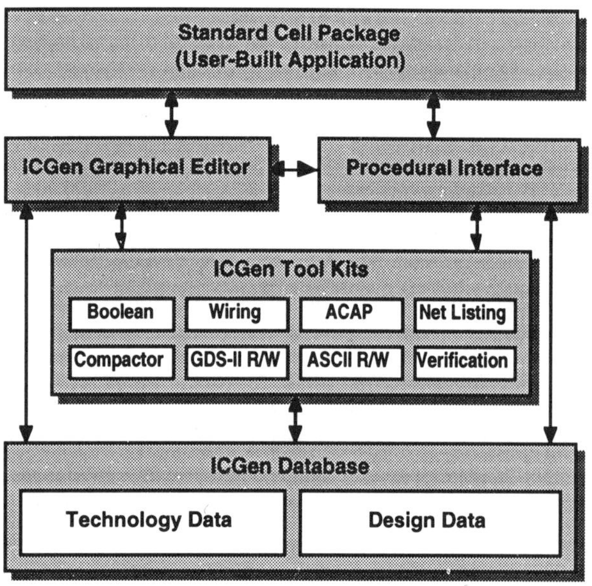

Fig. 1 - Layering the Standard Cell System on ICGen. Arrows Indicate Data Flow.

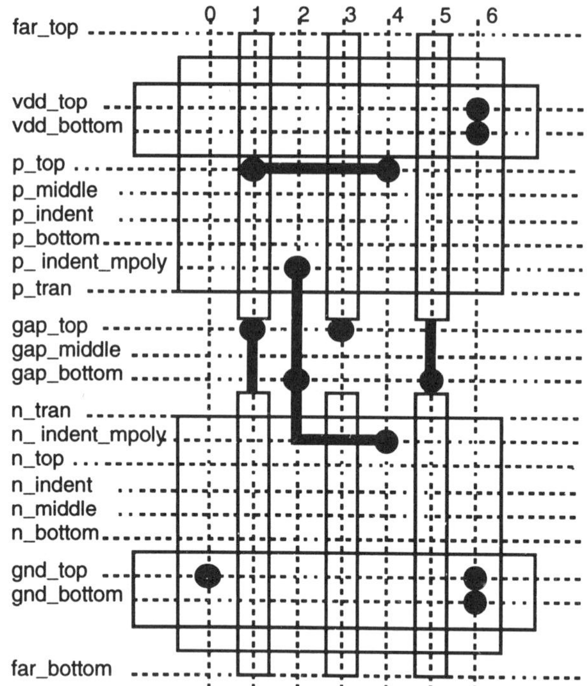

Fig.. 2 - Basic Layout Topology.

System Overview

Our system is layered on an extensible symbolic layout system, ICGen, as illustrated in Fig. 1 (6). We made heavy use of the tool kits exposed through procedural interfaces to build our layer of automation. We also took advantage of the underlying system's ability to define new parameterized transistor and contact types, know collectively as "elements", that worked efficiently with our automation layer to produce dense layout, as described below.

Our approach borrows some features from symbolic layout, virtual grid, and procedural layout approaches. Designers are given control of both the placement and the routing of the internals of the cells on a set of grid lines, as shown in Fig. 2. The basic cell layout topology defines a set of rules for object placement and boundary conditions, and rules for object placement relative to other objects in the layout. This structure acts as a target place holder for the elements declared for a particular cell, and defines the topology in terms of the cell pitch, position or placement of substrate/well contacts, and the positioning of I/O contacts. Designers express the layout in a simplified procedural language that we call Abstracted Procedural (AP). AP is optimized with only the expressions needed to describe the contents of the cells. Designers determine the placement and interconnection of the components to minimize the width in the horizontal direction. (The vertical height is typically fixed as part of the process definition.) We have found that giving the designers control of the object ordering, placement, and routing inside the cell allows them to use their expertise in making tradeoffs in layout, resulting in denser layouts from the system.

The system takes the AP generator description as input and produces a mask-ready cell as output. The internal steps are:

1. Parse the AP description. This step is similar to what is done in most procedural layout systems.
2. Build an object-ordering table. This and the next few steps are similar to what is done in most symbolic layout with compaction systems.
3. Build a wiring list.
4. Create a loose layout in the ICGen data base containing widely spaced elements (transistors and contacts).
5. Add wires to the ICGen data base to interconnect the elements already there.
6. Bend transistors and chamfer contacts as needed to optimize density as shown in Fig. 3. This step takes advantage of ICGen's extensible element capability to create transistors with optimally placed bends, and contacts with specially chamfered corners to maximize packing. This is where our method departs from other methods. This step extends the compaction of what is in the cell all the way to the actual contents of the elements themselves. Because we can build custom elements as needed during this step, it is like compacting the internals of the elements in addition to compacting around them. This helps significantly in optimizing the layout density.
7. Compact the layout in the horizontal direction using the "As Close As Possible" (ACAP) tool kit that is part of ICGen (7). The ACAP tool kit places pairs of objects at their minimal legal spacing. Here also we depart from other methods. Having access to the ACAP tool kit through its standard procedural interface allowed us to create a type of compaction that is tailored to the special needs of standard cell layout and that works well with the element customization of the previous step.
8. Process the layout to complete the mask. Here we go further than most other systems by using the Boolean tool kit in ICGen to automate the mask manipulation that is often done as separate post-processing. This step includes the creation of derived layers, such as the n+ and p+ implants and wells, as well as other kinds of mask processing, such as notch and gap filling.
9. Verify the layout. Too often "correct by construction" layout isn't really. To insure correctness and to provide feedback to the cell layout generator developer during debug, in this step we again used the Boolean tool kit to do some basic rule checking and verify completeness.

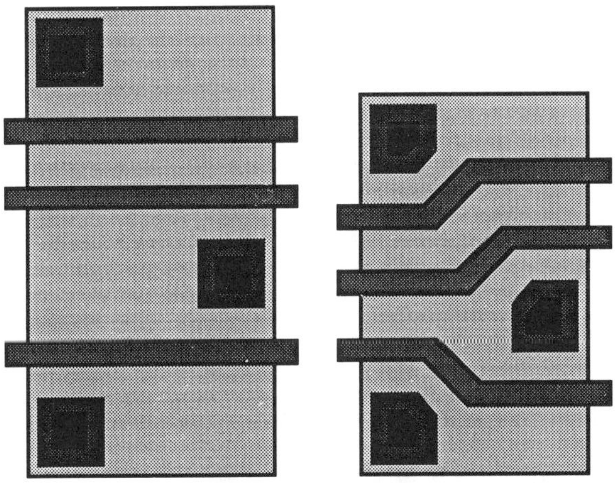

Fig. 3 - Area savings from bending transistors and chamfering contacts.

Table 1
Experimental Cell Sizes Compared to Hand-Crafted.

Number of Transistors	Automated / Hand-Crafted % Area
2 - 5	84.4%
6-10	99.3%
11-15	104.3%
16-20	103.1%
21-30	105.7%
31-40	95.7%
41-60	100%

Experimental Results

The system described here was used to create a set of commercial standard cells in 0.35μ CMOS. The experience demonstrated excellent results. First, the overall layout density of the cell library is <u>as good as previous generations that were produced by hand</u>. Quantitative results are shown in Table 1 which compares the layout from the system described here with a previous generation of 200 hand crafted equivalent cells. Because the two sets of cells being compared were fabricated in technologies with different minimum feature sizes, we compared them by expressing the cell sizes in terms of the minimal grid spacing for each technology. The overall average size of the cells produced by the system described here was only 0.7% larger than the hand-crafted cells. Because the system allowed layout designers the freedom to experiment with topologies and tradeoffs that they normally wouldn't have had time to try, in some cases the generated cells had better density than their manual counterparts. Secondly, because the system was able to accommodate process changes as they occurred, the library was developed in parallel with the fabrication process.

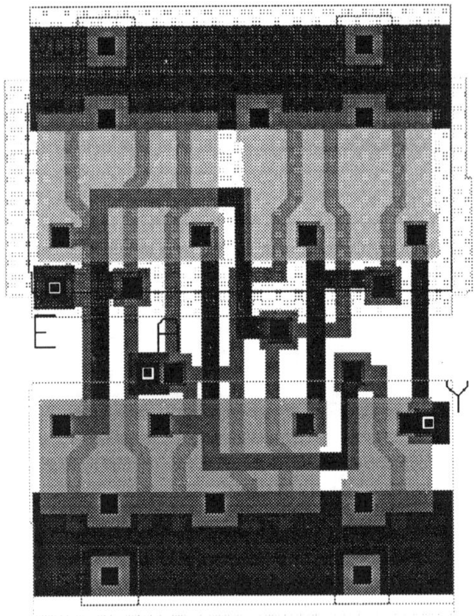

Fig. 4 - A tristate buffer produced by the system.

Normally designers had to wait until after the process stabilized to begin their layout, and libraries weren't completed for 3 to 6 months. Instead, a fully qualified library was available only one month after process freeze. In addition to these results, it is expected that the generator libraries that were developed will track the next generation of fabrication technology without significant modification.

Fig. 4 and Fig. 5 show sample layouts produced by the system.

Conclusions

We have shown that domain-specific layout automation built on top of a general-purpose extensible symbolic layout system can produce process-portable layout generators without sacrificing any layout density. Allowing layout designers to capture their expertise results in improved density over other types of automation. This type of automation can significantly shorten the time needed to produce cell layout libraries by enabling concurrent development of the fabrication process and the cell library, while also creating layout libraries that can port to future-generation processes.

References

(1) R. L. Maziasz and John P. Hayes, *Layout Minimization of CMOS Cells*, Kluwer Academic Publishers, Boston, MA, 1992.

(2) D.G.Boyer, "Symbolic Layout Compaction Review", *Proc. 25th Design Automation Conference*, pp. 383-389, 1988.

(3) J. Dao, N. Matsumoto, T. Hamai, C. Ogawa, and S. Mori, "A compaction method for full chip VLSI layouts", *Proc. 30th Design Automation Conference*, pp. 407-412, 1993.

(4) T. G. Matheson, C. Christensen, and M. R. Buric, "A software environment for building core-microcomputer compilers", *Proc. International Conference on Computer Design: VLSI in Computers (ICCD '85)*, pp. 221-4, 1985.

(5) J. C. Herbert, "An integrated design and characterization environment for the development of a standard cell library", *Proc. IEEE 1991 Custom Integrated Circuits Conference.*, pp. 25.6.1-25.6.5, 1991.

(6) T. G. Matheson, "Extensibility in a symbolic layout system", unpublished.

(7) R. A. Eesley and M. Tarsi, "ACAP - A system for the interactive graphical capture of generators", *Proc. IEEE 1991 Custom Integrated Circuits Conference.*, pp 22.1.1-22.1.4, 1991.

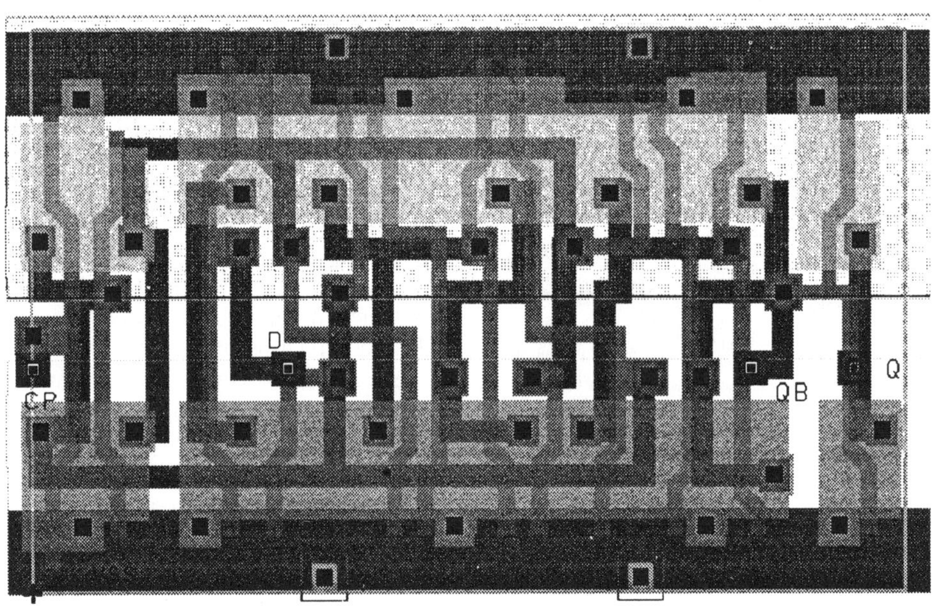

Fig. 5 - A D Flip Flop. produced by the system.

Efficient Standard Cell Generation When Diffusion Strapping Is Required

Bingzhong (David) Guan and Carl Sechen

Department of Electrical Engineering
University of Washington
Box 352500, Seattle, WA 98195-2500

Abstract

In [3], we proposed a single contact layout style (SC style) for CMOS standard cells with regular and compact structure, based on the assumption that a single diffusion contact is sufficient. In reality, the assumption is not always true. We therefore propose a partial strapping style (PS style) for use when diffusion strapping is required. The PS style keeps all the features of the SC style. The structure uses less area for individual cells, allows easy embedding of feedthroughs in the cell, and enables output pins to occur at any grid location. Using an exact algorithm to generate static CMOS cells with a minimum number of diffusion breaks ensures that the width of the cells is minimized. For the PS style, a constructive routing algorithm is used to perform the intra-cell routing. An exhaustive search among the minimum width cells produces the minimum height cell. Our results show that cells in the PS style have cell height very close to those in the SC style. Furthermore, cells using either layout style achieve significant area savings compared to cells using the traditional full strapping style.

Introduction

Layout generation, also known as silicon compilation, transforms the logic description of a system into physical masks for silicon fabrication. Since the problem is so vastly complicated, over the past two decades, the standard cell design methodology gained popularity because it solves the problem in a reasonable fashion with the divide-and-conquer approach. The layout generation process typically has been divided into subtasks of logic synthesis and technology mapping, cell generation, placement, global and detail routing, and compaction. The ultimate goal is to minimize chip area while satisfying performance requirements.

The building blocks of this approach are these standard cells. Layout minimization depends a lot on the structures of individual cells (cell layout style) and also the structure of the standard cell library (mainly, the content and the size of the library).

Since Uehara and vanCleemput [12] proposed a layout style (UvC78) for static CMOS cells in 1978, almost all cell layout styles have followed UvC78 with minor variations [1][7][9]. In the style of UvC78 and its variations, the power lines run parallel to the diffusion rows. However, the original style was targeted for a one metal layer process. This style is disadvantageous with respect to layout area [5]. In layout generator THEDA.P, a new layout style was introduced to target 2-layer metal CMOS processes [6]. However, in this style, the pins are not aligned and the metal-2 layer is not obstacle free.

In [3], we proposed a single contact layout style (SC style) for CMOS standard cells. In this paper, we propose a new partial strapping style (PS style). In designing our styles, the main goal was integration with place and route tools so that the total chip area is minimized after the cells are placed and routed.

New Layout Styles

The traditional fabrication process typically mandates that the drain and source regions be fully contacted (called full strapping, as shown in Figure 1(c)) to improve performance and ensure reliability. With advanced processing technology, self-aligned silicide

$O = \overline{a + b\,(c + d)\,(e + fg + hi)}$

Figure 1: *The layouts of the same complex logic function are shown in (a) single contact style, (b) partial strapping style, and, (c) traditional full strapping style. These cells have same width, same diffusion width, but different cell height and quite different metal 1 blockage.*

0-7803-3177-6 $5.00 © 1996 IEEE

IEEE 1996 CUSTOM INTEGRATED CIRCUITS CONFERENCE

(salicide) can make the sources and drains very low in resistance. With either a local interconnect being available to strap the diffusion areas or salicide being available, connecting to diffusion with a single contact does not effect performance much. One industry source, based on the latest 0.25μm processing technology, found that the performance degradation is minimal when moving from full strapping to a single contact style. The worst case degradation is less than 5%, when only having one minimum size contact driving a 10μm transistor from one end of the diffusion.

In [3], we proposed the SC style for CMOS standard cells (Figure 1(a)) based on the single contact assumption. The style is applicable to the processes where a single contact is sufficient due to salicide. It is also applicable to the processes where a local interconnect (LI) layer is available to accommodate full strapping of diffusion area. The SC style has a regular yet compact structure. The structure uses less area for individual cells, makes routing problems straightforward and allows easy embedding of feedthroughs in the cell. Using that new style, we developed a cell generator using an exact algorithm to minimize cell width and height.

Unfortunately, only a few semiconductor manufactures currently support such advanced fabrication processes (either salicide or LI). In this paper, we therefore propose a partial strapping style (PS style, Figure 1(b)). Although different from the SC style due to the strapping requirement, the PS style follows the same discipline of maintaining the cell structure as compact and regular as possible. The strapping requirement however creates a quite different intra-cell routing problem. We developed a constructive heuristic algorithm for the intra-cell routing problem. This algorithm has been integrated into our area minimizing cell generator.

Figure 1(a) shows a complex cell in the SC layout style. Figure 1(b) shows the same complex function cell in the new PS layout style. Both new layout styles only use metal-1 for intra-cell connections. They have same configuration for input and output pins or terminals. Pins are aligned in a row between the diffusion regions and are equally spaced between the poly gates. The differences between the two styles are power bus positions and intra-cell routing schemes. In the SC style, the power lines are over the diffusion area, running horizontally near the middle of the cell to facilitate cell abutment. In contrast, in the PS style, the power lines are partially overlapping the diffusion area on the top and bottom boundaries of the cell. In addition, in the SC style, the intra-cell routing scheme will use the horizontal tracks that are close to the center of the cell. Those regions over the diffusion area, but not used for intra-cell routing are outside the power buses and can be grouped together with the channel to form one routing region for inter-cell routing. In contrast, the intra-cell routing and partial strapping of diffusion area uses up all the area in between power buses in the metal-1 layer.

The features of these styles are: 1) The output pins can be at any grid position. The freedom of putting the output pin at any grid offers the potential to reduce the number of tracks needed to route all intra-cell nets. Reducing the number of tracks means the portion of the height of a cell used for intra-cell routing is reduced.

2) The pins are at the cell center, aligned in a row, on a uniformly spaced grid. This regularity made the inter-cell routing problem easier. The only obstacle for routing the inter-cell connections are the regions used for intra-cell routing in the metal-1 layer. Thus the routing region in the metal-1 layer is a rectilinear area between two rows, including the routing channel and the area over the cell which is not used for intra-cell connections in SC style. In PS style, the routing region will just be the channel area in the metal-1 layer. The metal-2 and metal-3 layers are obstacle free. Over-the-cell routing is simple with a uniform pin grid.

3) These styles provide a lot of built-in feedthrough positions. After place and route, the connection directions of pins in a cell are known. Two neighboring pins having opposite direction connections can be moved into one pin column, thus one column is free to be used as a feedthrough position. For example, in Figure 2, pin *f* only connects upward, and pin *e* connects downward. Pins *e* and *f* can share the same pin grid, while a straight feedthrough position is created. These layout styles also provide dog-leg feedthroughs.

Cell Generation

Functional cell generation is the process of translating a design from the transistor circuit level to the transistor layout level. The process of generating a minimum area cell has been proved to be NP-hard [2]. The primary goal in optimization is normally to minimize the cell area (width times height). One important aspect in reducing cell width is to utilize diffusion abutment. This abutment can be achieved when the source and drain diffusions of adjacent

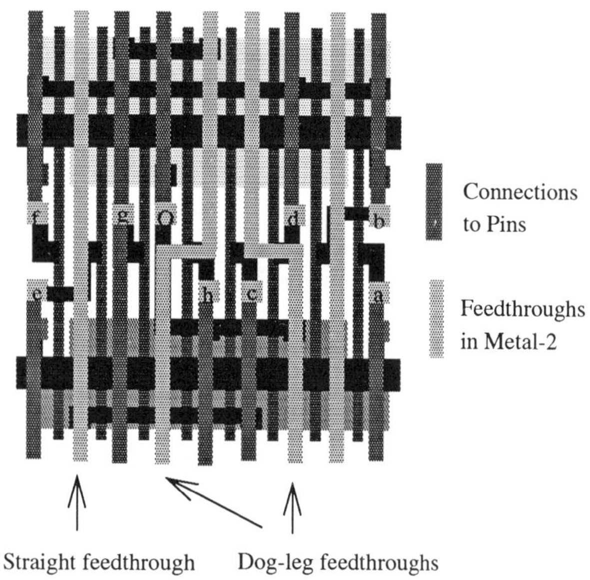

$O = \overline{ab + (c + d)(e + f) + gh}$

Figure 2: *The new layout style provides feedthroughs, both straight and dog-leg feedthroughs.*

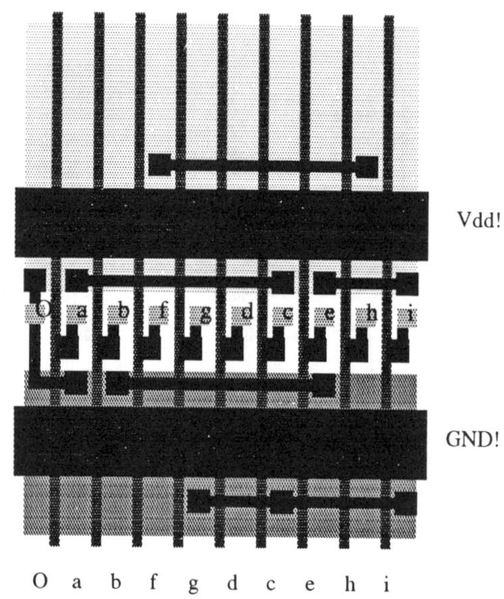

$O = \overline{a + b(c + d)(e + fg + hi)}$

Figure 3: *This complex cell is minimum height in our single contact layout style, while its corresponding partial strapping style cell is not minimum area.*

transistors are electrically equivalent. If they are not electrically connected to each other, a diffusion gap is needed to isolate these transistor terminals. In our layout styles, a diffusion gap forces the separation between neighboring poly gates to be twice as large as that needed by diffusion abutment. In fact, one of our goals is to maximize diffusion abutment.

Exact Algorithm to Minimize Cell Width and Height

In [8], an exact algorithm, HR-TrailTrace, was proposed to minimize cell width by minimizing the number of diffusion breaks. The algorithm utilizes the transistor reordering (also called delayed binding) technique. An exhaustive search among all minimum width cells produces a minimum height cell. The algorithm only counted the density of intra-cell nets as the height. The algorithm was shown to be feasible for all cells of practical size. We implemented that algorithm with a few modifications and extensions. Modifications include utilizing logical equivalence to drastically reduce the number of permutations we need to consider and increasing the efficiency of the algorithm [3]. This algorithm can handle series-parallel connected transistor netlists.

For the SC style, a modified left edge algorithm (LEA) has been implemented to route the intra-cell connections. The modification stems from the necessity of determining the output pin location and having the output net always on the first track from the middle. Among all these routed minimum width cells, those which have the least number of tracks are picked as the minimum height cells. As an example, corresponding to the complex function shown in Figure 1, the minimum height layout is shown in Figure 3. This area minimum cell only needs four routing tracks, while the layout shown in Figure 1(a), although also minimum width, needs five tracks.

Intra-Cell Routing for Partial Strapping Style

For the PS style, because of the strapping requirement, the LEA algorithm in itself cannot be applied. The routing problem becomes how to route all connections so that there are the fewest number of intersections between nets. If we don't have any (metal-1) intersections between nets, all routing can be placed over the diffusion regions. Otherwise, some nets will have to be routed in the area between the diffusion and the power bus, not only increasing the cell height, but also requiring some segments of poly wires to be used (as used in Figure 1(c)). For example, the corresponding PS style layout of Figure 3 will have the segment shown on top to have one intersection with one of the power connections. That intersection forces the top segment to be routed in the area between the power bus and the diffusion, which results in a taller cell than the one shown in Figure 1(b). This complication renders both HR-TrailTrace's density counting algorithm and LEA not applicable.

The intra-cell routing problem for each of the PMOS and NMOS sections actually is a single-row-planar-routing (SRPR [10]) problem with some constraints. One constraint is the power connections which block one side entirely, while the connection for the output pin blocks partially the other side. The other constraint is that vertical tracks between nodes are quite limited, and sometimes even do not exist.

We developed a heuristic to solve this routing problem. Listed in Figure 4 is the intra-cell routing algorithm for the PS style. Shown in Figure 5 is the step-by-step flow to generate intra-cell routing for the cell in Figure 1(b). The PS style routing starts from the results of the LEA (step 1). The track that contains the output net is assigned to the middle (step 2) and all other tracks are assigned to the outside (step 3). At this time, all violations (intersections), if any, are in the OT. Figure 5(b1) shows that the initial intersections are (*n1, n2*) and (*n2, n3*).

Step 4 calls the function *FindViolations&Flip* with the *AlwaysFlip* Flag to resolve all these violations in OT. This function first finds all segments which have violations (step 10) (nets *n1, n2,* and, *n3* in Figure 5(b1)) and then calculates (step 11) the costs associated with these segments: *CostOrig* is the number of intersections the segment has in its current location; *CostDest* is the number of intersections if the segment is flipped to the other side;

```
Algorithm DetailRouteOneMinimumWidthCell(Cell)

1  LEA(Cell);
2  Assign the output net track to MiddleTracks (MT);
3  Assign all other tracks to OutsideTracks (OT);
   /* Any intersections are in the OutsideTracks */

4  FindViolations&Flip (OT, AlwaysFlip);

5  While (ViolationsExist & NumIteration < Threshold) {
6    FindViolations&Flip (MT, CostReductionOnly);
7    FindViolations&Flip (OT, CostReductionOnly);
     /* Flip only the net not costing any more intersections */
   }

   /* Any remaining intersections need to be at the outside */
8  If (ViolationsExist) {
9    FindViolations&Flip (MT, AlwaysFlip);
   }

Function FindViolations&Flip(Tracks, FlipFlag) {
     While (ViolationsExist) {
10        Find all segments with violations;
11        Calculate costs for all these segments;
          /* Three costs are associated with a given configuration: CostOrig is the number of violations in its current location; CostDest is the number of violations if the segment is flipped to the other side; And CostDiff is the difference of the previous two (CostDiff= CostOrig - CostDest). */
12        Order these segments according to CostDiff, with CostOrig as the tie breaker;
13        if (FlipFlag is CostReductionOnly and the largest CostDiff is negative) return;
14        Flip the segment with the biggest CostDiff;
     }
}
```

Figure 4: *The intra-cell routing algorithm for the PS style.*

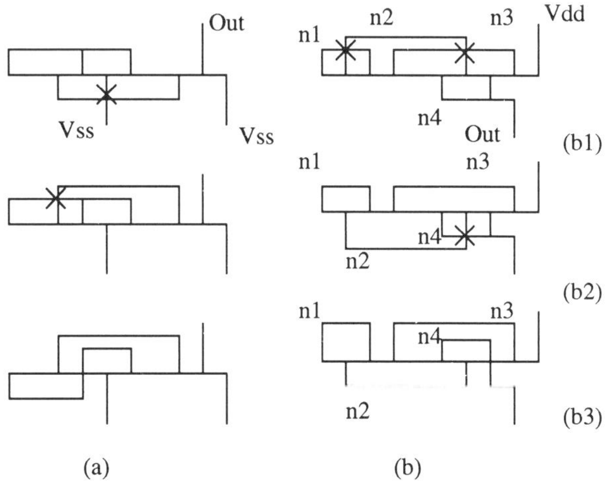

Figure 5: *Intra-cell routing flow for the cell layout in Figure 1(b). The cross indicates the intersection violation. (a) shows the routing steps for the NMOS section. (b) shows the routing steps for the PMOS section.*

And *CostDiff* is the difference of the previous two (*CostDiff* = *CostOrig* - *CostDest*). Step 12 orders these segments according to *CostDiff*, with *CostOrig* as the tie breaker. In Figure 5(b1), the ordered list of segments is (*n2, n1, n3*) with the corresponding (*CostDiff, CostOrig*) list as (1, 2), (1, 1), and (0,1). When the *FlipFlag* is *AlwaysFlip*, we will flip the segment with the largest *CostDiff* to the other side, even when it means an increase in the total number of intersections (step 14). The function returns when all violations are resolved on that side. When the *FlipFlag* is *CostReductionOnly*, the function will return when the cost cannot be reduced (steps 13 and 14). In Figure 5(b1), *n2* will be flipped to the middle tracks, and that resolves all violations in OT.

Afterwards, during the while loop in steps 5 through 7, the algorithm cycles through from side to side to resolve any violations we may have. The flipping is only done when the moving net will not be causing more violations. In Figure 5(b2), *n4* has the most *CostDiff*, and thus is flipped. This move resolves all violations as shown in Figure 5(b3). If there are any remaining intersections, they have to be placed outside (step 9). This concludes the intra-cell detail routing.

The intra-cell routing is performed over all minimum width cells of a given function, then the layouts which have the least intersections will be picked. Experiments have shown that the algorithm is very effective in routing nets over the diffusion area. The algorithm finds that all 87 cells having no more than 3 transistors in series in both the NMOS and PMOS sections have layout implementations in which all the intra-cell routing is over the diffusion.

Results

We studied the layout area and circuit performance as a result of utilizing a large library of standard cells [4]. We built libraries of all static CMOS cells having a chain length of up to 7, and this was made possible by our new cell generator's capability of generating any static CMOS cell with a user-specified limit on the number of series transistors. We used the TimberWolf place and route tools [11] to generate the actual chip layout. We used an industrial timing analyzer, which included wiring parasitics, and state-of-the-art 0.25 μm design rules, to provide performance information. We refer to a library of all possible cells having a chain length limit of *n* as s*n*. We experimented on 13 MCNC benchmark circuits, ranging from 124 to 2090 cells, 165 to 2135 nets. Our results show that compared to using library s2, in terms of chip area, s3, s4, s5, s6, and s7 saves 16%, 22%, 24%, 25%, and, 26% respectively. For larger designs, the area savings amounted to 50% when using s7. In the meantime, since a netlist which uses larger cells will have fewer cells on each critical path, we found that the average worst path delay is quite similar for libraries s3 through s7. We concluded that using a very large library (*e. g.*, s7) is optimal in terms of area and delay. Due to the sheer sizes of these large libraries, it is not possible to draw all those possible cells by hand. An efficient automatic standard cell generator is necessary to make this approach feasible and to realize the chip area savings.

As can be seen from Figure 1, while the SC, PS and FS style cells all have the same width, their height is quite different. Compared to the SC style cells, the PS style cells are taller due to partial strapping. For FS cells, intra-cell connections use some more area. Table I shows the cell height comparison between the three styles. The cells in the SC style are 12% shorter than those in the PS style, while the cells in the FS style are 30% taller than those in the PS style. Table I also shows the comparison of metal-1 blockage. Since the unblocked area can be used in the SC style, the height of the metal-1 blockage can be considered as the effective cell height. We can see from Table I that relaxing the partial strapping requirement could potentially reduce metal-1 usage by 43%.

In summary, the PS style is very close in matching the area savings offered by the SC style, while satisfying the strapping requirement. The area savings that a semiconductor manufacturer can gain from using a more advanced fabrication process (e. g., salicide or local interconnect) is quite clear. Nonetheless the PS style is very dense as can be seen by the very high metal-1 usage in the cells. Further, the PS style represents a significant area improvement over the conventional FS style.

Table I Cell height comparison, normalized to PS case.

Cell Style	Cell Height	Effective Cell Height (Metal-1 Blockage)
SC	0.88	0.57
PS	1	1
FS	1.30	1.30

In [3], we presented chip area comparison between the SC style and the Mississippi State University (MSU) library and an industrial library. Our results show that circuits using the SC style cells achieve significant area savings (as much as 50%) compared to the use of these manually laid out compact cells. Since the PS style is very close in matching the height of the SC style, similar area savings can be achieved.

Conclusions

We presented a partial strapping style for use when diffusion strapping is required. The PS style keeps all the features of the SC style. The structure uses less area for individual cells, allows easy embedding of feedthroughs in the cell, and enables output pins to occur at any grid location. Using an exact algorithm to generate static CMOS cells with a minimum number of diffusion breaks ensures that the width of the cells is minimized. We developed a constructive routing algorithm to perform the intra-cell routing. An exhaustive search among the minimum width cells produces the minimum height cell. Our results show that cells in the PS style have cell height very close to those in the SC style. Furthermore, cells using both layout styles achieve significant area savings compared to cells using the traditional full strapping style.

References

[1] S. Bhingarde, A. Panyam, and N. A. Sherwani, "Middle terminal cell models for efficient over-the-cell routing in high performance circuits," *IEEE Trans. on VLSI*, vol. 1, pp. 462-472, December 1993.

[2] S. Chakravarty, X. He, and S. S. Ravi, "Minimum area layout of series-parallel transistor networks is NP-hard," *IEEE Trans. Computer-Aided Design*, vol. 10, pp. 943-949, July 1991.

[3] B. Guan and C. Sechen, "An area minimizing layout generator for random logic blocks," *Proc. of CICC 1995*, Santa Clara, CA, May 1995.

[4] B. Guan and C. Sechen, "Large standard cell libraries and their impact on layout area and circuit performance," submitted to the 33rd Design Automation Conference, 1996.

[5] Y-C. Hsieh, C-Y. Hwang, Y-L. Lin, and Y-C. Hsu, "LiB: A CMOS cell compiler," *IEEE Trans. Computer-Aided Design*, vol. 10, pp. 994-1005, August 1991.

[6] C. Y. Hwang, Y-C. Hsieh, Y-L. Lin, and Y-C. Hsu, "An efficient layout style for two-metal CMOS leaf cells and its automatic synthesis", *IEEE Trans. Computer-Aided Design*, vol. 12, pp. 410-424, March 1993.

[7] S. M. Kang, "Metal-metal matrix (M^3) for high-speed VLSI layout," *IEEE Trans. Comp.-Aid. Design*, vol. 6, pp. 886-891, Sept. 1987.

[8] R. L. Maziasz and J. P. Hayes. *Layout Minimization of CMOS Cells.* Kluwer Academic Publishers, 1992.

[9] S. S. Sapatnekar and S. M. Kang. *Design Automation for Timing-Driven Layout Synthesis.* Kluwer Academic Publishers, 1993.

[10] N. Sherwani. *Algorithms for VLSI Physical Design Automation. 2nd Edition.* Kluwer Academic Publishers, Boston, 1995.

[11] TimberWolf Systems, Inc. *TimberWolf: Mixed Macro/Standard Cell Floorplanning, Placement and Routing Package (Version 1.0).* Obtained from Bill Swartz at TWS, Dallas, TX, 1994.

[12] T. Uehara and W. M. vanCleemput, "Optimal layout of CMOS functional arrays," *Proc. 16th ACM/IEEE DAC*, pp. 287-289, May 1978.

Efficient Area Minimization for Dynamic CMOS Circuits

Bulent Basaran and Rob A. Rutenbar

Department of Electrical and Computer Engineering
Carnegie Mellon University, Pittsburgh, PA 15213

Abstract

We present a new transistor ordering technique for the layout of dynamic CMOS leaf-cells which minimizes the cell area. The technique employs an Eulerian trail formulation which guarantees that the result is always of minimum width, i.e., the total diffusion area is optimum. A novel iterative improvement strategy minimizes the cell height as defined by the number of wiring tracks required to connect source and drain terminals. We were able to find a transistor ordering with the theoretical optimum area for many industrial circuits and standard test cases from the literature.

1 Introduction

In the layout of custom CMOS circuits, *transistor ordering* is the problem of placing transistors in a row and orienting them to maximize the sharing of source and drain regions [1]. This is crucial for reducing the total diffusion area and the cell width. Transistor ordering also has an impact on the number of tracks required to complete the routing which determines the cell height.

The focus of this paper is on *dynamic CMOS* circuits where transistor ordering is applied to the transistors in the discharge circuitry (the nFET logic block). For the row-based layout of dynamic CMOS circuits, *area minimization* is known to be NP-hard [2]. There are some techniques that minimize the cell width [3][4][5][6], but none of these techniques explicitly minimizes the total area.

On the other hand, for *static CMOS* (sCMOS) there are some exhaustive techniques that minimize the area, but they have exponential time complexity [7][8]. The relaxation approach of [9] is an efficient technique but it does not guarantee minimum cell width. The chaining technique of [10] does not consider the cell height during transistor placement.

In this paper, we present the *first* technique that minimizes both the width and the height for dynamic CMOS circuits. Our technique is efficient: it takes less than 2 seconds for typical cells. It *guarantees* optimum cell width, and is also very effective in reducing the cell height. The paper is organized as follows: Section 2 describes the transistor ordering problem. Section 3 introduces a new circuit representation based on Eulerian trails. Section 4 presents the iterative improvement technique. Results on circuits from industry and literature are given in Section 5. Finally, Section 6 offers some concluding remarks.

2 Transistor Ordering

Two MOSFET transistors that have a common drain or source node (e.g., two transistors in series have one common node, two transistors in parallel have two) can share a diffusion region[1]. Sharing a diffusion is advantageous both for saving die area and for reducing parasitic diffusion capacitances. Hence, circuit designers pay considerable attention to how to stack transistors best for compact layouts [11].

In row-based designs, transistors are located in a row to maximize possible diffusion sharing. Two parameters impact the quality of diffusion sharing: 1) order of transistors, 2) orientation of source and drain terminals. In what follows, we use the term *transistor ordering* to denote an order of transistors in a row with specified orientations for each transistor. In a given transistor ordering, if two adjacent transistors can not share a diffusion, then we have to leave a *diffusion break* (also called a *gap*) for correct DRC (See Figure 1). The number of gaps determine the width of the row, and in turn the cell width. Fewer gaps lead to smaller diffusion area and smaller cell width.

Figure 1. Effect of transistor ordering on diffusion sharing and the width of the cell.

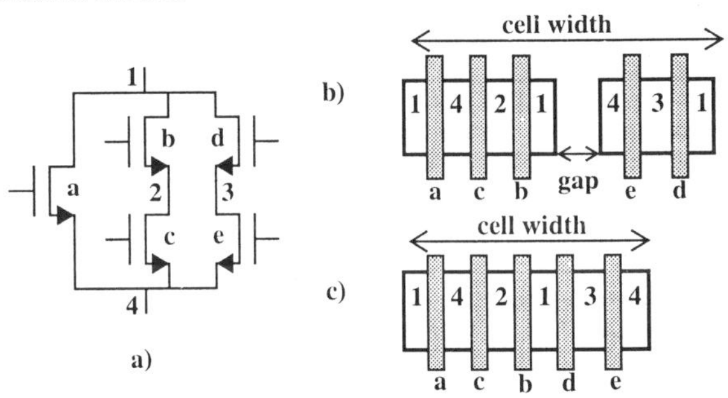

Transistor ordering also has an impact on the number of wiring tracks required to route the cell, which in turn determines the cell height (Figure 2). In what follows, we consider only the diffusion connections. For most practical dynamic circuits, poly connections can be done afterwards without increasing the cell height[2].

3 Circuit Representation

As introduced in [12], an Eulerian trail in a diffusion graph, if it exists, minimizes the diffusion area of sCMOS circuits.

1. With the assumption that they are of the same type (e.g., both nFET) and also have the same body potential.

2. The corresponding problem in static CMOS is considerably harder.

0-7803-3177-6 $5.00 © 1996 IEEE

IEEE 1996 CUSTOM INTEGRATED CIRCUITS CONFERENCE

Later [6] proposed an Eulerian trail method for dynamic CMOS circuits consisting of only one type of network (e.g., nFET logic and pFET pre-charge circuitry) which is not restricted to series-parallel circuits. We use a similar method to represent a dynamic circuit. Given a circuit, we first generate a so-called *diffusion graph, G* (undirected and possibly having parallel edges) corresponding to the nFET discharge circuitry. Each vertex in G corresponds to a diffusion node (source or drain) in the circuit, and each edge in G corresponds to a transistor (solid edges in Figure 3).

Figure 2. Effect of transistor ordering on routing and the height of the cell. The layout in b) has a height of 3, but c) has a height of 2.

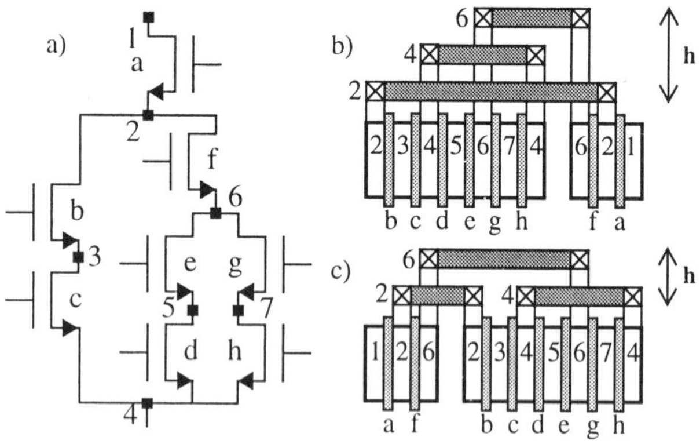

A *trail* on G is a set of adjacent edges which does not touch any edge in the graph twice [13]. We denote it as: $T=(v_0,e_0,v_1,e_1,v_2,\ldots,v_{k-1},e_{k-1},v_k)$ where $e_i=(v_i,v_{i+1})$ is an edge in the graph and $e_i \neq e_j$ for all $i \neq j$. If $v_k=v_0$, the trail is a *closed trail.* A closed trail which touches every edge of G is an *Eulerian* trail (*e-trail*). We may use the shorthand $(v_0,v_1,v_2,\ldots,v_k)$ for a trail, if we do not discriminate parallel edges. Note that a vertex v_i can appear at more than one position in the trail. Each such position is called a *terminal.* v_k and v_0 are called the *end terminals* of the trail. The *degree* of a vertex v, denoted $d(v)$, is the number of edges adjacent to it. A graph is *Eulerian* if an e-trail exists in it. It is well known that a graph is Eulerian if and only if all vertices in the graph have even degree. An e-trail for the graph in Figure 3 is $T_1=(s,2,3,4,5,6,7,4,s,6,2,1,s)$.

After G is created, we determine whether it is Eulerian or not. If it is not, we add a vertex, called the *super-vertex*, v_s, to G and make it Eulerian by adding a new edge (v_s, v_i), called a *super-edge,* for each odd-degreed v_i (dashed lines in Figure 3). We call the resulting graph *the modified diffusion graph.* On the modified diffusion graph with m edges, we can find an e-trail using a well known algorithm with $O(m^2)$ time complexity [13]. There is also a linear time recursive algorithm for finding an e-trail which can be modified to handle various constraints [14].

There are exponentially many e-trails in an Eulerian graph [15]. Another e-trail for the modified diffusion graph in Figure 3 is: $T_2=(s,1,2,6,s,2,3,4,5,6,7,4,s)$. There are 176 e-trails in this graph. The corresponding transistor orderings for T_1 and T_2 are shown in Figure 2 b) and c).

Figure 3. The diffusion graph, G, of the discharge circuitry in Figure 2. s is the super-vertex that makes G Eulerian.

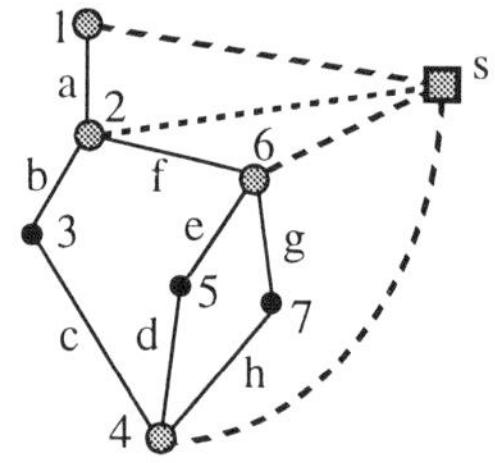

The main goal of transistor ordering is to minimize the total diffusion area of the layout. It is easy to see that any e-trail that starts with a super-vertex minimizes the diffusion area. Conversely, any transistor ordering that minimizes the diffusion area is an e-trail in the modified diffusion graph. This property is essential for the efficacy of our modified diffusion graph representation. On the other hand, some e-trails are better than others with respect to the cell height (See Figure 2), which is the subject of the next section.

4 Minimum-width Area Minimization

We propose an efficient technique that reduces the cell height significantly by searching *only* among the minimum-width transistor orderings. The modified diffusion graph (described in the previous section) is employed to model the circuit and the layout. As mentioned above, this representation ensures that the cell width is kept at a minimum and the total diffusion area is minimized. This is critical for circuit performance and for fixed-width cell methodologies (e.g., in data-path design [16]). The second goal, which is also very important for area minimization, is to find a transistor ordering that requires a minimum number of routing tracks. The following sub-sections explain the basic ingredients of our iterative improvement technique to perform this task.

A. Move generation

A *move* is an operation on an e-trail which converts it into another e-trail. In a good iterative improvement technique it is important that the *move set*, i.e., the set of moves employed, corresponds to a neighborhood structure which spans the whole search space. Otherwise, there will be some configurations (e-trails or transistor orderings) that will not be reachable, resulting in a sub-optimal solution. In the previous section we have shown that any minimum-width ordering can be represented with an e-trail in the modified diffusion graph. Here, we introduce two moves which are sufficient to ensure that it is possible to transform any e-trail into any other via a sequence of moves: *sub-trail modification* and *rotation*.

A *sub-trail* is a set of consecutive edges in an e-trail. The *sub-trail modification* move changes the order of a set of adjacent transistors. We first pick an arbitrary sub-trail of the e-trail. Next, we generate a sub-graph, G', that has *only* the vertices and the edges in that sub-trail (including the super-vertex and super-edges if any). Then, we find a *different* sub-trail in G'. Figure 4 shows an example. Note that G' is either Eulerian or it has at most 2 odd-degreed vertices. In either case, the algorithm given in [14] can find an e-trail in $O(m')$, where m' is the number of edges in the sub-trail.

Figure 4. Sub-trail modification move on the graph of Figure 3. Sub-trail *(6, s, 4, 7, 6, 2)* is replaced with *(6, 7, 4, s, 6, 2)*, where *s* corresponds to the gap.

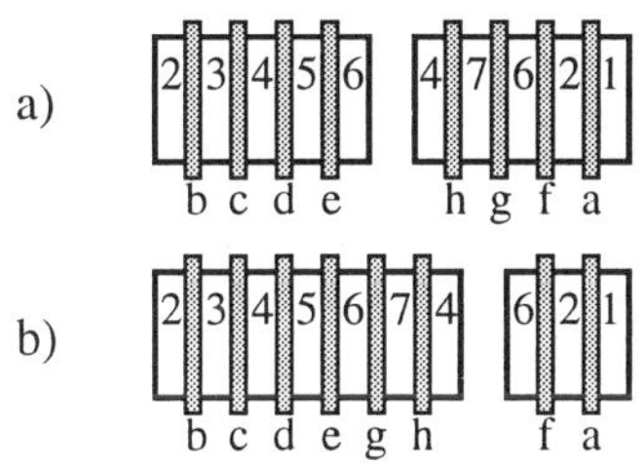

The *rotation* move rotates the transistors in the ordering cyclically without breaking the shared diffusions. Figure 5 shows an example.

Figure 5. Rotation move. The e-trail of Figure 4 b), *(s,2,3,4,5,6,7,4,s,6,2,1,s)*, is rotated to obtain a new e-trail: *(s,6,2,1,s,2,3,4,5,6,7,4,s)*.

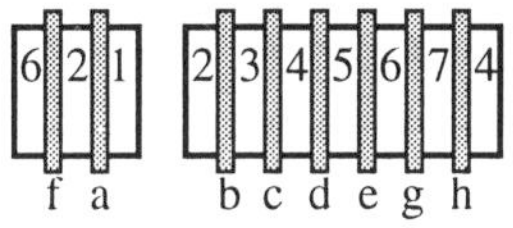

Note that neither of these two moves violates the Eulerian property of the trail: the trail generated by the move is also an e-trail. Therefore, the corresponding transistor ordering has minimum number of gaps i.e., minimum diffusion area is guaranteed.

B. Move evaluation

A move is evaluated with respect to the change in the cost of the transistor ordering as a result of the move. The cost of the transistor ordering is the final layout area. Since the moves in our move set do not change the cell width, we only need to look at the cell height to evaluate the new transistor ordering. The cell height is determined by the number of horizontal wiring tracks required to route the diffusion terminals. We denote it as the *routing height.* To find the routing height, we use the left-edge algorithm [17]. For a one-sided channel with n nets, the left-edge algorithm finds the optimal routing in $O(n)$. Figure 2 b) and c) show two routing examples generated by the left edge algorithm.

C. Iteration control

It is important that an iterative improvement scheme does not get stuck in local minima. We employed simulated annealing, which has been found to be very effective in optimizing many difficult layout problems [18][19]. In our implementation, we use a generic simulated annealing library that provides various effective iteration control mechanisms [20].

5 Results

The transistor ordering technique presented in this paper has been implemented in the C language on an IBM PowerPC 604 workstation running AIX 4.1. We have tested the algorithm on various dynamic CMOS circuits from the industry and literature. Table 1 lists some of these circuits and shows the results that were obtained.

Table 1. *m* is the number of edges in the modified diffusion graph. *m´* is the number of transistors in the circuit and *s* is the number of super-edges added. *H* is the number of routing tracks for the worst ordering. *h* is the result of our technique.

circuit	m=m´+s	H	h	cpu (sec.)
chak1	8+4	4	2	0.6
chak2	18+8	10	3	1.5
chak3	8+2	3	2	0.5
ck1	9+6	5	2	0.8
ck2	8+4	4	2	0.6
d1	11+2	5	2	0.7
d2	14+2	3	2	0.9
d3	34+2	3	2	2.1
dom3	7+2	3	2	0.5
uv1	16+8	9	3	1.4
uv1d	16+8	9	3	1.4
uv2	24+12	13	3	13.0
uv2d	24+12	13	3	14.0

Circuits *chak1-3* are from [6]; *ck1,2* are from [5]; *d1-3, dom3* are domino CMOS circuits from industry. The last four circuits are from [21]. For *all* of the circuits, the optimum area is obtained: the number of diffusion breaks (cell width) and the number of routing tracks (cell height) are minimum[1]. Figure 6 shows the minimum area result for *ck1*. Both the number of gaps (2) and the number of routing tracks (2) is less than that reported in [5] (3 gaps, 4 tracks)[2].

It is asserted in [21] that *uv1* (*uv1d* is the dual), is a "difficult" circuit for row-based layouts. Indeed, when there is no height minimization, the area can be proportional to $O(m^2)$, for m transistors, whereas the gate-matrix style can produce a layout with $O(m\log(m))$ area. However, with our area minimization technique the transistor ordering has an area of $O(m)$. This is a positive result for row-based layouts as compared to gate-matrix.

It is also worth noting that the enumeration technique of [8] for sCMOS (it was recently used in [22]) could also be applied to these circuits[3]. Unfortunately, for the last two circuits that exhaustive algorithm would require time proportional to

1. For verification, the optimum height as well as the worst case height for the last four circuits were found exploiting the special graph structure. For the other circuits we used a backtrack search technique for enumeration [15]. The full search takes around **16 hours** for *chak2*.

2. We should also mention that this is not a complete comparison, since [5] uses a different layout style, called *folding layout*.

3. The time complexity of that algorithm is super-exponential, $O(h^{3h*h})$, where h is the number of transistors in the longest discharge path [8].

$13^{3x13x13}$. However, our iterative improvement technique was able to find the *theoretical optimum* in less than 15 seconds.

Figure 6. Schematic of a dynamic CMOS circuit, *ck1*, and the optimum transistor ordering obtained for the nFET circuitry. For illustration purposes the routing is done in the channel, but it can also be done over-the-transistors to eliminate the routing footprint and save area.

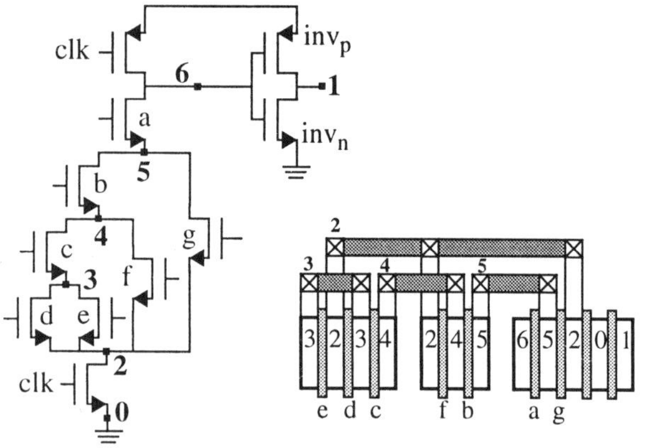

6 Conclusions

A new transistor ordering technique was presented for area minimization of dynamic CMOS circuits. It is able to find a minimum area transistor ordering for all the practical circuits we have tested it on. It is very efficient, finding the optimum layout in less then 2 seconds for realistic dynamic CMOS leaf-cells. We are currently working to reduce routing blockages and handle gate connections effectively. We are also extending the current one-row technique to multi-row and 2-D styles by integrating it into a macro-cell style placer for overall area efficiency.

We are also working on a performance-driven transistor ordering mechanism for parasitic sensitive dynamic CMOS circuits. We hope to extend analog leaf-cell layout techniques that control circuit parasitics in order to meet constraints on the performance of the circuit [14][23][24].

Acknowledgments

The authors would like to thank Prof. Ron Bianchini and Aykut Dengi (CMU) for helpful discussions and Dr. John Cohn (IBM) for providing some of the circuits. This work is supported in part by the Intel Corporation and the Semiconductor Research Corporation.

References

[1] A. Domic, "Layout synthesis of MOS digital cells", *Proceedings of the IEEE/ACM Design Automation Conference,* June 1990, pp. 241-5.

[2] S. Chakravarty, X. He, S.S. Ravi, "Minimum Area Layout of Series-Parallel Transistor Networks is NP-Hard", *IEEE Transactions on Computer-Aided Design,* Vol. 10, No. 7, July 1991, pp. 943-949.

[3] T. Lengauer and R. Muller, "Linear Algorithms for Optimizing the Layout of Dynamic CMOS Cells", *IEEE Transactions on Circuits and Systems,* Vol. 35, No. 3, March 1988, pp. 279-285.

[4] C.T. McMullen and R.H.J.M. Otten, "Minimum Length Linear Transistor Arrays in MOS", *IEEE International Symposium on Circuits and Systems,* August 1988, pp. 1783-1786.

[5] H.Y. Chen and S.M. Kang, "iCOACH: A Circuit Optimization aid for CMOS high-performance circuits", *Proceedings of the IEEE/ACM International Conference on Computer-Aided Design,* November 1988, pp. 372-375.

[6] S. Chakravarty, X. He, S.S. Ravi, "On Optimizing nMOS and Dynamic CMOS Functional Cells", *IEEE International Symposium on Circuits and Systems,* Vol. 3:, May 1990, pp. 1701-4.

[7] S. Wimer, R.Y. Pinter, J.A. Feldman, "Optimal Chaining of CMOS Transistors in a Functional Cell", *IEEE Transactions on Computer-Aided Design,* Vol. CAD-6, September 1987, pp. 795-801.

[8] R.L. Maziasz and J.P. Hayes, *Layout Minimization of CMOS Cells,* Kluwer Academic Publishers, Boston, 1992.

[9] A. Stauffer and R. Nair, "Optimal CMOS cell transistor placement: a relaxation approach", *Proceedings of the IEEE/ACM International Conference on Computer-Aided Design,* November 1988, pp. 364-7.

[10]C.Y. Hwang, Y.C. Hsieh, Y.-L. Lin, Y.-C. Hsu, "An Efficient Layout Style for Two-Metal CMOS Leaf Cells and its Automatic Synthesis", *IEEE Transactions on Computer-Aided Design,* Vol. 12, No. 3, March 1993, pp. 410-424.

[11]N.H.E. Weste and K. Eshraghian, *Principles of CMOS VLSI Design,* Addison-Wesley Publishing Company, 1993.

[12]T. Uehara and W. M. vanCleemput, "Optimal Layout of CMOS Functional Arrays", *IEEE Transactions on Computers,* Vol. C-30, No. 5, May 1981, pp. 305-312.

[13]J.A. Bondy and U.S.R. Murty, *Graph Theory with Applications,* Elsevier Science Publishing, New York, 1976.

[14]B. Basaran and R.A. Rutenbar, "An O(n) algorithm for optimum CMOS device stacking with analog constraints", *Research Report No. CMUCAD-95-56,* Dept. of Electrical and Computer Engineering, Carnegie Mellon University, 1995.

[15]A. Nijenhuis and H.S. Wilf, *Combinatorial Algorithms,* Academic Press, New York, 1975.

[16]Y. Tsujihashi, et al., "A High-Density Data-Path Generator with Stretchable Cells", *IEEE Journal of Solid-State Circuits,* Vol. 29, No. 1, Jan 1994, pp. 1-8.

[17]A. Hashimoto and J. Stevens, "Wire routing by optimizing channel assignment within large apertures", *Proceedings of the IEEE/ACM DAC,* June 1971, pp. 155-69.

[18]S. Kirkpatrick, C. D. Gelatt, and M. P. Vecchi, "Optimization by Simulated Annealing", *Science,* November 83, pp. 671-680.

[19]C. Sechen and K. Lee, "An Improved Simulated Annealing Algorithm for Row-Based Placement", *Proceedings of the IEEE/ACM ICCAD,* 1987, pp. 478-481.

[20]E. S. Ochotta, L. R. Carley, and R. A. Rutenbar, "Analog Circuit Synthesis for Large, Realistic Cells: Designing a Pipelined A/D Convert with ASTRX/OBLX", *IEEE Custom Integrated Circuits Conference,* May 1994, pp. 15.4/1-4.

[21]C.C. Su and M. Sarrafzadeh, "Optimal gate-matrix layout of CMOS functional cells", *Integration, the VLSI journal,* Vol. 9, 1990, pp. 3-23.

[22]B. Guan and C. Sechen, "An area minimizing layout generator for random logic blocks", *Proc. 1995 IEEE CICC,* May 1995, pp. 23.1/1-4.

[23]E. Charbon, E. Malavasi, D. Pandini, A. Sangiovanni-Vincentelli, "Imposing Tight Specifications On Analog ICs through Simultaneous Placement and Module Optimization", *Proc. IEEE CICC,* May 1994, pp. 525-28.

[24]B. Basaran, R. A. Rutenbar and L. R. Carley, "Latchup-Aware Placement and Parasitic-Bounded Routing of Custom Analog Cells", *Proceedings of the IEEE/ACM International Conference on Computer-Aided Design,* November 1993, pp. 415-421.

Gate-Array Library Design Using Local Interconnect

Larry Wissel, Douglas Stout, Nathan Buck

IBM Microelectronics
Essex Junction, Vermont 05452

Abstract

An ASIC gate-array library has been created in 0.4 µm CMOS technology using a local interconnect level. The gate-array cells in this library are denser than their counterparts in a library without local interconnect. The comparison of two benchmarks, including a 520K-gate ASIC routed with both libraries, further shows that the local interconnect allows higher density of ASIC designs due to more efficient use of the global interconnect layers.

Introduction

Advances in processing technology can improve the interconnect between circuits to increase ASIC gate density. Metal layers may be added to an ASIC, allowing a denser interconnect of the underlying gate-array cells; or a local interconnect layer can be included, which makes more global routing tracks available with the existing metallurgy through more efficient intracell routing [1].

The local interconnect profoundly impacts the design of gate-array cells. Access to this technology feature has prompted the definition of a new gate-array-cell template, and a new convention for intracell routing of the transistors within the gate-array cell. This paper concentrates on gate-array-cell design; standard-cell design types have been previously described [2].

The local interconnect used by some foundries allows direct connection to diffusion or polysilicon. It is typically a thin Ti-based film that is well suited for local interconnections, but too resistive to be freely used as a component of global interconnections [3, 4]. The local interconnect used by IBM Microelectronics is a robust damascene tungsten with resistance of only 0.3 ohms per square [5]. This structure may be used for not only local interconnects within a gate-array cell, but also as a component of global routing. This layer is required as the bottom component of a two-layer contact scheme between metal-1 and silicon (or polysilicon). Fig. 1 illustrates the structures through metal-1 of RAM cells with local interconnect.

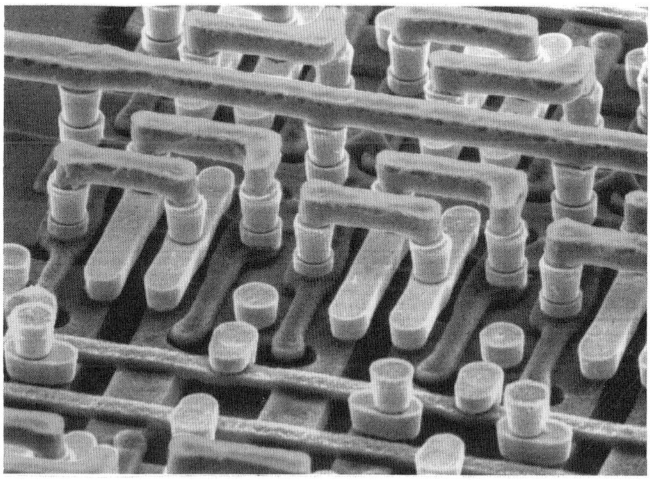

Figure 1. RAM cells with local interconnect.

In this SEM view of an array with dielectrics (and some metal-1 shapes) removed, the local interconnect connects PMOS and NMOS drains, and is the lower component of the two-layer contact to metal-1. Chemical-mechanical polishing has planarized this layer, as demonstrated by the uniform elevation of the top of the local interconnect layer over both diffusion and polysilicon [5]. The upper component of the two-layer contact is the uniform-size tungsten stud.

A selective etch allows the local interconnect tungsten to contact either diffusion or polysilicon without a border. The borderless contact allows the inclusion of features in the gate-array cell that would not be possible with processing requiring a traditional border.

Comparison of Cell Templates

Two technologies have been used to create the ASIC libraries to be compared: CMOS 5L without local interconnect, and CMOS 5S with local interconnect [6]. The technology features are summarized in Table 1. The diffusion-isolation layout rules are identical, as are the metallization rules from the contact level through all five metal layers with vias. From a density perspective, the only differing aspect is the availability of local interconnect and the accompanying borderless contact in CMOS 5S.

0-7803-3177-6 $5.00 © 1996 IEEE

IEEE 1996 CUSTOM INTEGRATED CIRCUITS CONFERENCE

Table 1. Technology characteristics.

	CMOS 5L	CMOS 5S
Leff	0.46 μm	0.36 μm
Tox	13.5 nm	9 nm
Local interconnect pitch		1.2 μm
Metal-1 pitch	1.4 μm	1.4 μm
Metal-2 - Metal-5 pitch	1.8 μm	1.8 μm

The cell templates for each gate-array library have the following characteristics:

1. sixteen 1.8μm routing tracks in height;
2. a continuous active area with PMOS-NMOS transistor pairs placed on a 1.8μm pitch;
3. metal-1 Vss and Vdd power busses which take up two horizontal routing tracks each;
4. boundary conditions at the vertical edges, which require that the diffusions be connected to the power supply to allow abutting gate-array cells to share common diffusion edges;
5. antenna-protection diodes to eliminate metal-etch induced threshold shifts on the gates connected to the input ports of the circuit built on the cell template;
6. an Nwell tap at the top edge of the cell.

The CMOS 5S cell template also has a p+ active area at the bottom edge to serve as a Pwell tap.

The basic unit of each gate-array template is two PMOS-NMOS pairs. Fig. 2 shows four such basic units placed side-by-side for each library.

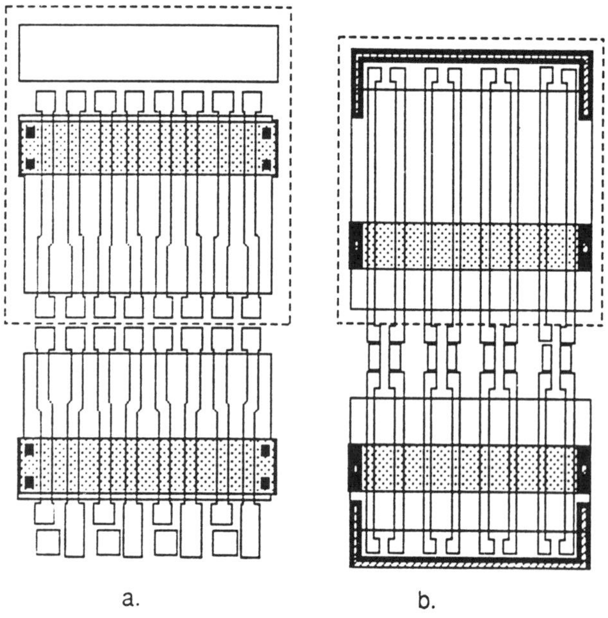

Figure 2. CMOS 5L (a) and CMOS 5S (b) cell templates.

CMOS 5L requires a border around the diffusion contacts, limiting the antenna-protection diodes to a 2.0μm pitch. Since the transistor pitch is 1.8μm, one antenna-protection diode is placed for every two transistor pairs.

Fig. 2(b) illustrates the gate-array template for the CMOS 5S ASIC library. CMOS 5S, with a local interconnect and no diffusion contact border requirement, can implement antenna-protection diodes at a 1.8μm pitch, allowing a diode to be placed for each transistor pair. Local interconnect is used to route the power supply potential at the cell-edge diffusions to the Nwell and Pwell taps, and also to strap the taps. In CMOS 5S, the Nwell tap and Pwell taps are each small due to the borderless feature.

Local interconnects, as the bottom component of the diffusion contact, can be placed on any diffusion. However, the local interconnect can directly connect PMOS-NMOS drains only between the second and third transistor pairs; between the fourth and fifth transistor pairs; and between the sixth and seventh transistor pairs without shorting to polysilicon.

Cell layouts

Fig. 3 shows a CMOS 5L implementation of AND3 with contact and metal-1 mask shapes on the gate-array template of Fig. 2(a). This cell is limited by antenna-protection density rather than by transistor density. The first two PMOS-NMOS pairs on the left are not required except that their antenna-protection diode is needed. Metal-1 is used to connect the gates to the diodes. Transistor pairs 3 through 5 (from left) are used to create the NAND3 input stage. Pair 6 is used to isolate the drains of the NAND3 from the output inverter with vertical tabs of metal-1 from the power busses to the gates. Transistor pairs 7 and 8 create the output inverter, completing the cell.

Metal-1 is also used to connect PMOS and NMOS polysilicon gates, and drains. For electromigration robustness, the metal-1 is expanded to cover two contacts over both drains of the output inverter. Metal-1 blocks 13 of the 16 global tracks of the cell.

To take fullest advantage of the second layer for cell design, a new intracell routing convention is used in CMOS 5S layouts: local interconnect is used primarily for vertical runs, and metal-1 is constrained to the horizontal orientation.

Local interconnect is used to isolate gates, to connect PMOS and NMOS gates and antenna-protection diodes together, and to make up as much of the vertical connection between PMOS and NMOS drains as possible.

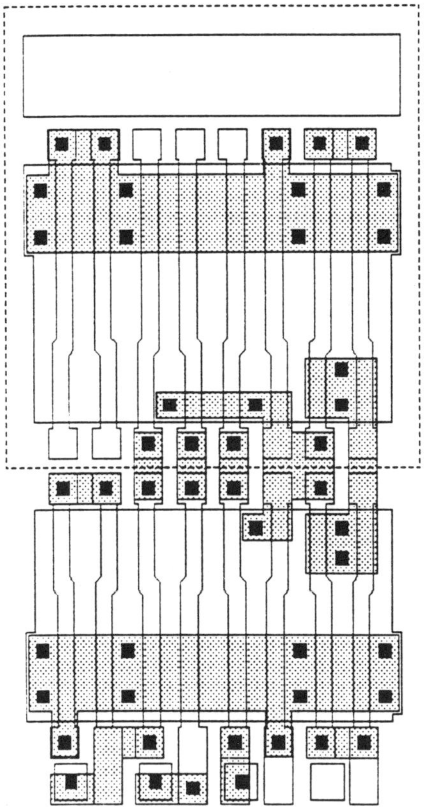

Figure 3. CMOS 5L AND3 gate-array cell.

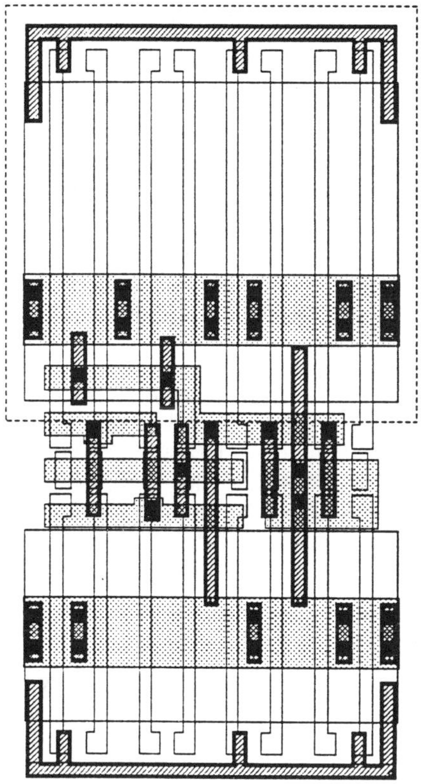

Figure 4. CMOS 5S AND3 gate-array cell.

Fig. 4 shows the CMOS 5S implementation of AND3. The first PMOS-NMOS pair is not used; pairs 2, 3, and 4 make up the NAND3 input, and pair 5 isolates the drains of the NAND3 from the output inverter. Pairs 6 and 7 are the output inverter, and pair 8 is not used. The local-interconnect drain-to-drain connection at pairs 6 and 7 has larger contact area to diffusions than even the contact pairs of Fig. 3 achieve, and the output port metal-1 has two contacts to the local interconnect for electromigration.

This cell could be implemented in just six transistor pairs by using pairs 1, 2, and 3 for the input nand, pair 4 for isolation, and pairs 5 and 6 for the output inverter. Since only the diffusions between pairs 2 and 3 and 4 and 5 could have PMOS-NMOS drain connections, the six-pair implementation would require metal-1 drain connections. This metal-1 would block two additional global tracks just to make the minimum contact from an electromigration perspective. Since ASIC designs are limited more by global interconnects than by library cell density, the six-pair implementation is less desirable.

In Fig. 4, only eight of the 16 global routing tracks are required by the cell metal-1. Considering that four tracks are required by the power busses in each library, the advantage of local interconnect is even more clear -- intracell routing requires nine tracks in CMOS 5L, and only four tracks in CMOS 5S.

The versatile intracell routing with local interconnect allows complex cells to be smaller; XNOR3 in CMOS 5L requires 18 transistor pairs; in CMOS 5S, only 12 transistor pairs are required.

Overall ASIC Density Benefit

A benchmark was created to demonstrate the superiority of the local-interconnect library from a cell density and routing perspective. A small design was synthesized to a netlist containing 2957 CMOS 5L library cells; the cells were then placed in an area constrained to 1.076 mm by 1.076 mm.

The CMOS 5L gate-array cells filled this area to a 94.8% utilization. To demonstrate the gate-array cell density of CMOS 5S with local interconnect, the CMOS 5L cells were replaced with their CMOS 5S equivalents; all placements were unchanged. The resulting utilization factor was 71.3%. The utilization ratio (94.8%:71.3%) is a valid indication of the average density advantage of the CMOS 5S gate-array library over the CMOS 5L library. The antenna-protection diode density and the better intracell routing with local interconnect allow the CMOS 5S cells to be smaller.

The benchmark contained 3065 nets interconnecting 9811 ports. Even though four metal layers were available, only 2842 nets could be successfully routed using the CMOS 5L library with its available routing tracks. The total wire length of the successfully routed nets was 1010.3mm, of which 54.0mm was on metal-1. The total Manhattan length of the 223 nets that could not be wired was 66.8mm.

As a contrast, the benchmark wired successfully with the CMOS 5S library. Metal-1 usage was up to 122.1mm. Typically, the CMOS 5L gate-array cells have three free metal-1 tracks (e.g., AND3 of Fig. 3), whereas the CMOS 5S gate-array cells have seven (AND3 of Fig. 4 has eight). The ratio of total metal-1 wire length in the benchmark (54.0mm:122.1mm) is almost identical to the ratio of the typical number of free tracks (3:7).

A second benchmark was created to verify this routing difference on a larger scale. The starting point for the second benchmark was an actual 520K-gate ASIC design in CMOS 5L. This ASIC is a four-metal-layer design containing 122,000 placeable objects. As with the first benchmark, the design was placed and routed in the CMOS 5L gate-array library, and then re-routed using the CMOS 5S library. Cell-for-cell replacements were again used for CMOS 5S without any placement modification.

The CMOS 5L netlist used in the second benchmark included custom-diffused register arrays, which are denser in CMOS 5S; the original arrays were left in the design to limit the variables in the benchmark.

The routing was done using CELL3 (Cadence Design Systems). The total routing time was 41.8 hours with the CMOS 5L library; the CMOS 5S library, with more available tracks, required only 24.8 hours. There are indications, however, that extra horizontal tracks available with the CMOS 5S library are even more important than the total routing time implies because the time required for the detailed routing step differed dramatically: 10.0 hours for CMOS 5L, and under one minute for CMOS 5S.

Wire lengths were also reduced. Total CMOS 5L wire length (all four layers) was 94.43m; total CMOS 5S wire length was 91.86m. The total vertical routing for both metal-2 and metal-4 differed little; likewise, the total horizontal routing for metal-1 and metal-3 did not greatly differ. But, as in the first benchmark, the total metal-1 contribution differed greatly. CMOS 5L metal-1 wire length was 5.85m; CMOS 5S metal-1 wire length was 14.46m; again, this ratio (5.85m:14.46m) is very close to the ratio of the typical number of free metal-1 tracks (3:7) of cells in the two libraries.

Summary

The IBM Microelectronics CMOS 5L and CMOS 5S technologies offer identical metal-1 through metal-5 layers. Thus, any difference in the ASIC routing efficiency between the CMOS 5L ASIC and CMOS 5S ASIC is attributable only to the lower usage of metal-1 for intracell routing in the CMOS 5S library because of the local-interconnect availability. A new intracell routing scheme has been developed to take advantage of local interconnect, freeing up metal-1 for additional global interconnects.

CMOS 5S gate-array cells are smaller than their CMOS 5L equivalents because of the local interconnect, which provides a density advantage in carefully placed portions of the ASIC design where the routing is not the limiting factor. Two benchmark experiments have demonstrated the benefits of gate-array cell design using local interconnect. Each benchmark has shown that the design is much more wirable because extra metal-1 routing tracks are made available by the local interconnect.

References

(1) C.W.Koburger III et al, "A Half-Micron CMOS Logic Generation", IBM J. Res. Dev.. V.39, No.1-2, Jan.-March 1995.

(2) C.Bachelu, M.Lefebvre, "A Study of the Use of Local Interconnect in CMOS Leaf Cell Design", IEEE Transactions on VLSI Systems, V.1, No.4, December 1993.

(3) A.I.Nasr, G.J.Grula, A.C.Berti, R.D.Jones,"CMOS-4 Technology for Fast Logic and Dense On-Chip Memory" Digital Technical Journal, V.4, No.2, Spring 1992.

(4) S.-P.Jeng, J.A.West, D.Wyke, "A Novel TiSi2/TiN Clad Local Interconnect Technology", 1993 Symposium on VLSI Technology Digest of Technical Papers.

(5) F.White et al. "Damascene Stud Local Interconnect in CMOS Technology", 1992 IEEE International Electron Devices Meeting Digest of Technical Papers, p.301.

(6) R.Gregor et al, "A One Million-Circuit CMOS ASIC Logic Family", 1993 IEEE Custom Integrated Circuits Conference Digest of Technical Papers.

Exploring Multiplier Architecture and Layout for Low Power

Pascal C. H. Meier, Rob A. Rutenbar and L. Richard Carley

Department of Electrical and Computer Engineering
Carnegie Mellon University, Pittsburgh, PA 15213
email: pascal@ece.cmu.edu rutenbar@ece.cmu.edu lrc@ece.cmu.edu

Abstract*

Multiplication represents a fundamental building block in all DSP tasks. Due to the large latency inherent in multiplication, schemes have been devised to minimize the delay. Two methods are common in current implementations: regular arrays and Wallace trees. Previous gate-level analyses have suggested that not only are Wallace trees faster than array schemes, they also consume much less power. However these analyses did not take wiring into account, resulting in optimistic timing and power estimates. We develop a simplified comparative layout methodology to analyze the effect of physical layout on these designs. Results for short bit-width (8, 16, 24 bit) DSP multipliers show that while wiring has a major impact on signal delay and power, Wallace trees still show roughly a 10% power advantage over array-based designs.

Introduction

There has been renewed interest in basic digital arithmetic in the last few years, driven originally by the migration of fast floating point hardware into standard microprocessors, and more recently by the demand for multimedia functionality in both desktop and portable systems. In the latter case, the demands of DSP-style systems for both high throughput (e.g., for voice, video) and low energy consumption has spawned new work in low-power circuit styles [1], DSP synthesis [2], and examination of basic tradeoffs in different arithmetic styles [3]. Our interest is in the basic building blocks of arithmetic circuits, in particular, short word width (8 - 24 bit) multipliers of the type that dominate in DSP applications.

For individual arithmetic blocks, different gate-level architectures have a substantial impact on hardware size, layout complexity, speed and power. For basic adder types (ripple, lookahead, skip, bypass, *etc.*), the power tradeoffs were clearly mapped by Callaway and Swartzlander in [4]. For example, ripple adders are slowest but use the least energy, whereas speculative styles such as bypass adders are fast but consume much more power, since they perform computations they later discard.

The tradeoffs for basic multipliers, however, are much less well characterized. Alternative designs focus on the manner in which the trapezoidal partial product array of individual bit-wise products are reduced (summed) to produce the final product (see Figure 1). *Array* styles [10] use a regular 2-D grid of adders for this reduction. Compact and easy to lay out, the arrays perform this reduction in gate depth that is linear in the bit width. At the other end of the spectrum, *Wallace tree* styles [3] use a log-depth tree network for this reduction. Faster, but irregular, they trade ease of layout for speed. Although the speed-size tradeoffs for these two styles are fairly well characterized, the power tradeoffs are not well understood.

For example, Bellaouar and Elmasry [5] suggest that Wallace tree styles are best avoided for low power applications, since the excess wiring is likely to consume extra power. On the hand, Callaway and Swartzlander [4] demonstrate quantitatively that switching activity within just the partial product reduction hardware is substantially better for the tree over the array---if one ignores the wires completely. Montoye [7] alludes to the difficulties in dealing with the tree wiring for one high-performance commercial processor (IBM RS/6000), but concludes that the speed gain is worth the trouble; unfortunately power was not a concern in this design.

In this paper, we attempt to shed some light on the question: *are Wallace tree style multipliers competitive with array styles for speed and power*? Intuition suggests that the log-versus-linear depth of the reduction network for the tree might well lead to shorter propagation paths and less power-consuming glitching. First order analysis of the average transition counts across sets of random vectors applied to both array and Wallace tree designs supports this point of view (see Figure 2.), as does the work of

Figure 1. Multiplier Architectures

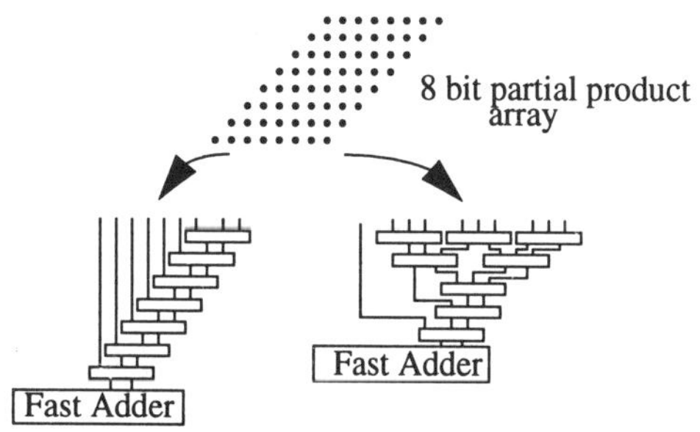

(a) Array architecture (b) Wallace Tree architecture

* This work was supported in part by the Semiconductor Research Corp, the National Science Foundation, and ARPA.

0-7803-3177-6 $5.00 © 1996 IEEE

IEEE 1996 CUSTOM INTEGRATED CIRCUITS CONFERENCE

Figure 2. Transition count comparison for multipliers.

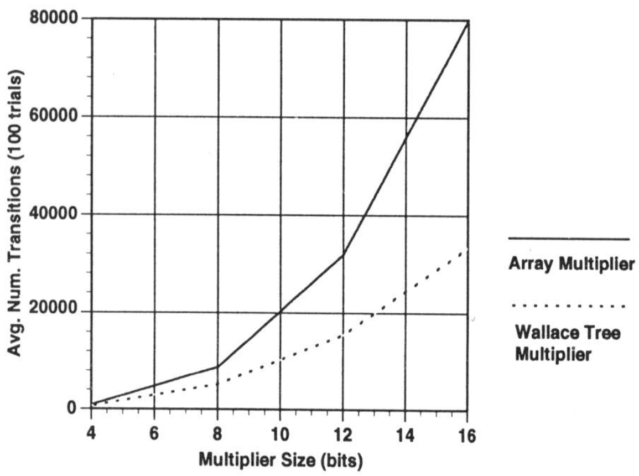

[4]. However, these analyses are incomplete due to the lack of any consideration of wiring effects, and are therefore unable to support or refute the claim of [5].

To remedy this, we suggest a more detailed evaluation model that considers not only the gate-level differences, but also the wiring effects due to layout. As we shall see, experiments suggest a posture of cautious optimism for the Wallace trees: they are neither as bad as [5] suggests, nor as good as [4] estimates, but they do appear roughly comparable within the limits of our model, with the tree style somewhat better on both energy consumption and on speed. The rest of the paper describes our modeling methodology, our layout strategy for comparing arrays and trees, and our experimental results.

Analysis Methodology

To fairly compare the array and Wallace tree multiplier styles, we need the following:

- **Cell-level generators for the multipliers' partial product reduction networks:** where "cell" here means the component (3,2) carry save adders and single-product AND gates needed in the design.
- **Cell-level generators for multipliers' final 2-level reducing adders:** each style requires a final addition, and each can use different adder styles to achieve different tradeoffs.
- **Cell characterization:** to understand power consumption in each of these primitive component cells, as implemented by conventional static CMOS circuits.
- **Layout model:** to allow exploration and first-order estimation of wiring, delay and area, we need a simple layout style that is easily automatable, yet not so detailed as to demand manual, full-custom layout.
- **Power and delay estimation:** to measure power and delay for the completed multiplier layouts–including the delay and capacitance of all the intercell wiring.

We summarize our strategy for each of these in this section.

For simplicity, we restrict ourselves to unsigned operands, and focus exclusively on the short word widths common in DSP applications: 8, 16 and 24 bits. Cell-level generation of the netlist for each multiplier is straightforward. The only degrees of freedom here are the detailed arrangement of the levels of carry save adders in the Wallace tree, for which we use the basic arrangement suggested by Wallace[3], and the final reducing adder at the bottom of each architecture. Since this adder is a large part of the overall delay, we evaluate with respect to three different adder styles for each: carry ripple, carry skip, and carry select addition. Carry select addition precomputes carries during the addition operation. Carry skip addition uses parallel evaluation of the criterion for carry propagation to allow the carry to traverse the adder faster. Ripple addition is simply a chain of full adders, and represents the most basic addition technique. See [11] for more details.

Given these netlists of basic cells, we next characterize the power and delay of each cell prior to analysis of the overall multiplier. We designed simple static CMOS circuits for each component in our multipliers, using parameters from the Hewlett-Packard 0.8μm CMOS process (CMOS26G); the designs assume a 3 V supply and are based on circuits presented in [10]. There are two components to power dissipation in these cells: a) the power dissipated in the circuit switching, and b) power delivered and removed from the capacitive load on each output. Both effects were measured using HSPICE simulations for all output transitions, at various capacitive loads. Cell delay was also calculated, in this case as the difference between the input and output signals at their 50% voltage points.

To evaluate complete multipliers, we require a common layout model for both the regular array and the Wallace tree. It is important here to avoid penalizing one style artificially at the expense of the other. We use a simple unit grid model, in which each component cell of the multiplier netlist occupies one slot in a set of standard cell-style rows (see Figure 3). We assume over-the-row routing of vertical wires connecting cells in different rows, but we conservatively estimate the impact on total height and wire length of the wires that must make horizontal jogs in each wiring channel. A post-process global router embeds wires into the placed model, and the maximum density of each channel is used to derive channel height and final wirelength estimates.

Figure 3. Multiplier layout model.

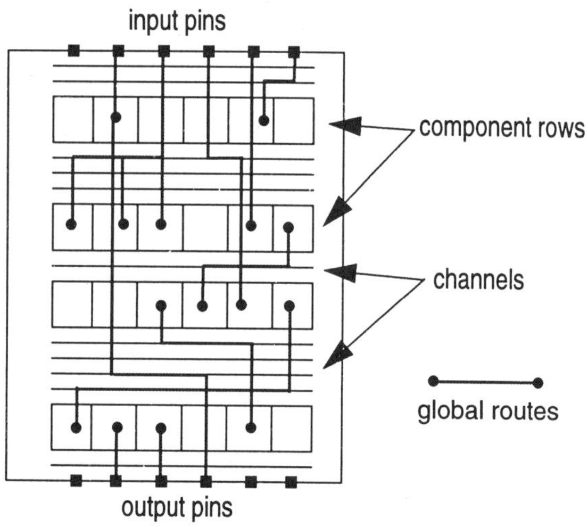

For the array, we procedurally tile the regular partial product reduction cells into the grid, with the final reduction adder carefully located at the bottom of the array. Since nearly all the connections in the array are nearest neighbor between cells in the partial product reduction network, the array fits well into this model. On the other hand, the Wallace tree requires constructive placement and routing of its partial product generation (AND gates), reduction network ((3,2) carry save adder tree) and final reduction adder. We use a simple annealing-based placement strategy [7] that strives to minimize the overall wirelength while densely packing the grid.

Finally, given a complete netlist and a real, although highly simplified placement, we can extract back per-net capacitance and determine estimates of both power and delay for each complete multiplier. Initially, we developed Verilog files for the various multiplier architectures, where power was computed by counting the number of transitions which occurred at each cell output. To accelerate this, we developed a simplified, custom logic simulator for these designs. The simulator is an event-driven evaluation engine, where drivers send signals to receivers. If the value of a receiver is unresolved when a new driving signal arrives, this corresponds to a potential glitch, and the previous driving signal may be preempted. Since the filtering of small glitches from the event queue can cause power estimation to be inaccurate (*i.e.,* too low), power consumed during glitches is also estimated. Comparisons between our custom logic simulator, operating on characterized cell-level netlists with backannotated wire delays, against transient device-level HSPICE simulation on small (4 bit) multipliers has shown this method to be acceptably accurate, typically within 5% of HSPICE estimates.

Determining the complete power dissipated in a multiplier requires sensitizing all possible combinations of inputs. This means for a circuit of n inputs, 2^{2n} input combinations have to be simulated. However, it is possible to estimate *average* power dissipation by running sets of test vectors, and performing mean estimation using statistical techniques. Such techniques have been shown to be effective in determining the actual power dissipation within a known tolerance [6]. In our approach, we run batches of vectors to insure 95% confidence intervals, *i.e.*, to estimate that the computed average power dissipation is within 2% of the actual mean, with a confidence of 95% [9].

To estimate delay, we use three strategies. For the smaller multipliers (up to 8 bit operands) we exhaustively simulate all pairs of inputs and use our custom logic simulator to measure the worst case delay. For larger designs, we use simple static timing analysis (topological longest path) to derive a *pessimistic* worst case for delay. We also simulate batches of random vectors to estimate an *optimistic* worst case delay. Together these two measurements bound–albeit loosely–the worst case delay. In the following section, we present results of all these analyses.

Experimental Results

We explored 18 different multiplier implementations in all: two different architectures (array versus Wallace tree), three different final reduction adders for each multiplier (carry select, carry skip, carry ripple), and three different word widths (8, 16 and 24 bits).

Table 1. and Table 2. show the area estimates for each multiplier Recall that wiring impacts the overall area because of the density estimates for each wiring channel. Unsurprisingly, the array multipliers are always smaller due to their much more local wiring. Also, the ripple-adder versions are always the smallest, again due to the smaller hardware and more local wiring in these adders. As a point of comparison here, [12] describes a 6 bit array multiplier as part of an 8-tap FIR filter that occupies roughly 0.4 mm^2 in a 0.8μm CMOS process; this suggests that our area estimates are reasonable.

Table 1. Array multipliers - estimated area (mm^2)

Adder Type	8 bit	16 bit	24 bit
Carry Sel.	0.668	2.195	4.579
Carry Skip	0.627	2.099	4.430
Ripple	0.532	1.913	4.156

Table 2. Wallace tree multipliers - estimated area (mm^2)

Adder type	8 bit	16 bit	24 bit
Carry Sel.	0.759	2.626	5.625
Carry Skip	0.736	2.537	5.576
Ripple	0.725	2.488	5.576

Figure 4. shows estimated average energy per multiply operation for each of the multipliers. Results suggest a fairly consistent 10% energy advantage for the Wallace trees, across the three bit widths examined. Despite the larger amount of irregular wiring, the shallower partial product reduction in the Wallace tree now appears to be advantageous for power. However, given the coarseness of our layout model, a more cautious conclusion is simply that the trees' energy use appears roughly competitive with that of arrays.

Figure 5. shows estimated delay for each multiplier. In gen-

Figure 4. Estimated average energy per multiply op.

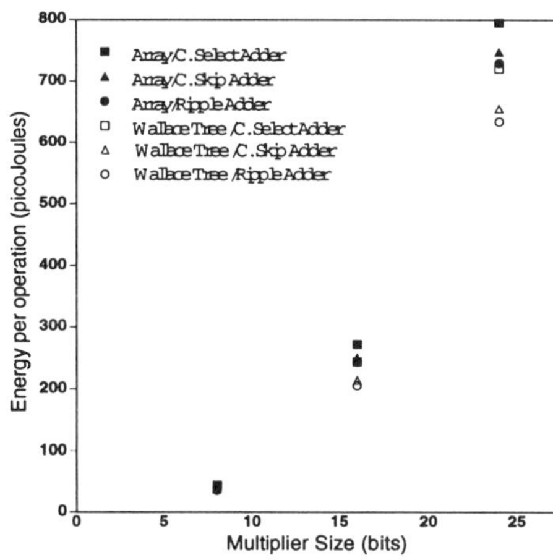

Table 3. 8 bit multiplier - estimated delay (ns)

	array		Wallace tree	
	exhaustive simulation	static timing	exhaustive simulation	static timing
Carry sel.	16.12	22.73	14.86	18.90
Carry skip	18.93	28.93	18.42	27.25
Ripple	18.79	27.94	18.81	26.47

eral, the delays are fairly long since we use a 0.8μm CMOS process, and because all devices are of minimum size. For the small 8 bit design, results of exhaustive simulation over all pairs of inputs appears in Table 3., along with more conservative static timing estimates for comparison. The static timing estimates are pessimistic by 30-50%. The greatest discrepancy occurs with the carry-skip adder, which has many false paths, due to carry propagate prediction circuitry. Note that the carry skip adder does not show a speed advantage at low bit width, due to the lookahead circuitry. Without this circuitry, the carry skip adder behaves like a ripple adder. Also note that the use of a ripple adder completely negates the advantage of using a Wallace tree, as expected.

Note that Figure 5. offers both a pessimistic (static timing) and an optimistic (worst case encountered during simulation of random patterns) timing estimate for each multiplier. The wide overlap of the array and Wallace tree timing intervals certainly suggests that the Wallace trees are at least competitive in delay. Indeed, the intervals for each Wallace tree cover smaller delays than the corresponding array interval, which also suggests that the trees are faster, again as expected.

Taken together, Figure 4. and Figure 5. suggest that the Wallace trees may indeed have an energy•delay advantage over the regular array multipliers. Wallace trees are not so bad as suggested by [5], nor as significantly superior as estimated by [4].

Figure 5. Estimated worst-case delay (ns).

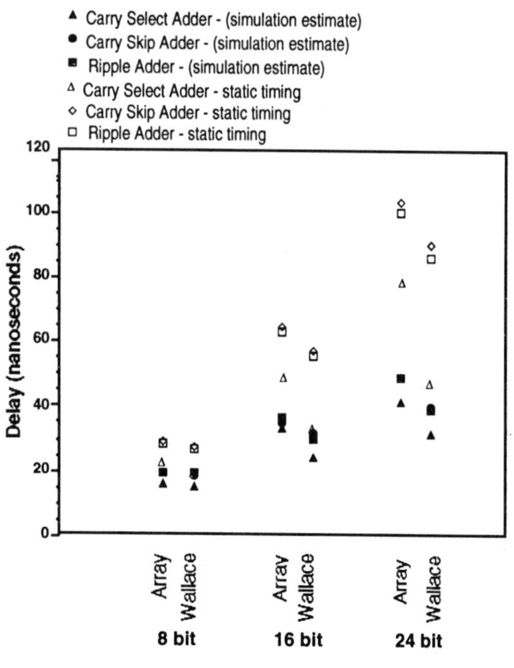

Conclusions

By introducing a simple unit grid layout model, we have been able to compare regular array and Wallace tree style unsigned multipliers over bit widths 8 to 24 bits, including first-order delay and area effects due to physical wiring. The model is clearly coarse, but capable of making basic predictions for area, for average power, and delay. Interestingly, the Wallace trees fare rather well, despite their irregularity and excess wiring. The smaller depth of their partial product reduction hardware seems to offset the power lost in the wiring, offering improved energy and delay. We believe this preliminary result justifies closer investigation with more refined models of problem, *e.g.*, use of more aggressive exact critical path analysis[11], and better layout optimization, to determine where the Wallace tree style may be useful.

Acknowledgments

We thank Dennis Ciplickas of Carnegie Mellon for his insights on circuit timing and power simulation issues.

References

[1] A.P. Chandrakasan, S. Sheng, and R.W. Brodersen, "Low Power Digital Design," *IEEE JSSC*, vol. 27, pp. 473-484, April 1992.

[2] A.P. Chandrakasan, M.Potkonjak, R. Mehra, J. Rabaey, R.W. Brodersen, "Optimizing Power Using Transformations," *IEEE Transactions on CAD*, Vol. 14, No.1 pp.12-31, Jan. 1995.

[3] C.S. Wallace, "Suggestions for a Fast Multiplier," *IEE Trans. Electron. Computers*, EC-13, pp. 14-17, 1964.

[4] T.K. Callaway, and E.E. Swartzlander Jr., "Low Power Arithmetic Components." in *Low Power Design Methodologies*, Rabey, J. and Pedram, M., eds., pp. 161-198, Norwell, Mass: Kluwer Academic Publishers, 1996.

[5] A. Bellaouar, and M.I. Elmasry, *Low-Power Digital VLSI Design, Circuits and Systems*, pp. 442-450, Norwell, Mass: Kluwer Academic Publishers, 1995.

[6] R. Burch, F. Najm, P. Yang, and T. Trick, "McPOWER: A Monte Carlo Approach to Power Estimation," *IEEE/ACM ICCAD*, pp.90-97, Santa Clara, CA, Nov. 8-12, 1992.

[7] R. Rutenbar, "Simulated Annealing Algorithms: An Overview," *IEEE Circuits and Devices Magazine*, pp. 19-26, Jan. 1989.

[8] R.K. Montoye E. Hokenek, S.L. Runyon, "Design of the IBM RISC System/6000 Floating-Point Execution Unit," *IBM Journal of Research and Development*, pp. 59-77, Vol. 34, No. 1, January 1990.

[9] N. Weste and K.Eshraghian, *Principles of CMOS VLSI Design*, p. 312, Addison-Wesley 1988.

[10] D. Goldberg, "Computer Arithmetic," in *Computer Architecture A Quantitative Approach*, J.L. Hennessy and D.A. Patterson, pp. A1-A66, San Mateo, CA: Morgan Kaufmann Publishers, Inc., 1990.

[11] M. Sivaraman and A.J. Strojwas, "Towards Incorporating Device Parameter Variations in Timing Analysis", *Proceedings of the European Design Conference*, 1994, pp. 338-342.

[12] L.E. Thon, P. Sutardja, F. Lai and G. Coleman, "A 240MHz 8-Tap Programmable FIR Filter for Disk-Drive Read Channels," *IEEE ISSCC*, 1995, pp. 82-83.

NOTES

NOTES

NOTES

NOTES

NOTES

NOTES